HAMMOND

CITATION

WORLD ATLAS

HAMMOND CITATION
WORLD

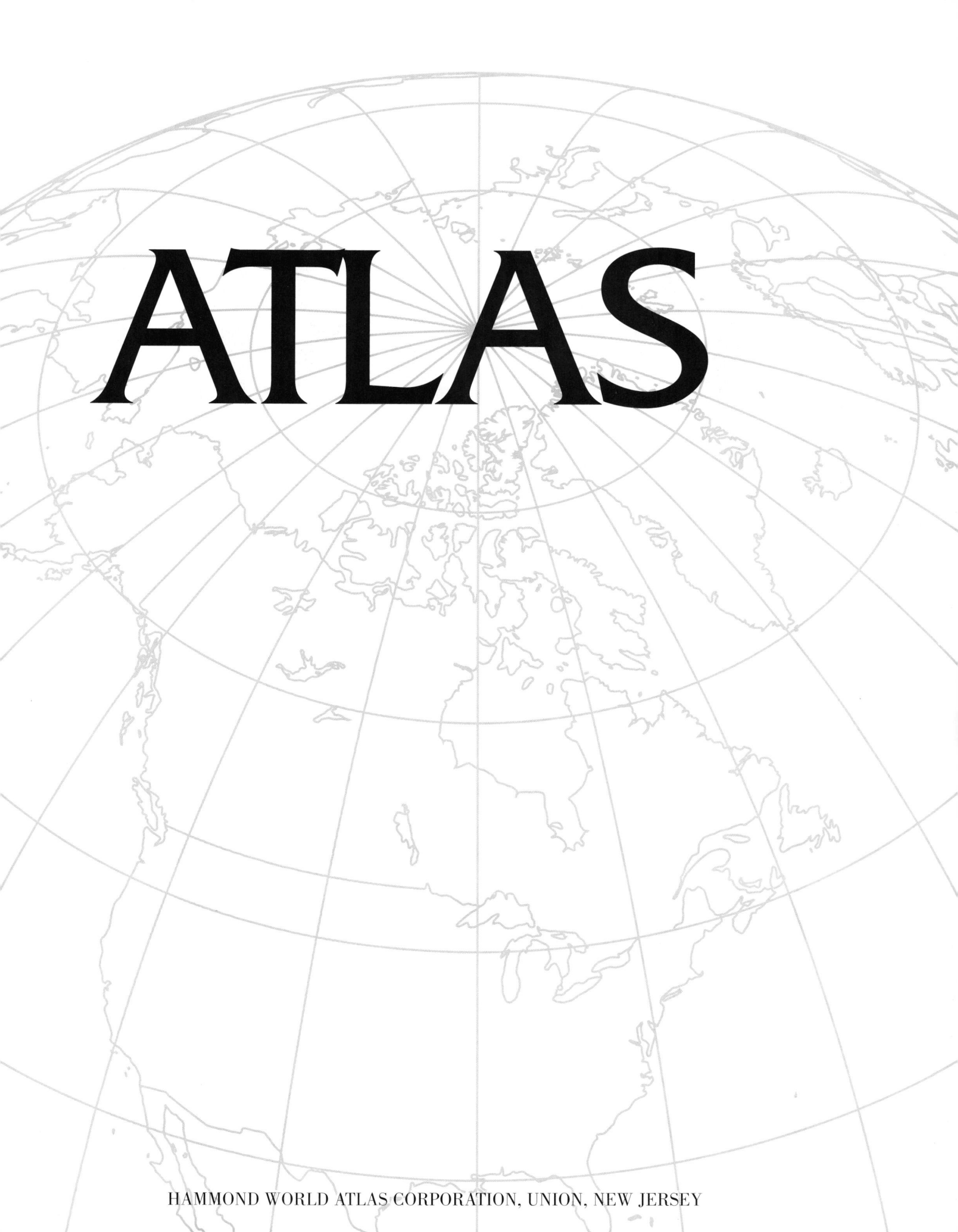

ATLAS

HAMMOND WORLD ATLAS CORPORATION, UNION, NEW JERSEY

Hammond Publications Advisory Board

Revised 2002 Edition
ENTIRE CONTENTS © COPYRIGHT 2000
BY HAMMOND WORLD ATLAS CORPORATION

PRINTED IN THE UNITED STATES OF AMERICA

Library of Congress Cataloging-in-Publication Data
Hammond World Atlas Corporation.
    Citation world atlas. -- Rev.
        p.   cm.
    At head of title: Hammond
    Includes indexes.
    ISBN 0-8437-1295-3 (softcover)
    ISBN 0-8437-1382-8 (hardcover)
    1. Atlases. I. Title. II. Title: Hammond citiation world atlas.
G1021. H2446   1998   <G&M>
912--DC21   98-12358
                               CIP
                              MAP

Contents

GAZETTEER-INDEX OF THE WORLD I   GLOSSARY OF ABBREVIATIONS VI

Part I—Terrain Maps of Land Forms and Ocean Floors

Contents IX   Index VII

Part II—Modern Maps and Indexes

Introduction

This unique Hammond World Atlas is organized to make the retrieval of information as pleasant and quick as possible. Our guiding principle is to present individual subjects on separate maps. In this manner, each map topic is shown with the greatest degree of clarity, unencumbered by extraneous information. Equally important is the use of separate atlas units to present all information on a given country or state. Thus, the basic reference map of an area is accompanied on adjacent pages by all supplementary information relating to that area. For example, the detailed index for any map always appears on the same page as, or on the pages immediately following, the reference map. This index provides population data for many cities, towns, and villages shown on the map. Pertinent statistics on the area, i.e. the total population and area, the capital, and the highest point, are found in the summary fact listing accompanying each unit. An adjacent locator map relates the subject area to the larger world, and a "three-dimensional" picture of the area is provided by a full-color topographic map. A separate economic map defines vital agricultural, industrial, and mineral resources. The flag of each independent nation or state also appears on the appropriate page. Finally, certain country units contain special subject maps dealing with the history, climate, demography, and vegetation of the area.

Another section features The Physical World - an outstanding series of terrain maps of land forms and ocean floors. These physical maps were originally produced as sculptured terrain models, thus simulating the earth's surface in a highly realistic manner. The three-dimensional effect is both informative and pleasing to the eye.

Of course, the maps have been thoroughly updated. These revisions reflect the new nations, and shifting international boundaries, and internal divisions of many countries. Even new communities generated by the tapping of resources in developing nations are recorded. Thorough research and worldwide contacts provide the most up-to-date geographical and demographic information available.

Uniquely designed, comprehensive, and easy-to-use, this World Atlas is the ideal reference for families, executives, students, travelers, or anyone who wants to be geographically informed about today's fast-changing world. Enjoy the adventure as you explore the pages of one of the world's finest atlases.

The Publisher

Introduction to the Maps and Indexes

The following notes have been added to aid the reader in making the best use of this atlas. Though the reader may be familiar with maps and map indexes, the publisher believes that a quick review of the material below will add to his enjoyment of this reference work.

Arrangement—*The Plan of the Atlas.* The atlas has been designed with maximum convenience for the user as its objective. Part I of the atlas is devoted to the physical world—terrain maps of land forms and the sea floor. Part II contains the general political reference maps, area by area. All geographically related information pertaining to a country or region appears on adjacent pages, eliminating the task of searching throughout the entire volume for data on a given area. Thus, the reader will find, conveniently assembled, political, topographic, economic and special maps of a political area or region, accompanied by detailed map indexes, statistical data, and illustrations of the national flags of the area.

The sequence of country units in this American-designed atlas is international in arrangement. Units on the world as a whole are followed by a section on the polar regions which, in turn, is followed by pages devoted to Europe and its countries. Every continent map is accompanied by special population distribution, climatic and vegetation maps of that continent. Following the maps of the European continent and its countries, the geographic sequence plan proceeds as follows: Asia, the Pacific and Australia, Africa, South America, North America, and ends with detailed coverage on the United States.

Political Maps—*The Primary Reference Tool.* The most detailed maps in each country unit are the *political maps*. It is our feeling that the reader is likely to refer to these maps more often than to any other in the book when confronted by such questions as—Where? How big? What is it near? Answering these common queries is the function of the political maps. Each political map stresses *political* phenomena—countries, internal political divisions, boundaries, cities and towns. The major political unit or units, shown on the map, are banded in distinctive colors for easy identification and delineation. First-order political subdivisions (states, provinces, counties on the state maps) are shown, scale permitting.

The reader is advised to make use of the *legend* appearing under the title on each political map. Map *symbols*, the special "language" of maps, are explained in the legend. Each variety of dot, circle, star or interrupted line has a special meaning which should be clearly understood by the user so that he may interpret the map data correctly.

Each country has been portrayed at a *scale* commensurate with its political, areal, economic or tourist importance. In certain cases, a whole map unit may be devoted to a single nation if that nation is considered to be of prime interest to most atlas users. In other cases, several nations will be shown on a single map if, as separate entities, they are of lesser relative importance. Areas of dense settlement and important significance within a country have been enlarged and portrayed in inset maps inserted on the margins of the main map. The scale of each map is indicated as a fractional representation (1:1,000,000). The reader is advised to refer to the linear or "bar" scale appearing on each map or map inset in order to determine the distance between points.

The *projection* system used for each map is noted near the title of the map. Map projections are the special graphic systems used by cartographers to render the curved three-dimensional surface of the globe on a flat surface. Optimum map projections determined by the attributes of the area have been used by the publishers for each map in the atlas.

A word here as to the choice of place names on the maps. Throughout the atlas names appear, with a few exceptions, in their local official spellings. However, conventional Anglicized spellings are used for major geographical divisions and for towns and topographic features for which English forms exist; i.e., "Spain" instead of "España" or "Munich" instead of "München." Names of this type are normally followed by the local official spelling in parentheses. As an aid to the user the indexes are cross-referenced for all current and most former spellings of such names.

Names of cities and towns in the United States follow the forms listed in the *Post Office Directory* of the United States Postal Service. Domestic physical names follow the decisions of the Board on Geographic Names, U.S. Department of the Interior, and of various state geographic name boards. It is the belief of the publishers that the boundaries shown in a general reference atlas should reflect current geographic and political realities. This policy has been followed consistently in the atlas. The presentation of *de facto* boundaries in cases of territorial dispute between various nations does not imply the political endorsement of such boundaries by the publisher, but simply the honest representation of boundaries as they exist at the time of the printing of the atlas maps.

Indexes—*Pinpointing a Location.* Each political map is accompanied by a comprehensive index of the place names appearing on the map. If you are unfamiliar with the location of a particular geographical place and wish to find its position within the confines of the subject area of the map, consult the map index as your first step. The name of the feature sought will be found in its proper alphabetical sequence with a key reference letter-number combination corresponding to its location on the map. After noting the key reference letter-number combination for the place name, turn to the map. The place name will be found within the square formed by the two lines of latitude and the two lines of longitude which enclose the coordinates—i.e., the marginal letters and numbers. The diagram below illustrates the system of indexing.

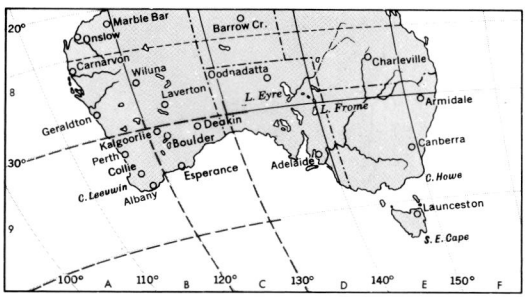

In the case of maps consisting entirely of insets, the place name is found near the intersection point of the imaginary lines connecting the coordinates at right angles. See below.

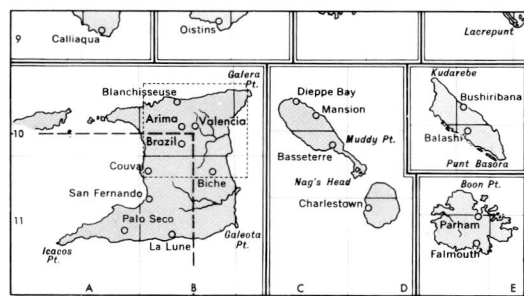

Where space on the map has not permitted giving the complete form of the place name, the complete form is shown in the index. Where a place is known by more than one name or by various spellings of the same name, the different forms have been included in the index. Physical features are listed under their proper names and not according to their generic terms; that is to say, Rio Negro will be found under Negro and not under Rio Negro. On the other hand, Rio Grande will be found under Rio Grande. Accompanying most index entries for cities and towns, and for other political units, are *population figures* for the particular entries. The large number of population figures in the atlas makes this work one of the most comprehensive statistical sources available to the public today. The population figures have been taken from the latest official censuses and estimates of the various nations.

Population and area figures for countries and major political units are listed in bold type *fact lists* on the margins of the index. In addition, the capital, largest city, highest point, monetary unit, principal languages and the prevailing religions of the country concerned are also listed. The Gazetteer-Index of the World on the following pages provides a quick reference index for countries and other important areas. Though population and area figures for each major unit are also found in the map section, the Gazetteer-Index provides a conveniently arranged statistical comparison contained in five pages.

Relief Maps. Accompanying each political map is a relief map of the area. The purpose of the relief map is to illustrate the surface configuration (TOPOGRAPHY) of the region. A shading technique in color simulates the relative ruggedness of the terrain — plains, plateaus, valleys, hills and mountains. Graded colors, ranging from greens for lowlands, yellows for intermediate elevations to brown in the highlands, indicate the height above sea level of each part of the land. A vertical scale at the margin of the map shows the approximate height in meters and feet represented by each color.

Economic Maps—Agriculture, Industry and Resources. One of the most interesting features that will be found in each country unit is the economic map. From this map one can determine the basic activities of a nation as expressed through its economy. A perusal of the map yields a full understanding of the area's economic geography and natural resources.

The agricultural economy is manifested in two ways: color bands and commodity names. The color bands express broad categories of dominant land use, such as cereal belts, forest lands, livestock range lands or nonagricultural wastes. The red commodity names, on the other hand, pinpoint the areas of production of *specific* crops, i.e., wheat, cotton, sugar beets, etc.

Major mineral occurrences are denoted by standard letter symbols appearing in blue. The relative size of the letter symbols signifies the relative importance of the deposit.

The manufacturing sector of the economy is presented by means of diagonal line patterns expressing the various *industrial* areas of consequence within a country.

The fishing industry is represented by names of commercial fish species appearing offshore in blue letters. Major waterpower sites are designated by blue symbols.

The publishers have tried to make this work the most comprehensive and useful atlas available, and it is hoped that it will prove a valuable reference work. Any constructive suggestions from the reader will be welcomed.

Sources and Acknowledgements

A multitude of sources goes into the making of a large-scale reference work such as this. To list them all would take many pages and would consume space better devoted to the maps and reference materials themselves. However, certain general sources were very useful in preparing this work and are listed below.

STATISTICAL OFFICE OF THE UNITED NATIONS.
Demographic Yearbook. New York. Issued annually.

STATISTICAL OFFICE OF THE UNITED NATIONS.
Statistical Yearbook. New York. Issued annually.

THE GEOGRAPHER, U.S. DEPARTMENT OF STATE.
International Boundary Study papers. Washington. Various dates.

THE GEOGRAPHER, U.S. DEPARTMENT OF STATE.
Geographic Notes. Washington. Various dates.

UNITED STATES BOARD ON GEOGRAPHIC NAMES.
Decisions on Geographic Names in the United States. Washington. Various dates.

UNITED STATES BOARD ON GEOGRAPHIC NAMES.
Official Standard Names Gazetteers. Washington. Various dates.

CANADIAN PERMANENT COMMITTEE ON GEOGRAPHICAL NAMES.
Gazetteer of Canada series. Ottawa. Various dates.

UNITED STATES POSTAL SERVICE.
National Five Digit ZIP Code and Post Office Directory. Washington. Issued annually.

UNITED STATES POSTAL SERVICE.
Postal Bulletin. Washington. Issued weekly.

UNITED STATES DEPARTMENT OF THE INTERIOR, BUREAU OF MINES.
Minerals Yearbook. 4 vols. Washington. Various dates.

UNITED STATES GEOLOGICAL SURVEY.
Elevations and distances in the United States. Reston, Va. 1990.

CARTACTUAL.
Cartactual — Topical Map Service. Budapest. Issues bi-monthly.

AMERICAN GEOGRAPHICAL SOCIETY.
Focus. New York. Issued ten times a year.

THE AMERICAN UNIVERSITY.
Foreign Area Studies. Washington. Various dates.

CENTRAL INTELLIGENCE AGENCY.
General reference maps. Washington. Various dates.

A sample list of sources used for specific countries follows:

Afghanistan
CENTRAL STATISTICS OFFICE.
Preliminary Results of the First Afghan Population Census 1979. Kabul.

Albania
DREJTORIA E STATISTIKES.
1979 Census. Tiranë.

Argentina
INSTITUTO NACIONAL DE ESTADISTICA Y CENSOS.
Censo Nacional de Población y Vivienda 1980. Buenos Aires.

Australia
AUSTRALIAN BUREAU OF STATISTICS.
Census of Population and Housing 1986. Canberra.

Brazil
FUNDACAO INSTITUTO BRASILEIRO DE GEOGRAFIA E ESTATISTICA.
IX Recenseamento Geral do Brasil 1980. Rio de Janeiro.

Canada
STATISTICS CANADA.
1986 Census of Canada. Ottawa.

Cuba
COMITE ESTATAL DE ESTADISTICAS.
Censo de Población y Viviendas 1981. Havana.

Hungary
HUNGARIAN CENTRAL STATISTICAL OFFICE.
1990 Census. Budapest.

Indonesia
BIRO PUSAT STATISTIK.
Sensus Penduduk 1980. Jakarta.

Kuwait
CENTRAL OFFICE OF STATISTICS.
1985 Census. Al Kuwait.

New Zealand
DEPARTMENT OF STATISTICS.
New Zealand Census of Population and Dwellings 1986. Wellington.

Panama
DIRECCION DE ESTADISTICA Y CENSO.
Censos Nacionales de 1990. Panamá.

Papua New Guinea
BUREAU OF STATISTICS.
National Population Census 1980. Port Moresby.

Philippines
NATIONAL CENSUS AND STATISTICS OFFICE.
1980 Census of Population. Manila.

Saint Lucia
CENSUS OFFICE.
1980 Population Census. Castries.

Singapore
DEPARTMENT OF STATISTICS.
Census of Population 1980. Singapore.

Russia
CENTRAL STATISTICAL ADMINISTRATION.
1989 Census. Moscow.

United States
BUREAU OF THE CENSUS.
1990 Census of Population. Washington.

Vanuatu
CENSUS OFFICE.
1979 Population Census. Port Vila.

Zambia
CENTRAL STATISTICAL OFFICE.
1980 Census of Population and Housing. Lusaka.

Gazetteer-Index of the World

This alphabetical list of continents, countries, states, possessions and other major geographical areas provides a quick reference to their area in square miles and square kilometers, population, capital or chief town, map page number and an alpha-numeric index reference. The index reference indicates the square on the respective page in which the name may be found. The population figures used in each case are the latest reliable figures obtainable. The government listings are based primarily on the nomenclature contained in the World Factbook published by the CIA of the United States Government. Those governments currently unsettled or in transition are indicated with a † symbol.

Country	Square Miles	Area Square Kilometers	Population	Capital or Chief Town	Page and Index Ref.	Government or Ownership
*Afghanistan	250,775	649,507	26,813,057	Kabul	68/A 2	authoritarian†
Africa	11,707,000	30,321,130	784,445,000	....	102/....	
Alabama, U.S.	51,705	133,916	4,447,100	Montgomery	195/....	state of the U.S.
Alaska, U.S.	591,004	1,530,700	626,932	Juneau	196/....	state of the U.S.
*Albania	11,100	28,749	3,510,484	Tiranë	45/E 5	emerging democracy†
Alberta, Canada	255,285	661,185	2,997,200	Edmonton	182/....	province of Canada
*Algeria	919,591	2,381,740	31,736,053	Algiers	106/D 3	republic
American Samoa	77	199	67,084	Pago Pago	87/J 7; 86/....	unincorporated, unorganized territory of the U.S.
*Andorra	188	487	67,627	Andorra la Vella	33/G 1	parliamentary democracy
*Angola	481,351	1,246,700	10,366,031	Luanda	114/C 6	transitional government†
Anguilla, U.K.	35	91	12,132	The Valley	156/F 3	dependent territory of the U.K.
Antarctica	5,500,000	14,245,000	....	....	5/....	
*Antigua and Barbuda	171	443	66,970	St. John's	161/E11; 156/G 3	constitutional monarchy
*Argentina	1,072,070	2,776,661	37,384,816	Buenos Aires	143/....	republic
Arizona, U.S.	114,000	295,260	5,130,632	Phoenix	198/....	state of the U.S.
Arkansas, U.S.	53,187	137,754	2,673,400	Little Rock	202/....	state of the U.S.
*Armenia	11,506	29,800	3,336,100	Yerevan	52/F 6	republic
Aruba, Netherlands	75	193	70,007	Oranjestad	161/E 9	autonomous member of the Netherlands realm
Ascension Island, St. Helena	34	88	719	Georgetown	102/A 5	part of St. Helena
Ashmore & Cartier Islands, Australia	61	159	....	(Canberra, Austr.)	88/C 2	territory of Australia
Asia	17,128,500	44,362,815	3,682,550,000	....	54/....	
*Australia	2,966,136	7,682,300	19,357,594	Canberra	88/....	federal parliamentary state
Australian Capital Territory	927	2,400	310,800	Canberra	96/E 4	territory of Australia
*Austria	32,375	83,851	8,150,835	Vienna	40/B 3	federal republic
*Azerbaijan	33,436	86,600	7,771,092	Baku	52/G 6	republic
Azores, Portugal	902	2,335	242,073	Ponta Delgada	32/....	autonomous region of Portugal
*Bahamas	5,382	13,939	297,852	Nassau	156/C 1	constitutional parliamentary democracy
*Bahrain	240	622	645,361	Manama	58/F 4	traditional monarchy
Baker Island, U.S.	1	2.6	....	....	87/J 5	unincorporated territory of the U.S.
Balearic Islands, Spain	1,936	5,014	796,483	Palma	33/H 3	autonomous community of Spain
*Bangladesh	55,126	142,776	131,269,860	Dhaka	68/G 4	republic
*Barbados	166	430	275,330	Bridgetown	161/B 8	parliamentary democracy
*Belarus	80,154	207,600	10,350,194	Minsk	52/C 4	republic
*Belgium	11,781	30,513	10,258,762	Brussels	27/E 7	constitutional monarchy
*Belize	8,867	22,966	256,062	Belmopan	154/C 2	parliamentary democracy
*Benin	43,483	112,620	6,590,782	Porto-Novo	106/E 6	republic
Bermuda, U.K.	21	54	63,503	Hamilton	156/H 3	dependent territory of the U.K.
*Bhutan	18,147	47,000	2,049,412	Thimphu	68/G 3	monarchy
*Bolivia	424,163	1,098,582	8,300,463	La Paz; Sucre	136/....	republic
Bonaire, Neth. Antilles	112	291	14,218	Kralendijk	161/E 9	part of Netherland Antilles
*Bosnia & Herzegovina	19,940	51,129	3,922,205	Sarajevo	45/C 3	emerging democracy†
*Botswana	224,764	582,139	1,586,119	Gaborone	119/C 4	parliamentary republic
Bouvet Island, Norway	22	57	....	....	5/D 1	territory of Norway
*Brazil	3,284,426	8,506,663	174,468,575	Brasília	132/....	federal republic
British Columbia, Canada	366,253	948,596	4,063,800	Victoria	184/....	province of Canada
British Indian Ocean Terr., U.K.	29	75	3,200	(London, U.K.)	54/L10	dependent territory of the U.K.
British Virgin Islands	59	153	19,615	Road Town	157/H 1	dependent territory of the U.K.
*Brunei	2,226	5,765	343,653	Bandar Seri Begawan	85/E 4	constitutional sultanate
*Bulgaria	42,823	110,912	7,707,495	Sofia	45/F 4	parliamentary democracy
*Burkina Faso	105,869	274,200	12,272,289	Ouagadougou	106/D 6	parliamentary
Burma, see Myanmar						
*Burundi	10,747	27,835	6,223,897	Bujumbura	114/E 4	republic
California, U.S.	158,706	411,049	33,871,648	Sacramento	204/.....	state of the U.S.
*Cambodia	69,898	181,036	12,491,501	Phnom Penh	72/E 4	constitutional monarchy†
*Cameroon	183,568	475,441	15,803,220	Yaoundé	114/B 2	one-party republic
*Canada	3,851,787	9,976,139	31,592,805	Ottawa	162/.....	confederation with parliamentary democracy
Canary Islands, Spain	2,808	7,273	1,630,015	Las Palmas; Santa Cruz	32/B 4	autonomous community of Spain
*Cape Verde	1,557	4,033	405,163	Praia	106/B 8	republic
Cayman Islands, U.K.	100	259	35,527	Georgetown	156/B 3	dependent territory of the U.K.
Celebes, Indonesia	72,986	189,034	7,732,383	Ujung Pandang	85/G 6	part of Indonesia
*Central African Republic	242,000	626,780	3,576,884	Bangui	114/C 2	republic
Central America	197,480	511,475	36,341,000	....	154/.....	
*Chad	495,752	1,283,998	8,707,078	N'Djamena	111/C 4	republic
Channel Islands, U.K.	75	194	147,000	St. Helier; St. Peter Port	13/E 8	part of the United Kingdom
*Chile	292,257	756,946	15,328,467	Santiago	138/....	republic
*China, People's Rep. of	3,705,386	9,596,960	1,273,111,290	Beijing	77/.....	communist party-led state
China, Republic of (Taiwan)	13,971	36,185	22,319,222	T'aipei	77/K 7	multiparty democratic
Christmas Island, Australia	52	135	2,564	The Settlement	54/M11	territory of Australia
Clipperton Island, France	2	5.2	....	....	146/H 8	possession of France
Cocos (Keeling) Islands, Australia	5.4	14	635	West Island	54/N11	territory of Australia

*Member of the United Nations

Gazetteer-Index of the World

Country	Square Miles	Area Square Kilometers	Population	Capital or Chief Town	Page and Index Ref.	Government or Ownership
*Colombia	439,513	1,138,339	40,349,388	Bogotá	126/.....	republic
Colorado, U.S.	104,091	269,596	4,301,261	Denver	208/.....	state of the U.S.
*Comoros	719	1,862	596,202	Moroni	119/G 2	republic
*Congo, Dem. Rep. of the	905,063	2,344,113	53,624,718	Kinshasa	114/D 4	dictatorship
*Congo, Rep. of the	132,046	342,000	2,894,336	Brazzaville	114/B 4	republic
Connecticut, U.S.	5,018	12,997	3,405,565	Hartford	210/.....	state of the U.S.
Cook Islands, New Zealand	91	236	20,611	Avarua	87/K 7	self-governing in free association with New Zealand
Coral Sea Islands, Australia	8.5	22	....	....	88/J 3	territory of Australia
Corsica, France	3,352	8,682	249,737	Ajaccio; Bastia	28/B 6	part of France
*Costa Rica	19,575	50,700	3,773,057	San José	154/E 5	democratic republic
Côte d'Ivoire, see Ivory Coast						
*Croatia	22,050	56,538	4,334,142	Zagreb	45/B 3	parliamentary democracy
*Cuba	44,206	114,494	11,184,023	Havana	158/.....	communist state
Curaçao, Neth. Antilles	178	462	151,448	Willemstad	161/G 7	part of Netherlands Antilles
*Cyprus	3,473	8,995	762,887	Nicosia	62/E 5	republic
*Czech Republic	30,449	78,863	10,264,212	Prague	41/C 2	parliamentary democracy
Delaware, U.S.	2,044	5,294	783,600	Dover	245/R 3	state of the U.S.
*Denmark	16,629	43,069	5,352,815	Copenhagen	21/.....	constitutional monarchy
District of Columbia, U.S.	69	179	572,059	Washington	244/F 5	district of the United States
*Djibouti	8,880	23,000	460,700	Djibouti	111/H 5	republic
*Dominica	290	751	70,786	Roseau	161/E 7	parliamentary democracy
*Dominican Republic	18,704	48,443	8,581,477	Santo Domingo	158/D 6	representative democracy
*Ecuador	109,483	283,561	13,183,978	Quito	128/C 3	republic
*Egypt	386,659	1,001,447	69,536,644	Cairo	110/E 2	republic
*El Salvador	8,260	21,393	6,237,662	San Salvador	154/C 4	republic
England, U.K.	50,516	130,836	49,997,100	London	13/.....	part of the United Kingdom
*Equatorial Guinea	10,831	28,052	486,060	Malabo	114/A 3	republic
*Eritrea	45,410	117,600	4,298,269	Asmara	110/G 4	transitional government†
*Estonia	17,413	45,100	1,423,316	Tallinn	53/.....	parliamentary democracy
*Ethiopia	426,366	1,104,300	65,891,874	Addis Ababa	110/G5.	federal republic
Europe	4,057,000	10,507,630	728,887,000	....	7/......	
Falkland Islands & Dependencies, U.K.	6,198	16,053	2,826	Stanley	120/E 8; 143/D 7	dependent territory of the U.K.
Faroe Islands, Denmark	540	1,399	45,661	Tórshavn	21/B 2	self-governing overseas administrative division of Denmark
*Fiji	7,055	18,272	844,330	Suva	87/H 8; 86/.....	republic
*Finland	130,128	337,032	5,175,783	Helsinki	18/O 6	republic
Florida, U.S.	58,664	151,940	15,982,378	Tallahassee	212/.....	state of the U.S.
*France	210,038	543,998	59,551,227	Paris	28/.....	republic
French Guiana	35,135	91,000	177,562	Cayenne	131/E 3	overseas department of France
French Polynesia	1,544	4,000	253,506	Papeete	87/L 8	overseas territory of France
*Gabon	103,346	267,666	1,221,175	Libreville	114/B 4	republic
*Gambia	4,127	10,689	1,411,205	Banjul	106/A 6	republic
Gaza Strip	139	360	1,178,119	Gaza	65/A 4	occupied by Israel
*Georgia	26,911	69,700	4,989,285	T'bilisi	52/F 6	republic
Georgia, U.S.	58,910	152,577	8,186,453	Atlanta	217/.....	state of the U.S.
*Germany	137,753	356,780	83,029,536	Berlin	22/.....	republic
*Ghana	92,099	238,536	19,894,014	Accra	106/D 7	constitutional democracy
Gibraltar, U.K.	2.28	5.91	27,649	Gibraltar	33/D 4	dependent territory of the U.K.
*Great Britain & Northern Ireland (United Kingdom)	94,399	244,493	57,236,000	London	10/.....	see United Kingdom
*Greece	50,944	131,945	10,623,835	Athens	45/F 6	presidential parliamentary republic
Greenland, Denmark	840,000	2,175,600	56,352	Nuuk (Godthåb)	4/B12	self-governing overseas administrative division of Denmark
*Grenada	133	344	89,227	St. George's	161/D 9; 156/G 4	constitutional monarchy
Guadeloupe & Dependencies, France	687	1,779	431,170	Basse-Terre	161/A 5; 156/F 4	overseas department of France
Guam, U.S.	209	541	157,557	Hagåtña	87/E 4; 86/.....	organized, unincorporated territory of the U.S.
*Guatemala	42,042	108,889	12,974,361	Guatemala	154/B 3	republic
*Guinea	94,925	245,856	7,613,870	Conakry	106/B 6	republic
*Guinea-Bissau	13,948	36,125	1,315,822	Bissau	106/A 6	republic
*Guyana	83,000	214,970	697,181	Georgetown	131/B 3	republic
*Haiti	10,694	27,697	6,964,549	Port-au-Prince	158/C 5	republic
Hawaii, U.S.	6,471	16,760	1,211,537	Honolulu	218/.....	state of the U.S.
Heard & McDonald Islands, Australia	113	293	....	....	2/N 8	territory of Australia
Holland, see Netherlands						
*Honduras	43,277	112,087	6,406,052	Tegucigalpa	154/D 3	republic
Hong Kong	422	1,092	7,210,505	Victoria	77/H 7; 78/.....	special administrative region of China
Howland Island, U.S.	1	2.6	....	....	87/J 5	unincorporated territory of the U.S.
*Hungary	35,919	93,030	10,106,017	Budapest	41/D 3	parliamentary democracy
*Iceland	39,768	103,000	277,906	Reykjavík	21/B 1	republic
Idaho, U.S.	83,564	216,431	1,293,953	Boise	220/.....	state of the U.S.
Illinois, U.S.	56,345	145,934	12,419,293	Springfield	222/.....	state of the U.S.
*India	1,269,339	3,287,588	1,029,991,145	New Delhi	68/D 4	federal republic
Indiana, U.S.	36,185	93,719	6,080,485	Indianapolis	227/.....	state of the U.S.
*Indonesia	788,430	2,042,034	228,437,870	Jakarta	85/D 7	republic
Iowa, U.S.	56,275	145,752	2,926,324	Des Moines	229/.....	state of the U.S.

Gazetteer-Index of the World

Country	Area Square Miles	Area Square Kilometers	Population	Capital or Chief Town	Page and Index Ref.	Government or Ownership
*Iran	636,293	1,648,000	66,128,965	Tehran	66/F 4	theocratic republic
*Iraq	172,476	446,713	23,331,985	Baghdad	66/C 4	republic
*Ireland	27,136	70,282	3,840,838	Dublin	17/.....	republic
Ireland, Northern, U.K.	5,452	14,121	1,697,800	Belfast	17/F 2	part of the United Kingdom
Isle of Man, U.K.	227	588	73,117	Douglas	13/C 3	part of the United Kingdom
*Israel	7,847	20,324	5,938,093	Jerusalem	65/B 4	parliamentary democracy
*Italy	116,303	301,225	57,679,825	Rome	34/.....	republic
*Ivory Coast (Côte d'Ivoire)	124,504	322,465	16,393,221	Yamoussoukro	106/C 7	republic
*Jamaica	4,411	11,424	2,665,636	Kingston	158/.....	parliamentary democracy
Jan Mayen, Norway	144	373	....	....	6/D 1	territory of Norway
*Japan	145,730	377,441	126,771,662	Tokyo	81/.....	constitutional monarchy
Jarvis Island, U.S.	1	2.6	....	....	87/K 6	unincorporated territory of the U.S.
Java, Indonesia	48,842	126,500	73,712,411	Jakarta	85/J 2	part of Indonesia
Johnston Atoll, U.S.	0.91	2.4	1,100	....	87/K 4	unincorporated territory of the U.S.
*Jordan	35,000	90,650	5,153,378	Amman	65/D 3	constitutional monarchy
Kansas, U.S.	82,277	213,097	2,688,418	Topeka	232/.....	state of the U.S.
*Kazakhstan	1,048,300	2,715,100	16,731,303	Astana	48/G 5	republic
Kentucky, U.S.	40,409	104,659	4,041,769	Frankfort	237/.....	state of the U.S.
*Kenya	224,960	582,646	30,765,916	Nairobi	115/G 3	republic
Kermadec Islands, New Zealand	13	33	5	....	87/J 9	part of New Zealand
Kingman Reef, U.S.	0.1	0.26	....	....	87/K 5	unincorporated territory of the U.S.
Kiribati	277	717	94,149	Tarawa	87/J 6	republic
*Korea, North	46,540	120,539	21,968,228	P'yŏngyang	80/D 3	authoritarian Socialist
*Korea, South	38,175	98,873	47,904,370	Seoul	80/D 5	republic
*Kuwait	6,532	16,918	2,041,961	Kuwait	58/E 4	constitutional monarchy
*Kyrgyzstan	76,641	198,500	4,753,003	Bishkek	48/H 5	republic
*Laos	91,428	236,800	5,635,967	Vientiane	72/D 3	communist
*Latvia	24,595	63,700	2,385,231	Riga	53/.....	parliamentary democracy
*Lebanon	4,015	10,399	3,627,774	Beirut	62/F 6	republic
*Lesotho	11,720	30,355	2,177,062	Maseru	119/D 5	constitutional monarchy
*Liberia	43,000	111,370	3,225,837	Monrovia	106/C 7	republic
*Libya	679,358	1,759,537	5,240,599	Tripoli	110/B 2	socialist people's (masses) state
*Liechtenstein	61	158	32,528	Vaduz	39/J 2	hereditary constitutional monarchy
*Lithuania	25,174	65,200	3,610,535	Vilnius	53/.....	parliamentary democracy
Louisiana, U.S.	47,752	123,678	4,468,976	Baton Rouge	238/.....	state of the U.S.
*Luxembourg	999	2,587	442,972	Luxembourg	27/J 9	constitutional monarchy
Macau	8	21	453,733	Macau	77/H 7	special administrative region of China
*Macedonia, Former Yugo. Rep. of	9,889	25,713	2,046,209	Skopje	45/E 5	emerging democracy
*Madagascar	226,657	587,041	15,982,563	Antananarivo	119/H 3	republic
Madeira Islands, Portugal	307	796	242,603	Funchal	32/A 2	autonomous region of Portugal
Maine, U.S.	33,265	86,156	1,274,923	Augusta	243/.....	state of the U.S.
*Malawi	45,747	118,485	10,548,250	Lilongwe	114/F 6	multiparty democracy
Malaya, Malaysia	50,806	131,588	17,740,609	Kuala Lumpur	72/D 6	part of Malaysia
*Malaysia	128,308	332,318	22,229,040	Kuala Lumpur	72/D 6; 85/E 4	constitutional monarchy
*Maldives	115	298	310,764	Male	54/L 9	republic
*Mali	464,873	1,204,021	11,008,518	Bamako	106/C 6	republic
*Malta	122	316	394,583	Valletta	34/E 7	parliamentary democracy
Manitoba, Canada	250,999	650,087	1,147,900	Winnipeg	179/.....	province of Canada
Marquesas Islands, French Polynesia	492	1,274	8,064	Atuona	87/N 6	part of French Polynesia
*Marshall Islands	70	181	70,822	Majuro	87/G 4	constitutional; free association with the U.S.
Martinique, France	425	1,101	418,454	Fort-de-France	161/D 5	overseas department of France
Maryland, U.S.	10,460	27,091	5,296,486	Annapolis	245/.....	state of the U.S.
Massachusetts, U.S.	8,284	21,456	6,349,097	Boston	249/.....	state of the U.S.
*Mauritania	419,229	1,085,803	2,747,312	Nouakchott	106/B 5	republic
*Mauritius	790	2,046	1,189,825	Port Louis	119/G 5	parliamentary democracy
Mayotte, France	144	373	163,366	Mamoutzou	119/G 2	territorial collectivity of France
*Mexico	761,601	1,972,546	101,879,171	Mexico City	150/.....	federal republic
Michigan, U.S.	58,527	151,585	9,938,444	Lansing	250/.....	state of the U.S.
*Micronesia, Federated States of	271	702	134,597	Palikir	87/E 5	constitutional; free association with the U.S.
Midway Islands, U.S.	1.9	4.9	....	....	87/J 3	unincorporated territory of the U.S.
Minnesota, U.S.	84,402	218,601	4,919,479	St. Paul	255/.....	state of the U.S.
Mississippi, U.S.	47,689	123,515	2,844,658	Jackson	256/.....	state of the U.S.
Missouri, U.S.	69,697	180,515	5,595,211	Jefferson City	261/.....	state of the U.S.
*Moldova	13,012	33,700	4,431,570	Chişinău	52/C 5	republic
*Monaco	368 acres	149 hectares	31,842		28/G 6	constitutional monarchy
*Mongolia	606,163	1,569,962	2,654,999	Ulaanbaatar	77/E 2	republic
Montana, U.S.	147,046	380,849	902,195	Helena	262/.....	state of the U.S.
Montserrat, U.K.	40	104	7,574	Plymouth	157/G 3	dependent territory of the U.K.
*Morocco	172,414	446,550	30,645,305	Rabat	106/C 2	constitutional monarchy
*Mozambique	303,769	786,762	19,371,057	Maputo	119/E 4	republic
*Myanmar (Burma)	261,789	678,034	41,994,678	Rangoon	72/B 2	military
*Namibia	317,827	823,172	1,797,677	Windhoek	118/B 3	republic
Nauru	7.7	20	12,088	Yaren (district)	87/G 6	republic
Navassa Island, U.S.	2	5	....	....	156/C 3	unincorporated territory of the U.S.
Nebraska, U.S.	77,355	200,349	1,711,263	Lincoln	264/.....	state of the U.S.
*Nepal	54,663	141,577	25,284,463	Kathmandu	68/E 3	parliamentary democracy
*Netherlands	15,892	41,160	15,981,472	The Hague; Amsterdam	27/F 5	constitutional monarchy
Netherlands Antilles	320	817	212,226	Willemstad	156/E 4	autonomous member of the Netherlands realm
Nevada, U.S.	110,561	286,353	1,998,257	Carson City	266/.....	state of the U.S.

Gazetteer-Index of the World

Country	Square Miles	Area Square Kilometers	Population	Capital or Chief Town	Page and Index Ref.	Government or Ownership
New Brunswick, Canada	28,354	73,437	756,600	Fredericton	170/.....	province of Canada
New Caledonia & Dependencies, France	7,335	18,998	204,863	Nouméa	87/G 8	overseas territory of France
Newfoundland, Canada	156,184	404,517	538,800	St. John's	166/.....	province of Canada
New Hampshire, U.S.	9,279	24,033	1,235,786	Concord	268/.....	state of the U.S.
New Jersey, U.S.	7,787	20,168	8,414,350	Trenton	273/.....	state of the U.S.
New Mexico, U.S.	121,593	314,926	1,819,046	Santa Fe	274/.....	state of the U.S.
New South Wales, Australia	309,498	801,600	6,428,700	Sydney	96/B 2	state of Australia
New York, U.S.	49,108	127,190	18,976,457	Albany	276/.....	state of the U.S.
*New Zealand	103,736	268,676	3,864,129	Wellington	100/.....	parliamentary democracy
*Nicaragua	45,698	118,358	4,918,393	Managua	154/D 4	republic
*Niger	489,189	1,267,000	10,355,156	Niamey	106/F 5	republic
*Nigeria	357,000	924,630	126,635,626	Abuja	106/F 6	republic
Niue, New Zealand	100	259	2,113	Alofi	87/K 7	self-governing territory in free association with New Zealand
Norfolk Island, Australia	13.4	34.6	1,892	Kingston	88/L 5	territory of Australia
North America	9,363,000	24,250,170	482,992,000	....	146/.....	
North Carolina, U.S.	52,669	136,413	8,049,313	Raleigh	281/.....	state of the U.S.
North Dakota, U.S.	70,702	183,118	642,200	Bismarck	282/.....	state of the U.S.
Northern Ireland, U.K.	5,452	14,121	1,663,300	Belfast	17/F 2	part of the United Kingdom
Northern Marianas, U.S.	184	477	74,612	Saipan	87/E 4	commonwealth associated with the U.S.
Northern Territory, Australia	519,768	1,346,200	193,400	Darwin	93/.....	territory of Australia
*North Korea	46,540	120,539	21,687,550	P'yŏngyang	80/D 3	authoritarian Socialist
Northwest Territories, Canada	519,731	1,346,106	42,100	Yellowknife	187/F 3	territory of Canada
*Norway	125,053	323,887	4,503,440	Oslo	18/F 7	constitutional monarchy
Nova Scotia, Canada	21,425	55,491	941,000	Halifax	168/.....	province of Canada
Nunavut, Canada	808,180	2,093,190	27,700	Iqaluit	187/J 3	territory of Canada
Oceania	3,292,000	8,526,280	30,393,000	....	87/.....	
Ohio, U.S.	41,330	107,045	11,353,140	Columbus	284/.....	state of the U.S.
Oklahoma, U.S.	69,956	181,186	3,450,654	Oklahoma City	288/.....	state of the U.S.
*Oman	120,000	310,800	2,622,198	Muscat	58/G 6	monarchy
Ontario, Canada	412,580	1,068,582	11,669,300	Toronto	175,177/	province of Canada
Oregon, U.S.	97,073	251,419	3,421,399	Salem	291/.....	state of the U.S.
Orkney Islands, Scotland	376	974	19,600	Kirkwall	15/E 1	part of the United Kingdom
*Pakistan	310,403	803,944	144,616,639	Islamabad	68/B 3	federal republic
*Palau	188	487	19,092	Koror	86/D 5	constitutional; free association with the U.S.
Palmyra Atoll, U.S.	12	31	....	....	87/K 5	unincorporated territory of the U.S.
*Panama	29,761	77,082	2,845,647	Panamá	154/G 6	constitutional democracy
*Papua New Guinea	183,540	475,369	5,049,055	Port Moresby	85/B 7; 87/E 6	parliamentary democracy
Paracel Islands, China	....	....	....	....	85/E 2	occupied by China; claimed by Taiwan and Vietnam
*Paraguay	157,047	406,752	5,734,139	Asunción	144/.....	republic
Pennsylvania, U.S.	45,308	117,348	12,281,054	Harrisburg	294/.....	state of the U.S.
*Peru	496,222	1,285,215	27,483,864	Lima	128/.....	republic
*Philippines	115,707	299,681	82,841,518	Manila	82/.....	republic
Pitcairn Islands, U.K.	18	47	54	Adamstown	87/O 8	dependent territory of the U.K.
*Poland	120,725	312,678	38,633,912	Warsaw	47/.....	republic
*Portugal	35,549	92,072	10,066,253	Lisbon	32/B 3	parliamentary democracy
Prince Edward Island, Canada	2,184	5,657	138,900	Charlottetown	168/E 2	province of Canada
Puerto Rico, U.S.	3,515	9,104	3,937,316	San Juan	161/.....	commonwealth associated with the U.S.
*Qatar	4,247	11,000	769,152	Doha	58/F 4	traditional monarchy
Québec, Canada	594,857	1,540,680	7,372,400	Québec	172,174/	province of Canada
Queensland, Australia	666,872	1,727,200	3,525,600	Brisbane	95/.....	state of Australia
Réunion, France	969	2,510	732,570	St-Denis	119/F 5	overseas department of France
Rhode Island, U.S.	1,212	3,139	1,048,319	Providence	249/H 5	state of the U.S.
*Romania	91,699	237,500	22,364,022	Bucharest	45/F 3	republic
*Russia	6,592,812	17,075,400	145,470,197	Moscow	48/D 4	federation
*Rwanda	10,169	26,337	7,312,756	Kigali	114/E 4	republic
Sabah, Malaysia	29,300	75,887	2,449,389	Kota Kinabalu	85/F 4	state of Malaysia
Saint Helena & Dependencies, U.K.	162	420	7,266	Jamestown	102/B 6	dependent territory of the U.K.
*Saint Kitts and Nevis	104	269	38,756	Basseterre	156/F 3; 161/C11	constitutional monarchy
*Saint Lucia	238	616	158,178	Castries	161/G 6	parliamentary democracy
Saint Pierre & Miquelon, France	93.5	242	6,928	Saint-Pierre	166/C 4	territorial collectivity of France
*Saint Vincent & the Grenadines	150	388	115,942	Kingstown	161/A 8; 157/G 4	parliamentary democracy
Sakhalin, Russia	29,500	76,405	632,000	Yuzhno-Sakhalinsk	48/P 4	part of Russia
*Samoa	1,133	2,934	179,058	Apia	87/J 7	constitutional monarchy
*San Marino	23.4	60.6	27,336	San Marino	34/D 3	republic
*São Tomé and Príncipe	372	963	165,034	São Tomé	106/F 8	republic
Sarawak, Malaysia	48,202	124,843	2,012,616	Kuching	85/E 5	state of Malaysia
Sardinia, Italy	9,301	24,090	1,450,483	Cagliari	34/B 4	region of Italy
Saskatchewan, Canada	251,699	651,900	1,023,600	Regina	181/.....	province of Canada
*Saudi Arabia	829,995	2,149,687	22,757,092	Riyadh	58/D 4	monarchy
Scotland, U.K.	30,414	78,772	5,114,600	Edinburgh	15/.....	part of the United Kingdom
*Senegal	75,954	196,720	10,284,929	Dakar	106/A 5	republic
*Seychelles	145	375	79,715	Victoria	119/H 5	republic
Shetland Islands, Scotland	552	1,430	22,740	Lerwick	15/G 2	part of the United Kingdom
Siam, see Thailand						
Sicily, Italy	9,926	25,708	5,087,794	Palermo	34/D 6	region of Italy
*Sierra Leone	27,925	72,325	5,426,618	Freetown	106/B 7	constitutional democracy
*Singapore	226	585	4,300,419	Singapore	72/F 6	republic
*Slovakia	18,924	49,014	5,414,937	Bratislava	41/E 2	parliamentary democracy

Gazetteer-Index of the World

Country	Square Miles	Area Square Kilometers	Population	Capital or Chief Town	Page and Index Ref.	Government or Ownership
*Slovenia	7,898	20,251	1,930,132	Ljubljana	45/A 3	republic
Society Islands, French Polynesia	677	1,753	167,398	Papeete	87/L 7	part of French Polynesia
*Solomon Islands	11,500	29,785	480,442	Honiara	87/G 6; 86/.....	parliamentary democracy
*Somalia	246,200	637,658	7,488,773	Mogadishu	115/H 3	no functioning government
*South Africa	455,318	1,179,274	43,586,097	Cape Town; Pretoria	118/C 5	republic
South America	6,875,000	17,806,250	345,782,000	....	120/.....	
South Australia, Australia	379,922	984,000	1,494,800	Adelaide	94/.....	state of Australia
South Carolina, U.S.	31,113	80,583	4,012,012	Columbia	296/.....	state of the U.S.
South Dakota, U.S.	77,116	199,730	754,844	Pierre	298/.....	state of the U.S.
*South Korea	38,175	98,873	47,350,529	Seoul	80/D 5	republic
*Spain	194,881	504,742	40,037,995	Madrid	33/.....	parliamentary monarchy
Spratly Islands	....	....	....	....	85/E 4	in dispute; claims by China, Malaysia, Philippines, Taiwan, Vietnam
*Sri Lanka	25,332	65,610	19,408,635	Colombo	68/E 7	republic
*Sudan	967,494	2,505,809	36,080,373	Khartoum	110/E 4	transitional†
Sumatra, Indonesia	164,000	424,760	19,360,400	Medan	84/B 5	see Indonesia
*Suriname	55,144	142,823	433,998	Paramaribo	131/C 3	constitutional democracy
Svalbard, Norway	23,957	62,049	2,416	Longyearbyen	18/C 2	territory of Norway
*Swaziland	6,705	17,366	1,104,343	Mbabane	119/E 5	monarchy
*Sweden	173,665	449,792	8,875,053	Stockholm	18/J 8	constitutional monarchy
Switzerland	15,943	41,292	7,283,274	Bern	39/.....	federal republic
*Syria	71,498	185,180	16,728,808	Damascus	62/G 5	military republic
Tahiti, French Polynesia	402	1,041	150,707	Papeete	87/L 7	see French Polynesia
Taiwan	13,971	36,185	22,370,461	T'aipei	77/K 7	multiparty democratic
*Tajikistan	55,251	143,100	6,578,681	Dushanbe	48/G 6	republic
*Tanzania	363,708	942,003	36,232,074	Dar es Salaam	114/F 5	republic
Tasmania, Australia	26,178	67,800	470,100	Hobart	99/.....	state of Australia
Tennessee, U.S.	42,144	109,153	5,689,283	Nashville	237/.....	state of the U.S.
Texas, U.S.	266,807	691,030	20,851,820	Austin	303/.....	state of the U.S.
*Thailand	198,455	513,998	61,797,751	Bangkok	72/D 3	constitutional monarchy
Tibet, China	463,320	1,200,000	1,790,000	Lhasa	76/C 5	part of China
*Togo	21,622	56,000	5,153,088	Lomé	106/E 7	republic†
Tokelau, New Zealand	3.9	10	1,458	Fakaofo	87/J 6	territory of New Zealand
Tonga	270	699	104,227	Nuku'alofa	87/J 8	hereditary constitutional monarchy
*Trinidad and Tobago	1,980	5,128	1,169,682	Port-of-Spain	157/G 5; 161/A10	parliamentary democracy
Tristan da Cunha, St. Helena	38	98	313	Edinburgh	2/J 7	see St. Helena
Tuamotu Archipelago, French Polynesia	341	883	15,370	Apataki	87/M 7	see French Polynesia
*Tunisia	63,378	164,149	9,705,102	Tunis	106/F 1	republic
*Turkey	300,946	779,450	66,493,970	Ankara	62/D 3	republican parliamentary democracy
*Turkmenistan	188,455	488,100	4,603,244	Ashgabat	48/F 6	republic
Turks and Caicos Islands, U.K.	166	430	18,122	Cockburn Town, Grand Turk	156/D 2	dependent territory of the U.K.
Tuvalu	9.78	25.33	10,991	Funafuti	87/H 6	democracy
*Uganda	91,076	235,887	23,985,712	Kampala	114/F 3	republic
*Ukraine	233,089	603,700	48,760,474	Kiev	52/D 5	republic
*United Arab Emirates	32,278	83,600	2,407,460	Abu Dhabi	58/F 5	federation of sheikdoms
*United Kingdom	94,399	244,493	59,647,790	London	10/.....	constitutional monarchy
*United States	3,623,420	9,384,658	278,058,881	Washington, D.C.	188/.....	federal republic
*Uruguay	72,172	186,925	3,360,105	Montevideo	145/.....	republic
Utah, U.S.	84,899	219,888	2,233,169	Salt Lake City	304/.....	state of the U.S.
*Uzbekistan	173,591	449,600	25,155,064	Tashkent	48/G 5	republic
*Vanuatu	5,700	14,763	192,910	Port-Vila	87/G 7	republic
Vatican City	108.7 acres	44 hectares	1,000	....	34/B 6	sacerdotal (priest-related) monarchy
*Venezuela	352,143	912,050	23,916,810	Caracas	124/.....	republic
Vermont, U.S.	9,614	24,900	608,827	Montpelier	268/.....	state of the U.S.
Victoria, Australia	87,876	227,600	4,726,600	Melbourne	96/B 5	state of Australia
Vietnam	128,405	332,569	79,939,014	Hanoi	72/E 3	communist state
Virginia, U.S.	40,767	105,587	7,078,515	Richmond	307/.....	state of the U.S.
Virgin Islands, British	59	153	20,812	Road Town	157/H 1	dependent territory of the U.K.
Virgin Islands, U.S.	132	342	101,809	Charlotte Amalie	161/A 4	organized, unincorporated territory of the U.S.
Wake Island, U.S.	2.5	6.5	302	Wake Islet	87/G 4	unincorporated territory of the U.S.
Wales, U.K.	8,017	20,764	2,946,200	Cardiff	13/D 5	part of the United Kingdom
Wallis and Futuna, France	106	275	15,435	Mata Utu	87/J 7	overseas territory of France
Washington, U.S.	68,139	176,480	5,894,121	Olympia	310/.....	state of the U.S.
West Bank	2,100	5,439	2,090,713	....	65/C 3	occupied by Israel
Western Australia, Australia	975,096	2,525,500	1,868,200	Perth	92/.....	state of Australia
Western Sahara	102,703	266,000	250,559	....	106/B 3	occupied by Morocco
West Virginia, U.S.	24,231	62,758	1,808,344	Charleston	312/.....	state of the U.S.
Wisconsin, U.S.	56,153	145,436	5,363,675	Madison	317/.....	state of the U.S.
World (land)	57,970,000	150,142,300	6,176,100,000	....	1,2/.....	
Wyoming, U.S.	97,809	253,325	493,782	Cheyenne	319/.....	state of the U.S.
*Yemen	188,321	487,752	18,078,035	Sanaa	58/D 7	republic
*Yugoslavia	38,989	102,173	11,210,243	Belgrade	45/C 3	republic
Yukon Territory, Canada	186,660	483,450	30,700	Whitehorse	186/E 3	territory of Canada
*Zambia	290,586	752,618	9,770,199	Lusaka	114/E 7	republic
*Zimbabwe	150,803	390,580	11,365,366	Harare	119/D 3	parliamentary democracy

Glossary of Abbreviations

A

A.A.F. — Army Air Field
Acad. — Academy
A.C.T. — Australian Capital Territory
adm. — administration; administrative
A.F.B. — Air Force Base
Afgh., Afghan. — Afghanistan
Afr. — Africa
Ala. — Alabama
Alb. — Albania
Alg. — Algeria
Alta. — Alberta
Amer. — American
Amer. Samoa — American Samoa
And. — Andorra
Ant., Antarc. — Antarctica
Ant. & Bar. — Antigua and Barbuda
Ar. — Arabia
arch. — archipelago
Arg. — Argentina
Ariz. — Arizona
Ark. — Arkansas
Arm. — Armenia
Aust. — Austria
Aust. Cap. Terr. — Australian Capital
    Territory
Austr., Austral. — Australian, Australia
aut. — autonomous
Aut. Obl. — Autonomous Oblast
Aut. Rep. — Autonomous
    Republic
Azer. — Azerbaijan

B

B. — Bay
Bah. — Bahamas
Barb. — Barbados
Battlef. — Battlefield
Bch. — Beach
Bel. — Belarus
Belg. — Belgium
Berm. — Bermuda
Bol. — Bolivia
Bos. — Bosnia & Hercegovina
Bots. — Botswana
Br. — Branch
Br. — British
Braz. — Brazil
Br. Col. — British Columbia
Br. Ind. Oc. Terr. — British Indian
    Ocean Territory
Bulg. — Bulgaria

C

C. — Cape
Calif. — California
Can. — Canada
can. — canal
cap. — capital
Cent. Afr. Rep. — Central African
    Republic
Cent. Amer. — Central America
C.G. Sta. — Coast Guard Station
C.H. — Court House
chan. — channel
Chan. Is. — Channel Islands
Chem. Ctr. — Chemical Center
co. — county
Col. — Colombia
Colo. — Colorado
comm. — commissary
Conn. — Connecticut
cont. — continent
cord. — cordillera (mountain range)
C. Rica — Costa Rica
Cro. — Croatia
C.S. — County Seat
C. Verde — Cape Verde
Czech. — Czech Republic

D

D.C. — District of Columbia
Del. — Delaware
Dem. — Democratic
Den. — Denmark
depr. — depression
dept. — department
des. — desert
dist., dist's — district, districts
div. — division
Dom. Rep. — Dominican Republic

E

E. — East
Ec., Ecua. — Ecuador
elec. div. — electoral division
El Salv. — El Salvador
Eng. — England

Equat. Guinea, Eq. Guin. — Equatorial
    Guinea
Erit. — Eritrea
escarp. — escarpment
est. — estuary
Est. — Estonia
Eth. — Ethiopia

F

Falk. Is. — Falkland Islands
Fin. — Finland
Fk., Fks. — Fork, Forks
Fla. — Florida
for. — forest
Fr. — France, French
Fr. Gui. — French Guiana
Fr. Poly. — French Polynesia
Ft. — Fort

G

G. — Gulf
Ga. — Georgia (state)
Game Res. — Game Reserve
Geo. — Georgia (nation)
Ger. — Germany
geys. — geyser
Gibr. — Gibraltar
glac. — glacier
gov. — governorate
Greenl. — Greenland
Gren. — Grenada
Gt. Brit. — Great Britain
Guad. — Guadeloupe
Guat. — Guatemala
Guinea-Biss. — Guinea-Bissau
Guy. — Guyana

H

har., harb., hbr. — harbor
hd. — head
highl. — highland, highlands
Hist. — Historic, Historical
Hond. — Honduras
Hts. — Heights
Hung. — Hungary

I

I., isl. — island, isle
I.C. — independent city
Ice., Icel. — Iceland
Ida. — Idaho
Ill. — Illinois
Ind. — Indiana
ind. city — independent city
Indon. — Indonesia
Ind. Res. — Indian Reservation
int. div. — internal division
inten. — intendency
Int'l — International
Ire. — Ireland
Is., isls. — islands
Isr. — Israel
isth. — isthmus
Iv. Coast — Ivory Coast

J

Jam. — Jamaica
Jct. — Junction

K

Kans. — Kansas
Kaz., Kazakh. — Kazakhstan
Ky. — Kentucky
Kyr. — Kyrgyzstan

L

L. — Lake, Loch, Lough
La. — Louisiana
Lab. — Laboratory
lag. — lagoon
Lat. — Latvia
ld. — land
Leb. — Lebanon
Les. — Lesotho
Liecht. — Liechtenstein
Lith. — Lithuania
Lux. — Luxembourg

M

Mac. — Macedonia
Mad., Madag. — Madagascar
Man. — Manitoba
Mart. — Martinique
Mass. — Massachusetts
Maur. — Mauritania
Md. — Maryland
met. area — metropolitan area

Mex. — Mexico
Mich. — Michigan
Minn. — Minnesota
Miss. — Mississippi
Mo. — Missouri
Mold. — Moldova
Mon. — Monument
Mong. — Mongolia
Mont. — Montana
Mor. — Morocco
Moz., Mozamb. — Mozambique
mt. — mount
mtn. — mountain

N

N., No. — North
N. Amer. — North America
Nam., Namib. — Namibia
N.A.S. — Naval Air Station
Nat'l — National
Nat'l Cem. — National Cemetery
Nat'l Mem. Park — National Memorial
    Park
Nat'l Mil. Park — National Military
    Park
Nat'l Pkwy. — National Parkway
Nav. Base — Naval Base
Nav. Sta. — Naval Station
N.B., N. Br. — New Brunswick
N.C. — North Carolina
N. Dak. — North Dakota
Nebr. — Nebraska
Neth. — Netherlands
Neth. Ant. — Netherlands Antilles
Nev. — Nevada
New Bruns. — New Brunswick
New Cal., New Caled. — New Caledonia
Newf. — Newfoundland
New Hebr. — New Hebrides
N.H. — New Hampshire
Nic. — Nicaragua
N. Ire. — Northern Ireland
N.J. — New Jersey
N. Mex. — New Mexico
Nor. — Norway, Norwegian
North. — Northern
North. Terr., No. Terr. — Northern
    Territory
    (Australia)
N.S. — Nova Scotia
N.S.W., N.S. Wales — New South Wales
N.W.T., N.W. Terrs. — Northwest
    Territories
    (Canada)
N.Y. — New York
N.Z., N. Zealand — New Zealand

O

Obl. — Oblast
Okla. — Oklahoma
Okr. — Okrug
Ont. — Ontario
Ord. Depot — Ordnance Depot
Oreg. — Oregon

P

Pa. — Pennsylvania
Pak. — Pakistan
Pan. — Panama
Papua N.G. — Papua New Guinea
Par. — Paraguay
par. — parish
passg. — passage
P.E.I. — Prince Edward Island
pen. — peninsula
Phil., Phil. Is. — Philippines
Pk. — Park
pk. — peak
plat. — plateau
P.N.G. — Papua New Guinea
Pol. — Poland
Port. — Portugal, Portuguese
Pr. Edward I. — Prince
    Edward Island
pref. — prefecture
P. Rico — Puerto Rico
prom. — promontory
prov. — province, provincial
pt. — point

Q

Que. — Québec
Queens. — Queensland

R

R. — River

ra. — range
Rec., Recr. — Recreation, Recreational
reg. — region
Rep. — Republic
res. — reservoir
Res. — Reservation, Reserve
R.I. — Rhode Island
riv. — river
Rom. — Romania

S

S. — South
sa. — sierra, serra
S. Afr., S. Africa — South Africa
salt dep. — salt deposit
salt des. — salt desert
S. Amer. — South America
São T. & Pr. — São Tomé
    and Príncipe
Sask. — Saskatchewan
Saudi Ar. — Saudi Arabia
S. Aust., S. Austral. — South Australia
S.C. — South Carolina
Scot. — Scotland
Sd. — Sound
S. Dak. — South Dakota
Sen. — Senegal
Seych. — Seychelles
Sing. — Singapore
S. Leone — Sierra Leone
Slvk. — Slovakia
Slvn. — Slovenia
S. Marino — San Marino
Sol. Is. — Solomon Islands
Sp. — Spanish
Spr., Sprs. — Spring, Springs
St., Ste. — Saint, Sainte
Sta. — Station
St. P. & M. — Saint Pierre and
    Miquelon
St. Vin. & Grens. — St. Vincent & The
    Grenadines
str., strs. — strait, straits
Sur. — Suriname
Swaz. — Swaziland
Switz. — Switzerland

T

Taj. — Tajikistan
Tanz. — Tanzania
Tas. — Tasmania
Tenn. — Tennessee
terr., terrs. — territory, territories
Tex. — Texas
Thai. — Thailand
trad. — traditional
Trin. & Tob. — Trinidad and Tobago
Tun. — Tunisia
Turk. — Turkmenistan
twp. — township

U

U.A.E. — United Arab Emirates
U.K. — United Kingdom
Ukr. — Ukraine
urb. area — urban area
Urug. — Uruguay
U.S. — United States
Uzb. — Uzbekistan

V

Va. — Virginia
Ven., Venez. — Venezuela
V.I. (U.K.) — Virgin Islands (U.K.)
V.I. (U.S.) — Virgin Islands (U.S.)
Vic. — Victoria
Viet. — Vietnam
Vill. — Village
vol. — volcano
Vt. — Vermont

W

W. — West, Western
Wash. — Washington
W. Aust., W. Austral. — Western
    Australia
W. Indies — West
    Indies
Wis. — Wisconsin
W. Va. — West Virginia
Wyo. — Wyoming

Y

Yugo. — Yugoslavia
Yukon — Yukon Territory

Z

Zim. — Zimbabwe

Index to Terrain Maps

on pages X through XXXII

This index contains only names of land and ocean physical features. Names of towns, internal divisions and countries are not included. The entry name is followed by a letter-number combination which refers to the area on the map in which the name will be found. The number following the map reference for the entry refers, not to the page on which the entry will be found, but to the map plate number.

Index Continued

THE PHYSICAL WORLD
Terrain Maps of Land Forms and Ocean Floors

CONTENTS

RELIEF MODELS BY ERNST G. HOFMANN, ASSISTED BY RAFAEL MARTINEZ

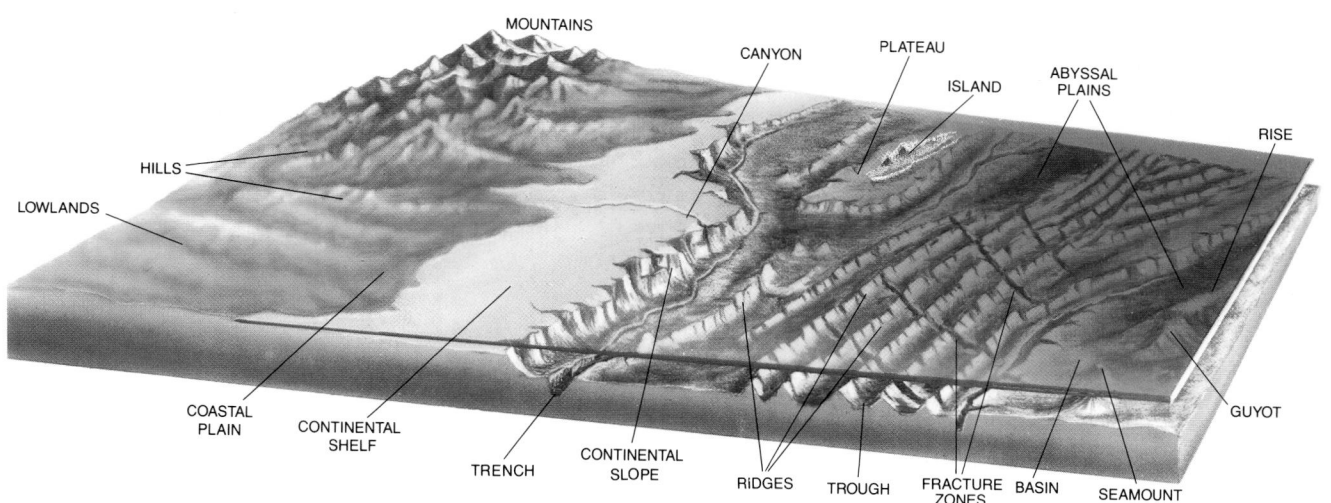

The oblique view diagram above is designed to provide a detailed view of the ocean floor as if seen through the depth of the sea. Graduating blue tones are used to contrast ocean floor depths: from light blue to represent shallow continental shelves to dark blues in the greater depths. Land relief is shown in conventional hypsometric tints.

In this dramatic collection of topographic maps of continents, oceans and major regions of the world, Hammond introduces a revolutionary new technique in cartography.

While most maps depicting terrain are created from painted artwork that is then photographed, Hammond now premiers the use of a remarkable sculptured model mapping technique created by one of our master cartographers.

The process begins with the sculpting of large scale three-dimensional models. Once physical details have been etched on the models and refinements completed, relief work is checked for accurate elevation based on a vertical scale exaggerated for visual effect.

Finished models are airbrushed and painted, then photographed using a single northwesterly light source to achieve a striking three-dimensional effect. The result is the dynamic presentation of mountain ranges and peaks on land, and canyons, trenches and seamounts on the ocean floor. Never before have maps conveyed such rich beauty while providing a realistic representation of the world as we know it.

A R C T I C O C E A N

SVALBARD
FRANZ JOSEF LAND
SEVERNAYA
ZEMLYA
NEW SIBERIAN IS.
Novaya
Zemlya
Laptev
Sea
Nordkapp
Kara
Barents
Sea
Sea
Wrangel
I.
Kjolen
Bering Sea
ALEUTIAN
BASIN
Baltic Sea
Ladoga
Ob
Yenisei
Lena
Aldan
Kamchatka
Pen.
ALEUTIAN ISLANDS
ALEUTIAN TRENCH
Volga
A   S   I   A
Irtysh
Ob
Angara
Lena
Sea
of
Okhotsk
KURIL KAMCHATKA TRENCH
EUROPE
Dnieper
Caspian Sea
L.
Balkhash
Amur
Sakhalin
NORTHWEST
Danube
Black Sea
Aral
Sea
Gobi
Sea of
Japan
PACIFIC
Euphrates
Honshu
JAPAN
TRENCH
BASIN
rranean Sea
Kunlun
Hwang
Japan
East
China
Sea
Nile
Himalaya
Mt.Everest
Chang
Taiwan
Tropic of Cancer
I C A
Ganges
Indus
Salween
PHILIPPINE
MARIANA
Arabian
Sea
Luzon
BASIN
MARIANA
TRENCH
MARSHALL IS.
CENTRAL
Mekong
ARABIAN
BASIN
Bay of
Bengal
South
China
Sea
Challenger
Deep
PACIFIC
BASIN
CHRISTMAS
RIDGE
C. Comorin
Ceylon
Mindanao
SOMALI
CEYLON
Borneo
CAROLINE IS
Victoria
CENTRAL
PLAIN
New Guinea
MELANESIAN
BASIN
Equator
O C E A N
Congo
Kilimanjaro
BASIN
Sumatra
Java
Celebes
INDIAN
RIDGE
Madagascar
Zambezi
Fiji Is.
INDIAN
Orange
BROKEN
PLATEAU
AUSTRALIA
Tropic of Capricorn
Tasman
Sea
North Cape
Good Hope
SOUTHEAST
C. Leeuwin
North I.
SIN
RIDGE
S. AUSTRALIA BASIN
Tasmania
South I.
SOUTHWEST INDIAN RIDGE
INDIAN
RIDGE
SOUTHEAST
KERGUELEN
PLATEAU
INDIAN
RIDGE
ENDERBY ABYSSAL PLAIN
AUSTRALIAN-ANTARCTIC BASIN
Antarctic Circle

A N T A R C T I C A

Amery
Ice Shelf
C. Adare
Ross
Sea

© Copyright 1987 by HAMMOND INCORPORATED, Maplewood, N.J.

LEGEND FOR TERRAIN MAPS

International Boundaries	—.—	Mountain Peaks	•
State and Provincial Boundaries	—.·—	National Capitals	⊛
Other Boundaries	– – –	Other Capitals	⊙
Boundaries Along Rivers	∿	Canals	⊱⊰

World | Plate 1

Plate 2 | **Europe**

Western Europe | **Plate 3**

0 100 200 300 400 500 MILES

0 100 200 300 400 500 KILOMETERS

GREENLAND
(Den.)

Plate 4 | Asia

| 0 | 300 | 600 | 900 | 1200 | 1500 MILES |

| 0 | 300 | 600 | 900 | 1500 KILOMETERS |

Southwest Asia | Plate 5

0   100   200   300   400   500 MILES
0   100   200   300   400   500 KILOMETERS

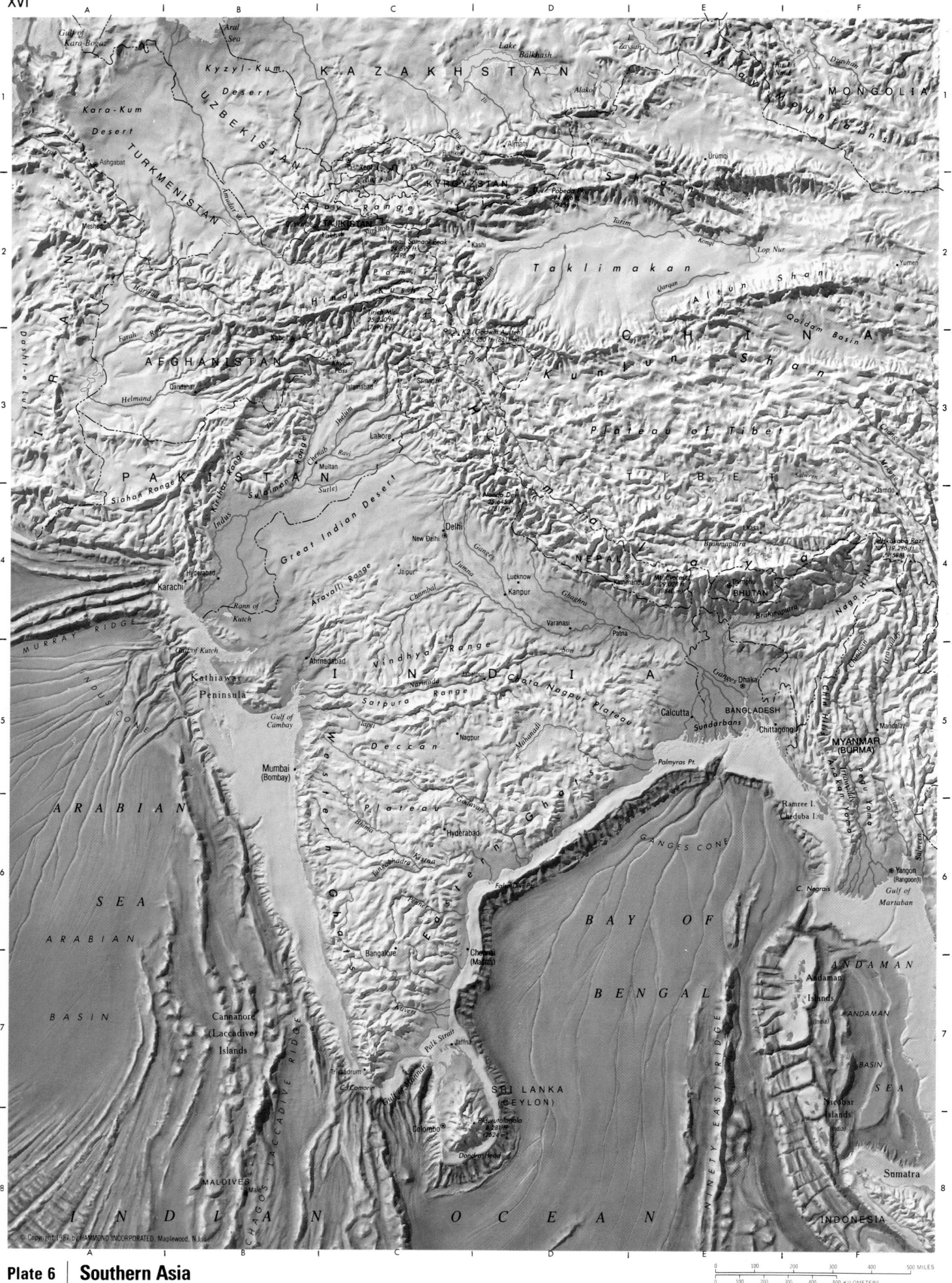

Plate 6 | **Southern Asia**

© Copyright 1997 by HAMMOND INCORPORATED, Maplewood, N.J.

| | 0 | 100 | 200 | 300 | 400 | 500 MILES |
| | 0 | 100 | 200 | 300 | 400 | 500 KILOMETERS |

East Asia | Plate 7

0 100 200 300 400 500 600 MILES
0 100 200 300 400 500 600 KILOMETERS

Plate 8 | Southeast Asia

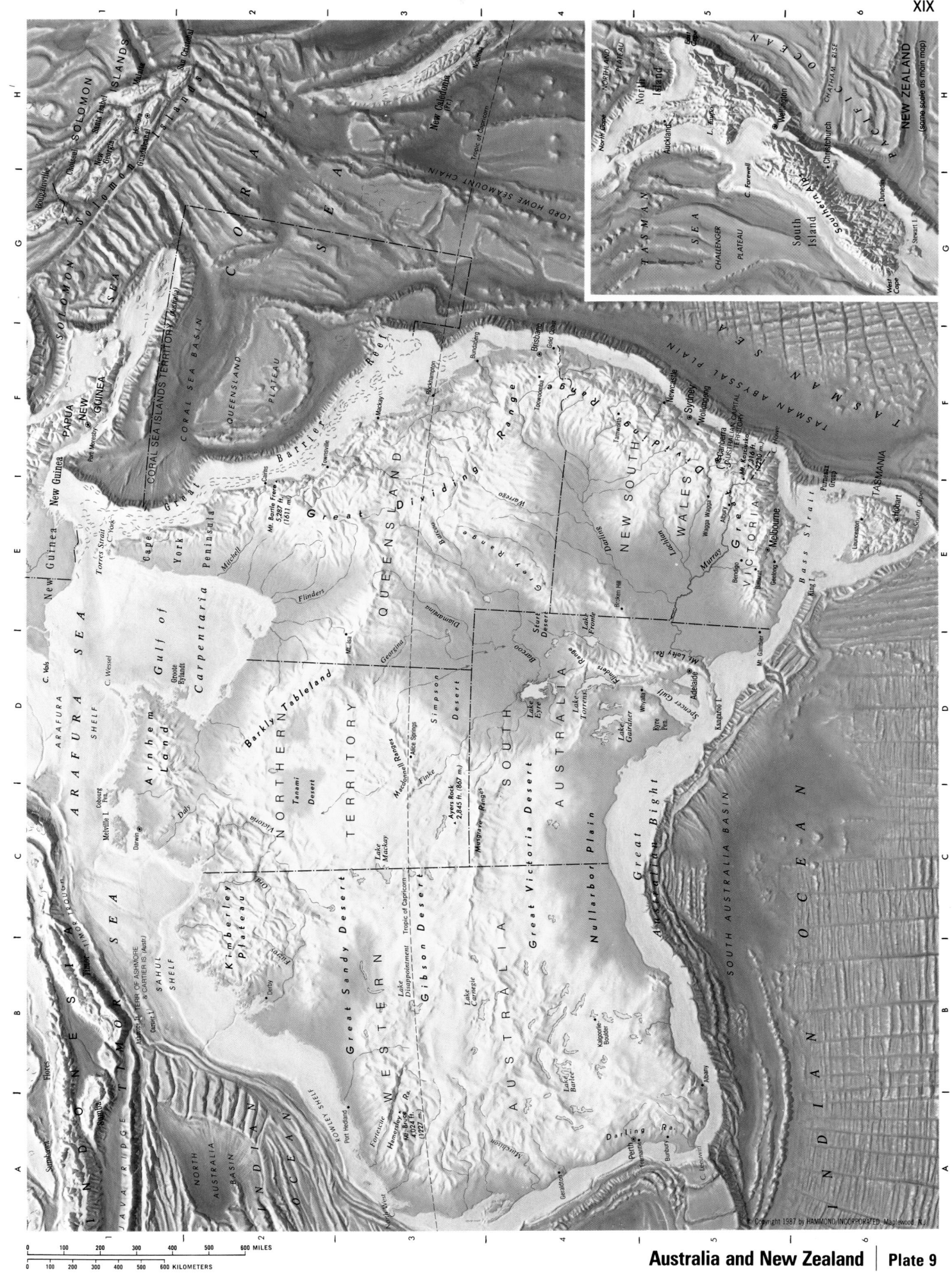

Australia and New Zealand | **Plate 9**

NEW ZEALAND
(same scale as main map)

© Copyright 1987 by HAMMOND INCORPORATED, Maplewood, N.J.

0   100   200   300   400   500   600 MILES

0   100   200   300   400   500   600 KILOMETERS

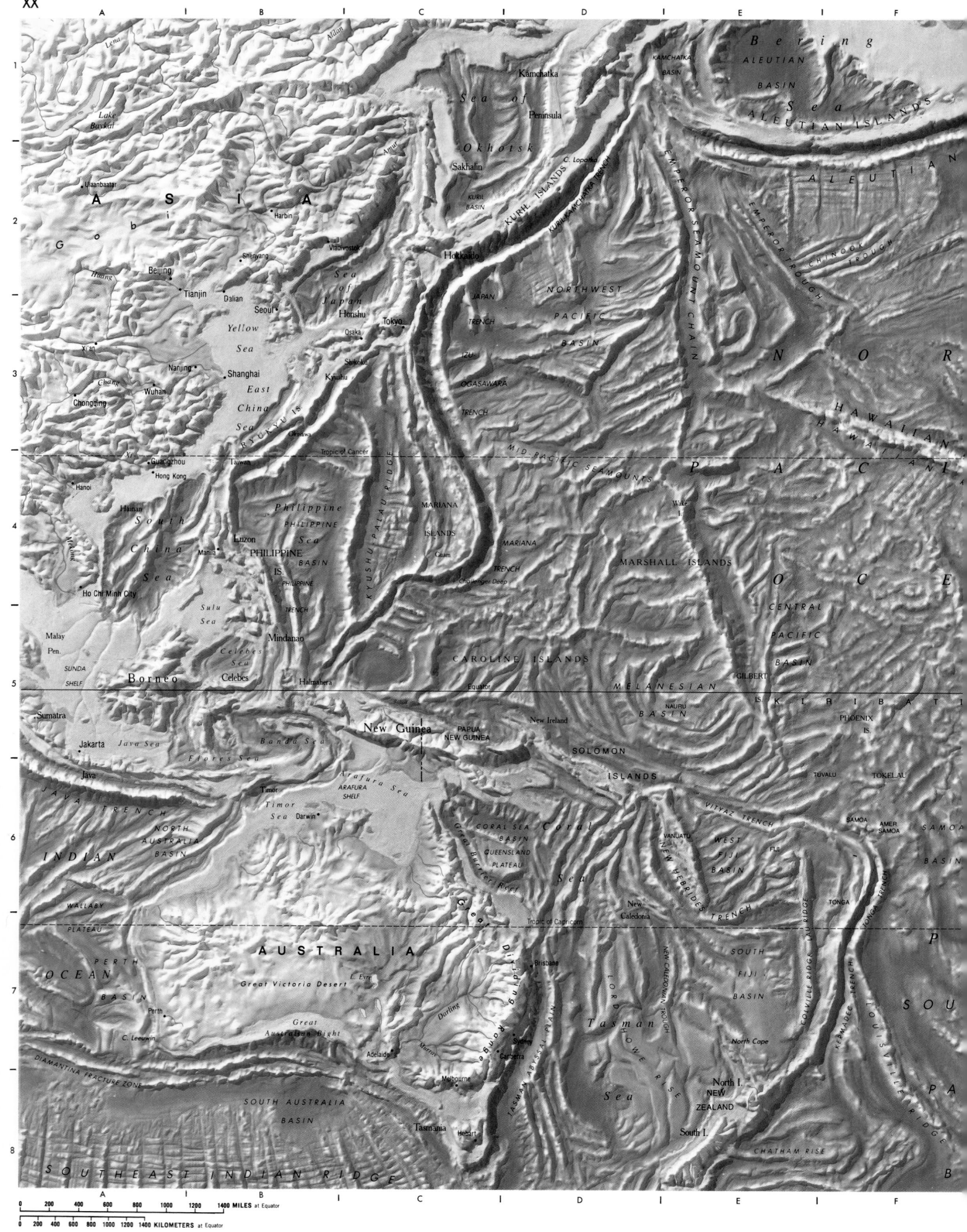

A   A   I   B   I   C   I   D   I   E   I   F   B

Bering

1   Lena   Aldan   Kamchatka   ALEUTIAN   BASIN
Lake   Sea of   Peninsula   KAMCHATKA   BASIN   Sea
Baykal   Okhotsk   ALEUTIAN ISLANDS

A   S   I   A   Amur   Sakhalin   C. Lopatka   ALEUTIAN
2   G   o   b   i   Harbin   KURIL   KURIL ISLANDS   EMPEROR SEAMOUNT CHAIN   CHINOOK
BASIN   Vladivostok   KURIL-KAMCHATKA TRENCH   EMPEROR TROUGH   TROUGH
Shenyang   Hokkaido   JAPAN   NORTHWEST
Huang   Beijing   Sea   TRENCH   PACIFIC   N   O   R
Tianjin   Dalian   of   IZU   BASIN
Xi'an   Seoul   Japan   OGASAWARA   H   A   W   A   I   I   A   N
3   Nanjing   Honshu   Tokyo   TRENCH
Chang   Wuhan   Shanghai   Osaka   HAWAIIAN
Chongqing   Yellow   Shikoku   OGASAWARA
Sea   Kyushu   Shenzhen   TRENCH
East   RYUKYU IS.   MID-PACIFIC SEAMOUNTS   P   A   C   I   C
China   Okinawa   Tropic of Cancer
Xi   Guangzhou   Sea   Taiwan   KYUSHU-PALAU RIDGE   Wake I.
Hanoi   Hong Kong   Philippine   MARIANA   Wake I.
Hainan   South   Sea   PHILIPPINE   ISLANDS   MARSHALL ISLANDS   O   C   E
4   China   Luzon   BASIN   MARIANA
Manila   PHILIPPINE   Guam   CENTRAL
Sea   IS.   TRENCH   PACIFIC
Ho Chi Minh City   PHILIPPINE   Challenger Deep   BASIN
Malay   Sulu   TRENCH   CAROLINE ISLANDS   GILBERT
Pen.   Sea   Mindanao   Equator   MELANESIAN   K   I   R   I   B   A   T   I
SUNDA   Celebes   Halmahera   NAURU   IS.   PHOENIX
5   SHELF   Borneo   Celebes   Sea   BASIN   IS.
Sumatra   New Ireland   TUVALU   TOKELAU
Jakarta   Java Sea   Banda Sea   New Guinea   PAPUA   SOLOMON   VITYAZ TRENCH   SAMOA   AMER.
Java   Flores Sea   NEW GUINEA   ISLANDS   VANUATU   WEST   SAMOA   SAMOA
JAVA   Timor   ARAFURA   Coral Sea   NEW HEBRIDES   FIJI   FIJI   BASIN
6   TRENCH   Timor   Sea   Darwin   CORAL SEA   Coral   BASIN   TRENCH
NORTH   Sea   BASIN   QUEENSLAND   Sea   New   SOUTH
INDIAN   AUSTRALIA   Great Barrier Reef   PLATEAU   Caledonia   FIJI
BASIN   Tropic of Capricorn   NEW CALEDONIA TROUGH   BASIN
WALLABY   AUSTRALIA   Brisbane   LORD   COLVILLE RIDGE
7   OCEAN   PLATEAU   Great Victoria Desert   L. Eyre   Sydney   HOWE   TONGA   NORFOLK RIDGE
PERTH   Darling   Canberra   Tasman   RISE   North Cape   KERMADEC TRENCH   LOUISVILLE RIDGE
BASIN   Perth   Great   Murray   Sea   SOU
C. Leeuwin   Australian Bight   Melbourne   NORTH I.   PA
Adelaide   DIAMANTINA FRACTURE ZONE   NEW
8   SOUTH AUSTRALIA   ZEALAND
BASIN   Tasmania   Hobart   South I.   CHATHAM RISE

S   O   U   T   H   E   A   S   T   I   N   D   I   A   N   R   I   D   G   E

A   A   I   B   I   C   I   D   I   E   I   F

0   200   400   600   800   1000   1200   1400 MILES at Equator

0   200   400   600   800   1000   1200   1400 KILOMETERS at Equator

Pacific Ocean | **Plate 10**

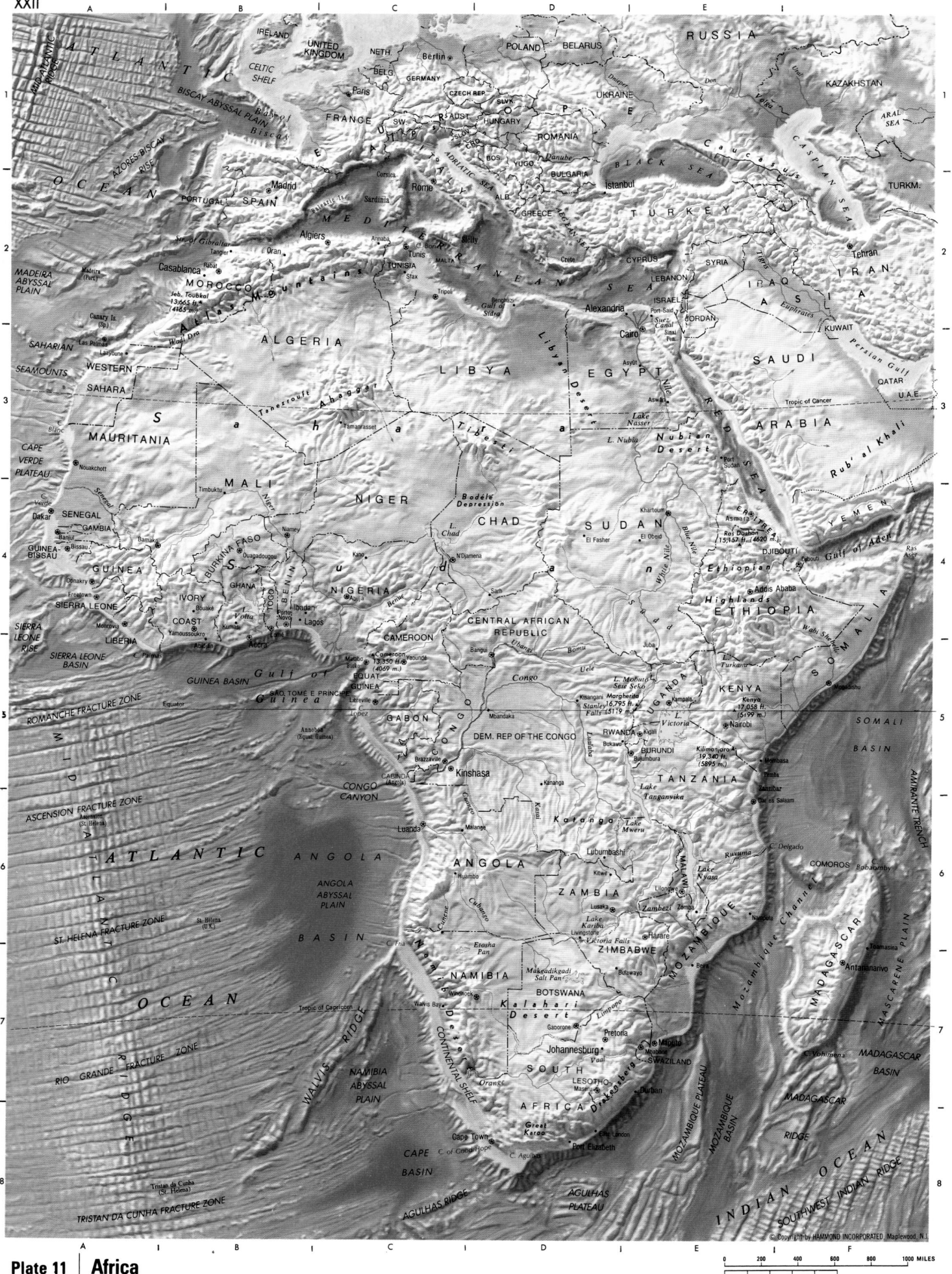

Plate 11 | **Africa**

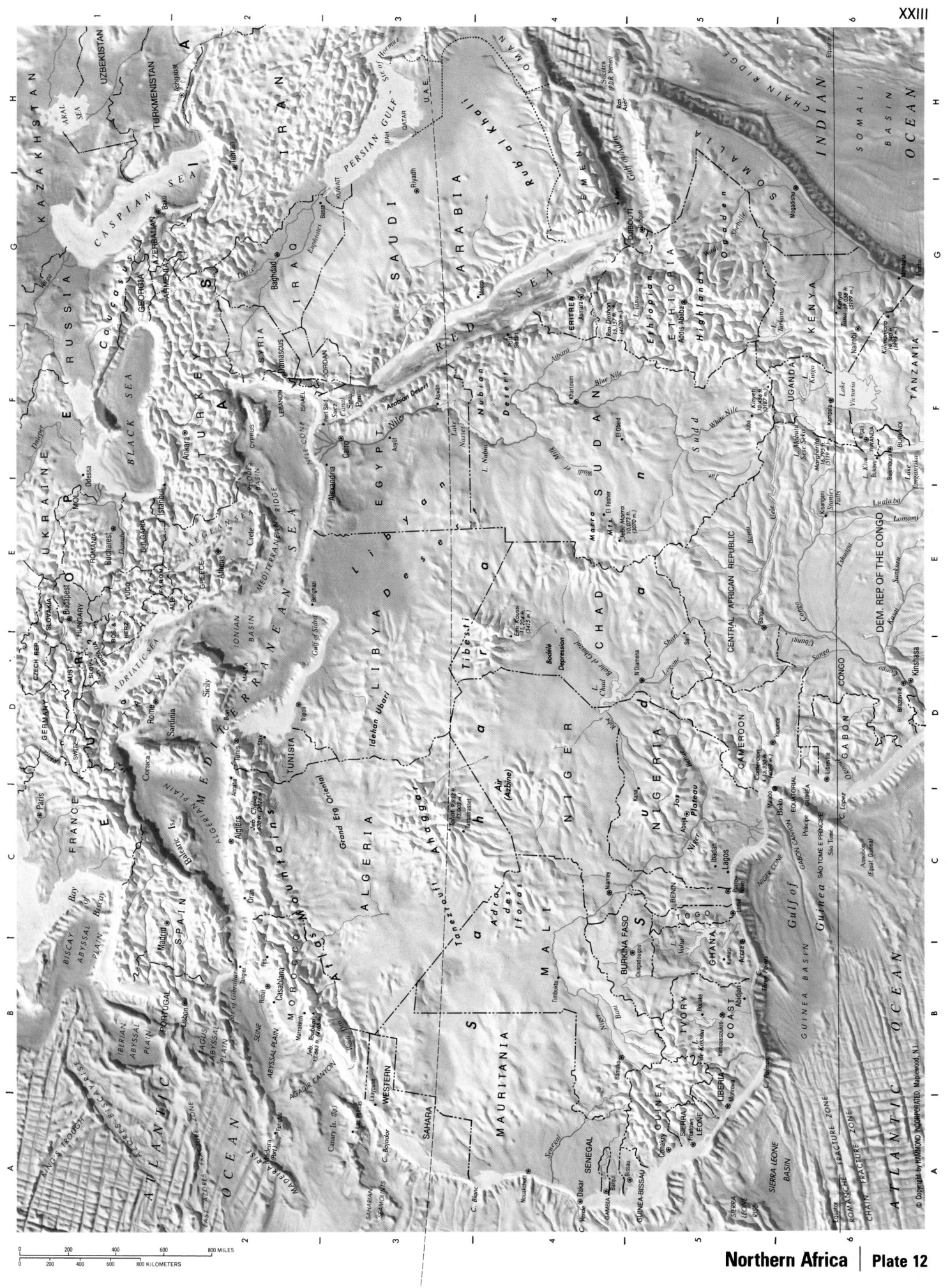

Northern Africa | **Plate 12**

© Copyright by HAMMOND INCORPORATED, Maplewood, N.J.

| 0 | 200 | 400 | 600 | 800 MILES |
| 0 | 200 | 400 | 600 | 800 KILOMETERS |

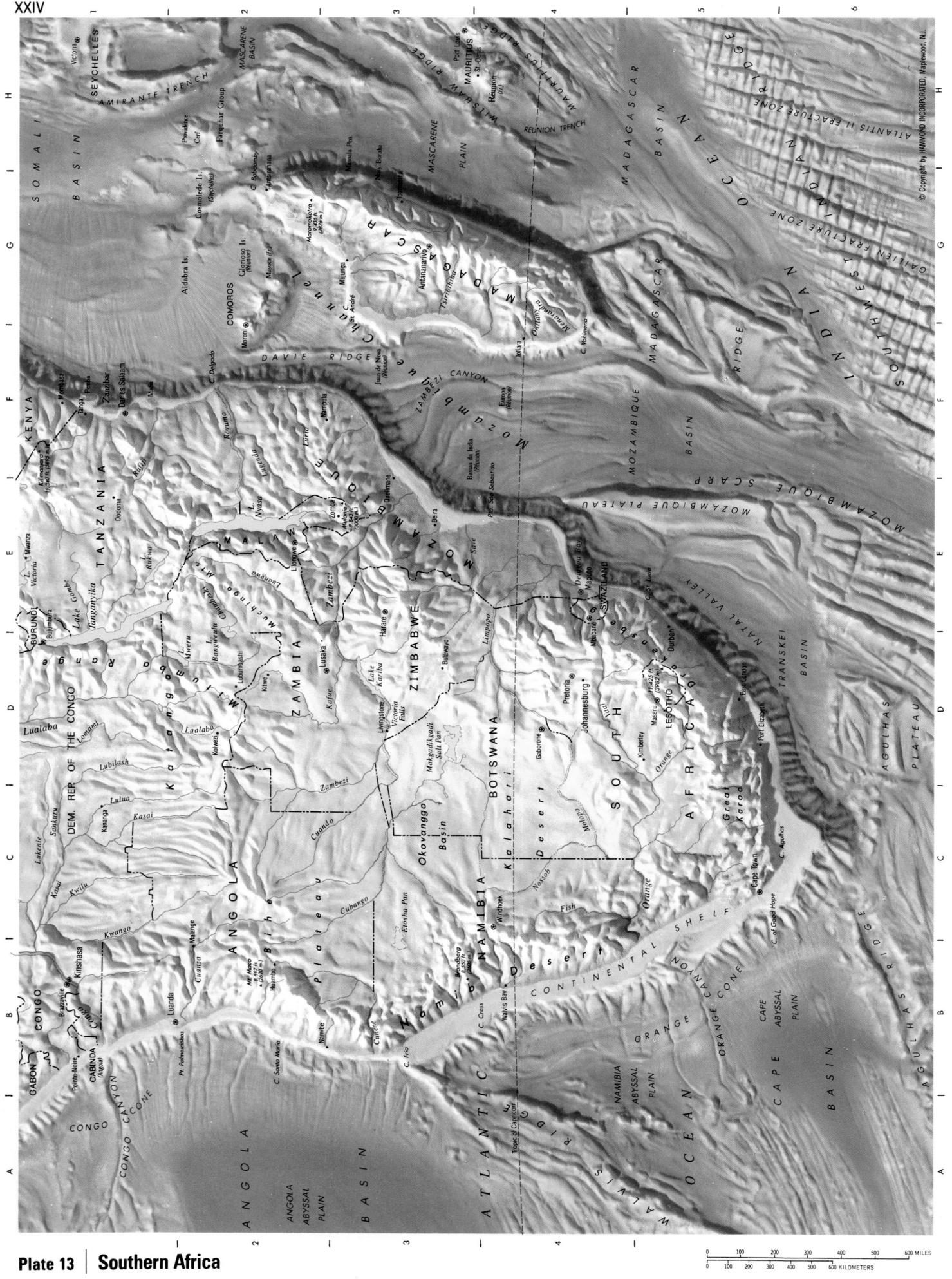

Plate 13 | Southern Africa

South America | Plate 14

© Copyright 1987 by HAMMOND INCORPORATED, Maplewood, N.J.

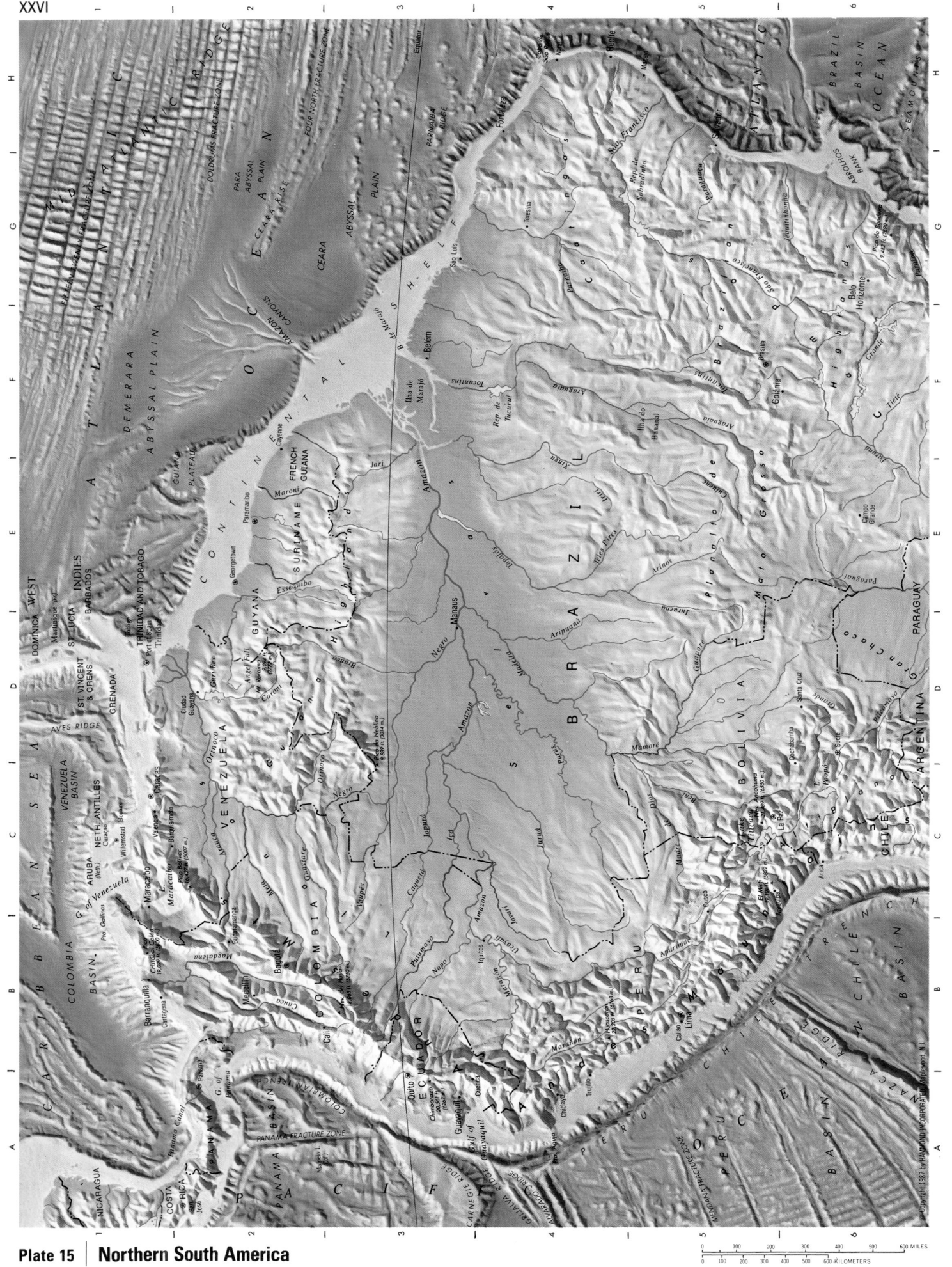

Plate 15 | **Northern South America**

| 0 | 100 | 200 | 300 | 400 | 500 | 600 MILES |

| 0 | 100 | 200 | 300 | 400 | 500 | 600 KILOMETERS |

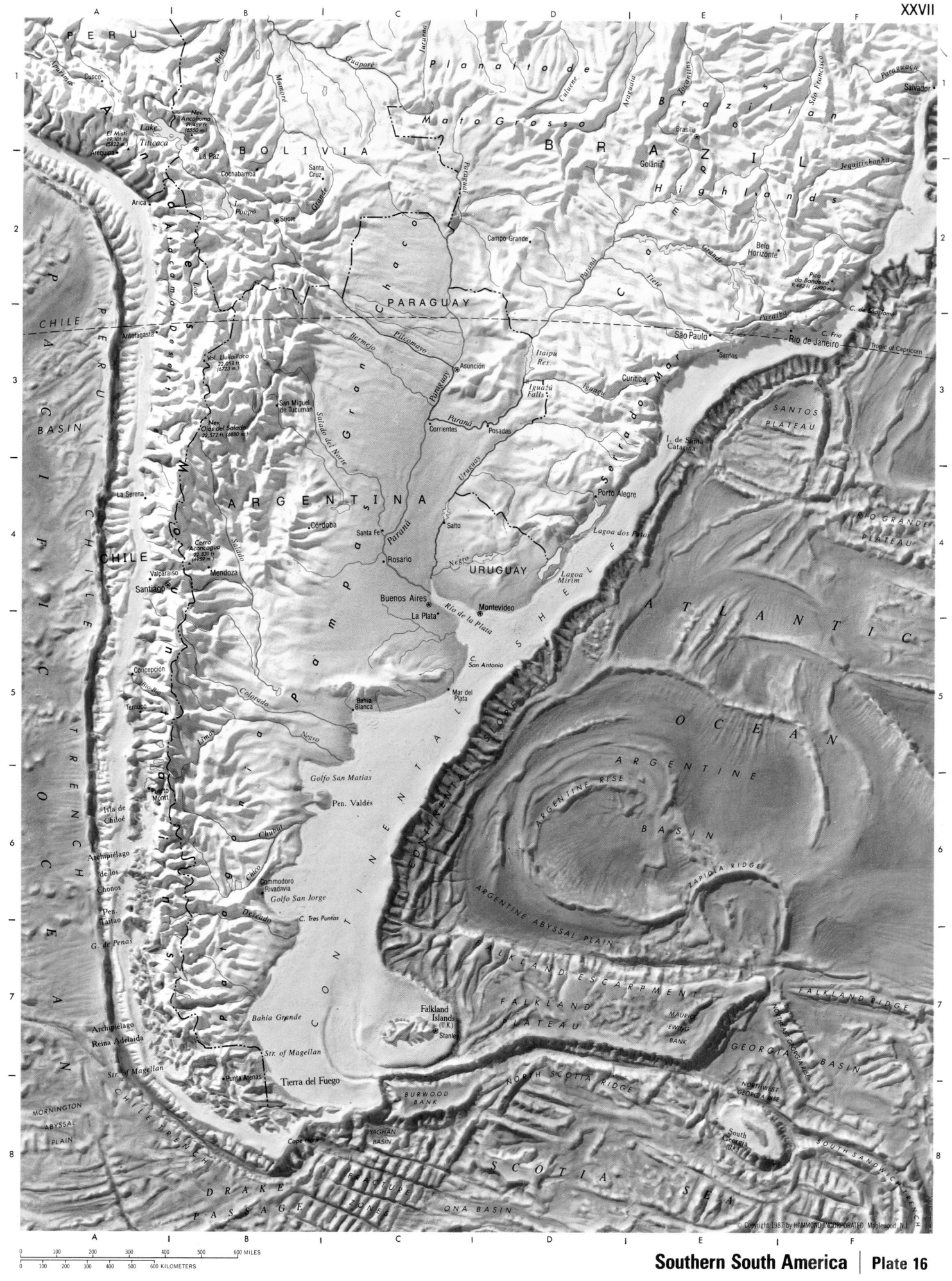

PERU

BRAZIL

Cusco

El Misti
19,101 ft.
(5822 m.)
Arequipa

Lake Titicaca

Nev. Ancohuma
(6550 m.)

La Paz

BOLIVIA

Cochabamba

Santa Cruz

Arica

L. Poopó

Sucre

Planalto de

Mato Grosso

Brasília

Goiânia

Brazilian

Highlands

Belo Horizonte

Pico da Bandeira
9,482 ft. (2890 m.)

Salvador

Jequitinhonha

CHILE

PERU

Antofagasta

Vol. Llullaillaco
22,057 ft.
(6723 m.)

Nev. Ojos del Salado
22,572 ft. (6880 m.)

San Miguel de Tucumán

Salado del Norte

PARAGUAY

Bermejo

Pilcomayo

Asunción

Paraguay

Gran

Chaco

Campo Grande

Itaipu Res.

Iguazú Falls

Iguaçu

Paraná

São Paulo

Santos

Curitiba

CHILE

La Serena

Cerro Aconcagua
22,831 ft.
(6959 m.)

Valparaíso

Santiago

Mendoza

ARGENTINA

Córdoba

Santa Fe

Rosario

Paraná

Corrientes

Posadas

Uruguay

Salto

Negro

URUGUAY

Montevideo

Buenos Aires

La Plata

Rio de la Plata

Porto Alegre

Lagoa dos Patos

Lagoa Mirim

Rio de Janeiro

C. Frio

Tropic of Capricorn

I. de Santa Catarina

SANTOS PLATEAU

RIO GRANDE PLATEAU

ATLANTIC

OCEAN

Concepción

Río Bío

Temuco

Colorado

Limay

Negro

C. San Antonio

Mar del Plata

Bahía Blanca

Puerto Montt

Isla de Chiloé

Golfo San Matías

Pen. Valdés

Chubut

ARGENTINE RISE

ARGENTINE

BASIN

Archipiélago de los Chonos

Pen. Taitao

G. de Penas

Chico

Commodoro Rivadavia

Golfo San Jorge

Deseado

C. Tres Puntas

ZAPIOLA RIDGE

ARGENTINE ABYSSAL PLAIN

FALKLAND ESCARPMENT

Archipiélago Reina Adelaida

Bahía Grande

Falkland Islands (U.K.)

Stanley

FALKLAND PLATEAU

MAURICE EWING BANK

FALKLAND RIDGE

GEORGIA BASIN

Str. of Magellan

Str. of Magellan

Punta Arenas

Tierra del Fuego

BURWOOD BANK

NORTH SCOTIA RIDGE

NORTHWEST GEORGIA RISE

South Georgia (U.K.)

SOUTH SANDWICH TRENCH

MORNINGTON ABYSSAL PLAIN

Cape Horn

YAGHAN BASIN

SCOTIA SEA

DRAKE PASSAGE

FRACTURE ZONE

ONA BASIN

© Copyright 1987 by HAMMOND INCORPORATED, Maplewood, N.J.

0 100 200 300 400 500 600 MILES

0 100 200 300 400 500 600 KILOMETERS

Southern South America | Plate 16

Plate 17 | **North America**

0   200   400   600   800   1000 MILES

0   200   400   600   800   1000 KILOMETERS

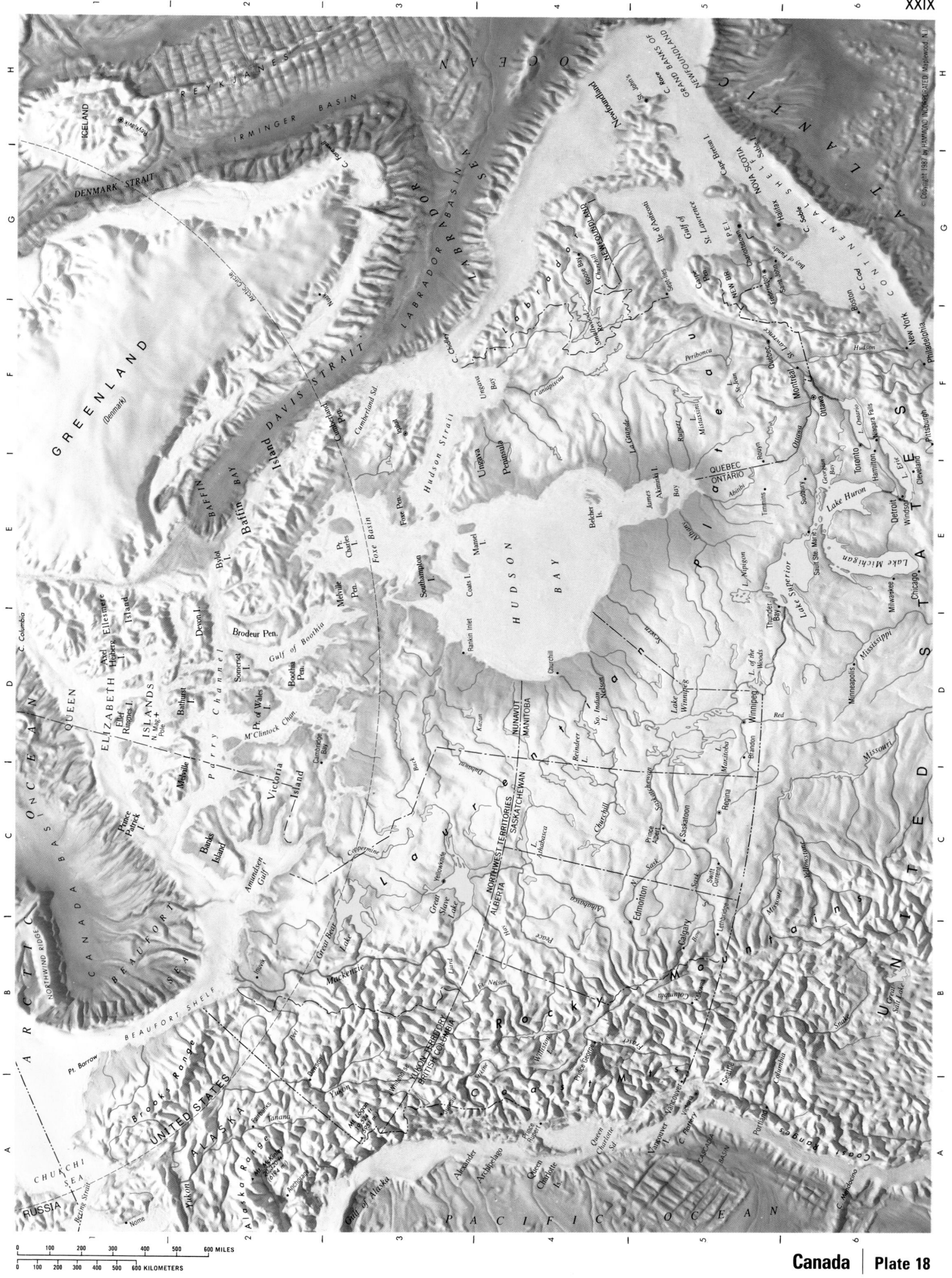

ICELAND

Reykjavik

REYKJANES

DENMARK STRAIT

IRMINGER BASIN

Arctic Circle

GREENLAND
(Denmark)

BAFFIN BAY

DAVIS STRAIT

BAFFIN Island

Cumberland Sd.

Cumberland Pen.

Iqaluit

LABRADOR SEA

ATLANTIC OCEAN

LABRADOR BASIN

NEWFOUNDLAND

GRAND BANKS OF NEWFOUNDLAND

C. Race
St. John's

Newfoundland

Gulf of St. Lawrence

Ile d'Anticosti

Cape Breton I.

NOVA SCOTIA

P.E.I.
Charlottetown
Halifax
Sable I.

NEW BR.
Fredericton

Bay of Fundy

Saint John

C. Cod
Boston

New York
Philadelphia
Pittsburgh

CONTINENTAL SHELF

C. Chidley

Goose Bay

Churchill

Schefferville

Smallwood Res.

Ungava Bay

Ungava Peninsula

Péribonca
Caniapiscau

La Grande
Rupert
Mistassini
Eastmain

St-Jean
Québec
Montréal
Ottawa

Roslyn

L. Ontario
Toronto
Niagara Falls
Hamilton

L. Erie
Cleveland

Hudson

QUÉBEC
ONTARIO

Abitibi

Timmins

Sudbury

Georgian Bay

Lake Huron

Sault Ste. Marie

Detroit
Windsor

UNITED STATES

HUDSON STRAIT

Foxe Pen.

Foxe Basin

Pt. Charles Pen.

Melville Pen.

Southampton I.

Mansel I.

Coats I.

Belcher Is.

James Bay

Akimiski I.

Albany

Moose

Lake Nipigon

Thunder Bay

L. Superior

Lake Michigan

Chicago
Milwaukee

Minneapolis

Mississippi

Missouri

Bylot I.

Brodeur Pen.

Devon I.

Ellesmere Island

C. Columbia

Axel Heiberg I.

Ellef Ringnes I.

Bathurst I.

N. Mag. Pole

QUEEN ELIZABETH ISLANDS

Prince Patrick I.

Melville I.

Banks Island

Somerset I.

Pr. of Wales I.

M'Clintock Chan.

Parry Channel

Gulf of Boothia

Boothia Pen.

Rankin Inlet

Churchill

NORTHWEST TERRITORIES

NUNAVUT

MANITOBA

SASKATCHEWAN

ALBERTA

Kazan

Thelon

Dubawnt

Back

Coppermine

Amundsen Gulf

Victoria Island

Cambridge Bay

Reindeer L.

So. Indian L.

L. Nelson

Nelson

Lake Winnipeg

L. of the Woods

Winnipeg

Brandon

L. Manitoba

L. Winnipegosis

Red

Regina

Saskatoon

Prince Albert

L. Athabasca

Athabasca

Sask.

Swift Current

Moose Jaw

Yellowknife

Great Slave Lake

Great Bear Lake

Inuvik

MacKenzie

Coppermine

ARCTIC OCEAN

CANADA BASIN

NORTHWIND RIDGE

BEAUFORT SEA

BEAUFORT SHELF

Pt. Barrow

CHUKCHI SEA

RUSSIA

Bering Strait

Nome

ALASKA

Fairbanks

Tanana

Mt. McKinley (6194)

Alaska Range

Yukon

Dawson

Whitehorse

YUKON TERRITORY
BRITISH COLUMBIA

Brooks Range

Peel

Liard

Ft. Nelson

Fort

Peace

Edmonton

Calgary

Lethbridge

Columbia

Kootenay

Bow

Yellowstone

Missouri

Snake

Columbia

Great Salt Lake

ROCKY Mountains

British Columbia

Prince George

Fraser

Williston L.

Prince Rupert

Queen Charlotte Is.

Queen Charlotte Sd.

Alexander Archipelago

Gulf of Alaska

PACIFIC OCEAN

Vancouver I.

Vancouver

Victoria

C. Flattery

Seattle

Portland

Coast Ranges

Juan de Fuca

LAURENTIAN

Coast Mts.

0   100   200   300   400   500   600 MILES

0   100   200   300   400   500   600 KILOMETERS

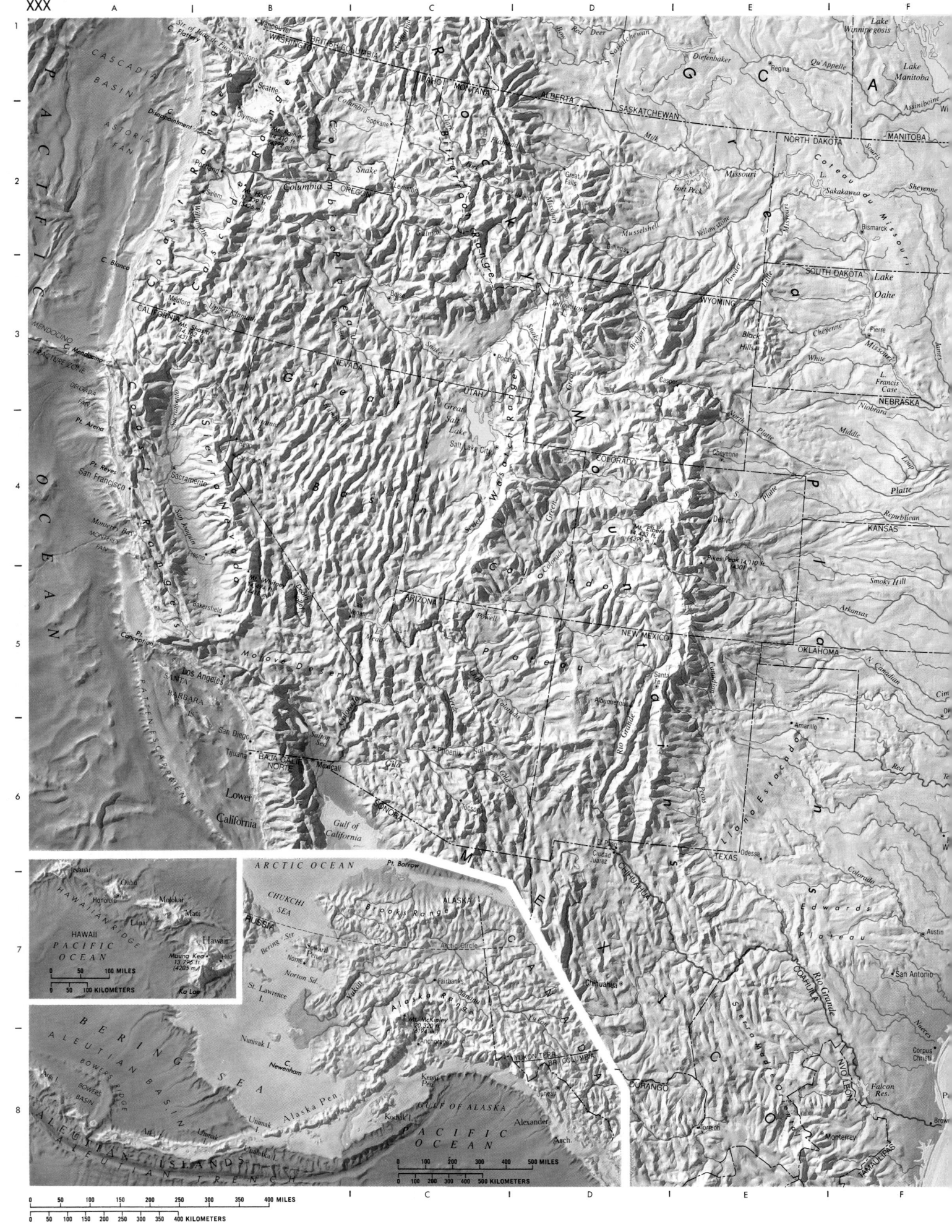

Copyright 1987 by HAMMOND INCORPORATED, Maplewood, N.J.

Plate 20 | **Middle America**

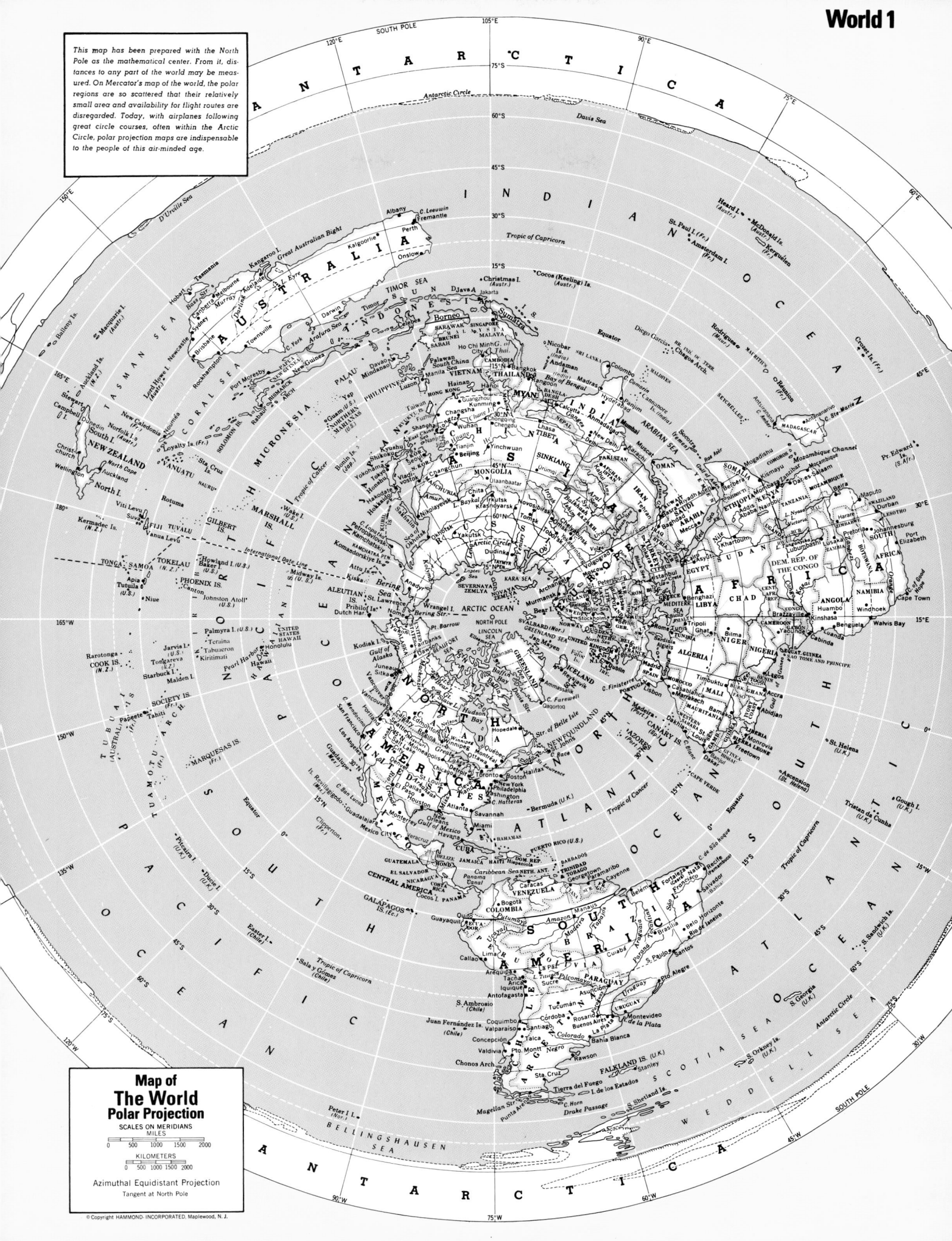

This map has been prepared with the North Pole as the mathematical center. From it, distances to any part of the world may be measured. On Mercator's map of the world, the polar regions are so scattered that their relatively small area and availability for flight routes are disregarded. Today, with airplanes following great circle courses, often within the Arctic Circle, polar projection maps are indispensable to the people of this air-minded age.

Map of The World Polar Projection

SCALES ON MERIDIANS

MILES

0   500   1000   1500   2000

KILOMETERS

0   500   1000 1500 2000

Azimuthal Equidistant Projection
Tangent at North Pole

The World

BRIESEMEISTER ELLIPTICAL
EQUAL-AREA PROJECTION

Capitals of Countries .... ⊛
Other Capitals .... ⊛
International Boundaries .... – – –

Scale 1:80,000,000

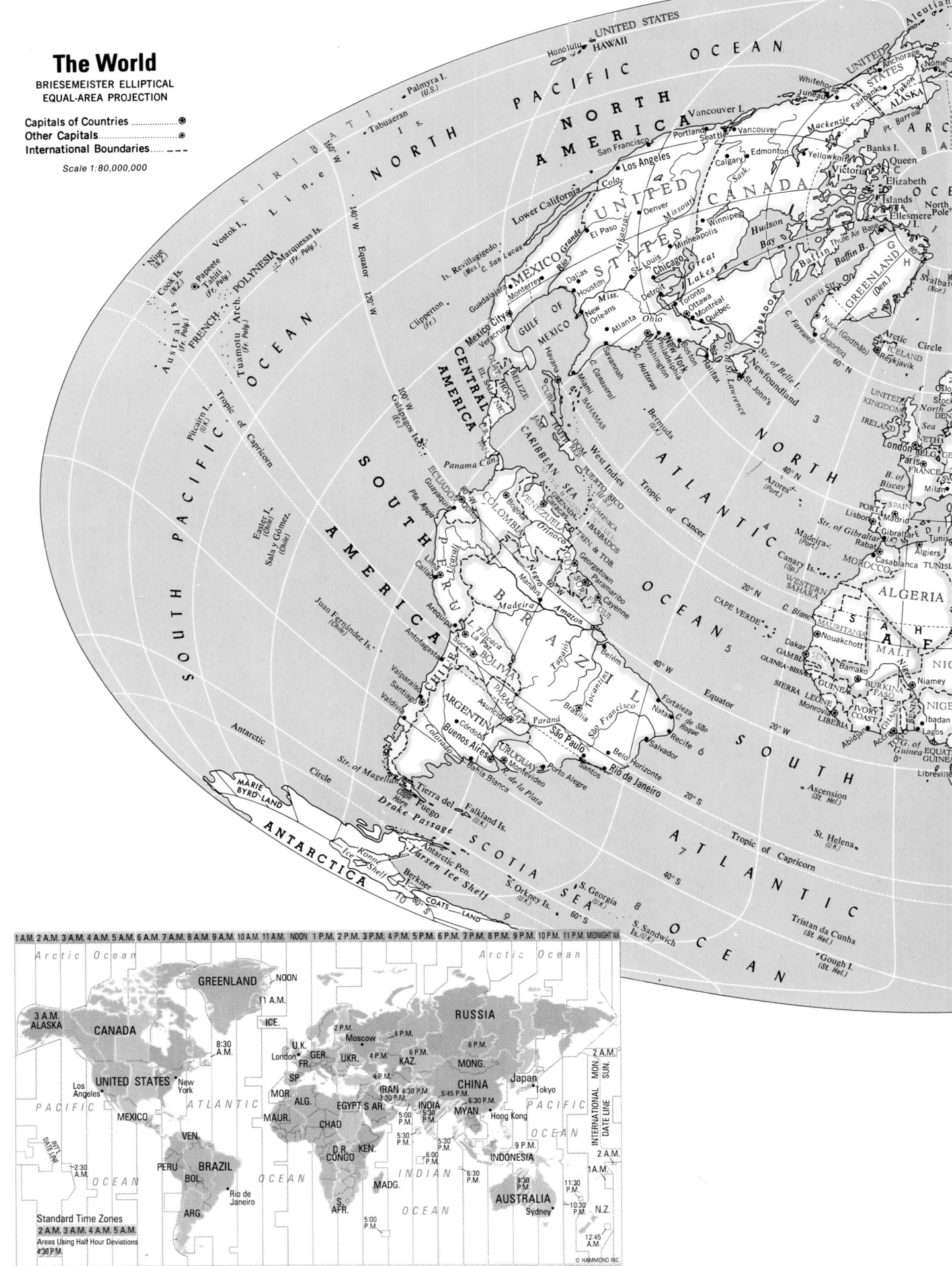

Standard Time Zones
2 A.M. 3 A.M. 4 A.M. 5 A.M.
Areas Using Half Hour Deviations
4:30 P.M.

© HAMMOND INC.

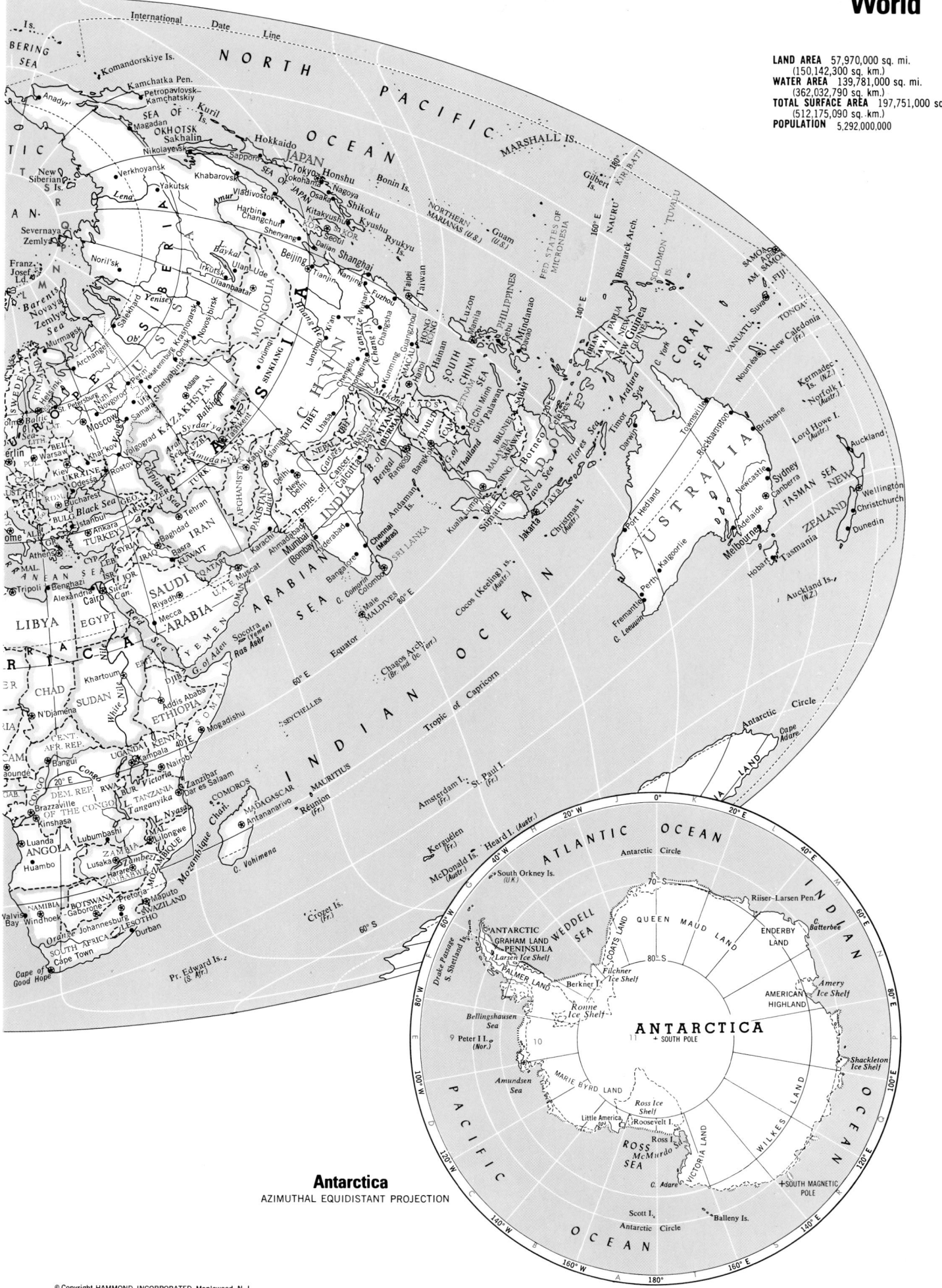

LAND AREA 57,970,000 sq. mi.
(150,142,300 sq. km.)
WATER AREA 139,781,000 sq. mi.
(362,032,790 sq. km.)
TOTAL SURFACE AREA 197,751,000 sq.mi.
(512,175,090 sq. km.)
POPULATION 5,292,000,000

Antarctica
AZIMUTHAL EQUIDISTANT PROJECTION

© Copyright HAMMOND INCORPORATED, Maplewood, N.J.

Arctic Ocean

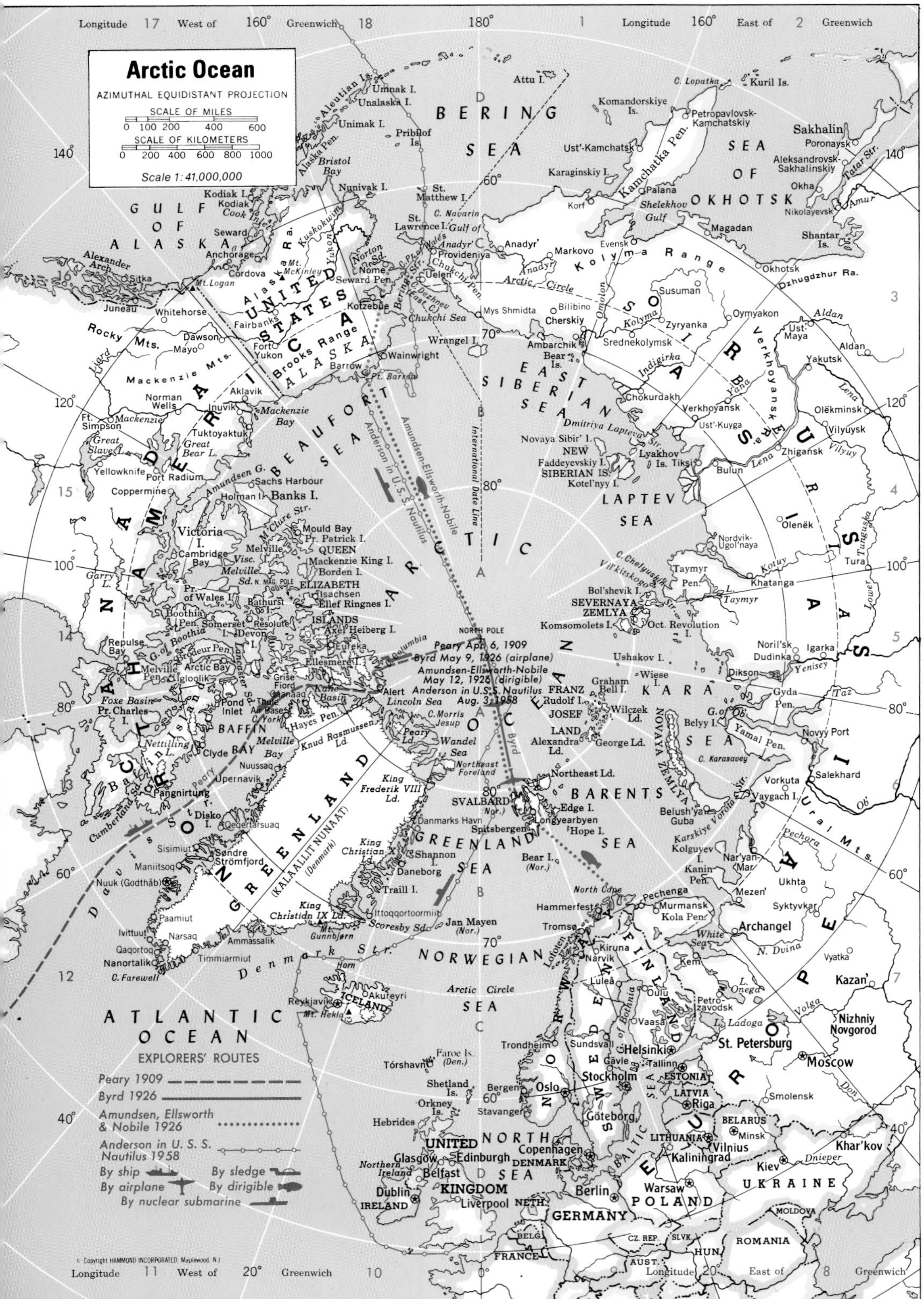

AZIMUTHAL EQUIDISTANT PROJECTION

SCALE OF MILES
0 100 200   400   600

SCALE OF KILOMETERS
0 200 400 600 800 1000

Scale 1: 41,000,000

EXPLORERS' ROUTES

Peary 1909
Byrd 1926
Amundsen, Ellsworth & Nobile 1926
Anderson in U.S.S. Nautilus 1958

By ship — By sledge
By airplane — By dirigible
By nuclear submarine

© Copyright HAMMOND INCORPORATED, Maplewood, N.J.

Antarctica
AZIMUTHAL EQUIDISTANT PROJECTION
SCALE OF MILES
0   200   400   600   800
KILOMETERS
0   200   400   600   800 1000

© Copyright HAMMOND INCORPORATED, Maplewood, N.J.

EXPLORERS' ROUTES

Palmer 1820
Amundsen 1910-12
Scott 1910-13
Byrd 1928-30
Fuchs 1957-58
By ship   By sledge   By airplane
By snow tractor

Amundsen Dec. 14, 1911
Scott Jan. 18, 1912
Byrd Nov. 29, 1929 (airplane)
Fuchs Jan. 19, 1958

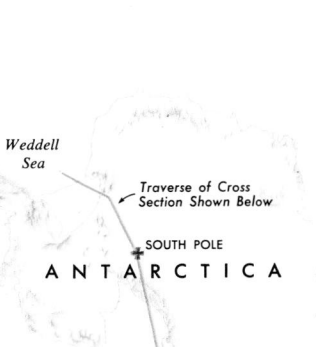

Weddell Sea

Traverse of Cross Section Shown Below

SOUTH POLE

ANTARCTICA

Ross Sea

Antarctic Cross Section: Weddell Sea to Ross Sea

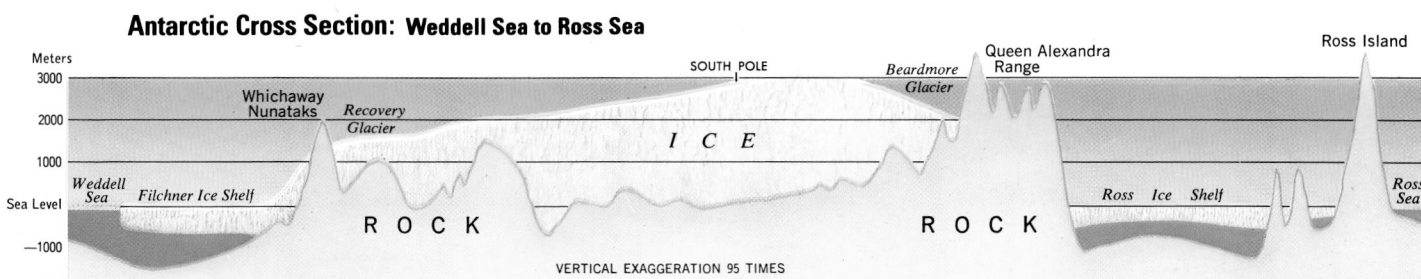

VERTICAL EXAGGERATION 95 TIMES

Information Based on American Geographical Society's "Antarctic Map Folio Series"

AREA 4,057,000 sq. mi.
(10,507,630 sq. km.)
POPULATION 689,000,000
LARGEST CITY Paris
HIGHEST POINT El'brus 18,510 ft.
(5,642 m.)
LOWEST POINT Caspian Sea -92 ft.
(-28 m.)

Population Distribution

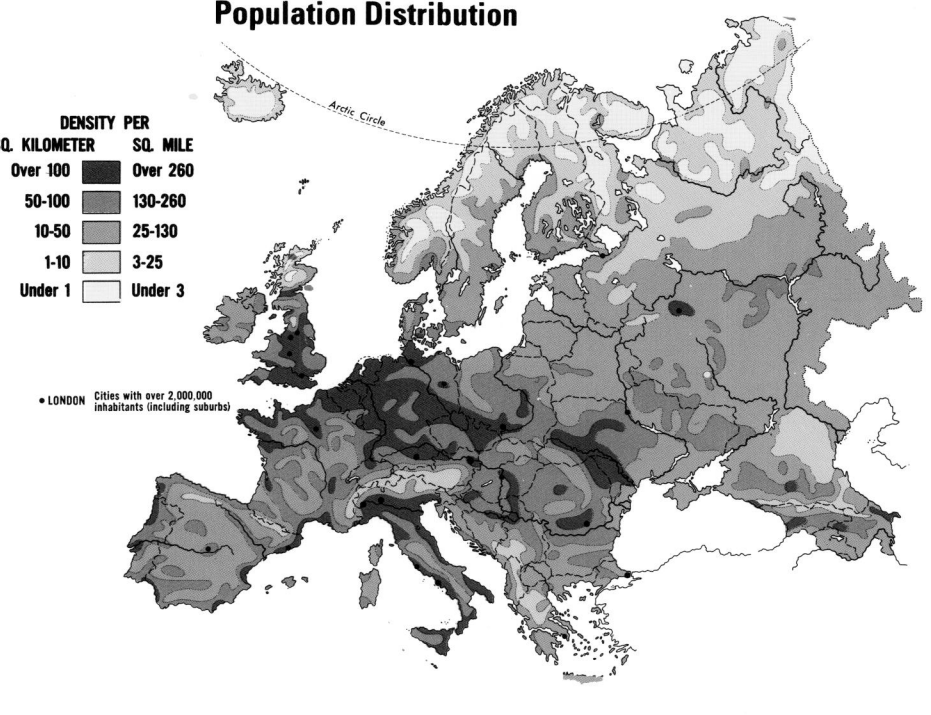

DENSITY PER

SQ. KILOMETER		SQ. MILE
Over 100		Over 260
50-100		130-260
10-50		25-130
1-10		3-25
Under 1		Under 3

• LONDON Cities with over 2,000,000 inhabitants (including suburbs)

Vegetation

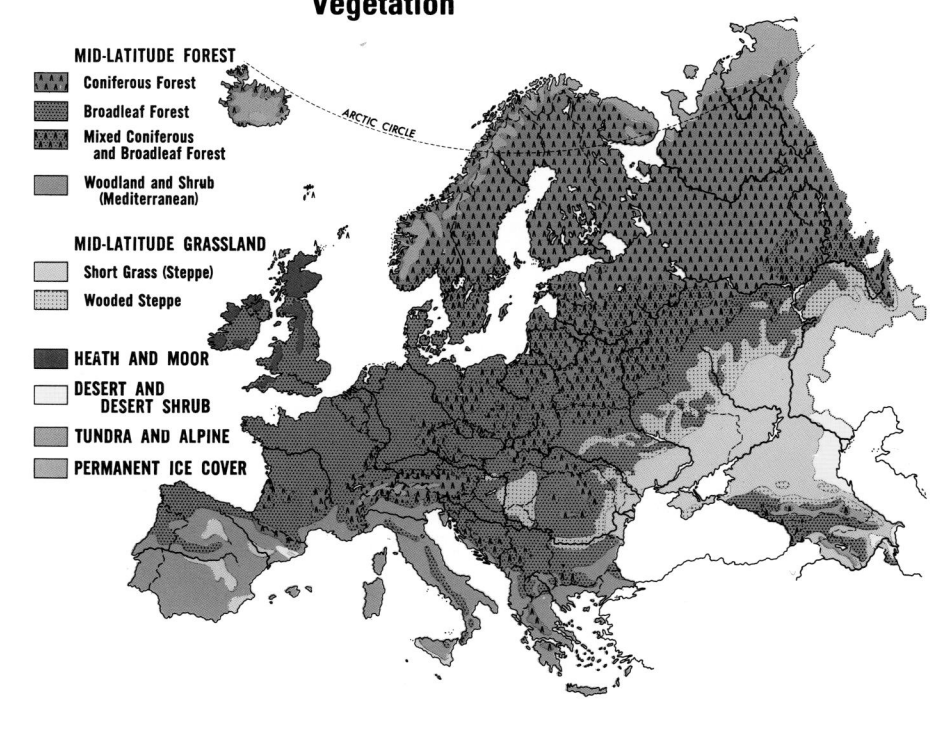

MID-LATITUDE FOREST
Coniferous Forest
Broadleaf Forest
Mixed Coniferous and Broadleaf Forest
Woodland and Shrub (Mediterranean)

MID-LATITUDE GRASSLAND
Short Grass (Steppe)
Wooded Steppe

HEATH AND MOOR
DESERT AND DESERT SHRUB
TUNDRA AND ALPINE
PERMANENT ICE COVER

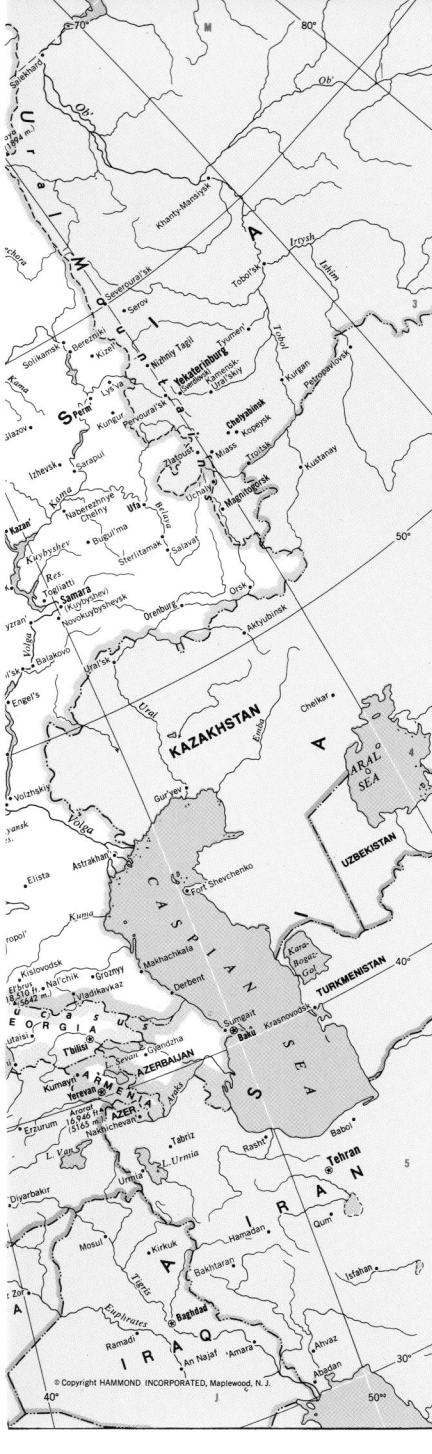

© Copyright HAMMOND INCORPORATED, Maplewood, N.J.

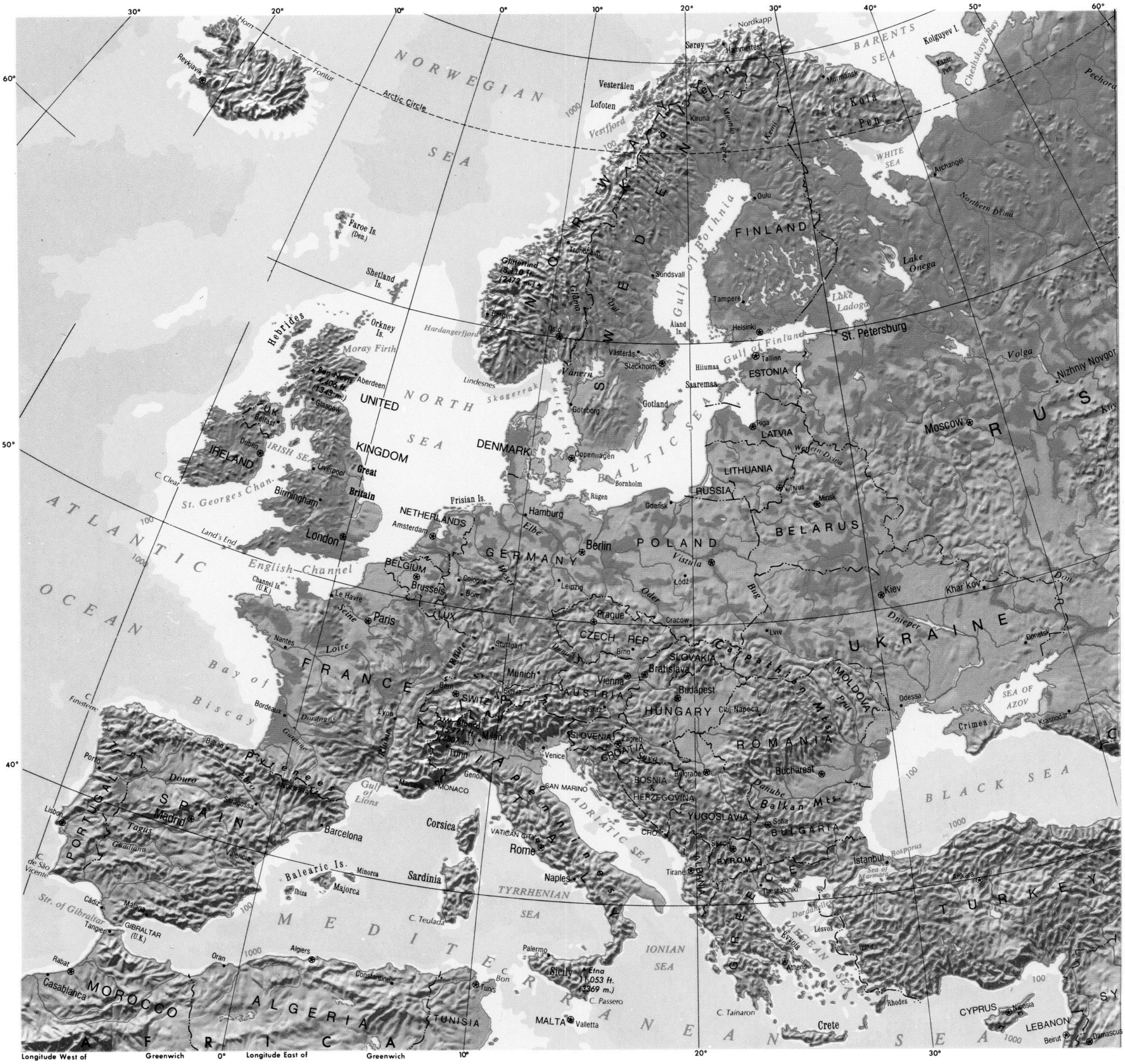

Vegetation/Relief

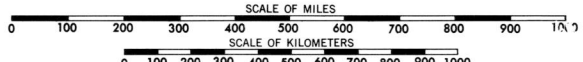

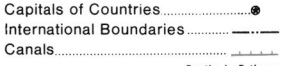

Capitals of Countries .... ⊛
International Boundaries .... —··—
Canals ....

Depths in Fathoms

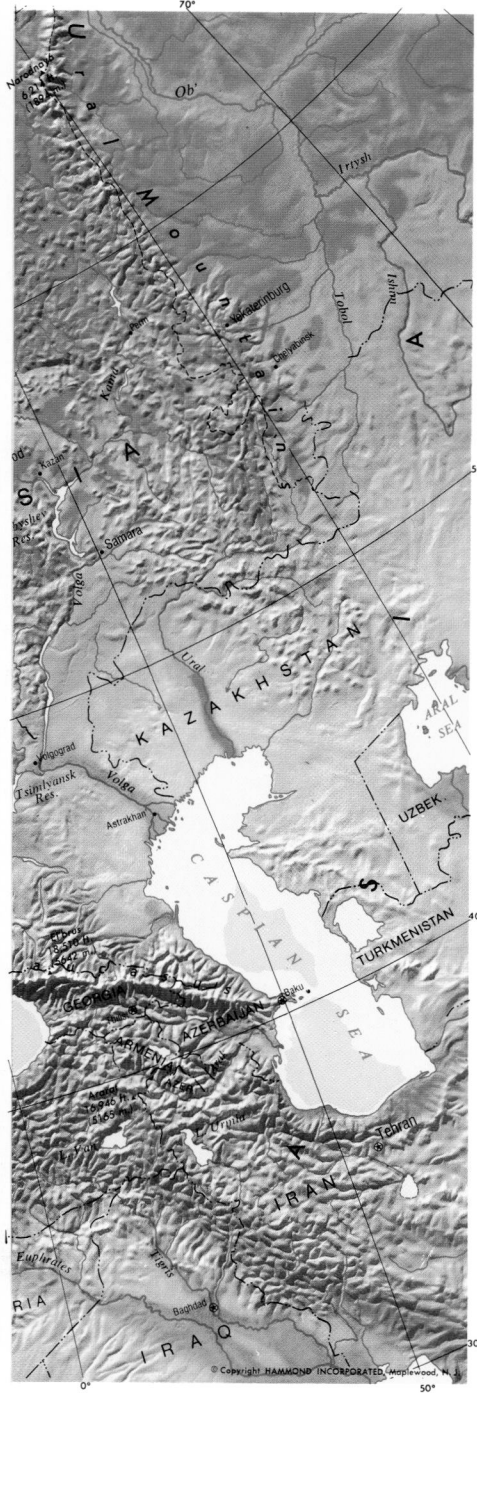

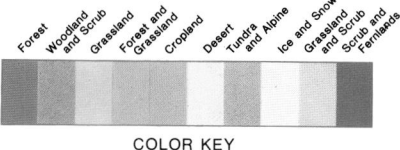

COLOR KEY

Forest | Woodland and Scrub | Grassland | Forest and Grassland | Cropland | Desert | Tundra and Alpine | Ice and Snow | Grassland and Scrub | Scrub and Fernlands

Rainfall

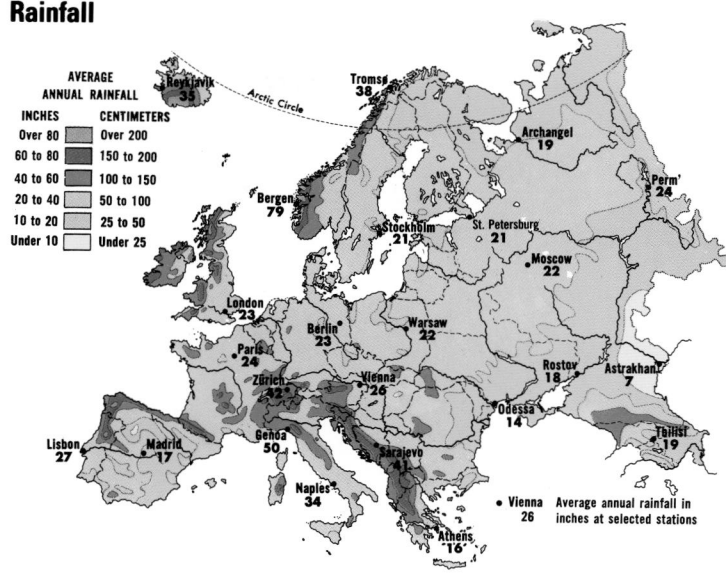

AVERAGE ANNUAL RAINFALL

INCHES	CENTIMETERS
Over 80	Over 200
60 to 80	150 to 200
40 to 60	100 to 150
20 to 40	50 to 100
10 to 20	25 to 50
Under 10	Under 25

• Vienna   Average annual rainfall in
  26   inches at selected stations

Reykjavík 35 · Tromsø 38 · Archangel 19 · Perm' 24 · Bergen 79 · Stockholm 21 · St. Petersburg 21 · Moscow 22 · London 23 · Berlin 23 · Warsaw 22 · Rostov 18 · Astrakhan 7 · Paris 24 · Zürich 42 · Vienna 26 · Odessa 14 · Tbilisi 19 · Lisbon 27 · Madrid 17 · Genoa 50 · Sarajevo 41 · Naples 34 · Athens 16

Average January Temperature

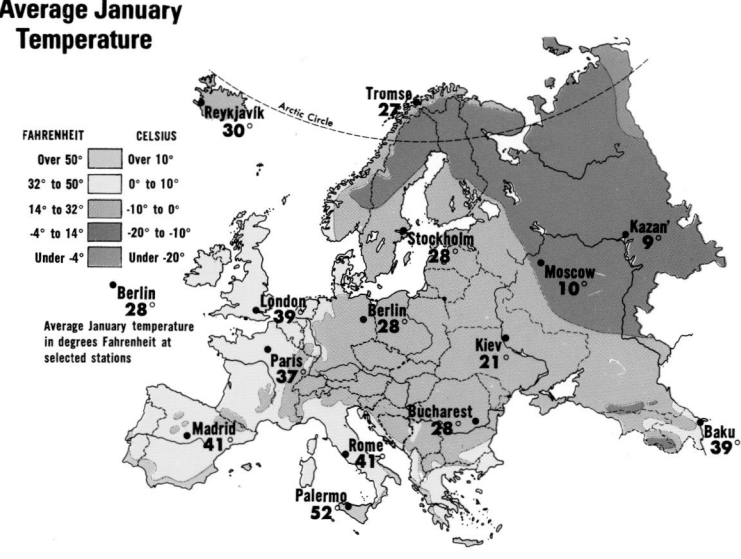

FAHRENHEIT	CELSIUS
Over 50°	Over 10°
32° to 50°	0° to 10°
14° to 32°	-10° to 0°
-4° to 14°	-20° to -10°
Under -4°	Under -20°

• Berlin   Average January temperature
  28   in degrees Fahrenheit at
           selected stations

Reykjavík 30° · Tromsø 27° · Stockholm 28° · Kazan' 9° · Moscow 10° · London 39° · Berlin 28° · Kiev 21° · Paris 37° · Madrid 41° · Rome 41° · Bucharest 28° · Baku 39° · Palermo 52°

Average July Temperature

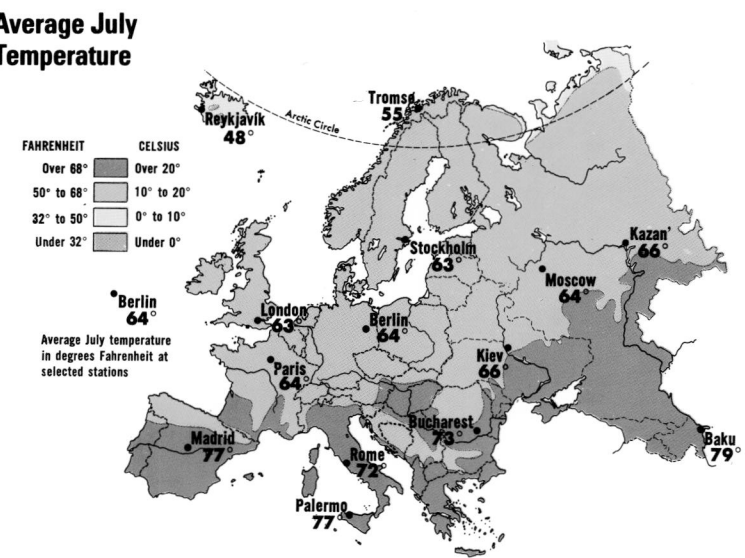

FAHRENHEIT	CELSIUS
Over 68°	Over 20°
50° to 68°	10° to 20°
32° to 50°	0° to 10°
Under 32°	Under 0°

• Berlin   Average July temperature
  64   in degrees Fahrenheit at
           selected stations

Reykjavík 48° · Tromsø 55° · Stockholm 63° · Kazan' 66° · Moscow 64° · London 63° · Berlin 64° · Kiev 66° · Paris 64° · Madrid 77° · Rome 72° · Bucharest 73° · Baku 79° · Palermo 77°

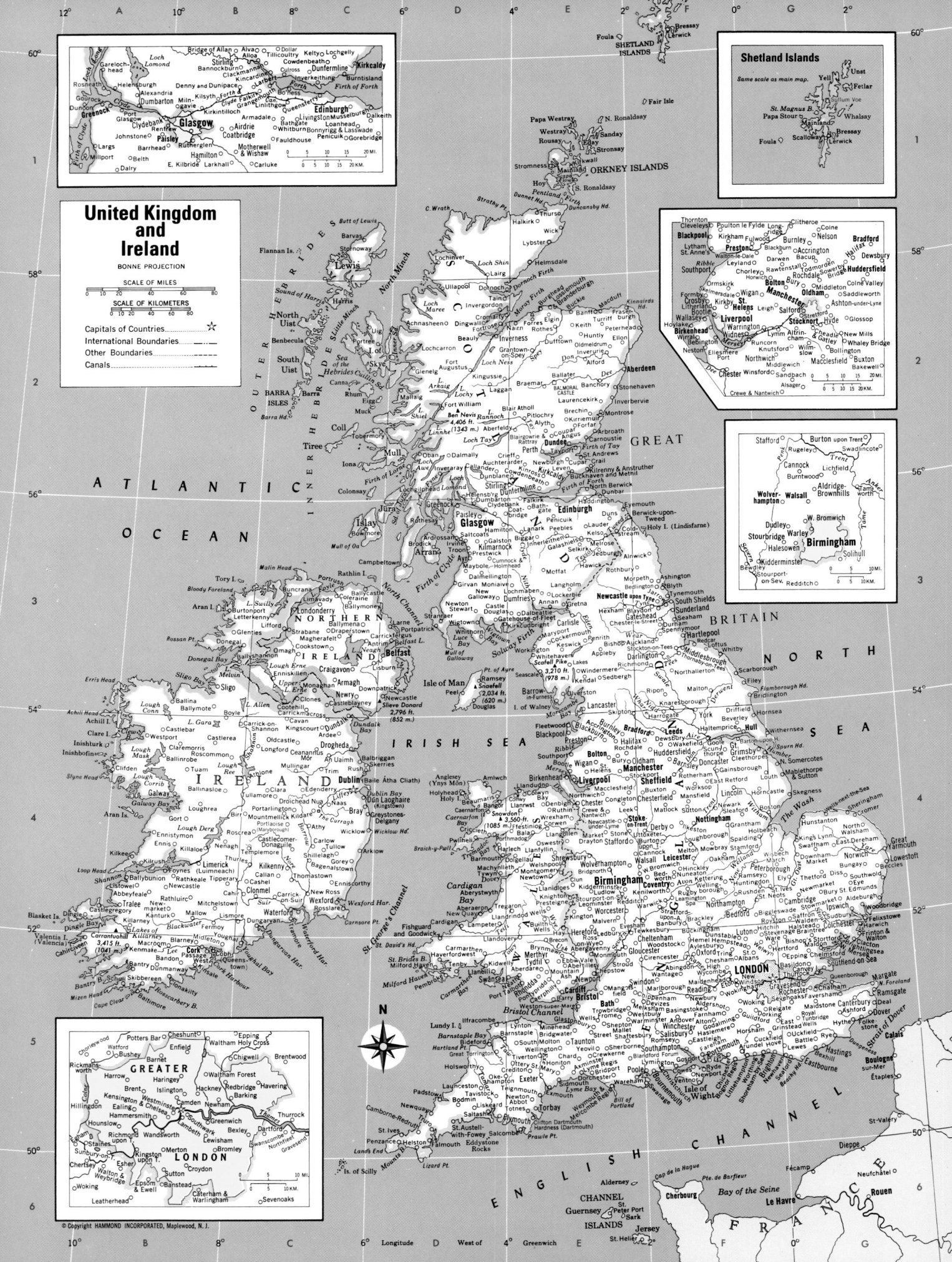

United Kingdom and Ireland

BONNE PROJECTION

SCALE OF MILES

SCALE OF KILOMETERS

Capitals of Countries..............☆
International Boundaries.........
Other Boundaries.................
Canals..................................

Shetland Islands

Same scale as main map.

© Copyright HAMMOND INCORPORATED, Maplewood, N.J.

UNITED KINGDOM

AREA 94,399 sq. mi. (244,493 sq. km.)
POPULATION 57,236,000
CAPITAL London
LARGEST CITY London
HIGHEST POINT Ben Nevis 4,406 ft. (1,343 m.)
MONETARY UNIT pound sterling
MAJOR LANGUAGES English, Gaelic, Welsh
MAJOR RELIGIONS Protestantism, Roman Catholicism

IRELAND

AREA 27,136 sq. mi. (70,282 sq. km.)
POPULATION 3,540,643
CAPITAL Dublin
LARGEST CITY Dublin
HIGHEST POINT Carrantuohill 3,415 ft. (1,041 m.)
MONETARY UNIT Irish pound
MAJOR LANGUAGES English, Gaelic (Irish)
MAJOR RELIGION Roman Catholicism

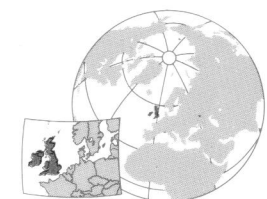

UNITED KINGDOM

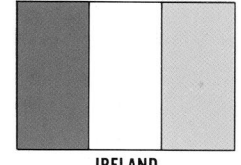
IRELAND

ENGLAND
(map on page 13)

COUNTIES

Avon 900,947	E6
Bedfordshire 502,164	G5
Berkshire 670,859	F6
Buckinghamshire 562,221	G6
Cambridgeshire 569,893	G5
Cheshire 921,623	E4
Cleveland 565,845	F3
Cornwall 418,631	C7
Cumbria 471,696	D3
Derbyshire 901,831	F5
Devon 930,112	D7
Dorset 578,993	E7
Durham 598,881	F3
East Sussex 641,016	H7
Essex 1,416,890	H6
Gloucestershire 493,166	E6
Hampshire 1,442,598	F6
Hereford and Worcester 624,393	E5
Hertfordshire 950,760	G6
Humberside 843,282	G4
Isle of Wight 114,879	F7
Isles of Scilly	A7
Kent 1,448,393	H6
Lancashire 1,362,801	E4
Leicestershire 835,647	F5
Lincolnshire 542,944	G4
London 6,608,598	H8
Manchester 2,575,407	H2

Merseyside 1,503,120	G2
Norfolk 685,232	H5
North Yorkshire 653,456	F3
Northamptonshire 524,967	G5
Northumberland 295,451	E2
Nottinghamshire 976,748	F4
Oxfordshire 507,230	F6
Shropshire 370,355	E5
Somerset 417,457	E6
South Yorkshire 1,292,020	F4
Staffordshire 1,005,641	E5
Suffolk 590,133	H5
Surrey 992,489	G6
Tyne and Wear 1,135,492	H3
Warwickshire 469,801	F5
West Midlands 2,628,419	F5
West Sussex 650,124	G7
West Yorkshire 2,021,707	J1
Wiltshire 512,635	E6
Yorkshire, North 653,456	F3
Yorkshire, South 1,292,029	F4
Yorkshire, West 2,021,707	J1

CITIES and TOWNS

Abingdon 29,130	F6
Accrington 36,459	H1
Adwickle Street 10,293	K2
Aldershot 53,665	G8
Aldridge 17,549	E5
Alfreton 21,284	F4
Alsager 12,944	E4
Alton 14,163	G6
Altrincham 39,528	H2

Amersham® 21,326	G7
Andover 30,632	F6
Arnold 37,721	F4
Ashford 45,198	H6
Ashington 27,786	F2
Ashton-under-Lyne 43,605	H2
Aylesbury 51,999	G7
Aylesford 21,017	J8
Bacup 14,082	H1
Banbury 37,463	F5
Banstead 35,360	H8
Barking 149,132	H8
Barnet 289,277	H7
Barnoldswick 10,125	H1
Barnsley 76,783	J2
Barnstaple 24,490	D6
Barrow-in-Furness 50,174	D3
Basildon 94,800	J8
Basingstoke 73,027	F6
Bath 84,283	E6
Batley 45,582	J1
Beaconsfield 13,397	G8
Bebington 62,618	G2
Beccles 10,677	J5
Bedford 75,632	G5
Bedlington 15,074	F2
Bedworth 29,192	F5
Beeston and Stapleford 64,785	F5
Benfleet 50,783	J8
Bentley with Arksey 34,273	F4
Berkhamsted 16,874	G7
Berwick-upon-Tweed 12,772	F2
Beverley 19,368	G4

Bexhill 34,625	H7
Bexley 213,215	H8
Bicester 15,946	F6
Bideford 13,826	C6
Biddulph 16,697	H2
Biggleswade 10,905	G5
Birkenhead 99,075	G2
Birmingham 1,013,995	F5
Bishop Auckland 23,560	E3
Bishop's Stortford 22,535	H6
Blackburn 109,564	H1
Blackpool 146,297	G1
Blaydon 16,719	H3
Blyth 35,101	F2
Bodmin 11,992	C7
Bognor Regis 50,323	G7
Boldon 11,639	J3
Bolsover 11,497	J2
Bolton 143,960	H2
Bootle 70,860	G5
Boston 33,908	G5
Bournemouth 142,829	F7
Bracknell 52,257	G8
Bradford 293,336	J1
Braintree 30,975	H6
Brent 251,238	H8
Brentwood 51,212	J8
Bridgnorth 10,332	E5
Bridgwater 30,782	E6
Bridlington 28,426	G3
Bridport 10,615	E7
Brighouse 32,597	J1
Brighton 134,581	G7
Bristol 413,861	E6

Broadstairs 21,551	J6
Bromley 280,525	H8
Bromsgrove 24,576	E5
Brownhills 18,200	E5
Buckingham 6,439	G6
Burgess Hill 23,577	G7
Burnham-on-Sea 17,022	D6
Burnley 76,365	H1
Burntwood 28,938	F5
Burton upon Trent 59,040	F5
Bury 61,785	H2
Bury Saint Edmunds 30,563	H5
Bushey 15,759	H7
Buxton 19,502	J2
Calne 10,235	F6
Camborne-Redruth 34,262	B7
Cambridge 87,111	G5
Camden 161,098	H8
Cannock 54,503	E5
Canterbury 34,546	H6
Canvey Island 35,243	J8
Carlisle 72,206	D3
Carlton 46,053	F5
Carterton 10,876	F6
Caterham and Warlingham 30,331	H8
Charlton Kings 10,786	F6
Chatham 65,835	J8
Cheadle 10,470	E5
Cheadle and Gatley 59,478	H2
Chelmsford 91,109	J7
Cheltenham 87,188	E6
Chertsey 10,195	G8
Chesham 20,883	G7
Cheshunt 49,616	H7
Chester 80,154	G2
Chester-le-Street 34,776	J3
Chesterfield 73,352	J2
Chichester 26,050	G7
Chippenham 21,325	E6
Chorley 33,465	G2
Christchurch 32,854	F7
Cirencester 13,491	E6
Clacton 39,618	J6
Clay Cross 22,635	J2
Cleethorpes 33,238	H4
Clevedon 17,875	D6
Clitheroe 13,671	H1
Coalville 28,831	F5
Colchester 87,476	H6
Colne 19,094	H1
Congleton 23,482	H2
Consett 22,409	H3
Corby 48,704	G5
Corsham 11,259	E6
Coventry 318,718	F5
Cowes 16,134	F7
Cranleigh 10,334	G6
Crawley 80,113	G6
Crewe 59,097	E4
Crosby 54,103	G2
Crowborough 17,008	H6
Croydon 298,794	H8
Darlington 85,519	F3
Dartford 62,032	J8
Darton 13,743	J2
Darwen 30,883	H1
Daventry 16,096	F5
Deal 26,311	J6
Dearne 13,391	K2
Denton 37,784	H2
Derby 218,026	F5
Devizes 12,430	F6
Dewsbury 49,612	J1
Didcot® 15,147	F6
Doncaster 74,727	F4
Dorchester 13,734	E7
Dorking 14,602	G8
Dover 33,461	J6
Droitwich 18,025	E5
Dronfield 22,641	J2
Dudley 186,513	E5
Dunstable 48,436	G6
Durham 38,105	J3
Ealing 278,677	H8
East Dereham 11,798	H5
East Grinstead 23,867	G6
East Retford 19,308	G4
Eastbourne 86,715	H7
Eastleigh 58,585	F7
Egham 21,810	G8
Ellesmere Port 65,829	G2

Enfield 257,154	H7
Epping 10,148	H7
Epsom and Ewell 65,830	G8
Esher 46,688	H8
Eston® 37,694	F3
Eton	G8
Evesham 15,069	E5
Exeter 88,235	D7
Exmouth 28,037	D7
Falmouth 17,810	B7
Fareham 55,563	F7
Farnborough 48,063	G8
Farnham 34,541	G8
Farnworth 25,591	H2
Faversham 15,914	H6
Felixstowe 24,207	J6
Felling 36,377	J3
Fleet 27,406	G8
Fleetwood 27,899	G1
Folkestone 42,949	J6
Formby 26,852	G2
Frinton and Walton 12,689	J6
Frome 16,371	E6
Gainsborough 20,326	G4
Gateshead 91,421	J3
Gillingham 92,531	J8
Glastonbury 6,751	E6
Glossop 29,923	J2
Gloucester 106,526	E6
Godalming 18,758	G8
Golborne 20,633	G2
Goole 19,394	G4
Gosport 69,664	F7
Grantham 30,700	G5
Gravesend 53,450	J8
Great Grimsby 91,532	H4
Great Harwood 10,968	H1
Great Malvern (Malvern) 30,153	E5
Great Yarmouth 54,777	J5
Greenwich 211,013	H8
Guildford 61,509	G8
Guisborough 19,242	F3
Hackney 179,529	H8

Hailsham 16,367	H7
Hale 16,362	H2
Halesowen 57,533	E5
Halifax 76,675	J1
Hammersmith 144,616	H8
Haringey 202,650	H8
Harlow 79,150	H7
Harrogate 63,637	J1
Harrow 195,292	G8
Hartlepool 91,749	F3
Harwich 17,245	J6
Haslemere 10,544	G6
Haslingden 14,347	H1
Hastings 74,979	H7
Hatfield 33,174	H7
Havant 50,098	G7
Haverhill 16,970	H5
Havering 238,335	J8
Haxby 11,415	F3
Hazel Grove and Bramhall 40,819	H2
Heanor 21,863	F4
Hebburn 20,098	J3
Hemel Hempstead 80,110	G7
Henley-on-Thames 10,910	G8
Hereford 48,277	E5
Hertford 21,350	H7
Hetton 14,529	J3
Heywood 29,639	H2
High Wycombe 69,575	G8
Hillingdon 226,659	G8
Hinckley 35,510	F5
Hitchin 33,480	G6
Hoddesdon 37,960	H7
Holmfirth 21,138	J2
Horley 17,700	H8
Horsham 38,356	G6
Horwich 16,758	G2
Houghton-le-Spring 35,337	J3
Hounslow 198,938	G8
Hove 65,587	G7
Hoylake 24,815	G2
Hoyland Nether 15,845	J2
Hucknall 27,463	F4

(continued on following page)

ENGLAND

AREA 50,516 sq. mi. (130,836 sq. km.)
POPULATION 46,220,955
CAPITAL London
LARGEST CITY London
HIGHEST POINT Scafell Pike 3,210 ft. (978 m.)

WALES

AREA 8,017 sq. mi. (20,764 sq. km.)
POPULATION 2,749,640
CAPITAL Cardiff
LARGEST CITY Cardiff
HIGHEST POINT Snowdon 3,560 ft. (1,085 m.)

SCOTLAND

AREA 30,414 sq. mi. (78,772 sq. km.)
POPULATION 5,130,735
CAPITAL Edinburgh
LARGEST CITY Glasgow
HIGHEST POINT Ben Nevis 4,406 ft. (1,343 m.)

NORTHERN IRELAND

AREA 5,452 sq. mi. (14,121 sq. km.)
POPULATION 1,543,000
CAPITAL Belfast
LARGEST CITY Belfast
HIGHEST POINT Slieve Donard 2,796 ft. (852 m.)

Topography

0   75   150 MI.

0   75   150 KM.

5,000 m. 16,404 ft. / 2,000 m. 6,562 ft. / 1,000 m. 3,281 ft. / 500 m. 1,640 ft. / 200 m. 656 ft. / 100 m. 328 ft. / Sea Level / Below

12 United Kingdom and Ireland

(continued)

Huddersfield 147,825 J2
Hugh Town⊙ A8
Hull 322,144 G4
Huntingdon 14,395 G5
Huyton-with-Roby 62,011 G2
Hyde 30,461 H2
Hythe 13,118 H6
Ilkeston 34,683 F5
Immingham⊙ 11,480 G4
Ipswich 129,661 J5
Islington 157,522 H8
Jarrow 31,345 J3
Kempston 15,454 G5
Kendal 23,710 E3
Kenilworth 18,782 F5
Kensington and Chelsea 125,892 G8
Kettering 44,758 G5
Kidderminster 50,385 E5
Kidsgrove 27,999 E4
King's Lynn 37,323 H5
Kingston upon Thames 130,829 H8
Kingswood 54,736 E6
Kirkby 52,825 G2
Knaresborough 12,910 F4
Knutsford 13,628 H2
Lambeth 244,143 H8
Lancaster 43,902 E3
Leamington Spa 56,552 F5
Leatherhead 42,399 G8
Leeds 451,841 J1
Leek 18,495 H2
Leicester 324,394 F5
Leigh 42,627 H2
Letchworth 31,146 G6
Lewes 14,499 H7
Lewisham 230,488 H8
Leyland 36,694 G1
Lichfield 25,408 F5
Lincoln 79,980 G4
Litherland 21,989 G2
Littlehampton 46,028 G7
Liverpool 538,809 G2
London (cap.) 7,566,620 H8
Long Eaton 42,285 F5
Longbenton 36,780 J3
Loughborough 44,895 F5
Louth 13,019 H4
Lowestoft 59,430 J5
Luton 163,209 G6
Lymington 11,614 F7
Lymm 10,036 H2
Lytham Saint Anne's 39,559 G1
Macclesfield 47,525 H2
Maidenhead 59,809 G8
Maidstone 86,067 J8
Maldon 14,638 H6
Malvern 30,153 E5
Manchester 448,604 H2
Mangotsfield 28,664 E6
Mansfield 71,325 K2
Mansfield Woodhouse 17,564 F4
March 14,155 H5
Margate 53,137 J6
Market Harborough 15,852 G5
Marlow 18,584 G8
Matlock 13,706 J2
Melksham 13,248 E6
Melton Mowbray 23,379 G5
Merton 165,102 H8
Middlesbrough 158,516 F3
Middleton 51,373 H2
Milton Keynes 93,305 G5
Morpeth 14,301 F2
Nantwich 11,867 E4
Nelson 30,449 H1
Neston 14,902 G2
New Romney 6,559 J7
Newark 33,143 G4
Newbury 31,488 F6
Newcastle upon Tyne 199,064 H3
Newcastle-under-Lyme 73,208 E4
Newham 209,128 H8
Newhaven 10,697 H7
Newmarket 15,861 H5
Newport, Isle of Wight 19,758 F7
Newport, Shropshire 10,339 E5
Newport Pagnell 10,733 G5
Newquay 13,905 B7
Newton Abbot 20,567 D7
Newton-le-Willows 19,466 H2
Northallerton 13,566 F3
Northampton 154,172 F5
Northfleet 21,400 J8
Northwich 32,664 H2
Norton-Radstock 17,668 E6
Norwich 169,814 J5
Nottingham 273,300 F5
Nuneaton 60,337 F5
Oadby 18,331 F5
Oldham 107,095 H2
Ormskirk 22,308 G2
Oswaldtwistle 11,188 H1
Oswestry 13,200 E5
Oxford 113,847 F6
Padiham 13,856 H1
Penrith 12,086 E3
Penzance 18,501 B7
Peterborough 113,404 G5
Peterlee⊙ 31,405 J3
Petersfield 10,078 F6
Plymouth 238,583 C7
Ponteland 10,215 H3
Poole 122,815 E7
Portishead 13,684 E6
Portslade 17,831 G7
Portsmouth 174,218 F7
Potters Bar 22,610 H7
Poulton-le-Fylde 18,477 G1
Preston 166,675 G1
Prestwich 31,854 H2
Prudhoe 11,140 H3
Radcliffe 27,664 H2
Ramsbottom 16,334 H2
Ramsgate 36,678 J6
Rawtenstall 21,247 H1
Rayleigh 28,574 J8
Reading 194,727 G8
Redbridge 226,977 H8
Redcar⊙ 35,373 F3
Redditch 61,639 E5
Reigate 48,241 H8
Richmond upon Thames 157,304 H8
Rickmansworth 15,960 G8
Ringwood 10,941 F7
Ripley 17,548 F4
Ripon 13,036 F3
Rochdale 97,292 H2
Rochester 23,840 J8
Romney (New Romney) 6,559 J7
Romsey 14,818 F6
Rotherham 122,374 K2
Royal Leamington Spa 56,552 F5
Royal Tunbridge Wells 57,699 H6
Royston 12,904 G5
Rugby 59,039 F5
Rugeley 23,751 E5
Runcorn 63,995 G2
Rushden 22,394 G5
Ryde 19,384 F7
Ryton 15,138 H3
Saffron Walden 11,879 H5
Saint Albans 76,709 H7
Saint Austell 20,267 C7
Saint Helens 114,397 G2
Saint Ives, Cambridgeshire 13,431 G5
Saint Ives, Cornwall 9,439 B7
Saint Neots 12,468 G5
Sale 57,872 H2
Salford 96,525 H2
Salisbury 36,890 F6
Saltash 12,486 C7
Sandbach 13,734 H2
Sandhurst 13,539 G8
Sandown-Shanklin 15,252 F7
Scarborough 36,665 G3
Scunthorpe 79,043 G4
Seaford 16,367 H7
Seaham 21,807 J3
Selby 12,224 F4
Sevenoaks 24,493 J8
Sheffield 470,685 J2
Shepshed 10,479 F5
Shildon 11,583 F3
Shoreham 20,562 G7
Shrewsbury 57,731 E5
Sidmouth 10,808 D7
Sittingbourne 35,893 H6
Skegness 12,645 H4
Skelmersdale 42,611 G2
Skipton 13,009 H1
Slough 106,341 G8
Solihull 93,940 F5
South Shields 86,488 J3
Southampton 211,321 F7
Southend-on-Sea 155,720 H6
Southport 88,596 G1
Southwark 209,735 H8
Southwick 11,364 G7
Sowerby Bridge 11,280 H1
Spalding 18,182 G5
Spennymoor 18,563 F3
Stafford 60,915 E5
Staines 51,949 G8
Stamford 16,127 G5
Standish 11,504 G2
Stanley 20,058 H3
Staveley 24,457 K2
Stevenage 74,757 G6
Stockport 135,489 H2
Stocksbridge 13,394 J2
Stockton-on-Tees 86,699 F3
Stoke-on-Trent 272,446 E4
Stone 12,119 E5
Stourbridge 55,136 E5
Stourport-on-Severn 17,880 E5
Stowmarket 10,913 J5
Stratford-upon-Avon 20,941 F5
Stretford 47,522 H2
Stroud 37,791 E6
Sudbury 17,723 H5
Sunbury 28,240 G8
Sunderland 195,064 J3
Sutton 165,323 H8
Sutton in Ashfield 39,536 K2
Swadlincote 33,667 F5
Swindon 127,348 F6
Tadley 13,668 F6
Tamworth 63,260 F5
Taunton 47,793 D6
Teignmouth 11,995 D7
Telford⊙ 28,645 E5
Tewkesbury 9,454 E6
Thatcham 14,940 F6
Thetford 19,529 H5
Thornaby⊙ 26,319 F3
Thornbury 11,948 E6
Thorne⊙ 16,662 F4
Thornton Cleveleys 26,697 G1
Tiverton 14,745 D7
Todmorden 11,936 H1
Tonbridge 34,407 H8
Torbay 93,995 D7
Tower Hamlets 139,996 H8
Tring 10,610 G7
Trowbridge 27,299 E6
Truro 17,852 B7
Tynemouth 17,877 J3
Uckfield 10,749 H7
Ulverston 11,976 D3
Urmston 43,706 H2
Uttoxeter 10,008 E5
Wakefield 74,764 J2
Wallasey 62,465 G2
Wallsend 44,542 J3
Walsall 177,923 E5
Waltham Forest 214,595 H8
Waltham Holy Cross 16,498 H7
Walton and Weybridge 50,031 G8
Wandsworth 252,240 H8
Ware 15,344 H7
Warminster 14,826 E6
Warrington 81,366 G2
Warsop 10,294 F4
Warwick 21,701 F5
Washington 48,856 J3
Waterloo 57,296 G7
Watford 109,503 H7
Wellingborough 38,598 G5
Wellington 8,980 D7
Welwyn 40,665 H7
West Bridgford 27,463 F5
West Bromwich 153,725 F5
Westminster 163,892 H8
Weston-super-Mare 60,821 D6
Weymouth 38,384 E7
Whickham 17,882 J3
Whitby 12,982 G3
Whitehaven 27,512 D3
Whitley Bay 36,040 J3
Widnes 55,973 G2
Wigan 88,725 G2
Wigston 32,373 F5
Wilmslow 28,827 H2
Wilton 4,002 F6
Wimborne Minster 14,193 E7
Winchester 34,127 F6
Windermere 6,835 E3
Windsor 30,832 G8
Winsford 26,548 G2
Wisbech 22,932 H5
Witham 21,875 H6
Witney 14,215 F6
Woking 92,667 G8
Wokingham 30,344 G8
Wolverhampton 263,501 E5
Wombwell 17,143 K2
Worcester 75,466 E5
Workington 25,978 D3
Worksop 34,551 F4
Worsbrough 10,821 J2
Worthing 90,687 G7
Yateley⊙ 14,121 G8
Yeovil 36,114 E7
York 123,126 F4

OTHER FEATURES

Aire (riv.) F4
Avon (riv.) F5
Barnstaple (bay) C6
Beachy (head) H7
Blackwater (riv.) H6
Bristol (chan.) C6
Cheviot (hills) E2
Chiltern (hills) G6
Cleveland (hills) F3
Colne (riv.) G8
Cotswold (hills) E6
Cross Fell (mt.) E3
Cumbrian (mts.) D3
Dart (riv.) D7
Dartmoor National Park C7
Dee (riv.) D4
Derwent (riv.) G3
Derwent (riv.) H3
Don (riv.) F4
Dove (riv.) J2
Dover (strait) J7
Dungeness (prom.) J7
Eddyston (rocks) C7
Eden (riv.) E3
English (chan.) D8
Esk (riv.) D2
Exe (riv.) D7
Exmoor National Park D6
Fens, The (reg.) G5
Flamborough (head) G3
Foulness Island (pen.) J6
Great Ouse (riv.) H5
Hartland (pt.) C6
Holderness (pen.) G4
Holy (isl.) F2
Humber (riv.) G4
Irish (sea) B4
Kennet (riv.) F6
Lake District National Park D3
Land's End (prom.) B7
Lea (riv.) G6
Lincoln Wolds (hills) G4
Lindisfarne (Holy) (isl.) F2
Lizard (pt.) B8
Lundy (isl.) C6
Lyme (bay) D7
Medway (riv.) H6
Mendip (hills) E6
Mersey (riv.) G2
Morecambe (bay) D3
Mounts (bay) B7
Naze, The (prom.) J6
Nene (riv.) H5
New (for.) F6
North (sea) J4
North Downs (hills) G6
North Foreland (prom.) J6
Northumberland National Park E2
North York Moors National Park G3
Ouse (riv.) G6
Ouse (riv.) G4
Peak District National Park F4
Peak, The (mt.) J2
Pennine Chain (range) E3
Portland, Bill of (pt.) E7
Purbeck, Isle of (pen.) F7
Ribble (riv.) E4
Saint Bees (head) D3
Saint Mary's (isl.) A8
Scafell Pike (mt.) D3
Scilly (isls.) A7
Severn (riv.) E6
Sheppey (isl.) J6
Sherwood (for.) F4
Solent (chan.) F7
Solway (firth) D3
South Downs (hills) G7
South Foreland (prom.) J6
Spithead (chan.) F7
Stonehenge (ruin) F6
Stour (riv.) H6
Stour (riv.) E7
Stour (riv.) J6
Swale (riv.) F3
Tees (riv.) F3
Thames (riv.) H6
Tintagel (head) C7
Trent (riv.) G4
Tweed (riv.) E2
Tyne (riv.) F3
Ure (riv.) F3
Walney, Isle of (isl.) D3
Wash, The (bay) H5
Weald, The (reg.) H6
Wear (riv.) F3
Welland (riv.) G5
Wey (riv.) G6
Wharfe (riv.) E3
Wight (isl.) 114,879 F7
Wirral (pen.) G2
Wolds, The (hills) G4
Wye (riv.) D5
Yare (riv.) J5
Yorkshire Dales National Park E3

CHANNEL ISLANDS

CITIES and TOWNS

Saint Helier (cap.), Jersey⊙ 27,549 E8
Saint Peter Port (cap.), Guernsey⊙ 16,085 E8
Saint Sampson's⊙ 7,475 E8

OTHER FEATURES

Alderney (isl.) 2,130 E8
Guernsey (isl.) 55,421 E8
Herm (isl.) 59 E8
Jersey (isl.) 82,809 E8
Sark (isl.) 560 E8

ISLE of MAN

CITIES and TOWNS

Castletown 2,788 C3
Douglas (cap.) 19,897 C3
Laxey 1,242 C3
Onchan 6,395 C3
Peel 3,295 C3
Port Erin 2,356 C3
Port Saint Mary 1,525 C3
Ramsey 5,372 C3

OTHER FEATURES

Ayre (pt.) C3
Calf of Man (isl.) C3
Langness (prom.) C3
Snaefell (mt.) C3

WALES

COUNTIES

Clwyd 385,581 D4
Dyfed 323,040 C6
Gwent 436,500 D6
Gwynedd 222,291 C4
Mid Glamorgan 533,770 D6
Powys 108,121 D5
South Glamorgan 376,718 A7
West Glamorgan 363,619 D6

CITIES and TOWNS

Abercarn 16,811 B6
Aberdare 31,617 A6
Abergavenny 13,880 B6
Abergele 12,264 D4
Abertillery and Brynmawr 28,239 B6
Aberystwyth 10,290 C5
Ammanford 10,735 C6
Bangor 12,244 C4
Barry 44,443 B7
Bethesda 3,558 D4
Brecknock (Brecon) 7,166 D6
Bridgend 31,008 A7
Brynmawr and Abertillery 28,239 B6
Buckley 16,693 G2
Caernarfon 9,271 C4
Caerphilly 28,681 B6
Caldicot 12,310 E6
Cardiff (cap.) 262,313 B7
Cardigan 3,815 C5
Carmarthen 13,860 C6
Chepstow 9,039 E6
Colwyn Bay 27,002 D4
Connah's Quay 14,785 G2
Cwmbran 44,592 B6
Denbigh 7,710 D4
Ebbw Vale 21,048 B6
Ffestiniog 4,507 D5
Flint 11,411 G2
Gelligaer 16,812 A6
Gwersyllt 13,374 E4
Harlech⊙ 1,292 C5
Haverfordwest 13,572 B6
Hawarden⊙ 22,361 G2
Holyhead 12,569 C4
Holywell 11,101 G2
Llandeilo 1,598 C6
Llandovery 1,676 D5
Llandrindod Wells 4,232 D5
Llandudno 13,202 D4
Llanelli 45,336 C6
Llanfairfechan 3,173 D4
Llangollen 2,546 D5
Llanidloes 2,392 D5
Llantrisant⊙ 8,317 A7
Llantwit Major 13,375 A7
Maesteg 21,821 D6
Menai Bridge 2,942 C4
Merthyr Tydfil 38,893 A6
Milford Haven 13,883 B6
Mold 8,487 G2
Monmouth 7,379 E6
Montgomery D5
Mountain Ash 23,520 A6
Narberth C6
Neath 48,687 D6
Nefyn⊙ 2,086 C5
Newport 115,896 B6
Newtown 8,906 D5
Neyland 3,095 B6
Ogmore 7,092 A6
Pembroke 8,235 C6
Penarth 22,467 B7
Pontypool 36,064 B6
Pontypridd 29,465 A6
Port Talbot 40,078 D6
Porthcawl 15,162 D6
Prestatyn 16,246 D4
Pwllheli 3,978 C5
Rhondda 70,980 A6
Rhosllanerchrugog 11,080 D4
Rhyl 23,130 D4
Risca 16,627 B6
Ruthin 4,417 D4
Saint David's⊙ 1,428 B6
Swansea 172,433 C6
Tenby 5,226 C6
Tredegar 16,188 B6
Welshpool 4,869 D5
Wrexham 39,929 E4
Ystradgynlais 10,406 D6

OTHER FEATURES

Anglesey (isl.) C4
Bardsey (isl.) C5
Brecon Beacons National Park D6
Bristol (chan.) C6
Caldy (isl.) C6
Cambrian (mts.) D5
Cardigan (bay) C5
Carmarthen (bay) C6
Conwy (bay) C4
Dee (riv.) D4
Gower (pen.) C6
Great Ormes (head) C4
Holy (isl.) C4
Irish (sea) B4
Lleyn (pen.) C5
Menai (strait) C4
Milford Haven (inlet) B6
Pembrokeshire Coast National Park C6
Radnor (for.) D5
Saint Brides (bay) B6
Saint George's (chan.) B5
Severn (riv.) E6
Snowdon (mt.) C4
Snowdonia National Park D4
Taff (riv.) B7
Teifi (riv.) C5
Towy (riv.) D6
Tremadoc (prom.) C5
Usk (riv.) B6
Wye (riv.) D5
Ynys Môn (Anglesey) (isl.) C4

⊙ Population of parish.

SCOTLAND

(map on page 15)

REGIONS

Borders 99,784 E5
Central 273,391 D4
Dumfries and Galloway 145,139 E5
Fife 327,362 E4
Grampian 471,942 F3
Highland 200,150 D3
Lothian 738,372 E5
Orkney (islands area) 19,056 E1
Shetland (islands area) 27,277 F2
Strathclyde 2,404,532 C4
Tayside 391,846 E4
Western Isles (islands area) 31,884 A3

CITIES and TOWNS

Aberchirder 1,021 F3
Aberdeen 190,465 F3
Aberfeldy 1,613 E4
Aberfoyle 793 D4
Abernethy 776 E4
Aboyne 1,529 F3
Achiltibuie⊙ 1,564 C3
Achnasheen⊙ 1,078 C3
Airdrie 45,747 C2
Alexandria 26,329 A1
Alford 764 F3
Alloa 26,428 C1
Alness 6,289 D3
Altnaharra⊙ 1,227 D2
Alva 4,874 C1
Alyth 2,289 E4
Annan 8,314 E6
Annbank Station 3,223 D5
Arbroath 24,119 F4
Ardrishaig 1,325 C4
Ardrossan 11,421 D5
Armadale 9,527 C2
Auchinleck 4,463 D5
Auchterarder 2,904 E4
Auchtermuchty 1,646 E4
Aviemore 1,224 E3
Ayr 49,522 D5
Baillieston 7,671 B2
Balerno 3,576 D2
Balfron 1,127 B1
Ballantrae 262 C5
Ballater 1,218 F3
Ballingry 7,021 D1
Balloch 1,484 B1
Banchory 4,890 F3
Banff 3,938 F3
Bankhead 1,492 F3
Bannockburn 5,889 C1
Barrhead 18,418 B2
Bathgate 14,477 C2
Bearsden 27,183 B2
Beauly 1,148 D3
Beith 5,742 D5
Bellsbank 2,482 D5
Bellshill 39,676 C2
Berriedale⊙ 1,927 E2
Bieldside 1,137 F3
Biggar 1,938 E5
Bishopbriggs 23,501 B2
Bishopton 5,283 B2
Blackburn 5,785 C2
Blair Atholl 437 E4
Blairgowrie and Rattray 7,184 E4
Blantyre 19,948 B2
Bo'ness 14,641 C1
Boddam 1,367 G3
Bonhill 4,385 B1
Bonnybridge 5,701 C1
Bonnyrigg and Lasswade 14,399 D2
Brechin 7,692 F4
Bridge of Allan 4,694 C1
Bridge of Don 4,086 F3
Bridge of Weir 4,724 A2
Brightons 3,106 C1
Brora 1,736 E2
Broxburn 12,032 D1
Buckhaven and Methil 18,265 F4
Buckie 7,839 E3
Bucksburn 6,567 F3
Burghead 1,380 E3
Burntisland 5,865 D1
Callander 2,520 D4
Cambuslang 14,607 B2
Campbeltown 6,098 C5
Caol 3,719 C4
Cardenden 5,898 D1
Carluke 11,674 E5
Carnoustie 9,225 F4
Carnwath 1,374 E5
Carron 2,626 C1
Castle Douglas 3,521 E6
Catrine 2,790 D5
Cawdor 111 E3
Chirnside 1,263 F5
Chryston 11,067 C2
Clackmannan 3,258 C1
Clarkston 8,404 B2
Clydebank 51,854 B2
Coalburn 1,241 E5
Coatbridge 50,957 C2
Cockenzie and Port Seton 3,760 D1
Coldstream 1,645 F5
Comrie 1,477 E4
Cononbridge 2 187 D3
Corpach 1,296 C4
Coupar Angus 2,186 E4
Cove and Kilcreggan 1,220 A1
Cove Bay 2,840 F3
Cowdenbeath 12,272 D1
Cowie 2,513 C1
Crail 1,181 F4
Creetown 769 D6
Crieff 5,477 E4
Crimond 1,002 G3
Cromarty 492 E3
Cruden Bay 1,453 G3
Cullen 1,414 F3
Culross 504 C1
Cults 3,336 F3
Cumbernauld 47,901 C1
Cumnock and Holmhead 9,650 D5
Cupar 6,637 E4
Currie 6,764 D2
Dailly 1,098 D5
Dalbeattie 3,917 E6
Dalkeith 11,255 D2
Dalmellington 1,425 D5
Dalry 5,856 D5
Dalrymple 1,237 D5
Darvel 3,461 D5
Denny and Dunipace 23,158 C1
Dervaig⊙ 1,081 B4
Dingwall 4,842 D3
Dollar 2,486 E4
Dornoch 880 D3
Douglas 1,727 E5
Doune 1,046 D4
Drongan 3,129 D5
Dufftown 1,643 E3
Dumbarton 23,430 B1
Dumfries 32,100 E5
Dunbar 6,035 F4
Dunblane 6,855 E4
Dundee 174,345 F4
Dundonald 2,669 D5
Dunfermline 52,227 D1
Dunoon 9,369 A2
Duns 2,253 F5
Duntocher 3,532 B2
Dyce 7,039 F3
Eaglesham 3,166 D5
Earlston 1,610 F5
East Calder 5,112 D2
East Kilbride 70,676 B2
East Linton 1,206 F5
Eastriggs 1,845 E5
East Wemyss 1,782 D1
Edinburgh (cap.) 420,169 D1
Elderslie 5,204 B2
Elgin 18,908 E3
Ellon 6,319 F3
Errol 762 E4
Eyemouth 3,398 F5
Fairlie 1,326 D5
Falkirk 36,880 C1
Falkland 998 E4
Fallin 2,663 C1
Fauldhouse 5,036 C2
Findhorn 664 E3
Findochty 1,019 E3
Fochabers 1,483 E3
Forfar 12,770 F4
Forres 8,354 E3
Forth 2,890 C2
Fortrose 1,332 D3
Fort William 11,061 C4
Fraserburgh 12,512 G3
Gairloch 125 C3
Galashiels 12,244 F5
Galston 5,311 D5
Garelochhead 2,072 A1
Gatehouse-of-Fleet 835 D6
Giffnock 33,634 B2
Girvan 7,795 D5
Glamis 190 F4
Glasgow 765,030 B2
Glenbarr⊙ 691 C5
Glencoe 195 C4
Glenelg⊙ 1,468 C3
Glenrothes 32,971 E4
Golspie 1,491 E3
Gorebridge 6,036 D2
Gourock 11,203 A1
Grangemouth 21,599 C1
Grantown-on-Spey 2,034 E3
Greenock 59,016 A2
Gretna 2,811 E5
Gullane 2,232 F4
Haddington 8,139 F5
Halkirk 679 E2
Hamilton 51,718 C2
Harthill 4,161 C2
Hawick 16,364 F5
Heathhall 1,365 E5
Helensburgh 16,621 A1
Helmsdale 727 E2
Hillside 1,233 F4
Hillswick⊙ 696 G2
Hopeman 1,398 E3
Huntly 3,952 F3
Hurlford 4,294 D5
Inchnadamph⊙ 833 D2
Innerleithen 2,468 E5
Insch 1,256 F3
Inveraray 473 C4
Inverbervie 1,799 F4
Invercassley⊙ 1,067 D3
Invergordon 4,067 D3
Invergowrie 1,389 E4
Inverie⊙ 1,468 C3
Inverkeithing 5,770 D1
Inverness 40,010 D3
Inverurie 7,680 F3
Irvine 32,968 D5
Jedburgh 4,069 F5
John O'Groats 195 E2
Johnstone 42,669 B2
Keith 4,407 F3
Kelso 5,648 F5
Kelty 5,623 D1
Kemnay 3,034 F3
Kilbarchan 2,669 A2
Kilbirnie 8,710 A2
Kilchoan⊙ 764 B4
Kildonan⊙ 1,105 E2
Killearn 1,771 B1
Kilmacolm 3,676 A2
Kilmarnock 52,083 D5
Kilmaurs 2,738 D5
Kilrenny and Anstruther 2,951 F4
Kilsyth 10,538 B1
Kilwinning 16,266 D5
Kinbrace⊙ 1,105 E2
Kincardine 3,166 C1
Kinghorn 2,698 D1
Kingussie 1,229 D3
Kinlochewe⊙ 1,794 C3
Kinlochleven 1,047 D4
Kinloss 2,813 E3
Kinross 3,496 E4
Kintore 1,644 F3
Kirkcaldy 46,522 D1
Kirkconnel 2,656 E5
Kirkcudbright 3,427 E6
Kirkintilloch 33,148 B2
Kirkmuirhill 3,624 E5
Kirkwall 5,995 E2
Kirriemuir 5,326 E4
Kyle of Lochalsh 687 C3
Kylestrome⊙ 745 D2
Ladybank 1,355 E4
Lairg 572 D2
Lanark 9,806 E5
Langholm 2,615 E5
Larbert 4,922 C1
Largs 9,905 A2
Larkhall 16,216 E5
Lauder 639 F5
Laurencekirk 1,329 F4
Lennoxtown 4,829 B1
Lerwick 7,561 G2
Leslie 3,551 E4
Lesmahagow 3,408 E5
Letham 804 F4
Leuchars 2,244 F4
Leven 8,624 F4
Lhanbryde 1,811 E3
Limekilns 1,444 D1
Linlithgow 9,582 C1
Linwood 10,510 B2
Livingston 38,954 C2
Loanhead 6,159 D2
Lochailort⊙ 673 C4
Locharbriggs 4,230 E5
Lochgelly 7,334 D1
Lochgilphead 2,461 C4
Lochinver 283 C2
Lochmaben 1,713 E5
Lochore 2,994 D1
Lochwinnoch 2,273 A2
Lockerbie 3,561 E5

(continued)

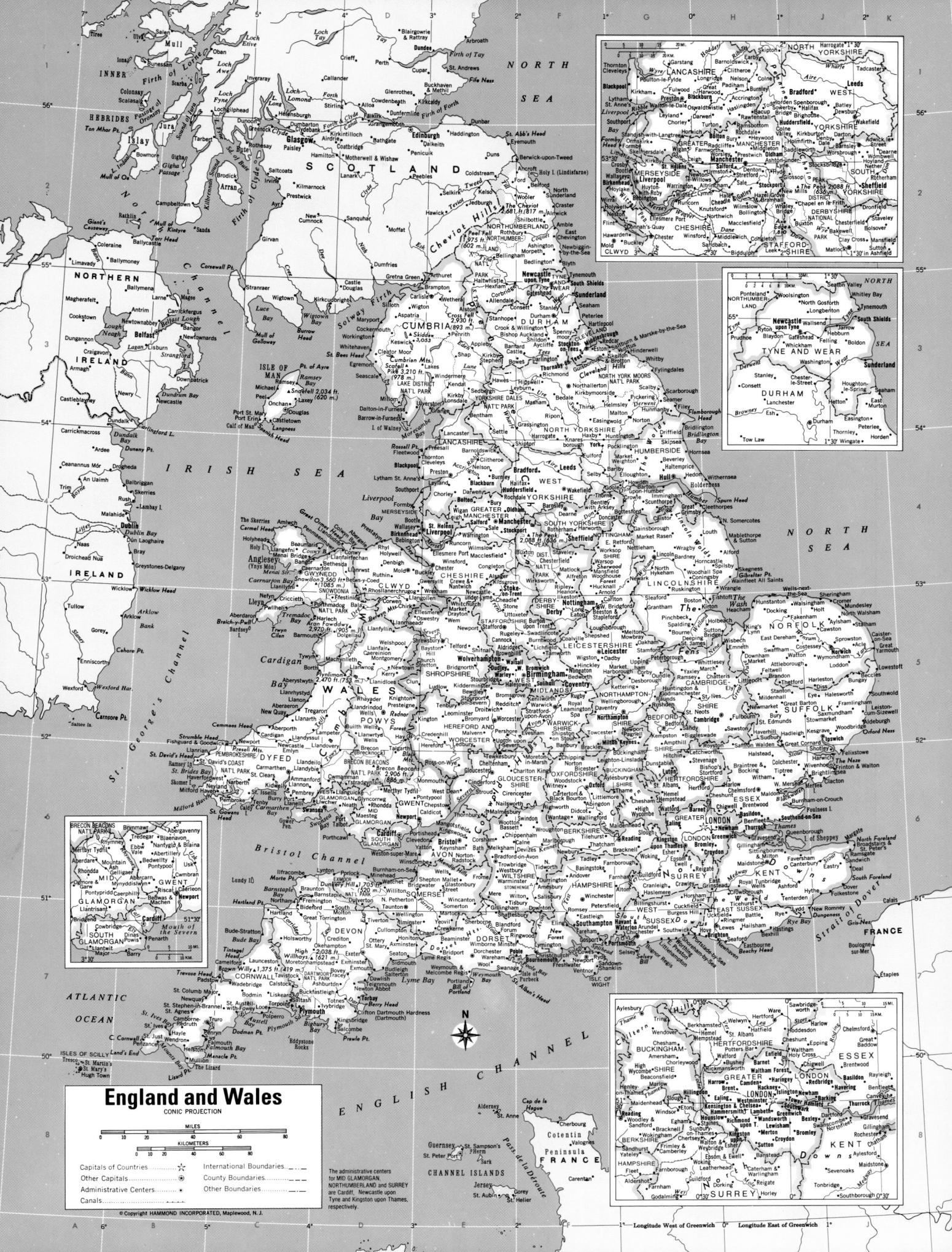

England and Wales

CONIC PROJECTION

MILES
0 10 20   40   60   80

KILOMETERS
0 10 20   40   60   80

Capitals of Countries .... ✪
Other Capitals .... ⊛
Administrative Centers .... •
Canals

International Boundaries
County Boundaries
Other Boundaries

The administrative centers for MID GLAMORGAN, NORTHUMBERLAND and SURREY are Cardiff, Newcastle upon Tyne and Kingston upon Thames, respectively.

© Copyright HAMMOND INCORPORATED, Maplewood, N.J.

Longitude West of Greenwich 0°   Longitude East of Greenwich 1°

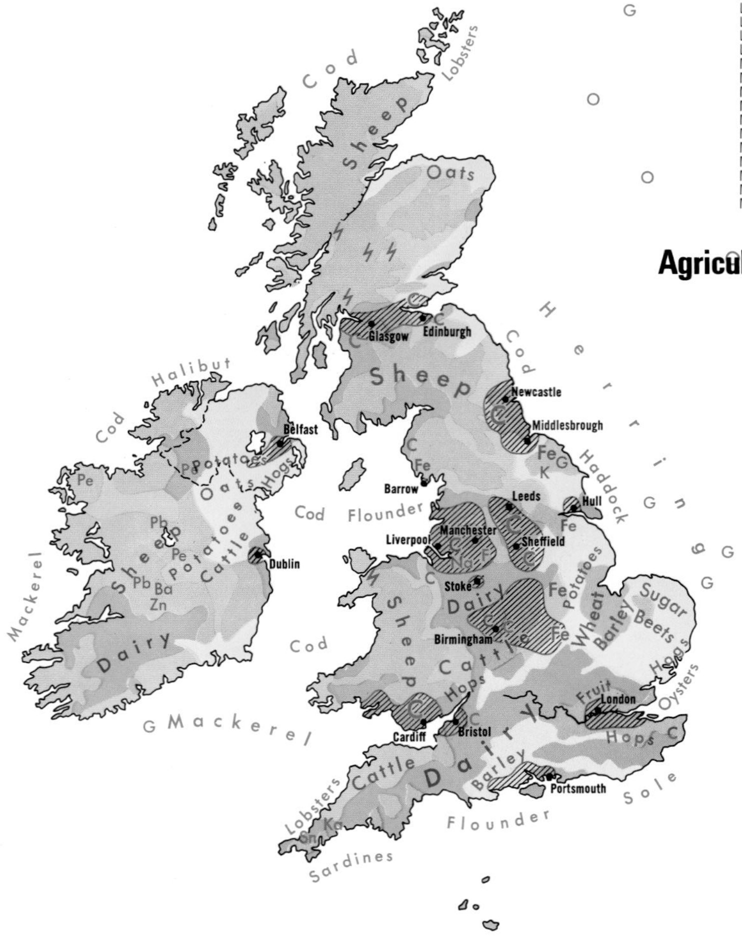

Agriculture, Industry and Resources

DOMINANT LAND USE

- Cereals (chiefly oats, barley)
- Truck Farming, Horticulture
- Dairy, Mixed Farming
- Livestock, Mixed Farming
- Pasture Livestock

MAJOR MINERAL OCCURRENCES

Ba	Barite	Na	Salt
C	Coal	O	Petroleum
F	Fluorspar	Pb	Lead
Fe	Iron Ore	Pe	Peat
G	Natural Gas	Sn	Tin
K	Potash	Zn	Zinc
Ka	Kaolin (china clay)		

Water Power

Major Industrial Areas

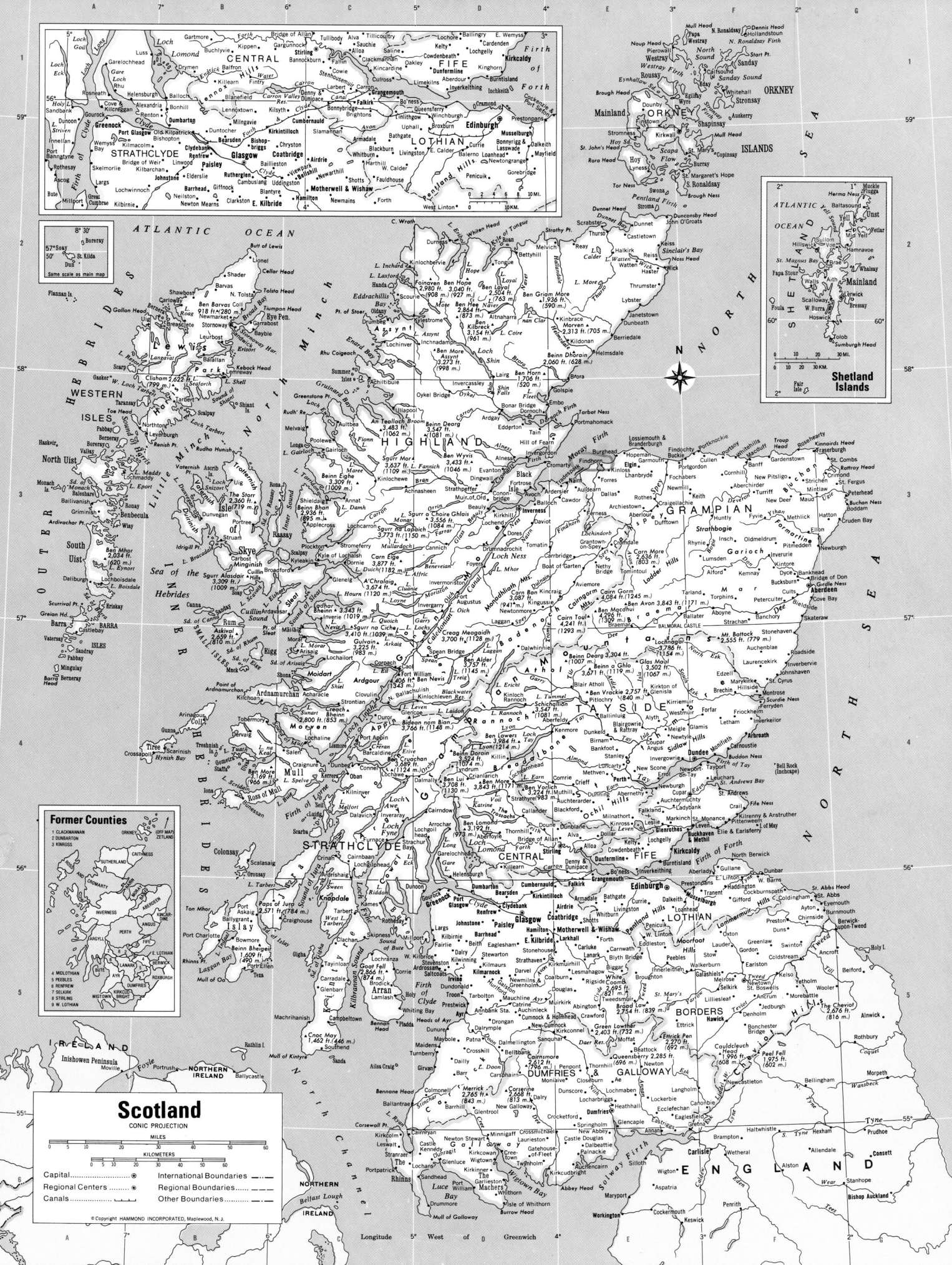

Scotland

CONIC PROJECTION

MILES
KILOMETERS

Capital..............................⊛
Regional Centers.........⊛
Canals................................

International Boundaries....._.._.._
Regional Boundaries......._._._._
Other Boundaries.............

© Copyright HAMMOND INCORPORATED, Maplewood, N.J.

Former Counties

1 CLACKMANNAN
2 DUNBARTON
3 KINROSS
4 MIDLOTHIAN
5 PEEBLES
6 RENFREW
7 SELKIRK
8 STIRLING
9 W. LOTHIAN

Longitude 5° West of Greenwich

IRELAND

COUNTIES

Carlow 40,988 H6
Cavan 53,965 G4
Clare 91,344 D6
Cork 412,735 D7
Donegal 129,664 F2
Dublin 1,021,449 J5
Galway 178,552 D5
Kerry 124,159 B7
Kildare 116,247 H5
Kilkenny 73,186 G6
Laois 53,284 G6
Leitrim 27,035 E3
Leix (Laois) 53,284 G6
Limerick 164,569 D7
Longford 31,496 F4
Louth 91,810 J4
Mayo 115,184 C4
Meath 103,881 H4
Monaghan 52,379 H3
Offaly 59,835 F5
Roscommon 54,592 E4
Sligo 56,046 D3
Tipperary 136,619 F6
Waterford 91,151 F7
Westmeath 63,379 G5
Wexford 102,552 H7
Wicklow 94,542 J6

CITIES and TOWNS

Abbeyfeale 1,483 C7
Abbeyleix 1,468 G6
Adare 792 D6
Aghada-Farsid-Rostellan 818 E8
An Uaimh 3,660 H4
Ardee 3,253 H4
Ardfinnan 827 F7
Ardmore 343 F8
Arklow 8,388 J6
Ashford 782 J5
Askeaton 951 D6
Athboy 1,055 H4
Athenry 1,642 D5
Athlone 8,815 F5
Athy 4,734 H6
Aughrim 756 J6
Avoca 490 J6
Bagenalstown (Muinebeag) 2,653 H6
Baile Atha Cliath (Dublin) (cap.) 502,749 K5
Bailieborough 1,645 G4
Balbriggan 5,680 J4
Ballaghaderreen 1,366 E4
Ballina, Mayo 6,714 C3
Ballina, Tipperary 507 E6
Ballinamore 810 F3
Ballinasloe 6,125 E5
Ballincollig-Carrigrohane 7,231 D8
Ballineen 592 D8
Ballinrobe 1,270 C4
Ballybofey-Stranorlar 2,964 F2
Ballybunion 1,452 B7
Ballycastle 219 C3
Ballyconnell 466 F3
Ballygar 472 E4
Ballygeary 891 J7
Ballyhaunis 1,338 D4
Ballyheigue 660 B7
Ballyjamesduff 842 G4
Ballylanders 343 E7
Ballylongford 523 B6
Ballymahon 859 F4
Ballymore Eustace 575 J5
Ballymote 1,064 D3
Ballyragget 833 G6
Ballyshannon 2,573 E3
Baltinglass 1,089 H6
Banagher 1,465 F5
Bandon 1,943 D8
Bantry 2,811 C8
Belmullet 1,033 B3
Belturbet 1,228 G3
Bennettsbridge 601 G6
Birr 3,417 F5
Blanchardstown H5
Blarney 1,952 D8
Blessington 1,322 J5
Borrisokane 837 E6
Borrisoleigh 624 E6
Boyle 1,859 E4
Bray 24,686 K5
Bri Chualann (Bray) 24,686 K5
Bruff 819 D7
Bunbeg-Derrybeg 1,469 E1
Bunclody-Carrickduff 1,423 H6
Buncrana 3,106 G1
Bundoran 1,535 E3
Buttevant 1,133 D7
Cahir 2,118 F7
Cahirciveen 1,310 A8
Callan 1,266 G7
Cappamore 765 E6
Cappoquin 920 F7
Carlingford 635 J3
Carlow 11,509 H6
Carndonagh 1,600 G1
Carnew 723 H6
Carrickmacross 1,815 H4
Carrick-on-Shannon 1,984 F4
Carrick-on-Suir 5,353 F7
Carrigaline 5,893 E8
Carrigtwohill 1,272 E8
Cashel 2,458 F7
Castlebar 6,349 C4
Castlebellingham 848 J4
Castleblayney 2,157 H3
Castlebridge 655 J7
Castlecomer-Donaguile 1,490 G6
Castledermot 792 H6
Castlefin 694 F2
Castleisland 2,281 B7
Castlemartyr 585 E8
Castlepollard 803 G4
Castlerea 1,840 D4
Castletown 303 F6
Castletownbere 905 B8
Castletownroche 474 D7
Cavan 3,381 G3
Ceanannus Mór 2,413 G4
Celbridge 7,135 H5
Charlestown-Bellahy 754 D4
Charleville (Rathluirc) 2,814 D7
Clara 2,736 F5
Claremorris 1,992 C4
Clifden 896 B5
Cloghan 496 F5
Clogh-Chatsworth 319 G6
Clogheen 502 F7
Clogherhead 765 J4
Clonakilty 2,567 D8
Clones 2,280 G3
Clonfert E5
Clonmel 11,759 F7
Cloughjordan 499 E6
Cloyne 721 E8
Cóbh 6,369 E8
Coill Dubh 772 H5
Collooney 705 E3
Convoy 891 F2
Coolgreany 352 J6
Cootehill 1,487 G3
Cork 133,271 E8
Corofin 391 C6
Courtown Harbour 317 J6
Creeslough 340 F1
Croom 1,024 D6
Crosshaven 1,362 E8
Crossmolina 1,250 C3
Daingean 659 G5
Delvin 309 G4
Dingle 1,253 A7
Donabate 599 J5
Donegal 2,242 F2
Doneraile 846 D7
Dooagh-Keel 650 A4
Doon 308 E6
Drimoleague 381 C8
Drogheda 24,086 J4
Droichead Nua 5,983 H5
Dromahair 353 E3
Drumconrath 334 H4
Drumshanbo 622 E3
Dublin (cap.) 502,749 K5
Duleek 1,679 J4
Duncannon 388 H7
Dundalk 26,669 H3
Dunfanaghy 314 F1
Dungarvan 6,849 F7
Dungloe 940 E2
Dún Laoghaire 54,715 K5
Dunkineely 442 E2
Dunlavin 734 H5
Dunleer 1,184 J4
Dunmanway 1,382 C8
Dunmore 445 D4
Dunmore East 1,041 G7
Dunshaughlin 878 H5
Durrow 707 G6
Edenderry 3,539 G5
Elphin 513 E4
Emyvale 464 G3
Ennis 5,917 D6
Enniscorthy 4,483 J7
Enniskerry 1,229 J5
Ennistymon 1,039 C6
Eyrecourt 351 E5
Fahan 367 G1
Falcarragh 996 E1
Fenit 401 B7
Ferbane 1,374 F5
Fermoy 2,872 E7
Ferns 811 J6
Fethard 982 F7
Foxford 1,033 C4
Foynes 707 C6
Frankford (Kilcormac) 1,118 F5
Freshford 700 G6
Galbally 248 E7
Galway 47,104 C5
Geashill 339 G5
Glanworth 379 E7
Glenamaddy 369 D4
Glenties 914 E2
Glin 569 C6
Golden 295 F7
Gorey 2,445 J6
Gormanston 870 J4
Gort 1,021 D5
Gowran 517 G6
Graiguenamanagh-Tinnahinch 1,485 H6
Granard 1,338 F4
Greencastle 584 H1
Greystones 8,455 K5
Hacketstown 710 H6
Headford 675 C5
Holycross 274 F6
Hospital 751 E7
Inniscrone 633 C3
Johnstown 408 G6
Kanturk 1,870 D7
Keel-Dooagh 650 A4
Kells (Ceanannus Mór) 2,413 G4
Kenmare 1,130 B8
Kilbeggan 603 G5
Kilcar 345 D2
Kilcock 1,414 H5
Kilcoole 2,335 K5
Kilcormac 1,118 F5
Kilcullen 1,693 H5
Kildare 4,268 H5
Kildysart 347 C6
Kilfinane 788 D7
Kilkee 1,448 B6
Kilkenny 8,969 G6
Killala 674 C3
Killaloe 1,033 D6
Killarney 7,837 C7
Killenaule 717 F6
Killeshandra 455 F3
Killorglin 1,304 B7
Killucan-Rathwire 353 G4
Killybegs 1,632 E2
Kilmacrennan 412 F1
Kilmacthomas 648 G7
Kilmallock 1,424 D7
Kilmihill 338 C6
Kilmore Quay 458 H7
Kilnaleck 321 G4
Kilronan 282 B5
Kilrush 2,961 C6
Kiltimagh 982 C4
Kilworth 411 E7
Kingscourt 1,242 H4
Kingstown (Dún Laoghaire) 54,715 K5
Kinnegad 433 G5
Kinnitty 261 F5
Kinsale 1,811 D8
Kinvara 425 D5
Knightstown 204 A8
Knock 332 D4
Knocklong 273 D7
Lahinch 511 C6
Lanesborough-Ballyleague 1,058 E4
Laytown-Bettystown-Mornington 3,321 J4
Leighlinbridge 540 H6
Leitrim F3
Leixlip 11,938 H5
Letterkenny 6,691 F2
Lifford 1,478 F2
Limerick 56,279 D6
Lisdoonvarna 648 C5
Lismore 703 F7
Listowel 3,494 C7
Littleton 566 F6
Longford 6,457 F4
Loughrea 3,360 E5
Louisburgh 209 B4
Louth 435 J4
Lucan 12,259 J5
Luimneach (Limerick) 56,279 D6
Lusk 1,831 J4
Macroom 2,449 C8
Malahide 9,940 J5
Mallow 6,488 D7
Manorhamilton 1,031 E3
Maryborough (Portlaoise) 3,773 G5
Maynooth 4,768 H5
Meathas Truim 806 G4
Midleton 3,111 E8
Milford 981 F1
Millstreet 1,330 D7
Milltown 347 A7
Miltownmalbay 719 C6
Mitchelstown 3,210 E7
Moate 1,659 F5
Mohill 930 F4
Monaghan 6,075 G3
Monasterevan 2,143 H5
Moneygall 346 F6
Mooncoin 868 G7
Mount Bellew 519 D5
Mountcharles 480 E2
Mountmellick 2,789 G5
Mountrath 1,402 F5
Moville 1,331 G1
Moycullen 366 C5
Muinebeag 2,653 H6
Mullagh 462 H4
Mullinahone 385 F7
Mullinavat 355 G7
Mullingar 8,077 G4
Naas 10,017 H5
Navan (An Uaimh) 3,660 H4
Nenagh 5,483 E6
Newbliss 293 G3
Newbridge (Droichead Nua) 5,983 H5
Newcastle 3,370 J5
Newmarket 1,022 C7
Newmarket-on-Fergus 1,678 D6
Newport, Mayo 492 C4
Newport, Tipperary 857 E6
New Ross 5,343 H7
Newtown Forbes 393 F4
Newtownmountkennedy 2,183 J5
Newtownsandes 357 C6
O'Briensbridge-Montpelier 385 D6
Oldcastle 869 G4
Oola 451 E7
Oranmore 1,064 D5
Oughterard 682 C5
Passage East 563 G7
Passage West 3,511 E8
Patrickswell 905 D6
Piltown 691 G7
Portarlington 3,295 G5
Portlaoise 3,773 G5
Portlaw 1,260 G7
Portmarnock 9,055 J5
Portumna 1,062 E5
Queenstown (Cóbh) 6,369 E8
Ramelton 989 F1
Raphoe 1,027 F2
Rathangan 1,270 G5
Rathcoole 2,991 J5
Rathdowney 1,095 F6
Rathdrum 1,307 J6
Rathkeale 1,815 D7
Rathluirc 2,814 D7
Rathmore 548 J5
Rathmullen 554 F1
Rathnew 1,389 J6
Rathvilly 512 H6
Ratoath 551 J5
Riverstown 1,416 E8
Roscommon 1,363 E4
Roscrea 4,378 F6
Rosscarbery 425 C8
Rosses Point 598 D3
Rosslare 704 J7
Rosslare Harbour (Ballygeary) 891 J7
Roundwood 371 J5
Rush 4,513 J4
Saint Johnston 468 F2
Scarriff 847 E6
Schull 509 B8
Shanagolden 402 C6
Shannon 8,005 D6
Shannon Bridge 310 F5
Shercock 406 G4
Shillelagh 334 J6
Shinrone 479 F5
Sixmilebridge 1,182 D6
Skerries 6,864 J4
Skibbereen 1,999 C8
Slane 689 H4
Sligo 17,259 E3
Sneem 309 B8
Stepaside 748 J5
Stradbally, Laois 1,046 G5
Stradbally, Waterford 255 F7
Strokestown 620 E4
Swinford 1,197 C4
Swords 15,312 J5
Taghmon 607 H7
Tallow 867 F7
Tarbert 683 C6
Templemore 2,258 F6
Templetuohy 242 F6
Termonfeckin 741 J4
Thomastown 1,465 G7
Thurles 7,049 F6
Timoleague 330 D8
Tinahely 594 H6
Tipperary 5,033 E7
Toomevara 428 E6
Tralee 17,109 B7
Tramore 5,999 G7
Trim 1,967 H4
Tuam 4,109 D4
Tubbercurry 1,250 D3
Tulla 403 D6
Tullamore 8,484 G5
Tullow 2,324 H6
Tyrrellspass 328 G5
Urlingford 676 F6
Virginia 699 G4
Waterford 39,529 G7
Waterville-Spunkane 475 A8
Westport 3,456 C4
Wexford 10,336 H7
Wicklow 5,304 K6
Woodford 242 E5
Youghal 5,706 F8

OTHER FEATURES

Achill (head) A4
Achill (isl.) A4
Allen (lake) E3
Allen, Bog of (marsh) H5
Annalee (riv.) G3
Aran (isls.) B5
Aran (isl.) D2
Arrow (lake) E3
Ballinskelligs (bay) A8
Ballyhoura (hills) E7
Ballyteige (bay) H7
Bandon (riv.) D8
Bantry (bay) B8
Barrow (riv.) H7
Baurtregaum (mt.) A7
Bear (isl.) B8
Ben Dash (hill) C6
Bertraghboy (bay) A5
Black (head) C5
Blacksod (bay) A3
Blackstairs (mt.) H6
Blackwater (riv.) H4
Blackwater (riv.) D7
Blasket (isls.) A7
Bloody Foreland (prom.) E1
Blue Stack (mts.) E2
Boderg (lake) E4
Boggeragh (mts.) D7
Bolus (head) A8
Boyne (riv.) J4
Brandon (bay) A7
Brandon (head) A7
Brandon (mt.) A7
Bray (head) A8
Bride (riv.) E7
Broad Haven (harb.) B3
Brosna (riv.) F5
Bull, The (isl.) A8
Caha (mts.) B8
Cahore (pt.) J6
Cark (mt.) F2
Carlingford (inlet) J3
Carnsore (pt.) J7
Carra (lake) C4
Carrantuohill (mt.) B7
Carrowmore (lake) B3
Clare (isls.) A4
Clare (riv.) D5
Clear (cape) B9
Clear (isl.) C9
Clew (bay) B4
Clonakilty (bay) D8
Comeragh (mts.) F7
Conn (lake) C3
Connacht (prov.) 431,409 C4
Connemara (dist.) B5
Cork (harb.) E8
Corrib (lake) C5
Croagh Patrick (mt.) C4
Cuilcagh (mt.) F3
Cullin (lake) C4
Curragh, The (plain) H5
Dash, Ben (hill) C6
Dee (riv.) H4
Deel (riv.) C3
Deel (riv.) C7
Deel (riv.) G4
Deele (riv.) F2
Derg (lake) F2
Derg (lake) E6
Derravaragh (lake) G4
Derryveagh (mts.) E2
Devilsbit (mt.) E6
Dingle (bay) A7
Donegal (bay) D3
Donegal (pt.) B6
Doulus (head) A8
Downpatrick (head) C3
Drum (hills) F7
Dublin (bay) J5
Dunany (pt.) J4
Dundalk (bay) J4
Dungarvan (harb.) G7
Dunkellin (riv.) D5
Dunmanus (bay) B8
Dursey (isl.) A8
Eask (lake) E2
Ennell (lake) G5
Erne (riv.) E3
Errigal (mt.) E1
Erris (head) A3
Fanad (head) F1
Fastnet Rock (isl.) B9
Feale (riv.) C7
Feeagh (lake) B4
Fergus (riv.) D6
Finn (riv.) F2
Finn (riv.) G3
Foul (sound) B5
Foyle (inlet) G1
Foyle (riv.) G2
Galley (head) D9
Galtee (mts.) E7
Galtymore (mt.) E7
Galway (bay) C5
Gara (lake) D4
Garadice (lake) F3
Gill (lake) E3
Gola (isl.) E1
Golden Vale (plain) E7
Gorumna (isl.) B5
Gowna (lake) G4
Grand (canal) G5
Great Blasket (isl.) A7
Gregory's (sound) B5
Gweebarra (bay) D2
Hags (head) B6
Hook (head) H7
Horn (head) E1
Iar Connacht (dist.) C5
Inishbofin (isl.) A4
Inisheer (isl.) B5
Inishmaan (isl.) C5
Inishmore (isl.) B5
Inishmurray (isl.) D3
Inishowen (head) H1
Inishowen (pen.) G1
Inishshark (isl.) A4
Inishtrahull (isl.) G1
Inishtrahull (sound) G1
Inishturk (isls.) A4
Inny (riv.) F4
Ireland's Eye (isl.) K5
Irish (sea) K4
Joyce's Country (dist.) B4
Keeper (hill) E6
Kenmare (riv.) A8
Kerry (head) A7
Key (lake) E3
Kilkieran (bay) B5
Killala (bay) C3
Killary (harb.) A4
Kinsale (harb.) E8
Kinsale, Old Head of (head) E8
Kippure (mt.) J5
Knockanefune (mt.) C7
Knockboy (mt.) B8
Knockmealdown (mts.) F7
Lambay (hill) K4
Laune (riv.) B7
Leane (lake) G4
Lee (riv.) D8
Leinster (mt.) H6
Leinster (prov.) 1,852,649 G5
Lettermullan (isl.) B5
Liffey (riv.) H5
Liscannor (bay) B6
Loop (head) A6
Loughros More (bay) D2
Lugnaquillia (mt.) J5
Lung (riv.) D4
Macgillicuddy's Reeks (mts.) B7
Macnean (lake) F3
Maigue (riv.) D6
Malin (head) F1
Mangerton (mt.) C8
Mask (lake) C4
Maumakeogh (mt.) C3
Maumturk (mts.) B5
Melvin (lake) E3
Mine (head) F8
Mizen (head) B9
Mizen (head) K6
Moher (cliff) B6
Monavullagh (mts.) F7
Moy (riv.) C3
Mullaghareirk (mts.) C7
Mulroy (bay) F1
Munster (prov.) 1,020,577 D7
Mutton (isl.) B6
Mweelrea (mt.) B4
Nagles (mts.) E7
Nephin (mt.) C3
Nephin Beg (mt.) B3
Nore (riv.) G7
North (sound) B5
North Inishkea (isl.) A3
Oughter (lake) G3
Ovoca (riv.) J6
Owenmore (riv.) B3
Owenmore (riv.) D3
Paps, The (mt.) C7
Partry (mts.) C4
Pollaphuca (res.) J5
Puffin (isl.) A8
Punchestown H5
Ramor (lake) G4
Rathlin O'Birne (isl.) C2
Ree (lake) F5
Rinn (lake) F4
Roaringwater (bay) B9
Rosscarbery (bay) D9
Rosskeeragh (pt.) D3
Royal (canal) G4
Saint Finan's (bay) A8
Saint George's (chan.) K7
Saint John's (pt.) D2
Saltee (isls.) H7
Scarriff (isl.) A8
Seven Hogs, The (isls.) A7
Shannon (riv.) E6
Shannon, Mouth of the (delta) B6
Sheeffry (hills) B4
Sheelin (lake) G4
Sheep Haven (harb.) F1
Sheeps (head) B8
Shehy (mts.) C8
Sherkin (isl.) C9
Silvermine (mts.) E6
Slaney (riv.) H7
Slieve Anierin (mt.) F3
Slieve Aughty (mts.) D5
Slieve Bernagh (mt.) D6
Slieve Bloom (mts.) F5
Slieve Callan (mt.) C6
Slieve Car (mt.) B3
Slieve Elva (mt.) C5
Slieve Gamph (Ox) (mts.) D3
Slievefelim (mts.) E6
Slievenaman (mt.) F7
Sligo (bay) D3
Slyne (head) A5
Smerwick (harb.) A7
South (sound) C5
Stacks (mts.) B7
Suck (riv.) E5
Sugarloaf (mt.) B8
Suir (riv.) G7
Swilly (inlet) F1
Tara (hill) H4
Toe (head) C9
Tory (isl.) E1
Tory (sound) E1
Tralee (bay) B7
Tramore (bay) G7
Truskmore (mt.) E3
Twelve Pins (mt.) B4
Ulster (part) (prov.) 236,008 G2
Valencia (Valentia) (isl.) A8
Valentia (isl.) A8
Waterford (harb.) G7
Wexford (harb.) J7
Wexford (bay) J7
Wicklow (mts.) J6
Youghal (bay) F8

NORTHERN IRELAND

DISTRICTS

Antrim 44,384 J2
Ards 57,626 K2
Armagh 47,618 H3
Ballymena 54,426 J2
Ballymoney 22,873 J1
Banbridge 29,885 J3
Belfast 295,223 J2
Carrickfergus 28,458 K2
Castlereagh 60,757 K2
Coleraine 46,272 H1
Cookstown 26,624 H2
Craigavon 71,202 J3
Down 52,869 K3
Dungannon 41,073 H3
Fermanagh 51,008 F3
Larne 28,929 K2
Limavady 26,270 H1
Lisburn 82,091 J2
Londonderry 83,384 G2
Magherafelt 30,825 H2
Mourne (Newry and Mourne) 72,243 J3
Moyle 14,252 J1
Newtownabbey 71,631 J2
North Down 65,849 K2
Omagh 41,159 G2
Strabane 35,028 G2

CITIES and TOWNS

Annalong 1,823 K3
Antrim 22,342 J2
Armagh 12,700 H3
Augher 1,874 G3
Aughnacloy 1,659 H3
Ballycarry 1,652 K2
Ballycastle 3,284 J1
Ballyclare 6,159 J2
Ballygawley 2,099 G3
Ballymena 28,166 J2
Ballymoney 5,679 J1
Ballynahinch 3,721 J3
Banbridge 9,650 J3
Bangor 46,585 K2
Belfast (cap.) 295,223 J2
Bellaghy 1,854 H2
Belleek and Boa 2,469 E3
Beragh 2,028 G2
Bessbrook 2,756 J3
Brookeborough 2,250 G3
Broughshane 1,503 J2
Bushmills 1,381 J1
Caledon 1,633 H3
Carnlough 1,462 K2
Carrickfergus 17,633 K2
Carrowdore 3,019 K2
Carryduff 2,666 K2
Castledawson 1,460 H2
Castlederg 1,730 F2
Castlewellan 2,105 K3
Claudy 2,516 G2
Clogher 1,792 G3
Cloughmills 1,558 J2
Coalisland 3,324 H2
Coleraine 15,967 H1
Comber 7,600 K2
Cookstown 7,649 H2
Craigavon 10,195 J3
Crumlin 1,708 J2
Cullybackey 2,098 J2
Derrygonnelly 2,627 F3
Donaghadee 3,874 K2
Downpatrick 8,245 K3
Dromore, Banbridge 3,089 J3
Dromore, Omagh 2,286 G3
Drumquin 1,865 F2
Dundrum 2,295 K3
Dungannon 8,295 H3
Dungiven 2,249 H2
Dunloy 1,593 J1
Dunnamanagh 2,191 G2
Ederney, Kesh and Lark 2,607 F2
Enniskillen 10,429 F3
Feeny 1,402 H2
Fintona 1,353 G3
Fivemiletown 1,758 G3
Garvagh 2,222 H2
Gilford 1,512 J3
Glenarm 1,533 J2
Glenavy 2,402 J2
Glynn 1,689 K2
Gortin 1,877 G2
Greenisland 5,103 K2
Grey Abbey 2,945 K2
Groomsport 3,870 K2
Holywood 9,462 K2
Irvinestown 1,827 F3
Keady 2,561 H3
Kells 2,564 J2
Kesh, Ederney and Lark 2,607 F3
Kilkeel 6,036 K3
Killough 3,104 K3
Killyclogher 5,557 G2
Killyleagh 2,094 K3
Kilrea 1,320 H2
Lambeg J2
Larne 18,224 K2
Limavady 8,015 H1
Lisbellaw 2,395 K2
Lisburn 40,391 J2
Lisnaskea 1,568 G3
Londonderry (Derry) 62,692 G2
Loughbrickland 2,244 J3
Lurgan 20,991 J3
Macosquin 2,267 H1
Maghera 1,953 H2
Magherafelt 5,044 H2
Millisle 1,373 K2
Moy 2,163 H3
Newcastle 6,246 J3
Newry 19,426 J3
Newtownabbey 56,149 K2
Newtownards 20,531 K2
Newtownbutler 2,632 G3
Newtownhamilton 1,654 H3
Newtownstewart 1,425 G2
Omagh 14,627 G2
Pomeroy 1,638 H2
Portadown 21,333 H3
Portaferry 2,148 K3
Portglenone 2,017 H2
Portrush 5,114 H1
Portstewart 5,312 H1
Randalstown 3,591 J2
Rathfriland 2,243 J3
Richhill 1,728 H3
Rostrevor 1,852 J3
Sion Mills 1,771 G2
Sixmilecross 1,613 G2
Stewartstown 1,554 H2
Strabane 9,413 G2
Strangford 2,062 K3
Strathfoyle 2,050 G1
Tandragee 2,224 J3
Tempo 2,149 G3
Trillick 2,017 G3
Warrenpoint 4,798 J3
Whitehead 3,546 K2

OTHER FEATURES

Arney (riv.) F3
Bann (riv.) H2
Beg (lake) J2
Belfast (inlet) K2
Blackwater (riv.) H3
Bush (riv.) H1
Copeland (isl.) K2
Derg (riv.) F2
Divis (mt.) J2
Dundrum (bay) K3
Erne, Lough (lake) F3
Fair (head) J1
Foyle (riv.) G2
Foyle (inlet) G1
Garron (pt.) K1
Giant's Causeway H1
Lagan (riv.) K2
Larne (inlet) K2
Macnean (lake) F3
Magee, Island (pen.) K2
Main (riv.) J2
Mourne (riv.) G2
Mourne (mts.) J3
Neagh (lake) J2
North (chan.) K1
Owenkillew (riv.) G2
Rathlin (sound) J1
Rathlin (isl.) J1
Red (bay) K1
Roe (riv.) H1
Saint John's (pt.) K3
Slieve Beagh (mt.) G3
Slieve Donard (mt.) K3
Sperrin (mts.) G2
Strangford (inlet) K3
Trostan (mt.) J1
Ulster (part) (prov.) G2
Upper Lough Erne (lake) F3

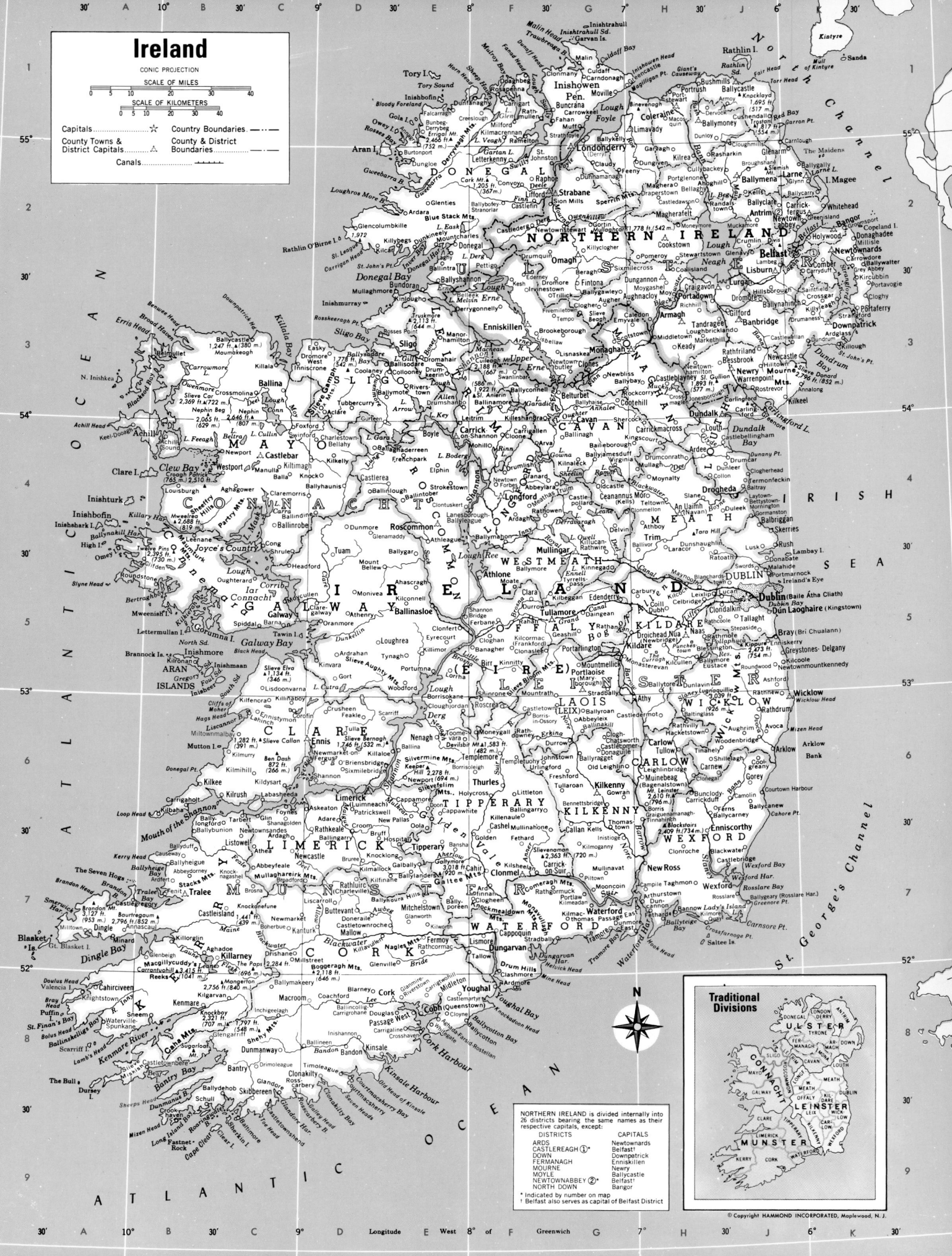

Ireland

CONIC PROJECTION

SCALE OF MILES

SCALE OF KILOMETERS

Capitals..................☆ Country Boundaries.----·-
County Towns & County & District
District Capitals........△ Boundaries...............-··-
Canals.....................

Traditional Divisions

NORTHERN IRELAND is divided internally into
26 districts bearing the same names as their
respective capitals, except:

DISTRICTS	CAPITALS
ARDS	Newtownards
CASTLEREAGH ① *	Belfast†
DOWN	Downpatrick
FERMANAGH	Enniskillen
MOURNE	Newry
MOYLE	Ballycastle
NEWTOWNABBEY ② *	Belfast†
NORTH DOWN	Bangor

* Indicated by number on map
† Belfast also serves as capital of Belfast District

© Copyright HAMMOND INCORPORATED, Maplewood, N.J.

Norway, Sweden, Finland and Denmark

CONIC PROJECTION

SUBDIVISIONS
Indicated by Numbers

Counties in NORWAY
1 Akershus   G 6
2 Vestfold   G 7
3 Østfold   G 7
4 Oslo   G 7

Oslo is the administrative
center for Akershus and
Oslo County.

Counties in SWEDEN
5 Göteborg och
  Bohus   G 7
6 Västmanland   K 7
7 Södermanland K 7
8 Östergötland H 7
9 Malmohus   H 9
10 Kristianstad J 9

SCALE OF MILES
0   50   100   150

SCALE OF KILOMETERS
0   50   100   150   200

Capitals of Countries .... ☆
Administrative Centers .... △
International Boundaries .... —·—·—
Internal Boundaries .... ———
Canals .... ·······

AREA 125,053 sq. mi.
(323,887 sq. km.)
POPULATION 4,242,000
CAPITAL Oslo
LARGEST CITY Oslo
HIGHEST POINT Glittertinden
8,110 ft. (2,472 m.)
MONETARY UNIT krone
MAJOR LANGUAGE Norwegian
MAJOR RELIGION Protestantism

AREA 173,665 sq. mi.
(449,792 sq. km.)
POPULATION 8,541,000
CAPITAL Stockholm
LARGEST CITY Stockholm
HIGHEST POINT Kebnekaise 6,946 ft.
(2,117 m.)
MONETARY UNIT krona
MAJOR LANGUAGE Swedish
MAJOR RELIGION Protestantism

AREA 130,128 sq. mi.
(337,032 sq. km.)
POPULATION 4,973,000
CAPITAL Helsinki
LARGEST CITY Helsinki
HIGHEST POINT Haltiatunturi
4,343 ft. (1,324 m.)
MONETARY UNIT markka
MAJOR LANGUAGES Finnish, Swedish
MAJOR RELIGION Protestantism

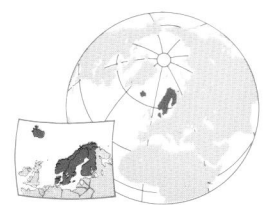

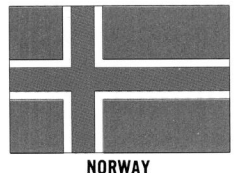

NORWAY

SWEDEN

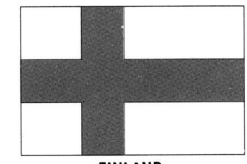

FINLAND

FINLAND

PROVINCES

Ahvenanmaa 23,591	L6
Åland (Ahvenanmaa) 23,591	L6
Häme 677,750	O6
Keski-Suomi 247,693	O5
Kuopio 256,036	P5
Kymi 340,665	Q6
Lappi 200,943	P3
Mikkeli 239,029	P6
Oulu 432,141	P4
Pohjois-Karjala 177,567	Q5
Turku ja Pori 713,050	N6
Uusimaa 1,187,851	O6
Vaasa 444,348	N5

CITIES and TOWNS

Abo (Turku) 161,398	N6
Alavus 10,701	N5
Äänekoski 11,447	O5
Anjalamkoski 19,703	P6
Borga 19,513	O6
Espoo 156,778	O6
Forssa 20,074	N6
Haapajärvi 8,454	O5
Hämeenlinna 42,382	O6
Hamina 10,313	P6
Hangö 12,071	N7
Hanko (Hangö) 12,071	N7
Harjavalta 8,955	M6
Heinola 16,112	P6
Helsinki (cap.) 485,795	O6
Hyvinkää 38,742	O6
Iisalmi 23,612	P5
Ikaalinen 8,184	N6
Imatra 35,085	Q6
Jakobstad 20,458	N5
Jämsä 12,498	O6
Järvenpää 27,220	O6
Joensuu 46,850	R5
Jyväskylä 65,282	O5
Kajaani 36,020	P4
Kankaanpää 13,652	M6
Karis (Karjaa)	N6
Karkkila 8,355	N6
Kauniainen 7,746	O6
Kemi 26,421	O4
Kemijärvi 12,762	P3
Kerava 26,207	O6
Kokemäki 9,741	N6
Kokkola 34,489	N5
Kotka 58,956	P6
Kouvola 31,829	P6
Kristiinankaupunki	
(Kristinestad) 9,081	N5
Kristinestad 9,081	N5
Kuopio 78,124	Q5
Kurikka 11,512	N5
Kuusankoski 22,089	P6
Lahti 94,447	O6
Lappeenranta 54,102	Q6
Lapua 14,644	N5
Lieksa 18,588	R5
Loimaa 7,053	N6
Lovisa 8,697	P6
Maarianhamina	
(Mariehamn) 9,829	M7
Mänttä 8,092	O6
Mariehamn 9,829	M7
Mikkeli 31,636	P6
Naantali 10,246	M6
Nokia 24,325	N6
Nurmes 11,419	Q5
Nykarleby 7,768	N5
Oulainen 8,225	O4
Oulu 97,297	O4
Outokumpu 9,678	Q5
Parainen 11,618	M6
Parkano 8,692	N6
Pieksämäki 14,372	P5
Pietarsaari	
(Jakobstad) 20,458	N5
Pori 78,376	M6
Pudasjärvi 11,453	P4
Raahe 18,932	O4
Raisio 19,671	M6
Rauma 30,921	M6
Riihimäki 24,366	O6
Rovaniemi 32,782	O3
Salo 20,495	N6
Savonlinna 28,667	Q6
Seinäjoki 26,257	N5
Suonenjoki 8,981	P5
Tampere 169,026	N6
Toijala 8,046	N6
Tornio 22,328	O4
Turku 161,398	N6
Utsjoki 1,548	P2
Uusikaarlepyy	
(Nykarleby) 7,768	N5
Uusikaupunki 14,026	M6
Vaasa 54,333	M5
Valkeakoski 22,582	N6
Vammala 16,024	N6
Vantaa 143,844	O6
Varkaus 24,856	Q5
Vasa (Vaasa) 54,333	M5
Virrat 9,391	N5
Ylivieska 12,559	O4

OTHER FEATURES

Åland (isls.)	L6
Baltic (sea)	K9
Bothnia (gulf)	M5
Finland (gulf)	P7
Hailuoto (isl.)	O4
Haltiatunturi (mt.)	M2
Haukivesi (lake)	Q5
Iijoki (riv.)	O4
Inari (lake)	P2
Ivalojoki (riv.)	P2
Kallavesi (lake)	P5
Karlö (Hailuoto) (isl.)	O4
Keitele (lake)	O5
Kemijärvi (lake)	Q3
Kemijoki (riv.)	Q3
Lapland (reg.)	O2
Lappajärvi (lake)	O5
Lapuanjoki (riv.)	N5
Lokka (reg.)	Q3
Muojärvi (lake)	R4
Muonio (riv.)	M2
Näsijärvi (lake)	O6
Orihvesi (lake)	Q5
Oulujärvi (lake)	O4
Oulujoki (riv.)	O4
Ounasjoki (riv.)	O3
Päijänne (lake)	O6
Pielinen (lake)	Q5
Porkkala (pen.)	O7
Puruvesi (lake)	Q6
Saimaa (lake)	Q6
Tana (riv.)	P2
Tornionjoki (riv.)	O3
Ylikitka (lake)	Q3

NORWAY

COUNTIES

Akershus 399,797	G6
Aust-Agder 95,475	E7
Buskerud 221,384	F6
Finnmark 74,690	O2
Hedmark 186,305	G6
Hordaland 402,343	E6
Møre og Romsdal 237,489	E5
Nordland 241,048	G3
Nord-Trøndelag 126,648	H4
Oppland 181,620	F6
Oslo (city) 449,220	D3
Østfold 235,813	G7
Rogaland 326,611	D6
Sogn og Fjordane 105,466	E6
Sør-Trøndelag 247,354	G5
Telemark 162,595	F7
Troms 146,595	L2
Vest-Agder 141,284	E7
Vestfold 192,934	D4

CITIES and TOWNS

Ålesund 40,868	D5
Ålgård 2,322	D7
Alta 5,582	N2
Åndalsnes 2,574	F5
Årdalstangen 2,360	F6
Arendal 11,701	F7
Årnes 2,267	G6
Askim 8,413	E4
Bamble† 7,031	F7
Bergen 213,434	D6
Bodø 31,077	J3
Borge† 3,294	H2
Brate 2,107	G7
Brønnøysund 3,130	G4
Drammen 50,777	C4
Drøbak 4,538	D4
Eidsvoll 2,906	G6
Eigersund 11,379	D7
Elverum 7,391	G6
Farsund 8,908	E7
Flekkefjord 8,750	E7
Flora 8,822	D6
Fredrikstad 29,024	D4
Gjøvik 25,963	G6
Grimstad 13,091	F7
Halden 27,087	G7
Hamar 16,418	G6
Hammerfest 7,610	N1
Harstad 21,125	K2
Hauge 2,079	E7
Haugesund 27,386	D7
Holmestrand 8,246	C4
Honningsvag 3,780	O1
Horten 13,746	D4
Kirkenes 4,466	Q2
Kongsberg 19,854	F7
Kongsvinger 16,146	H6
Kopervik 4,221	D7
Kornsjø† 6,079	G7
Kragerø 5,249	F7
Kristiansand 59,488	F8
Kristiansund 18,847	E5
Kvinnherad† 2,898	E6
Larvik 9,097	C4
Lenvik† 11,098	L2
Levanger 5,066	G5
Lillehammer 21,248	F6
Lillestrøm† 11,550	E3
Lodingen 1,840	J2
Longyearbyen	D2
Lysaker† 81,612	D3
Mandal 11,579	E7
Meråker† 2,907	G5
Mo 21,033	J3
Molde 20,334	E5
Mosjøen 9,341	H4
Moss 25,786	D4
Mysen 3,760	G7
Namsos 11,452	G4
Narvik 19,582	K2
Nesttun† 11,519	D6
Nittedal† 8,889	D3
Notodden 12,970	F7
Nøtterøy 11,944	D4
Odda 7,401	E6
Oppdal 2,173	F5
Oranger 3,685	F5
Oslo (cap.) 462,732	D3
Oslo* 645,413	D3
Porsgrunn 31,709	G7
Rakkestad 2,392	G7
Ringerike 30,156	C3
Risør 6,560	F7
Rjukan 5,334	F7
Røros 3,041	G5
Saetermoen 2,114	L2
Sandefjord 33,350	C4
Sandnes 33,934	D7
Sandvika† 34,337	C3
Sarpsborg 12,889	D4
Seljet 3,386	D5
Ski 9,081	D4
Skien 47,105	F7
Skudeneshavn 2,206	D7
Stavanger 86,639	D7
Stavern 2,604	D4
Steinkjer 20,553	G4
Stor-Elvdal† 2,993	G6
Stordalsøra 5,114	F5
Svelvik 2,256	D4
Svolvaer 3,942	J2
Tana 1,893	Q1
Tønsberg 9,964	D4
Tromsø 43,830	L2
Trondheim 134,910	F5
Tvedestrand 1,689	F7
Ullensvang† 2,326	E6
Vadsø 6,019	Q1
Vardø 3,875	R1
Vanylven 1,966	E5
Vik 1,019	E6
Volda 3,511	E5
Voss 5,944	E6

OTHER FEATURES

Andøya (isl.)	J2
Barentsøya (isl.)	D2
Bjornøya (isl.)	D3
Boknafjord (fjord)	D7
Dovrefjell (hills)	F5
Edgeøya (isl.)	E2
Femundsjø (lake)	G6
Folda (fjord)	J3
Folda (fjord)	H3
Frohavet (bay)	F5
Frøya (isl.)	F5
Glittertinden (mt.)	F6
Greenland (sea)	C3
Hadselfjorden (fjord)	J2
Haltiatunturi (mt.)	M2
Hardangerfjord (fjord)	D7
Hardangervidda (plat.)	E6
Hinlopenstreten (strait)	C1
Hinnøya (isl.)	J2
Hitra (isl.)	F5
Hortensfjord (fjord)	G4
Isfjorden (fjord)	C2
Kjølen (mts.)	K3
Kvaenangen (fjord)	M2
Kvaløy (isl.)	K2
Kvaløya (isl.)	O1
Lakselfjorden (fjord)	P1
Langøya (isl.)	J2
Lapland (reg.)	K2
Lindesnes (cape)	E8
Lofoten (isls.)	H2
Lopphavet (bay)	M1
Magerøya (isl.)	P1
Moskenesøya (isl.)	H3
Namsen (riv.)	H4
Nordaustlandet (isl.)	D1
Nordfjord (fjord)	E6
Nordkapp (pt.)	C1
North Cape	
(Nordkapp) (cape)	P1
Norwegian (sea)	F3
Ofotfjorden (fjord)	K2
Oslofjord (fjord)	D4
Otra (riv.)	E7
Pasvikelv (riv.)	Q2
Porsangen (fjord)	O1
Prins Karls Forland (isl.)	B2
Rana (fjord)	H3
Rauma (riv.)	F5
Ringvassøy (isl.)	L2
Romsdalsfjorden (fjord)	E5
Saltfjorden (fjord)	J3
Senja (isl.)	K2
Seiland (isl.)	N1
Skagerrak (strait)	C7
Sognafjorden (fjord)	D6
Sørkapp (pt.)	C2
Sørøya (isl.)	N1
Spitsbergen (isl.)	D2
Steinneset (cape)	E2
Storfjorden (fjord)	E5
Sulitjelma (mt.)	J3
Svalbard (isls.)	C3
Tana (riv.)	P1
Tanafjord (fjord)	Q1
Trondheimsfjorden (fjord)	G5
Tyrifjord (lake)	F6
Vannøy (isl.)	L1
Varangerfjord (fjord)	Q1
Varangerhalvøya (pen.)	Q1
Vegafjorden (fjord)	G4
Vesterålen (isls.)	J2
Vestfjord (fjord)	H3
Vestvågøya (isl.)	H3
Vikna (isls.)	G4

(continued on following page)

Horn
Fontur
Nordkapp
(North Cape)
Varangerfjord
VATNA-
JÖKULL
VESTER-
ÅLEN
Haltiatunturi
4,343 ft.
(1324m.)
Inari
Faxaflói
Hvannadal-
Hekla   snnúkur
3,891 ft.   6,946 ft.
(1491 m.)   (2117 m.)
Reykjavík
LOFOTEN
Iceland
Kebnekaise
6,946 ft.
(2117 m.)
Lapland
Uddjaur
Ylikitka
Trondheim
fjorden
Storsjön
Oulujärvi
GULF OF BOTHNIA
Glittertinden
8,110 ft.
(2472 m.)
Bergen
Mjøsa
Hardanger
fjord
Oslo
ÅLAND
Helsinki
IS.
Nordfjord
Vänern
Stockholm
Lindesnes
Vättern
Göta
Canal
Topography
Skagerrak
Göteborg
Gotland
0   100   200 MI.
0   100   200 KM.
Yding
Skovhøj
568 ft.
(173 m.)
Öland
Copenhagen
Fyn
Below Sea 100 m.   200 m.   500 m. 1,000 m. 2,000 m. 5,000 m.
Lolland   Level 328 ft. 656 ft. 1,640 ft. 3,281 ft. 6,562 ft. 16,404 ft.
Bornholm

Agriculture, Industry and Resources

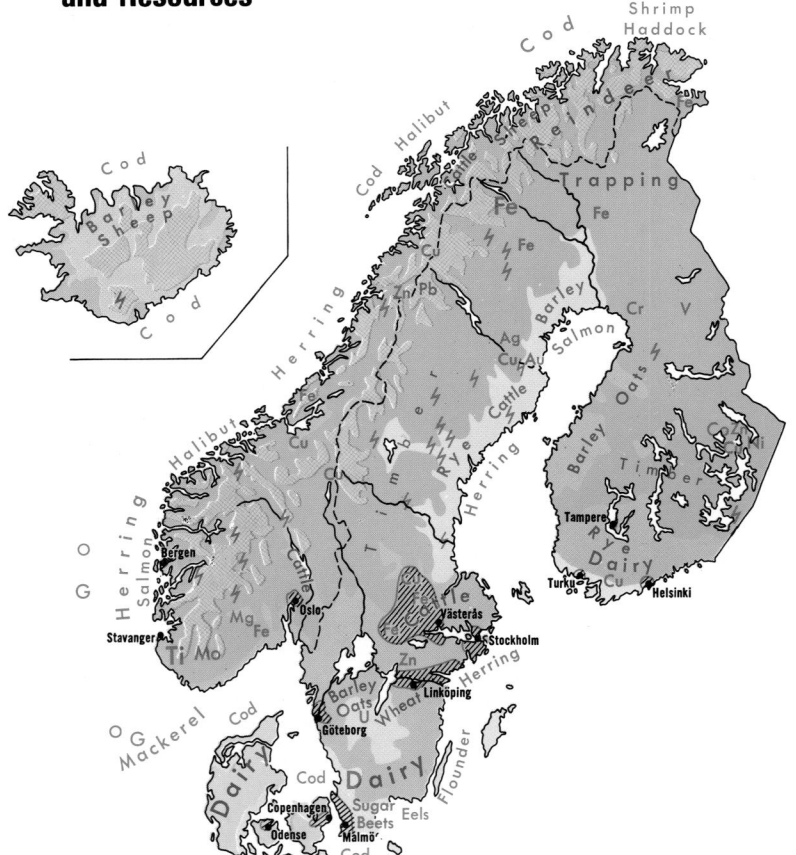

DOMINANT LAND USE

- Cash Cereals, Dairy
- Dairy, Cattle, Hogs
- Dairy, General Farming
- General Farming (chiefly cereals)
- Nomadic Sheep Herding
- Forests, Limited Mixed Farming
- Nonagricultural Land

MAJOR MINERAL OCCURRENCES

Ag	Silver	Ni	Nickel
Au	Gold	O	Petroleum
Co	Cobalt	Pb	Lead
Cr	Chromium	Ti	Titanium
Cu	Copper	U	Uranium
Fe	Iron Ore	V	Vanadium
Mg	Magnesium	Zn	Zinc
Mo	Molybdenum		

⚡ Water Power

▨ Major Industrial Areas

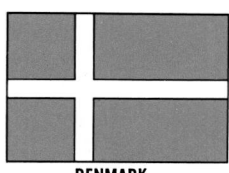

DENMARK

ICELAND

DENMARK

AREA 16,629 sq. mi. (43,069 sq. km.)
POPULATION 5,135,000
CAPITAL Copenhagen
LARGEST CITY Copenhagen
HIGHEST POINT Yding Skovhøj
568 ft. (173 m.)
MONETARY UNIT krone
MAJOR LANGUAGE Danish
MAJOR RELIGION Protestantism

ICELAND

AREA 39,768 sq. mi. (103,000 sq. km.)
POPULATION 250,000
CAPITAL Reykjavík
LARGEST CITY Reykjavík
HIGHEST POINT Hvannadalshnúkur
6,952 ft. (2,119 m.)
MONETARY UNIT króna
MAJOR LANGUAGE Icelandic
MAJOR RELIGION Protestantism

Ringe 4,440	D7
Ringkøbing 7,907	A5
Ringsted 16,564	E7
Rødby 2,572	E8
Rødding 2,291	B7
Rødekro 4,195	C7
Rønde 1,745	D5
Ronne 14,143	F9
Roskilde 39,659	E6
Rudkøbing 4,667	D8
Ry 3,648	C5
Ryomgård 1,578	D5
Saeby 7,464	D3
Sakskøbing 7,518	E8
Silkeborg 33,304	C5
Sindal 2,665	D3
Skaelskør 5,574	E7
Skaerbaek 2,946	B7
Skagen 11,743	D2
Skals 1,410	C4
Skanderborg 11,094	D5
Skibby 2,259	E6
Skive 19,034	B4
Skjern 6,351	B6
Skørping 2,016	C4
Slagelse 28,539	E7
Slangerup 5,218	E6
Søllested 1,431	E8
Sønderborg 25,885	C8
Sønder Omne 1,751	B6
Sønderso 2,667	D7
Sorø 6,067	E7
Stege 3,906	F8
Stenlille 1,370	E6
Stenstrup 1,501	D7
Stoholm 1,830	C5
Store Heddinge 2,881	F7
Støvring 4,615	C4
Strandby 2,264	D3
Struer 10,973	B5
Stubbekøbing 2,255	F8
Svendborg 23,847	D7
Svinninge 2,281	E6
Tarm 4,008	B6
Tårnby 41,517	F6
Them 1,382	C5
Thisted 12,469	B4
Thyborøn 2,766	A4
Tinglev 2,508	C8
Toftlund 3,388	B7
Tølløse 2,279	E6
Tommerup 1,890	D7
Tønder 7,914	B8
Tørring 1,922	C6
Ulfborg 1,781	B5
Vamdrup 3,960	C7
Varde 10,888	B6
Vejen 7,412	C7
Vejle 43,300	C6
Vemb 1,261	B5
Viborg 28,659	C5
Viby 2,942	F6
Videbaek 3,440	B5
Vildbjerg 2,703	B5
Vinderup 2,865	B5
Vojens 6,792	C7
Vordingborg 8,706	E7
Vrå 2,312	C3

OTHER FEATURES

AEro (isl.)	D8
Alborg (bay)	D4
Als (isl.)	C8
Amager (isl.)	F6
Anholt (isl.)	E4
Baltic (sea)	E9
Blavands Huk (pt.)	A6
Bornholm (isl.)	F9
Dovns Klint (cliff)	D8
Endelave (isl.)	D6
Fakse (bay)	F7
Falster (isl.)	F8
Fanø (isl.)	B7
Fehmarn (strait)	E8
Fejø (isl.)	E8
Femø (isl.)	E8
Frisian, North (isls.)	B7
Fyn (isl.)	D7
Fyns Hoved (pt.)	D6
Gedser Odde (pt.)	E8
Gelså (riv.)	C7
Gudenå (riv.)	C5
Isefjord (fjord)	E6
Jammerbugt (bay)	C3
Jutland (pen.)	C5
Jylland (Jutland) (pen.)	C5
Kattegat (strait)	E4
Knosen (mt.)	D3
Koge (bay)	F7

Laesø (isl.)	D3
Langeland (isl.)	D8
Langelands Baelt (chan.)	D8
Lille Baelt (chan.)	C7
Lillea (riv.)	B5
Limfjorden (fjord)	A4
Limfjorden (fjord)	D4
Løgstør Bredning (fjord)	C4
Lolland (isl.)	E8
Mariager (fjord)	D4
Møn (isl.)	F8
Mons Klint (cliff)	F8
Mors (isl.)	B4
Nissum (fjord)	A5
North (sea)	B9
North Frisian (isls.)	B7
Omme (riv.)	B6
Ømø (isl.)	E7
Øresund (sound)	F6
Ringkobing (fjord)	B6
Rømø (isl.)	B7
Rosnaes (pen.)	D6
Sams Baelt (chan.)	D6
Samsø (isl.)	D6
Sejerø (isl.)	E6
Sjaelland (isl.)	E6
Sjaellands Odde (pen.)	E5
Skagens Odde (cape)	D2
Skagerrak (strait)	C2
Skaw, The (Skagens Odde) (cape)	D2
Skive (riv.)	C5
Stevns Klint (cliff)	F7
Storå (riv.)	B5
Store Baelt (chan.)	D7
The Skaw (Skagens Odde) (cape)	D2
Varde (riv.)	B6
Vejle (fjord)	C6
Vorgod (riv.)	B6
Yding Skovhøj (mt.)	C6

FAROE ISLANDS

CITIES and TOWNS

Klaksvík 4,536	B2
Tórshavn (cap.), Faroe Is. 11,618	A3

OTHER FEATURES

Faroe (isls.)	B2
Sandoy (isl.)	B3
Streymoy (isl.)	B3
Sudhuroy (isl.)	B3

ICELAND

CITIES and TOWNS

Akranes 5,404	B1
Akureyri 13,972	C1
Hafnarfjördhur 14,199	B2
Húsavík 2,499	C1
Ísafjördhur 3,458	B1
Keflavík 7,305	B1
Kópavogur 15,551	B1
Olafsfjördhur 1,179	C1
Reykjavík (cap.) 95,811	B1
Saudhárkrókur 2,478	C1
Vestmannaeyjar 4,743	B2

OTHER FEATURES

Bjargtangar (pt.)	A1
Breidhafjördhur (fjord)	B1
Faxaflói (bay)	B1
Fontur (pt.)	D1
Gerpir (cape)	D1
Grímsey (isl.)	C1
Hekla (vol.)	C1
Hofsjökull (glacier)	C1
Horn (cape)	B1
Húnaflói (bay)	C1
Hvannadalshnúkur (mt.)	C1
Jökulsá (riv.)	C1
Lagarfljót (stream)	D1
Langjökull (glacier)	B1
North (Horn) (cape)	B1
Reykjanestá (cape)	A2
Rifstangi (cape)	C1
Skagata (cape)	B1
Skjálfandafljót (stream)	C1
Surtsey (isl.)	B2
Thjórsá (riv.)	C1
Vatnajökull (glacier)	C1
Vopnafjördhur (fjord)	D1

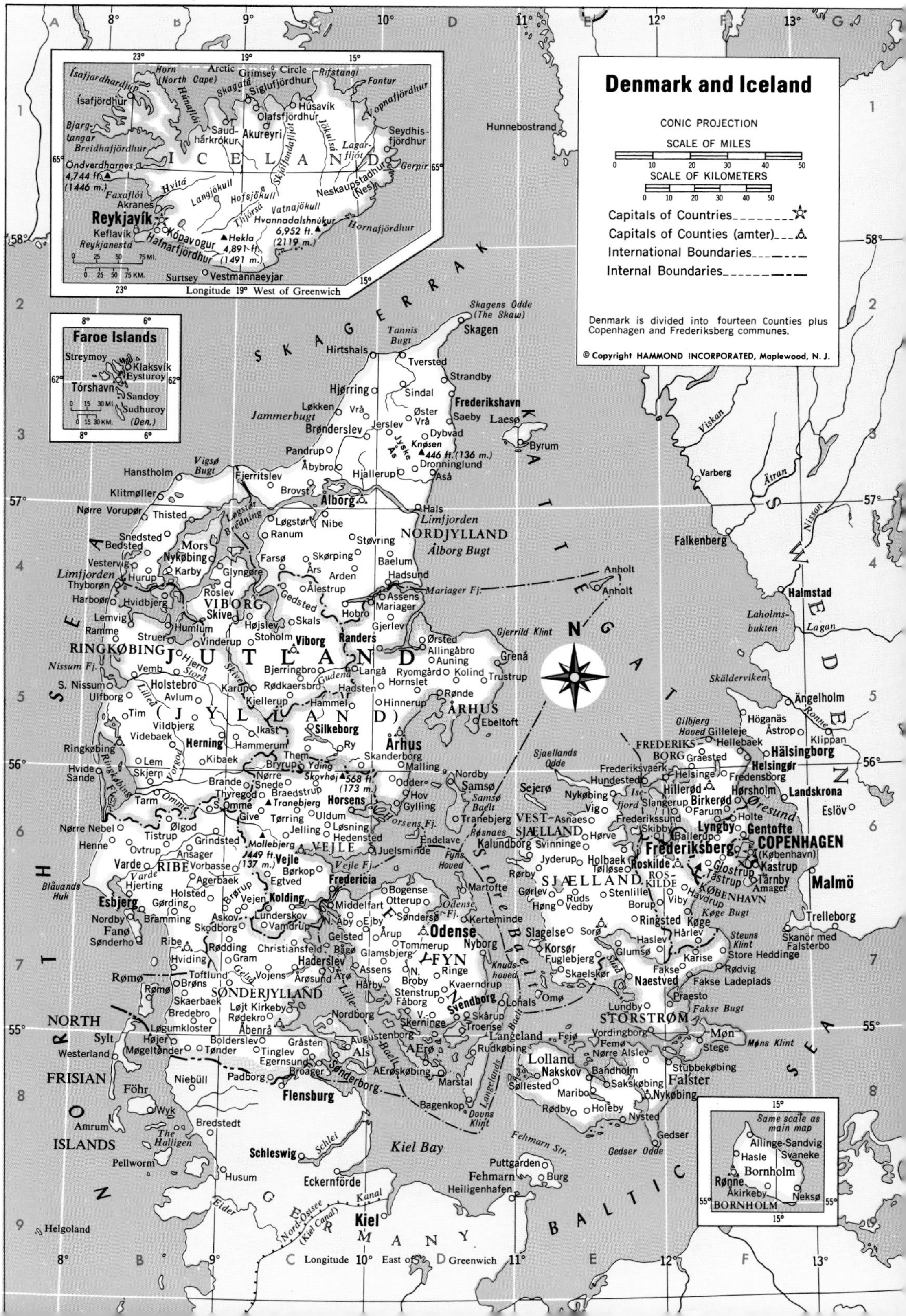

Denmark and Iceland

CONIC PROJECTION

SCALE OF MILES

SCALE OF KILOMETERS

Capitals of Countries ········· ☆
Capitals of Counties (amter) ··· △
International Boundaries ········
Internal Boundaries ···········

Denmark is divided into fourteen Counties plus Copenhagen and Frederiksberg communes.

AREA 137,753 sq. mi. (356,780 sq. km.)
POPULATION 78,890,000
CAPITAL Berlin
LARGEST CITY Berlin
HIGHEST POINT Zugspitze 9,718 ft. (2,962 m.)
MONETARY UNIT Deutsche mark
MAJOR LANGUAGE German
MAJOR RELIGIONS Protestantism, Roman Catholicism

GERMANY

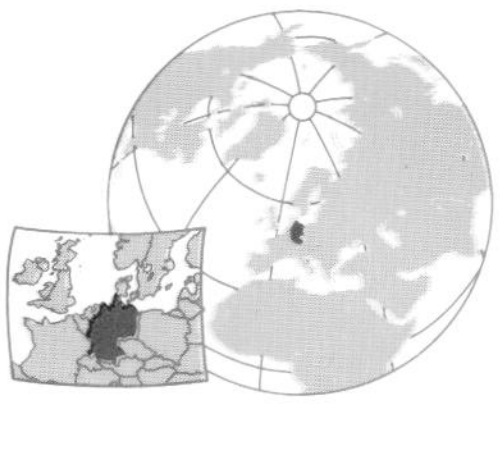

Topography

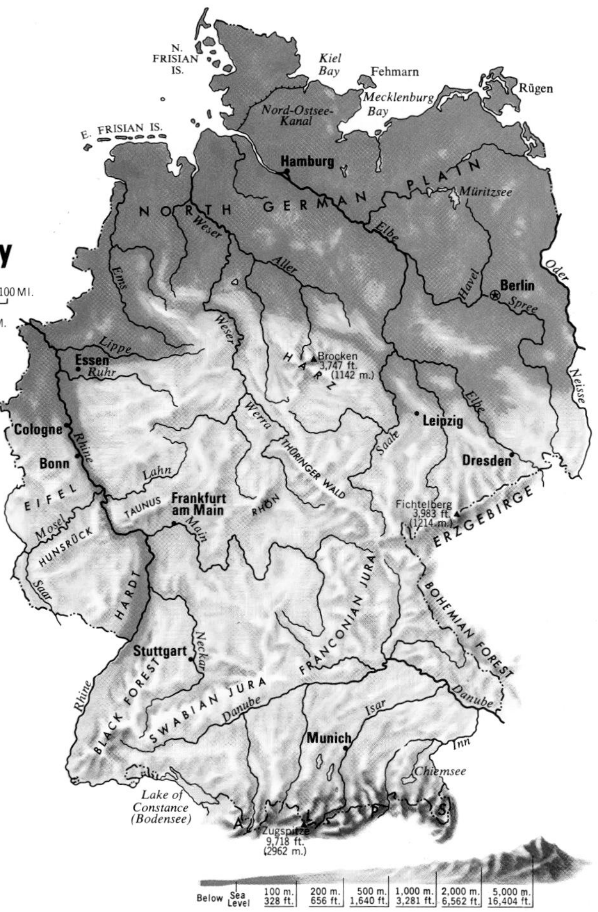
N. FRISIAN IS.
Kiel Bay
Fehmarn
Rügen
Mecklenburg Bay
Nord-Ostsee-Kanal
E. FRISIAN IS.
Hamburg
NORTH GERMAN PLAIN
Müritzsee
Weser
Aller
Elbe
Havel
Oder
Berlin
Spree
Ems
Lippe
Essen
Ruhr
Brocken 3,747 ft. (1142 m.)
HARZ
Neisse
Cologne
Werra
Saale
Leipzig
Bonn
Rhine
Lahn
THÜRINGER WALD
Dresden
EIFEL
TAUNUS
Frankfurt am Main
RHÖN
Main
Fichtelberg 3,983 ft. (1214 m.)
ERZGEBIRGE
Mosel
HUNSRÜCK
Saar
HARDT
FRANCONIAN JURA
BOHEMIAN FOREST
Stuttgart
Neckar
SWABIAN JURA
Danube
Isar
BLACK FOREST
Munich
Inn
Chiemsee
Lake of Constance (Bodensee)
ALPS
Zugspitze 9,718 ft. (2962 m.)
0 50 100 MI.
0 50 100 KM.
Below Sea Level
100 m. 328 ft.
200 m. 656 ft.
500 m. 1,640 ft.
1,000 m. 3,281 ft.
2,000 m. 6,562 ft.
5,000 m. 16,404 ft.

GERMANY

STATES

Baden-Württemberg 9,432,709 C4
Bavaria 11,049,263 D4
Berlin 3,304,561 E4
Brandenberg* E2
Bremen 661,992 C2
Hamburg 1,603,070 D2
Hesse 5,568,892 C3
Lower Saxony 7,184,943 C2
Mecklenburg-Western Pomerania* E2
North Rhine-Westphalia 16,874,059 B3
Rhineland-Palatinate 3,653,155 B4
Saarland 1,054,142 B4
Saxony* E3
Saxony-Anhalt* D3
Schleswig-Holstein 2,564,565 C1
Thuringia* D3

*East German States 15,611,488 D-E 2-3

CITIES and TOWNS

Aachen 233,255 B3
Aalen 62,812 D4
Ahaus 30,180 B2
Ahlen 52,836 B3
Ahrensburg 27,174 D2
Alfeld 21,986 C2
Alsdorf 46,328 B3
Alsfeld 16,686 C3
Altena 23,301 B3
Altenburg 53,602 E3
Amberg 42,246 D4
Andernach 27,171 B3
Anklam 19,946 E2
Annaberg-Buchholz 26,002 E3
Ansbach 36,912 D4
Apolda 28,230 D3
Arnsberg 73,912 C3
Arnstadt 30,207 D3
Aschaffenburg 62,048 C4
Aschersleben 34,166 D3
Aue 27,935 E3
Auerbach 22,324 E3
Augsburg 247,731 D4
Aurich 36,063 B2
Backnang 30,583 C4
Bad Berleburg 20,080 C3
Bad Driburg 16,698 C3
Bad Dürkheim 16,670 C4
Baden-Baden 50,761 C4
Bad Harzburg 23,079 D3
Bad Hersfeld 28,214 C3
Bad Homburg vor der Höhe 51,035 C3
Bad Honnef 21,812 B3
Bad Kissingen 20,237 D3
Bad Kreuznach 39,400 B4
Bad Langensalza 17,027 D3
Bad Mergentheim 19,801 C4
Bad Münstereifel 15,232 B3
Bad Nauheim 27,561 C3
Bad Neuenahr-Ahrweiler 24,610 B3
Bad Oldesloe 20,473 D2
Bad Pyrmont 20,437 C3
Bad Reichenhall 16,365 E5
Bad Salzuflen 50,875 C2
Bad Salzungen 21,387 D3
Bad Schwartau 19,960 D2
Bad Vilbel 24,567 C3
Bad Zwischenahn 23,348 B2
Balingen 30,615 C4
Bamberg 69,809 D4
Barsinghausen 37,792 C2
Bautzen 52,354 F3
Bayreuth 70,933 D4
Bensheim 34,241 C4
Berchtesgaden 7,644 E5
Bergen 16,713 E1
Bergisch Gladbach 101,983 B3
Berleburg (Bad Berleburg) 20,080 C3
Berlin (cap.) 3,304,561 E4
Bernau bei Berlin 19,919 E2
Bernburg 40,834 D3
Biberach an der Riss 28,319 C4
Bielefeld 311,946 C2
Bietigheim-Bissingen 37,573 C4
Bingen 23,141 B4
Bitburg 10,758 B4
Bitterfeld 20,869 E3
Blankenburg am Harz 19,279 D3
Böblingen 43,400 C4
Bocholt 67,565 B3
Bochum 389,087 B3
Bonn 282,190 B3
Borghorst 17,238 B2
Borken 34,710 B3
Borna 24,397 E3
Bornheim 34,536 B3
Bottrop 116,363 B3
Brake 16,069 C2
Bramsche 24,653 B2
Brandenburg 94,755 E2
Braunschweig 253,794 D2
Bremen 535,058 C2
Bremerhaven 126,934 C2
Bremervörde 17,629 C2
Bretten 23,894 C4
Brilon 24,341 C3
Bruchsal 36,831 C4
Brühl 40,710 B3
Buchholz in der Nordheide 30,523 C2
Bückeburg 19,758 C2
Büdingen 17,013 C3
Bühl 23,470 C4
Bünde 39,103 C2
Büren 17,720 C3
Burg bei Magdeburg 28,359 D2
Burghausen 16,767 E4
Burgsteinfurt 31,367 B2
Butzbach 21,095 C3
Buxtehude 31,132 C2
Castrop-Rauxel 77,660 B3
Celle 71,050 D2
Cham 16,641 E4
Chemnitz 313,799 E3
Clausthal-Zellerfeld 16,069 D3
Cloppenburg 22,536 B2
Coburg 43,233 D3
Coesfeld 31,979 B3
Cologne 937,482 B3
Coswig 27,590 E3
Cottbus 126,592 F3
Crailsheim 26,678 D4
Crimmitschau 24,440 E3
Cuxhaven 55,249 C2
Dachau 34,183 D4
Darmstadt 136,067 C4
Deggendorf 28,680 E4
Delitzsch 27,636 E3
Delmenhorst 72,901 C2
Demmin 16,992 E2
Dessau 103,538 E3
Detmold 66,809 C3
Dillenburg 23,672 C3
Dillingen 21,358 B4
Döbeln 27,706 E3
Donaueschingen 18,296 C5
Donauwörth 17,420 D4
Dorsten 75,518 B3
Dortmund 587,328 B3
Dresden 519,810 E3
Duderstadt 22,265 D3
Duisburg 527,447 B3
Dülmen 39,344 B3
Düren 83,120 B3
Düsseldorf 569,641 B3
Eberswalde-Finow 54,566 E2
Eckernförde 22,197 C1
Ehingen 22,580 C4
Eilenburg 21,931 E3
Einbeck 25,813 C3
Eisenach 49,534 D3
Eisenhüttenstadt 51,729 F2
Eisleben 26,484 D3
Ellwangen 21,857 D4
Elmshorn 42,784 C2
Emden 49,803 B2
Emmendingen 22,959 B4
Emmerich 27,906 B3
Emsdetten 31,063 B2
Erfurt 217,134 D3
Erkelenz 36,525 B3
Erlangen 100,583 D4
Eschwege 21,527 C3
Eschweiler 53,516 B3
Espelkamp 23,868 C2
Essen 620,594 B3
Esslingen am Neckar 90,537 C4
Ettlingen 37,269 C4
Euskirchen 47,756 B3
Eutin 16,567 D1
Falkensee 23,024 E3
Fellbach 39,612 C4
Finsterwalde 23,857 E3
Flensburg 85,830 C1
Forchheim 28,784 D4
Forst 26,501 F3
Frankenberg-Eder 16,283 C3
Frankenthal 45,408 C4
Frankfurt am Main 625,258 C3
Frankfurt an der Oder 86,441 F2
Frechen 42,516 B3
Freiberg 50,415 E3
Freiburg im Breisgau 183,979 B5
Freising 35,201 D4
Freital 43,092 E3
Freudenstadt 21,355 C4
Friedberg 24,279 C3
Friedrichshafen 52,295 C5
Fulda 54,320 C3
Fürstenfeldbruck 30,313 D4
Fürstenwalde 35,282 F2
Fürth 98,832 D4
Füssen 13,173 D5
Gaggenau 28,182 C4
Garbsen 59,225 C2
Garmisch-Partenkirchen 25,908 D5
Geesthacht 25,054 D2
Geislingen an der Steige 26,176 C4
Geldern 28,465 B3
Gelnhausen 18,866 C3
Gelsenkirchen 287,255 B3
Genthin 17,347 E2
Georgsmarienhütte 30,880 B2
Gera 132,319 E3
Geretsried 21,081 D5
Gifhorn 35,697 D2
Glauchau 28,309 E3
Goch 29,592 B3
Göppingen 52,873 C4
Görlitz 78,856 F3
Goslar 45,614 D3
Gotha 57,423 D3
Göttingen 118,073 D3
Greifswald 67,298 E1
Greiz 34,858 E3
Greven 29,671 B2
Grevenbroich 59,204 B3
Griesheim 20,531 C4
Grimma 17,812 E3
Gronau 39,397 B2
Guben 34,665 F3
Gummersbach 49,017 B3
Günzburg 18,303 D4
Güstrow 38,971 E2
Gütersloh 83,407 C3
Haar 16,553 D4
Hagen 210,640 B3
Halberstadt 47,017 D3
Haldensleben 20,369 D2
Halle 236,148 D3
Halle-Neustadt 93,477 D3
Haltern 33,093 B3
Hamburg 1,603,070 D2
Hameln 57,642 C2
Hamm 173,611 B3
Hanau 84,300 C3
Hannover 498,495 C2
Hassloch 18,646 C4
Heide 19,909 C1
Heidelberg 131,429 C4
Heidenau 19,133 E3
Heidenheim an der Brenz 48,497 D4
Heilbronn 112,279 C4
Helmstedt 26,554 D2
Hennef 30,516 B3
Hennigsdorf bei Berlin 26,574 E3
Herborn 20,409 C3
Herford 61,700 C2
Herne 174,664 B3
Hettstedt 21,861 D3
Hildesheim 103,512 D2
Hof 50,938 D3
Holzminden 20,877 C3
Homburg 41,888 B4
Höxter 31,925 C3
Hoyerswerda 69,113 F3
Hückelhoven 33,841 B3
Hürth 49,094 B3
Husum 20,649 C1
Ibbenbüren 43,424 B2
Idar-Oberstein 33,227 B4
Ilmenau 29,338 D3
Ingolstadt 97,702 D4
Iserlohn 93,337 B3
Itzehoe 32,342 C2
Jena 107,610 D3
Jülich 30,496 B3
Kaiserslautern 96,990 B4
Kamenz 18,323 F3
Karlsruhe 265,100 C4
Kassel 189,156 C3
Kaufbeuren 39,192 D5
Kehl 28,902 B4
Kempten 60,052 D5
Kevelaer 22,633 B3
Kiel 240,675 D1
Kirchheim unter Teck 34,534 C4
Kitzingen 19,085 C4
Koblenz 107,286 B3
Köln (Cologne) 937,482 B3
Königs Wusterhausen 19,085 E2
Königswinter 34,136 B3
Konstanz 72,862 C5
Köpenick 118,059 F4
Korbach 21,406 C3
Kornwestheim 28,519 C4
Köthen 34,617 E3
Krefeld 235,423 B3
Kreuztal 29,716 C3
Kronach 18,246 D3
Kulmbach 27,116 D3
Lage 32,612 C3
Lahnstein 17,972 B3
Lahr 33,369 B4
Lampertheim 30,263 C4
Landau in der Pfalz 36,297 C4
Landsberg am Lech 19,808 D4
Landshut 57,194 E4
Langen 31,206 C4
Langenhagen 46,298 C2
Lauchhammer 24,391 E3
Lauenburg an der Elbe 10,786 D2
Lauf an der Pegnitz 22,593 D4
Leer 31,292 B2
Lehrte 39,600 D2
Leipzig 550,641 E3
Lemgo 38,351 C2
Lengerich 20,235 B2
Leverkusen 157,358 B3
Lichtenberg 95,426 F4
Lichtenfels 20,252 D3
Limburg an der Lahn 29,196 C3
Limbach-Oberfrohna 22,059 E3
Lindau 23,699 C5
Lingen 47,837 B2
Lippstadt 60,396 C3
Löbau 18,492 F3
Löhne 36,882 C2
Lörrach 41,087 B5
Lübbenau 20,815 F3
Lübeck 210,681 D2
Luckenwalde 26,761 E2
Lüdenscheid 76,118 B3
Ludwigsburg 79,342 C4
Ludwigshafen am Rhein 158,478 C4
Lüneburg 60,053 D2
Lünen 85,584 B3
Magdeburg 288,975 D2
Mainz 174,828 C4
Mannheim 300,468 C4
Marburg 70,905 C3
Markkleeberg 19,240 E3
Marktredwitz 18,605 E4
Marl 89,601 B3
Mayen 18,427 B3
Mechernich 21,986 B3
Meerane 21,879 E3
Meiningen 25,823 D3
Meissen 37,757 E3
Melle 40,490 C2
Memmingen 37,942 D5
Meppen 29,900 B2
Merseburg 46,188 D3
Merzig 29,312 B4
Meschede 30,853 C3
Metzingen 19,895 C4
Minden 75,169 C2
Mittenwald 7,998 D5
Mittweida 18,469 E3
Mönchengladbach 252,910 B3
Mosbach 23,897 C4
Mülhausen 43,046 D3
Mülheim an der Ruhr 175,454 B3
München (Munich) 1,211,617 D4
Münden 24,794 C3
Munich 1,211,617 D4
Münster 248,919 B3
Nagold 20,405 C4
Naumburg 32,100 D3
Neckarsulm 21,765 C4
Neubrandenburg 87,235 E2
Neuburg an der Donau 24,502 D4
Neu-Isenburg 34,896 C3
Neumarkt in der Oberpfalz 33,603 D4
Neumünster 79,574 C1
Neunkirchen 50,784 B4
Neuruppin 26,934 E2
Neuss 143,976 B3
Neustadt an der Weinstrasse 50,453 B4
Neustadt bei Coburg 16,211 D3
Neustrelitz 27,300 E2
Neu-Ulm 45,116 D4
Neuwied 60,665 B3
Nienburg 29,545 C2
Norden 23,655 B2
Nordenham 28,393 C2
Norderstedt 66,747 D2
Nordhausen 47,681 D3
Nordhorn 48,556 B2
Nördlingen 18,278 D4
Northeim 30,349 C3
Nuremberg 480,078 D4
Nürnberg (Nuremberg) 480,078 D4
Nürtingen 36,807 C4
Oberammergau 4,980 D5
Oberhausen 221,017 B3
Oberursel 39,105 C3
Offenbach am Main 112,450 C3
Offenburg 51,730 B4
Oldenburg 140,785 C2
Oranienburg 28,667 E2
Oschatz 19,100 E3
Oschersleben 16,976 D2
Osnabrück 154,594 C2
Osterholz-Scharmbeck 24,205 C2
Osterode am Harz 26,631 D3
Paderborn 114,148 C3
Pankow 62,847 F3
Papenburg 29,237 B2
Parchim 23,454 D2
Passau 49,137 E4
Peenemünde E1
Peine 45,522 D2
Pfaffenhofen an der Ilm 18,335 D4
Pforzheim 108,887 C4
Pinneberg 36,583 C2
Pirmasens 47,102 B4
Pirna 46,991 E3
Plauen 77,514 E3
Plettenberg 28,113 C3
Pössneck 17,895 D3
Potsdam 141,231 E2
Prenzlau 23,642 E2
Quedlinburg 29,168 D3
Radeberg 15,702 E3
Radebeul 33,757 E3
Radolfzell 25,712 C5
Rastatt 40,909 C4
Rastede 18,191 C2
Rathenow 31,302 E2
Ratingen 89,880 B3
Ravensburg 44,146 C5
Recklinghausen 121,666 B3
Regensburg 119,078 E4

(continued on following page)

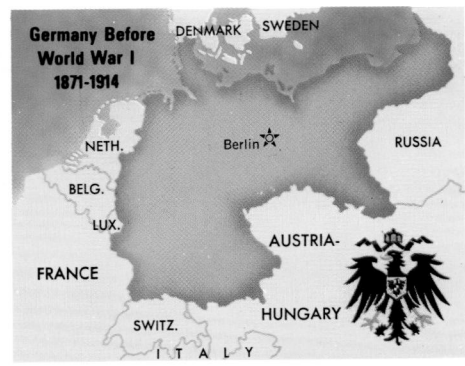

Germany Before World War I 1871-1914

Germany Between Wars 1919-1937

Occupied Germany 1945-1949

Reichenbach 24,749 ....E3
Remagen 14,375 ....B3
Remscheid 120,979 ....B3
Rendsburg 30,752 ....C1
Reutlingen 100,400 ....C4
Rheda-Wiedenbrück 36,990 ..C3
Rheine 69,324 ....B2
Rheinfelden 27,711 ....B5
Ribnitz-Damgarten 17,512 ..E1
Riesa 49,108 ....E3
Rietberg 23,058 ....C3
Rinteln 26,120 ....C2
Rosenheim 54,304 ....D5
Rostock 249,349 ....E1
Rotenburg 18,392 ....C2
Roth bei Nürnberg 20,288 ..D4
Rothenburg ob der Tauber
11,071 ....D4
Rottenburg am Neckar
33,907 ....C4
Rottweil 23,080 ....C4
Rudolstadt 32,264 ....D3
Rüsselsheim 58,426 ....C4
Saalfeld 33,453 ....D3
Saarbrücken 188,467 ....B4
Saarlouis 37,662 ....B4
Salzgitter 111,674 ....D2
Salzwedel 23,163 ....D2
Sangerhausen 33,604 ....D3
Sankt Ingbert 40,527 ....B4
Sankt Wendel 26,649 ....B4
Saulgau 14,864 ....C5
Schleswig 26,648 ....C1
Schmalkalden 17,409 ....D3
Schneeberg 22,105 ....E3
Schönebeck 45,155 ....D2
Schramberg 18,208 ....C4
Schwabach 34,217 ....D4
Schwäbisch Gmünd 57,861 ....C4
Schwäbisch Hall 31,375 ....C4

Schwalmstadt 17,371 ....C3
Schwandorf im Bayern
25,874 ....E4
Schwedt 51,753 ....F2
Schweinfurt 52,818 ....D3
Schwelm 29,564 ....B3
Schwetzingen 18,029 ....C4
Seesen 21,604 ....D2
Selb 19,275 ....E3
Senftenberg 32,428 ....F3
Siegburg 34,402 ....B3
Siegen 106,160 ....C3
Sigmaringen 15,270 ....C4
Sindelfingen 57,524 ....C4
Singen 42,605 ....C5
Soest 40,775 ....C3
Solingen 160,824 ....B3
Soltau 19,115 ....C2
Sömmerda 23,398 ....D3
Sondershausen 24,178 ....D3
Sonneberg 28,512 ....D3
Sonthofen 20,037 ....D5
Spandau E3
Speyer 45,089 ....C4
Spremberg 24,815 ....F3
Springe 29,209 ....C2
Stade 41,223 ....C2
Stadthagen 22,218 ....C2
Starnberg 19,845 ....D4
Stassfurt 27,372 ....D3
Stendal 47,880 ....D2
Stolberg 56,182 ....B3
Stralsund 75,857 ....E1
Straubing 40,612 ....E4
Strausberg 27,527 ....F2
Stuttgart 562,658 ....C4
Suhl 55,295 ....D3
Sulzbach 19,753 ....C4
Sulzbach-Rosenberg 18,134 ..D4

Telgte 16,834 ....B3
Tempelhof F4
Thale 16,605 ....D3
Torgau 22,749 ....E3
Traunstein 17,145 ....E5
Treptow 58,938 ....F4
Treuchtlingen 12,314 ....D4
Triberg im Schwarzwald
5,697 ....C4
Trier 95,692 ....B4
Troisdorf 62,011 ....B3
Tübingen 76,046 ....C4
Tuttlingen 31,752 ....C5
Übach-Palenberg 23,005 ....B3
Überlingen 18,043 ....C5
Ueckermünde 12,304 ....F2
Uelzen 34,891 ....D2
Uetersen 17,218 ....C2
Ulm 106,508 ....C4
Varel 23,718 ....C2
Vechta 22,759 ....C2
Verden 23,770 ....C2
Viersen 76,163 ....B3
Villingen-Schwenningen
76,258 ....C4
Völklingen 42,916 ....B4
Waldheim 10,316 ....E3
Waldkirch 18,893 ....B4
Waldkraiburg 23,177 ....E4
Waldshut-Tiengen 21,372 ....C5
Walsrode 22,232 ....C2
Waltershausen 14,127 ....D3
Wangen im Allgäu 23,822 ....C5
Warburg 21,802 ....C3
Waren 24,318 ....E2
Warendorf 33,891 ....B3
Wedel 30,158 ....C2
Weida 10,602 ....D3
Weiden in der Oberpfalz
41,539 ....D4

Weilheim im Oberbayern
17,602 ....D5
Weimar 63,910 ....D3
Weingarten 21,522 ....C5
Weinheim 41,876 ....C4
Weissenburg im Bayern
17,318 ....D4
Weissenfels 38,763 ....D3
Weissensee 31,858 ....F3
Weisswasser 36,472 ....F3
Werdau 19,451 ....E3
Wernigerode 36,499 ....D3
Wertheim 20,457 ....C4
Wesel 57,986 ....B3
Westerstede 18,184 ....B2
Wiehl 21,897 ....B3
Wiesbaden 254,209 ....C4
Wiesmoor 10,827 ....B2
Wilhelmshaven 89,892 ....B2
Winsen 26,139 ....D2
Wismar 58,066 ....D1
Witten 103,637 ....B3
Wittenberg 53,670 ....E3
Wittenberge 30,389 ....D2
Wolfen 43,606 ....E3
Wolfenbüttel 50,960 ....D2
Wolfsburg 125,831 ....D2
Worms 74,809 ....C4
Wunstorf 37,115 ....C2
Wuppertal 371,283 ....B3
Würzburg 125,589 ....C4
Xanten 16,097 ....B3
Zeitz 42,985 ....E3
Zerbst 18,717 ....E3
Zeulenroda 14,409 ....D3
Zirndorf 21,608 ....D4
Zittau 39,305 ....F3
Zwickau 120,923 ....E3

OTHER FEATURES

Aller (riv.) C2
Allgäu (reg.) D5
Altmark (reg.) D2
Ammersee (lake) D4
Amrum (isl.) C1
Arkona (cape) E1
Baltic (sea) E4
Bavarian (forest) E4
Bavarian Alps (range) E5
Bayerischer Wald Nat'l Park ..E4
Black (forest) C4
Black Elster (riv.) E3
Bodensee (Constance) (lake) ..C5
Bohemian (forest) E4
Borkum (isl.) B2
Breisgau (reg.) B5
Brocken (mt.) D3
Chiemsee (lake) E5
Constance (lake) C5
Danube (riv.) C4
Donau (Danube) (riv.) E4
East Friesland (reg.) B2
Eder (riv.) C3
Elbe (riv.) D2
Elde (riv.) D2
Ems (riv.) B2
Erzgebirge (mts.) E3
Fehmarn (isl.) D1
Feldberg (mt.) C5
Fichtelberg (mt.) E3
Fichtelgebirge (range) E3
Föhr (isl.) C1
Franconian Jura (range) D4
Frisian, East (isls.) B2
Frisian, North (isls.) B1
Fulda (riv.) C3
Grosser Arber (mt.) E4
Harz (mts.) D3

Havel (riv.) E2
Hegau (reg.) C5
Helgoland (bay) C1
Helgoland (isl.) B1
Hunsrück (mts.) B4
Iller (riv.) D4
Ilmenau (riv.) D2
Inn (riv.) E4
Isar (riv.) E4
Jade (bay) C2
Juist (isl.) B2
Kaiserstuhl (mt.) B4
Kiel (bay) D1
Kiel (Nord-Ostsee) (canal) C1
Königssee (lake) E5
Lahn (riv.) C3
Langeoog (isl.) B2
Lech (riv.) D4
Leine (riv.) C2
Lippe (riv.) B3
Lüneburger Heide (dist.) C2
Lusatia (reg.) F3
Main (riv.) C4
Mecklenburg (bay) D1
Mosel (riv.) B3
Mulde (riv.) E3
Müritzsee (lake) E2
Naab (riv.) E4
Neckar (riv.) C4
Neisse (riv.) F3
Norderney (isl.) B2
Nord-Ostsee (canal) C1
Nordstrand (isl.) C1
North (sea) C1
North Friesland (reg.) C1
Oder (riv.) F2
Oder-Haff (lag.) F2
Our (riv.) B3
Peene (riv.) E2

Pellworm (isl.) C1
Plauersee (lake) E2
Pomeranian (bay) F1
Regnitz (riv.) D4
Rhine (riv.) B3
Rhön (mts.) D3
Rügen (isl.) E1
Ruhr (riv.) B3
Saale (riv.) D3
Saar (riv.) B4
Salzach (riv.) E5
Sauer (riv.) B4
Sauerland (reg.) B3
Schwarzwald (Black) (forest) ..C4
Schwerinersee (lake) D2
Spessart (range) C4
Spiekeroog (isl.) B2
Spree (riv.) F3
Spreewald (forest) F3
Starnbergersee (lake) D5
Swabian Jura (range) C4
Sylt (isl.) C1
Taunus (range) C4
Tegernsee (lake) D5
Teutoburger Wald (forest) C2
Thüringer Wald (forest) D3
Unstrut (riv.) D3
Usedom (isl.) F1
Vechte (riv.) B2
Vogelsberg (mts.) C3
Walchensee (lake) D5
Wasserkuppe (mt.) C3
Watzmann (mt.) E5
Werra (riv.) C3
Weser (riv.) C2
Westerwald (forest) B3
White Elster (riv.) E3
Würmsee (Starnbergersee)
(lake) D5
Zugspitze (mt.) D5

Agriculture, Industry and Resources

DOMINANT LAND USE

- Wheat, Sugar Beets
- Cereals (chiefly rye, oats, barley)
- Potatoes, Rye
- Dairy, Livestock
- Mixed Cereals, Dairy
- Truck Farming
- Grapes, Fruit
- Forests

MAJOR MINERAL OCCURRENCES

Ag	Silver	K	Potash
Ba	Barite	Lg	Lignite
C	Coal	Na	Salt
Cu	Copper	O	Petroleum
Fe	Iron Ore	Pb	Lead
G	Natural Gas	U	Uranium
Gr	Graphite	Zn	Zinc

⚡ Water Power

▨ Major Industrial Areas

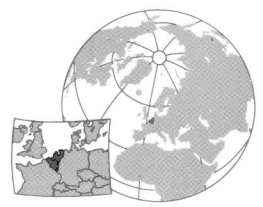

AREA 15,892 sq. mi. (41,160 sq. km.)
POPULATION 14,906,000
CAPITALS The Hague, Amsterdam
LARGEST CITY Amsterdam
HIGHEST POINT Vaalserberg 1,056 ft. (322 m.)
MONETARY UNIT guilder (florin)
MAJOR LANGUAGE Dutch
MAJOR RELIGIONS Protestantism, Roman Catholicism

AREA 11,781 sq. mi. (30,513 sq. km.)
POPULATION 9,883,000
CAPITAL Brussels
LARGEST CITY Brussels (greater)
HIGHEST POINT Botrange 2,277 ft. (694 m.)
MONETARY UNIT Belgian franc
MAJOR LANGUAGES French (Walloon), Flemish
MAJOR RELIGION Roman Catholicism

AREA 999 sq. mi. (2,587 sq. km.)
POPULATION 378,000
CAPITAL Luxembourg
LARGEST CITY Luxembourg
HIGHEST POINT Ardennes Plateau 1,825 ft. (556 m.)
MONETARY UNIT Luxembourg franc
MAJOR LANGUAGES Luxembourgeois (Letzeburgisch), French, German
MAJOR RELIGION Roman Catholicism

NETHERLANDS

BELGIUM

LUXEMBOURG

BELGIUM

PROVINCES

Antwerp 1,569,876 .... F6
Brabant 2,221,222 ....
East Flanders 1,331,192 .... D7
Hainaut 1,301,477 .... D7
Liège 999,413 .... H7
Limburg 716,888 .... G7
Luxembourg 221,926 .... G9
Namur 407,400 .... H8
West Flanders 1,079,253 ... B7

CITIES and TOWNS

Aalst 78,938 .... D7
Aalter 15,554 .... C6
Aarlen (Arlon) 22,279 ...... H9
Aarschot 25,168 .... F7
Alken 9,563 .... G7
Amay 12,725 .... G7
Andenne 22,341 .... G8
Anderlecht 94,764 .... B9

Anderlues 11,700 .... E8
Ans 26,016 .... H7
Antoing 7,970 .... C7
Antwerp 185,897 .... E6
Antwerp* 918,144 .... E6
Antwerpen (Antwerp) 185,897 .... E6
Ardooie 9,458 .... C7
Arendonk 10,561 .... G6
Arlon 22,279 .... H9
Asse 26,425 .... E7
Assenede 13,353 .... D6
Aubange 14,696 .... H9
Audenarde (Oudenarde) 26,615 .... D7
Auderghem 30,435 .... C9
Aywaille 8,194 .... H8
Baerle-Hertog 2,111 .... F6
Balen 18,162 .... G6
Bastenaken (Bastogne) 11,386 .... H9
Bastogne 11,386 .... H9
Beauraing 7,641 .... F8
Beernem 13,526 .... C6

Beloeil 13,553 .... D7
Berchem 45,423 .... F6
Berchem-Sainte-Agathe 18,719 .... B9
Bergen (Mons) 94,417 ...... E8
Beringen 34,254 .... G6
Bertrix 7,244 .... G9
Beveren 40,857 .... E6
Bilzen 25,683 .... G7
Binche 33,651 .... E8
Blankenberge 14,832 ...... C6
Bocholt 10,142 .... H6
Boom 14,827 .... E6
Borgerhout 43,521 .... E6
Borgworm (Waremme) 11,907 .... G7
Bourg-Léopold (Leopoldsburg) 9,593 .... G6
Boussu 21,558 .... D8
Braine-l'Alleud 30,028 .... E7
Braine-le-Comte 16,475 ... D7
Brecht 16,391 .... F6
Bredene 10,538 .... B6
Bree 13,345 .... H6

Bruges 118,020 .... C6
Brugge (Bruges) 118,020 .. C6
Brussels (cap.)* 997,293 .. C9
Bruxelles (Brussels) (cap.)* 997,293 ...... C9
Charleroi 222,343 .... E8
Charleroi* 443,832 .... E8
Châtelet 38,506 .... F8
Chimay 9,273 .... E8
Ciney 13,330 .... G8
Comines 18,034 .... B7
Courcelles 29,757 .... E8
Courtrai (Kortrijk) 75,917 . C7
Couvin 12,909 .... F8
Damme 9,881 .... C6
De Haan 8,655 .... C6
Deinze 24,871 .... C7
Denderleeuw 16,497 ..... E7
Dendermonde 22,119 .... E6
De Panne 9,507 .... B6
Dessel 8,074 .... G6
Destelbergen 15,741 ..... D6
Deurne 77,635 .... F6
Diest 20,491 .... F7

Diksmuide 15,347 .... B6
Dilbeek 35,050 .... B9
Dilsen 15,910 .... H6
Dinant 12,105 .... G8
Dison 14,225 .... H7
Dixmude (Diksmuide) 15,347 B6
Doornik (Tournai) 67,906 .. C7
Dour 17,737 .... D8
Duffel 14,684 .... F6
Durbuy 7,729 .... H8
Ecaussinnes 9,739 ...... E7
Edingen (Enghien) 10,095 .. D7
Eeklo 19,637 .... D6
Eghezée 10,683 .... F7
Eigenbrakel (Braine-l'Alleud) 30,028 .. E7
Ekeren 30,294 .... E6
Enghien 10,095 .... D7
Erquelinnes 10,029 .... E8
Esneux 11,559 .... H7
Essen 12,505 .... F6
Estampuis 9,601 ...... C7
Etterbeek 44,218 ...... B9
Eupen 16,847 .... J7

Evere 30,520 .... C9
Evergem 28,974 .... D6
Farciennes 12,205 .... E8
Flémalle 28,217 ...... G7
Fleurus 22,574 .... E8
Florennes 10,537 ..... F8
Forest 50,607 .... B9
Fosses-La-Ville 7,678 . F8
Frameries 21,470 ..... D8
Frasnes-lez Anvaing 10,751 D7
Furnes (Veurne) 11,253 B6
Ganshoren 21,445 .... B9
Geel 31,463 .... G6
Geldenaken (Jodoigne) 8,983 F7
Gembloux-sur-Orneau 17,636 F7
Genk 61,502 .... H7
Gent (Ghent) 239,256 . D6
Geraardsbergen 17,533 . D7
Gerpinnes 10,808 .... F8
Ghent 239,256 .... D6
Ghent* 485,565 ...... D6
Gistel 9,531 .... B6
Grammont (Geraardsbergen) 17,533 ..D7
Grez-Doiceau 8,795 .. F7
Grimbergen 32,038 ... F7
Haacht 11,285 .... F7
Hal (Halle) 15,293 ... E7
Halen 7,865 .... G7
Halle 15,293 .... E7
Hamme 22,790 .... E6
Hamont-Achel 11,939 . H6
Hannuit (Hannut) 11,527 G7
Hannut 11,527 ...... G7
Harelbeke 25,214 .... C7
Hasselt 64,613 ...... G7
Heist-op-den-Berg 34,617 F6
Hensies 6,806 .... D8
Herentals 23,797 .... F6
Herselt 11,340 ...... F6
Herstal 38,592 ...... H7
Herve 14,276 .... H7
Heuvelland 8,540 .... B7
Hoboken 34,563 ..... E6
Hoei (Huy) 17,331 ... G8
Hoeselt 8,497 ...... G7
Hoogstraten 14,368 .. F6
Huy 17,331 .... G8
Ichtegem 12,259 .... B6
Ieper 34,425 .... B7
Ingelmunster 10,434 . C7
Ixelles 75,723 ...... C9
Izegem 26,410 ...... C7
Jabbeke 10,629 ..... C6
Jemappes 18,632 .... D8
Jemeppe-sur-Sambre 17,120 F8
Jette 40,109 .... B9
Jodoigne 8,983 ..... F7
Kalmthout 14,960 ... F6
Kapellen 14,536 .... E6
Kasterlee 14,612 ... F6
Kinrooi 10,138 ..... H6
Knokke-Heist 28,868 . C6
Koekelare 7,606 .... B6
Koekelberg 16,643 .. B9
Koksijde 13,679 .... B6
Kontich 17,878 ..... E6
Kortemark 12,580 ... C6
Kortrijk 75,917 .... C7
Kraainem 11,780 .... C9
La Louvière 77,326 .. E8
Lanaken 20,272 .... H7
Landen 14,081 ..... G7
Langemark-Poelkapelle 7,097 B7
Lasne 10,919 ...... F7
Lede 17,249 .... D7
Lens 3,726 .... D7
Leopoldsburg 9,593 . G6
Le Roeulx 7,754 ... E8
Lessen (Lessines) 16,553 D7
Lessines 16,553 ... D7
Leuven 85,076 .... F7
Leuze-en-Hainaut 12,863 C7
Libramont-Chevigny 7,859 G9
Lichtervelde 7,459 . C7
Liedekerke 11,609 . D7
Liège 214,119 ..... H7
Liège* 605,123 .... H7

Lier 31,261 .... F6
Lierre (Lier) 31,261 . F6
Limbourg 5,350 ... J7
Limburg (Limbourg) 5,350 . J7
Linter 6,568 ..... G7
Lochristi 16,125 .. D6
Lokeren 33,369 .. D6
Lommel 25,412 .. G6
Louvain (Leuven) 85,076 . F7
Luik (Liège) 214,119 . H7
Lummen 11,793 ... G7
Maaseik 20,056 .. H6
Maasmechelen 33,618 . H7
Machelen 11,273 . C9
Malines (Mechelen) 77,269 . F6
Malmédy 10,036 .. J8
Marche-en-Famenne 14,115 G8
Mechelen 77,269 . F6
Meerhout 8,613 .. G6
Meise 15,078 .... E7
Menen 33,542 ... C7
Menin (Menen) 33,542 . C7
Merchtem 12,972 . E7
Merelbeke 19,773 . D7
Merksem 41,600 .. E6
Merksplas 6,136 . F6
Mettet 9,958 .... F8
Meulebeke 10,471 . C7
Middelkerke 14,168 . B6
Moeskroen (Mouscron) 54,590 .. C7
Mol 29,798 ...... G6
Molenbeek-Saint-Jean 70,850 .... B9
Mons 94,417 .... E8
Montigny-le-Tilleul 9,726 . E8
Moorslede 10,974 . B7
Mortsel 26,746 .. E6
Mouscron 54,590 . C7
Namen (Namur) 102,321 . F8
Namur 102,321 .. F8
Nazareth 9,248 .. D7
Neerpelt 12,779 .. G6
Neufchâteau 6,039 . G9
Nevele 10,471 ... D6
Nieuport (Nieuwpoort) 8,195 B6
Nieuwpoort 8,195 . B6
Nijvel (Nivelles) 21,580 . E7
Ninove 33,393 ... D7
Nivelles 21,580 .. E7
Oostende (Ostend) 68,915 B6
Oostkamp 19,747 . C6
Opwijk 11,451 ... E7
Ostend 68,915 ... B6
Oudenaarde 26,615 . D7
Oudenburg 8,138 . B6
Oud-Turnhout 10,733 . F6
Oupeye 22,453 .. H7
Overijse 21,428 .. F7
Overpelt 11,233 .. G6
Peer 12,099 .... G6
Péruwelz 16,664 . D8
Philippeville 6,916 . E8
Poelkapelle-Langemark 7,097 B7
Pont-à-Celles 15,444 . E8
Poperinge 19,886 . B7
Profondeville 8,724 . F8
Putte 14,017 .... F6
Quaregnon 20,071 . D8
Quévy 7,391 .... D8
Quiévrain 6,945 .. D8
Raeren 8,046 ... J7
Ravels 10,328 ... G6
Rebecq 8,891 ... E7
Renaix (Ronse) 25,056 . D7
Retie 8,359 .... G6
Rochefort 4,357 . G8
Roeselare 51,984 . C7
Ronse (Renaix) 25,056 . D7
Roulers (Roeselare) 51,984 C7
Saint-Gilles 46,076 . B9
Saint-Josse-ten-Noode 20,381 ..... C9
Saint-Nicolas 25,755 . G7
Saint-Trond (Sint-Truiden) 36,374 . G7
Saint-Vith (Sankt Vith) 8,434 . J8

(continued on following page)

Agriculture, Industry and Resources

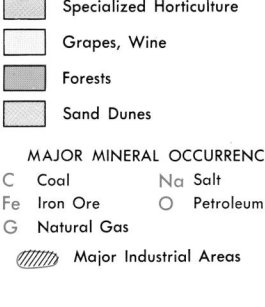

DOMINANT LAND USE

- Dairy, Truck Farming
- Cash Crops, Livestock
- Mixed Cereals, Dairy
- Specialized Horticulture
- Grapes, Wine
- Forests
- Sand Dunes

MAJOR MINERAL OCCURRENCES

C   Coal
Fe Iron Ore
G   Natural Gas
Na Salt
O   Petroleum

////// Major Industrial Areas

26 Netherlands, Belgium and Luxembourg

(continued)

Sankt Vith 8,434 .... J8
Schaerbeek 106,754 .... C9
Schoten 31,128 .... F6
Seraing 64,543 .... G7
's-Gravenbrakel
(Braine-le-Comte) 16,475 ... D7
Sint-Laureins 6,620 .... D6
Sint-Niklaas 67,992 .... E6
Sint-Pieters-Leeuw 27,968 ..... B9
Sint-Truiden 36,374 .... G7
Soignies 23,352 .... D7
Spa 9,619 .... H8
Sprimont 9,660 .... H8
Staden 11,135 .... B7
Steenokkerzeel 9,638 .... C9
Stekene 14,125 .... E6
Tamise (Temse) 23,525 .... E6
Temse 23,525 .... E6
Termonde
(Dendermonde) 22,119 .... E6
Tessenderlo 13,800 .... G6
Theux 9,167 .... H8
Thuin 13,757 .... E8
Tielt 19,103 .... C7
Tielt-Winge 8,237 .... F7
Tienen 32,620 .... F7
Tirlemont 32,620 .... F7
Tongeren 29,603 .... G7
Tongres (Tongeren) 29,603 .... G7
Torhout 17,165 .... C6
Tournai 67,906 .... C7
Tubeke (Tubize) 19,827 .... E7
Tubize 19,827 .... E7
Turnhout 37,453 .... F6
Uccle 76,004 .... B9
Ukkel (Uccle) 76,004 .... B9
Verviers 55,371 .... H7
Veurne 11,253 .... B6
Vielsalm 6,731 .... H8
Vilvoorde 33,264 .... F7
Vilvorde (Vilvoorde) 33,264 .... F7
Viroinval 5,589 .... F8
Virton 10,490 .... H9
Visé 16,469 .... H7
Vorst (Forest) 50,607 .... B9
Waarschoot 7,574 .... D6
Wachtebeke 6,951 .... D6
Waimes (Weismes) 5,713 .... J8
Walcourt 14,866 .... F8
Waregem 32,810 .... C7
Waremme 11,907 .... G7
Waterloo 24,755 .... E7
Watermael-Boitsfort 24,880 .... C9
Watermael-Bosvoorde
(Watermael-Boitsfort)
24,880 .... C9
Waver (Wavre) 25,153 .... F7
Wavre 25,153 .... F7
Wemmel 13,547 .... B9
Wervik 18,086 .... B7
Westerlo 19,459 .... F6
Wetteren 23,460 .... D7
Wezembeek-Oppem 12,006 .... D9
Wezet (Visé) 16,469 .... H7
Willebroek 22,265 .... E6
Wilrijk 42,328 .... E6
Wingene 12,188 .... C6
Woluwe-Saint-Lambert
48,801 .... C9
Woluwe-Saint-Pierre 40,686 ... C9
Ypres (Ieper) 34,425 .... B7
Yvoir 6,527 .... F8
Zaventem 25,393 .... C9
Zedelgem 19,198 .... C6
Zele 19,631 .... E6
Zelzate 12,934 .... D6
Zemst 17,167 .... E7
Zinnik (Soignies) 23,352 .... D7
Zonhoven 15,965 .... G6
Zottegem 25,109 .... D7

OTHER FEATURES

Albert (canal) F6
Ardennes (forest) F9
Botrange (mt.) J8
Dender (riv.) D7
Deûle (riv.) B7
Dyle (riv.) F7
Hohe Venn (plat.) H8
Lys (riv.) B7
Mark (riv.) F6
Meuse (riv.) F8
Nethe (riv.) F6
North (sea) D4
Ourthe (riv.) G8
Rupel (riv.) F7
Sambre (riv.) D8
Schelde (Scheldt) (riv.) C7
Scheldt (riv.) C7
Semois (riv.) G9
Senne (riv.) E7
Vaalserberg (mt.) J7
Vesdre (riv.) H7
Yser (riv.) B7

LUXEMBOURG

CITIES and TOWNS

Bascharage 4,870 .... H9
Diekirch† 5,470 .... J9
Differdange 15,940 .... H9
Dudelange† 14,070 .... J1
Echternach† 4,290 .... J9
Esch-sur-Alzette† 23,800 .... H9
Ettelbruck† 6,600 .... J9
Grevenmacher† 2,940 .... J9
Hesperange 9,470 .... J9
Luxembourg (cap.) 75,540 ...... J9
Mamer 6,090 .... H9
Mersch 5,560 .... J9
Mertert 3,000 .... J9
Pétange 11,800 .... H9
Remich 2,430 .... J9
Troisvierges 1,890 .... J9
Viandent† 1,510 .... J9
Wasserbillig 2,097 .... J9
Wiltz 3,850 .... H9

OTHER FEATURES

Alzette (riv.) J9
Clerf (riv.) J8
Mosel (riv.) J9
Our (riv.) J9
Sauer (riv.) J9

NETHERLANDS

PROVINCES

Drenthe 439,066 .... K3
Flevoland 202,678 .... G4
Friesland 599,190 .... H2
Gelderland 1,794,678 .... H4
Groningen 555,200 .... K2
Limburg 1,099,622 .... H6
North Brabant 2,172,604 .... F5
North Holland 2,365,160 .... F3
Overijssel 1,014,949 .... J4
South Holland 3,200,408 .... E5
Utrecht 1,004,632 .... G4
Zeeland 355,585 .... D6

CITIES and TOWNS

Aalsmeer 21,984 .... F4
Aalten 18,202 .... K5
Alkmaar 88,571 .... F3
Almelo 62,008 .... K4
Almere 63,785 .... G4
Alphen aan de Rijn 59,586 ...... F4
Amersfoort 96,072 .... G4
Amstelveen 69,505 .... B5
Amsterdam (cap.) 694,680 ..... B4
Apeldoorn 147,270 .... H4
Appingedam 12,668 .... K2
Arnhem 128,946 .... H4
Assen 49,398 .... K3
Asten 14,965 .... H6
Axel 12,219 .... D6
Baarn 24,897 .... G4
Barneveld 41,649 .... H4
Beilen 14,057 .... K3
Bemmel 15,842 .... H5
Bergen 14,075 .... F3
Bergen op Zoom 46,842 .... E5
Berkel 15,690 .... F5
Beverwijk 35,126 .... F4
Bloemendaal 8,977 .... E4
Bodegraven 17,720 .... F4
Bolsward 9,799 .... H2
Borculo 10,057 .... J4
Borger 12,730 .... K3
Borne 21,261 .... K4
Boskoop 14,524 .... F4
Boxmeer 14,363 .... H5
Boxtel 24,951 .... G5
Breda 121,362 .... F5
Brielle 14,973 .... E5
Brummen 20,802 .... J4
Brunssum 29,799 .... J7
Bussum 31,988 .... G4
Capelle 57,423 .... F5
Castricum 22,433 .... F3
Coevorden 14,344 .... K3
Culemborg 21,116 .... G5
De Bilt 31,729 .... G4
Delft 88,135 .... E4
Delfzijl 23,472 .... K2
Denekamp 12,206 .... L4
Den Helder 62,094 .... F3
Deurne 29,308 .... H6
Deventer 66,398 .... J4
Didam 16,036 .... J5
Diemen 18,083 .... C5
Dinxperlo 8,133 .... K5
Dirksland 7,341 .... E5
Doesburg 10,578 .... J4
Doetinchem 41,260 .... J5
Dongen 21,124 .... F5
Doorn 10,419 .... G4
Dordrecht 108,519 .... F5
Driebergen 18,294 .... G4
Dronten 24,281 .... H3
Druten 14,630 .... H5
Echt 16,927 .... H6
Edam-Volendam 24,572 .... G4
Ede 92,293 .... H4
Egmond aan Zee 11,163 .... E3
Eindhoven 190,736 .... G6
Elst 17,654 .... H5
Emmen 92,422 .... K3
Enkhuizen 15,939 .... G3
Enschede 145,223 .... K4
Epe 33,872 .... H4
Ermelo 25,644 .... H4
Etten-Leur 32,010 .... F5
Flushing 44,022 .... C6
Geertruidenberg 6,645 .... F5
Geldermalsen 22,017 .... G5
Geldrop 25,817 .... H6
Geleen 33,756 .... H7
Gemert 17,613 .... H5
Gendringen 20,186 .... J5
Genemuiden 7,545 .... H3
Gennep 16,264 .... H5
Giessendam 16,722 .... F5
Gilze 22,577 .... F5
Goes 31,815 .... D6
Goirle 18,852 .... G5
Goor 11,804 .... K4
Gorinchem 28,222 .... G5
Gouda 63,232 .... F4
Gramsbergen 6,080 .... K3
Grave 10,447 .... H5
Groenlo 8,895 .... K4
Groesbeek 18,221 .... H5
Groningen 167,788 .... K2
Haaksbergen 22 690 .... K4
Haarlem 149,198 .... F4
Haarlemmermeer
(Hoofddorp) 93,427 .... F4
Hague, The (cap.) 443,845 ...... E4
Hardenberg 32,065 .... J3
Harderwijk 34,600 .... H4
Hardinxveld-Giessendam
16,722 .... G5
Harlingen 15,727 .... G2
Hasselt 6,871 .... J3
Hattem 11,571 .... H4
Heemskerk 32,910 .... F3
Heemstede 26,308 .... F4
Heerde 18,171 .... H4
Heerenveen 37,700 .... H3
Heerhugowaard 35,522 .... F3
Heerlen 94,149 .... J7
Heesch 11,309 .... G5
Heiloo 20,467 .... F3
Hellendoorn 34,287 .... J4
Hellevoetsluis 34,276 .... E5
Helmond 66,791 .... H6
Hengelo 76,175 .... J4
's Hertogenbosch 90,584 .... G5
Heusden 5,761 .... G5
Hillegom 20,001 .... E4
Hilvarenbeek 9,975 .... G6
Hilversum 84,983 .... G4
Hoek van Holland
(Hook of Holland) D4
Hoofddorp
(Haarlemmermeer) 93,427 .. F4
Hoogeveen 45,601 .... J3
Hoogezand-Sappemeer
34,618 .... K2
Hook of Holland D4
Hoorn 56,474 .... G3
Horst 17,614 .... H6
Huissen 15,544 .... H5
Huizen 20,501 .... G4
Hulst 18,575 .... E6
IJsselstein 19,516 .... F4
Kampen 32,769 .... H3
Katwijk aan Zee 39,441 .... E4
Kerkrade 52,994 .... J7
Kesteren 9,389 .... G5
Krimpen aan den IJssel
27,638 .... F5
Landsmeer 9,121 .... C4
Laren 11,643 .... G4
Leek 17,743 .... J2
Leerdam 19,015 .... G5
Leeuwarden 85,296 .... H2
Leiden 109,254 .... E4
Lelystad 58,125 .... H3
Lisse 20,826 .... F4
Lith 6,115 .... G5
Lochem 18,295 .... J4
Loon op Zand 21,372 .... G5
Losser 22,526 .... L4
Maarssen 37,629 .... F4
Maasbree 11,752 .... H6
Maassluis 33,155 .... E5
Maastricht 116,380 .... H7
Margraten 13,365 .... H7
Medemblik 6,876 .... G3
Meerssen 20,462 .... H7
Meppel 23,492 .... J3
Middelburg 39,462 .... C6
Middelharnis 15,480 .... E5
Millingen aan den Rijn 5,287 ... J5
Monnickendam 9,953 .... C4
Montfoort 12,397 .... G4
Muiden 6,772 .... G4
Muntendam 5,022 .... K2
Naaldwijk 27,683 .... E4
Naarden 16,101 .... G4
Neede 10,982 .... K4
Nieuwegein 58,316 .... G4
Nieuwkoop 10,723 .... F4
Nijkerk 25,613 .... H4
Nijmegen 145,405 .... H5
Noordwijk 24,996 .... E4
Norg 6,595 .... J2
Nunspeet 24,573 .... H4
Odoorn 12,225 .... K3
Oisterwijk 18,177 .... G5
Oldenzaal 29,680 .... K4
Olst 9,039 .... J4
Ommen 17,957 .... J3
Oostburg 18,145 .... C6
Oosterhout 48,157 .... F5
Oostzaan 7,292 .... C4
Oss 50,987 .... H5
Oud-Beijerland 20,385 .... E5
Oude-Pekela 8,028 .... K2
Oudenbosch 12,576 .... E5
Oudewater 9,410 .... F4
Purmerend 56,233 .... F4
Putten 20,898 .... H4
Raalte 26,883 .... J4
Renkum 33,841 .... H5
Reusel 7,813 .... G6
Rheden 46,088 .... J4
Rhenen 16,613 .... H5
Ridderkerk 46,163 .... F5
Rijnsburg 13,412 .... F4
Rijssen 23,927 .... J4
Rijswijk 48,189 .... E4
Roden 18,331 .... J2
Roermond 38,486 .... J6
Roosendaal 59,237 .... F5
Rotterdam 576,232 .... E5
Ruurlo 7,418 .... J4
Sappemeer-Hoogezand
34,618 .... K2
Schagen 16,759 .... F3
Schiedam 69,438 .... E5
Schijndel 21,397 .... G5
Schoonebeek 7,740 .... K3
Schoonhoven 11,231 .... F5
's Gravendeel 8,424 .... E5
's Gravenhage (The Hague)
(cap.) 443,845 .... E4
's Gravenzande 18,453 .... E4
Simpelveld 11,882 .... J7
Sittard 44,894 .... H6
Sliedrecht 22,833 .... F5
Slochteren 13,958 .... K2
Sloten H3
Sluis 2,882 .... C6
Smilde 9,212 .... K3
Sneek 29,408 .... H2
Soest 41,598 .... G4
Stadskanaal 33,047 .... L3
Staphorst 13,580 .... J3
Staveren G3
Steenbergen 13,826 .... E5
Steenwijk 20,907 .... J3
Stiens H2
Tegelen 18,991 .... J6
Ter Apel L3
Termunten 4,378 .... K2
Terneuzen 35,043 .... D6
The Hague (cap.) 443,845 .... E4
Tholen 19,019 .... E5
Tiel 31,394 .... G5
Tilburg 155,110 .... G5
Twello J4
Uden 35,057 .... H5
Uithoorn 22,205 .... F4
Uithuizen K2
Ulrum 3,657 .... J2
Urk 12,728 .... H3
Utrecht 230,634 .... G4
Vaals 10,639 .... H7
Valkenswaard 29,811 .... H6
Veendam 28,234 .... K2
Veenendaal 47,258 .... G4
Veere 4,836 .... D5
Veghel 25,701 .... H5
Veldhoven 38,644 .... G6
Velsen 57,608 .... F4
Venlo 63,607 .... J6
Venraij 34,172 .... H6
Vianen 18,704 .... G5
Vlaardingen 74,480 .... E5
Vlagtwedde 16,181 .... L3
Vlijmen 15,655 .... G5
Vlissingen (Flushing) 44,022 .. C6
Volendam-Edam 24,572 .... G4
Voorburg 40,455 .... E4
Voorst 23,678 .... J4
Vorden 8,282 .... J4
Vriezenveen 18,601 .... K4
Vught 23,718 .... G5
Waalre 15,126 .... G6
Waalwijk 28,674 .... F5
Wageningen 32,370 .... H5
Warmenhuizen 4,765 .... F3
Weert 40,068 .... H6
Weesp 18,362 .... C5
Westkapelle 2,666 .... C5
Wierden 22,200 .... K4
Wijhe 7,155 .... J4
Wijk bij Duurstede 15,401 ...... G5
Willemstad 3,357 .... F5
Winschoten 19,680 .... L2
Winsum 6,583 .... K2
Winterswijk 28,024 .... K5
Woensdrecht 10,077 .... E6
Woerden 34,166 .... F4
Wolvega J3
Workum G3
Zaandam (Zaanstad) 129,653 . B4
Zaltbommel 9,534 .... G5
Zandvoort 15,428 .... E4
Zeewolde 5,930 .... G4
Zeist 59,431 .... G4
Zevenaar 26,848 .... J5
Zevenbergen 15,562 .... E5
Zierikzee 9,804 .... D5
Zundert 13,385 .... F6
Zutphen 31,144 .... J4
Zwartsluis 4,465 .... H3
Zwijndrecht 41,357 .... E5
Zwolle 92,517 .... J3

OTHER FEATURES

Alkmaardermeer (lake) F3
Ameland (isl.) H2
Beulaker Wijde (lake) H3
Borndiep (chan.) H2
De Fluessen (lake) G3
De Honte (bay) D6
De Peel (reg.) H6
De Twente (reg.) K4
De Zaan (riv.) B4
Dollard (bay) L2
Dommel (riv.) H6
Duiveland (isl.) D5
Eems (riv.) K2
Eijerlandsche Gat (strait) F2
Flevoland Polders G4
Frisian, West (isls.) G2
Goeree (isl.) D5
Grevelingen (strait) E5
Griend (isl.) G2
Groninger Wad (sound) J2
Groote IJ Polder B4
Haarlemmermeer Polder B5
Haringvliet (strait) E5
Het IJ (est.) C4
Hoek van Holland (cape) D5
Houtrak Polder A4
Hunse (riv.) K3
IJmeer (bay) C4
IJssel (riv.) J4
IJsselmeer (lake) G3
Lauwers (chan.) J1
Lauwers Zee (bay) J2
Lek (riv.) F5
Lower Rhine (riv.) H5
Maas (riv.) G5
Marken (isl.) G4
Markerwaard Polder G3
Marsdiep (chan.) F3
North (sea) E3
North Beveland (isl.) D5
North East Polder H3
North Holland (canal) C4
North Sea (canal) F4
Old Rhine (riv.) E4
Oostzaan Polder B4
Orange (canal) K3
Overflakkee (isl.) E5
Rhine (riv.) J5
Roer (riv.) J6
Scheldt, Eastern (est.) D5
Scheldt, Western
(De Honte) (bay) D6
Schiermonnikoog (isl.) J1
Schouwen (isl.) D5
Slotermeer (lake) H3
Sneekermeer (lake) H2
South Beveland (isl.) D6
Terschelling (isl.) G2
Texel (isl.) F2
Tjeukemeer (lake) H3
Vaalserberg (mt.) J7
Vecht (riv.) F4
Vechte (riv.) J3
Veersche Meer (lake) D5
Veluwe (reg.) H4
Vlieland (isl.) F2
Vliestroom (strait) G2
Voorne (isl.) D5
Waal (riv.) G5
Waddenzee (sound) G2
Walcheren (isl.) C5
West Frisian (isls.) F2
Wester Eems (chan.) K1
Western Scheldt
(De Honte) (bay) D6
Wieringermeer Polder G3
Wilhelmina (canal) G6
Willems (canal) G5

* City and suburbs.
† Population of urban area.

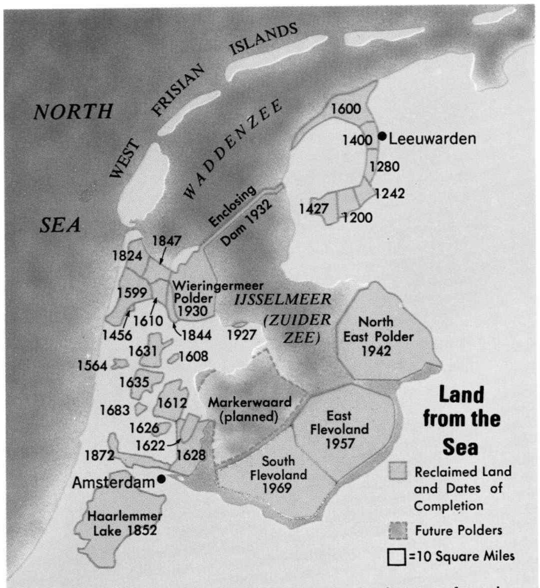

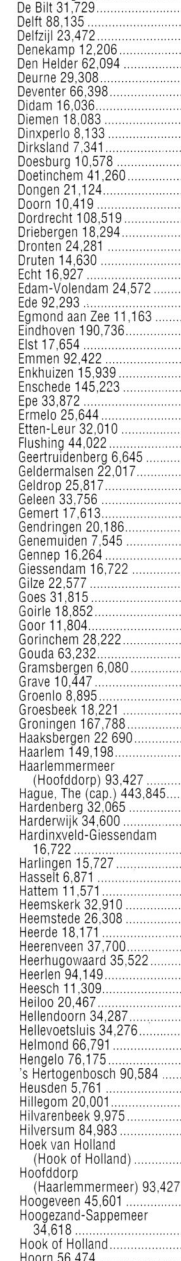

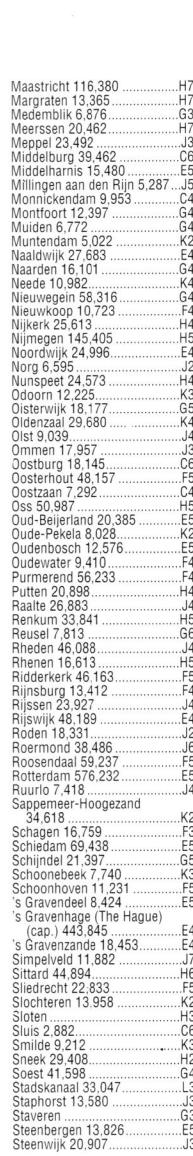

WEST FRISIAN ISLANDS
NORTH
SEA
WADDENZEE
Enclosing Dam 1932
Leeuwarden
1600
1400
1280
1242
1427
1200
1847
1824
Wieringermeer Polder 1930
IJSSELMEER (ZUIDER ZEE)
North East Polder 1942
1599
1610
1456
1844
1927
1631
1608
1564
1635
1683
1612
Markerwaard (planned)
East Flevoland 1957
1626
1622
1628
1872
South Flevoland 1969
Amsterdam
Haarlemmer Lake 1852
Land from the Sea
Reclaimed Land and Dates of Completion
Future Polders
=10 Square Miles

For centuries the Dutch have been renowned for the drainage of marshes and the construction of polders, i.e., arable land reclaimed from the sea. Future projects will convert much of the present IJsselmeer to agricultural land.

Topography

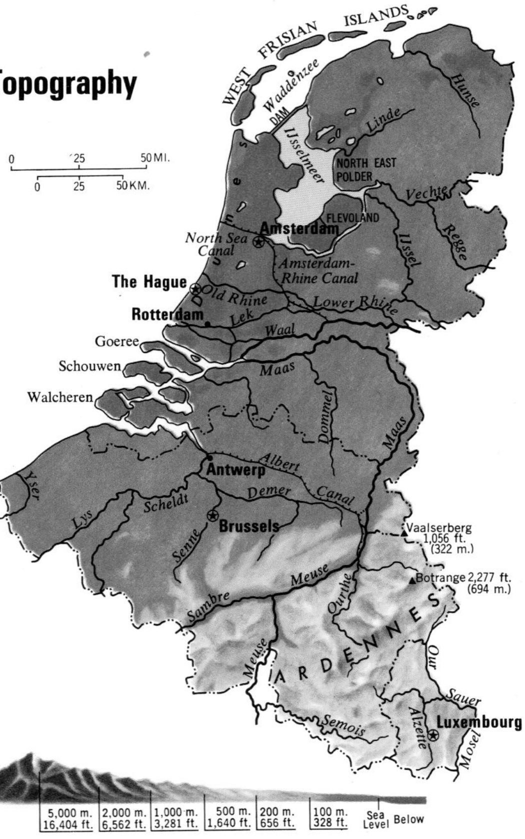

WEST FRISIAN ISLANDS
Waddenzee
DAM
IJsselmeer
NORTH EAST POLDER
FLEVOLAND
Linde
Hunse
Vechte
Regge
IJssel
Amsterdam
North Sea Canal
Amsterdam-Rhine Canal
The Hague
Old Rhine
Lek
Lower Rhine
Rotterdam
Waal
Goeree
Schouwen
Walcheren
Maas
Dommel
Albert
Antwerp
Canal
Scheldt
Demer
Yser
Lys
Brussels
Senne
Vaalserberg 1,056 ft. (322 m.)
Botrange 2,277 ft. (694 m.)
Meuse
Ourthe
Sambre
ARDENNES
Our
Sauer
Alzette
Semois
Mosel
Luxembourg
0 25 50 MI.
0 25 50 KM.
5,000 m. 16,404 ft.
2,000 m. 6,562 ft.
1,000 m. 3,281 ft.
500 m. 1,640 ft.
200 m. 656 ft.
100 m. 328 ft.
Sea Level
Below

France
CONIC PROJECTION
SCALE OF MILES
SCALE OF KILOMETERS

Capitals of Countries
Capitals of Departments
International Boundaries
Department Boundaries
Canals

Paris and Environs

Corsica
Same Scale as Main Map

© Copyright HAMMOND INCORPORATED, Maplewood, N.J.

DEPARTMENTS

Ain 418,516 F 4
Aisne 533,970 E 3
Allier 369,580 E 4
Alpes-de-Haute-Provence 119,068 G 5
Alpes-Maritimes 881,198 G 6
Ardèche 267,970 F 5
Ardennes 302,338 F 3
Ariège 135,725 D 6
Aube 289,300 E 3
Aude 280,686 E 6
Aveyron 278,654 E 5
Bas-Rhin 915,676 G 3
Belfort 131,999 G 4
Bouches-du-Rhône 1,724,199 F 6
Calvados 589,559 C 3
Cantal 162,838 E 5
Charente 340,770 D 5
Charente-Maritime 513,220 C 5
Cher 320,174 E 4
Corrèze 241,448 D 5
Corse du Sud 108,604 B 6
Côte-d'Or 473,548 F 4
Côtes-du-Nord 538,869 B 3
Creuse 139,968 D 4
Deux-Sèvres 342,812 C 4
Dordogne 377,356 D 5
Doubs 477,163 G 4
Drôme 389,781 F 5
Essonne 988,000 E 3
Eure 462,323 D 3
Eure-et-Loir 362,813 D 3
Finistère 828,364 A 3
Gard 530,478 F 6
Gers 174,154 D 6
Gironde 1,127,546 C 5
Haute-Corse 131,574 B 6
Haute-Garonne 824,501 D 6
Haute-Loire 205,895 E 5
Haute-Marne 210,670 F 3
Hautes-Alpes 105,070 G 5
Haute-Saône 231,962 G 4
Haute-Savoie 494,505 G 5
Hautes-Pyrénées 227,922 D 6
Haute-Vienne 355,737 D 5
Haut-Rhin 650,372 G 4
Hauts-de-Seine 1,387,039 A 2
Hérault 706,499 E 6
Ille-et-Vilaine 749,764 C 3
Indre 243,191 D 4
Indre-et-Loire 506,097 D 4
Isère 936,771 F 5
Jura 242,925 F 4
Landes 297,424 C 5
Loire 739,521 F 5
Loire-Atlantique 995,498 C 4
Loiret 535,669 E 4
Loir-et-Cher 296,220 D 4
Lot 154,533 D 5
Lot-et-Garonne 298,522 D 5
Lozère 74,294 E 5
Maine-et-Loire 675,321 C 4
Manche 465,948 C 3
Marne 543,627 F 3
Mayenne 271,784 C 3
Meurthe-et-Moselle 716,846 G 3
Meuse 200,101 F 3
Morbihan 590,889 B 4
Moselle 1,007,189 G 3
Nièvre 239,635 E 4
Nord 2,520,526 E 2
Oise 661,781 E 3
Orne 295,472 C 3
Paris 2,188,918 B 2
Pas-de-Calais 1,412,413 E 2
Puy-de-Dôme 594,365 E 5
Pyrénées-Atlantiques 555,696 C 6
Pyrénées-Orientales 334,557 E 6
Rhône 1,445,208 F 5
Saône-et-Loire 571,852 F 4
Sarthe 504,768 D 3
Savoie 323,675 G 5
Seine-et-Marne 887,112 E 3
Seine-Maritime 1,324,301 D 3
Seine-Saint-Denis 1,324,301 C 1
Somme 544,570 E 3
Tarn 339,345 E 6
Tarn-et-Garonne 190,485 D 5
Val-de-Marne 1,193,655 C 1
Val-d'Oise 920,598 E 3
Var 708,331 G 6
Vaucluse 427,343 F 6
Vendée 483,027 C 4
Vienne 371,428 D 4
Vosges 395,769 G 3
Yonne 311,019 E 4
Yvelines 1,196,111 D 3

CITIES and TOWNS

Aigues-Mortes 4,106 F 6
Aix-en-Provence 100,221 F 6
Aix-les-Bains 22,331 G 5
Ajaccio 48,324 B 7
Alençon 30,952 D 3
Amboise 10,823 D 4
Amiens 130,302 E 3
Angers 135,293 C 4
Angoulême 45,495 D 5
Annecy 49,753 G 5
Antibes 62,427 G 6
Argenteuil 94,826 A 1
Arles 37,554 F 6
Armentières 22,849 E 2
Arras 41,376 E 2
Asnières-sur-Seine 71,058 A 1
Aubervilliers 67,684 B 1
Aubusson 5,326 E 4
Aulnay-sous-Bois 75,543 B 1
Aurignac 772 D 6
Avignon 75,178 F 6
Ax-les-Thermes 1,283 D 6
Bagnolet 32,556 B 2
Barbizon 478 E 3
Barcelonnette 2,674 G 5
Barfleur 617 C 3
Bastia 43,502 B 6
Bayeux 14,568 C 3
Bayonne 40,088 C 6
Beaucaire 10,622 F 6
Beaune 19,110 F 4
Beauvais 51,542 E 3
Belfort 51,034 G 4
Bergerac 24,604 D 5
Besançon 112,023 G 4
Bessèges 4,352 F 5
Béziers 74,114 E 6
Biarritz 26,579 C 6
Blois 46,925 D 4
Bobigny 42,630 B 1
Bonifacio 1,727 B 7
Bordeaux 201,965 C 5
Boulogne-Billancourt 102,582 A 2
Boulogne-sur-Mer 47,482 D 2
Bourg-en-Bresse 37,582 F 4
Bourges 74,622 E 4
Brest 154,110 A 3
Brignoles 8,529 G 6
Brive-la-Gaillarde 50,898 D 5
Bruay-en-Artois 22,502 E 2
Caen 112,332 C 3
Calais 76,206 D 2
Caluire-et-Cuire 41,864 F 5
Cambrai 35,070 E 2
Cannes 71,888 G 6
Carcassonne 38,379 E 6
Castres 39,216 E 6
Châlons-sur-Marne 49,941 F 3

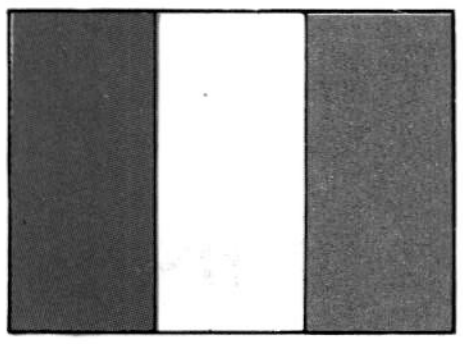

AREA 210,038 sq. mi. (543,998 sq. km.)
POPULATION 56,160,000
CAPITAL Paris
LARGEST CITY Paris
HIGHEST POINT Mont Blanc 15,771 ft. (4,807 m.)
MONETARY UNIT franc
MAJOR LANGUAGE French
MAJOR RELIGION Roman Catholicism

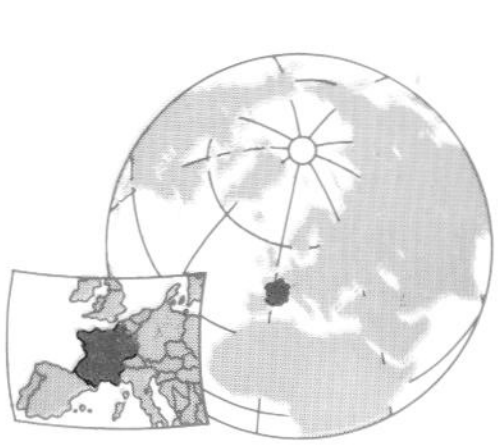

Topography

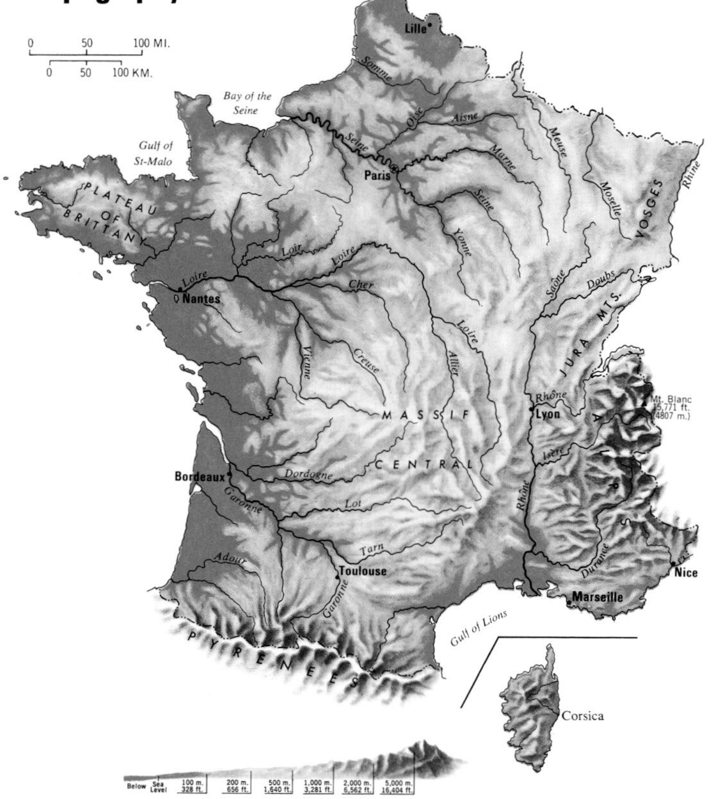

Historic Provinces

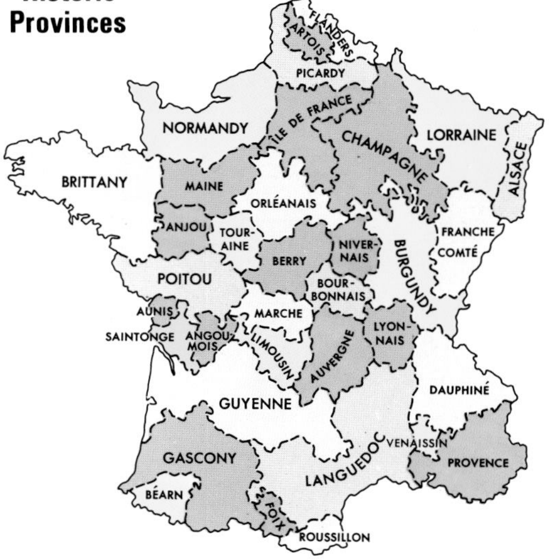

A resident of the city of Caen thinks of himself as a Norman rather than as a citizen of the modern department of Calvados. In spite of the passing of nearly two centuries, the historic provinces which existed before 1790 command the local patriotism of most Frenchmen.

Chalon-sur-Saône 53,893 F 4
Chambéry 49,465 F 5
Chambord 159 D 4
Chamonix-Mont-Blanc 7,406 G 5
Champigny-sur-Marne 76,039 C 2
Chantilly 10,065 E 3
Charleville-Mézières 7,814 F 3
Chartres 36,706 D 3
Châteaudun 15,905 D 3
Châteauneuf-sur-Loire 5,630 E 4
Châteauroux 51,744 D 4
Château-Thierry 14,427 E 3
Chatou 28,435 A 1
Cherbourg 28,324 C 3
Chinon 6,030 D 4
Choisy-le-Roi 35,443 B 2
Cholet 51,620 C 4
Clamart 48,210 A 2
Clermont-Ferrand 145,901 E 5
Clichy 46,830 B 1
Cluny 4,133 F 4
Cognac 20,247 C 5
Colmar 61,560 G 3
Colombes 78,485 A 1
Compiègne 39,909 E 3
Courbevoie 59,821 A 1
Creil 34,332 E 3
Créteil 71,559 B 2
Deauville 4,682 C 3
Dieppe 35,659 D 3
Digne 12,540 G 5
Dijon 139,188 F 4
Dinard 9,562 B 3
Domrémy-la-Pucelle 162 F 3
Douai 41,576 E 2
Drancy 60,122 B 1
Dunkirk 71,756 E 2
Ernée 5,253 C 3
Évreux 45,215 D 3
Falaise 8,424 C 3
Fécamp 21,212 D 3
Foix 9,212 D 6
Fontainebleau 14,687 E 3
Fontenay-sous-Bois 52,397 C 2
Gex 4,776 G 4
Grasse 24,257 G 6
Grenoble 156,437 F 5
Guise 6,179 E 3
Harfleur 9,470 D 3
Hazebrouck 19,266 E 2
Hendaye 10,492 C 6
Héricourt 9,239 G 4
Honfleur 8,125 D 3
Issy-les-Moulineaux 45,702 A 2
Istres 21,286 F 6
Ivry-sur-Seine 55,682 B 2
La Baule-Escoublac 13,151 B 4
La Courneuve 33,525 B 1
Langres 9,718 F 4
Lapalisse 3,173 E 4
La Rochelle 74,728 C 4
La Roche-sur-Yon 42,026 C 4
Laval 53,582 C 3
Le Bourget 11,020 B 1
Le Creusot 32,013 F 4
Le Havre 198,700 C 3
Le Mans 145,976 C 3
Le Puy 22,806 F 5
Le Tréport 6,330 D 2
Levallois-Perret 53,485 B 1
Lille 167,791 E 2
Limoges 137,809 D 5
Lisieux 24,454 D 3
Lorient 62,207 B 4
Lourdes 17,252 C 6
Lunéville 21,200 G 3
Lyon 410,455 F 5
Mâcon 36,517 F 4
Maisons-Alfort 51,041 B 2
Maisons-Laffitte 22,565 A 1
Mantes-la-Jolie 43,551 D 3
Marmande 14,264 C 5
Marseille 868,435 F 6
Maubeuge 35,424 F 2
Mayenne 12,156 C 3
Meaux 44,386 E 3
Melun 34,379 E 3
Mende 10,520 E 5
Menton 22,234 G 6
Metz 113,236 G 3
Meudon 29,356 A 2
Montauban 36,122 D 5
Montbéliard 31,174 G 4
Montceau-les-Mines 26,877 F 4
Mont-de-Marsan 25,896 C 6
Mont-Dore 2,091 E 5
Montfort 4,029 C 3
Montluçon 49,737 E 4
Montmédy 1,880 F 3
Montpellier 190,423 E 6
Montreuil 96,441 B 2
Mont-Saint-Michel 65 C 3
Mulhouse 111,742 G 4
Nancy 95,654 G 3
Nanterre 88,567 A 1
Nantes 237,789 C 4
Narbonne 38,222 E 6
Nemours 11,624 E 3
Neufchâtel-en-Bray 5,452 D 3
Neuilly-sur-Seine 64,093 A 1
Nice 331,165 G 6
Nîmes 120,515 F 6
Niort 56,256 C 4
Nogent-le-Rotrou 11,963 D 3
Noisy-le-Sec 36,821 B 1
Nontron 3,407 D 5
Noyon 13,949 E 3
Nyons 5,219 F 5
Orléans 81,615 D 3
Orly 23,729 B 2
Oyonnax 22,516 F 4
Paris (cap.) 2,165,892 B 2
Paris *10,073,059 B 2
Pau 82,186 C 6
Périgueux 32,632 D 5
Perpignan 107,812 E 6
Pessac 49,019 C 5
Poitiers 76,793 D 4
Pontoise 27,885 E 3
Port-Vendres 4,871 E 6
Privas 9,253 F 5
Quimper 52,335 A 4
Rambouillet 21,136 D 3
Redon 9,071 C 4
Reims 176,419 E 3
Rennes 190,861 C 3
Roanne 48,574 E 4
Rochefort 25,392 C 4
Roubaix 101,488 E 2
Rouen 100,696 D 3
Rueil-Malmaison 63,310 A 2
Saint-Brieuc 48,259 B 3
Saint-Cloud 28,561 A 2
Saint-Denis 90,686 B 1
Saint-Dizier 34,074 F 3
Sainte-Mère-Église 1,205 C 3
Saint-Étienne 193,938 F 5
Saint-Germain-en-Laye 36,585 D 3
Saint-Jean-d'Angély 3,268 C 4

(continued on following page)

Wine Regions

Climate, soil and variety of grape planted determine the quality of wine. Long, hot and fairly dry summers with cool, humid nights constitute an ideal climate. The nature of the soil is such a determining influence that identical grapes planted in Bordeaux, Burgundy and Champagne, will yield wines of widely different types.

Agriculture, Industry and Resources

DOMINANT LAND USE

- Cereals (chiefly wheat)
- Cereals (chiefly rye, oats, barley)
- Dairy
- Pasture Livestock
- Truck Farming, Horticulture
- Grapes, Wine
- Forests

MAJOR MINERAL OCCURRENCES

Ab	Asbestos	Na	Salt
Al	Bauxite	O	Petroleum
C	Coal	Pb	Lead
F	Fluorspar	U	Uranium
Fe	Iron Ore	W	Tungsten
G	Natural Gas	Zn	Zinc
K	Potash		

⚡ Water Power

▨ Major Industrial Areas

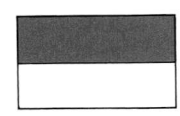

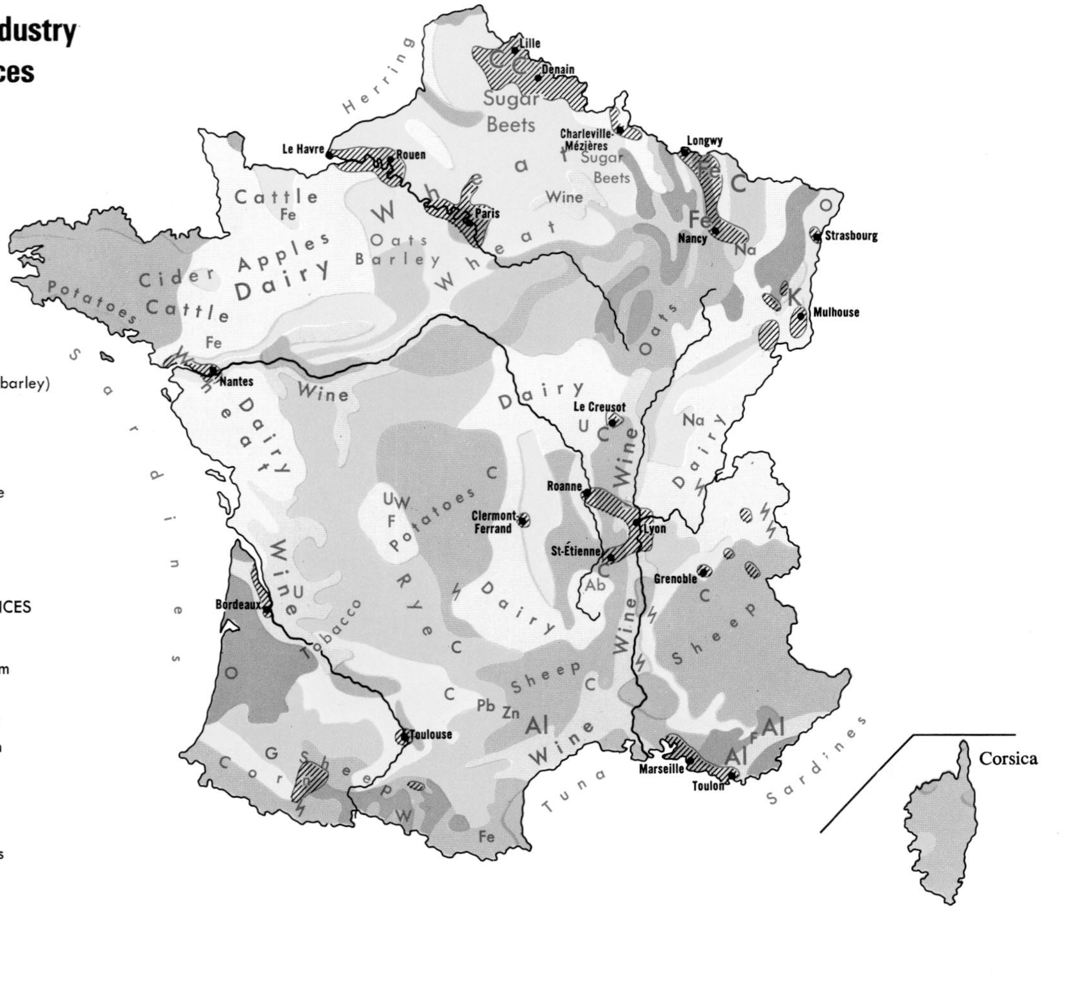

Corsica

ANDORRA

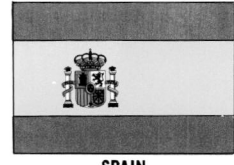
SPAIN

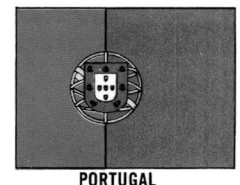
PORTUGAL

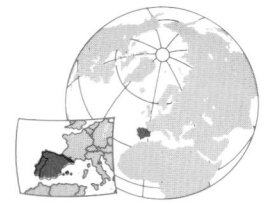

SPAIN

SPAIN

AREA 194,881 sq. mi. (504,742 sq. km.)
POPULATION 39,328,000
CAPITAL Madrid
LARGEST CITY Madrid
HIGHEST POINT Pico de Teide 12,172 ft. (3,710 m.)
(Canary Is.); Mulhacén 11,411 ft. (3,478 m.)
(mainland)
MONETARY UNIT peseta
MAJOR LANGUAGES Spanish, Catalan, Basque,
Galician, Valencian
MAJOR RELIGION Roman Catholicism

ANDORRA

AREA 188 sq. mi. (487 sq. km.)
POPULATION 32,000
CAPITAL Andorra la Vella
MONETARY UNITS French franc, Spanish peseta
MAJOR LANGUAGE Catalan
MAJOR RELIGION Roman Catholicism

PORTUGAL

AREA 35,549 sq. mi. (92,072 sq. km.)
POPULATION 10,467,000
CAPITAL Lisbon
LARGEST CITY Lisbon
HIGHEST POINT Malhão da Estrela
6,532 ft. (1,991 m.)
MONETARY UNIT escudo
MAJOR LANGUAGE Portuguese
MAJOR RELIGION Roman Catholicism

GIBRALTAR

AREA 2.28 sq. mi. (5.91 sq. km.)
POPULATION 31,000
CAPITAL Gibraltar
MONETARY UNIT pound sterling
MAJOR LANGUAGES English, Spanish
MAJOR RELIGION Roman Catholicism

(continued on following page)

Agriculture, Industry and Resources

DOMINANT LAND USE

- Cereals (chiefly wheat)
- Livestock (chiefly sheep, goats)
- Mixed Cereals, Livestock
- Olives, Fruit
- Grapes, Fruit, Nuts, Mixed Cereals
- Forests
- Nonagricultural Land

MAJOR MINERAL OCCURRENCES

Ag	Silver	Na	Salt
C	Coal	O	Petroleum
Cu	Copper	Pb	Lead
Fe	Iron Ore	Py	Pyrites
G	Natural Gas	Sb	Antimony
Hg	Mercury	Sn	Tin
K	Potash	U	Uranium
Lg	Lignite	W	Tungsten
Mg	Magnesium	Zn	Zinc

⚡ Water Power

▨ Major Industrial Areas

(continued on following page)

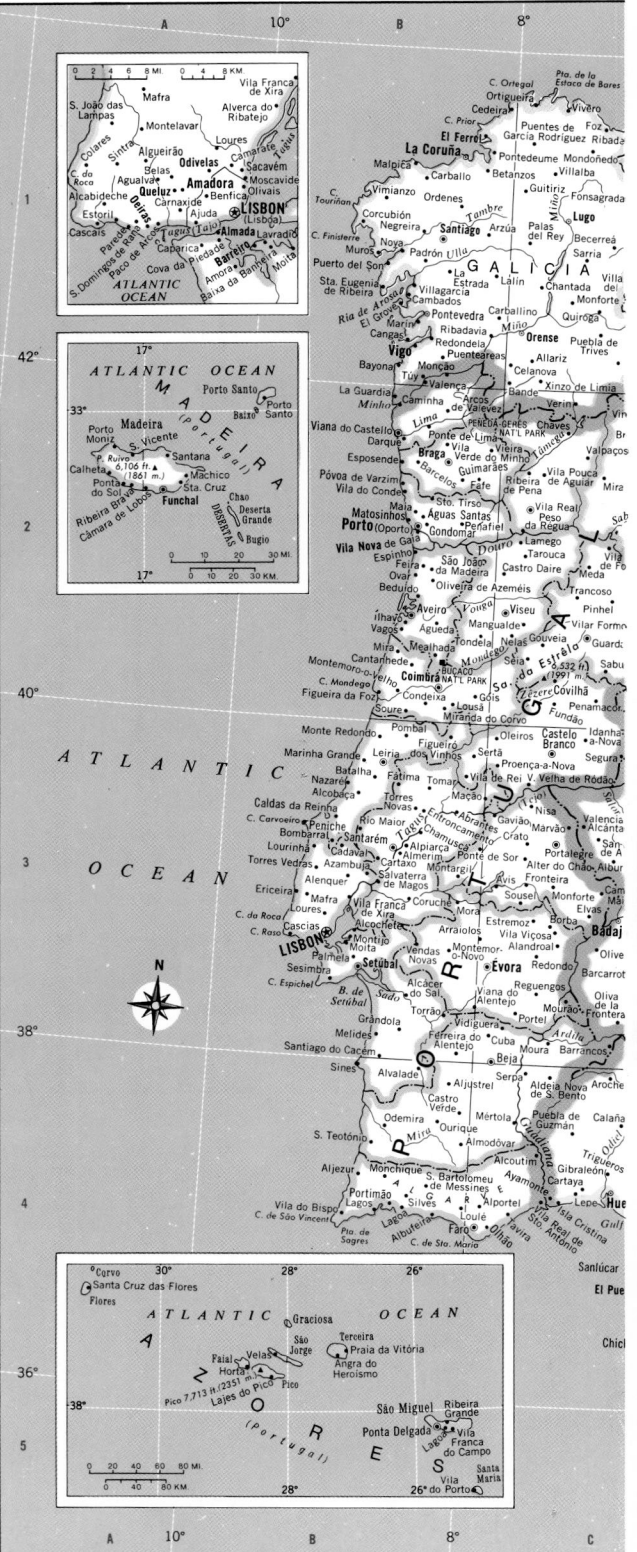

Topography

VATICAN CITY

AREA 108.7 acres (44 hectares)
POPULATION 1,000

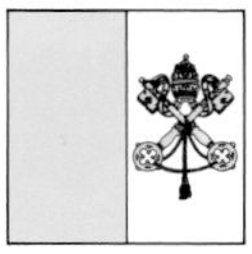

SAN MARINO

AREA 23.4 sq. mi. (60.6 sq. km.)
POPULATION 23,000

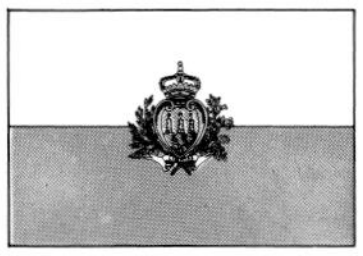

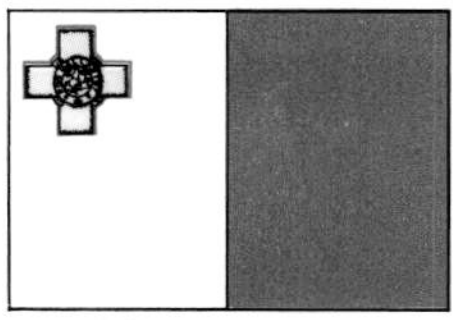

MALTA

AREA 122 sq. mi. (316 sq. km.)
POPULATION 353,000
CAPITAL Valletta
LARGEST CITY Sliema
HIGHEST POINT 787 ft. (240 m.)
MONETARY UNIT Maltese lira
MAJOR LANGUAGES Maltese, English
MAJOR RELIGION Roman Catholicism

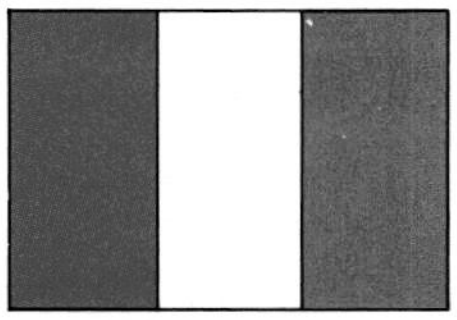

ITALY

AREA 116,303 sq. mi. (301,225 sq. km.)
POPULATION 57,574,000
CAPITAL Rome
LARGEST CITY Rome
HIGHEST POINT Dufourspitze (Mte. Rosa) 15,203 ft. (4,634 m.)
MONETARY UNIT lira
MAJOR LANGUAGE Italian
MAJOR RELIGION Roman Catholicism

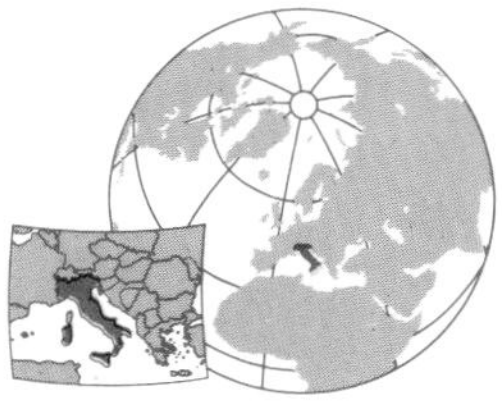

ITALY

REGIONS

Abruzzi 1,217,791 D3
Aosta 112,353 A2
Apulia (Puglia) 3,871,617 F4
Basilicata 610,186 F4
Calabria 2,061,182 F5
Campania 5,463,134 E4
Emilia-Romagna 3,957,513 C2
Friuli-Venezia Giulia 1,233,984 D1
Latium (Lazio) 5,001,684 D3
Liguria 1,807,893 B2
Lombardy 8,891,652 B2
Marche 1,412,404 D3
Molise 328,371 E4
Piedmont 4,479,031 A2
Sardinia 1,594,175 B4
Sicily 4,906,878 D6
Trentino-Alto Adige 873,413 C1
Tuscany 3,581,051 C3
Umbria 807,552 D3
Veneto 4,345,047 C2

PROVINCES

Agrigento 466,495 D6
Alessandria 466,102 B2
Ancona 433,417 D3
Arezzo 313,157 C3
Ascoli Piceno 352,567 D3
Asti 215,382 B2
Avellino 434,021 E4
Bari 1,464,627 F4
Belluno 220,335 D1
Benevento 289,143 E4
Bergamo 896,117 B2
Bologna 930,284 C2
Bolzano-Bozen 430,568 C1
Brescia 1,017,093 C2
Brindisi 391,064 G4
Cagliari 730,473 B5
Caltanissetta 285,829 D6
Campobasso 235,847 E4
Caserta 755,628 E4
Catania 1,005,577 E6
Catanzaro 744,834 F5
Chieti 370,534 E3
Como 755,979 B2
Cosenza 743,255 F5
Cremona 332,236 B2
Cuneo 548,452 A2
Enna 190,939 E6
Ferrara 381,118 C2
Florence 1,202,013 C3
Foggia 681,595 E4
Forlì 599,420 D2
Frosinone 460,395 D4
Genoa 1,045,109 B2
Gorizia 144,726 D2
Grosseto 220,905 C3
Imperia 223,738 B3
Isernia 92,524 E4
L'Aquila 291,742 D3
La Spezia 241,371 B2
Latina 434,086 D4
Lecce 762,017 G4
Livorno (Leghorn) 346,657 C3
Lucca 385,876 C3
Macerata 292,932 D3
Mantua 377,158 C2
Massa-Carrara 203,530 C2
Matera 203,570 F4
Messina 669,323 E5
Milan 4,018,108 B2
Modena 596,025 C2
Naples 2,970,563 E4
Novara 507,367 B2
Nuoro 274,817 B4
Padua 809,667 C2
Palermo 1,198,575 D5
Parma 400,192 C2
Pavia 512,895 B2
Perugia 580,988 D3
Pesaro e Urbino 333,488 D3
Pescara 286,240 E3
Piacenza 278,424 B2
Pisa 388,800 C3
Pistoia 264,995 C3
Pordenone 275,888 D2
Potenza 406,616 E4
Ragusa 274,583 E6
Ravenna 358,654 D2
Reggio di Calabria 573,093 E5
Reggio nell'Emilia 413,396 C2
Rieti 142,794 D3
Rome 3,695,961 F6
Rovigo 253,508 C2
Salerno 1,013,779 E4
Sassari 433,842 B4
Savona 297,675 B2
Siena 255,118 C3
Sondrio 174,009 B1
Syracuse 394,692 E6
Taranto 572,314 F4
Teramo 269,275 D3
Terni 807,552 D3
Trapani 420,865 D5
Trento 442,845 C1
Treviso 720,580 D2
Trieste 283,641 D2
Turin 2,345,771 A2
Udine 529,729 D1
Varese 788,057 B2
Venice 838,794 D2
Vercelli 395,957 B2
Verona 775,745 C2
Vicenza 726,418 C2
Viterbo 268,448 C3

CITIES and TOWNS

Acireale 46,711 E6
Acqui Terme 20,951 B2
Adrano 32,865 E6
Agrigento 38,681 D6
Alba 25,853 B2
Albano Laziale 27,796 F7
Alcamo 41,626 D6
Alessandria 79,552 B2
Alghero 32,519 B4
Altamura 50,539 F4
Amalfi 4,423 E4
Ancona 97,118 D3
Andria 84,070 F4
Anzio 25,932 D4
Aosta 36,649 A2
Aprilia 31,604 D4
Arezzo 74,477 C3
Ascoli Piceno 44,411 D3
Assisi 4,683 D3
Asti 65,483 B2
Augusta 37,162 E6
Avellino 50,894 E4
Aversa 55,788 E4
Avezzano 30,227 D3
Avola 30,360 E6
Bagheria 39,869 D5
Barcellona Pozzo di Gotto 33,404 E5
Bari 369,444 F4
Barletta 82,290 F4
Bassano del Grappa 33,724 C2
Belluno 28,468 D1
Benevento 51,831 E4
Bergamo 121,389 B2
Biancavilla 20,047 E6
Biella 52,587 B2
Bisceglie 46,209 F4
Bitonto 46,538 F4
Bologna 454,897 C2
Bolzano (Bolzen) 103,241 C1
Borgomanero 18,701 B2
Bra 21,304 A2
Brescia 202,539 C2
Brindisi 84,887 G4
Bronte 17,477 E6
Busto Arsizio 79,321 B2
Cagliari 219,423 B5
Caltagirone 32,860 E6
Caltanissetta 57,704 D6
Camaiore 24,284 C3
Campobasso 41,687 E4
Canicattì 31,726 D6
Canosa di Puglia 30,555 E4
Cantù 35,644 B2
Capannori 39,717 C3
Carbonia 25,140 B5
Carmagnola 19,581 A2
Carpi 49,370 C2
Carrara 61,709 C2
Casale Monferrato 37,157 B2
Cascina-Navacchio 32,570 C3
Caserta 59,185 E4
Cassino 22,406 D4
Castel Gandolfo 6,176 F7
Castelfranco Veneto 20,196 D2
Castellammare di Stabia 70,507 E4
Castelvetrano 29,503 D6
Castrovillari 18,648 F5
Catania 379,754 E6
Catanzaro 96,930 F5
Cava de'Tirreni 47,007 E4
Cecina 22,264 C3
Ceglie Messapico 17,915 F4
Cerignola 48,105 E4
Cesena 72,145 D2
Cesenatico 15,634 D2
Chiavari 29,171 B2
Chieri 28,296 A2
Chieti 49,267 E3
Chioggia 46,728 D2
Chivasso 22,230 A2
Ciampino 31,981 F7
Città di Castello 21,492 C3
Civitavecchia 46,465 C3
Comiso 25,469 E6
Como 94,167 B2
Conegliano 32,406 D2
Conversano 18,518 F4
Corato 41,078 F4
Cosenza 101,144 F5
Crema 33,901 B2
Cremona 74,341 B2
Crotone 51,204 F5
Cuneo 47,836 A2
Desenzano del Garda 17,296 C2
Domodossola 19,825 A1
Eboli 24,152 E4
Empoli 34,066 C3
Enna 26,760 E6
Fabriano 21,155 D3
Faenza 40,635 D2
Fano 42,440 D3
Fasano 22,918 F4
Favara 30,031 D6
Fermo 17,603 D3
Ferrara 117,590 C2
Fidenza 19,482 B2
Fiesole 3,711 C3
Firenze (Florence) 442,721 C3
Fiumicino 21,167 F7
Florence 442,721 C3
Floridia 17,790 E6
Foggia 150,480 E4
Foligno 41,696 D3
Fondi 19,580 D4
Forlì 91,366 D2
Formia 29,147 D4
Fossano 17,116 A2
Francavilla Fontana 31,371 F4
Frascati 18,356 F7
Frosinone 42,626 D4
Gaeta 23,190 D4
Galatina 22,611 G4
Gallarate 47,259 B2
Gela 74,077 E6
Genoa 755,389 B2
Genova (Genoa) 787,011 B2
Giarre 23,377 E6
Gioia del Colle 23,868 F4
Giovinazzo 18,832 F4
Giulianova 20,189 E3
Gorizia 40,679 D2
Gravina in Puglia 35,891 F4
Grosseto 55,569 C3
Grottaglie 27,140 F4
Iglesias 26,313 B5
Imola 47,365 C2
Imperia 39,151 B3
Isernia 16,919 E4
Ivrea 26,446 B2
Jesi 37,075 D3
L'Aquila 40,467 D3
La Spezia 110,632 B2
Lanciano 25,828 E3
Latina 64,529 D4
Lecce 80,127 G4
Lecco 51,160 B2
Leghorn (Livorno) 171,811 C3
Legnago 23,232 C2
Lentini 30,950 E6
Leonforte 15,745 E6
Licata 40,309 D6
Lido di Ostia 85,043 F7
Lido di Venezia 20,863 D2
Livorno 171,811 C3
Lodi 41,338 B2
Lucca 84,836 C3
Lucera 31,252 E4
Lugo 21,593 D2
Macerata 34,409 D3
Manduria 28,112 F4
Manfredonia 52,162 F4
Mantua 52,477 C2
Marino 30,261 F7
Marsala 76,843 D6
Martina Franca 34,911 F4
Massa 60,810 C2
Massafra 26,172 F4
Matera 48,226 F4
Mazara del Vallo 42,320 D6
Merano 31,854 C1
Mesagne 29,770 G4
Messina 240,121 E5
Mestre 197,952 D2
Milan 1,601,797 B2
Milazzo 29,868 E5
Minturno 15,795 D4
Mira Taglio 26,031 D2
Modena 164,529 C2
Modica 34,488 E6
Mola di Bari 25,744 F4
Molfetta 64,738 F4
Moncalieri 59,344 A2
Monfalcone 29,960 D2
Monopoli 33,928 F4
Monreale 18,168 D5
Monte Sant'Angelo 16,491 F4
Montebelluna 19,708 D2
Monterotondo 25,383 F6
Montevarchi 17,110 C3
Monza 122,541 B2
Naples 1,210,365 E4
Nardò 27,384 F4
Nettuno 27,929 D4
Nicastro-Sambiase 49,325 F5
Niscemi 25,677 E6
Nocera Inferiore 43,879 E4
Noto 20,609 E6
Novara 94,477 B2
Novi Ligure 28,756 B2
Nuoro 35,491 B4
Olbia 26,702 B4
Oristano 23,938 B5
Orvieto 7,509 D3
Ostia Antica 3,939 F7
Ostuni 27,948 F4
Otranto 4,334 G4
Pachino 20,631 E6
Padua 228,333 C2
Palermo 698,481 D5
Palma di Montechiaro 23,918 D6
Palmi 16,394 E5
Pantelleria 3,454 C6
Parma 160,374 C2
Partinico 27,479 D6
Paterno 42,916 E6
Pavia 82,629 B2
Perugia 103,542 D3
Pesaro 78,550 D3
Pescara 131,016 E3
Piacenza 103,584 B2
Piazza Armerina 20,119 E6
Pietrasanta 20,404 B3
Pinerolo 33,176 A2
Piombino 35,312 C3
Pisa 95,015 C3
Pistoia 78,105 C3
Poggibonsi 22,644 C3
Pomezia 19,453 F7
Pordenone 51,270 D2
Porto Empedocle 16,126 D6
Porto Torres 20,233 B4
Portocivitanova 28,155 D3
Portoferraio 8,108 C3
Portofino 615 B2
Potenza 55,175 E4
Pozzuoli 61,856 D4
Prato 156,894 C3
Putignano 22,361 F4
Quartu Sant'Elena 40,506 B5
Ragusa 60,871 E6
Rapallo 26,457 B2
Ravenna 87,582 D2
Reggio di Calabria 159,416 E5
Reggio nell'Emilia 107,484 C2
Rho 50,373 B2
Rieti 33,614 D3
Rimini 111,991 D2
Rome (cap.) 2,605,441 F6
Rovereto 31,286 C2
Rovigo 41,050 C2
Ruvo di Puglia 23,510 F4
Salerno 150,252 E4
Saluzzo 13,078 A2
San Benedetto del Tronto 43,189 E3
San Cataldo 20,694 D6
San Giovanni in Fiore 19,391 F5
Sannicandro Garganico 18,652 E4
San Remo 59,872 A3
San Severo 53,948 E4
Santa Maria Capua Vetere 32,129 E4
Santeramo in Colle 21,154 F4
San Vito dei Normanni 18,366 F4
Saronno 36,732 B2
Sassari 104,334 B4
Sassuolo 37,515 C2
Savona 65,040 B2
Schio 30,738 C2
Sciacca 35,063 D6
Scicli 18,419 E6
Senigallia 27,474 D3
Sesto Fiorentino 43,307 C3
Sestri Levante 19,672 B2
Siena 54,982 C3
Siracusa (Syracuse) 109,038 E6
Sondrio 19,955 B1
Sora 20,380 D4
Sorrento 15,747 E4
Spoleto 21,625 D3
Stresa 4,290 B2
Sulmona 21,504 D3
Syracuse 109,038 E6
Taranto 231,441 F4
Teramo 35,142 D3
Termini Imerese 24,252 D6

(continued on following page)

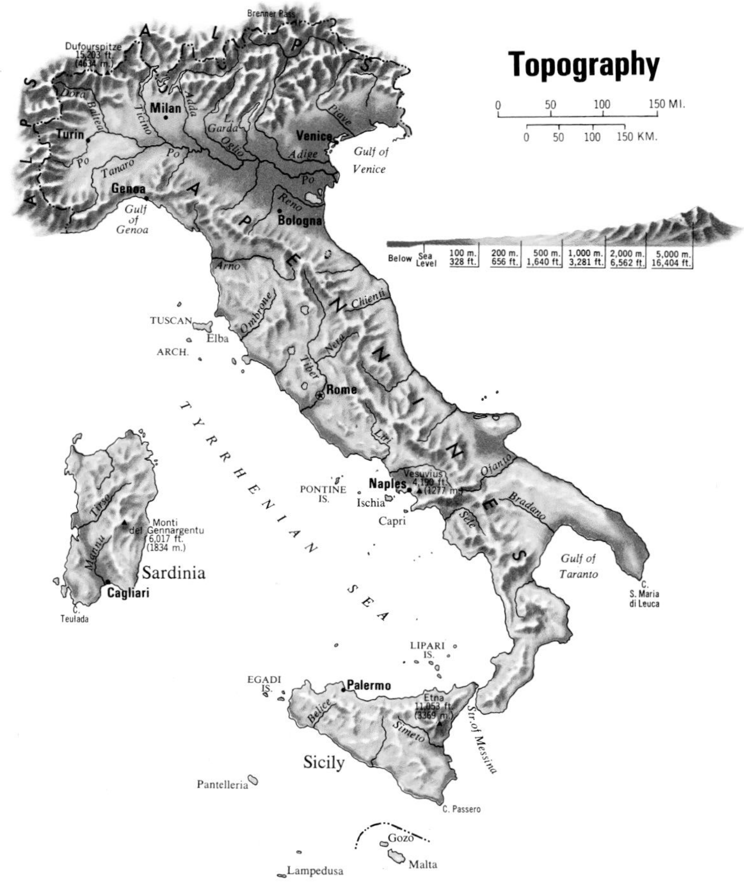
Topography
0 50 100 150 MI.
0 50 100 150 KM.
Below Sea Level
100 m. 328 ft.
200 m. 656 ft.
500 m. 1,640 ft.
1,000 m. 3,281 ft.
2,000 m. 6,562 ft.
5,000 m. 16,404 ft.
Brenner Pass
Dufourspitze 15,203 ft. (4634 m.)
ALPS
Milan
Turin
Venice
Gulf of Venice
Po
Tanaro
Ticino
Adda
Garda
Oglio
Adige
Piave
Genoa
Gulf of Genoa
Bologna
Reno
Arno
Ombrone
Chienti
Tiber
Nera
Rome
APENNINES
TUSCAN ARCH.
Elba
TYRRHENIAN SEA
PONTINE IS.
Naples
Vesuvius 4,190 ft. (1277 m.)
Ischia
Capri
Ofanto
Bradano
Sele
Gulf of Taranto
C. S. Maria di Leuca
Monti del Gennargentu 6,017 ft. (1834 m.)
Tirso
Sardinia
Cagliari
C. Teulada
LIPARI IS.
EGADI IS.
Palermo
Etna 11,053 ft. (3369 m.)
Str. of Messina
Simeto
Belice
Sicily
C. Passero
Pantelleria
Gozo
Malta
Lampedusa

DOMINANT LAND USE

- Wheat, Rice, Dairy
- Pasture Livestock
- Cereals, Livestock
- Fruit, Truck and Mixed Farming
- Grapes, Wine
- Forests
- Nonagricultural Land

MAJOR MINERAL OCCURRENCES

Ab	Asbestos	K	Potash	Pb	Lead
Al	Bauxite	Lg	Lignite	Py	Pyrites
C	Coal	Mr	Marble	S	Sulfur
Fe	Iron Ore	Na	Salt	Sb	Antimony
G	Natural Gas	O	Petroleum	Zn	Zinc
Hg	Mercury				

⚡ Water Power
▨ Major Industrial Areas

Agriculture, Industry and Resources

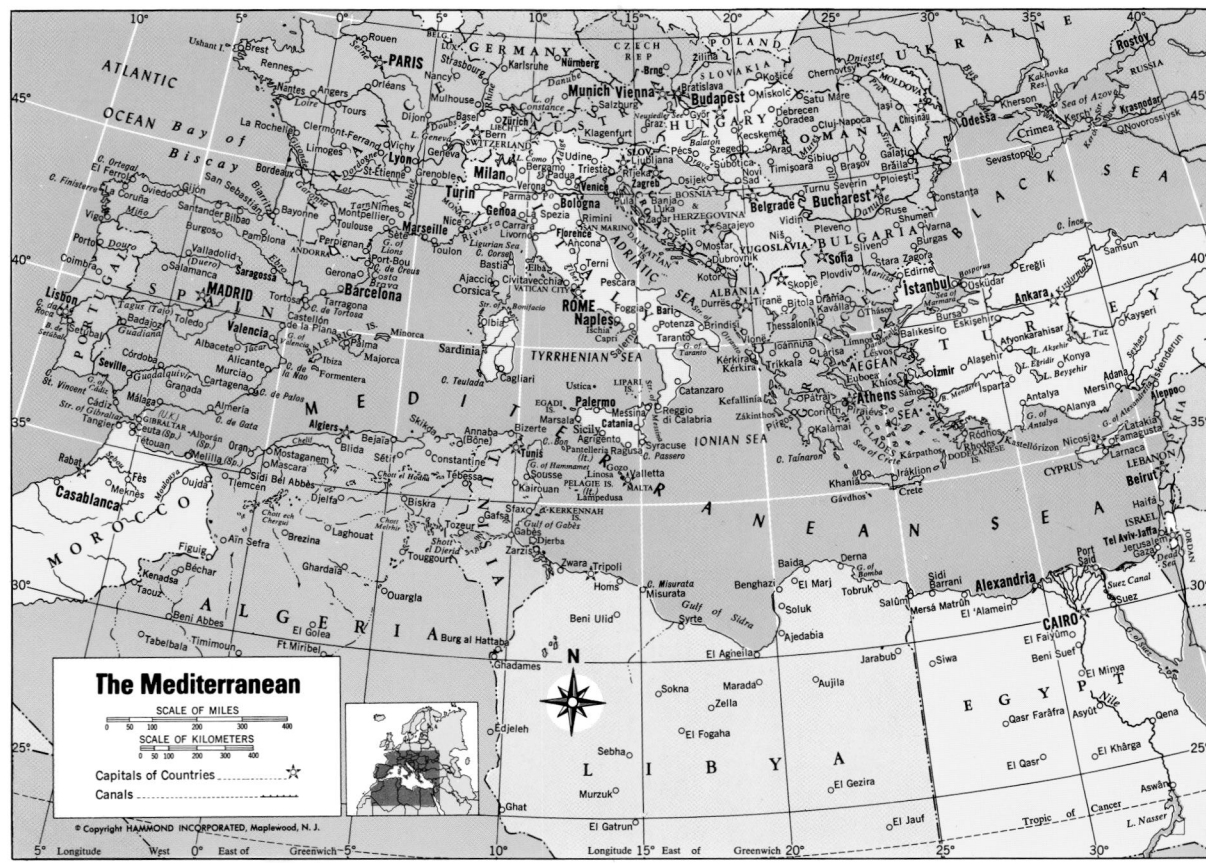

The Mediterranean

SCALE OF MILES
0 50 100 200 300 400

SCALE OF KILOMETERS
0 50 100 200 300 400

Capitals of Countries..........★
Canals.........................

© Copyright HAMMOND INCORPORATED, Maplewood, N.J.

SWITZERLAND
AREA 15,943 sq. mi. (41,292 sq. km.)
POPULATION 6,647,000
CAPITAL Bern
LARGEST CITY Zürich
HIGHEST POINT Dufourspitze
(Mte. Rosa) 15,203 ft. (4,634 m.)
MONETARY UNIT Swiss franc
MAJOR LANGUAGES German, French,
Italian, Romansch
MAJOR RELIGIONS Protestantism,
Roman Catholicism

LIECHTENSTEIN
AREA 61 sq. mi. (158 sq. km.)
POPULATION 28,000
CAPITAL Vaduz
LARGEST CITY Vaduz
HIGHEST POINT Grauspitze 8,527 ft.
(2,599 m.)
MONETARY UNIT Swiss franc
MAJOR LANGUAGE German
MAJOR RELIGION Roman Catholicism

SWITZERLAND

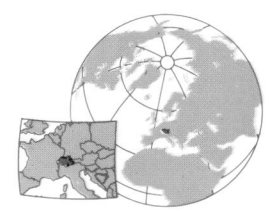

LIECHTENSTEIN

Languages

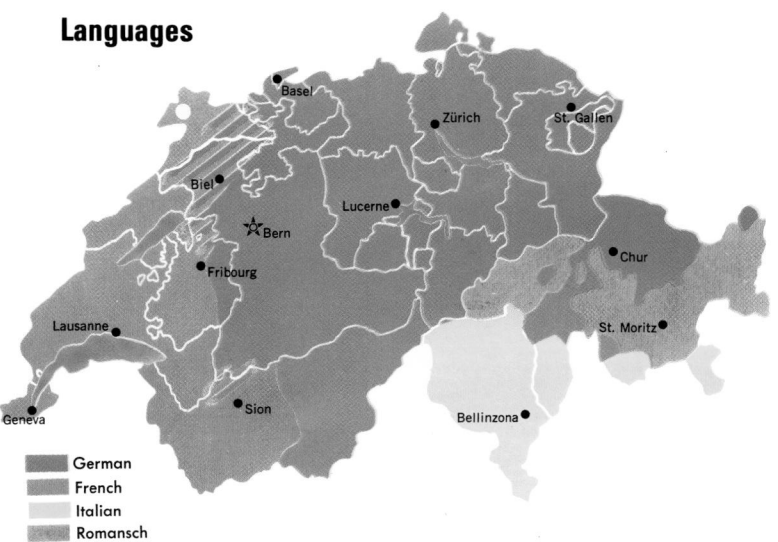

German
French
Italian
Romansch

Switzerland is a multilingual nation with four
official languages. 70% of the people speak
German, 19% French, 10% Italian and 1% Romansch.

Agriculture, Industry and Resources

DOMINANT LAND USE

Cereals, Dairy

Pasture Livestock

General Farming, Livestock

Fruit, Truck, Mixed Farming

Forests

Nonagricultural Land

⚡ Water Power

▨ Major Industrial Areas

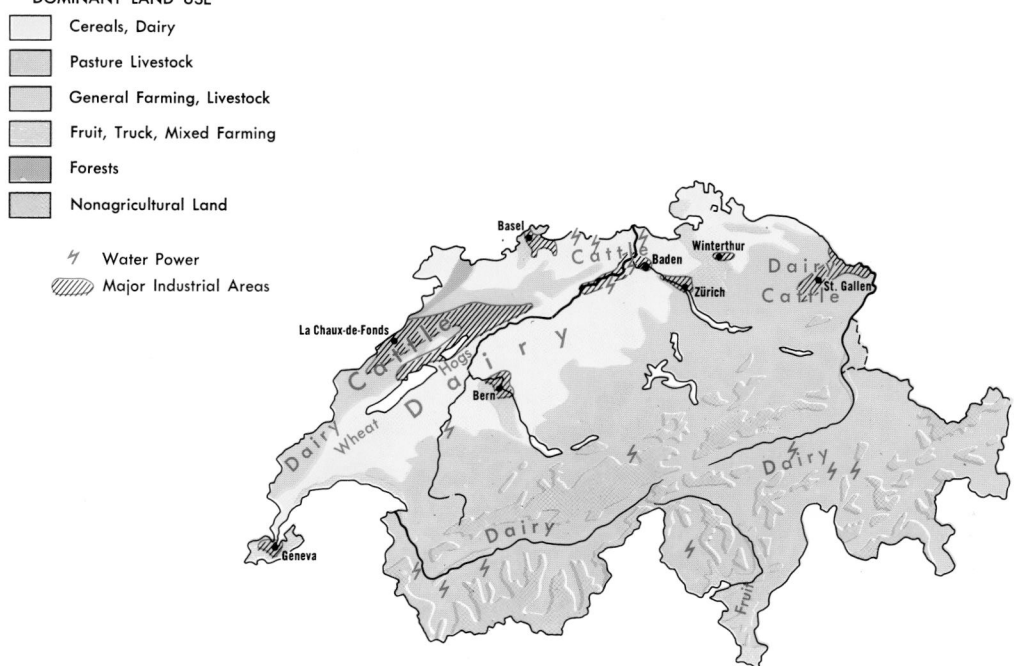

SWITZERLAND

CANTONS

Aargau 453,442	F2
Appenzell, Ausser Rhoden 47,611	H2
Appenzell, Inner Rhoden 12,844	H2
Baselland 219,822	E2
Baselstadt 203,915	E1
Bern 912,022	D2
Fribourg 185,246	D3
Geneva (Genève) 349,040	B4
Glarus 36,718	H3
Graubünden (Grisons) 164,641	H3
Jura 64,986	D2
Lucerne (Luzern) 296,159	F2
Luzern 296,159	F2
Neuchâtel 158,368	C3
Nidwalden 28,617	F3
Obwalden 25,865	F3
Sankt Gallen 391,995	H2
Schaffhausen 69,413	G1
Schwyz 97,354	G2
Soleure (Solothurn) 218,102	E2
Solothurn 218,102	E2
Thurgau 183,795	H1
Ticino 265,899	G4
Uri 33,883	G3
Valais 218,707	D4
Vaud 528,747	B3

Zug 75,930	G2
Zürich 1,122,839	G2

CITIES and TOWNS

Aadorf 3,257	G2
Aarau 15,788	F2
Aarberg 3,212	D2
Aarburg 5,354	E2
Adelboden 3,276	E3
Adliswil 16,418	F2
Affoltern am Albis 8,064	F2
Aigle 6,233	C4
Allschwil 17,952	D1
Alpnach 3,556	F3
Altdorf 8,230	G3
Altstätten 9,260	J2
Amriswil 8,790	H1
Appenzell 4,781	H2
Arbedo-Castione 3,058	G4
Arbon 11,333	H1
Arosa 2,782	J3
Arth 7,795	F2
Ascona 4,722	G4
Au 5,434	J2
Avenches 2,177	D3
Baar 15,196	F2
Bad Ragaz 3,721	H2
Baden 13,870	F2
Balerna 3,455	G5
Balsthal 5,090	E2
Bäretswil 3,145	G2
Basel 182,143	E1

Basel 364,813	E1
Bassecourt 2,942	D2
Bauma 3,010	G2
Bellinzona 16,743	H4
Belp 7,578	D3
Bern (cap.) 145,254	D3
Bettlach 3,851	D2
Bex 4,843	D4
Biasca 5,447	H4
Biberist 7,519	D2
Biel 53,793	D2
Binningen 14,195	D1
Bischofszell 3,390	H1
Bolligen 32,312	E3
Boudry 4,488	C3
Breitenbach 2,518	E2
Bremgarten 4,815	F2
Brienz 2,759	F3
Brig 9,608	F4
Brittnau 2,822	E2
Brugg 8,911	F2
Bubikon 3,601	G2
Buchs 9,066	H2
Bülach 12,292	G1
Bulle 7,595	D3
Buochs 3,742	F3
Büren an der Aare 2,761	D2
Burgdorf 15,379	E2
Bürglen 3,456	G3
Bussigny-près-Lausanne 4,909	B3
Bütschwil 3,423	H2
Carouge 13,100	B4
Castagnola 4,430	G4
Cham 9,275	G2
Château-d'Oex 2,872	D4
Châtel-Saint-Denis 3,141	C3
Chêne-Bougeries 9,068	B4
Chiasso 8,583	G5
Chur 32,037	J3
Collombey-Muraz 2,982	C4
Collonge-Bellerive 4,531	B4
Conthey 4,828	D4
Courrendlin 2,435	D2
Couvet 2,627	C3
Davos 10,468	J3
Degersheim 3,269	H2
Delémont 11,682	D2
Derendingen 4,675	E2
Dielsdorf 3,767	F1
Diepoldsau 3,562	J2
Diessenhofen 2,535	G1
Dietikon 21,765	F2
Disentis-Muster 2,320	G3
Domat-Ems 6,266	H3
Dornach 5,442	E2
Döttingen 3,264	F1
Dübendorf 20,683	G2
Düdingen 5,572	D3
Dürnten 4,927	G2
Ebnat-Kappel 4,950	H2
Echallens 2,163	C3
Ecublens 7,615	B3
Effretikon 14,788	G2
Egg 6,074	G2
Eggiwil 2,323	E3
Egnach 3,397	H1
Einsiedeln 9,629	G2
Elgg 3,041	G2
Emmen 22,392	F2
Engelberg 2,963	F3
Ennenda 2,512	H2
Entlebuch 3,238	E3
Erstfeld 4,158	G3
Eschenbach 3,661	G2
Escholzmatt 3,033	E3
Estavayer-le-Lac 3,662	C3
Feuerthalen 2,920	G1
Flawil 8,575	H2
Fleurier 3,573	C3
Flims 2,136	H3
Flums 4,228	H2
Frauenfeld 18,607	H1
Freienbach 9,912	G2
Fribourg 37,400	D3
Frick 3,116	E1
Frutigen 5,779	E3
Fully 3,926	D4
Gais 2,388	H2
Gelterkinden 4,954	E2

(continued on following page)

Topography

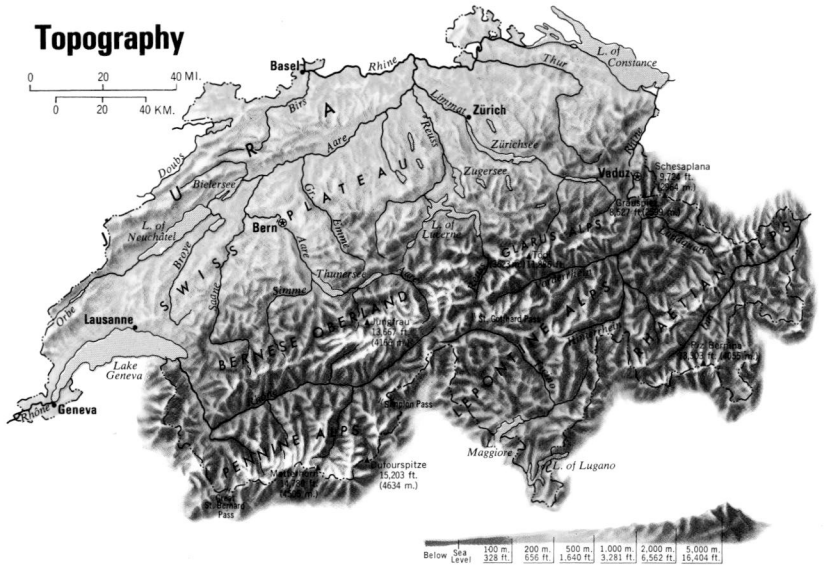

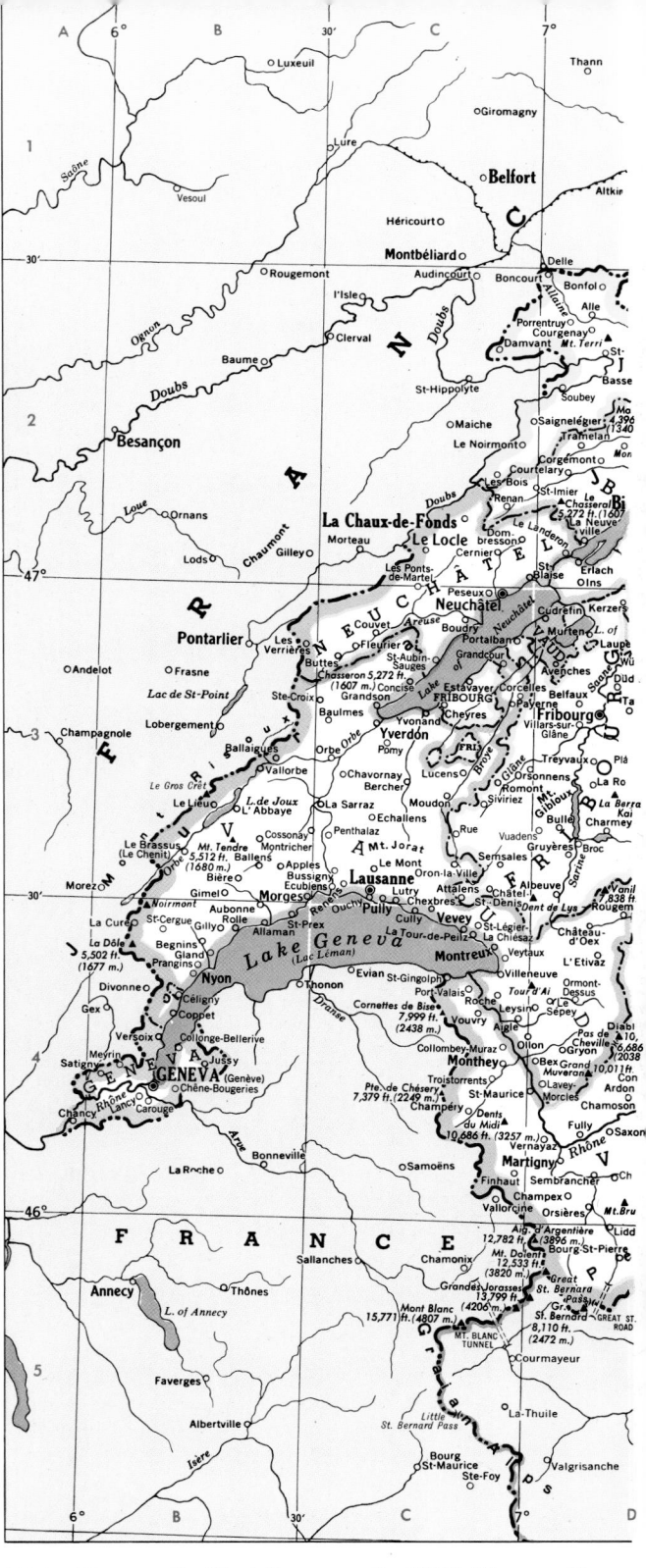

Geneva (Genève) 156,505......B4
Giswil 2,595......F3
Giubiasco 6,585......G4
Gland 4,906......B4
Glarus 5,969......H2
Glattfelden 2,753......F1
Glis 3,389......G4
Gordola 2,956......G4
Gossau 14,584......H2
Grabs 4,844......H2
Grenchen 16,800......C2
Grindelwald 3,555......E3
Grosswangen 2,235......F2
Gstaad 3,620......D4
Heiden 3,620......H1
Heimberg 4,107......E3
Hergiswil 4,254......F3
Herisau 14,160......H2
Herzogenbuchsee 5,107......E2
Hilterfingen 3,600......E3
Hinwil 7,554......G2
Hochdorf 6,034......F2
Horgen 16,577......G2
Huttwil 4,612......E2
Igis 5,392......J3
Ingenbohl 6,232......G2
Ins 2,608......D2
Interlaken 4,852......E3
Jegenstorf 3,541......D2
Jona 12,156......G2
Kaltbrunn 2,735......G2
Kerns 4,200......F3
Kerzers 2,658......D3
Kirchberg, Bern 3,966......E2
Kirchberg, St. Gallen 6,398......G2
Klingnau 2,433......F1
Klosters-Serneus 3,487......J3
Kloten 15,845......G2
Kölliken 3,080......F2
Köniz 33,441......D3
Konolfingen 4,360......E2
Kreuzlingen 16,101......H1
Kriens 21,097......F2
Küsnacht 12,766......G2
Küssnacht am Rigi 8,091......F2
Küttigen 4,356......F2
La Chaux-de-Fonds 37,234......C2
Lachen 5,352......G2
Lancy 20,877......B4
La Neuveville 3,519......D2
Langenthal 13,408......E2
Langnau am Albis 6,694......G2
Langnau in Emmental 8,821......E3
La Tour-de-Peilz 9,411......C4
Laufen 4,444......D2
Laupen 2,261......D3
Lauperswil 2,482......E2
Lausanne 127,349......C3
Lauterbrunnen 3,077......E3
Le Brassus 4,359......B3
Le Châble 4,541......C4
Le Chenit (Le Brassus) 4,359 B3
Le Landeron 3,287......C2
Le Locle 12,039......C2
Le Mont-sur-Lausanne 3,664 C3
Lengnau 4,317......D2
Lenk 2,089......D4
Lens 2,412......D4
Lenzburg 7,585......F2
Leuk 2,983......E4
Leukerbad 1,070......E4
Liestal 12,158......E1
Liestal-Sissach 40,800......E1
Littau 14,996......F2
Locarno 14,103......G4
Lucerne 63,278......F2
Lugano 27,815......G4
Lutry 5,884......C3
Lützelflüh 3,770......E2
Luzern (Lucerne) 63,278......F2
Lyss 8,723......D2
Malters 4,900......F2
Männedorf 7,833......G2
Martigny 11,309......C4

Meilen 10,430......G2
Meiringen 4,072......F3
Mellingen 3,285......F2
Mels 6,235......H2
Mendrisio 6,590......G5
Menzingen 3,564......G2
Menznau 2,248......E2
Meyrin 18,808......B4
Minusio 5,602......G4
Möhlin 6,360......E1
Mollis 2,621......H2
Monthey 11,285......C4
Montreux 19,685......C4
Morges 13,057......B3
Moudon 3,805......C3
Moutier 7,959......D2
Mümliswil-Ramiswil 2,386......E2
Münchenbuchsee 8,395......D2
Münsingen 9,340......E3
Muotathal 2,896......G3
Muri 5,399......F2
Muri bei Bern 12,285......D3
Murten 4,558......D3
Muttenz 16,911......E1
Näfels 3,766......H2
Naters 6,662......E4
Nendaz 4,372......D4
Netstal 2,642......H2
Neuchâtel 34,428......C3
Neuenegg 3,727......D3
Neuhausen am Rheinfall
  10,662......G1
Niederbipp 3,165......E2
Niederurnen 3,438......H2
Nyon 12,842......B4
Oberägeri 3,563......G2
Oberburg 2,869......E2
Oberdiessbach 2,319......E3
Oberriet 6,222......J2
Obersiggenthal 7,442......F1
Oberuzwil 4,616......H2
Oensingen 3,543......E2
Oftringen 9,006......F2
Ollon 4,429......D4
Olten 18,991......E2
Opfikon 11,444......G2
Orbe 3,985......C3
Orsières 2,357......C4
Paradiso 3,261......G5
Payerne 6,713......C3
Peseux 5,212......C3
Pfäffikon 8,306......G2
Pfaffnau 2,453......E2
Pieterlen 3,127......D2
Porrentruy 7,039......C2
Poschiavo 3,294......J4
Prangins 2,028......B4
Pratteln 15,751......E1
Pully 14,988......C4
Rafz 2,325......G1
Rapperswil 7,828......G2
Regensdorf 12,300......F1
Reichenbach im Kandertal
  2,948......E3
Reiden 3,363......F2
Reinach in Aargau 5,696......F2
Reinach in Baselland 17,813...E2
Renens 16,977......C3
Rheineck 3,037......J2
Rheinfelden 9,456......E1
Richterswil 8,672......G2
Riehen 20,611......E1
Riggisberg 2,196......D3
Roggwil 3,333......E2
Rolle 3,409......B4
Romanshorn 7,893......H1
Romont 3,495......C3
Rorschach 9,878......H1
Rothrist 6,015......E2
Rüti, Zürich 9,331......G2
Rumlang 5,055......G2
Ruswil 4,870......F2

Saint-Blaise 2,788......D2
Sainte-Croix 4,543......B3
Saint-Imier 5,430......C2
Saint-Légier-La Chiésaz 2,787 C4
Saint-Maurice 3,458......C4
Saint-Moritz 5,900......J3
Saint Niklaus 2,036......E4
Saint-Prex 2,937......B4
Samedan 2,553......J3
Sankt Gallen 75,847......H2
Sankt Margrethen 4,935......J2
Sargans 4,267......H2
Sarnen 7,372......F3
Savièse 4,097......D4
Saxon 2,394......D4
Schänis 2,426......H2
Schaffhausen 34,250......G1
Schattdorf 4,516......G3
Schiers 2,253......J3
Schlieren 12,891......F2
Schönenwerd 4,746......E2
Schübelbach 4,720......G2
Schüpfheim 3,537......F3
Schwanden 2,519......H2
Schwyz 12,100......G2
Sempach 2,237......F2
Seon 3,826......F2
Seuzach 4,659......G1
Sevelen 2,839......H2
Sierre 13,050......D4
Signau 2,606......E3
Sigriswil 3,536......E3
Liechen 2,115......G3
Simplon 328......F4
Sins 2,625......F2
Sion 22,877......D4
Sirnach 4,170......G2
Sissach 4,564......E2
Solothurn (Soleure) 15,778......E2
Spiez 9,800......E3
Stäfa 10,558......G2
Stans 5,681......F3
Steckborn 3,232......G1
Steffisburg 12,539......E3
Stein am Rhein 2,507......G1
Suhr 7,366......F2
Sumiswald 5,070......E2
Sursee 7,645......F2
Tafers 2,263......D3
Tavannes 3,336......D2
Teufen 5,027......H2
Thal 4,725......J2
Thalwil 15,412......G2
Thayngen 3,751......G1
Therwil 7,311......E1
Thun 36,891......E3
Thunstetten 2,567......E2
Thusis 2,525......H3
Tramelan 4,733......D2
Turbenthal 2,975......G2
Uetendorf 4,538......E3
Unterägeri 5,371......G2
Unterkulm 2,558......F2
Unterseen 4,558......E3
Uster 23,702......G2
Utzenstorf 3,141......E2
Uznach 4,269......H2
Uzwil 9,614......H2
Vallorbe 3,375......B3
Vechigen 4,036......E3
Versoix 7,483......B4
Vevey 16,139......C4
Vevey-Montreux 60,558......C4
Villars-sur-Glâne 5,788......D3
Villeneuve 3,573......C4
Visp 6,383......E4
Wädenswil 18,485......G2
Wängi 2,909......G1
Wahlern 5,104......D3
Wald 7,447......G2
Waldkirch 2,622......H2
Walenstadt 3,605......H2
Wallisellen 10,887......G2
Wartau 3,692......H2

Wattwil 7,874......H2
Weinfelden 8,793......H1
Wettingen 18,377......F2
Wetzikon 15,859......G2
Wil 16,245......H2
Willisau 2,639......F2
Windisch 7,598......F2
Winterthur 86,758......G1
Wohlen 12,024......F2
Wohlen 15,746......F2
Wohlen bei Bern 7,666......D3
Wolhusen 3,670......E3
Worb 11,080......E3
Wünnewil 3,774......D3
Yverdon 20,802......C3
Zell 4,138......G2
Zermatt 3,548......E4
Zofingen 8,643......E2
Zollikofen 8,717......D3
Zollikon 12,134......G2
Zug 21,609......G2
Zürich 369,522......F2
Zurzach 3,068......F1
Zweisimmen 2,852......D3

OTHER FEATURES

Aa (riv.)......F3
Aare (riv.)......D2
Ägerisee (lake)......G2
Aiguille d'Argentière (mt.)......C5
Albristhorn (mt.)......D4
Aletschhorn (mt.)......E4
Allaine (riv.)......D2
Areuse (riv.)......C3
Aroser Rothorn (mt.)......J3
Ault (peak)......H3
Baldeggersee (lake)......F2
Balmhorn (mt.)......E4
Bärenhorn (mt.)......H3
Basodino (peak)......G4
Bernese Oberland (reg.)......E3
Bernina (mts.)......K4
Bernina (pass)......K4
Bernina (peak)......J4
Bernina (riv.)......J4
Bielersee (lake)......D2
Bietschhorn (mt.)......E4
Birs (riv.)......D2
Blas (peak)......H3
Blinnenhorn (mt.)......F4
Blümlisalp (mt.)......E4
Bodensee (Constance) (lake)..H1
Borgne (riv.)......D4
Breithor (peak)......E5
Breithorn (mt.)......E4
Brienzer Rothorn (mt.)......F3
Brienzersee (lake)......E3
Broye (riv.)......C3
Brule (riv.)......D2
Buchegg (mts.)......D2
Buin (peak)......K3
Burkelkopf (mt.)......K3
Bütschelegg (mt.)......D3
Calancasca (riv.)......H4
Campo Tencia (peak)......G4
Ceneri (pass)......G4
Chasseron (mt.)......C3
Chésery, Pointe de (mt.)......C4
Cheville (pass)......D4
Churfirsten (mts.)......H2
Clariden (mt.)......G3
Collon (mt.)......D5
Constance (Bodensee) (lake)..H1
Cornettes de Bise (mts.)......C4
Dammastock (mt.)......F3
Danis (valley)......H3
Dent Blanche (mt.)......E4
Dent de Lys (mt.)......C4
Dent de Ruth (mt.)......D4
Dent d'Hérens (mt.)......E5
Dents du Midi (mt.)......C4
Diablerets (mt.)......D4

Doldenhorn (mt.)......E4
Dolent (mt.)......C5
Dom (mt.)......E4
Doubs (riv.)......D4
Drance (riv.)......C4
Dufourspitze (mt.)......E5
Emmental (riv.)......E3
Engadine (valley)......K3
Err (peak)......J3
Finsteraarhorn (mt.)......F3
Finstermünz (pass)......K3
Fletschhorn (mt.)......E4
Fluchthorn (mt.)......K3
Flüela (pass)......J3
Fluhberg (mt.)......G2
Fort (mt.)......D4
Frienisberg (mt.)......D2
Furka (pass)......F3
Gelgia (riv.)......J3
Generoso (mt.)......G5
Geneva (lake)......C4
Giacomo (pass)......G4

Gibloux (mt.)......D3
Glâne (riv.)......C3
Glärnisch (mt.)......H2
Glarus Alps (mts.)......H3
Glatt (riv.)......G2
Goms (valley)......F4
Grand Combin (mt.)......D5
Grand Muveran (mt.)......D4
Grande Dixence (dam)......D4
Grauhörner (mts.)......H3
Great Saint Bernard (pass)......D5
Great Saint Bernard (tunnel)...D5
Greifensee (lake)......G2
Greina (pass)......H3
Gridone (mt.)......G4
Grimsel (pass)......F3
Gross Emme (riv.)......E2
Gross Litzner (mt.)......K3
Hallwilersee (lake)......F2
Hausstock (mt.)......H3
Helsenhorn (mt.)......F4

Hinterrhein (riv.)......H3
Hochwang (mt.)......J3
Hohenstollen (mt.)......F3
Honegg (mt.)......E3
Hörnli (mt.)......G2
Ilfis (riv.)......E3
Inn (riv.)......K3
Joch (pass)......F3
Jorat (mt.)......C3
Joux (lake)......B3
Jungfrau (mt.)......E3
Jura (mts.)......B3
Kaiseregg (mt.)......D3
Kesch (peak)......J3
Kisten (pass)......H3
Klausen (pass)......G3
Kleine Emme (riv.)......F3
La Berra (mt.)......D3
La Dôle (mt.)......B4
Landquart (riv.)......J3
Le Chasseral (mt.)......D2
Le Gros Crêt (mt.)......B3

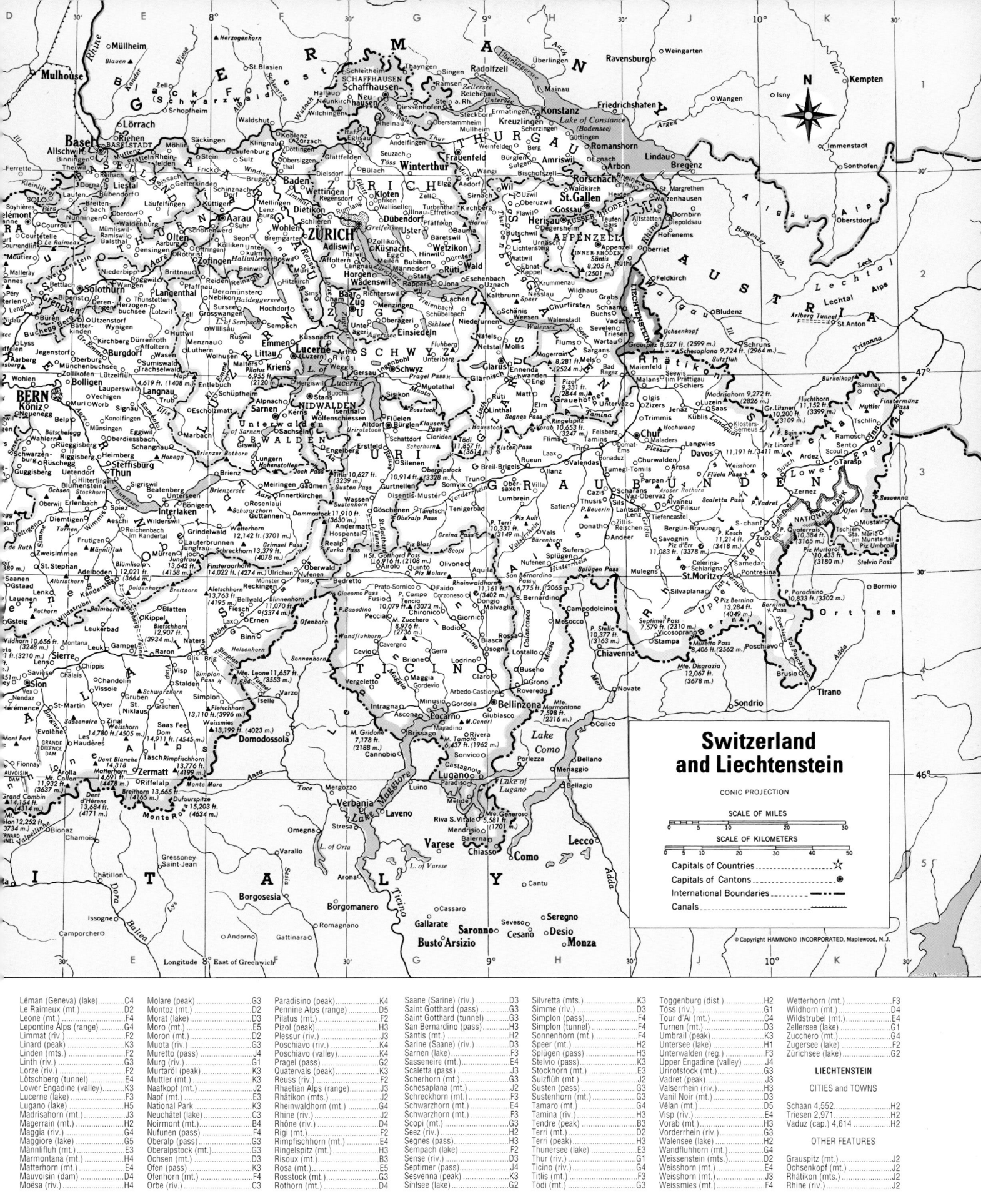

AUSTRIA

PROVINCES

Burgenland 272,274 ....D3
Carinthia 536,727 ....C3
Lower Austria 1,439,137 ....C2
Salzburg 441,842............C3
Styria 1,187,512 ....C3
Tirol 586,139 ....A3
Upper Austria 1,270,426.....B2
Vienna (city) 1,515,666......D2
Vorarlberg 305,615..........A3

CITIES and TOWNS†

Altheim 4,702 ....B2
Althofen 4,274 ....C3
Amstetten 22,015 ....C2
Arnoldstein 6,641 ....C3
Attnang-Puchheim 8,058 .....B2
Bad Aussee 5,047............B3
Bad Goisern 6,500 ....B3
Bad Hofgastein 5,960 ....B3
Bad Ischl 13,027 ....B3
Bad Sankt-Leonhard im
  Lavanttal 5,008 ....C3
Baden 23,235 ....C2
Badgastein 5,600 ....B3
Berndorf 8,189 ....C2
Bischofshofen 9,520 ....B3
Bludenz 12,893 ....A3
Bramberg am Wildkogel
  3,410 ....B3
Braunau am Inn 16,192 .....B2
Bregenz 24,683 ....A3
Bruck an der Leitha 7,170 ...D2
Bruck an der Mur 15,086.....C3
Deutsch Feistritz 3,719 .....C3
Deutschkreutz 3,563........D3
Deutsch Landsberg 7,623....C3

Hermagor-Preseggersee
  7,116 ....B3
Herzogenburg 7,313 ....C2
Hohenems 12,669 ....A3
Hollabrunn 10,254 ....D2
Hopfgarten in Nordtirol
  4,956 ....B3
Horn 6,319 ....C2
Imst 6,691 ....A3
Innsbruck 116,110 ....A3
Jenbach 5,725 ....A3
Jennersdorf 4,131 ....D3
Judenburg 11,199 ....C3
Kapfenberg 25,719 ....C3
Kaprun 2,764 ....B3
Kindberg 6,269 ....C3
Kirchdorf an der Krems
  3,708 ....C3
Kitzbühel 7,872.............B3
Klagenfurt 86,303 ....C3
Klosterneuburg 23,307 ......D2
Knittelfeld 14,153...........C3
Köflach 12,009..............C3
Korneuburg 9,132 ....D2
Kötschach-Mauthen 3,633....B3
Krems an der Donau 23,123..C2
Kufstein 13,125 ....B3
Laa an der Thaya 6,485 .....D2
Laakirchen 7,670 ....B3
Landeck 7,325 ....A3
Landskron 10,429...........B3
Langenlois 6,474 ....C2
Langenwang 4,187 ....C3
Lavamünd 3,824 ....C3
Leibnitz 6,659 ....C3
Lenzing 5,079 ....B3
Leoben 32,006 ....C3
Lienz 11,699 ....B3
Liezen 7,021 ....C3
Lilienfeld 3,030 ....C3
Linz 197,962 ....C2

Salzburg 138,213............B3
Sankt Johann in Tirol 6,495..B3
Sankt Michael im Lungau
  3,246 ....B3
Sankt Michael in
  Obersteiermark 3,604 .....C3
Sankt Paul in Lavanttal
  5,770 ....C3
Sankt Pölten 51,102 ....C2
Sankt Valentin 8,759........C2
Sankt Veit an der Glan 12,021.C3
Schärding 5,784 ....B2
Scheibbs 4,537 ....C2
Schladming 3,930 ....B3
Schrems 6,010 ....C2
Schwarzach im Pongau
  3,607 ....B3
Schwaz 10,936 ....A3
Schwechat 14,844 ....C2
Schwertberg 4,385 ....C2
Sierning 7,891 ....C2
Solbad Hall in Tirol 12,622..A3
Spittal an der Drau 14,769...B3
Steyr 38,967 ....C2
Stockerau 12,692 ....C2
Tamsweg 5,256 ....B3
Telfs 7,749 ....A3
Ternitz 16,154 ....C3
Traiskirchen 14,102 ....C2
Traun 21,524 ....C2
Trieben 4,241 ....C3
Trofaiach 8,959.............C3
Tulln 11,287 ....C2
Velden am Wörthersee 7,458..C3
Vienna (cap.) 1,515,666.....D2
Villach 52,744 ....B3
Vöcklabruck 11,053 ....B2
Voitsberg 110,951 ....C3
Völkermarkt 10,900 ....C3
Waidhofen an der Thaya
  5,401 ....C2

Waidhofen an der Ybbs
  11,339 ....C3
Weitensfeld-Flattnitz 5,158...C3
Weiz 8,418..................C3
Wels 51,024.................C2
Wien (Vienna) (cap.)
  1,515,666.................D2
Wiener Neustadt 35,050 .....D3
Wilhelmsburg 6,339 ....C2
Wolfsberg 28,182 ....C3
Wörgl 8,644 ....B3
Ybbs an der Donau 5,983 ....C2
Zell am See 7,959 ....B3
Zeltweg 8,722...............C3
Zistersdorf 5,814 ....D2
Zwettl-Niederösterreich
  11,579 ....C2

OTHER FEATURES

Allgäu Alps (mts.) A3
Atter See (lake) A3
Bavarian Alps (mts.) A3
Bodensee (Constance) (lake)..A3
Brenner (pass)...............A3
Carnic Alps (mts.) A3
Coglians (mt.) A3
Constance (lake) A3
Danube (riv.) D2
Donau (Danube) (riv.) C2
Drau (riv.) C3
Enns (riv.) C2
Greiner Wald (mts.) C2
Grosser Peilstein (mt.) B3
Grossglockner (mt.) B3
Hochgolling (mt.) B3
Hohe Tauern (range) B3
Hohe Warte (Coglians) (mt.)..B3
Inn (riv.) B2
Kamp (riv.) C2
Karawanken (range) C3

Lafnitz (riv.) D3
March (riv.) D2
Mühlviertel (reg.) C2
Mur (riv.) C3
Mürz (riv.) C3
Neusiedler See (lake) D3
Niedere Tauern (range) B3
Olsa (riv.) A3
Ötztal Alps (mts.) A3
Parseierspitze (mt.) A3
Raab (riv.) D3
Rhine (riv.) A3
Salzach (riv.) B2
Salzkammergut (reg.) B3
Semmering (pass) C3
Thaya (riv.) C2
Traun (riv.) B3
Traun See (lake) B3
Wildspitze (mt.) A3
Zugspitze (mt.) A3

CZECH REPUBLIC

REGIONS

Jihočeský 689,229 ....C2
Jihomoravský 2,040,903 .....D2

Praha (city) 1,182,186........C1
Severočeský 1,167,231 ......C1
Severomoravský 1,932,722...D2
Středočeský 1,151,265 ......C2
Východočeský 1,248,466 .....C1
Západočeský 879,925........B2

CITIES and TOWNS

Aš 13,551 ....B1
Austerlitz (Slavkov) 6,316....D2
Benešov 15,172 ....C2
Beroun 23,580 ....C2
Bílina 18,836 ....B1
Blansko 19,508 ....D2
Blatná 7,264 ....B2
Boskovice 12,025 ....D2
Brandýs nad Labem-Stará
  Boleslav 15,071 ....C1
Břeclav 23,978 ....D2
Brno 371,463 ....D2
Broumov 7,834 ....D1
Bruntál 17,062 ....D2
Bystřice nad Pernštejnem
  10,044 ....D2
Bystřice pod Hostýnem
  10,359 ....D2

Čáslav 9,950 ....C2
Česká Kamenice 7,272 ....C1
Česká Lípa 24,924...........C1
Česká Třebová 17,136 ....D2
České Budějovice 90,415.....C2
Český Krumlov 13,776 ....C2
Český Těšín 23,389 ....E2
Cheb 31,039 ....B1
Chodov 14,704 ....B1
Chomutov 51,769 ....B1
Chotěboř 8,744 ....C2
Chrastava 7,022 ....C1
Chrudim 20,517 ....C2
Dačice 7,443 ....C2
Děčín 49,682 ....B1
Dobříš 7,466 ....C2
Domažlice 11,461 ....B2
Duchcov 10,554 ....B1
Dvůr Králové nad Labem
  17,270 ....C1
Frenštát pod Radhoštěm
  10,434 ....E2
Frýdek-Místek 59,430 ....E2
Frýdlant nad
  Ostravicí 14,065...........E2
Frýdlant v. Čechách 7,418....C1
(continued)

Topography

Deutsch Wagram 5,111 ....D2
Dornbirn 38,663 ....A3
Ebensee 9,005 ....B3
Eggenburg 3,729 ....C2
Eisenerz 10,074 ....C3
Eisenkappel-Vellach 3,520....C3
Eisenstadt 10,150 ....D3
Enns 9,731 ....C2
Feldbach 4,073 ....C3
Feldkirch 23,876 ....A3
Feldkirchen in Kärnten
  12,181 ....B3
Ferlach 7,658 ....B3
Fieberbrunn 3,926 ....B3
Fohnsdorf 10,360 ....C3
Frankenmarkt 3,166 ....B2
Freistadt 6,289 ....C2
Frohnleiten 5,061 ....C3
Fürstenfeld 6,040 ....C3
Gaming 4,099 ....C3
Gänserndorf 4,948 ....D2
Gleisdorf 5,078 ....C3
Gloggnitz 6,290 ....C3
Gmünd 6,457 ....B3
Gmunden 12,720............B3
Golling an der Salzach 3,409..B3
Götzis 8,740 ....A3
Graz 243,405 ....C3
Grieskirchen 4,813 ....B2
Grosssiegharts 3,374 ....C2
Grünburg 3,630 ....C3
Güssing 3,895 ....D3
Haag 5,095 ....C2
Hainburg an der Donau
  5,749 ....D2
Hainfeld 3,735 ....C2
Hallein 15,404 ....B3
Hartberg 6,048 ....C3
Heidenreichstein 5,351 ......C2
Heiligenblut 1,334 ....B3

Lustenau 17,404 ....A3
Mannersdorf am
  Leithagebirge 3,878 ....D3
Marchegg 2,676 ....D2
Matrei in Osttirol 4,298 .....B3
Mattersburg 5,682 ....D3
Mattighofen 4,566 ....B2
Mauthausen 4,353 ....C2
Mauthen-Kötschach 3,633....B3
Mayrhofen 3,274 ....A3
Melk 5,074 ....C2
Mistelbach an der Zaya
  10,300 ....D2
Mittersill 5,033 ....B3
Mödling 19,333 ....C2
Mürzzuschlag 10,770 ....C3
Neumarkt am Wallersee
  3,703 ....B2
Neunkirchen 10,780 ....D3
Neusiedl am See 4,154 ......D3
Ober Grafendorf 4,475 ......C2
Oberndorf bei Salzburg 3,838.B2
Oberwart 5,973 ....D3
Oberwölz 9,510 ....B3
Paternion 5,914 ....B3
Perg 5,226 ....C2
Pinkafeld 4,802 ....D3
Pöchlarn 3,637 ....C2
Poysdorf 5,658 ....D2
Pregarten 3,823 ....C2
Raabs an der Thaya 3,839 ...C2
Radenthein 7,083...........B3
Radstadt 3,994 ....B3
Rankweil 9,929 ....A3
Reichenau an der Rax 3,601..C3
Retz 4,373 ....C2
Reutte 5,145 ....A3
Ried im Innkreis 10,952 .....B2
Rottenmann 5,425 ....C3
Saalfelden am Steinernen
  Meer 11,436 ....B3

AUSTRIA
AREA 32,375 sq. mi. (83,851 sq. km.)
POPULATION 7,666,000
CAPITAL Vienna
LARGEST CITY Vienna
HIGHEST POINT Grossglockner 12,457 ft. (3,797 m.)
MONETARY UNIT schilling
MAJOR LANGUAGE German
MAJOR RELIGION Roman Catholicism

CZECH REPUBLIC
AREA 30,449 sq. mi. (78,863 sq. km.)
POPULATION 10,291,927
CAPITAL Prague
LARGEST CITY Prague
HIGHEST POINT Sněžka 5,256 ft. (1,602 m.)
MONETARY UNIT Czech koruna
MAJOR LANGUAGE Czech
MAJOR RELIGIONS Roman Catholicism, Protestantism

HUNGARY
AREA 35,919 sq. mi. (93,030 km.)
POPULATION 10,558,000
CAPITAL Budapest
LARGEST CITY Budapest
HIGHEST POINT Kékes 3,330 ft. (1,015 m.)
MONETARY UNIT forint
MAJOR LANGUAGE Hungarian
MAJOR RELIGIONS Roman Catholicism, Protestantism

SLOVAKIA
AREA 18,924 sq. mi. (49,014 sq. km.)
POPULATION 4,991,168
CAPITAL Bratislava
LARGEST CITY Bratislava
HIGHEST POINT Gerlachovky Štit 8,707 ft. (2,654 m.)
MONETARY UNIT Slovak koruna
MAJOR LANGUAGE Slovak
MAJOR RELIGIONS Roman Catholicism, Protestantism

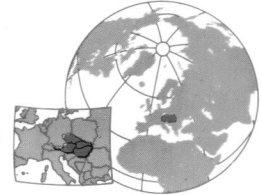

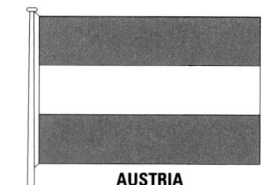
AUSTRIA

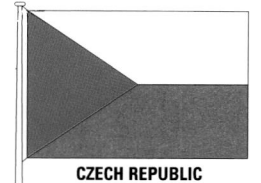
CZECH REPUBLIC

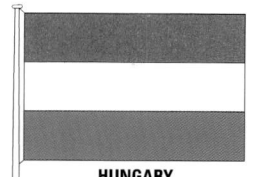
HUNGARY

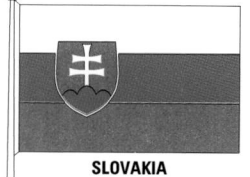
SLOVAKIA

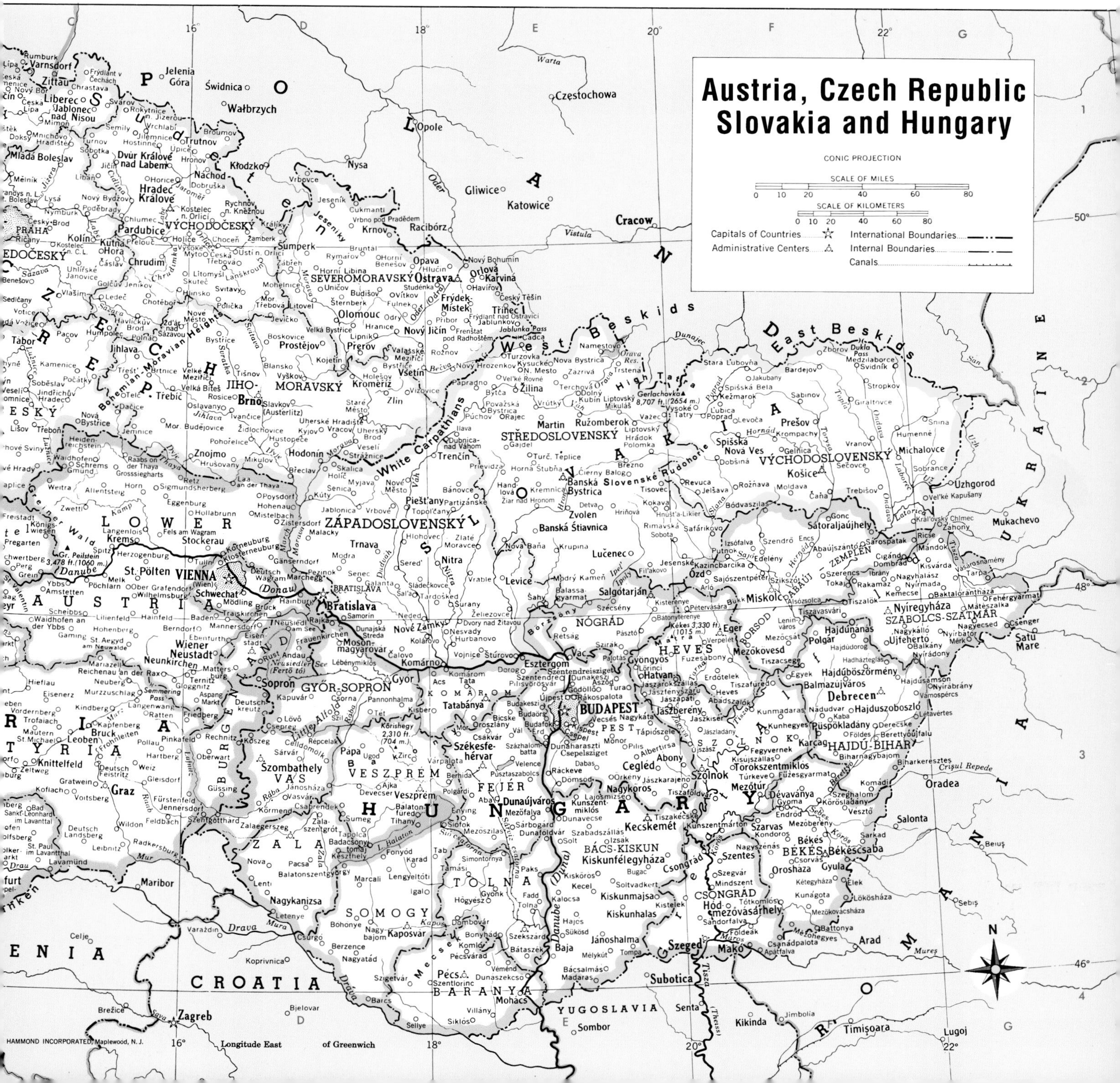

Austria, Czech Republic Slovakia and Hungary

CONIC PROJECTION

SCALE OF MILES
0 10 20 40 60 80

SCALE OF KILOMETERS
0 10 20 40 60 80

Capitals of Countries ☆
Administrative Centers △
International Boundaries _____
Internal Boundaries _____
Canals _____

42 Austria, Czech Republic, Slovakia and Hungary

(continued)

Fulnek 8,214 ..... D2
Havířov 89,920 ..... E2
Havlíčkuv Brod 24,550 ..... C2
Hlinsko 10,635 ..... C2
Hlučín 22,581 ..... E2
Hodonín 25,485 ..... D2
Holešov 13,323 ..... D2
Hořice v Podkrkonoší 9,251 ..... C1
Hradec Králové 95,588 ..... C1
Hranice 18,099 ..... D2
Hronov 9,609 ..... D1
Humpolec 10,042 ..... C2
Ivančice 9,746 ..... D2
Jablonec nad Nisou 42,179 ..... C1
Jablunkov 15,962 ..... E2
Jaroměř 11,562 ..... C1
Jeseník 14,314 ..... D1
Jičín 16,440 ..... C1
Jihlava 51,144 ..... C2
Jindřichuv Hradec 20,096 ..... C2
Jiřkov 11,980 ..... B1
Kadăn 18,420 ..... B1
Karlovy Vary 60,950 ..... B1
Karviná 78,334 ..... E2
Kladno 71,141 ..... B1
Klatovy 21,782 ..... B2
Kojetín 8,881 ..... D2
Kolín 30,921 ..... C1
Kralupy nad Vltavou 17,528 ..... C1
Kraslice 7,371 ..... B1
Krnov 25,678 ..... D1
Kroměříž 25,887 ..... D2
Krupka 9,336 ..... B1
Kutná Hora 20,927 ..... C2
Kyjov 12,632 ..... D2
Lanškroun 10,620 ..... D2
Liberec 97,474 ..... C1
Lidice C1
Lipník nad Bečvou 9,961 ..... D2
Litoměřice 23,835 ..... C1
Litomyšl 10,079 ..... D2
Prague (Praha) (cap.) 1,182,186 ..... C1
Přelouč 8,561 ..... C1
Přerov 50,265 ..... D2
Příbor 12,711 ..... E2
Příbram 37,854 ..... C2
Prostějov 49,599 ..... D2
Rakovník 16,233 ..... B1
Říčany u Prahy 10,703 ..... C2
Rokycany 15,041 ..... B2
Roudnice nad Labem 13,956 ..... C1
Rožnov pod Radhoštěm 15,468 ..... E2
Rumburk 10,255 ..... C1
Rychnov nad Kněžnou 8,955 ..... D1
Rýmařov 9,927 ..... D2
Sedlčany 7,453 ..... C2
Semily 8,464 ..... C1
Slaný 14,705 ..... C1
Slavkov 6,316 ..... D2
Soběslav 8,406 ..... C2
Sokolov 28,523 ..... B1
Staré Město 6,293 ..... D2
Šternberk 16,342 ..... D2
Strakonice 22,611 ..... B2
Stříbro 8,169 ..... B2
Studénka 12,497 ..... E2
Šumperk 31,873 ..... D2
Sušice 11,400 ..... B2
Svitavy 19,075 ..... D2
Tábor 31,867 ..... C2
Tachov 12,798 ..... B2
Teplice 53,964 ..... B1
Tišnov 12,179 ..... D2
Třebíč 30,246 ..... C2
Třeboň 8,878 ..... C2
Třinec 44,739 ..... E2
Trutnov 27,648 ..... C1
Turnov 13,906 ..... C1
Ústí nad Labem 87,909 ..... C1
Jihlava (riv.) D2
Jizera (riv.) C1
Krušné Hory (Erzgebirge) (mts.) B1
Labe (riv.) C1
Lipno (res.) C2
Lužnice (riv.) C2
Moldau (Vltava) (riv.) C2
Morava (riv.) D2
Mže (riv.) B2
Oder (Odra) (riv.) D2
Ohře (riv.) B1
Ondava (riv.) F2
Orlice (riv.) D1
Orlickă (res.) C2
Otava (riv.) B2
Radbuza (riv.) B2
Sázava (riv.) C2
Sudeten (mts.) C1
Svitava (riv.) D2
Svratka (riv.) D2
Úhlava (riv.) B2
Vltava (riv.) C2

HUNGARY

COUNTIES

Bács-Kiskun 553,000 ..... E3
Baranya 434,000 ..... E4
Békés 416,000 ..... F3
Borsod-Abaúj-Zemplén 779,000 ..... F2
Budapest (city) 2,104,000 ..... E3
Csongrád 457,000 ..... F3
Fejér 426,000 ..... E3
Győr-Sopron 426,000 ..... D3
Hajdú-Bihar 549,000 ..... F3
Heves 338,000 ..... F3
Komárom 320,000 ..... E3
Nógrád 229,000 ..... E3
Pest 988,000 ..... E3
Csorna 13,000 ..... D3
Dabas 13,075 ..... E3
Debrecen 217,000 ..... F3
Derecske 9,579 ..... F3
Dévaványa 11,208 ..... F3
Dombóvár 21,000 ..... E3
Dorog 13,000 ..... E3
Dunaföldvár 10,318 ..... E3
Dunaharaszti 15,788 ..... E3
Dunakeszi 29,000 ..... E3
Dunaújváros 62,000 ..... E3
Edelény 12,000 ..... F2
Eger 67,000 ..... F3
Egyek 7,956 ..... F3
Endrőd 8,136 ..... F3
Enying 7,518 ..... E3
Érd 44,904 ..... E3
Esztergom 30,476 ..... E3
Fegyvernek 8,421 ..... F3
Fehérgyarmat 9,000 ..... G3
Füzesgyarmat 7,097 ..... F3
Gödöllő 30,000 ..... E3
Gyöngyös 36,000 ..... E3
Gyoma 10,392 ..... F3
Győr 131,000 ..... D3
Gyula 36,000 ..... F3
Hadházteglás 13,626 ..... F3
Hajdúboszormény 31,000 ..... F3
Hajdúdorog 10,118 ..... F3
Hajdúnánás 18,000 ..... F3
Hajdúsámson 7,492 ..... F3
Hajdúszoboszló 24,000 ..... F3
Hatvan 25,000 ..... E3
Heves 11,000 ..... F3
Hódmezővásárhely 54,000 ..... F3
Izsák 7,686 ..... E3
Jánoshalma 12,534 ..... E3
Jászapáti 10,424 ..... F3
Jászárokszállás 10,139 ..... E3
Jászberény 30,000 ..... E3
Jászladány 7,823 ..... F3
Kalocsa 20,000 ..... E3
Mezőtúr 21,000 ..... F3
Mindszent 8,730 ..... F3
Miskolc 210,000 ..... F2
Mohács 21,000 ..... E4
Monor 16,838 ..... E3
Mór 12,066 ..... E3
Mosonmagyaróvár 30,000 ..... D3
Nádudvar 9,447 ..... F3
Nagyatád 15,000 ..... D3
Nagyecsed 8,225 ..... G3
Nagykálló 11,282 ..... F3
Nagykanizsa 55,000 ..... D3
Nagykáta 11,922 ..... E3
Nagykőrös 27,000 ..... E3
Nagyszénás 7,124 ..... F3
Nyíradony 7,146 ..... F3
Nyírbátor 14,000 ..... G3
Nyíregyháza 119,000 ..... F3
Orosháza 36,000 ..... F3
Oroszlány 22,000 ..... E3
Ózd 45,000 ..... F2
Paks 26,000 ..... E3
Pápa 35,000 ..... D3
Pásztó 12,000 ..... E3
Pécs 182,000 ..... E3
Pilis 9,055 ..... E3
Pilisvörösvár 10,217 ..... E3
Polgár 9,429 ..... F3
Püspökladány 16,000 ..... F3
Putnok 7,103 ..... F2
Ráckeve 7,534 ..... E3
Rákospalota 60,983 ..... E3
Sajószentpéter 13,992 ..... F2
Salgótarján 49,000 ..... E2
Sárbogárd 13,000 ..... E3
Sarkad 11,937 ..... F3
Sárospatak 15,000 ..... F2
Sárvár 16,000 ..... D3
Sátoraljaújhely 20,000 ..... F2
Siklós 11,000 ..... E4
Siófok 24,000 ..... E3
Soltvadkert 7,934 ..... E3

OTHER FEATURES

Bakony (mts.) D3
Balaton (lake) D3
Berettyó (riv.) F3
Börzsöny (mts.) E3
Bükk (mts.) F2
Csepelsziget (isl.) E3
Danube (riv.) E3
Dráva (riv.) D3
Duna (Danube) (riv.) E3
Fertő tó (Neusiedler See) (lake) D3
Great Alföld (plain) F3
Hernád (riv.) F2
Ipoly (riv.) E2
Kapos (riv.) D3
Kékes (mt.) E3
Korishegy (mt.) D3
Körös (riv.) F3
Little Alföld (plain) D3
Maros (riv.) F3
Mátra (mts.) E3
Mecsek (mts.) D3
Mura (riv.) D3
Rába (riv.) D3
Sajó (riv.) F2
Sárviz csatorna (canal) E3
Sebes Körös (riv.) F3
Sió csatorna (canal) E3
Szentendreisziget (isl.) E3
Tarna (riv.) F3
Tisza (riv.) F3
Zagyva (riv.) F3
Zala (riv.) D3

SLOVAKIA

REGIONS

Bratislava (city) 380,259 ..... D2
Středoslovenský
Liptovský Mikuláš 24,520 ..... E2
Lučenec 26,399 ..... E2
Malacky 15,218 ..... D2
Martin 56,208 ..... E2
Michalovce 29,765 ..... F2
Modra 7,679 ..... D2
Myjava 11,668 ..... D2
Nitra 76,663 ..... E2
Nová Baňa 8,321 ..... E2
Nové Město nad Váhom 18,170 ..... D2
Nové Zámky 34,147 ..... D3
Partizánske 23,266 ..... E2
Pezinok 17,116 ..... D2
Piešt'any 30,487 ..... D2
Poprad 38,077 ..... F2
Považská Bystrica 30,444 ..... E2
Prešov 71,500 ..... F2
Prievidza 40,813 ..... E2
Púchov 17,554 ..... E2
Revúca 11,881 ..... F2
Rimavská Sobota 19,699 ..... E2
Rožňava 18,039 ..... F2
Ružomberok 26,396 ..... E2
Sabinov 7,008 ..... F2
Šafárikovo 7,021 ..... F2
Šahy 8,034 ..... E2
Šal'a 19,167 ..... D2
Samorín 9,677 ..... D2
Senec 10,772 ..... D2
Senica 15,515 ..... D2
Sered' 16,071 ..... D2
Skalica 13,833 ..... D2
Snina 13,347 ..... G2
Spišská Nová Ves 31,917 ..... F2
Stropkov 7,405 ..... F2
Štúrovo 12,807 ..... E3
Šurany 11,320 ..... E2
Svidník 7,538 ..... F2
Topol'čany 31,340 ..... E2
Trebišov 14,961 ..... F2
Trenčín 47,887 ..... E2

Agriculture, Industry and Resources

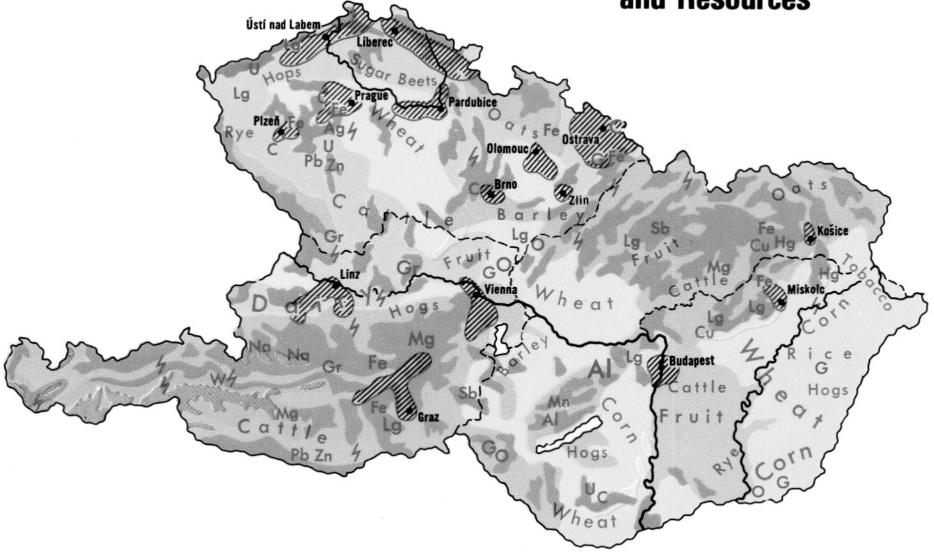

Ústí nad Labem
Liberec
Prague
Pardubice
Plzeň
Olomouc
Ostrava
Brno
Zlín
Košice
Linz
Vienna
Miskolc
Budapest
Graz
Sugar Beets
Hops
Oats
Wheat
Barley
Cattle
Fruit
Hogs
Corn
Rice
Rye
Tobacco
DOMINANT LAND USE
Cereals (chiefly wheat, corn)
Other Cereals, Livestock, Dairy
General Farming, Livestock
General Farming, Truck Farming
Pasture Livestock
Grapes, Wine
Forests
Nonagricultural Land
MAJOR MINERAL OCCURRENCES
Ag Silver
Al Bauxite
C Coal
Cu Copper
Fe Iron Ore
G Natural Gas
Gr Graphite
Hg Mercury
Lg Lignite
Mg Magnesium
Mn Manganese
Na Salt
O Petroleum
Pb Lead
Sb Antimony
U Uranium
W Tungsten
Zn Zinc
Water Power
Major Industrial Areas

Litovel 12,454 ..... D2
Litvínov 22,624 ..... B1
Louny 20,436 ..... B1
Lovosice 11,456 ..... C1
Lysá nad Labem 9,113 ..... C1
Mariánské Lázně 17,932 ..... B2
Mělník 18,941 ..... C1
Mikulov 8,472 ..... D2
Milevsko 8,852 ..... C2
Mimoň 7,437 ..... C1
Mladá Boleslav 45,896 ..... C1
Mnichovo Hradiště 7,340 ..... C1
Mohelnice 9,405 ..... D2
Moravská Třebová 11,543 ..... D2
Moravské Budějovice 8,943 ..... C2
Most 60,119 ..... B1
Náchod 19,892 ..... D1
Nejdek 9,768 ..... B1
Nové Město na Moravě 11,330 ..... D2
Nový Bohumín 16,700 ..... E2
Nový Bor 10,493 ..... C1
Nový Bydžov 9,317 ..... C1
Nový Jičín 31,506 ..... D2
Nymburk 14,033 ..... C1
Odry 10,032 ..... D2
Olomouc 102,112 ..... D2
Opava 59,384 ..... D2
Orlová 31,190 ..... E2
Ostrava 322,073 ..... E2
Ostrov 19,618 ..... B1
Pardubice 91,855 ..... C1
Písek 28,104 ..... C2
Plzeň 170,701 ..... B2
Poděbrady 13,782 ..... C1
Pohořelice 5,125 ..... D2
Polička 8,972 ..... D2
Prachatice 10,354 ..... B2
Ustí nad Orlicí 15,945 ..... D2
Uherské Hradiště 36,756 ..... D2
Uherský Brod 17,459 ..... D2
Uničov 12,507 ..... D2
Valašské Meziříčí 26,531 ..... D2
Varnsdorf 16,356 ..... C1
Velké Meziřičí 14,073 ..... D2
Veselí nad Moravou 12,464 ..... D2
Vimperk 7,257 ..... B2
Vítkov 7,543 ..... D2
Vlašim 13,284 ..... C2
Vodňany 6,989 ..... C2
Vrbno pod Pradědem 6,912 ..... D1
Vrchlabí 12,419 ..... C1
Vsetín 29,927 ..... D2
Vyškov 18,330 ..... D2
Vysoké Mýto 10,887 ..... D2
Zábřeh 15,184 ..... D2
Žatec 19,529 ..... B1
Žd'ár nad Sázavou 25,015 ..... C2
Zlin 83,983 ..... D2
Znojmo 39,271 ..... D2

OTHER FEATURES

Bečva (riv.) E2
Berounka (riv.) C2
Bohemian (for.) B2
Bohemian-Moravian Heights (hills) C2
Chrudimka (riv.) C2
Cidlina (riv.) C1
Danube (riv.) C2
Dyje (riv.) D2
Erzgebirge (mts.) B1
Jablunka (pass) E2
Jeseniky (mts.) D1
Somogy 349,000 ..... D3
Szabolcs-Szatmár 570,000 ..... G3
Szolnok 428,000 ..... F3
Tolna 263,000 ..... E3
Vas 277,000 ..... D3
Veszprém 387,000 ..... D3
Zala 311,000 ..... D3

CITIES and TOWNS

Abádszalók 6,386 ..... F3
Abaújszántó 4,209 ..... F2
Abony 15,624 ..... E3
Ács 8,423 ..... E3
Ajka 34,000 ..... D3
Albertirsa 11,252 ..... E3
Alsózsolca 5,045 ..... F2
Bácsalmás 8,000 ..... E3
Baja 41,000 ..... E3
Balassagyarmat 20,000 ..... E2
Balatonfüred 15,000 ..... D3
Balkány 7,667 ..... F3
Balmazújváros 17,371 ..... F3
Barcs 12,000 ..... D4
Bátaszék 7,274 ..... E3
Battonya 9,324 ..... F3
Békés 22,000 ..... F3
Békéscsaba 71,000 ..... F3
Berettyóújfalu 18,000 ..... F3
Bicske 13,000 ..... E3
Bonyhád 15,000 ..... E3
Budafok 40,623 ..... E3
Budakeszi 10,429 ..... E3
Budaörs 22,000 ..... E3
Budapest (cap.) 2,104,000 ..... E3
Cegléd 40,000 ..... E3
Celldömölk 12,000 ..... D3
Csepel 71,693 ..... E3
Csongrád 21,000 ..... F3
Kaposvár 74,000 ..... D3
Kapuvár 11,000 ..... D3
Karcag 25,000 ..... F3
Kazincbarcika 39,000 ..... F2
Kecel 10,493 ..... E3
Kecskemét 105,000 ..... E3
Keszthely 23,000 ..... D3
Kisbér 8,000 ..... D3
Kiskőrös 15,000 ..... E3
Kiskunfélegyháza 35,000 ..... E3
Kiskunhalas 32,000 ..... E3
Kiskunmajsa 14,439 ..... E3
Kispest 65,106 ..... E3
Kistelek 8,544 ..... E3
Kisújszállás 13,000 ..... F3
Kisvárda 17,828 ..... G2
Komádi 8,765 ..... F3
Komárom 19,955 ..... E3
Komló 30,301 ..... E3
Kondoros 7,319 ..... F3
Körmend 12,000 ..... D3
Kőszeg 14,000 ..... D3
Kunhegyes 10,116 ..... F3
Kunmadaras 7,343 ..... F3
Kunszentmárton 12,000 ..... F3
Kunszentmiklós 7,952 ..... E3
Lajosmizse 12,872 ..... E3
Leninváros 19,000 ..... F3
Lenti 9,000 ..... D3
Létavértes 9,106 ..... F3
Lőrinci 10,679 ..... E3
Makó 29,000 ..... F3
Marcali 13,000 ..... D3
Mátészalka 20,000 ..... G3
Mélykút 7,640 ..... E3
Mezőberény 12,702 ..... F3
Mezőhegyes 8,631 ..... F3
Mezőkovácsháza 7,000 ..... F3
Mezőkövesd 18,000 ..... F3
Sopron 57,000 ..... D3
Szabadszállás 8,223 ..... E3
Szarvas 19,000 ..... F3
Százhalombatta 18,000 ..... E3
Szeged 188,000 ..... F3
Szeghalom 10,000 ..... F3
Székesfehérvár 113,000 ..... E3
Szekszárd 39,000 ..... E3
Szentendre 20,000 ..... E3
Szentes 35,000 ..... F3
Szentgotthárd 8,000 ..... D3
Szerencs 10,000 ..... F2
Szigetvár 13,000 ..... D3
Szolnok 81,000 ..... F3
Szombathely 87,000 ..... D3
Tamási 10,000 ..... E3
Tapolca 18,000 ..... D3
Tata 26,000 ..... E3
Tatabánya 76,000 ..... E3
Tiszaföldvár 12,560 ..... F3
Tiszafüred 14,000 ..... F3
Tiszakécske 12,000 ..... F3
Tiszavasvári 14,000 ..... F3
Tolna 8,997 ..... E3
Törökszentmiklós 24,000 ..... F3
Tótkomlós 8,803 ..... F3
Tura 8,235 ..... E3
Túrkeve 11,000 ..... F3
Újfehértó 14,412 ..... F3
Újpest 80,384 ..... E3
Vác 36,000 ..... E3
Várpalota 28,000 ..... E3
Vásárosnamény 9,000 ..... G2
Vecsés 19,193 ..... E3
Veszprém 66,000 ..... D3
Vészto 9,815 ..... F3
Zalaegerszeg 63,000 ..... D3
Zalaszentgrót 8,000 ..... D3
Zirc 11,000 ..... D3
1,524,766 ..... E2
Východoslovenský 1,402,252 ..... F2
Západoslovenský 1,683,891 ..... D2

CITIES and TOWNS

Bánovce nad Bebravou 15,342 ..... E2
Banská Bystrica 66,412 ..... E2
Banská Štiavnica 9,180 ..... E2
Bardejov 23,741 ..... F2
Bratislava (cap.) 380,259 ..... D2
Brezno 17,872 ..... E2
Bytča 11,789 ..... E2
Čadca 19,319 ..... E2
Čalovo 8,063 ..... D3
Detva 14,261 ..... E2
Dolný Kubín 13,971 ..... E2
Dubnica nad Váhom 15,580 ..... E2
Dunajská Streda 18,715 ..... D3
Fil'akovo 10,497 ..... E2
Galanta 15,477 ..... D2
Handlová 17,777 ..... E2
Hlohovec 21,148 ..... D2
Holíč 8,741 ..... D2
Hrĭnová 8,485 ..... E2
Humenné 27,285 ..... F2
Hurbanovo 7,613 ..... E3
Kežmarok 17,570 ..... F2
Kolárovo 11,295 ..... D3
Komárno 32,520 ..... D3
Košice 202,368 ..... F2
Kremnica 7,168 ..... E2
Krupina 7,337 ..... E2
Kysucké Nové Mesto 14,083 ..... E2
Levice 26,132 ..... E2
Levoča 11,025 ..... F2
Liptovský Hrádok 9,197 ..... E2
Trnava 64,062 ..... D2
Turzovka 6,962 ..... E2
Vel'ké Kapušany 8,459 ..... G2
Vráble 7,586 ..... E2
Vranov nad Teplou 18,423 ..... F2
Žiar nad Hronom 19,098 ..... E2
Žilina 83,016 ..... E2
Zlaté Moravce 14,119 ..... E2
Zvolen 36,538 ..... E2

OTHER FEATURES

Beskids, East (mts.) F2
Beskids, West (mts.) E2
Dudvá (riv.) D2
Dukla (pass) F2
Dunajec (riv.) F2
Gerlachovka (mt.) F2
Hornád (riv.) F2
Hron (riv.) E2
Ipel' (riv.) E2
Laborec (riv.) F2
Latorica (riv.) F2
Nitra (riv.) E2
Orava (riv.) E2
Poprad (riv.) F2
Slaná (riv.) F2
Slovenské Rudohorie (mts.) E2
Tatra, High (riv.) E2
Topl'a (riv.) F2
Torysa (riv.) F2
Už (riv.) G2
Váh (riv.) D2
White Carpathians (mts.) E2

†Population of Austrian cities are communes.

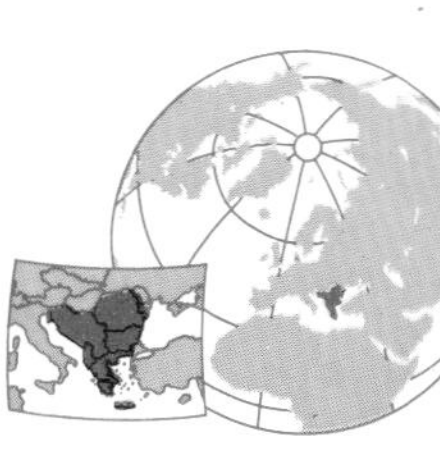

ALBANIA

AREA 11,100 sq. mi. (28,749 sq. km.)
POPULATION 3,401,126
CAPITAL Tiranë
LARGEST CITY Tiranë
HIGHEST POINT Korab 9,026 ft. (2,751 m.)
MONETARY UNIT lek
MAJOR LANGUAGE Albanian
MAJOR RELIGIONS Islam, Eastern Orthodoxy, Roman Catholicism

BOSNIA AND HERZEGOVINA

AREA 19,940 sq. mi. (51,129 sq. km.)
POPULATION 3,591,618
CAPITAL Sarajevo
LARGEST CITY Sarajevo
HIGHEST POINT Pločna 7,310 ft. (2,228 m.)
MONETARY UNIT dinar
MAJOR LANGUAGE Serbo-Croatian
MAJOR RELIGIONS Islam, Roman Catholicism, Eastern Orthodoxy,

BULGARIA

AREA 42,823 sq. mi. (110,912 sq. km.)
POPULATION 8,155,828
CAPITAL Sofia
LARGEST CITY Sofia
HIGHEST POINT Musala 9,597 ft. (2,925 m.)
MONETARY UNIT lev
MAJOR LANGUAGE Bulgarian
MAJOR RELIGION Eastern Orthodoxy

CROATIA

AREA 22,050 sq. mi. (56,538 sq. km.)
POPULATION 4,681,015
CAPITAL Zagreb
LARGEST CITY Zagreb
HIGHEST POINT Mali Rajinac 5,574 ft. (1,699 m.)
MONETARY UNIT Croatian kuna
MAJOR LANGUAGE Serbo-Croatian
MAJOR RELIGIONS Roman Catholicism, Eastern Orthodoxy

FORMER YUGOSLAV REP. OF MACEDONIA

AREA 9,889 sq. mi. (25,713 sq. km.)
POPULATION 2,035,044
CAPITAL Skopje
LARGEST CITY Skopje
HIGHEST POINT Solunska Glava 8,333 ft. (2,540 m.)
MONETARY UNIT denar
MAJOR LANGUAGES Macedonian, Serbo-Croatian, Albanian
MAJOR RELIGIONS Eastern Orthodoxy, Islam, Roman Catholicism

GREECE

AREA 50,944 sq. mi. (131,945 sq. km.)
POPULATION 10,750,705
CAPITAL Athens
LARGEST CITY Athens
HIGHEST POINT Olympus 9,570 ft. (2,917 m.)
MONETARY UNIT drachma
MAJOR LANGUAGE Greek
MAJOR RELIGION Eastern (Greek) Orthodoxy

ROMANIA

AREA 91,699 sq. mi. (237,500 sq. km.)
POPULATION 22,291,200
CAPITAL Bucharest
LARGEST CITY Bucharest
HIGHEST POINT Molodoveanul 8,343 ft. (2,543 m.)
MONETARY UNIT leu
MAJOR LANGUAGES Romanian, Hungarian
MAJOR RELIGION Eastern Orthodoxy

SLOVENIA

AREA 7,898 sq. mi. (20,251 sq. km.)
POPULATION 1,970,056
CAPITAL Ljubljana
LARGEST CITY Ljubljana
HIGHEST POINT Triglav 9,393 ft. (2,863 m.)
MONETARY UNIT tolar
MAJOR LANGUAGES Slovenian, Serbo-Croatian
MAJOR RELIGIONS Roman Catholicism, Eastern Orthodoxy

YUGOSLAVIA

AREA 38,989 sq. mi. (102,173 sq. km.)
POPULATION 11,210,243
CAPITAL Belgrade
LARGEST CITY Belgrade
HIGHEST POINT Daravica 8,714 ft. (2,656 m.)
MONETARY UNIT Yugoslav new dinar
MAJOR LANGUAGES Serbo-Croatian, Slovenian, Montenegrin, Albanian
MAJOR RELIGIONS Eastern Orthodoxy, Roman Catholicism

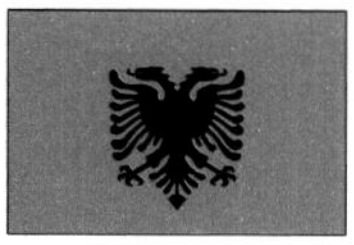
ALBANIA

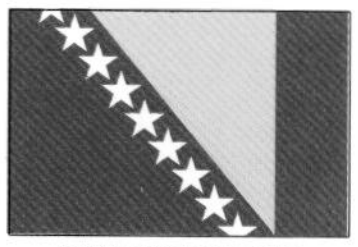
BOSNIA AND HERZEGOVINA

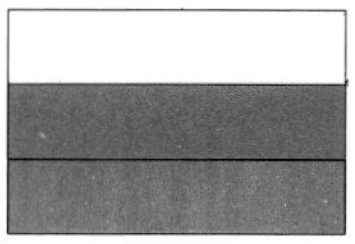
BULGARIA

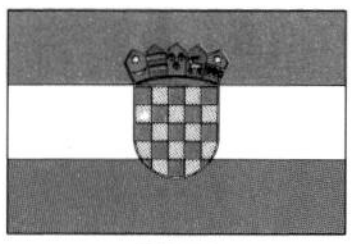
CROATIA

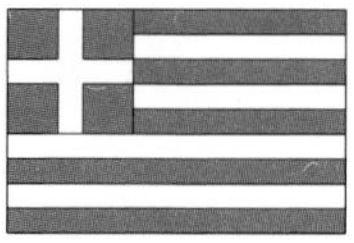
GREECE

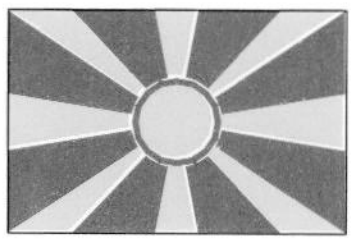
MACEDONIA

ROMANIA

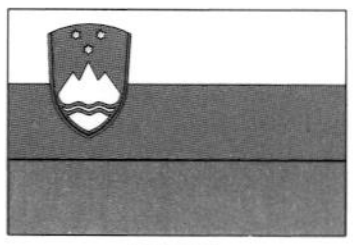
SLOVENIA

YUGOSLAVIA

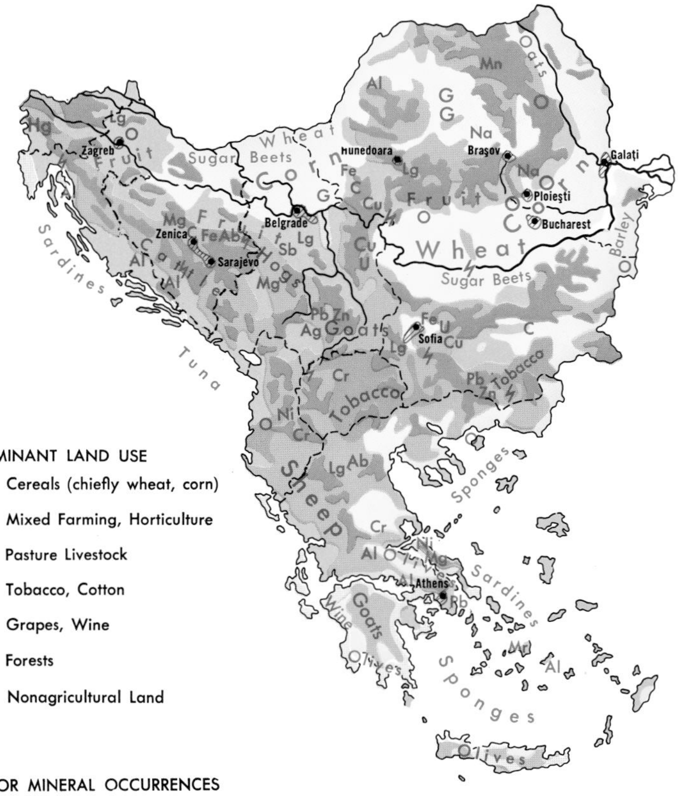

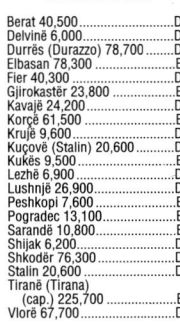

DOMINANT LAND USE

- Cereals (chiefly wheat, corn)
- Mixed Farming, Horticulture
- Pasture Livestock
- Tobacco, Cotton
- Grapes, Wine
- Forests
- Nonagricultural Land

MAJOR MINERAL OCCURRENCES

Ab	Asbestos	Mg	Magnesium
Ag	Silver	Mn	Manganese
Al	Bauxite	Mr	Marble
C	Coal	Na	Salt
Cr	Chromium	Ni	Nickel
Cu	Copper	O	Petroleum
Fe	Iron Ore	Pb	Lead
G	Natural Gas	Sb	Antimony
Hg	Mercury	U	Uranium
Lg	Lignite	Zn	Zinc

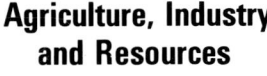

Agriculture, Industry and Resources

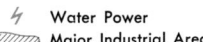

Water Power
Major Industrial Areas

ALBANIA

CITIES and TOWNS

Berat 40,500 .... D5
Delvinë 6,000 .... D6
Durrës (Durazzo) 78,700 .... D5
Elbasan 78,300 .... E5
Fier 40,300 .... D5
Gjirokastër 23,800 .... E5
Kavajë 24,200 .... D5
Korçë 61,500 .... E5
Krujë 9,600 .... D5
Kuçovë (Stalin) 20,600 .... D5
Kukës 9,500 .... E4
Lezhë 6,900 .... D5
Lushnjë 26,900 .... D5
Peshkopi 7,600 .... E5
Pogradec 13,100 .... E5
Sarandë 10,800 .... E6
Shijak 6,200 .... D5
Shkodër 76,300 .... D4
Stalin 20,600 .... D5
Tiranë (Tirana) (cap.) 225,700 .... E5
Vlorë 67,700 .... D5

OTHER FEATURES

Adriatic (sea) B4
Drin (riv.) E4
Korab (mt.) E4
Ohrid (lake) E5
Otranto (str.) D5
Prespa (lake) E5
Sazan (isl.) D5
Scutari (lake) D4
Vijosë (riv.) D5

BOSNIA and HERZEGOVINA

CITIES and TOWNS

Banja Luka 183,618 .... C3
Bihać 65,544 .... B3
Bijeljina 92,808 .... D3
Bileca 13,199 .... D4
Bosanska Dubica 30,867 .... C3
Bosanska Gradiška 58,095 .... C3
Bosanska Krupa 55,229 .... C3
Bosanski Brod 32,286 .... D3
Bosanski Novi 42,142 .... C3
Bosanski Petrovac 16,095 .... C3
Bosanski Šamac 32,320 .... D3
Brčko 82,768 .... D3
Bugojno 39,969 .... C3
Čapljina 26,032 .... C4
Cazin 57,110 .... B3
Derventa 57,010 .... C3
Doboj 99,548 .... C3
Drvar 17,983 .... C3
Donji Vakuf 22,606 .... C3
Foča 44,661 .... D4
Gacko 10,729 .... D4
Glamoč 14,120 .... C3
Gornji Vakuf 22,432 .... C4
Gračanica 54,311 .... D3
Gradačac 54,281 .... D3
Jajce 41,197 .... C3
Kladanj 15,641 .... D3
Ključ 40,008 .... C3
Konjic 43,677 .... C4
Livno 40,438 .... C4
Ljubinje 4,516 .... D4
Ljubuški 27,603 .... C4
Maglaj 42,160 .... D3
Modriča 34,541 .... D3
Mostar 110,377 .... C4
Nevesinje 16,326 .... D4
Prijedor 108,868 .... C3
Prozor 19,108 .... C4
Rogatica 23,578 .... D4
Sanski Most 62,467 .... C3
Sarajevo (cap.) 448,500 .... D4
Srebrenica 36,292 .... D3
Stolac 18,910 .... C4
Teslić 60,434 .... C3
Travnik 64,100 .... C3
Trebinje 30,372 .... D4
Tuzla 121,717 .... D3
Vareš 22,822 .... D3
Višegrad 23,201 .... D4
Visoko 40,901 .... D4
Vlasenica 30,498 .... D3
Zenica 132,733 .... C3
Žepče 19,754 .... D3
Zvornik 73,845 .... D3

OTHER FEATURES

Adriatic (sea) B4
Bosna (riv.) D3
Dinaric Alps (mts.) B3
Drina (riv.) D3
Neretva (riv.) D4
Tara (riv.) D4
Una (riv.) C3
Vrbas (riv.) C3

BULGARIA

CITIES and TOWNS

Asenovgrad 47,159 .... G5
Aytos 23,124 .... H4
Balchik 12,764 .... J4
Bansko 10,025 .... F5
Berkovitsa 16,340 .... F4
Blagoevgrad 65,481 .... F5
Botevgrad 22,659 .... F4
Burgas 182,856 .... H4
Byala 11,017 .... G4
Byala Slatina 16,034 .... F4
Chirpan 20,440 .... G4
Dimitrovgrad 54,056 .... G4
Dobrich (Tolbukhin) 109,170 .... H4
Dryanovo 10,306 .... G4
Elkhovo 13,655 .... H4
Gabrovo 81,629 .... G4
Gorna Oryakhovitsa 40,895 .... G4
Gotse Delchev 19,836 .... F5
Grudovo 10,736 .... H4
Ikhtiman 13,001 .... F4
Isperikh 11,235 .... H4
Karlovo 28,403 .... G4
Karnobat 22,536 .... H4
Kavarna 12,024 .... J4
Kazanlŭk 61,396 .... G4
Kharmanli 21,050 .... G5
Khaskovo 87,847 .... G5
Kubrat 10,758 .... H4
Kŭrdzhali 55,201 .... G5
Kyustendil 53,498 .... F4
Lom 32,307 .... F4
Lovech 48,992 .... G4
Lukovit 10,645 .... G4
Mikhaylovgrad 51,714 .... F4
Momchilgrad 10,189 .... G5
Nesebŭr 8,130 .... H4
Nova Zagora 25,327 .... H4
Novi Pazar 16,314 .... H4
Omurtag 9,505 .... H4
Oryakhovo 14,012 .... F4
Panagyurishte 22,034 .... G4
Pazardzhik 77,603 .... G4
Pernik 94,460 .... F4
Peshtera 18,763 .... G4
Petrich 26,451 .... F5
Pirdop 8,248 .... G4
Pleven 129,863 .... G4
Plovdiv 343,064 .... G4
Pomorie 13,507 .... H4
Popovo 21,236 .... H4
Provadiya 15,762 .... H4
Radomir 16,733 .... F4
Razgrad 49,582 .... H4
Razlog 14,010 .... F5
Rositsa 185,485 .... H4
Ruse 185,485 .... H4
Samokov 27,485 .... F4
Sandanski 24,629 .... F5
Sevlievo 26,560 .... G4
Shumen 100,125 .... H4
Silistra 53,537 .... H3
Sliven 9,037 .... H4
Smolyan 31,456 .... G5
Sofia (cap.) 1,121,763 .... F4
Stanke Dimitrov 41,897 .... F4
Stara Zagora 151,163 .... G4
Svilengrad 17,472 .... H5
Svishtov 30,555 .... G4
Teteven 12,784 .... G4
Tolbukhin 109,170 .... H4
Troyan 26,179 .... G4
Tŭrgovishte 46,043 .... H4
Tutrakan 12,153 .... H4
Varna 302,816 .... H4
Veliko Tŭrnovo 69,173 .... G4
Vidin 62,541 .... F4
Vratsa 75,180 .... F4
Yambol 90,019 .... H4
Zlatograd 8,780 .... G5

OTHER FEATURES

Arda (riv.) G5
Balkan (mts.) G4
Black (sea) J4
Danube (riv.) H4
Dunav (Danube) (riv.) H4
Emine (cape) J4
Iskŭr (riv.) G4
Kaliakra (cape) J4
Maritsa (riv.) G4
Mesta (riv.) F5
Midzhur (mt.) F4
Musala (mt.) F4
Osŭm (riv.) G4
Rhodope (mts.) G5
Rujen (mt.) F4
Struma (riv.) F5
Timok (riv.) F3
Tundzha (riv.) G4
Vit (riv.) G4

CROATIA

CITIES and TOWNS

Beli Manastir 53,409 .... D3
Biograd 15,865 .... B4
Bjelovar 66,553 .... C3
Čakovec 116,825 .... C2
Daruvar 31,424 .... C3
Djakovo 52,349 .... D3
Dubrovnik 66,131 .... D4
Fiume (Rijeka) 193,044 .... B3
Gospić 31,263 .... B3
Gračac 11,863 .... B3
Karlovac 78,363 .... B3
Knin 43,731 .... C3
Koprivnica 61,166 .... C2
Kostajnica 15,548 .... C3
Križevci 41,316 .... C2
Krk 13,334 .... B3
Kutina 38,597 .... C3
Makarska 17,819 .... C4
Našice 38,938 .... D3
Nova Gradiška 61,267 .... C3
Novska 24,530 .... C3
Ogulin 31,076 .... B3
Omiš 24,082 .... C4
Opatija 29,274 .... B3
Osijek 158,790 .... D3
Pag 7,076 .... B3
Petrinja 33,570 .... C3
Plŏce (Kardeljevŏ) 11,328 .... C4
Pola (Pula) 77,278 .... A3

(continued on following page)

(continued)

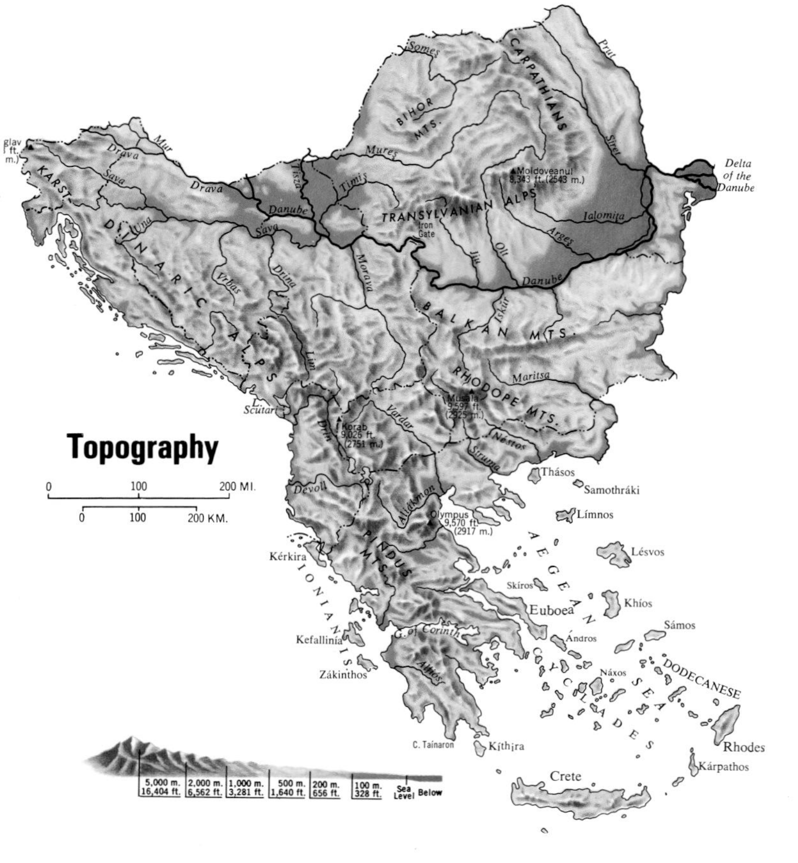

Topography
0 100 200 MI.
0 100 200 KM.
5,000 m. 16,404 ft.
2,000 m. 6,562 ft.
1,000 m. 3,281 ft.
500 m. 1,640 ft.
200 m. 656 ft.
100 m. 328 ft.
Sea Level
Below
CARPATHIANS
BIHOR MTS.
TRANSYLVANIAN ALPS
Moldoveanul 8,343 ft. (2543 m.)
Delta of the Danube
DINARIC ALPS
BALKAN MTS.
RHODOPE MTS.
Korab 9,026 ft. (2751 m.)
Musala 9,596 ft. (2925 m.)
Olympus 9,570 ft. (2917 m.)
PINDUS MTS.
AEGEAN SEA
IONIAN SEA
DODECANESE
CYCLADES
Danube
Drava
Sava
Mur
Mureş
Someş
Prut
Siret
Ialomiţa
Argeş
Olt
Jiu
Timiş
Tisza
Morava
Drina
Vrbas
Una
Iskăr
Maritsa
Vardar
Strimon
Nestos
Aliákmon
Devoll
Drin
Lim
Ibar
Iron Gate
Scutari
Thásos
Samothráki
Límnos
Lésvos
Khíos
Sámos
Skíros
Euboea
Ándros
Náxos
Kérkira
Kefallinía
Zákinthos
G. of Corinth
Athens
C. Taínaron
Kíthira
Crete
Rhodes
Kárpathos

Poreč 19,946 ... A3
Pula 77,278 ... A3
Rab 8,877 ... B3
Ragusa (Dubrovnik) 66,131 ... C4
Rijeka 193,044 ... B3
Rovinj 18,277 ... A3
Samobor 43,855 ... B3
Senj 9,582 ... B3
Šibenik 80,148 ... C4
Sinj 59,298 ... C4
Sisak 84,756 ... C3
Slavonska Požega 71,286 ... C3
Slavonski Brod 106,400 ... D3
Split 235,922 ... C4
Trogir 19,856 ... C4
Varaždin 90,729 ... C2
Vinkovci 95,245 ... D3
Virovitica 47,417 ... C3
Vukovar 81,203 ... D3
Zadar 116,174 ... B3
Zagreb (cap.) 681,173 ... C3
Zara (Zadar) 116,174 ... B3

OTHER FEATURES

Adriatic (sea) ... B4
Brač (isl.) ... C4
Cazma (riv.) ... C3
Cres (isl.) ... B3
Dalmatia (reg.) ... C4
Danube (riv.) ... E3
Dinaric Alps (mts.) ... B3
Drava (riv.) ... C3
Dugi Otok (isl.) ... B3
Hvar (isl.) ... C4
Istria (pen.) ... A3
Kamenjak (cape) ... A3
Korčula (isl.) ... C4
Kornat (isl.) ... B4
Krk (isl.) ... B3
Kupa (riv.) ... B3
Kvarner (gulf) ... B3
Lastovo (Lagosta) (isl.) ... C4
Lošinj (isl.) ... B3
Mljet (isl.) ... C4
Pag (isl.) ... B3
Palagruža (Pelagosa) (isl.) ... C4
Rab (isl.) ... B3
Sava (riv.) ... D3
Slavonia (reg.) ... C3
Šolta (isl.) ... C4
Una (riv.) ... C3
Vis (isl.) ... C4
Žirje (isl.) ... B4

FORMER YUGOSLAV REP. OF MACEDONIA

CITIES and TOWNS

Berovo 20,226 ... F5
Bitola 137,835 ... E5
Debar 22,506 ... E5
Gevgelija 32,023 ... F5
Gostivar 101,188 ... E5
Kavadarci 39,738 ... F5
Kičevo 51,452 ... E5
Kočani 47,976 ... F5
Kumanovo 126,368 ... E4
Ohrid 64,316 ... E5
Prilep 99,941 ... E5
Radoviš 28,574 ... F5
Skopje (cap.) 506,547 ... E5
Štip 46,651 ... F5
Struga 54,489 ... E5
Strumica 87,446 ... F5
Tetovo 162,414 ... E5
Titov Veles 64,901 ... E5

OTHER FEATURES

Korab (mt.) ... E5
Ohrid (lake) ... E5
Prespa (lake) ... E5
Rujen (mt.) ... F4
Vardar (riv.) ... E5

GREECE

REGIONS

Aegean Islands 417,813 ... G6
Athens, Greater 3,027,331 ... F7
Áyion Óros (aut. dist.) 1,732 ... G5
Central Greece and Euboea 1,099,841 ... F6
Crete 502,165 ... G8
Epirus 324,541 ... E6
Ionian Islands 182,651 ... D6
Macedonia 2,121,953 ... F5
Pelopónnisos 1,012,528 ... F7
Thessaly 695,654 ... F6
Thrace 345,220 ... G5

CITIES and TOWNS

Agrínion 34,328 ... E6
Aíyion 20,824 ... F6
Alexandroúpolis 34,535 ... G5
Amaliás 14,698 ... E7
Árgos 20,702 ... F7
Árta 18,283 ... E6
Atalándi 5,456 ... F6
Athens (cap.) 885,737 ... F7
Áyios Nikólaos 8,130 ... G8
Candia (Iráklion) 101,634 ... G8
Canea (Khaniá) 40,564 ... G8
Corinth 22,658 ... F7
Dhidhimótikhon 8,374 ... H5
Dráma 36,109 ... G5
Édhessa 16,054 ... F5
Ermoúpolis 13,876 ... G7
Flórina 12,562 ... E5
Grevená 7,433 ... E5
Ierápetra 8,575 ... G8
Ioánnina 44,829 ... E6
Iráklion 101,634 ... G8
Itháki 2,037 ... E6
Kalámai 41,911 ... F7
Kálimnos 10,118 ... H7
Kardhítsa 27,291 ... E6
Kastoría 17,133 ... E5
Kateríni 38,016 ... F5
Kavála 56,375 ... G5
Kérkira 33,561 ... D6
Khalkís 44,867 ... F6
Khaniá 40,564 ... G8
Khíos 24,070 ... G6
Kiáton 7,392 ... F6
Kilkís 11,148 ... F5
Komotiní 34,051 ... G5
Koropí 11,214 ... G7
Kos 11,851 ... H7
Kozáni 30,994 ... F5
Lamía 41,667 ... F6
Lárisa 102,048 ... F6
Lávrion 8,921 ... G7
Levádhia 16,864 ... F6
Marathón 2,052 ... G6
Mégara 17,719 ... F6
Mesolóngion 10,164 ... E6
Mitilíni 24,115 ... H6
Náousa 19,383 ... F5
Návpaktos 9,012 ... F6
Návplion 10,609 ... F7
Náxos 3,735 ... G7
Orestiás 12,685 ... H5
Pátrai 141,529 ... E6
Piraiévs (Piraeus) 196,389 ... F7
Pírgos 21,958 ... E7
Préveza 12,662 ... E6
Psakhná 5,320 ... F6
Ptolemaís 22,109 ... E5
Réthimnon 17,736 ... G8
Rhodes (Ródhos) 40,392 ... J7
Salamís 20,437 ... F6
Salonika (Thessaloníki) 406,413 ... F5
Sámos 5,575 ... H7
Samothráki 941 ... G5
Sérrai 45,213 ... F5
Sparta 11,911 ... F7
Thásos 2,300 ... G5
Thessaloníki 406,413 ... F5
Thívai 18,712 ... F6
Tirnávos 10,965 ... F6
Trikkála 40,857 ... E6
Trípolis 21,311 ... F7
Vérria 37,087 ... F5
Vólos 71,378 ... F6
Vónitsani 3,627 ... E6
Xánthi 31,541 ... G5
Yiannitsá 21,082 ... F5
Zante (Zákinthos) 9,764 ... E7

OTHER FEATURES

Aegean (sea) ... G6
Akrí (cape) ... E7
Aktí (pen.) ... G5
Amorgós (isl.) ... G7
Anáfi (isl.) ... G7
Andikíthira (isl.) ... F8
Ándros (isl.) ... G7
Arda (riv.) ... G5
Argolis (gulf) ... F7
Astipálaia (isl.) ... H7
Áthos (mt.) ... G5
Áyios Evstrátios (isl.) ... G6
Áyios Yeóryios (cape) ... G5
Cephalonia (Kefallinía) (isl.) ... E6
Corfu (Kérkira) (isl.) ... D6
Corinth (gulf) ... F6
Crete (isl.) ... G8
Crete (sea) ... G8
Cyclades (isls.) ... G7
Día (isl.) ... G8
Dodecanese (isls.) ... H8
Euboea (Évvoia) (isl.) ... G6
Évros (riv.) ... H5
Gávdhos (isl.) ... F8
Ídhi (mt.) ... G8
Ikaría (isl.) ... G7
Ionian (sea) ... D7
Íos (isl.) ... G7
Itháki (Ithaca) (isl.) ... E6
Kafirévs (cape) ... G6
Kálimnos (isl.) ... H7
Kárpathos (isl.) ... H8
Kásos (isl.) ... H8
Kassándra (pen.) ... F6
Kéa (isl.) ... G7
Kefallinía (isl.) ... E6
Kérkira (isl.) ... D6
Khálki (isl.) ... H7
Khaniá (gulf) ... G8
Khíos (isl.) ... G6
Kímilos (isl.) ... G7
Kiparissía (gulf) ... E7
Kíthira (isl.) ... F7
Kíthnos (isl.) ... G7
Kos (isl.) ... H7
Kriós (cape) ... F8
Kríti (Crete) (isl.) ... G8
Lakonía (gulf) ... F7
Léros (isl.) ... H7
Lésvos (isl.) ... G6
Levítha (isl.) ... H7
Levkás (isl.) ... E6
Límnos (isl.) ... G6
Maléa (cape) ... F7
Matapan (Taínaron) (cape) ... F7
Merabéllou (gulf) ... H8
Mesará (gulf) ... G8
Messíni (gulf) ... E7
Míkinos (isl.) ... G7
Mílos (isl.) ... G7
Mirtóön (sea) ... F7
Náxos (isl.) ... G7
Néstos (riv.) ... G5
Nísiros (isl.) ... H7
Northern Sporades (isls.) ... F6
Olympia (isls.) ... E7
Olympus (mt.) ... F5
Parnassus (mt.) ... F6
Páros (isl.) ... G7
Pátmos (isl.) ... H7
Paxoí (isl.) ... D6
Pindus (mts.) ... E6
Piniós (riv.) ... E6
Prespa (lake) ... E5
Psará (isl.) ... G6
Psevdhókavos (cape) ... G6
Rhodes (isl.) ... H7
Rhodope (mts.) ... G5
Salonika (Thermaic) (gulf) ... F6
Sámos (isl.) ... H7
Samothráki (isl.) ... G5
Saría (isl.) ... H8
Saronic (gulf) ... F7
Sérifos (isl.) ... G7
Sídheros (cape) ... H8
Sífnos (isl.) ... G7
Sími (isl.) ... H7
Síros (isl.) ... G7
Sithoniá (pen.) ... F5
Skíros (isl.) ... G6
Spátha (cape) ... F8
Strimón (gulf) ... G5
Strofádhes (isls.) ... E7
Taínaron (cape) ... F7
Thásos (isls.) ... G5
Thermaic (gulf) ... F5
Thíra (isl.) ... G7
Tílos (isl.) ... H7
Tínos (isl.) ... G7
Toronaic (gulf) ... F5
Vardar (riv.) ... F5
Vólvi (lake) ... F5
Voïviis (lake) ... F6
Voúxa (cape) ... F8
Zákinthos (Zante) (isl.) ... E7

ROMANIA

CITIES and TOWNS

Aiud 27,600 ... F2
Alba Iulia 53,000 ... F2
Alexandria 43,700 ... G3
Anina 11,300 ... E3
Arad 182,000 ... E2
Babadag 9,000 ... J3
Bacău 156,200 ... H2
Baia Mare 123,300 ... F2
Băileşti 21,500 ... F3
Balş 17,300 ... G3
Beiuş 10,100 ... F2
Bicaz 9,300 ... G2
Bîrlad 63,800 ... H2
Bistrita 59,800 ... G2
Blaj 22,200 ... F2
Borşa 25,287 ... F2
Botoşani 84,900 ... H2
Brad 18,600 ... F2
Brăila 219,200 ... H3
Braşov 320,200 ... G3
Bucharest (Bucureşti) (cap.) 1,929,400 ... G3
Buhuşi 20,300 ... H2
Buzău 116,300 ... H3
Buziaş 8,700 ... E3
Calafat 17,100 ... F3
Călăraşi 58,000 ... H3
Caracal 33,600 ... G3
Caransebeş 28,800 ... F3
Carei 25,500 ... F2
Cernavodă 15,000 ... J3
Chişineu Criş 9,600 ... E2
Cîmpia Turzii 25,300 ... F2
Cîmpina 35,300 ... G3
Cîmpulung 37,400 ... G3
Cîmpulung Moldovenesc 20,500 ... G2
Cisnădie 21,100 ... G3
Cluj-Napoca 289,800 ... F2
Comaneşti 18,500 ... H2
Constanţa 293,900 ... J3
Corabia 20,300 ... G4
Costeşti 10,900 ... G3
Craiova 239,700 ... F3
Curtea de Argeş 26,900 ... G3
Darabani 11,500 ... H1
Dej 36,500 ... F2
Deva 73,300 ... F3
Dorohoi 25,700 ... H2
Drăgăneşti Olt 11,800 ... G3
Drăgăşani 17,300 ... G3
Drobeta-Turnu Severin 86,600 ... F3
Făgăraş 37,200 ... G3
Fălticeni 24,000 ... H2
Feteşti 29,600 ... H3
Focşani 70,700 ... H3
Găeşti 14,000 ... G3
Galati 268,000 ... H3
Gheorghe Gheorghiu-Dej 46,100 ... H2
Gheorghieni 21,800 ... G2
Gherla 20,700 ... G2
Giurgiu 57,000 ... G4
Hateg 10,200 ... F3
Hîrlău 8,900 ... H2
Hîrşova 9,000 ... J3
Huedin 8,700 ... F2
Hunedoara 85,700 ... F3
Huşi 26,000 ... J2
Iaşi 279,800 ... H2
Ineu 10,800 ... E2
Jimbolia 14,600 ... E3
Lipova 12,900 ... E2
Luduş 16,000 ... G2
Lugoj 50,000 ... E3
Lupeni 29,100 ... F3
Mangalia 31,100 ... J4
Medgidia 45,300 ... J3
Mediaş 69,000 ... G2
Miercurea Ciuc 40,400 ... G2
Mizil 15,200 ... H3
Moineşti 21,200 ... H2
Moldova Nouă 17,800 ... E3
Moreni 18,900 ... G3
Ocna Mureş 16,200 ... G2
Odorheiu Secuiesc 36,200 ... G2
Oltenita 26,800 ... H3
Oradea 192,600 ... E2
Orăştie 19,900 ... F3
Oravita 114,300 ... E3
Orşova 115,800 ... F3
Panciu 77,900 ... H3
Paşcani 229,500 ... H2
Petrila 25,900 ... F3
Petroşeni 45,600 ... F3
Piatra Neamt 93,300 ... H2
Piteşti 143,600 ... G3
Ploieşti 219,900 ... H3
Pucioasa 14,100 ... G3
Rădăuti 26,000 ... G2
Reghin 33,600 ... G2
Reşita 96,800 ... E3
Rîmnicu Sărat 32,400 ... H3
Rîmnicu Vîlcea 78,900 ... G3
Roman 62,700 ... H2
Roşiori de Vede 31,700 ... G3
Săcele 33,900 ... G3
Salonta 20,400 ... E2
Satu Mare 115,600 ... F2
Sebeş 29,500 ... F3
Segarcea 8,700 ... F3
Sfîntu Gheorghe 57,900 ... G3
Sibiu 164,200 ... G3
Sighetu Marmaţiei 40,500 ... F2
Sighişoara 33,000 ... G2
Şimleul Silvaniei 15,100 ... F2
Sinaia 14,700 ... G3
Sînnicolaul Mare 13,600 ... E2
Slatina 62,800 ... G3
Slobozia 39,400 ... H3
Sovata 11,200 ... G2
Strehaia 11,800 ... F3
Suceava 76,500 ... H2
Tăşnad 10,400 ... F2
Techirghiol 11,800 ... J3
Tecuci 40,300 ... H3
Timişoara 288,200 ... E3
Tîrgovişte 77,500 ... G3
Tîrgu Jiu 75,200 ... F3
Tîrgu Mureş 141,300 ... G2
Tîrgu Neamt 16,600 ... H2
Tîrgu Ocna 12,800 ... H2
Tîrgu Secuiesc 19,800 ... H2
Tîrnăveni 27,900 ... G2
Toplita 15,200 ... G2
Tulcea 73,600 ... J3
Turda 58,700 ... F2
Turnu Măgurele 33,000 ... G4
Urlata 11,200 ... H3
Urziceni 14,300 ... H3
Vaslui 50,100 ... H2
Vatra Dornei 17,800 ... G2
Videle 11,500 ... G3
Vişeul de Sus 20,800 ... G2
Zalău 43,300 ... F2
Zărneşti 25,000 ... G3
Zimnicea 16,400 ... G4

OTHER FEATURES

Argeş (riv.) ... G3
Bîrlad (riv.) ... H2
Black (sea) ... J4
Brăila (marshes) ... H3
Buzău (riv.) ... H3
Carpathian (mts.) ... G2
Crişul Alb (riv.) ... F2
Crişul Repede (riv.) ... F2
Danube (delta) ... J3
Danube (riv.) ... H4
Ialomiţa (marshes) ... J3
Ialomiţa (riv.) ... H3
Jijia (riv.) ... H2
Jiu (riv.) ... F3
Moldoveanul (mt.) ... G3
Mureş (riv.) ... E2
Olt (riv.) ... G3
Peleaga (mt.) ... F3
Pietrosul (mt.) ... G2
Prut (riv.) ... J2
Siret (riv.) ... H2
Someş (riv.) ... F2
Timiş (riv.) ... E3
Tîrnava Mare (riv.) ... G2
Transylvanian Alps (mts.) ... G3

SLOVENIA

CITIES and TOWNS

Bled 4,710 ... A2
Brežice 25,238 ... C3
Celje 63,877 ... B2
Jesenice 31,094 ... A2
Kočevje 18,139 ... B3
Koper 41,843 ... A3
Kranj 66,879 ... B2
Krško 27,774 ... B3
Ljubljana (cap.) 305,211 ... B3
Maribor 185,699 ... B2
Murska Sobota 64,299 ... C2
Nova Gorizia 56,758 ... A3
Novo Mesto 55,584 ... B3
Piran 15,235 ... A3
Postojna 19,892 ... B3
Ptuj 67,754 ... B2
Ravne na Koroškem 25,907 ... B2
Škofja Loka 35,276 ... B2
Trbovlje 18,786 ... B2
Tržič 14,014 ... B2
Velenje 38,041 ... B2

OTHER FEATURES

Adriatic (sea) ... B4
Drava (riv.) ... C3
Kupa (riv.) ... B3
Mur (riv.) ... B2
Triglav (mt.) ... A2

YUGOSLAVIA

INTERNAL DIVISIONS

Kosovo (aut. reg.) 1,240,919 ... E4
Montenegro (rep.) 527,207 ... D4
Serbia (rep.) 8,401,673 ... E3
Vojvodina (aut. prov.) 1,953,980 ... D3

CITIES and TOWNS

Aleksinac 67,286 ... E4
Apatin 33,843 ... D3
Arendjelovac 46,803 ... E3
Bačka Topola 41,889 ... D3
Bar 32,535 ... D4
Bečej 44,243 ... E3
Bela Crkva 25,690 ... E3
Belgrade (cap.) 1,470,073 ... E3
Beograde (Belgrade) (cap.) 1,470,073 ... E3
Bijelo Polje 55,634 ... D4
Bor 56,486 ... E3
Čačak 110,676 E4
Caribrod (Dimitrovgrad) 15,158 ... F4
Cetinje 20,213 ... D4
Ćuprija 38,841 ... E4
Dimitrovgrad 15,158 ... F4
Djakovica 92,203 ... E4
Gnjilane 84,085 ... E4
Gornji Milanovac 50,651 ... E3
Herceg Novi 23,258 ... D4
Ivangrad 49,772 ... D4
Kanjiža 32,709 ... D2
Kikinda 69,854 ... E3
Knjaževac 48,789 ... F4
Kosovska Mitrovica 105,353 ... E4
Kotor 20,455 ... D4
Kragujevac 164,823 ... E3
Kraljevo 121,622 ... E4
Kruševac 132,972 ... E4
Leskovac 159,001 ... E4
Loznica 84,180 ... D3
Negotin 63,973 ... F3
Nikšić 72,299 ... D4
Niš 230,711 ... E4
Novi Pazar 74,000 ... E4
Novi Sad 257,685 ... D3
Pančevo 123,791 ... E3
Paraćin 64,718 ... E4
Peć 111,071 ... E4
Pirot 69,653 ... F4
Plav 19,560 ... D4
Pljevlja 43,316 ... D4
Podgorica 132,290 ... D4
Požarevac 81,123 ... E3
Preševo 33,948 ... E4
Priboj 35,200 ... D4
Prijedor 108,868 ... C3
Prijepolje 46,902 ... D4
Priština 210,040 ... E4
Prizren 134,526 ... E4
Prokuplje 56,256 ... E4
Ruma 55,083 ... D3
Šabac 119,668 ... D3
Senta 30,519 ... D3
Šid 37,459 ... D3
Sjenica 35,570 ... E4
Smederevo 107,366 ... E3
Smederevska Palanka 60,945 ... E3
Sombor 99,168 ... D3
Sremska Mitrovica 85,129 ... D3
Subotica 154,611 ... D2
Surdulica 27,029 ... E4
Svetozarevo 76,460 ... E4
Svilajnac 34,888 ... E3
Titovo Užice 77,049 ... D4
Trstenik 53,695 ... E4
Ub 36,259 ... E3
Ulcinj 21,575 ... D5
Uroševac 113,680 ... E4
Valjevo 95,449 ... D3
Velika Plana 52,619 ... E3
Veliki Bečkerek (Zrenjanin) 139,000 ... E3
Vranje 82,527 ... E4
Vrbas 45,755 ... D3
Vršac 61,005 ... E3
Vučitrn 65,512 ... E4
Zaječar 76,681 ... F4
Zrenjanin 139,000 ... E3

OTHER FEATURES

Adriatic (sea) ... B4
Bobotov Kuk (mt.) ... D4
Danube (riv.) ... E3
Drina (riv.) ... D3
Ibar (riv.) ... E4
Lim (riv.) ... D4
Midzhur (mt.) ... F4
Morava (riv.) ... E3
Sava (riv.) ... D3
Scutari (lake) ... D4
Timok (riv.) ... F3
Tisa (riv.) ... E3

The
Balkan States

CONIC PROJECTION

SCALE OF MILES

SCALE OF KILOMETERS

Capitals of Countries -------- ☆
Administrative Centers -------- △
International Boundaries -------
Major Internal Boundaries -------
Minor Internal Boundaries -------
Canals -------

* Former Yugoslav Republic of Macedonia

BULGARIA and GREECE are divided into regions and departments, respectively. Because of the scale no attempt has been made to delimit and name these subdivisions; their administrative centers have, however, been designated.
The larger divisions named in Greece are well-known geographical regions, without administrative function.
ROMANIA consists of thirty-nine counties and three cities of regional status, Bucharest, Constanța and Petroşeni. Scale does not permit delimiting these counties.
ALBANIA is divided into twenty-seven districts. Scale does not permit the delimitation of these divisions.

© Copyright HAMMOND INCORPORATED, Maplewood, N. J.

Topography

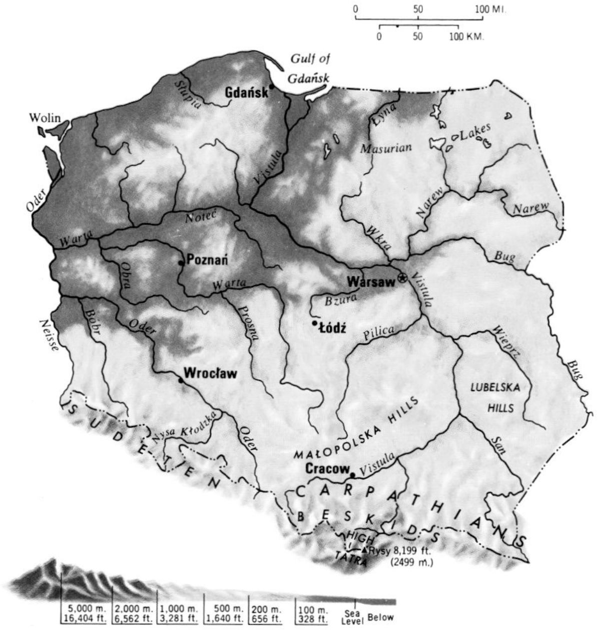

Gulf of Gdańsk

Wolin

Gdańsk

Oder

Poznań

Warta

Wrocław

Warsaw

Łódź

SUDETEN

MAŁOPOLSKA HILLS

Cracow

CARPATHIANS

BESKIDS

HIGH TATRA

Rysy 8,199 ft. (2499 m.)

LUBELSKA HILLS

Masurian Lakes

Narew

Bug

5,000 m. 16,404 ft. | 2,000 m. 6,562 ft. | 1,000 m. 3,281 ft. | 500 m. 1,640 ft. | 200 m. 656 ft. | 100 m. 328 ft. | Sea Level | Below

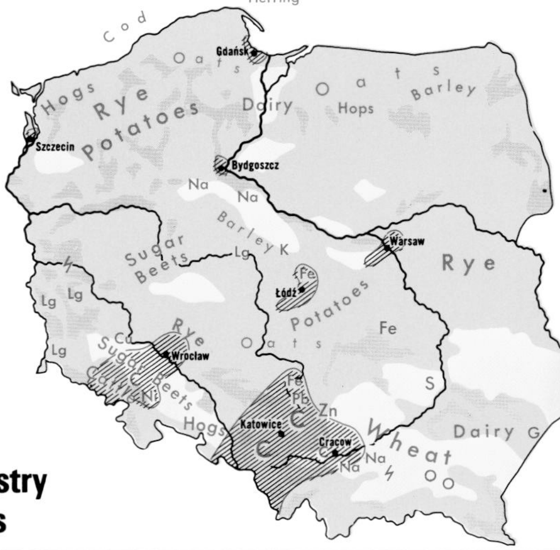

MAJOR MINERAL OCCURRENCES

Ag Silver
C Coal
Cu Copper
Fe Iron Ore
G Natural Gas
K Potash
Lg Lignite

Na Salt
Ni Nickel
O Petroleum
Pb Lead
S Sulfur
Zn Zinc

⚡ Water Power

▨ Major Industrial Areas

DOMINANT LAND USE

▢ Cereals (chiefly wheat)

▢ Rye, Oats, Barley, Potatoes

▢ General Farming, Livestock

▢ Forests

PROVINCES

Biała Podlaska 304,028 ....F3
Białystok 687,806 ....F2
Bielsko 895,357 ....D4
Bydgoszcz 1,104,048 ....C2
Chełm 245,484 ....F3
Ciechanów 425,608 ....E2
Cracow (Kraków) 1,223,137 ..E3
Cracow (city) 651,300 ......E3
Częstochowa 773,365 ......D3
Elbląg 475,862 ....D1
Gdańsk 1,417,801 ....D1
Gorzów 497,342 ....B2
Jelenia Góra 514,947 ......B3
Kalisz 706,514 ....D3
Katowice 3,953,769 ....D3
Kielce 1,123,691 ....E3
Konin 465,928 ....D2
Koszalin 502,750 ....C1
Krosno 491,471 ....E4
Legnica 510,000 ....C3
Leszno 383,315 ....C3
Łódź 777,800 ....D3

Łódź (city) 1,139,379 ......D3
Łomza 344,518 ....F2
Lublin 1,010,641 ....F3
Nowy Sącz 690,737 ....E4
Olsztyn 746,185 ....E2
Opole 1,010,416 ....C3
Ostrołęka 393,427 ....E2
Piła 475,953 ....C2
Piotrków 638,948 ....D3
Płock 512,626 ....D2
Poznań 1,323,368 ....C2
Przemyśl 404,200 ....F4
Radom 745,374 ....E3
Rzeszów 716,317 ....F4
Siedlce 648,111 ....F2
Sieradz 408,082 ....D3
Skierniewice 416,690 ....E3
Słupsk 410,049 ....C1
Suwałki 467,048 ....F1
Szczecin 964,298 ....B2
Tarnobrzeg 594,255 .....E3
Tarnów 664,953 ....E4
Toruń 656,421 ....D2
Wałbrzych 738,092 .....C3

Warsaw 2,415,950 ....E2
Warsaw (city) 1,377,100 ..E2
Włocławek 427,418 .....D2
Wrocław 1,122,806 .....C3
Zamość 488,193 ....F3
Zielona Góra 655,146 ...B3

CITIES and TOWNS

Aleksandrów Łódzki 19,711 ..D3
Allenstein (Olsztyn) 160,956 ..E2
Andrychów 22,387 ....D4
Augustów 28,307 ....F1
Auschwitz (Oświęcim) 45,402 ..D3
Bartoszyce 25,195 ....E1
Bedzin 76,883 ....B3
Belchatów 55,632 ....D3
Beuthen (Bytom) 229,991 ..A3
Biala Podlaska 52,119 ....F3
Bialogard 23,973 ....C1
Bialystok 267,670 ....F2
Bielawa 34,224 ....C3
Bielsk Podlaski 26,145 ...F2

Bielsko-Biala 181,072 .....D4
Bilgoraj 25,542 ....F3
Bochnia 28,846 ....E4
Bogatynia 18,616 ....B3
Boguszów-Gorce 19,452 ..B3
Bolesławiec 43,076 .....B3
Braniewo 17,594 ....D1
Breslau (Wrocław) 640,557 ..C3
Brieg (Brzeg) 38,504 ......C3
Brodnica 26,056 ....D2
Brzeg 38,504 ....C3
Busko Zdrój 17,675 .....E3
Bydgoszcz 380,426 .....C2
Bytom 229,991 ....A3
Bytów 16,720 ....C1
Chelm 64,683 ....F3
Chelmno 21,506 ....D2
Chodzież 19,831 ....C2
Chojnice 37,733 ....C2
Chorzów 131,850 ....B4
Chrzanów 42,195 ....B4
Ciechanów 43,068 ......E2
Cieszyn 36,682 ....D4
Cracow 745,568 ....E3

Agriculture, Industry and Resources

Cod
Herring
Oats
Gdańsk
Hogs
Rye
Oats
Dairy
Oats
Barley
Hops
Szczecin
Potatoes
Na
Bydgoszcz
Na
Sugar Beets
Barley
K
Warsaw
Lg
Lg
Rye
Łódź
Potatoes
Rye
Wrocław
Oats
Fe
Fe
Sugar Beets
Hogs
Zn
Katowice
Na
Cracow
Wheat
Dairy
G
S

Former Republics of Yugoslavia

CONIC PROJECTION

MILES
0 25 50 75 100

KILOMETERS
0 25 50 75 100

Capitals

⊛ National
★ Federal Republics
⊙ Autonomous Provinces

Boundaries

Canals

© Copyright HAMMOND INCORPORATED, Maplewood, N. J.

AUSTRIA

HUNGARY

ITALY

SLOVENIA

Ljubljana

Zagreb

CROATIA

VOJVODINA

Novi Sad

ROMANIA

BOSNIA

AND

HERZEGOVINA

Sarajevo

Belgrade (Beograd)

SERBIA

MONTENEGRO

Podgorica

KOSOVO

BULGARIA

Skopje

ALBANIA

Tirane

F.Y.R.O.M.*

GREECE

ADRIATIC SEA

IONIAN SEA

* Former Yugoslav Republic of Macedonia

Longitude 18° East of Greenwich

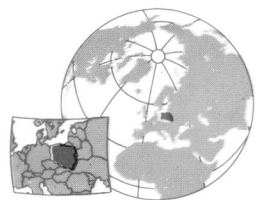

AREA 120.725 sq. mi. (312,678 sq. km.)
POPULATION 37,931,000
CAPITAL Warsaw
LARGEST CITY Warsaw
HIGHEST POINT Rysy 8,199 ft.
(2,499 m.)
MONETARY UNIT zloty
MAJOR LANGUAGE Polish
MAJOR RELIGION Roman Catholicism

Czechowice-Dziedzice 35,194	D4
Czeladź 37,569	B4
Częstochowa 256,578	D3
Dabrowa Górnicza 134,934	B3
Danzig (Gdańsk) 462,076	D1
Debica 44,966	E3
Deblin 18,763	E3
Działdowo 19,295	E2
Dzierzoniów 37,908	C3
Elbing (Elbląg) 125,778	D1
Elbląg 125,778	D1
Ełk 51,274	F2
Gdańsk 462,076	D1
Gdynia 251,303	D1
Gizycko 28,918	E1
Kraków (Cracow) 745,568	E3
Krapkowice 19,452	C3
Kraśnik Fabryczny 36,202	F3
Krasnystaw 19,832	F3
Krosno 49,094	E4
Krotoszyn 27,807	C3
Kutno 49,753	D2
Kwidzin 36,409	D2
Landsberg (Gorzów Wielkopolski) 123,222	B2
Lask 19,569	D3
Laziska Górne 19,569	A4
Lebork 33,981	C1
Leczyca 16,491	D2
Legionowo 50,577	E2
Legnica 103,949	C3
Leszczyna 29,746	A4
Leszno 57,673	C3
Liegnitz (Legnica) 103,949	C3
Łódź 849,204	D3
Lomza 57,976	F2
Lowicz 30,322	D2
Lubań 23,708	B3
Lubartów 22,117	F3
Lubin 80,757	C3
Lubliniec 33,084	D3
Luboń 20,255	C2
Łuków 30,536	F3
Malbork (Marienburg) 39,018	D1
Miedzyrzecz 19,805	B2
Mielec 60,187	F3
Mikolow 37,022	B4
Mińsk Mazowiecki 33,913	E2
Plock 120,933	D2
Plońsk 20,956	E2
Mragowo 21,674	E2
Mysłowice 92,009	B4
Kęty 18,731	D4
Kielce 213,012	E3
Klodzko 30,261	C3
Kluczbork 25,988	D3
Knurów 44,468	A4
Kolberg (Kołobrzeg) 44,426	B1
Kolo 22,807	D2
Kołobrzeg 44,426	B1
Konin 79,315	D2
Końskie 21,548	E3
Kościan 23,554	C2
Kościerzyna 22,287	C1
Kösslin (Koszalin) 107,592	C1
Koszalin 107,592	C1
Kozienice 20,557	E3
Myszków 33,084	D3
Naklo nad Notecia 20,056	C2
Prudnik 24,598	C3
Pruszcz Gdanski 20,575	D1
Pruszków 53,889	E2
Przemyśl 68,121	F4
Puławy 52,624	F3
Pultusk 17,619	E2
Nowa Ruda 27,507	C3
Nowa Sól 43,053	B3
Nowy Dwór Mazowiecki 26,842	E2
Nowy Sącz 76,658	E4
Nowy Targ 32,143	E4
Nysa 46,686	C3
Olawa 31,188	C3
Oleśnica 37,767	C3
Olkusz 40,571	D3
Olsztyn 160,956	E2
Opoczno 26,878	E3
Opole (Oppeln) 126,962	C3
Orzesze 18,100	A4
Ostróda 33,641	D2
Ostrołeka 49,032	E2
Ostrów Mazowiecka 20,367	E2
Ostrów Wielkopolski 72,085	C3
Ostrowiec Świetokrzyski 77,466	E3
Oświęcim 45,402	D4
Otwock 44,488	E2
Ozorków 21,676	D3
Pabianice 74,755	D3
Piaseczno 24,359	E2
Piekary Śląskie 68,274	B4
Pila 71,109	C2
Pionki 20,701	E3
Piotrków Trybunalski 80,598	D3
Pisz 18,286	F2
Pleszew 17,712	C3
Police 33,478	B2
Polkowice 20,823	B3
Poznań 586,908	C2
Radom 226,025	E3
Radomsko 50,059	D3
Raciborz 62,733	D3
Ratibor (Raciborz) 62,733	D3
Rawa Mazowiecka 17,428	E3
Rawicz 20,548	C3
Ruda Slaska 169,017	B4
Rumia 36,762	D1
Rybnik 142,059	D4
Rzeszów 150,702	F4
Sandomierz 23,815	E3
Sanok 39,163	F4
Schneidemühl (Piła) 71,109	C2
Schweidnitz (Świdnica) 62,424	C3
Siedlce 70,529	F2
Siemianowice Śląskie 80,412	B4
Sieradz 42,041	D3
Sierpc 18,286	D2
Skarzysko-Kamienna 50,455	E3
Skawina 23,487	D4
Skierniewice 43,963	E3
Słupsk 100,127	C1
Sochaczew 38,170	E2
Sokółka 19,059	F2
Sopot 46,874	D1
Sosnowiec 259,318	B4
Śrem 27,719	C3
Środa Wielkopolska 20,053	C2
Stalowa Wola 68,856	F3
Starachowice 55,996	E3
Stargard Szczeciński 69,852	B2
Stargard Gdanski 47,142	D2
Stettin (Szczecin) 411,275	B2
Stolp (Słupsk) 100,127	C1
Strzegom 17,185	C3
Strzelce Opolskie 21,238	D3
Suwalki 59,684	F1
Swarzedz 22,022	C2
Świdnica 62,424	C3
Świdnik 39,651	F3
Świebodzice 24,392	C3
Świebodzin 21,617	B2
Świecie 25,415	D2
Świetochlowice 58,770	A4
Świnoujście (Swinemünde) 42,932	B1
Szamotuly 17,970	C2
Szczecin 411,275	B2
Szczecinek 40,428	C2
Szczytno 26,878	E2
Tarnobrzeg 45,702	E3
Tarnów 120,639	E4
Tarnowskie Góry 73,460	A3
Tczew 58,887	D1
Tomaszów Lubelski 20,011	F3
Tomaszów Mazowiecki 69,579	E3
Toruń 201,860	D2
Trzebinia-Siersza 20,376	C4
Turek 28,559	D2
Tychy 189,816	B4
Wadowice 18,691	D4
Wagrowiec 22,736	C2
Wałbrzych 141,504	C3
Walcz 26,367	C2
Waldenburg (Wałbrzych) 141,504	C3
Warsaw (Warszawa) (cap.) 1,651,225	E2
Wejherowo 46,465	D1
Wieliczka 17,775	D4
Wieluń 24,061	D3
Włocławek 120,680	D2
Wodzisław Slaski 111,099	D4
Wolomin 36,762	E2
Wrocław 640,557	C3
Września 27,449	C2
Wyszków 23,411	E2
Zabrze 202,824	A4
Zagań 27,333	B3
Zakopane 28,417	D4
Zambrów 22,175	F2
Zamość 60,565	F3
Zary 39,172	B3
Zawiercie 56,017	D3
Zduńska Wola 44,686	D3
Zgierz 58,836	D3
Zgorzelec 36,008	B3
Zielona Góra 113,108	B3
Zlotów 17,533	C2
Zyrardów 33,196	D2
Żywiec 30,572	D4

OTHER FEATURES

Baltic (sea)	B1
Beskids (range)	D4
Biebrza (riv.)	F2
Bobr (riv.)	B3
Brda (riv.)	C2
Brynica (riv.)	B4
Bug (riv.)	F2
Danzig (Gdańsk) (gulf)	D1
Dukla (pass)	E4
Dunajec (riv.)	E4
Frisches Haff (lag.)	D1
Gwda (riv.)	C2
Hel (pen.)	D1
High Tatra (range)	D4
Jezioro Sniardwy (lake)	E2
Kłodnica (riv.)	A4
Łyna (riv.)	E1
Mamry, Jezioro (lake)	E1
Masurian (lakes)	E2
Narew (riv.)	E2
Neisse (riv.)	B3
Noteć (riv.)	B2
Nysa Łużycka (Neisse) (riv.)	B3
Nysa Kłodzka (riv.)	C3
Oder (riv.)	B2
Oder-Haff (lag.)	B1
Orava (riv.)	D4
Pilica (riv.)	D3
Plonia (riv.)	B2
Pomerania (bay)	B1
Prosna (riv.)	C3
Przemsza (riv.)	B4
Rega (riv.)	B2
Rysy (mt.)	D4
San (riv.)	F3
Słupia (riv.)	C1
Sudeten (range)	B3
Uznam (Usedom) (isl.)	B1
Vistula (riv.)	D2
Vistula (spit)	D1
Warmia (reg.)	D1
Warta (riv.)	B2
Wieprz (riv.)	F3
Wisła (Vistula) (riv.)	D2
Włocławske (lake)	D2
Wolin (Wollin) (isl.)	B2
Zegrzyńske (lake)	E2

Map

Poland — Conic Projection

SCALE OF MILES / SCALE OF KILOMETERS
0 10 20 40 60 80

Capitals of Countries
Other Capitals
International Boundaries
Internal Boundaries
Canals

Poland is divided into 49 provinces (bearing the same name as their capitals) and the autonomous cities of Warsaw, Łódź and Cracow.

(continued)

Russia and Neighboring Countries

CONIC PROJECTION

SCALE OF MILES
0 100 200 300 400 500 600

SCALE OF KILOMETERS
0 100 200 300 400 500 600

Capitals   Boundaries
★ National
◉ Autonomous Republic
◎ Autonomous Oblast
○ Autonomous Okrug

ADMINISTRATIVE DIVISIONS NOT NAMED ON MAP			
Division	Ref.	Division	Ref.
1 Abkhaz Aut. Rep.	E5	13 Khakass Aut Oblast	J4
2 Adygey Aut. Oblast	D5	14 Komi-Permyak Aut. Okrug	F4
3 Adzhar Aut. Rep.	E5	15 Mari Aut. Rep.	F4
4 Aginsk Buryat		16 Mordovian Aut. Rep.	F4
Autonomous Okrug	M4	17 Nagorno-Karabakh Aut. Oblast	E5
5 Checken-Ingush Aut. Rep.	E5	18 Nakhichevan Aut. Rep.	E6
6 Chuvash Aut. Rep.	F4	19 North Ossetian Aut. Rep.	E5
7 Gorno-Altay Aut. Oblast	J4	20 South Ossetian Aut. Oblast	E5
8 Gorno-Badakhshan Aut. Oblast	H6	21 Tatar Aut Rep	F4
9 Jewish Aut Oblast	O5	22 Tuvinian Aut. Rep.	K4
10 Kabardin-Balkar Aut. Rep.	E5	23 Udmurt Aut. Rep.	F4
11 Karachay-Cherkess Aut. Oblast	E5	24 Ust-Ordynsk Buryat	
12 Karakalpak Aut. Rep.	G5	Autonomous Okrug	L4

ARMENIA

AZERBAIJAN

BELARUS

GEORGIA

KAZAKHSTAN

KYRGYZSTAN

MOLDOVA

RUSSIA

TAJIKISTAN

TURKMENISTAN

UKRAINE

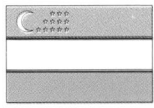
UZBEKISTAN

ARMENIA

AREA 11,506 sq. mi. (29,800 sq. km.)
POPULATION 3,283,000
CAPITAL Yerevan
LARGEST CITY Yerevan
HIGHEST POINT Alagez 13,435 ft. (4,095 m.)
MAJOR LANGUAGES Armenian, Azerbaijani, Kurdish, Russian
MAJOR RELIGIONS Eastern (Armenian Apostolic) Orthodoxy, Islam

AZERBAIJAN

AREA 33,436 sq. mi. (86,600 sq. km.)
POPULATION 7,029,000
CAPITAL Baku
LARGEST CITY Baku
HIGHEST POINT Bazardyuzyu 14,653 ft. (4,466 m.)
MAJOR LANGUAGES Azerbaijani, Russian, Armenian
MAJOR RELIGIONS Islam, Eastern (Russian) Orthodoxy

BELARUS

AREA 80,154 sq. mi. (207,600 sq. km.)
POPULATION 10,200,000
CAPITAL Minsk
LARGEST CITY Minsk
HIGHEST POINT Dzerzhinskaya 1,135 ft. (346 m.)
MAJOR LANGUAGES Belorussian, Russian, Polish, Ukrainian, Yiddish
MAJOR RELIGIONS Eastern (Russian) Orthodoxy, Roman Catholicism, Judaism

GEORGIA

AREA 26,911 sq. mi. (69,700 sq. km.)
POPULATION 5,449,000
CAPITAL T'bilisi
LARGEST CITY T'bilisi
HIGHEST POINT Kazbek 16,558 ft. (5,047 m.)
MAJOR LANGUAGES Georgian, Armenian, Russian, Azerbaijani, Abkhazian, Ossetian
MAJOR RELIGIONS Eastern (Georgian) Orthodoxy, Islam

KAZAKHSTAN

AREA 1,048,300 sq. mi. (2,715,100 sq. km.)
POPULATION 16,538,000
CAPITAL Astana
LARGEST CITY Almaty
HIGHEST POINT Khan-Tengri 22,951 ft. (6,995 m.)
MAJOR LANGUAGES Kazakh, Russian, German, Ukrainian, Uzbek, Tatar
MAJOR RELIGIONS Islam, Eastern (Russian) Orthodoxy

KYRGYZSTAN

AREA 76,641 sq. mi. (198,500 sq. km.)
POPULATION 4,291,000
CAPITAL Bishkek (Frunze)
LARGEST CITY Bishkek (Frunze)
HIGHEST POINT Pobeda Peak 24,406 ft. (7,439 m.)
MAJOR LANGUAGES Kirgiz, Russian, Uzbek, Ukrainian, German, Tatar
MAJOR RELIGIONS Islam, Eastern (Russian) Orthodoxy

MOLDOVA

AREA 13,012 sq. mi. (33,700 sq. km.)
POPULATION 4,341,000
CAPITAL Chişinău
LARGEST CITY Chişinău
HIGHEST POINT 1,408 ft. (429 m.)
MAJOR LANGUAGES Moldavian (Romanian), Ukrainian, Russian, Gagauzi, Yiddish
MAJOR RELIGIONS Eastern (Romanian) Orthodoxy, Judaism

RUSSIA

AREA 6,592,812 sq. mi. (17,075,400 sq. km.)
POPULATION 147,386,000
CAPITAL Moscow
LARGEST CITY Moscow
HIGHEST POINT El'brus 18,510 ft. (5,642 m.)
MONETARY UNIT ruble
MAJOR LANGUAGES Russian, Tatar, Ukrainian, Chuvash, Bashkir, Belorussian, Mordvinian, German, Kazakh, Yiddish, Chechen, Udmurt, Ossetian, Buryat, Yakut, Ingush, Tuvan
MAJOR RELIGIONS Eastern (Russian) Orthodoxy, Roman Catholicism, Islam, Judaism, Lamaism, Buddhism, Animism

TAJIKISTAN

AREA 55,251 sq. mi. (143,100 sq. km.)
POPULATION 5,112,000
CAPITAL Dushanbe
LARGEST CITY Dushanbe
HIGHEST POINT Ismail Samani Peak 24,590 ft. (7,495 m.)
MAJOR LANGUAGES Tajik, Uzbek, Russian, Tatar, Kirgiz
MAJOR RELIGIONS Islam, Eastern (Russian) Orthodoxy

TURKMENISTAN

AREA 188,455 sq. mi. (488,100 sq. km.)
POPULATION 3,534,000
CAPITAL Ashgabat
LARGEST CITY Ashgabat
HIGHEST POINT Rize 9,653 ft. (2,942 m.)
MAJOR LANGUAGES Turkmenian, Russian, Uzbek, Kazakh, Tatar
MAJOR RELIGIONS Islam, Eastern (Russian) Orthodoxy

UKRAINE

AREA 233,089 sq. mi. (603,700 sq. km.)
POPULATION 51,704,000
CAPITAL Kiev
LARGEST CITY Kiev
HIGHEST POINT Goverla 6,762 ft. (2,061 m.)
MAJOR LANGUAGES Ukrainian, Russian, Yiddish, Belorussian, Moldavian (Romanian), Polish, Tatar
MAJOR RELIGIONS Eastern (Ukrainian) Orthodoxy, Roman (Ukrainian Uniate) Catholicism, Judaism

UZBEKISTAN

AREA 173,591 sq. mi. (449,600 sq. km.)
POPULATION 19,906,000
CAPITAL Tashkent
LARGEST CITY Tashkent
HIGHEST POINT Khodzha-Pir'yakh 14,515 ft. (4,424 m.)
MAJOR LANGUAGES Uzbek, Russian, Tajik, Kazakh, Tatar, Karakalpak, Kirgiz, Ukrainian, Turkmenian
MAJOR RELIGIONS Islam, Eastern (Russian) Orthodoxy

Topography

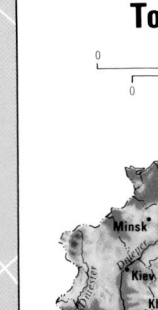

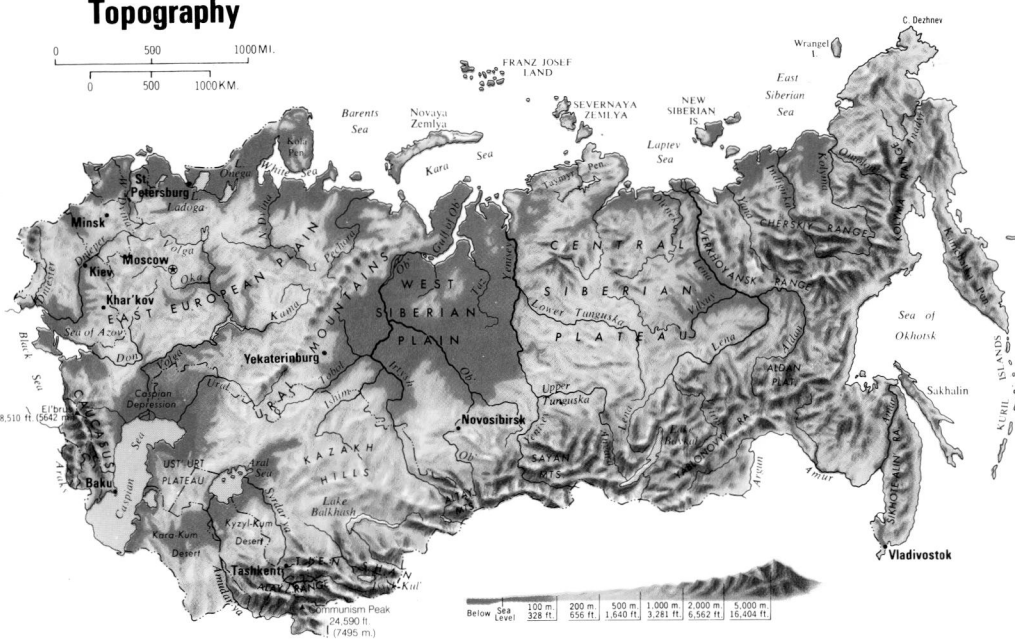

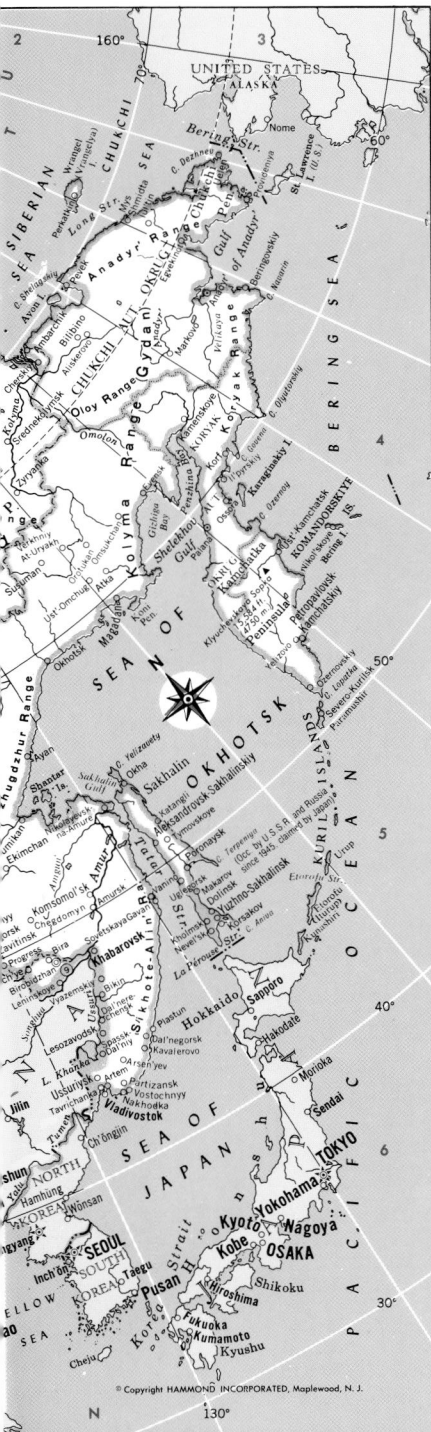

(continued)

Agriculture, Industry and Resources
(Eastern Europe)

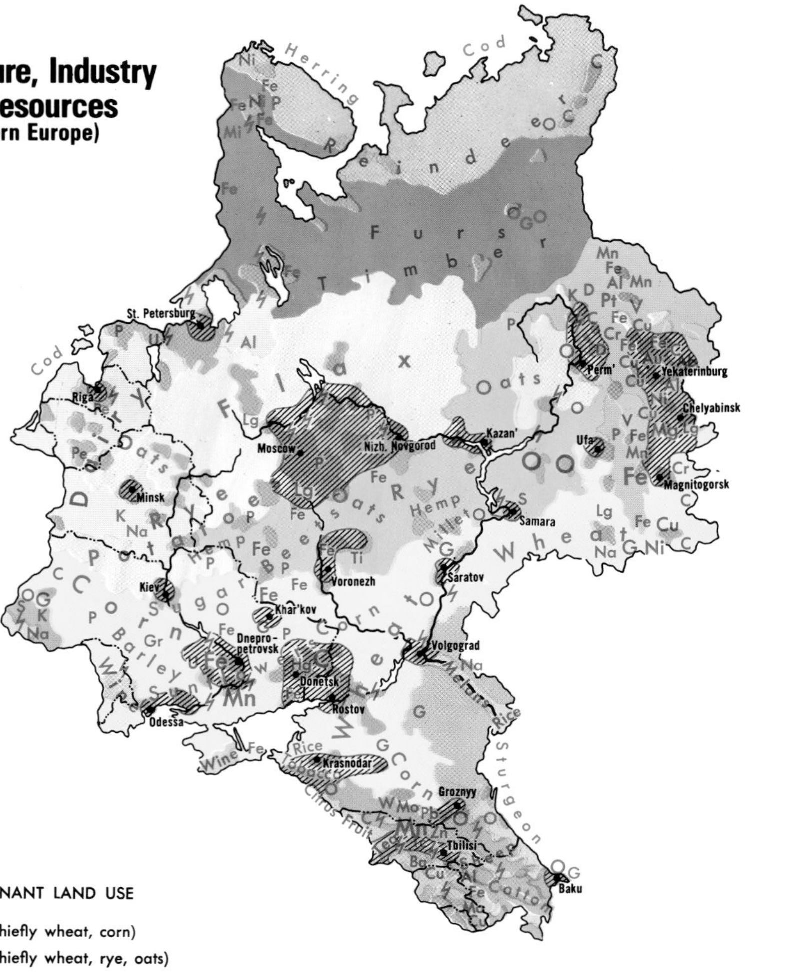

DOMINANT LAND USE

- Cereals (chiefly wheat, corn)
- Cereals (chiefly wheat, rye, oats)
- Dairy, Hogs, Livestock
- Livestock, Dairy
- Pasture Livestock
- Truck Farming, Potatoes, Vegetables, Dairy
- Flax, Dairy, Potatoes
- Cotton
- Vineyards, Orchards, Horticulture
- Sheep Herding, Limited Agriculture
- Forests
- Nonagricultural Land

MAJOR MINERAL OCCURRENCES

Ab	Asbestos	Hg	Mercury	Pb	Lead
Al	Bauxite	K	Potash	Pe	Peat
Au	Gold	Lg	Lignite	Pt	Platinum
Ba	Barite	Mg	Magnesium	S	Sulfur, Pyrites
C	Coal	Mi	Mica	Tc	Talc
Cr	Chromium	Mn	Manganese	Ti	Titanium
Cu	Copper	Mo	Molybdenum	U	Uranium
D	Diamonds	Na	Salt	V	Vanadium
Fe	Iron Ore	Ni	Nickel	W	Tungsten
G	Natural Gas	O	Petroleum	Zn	Zinc
Gr	Graphite	P	Phosphates		

⚡ Water Power   ▨ Major Industrial Areas

Agriculture, Industry and Resources
(Northern Asia)

DOMINANT LAND USE

- Cereals (chiefly wheat, corn)
- Livestock, Dairy
- Truck Farming, Potatoes, Vegetables, Dairy
- Cotton
- Sheep Herding, Limited Agriculture
- Forests
- Nonagricultural Land

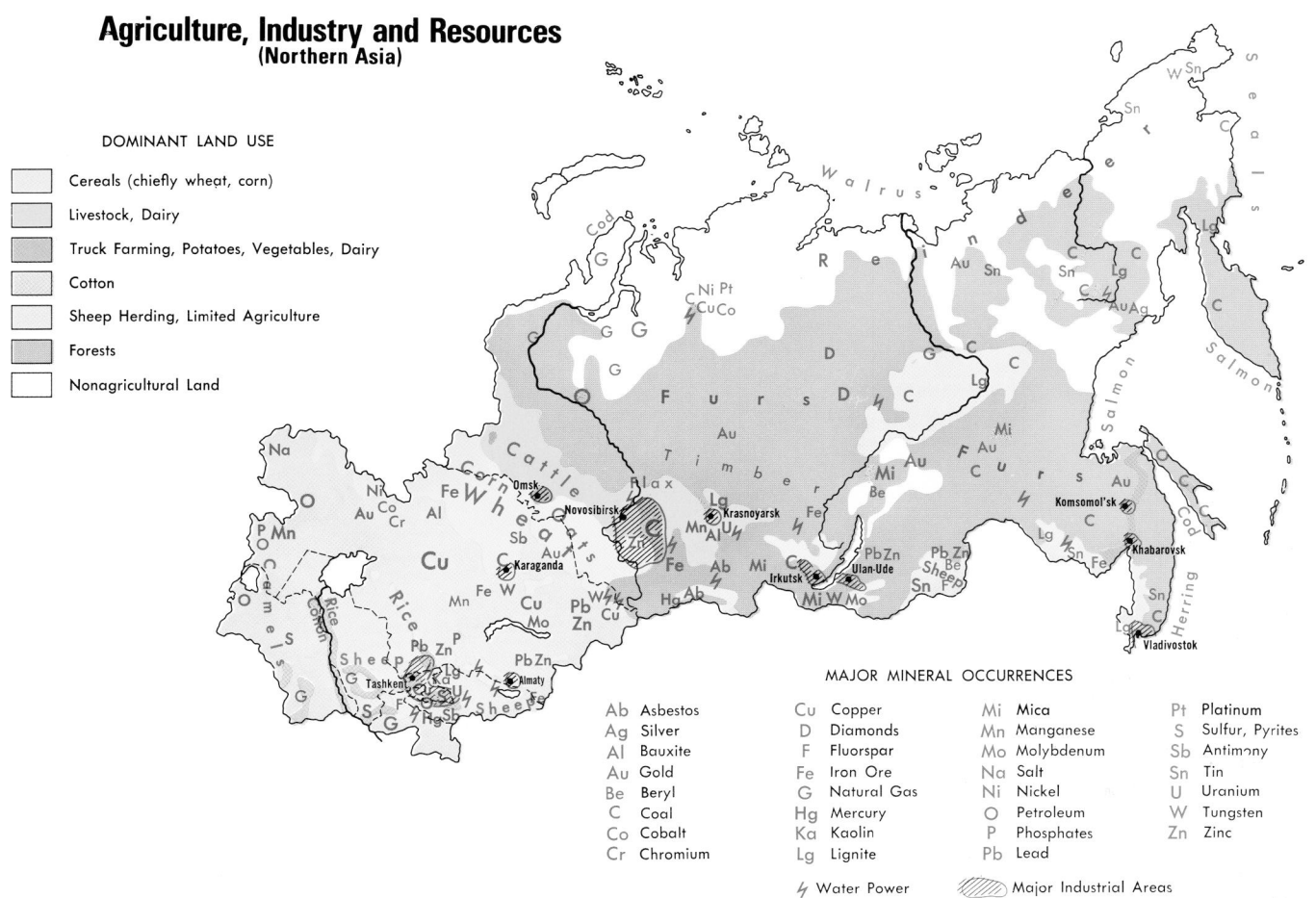

MAJOR MINERAL OCCURRENCES

Ab	Asbestos	Cu	Copper	Mi	Mica	Pt	Platinum
Ag	Silver	D	Diamonds	Mn	Manganese	S	Sulfur, Pyrites
Al	Bauxite	F	Fluorspar	Mo	Molybdenum	Sb	Antimony
Au	Gold	Fe	Iron Ore	Na	Salt	Sn	Tin
Be	Beryl	G	Natural Gas	Ni	Nickel	U	Uranium
C	Coal	Hg	Mercury	O	Petroleum	W	Tungsten
Co	Cobalt	Ka	Kaolin	P	Phosphates	Zn	Zinc
Cr	Chromium	Lg	Lignite	Pb	Lead		

⚡ Water Power   ▨ Major Industrial Areas

Russia and Neighboring Countries
European Part

CONIC PROJECTION
SCALE OF MILES

SCALE OF KILOMETERS

National Capitals	★
Administrative Centers	△
International boundaries	
Aut. Rep. Oblast, Kray boundaries	
Autonomous Oblast boundaries	
Autonomous Okrug boundaries	

© Copyright HAMMOND INCORPORATED, Maplewood, N.J.

Administrative Divisions bear same names as their respective Capitals or Centers, except:

Abkhaz Aut. Rep.	Sukhumi	F6
Adygey Aut. Oblast	Maykop	F6
Adzhar Aut. Rep.	Batumi	F6
Bashkir Aut. Rep.	Ufa	J4
Chechen-Ingush Aut. Rep.	Groznyy	G6
Chuvash Aut. Rep.	Cheboksary	G3
Crimean Oblast	Simferopol'	D6
Dagestan Aut. Rep.	Makhachkala	G6
Kabardin-Balkar Aut. Rep.	Nal'chik	F6
Kalmuck Aut. Rep.	Elista	F5
Karachay-Cherkess Aut. Obl.	Cherkessk	F6
Karelian Aut. Rep.	Petrozavodsk	D2
Komi Aut. Rep.	Syktyvkar	H2
Komi-Permyak Aut. Okrug	Kudymkar	H3
Mari Aut. Rep.	Yoshkar-Ola	G3
Mordvinian Aut. Rep.	Saransk	G4
Nagorno-Karabakh Aut. Obl.	Stepanakert	G7
Nenets Aut. Okrug	Nar'yan-Mar	H1
North Ossetian Aut. Rep.	Vladikavkaz	F6
South Ossetian Aut. Obl.	Tskhinvali	F6
Tatar Aut. Rep.	Kazan	G4
Trans-Carpathian Oblast	Uzhgorod	B5
Udmurt Aut. Rep.	Izhevsk	H3
Volyn Oblast	Lutsk	C4

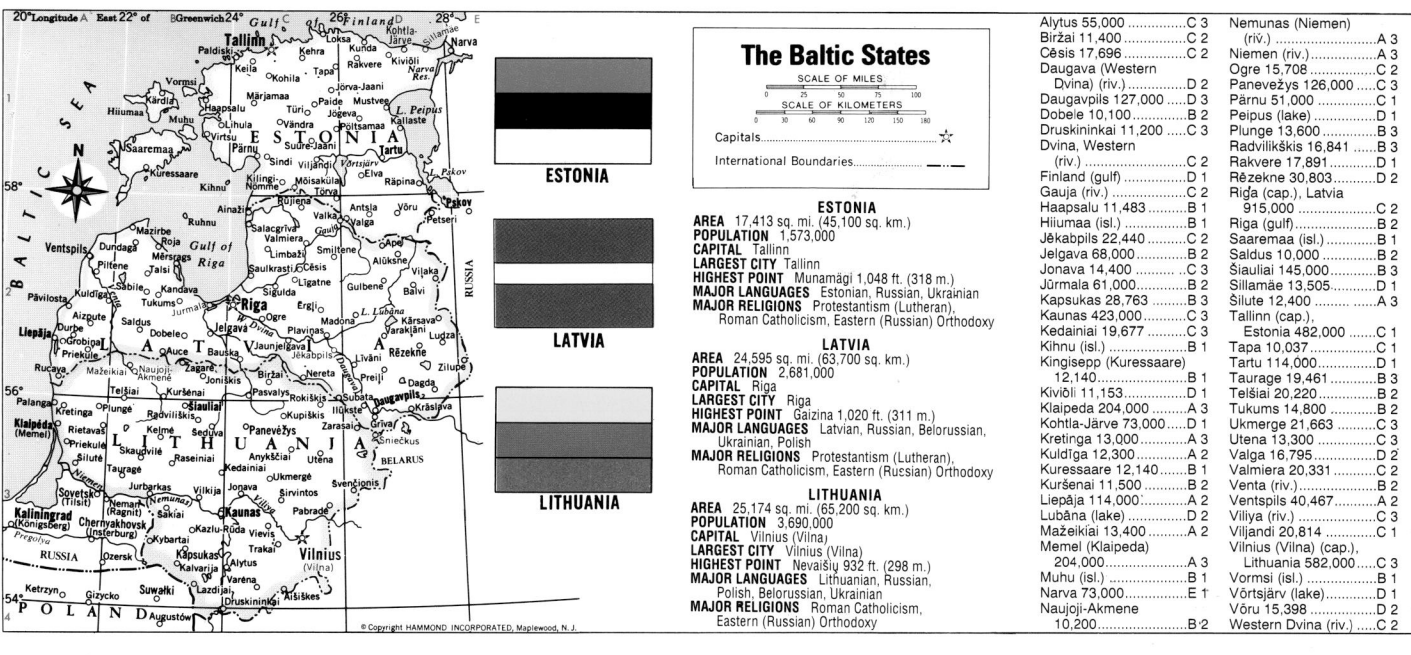

The Baltic States

SCALE OF MILES
0   25   50   75   100

SCALE OF KILOMETERS
0   30   60   90   120   150   180

Capitals..................................☆

International Boundaries.................. — ▪ —

ESTONIA
AREA 17,413 sq. mi. (45,100 sq. km.)
POPULATION 1,573,000
CAPITAL Tallinn
LARGEST CITY Tallinn
HIGHEST POINT Munamägi 1,048 ft. (318 m.)
MAJOR LANGUAGES Estonian, Russian, Ukrainian
MAJOR RELIGIONS Protestantism (Lutheran),
  Roman Catholicism, Eastern (Russian) Orthodoxy

LATVIA
AREA 24,595 sq. mi. (63,700 sq. km.)
POPULATION 2,681,000
CAPITAL Riga
LARGEST CITY Riga
HIGHEST POINT Gaizina 1,020 ft. (311 m.)
MAJOR LANGUAGES Latvian, Russian, Belorussian,
  Ukrainian, Polish
MAJOR RELIGIONS Protestantism (Lutheran),
  Roman Catholicism, Eastern (Russian) Orthodoxy

LITHUANIA
AREA 25,174 sq. mi. (65,200 sq. km.)
POPULATION 3,690,000
CAPITAL Vilnius (Vilna)
LARGEST CITY Vilnius (Vilna)
HIGHEST POINT Nevaišių 932 ft. (298 m.)
MAJOR LANGUAGES Lithuanian, Russian,
  Polish, Belorussian, Ukrainian
MAJOR RELIGIONS Roman Catholicism,
  Eastern (Russian) Orthodoxy

© Copyright HAMMOND INCORPORATED, Maplewood, N. J.

Population Distribution

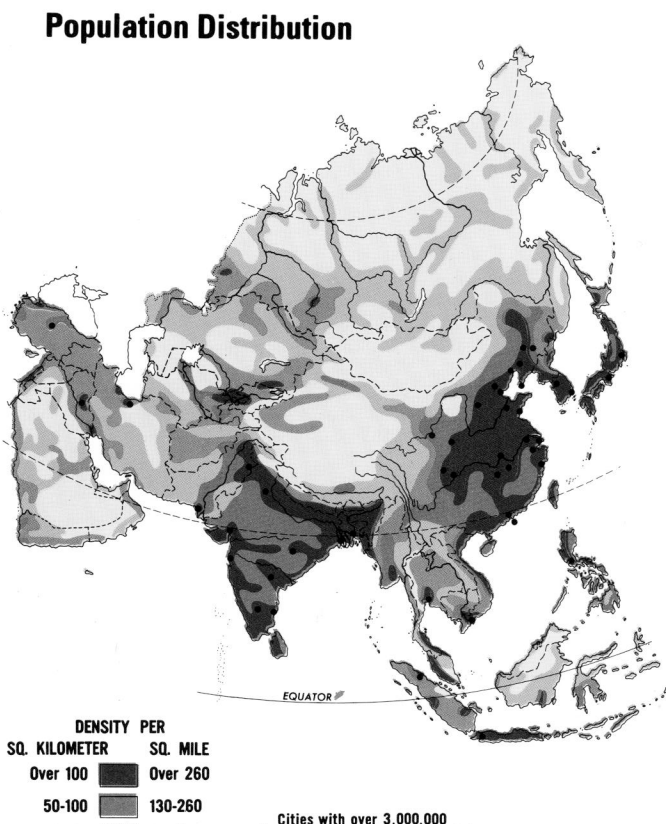

AREA 17,128,500 sq. mi.
(44,362,815 sq. km.)
POPULATION 3,176,000,000
LARGEST CITY Tokyo
HIGHEST POINT Mt. Everest 29,028 ft.
(8,848 m.)
LOWEST POINT Dead Sea -1,296 ft.
(-395 m.)

Vegetation

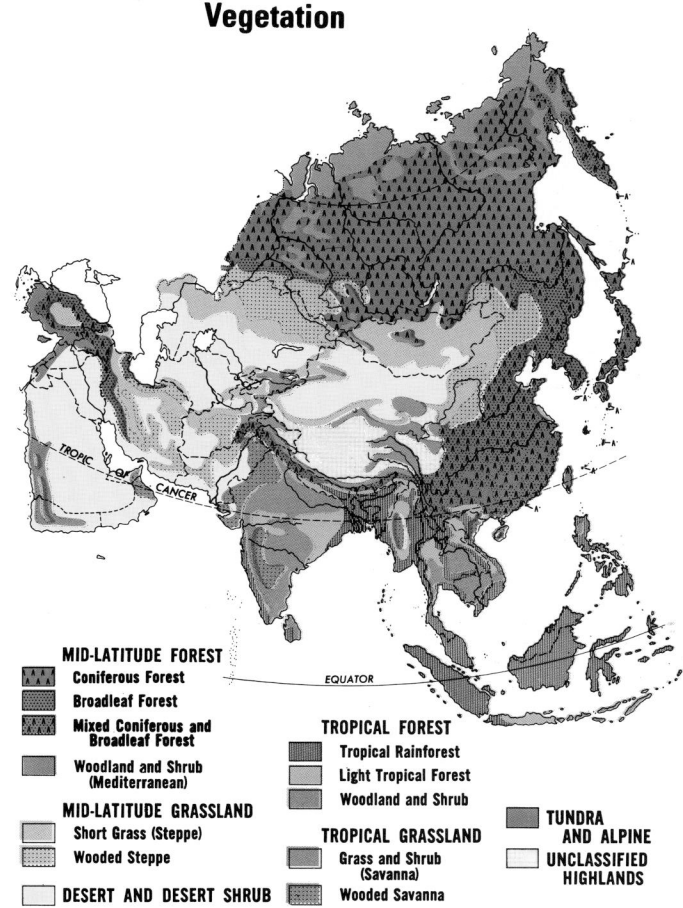

DENSITY PER

SQ. KILOMETER	SQ. MILE
Over 100	Over 260
50-100	130-260
10-50	25-130
1-10	3-25
Under 1	Under 3

• Cities with over 3,000,000 inhabitants (including suburbs)

MID-LATITUDE FOREST
Coniferous Forest
Broadleaf Forest
Mixed Coniferous and Broadleaf Forest
Woodland and Shrub (Mediterranean)

MID-LATITUDE GRASSLAND
Short Grass (Steppe)
Wooded Steppe

DESERT AND DESERT SHRUB

TROPICAL FOREST
Tropical Rainforest
Light Tropical Forest
Woodland and Shrub

TROPICAL GRASSLAND
Grass and Shrub (Savanna)
Wooded Savanna

TUNDRA AND ALPINE

UNCLASSIFIED HIGHLANDS

Average January Temperature

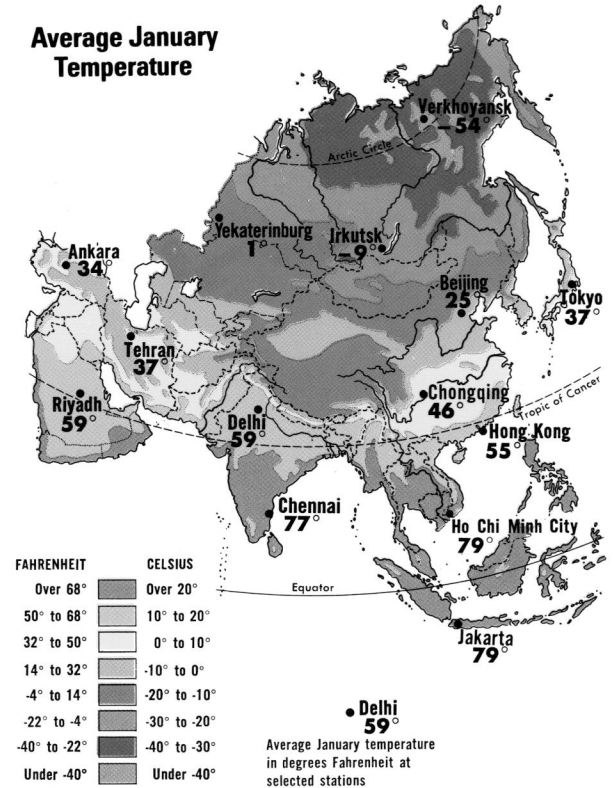

Verkhoyansk 54°
Yekaterinburg 1° Irkutsk -9°
Ankara 34°
Beijing 25°
Tokyo 37°
Tehran 37°
Chongqing 46°
Riyadh 59°
Delhi 59°
Hong Kong 55°
Chennai 77°
Ho Chi Minh City 79°
Jakarta 79°

FAHRENHEIT / **CELSIUS**

FAHRENHEIT	CELSIUS
Over 68°	Over 20°
50° to 68°	10° to 20°
32° to 50°	0° to 10°
14° to 32°	-10° to 0°
-4° to 14°	-20° to -10°
-22° to -4°	-30° to -20°
-40° to -22°	-40° to -30°
Under -40°	Under -40°

• Delhi 59°
Average January temperature in degrees Fahrenheit at selected stations

Average July Temperature

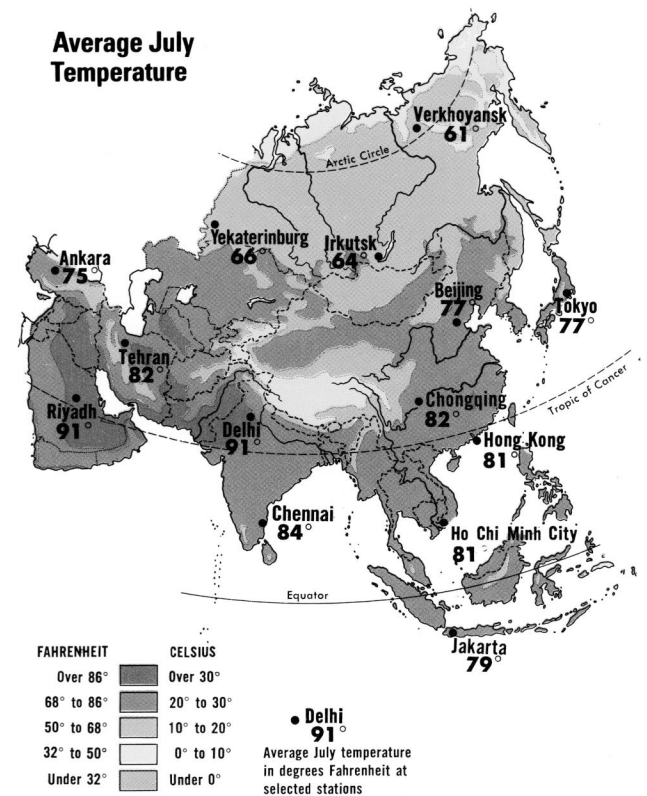

Verkhoyansk 61°
Yekaterinburg 66° Irkutsk 64°
Ankara 75°
Beijing 77°
Tokyo 77°
Tehran 82°
Chongqing 82°
Riyadh 91°
Delhi 91°
Hong Kong 81°
Chennai 84°
Ho Chi Minh City 81°
Jakarta 79°

FAHRENHEIT	CELSIUS
Over 86°	Over 30°
68° to 86°	20° to 30°
50° to 68°	10° to 20°
32° to 50°	0° to 10°
Under 32°	Under 0°

• Delhi 91°
Average July temperature in degrees Fahrenheit at selected stations

Rainfall

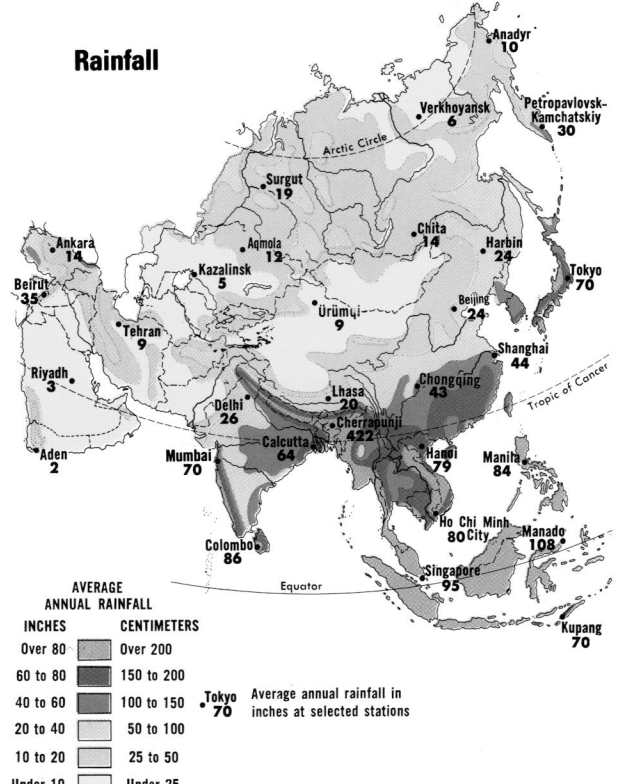

Anadyr 10
Verkhoyansk 6
Petropavlovsk-Kamchatskiy 30
Surgut 19
Chita 14
Harbin 24
Ankara 14
Aqmola 12
Tokyo 70
Beirut 35
Kazalinsk 5
Ürümqi 9
Beijing 24
Tehran 9
Shanghai 44
Riyadh 3
Lhasa 20
Chongqing 43
Delhi 26
Cherrapunji 422
Calcutta 64
Aden 2
Mumbai 70
Hanoi 79
Manila 84
Ho Chi Minh City 80
Manado 108
Colombo 86
Singapore 95
Kupang 70

AVERAGE ANNUAL RAINFALL

INCHES	CENTIMETERS
Over 80	Over 200
60 to 80	150 to 200
40 to 60	100 to 150
20 to 40	50 to 100
10 to 20	25 to 50
Under 10	Under 25

• Tokyo 70
Average annual rainfall in inches at selected stations

Vegetation/Relief

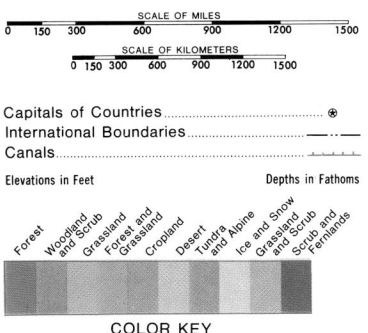

SCALE OF MILES
0 150 300 600 900 1200 1500

SCALE OF KILOMETERS
0 150 300 600 900 1200 1500

Capitals of Countries .... ⊛
International Boundaries ....
Canals ....

Elevations in Feet   Depths in Fathoms

Forest
Woodland and Scrub
Grassland
Forest and Grassland
Cropland
Desert
Tundra and Alpine
Ice and Snow
Grassland and Scrub
Scrub and Fernlands

COLOR KEY

Longitude 70° East of Greenwich

SAUDI ARABIA

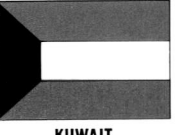
KUWAIT

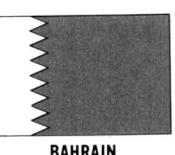
YEMEN

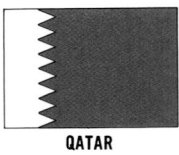
BAHRAIN

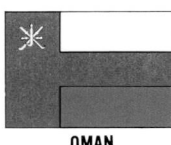
QATAR

OMAN

AFGHANISTAN

CITIES and TOWNS

Andkhvoy 13,137	H2
Aybak 33,016	J2
Baghlan 75,130	J2
Bamian 7,355	J3
Chaghcharan 2,974	J3
Charikar 25,093	J2
Farah 18,797	H3
Feyzabad 10,142	K2
Gardez 11,415	J3
Ghazni 30,425	J3
Ghurian 12,404	H3
Herat 163,960	H3
Jalalabad 56,384	J3
Kabul (cap.) 905,108	J3

Kalat (Qalat) 5,946	J3
Kandahar (Qandahar) 178,409	J3
Khanabad 26,803	J2
Kholm 28,078	J2
Khowst	J3
Kuhestan	H3
Landay	H3
Lashkar Gah 26,646	H3

Mazar-e Sharif 122,567	J2
Meymaneh 54,954	H2
Pol-e Khomri 31,101	J2
Qalat 5,946	J3
Qal'eh-ye Now 5,340	H3
Qandahar 178,409	J3
Qonduz 107,191	J2
Sar-e Pol 15,699	J2
Sheberghan 54,870	H2

Taloqan 46,202	J2
Zaranj 6,477	H3

OTHER FEATURES

Farah Rud (riv.)	H3
Gowd-e Zerreh (depr.)	H4
Harirud (riv.)	H3
Helmand (riv.)	J3

Hindu Kush (mts.)	J2
Kabul (riv.)	K3
Konar (riv.)	K2
Lurah (riv.)	J3
Margow, Dasht-e (des.)	H3
Murghab (riv.)	H2
Namaksar (salt lake)	H3
Nurestan (reg.)	K2
Paropamisus (mts.)	H3

Qonduz (riv.)	J2
Rigestan (reg.)	H3
Vakhan (reg.)	K2

BAHRAIN

CITIES and TOWNS

Manama (cap.) 88,785	F4

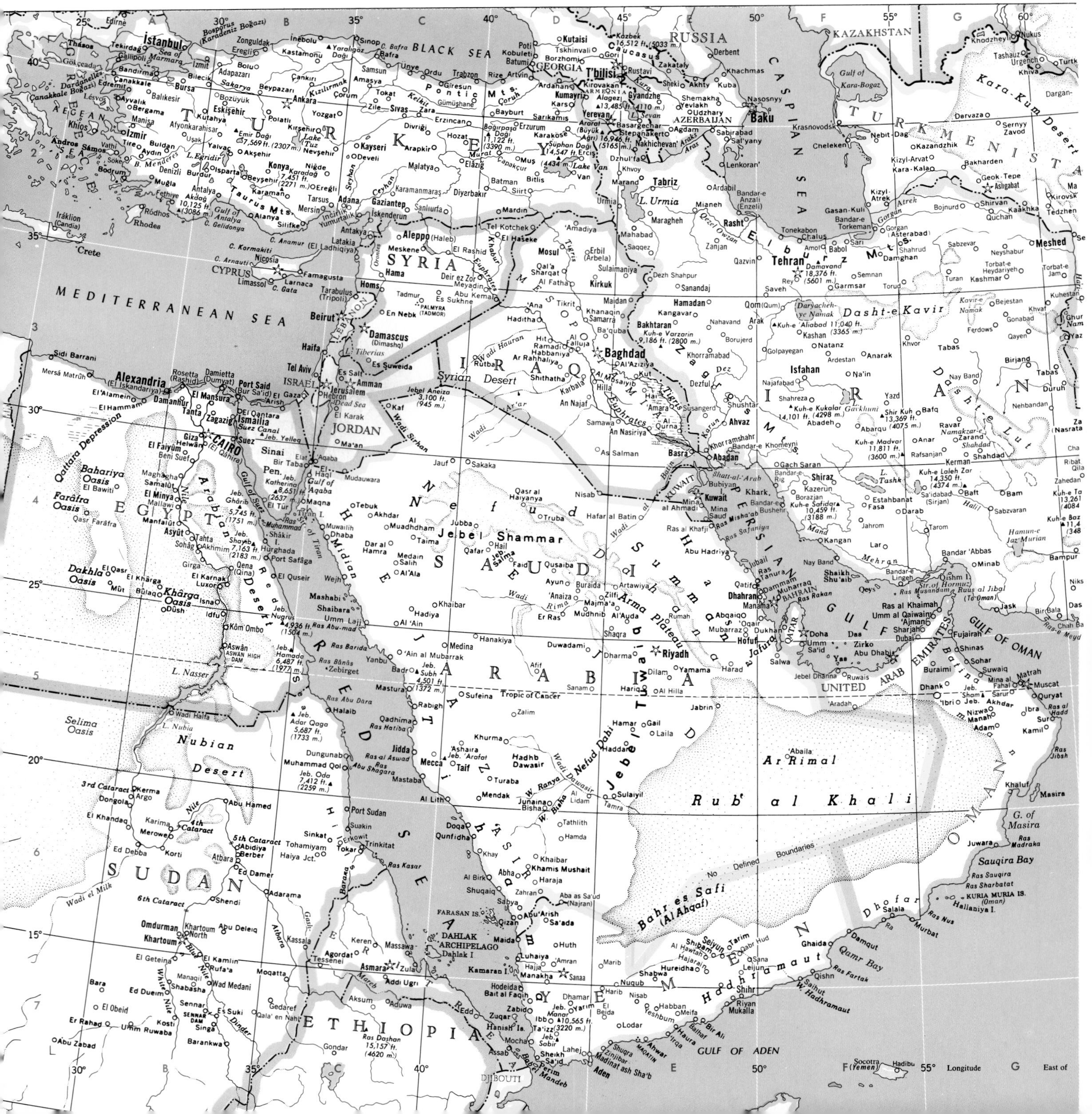

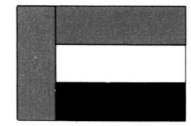

UNITED ARAB EMIRATES

Muharraq 37,732 ....F4

GAZA STRIP

CITIES and TOWNS

Gaza*
118,272 ....B3

IRAN

CITIES and TOWNS

Abadan 296,081 ....E3	Anarak 2,038 ....F3
Abadeh 16,000 ....F3	Arak 114,507 ....E3
Abarqu 8,000 ....F3	Ardabil 147,404 ....E2
Ahvaz 329,006 ....E3	Ardestan 5,868 ....F3
Amol 68,782 ....F2	Asterabad (Gorgan) 88,348...F2
	Babol 67,790 ....F2
	Bafq 5,000 ....G3
	Baft 6,000 ....G4
	Bakhtaran 290,861 ....E3

Bam 22,000 ....G4	Bejestan 3,823 ....G3
Bandar 'Abbas 89,103 ......G4	Birjand 25,854 ....G3
Bandar-e Anzali (Enzeli)	Bojnurd 31,248 ....G2
55,978...................E2	Borazjan 20,000 ....F4
Bandar-e Bushehr 57,681 ...F4	Borujerd 100,103 ....E3
Bandar-e Khomeyni 6,000...E3	Chalus 15,000 ....F2
Bandar-e Lengeh 4,920 .....F4	Damghan 13,000 ....F2
Bandar-e Rig 1,889 ....F4	Darab 13,000 ....G4
Bandar-e Torkeman 13,000...F2	Dezful 110,287 ....E3

(continued on following page)

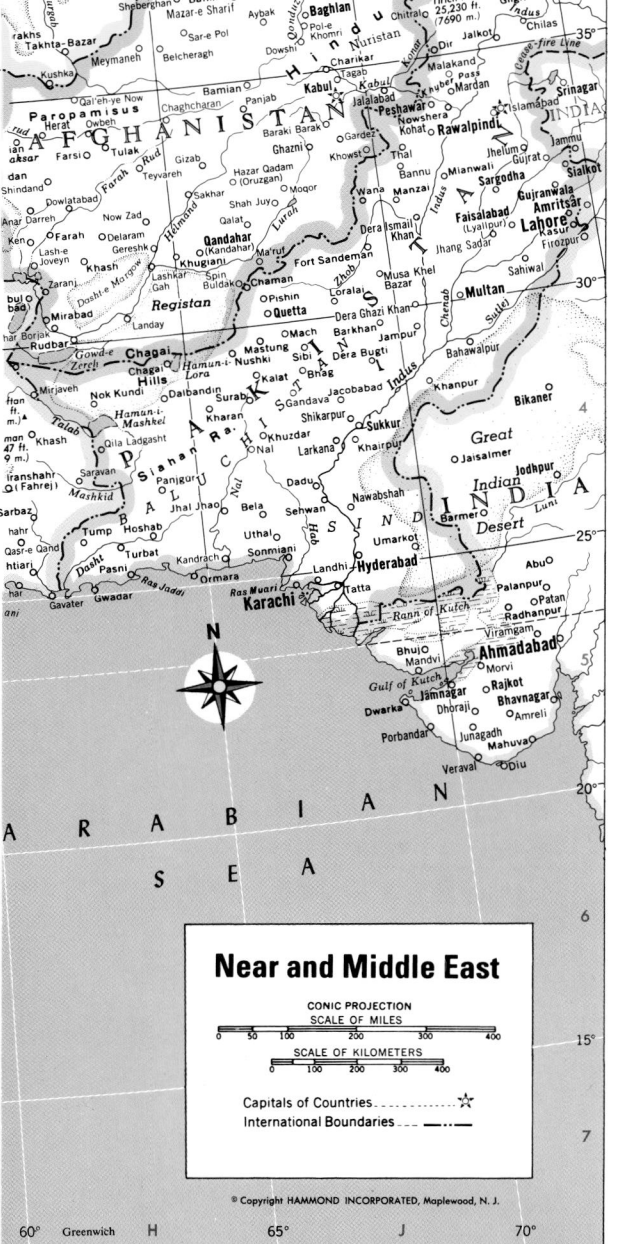

SAUDI ARABIA

AREA	829,995 sq. mi.
	(2,149,687 sq. km.)
POPULATION	14,435,000
CAPITAL	Riyadh
MONETARY UNIT	Saudi riyal
MAJOR LANGUAGE	Arabic
MAJOR RELIGION	Islam

YEMEN

AREA	188,321 sq. mi. (487,792 sq. km.)
POPULATION	10,183,000
CAPITAL	Sanaa
MONETARY UNIT	Yemeni rial
MAJOR LANGUAGE	Arabic
MAJOR RELIGION	Islam

QATAR

AREA	4,247 sq. mi. (11,000 sq. km.)
POPULATION	422,000
CAPITAL	Doha
MONETARY UNIT	Qatari riyal
MAJOR LANGUAGE	Arabic
MAJOR RELIGION	Islam

KUWAIT

AREA	6,532 sq mi. (16,918 sq. km.)
POPULATION	2,048,000
CAPITAL	Kuwait
MONETARY UNIT	Kuwaiti dinar
MAJOR LANGUAGE	Arabic
MAJOR RELIGION	Islam

BAHRAIN

AREA	240 sq. mi. (622 sq. km.)
POPULATION	489,000
CAPITAL	Manama
MONETARY UNIT	Bahraini dinar
MAJOR LANGUAGE	Arabic
MAJOR RELIGION	Islam

OMAN

AREA	120,000 sq. mi. (310,800 sq. km.)
POPULATION	2,000,000
CAPITAL	Muscat
MONETARY UNIT	Omani rial
MAJOR LANGUAGE	Arabic
MAJOR RELIGION	Islam

UNITED ARAB EMIRATES

AREA	32,278 sq. mi. (83,600 sq. km.)
POPULATION	1,206,000
CAPITAL	Abu Dhabi
MONETARY UNIT	dirham
MAJOR LANGUAGE	Arabic
MAJOR RELIGION	Islam

Topography

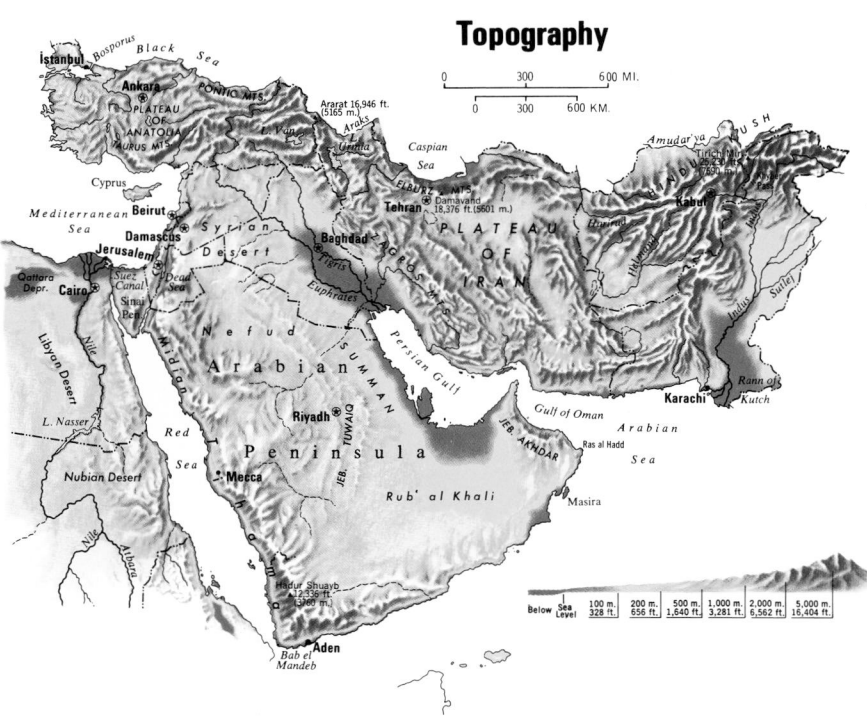

® Copyright HAMMOND INCORPORATED, Maplewood, N.J.

Agriculture, Industry and Resources

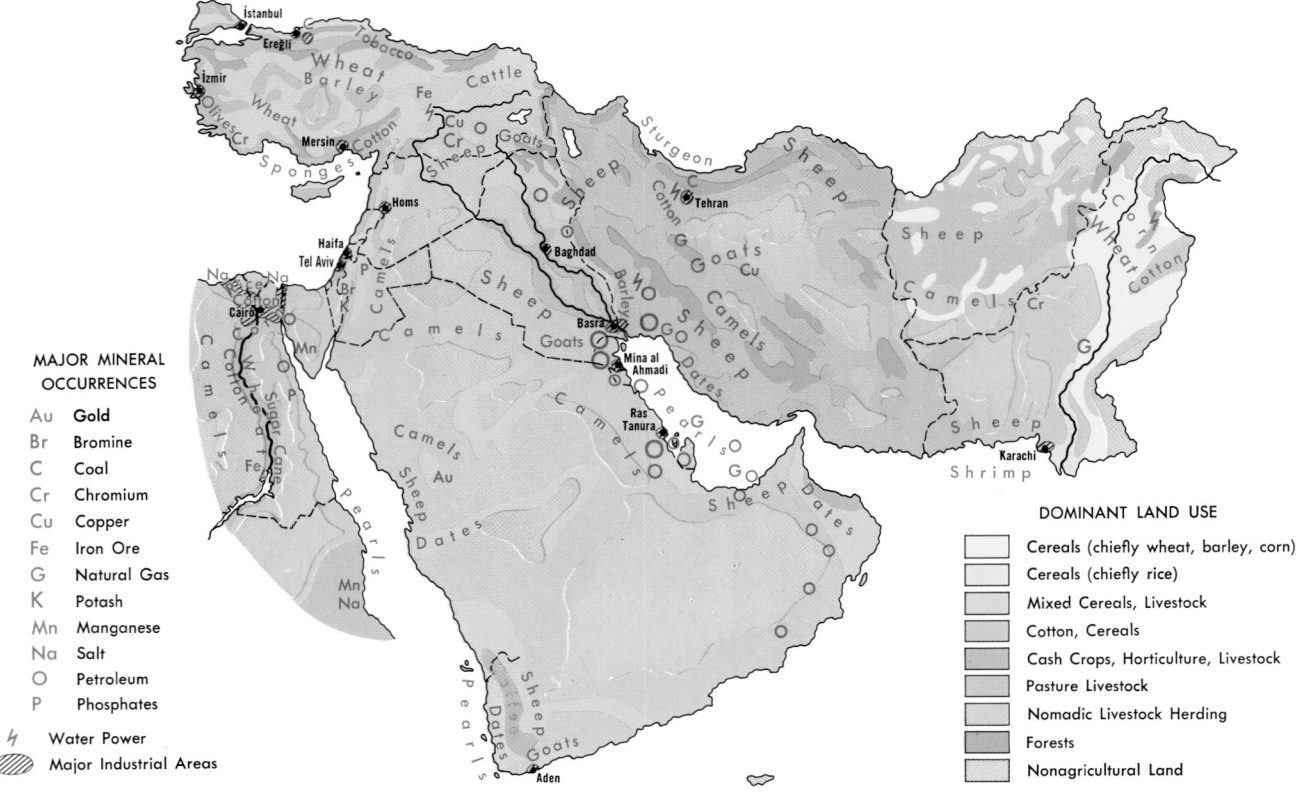

MAJOR MINERAL OCCURRENCES

Au Gold
Br Bromine
C Coal
Cr Chromium
Cu Copper
Fe Iron Ore
G Natural Gas
K Potash
Mn Manganese
Na Salt
O Petroleum
P Phosphates
⚡ Water Power
▨ Major Industrial Areas

DOMINANT LAND USE

Cereals (chiefly wheat, barley, corn)
Cereals (chiefly rice)
Mixed Cereals, Livestock
Cotton, Cereals
Cash Crops, Horticulture, Livestock
Pasture Livestock
Nomadic Livestock Herding
Forests
Nonagricultural Land

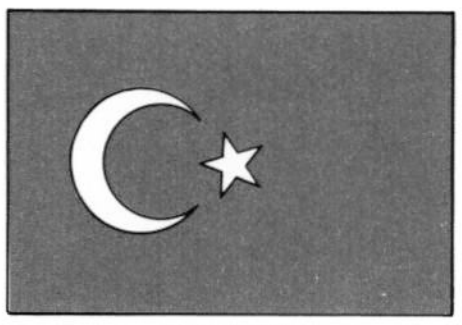

TURKEY

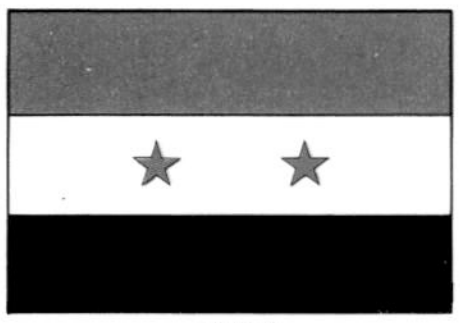

SYRIA

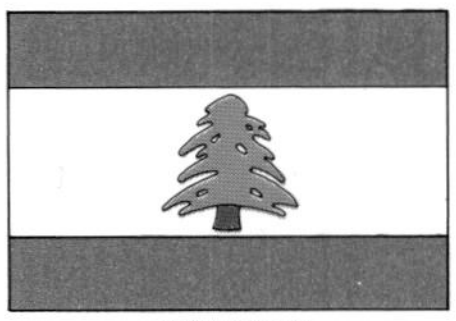

LEBANON

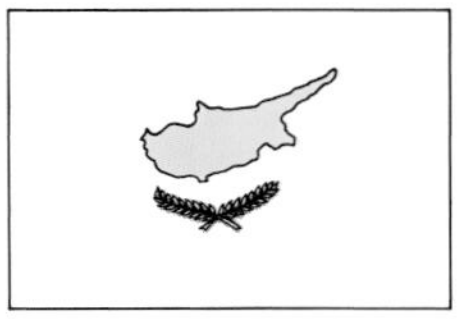

CYPRUS

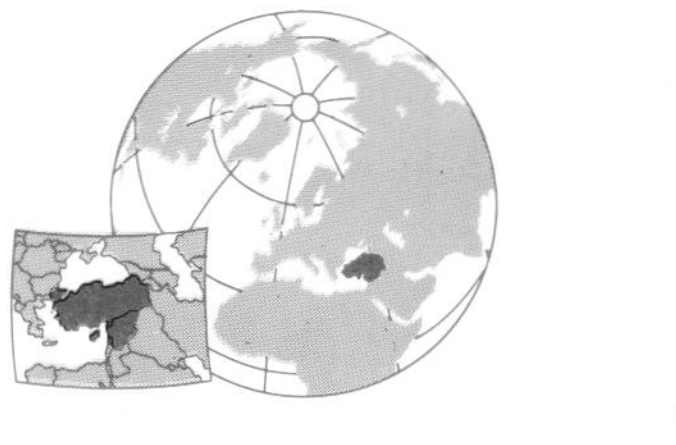

AREA 300,946 sq. mi. (779,450 sq. km.)
POPULATION 56,741,000
CAPITAL Ankara
LARGEST CITY Istanbul
HIGHEST POINT Ararat 16,946 ft. (5,165 m.)
MONETARY UNIT Turkish lira
MAJOR LANGUAGE Turkish
MAJOR RELIGION Islam

AREA 71,498 sq. mi. (185,180 sq. km.)
POPULATION 11,719,000
CAPITAL Damascus
LARGEST CITY Damascus
HIGHEST POINT Hermon 9,232 ft. (2,814 m.)
MONETARY UNIT Syrian pound
MAJOR LANGUAGES Arabic, French, Kurdish, Armenian
MAJOR RELIGIONS Islam, Christianity

AREA 4,015 sq. mi. (10,399 sq. km.)
POPULATION 2,897,000
CAPITAL Beirut
LARGEST CITY Beirut
HIGHEST POINT Qurnet es Sauda 10,131 ft. (3,088 m.)
MONETARY UNIT Lebanese pound
MAJOR LANGUAGES Arabic, French
MAJOR RELIGIONS Christianity, Islam

AREA 3,473 sq. mi. (8,995 sq. km.)
POPULATION 699,000
CAPITAL Nicosia
LARGEST CITY Nicosia
HIGHEST POINT Troödos 6,406 ft. (1,953 m.)
MONETARY UNIT Cypriot pound
MAJOR LANGUAGES Greek, Turkish, English
MAJOR RELIGIONS Eastern (Greek) Orthodoxy,

CYPRUS

CITIES and TOWNS

Famagusta 38,960 F5
Kyrenia 3,892 E5
Kythrea 3,400 E5
Lapithos 3,600 E5
Larnaca 19,608 E5
Lefka 3,650 E5
Limassol 79,641 E5
Morphou 9,040 E5
Nicosia (cap.) 115,718 E5
Paphos 8,984 E5
Polis 2,200 E5
Rizokarpasso 3,600 F5
Yialousa 2,750 E5

OTHER FEATURES

Andreas (cape) F5
Arnauti (cape) E5
Famagusta (bay) F5
Gata (cape) E5
Greco (cape) F5
Klides (isls.) F5
Kormakiti (cape) E5
Larnaca (bay) E5
Morphou (bay) E5
Pomos (pt.) E5
Troodos (mt.) E5

LEBANON

CITIES and TOWNS

'Aleih 18,630 F6
Amyun 7,926 F5
Ba'albek 15,560 G5
Beirut (cap.) 474,870 F6
Beirut* 938,940 F6
Merj 'Uyun 9,318 F6
Rasheiya 6,731 F6
Saida 32,200 F6
Sidon (Saida) 32,200 F6
Sur 16,483 F6
Tarabulus 127,611 F5
Tripoli (Tarabulus) 127,611 F5
Tyre (Sur) 16,483 F6
Zahle 53,121 F6
Zegharta 18,210 G5

OTHER FEATURES

Lebanon (mts.) F6
Leontes (Litani) (riv.) F6
Litani (riv.) F6
Sauda, Qurnet es (mt.) G5

SYRIA

PROVINCES

Aleppo 1,316,872 G4
Damascus 1,457,934 G6
Deir ez Zor 292,780 H5
Der'a 230,481 G6
El Quneitra 16,490 F6
Es Suweida 139,650 G6
Hama 514,748 G5
Haseke 468,506 J4
Homs 546,176 G5
Idlib 383,695 G5
Latakia 389,552 G5
Rashid 243,736 H5
Tartus 302,065 G5

CITIES and TOWNS

Abu Kemal 6,907 J5
'Ain el 'Arab 4,529 H4
Aleppo 639,428 G4
Azaz 13,923 G4
Baniyas 8,537 F5
Damascus (cap.) 836,668 G6
Damascus* 923,253 G6
Deir ez Zor 66,164 H5
Der'a 27,651 G6
Dimashq (Damascus) (cap.) 836,668 G6
Duma 30,050 G6
El Bab 27,366 G4
El Haseke 32,746 J4
El Ladhiqya (Latakia) 125,716 F5
El Qaryatein G5
El Quneitra 17,752 F6
El Rashid 37,151 H5
En Nebk 16,334 G5
Es Suweida 29,524 G6
Et Tell el Abyad H4
Haffe 4,656 G5
Haleb (Aleppo) 639,428 G4
Hama 137,421 G5
Harim 6,837 G4
Homs 215,423 G5
Idlib 34,515 G5
Izra 3,226 G6
Jeble 15,715 F5
Jerablus 8,610 G4
Jisr esh Shughur 13,131 G5
Khan Sheikhun G5
Latakia 125,716 F5
Masyaf 7,058 G5
Membij 13,796 G4
Meskene H5
Meyadin 12,515 J5
Qal'at es Salihiye J5
Qamishliye 31,448 J4
Quteife 4,993 G6
Raqqa (El Rashid) 37,151 H5
Safita 9,650 G5
Selemiya 21,677 G5
Tadmur 10,670 H5
Tartus 29,842 F5
Telkalakh 6,242 G5
Zebdani 10,010 G6

OTHER FEATURES

Abdul 'Aziz, Jebel (mts.) J4
'Amrit (ruins) F5
Arwad (Ruad) (isl.) F5
'Asi (Orontes) (riv.) G5
Bahrat Assad (lake) H4
Druz, Jebel ed (mts.) G6
El Furat (riv.) H4
Euphrates (El Furat) (riv.) H4
Hermon (mt.) F6
Khabur (riv.) J5
Orontes (riv.) G5
Palmyra (Tadmor) (ruins) H5
Tigris (riv.) K4

TURKEY

PROVINCES

Adana 1,485,743 F4
Adıyaman 367,595 H4
Afyonkarahisar 597,516 D3
Ağri 368,009 K3
Amasya 341,287 F2
Ankara 2,854,689 E3
Antalya 748,706 D4
Artvin 228,997 J2
Aydın 652,488 B4
Balıkesir 853,177 B3
Bilecik 147,001 D2
Bingöl 228,702 J3
Bitlis 257,908 J3
Bolu 471,751 D2
Burdur 235,009 D4
Bursa 1,148,492 C2
Çanakkale 338,091 B2
Çankırı 258,436 E2
Çorum 571,831 F2
Denizli 603,338 C4
Diyarbakır 778,150 H4
Edirne 363,286 B2
Elâzığ 440,808 H3
Erzincan 282,022 H3
Erzurum 801,809 J3
Eskişehir 543,802 D3
Gaziantep 808,697 G4
Giresun 480,083 H2
Gümüşhane 275,191 H2
Hakkâri 155,463 K4
Hatay 856,271 G4
İçel 842,817 F4
İsparta 301,166 D4
İstanbul 3,264,393 C2
İzmir 1,976,763 B3
Kahramanmaraş 738,032 G4
Kars 700,238 K2
Kastamonu 450,946 E2
Kayseri 778,383 F3
Kırklareli 283,408 B2
Kırşehir 240,497 F3
Kocaeli 596,899 C2
Konya 1,562,139 E4
Kütahya 497,089 C3
Malatya 669,962 H3
Manisa 941,941 B3
Mardin 564,967 J4
Muğla 438,145 C4
Muş 302,406 J3
Nevşehir 256,933 F3
Niğde 512,071 F4
Ordu 713,535 G2
Rize 361,258 J2
Sakaryâ 548,747 D2
Samsun 1,008,113 F2
Siirt 445,483 J4
Sinop 276,242 F2
Sivas 750,144 G3
Tekirdağ 360,742 B2
Tokat 624,508 G2
Trabzon 731,045 H2
Tunceli 157,974 H3
Urfa 602,736 H4
Uşak 247,224 C3
Van 468,646 K3
Yozgat 504,433 F3
Zonguldak 972,856 D2

CITIES and TOWNS

Adalia (Antalya) 176,446 D4
Adana 842,845 F4
Adapazarı 131,400 D2
Adilcevaz 10,342 K3
Adıyaman 116,986 H4
Afşin 20,084 G3
Afyonkarahisar 597,516 D3
Ağri (Karaköse) 41,103 K3
Ahlat 10,422 K3
Akçaabat 13,384 H2
Akçadağ 8,015 G3
Akçakale 11,184 H4
Akçakoca 9,639 D2
Akdağmadeni 10,192 F3
Akhisar 61,491 B3
Aksaray 62,927 F3
Akşehir 40,312 D3
Akseki 6,815 D4
Akyazı 14,795 D2
Alaca 15,649 F2
Alaçam 11,402 F2
Alanya 22,190 D4
Alaşehir 25,611 C3
Alexandretta (Iskenderun) 120,985 G4
Alibeyköyü 33,387 D6
Altındağ 608,689 E2
Altınova 6,980 B3
Alucra 8,795 H2
Amasya 48,010 F2
Anamur 23,025 E4
Andırın 6,045 G4
Ankara (cap.) 2,203,729 E3
Antakya 99,551 F4
Antalya 176,446 D4
Antioch (Antakya) 99,551 F4
Arapkir 8,816 H3
Ardahan 14,912 K2
Ardeşen 9,582 J2
Arhavi 6,801 J2
Arsin 6,892 H2
Artvin 14,203 J2
Aşkale 12,045 J3
Avanos 8,927 F3
Ayancık 8,257 F1
Aybastı 13,517 G2
Aydın 37,696 B4
Aydıncık 19,371 E4
Ayvalık 19,371 B3
Babaeski 18,145 B2
Bafra 50,213 F2
Bahçe 12,366 G4
Bakırköy 234,226 D6
Balıkesir 124,122 B3
Banaz 8,356 C3
Bandırma 53,497 B2
Bartın 20,728 E2
Başkale 9,770 K3
Batman 86,172 J4
Bayat 5,366 F2
Bayburt 22,578 J2
Bayındır 12,440 B3
Bayramiç 7,854 B3
Bernama 34,716 B3
Beşiktaş 188,117 D6
Besni 15,833 G4
Beykoz 94,101 D5
Beyoğlu 223,360 D6
Beypazarı 16,971 D2
Beyşehir 15,845 D4
Biga 16,359 B2
Bigadiç 8,955 C3
Bilecik 15,108 D2
Bingöl (Çapakçur) 27,904 J3
Birecik 20,412 H4
Bismil 19,059 J4
Bitlis 27,114 J3
Bodrum 32,517 B4
Boğazlıyan 10,827 F3
Bolu 38,400 D2
Bolvadin 30,599 D3
Bor 45,480 F4
Bornova 60,397 B3
Boyabat 14,797 F2
Bozdoğan 7,682 C4
Bozova 5,510 H4
Bozüyük 18,052 C3
Bucak 18,852 D4
Bulancak 16,089 H2
Bulanık 9,140 K3
Buldan 10,939 C3

(continued on following page)

Agriculture, Industry and Resources

DOMINANT LAND USE

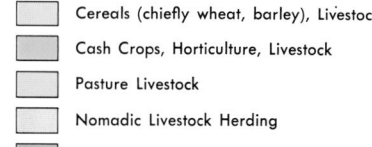

- Cereals (chiefly wheat, barley), Livestock
- Cash Crops, Horticulture, Livestock
- Pasture Livestock
- Nomadic Livestock Herding
- Forests
- Nonagricultural Land

MAJOR MINERAL OCCURRENCES

Ab	Asbestos	Na	Salt
Al	Bauxite	O	Petroleum
C	Coal	P	Phosphates
Cr	Chromium	Pb	Lead
Cu	Copper	Py	Pyrites
Fe	Iron Ore	Sb	Antimony
Hg	Mercury	Zn	Zinc
Mg	Magnesium		

Water Power
Major Industrial Areas

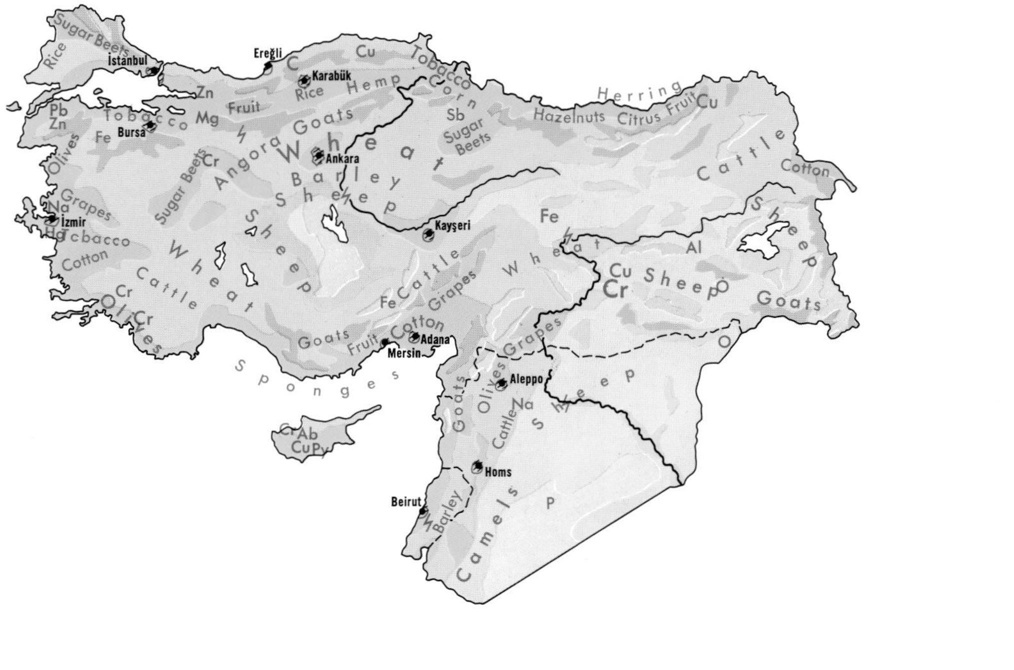

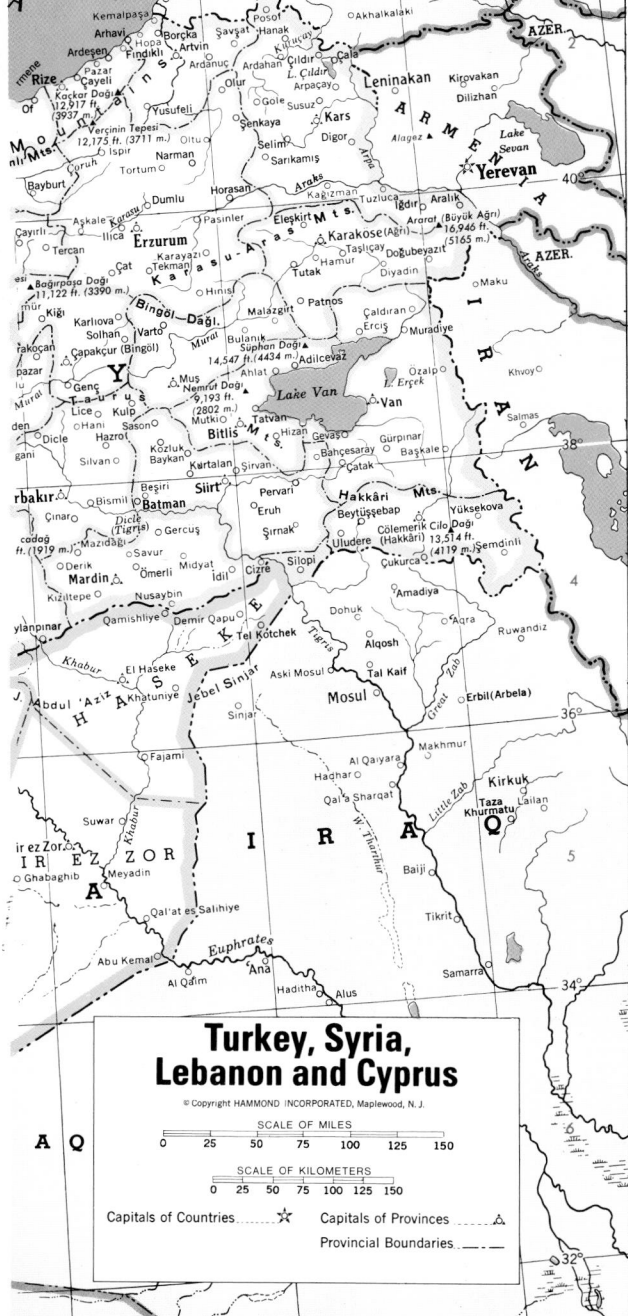

Topography

0 — 100 — 200 MI.

0 — 100 — 200 KM.

Below Sea Level	100 m. 328 ft.	200 m. 656 ft.	500 m. 1,640 ft.	1,000 m. 3,281 ft.	2,000 m. 6,562 ft.	5,000 m. 16,404 ft.

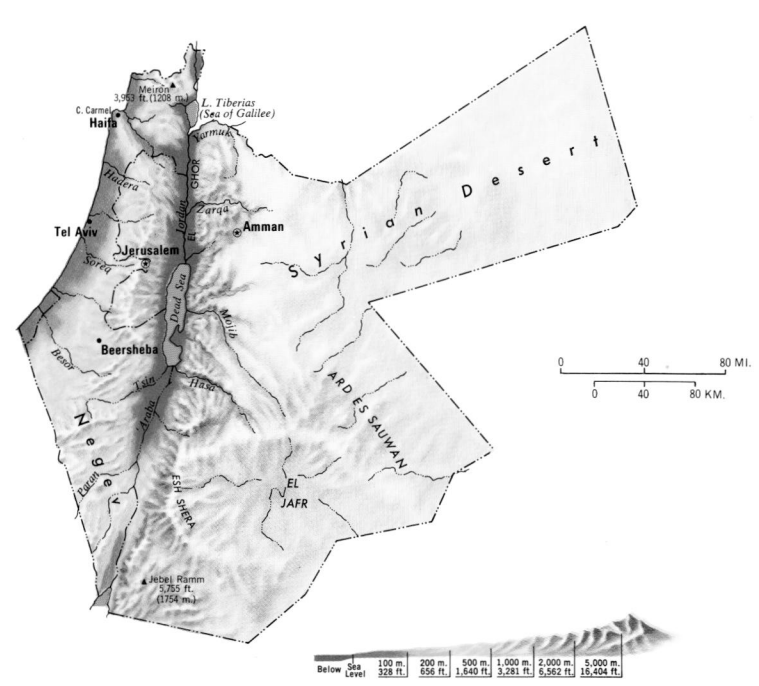

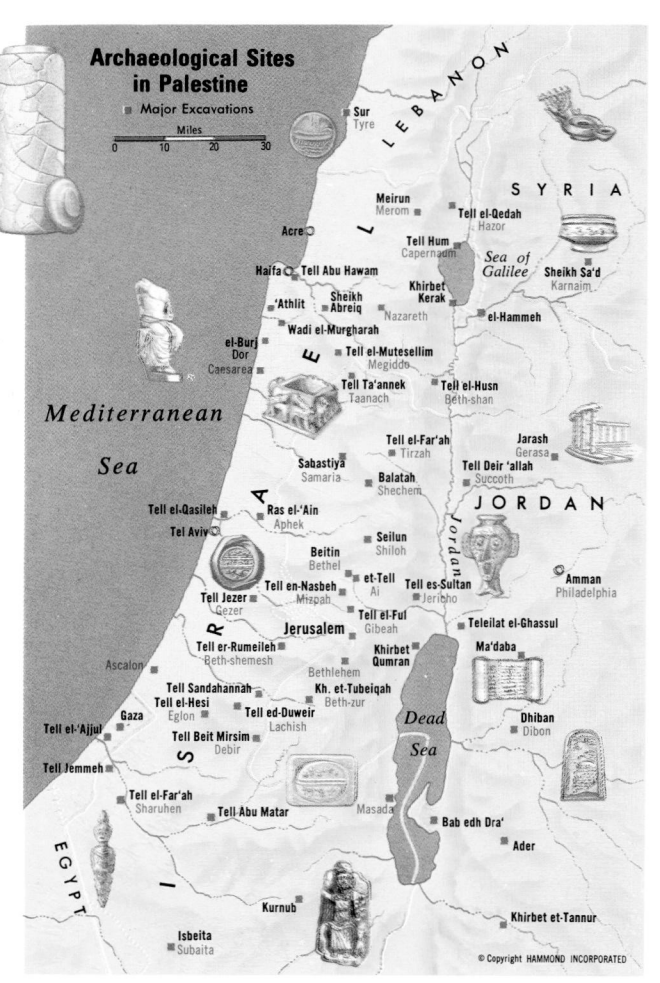

Archaeological Sites in Palestine

■ Major Excavations

Agriculture, Industry and Resources

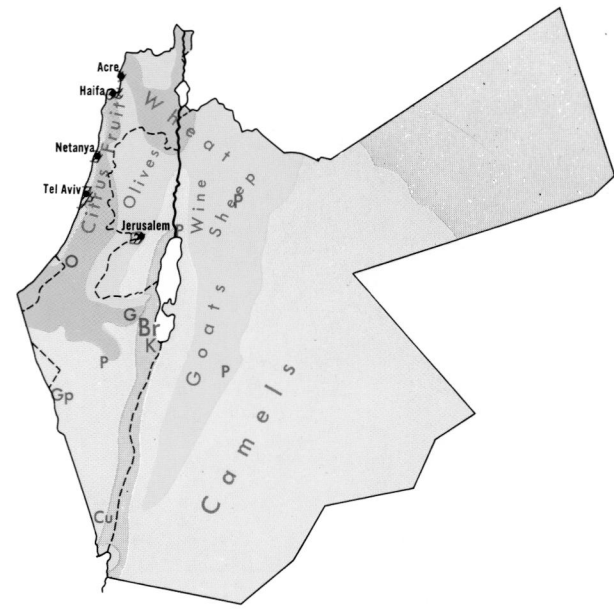

DOMINANT LAND USE

Cereals, Livestock

Cash Crops, Horticulture

Nomadic Livestock Herding

Nonagricultural Land

MAJOR MINERAL OCCURRENCES

Br Bromine
Cu Copper
G Natural Gas
Gp Gypsum

K Potash
O Petroleum
P Phosphates

////// Major Industrial Areas

© Copyright HAMMOND INCORPORATED

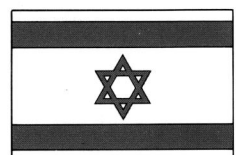

ISRAEL

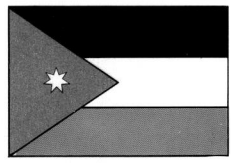

JORDAN

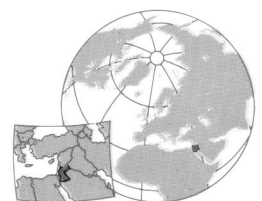

ISRAEL

AREA 7,847 sq. mi. (20,324 sq. km.)
POPULATION 4,625,000
CAPITAL Jerusalem
LARGEST CITY Tel Aviv
HIGHEST POINT Meiran 3,963 ft.
  (1,208 m.)
MONETARY UNIT shekel
MAJOR LANGUAGES Hebrew, Arabic
MAJOR RELIGIONS Judaism, Islam,
  Christianity

JORDAN

AREA 35,000 sq. mi.
  (90,650 sq. km.)
POPULATION 2,779,000
CAPITAL Amman
LARGEST CITY Amman
HIGHEST POINT Jeb. Ramm 5,755 ft.
  (1,754 m.)
MONETARY UNIT Jordanian dinar
MAJOR LANGUAGE Arabic
MAJOR RELIGION Islam

GAZA STRIP

CITIES and TOWNS

'Abasan 1,481 ....A5
Bani Suheila 7,561 ....A5
Beit Hanun 4,756 ....A4
Deir el Balah 10,854 ....A5
Deir el Balah* 18,118............A5
Gaza 87,793 ....A5
Gaza* 118,272 ....A4
Jabaliya 10,508 ....A4
Jabaliya* 43,604 ....A5
Khan Yunis 29,522 ....A5
Khan Yunis* 52,997 ....A5
Rafah 10,812 ....A5
Rafah* 49,812 ....A5

WEST BANK

CITIES and TOWNS

'Ajja 1,322 ....C3
'Anabta 3,426 ....C3
Anin 914 ....C2
'Aqqaba 1,127 ....C3
'Aqraba 2,501 ....C3
Ariha (Jericho) 5,312 ....C4
'Arraba 4,231 ....C3
'Arura 849 ....C3
'Attil 3,808 ....C3
Beit Fajjar 2,474 ....C4
Beit Hanina 1,177 ....C4
Beit Jala 6,041 ....C4
Beit Lahm
  (Bethlehem) 14,439............C4
Beit Sahur 5,380 ....C4
Bethlehem 14,439................C4
Biddu 1,259 ....C4
Birqin 2,477 ....C3
Bir Zeit 2,311 ....C4
Burqa 2,477 ....C3
Deir Ballut 1,058 ....C3
Deir Sharaf 973 ....C3
Dhahiriya 4,875 ....B5
Dura 4,954 ....C4
El Bira 9,674 ....C4
El Bira* 13,037 ....C4
El Khalil (Hebron) 38,309......C4
Er Rihiya 679 ....C4
Ez Zababida 1,474 ....C3
Halhul 6,041 ....C4
Hebron 38,309 ....C4
Idna 3,713 ....B4
Jaba 2,817 ....C3
Jalama 784 ....C3
Jalbun 914 ....C3
Jenin 8,346 ....C3
Jenin* 13,365 ....C3
Jericho 5,312 ....C4
Jericho* 6,931 ....C4
Jifna 655 ....C4
Kharas 1,364 ....C4
Nablus (Nabulus) 41,799......C3
Nahhalin 1,109 ....C4
Ni'lin 1,227 ....C4
Qabalan 1,970 ....C3
Qabatiya 6,005 ....C3
Qaffin 2,480 ....C3
Qalqiliya 8,926 ....C4
Qibya 926 ....C3
Rafidiya 1,123 ....C3
Ramallah 12,134...................C4
Rammin 1,198 ....C4
Rantis 897 ....C3
Salfit 3,201 ....C3
Samu 3,784 ....C4
Shu'fat 14,000 ....C4
Shuweika 2,332 ....C3
Silat Dhahr 2,104 ....C3
Sinjil 1,823 ....C3
Siris 1,285 ....C3
Tammun 2,952 ....C3
Tarqumiya 2,412...................C4
Tubas 5,262 ....C3
Tulkarm 10,255 ....C3
Tulkarm* 15,275 ....C3
Tur 12,200 ....C4
Ya'bad 4,857 ....C3
Yamun 4,384 ....C3
Yatta 7,281 ....C5

OTHER FEATURES

Ebal (mt.) C3
Golan Heights (reg.) D1
Judaea (reg.) C4
Khirbet Qumran (site) C4
Mashash, Wadi (riv.) C4
Samaria (reg.) C4
Tell 'Asur (mt.) C4
West Bank (reg.) C3

JORDAN

GOVERNORATES

Amman 1,000,000.................D4
El Balqa 113,000..................D4
El Karak 93,000...................E5
Irbid 506,000........................D3
Ma'an 62,000.......................D5

CITIES and TOWNS

'Ajlun⊙ 42,000 ....D3
Amman (cap.) 711,850..........D4
'Anjara 3,163 ....D3
Aqaba 15,000 ....D6
Bal'ama 769 ....E3
Baqura 3,042 ....D2
Damiya 483 ....D3
Dana 844 ....E5
Deir Abu Sa'id 1,927............D3
El 'Al 492 ....D4
El Husn 3,728 ....D3
El Karak 10,000...................E4
El Kitta 987 ....D3
El Mafraq 15,500..................E3
Er Ramtha 19,000................E2
Er Ruseifa 6,200 ....E3
Es Salt 24,000 ....D4
Es Sukhna 649 ....E3
Esh Shaubak 4,634..............D5
Et Tafila 17,000....................E5
Ez Zarqa' 263,400................E3
Harima 635 ....D3
Hawara 2,342 ....D3
Hisban 718 ....D4
'Ibbin 1,364 ....D3
Irbid 136,770........................D3
Jarash 29,000......................D3
Kitim 1,026...........................D3
Kufrinja 3,922 ....D3
Ma'da 125.............................D2
Ma'an 9,500.........................E5
Ma'daba 22,600....................D4
Ma'in 1,271 ....D4
Mazra'..................................D5
Na'ur 2,382...........................D4
Qumeim 955.........................D2
Ra's en Naqb 225.................E5
Safi.......................................E5
Safut 4,210 ....D3
Samar 716 ....D2
Sarih 3,390 ....D2
Subeihi 514 ....D3
Suweilih 3,457 ....D4
Suweima 315........................D4
Um Jauza 582 ....D3
Wadi es Sir 4,455.................D4
Wadi Musa 654 ....E5
Waqqas 2,321 ....D3

OTHER FEATURES

'Ajlun, Jebel (range) D3
Aqaba (gulf) D6
'Araba, Wadi (valley) D5
Dead (sea) C4
Hasa, Wadi el (dry riv.) E5
Jordan (riv.) D3
Nebo (mt.) D4
Petra (ruins) D5
Ramm, Jebel (mt.) D5
Shallala, Wadi esh (dry riv.)..D2
Shu'eib, Wadi (dry riv.) D4
Zarqa' (riv.) D3

*City and suburbs
⊙ Population of subdivision

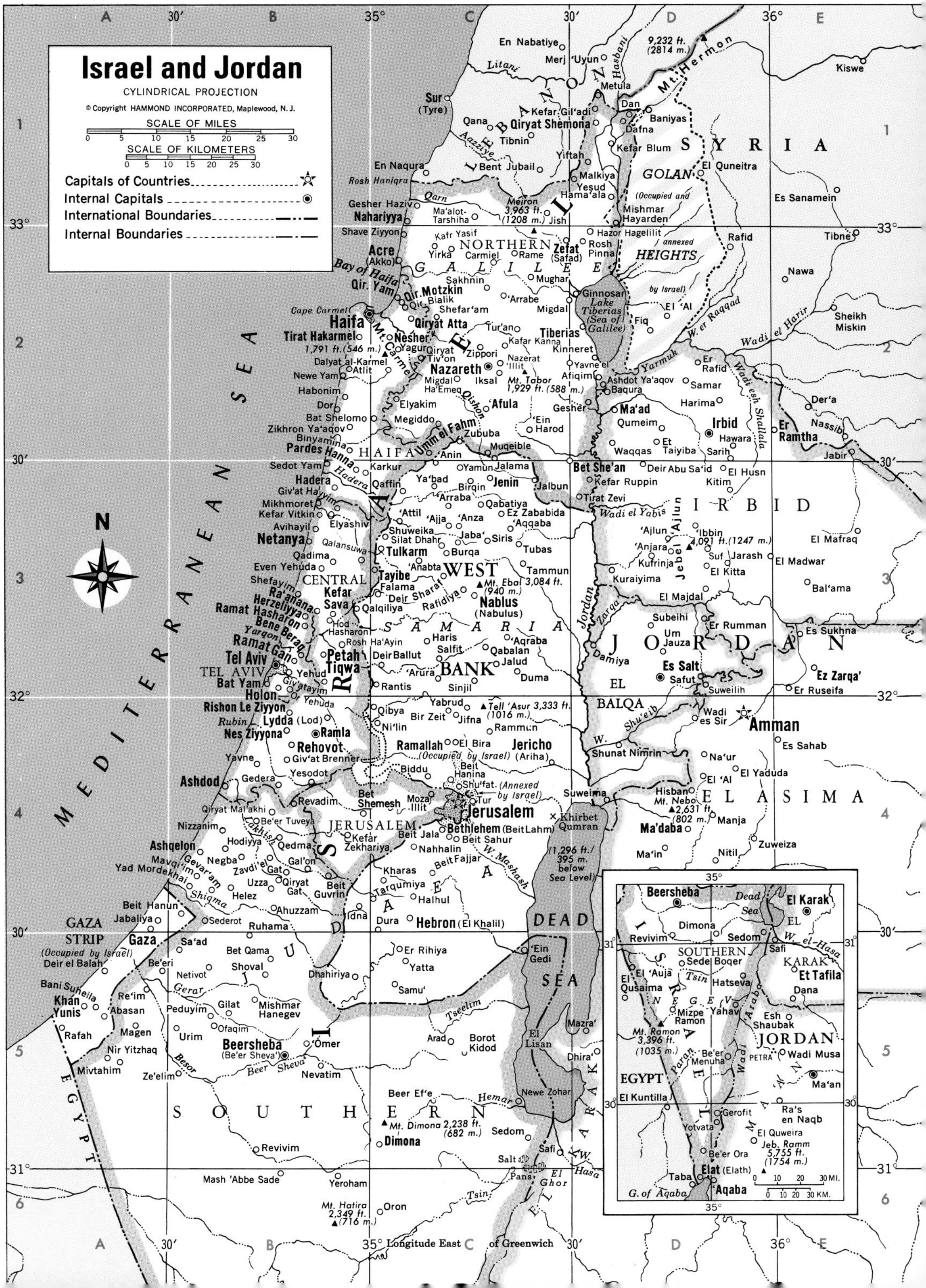

IRAN

INTERNAL DIVISIONS

Azerbaijan, East (prov.)
3,194,543................E1
Azerbaijan, West (prov.)
1,404,875................D1
Bakhtaran (prov.) 1,016,199...E3
Bakhtiari (governorate)
394,300................F4
Bushehr (prov.) 345,427......G6
Central (Markazi) (prov.)
6,921,283................G3
Esfahan (Isfahan) (prov.)
1,974,938................H4
Fars (prov.) 2,020,947......H6
Gilan (prov.) 1,577,800......F2
Hamadan (governorate)
1,086,512................F3
Hormozgan (prov.) 463,419...J7
Ilam (governorate) 244,222...E4
Isfahan (prov.) 1,974,938...H4
Kerman (prov.) 1,088,045...K6
Khorasan (prov.) 3,266,650...K3
Khuzestan (prov.) 2,176,612...F5
Kohkiluyeh and Boyer
Ahmediyeh (governorate)
244,750................G5
Kordestan (Kurdistan) (prov.)
781,889................E3

Lorestan (Luristan)
(governorate) 924,848....F4
Mazandaran (prov.)
2,384,226................H2
Semnan (governorate)
485,875................J3
Sistan and Baluchestan
(prov.) 659,297........M6
Yazd (governorate) 356,218...J5
Zanjan (governorate) 579,000..F2

CITIES and TOWNS

Abadan 296,081.............F5
Abadeh 16,000.............H5
Abhar 24,000.............F2
Agha Jari 24,195.............F5
Ahar 24,000.............F1
Ahvaz (Ahwaz) 329,006....F5
Amol 68,782.............H2
Andimeshk 16,000.............F4
Arak 114,507.............F3
Ardabil 147,404.............F1
Asterabad (Gorgan) 88,348...J2
Babol 67,790.............H2
Bakhtaran (Kermanshah)
290,861................E3
Bam 22,000.............L6
Bandar 'Abbas 89,103......J7
Bandar Behesti (Bahar)
1,800................M8

Bandar-e Anzali (Enzeli)
55,978................F2
Bandar-e Bushehr (Bushire)
57,681................G6
Bandar-e Khomeyni 6,000...F5
Bandar-e Lingeh 4,920......J7
Bandar-e Ma'shur 17,000...F5
Bandar-e Torkeman 13,000..H2
Behbehan 39,874.............F5
Behshahr 26,032.............H2
Birjand 25,854.............L4
Bojnurd 31,248.............K2
Borazjan 20,000.............G6
Borujerd 100,103.............F4
Bostan 4,619.............F5
Chalus 15,000.............G2
Damghan 13,000.............J2
Dasht-e Azadegan
(Susangerd) 21,000......F5
Dizful (Dezful) 110,287....F4
Duzdab (Zahedan) 92,628...M6
Emamshahr (Shahrud)
30,767................J2
Enzeli 55,978.............F2
Esfahan (Isfahan) 671,825...G4
Eslamabad 12,000.............E3
Estahbanat 18,187.............H6
Fahrej (Iranshahr) 5,000...M7
Fasa 19,000.............H6
Ganaveh 9,000.............G6
Garmsar 4,723.............H3

Ghaemshahr 63,289.........H2
Golpayegan 20,515.........G4
Golshan (Tabas) 10,000....K4
Gonbad-e Kavus 59,868....J2
Gorgan (Gurgan) 88,348....J2
Hamadan 155,846.........F3
Hormoz 2,569.............K7
Iranshahr 5,000.............M7
Isfahan 671,825.............G4
Jahrom 38,236.............H6
Karaj 138,774.............G3
Kashan 84,545.............G3
Kashmar 17,000.............K3
Kazerun 51,309.............G6
Kazvin (Qazvin) 138,527...F2
Kerman 140,309.............K5
Khomeinishar 46,836.........G4
Khorramabad 104,928......F4
Khorramshahr 146,709......F5
Khvoy (Khoi) 70,040......D1
Lahijan 25,725.............F2
Lar 22,000.............J7
Mahabad 28,610.............D2
Malayer 28,434.............F3
Marageh 60,820.............E2
Marand 24,000.............D1
Marv Dasht 25,498.........H6
Masjed Soleyman 77,161...F5
Mashhad (Meshed) 670,180..L2
Miandowab 19,000.........E2
Mianeh 28,447.............E2

Nahavand 24,000.............F3
Najafabad 76,236.............G4
Nasratabad (Zabol) 20,000..M5
Neyriz 16,114.............J6
Nishapur (Neyshabur) 59,101..L2
Nosratabad 20,000.........M5
Orumiyeh (Urmia) 163,991...D2
Pahlevi (Enzeli) 55,978....F2
Qayen 6,000.............L4
Qazvin 138,527.............F2
Quchan 29,133.............L2
Qum (Qom) 246,831......G3
Rafsanjan 21,000.............K5
Resht (Rasht) 187,203......F2
Rey 102,825.............G3
Reza'iyeh (Urmia) 163,991...D2
Sa'idabad 20,000.............J6
Sabzevar 69,174.............K2
Sakht-Sar 12,000.............G2
Salmas 13,161.............D1
Sanandaj 95,834.............E3
Saqqez 17,000.............E3
Sari 70,936.............H2
Savanat (Estahbanat) 18,187..J6
Saveh 17,565.............G3
Semnan 31,058.............H3
Shahr Kord 24,000.........G4
Shahreza 34,220.............H4
Shahrud 30,767.............J2
Shiraz 416,408.............H6
Shushtar 24,000.............F4

Sinneh (Sanandaj) 95,834...E3
Sirjan (Sai'dabad) 20,000...J6
Sultanabad (Kashmar)
17,000................K3
Susangerd 21,000.............F5
Tabas 10,000.............K4
Tabriz 598,576.............D2
Tajrish 157,486.............G3
Tehran (cap.) 4,496,159...G3
Torbat-e Heydariyeh 30,106...L3
Urmia 163,991.............D2
Yazd (Yezd) 135,978......J5
Zabol 20,000.............M5
Zahedan 92,628.............M6
Zenjan (Zanjan) 99,967...F2

OTHER FEATURES

'Arabi (isl.)...............G7
Araks (Aras) (riv.).......E1
Atrak (Atrek) (riv.).......J2
Azerbaijan (reg.).........D1
Bakhtegan (lake).........J6
Baluchistan (reg.).........M7
Bampur (riv.).............M7
Behistun (ruins).........E3
Caspian Sea (sea).........G1
Damavand (Demavend)
(mt.)................H2
Daryacheh-ye Namak
(salt lake)...........G3

Daryacheh-ye Sistan
(salt lake)...........M5
Dasht-e Kavir (salt des.)...J3
Dasht-e Lut (des.).........L5
Dez (riv.)...............F5
Elburz (mts.).............G2
Farsi (isl.).............G7
Gabrik (riv.).............L7
Gamas Ab (riv.).........E3
Gavkhuni (marsh).........H4
Gorgan (riv.).............J2
Hamun-e Saberi (lake)...M5
Harirud (riv.)...........M3
Hormoz (riv.).............K7
Hormuz (str.)...........K7
Jaba Rud (riv.).........L2
Joveyn (riv.).............K2
Kabir Kuh (mts.).........E4
Karkheh (riv.)...........E4
Karun (riv.).............F5
Khark (Kharg) (isl.)......G6
Khusf Rud (riv.).........L4
Khvojeh Lak, Kuh-e (mt.)...E2
Kor (riv.)...............H6
Laristan (reg.)...........H6
Makran (reg.)...........M8
Mand (riv.).............H7
Mand (riv.).............G6
Mashkid (riv.)...........N7
Mehran (riv.)...........J7
Nahang (riv.)...........N7

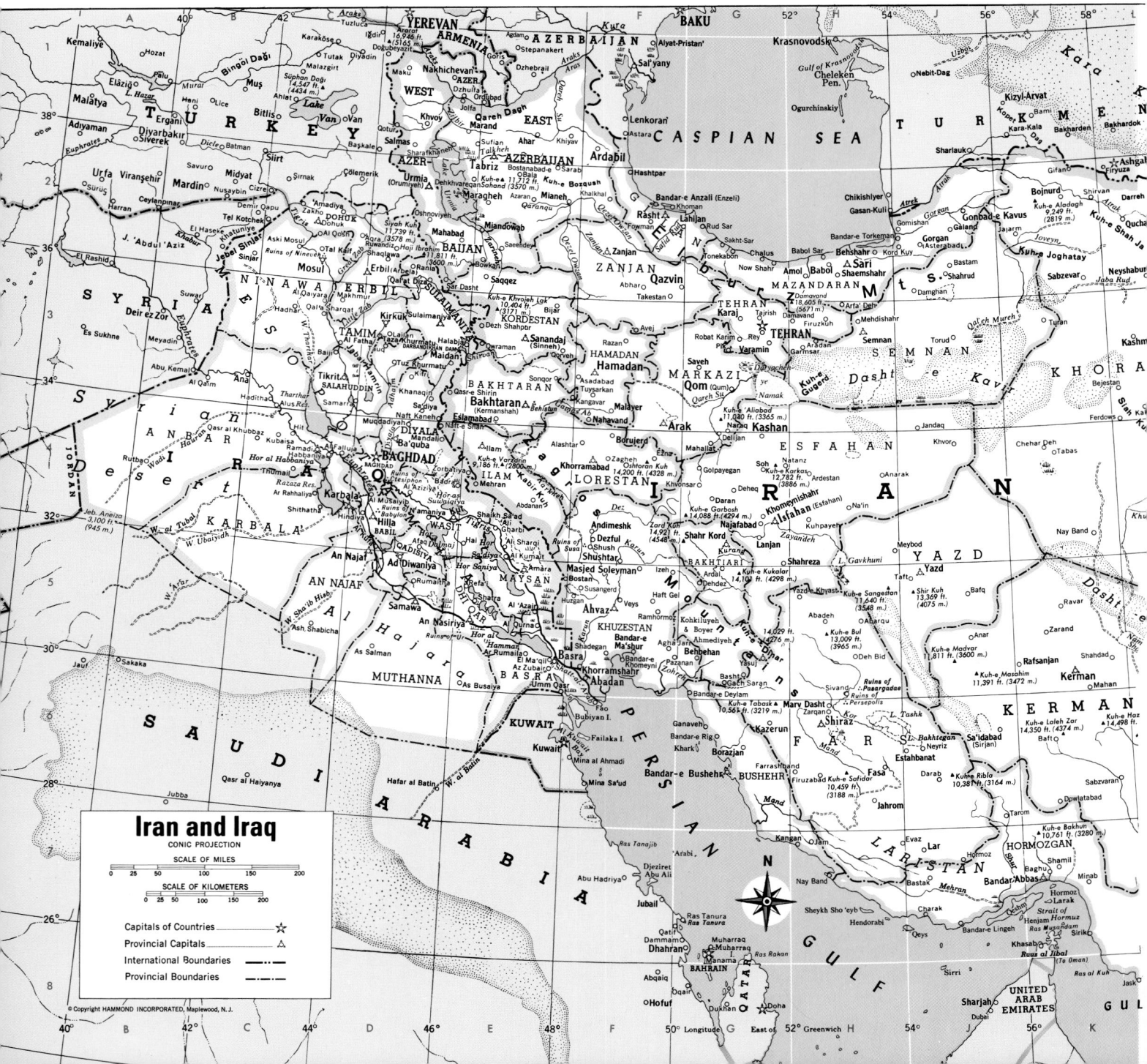

© Copyright HAMMOND INCORPORATED, Maplewood, N.J.

Iran and Iraq
CONIC PROJECTION
SCALE OF MILES
0 25 50 100 150 200
SCALE OF KILOMETERS
0 25 50 100 150 200

Capitals of Countries.........★
Provincial Capitals...........△
International Boundaries......_____
Provincial Boundaries........_ _ _ _

IRAN

IRAQ

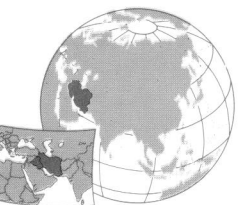

AREA 636,293 sq. mi. (1,648,000 sq. km.)
POPULATION 55,208,000
CAPITAL Tehran
LARGEST CITY Tehran
HIGHEST POINT Damavand 18,605 ft. (5,671 m.)
MONETARY UNIT Iranian rial
MAJOR LANGUAGES Persian, Azerbaijani, Kurdish
MAJOR RELIGION Islam

AREA 172,476 sq. mi. (446,713 sq. km.)
POPULATION 16,335,000
CAPITAL Baghdad
LARGEST CITY Baghdad
HIGHEST POINT Haji Ibrahim 11,811 ft.
  (3,600 m.)
MONETARY UNIT Iraqi dinar
MAJOR LANGUAGES Arabic, Kurdish
MAJOR RELIGION Islam

Topography

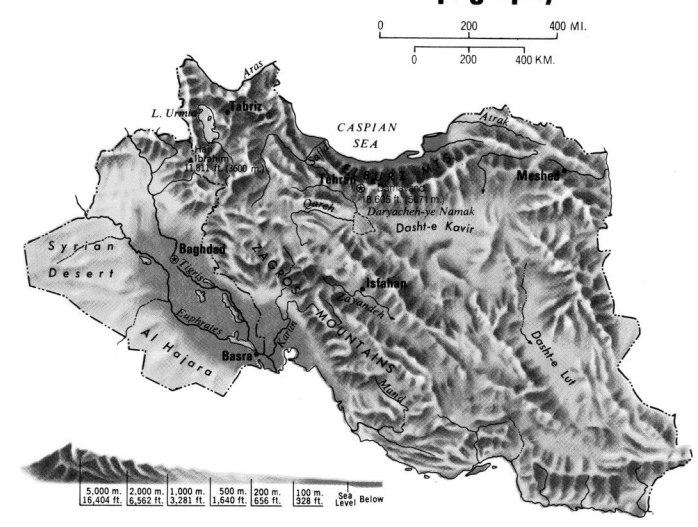

Agriculture, Industry and Resources

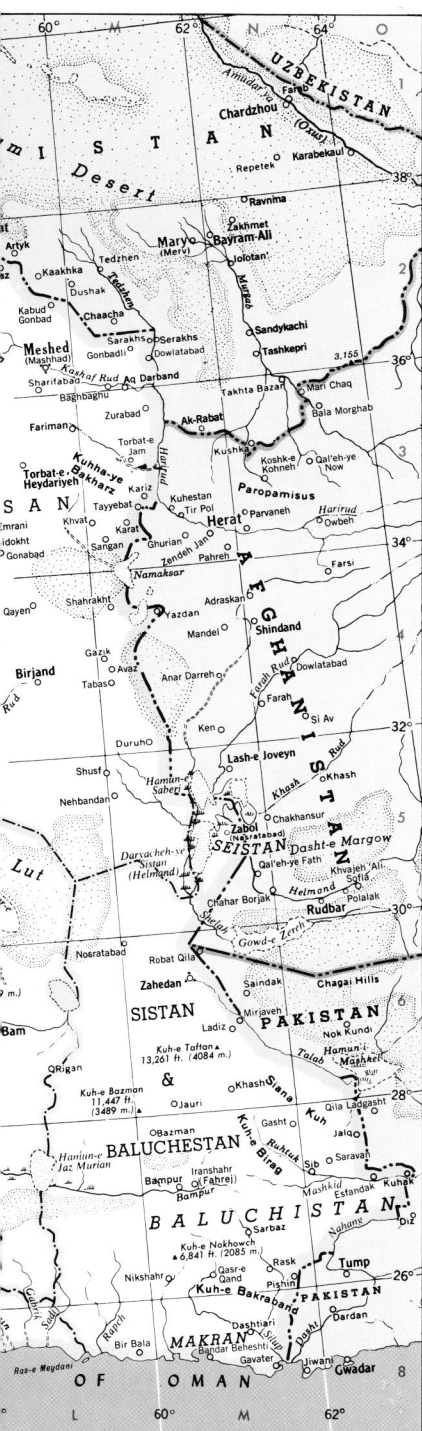

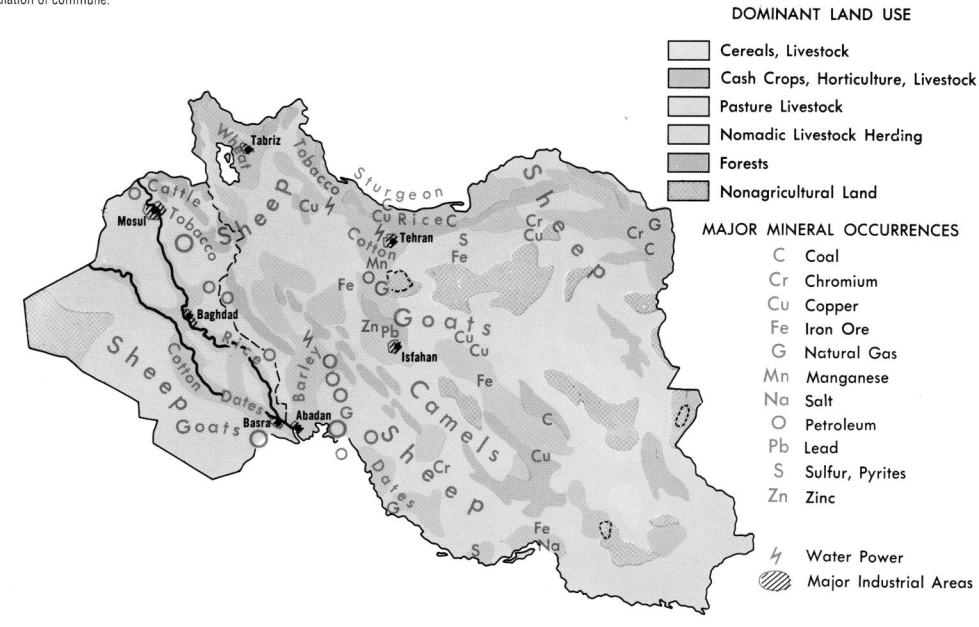

DOMINANT LAND USE

Cereals, Livestock
Cash Crops, Horticulture, Livestock
Pasture Livestock
Nomadic Livestock Herding
Forests
Nonagricultural Land

MAJOR MINERAL OCCURRENCES

C   Coal
Cr   Chromium
Cu   Copper
Fe   Iron Ore
G   Natural Gas
Mn   Manganese
Na   Salt
O   Petroleum
Pb   Lead
S   Sulfur, Pyrites
Zn   Zinc

Water Power
Major Industrial Areas

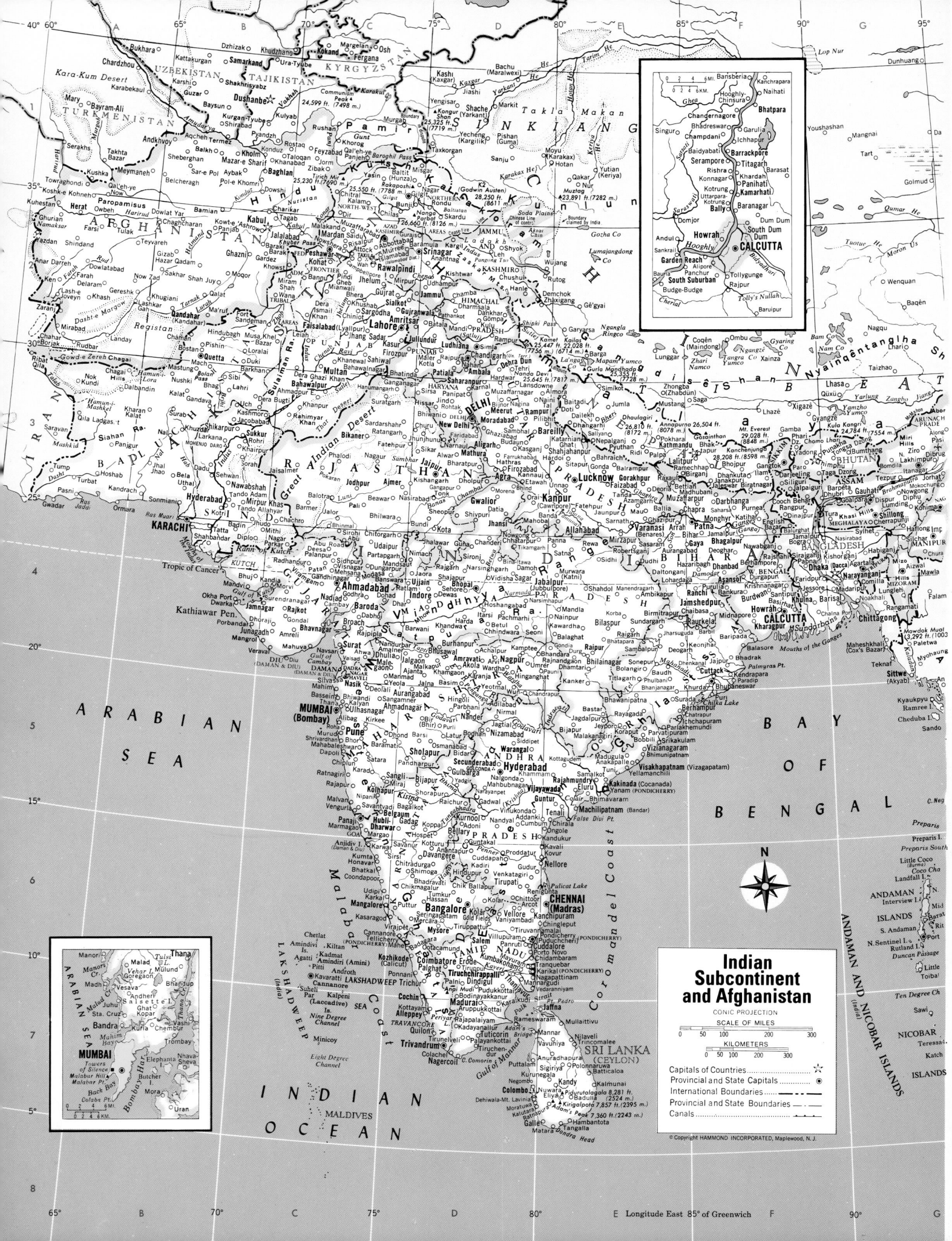

Indian Subcontinent and Afghanistan

CONIC PROJECTION

SCALE OF MILES

0 50 100   200   300

KILOMETERS

0 50 100 200 300

Capitals of Countries............................☆

Provincial and State Capitals..............◉

International Boundaries..............━ ━ ━

Provincial and State Boundaries..━ ·· ━

Canals...━━━━

© Copyright HAMMOND INCORPORATED, Maplewood, N.J.

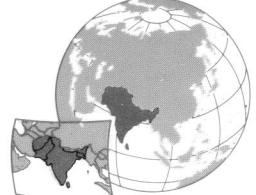

INDIA

AREA 1,269,339 sq. mi. (3,287,588 sq. km.)
POPULATION 843,930,861
CAPITAL New Delhi
LARGEST CITY Calcutta (greater)
HIGHEST POINT Nanda Devi 25,645 ft. (7,817 m.)
MONETARY UNIT Indian rupee
MAJOR LANGUAGES Hindi, English, Bengali, Telugu, Marathi, Tamil, Urdu, Gujarati, Malayalam, Kannada, Oriya, Punjabi, Assamese, Kashmiri, Sindhi
MAJOR RELIGIONS Hinduism, Islam, Christianity, Sikhism, Buddhism, Jainism, Zoroastrianism, Animism

PAKISTAN

AREA 310,403 sq. mi. (803,944 sq. km.)
POPULATION 112,050,000
CAPITAL Islamabad
LARGEST CITY Karachi
HIGHEST POINT K2 (Godwin Austen) 28,250 ft. (8,611 m.)
MONETARY UNIT Pakistani rupee
MAJOR LANGUAGES Urdu, English, Punjabi, Pushtu, Sindhi, Baluchi, Brahui
MAJOR RELIGIONS Islam, Hinduism, Sikhism, Christianity, Buddhism

SRI LANKA (CEYLON)

AREA 25,332 sq. mi. (65,610 sq. km.)
POPULATION 16,806,000
CAPITAL Colombo
LARGEST CITY Colombo
HIGHEST POINT Pidurutalagala 8,281 ft. (2,524 m.)
MONETARY UNIT Sri Lanka rupee
MAJOR LANGUAGES Sinhala, Tamil, English
MAJOR RELIGIONS Buddhism, Hinduism, Christianity, Islam

AFGHANISTAN

AREA 250,775 sq. mi. (649,507 sq. km.)
POPULATION 15,814,000
CAPITAL Kabul
LARGEST CITY Kabul
HIGHEST POINT Nowshak 24,557 ft. (7,485 m.)
MONETARY UNIT afghani
MAJOR LANGUAGES Pushtu, Dari, Uzbek
MAJOR RELIGION Islam

NEPAL

AREA 54,663 sq. mi. (141,577 sq. km.)
POPULATION 18,442,000
CAPITAL Kathmandu
LARGEST CITY Kathmandu
HIGHEST POINT Mt. Everest 29,028 ft. (8,848 m.)
MONETARY UNIT Nepalese rupee
MAJOR LANGUAGES Nepali, Maithili, Tamang, Newari, Tharu
MAJOR RELIGIONS Hinduism, Buddhism

MALDIVES

AREA 115 sq. mi. (298 sq. km.)
POPULATION 206,000
CAPITAL Male
LARGEST CITY Male
HIGHEST POINT 20 ft. (6 m.)
MONETARY UNIT Maldivian rufiyaa
MAJOR LANGUAGE Divehi
MAJOR RELIGION Islam

BHUTAN

AREA 18,147 sq. mi. (47,000 sq. km.)
POPULATION 1,483,000
CAPITAL Thimphu
LARGEST CITY Thimphu
HIGHEST POINT Kula Kangri 24,784 ft. (7,554 m.)
MONETARY UNIT ngultrum
MAJOR LANGUAGES Dzongka, Nepali
MAJOR RELIGIONS Buddhism, Hinduism

BANGLADESH

AREA 55,126 sq. mi. (142,776 sq. km.)
POPULATION 106,507,000
CAPITAL Dhaka
LARGEST CITY Dhaka
HIGHEST POINT Keokradong 4,034 ft. (1,230 m.)
MONETARY UNIT taka
MAJOR LANGUAGES Bengali, English
MAJOR RELIGIONS Islam, Hinduism Christianity

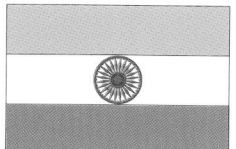

INDIA

PAKISTAN

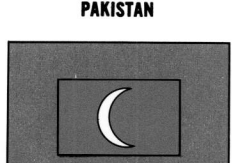

SRI LANKA (CEYLON)

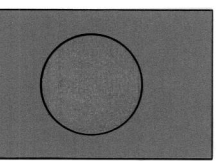

BHUTAN

AFGHANISTAN   **MALDIVES**   **BANGLADESH**   **NEPAL**

AFGHANISTAN

CITIES and TOWNS

Andkhvoy 13,137 ....A1
Aybak 33,016 ....B1
Baghlan 75,130 ....B1
Bamian 7,355 ....B2
Chaghcharan 2,974 ....B2
Charikar 25,093 ....B1
Farah 18,797 ....A2
Feyzabad 10,142 ....C1
Gardez 11,415 ....B2
Gereshk A2
Ghazni 30,425 ....B2
Ghurian 12,404 ....A2
Hazar Qadam B2
Herat 163,960 ....A2
Jalalabad 56,384 ....B2
Kabul (cap.) 905,108 ....B2
Kalat (Qalat) 5,946 ....B2
Kandahar (Qandahar) 178,409 ....B2
Khanabad 26,803 ....B1
Khash................................A2
Kholm 28,078 ....B1
Khowst B2
Konduz 107,191 ....B1
Kuhestan..........................A2
Lashkar Gah 26,646 ....A2
Mazar-e Sharif 122,567 ..B1
Meymaneh 54,954 ....A1
Mirabad A2
Oruzgan (Hazar Qadam)...B2
Panjab..............................B2
Pol-e Khomri 31,101 ....B1
Qalat 5,946 ....B2
Qal'eh-ye Now 5,340 ....A1
Qandahar 178,409 ....B2
Qonduz (Konduz) 107,191..B1
Sakhar B2
Sar-e Pol 15,699 ....B1
Sheberghan 54,870 ....A1
Shindand A2
Tagab................................B2
Taloqan 46,202 ....B1
Zaranj 6,477 ....A2

OTHER FEATURES

Baroghil (pass) C1
Chagai (hills)....................A3
Margow, Dasht-e (des.)....A2
Farah Rud (riv.) A2
Gowd-e Zereh (depr.) A3
Harirud (riv.) A1
Helmand (riv.) B2
Hindu Kush (mts.) B1
Kabul (riv.) C1
Konar (riv.) C1
Konduz (riv.) B1
Lurah (riv.) B2
Namaksar (salt lake) A2
Nuristan (reg.) C1

Panj (riv.)...........................C1
Paropamisus (range) A2
Qonduz (Konduz) (riv.) B1
Registan (reg.)...................A2
Tarnak (riv.).......................B2

BANGLADESH

CITIES and TOWNS

Barisal 159,298 ....G4
Bogra 68,237 ....F4
Chalna Port 14,590 ....F4
Chittagong 1,388,476 ......G4
Comilla 126,130 ....G4
Cox's Bazar G4
Dhaka (cap.) 3,458,602 ...G4
Dinajpur 96,348 ....F3
Faridpur 66,911 ....F4
Habiganj 16,281 ....G4
Jamalpur 89,847 ....F4
Jessore 149,426 ....F4
Khulna 623,184 ....F4
Kishorganj 52,081 ....G4
Madaripur 58,645 ....G4
Maheshkhali 29,530 ....G4
Mymensingh (Nasirabad) 107,863 ....G4
Narayanganj 196,139.......G4
Nasirabad 107,863 ....G4
Nawabganj 65,286 ....F4
Noakhali 32,490...............G4
Pabna 101,080 ....F4
Rajshahi 171,600 ....F4
Rangamati 36,490 ....G4
Rangpur 72,829 ....F3
Sirajganj 74,457 ....F4
Sylhet 59,546 ....G4

OTHER FEATURES

Bengal, Bay of (bay) F5
Brahmaputra (riv.) G3
Ganges (riv.) F3
Ganges, Mouths of the (delta).F3
Mowdok Mual (mt.) G4
Sundarbans (reg.) F4

BHUTAN

CITIES and TOWNS

Bumthang 10,000 ....G3
Paro 35,000 ....F3
Punakha 12,000...............G3
Taga Dzong 18,000..........G3
Thimphu (cap.) 50,000......G3
Tongsa Dzong 2,500.........G3

OTHER FEATURES

Chomo Lhari (mt.)..............F3
Himalaya (mts.) E2
Kula Kangri (mt.) G3

(continued on following page)

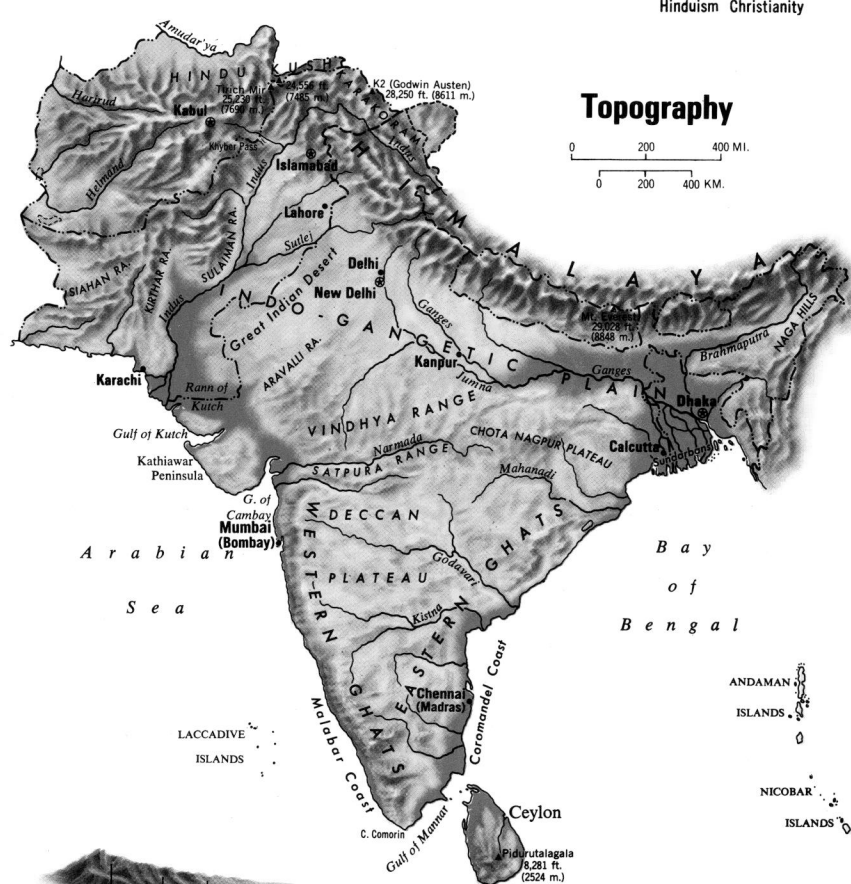

Topography

0   200   400 MI.
0   200   400 KM.

5,000 m. | 2,000 m. | 1,000 m. | 500 m. | 200 m. | 100 m. | Sea
16,404 ft. | 6,562 ft. | 3,281 ft. | 1,640 ft. | 656 ft. | 328 ft. | Level Below

(continued)

INDIA

INTERNAL DIVISIONS

Andaman and Nicobar Isls. (terr.) 188,741 ... G6
Andhra Pradesh (state) 53,549,673 ... D5
Arunachal Pradesh (state) 631,839 ... G3
Assam (state) 19,902,826 ... G3
Bihar (state) 69,914,734 ... F4
Chandigarh (terr.) 451,610 ... D2
Dādra and Nagar Haveli (terr.) 103,676 ... C4
Daman and Diu (state) ... C4
Delhi (terr.) 6,220,406 ... D3
Goa (state) ... C4
Gujarāt (state) 34,085,799 ... C4
Haryana (state) 12,922,618 ... D3
Himachal Pradesh (state) 4,280,818 ... D2
Jammu and Kashmir (state) 5,987,389 ... D2
Karnataka (state) 37,135,714 ... D6
Kerala (state) 25,453,680 ... D6
Lakshadweep (terr.) 40,249 ... C6
Madhya Pradesh (state) 52,178,844 ... D4
Mahārāshtra (state) 62,784,171 ... C5
Manipur (state) 1,420,953 ... G4
Meghalaya (state) 1,335,819 ... G3
Mizoram (state) 493,757 ... G4
Nagaland (state) 774,930 ... G3
Orissa (state) 26,370,271 ... E5
Pondicherry (terr.) 604,471 ... E6
Punjab (state) 16,788,915 ... D2
Rajasthan (state) 34,261,862 ... C3
Sikkim (state) 316,385 ... F3
Tamil Nadu (state) 48,408,077 ... D6
Tripura (state) 2,053,058 ... G4
Uttar Pradesh (state) 110,862,013 ... D3
West Bengal (state) 54,580,647 ... F4

CITIES and TOWNS

Abu 9,840 ... C4
Achalpur 42,326 ... D4
Adoni 108,905 ... D5
Agartala 131,513 ... G4
Agra 770,352 ... D3
Ahmadabad 2,515,195 ... C4
Ahmadnagar 181,239 ... C5
Ajmer 374,350 ... C3
Akola 225,402 ... D4
Alibag 11,913 ... C5
Aligarh 319,981 ... D3
Allahabad 642,420 ... E3
Alleppey-Cochin 160,166 ... D7
Almora 19,671 ... D3
Alwar 139,973 ... D3
Amalner 55,544 ... C4
Ambala 121,135 ... D2
Ambikapur 23,087 ... E4
Amravati 261,387 ... D4
Amreli 39,520 ... C4
Amritsar 589,229 ... C2
Anakapalle 57,273 ... E5
Anantapur 119,536 ... D6
Arrah 124,614 ... E3
Aruppukkottai 62,223 ... D7
Asansol 366,371 ... F4
Azamgarh 40,963 ... E3
Badagara 53,938 ... D6
Bagalkot 51,746 ... D5
Bahraich 102,580 ... E3
Baidyabati 54,130 ... F1
Balasore 46,239 ... F4
Ballia 47,101 ... E3
Bally 38,892 ... F1
Balurghat 67,088 ... F3
Banda 50,575 ... D3
Bandar (Machilipatnam) 138,525 ... E5
Bandra ... B7
Bangalore 4,100,000 ... D6
Bankura 79,129 ... F4
Bansberia 61,748 ... F1
Baranagar 136,842 ... F1
Barasat 42,642 ... F1
Bareilly 437,801 ... D3
Barmer 38,630 ... C3
Baroda (Vadodara) 744,043 ... C4
Barrackpore 96,889 ... F1
Barsi 62,374 ... D5
Barwani 22,099 ... D4
Basirhat 63,816 ... F4
Batala 100,790 ... D2
Beawar 66,114 ... C3
Belgaum 300,290 ... C5
Bellary 201,014 ... D5
Benares (Varanasi) 793,542 ... E3
Berhampore 100,150 ... F4
Berhampur 162,407 ... F5
Bettiah 51,018 ... E3
Bhadrak 40,487 ... F4
Bhadravati 130,459 ... D6
Bhadreswar 45,586 ... F1
Bhagalpur 221,276 ... F4
Bhandara 39,423 ... E4
Bharuch 91,589 ... C4
Bharatpur 105,239 ... D3
Bhatinda 127,450 ... C2
Bhatpara 204,750 ... F1
Bhavnagar 308,194 ... C4
Bhawanipatna 22,808 ... E5
Bhilainagar 157,173 ... E4
Bhilwara 122,338 ... C3
Bhimavaram 101,940 ... E5
Bhind 42,371 ... D3
Bhir (Bir) 49,965 ... D5
Bhiwandi 115,256 ... C5
Bhiwani 101,263 ... D3
Bhopal 672,329 ... D4
Bhubaneswar 219,419 ... F4
Bhuj 52,177 ... B4
Bhusawal 132,146 ... D4
Bidar 50,670 ... D5
Bihar 151,308 ... F3
Bijapur 146,808 ... D5
Bijnor 43,290 ... D3
Bikaner 280,356 ... C3
Bilaspur 186,885 ... E4
Bir 49,965 ... D5
Bodhan 37,589 ... D5
Bodinayakkanur 54,176 ... D6
Bolangir 35,748 ... E4
Broach (Bharuch) 91,589 ... C4
Budaun 72,204 ... D3
Budge-Budge 51,039 ... F2
Bundi 34,279 ... D3
Burdwan 143,318 ... F4
Burhanpur 141,142 ... D4
Calcutta 10,860,000 ... F2
Calicut (Kozhikode) 333,979 ... D6
Cambay 62,097 ... C4
Cannanore 157,777 ... C6
Cawnpore (Kanpur) 1,688,242 ... E3
Chaibasa 35,386 ... F4
Chamba 11,814 ... D2
Champdani 58,596 ... F1
Chanderi 10,294 ... D4
Chandernagore 75,238 ... F1
Chandigarh 421,256 ... D2
Chandrapur 115,352 ... D5
Chapra 111,461 ... F3
Chennai (Madras) 5,360,000 ... E6
Cherrapunji▲ 83,987 ... G3
Chhatarpur 32,271 ... D4
Chhindwara 53,492 ... D4
Chidambaram 48,811 ... E6
Chikmagalur 41,639 ... D6
Chinglept 38,419 ... E6
Chirala 54,487 ... E5
Chitradurga 50,254 ... D6
Chittoor 63,035 ... D6
Churachandpur 8,706 ... G4
Churu 52,502 ... D3
Cocanada (Kakinada) 226,642 ... E5
Cochin-Alleppey 439,066 ... D6
Coimbatore 917,155 ... D6
Colachel 18,819 ... D7
Cooch Behar 53,684 ... F3
Cuddalore 127,569 ... E6
Cuddapah 103,146 ... D6
Cuttack 326,468 ... F4
Dabhoi 37,892 ... C4
Damoh 59,489 ... D4
Darbhanga 175,879 ... F3
Darjeeling 42,873 ... F3
Datia 36,439 ... D3
Davangere 196,481 ... D6
Dehra Dun 293,628 ... D2
Delhi 8,380,000 ... D3
Deoghar 40,356 ... F4
Deolali 55,436 ... C5
Deoria 38,161 ... E3
Dewas 51,545 ... D4
Dhanbad 676,736 ... F4
Dhar 36,172 ... C4
Dharmsala 10,939 ... D2
Dharwar-Hubli 379,166 ... C5
Dhoraji 59,773 ... C4
Dhubri 36,503 ... G3
Dhulia 210,927 ... C4
Dibrugarh 80,348 ... G3
Dindigul 170,196 ... D6
Diphu 10,200 ... G3
Diu 6,214 ... C4
Dungarpur 19,773 ... C4
Durg 67,892 ... E4
Durg-Bhilainagar 490,158 ... E4
Durgapur 305,838 ... F4
Dwarka 17,801 ... B4
Eluru 168,148 ... E5
English Bazar 61,335 ... F3
Erode 275,103 ... D6
Etawah 112,426 ... D3
Faizabad-cum-Ayodhya 102,835 ... E3
Faridabad 326,968 ... D3
Farrukhabad-cum-Fatehgarh 160,927 ... D3
Firozabad 202,837 ... D3
Firozpur 49,545 ... C2
Gadag 116,596 ... D5
Ganganagar 121,516 ... C3
Gangtok 12,000 ... F3
Garden Reach 154,913 ... F2
Garulia 44,271 ... F1
Gauhati 123,783 ... G3
Gaya 246,778 ... F4
Ghaziabad 291,995 ... D3
Ghazipur 45,635 ... E3
Godhra 66,403 ... C4
Gonda 52,662 ... E3
Gondal 54,928 ... C4
Gondia 100,342 ... E4
Gorakhpur 306,399 ... E3
Gulbarga 218,621 ... D5
Guna 40,006 ... D4
Guntakal 66,320 ... D5
Guntur 367,219 ... D5
Gwalior 38,472 ... D3
Haflong 5,197 ... G3
Hanumangarh 30,017 ... C3
Hardoi 46,639 ... E3
Hardwar 146,186 ... D2
Hassan 51,325 ... D6
Hathras 74,349 ... D3
Hazaribagh 54,818 ... F4
Hindupur 42,959 ... D6
Hinganghat 44,349 ... D4
Hissar 137,254 ... D3
Honavar 12,444 ... C6
Hooghly-Chinsura 105,241 ... F1
Hospet 114,711 ... D5
Howrah 737,877 ... F2
Hubli-Dharwar 526,493 ... C5
Hyderabad 4,270,000 ... D5
Ichchapuram 15,850 ... F5
Imphal 155,639 ... G4
Indore 827,021 ... D4
Itanagar▲ 18,787 ... G3
Itarsi 44,191 ... D4
Jabalpur 757,726 ... D4
Jaipur 1,004,669 ... D3
Jaisalmer 16,578 ... C3
Jajpur 16,707 ... F4
Jalgaon 145,254 ... D4
Jalna 122,246 ... D4
Jalor 15,478 ... C3
Jalpaiguri 55,159 ... F3
Jamalpur 61,731 ... F3
Jammu 155,338 ... D2
Jamnagar 317,037 ... B4
Jamshedpur 669,984 ... F4
Jaora 37,235 ... D4
Jaunpur 104,994 ... E3
Jhansi 231,332 ... D3
Jind 38,161 ... D3
Jodhpur 493,604 ... C3
Jubbulpore (Jabalpur) 757,726 ... D4
Jullundur 296,106 ... D2
Junagadh 120,072 ... B4
Kadayanallur 50,295 ... D7
Kakinada 226,642 ... E5
Kalyan 99,547 ... C5
Kamarhati 169,404 ... F1
Kamptee 53,412 ... D4
Kanchipuram 145,329 ... E6
Kanchrapara 78,768 ... F1
Kandla 17,995 ... C4
Kanker 9,278 ... E4
Kanpur 1,688,242 ... E3
Karad 42,329 ... C5
Karaikudi 100,187 ... D7
Kargil 2,390 ... D2
Karnal 132,067 ... D3
Kasganj 46,467 ... D3
Katarnian Ghat ... E3
Katihar 121,693 ... F3
Katni (Murwara) 125,096 ... E4
Kavaratti 4,420 ... C6
Kendrapara 20,079 ... F4
Khamgaon 53,692 ... D4
Khamman 56,919 ... D5
Khandwa 114,463 ... D4
Kharagpur 234,931 ... F4
Kirkee 65,497 ... C5
Kishangarh 37,405 ... D3
Kishtwar 5,276 ... D2
Kohima 21,545 ... G3
Kolar 43,418 ... D6
Kolar Gold Fields 144,406 ... D6
Kolhapur 351,073 ... C5
Koraput 21,505 ... E5
Korba 30,963 ... E4
Kota 346,928 ... D3
Kottaguden 75,542 ... E5
Kottayam 59,714 ... D7
Kozhikode 333,979 ... D6
Krishnanagar 85,923 ... F4
Kumbakonam 141,639 ... D6
Kumta 19,112 ... C6
Kurnool 206,661 ... D5
Latur 111,961 ... D5
Leh 5,519 ... D2
Lucknow 1,006,538 ... E3
Ludhiana 606,250 ... D2
Machilipatnam 138,525 ... E5
Madugula 8,376 ... E5
Madurai 904,362 ... D7
Mahabaleshwar 7,318 ... C5
Mahbubnagar 51,756 ... D5
Mahe 8,972 ... D6
Mahuva 39,497 ... C4
Malegaon 245,769 ... C4
Maler Kotla 48,536 ... D2
Malkapur 35,476 ... D4
Malvan 17,579 ... C5
Mandi 16,849 ... D2
Mandla 24,406 ... E4
Mandsaur 52,347 ... C4
Mangalore 3,055,113 ... C6
Mannargudi 42,783 ... E6
Margao 41,655 ... C5
Marmagao 44,065 ... C5
Mathura 160,995 ... D3
Mau 64,058 ... E3
Mayuram 60,195 ... D6
Meerut 538,461 ... D3
Mehsana 51,598 ... C4
Mercara 19,357 ... D6
Mhow 59,037 ... D4
Midnapore 71,326 ... F4
Miraj 77,606 ... D5
Mirzapur-cum-Vindhyachal 128,179 ... E4
Monghyr 102,474 ... F3
Moradabad 347,983 ... D3
Morena 44,901 ... D3
Morvi 60,976 ... C4
Mumbai (Bombay) (Greater)* 8,227,332 ... B7
Murwara 125,096 ... E4
Muzaffarnagar 172,435 ... D3
Muzaffarpur 189,765 ... E3
Mysore 476,446 ... D6
Nadiad 142,279 ... C4
Nagapattinam 68,026 ... E6
Nagaur 36,448 ... C3
Nagercoil 171,641 ... D7
Nagina 37,066 ... D3
Nagpur 1,297,977 ... D4
Nahan 16,017 ... D2
Naihati 82,080 ... F1
Naini Tal 23,986 ... D3
Nander 190,819 ... D5
Nandurbar 54,070 ... C4
Nandyal 63,193 ... D5
Nasik 428,778 ... C5
Navsari 129,122 ... C4
Nellore 236,225 ... E6
New Delhi (cap.) 301,801 ... D3
Nimach 47,113 ... D4
Nipani 35,116 ... C5
Nizamabad 183,135 ... D5
Nova Goa (Panaji) 34,953 ... C5
Nowgong 56,537 ... G3
Okha Port 10,687 ... B4
Ongole 53,330 ... E5
Ootacamund 63,310 ... D6
Orai 42,513 ... D3
Pachmarhi 1,212 ... D4
Palanpur 42,114 ... C4
Palayankottai 70,070 ... D7
Palghat 117,961 ... D6
Pali 49,834 ... C3
Palni 49,575 ... D6
Panaji 34,953 ... C5
Panchur 59,021 ... F2
Pandharpur 53,638 ... D5
Panihati 148,046 ... F1
Panipat 137,953 ... D3
Panna 22,316 ... E4
Parbhani 109,328 ... D5
Pasighat 5,116 ... G3
Patan 105,191 ... C4
Pathankot 108,777 ... D2
Patiala 205,849 ... D2
Patna 916,102 ... F3
Pilibhit 68,273 ... D3
Pondicherry 251,471 ... E6
Ponnani 35,723 ... D6
Porbandar 133,545 ... B4
Port Blair 26,218 ... G6
Porto Novo 17,412 ... E6
Proddatur 107,068 ... D6
Puduchcheri (Pondicherry) 251,471 ... E6
Pudukkottai 66,384 ... D6
Pune 1,135,034 ... C5
Puri 101,089 ... F5
Purli 31,078 ... D5
Purnea 109,649 ... F3
Purulia 57,708 ... F4
Quilon 167,583 ... D7
Raichur 124,600 ... D5
Raigarh 46,745 ... E4
Raipur 338,973 ... E4
Rajahmundry 267,749 ... E5
Rajapalaiyam 101,633 ... D7
Rajapur 9,017 ... C5
Rajkot 444,156 ... C4
Rajnandgaon 41,183 ... E4
Rajpur 34,393 ... F2
Rameswaram 16,755 ... D7
Ranchi 500,593 ... F4
Ratlam 156,490 ... C4
Ratnagiri 37,551 ... C5
Raurkela 321,326 ... F4
Raxaul 12,064 ... E3
Rewa 100,519 ... E4
Rishra 63,486 ... F1
Rohtak 166,631 ... C5
Sadiya▲ 64,252 ... H3
Sagar 207,401 ... D4
Saharanpur 294,391 ... D3
Salem 515,021 ... D6
Sambalpur 162,190 ... E4
Sambhal 108,379 ... D3
Sangli 268,962 ... C5
Santipur 61,166 ... F4
Sardarshahr 37,703 ... C3
Sasaram 48,282 ... E4
Satara 66,433 ... C5
Satna 57,531 ... E4
Sehore 35,657 ... D4
Seoni 38,396 ... D4
Serampore 102,023 ... F1
Seringapatam 14,100 ... D6
Shahjahanpur 205,325 ... E3
Shillong 173,064 ... G3
Shimoga 151,562 ... D6
Shivpuri 42,120 ... D3
Sholapur 514,461 ... D5
Sidhi 8,341 ... E4
Sidhpur 40,521 ... C4
Sikar 102,946 ... D3
Silchar 52,596 ... G4
Siliguri 153,825 ... F3
Simla 55,368 ... D2
Sirohi 18,774 ... C4
Sirsa 48,808 ... D3
Sitapur 66,715 ... E3
South Dum Dum 174,538 ... F2
South Suburban 272,600 ... F2
Srikakulam 45,179 ... E5
Srinagar 403,413 ... D2
Sundargarh 17,244 ... E4
Surat 912,568 ... C4
Surendranagar 66,667 ... C4
Tanda 41,611 ... E3
Tellicherry 68,759 ... C6
Tenali 119,216 ... E5
Tezpur 39,870 ... G3
Thana 388,577 ... B6
Thanjavur 183,464 ... D6
Tinsukia 54,911 ... H3
Tiruchchirappalli 607,815 ... D6
Tiruchendur 18,126 ... D7
Tirunelveli 324,034 ... D7
Tirupati 115,244 ... D6
Tiruppattur 40,357 ... D6
Tiruppur 215,743 ... D6
Tiruvannamalai 61,370 ... D6
Titagarh 88,218 ... F1
Tonk 55,866 ... D3
Tranquebar 17,318 ... E6
Trichur 170,093 ... D6
Trivandrum 519,766 ... D7
Tumkur 109,231 ... D6
Tura 15,489 ... G3
Tuticorin 250,673 ... D7
Udaipur 229,762 ... C4
Udhampur 16,392 ... D2
Ujjain 231,878 ... D4
Ulhasnagar 648,149 ... C5
Unnao 38,195 ... E3
Uttarpara-Kotrung 67,568 ... F1
Vadodara 744,043 ... C4
Valsad 43,254 ... C4
Vaniyambadi 51,810 ... D6
Varanasi 793,542 ... E3
Vellore 246,937 ... D6
Vengurla 11,805 ... C5
Veraval 58,771 ... C4

British India

RUSSIA
AFGHANISTAN
GILGIT AGENCY
KASHMIR & JAMMU
N.W. FRONTIER PROV.
PUNJAB
BALUCHISTAN
Indus
PUNJAB STATES
BAHAWALPUR (PUNJ. ST.)
IRAN
PUNJ. ST.
RAJPUTANA
AJMER-MERWARA
SIND
Gwadar (Oman)
DELHI
RAMPUR
UNITED PROVINCES
TIBET
CHINA
NEPAL
SIKKIM
BHUTAN
Brahmaputra
E. ST.
KHASI HILLS
ASSAM
MANIPUR
Ganges
GWALIOR
BENARES
BIHAR
CENTRAL INDIA
EASTERN STATES
BENGAL
TRIPURA (E. ST.)
Chandernagore (Fr.)
BURMA
WESTERN INDIA
Arabian Sea
GUJARAT ST.
BOMBAY
CENTRAL PROVINCES
BERAR
E. ST.
ORISSA
Diu (Port.)
Damão (Port.)
EASTERN STATES
HYDERABAD
DECCAN STATES
Yanaon (Fr.)
Bay of Bengal
Gôa (Port.)
MADRAS
MYSORE
Bangalore (Br.)
COORG
Mahé (Fr.)
Pondichéry (Fr.)
Karikal (Fr.)
M. ST.
Laccadive Islands (Madras)
Cochin (Br.)
MADRAS STATES
CEYLON
Andaman Islands (Br.)
Nicobar Islands (Br.)
British India. The provinces of British India were directly administered by Britain. A few areas were leased from the Indian princes.
Indian States. The Indian States, sometimes referred to as the "Native" or "Princely States," were under the nominal control of maharajas or other hereditary princes.
Possessions of Other Countries in India
State or Provincial Boundaries
Other Internal Boundaries

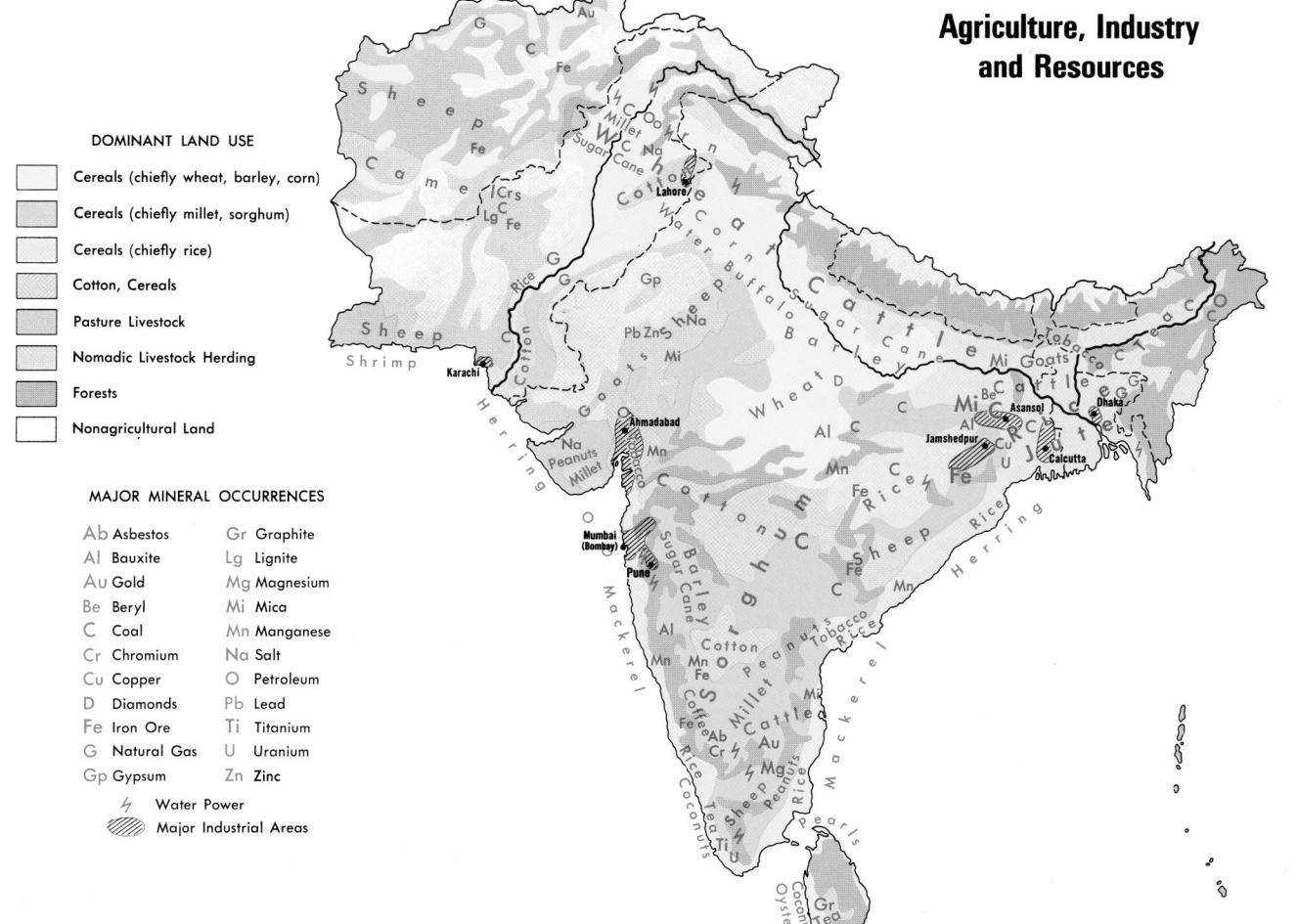

Agriculture, Industry and Resources

DOMINANT LAND USE

Cereals (chiefly wheat, barley, corn)
Cereals (chiefly millet, sorghum)
Cereals (chiefly rice)
Cotton, Cereals
Pasture Livestock
Nomadic Livestock Herding
Forests
Nonagricultural Land

MAJOR MINERAL OCCURRENCES

Ab Asbestos
Al Bauxite
Au Gold
Be Beryl
C Coal
Cr Chromium
Cu Copper
D Diamonds
Fe Iron Ore
G Natural Gas
Gp Gypsum

Gr Graphite
Lg Lignite
Mg Magnesium
Mi Mica
Mn Manganese
Na Salt
O Petroleum
Pb Lead
Ti Titanium
U Uranium
Zn Zinc

⚡ Water Power
▨ Major Industrial Areas

Burma, Thailand, Indochina and Malaya

CONIC PROJECTION

SCALE OF MILES

SCALE OF KILOMETERS

International Boundaries	---------
Division and State Boundaries	---------
Capitals of Countries	☆
Division and State Capitals	◉

© Copyright HAMMOND INCORPORATED, Maplewood, N.J.

Longitude East 96° of Greenwich

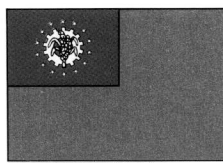

MYANMAR (BURMA)

THAILAND

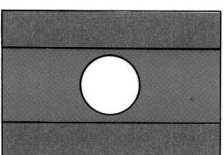

LAOS

CAMBODIA

VIETNAM

MALAYSIA

SINGAPORE

CAMBODIA

AREA 69,898 sq. mi. (181,036 sq. km.)
POPULATION 11,918,865
CAPITAL Phnom Penh
LARGEST CITY Phnom Penh
HIGHEST POINT 5,948 ft. (1,813 m.)
MONETARY UNIT new riel
MAJOR LANGUAGE Khmer (Cambodian)
MAJOR RELIGIONS Buddhism

MALAYSIA

AREA 128,308 sq. mi. (332,318 sq. km.)
POPULATION 21,820,143
CAPITAL Kuala Lumpur
LARGEST CITY Kuala Lumpur
HIGHEST POINT Mt. Kinabalu 13,455 ft. (4,101 m.)
MONETARY UNIT ringgit
MAJOR LANGUAGES Malay, Chinese, English, Tamil, Dayak, Kadazan
MAJOR RELIGIONS Islam, Confucianism, Buddhism, tribal religions, Hinduism, Taoism, Christianity, Sikhism

SINGAPORE

AREA 226 sq. mi. (585 sq. km.)
POPULATION 3,571,710
CAPITAL Singapore
LARGEST CITY Singapore
HIGHEST POINT Bukit Timah 581 ft. (177 m.)
MONETARY UNIT Singapore dollar
MAJOR LANGUAGES Chinese, Malay, Tamil, English, Hindi
MAJOR RELIGIONS Confucianism, Buddhism, Taoism, Hinduism, Islam, Christianity

LAOS

AREA 91,428 sq. mi. (236,800 sq. km.)
POPULATION 5,556,821
CAPITAL Vientiane
LARGEST CITY Vientiane
HIGHEST POINT Phou Bia 9,252 ft. (2,820 m.)
MONETARY UNIT new kip
MAJOR LANGUAGE Lao
MAJOR RELIGIONS Buddhism, tribal religions

MYANMAR (BURMA)

AREA 261,789 sq. mi. (678,034 sq. km.)
POPULATION 48,852,098
CAPITAL Yangon (Rangoon)
LARGEST CITY Yangon (Rangoon)
HIGHEST POINT Hkakabo Razi 19,296 ft. (5,881 m.)
MONETARY UNIT kyat
MAJOR LANGUAGES Burmese, Karen, Shan, Kachin, Chin, Kayah, English
MAJOR RELIGIONS Buddhism, tribal religions

THAILAND

AREA 198,455 sq. mi. (513,998 sq. km.)
POPULATION 61,163,833
CAPITAL Bangkok
LARGEST CITY Bangkok
HIGHEST POINT Doi Inthanon 8,452 ft. (2,576 m.)
MONETARY UNIT baht
MAJOR LANGUAGES Thai, Lao, Chinese, Khmer, Malay
MAJOR RELIGIONS Buddhism, tribal religions

VIETNAM

AREA 128,405 sq. mi. (332,569 sq. km.)
POPULATION 78,349,503
CAPITAL Hanoi
LARGEST CITY Ho Chi Minh City
HIGHEST POINT Fan Si Pan 10,308 ft. (3,142 m.)
MONETARY UNIT new dong
MAJOR LANGUAGES Vietnamese, Thai, Muong, Meo, Yao, Khmer, French, Chinese, Cham
MAJOR RELIGIONS Buddhism, Taoism, Confucianism, Roman Catholicism, Cao-Dai

Topography

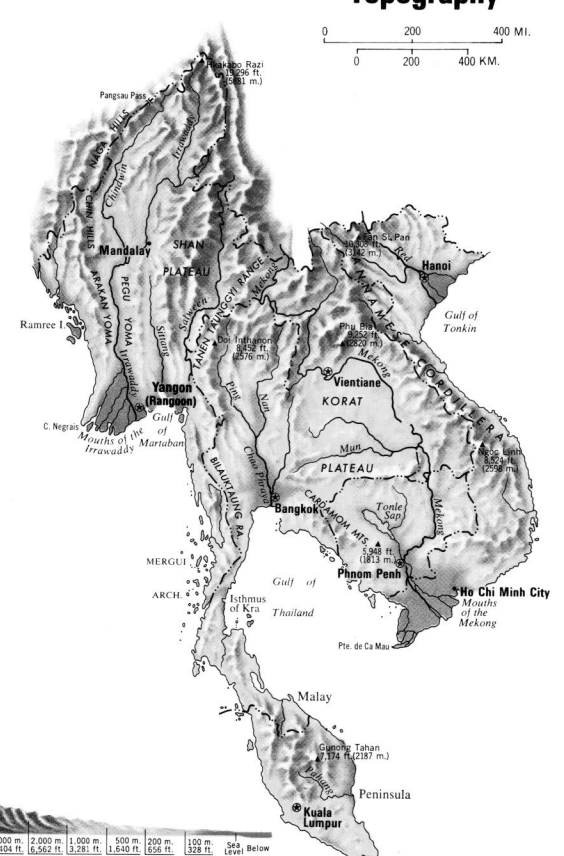

MYANMAR (BURMA)

INTERNAL DIVISIONS

CITIES and TOWNS

OTHER FEATURES

(continued on following page)

Agriculture, Industry and Resources

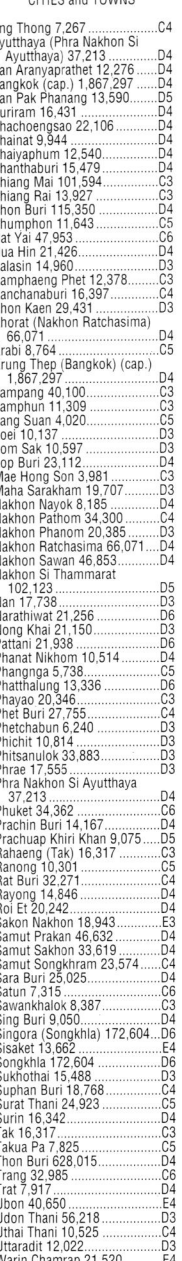

DOMINANT LAND USE

- Rice
- Diversified Tropical Crops
- Livestock Grazing, Limited Agriculture
- Tropical Forests

MAJOR MINERAL OCCURRENCES

Ag	Silver	Cu	Copper	O	Petroleum	Sn	Tin
Al	Bauxite	Fe	Iron Ore	P	Phosphates	Ti	Titanium
Au	Gold	G	Natural Gas	Pb	Lead	W	Tungsten
C	Coal	Mn	Manganese	Sb	Antimony	Zn	Zinc
Cr	Chromium				

⚡ Water Power   Major Industrial Areas

CHINA (MAINLAND)
AREA 3,705,386 sq. mi. (9,596,960 sq. km.)
POPULATION 1,256,167,701
CAPITAL Beijing
LARGEST CITY Shanghai
HIGHEST POINT Mt. Everest 29,028 ft. (8,848 m.)
MONETARY UNIT yuan
MAJOR LANGUAGES Chinese, Chuang, Uiguar, Yi, Tibetan, Maio, Mongol, Kazakh
MAJOR RELIGIONS Confucianism, Buddhism, Taoism, Islam

CHINA (TAIWAN)
AREA 13,971 sq. mi. (36,185 sq. km.)
POPULATION 22,319,222
CAPITAL T'aipei
LARGEST CITY T'aipei
HIGHEST POINT Yü Shan 13,115 ft. (3,997 m.)
MONETARY UNIT new Taiwan dollar
MAJOR LANGUAGES Chinese, Formosan
MAJOR RELIGIONS Confucianism, Buddhism, Taoism, Christianity, tribal religions

MONGOLIA
AREA 606,163 sq. mi. (1,569,962 sq. km.)
POPULATION 2,538,211
CAPITAL Ulaanbaatar
LARGEST CITY Ulaanbaatar
HIGHEST POINT Tabun Bogdo 14,288 ft. (4,355 m.)
MONETARY UNIT tughrik
MAJOR LANGUAGES Khalkha Mongolian, Kazakh (Turkic)
MAJOR RELIGION Buddhism

MONGOLIA
AREA 606,163 sq. mi. (1,569,962 sq. km.)
POPULATION 2,654,572
CAPITAL Ulaanbaatar
LARGEST CITY Ulaanbaatar
HIGHEST POINT Tabun Bogdo 14,288 ft. (4,355 m.)
MONETARY UNIT tughrik
MAJOR LANGUAGES Khalkha Mongolian, Kazakh (Turkic)
MAJOR RELIGION Buddhism

MACAU
AREA 8 sq. mi. (21 sq. km.)
POPULATION 429,152
CAPITAL Macau
MONETARY UNIT pataca
MAJOR LANGUAGES Chinese, Portuguese
MAJOR RELIGIONS Confucianism, Buddhism, Taoism, Christianity

CHINA (MAINLAND)   **CHINA (TAIWAN)**   **MONGOLIA**

CHINA
PROVINCES

Anhui (Anhwei) 49,665,724 ....J5
Chekiang (Zhejiang)
 38,884,603 ....K6
Fujian (Fukien) 25,931,106.....J6
Gansu (Kansu) 19,569,261 .....E3
Guangdong (Kwangtung)
 59,299,220 ....H7
Guangxi Zhuangzu
 (Kwangsi Chuang Aut. Reg.)
 36,420,960 ....G7
Guizhou (Kweiichow)
 28,552,997 ....G6
Hainan H8
Hebei (Hopei) 53,005,875 .....J4
Heilongjiang (Heilungkiang)
 32,665,546 ....K2
Henan (Honan) 74,422,739 ....H5
Hubei (Hupei) 47,804,150 .....H5
Hunan 54,008,851 ....H6
Inner Mongolian Aut. Reg.
 (Nei Monggol) 19,274,279 .H3
Jiangsu (Kiangsu)
 60,521,114 ....K5
Jiangxi (Kiangsi) 33,184,827 ..J6
Jilin (Kirin) 22,560,053 ....L3
Kansu (Gansu) 19,569,261 .....E3
Kiangsi (Jiangxi) 33,184,827 ...J6
Kiangsu (Jiangsu)
 60,521,114 ....K5
Kirin (Jilin) 22,560,053 ....L3
Kwangsi Chuang Aut. Reg.
 (Guangxi Zhuang)
 36,420,960 ....G7

Kwangtung (Guangdong)
 59,299,220 ....H7
Kweiichow (Guizhou)
 28,552,997 ....G6
Liaoning 35,721,693 ....K3
Nei Monggol
 (Inner Mongolian Aut. Reg.)
 19,274,279 ....H3
Ningxia Huizu
 (Ningsia Hui Aut. Reg.)
 3,895,578 ....F3
Qinghai (Tsinghai) 3,895,706 ..E4
Shaanxi (Shensi) 28,904,423 .G5
Shandong (Shantung)
 74,419,054 ....J4
Shanxi (Shansi) 25,291,389 ...H4
Sichuan (Szechwan)
 99,713,310 ....F5
Sinkiang-Uigur Aut. Reg.
 (Xinjiang Uygur) 13,081,631 ..B3
Taiwan 21,665,515 ....K7
Tibet Aut. Reg. (Xizang)
 1,892,393 ....B5
Tsinghai (Qinghai) 3,895,706 ..E4
Xinjian Uygur
 (Sinkiang-Uigur Aut. Reg.)
 13,081,631 ....B3
Xizang (Tibet Aut. Reg.)
 1,892,393 ....B5
Yunnan 32,553,817 ....F7
Zhejiang (Chekiang)
 38,884,603 ....K6

SPECIAL ADMINISTRATIVE REGIONS

Hong Kong 6,966,929 ....H7
Macau 445,427 ....H7

CITIES and TOWNS

Aihui (Aigun) (Heihe) 73,660 ..L1
Amoy (Xiamen) 507,390 ....J7
Anqing (Anking) 449,310 ......J5
Anshan 1,195,580.................K3
Anshun 200,680 ....G6
Anyang 501,390 ....H4
Baicheng, Jilin 276,420 ....K2
Baoding (Paoting) 495,140 ....J4
Baoji (Paoki) 341,240 ....G5
Baotou (Paotow) 1,075,920...G3
Beihai (Pakhoi) 173,740 ....G7
Beijing (Peking) (cap.)
 5,715,368 ....J3
Bengbu (Pengpu) 550,360 .....J5
Canton (Guangzhou,
 Kwangzhou) 3,181,510 .....H7
Changchih (Changzhi)
 450,320 ....H4
Changchow (Changzhou)
 533,940 ....J5
Changchow (Zhangzhou)
 283,490 ....J7
Changchun 1,747,410.............K3
Changde (Changteh) 213,890..H6
Changhua 185,816 ....K7
Changsha 1,066,030..............H6
Changteh (Changde) 213,890..H6
Changzhi (Changchih)
 450,320 ....H4
Changzhou (Changchow)
 533,940 ....K5
Chankiang (Zhanjiang)
 853,970 ....H7
Chaotung (Zhaotung)
 133,080 ....F6
Chaoyang, Liaoning 206,700 ..J3
Chefoo (Yantai) 385,180.........K4
Chengchow (Zhengzhou)
 1,404,050 ....H5
Chengde (Chengteh) 326,910 .J3
Chengdu (Chengtu)
 2,499,000 ....F5
Chiai 251,840.......................K7
Chifeng 293,460....................J3
Chinchow (Jinzhou) 599,490..K3
Chinkiang (Zhenjiang)
 345,560 ....J5
Chinwangtao (Qinhuangdao)
 374,210 ....K4
Chongqing (Chungking)
 2,673,170 ....G6
Chüanchow (Quanzhou)
 403,180 ....J7
Chuchow (Zhuzhou) 382,950..H6
Chumatien (Zhumadian)
 150,440 ....H5
Chungking (Chongqing)
 2,673,170 ....G6
Chungshan (Zhongshan)
 135,000 ....H7
Conghua 280,250..................H7
Dafang 962,470.....................G6
Dalian 1,480,240...................K4
Dandong (Tantung) 545,180...K3
Daqing 758,430 ....L2
Datong (Tatung), Shanxi
 962,470 ....H3

(continued on following page)

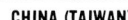

China and Mongolia Transportation

Railroads	————
Under Construction	– – – –
Connecting Roads	————
Navigable Rivers	
Canals	
Major Seaports	⚓

© Copyright HAMMOND INCORPORATED, Maplewood, N. J.

Da Xian 193,490 ....G5
Dezhou (Tehchow) 258,860 ....J4
Dukou 497,330 ....F6
Fatshan (Foshan) 273,840 ....H7
Fengcheng 995,900 ....K3
Foochow (Fuzhou) 1,111,550 ....J6
Foshan (Fatshan) 273,840 ....H7
Fowyang (Fuyang) 177,850 ....J5
Fushun 1,184,940 ....K3
Fuyang (Fowyang) 177,850 ....J5
Fuzhou (Foochow), Fujian
 1,111,550 ....J6
Fuzhou, Jiangxi 158,300 ....J6
Ganzhou (Kanchow) 362,880 ....H6
Gejiu (Kokiu) 352,980 ....F7
Guangzhou (Canton)
 3,181,510 ....H7
Guilin (Kweilin) 432,410 ....G6
Guiyang (Kweiyang), Guizhou
 1,350,190 ....G6
Gulja (Yining) 257,280 ....B3
Haikou (Hoihow) 263,280 ....H7
Hailar 157,490 ....K1
Hanchung (Hanzhong)
 374,270 ....G5
Handan (Hantan) 929,530 ....H4
Hangzhou (Hangchow)
 1,171,450 ....J5
Hanton (Handan) 929,530 ....H4
Hanzhong (Hanchung)
 374,270 ....G5
Harbin 2,519,120 ....L2
Hebi 336,430 ....H4
Hefei (Hofei) 795,420 ....J5
Hegang (Hokang) 592,470 ....L2
Heihe (Aigun, Aihui) 73,660 ....L1
Hengshui 101,260 ....H4
Hengyang 531,730 ....H6
Hofei (Hefei) 795,420 ....J5
Hohhot (Huhehot) 754,120 ....H3
Hoihow (Haikou) 263,280 ....H7
Hokang (Hegang) 592,470 ....L2
Horqin Youyi Qianqi
 (Ulanhot) 174,050 ....K2
Houma 144,460 ....H4
Hsüchang (Xuchang)
 218,960 ....H5
Huaibei 444,820 ....J5
Huainan 1,029,220 ....J5
Huangshi 375,640 ....J5
Huize 158,380 ....F6
Hunjiang 694,160 ....L3
Huzhou (Wuxing) 925,900 ....K5
Hwainan (Huainan)
 1,029,220 ....J5

Hwangshih (Huangshi)
 375,640 ....J5
Ichang (Yichang) 365,000 ....H5
Ichun (Yichun) 755,830 ....L2
Ipin (Yibin) 245,240 ....F6
Jiamusi (Kiamusze) 540,190 ....M2
Ji'an (Kian) 167,550 ....J6
Jiangmen (Kongmoon)
 212,450 ....H7
Jiaozuo (Tsiaotso) 484,370 ....H4
Jiaxing (Kashing) 656,130 ....K5
Jilin (Kirin) 1,808,420 ....L3
Jinan (Tsinan) 1,359,130 ....J4
Jingdezhen (Kingtehchen)
 611,030 ....J6
Jinhua (Kinhwa) 869,490 ....J6
Jining (Tsining), Nei Monggol
 158,570 ....H3
Jining (Tsining), Shandong
 190,420 ....J4
Jinzhou (Chinchow) 599,490 ....K3
Jiujiang (Kiukiang) 350,910 ....J6
Jixi (Kisi) 781,800 ....M2
Kaifeng 602,230 ....H5
Kaiyuan, Yunnan 223,420 ....F7
Kalgan (Zhangjiakou)
 617,120 ....J3
Kanchow (Ganzhou) 362,880 ....H6
Kaohsiung 1,227,454 ....J7
Karamay 156,970 ....B2
Kashi 256,890 ....A4
Kashing (Jiaxing) 656,130 ....K5
Keelung 347,828 ....K6
Kiamusze (Jiamusi) 540,190 ....M2
Kian (Ji'an) 167,550 ....J6
Kingtehchen (Jingdezhen)
 611,030 ....J6
Kinhwa (Jinhua) 869,490 ....J6
Kirin (Jilin) 1,808,420 ....L3
Kisi (Jixi) 781,800 ....M2
Kiukiang (Jiujiang) 350,910 ....J6
Kokiu (Gejiu) 352,980 ....F7
Kongmoon (Jiangmen)
 212,450 ....H7
Korla 117,690 ....C3
Kowloon 2,450,187 ....H7
Kuldja (Yining) 257,280 ....B3
Kunming 1,418,640 ....F6
Kuytun 239,870 ....C3
Kwangchow (Canton)
 3,181,510 ....H7
Kweilin (Guilin) 432,410 ....G6
Kweisui (Hohhot) 754,120 ....H3
Kweiyang (Guiyang)
 1,350,190 ....G6
Lanzhou (Lanchow)
 1,364,480 ....F4

Lengshuijiang 254,590 ....H6
Leshan (Loshan) 958,360 ....F6
Lhasa 83,540 ....D6
Lianyungang (Lienyünkang)
 397,090 ....J5
Liaoyang 646,580 ....K3
Liaoyuan 771,510 ....K3
Linfen 208,210 ....H4
Liuzhou (Luchow) 581,940 ....G7
Loho (Luohe) 157,670 ....H5
Longyan 346,700 ....J6
Loshan (Leshan) 958,360 ....F6
Loyang (Luoyang) 951,610 ....H5
Lu'an 145,880 ....J5
Luchow (Luzhou) 305,220 ....G6
Lüda (Dalian) 1,480,240 ....K4
Luohe 157,670 ....H5
Luoyang (Loyang) 951,610 ....H5
Luzhou (Luchow) 305,220 ....G6
Ma'anshan 351,880 ....J5
Macau (Macao) 238,413 ....H7
Manchouli (Manzhouli)
 104,220 ....J2
Maoming (Mowming)
 412,540 ....H7
Mianyang, Sichuan 768,500 ....G5
Mowming (Maoming)
 412,540 ....H7
Mudanjiang (Mutankiang)
 581,300 ....M3
Mukden (Shenyang)
 3,944,240 ....K3
Nanchang 1,075,710 ....J6
Nanchong (Nanchung)
 228,340 ....G5
Nanjing (Nanking) 2,091,400 ....J5
Nanning 889,790 ....G7
Nanping 407,810 ....J6
Nantong 402,990 ....K5
Nanyang 288,300 ....H5
Neijiang (Neikiang) 270,950 ....G6
Ningbo (Ningpo) 478,940 ....K6
Ningpo (Ningbo) 478,940 ....K6
Ningsia (Yinchuan,
 Yinchuan) 354,100 ....G4
Paicheng (Baicheng)
 276,420 ....K2
Pakhoi (Beihai) 173,740 ....G7
Paoki (Baoji) 341,240 ....G5
Paoting (Baoding) 495,140 ....J4
Paotow (Baotou) 1,075,920 ....G3
Peking (Beijing) (cap.)
 5,715,368 ....J3
Pingtung 189,347 ....K7
Pingxiang, Guangxi
 1,189,030 ....G7

Pingxiang, Jiangxi 76,260 ....H6
Qingdao (Tsingtao)
 1,172,370 ....K4
Qingjiang 234,750 ....J5
Qinhuangdao (Chinwangtao)
 374,210 ....K4
Qinzhou 981,280 ....G7
Qiqihar (Tsitsihar) 1,209,180 ....K2
Qitaihe 283,420 ....M2
Quanzhou (Chüanchow)
 403,180 ....J6
Sanmenxia 147,050 ....H5
Sanming 199,230 ....J6
Shanghai 7,551,236 ....K5
Shangqiu (Shangkiu)
 186,760 ....J5
Shangrao (Shangjao)
 135,160 ....J6
Shantou (Swatow) 717,620 ....J7
Shaoguan (Shiukwan)
 370,550 ....H7
Shaoxing (Shaohing)
 1,091,170 ....K5
Shaoyang 396,600 ....H6
Shashi 238,960 ....H5
Shenyang (Mukden)
 3,944,240 ....K3
Shenzhen 98,060 ....H7
Shihezi (Shihhotzu) 563,740 ....C3
Shijiazhuang (Shihkiachwang)
 1,068,720 ....J4
Shiukwan (Shaoguan)
 370,550 ....H7
Shiyan 306,830 ....H5
Shuangyashan 400,050 ....M2
Siakwan (Xiaguan) 117,190 ....E6
Sian (Xi'an) 2,185,040 ....G5
Siangfan (Xiangfan) 323,000 ....H5
Siangtan (Xiangtan) 493,040 ....H6
Sienyang (Xianyang)
 501,810 ....G5
Sinchu 208,038 ....K7
Singtai (Xingtai) 334,210 ....H4
Sining (Xining) 566,650 ....F4
Sinsiang (Xinxiang) 525,280 ....H4
Sinyang (Xinyang) 240,000 ....H5
Siping (Szeping) 333,850 ....K3
Soochow (Suzhou) 191,710 ....K5
Süchow (Xuzhou) 776,770 ....J5
Suizhong 669,940 ....K3
Suzhou (Soochow) 191,710 ....K5
Swatow (Shantou) 717,620 ....J7
Szeping (Siping) 333,850 ....K3
Tai'an 1,274,770 ....J4
Taichow (Taizhou) 161,200 ....K5
Taichung 565,255 ....K6
Tainan 541,390 ....J7

T'aipei 2,108,193 ....K7
Taiyuan 1,745,820 ....H4
Taizhou (Taichow) 161,200 ....K5
Tangshan 1,407,840 ....J4
Tantung (Dandong) 545,180 ....K3
Taoyuan 105,841 ....K6
Tatung (Datong) 962,470 ....H3
Tehchow (Dezhou) 258,860 ....J4
Tianjin (Tientsin) 4,521,266 ....J4
Tianshui 185,230 ....F5
Tieling 220,850 ....K3
Tientsin (Tianjin) 4,521,266 ....J4
Tienshui (Tianshui) 185,230 ....F5
Tongchuan (Tungchwan)
 353,520 ....G5
Tongliao 213,470 ....K3
Tongling 184,060 ....J5
Tsiaotso (Jiaozuo) 484,370 ....H4
Tsinan (Jinan) 1,359,130 ....J4
Tsingkiang (Qingjiang)
 234,750 ....J5
Tsingtao (Qingdao)
 1,172,370 ....K4

Tsining (Jining), Nei Monggol
 158,570 ....H3
Tsining (Jining), Shandong
 190,420 ....J4
Tsitsihar (Qiqihar) 1,209,180 ....K2
Tsunyi (Zunyi) 250,670 ....G6
Tungchuan (Tongchuan)
 353,520 ....G5
Tunghwa (Tonghua)
 359,960 ....L3
Tungliao (Tongliao) 213,470 ....K3
Tunxi (Tunki) 103,560 ....J6
Tzekung (Zigong) 866,020 ....F6
Tzepo (Zibo) 2,197,668 ....J4
Ulanhot (Horquin Youyi
 Qianqi) 174,050 ....K2
Ürümqi (Urumchi)
 961,240 ....C3
Victoria 1,183,621 ....H7
Wanxian (Wanhsien)
 267,000 ....G5
Weifang 393,410 ....J4
Weihai (Weihaiwei) 205,010 ....K4

Wenchow (Wenzhou)
 515,650 ....J6
Wenzhou 515,650 ....J6
Wuchow (Wuzhou) 245,250 ....H7
Wuchung (Wuzhong)
 245,250 ....G4
Wuhan 3,287,720 ....H5
Wuhu 449,070 ....J5
Wusih (Wuxi) 798,310 ....K5
Wuxi (Wusih) 798,310 ....K5
Wuxing 925,900 ....K5
Wuzhou (Wuchow) 245,250 ....H7
Xiaguan (Siakwan) 117,190 ....E6
Xiamen (Amoy) 507,390 ....J7
Xi'an (Sian) 2,185,040 ....G5
Xiangfan (Siangfan) 323,000 ....H5
Xiangtan (Siangtan) 493,040 ....H6
Xianyang (Sienyang) 501,810 ....G5
Xingtai (Singtai) 334,210 ....H4
Xining (Sining) 566,650 ....F4
Xinxiang (Sinsiang) 525,280 ....H4

Topography

0 300 600 MI.

0 300 600 KM.

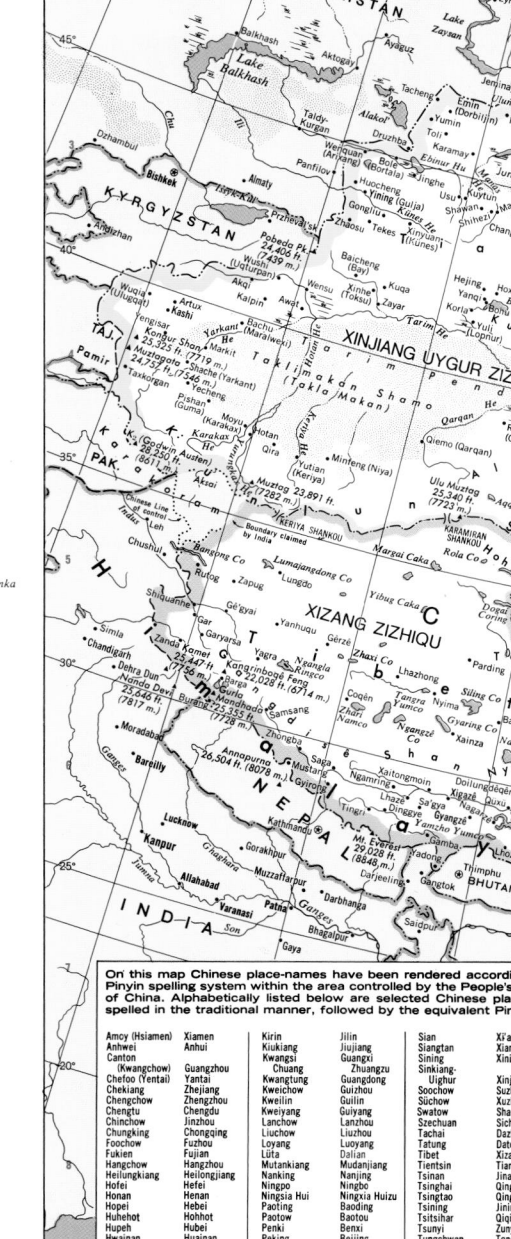

On this map Chinese place-names have been rendered according to the Pinyin spelling system within the area controlled by the People's Republic of China. Alphabetically listed below are selected Chinese place-names spelled in the traditional manner, followed by the equivalent Pinyin form.

Amcy (Hsiamen)	Xiamen	Kirin	Jilin	Sian	Xi'an
Anhwei	Anhui	Kiukiang	Jiujiang	Siangtan	Xiangtan
Canton		Kwangsi	Guangxi	Sining	Xining
(Kwangchow)	Guangzhou	Chuang	Zhuangzu	Sinkiang-	
Chefoo (Yentai)	Yantai	Kwangtung	Guangdong	Uighur	Xinjiang Uygur
Chekiang	Zhejiang	Kweichow	Guizhou	Soochow	Suzhou
Chengchow	Zhengzhou	Kweilin	Guilin	Süchow	Xuzhou
Chengtu	Chengdu	Kweiyang	Guiyang	Swatow	Shantou
Chinchow	Jinzhou	Lanchow	Lanzhou	Szechwan	Sichuan
Chungking	Chongqing	Liuchow	Liuzhou	Tachai	Dazhai
Foochow	Fuzhou	Loyang	Luoyang	Tatung	Datong
Fukien	Fujian	Lüta	Dalian	Tibet	Xizang
Hangchow	Hangzhou	Mutankiang	Mudanjiang	Tientsin	Tianjin
Heilungkiang	Heilongjiang	Nanking	Nanjing	Tsinan	Jinan
Hofei	Hefei	Ningpo	Ningbo	Tsinghai	Qinghai
Honan	Henan	Ningsia Hui	Ningxia Huizu	Tsingtao	Qingdao
Hopei	Hebei	Paoting	Baoding	Tsining	Jining
Huhehot	Hohhot	Paotow	Baotou	Tsitsihar	Qiqihar
Hupeh	Hubei	Penki	Benxi	Tsunyi	Zunyi
Hwainan	Huainan	Peking	Beijing	Tungchuan	Tongchuan
Inner Mongolia	Nei Monggol	Pengpu	Bengbu	Tzepo	Zibo
Kansu	Gansu	Shansi	Shanxi	Urumchi	Ürümqi
Kiangsi	Jiangxi	Shantung	Shandong	Wusih	Wuxi
Kiangsu	Jiangsu	Shensi	Shaanxi	Yenan	Yan'an
Kingtehchen	Jingdezhen	Shihkiachwang	Shijiazhuang	Yinchwan	Yinchuan

Yanji (Yenki) 176,000 ....L3	Zhaoqing 172,080 ....H7	Argun' (Ergun He) (riv.) K1	Gangdisê Shan (range) B5
Yantai (Chefoo) 385,180 ......K4	Zhaotong (Chaotung)	Ayakkum Hu (lake) C4	Gan He (riv.) K2
Yenki (Yanji) 176,000 ....L3	133,080 ....F6	Bagrax (Bosten Hu) (lake) C3	Gaoyou Hu (lake) J5
Yibin (Ipin) 245,240 ....F6	Zhengzhou (Chengchow)	Bangong Co (lake) A5	Ghenghis Khan Wall (ruin) H2
Yichang (Ichang) 365,000 ......H5	1,404,050 ....H5	Bashi (chan.) K7	Gobi (des.) G3
Yichun, Heilongjiang 755,830 ..L2	Zhenjiang (Chinkiang)	Bayan Har Shan (range) E5	Gongga Shan (mt.) F6
Yichun, Jiangxi 171,720 ......H6	345,560 ....J5	Bo Hai (gulf) J4	Grand (canal) H5
Yinchuan Ningsia 354,100 ....G4	Zhongshan (Chungshan)	Bosten Hu (Bagrax) (lake) C3	Great Wall (ruins) G4
Yingkou 422,590 ....K3	135,000 ....H7	Chang Jiang (Yangtze) (riv.) ..J6	Gurla Mandhada (mt.) B5
Yining 257,280 ....B3	Zhumadian (Chumatien)	Daba Shan (range) G5	Gyaring Co (lake) B5
Yiyang 165,040 ....H6	150,440 ....H5	Da Hingan Ling (range) J3	Gyaring Hu (lake) E5
Yuci (Yütze) 270,890 ....H6	Zhuzhou (Chuchow) 382,950 ..H6	Dian Chi (lake) F7	Hailar He (riv.) K2
Yueyang 971,790 ....H6	Zibo (Tzepo) 2,197,668 ......J4	Dogai Coring (lake) C5	Hainan (isl.) H8
Yumen 195,290 ....E4	Zigong (Tzekung) 866,020 ....F6	Dongsha (isl.) J7	Hangzhou Wan (bay) J5
Yungkia (Wenzhou) 515,650 ..J6	Zunyi (Tsunyi) 350,670 ......G6	Dongting Hu (lake) H6	Han Shui (riv.) H5
Yütze (Yuci) 270,890 ....H6		East China (sea) L6	Har Hu (lake) D4
Zaozhuang 1,244,020 ....J5	**OTHER FEATURES**	Ebinur Hu (lake) B2	Heilong Jiang (Amur) (riv.) L2
Zhangjiakou (Kalgan)	Altun Shan (range) C4	Ergun He (Argun') (riv.) K1	Hengduan Shan (mts.) C6
617,120 ....J3	Alxa Shamo (des.) F4	Everest (mt.) C6	Himalaya (mts.) C6
Zhangzhou (Changchow)	Amur (Heilong Jiang) (riv.) L2	Fen He (riv.) H4	Hoh Xil Shan (range) C5
283,490 ....J7	A'nyêmaqên Shan (mts.) E5	Formosa (Taiwan) (isl.) K7	Hongshui He (riv.) G7
Zhanjiang (Chankiang)	Aqqikkol Hu (lake) C4	Formosa (Taiwan) (str.) J7	Hongze Hu (lake) J5
853,970 ....H7			Hotan He (riv.) B4

Huang He (Yellow) (riv.) F5	Liaodong Bandao (pen.) K3	Ngoring Hu (lake) E5
Hulun Nur (lake) J2	Lop Nor (Lop Nur) (lake) D3	Nu Jiang (riv.) E6
Huma (riv.) K1	Lumajangdong Co (lake) B5	Nyainqêntanglha Shan
Inner Mongolia (reg.) H3	Manas He (riv.) C3	(range) D5
Jinmen (Quemoy) (isl.) J7	Manas Hu (lake) C3	Ordos (reg.) G4
Jinsha Jiang (Yangtze) (riv.) ..E5	Margai Caka (lake) C4	Penghu (Pescadores) (isls.) ..J7
Junggar Pendi (desert basin) ..C2	Mazu (Matsu) (isl.) K6	Pingtan (isl.) J7
Kangrinboqê Feng (mt.) B5	Mekong (Lancang Jiang) F7	Pobeda (peak) A3
Karakax He (riv.) A4	Min Jiang (riv.) J6	Poyang Hu (lake) J6
Karakhoto (ruins) F3	Moron Us He (riv.) D5	Pratas (Dongsha) (isl.) J7
Karamiran Shankou (pass) C4	Mudan Jiang (riv.) L3	Qaidam Pendi (basin) D4
Keriya He (riv.) B4	Mu Us Shamo (des.) G4	Qarqan He (riv.) C4
Keriya Shankou (pass) B4	Muztag (mt.) B4	Qilian Shan (range) E4
Khanka (lake) M3	Muztagata (mt.) A4	Qinghai Hu (lake) E4
Kongur Shan (mt.) A4	Nam Co (lake) D5	Qumar He (riv.) D4
Konqi He (riv.) C3	Namzha Parwa (mt.) E6	Rola Co (lake) C5
Künes (riv.) B3	Nan Ling (mts.) H6	Salween (Nu Jiang) (riv.) E6
Kunlun Shan (range) C3	Nan Shan (range) F7	Siling Co (lake) C5
Kuruktag Shan (range) C3	Nen Jiang (riv.) K2	Songhua Hu (lake) L3
Lancang Jiang (riv.) F7	Ngangze Co (lake) B5	Songhua Jiang (Sungari) M2
Laoha He (riv.) J3	Ngangzê Co (lake) C5	Tai Hu (lake) J5
Leizhou Bandao (pen.) G7	Ngom Qu (riv.) E5	Taiwan (Formosa) (isl.) K7
Liao He (riv.) K3		(continued on following page)

† Populations of mainland cities, excluding Peking (Beijing), Shanghai and Tianjin (Tientsin), courtesy of Kingsley Davis.
Office of Int'l Pop. and Research, Inst. of Int'l Studies Univ. of California.

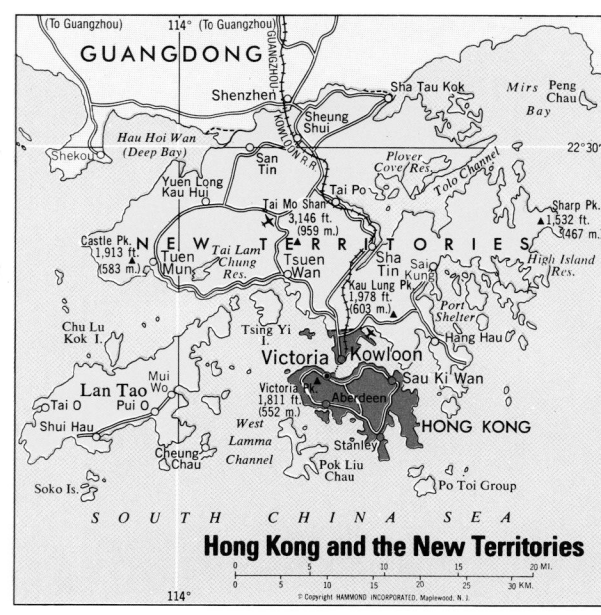

Hong Kong and the New Territories

Agriculture, Industry and Resources

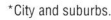

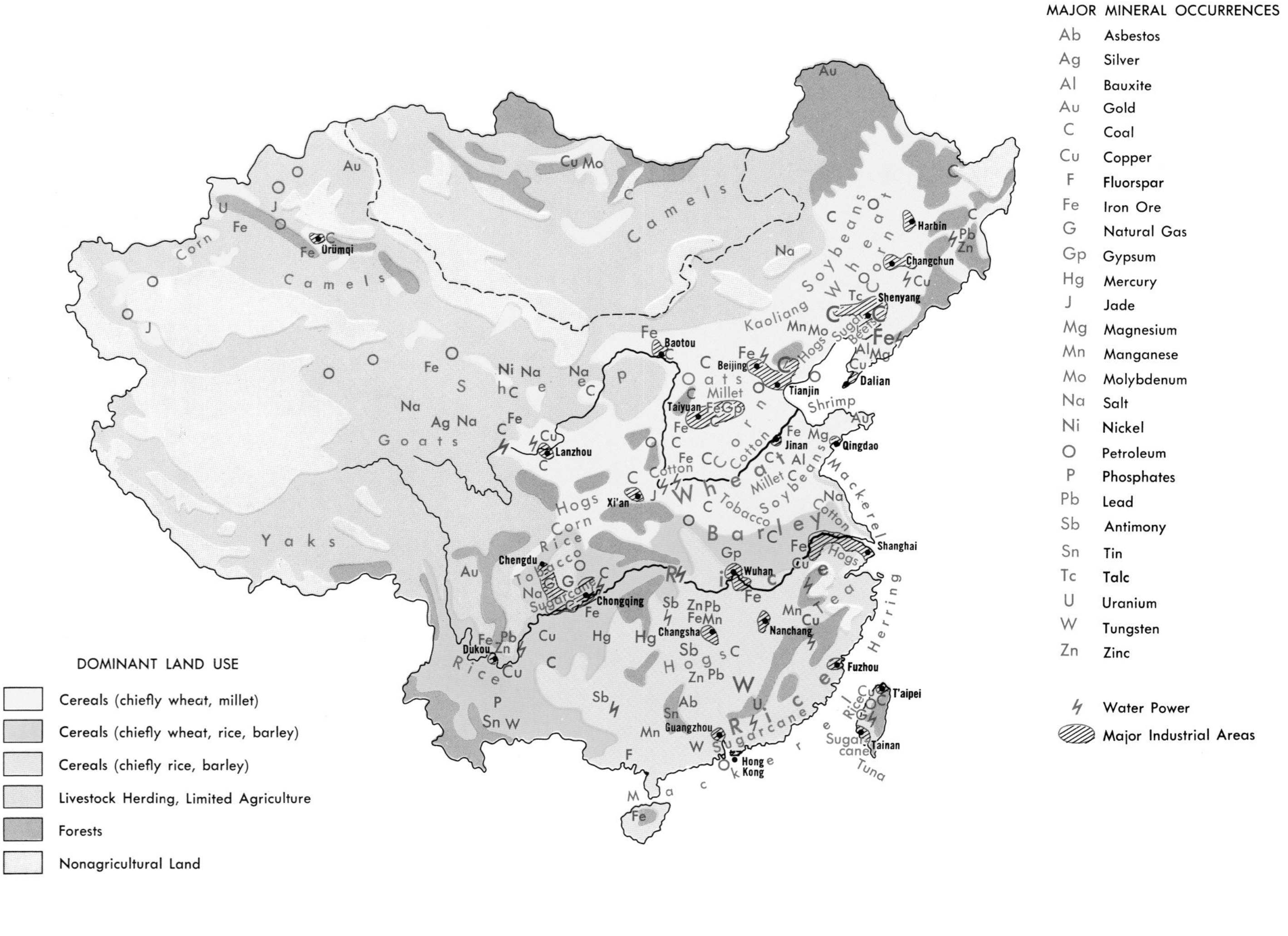

DOMINANT LAND USE

Cereals (chiefly wheat, millet)

Cereals (chiefly wheat, rice, barley)

Cereals (chiefly rice, barley)

Livestock Herding, Limited Agriculture

Forests

Nonagricultural Land

MAJOR MINERAL OCCURRENCES

Ab   Asbestos
Ag   Silver
Al   Bauxite
Au   Gold
C   Coal
Cu   Copper
F   Fluorspar
Fe   Iron Ore
G   Natural Gas
Gp   Gypsum
Hg   Mercury
J   Jade
Mg   Magnesium
Mn   Manganese
Mo   Molybdenum
Na   Salt
Ni   Nickel
O   Petroleum
P   Phosphates
Pb   Lead
Sb   Antimony
Sn   Tin
Tc   Talc
U   Uranium
W   Tungsten
Zn   Zinc

Water Power

Major Industrial Areas

AREA 145,730 sq. mi. (377,441 sq. km.)
POPULATION 123,116,000
CAPITAL Tokyo
LARGEST CITY Tokyo
HIGHEST POINT Fuji 12,389 ft. (3,776 m.)
MONETARY UNIT yen
MAJOR LANGUAGE Japanese
MAJOR RELIGIONS Buddhism, Shintoism

AREA 46,540 sq. mi. (120,539 sq. km.)
POPULATION 22,419,000
CAPITAL P'yŏngyang
LARGEST CITY P'yŏngyang
HIGHEST POINT Paektu 9,003 ft. (2,744 m.)
MONETARY UNIT won
MAJOR LANGUAGE Korean
MAJOR RELIGIONS Confucianism, Buddhism, Ch'ondogyo

AREA 38,175 sq. mi. (98,873 sq. km.)
POPULATION 42,793,000
CAPITAL Seoul
LARGEST CITY Seoul
HIGHEST POINT Halla 6,398 ft. (1,950 m.)
MONETARY UNIT won
MAJOR LANGUAGE Korean
MAJOR RELIGIONS Confucianism, Buddhism, Ch'ondogyo, Christianity

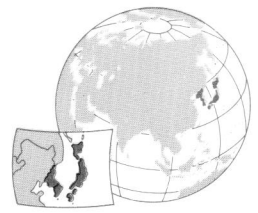

JAPAN

NORTH KOREA

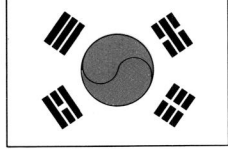

SOUTH KOREA

JAPAN

PREFECTURES

Aichi 6,221,638 ....H6
Akita 1,256,745 ....J4
Aomori 1,523,907 ....K3
Chiba 4,735,424 ....P2
Ehime 1,506,637 ....F7
Fukui 794,354 ....G5
Fukuoka 4,553,461 ....D7
Fukushima 2,035,272 ....K5
Gifu 1,960,107 ....H6
Gumma 1,848,562 ....J5
Hiroshima 2,739,161 ....E6
Hokkaido 5,575,989 ....K2
Hyogo 5,144,892 ....H7
Ibaraki 2,558,007 ....K5
Ishikawa 1,119,304 ....H5
Iwate 1,421,927 ....K4
Kagawa 999,864 ....G6
Kagoshima 1,784,623 ....E8
Kanagawa 6,924,348 ....O2
Kochi 831,275 ....F7
Kumamoto 1,790,327 ....E7
Kyoto 2,527,330 ....J7
Mie 1,686,936 ....H6
Miyagi 2,082,320 ....K4
Miyazaki 1,151,587 ....E7
Nagano 2,083,934 ....J5

Nagasaki 1,590,564 ....D7
Nara 1,209,365 ....J8
Niigata 2,451,357 ....J5
Oita 1,228,913 ....E7
Okayama 1,871,023 ....F6
Okinawa 1,106,559 ....N6
Osaka 8,473,446 ....J8
Saga 865,574 ....E7
Saitama 5,420,480 ....O2
Shiga 1,079,898 ....J7
Shimane 784,795 ....F6
Shizuoka 3,446,804 ....H6
Tochigi 1,792,201 ....K5
Tokushima 825,261 ....G7
Tokyo 11,618,281 ....O2
Tottori 604,221 ....G6
Toyama 1,103,459 ....H5
Wakayama 1,087,012 ....G6
Yamagata 1,251,917 ....K4
Yamaguchi 1,587,079 ....E6
Yamanashi 804,256 ....J6

CITIES and TOWNS

Abashiri 44,777 ....M1
Ageo 166,243 ....O2
Aizuwakamatsu 114,528 ....J5
Akashi 254,869 ....H8
Akita 284,863 ....J4
Amagasaki 523,650 ....H8

Amagi 42,863 ....E7
Anan 61,253 ....G7
Aomori 287,594 ....K3
Asahi 35,721 ....K6
Asahikawa 352,619 ....L2
Ashikaga 165,756 ....J5
Ashiya 81,745 ....H8
Atami 50,082 ....J6
Atsugi 145,392 ....O2
Ayabe 42,552 ....G6
Beppu 136,485 ....E7
Chiba 746,430 ....P2
Chichibu 61,285 ....J5
Chigasaki 171,016 ....O3
Chitose 66,788 ....K2
Chofu 180,548 ....O2
Choshi 89,416 ....K6
Daito 116,635 ....J8
Ebetsu 86,349 ....K2
Eniwa 42,911 ....K2
Fuchu, Hiroshima 49,026 ......F6
Fuchu, Tokyo 192,198 ....O2
Fuji 205,751 ....J6
Fujieda 103,225 ....J6
Fujisawa 300,248 ....O3
Fukagawa 35,376 ....L2
Fukuchiyama 63,788 ....G6
Fukui 240,962 ....G5
Fukuoka 1,088,588 ....D7

Fukushima 262,837 ....K5
Fukuyama 346,030 ....F6
Funabashi 479,439 ....P2
Furukawa 57,060 ....K4
Gifu 410,357 ....H6
Goshogawara 50,632 ....K3
Habikino 103,181 ....J8
Hachinohe 238,179 ....K3
Hachioji 387,178 ....O2
Hadano 123,133 ....O2
Hagi 53,693 ....E6
Hakodate 320,154 ....K3
Hamada 50,799 ....E6
Hamamatsu 490,824 ....H6
Hanamaki 68,873 ....K4
Hanno 61,179 ....O2
Haramachi 46,052 ....K4
Higashiosaka 521,558 ....J8
Hikone 89,701 ....H6
Himeji 446,256 ....G6
Himi 62,413 ....H5
Hino 145,448 ....O2
Hirakata 353,358 ....J7
Hiratsuka 214,293 ....O3
Hirosaki 175,330 ....K3
Hiroshima 899,399 ....E6
Hitachi 204,596 ....K5
Hitoyoshi 42,236 ....E7
Hofu 111,468 ....E6
Hondo 42,460 ....E7

Honjo 42,962 ....J4
Hyuga 58,347 ....E7
Ibaraki 234,062 ....J7
Ichihara 216,394 ....P3
Ichikawa 364,244 ....P2
Ichinomiya 253,139 ....H6
Ichinoseki 60,214 ....K4
Iida 78,515 ....H6
Iizuka 80,288 ....E7
Ikeda 101,121 ....H7
Ikoma 70,461 ....J8
Imabari 123,234 ....F6
Imari 61,243 ....D7
Ina 56,086 ....H6
Isahaya 83,723 ....D7
Ise 105,621 ....H6
Ishinomaki 120,699 ....K4
Ishioka 47,829 ....K5
Itami 178,228 ....H7
Ito 69,638 ....J6
Itoman 42,239 ....N6
Iwaki 342,074 ....K5
Iwakuni 112,525 ....E6
Iwamizawa 78,311 ....L2
Iwata 75,810 ....H6
Iwatsuki 94,696 ....O2
Izumi 124,323 ....J8
Izumiotsu 67,474 ....J8
Izumisano 90,684 ....G6
Izumo 77,303 ....F6

Joetsu 127,842 ....H5
Joyo 74,350 ....J7
Kadoma 138,902 ....J7
Kaga 65,282 ....H5
Kagoshima 505,360 ....E8
Kaizuka 81,162 ....H8
Kakogawa 212,233 ....G6
Kamaishi 65,250 ....L4
Kamakura 172,629 ....O3
Kameoka 69,410 ....J7
Kanazawa 417,684 ....H5
Kanonji 44,927 ....F6
Kanoya 73,242 ....E8
Kanuma 85,159 ....J5
Karatsu 77,710 ....D7
Kaseda 25,392 ....D8
Kashihara 107,316 ....J8
Kashima 239,198 ....P2
Kashiwara 69,836 ....J8
Kashiwazaki 83,499 ....J5
Kasugai 244,119 ....H6
Kasukabe 155,555 ....O2
Katsuta 92,621 ....K5
Kawachinagano 78,572 ....J8
Kawagoe 259,314 ....O2
Kawaguchi 379,360 ....J6
Kawanishi 129,834 ....H7
Kawasaki 1,040,802 ....O2
Kesennuma 68,551 ....K4
Kimitsu 77,286 ....O3

Kiryu 132,889 ....J5
Kisarazu 110,711 ....P3
Kishiwada 180,317 ....J8
Kitaibaraki 47,670 ....K5
Kitakami 53,647 ....K4
Kitakyushu 1,065,078 ....E6
Kitami 102,915 ....L2
Kobayashi 40,033 ....E8
Kobe 1,367,390 ....H7
Kochi 300,822 ....F7
Kodaira 154,610 ....O2
Kofu 199,262 ....J6
Koga 56,657 ....J5
Koganei 102,456 ....O2
Komatsu 104,329 ....H5
Koriyama 286,451 ....K5
Koshigaya 223,241 ....P2
Kuki 54,410 ....O2
Kumagaya 136,806 ....J5
Kumamoto 525,662 ....E7
Kurashiki 403,785 ....F6
Kurayoshi 52,270 ....F6
Kure 234,549 ....F6
Kuroiso 46,574 ....K5
Kurume 216,972 ....E7
Kushiro 214,694 ....M2
Kyoto 1,473,065 ....J7
Machida 295,405 ....O2
Maebashi 265,169 ....J5
Maizuru 97,578 ....G6
Masuda 52,756 ....E6
Matsubara 135,849 ....H8
Matsudo 400,863 ....P2
Matsue 135,568 ....F6
Matsumoto 192,085 ....H5
Matsusaka 113,481 ....H6
Matsuto 43,766 ....H5
Matsuyama 401,703 ....F7
Mihara 84,450 ....F6
Miki 70,201 ....H7
Minoo 104,112 ....J7
Mitaka 164,526 ....O2
Mito 215,566 ....K5
Mitsukaido 40,435 ....P2
Miura 48,687 ....O3
Miyako 62,478 ....L4
Miyakonojo 129,009 ....E8
Miyazaki 264,855 ....E8
Mizusawa 55,226 ....K4
Mobara 71,521 ....K6
Mooka 52,764 ....K5
Moriguchi 165,630 ....J7
Morioka 229,114 ....K4
Muko 50,604 ....J7
Muroran 150,199 ....K2
Musashino 136,910 ....O2
Mutsu 47,610 ....K3
Nagahama 54,935 ....H6
Nagano 324,360 ....J5
Nagaoka 180,259 ....J5
Nagaokakyo 71,445 ....J7
Nagasaki 447,091 ....D7
Nago 45,991 ....N6
Nagoya 2,087,902 ....H6
Naha 295,778 ....N6
Nakatsu 63,941 ....E7
Nanao 50,394 ....H5
Nankoku 44,866 ....F7
Nara 297,953 ....J8
Narashino 125,155 ....P2
Naze 49,021 ....O5
Nemuro 42,880 ....M2
Neyagawa 255,859 ....J7
Nichinan 52,949 ....E8
Niigata 457,785 ....J5
Niihama 132,339 ....F6
Niitsu 62,282 ....J5
Nishinomiya 410,329 ....H8
Nobeoka 136,598 ....E7
Noboribetsu 56,503 ....K2
Noda 93,958 ....P2
Nogata 62,595 ....E7
Noshiro 60,674 ....J3
Noto 15,480 ....H5
Numata 47,150 ....J5
Numazu 203,695 ....J6
Obihiro 153,861 ....L2
Oda 38,026 ....F6

(continued on following page)

Agriculture, Industry and Resources

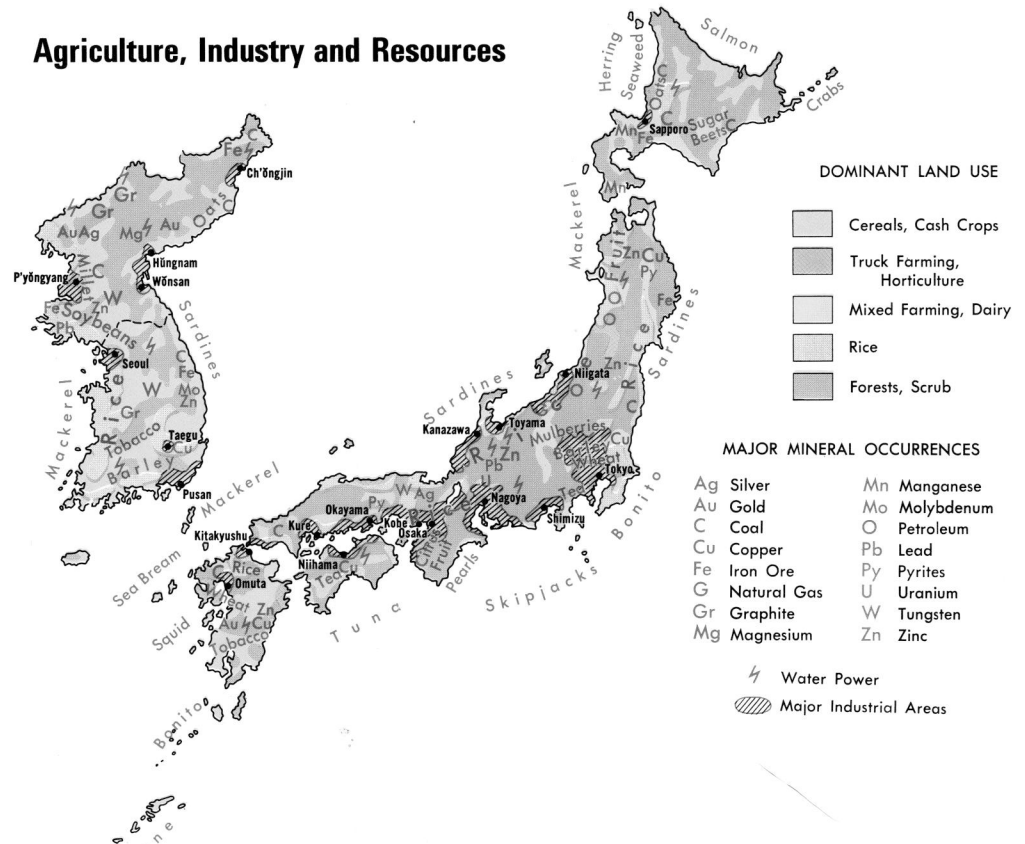

DOMINANT LAND USE

- Cereals, Cash Crops
- Truck Farming, Horticulture
- Mixed Farming, Dairy
- Rice
- Forests, Scrub

MAJOR MINERAL OCCURRENCES

Ag Silver
Au Gold
C Coal
Cu Copper
Fe Iron Ore
G Natural Gas
Gr Graphite
Mg Magnesium

Mn Manganese
Mo Molybdenum
O Petroleum
Pb Lead
Py Pyrites
U Uranium
W Tungsten
Zn Zinc

⚡ Water Power
▨ Major Industrial Areas

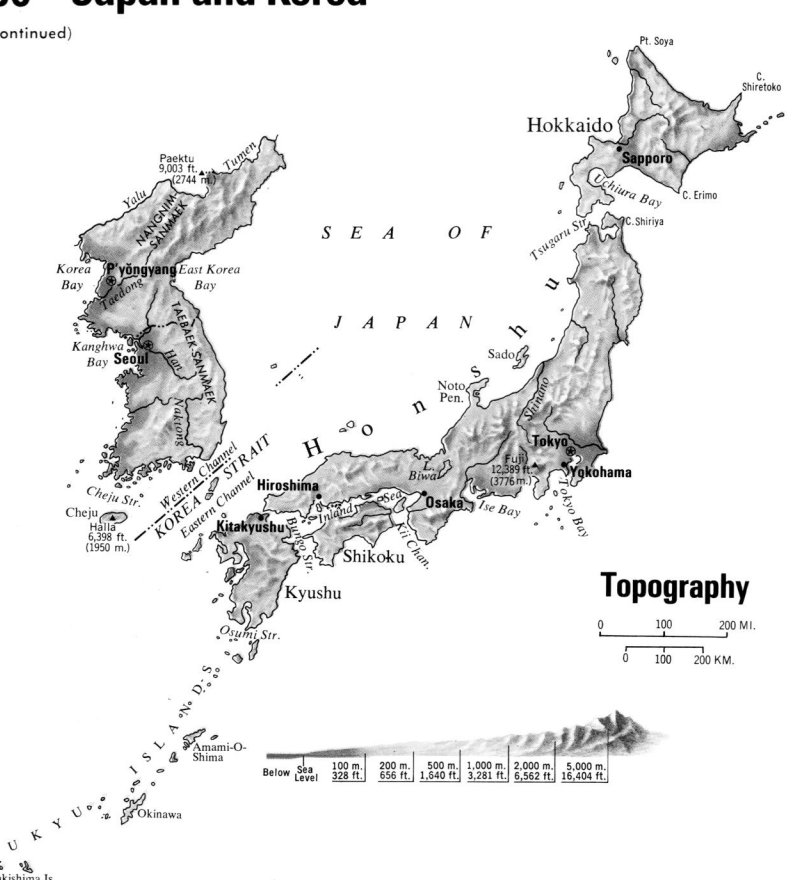

Topography

Below Sea Level	100 m. 328 ft.	200 m. 656 ft.	500 m. 1,640 ft.	1,000 m. 3,281 ft.	2,000 m. 6,562 ft.	5,000 m. 16,404 ft.

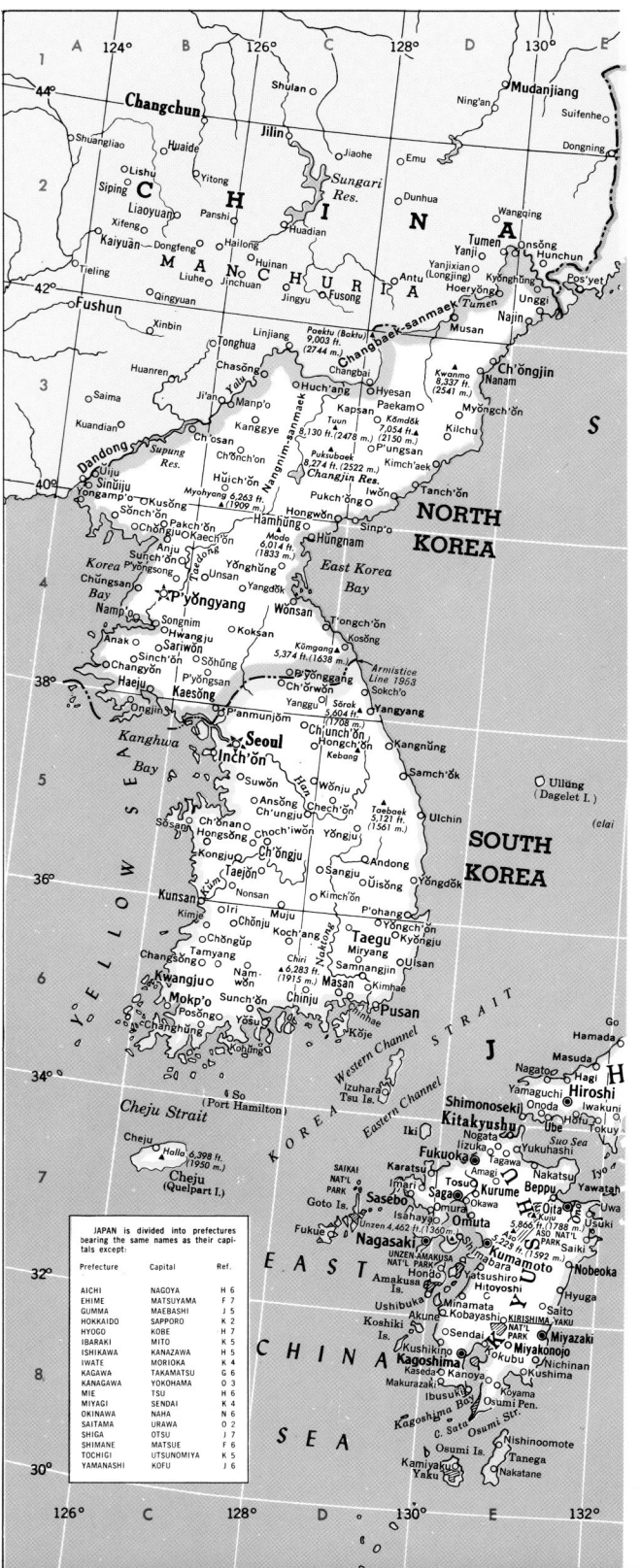

Yoshino (riv.) G6	Kaesŏng 175,000 ....C4	Sinŭiju 300,000 ....B3	East Korea (bay) D4	Cheju 167,546 ....C7	Masan 386,773 ....D6	Yŏngch'ŏn 50,765 ....D6	
KOREA (NORTH)	Kapsan C3	Sŏhŭng C4	Japan (sea) G4	Chinhae 112,098 ....D6	Miryang 42,951 ....D6	Yŏngju 77,890 ....D5	
	Kimch'aek 100,000 ....C4	Sŏnch'ŏn B4	Kanghwa (bay) B5	Chinju 202,753 ....D6	Mokp'o 221,856 ....C6	Yŏsu 161,009 ....C6	
CITIES and TOWNS	Koksan C4	Sognim B4	Korea (bay) B4	Ch'ŏnan 120,618 ....C5	Namwŏn 50,857 ....C6		
	Kosŏng D4	Sunch'ŏn D3	Kwanmo (mt.) D3	Ch'ŏngju 252,985 ....C5	Nonsan 226,429 ....C5	**OTHER FEATURES**	
Anju B4	Kusŏng B3	Tanch'ŏn D3	Nangnim-sanmaek (range)..C3	Chŏngŭp 54,864 ....C6	P'anmunjŏm C5		
Ch'ŏngjin 306,000 ....E3	Manp'o C3	T'ongch'ŏn C4	Paektu (mt.) C3	Chŏnju 366,997 ....C6	P'ohang 201,355 ....D5	Cheju (isl.) C7	
Ch'osan C3	Musan D2	Ŭiju B3	Puksubaek (mt.) C3	Ch'unch'ŏn 155,247 ....C5	Pusan 3,160,276 ....D6	Dagelet (Ullŭng) (isl.) E5	
Changyŏn B4	Najin E2	Unggi E2	Supung (res.) B3	Ch'ungju 113,138 ....C5	Samch'ŏk 42,526 ....D5	East China (sea) C8	
Chasŏng C3	Nampo 140,000 ....B4	Unsan C4	Taedong (riv.) C4	Inch'ŏn 1,084,730 ....C5	Sangju 52,839 ....D5	Halla (mt.) C7	
Chŏngju B4	Nanam D3	Wŏnsan 275,000 ....C4	Tumen (riv.) D2	Iri 145,358 ....C6	Seoul (cap.) 8,366,756 ...C5	Han (riv.) C5	
Haeju 140,000 ....B4	Ongjin B4	Yangdŏk C4	Yalu (riv.) C3	Kangnŭng 116,903 ....D5	Sokch'o 65,798 ....D4	Japan (sea) G4	
Hamhŭng 484,000 ....C4	Onsŏng E2	Yŏnghŭng C4	Yellow (sea) B6	Kimch'ŏn 72,229 ....D5	Sŏsan 38,081 ....C6	Kanghwa (bay) B5	
Heiju (P'yŏngyang) (cap.)	P'anmunjŏm C5	Yongamp'o B4		Kimhae 203,428 ....D6	Sunch'ŏn 114,223 ....C6	Kŏje (isl.) D6	
1,250,000 ....C4	P'ungsan D3		**KOREA (SOUTH)**	Kimje 221,414 ....C6	Suwŏn 310,757 ....C5	Korea (strait) D6	
Hongwŏn C4	P'yŏngyang (cap.)	**OTHER FEATURES**		Kŏhŭng 217,446 ....C6	Taegu 1,607,458 ....D6	Naktong (riv.) D6	
Hŭich'ŏn C3	1,250,000 ....C4		**CITIES and TOWNS**	Kongju 39,756 ....C5	Taejŏn 651,642 ....C6	Quelpart (Cheju) (isl.) C7	
Hŭngnam C4	Pukch'ŏng C4	Baktu (Paektu) (mt.) C3		Kunsan 165,318 ....C6	Ulsan 418,415 ....D6	Ullŭng (isl.) E5	
Hyesan D3	Sariwŏn B4	Changbaek-sanmaek (mts.)..D2	Andong 102,024 ....D5	Kwangju 727,627 ....C6	Wŏnju 136,961 ....D5	Yellow (sea) B6	
	Sinp'o D4	Changjin (res.) C3	Chech'ŏn 74,239 ....D5	Kyŏngju 122,038 ....D6	Yanggu 277,986 ....C4		

RUSSIA

SEA OF OKHOTSK

HOKKAIDŌ

Sapporo

TOKYO

Tokyo

KAWASAKI

Yokohama

Yoko-suka

NAMPO-SHOTO

Same scale as main map

BONIN ISLANDS
(OGASAWARA-GUNTO)

VOLCANO ISLANDS
(KAZAN-RETTO)

KYUSHU

PACIFIC OCEAN

EAST CHINA SEA

RYUKYU ISLANDS

OKINAWA ISLANDS

Naha

Kyoto

Otsu

Kobe

Osaka

Sakai

Nara

Higashiosaka

Nishinomiya

Same scale as main map

Tropic of Cancer

Japan and Korea

CONIC PROJECTION

SCALE OF MILES

0   50   100   150

SCALE OF KILOMETERS

0   50   100   150   200   250   300

Capitals of Countries _____ ⭐
Capitals of Prefectures _____ ◉
International Boundaries ___ ___ ___

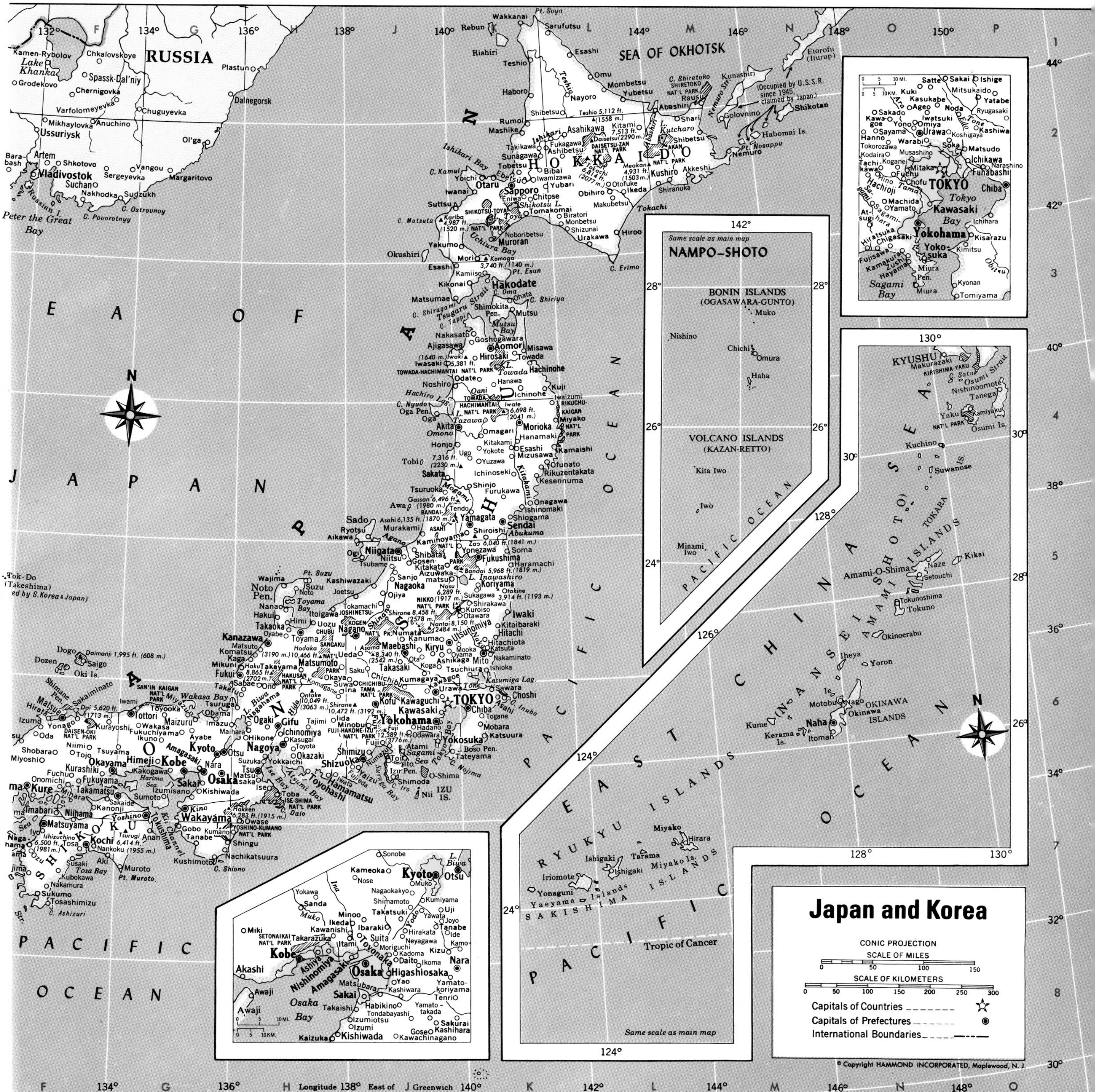

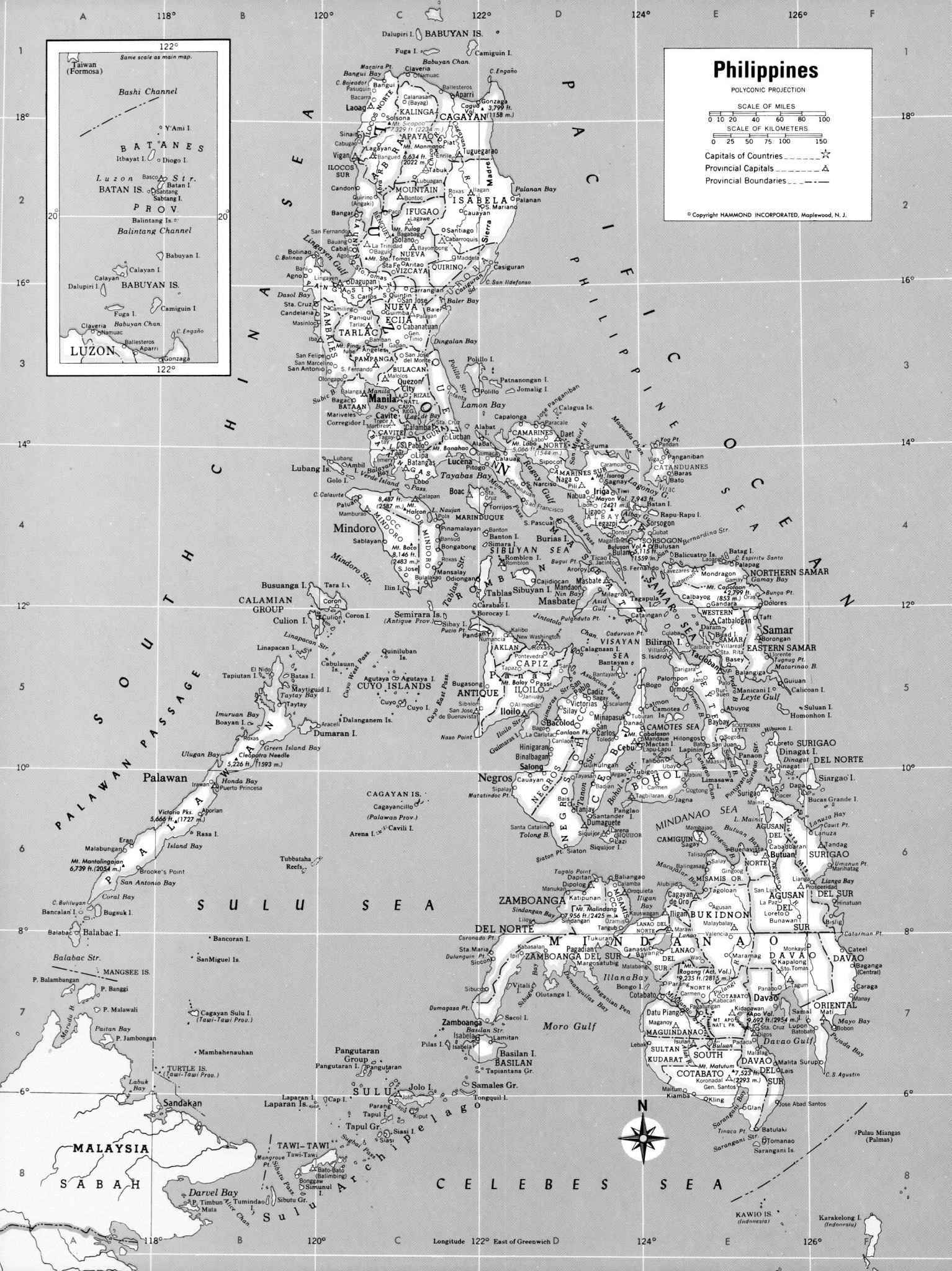

Philippines

POLYCONIC PROJECTION

SCALE OF MILES

0 10 20 40 60 80 100

SCALE OF KILOMETERS

0 25 50 75 100 150

Capitals of Countries _____ ☆

Provincial Capitals _____ △

Provincial Boundaries ___._.

© Copyright HAMMOND INCORPORATED, Maplewood, N.J.

PROVINCES

Abra 160,198 C2
Agusan del Norte 365,421 E6
Agusan del Sur 631,634 E6
Aklan 324,563 D5
Albay 809,177 D4
Antique 344,879 D5
Aurora 107,145 C3
Basilan 201,407 D7
Bataan 323,254 C3
Batanes 12,091 A2
Batangas 1,174,201 C4
Benguet 354,751 C2
Bohol 806,031 E6
Bukidnon 631,634 E6
Bulacan 1,098,046 C3
Cagayan 711,476 C1
Camarines Norte 368,007 D3
Camarines Sur 1,099,346 D4
Camiguin 57,126 E6
Capiz 492,231 D5
Catanduanes 175,247 E4
Cavite 771,320 C3
Cebu 2,091,602 D5
Davao 725,153 E7
Davao del Sur 1,133,599 E7
Davao Oriental 339,931 F7
Eastern Samar 320,637 E5
Ifugao 111,368 C2
Ilocos Norte 390,666 C1
Ilocos Sur 443,591 C2
Iloilo 1,433,641 D5
Isabela 870,604 C2
Kalinga-Apayao 185,063 C1
Laguna 973,104 C3
Lanao del Norte 461,049 E6
Lanao del Sur 404,971 E7
La Union 452,578 C2
Leyte 1,302,648 E5
Maguindanao 536,546 E7
Manila 5,925,884 C3
Marinduque 173,715 C4
Masbate 584,526 D4
Misamis Occidental 386,328 D6
Misamis Oriental 690,032 E6
Mountain 103,052 C2
National Capital Region (Manila) 5,925,884 C3
Negros Occidental 1,930,301 D6
Negros Oriental 819,399 D6
North Cotabato 564,599 E7
Northern Samar 378,516 E4
Nueva Ecija 1,069,409 C3
Nueva Vizcaya 241,690 C2
Occidental Mindoro 222,431 C4
Oriental Mindoro 448,938 C4
Palawan 371,782 B6
Pampanga 1,181,590 C3
Pangasinan 1,636,057 C3
Quezon 1,129,277 C3
Quirino 83,230 C2
Rizal 555,533 C3
Romblon 193,174 D4
Siquijor 70,300 D6
Sorsogon 500,685 E4
South Cotabato 770,473 E7
Southern Leyte 298,294 E5
Sultan Kudarat 303,784 E7
Sulu 360,588 C7
Surigao del Norte 363,414 F5
Surigao del Sur 377,647 F6
Tarlac 638,457 C3
Tawi-Tawi 194,651 B8
Western Samar 501,439 E5
Zambales 444,037 C3
Zamboanga del Norte 588,015 D6
Zamboanga del Sur 1,183,845 D7

CITIES and TOWNS

Angeles 188,834 C3
Aparri 45,070 C1
Bacolod 262,415 D5
Bagac 13,109 C3
Bago 99,631 D5
Baguio 119,009 C2
Balanga 39,132 C3
Baler 18,349 C3
Balimbing (Bato-Bato) 22,189 C8
Bamban 26,072 C3
Basco 4,341 A2
Batangas 143,570 C4
Bato-Bato 22,189 C8
Baybay 74,640 E5
Bislig 81,615 F6
Boac 37,005 C4
Bontoc 17,091 C2
Burauen 48,058 E5
Butuan 172,489 E6
Cabanatuan 138,298 C3
Cabarroquis 17,450 C2
Cadiz 129,632 D5
Cagayan de Oro 227,312 E6
Calamba 121,175 C3
Calbayog 106,719 E4
Carigara 34,377 E5
Cauayan 70,017 D6
Cavite 87,666 C3
Cebu 490,281 D5
Cotabato 83,871 D7
Dagupan 98,344 C2
Davao 610,375 E7
Digos 70,065 E7
Escalante 71,293 D5
General Santos 149,396 E7
Gingoog 79,937 E6
Guihulngan 84,156 D5
Guimba 58,847 C3
Iba 22,791 B3
Ilagan 79,336 C2
Iligan 167,358 E6
Iloilo 244,827 D5
Infanta 27,914 C3
Jaro 29,739 E5
Jolo 52,429 C8
Koronadal 80,566 E7
Lagawe 15,075 C2
Lapu-Lapu 98,723 E5
Legazpi 99,766 D4
Ligao 69,860 D4
Lingayen 65,187 C2
Lipa 121,166 C4
Lucena 107,880 C4
Maganoy 45,845 E7
Mainit 18,078 E6
Malabang 18,955 D7
Malolos 95,699 C3
Mandaue 110,590 E5
Manila (cap.) 1,630,485 C3
Mariveles 48,594 C3
Mati 78,178 F7
Naga 90,712 D4
Olongapo 156,430 C3
Ormoc 104,978 E5
Ozamis 77,832 D6
Pagadian 80,861 D7
Palo 31,124 E5
Palompon 40,242 E5
Panabo 71,098 E7
Prosperidad 33,824 F6
Puerto Princesa 60,234 B6
Quezon City 1,165,865 C3
Romblon 24,251 D4
Roxas 81,183 D5
Sagay 99,118 D5
San Antonio 42,969 B3
San Carlos, Negros Occ. 91,627 D5
San Carlos Pangasinan 101,243 C3
San Fernando, La Union 68,410 C2
San Fernando, Pampanga 110,891 C3
San Jose 64,254 C3
San Jose del Monte 90,732 C3
San Pablo 131,655 C4
Santa Fe 6,338 C2
Santiago 69,877 C2
Silay 111,131 D5
Siquijor 17,533 D6
Surigao 79,745 E6
Tacloban 102,523 E5
Tagaytay 16,322 C3
Tagum 86,201 E7
Tarlac 175,691 C3
Toledo 91,668 D5
Tuguegarao 73,507 C2
Zamboanga 343,722 C7

OTHER FEATURES

Agusan (riv.) E6
Alabat (isl.) D3
Apo (vol.) E7
Babuyan (isl.) B2
Balabac (isl.) A7
Balayan (bay) C4
Balintang (chan.) A2
Baloy (mt.) D5
Bantayan (isl.) D5
Banton (isl.) D4
Bashi (chan.) A1
Basilan (isl.) D7
Batan, Albay (isl.) E4
Batan, Batanes (isl.) B2
Batan (isls.) A2
Bay, Laguna de (lake) C3
Biliran (isl.) E5
Bohol (isl.) E6
Bojeador (cape) C1
Borocay (isl.) D5
Bucas Grande (isl.) F6
Bugsuk (isl.) A6
Bulíluyan (cape) A6
Bunga (pt.) E4
Burias (isl.) D4
Busuanga (isl.) B4
Cabalasan (mt.) E5
Cabulauan (isls.) C5
Cagayan (isls.) C6
Cagayan (riv.) C2
Cagayan Sulu (isl.) B7
Cagua (vol.) D1
Calagua (isls.) D3
Calamian Group (isls.) B4
Calayan (isl.) A2
Calicoan (isl.) E5
Camiguin, Cagayan (isl.) B3
Camiguin, Camiguin (isl.) E6
Camotes (isls.) E5
Camotes (sea) E5
Canigao (chan.) E5
Canlaon (peak) D5
Capotoan (mt.) E4
Carabao (isl.) D4
Catanduanes (isl.) E4
Cebu (isl.) D5
Celebes (sea) D8
Cleopatra Needle (mt.) B5
Coron (isl.) C5
Corregidor (isl.) C3
Culion (isl.) B5
Cuyo (isl.) C5
Cuyo (isls.) C5
Daram (isl.) E5
Davao (gulf) E7
Dinagat (isl.) E5
Diuata (mts.) E6
Dumanquilas (bay) D7
Dumaran (isl.) C5
Engaño (cape) D1
Espiritu Santo (cape) E4
Fuga (isl.) A3
Guimaras (isl.) D5
Halcon (mt.) C4
Hibuson (isl.) E5
Homonhon (isl.) E5
Honda (bay) B6
Iligan (bay) E6
Ilin (isl.) C4
Illana (bay) D7
Imuruan (bay) B5
Island (bay) B6
Itbayat (isl.) A2
Jintotolo (chan.) D5
Jolo (isl.) C7
Jomalig (isl.) D3
Lagonoy (gulf) E4
Lamon (bay) C3
Lanao (lake) E7
Laparan (isls.) B8
Lapinin (isl.) E5
Leyte (gulf) E5
Leyte (isl.) E5
Limasawa (isl.) E6
Linapacan (isl.) B5
Lingayen (gulf) C2
Lubang (isls.) B4
Luzon (isl.) C3
Luzon (str.) A2
Macajalar (bay) E6
Malindang (mt.) D6
Mangsee (isls.) A7
Manila (bay) C3
Mantalingajan (mt.) A6
Maqueda (chan.) D3
Maraira (pt.) C1
Marinduque (isl.) C4
Masbate (isl.) D4
Mayon (vol.) D4
Maytiguid (isl.) B5
Mindanao (isl.) D7
Mindanao (riv.) E7
Mindoro (isl.) C4
Mindoro (str.) C4
Mompog (passg.) D4
Moro (gulf) D7
Mount Apo National Park E7
Naso (pt.) C5
Negros (isl.) D6
Olutanga (isl.) D7
Pacsan (mt.) C2
Palawan (isl.) B6
Palawan (passg.) A6
Panaon (isl.) E5
Panay (isl.) D5
Panglao (isl.) D6
Pangutaran (isl.) C7
Pangutaran Group (isls.) C7
Patnanongan (isl.) D3
Philippine (sea) D3
Pilas (isl.) C7
Pinatubo (mt.) C3
Polillo (isl.) C3
Pujada (bay) F7
Pulangi (riv.) E7
Ragang (vol.) E7
Ragay (gulf) D4
Rapu-Rapu (isl.) E4
Romblon (isl.) D4
Sabtang (isl.) B2
Sacol (isl.) D7
Samal (isl.) E7
Samales Group (isls.) D7
Samar (isl.) E5
Samar (sea) E4
San Agustin (cape) F7
San Bernardino (str.) E4
San Miguel (bay) D3
San Pedro (bay) E5
Santo Tomas (mt.) C2
Semirara (isls.) C5
Siargao (isl.) F6
Sibay (isl.) C5
Sibuguey (bay) D7
Sibutu Group (isls.) B8
Sibuyan (isl.) D4
Sibuyan (sea) D4
Sierra Madre (mt.) D2
Simunul (isl.) B8
Siquijor (isl.) D6
South China (sea) B3
Subic (bay) C3
Sulu (arch.) B8
Sulu (sea) B6
Suluan (isl.) F5
Surigao (str.) E6
Taal (lake) C4
Tablas (isl.) D4
Tablas (str.) C4
Tagapula (isl.) E4
Tagolo (pt.) D6
Tanon (str.) D6
Tapul (isl.) C8
Tapul Group (isls.) C8
Tara (isl.) C4
Tawi-Tawi (isl.) B8
Tayabas (bay) C4
Ticao (isl.) D4
Tinaca (pt.) E8
Tongquil (isl.) D8
Tumindao (isl.) B8
Turtle (isls.) B7
Verde Island (passg.) C4
Victoria (peaks) B6
Visayan (sea) D5

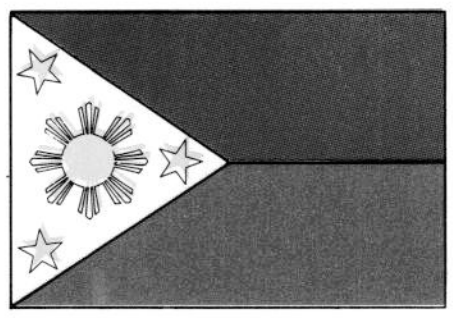

AREA 115,707 sq. mi. (299,681 sq. km.)
POPULATION 60,097,000
CAPITAL Manila
LARGEST CITY Manila
HIGHEST POINT Apo 9,692 ft. (2,954 m.)
MONETARY UNIT peso
MAJOR LANGUAGES Pilipino (Tagalog), English, Spanish, Bisayan, Ilocano, Bikol
MAJOR RELIGIONS Roman Catholicism, Islam, Protestantism, tribal religions

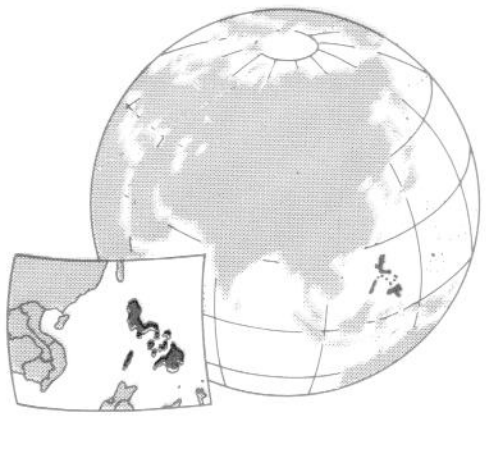

Topography

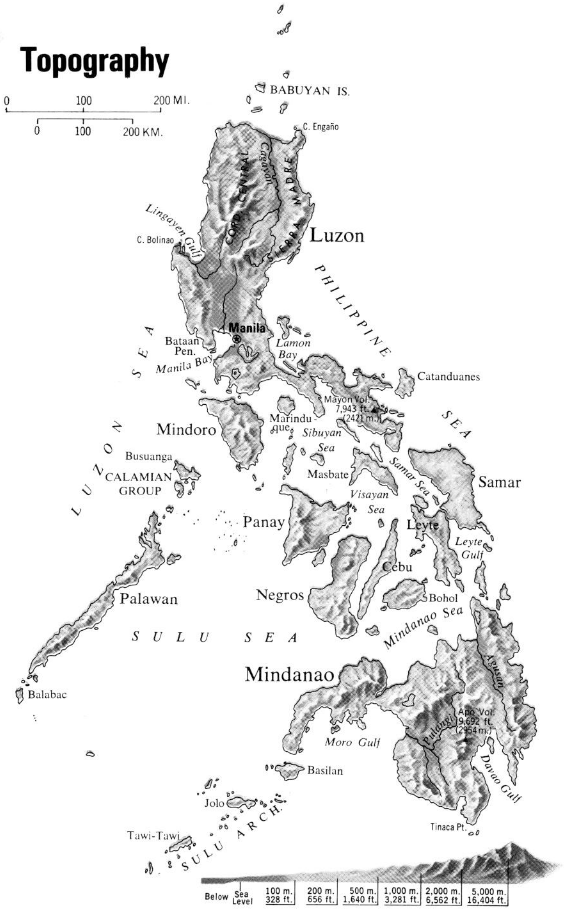

0 100 200 MI.
0 100 200 KM.
BABUYAN IS.
C. Engaño
Luzon
C. Bolinao
Lingayen Gulf
CORDILLERA CENTRAL
SIERRA MADRE
Cagayan
PHILIPPINE SEA
Manila
Bataan Pen.
Manila Bay
Lamon Bay
Catanduanes
Mayon Vol. 7,943 ft. (2421 m.)
Mindoro
Marinduque
Sibuyan Sea
Busuanga
CALAMIAN GROUP
Masbate
Samar Sea
Samar
Visayan Sea
Panay
Leyte
Leyte Gulf
Cebu
Negros
Bohol
Palawan
Mindanao Sea
SULU SEA
Mindanao
Agusan
Balabac
Apo Vol. 9,692 ft. (2954 m.)
Pulangi
Moro Gulf
Davao Gulf
Basilan
Jolo
Tinaca Pt.
Tawi-Tawi
SULU ARCH.
LUZON SEA
Below Sea Level
100 m. 328 ft.
200 m. 656 ft.
500 m. 1,640 ft.
1,000 m. 3,281 ft.
2,000 m. 6,562 ft.
5,000 m. 16,404 ft.

Agriculture, Industry and Resources

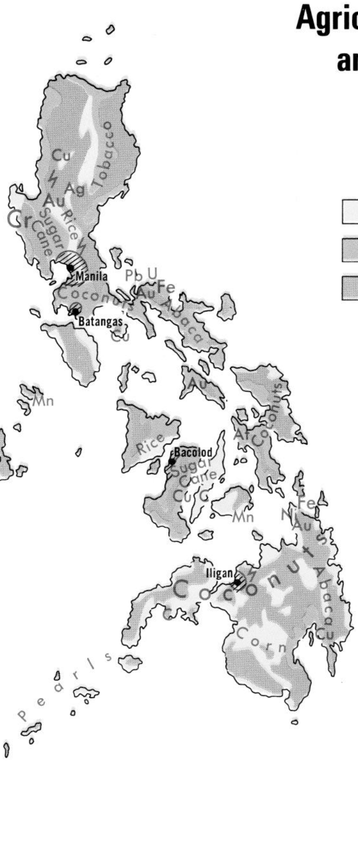

Cu
Ag
Au
Cr
Tobacco
Rice
Sugar Cane
Manila
Pb U
Coconuts
Fe
Batangas
Abaca
Cu
Mn
O
Hg
Rice
Bacolod
Sugar Cane
Cu
Coconuts
Mn
Fe
Ni
Au
Iligan
Coconuts
Abaca
Corn
Pearls

DOMINANT LAND USE

- Cereals (chiefly rice, corn)
- Cash Crops
- Tropical Forests

MAJOR MINERAL OCCURRENCES

Ag Silver
At Asphalt
Au Gold
C Coal
Cr Chromium
Cu Copper
Fe Iron
Hg Mercury
Mn Manganese
Ni Nickel
O Petroleum
Pb Lead
U Uranium

Water Power
Major Industrial Areas

BRUNEI

CITIES and TOWNS

Bandar Seri Begawan 63,868 ... E4
Seria 23,511 ... E5

INDONESIA

CITIES and TOWNS

Adaut ... J7
Agats ... K7
Ambon (Amboina) 208,898 ... H6
Amuntai ... F6
Amurang ... G5
Atambua ... G7
Aubá ... H7
Baa ... G8
Bagansiapiapi ... C5
Balikpapan 280,675 ... F6
Banda Aceh 72,090 ... A4
Bandanaira ... H6
Bandung 1,462,637 ... H2
Banggai ... G6
Banjarmasin 381,286 ... E6
Banyumas ... J2
Batang ... J2
Batavia (Jakarta) (cap.)
6,503,449 ... H1
Baukau ... H7
Bekasi ... H2
Belawan ... B5
Bengkulu 64,783 ... C6
Beo ... H5
Biak ... K6
Binjai 76,464 ... B5
Bintuhan ... C6
Blitar 78,503 ... K2
Bogor 247,409 ... H2
Bojonegoro ... J2
Bukittinggi 70,771 ... B6
Bula ... J6
Bulukumba ... F7
Buntok ... F6
Cianjur ... H2
Cimahi ... H2
Cirebon 223,776 ... H2
Demta ... L6
Denpasar ... F7
Dili ... H7
Djambi (Jambi) 230,373 ... C6
Djokjakarta (Yogyakarta)
398,727 ... J2
Dobo ... J7
Donggala ... F6
Enaratoli ... K6
Ende ... G7
Fakfak ... J6
Garut ... H2

Gorontalo 97,628 ... G5
Hollandia (Jayapura) ... K6
Indramayu ... H2
Jailolo ... H5
Jakarta (cap.) 6,503,449 ... H1
Jambi 230,373 ... D6
Jayapura (Hollandia) ... K6
Jogjakarta (Yogyakarta)
398,727 ... J2
Jombang ... J2
Kaimana ... J6
Kampung Baru (Tolitoli) ... G5
Kediri 221,820 ... K2
Kendari ... K7
Kepi ... K7
Ketapang ... E6
Kokonau ... K6
Kolonodale ... G6
Kotabaharu ... G6
Kotabaru ... F6
Kotawaringin ... E6
Kragen ... K2
Kupang ... G8
Kutaraja (Banda Aceh)
72,090 ... A4
Labuha ... H6
Labuhan ... H2
Laiwui ... H6
Larantuka ... G7
Lekitobi ... H6
Longiram ... F5
Madiun 150,562 ... K2
Magelang 123,484 ... J2
Majalengka ... J2
Makassar (Ujung Pandang)
709,038 ... F7
Malang 511,780 ... K2
Malili ... G6
Manado 217,159 ... G5
Manokwari ... J6
Maumere ... G7
Medan 1,378,955 ... B5
Menggala ... D6
Merauke ... L7
Mindiptana ... L7
Mojokerto 68,849 ... K2
Muarasiberut ... B6
Nangatayap ... E6
Pacitan ... J2
Padang 480,922 ... B6
Padangpanjang 34,517 ... B6
Padangsidempuan ... B5
Pakanbaru 186,262 ... C6
Palangkaraya 60,447 ... E6
Palembang 787,187 ... C6
Pangkalanbuun ... E6
Pangkalpinang 90,096 ... D6
Parepare 86,450 ... F6
Pasangkayu ... F6
Pasuruan 95,864 ... K2

Payakumbuh 78,836 ... C6
Pekalongan 132,558 ... J2
Pemalang ... J2
Pematangsiantar 150,376 ... B5
Pinrang ... F6
Plaju ... D6
Pontianak 304,778 ... D6
Probolinggo 100,296 ... K2
Purbolinggo ... J2
Purwokerto ... H2
Raha ... G6
Rantauprapat ... C5
Rembang ... K2
Sabang, Celebes ... F5
Sabang, Weh 23,821 ... B4
Salatiga 85,849 ... J2
Samarinda 264,718 ... F6
Sampit ... E6
Sarmi ... K6
Sawahlunto 13,561 ... C6
Seba ... G8
Semarang 1,026,671 ... J2
Semitau ... E5
Serui ... K6
Sibolga 59,897 ... B5
Sigli ... B4
Sinabang ... B5
Singaraja ... F7
Solo (Surakarta) 469,888 ... J2
Solok 31,724 ... C6
Sorong ... J6
Sragen ... J2
Subang ... H2
Sukabumi 109,994 ... H2
Sumbawa Besar ... F7
Sumedang ... H2
Surabaya 2,027,913 ... K2
Surakarta 469,888 ... J2
Tanahmerah ... K7
Tanjungbalai 41,894 ... C5
Tanjungkarang 284,275 ... D7
Tanjungpandan ... C5
Tanjungselor ... F5
Tarakan ... F5
Tebingtinggi 92,087 ... B5
Tegal 131,728 ... J2
Telukbayur ... C6
Tepa ... H7
Terempa ... D5
Tjilatjap (Cilacap) ... H2
Tjirebon (Cirebon) 223,776 ... H2
Tolitoli ... G5
Tuban ... K2
Ujung Pandang 709,038 ... F7
Vikeke ... H7
Wahai ... H6
Waigama ... H5
Wajabula ... H5
Waren ... K6
Weda ... H5
Wonreli ... H7

Yogyakarta 398,727 ... J2

OTHER FEATURES

Anambas (isls.) 29,572 ... D5
Arafura (sea) ... J8
Aru (isls.) 34,195 ... K7
Babar (isl.) ... H7
Bali (isl.) 2,074,438 ... F7
Banda (sea) ... H7
Banggai (arch.) 169,025 ... G6
Bangka (isl.) 298,017 ... D6
Banyak (isls.) 1,980 ... B5
Barisan (mts.) ... C6
Barito (riv.) ... E6
Batu (isls.) 16,390 ... B6
Bawean (isl.) 64,551 ... K1
Belitung (Billiton) (isl.)
128,694 ... D6
Berau (bay) ... J6
Biak (isl.) ... K6
Billiton (isl.) 128,694 ... D6
Binongko (isl.) 11,549 ... G7
Bone (gulf) ... G6
Borneo (isl.) ... E5
Bosch, van den (cape) ... J6
Bunguran (Great Natuna)
(isl.) ... D5
Buru (isl.) 23,034 ... H6
Butung (isl.) 188,173 ... G6
Celebes (Sulawesi) (isl.)
7,732,383 ... G6
Celebes (sea) ... G5
Cenderawasih (bay) ... K6
Dampier (str.) ... J6
Digul (riv.) ... K7
Doberai (pen.) ... J6
Enggano (isl.) 1,082 ... C7
Ewab (Kai) (isls.) 108,328 ... J7
Flores (isl.) 860,328 ... G7
Flores (sea) ... F7
Frederik Hendrik (Kolepom)
(isl.) ... K7
Geelvink (Cenderawasih)
(bay) ... K6
Great Kai (isl.) 38,748 ... J7
Halmahera (isl.) 122,521 ... H5
Irian Jaya (reg.) 923,440 ... K6
Jambuair (cape) ... B4
Jamursba (cape) ... J6
Java (head) ... C7
Java (isl.) 73,712,411 ... J2
Java (sea) ... D6
Jaya, Puncak (mt.) ... K6
Jayawijaya (range) ... K6
Jemaja (isl.) 5,628 ... D5
Kabaena (isl.) ... G7
Kai (isls.) 108,328 ... J7
Kalao (isl.) ... G7
Kalaotoa (isl.) ... G5

Kalimantan (reg.) 4,956,865 ... E5
Kangean (isls.) ... F7
Kapuas (riv.) ... D6
Karakelong (isl.) ... H5
Karimata (arch.) 9,398 ... D6
Karimunjawa (isls.) 5,025 ... J1
Kerinci (mt.) ... C6
Kisar (isl.) ... H7
Komodo (isl.) 30,407 ... F7
Krakatau (Rakata) (isl.) ... C7
Laut (isl.) 55,711 ... F6
Leuser (mt.) ... B5
Lingga (arch.) 46,658 ... D5
Lingga (isl.) 18,027 ... D6
Lombok (isl.) 1,581,193 ... F7
Madura (isl.) 1,509,774 ... K2
Mahakam (riv.) ... F6
Makassar (str.) ... F6
Malacca (str.) ... C5
Mamberamo (riv.) ... K6
Maoke (mts.) ... K6
Mapia (isl.) ... J5
Mentawai (isls.) 30,107 ... B6
Misool (isl.) ... J6
Molucca (sea) ... H6
Moluccas (isls.) 944,240 ... H6
Morotai (isl.) 27,333 ... H5
Muli (str.) ... K7
Müller (mts.) ... E5
Muna (isl.) 156,186 ... G7
Musi (riv.) ... C6
Natuna (isls.) 23,893 ... D5
Ngunju (cape) ... F8
Nias (isl.) 356,093 ... B5
Numfoor (isl.) ... J6
Obi (isls.) 12,437 ... H6
Ombai (str.) ... H7
Pantar (isl.) 28,259 ... G7
Perkam (cape) ... K6
Puting, Borneo (cape) ... E6
Puting, Sumatra (cape) ... C7
Raja Ampat Group (isls.) ... H6
Rakata (isl.) ... C7
Rantekombola (mt.) ... F6
Raya (mt.) ... E6
Riau (arch.) 483,230 ... C5
Rokan (riv.) ... C5
Roti (isl.) 76,270 ... G8
Salawati (isl.) ... J6
Sanghie (isl.) ... H5
Sanghie (isls.) 183,000 ... H5
Sawu (isls.) 51,002 ... G8
Sawu (sea) ... G7
Schouten (isls.) 110,148 ... K6
Schwaner (mts.) ... E6
Sebuku (bay) ... F5
Selatan (cape) ... E6
Selayar (isl.) 92,342 ... G7
Semeru (mt.) ... K2
Siau (isl.) 46,801 ... H5

Siberut (str.) ... B6
Simeulue (isl.) 29,147 ... A5
Singkep (isl.) 28,631 ... D6
Sipura (isl.) 6,051 ... B6
Slamet (mt.) ... J2
Sorikmerapi (mt.) ... B5
South Natuna (isls.) ... D5
Sudirman (range) ... K6
Sula (isls.) 36,922 ... H6
Sulawesi (isl.) 7,732,383 ... G6
Sumatra (isl.) 19,360,400 ... B5
Sumba (isl.) 291,190 ... F7
Sumba (str.) ... F7
Sunda (isl.) 621,140 ... F7
Sunda (str.) ... C7
Tahulandang (isl.) 21,493 ... H5
Talaud (isls.) 46,395 ... H5
Taliabu (isl.) 18,303 ... G6
Tambelan (isls.) 4,032 ... D5
Tanimbar (isls.) 55,405 ... J7
Tariku (riv.) ... K6
Tidore (isl.) 28,655 ... H5
Timor (reg.) 1,435,527 ... H7
Timor (isl.) ... H7
Toba (lake) ... B5
Tolo (gulf) ... G6
Tomini (gulf) ... G6
Tukangbesi (isls.) 73,106 ... G7
Vals (cape) ... K7
Vogelkop (Doberai) (pen.) ... J6
Waigeo (isl.) ... J5

MALAYSIA

STATES

North Borneo (Sabah)
1,002,608 ... F3
Sarawak 1,294,753 ... E5

CITIES and TOWNS

Beaufort 2,709 ... F4
Bintulu 4,424 ... E5
Kabong ... E5
Kampong Sibuti ... E5
Kapit 1,929 ... E5
Keningau 2,037 ... F4
Kota Kinabalu 40,939 ... F4
Kuching 63,535 ... E5
Kudat 5,089 ... F4
Labuan 7,216 ... F4
Lahad Datu 5,169 ... F4
Lamag ... F4
Marudi 4,700 ... E5
Miri 35,702 ... E5
Mukah 1,717 ... E5

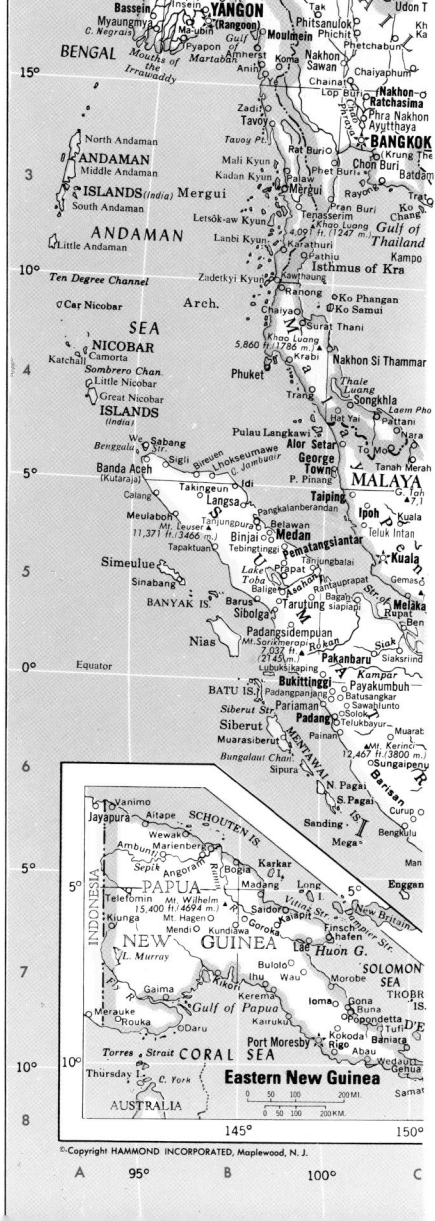

Topography

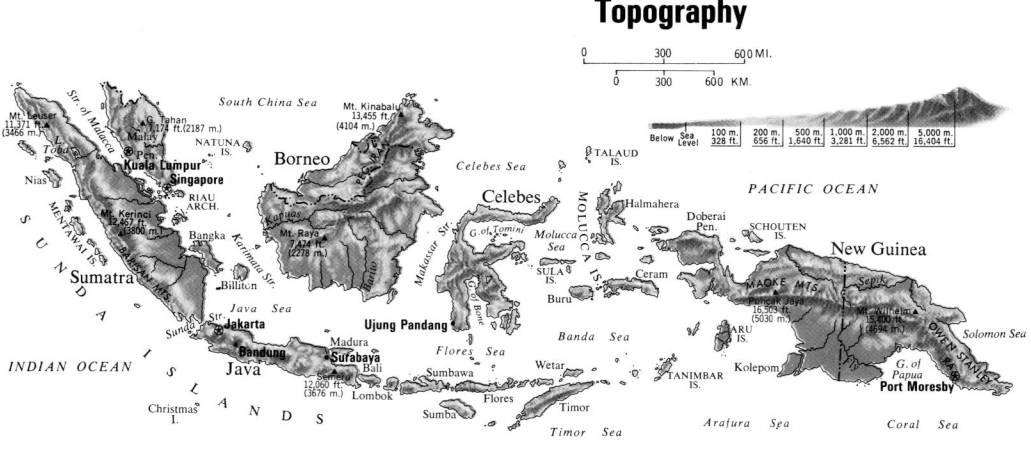

Agriculture, Industry and Resources

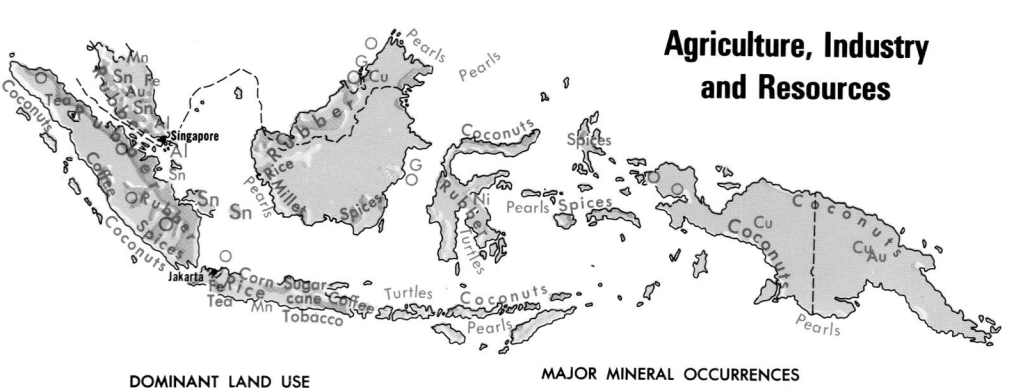

DOMINANT LAND USE

Cereals (chiefly rice, corn)

Diversified Tropical Crops

Forests

MAJOR MINERAL OCCURRENCES

Al Bauxite
Au Gold
C Coal
Cu Copper
Fe Iron Ore
G Natural Gas
Mn Manganese
Ni Nickel
O Petroleum
Sn Tin

Major Industrial Areas

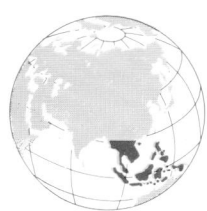

Papar 1,855 . . . . F4
Ranau 2,024 . . . . F4
Sandakan 42,413 . . . . F4
Sematan D5
Semporna 3,371 . . . . F5
Serian 2,209 . . . . E5
Sibu 50,635 . . . . E5
Simanggang 8,445 . . . E5
Suai E5
Tawau 24,247 . . . . . . F5
Weston F4

OTHER FEATURES

Balambangan (isl.) F4
Banggi (isl.) F4
Iran (mts.) E5
Kinabalu (mt.) F4
Labuan (isl.) 17,189 . . F4
Labuk (bay) F4
Rajang (riv.) E5
Sirik (cape) E5

PAPUA NEW GUINEA

CITIES and TOWNS

Abau C7
Aitape 3,368 . . . . B6
Ambunti 1,035 . . . . . . B6
Angoram 1,846 . . . . . . B6

Baniara C7
Bogia 755 . . . . B6
Bulolo 6,730 . . . . B7
Buna C7
Daru 7,127 . . . . B7
Finschhafen 756 . . . . C7
Gaima C8
Gehua C8
Gona C7
Goroka 18,511 . . . . . . B7
Ihu 541 . . . . B7
Ioma C7
Kaiapit 515 . . . . B7
Kairuku C7
Kerema 3,389 . . . . . . B7
Kikori 763 . . . . B7
Kiunga 1,407 . . . . B7
Kokoda C7
Kundiawa 4,299 . . . . . B7
Lae 61,617 . . . . B7
Madang 21,335 . . . . . B7
Marienberg B6
Mendi 4,130 . . . . B7
Morobe C7
Mount Hagen 13,441 . . B7
Popondetta 6,429 . . . . C7
Port Moresby
 (cap.) 123,624 . . . . . B7
Rouka B7
Saidor 500 . . . . B7
Samarai 864 . . . . C8

Telefomin B7
Vanimo 3,071 . . . . . . B6
Wau 2,349 . . . . B7
Wedau C7
Wewak 19,890 . . . . . . B6

OTHER FEATURES

Dampier (str.) C7
D'Entrecasteaux (isls.) . C7
Fly (riv.) A7
Huon (gulf) C7
Karkar (isl.) B6
Kiriwina (isl.) C7
Long (isl.) B7
Louisiade (arch.) D8
Milne (bay) C8
Misima (isl.) C8
New Britain (isl.) 148,773 . C7
Ramu (riv.) B7
Rossel (isl.) D8
Schouten (isls.) B6
Sepik (riv.) B6
Solomon (sea) C7
Tagula (isl.) C8
Torres (str.) B7
Trobriand (isls.) C7
Vitiaz (str.) B7
Woodlark (isl.) C7

★See page 74 for other
Malaysian entries.

INDONESIA
AREA 788,430 sq. mi. (2,042,034 sq. km.)
POPULATION 179,136,000
CAPITAL Jakarta
LARGEST CITY Jakarta
HIGHEST POINT Puncak Jaya 16,503 ft.
 (5,030 m.)
MONETARY UNIT rupiah
MAJOR LANGUAGES Bahasa Indonesia,
 Indonesian and Papuan languages,
 English
MAJOR RELIGIONS Islam, tribal religions,
 Christianity, Hinduism

PAPUA NEW GUINEA
AREA 183,540 sq. mi. (475,369 sq. km.)
POPULATION 3,593,000
CAPITAL Port Moresby
LARGEST CITY Port Moresby
HIGHEST POINT Mt. Wilhelm 15,400 ft.
 (4,694 m.)
MONETARY UNIT kina
MAJOR LANGUAGES pidgin English,
 Hiri Motu, English
MAJOR RELIGIONS Tribal religions,
 Christianity

BRUNEI
AREA 2,226 sq. mi. (5,765 sq. km.)
POPULATION 249,000
CAPITAL Bandar Seri Begawan
LARGEST CITY Bandar Seri Begawan
HIGHEST POINT Pagon 6,070 ft.
 (1,850 m.)
MONETARY UNIT Brunei Dollar
MAJOR LANGUAGES Malay, English,
 Chinese
MAJOR RELIGIONS Islam, Buddhism,
 Christianity, tribal religions

INDONESIA

PAPUA NEW GUINEA

BRUNEI

FIJI

AREA 7,055 sq. mi. (18,272 sq. km.)
POPULATION 792,441
CAPITAL Suva
LARGEST CITY Suva
HIGHEST POINT Tomaniivi 4,341 ft.
 (1,323 m.)
MONETARY UNIT Fijian dollar
MAJOR LANGUAGES Fijian, Hindi, English
MAJOR RELIGIONS Protestantism, Hinduism

KIRIBATI

AREA 277 sq. mi. (717 sq. km.)
POPULATION 82,449
CAPITAL Tarawa
HIGHEST POINT (on Banaba I.) 285 ft. (87 m.)
MONETARY UNIT Australian dollar
MAJOR LANGUAGES I-Kiribati, English
MAJOR RELIGIONS Protestantism, Roman
 Catholicism

NAURU

AREA 7.7 sq. mi. (20 sq. km.)
POPULATION 10,390
CAPITAL Yaren (district)
MONETARY UNIT Australian dollar
MAJOR LANGUAGES Nauruan, English
MAJOR RELIGION Protestantism

MARSHALL ISLANDS

AREA 70 sq. mi. (181 sq. km.)
POPULATION 60,652
CAPITAL Majuro
MONETARY UNIT U.S. dollar
MAJOR LANGUAGES English, Marshallese,
 Japanese
MAJOR RELIGION Protestantism

SOLOMON ISLANDS

AREA 11,500 sq. mi. (29,785 sq. km.)
POPULATION 386,000
CAPITAL Honiara
LARGEST CITY Honiara
HIGHEST POINT Mount Popomanatseu
 7,647 ft. (2,331 m.)
MONETARY UNIT Solomon Islands dollar
MAJOR LANGUAGES English,
 pidgin English, Melanesian dialects
MAJOR RELIGIONS Tribal religions,
 Protestantism, Roman Catholicism

TONGA

AREA 289 sq. mi. (748 sq. km.)
POPULATION 105,000
CAPITAL Nuku'alofa
HIGHEST POINT Kao Island 3,389 ft. (1,033 m.)
MONETARY UNIT pa'anga
MAJOR LANGUAGES Tongan, English
MAJOR RELIGION Protestantism

TUVALU

AREA 9.78 sq. mi. (25.33 sq. km.)
POPULATION 10,000
CAPITAL Funafuti
MONETARY UNIT Australian dollar
MAJOR LANGUAGES English, Tuvaluan
MAJOR RELIGION Protestantism

MICRONESIA

AREA 271 sq. mi. (702 sq. km.)
POPULATION 122,950
CAPITAL Palikir
MONETARY UNIT U.S. dollar
MAJOR LANGUAGES English, Trukese,
 Pohnpeian, Yapese, Kosrean
MAJOR RELIGIONS Roman Catholicism,
 Protestantism

Abaiang (atoll) 3,296	H 5
Abemama (atoll) 2,300	H 5
Adamstown (cap.), Pitcairn Is. 54	N 8
Admiralty (isls.)	E 6
Agrihan (isl.)	E 4
Ailinglapalap (atoll) 1,385	G 5
Ailuk (atoll) 413	H 4
Aitutaki (atoll) 2,348	K 7
Alofi (cap.), Niue 960	K 7
Alotau 4,310	E 7
Ambrym (isl.) 6,324	G 7
American Samoa 32,297	J 7
Anaa (atoll) 444	M 7
Angaur (isl.) 243	D 5
Apataki (atoll)	M 7
Apia (cap.), Samoa 33,100	J 7
Arno (atoll) 1,487	H 5
Arorae (atoll) 1,626	H 6
Atafu (atoll) 577	J 6
Atiu (isl.) 1,225	L 8
Austral (isls.) 5,208	L 8
Avarua (cap.), Cook Is.	L 8
Babelthuab (isl.) 10,391	D 5
Baker (isl.)	J 5
Banaba (isl.) 2,314	G 6
Banks (isls.) 3,158	G 7
Belep (isls.) 624	G 8
Bellona (reefs)	G 8
Beru (atoll) 2,318	H 6
Bikini (atoll)	G 4
Bismarck (arch.) 218,339	E 6
Bonin (isls.) 1,879	E 3
Bora-Bora (isl.) 2,572	L 7
Bougainville (isl.) 71,761	F 6
Bounty (isls.)	H10
Bourail 3,149	G 8
Butaritari (atoll) 2,971	H 5
Caroline (isl.)	M 7
Caroline (isls.)	E 5
Chichi (isl.) 1,879	E 3
Choiseul (isl.) 10,349	F 6
Christmas (Kiritimati) (isl.) 674	L 5
Cook (isls.) 17,695	K 7
Coral (sea)	F 7
Danger (Pukapuka) (atoll) 797	K 7
Daru 7,127	E 6
Disappointment (isls.) 373	N 7
Ducie (isl.)	O 8
Easter (isl.) 1,598	Q 8
Ebon (atoll) 887	G 5
Efate (isl.) 18,038	G 7
Enderbury (atoll)	J 6
Enewetak (Eniwetok) (atoll) 542	G 4
Erromanga (isl.) 945	H 7
Espíritu Santo (isl.) 16,220	G 7
Fais (isl.) 207	E 5
Fakaofo (atoll) 654	J 6
Fanning (Tabuaeran) (isl.), 340	L 5
Faraulep (atoll) 132	E 5
Fatuhiva (isl.) 386	N 7
Fiji 792,441	H 8
Flint (isl.)	L 7
Fly (riv.)	E 6
Funafuti (cap.), Tuvalu	H 6
French Polynesia 137,382	L 8
Funafuti (atoll) 2,120	H 6
Futuna (Hoorn) (isls.) 3,173	J 7
Gambier (isls.) 556	N 8
Gardner (Nukumaroro) (isl.)	J 6
Gilbert (isls.) 47,711	H 6
Greenwich (Kapingamarangi) (atoll) 508	F 5
Guadalcanal (isl.) 46,619	F 7
Guam (isl.) 105,979	E 4
Hagåtña (cap.) Guam 896	E 4
Hall (isls.) 647	F 5
Hawaiian (isls.) 964,691	J 3
Henderson (isl.)	O 8
Hivaoa (isl.) 1,445	N 6
Honiara (cap.), Solomon Is. 14,942	F 6
Hoorn (isls.) 3,173	J 7
Howland (isl.)	J 5
Huahine (isl.) 3,140	L 7
Hull (Orona) (isl.)	J 6
Huon (Gulf)	E 6
Ifalik (atoll) 389	E 5
Iwo (isl.)	E 3
Jaluit (atoll) 1,450	G 5
Jarvis (isl.)	K 6
Johnston (atoll) 327	K 4
Kadavu (Kandavi) (isl.) 8,699	H 7
Kanton (isl.)	J 6
Kapingamarangi (atoll) 508	F 5
Kavieng 4,633	E 6
Kermadec (isls.) 5	J 9
Kieta 3,491	F 6
Kimbe 4,662	E 6
Kingman (reef)	K 5
Kiribati 82,449	H 6
Kiritimati (isl.) 674	L 5
Koror (cap.), Palau 6,222	D 5
Kosrae (isl.) 5,491	G 5
Kwajalein (atoll) 6,624	G 5
Lae 61,617	E 6
Lau Group (isls.) 14,452	J 7
Lavongai (isl.)	F 6
Lifu (isl.) 7,585	G 8
Line (isls.)	K 5
Little Makin (isl.) 1,445	H 5
Lord Howe (Ontong Java) (isl.) 1,082	F 6
Lord Howe (isl.) 287	G 9
Lorengau 3,986	E 6
Louisiade (arch.)	F 7
Loyalty (isls.) 14,518	G 8
Luganville 4,935	G 7
Madang 21,335	E 6

Majuro (atoll) (cap.), Marshall Is. 8,583	H 5
Makin (Butaritari) (atoll) 2,971	H 5
Malaita (isl.) 50,912	G 6
Malden (isl.)	L 6
Malekula (isl.) 15,931	G 7
Maloelap (atoll) 763	H 5
Mangaia (isl.) 1,364	L 8
Mangareva (isl.) 556	N 8
Manihiki (atoll) 405	K 7
Manua (isl.) 1,459	K 7
Manus (isl.) 25,844	E 6
Marcus (isl.)	F 3
Maré (isl.) 4,156	G 8
Marianas, Northern 16,780	E 4
Mariana Trench	E 4
Marquesas (isls.) 5,419	N 6
Marshall Islands 60,652	G 4
Marutea (atoll)	M 7
Mata Utu (cap.), Wallis and Futura 558	J 7
Mauke (isl.) 684	L 8
Melanesia (reg.)	E 5
Micronesia (reg.)	E 4
Micronesia Federated States, of 122,950	J 7
Midway (isls.) 453	J 3
Mili (atoll) 763	H 5
Moen (isl.) 10,351	F 5
Moorea (isl.) 5,788	L 7
Mururoa (isl.)	M 8
Nadi 6,938	H 7
Namonuito (atoll) 783	E 5
Namorik (atoll) 617	G 5
Nanumea (atoll) 844	H 6
Nauru 10,390	G 8
Ndeni (isl.) 4,854	G 7
New Britain (isl.) 148,773	F 6
New Caledonia 133,233	G 8
New Caledonia (isl.) 118,715	G 8
New Georgia (isl.) 16,472	F 6
New Guinea (isl.)	E 6
New Ireland (isl.) 65,657	F 6
Ngatik (atoll) 560	F 5
Ngulu (atoll) 21	D 5
Niuatoputapu (isl.) 1,650	J 7
Niue (isl.) 3,578	K 7
Niutao (atoll) 866	H 6
Nomoi (isls.) 1,879	F 5
Nonouti (atoll) 2,223	H 6
Norfolk Island (terr.) 2,175	G 8
Northern Marianas 116,780	E 4
Nouméa (cap.), New Caled. 56,078	G 8
Nouméa *74,335	G 8
Nui (atoll) 603	H 6
Nuku'alofa (cap.) Tonga 18,356	J 8
Nukuhiva (isl.) 1,484	M 6

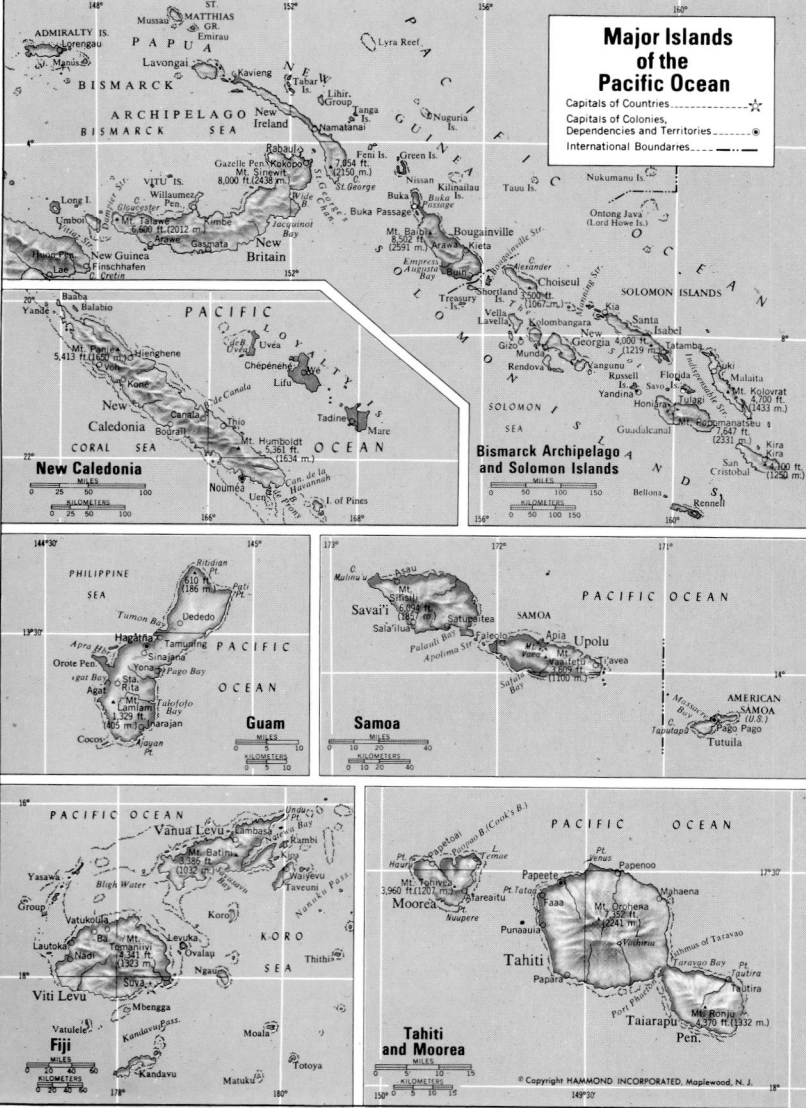

Major Islands of the Pacific Ocean

Capitals of Countries .... ☆
Capitals of Colonies, Dependencies and Territories ...... ◉
International Boundaries ...... ——— ·—·—

New Caledonia

Bismarck Archipelago and Solomon Islands

Guam

Samoa

Fiji

Tahiti and Moorea

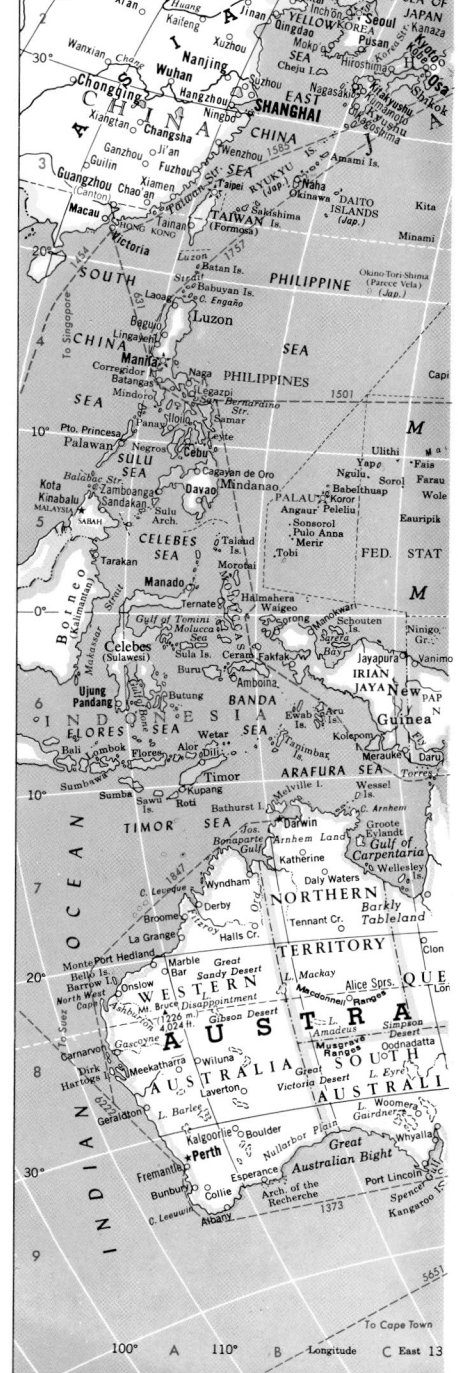

VANUATU

AREA 5,700 sq. mi. (14,763 sq. km.)
POPULATION 170,000
CAPITAL Port-Vila
LARGEST CITY Port-Vila
HIGHEST POINT Mt. Tabwemasana
  6,165 ft. (1,879 m.)
MONETARY UNIT Vatu
MAJOR LANGUAGES Bislama, English,
  French
MAJOR RELIGIONS Christian, animist

SAMOA

AREA 1,133 sq. mi. (2,934 sq. km.)
POPULATION 204,000
CAPITAL Apia
LARGEST CITY Apia
HIGHEST POINT Mt. Silisili 6,094 ft.
  (1,857 m.)
MONETARY UNIT tala
MAJOR LANGUAGES Samoan, English
MAJOR RELIGIONS Protestantism,
  Roman Catholicism

PALAU

AREA 177 sq. mi. (458 sq. km.)
POPULATION 15,122
CAPITAL Koror
HIGHEST POINT Mt. Makelulu 804 ft.
  (242 m.)
MONETARY UNIT U.S. dollar
MAJOR LANGUAGES English,
  Sonsorolese, Angaur, Japanese,
  Tobi, Palauan
MAJOR RELIGIONS Christian,
  Modekngei

*City and suburbs.
•Population of urban area.

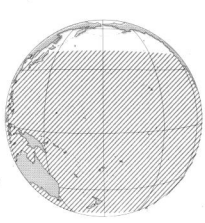

FIJI
TONGA
KIRIBATI
TUVALU
NAURU
VANUATU
SOLOMON ISLANDS
SAMOA
MARSHALL ISLANDS
MICRONESIA
PALAU

Pacific Ocean

LAMBERT AZIMUTHAL EQUAL-AREA PROJECTION
©Copyright HAMMOND INCORPORATED, Maplewood, N.J.

NAUTICAL MILES
0 200 400 600 800 1000 1200
STATUTE MILES
0 200 400 600 800 1000 1200
KILOMETERS
0 200 400 600 800 1000 1200

Capitals of Countries . . . . ★
Capitals of Colonies,
  Dependencies, States and Territories . ★
Administrative Centers . . . . ⊛
International Boundaries _ _ _ _ _ _ _
Internal Boundaries _ _ _ _ _ _ _ _ _
Railroads _ _ _ _ _ _ _ _ _ _ _
Distances Between Points _ _ _ 5444
                        (nautical miles)

Scale 1:50,000,000

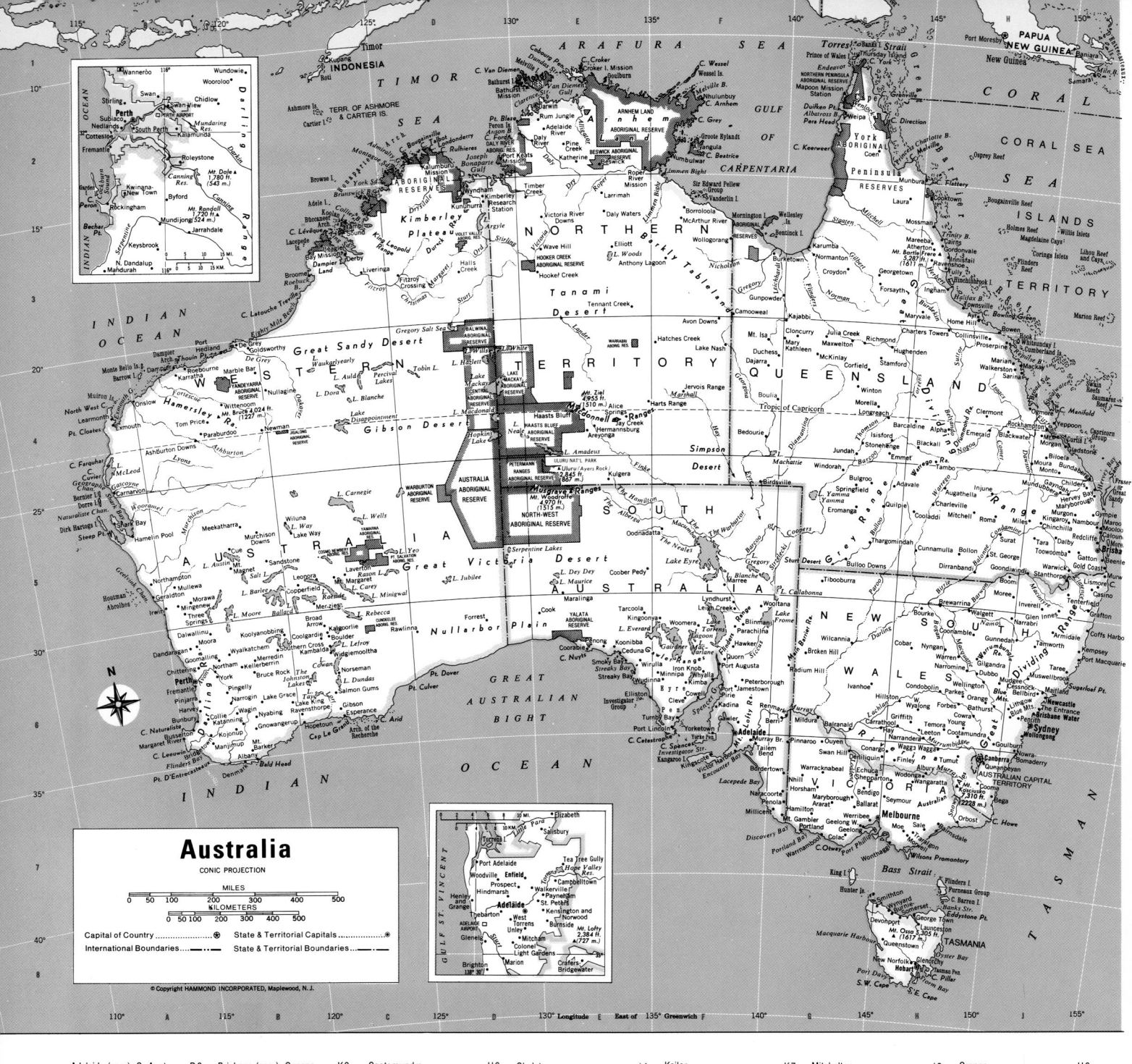

Australia

CONIC PROJECTION

MILES
0 50 100 200 300 400 500

KILOMETERS
0 50 100 200 300 400 500

Capital of Country .... ⊛   State & Territorial Capitals .... ⊛
International Boundaries ....   State & Territorial Boundaries ....

© Copyright HAMMOND INCORPORATED, Maplewood, N.J.

AREA 2,966,136 sq. mi. (7,682,300 sq. km.)
POPULATION 15,602,156
CAPITAL Canberra
LARGEST CITY Sydney
HIGHEST POINT Mt. Kosciusko 7,310 ft. (2,228 m.)
LOWEST POINT Lake Eyre -39 ft. (-12 m.)
MONETARY UNIT Australian dollar
MAJOR LANGUAGE English
MAJOR RELIGIONS Protestantism, Roman Catholicism

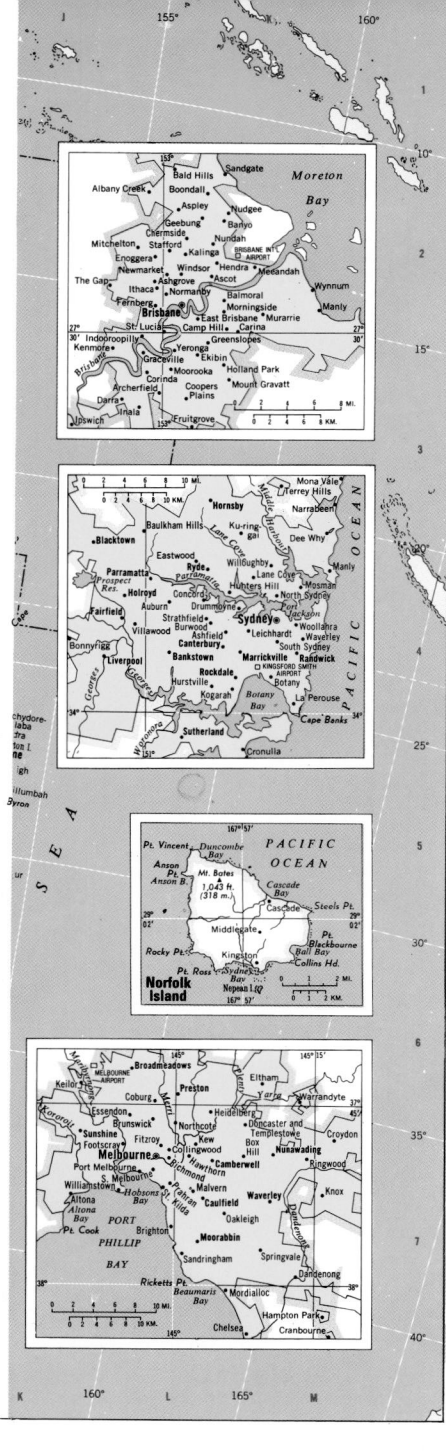

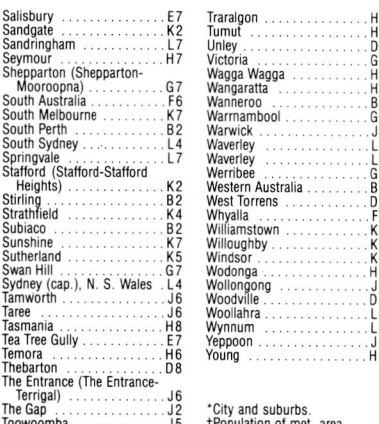

Population Distribution

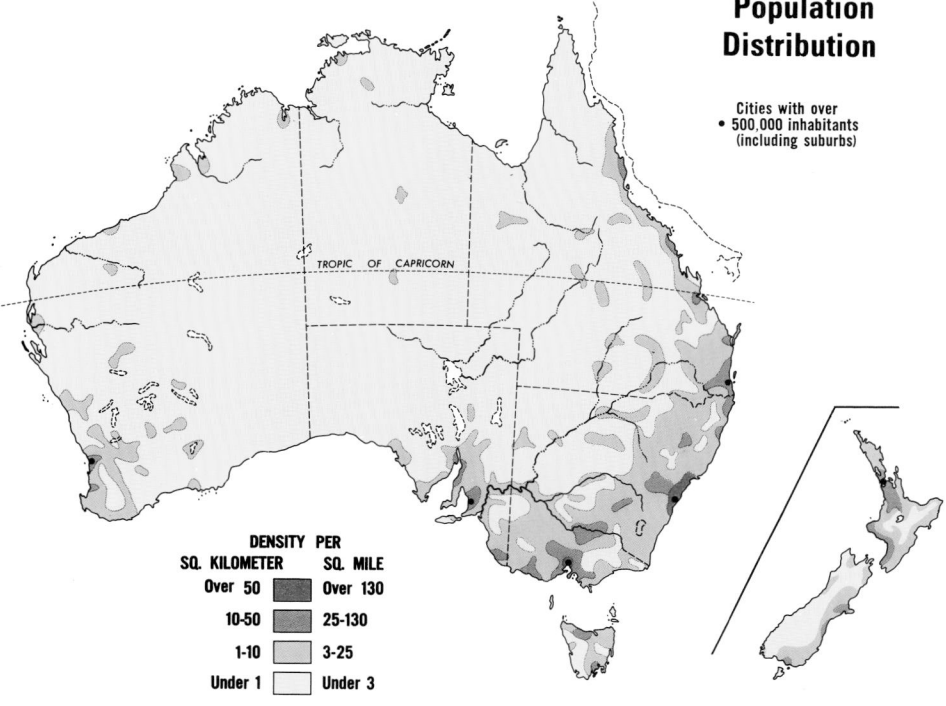

• Cities with over 500,000 inhabitants (including suburbs)

DENSITY PER

SQ. KILOMETER	SQ. MILE
Over 50	Over 130
10-50	25-130
1-10	3-25
Under 1	Under 3

Vegetation

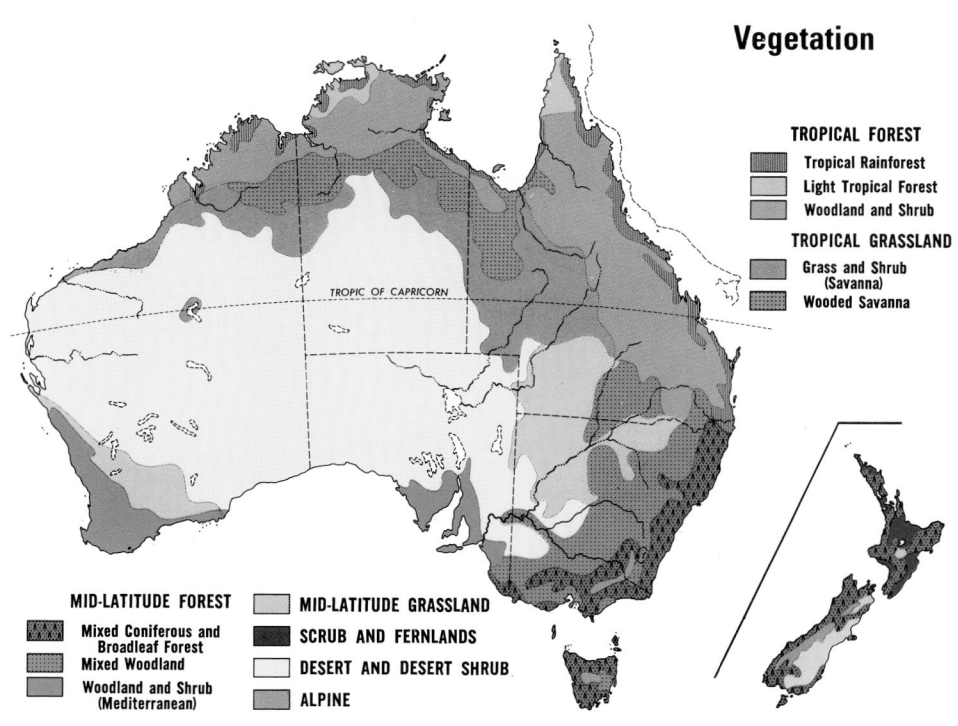

TROPICAL FOREST
- Tropical Rainforest
- Light Tropical Forest
- Woodland and Shrub

TROPICAL GRASSLAND
- Grass and Shrub (Savanna)
- Wooded Savanna

MID-LATITUDE FOREST
- Mixed Coniferous and Broadleaf Forest
- Mixed Woodland
- Woodland and Shrub (Mediterranean)

MID-LATITUDE GRASSLAND

SCRUB AND FERNLANDS

DESERT AND DESERT SHRUB

ALPINE

Average January Temperature

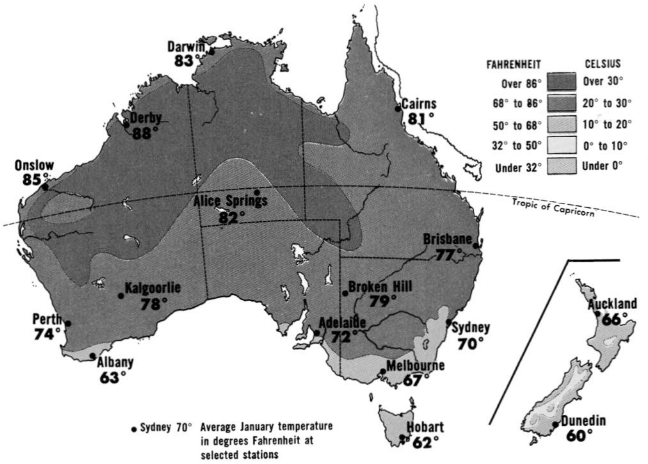
FAHRENHEIT
CELSIUS
Over 86° Over 30°
68° to 86° 20° to 30°
50° to 68° 10° to 20°
32° to 50° 0° to 10°
Under 32° Under 0°
Darwin 83°
Derby 88°
Cairns 81°
Onslow 85°
Alice Springs 82°
Tropic of Capricorn
Brisbane 77°
Kalgoorlie 78°
Broken Hill 79°
Perth 74°
Adelaide 72°
Sydney 70°
Albany 63°
Melbourne 67°
Hobart 62°
Auckland 66°
Dunedin 60°
• Sydney 70° Average January temperature in degrees Fahrenheit at selected stations

Average July Temperature

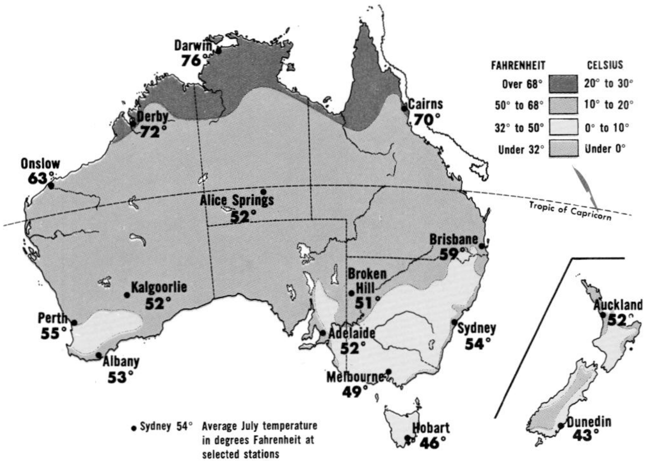
FAHRENHEIT
CELSIUS
Over 68° 20° to 30°
50° to 68° 10° to 20°
32° to 50° 0° to 10°
Under 32° Under 0°
Darwin 76°
Derby 72°
Cairns 70°
Onslow 63°
Alice Springs 52°
Tropic of Capricorn
Brisbane 59°
Broken Hill 51°
Kalgoorlie 52°
Perth 55°
Adelaide 52°
Sydney 54°
Albany 53°
Melbourne 49°
Hobart 46°
Auckland 52°
Dunedin 43°
• Sydney 54° Average July temperature in degrees Fahrenheit at selected stations

Rainfall

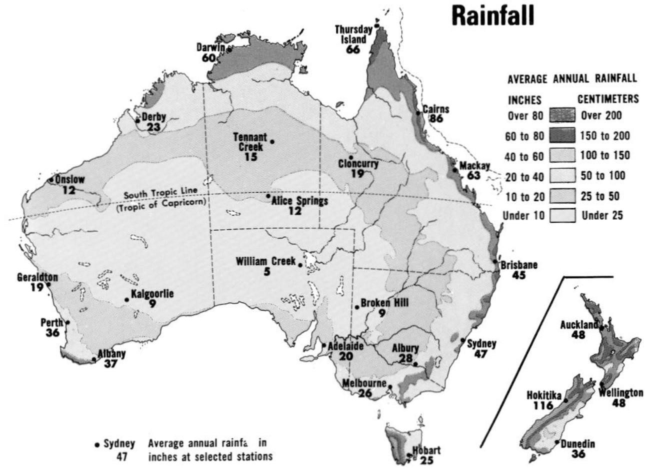
AVERAGE ANNUAL RAINFALL
INCHES CENTIMETERS
Over 80 Over 200
60 to 80 150 to 200
40 to 60 100 to 150
20 to 40 50 to 100
10 to 20 25 to 50
Under 10 Under 25
Thursday Island 66
Darwin 60
Derby 23
Cairns 86
Tennant Creek 15
Cloncurry 19
Mackay 63
Onslow 12
South Tropic Line (Tropic of Capricorn)
Alice Springs 12
William Creek 5
Brisbane 45
Geraldton 19
Kalgoorlie 9
Broken Hill 9
Perth 36
Adelaide 20
Albury 28
Sydney 47
Albany 37
Melbourne 26
Hobart 25
Auckland 48
Hokitika 116
Wellington 48
Dunedin 36
• Sydney 47 Average annual rainfall in inches at selected stations

DOMINANT LAND USE

- Cereals (chiefly wheat), Livestock
- Dairy, Truck Farming
- Cash Crops, Horticulture, Fruit
- Pasture Livestock
- Range Livestock
- Forests
- Nonagricultural Land

MAJOR MINERAL OCCURRENCES

Ab	Asbestos	Na	Salt
Ag	Silver	Ni	Nickel
Al	Bauxite	O	Petroleum
Au	Gold	Op	Opals
C	Coal	P	Phosphates
Cu	Copper	Pb	Lead
D	Diamonds	S	Sulfur, Pyrites
Fe	Iron Ore	Sb	Antimony
G	Natural Gas	Sn	Tin
Gp	Gypsum	Ti	Titanium
Lg	Lignite	U	Uranium
Ls	Limestone	W	Tungsten
Mg	Magnesium	Zn	Zinc
Mi	Mica	Zr	Zirconium
Mn	Manganese		

Water Power

Major Industrial Areas

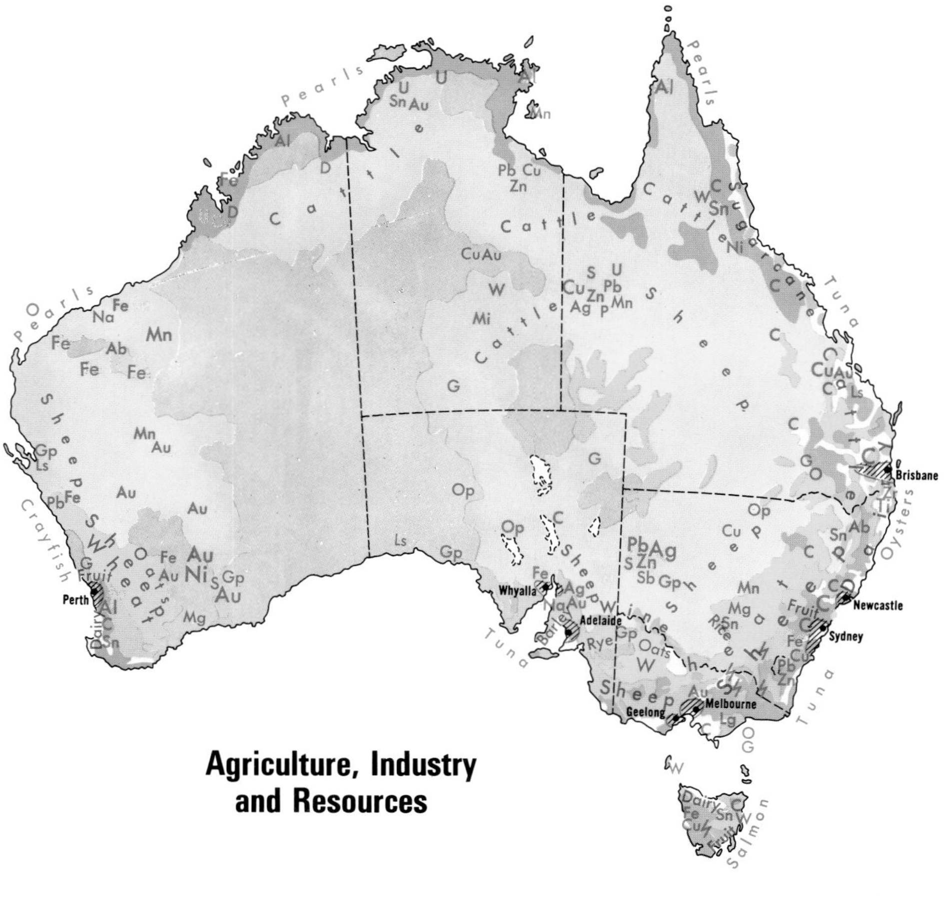
Pearls
Cattle
Sheep
Sugarcane
Tuna
Oysters
Crayfish
Fruit
Dairy
Wheat
Oats
Rye
Rice
Salmon
Perth
Whyalla
Adelaide
Geelong
Melbourne
Sydney
Newcastle
Brisbane

Agriculture, Industry and Resources

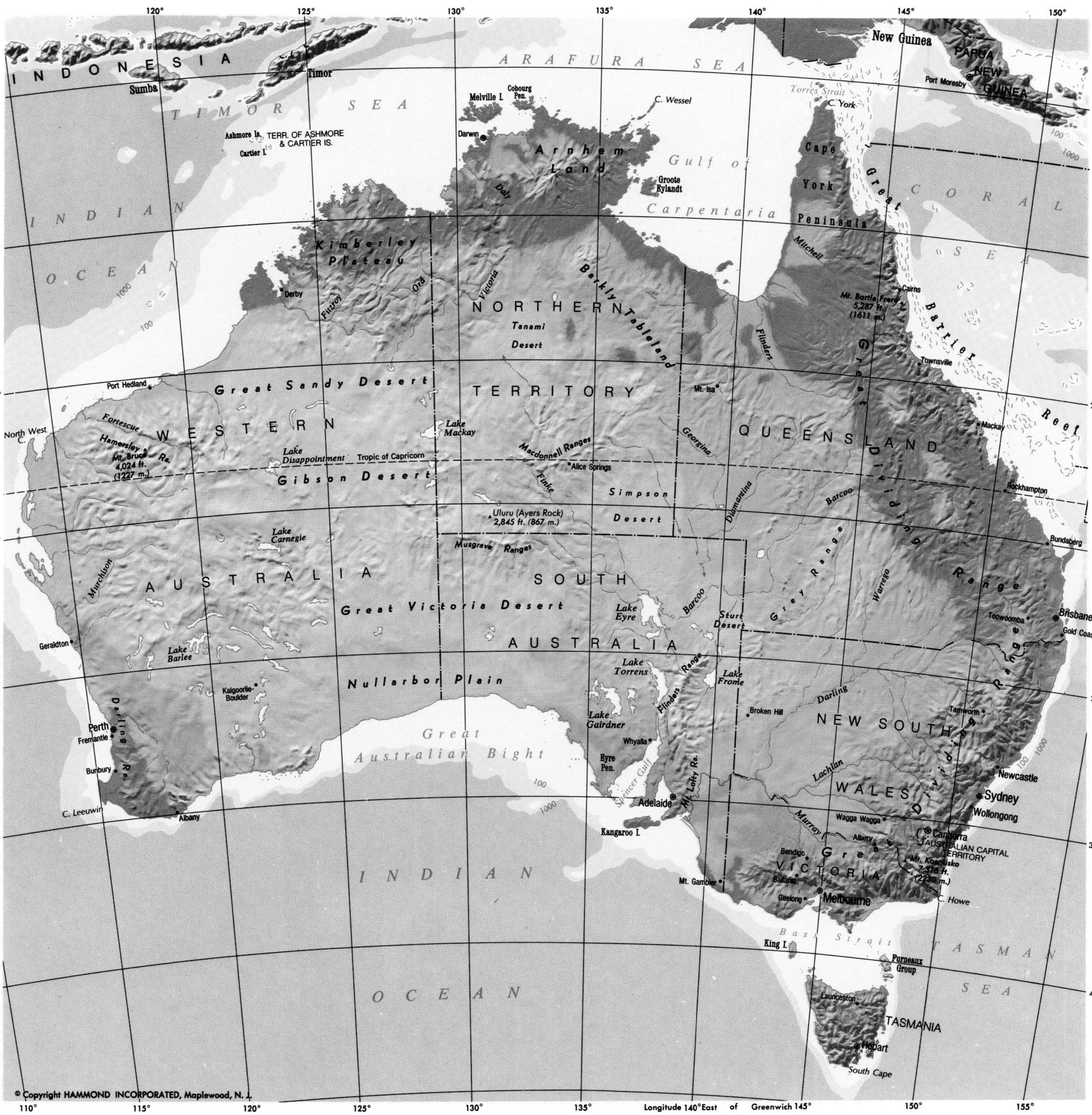

Vegetation/Relief

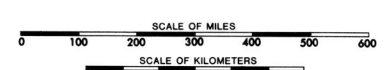

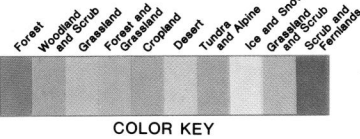

Capital of Country .... ⊕
State and Territorial Capitals .... ●
International Boundaries .... ——
State and Territorial Boundaries .... —·—

Elevations in Feet   Depths in Fathoms

Forest | Woodland and Scrub | Grassland | Forest and Grassland | Grassland and Cropland | Desert | Tundra and Alpine | Ice and Snow | Grassland and Scrub | Scrub and Fernlands

COLOR KEY

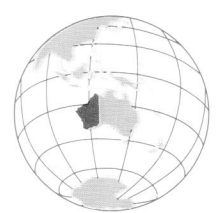

AREA 975,096 sq. mi.
(2,525,500 sq. km.)
POPULATION 1,406,929
CAPITAL Perth
LARGEST CITY Perth
HIGHEST POINT Mt. Bruce 4,024 ft.
(1,227 m.)

Topography

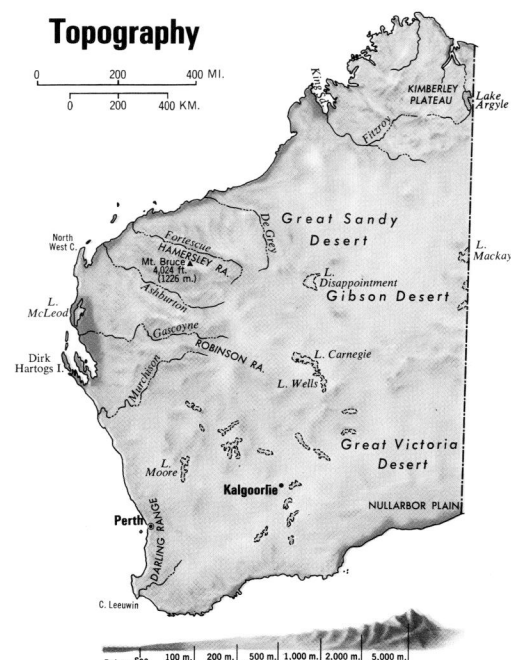

CITIES and TOWNS

Albany 15,222 . . . . B6
Augusta 588 . . . . A6
Australind 1,681 . . . . A2
Balladonia D6
Beverley 756 . . . . B1
Boddington 367 . . . . B6
Boulder-Kalgoorlie 19,848 . . C5
Boyanup 365 . . . . A2
Bridgetown 1,521 . . . . B6
Brookton 595 . . . . B6
Broome 3,666 . . . . C2
Bruce Rock 565 . . . . B5
Brunswick Junction 889 . . . . A2
Bunbury 21,749 . . . . A6
Busselton 6,463 . . . . A6
Canning 52,816 . . . . A1
Capel 680 . . . . A2
Carnamah 422 . . . . A5
Carnarvon 5,053 . . . . A4
Collie 7,667 . . . . B6
Coolgardie 891 . . . . C5

Coorow 226 . . . . B5
Corrigin 841 . . . . B6
Cranbrook 316 . . . . B6
Cuballing ○647 . . . . B6
Cue 320 . . . . B4
Cunderdin 731 . . . . B5
Dalwallinu 639 . . . . B5
Dampier 2,471 . . . . B3
Dandaragan ○1,748 . . . . . . A5
Darkan 242 . . . . B6
Denham 402 . . . . A4
Denmark 985 . . . . B6
Derby 2,933 . . . . C2
Dongara-Port Denison 1,155 A5
Donnybrook 1,197 . . . . A2
Dwellingup 453 . . . . A2
Esperance 6,375 . . . . C6
Eucla E2
Exmouth 2,583 . . . . A3
Fitzroy Crossing D2
Fremantle 22,484 . . . . A1
Geraldton 20,895 . . . . A5
Gingin 382 . . . . A1
Gnowangerup 872 . . . . B6

Goldsworthy 923 . . . . B3
Goomalling 600 . . . . B1
Halls Creek 966 . . . . D2
Harvey 2,479 . . . . A2
Hopetoun C6
Hyden B6
Jarrahdale 315 . . . . B2
Kalbarri 820 . . . . A4
Kalgoorlie 9,145 . . . . C5
Kalgoorlie-Boulder 19,848 . . C5
Kambalda 4,463 . . . . C5
Karratha 8,341 . . . . B3
Katanning 4,413 . . . . B6
Kellerberrin 1,091 . . . . B5
Kojonup 544 . . . . B6
Koolyanobbing 277 . . . . . . B5
Kununurra E2
Kwinana New Town 12,355 . . A1
Lake Grace 575 . . . . B6
Laverton 872 . . . . C5
Learmonth A3
Leonora 524 . . . . C5
Madura D5
Mandurah 10,978 . . . . A2

Manjimup 4,150 . . . . B6
Marble Bar 357 . . . . C3
Margaret River 798 . . . . . . A6
Meekatharra 989 . . . . B4
Melville 61,211 . . . . A1
Menzies 232 . . . . C5
Merredin 3,520 . . . . B5
Mingenew 368 . . . . A5
Moora 1,677 . . . . B5
Morawa 694 . . . . B5
Mount Barker 1,519 . . . . . . B6
Mount Magnet 618 . . . . B5
Mukinbudin 370 . . . . B5
Mullewa 918 . . . . A5
Mundijong 356 . . . . A2
Nannup 552 . . . . B6
Narrogin 4,969 . . . . B2
Nedlands 20,257 . . . . A1
New Norcia A5
Newman 5,466 . . . . B3
Norseman 1,895 . . . . C6
Northam 6,791 . . . . B1
Northampton 750 . . . . A5
Northcliffe B6
Nungarin ○332 . . . . B5
Onslow 594 . . . . A3
Pannawonica 1,170 . . . . . . B3
Paraburdoo 2,357 . . . . B3
Pardoo B3
Pemberton 871 . . . . A6
Perenjori 257 . . . . B5
Perth (cap.) 809,035 . . . . . A1
Perth *898,918 . . . . A1
Pingelly 937 . . . . B2
Pinjarra 1,336 . . . . A2
Port Denison-Dongara 1,155 A5
Port Hedland 12,948 . . . . . B3
Quairading 741 . . . . B1
Ravensthorpe 327 . . . . B6
Rockingham 24,932 . . . . . . A2
Roebourne 1,688 . . . . B3

Sandstone ○133 . . . . B4
Shay Gap 853 . . . . C3
Southern Cross 798 . . . . . . B5
South Perth 31,524 . . . . . . A1
Stirling 161,858 . . . . A1
Three Springs 638 . . . . A5
Tom Price 3,540 . . . . B3
Toodyay 560 . . . . B1
Turkey Creek 212 . . . . E2
Wagin 1,488 . . . . B2
Walpole 291 . . . . B6
Wandering ○470 . . . . B2
Wanneroo 6,745 . . . . A1
Waroona 1,462 . . . . A2
Wickepin 267 . . . . B2
Wickham 2,387 . . . . B3
Williams 453 . . . . B2
Wiluna 221 . . . . C4
Wittenoom 247 . . . . B3
Wongan Hills 947 . . . . B1
Wundowie 720 . . . . B1
Wyalkatchem 453 . . . . B5
Wyndham 1,509 . . . . E1
Yalgoo ○315 . . . . B5
Yampi Sound C2
York 1,136 . . . . B1

OTHER FEATURES

Adele (isl.) C1
Admiralty (gulf) D1
Aloysius (mt.) E4
Argyle (lake) E2
Arid (cape) C6
Ashburton (riv.) A3
Augustus (mt.) B4
Austin (lake) B4
Australia Aboriginal Res. . . . E4
Bald (head) B6
Balwina Aboriginal Res. . . . E3
Barlee (lake) B5
Barrow (isl.) A3
Beaglebay Aboriginal Res. . . C2
Bluff Knoll (mt.) B6
Bonaparte (arch.) D1
Bougainville (cape) D1
Brassey (range) C4
Bruce (mt.) B3
Brunswick (bay) D1
Buccaneer (arch.) C2
Carey (lake) C5
Carnegie (lake) C4
Central Aboriginal Res. E3
Churchman (mt.) B5
Collier (bay) C1
Cosmo Newbery Aboriginal
  Res. C5
Cowan (lake) C5
Cundeelee Aboriginal Res. . . C5
Dale (mt.) B1
Dampier (arch.) B3
Dampier Land (reg.) C2
Darling (range) A1
De Grey (riv.) B3
D'Entrecasteaux (pt.) A6
Dirk Hartogs (isl.) A4
Disappointment (lake) C3
Drysdale (riv.) D1
Dundas (lake) C6
Egerton (mt.) B4
Eighty Mile (beach) B3
Enid (mt.) B3
Esperance (bay) C6

Exmouth (gulf) A3
Fitzroy (riv.) D2
Flinders (bay) A6
Forrest River Aboriginal Res. D1
Fortescue (riv.) B3
Garden (isl) A1
Gascoyne (riv.) B4
Geelvink (chan.) A5
Geographe (bay) A6
Geographe (chan.) A4
Gibson (des.) D3
Great Australian (bight) E6
Great Sandy (des.) C3
Great Victoria (des.) D5
Hamersley (range) B3
Hann (mt.) D1
Hopkins (lake) E4
Houtman Abrolhos (isls.) . . . A5
Indian Ocean A5
Johnston, The (lakes) C6
Joseph Bonaparte (gulf) . . . E1
Kimberley (plat.) D2
King (sound) C2
King Leopold (range) D2
Koolan (isl.) C1
Koolan (isl.) C1
Le Grand (cape) C6
Lévêque (cape) C2
Londonderry (cape) C1
Lyons (riv.) A4
Macdonald (lake) E3
Mackay (lake) E3
McLeod (lake) A4
Minigwal (lake) C5
Monte Bello (isls.) A3
Moore (lake) B5
Murchison (riv.) B4
Murray (riv.) A2
Naturaliste (cape) A6
Naturaliste (chan.) A4
North West (cape) A3
North-West Aboriginal Res. . E4
Nullarbor (plain) D5
Oakover (riv.) C3
Ord (mt.) D1
Ord (riv.) E2
Percival (lakes) D3
Peron (pen.) A4
Petermann (ranges) E4
Rason (lake) D5
Rebecca (lake) C5
Recherche (arch.) C6
Robinson (range) B4
Roebuck (bay) C2
Rottnest (isl.) A1
Saint George (ranges) D2
Shark (bay) A4
Southesk Tablelands D3
Sturt (creek) D2
Swan (riv.) A1
Timor (sea) D1
Tomkinson (ranges) E4
Wanna (lake) D4
Warburton Aboriginal Res. . . D4
Way (lake) C4
Weld (range) B4
Wells (lake) C4
Whaleback (mt.) B3
Wooramel (riv.) A4
York (sound) D1

○ Population of district.
*Population of met. area.

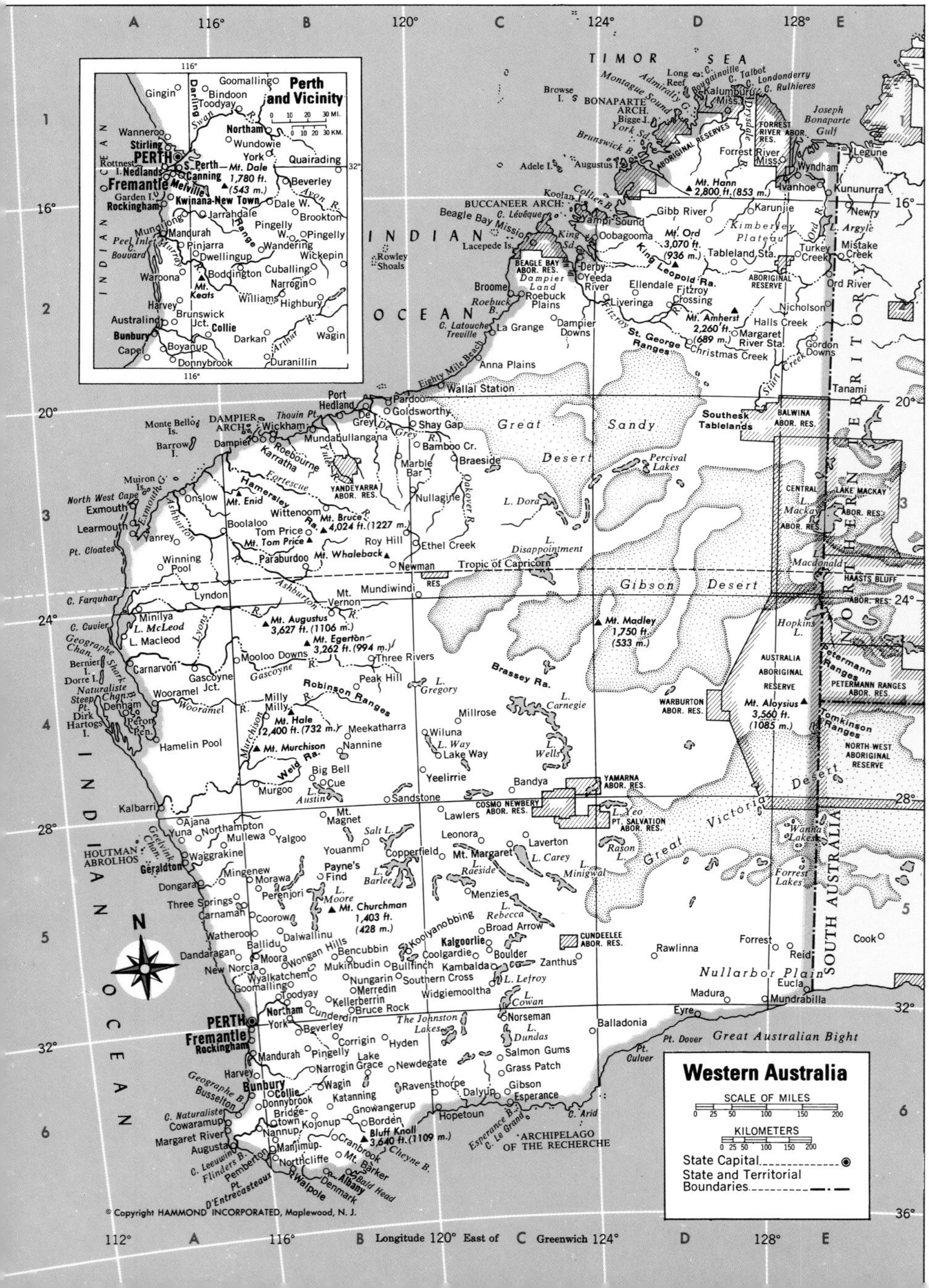

Western Australia

SCALE OF MILES

KILOMETERS

State Capital . . . . ◉
State and Territorial
Boundaries . . . .

© Copyright HAMMOND INCORPORATED, Maplewood, N. J.

CITIES and TOWNS

Adelaide River	B2
Aileron	C7
Alice Springs 18,395	D7
Alyangula 1,181	E2
Angurugu 597	E3
Anthony Lagoon	D4
Areyonga	C8
Arltunga	D7
Avon Downs	E5
Bamyili-Beswick 685	C3
Banka Banka	C5
Barrow Creek	D6
Batchelor	B2
Bathurst Island 1,032	B1
Birdum	C3
Birrimbah	C3
Birrindudu	A5
Borroloola 420	E4
Bundooma	D8
Burramurra	E6
Charlotte Waters	D8
Claravale	B3
Coniston	C7
Coolibah	B3
Creswell Downs	E4
Croker Island Mission	C1
Daly River	B2
Daly Waters	C4
Darwin (cap.) 56,482	B2
Docker River 217	A8
Elliott	C4
Epenarra	D6
Erldunda	C8
Eva Downs	D5
Ewaninga	D7
Goulburn Island 277	C1
Gove (Nhulunbuy) 3,879	E2
Harts Range	D7
Hatches Creek	D6
Helen Springs	C5
Henbury	C8
Hermannsburg 541	C7
Hooker Creek 671	B5
Humpty Doo	B2
Katherine 3,737	B3
Kildurk	A4
Koolpinyah	B2
Kulgera	C8
Kurundi	D6
Lake Nash	E6
Larrimah	C3
Legune	A3
Limbunya	B4
Lucy Creek	E7
Mainoru	C3
Maningrida 702	D2
Mataranka	C3
Milingimbi 564	D2
Mistake Creek	A4
Montejinni	C4
Mount Cavenagh	C8
Mount Doreen	B7
Murray Downs	D6
Napperby	C7
Newcastle Waters	C4
Nhulunbuy 3,879	E2
Numbulwar 422	D3
Oenpelli 452	C2
O. T. Downs	D4
Papunya 635	C7
Pine Creek 214	C2
Plenty River Mine	D7
Port Keats 819	A3
Powell Creek	C5
Rankine Store	E5
Robinson River	E4
Rockhampton Downs	D5
Rodinga	D8
Rum Jungle	B2
Santa Teresa 479	D8
Soudan	E6
Stirling Station	C6
Tanami	A5
Tarlton Downs	E7
Tea Tree Well	C7
Tempe Downs	C8
Tennant Creek 3,118	C5
The Granites	B6
Top Springs	C4
Ucharonidge	D4
Umbakumba 247	E3
Umbeara	C8
Urapunga	D3
Utopia	D7
Victoria River Downs	B4
Warrabri 459	D6
Warrego 991	C5
Wave Hill	B4
White Quartz Hill	D7
Willeroo	B3
Willowra	C6
Wollogorang	F4
Yambah	C7
Yirrkala 543	E2
Yuendumu 687	B7

OTHER FEATURES

Amadeus (lake)	B8
Arafura (sea)	D1
Arnhem (cape)	E2
Arnhem Land (reg.)	D2
Arnhem Land Aboriginal Res.	C2
Arnold (riv.)	D3
Barkly Tableland	D4
Bathurst (isl.)	A1
Beagle (gulf)	A2
Beatrice (cape)	E3
Bennett (lake)	B7
Beswick Aboriginal Res.	C3
Bickerton (isl.)	E2
Blaze (pt.)	A2
Carpentaria (gulf)	E3
Central Wedge (mt.)	C7
Clarence (str.)	B2
Cobourg (pen.)	C1
Conner (mt.)	B8
Croker (cape)	C1
Daly (riv.)	B2
Daly River Aboriginal Res.	A2
Davenport (mt.)	B7
Dundas (str.)	B1
East Alligator (riv.)	C2
Ehrenberg (range)	B7
Elcho (isl.)	D1
Finke (riv.)	C8
Fitzmaurice (riv.)	B3
Ford (cape)	A2
Georgina (riv.)	E6
Goulburn (isls.)	C1
Goyder (riv.)	D2
Groote Eylandt (isl.) 2,230	E3
Haasts Bluff Aboriginal Res.	B7
Hale (riv.)	D8
Hanson (riv.)	C6
Hay (dry riv.)	E7
Hogarth (mt.)	E6
Hopkins (lake)	A8
Joseph Bonaparte (gulf)	A3
Kata Tjuta (Olga) (mt.)	B8
Katherine (riv.)	C3
Lake MacKay Aboriginal Res.	A6
Lander (riv.)	C6
Leisler (mt.)	A7
Limmen Bight (riv.)	D4
Macdonald (lake)	B7
Macdonnell (ranges)	C7
MacKay (lake)	A7
Mann (riv.)	D2
Marshall (riv.)	D7
Melville (bay)	E2
Melville (isl.)	B1
Murchison (range)	D6
Napier (mt.)	A4
Neale (lake)	A8
Newcastle (creek)	C4
Nicholson (riv.)	E5
Peron (isls.)	A2
Petermann (ranges)	A8
Petermann Ranges Aboriginal Res.	A8
Port Darwin (inlet)	B2
Ranken (riv.)	E6
Robinson (riv.)	E4
Roper (riv.)	C3
Sandover (riv.)	D6
Simpson (des.)	E8
Singleton (mt.)	B6
Sir Edward Pellew Group (isls.)	E3
South Alligator (riv.)	C2
Stanley (mt.)	B7
Stewart (cape)	D1
Stirling (creek)	A4
Sturt (plain)	C4
Tanami (des.)	C5
Timor (sea)	A2
Todd (riv.)	D8
Uluru Nat'l Park	B8
Vanderlin (isl.)	E3
Van Diemen (cape)	A1
Van Diemen (gulf)	B1
Victoria (riv.)	B3
Wagait Aboriginal Res	B2
Warwick (chan.)	E3
Wessel (cape)	E1
Wessel (isls.)	E1
West Baines (riv.)	A4
White (lake)	A6
Woods (lake)	C4
Young (mt.)	D3
Ziel (mt.)	C7

AREA 519,768 sq. mi.
(1,346,200 sq. km.)
POPULATION 154,848
CAPITAL Darwin
LARGEST CITY Darwin
HIGHEST POINT Mt. Ziel 4,955 ft.
(1,510 m.)

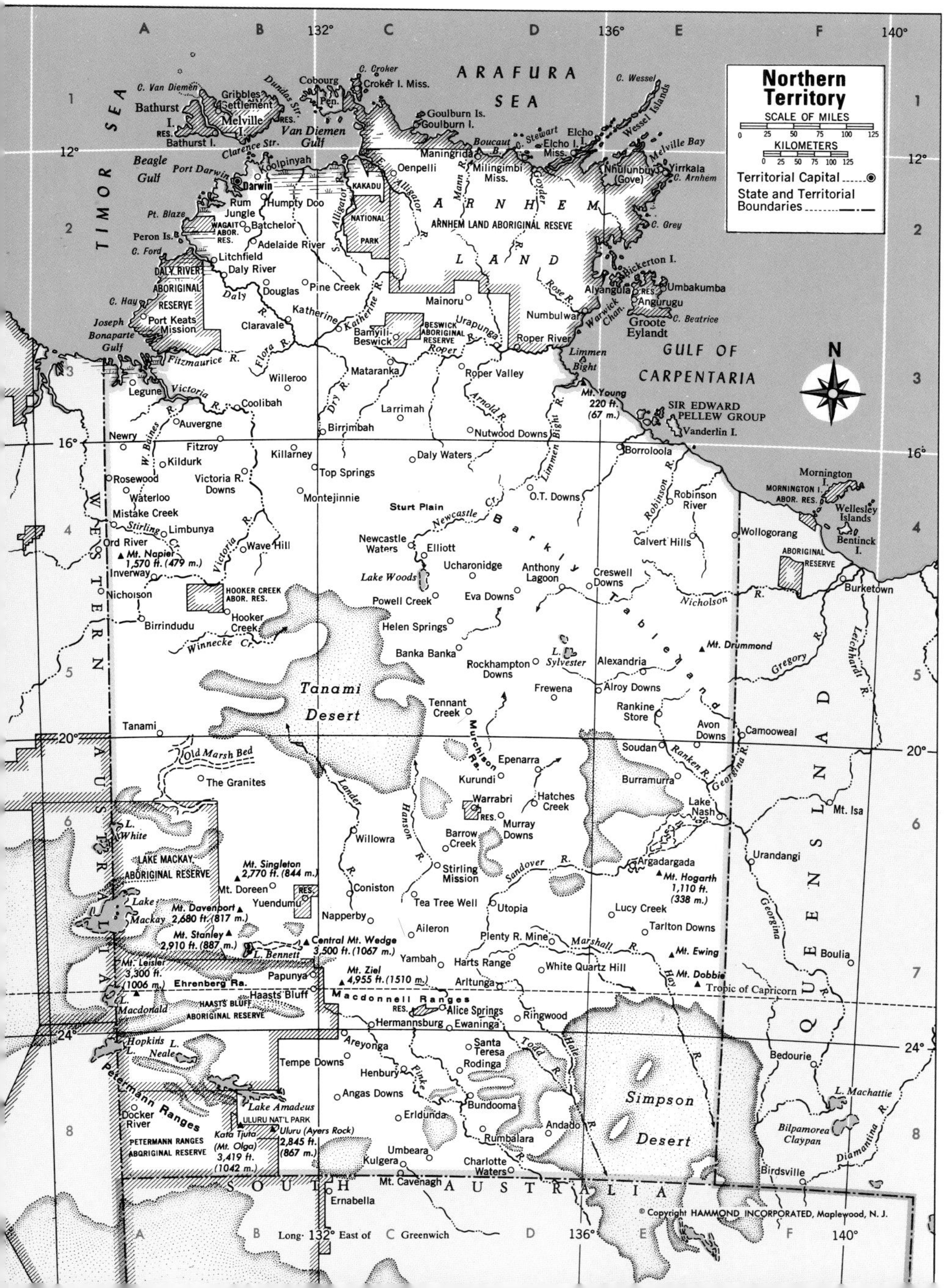

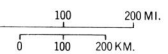

Topography

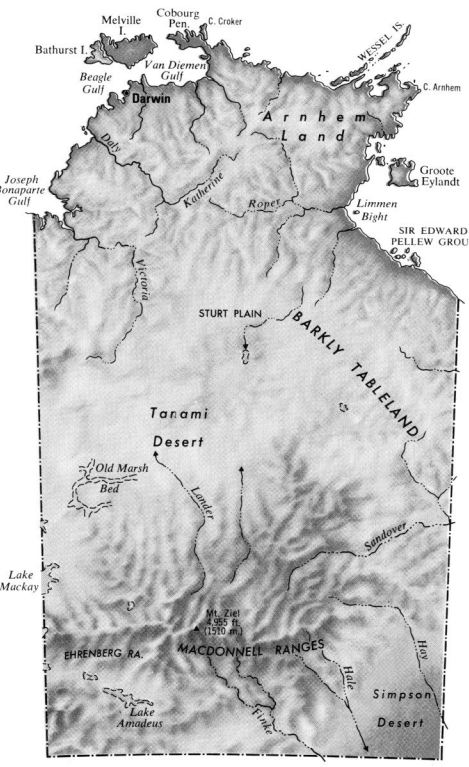

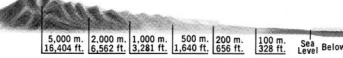

© Copyright HAMMOND INCORPORATED, Maplewood, N.J.

AREA 379,922 sq. mi. (984,000 sq. km.)
POPULATION 1,345,945
CAPITAL Adelaide
LARGEST CITY Adelaide
HIGHEST POINT Mt. Woodroffe 4,970 ft.
(1,515 m.)

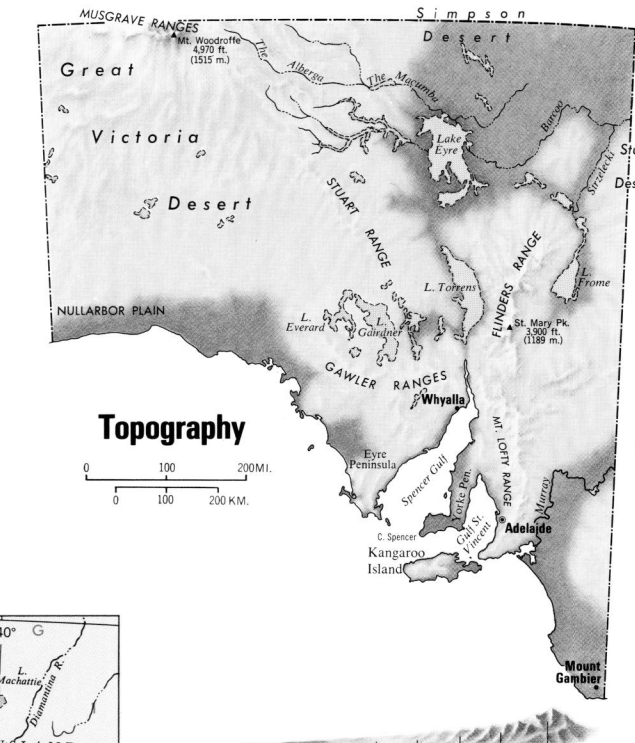

Topography

CITIES and TOWNS

Adelaide (cap.) 882,520 . . . . B6	Crafters-Bridgewater 9,764 . . B8	Kingston 1,325 . . . . G7
Adelaide *931,886 . . . . B6	Crystal Brook 1,240 . . . . . E5	Lameroo 599 . . . . G6
Andamooka 402 . . . . E4	Cummins 767 . . . . D6	Laura 504 . . . . F5
Angaston 1,753. . . . . F6	Edithburgh 359 . . . . E6	Leigh Creek 1,635 . . . . . F4
Balaklava 1,306 . . . . F6	Elizabeth 32,608 . . . . . . B7	Lobethal 1,522 . . . . C7
Barmera 2,014 . . . . G6	Elliston ○1,345 . . . . D5	Lock 213 . . . . D5
Beachport 357 . . . . F7	Enfield 66,797 . . . . B7	Loxton 3,100 . . . . G6
Berri 3,419 . . . . G6	Gawler 9,433 . . . . B6	Lyndoch 539 . . . . C6
Birdwood 397 . . . . C7	Gladstone 680 . . . . F5	Maitland 1,085 . . . . E6
Blinman F4	Glenelg 13,306 . . . . A8	Mannum 1,984 . . . . F6
Bordertown 2,138 . . . . . . G7	Gumeracha 387 . . . . C7	Marion 66,580 . . . . A8
Brighton 19,441 . . . . A8	Hahndorf 1,274 . . . . C8	Marree E3
Burnside 37,593 . . . . B8	Hawker 351 . . . . F4	Meadows 388 . . . . B8
Burra 1,222 . . . . F5	Hindmarsh 7,593 . . . . . . A7	Meningie 807 . . . . F6
Campbelltown 43,084 . . . . B7	Iron Knob 398 . . . . E5	Millicent 5,255 . . . . F7
Ceduna 2,794 . . . . D5	Jamestown 1,384 . . . . . . F5	Minlaton 865 . . . . E6
Clare 2,381 . . . . F5	Kadina 2,943 . . . . E5	Mitcham 60,309 . . . . . . A8
Cleve 827 . . . . E5	Kapunda 1,340 . . . . F6	Moonta 1,751 . . . . E5
Coober Pedy 2,078 . . . . . D3	Keith 1,147 . . . . G7	Mount Barker 4,190 . . . . C8
Cowell 626 . . . . E5	Kensington and Norwood	Mount Gambier 18,193 . . . G7
	8,950 . . . . B8	Murray Bridge 8,664 . . . . F6
	Kimba 862 . . . . E5	Nairne 706 . . . . C7
	Kingscote 1,236 . . . . . . E6	Nangwarry 758 . . . . G7

Naracoorte 4,758 . . . . G7	Flinders (range) F4
Noarlunga 60,928 . . . . A8	Frome (lake) G4
Nuriootpa 2,851 . . . . F6	Gairdner (lake) D4
Oodnadatta D2	Gawler (ranges) E5
Ororoo 604 . . . . F5	Gawler (riv.) B6
Payneham 16,502 . . . . B7	Gilles (lake) E5
Penola 1,205 . . . . G7	Goyders (lag.) F2
Peterborough 2,575 . . . . . . F5	Great Australian (bight) A5
Pinnaroo 731 . . . . G6	Great Victoria (des.) B3
Port Adelaide 35,407 . . . . . A7	Gregory (lake) E3
Port Augusta 15,566 . . . . . E5	Hack (mt.) F4
Port Broughton 587 . . . . . . F5	Hamilton, The (riv.) D2
Port Lincoln 9,846 . . . . . . E6	Harris (lake) D4
Port Pirie 14,695 . . . . E5	Head of Bight (bay) B4
Prospect 18,591 . . . . B7	Indian Ocean E7
Quorn 1,049 . . . . F5	Investigator (str.) E6
Renmark 3,475 . . . . G5	Investigator Group (isls.) . . . D5
Robe 590 . . . . F7	Island (lag.) E4
Salisbury 86,451 . . . . B7	Jaffa (cape) F7
Snowtown 492 . . . . E5	Kangaroo (isl.) 3,515 . . . . . . E6
Strathalbyn 1,756 . . . . F6	Lacepede (bay) F7
Streaky Bay 985 . . . . D5	Lofty (mt.) B8
Tailem Bend 1,677 . . . . . . F6	Macfarlane (lake) E5
Tanunda 2,621 . . . . C6	Macumba, The (riv.) D2
Tea Tree Gully 67,237 . . . . B7	Maurice (lake) B3
Thebarton 9,208 . . . . A7	Meramangye (lake) C3
Tumby Bay 933 . . . . E6	Morris (mt.) B2
Unley 35,844 . . . . B8	Murray (res.) F6
Uraidla 303 . . . . B8	Musgrave (ranges) B2
Victor Harbor 4,522 . . . . . . F6	Neales, The (riv.) E3
Virginia 353 . . . . B7	Northumberland (cape) F8
Waikerie 1,629 . . . . F6	Nukey Bluff (mt.) D5
Wallaroo 2,043 . . . . E5	Nullarbor (plain) A4
West Torrens 45,099 . . . . . A8	Nuyts (arch.) C5
Whyalla 30,518 . . . . E5	Nuyts (cape) C5
Williamstown 495 . . . . C7	Peera Peera Poolanna (lake) . F2
Willunga 667 . . . . F6	Saint Mary (peak) F4
Wilmington 227 . . . . F6	Saint Vincent (gulf) E6
Woodside 724 . . . . C8	Serpentine (lakes) A3
Woodville 77,634 . . . . A7	Simpson (des.) E1
Woomera 1,658 . . . . E4	Sir Joseph Banks Group
Wudinna 572 . . . . D5	(isls.) E6
Yorketown 713 . . . . E6	Spencer (cape) E6
	Spencer (gulf) E6
	Stevenson, The (riv.) D2
OTHER FEATURES	Streaky (bay) C5
	Strzelecki (creek) G3
Acraman (lake) D5	Stuart (range) D3
Alberga, The (riv.) D2	Sturt (des.) G3
Alexandrina (lake) F6	The Alberga (riv.) D2
Anxious (bay) D5	The Coorong (lag.) F6
Arckaringa (creek) D2	The Hamilton (riv.) D2
Barcoo (creek) F3	The Macumba (riv.) E2
Birksgate (range) A2	The Neales (riv.) E3
Blanche (lake) F3	The Stevenson (riv.) D2
Brady (mt.) D3	The Warburton (riv.) E2
Cadibarrawirracanna (lake) . D3	Thistle (isl.) E6
Callabonna (lake) F3	Torrens (lake) E4
Catastrophe (cape) D6	Torrens (riv.) C7
Coffin (bay) D6	Warburton, The (riv.) E2
Coffin Bay (pen.) D6	Wilkinson (lakes) C3
Coopers (Barcoo) (creek) . . F3	Woodroffe (mt.) C2
Coorong, The (lag.) F6	Yalata Aboriginal Res. B4
Dey Dey (lake) B3	Yarle (lakes) B4
Encounter (bay) F6	Yorke (pen.) E6
Everard (lake) D4	
Everard (ranges) C2	
Eyre (pen.) D5	○ Population of district.
Eyre North (lake) E3	*Population of met. area.
Eyre South (lake) E3	
Finke (riv.) C1	

Adelaide and Vicinity

South Australia

SCALE OF MILES

KILOMETERS

State Capital
State and Territorial
Boundaries

® Copyright HAMMOND INCORPORATED, Maplewood, N.J.

Longitude D East 136° of E Greenwich

CITIES and TOWNS

Aramac 428 .... C4
Archerfield 785 .... D3
Ascot 4,298 .... E2
Atherton 4,196 .... C3
Ayr 8,787 .... C3
Balmoral 2,915 .... E2
Barcaldine 1,432 .... C4
Beaudesert 3,780 .... E6
Biloela 4,643 .... D5
Birdsville A5
Blackall 1,609 .... C5
Blackwater 5,434 .... D4
Boulia 292 .... A4
Bowen 7,663 .... D3
Brisbane (cap.) 689,378 .... D2
Brisbane *1,028,527 .... D2
Bucasia 1,356 .... D4
Bundaberg 32,560 .... D5
Burketown 210 .... A3
Cairns 48,557 .... C3
Caloundra 16,758 .... E5
Camooweal 251 .... A3
Camp Hill 8,999 .... E3
Capella 660 .... D4
Cardwell 1,249 .... C3
Charleville 3,523 .... C5
Charters Towers 6,823 .... C4
Cherbourg 963 .... D5
Chermside 6,892 .... D2
Clermont 1,659 .... D4
Cloncurry 1,961 .... B4
Collinsville 2,756 .... C4
Cooktown 913 .... C3
Coopers Plains 4,492 .... E3
Corinda 4,894 .... D3
Croydon ○255 .... B3
Cunnamulla 1,627 .... C5
Dalby 8,784 .... D5
Dirranbandi 480 .... D6
East Brisbane 4,853 .... E3
Eidsvold 613 .... D5
Emerald 4,628 .... C4
Esk 676 .... E5
Gatton 4,190 .... D5
Gayndah 1,708 .... D5
Geebung 4,850 .... E2
Georgetown 319 .... B3
Gladstone 22,083 .... D4
Gold Coast 135,437 .... E6
Goondiwindi 3,576 .... D6
Gordonvale 2,375 .... C3
Greenslopes 7,219 .... E3
Gympie 10,768 .... E5

Hervey Bay 13,569 .... E5
Holland Park 7,363 .... E3
Home Hill 3,138 .... C3
Hughenden 1,657 .... B4
Inala 17,383 .... D3
Indooroopilly 7,959 .... D3
Ingham 5,598 .... C3
Injune 407 .... D5
Innisfail 7,933 .... C3
Ipswich 68,297 .... E5
Isisford ○605 .... C4
Jandowae 781 .... D5
Jericho ○1,177 .... C4
Julia Creek 602 .... B4
Karumba 670 .... B3
Kilcoy 1,257 .... E5
Kingaroy 5,134 .... D5
Longreach 2,971 .... B4
Mackay 35,361 .... D4
Mareeba 6,309 .... C3
Marian 796 .... D4
Maroochydore-Mooloolaba
  17,460 .... E5
Maryborough 20,111 .... E5
Mary Kathleen 830 .... A4
McKinlay ○1,477 .... B4
Millmerran 1,107 .... D5
Mitchell 1,171 .... C5
Mitchelton 5,810 .... D2
Monto 1,397 .... D5
Moorooka 8,740 .... D3
Moranbah 4,362 .... C9
Mossman 1,614 .... C3
Mount Isa 23,679 .... A4
Moura 2,871 .... D5
Murgon 2,327 .... D5
Nambour 7,965 .... E5
Newmarket 3,520 .... D2
Normanton 926 .... B3
Nundah 7,358 .... E2
Proserpine 3,058 .... D4
Quilpie 694 .... C5
Ravenshoe 915 .... C3
Redcliffe 42,223 .... E5
Richmond 784 .... B4
Rockhampton 50,146 .... D4
Roma 5,706 .... D5
Saint George 2,204 .... D5
Saint Lucia 6,075 .... D3
Sandgate 6,776 .... D2
Sarina 2,815 .... D4
Springsure 774 .... D5
Stafford (Stafford Heights)
  13,731 .... D2
Stanthorpe 3,966 .... D6
Tara 864 .... D5

Taroom 688 .... D5
Tewantin-Noosa 9,965 .... E5
Theodore 643 .... D5
Thursday Island 2,283 .... B1
Toowoomba 63,401 .... D5
Townsville 86,112 .... C3
Tully 2,728 .... C3
Walkerston 1,277 .... D4
Warwick 8,853 .... D6
Weipa 2,433 .... B2
Windsor 6,119 .... D2
Winton 1,259 .... B4
Wynnum 10,794 .... E5
Yeppoon 6,447 .... D4
Yeronga 4,579 .... D3

OTHER FEATURES

Albatross (bay) B2
Archer (riv.) B2
Balonne (riv.) D6
Banks (isl.) B1
Barcoo (creek) C5
Barkly Tableland A4
Bartle Frere (mt.) C3
Beal (range) B5

AREA 666,872 sq. mi. (1,727,200 sq. km.)
POPULATION 2,587,315
CAPITAL Brisbane
LARGEST CITY Brisbane
HIGHEST POINT Mt. Bartle Frere 5,287 ft.
  (1,611 m.)

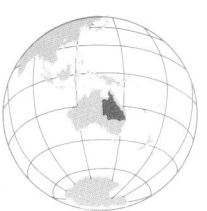

Topography

Belyando (riv.) C4
Broad (sound) D4
Bulloo (lake) B6
Bulloo (riv.) B6
Bunker Group (isls.) E4
Burdekin (riv.) C3
Cape York (pen.) C2
Capricorn (chan.) D4
Capricorn Group (isls.) E4
Carnarvon (range) D5
Carpentaria (gulf) A2
Cloncurry (riv.) B4
Coopers (Barcoo) (creek) B5
Coral (sea) C1
Culgoa (riv.) C6
Cumberland (isls.) D4
Curtis (isl.) D4
Darling Downs D5
Dawson (riv.) D5
Diamantina (riv.) B4
Drummond (range) C4
Duifken (pt.) B2
Endeavour (str.) B1

Fitzroy (riv.) D4
Flinders (riv.) B3
Fraser (isl.) E5
Georgina (riv.) A4
Gilbert (riv.) B3
Great Dividing (range) C4
Gregory (range) B3
Gregory (riv.) A3
Grey (range) B5
Hamilton (riv.) A4
Hervey (bay) E5
Hinchinbrook (isl.) C3
Hook (isl.) D4
Leichhardt (riv.) A3
Machattie (lake) B5
Macintyre (riv.) C6
Maranoa (riv.) C5
Mary (riv.) E5
Melville (cape) C2
Mitchell (riv.) B2
Moreton (bay) E5
Moreton (isl.) E5
Mornington (isl.) A3

Norman (riv.) B3
Northern Peninsula
  Aboriginal Res. B1
Prince of Wales (isl.) B1
Princess Charlotte (bay) C2
Sandy (cape) E5
Selwyn (range) B4
Simpson (des.) A5
Sturt (des.) B3
Suttor (riv.) C4
Swain (reefs) E4
Thompson (riv.) B5
Torres (str.) B1
Warrego (range) C5
Warrego (riv.) C5
Wellesley (isls.) A3
Whitsunday (isl.) D4
Willies (isls.) C6
Yamma Yamma (lake) B6
York (cape) B1

○ Population of district.
*Population of met. area.

96 New South Wales and Victoria

Cobbadah F2
Coffs Harbour 16,020 . . . . . . G2
Collarenebri 602 . . . . E1
Collie E2
Comboyne G2
Concord 23,926 . . . . J3
Condobolin 3,355 . . . . D3
Conoble C3
Coogee K3
Coolah 878 . . . . E2
Coolamon 1,088 . . . . D4
Cooma 7,978 . . . . E5
Coonabarabran 3,001 . . . . . . E2
Coonamble 3,090 . . . . E2
Cootamundra 6,540 . . . . D4
Copmanhurst ○2,857 . . . . . G1
Coraki 895 . . . . G1
Coramba 202 . . . . G2
Corowa-Wahgunyah 3,390 . . D4
Cowra 7,900 . . . . E3
Crescent Head 944 . . . . G2
Cronulla J4
Crookwell 2,063 . . . . E4
Cudal 400 . . . . E3
Cullburra-Orient Point 2,068 . F4
Culcairn 1,027 . . . . D4
Cullen Bullen 231 . . . . F3
Cumnock 252 . . . . E3
Curlewis 487 . . . . F2
Dareton 612 . . . . B4
Darlington Point 599 . . . . C4
Deepwater 260 . . . . F1
Dee Why K3
Delegate 347 . . . . E5
Delungra 332 . . . . F1
Deniliquin 7,354 . . . . C4
Denman 1,122 . . . . F3
Dorrigo 1,192 . . . . G2
Drummoyne 30,961 . . . . J3
Dubbo 23,986 . . . . E3
Dunedoo 836 . . . . E3
Dungalear Station D1
Dungog 2,126 . . . . F3
Eden 3,107 . . . . E5
Emmaville 503 . . . . F1
Ermeran Station D3
Eugowra 577 . . . . E3
Evans Head 1,802 . . . . G1

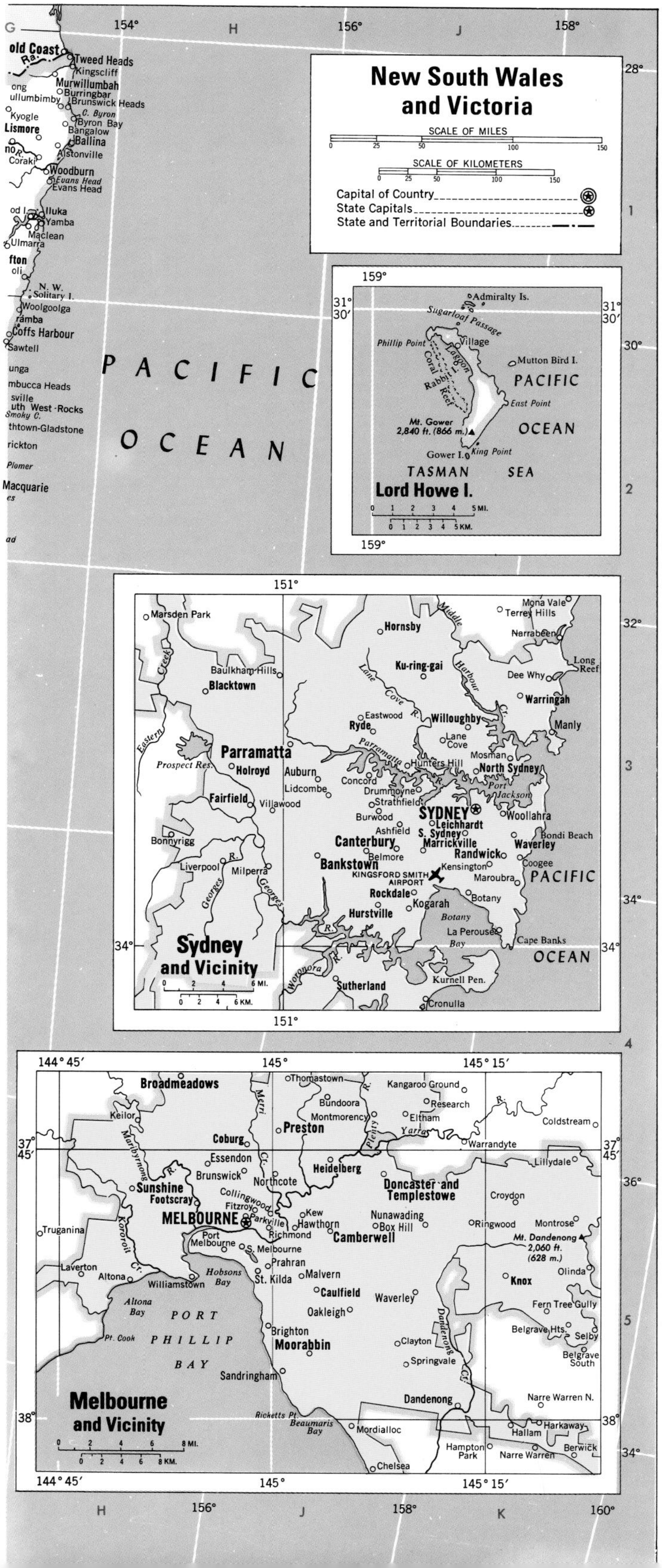

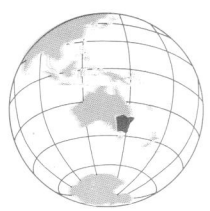

NEW SOUTH WALES

AREA 309,498 sq. mi. (801,600 sq. km.)
POPULATION 5,401,881
CAPITAL Sydney
LARGEST CITY Sydney
HIGHEST POINT Mt. Kosciusko 7,310 ft. (2,228 m.)

VICTORIA

AREA 87,876 sq. mi. (227,600 sq. km.)
POPULATION 4,019,478
CAPITAL Melbourne
LARGEST CITY Melbourne
HIGHEST POINT Mt. Bogong 6,508 ft. (1,984 m.)

Topography

Fairfield 129,557 . . . . H3
Finley 2,193 . . . . C4
Forbes 8,029 . . . . E3
Forest Hill 1,977 . . . . D4
Forster-Tuncurry 9,261 . . . . . G3
Frederickton 616 . . . . G2
Ganmain 650 . . . . D4
Gerringong 1,775 . . . . F4
Geurie 290 . . . . E3
Gilgai 257 . . . . F1
Gilgandra 2,700 . . . . E2
Gilgunnia C3
Glen Innes 6,052 . . . . F1
Glenreagh 233 . . . . G2
Gloucester 2,488 . . . . F2
Goodooga 248 . . . . D1
Goolgowi 245 . . . . C3
Gooloogong 221 . . . . E3
Goombalie C1
Goulburn 21,755 . . . . E4
Grafton 17,005 . . . . G1
Grenfell 2,070 . . . . E3
Greta-Branxton 2,849 . . . . . . F3
Greta East 398 . . . . F3
Griffith 13,187 . . . . C4
Gulargambone 457 . . . . E2
Gulgong 1,740 . . . . E3
Gundagai 2,308 . . . . D4
Gunnedah 8,909 . . . . F2
Gunning 438 . . . . E4
Guyra 1,840 . . . . F2
Hanwood 306 . . . . C4
Harrington 1,183 . . . . G2
Hay 2,958 . . . . C4
Henty 883 . . . . D4
Hillston 999 . . . . C3
Holbrook 1,276 . . . . D4
Holroyd 80,116 . . . . H3
Hornsby 111,081 . . . . J3
Howlong 1,072 . . . . D4
Hunters Hill 12,537 . . . . J3
Hurstville 64,910 . . . . J4
Huskisson 2,296 . . . . F4
Iluka 1,362 . . . . G1
Inverell 9,734 . . . . F1
Ivanhoe 517 . . . . C3
Jerilderie 1,075 . . . . C4
Jindabyne 1,602 . . . . E5
Junee 3,993 . . . . D4
Kandos 1,626 . . . . F3
Karpakora B3
Katoomba-Wentworth Falls 13,942 . . . . F3
Keewong C3
Kempsey 9,037 . . . . G2
Kendall 522 . . . . G2
Khancoban 515 . . . . E5
Kiama 7,717 . . . . F4
Kinalung B3
Kogarah 46,322 . . . . J4
Koorawatha 262 . . . . E4
Kootingal 731 . . . . F2
Ku-ring-gai 101,501 . . . . J3
Kurri Kurri-Weston 12,795 . . F3
Kyogle 3,070 . . . . G1
Lake Cargellico 1,240 . . . . . . D3
Lane Cove 29,113 . . . . J3
Leeton 6,498 . . . . D4
Leichhardt 57,332 . . . . J3
Lette B4
Lidcombe J3
Lightning Ridge 1,112 . . . . . E1
Lismore 24,033 . . . . G1
Lithgow 12,793 . . . . F3
Liverpool 92,715 . . . . H4
Lockhart 923 . . . . D4
Macksville 2,352 . . . . G2
Maclean 2,593 . . . . G1
Maitland 38,865 . . . . F3
Manildra 520 . . . . E3
Manilla 1,884 . . . . F2
Manly 37,080 . . . . K3
Maroubra K3
Marrickville 83,448 . . . . J3
Marsden D3
Marsden Park 518 . . . . H3
Marulan 330 . . . . E4
Mathoura 582 . . . . C4
Melrose D3
Mendooran 325 . . . . E2
Menindee 455 . . . . B3
Merimbula 2,899 . . . . F5
Merriwa 943 . . . . F3
Millthorpe 607 . . . . E5
Milpa B2
Milperra H4
Milton 740 . . . . F4
Mittagong 4,266 . . . . F4
Moama C5
Mogil Mogil E1
Molong 1,374 . . . . E3
Mona Vale K2
Moree 10,459 . . . . E1
Morisset 1,593 . . . . F3
Moruya 2,003 . . . . F4
Mosman 26,200 . . . . J3
Moss Vale 4,415 . . . . F4
Mossgiel C4
Moulamein 396 . . . . C4
Mount Arrowsmith A2
Mount Drysdale C2
Mount Hope C3
Mudgee 6,015 . . . . E3
Mullumbimby 2,234 . . . . G1
Mungindi 707 . . . . E1
Murrumburrah 2,070 . . . . . . E4
Murrurundi 861 . . . . F2
Murwillumbah 7,807 . . . . . . G1
Muswellbrook 8,548 . . . . F3
Nabiac 363 . . . . G3
Nambucca Heads 4,053 . . . . G2
Narellan 2,104 . . . . F3
Narooma 2,758 . . . . F5
Narrabeen K3
Narrabri 7,296 . . . . E2
Narrandera 5,013 . . . . D4
Narromine 2,994 . . . . E3
New Angledool E1
Newcastle 135,207 . . . . F3
Newcastle †389,237 . . . . F3
Nimmitabel 257 . . . . E5
North Sydney 48,500 . . . . . . J3
Nowra-Bomaderry 17,887 . . . F4
Nundle 235 . . . . F2
Nymboida ○2,044 . . . . G1
Nyngan 2,485 . . . . D2
Oaklands 283 . . . . D4
Oberon 1,937 . . . . E3
Old Bar 970 . . . . G2
Orange 27,626 . . . . E3
Pallamallawa 340 . . . . F1
Pambula 604 . . . . E5
Parkes 9,047 . . . . E3
Parramatta 130,943 . . . . H3
Peak Hill 1,037 . . . . E3
Penrith 108,720 . . . . F3
Perthville 639 . . . . E3
Picton 1,817 . . . . F4
Pilliga 206 . . . . E2
Popiltah A3
Portland 1,980 . . . . E3
Port Kembla F4
Port Macquarie 19,581 . . . . . G2
Queanbeyan 19,383 . . . . E4
Quirindi 2,851 . . . . F2
Randwick 116,202 . . . . J3
Raymond Terrace 7,548 . . . . F3
Richmond-Windsor 15,491 . . F3
Rockdale 83,719 . . . . J4

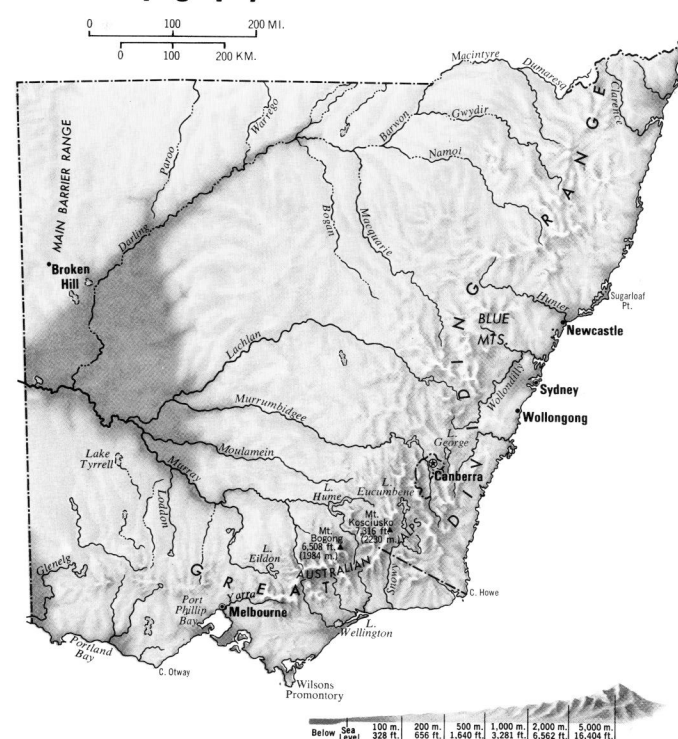

(continued on following page)

(continued)

Ryde 88,948 J3
Rylstone 651 E3
Salisbury Downs B1
Sawtell 5,970 G2
Scone 3,949 F3
Shellharbour 41,790 F4
Singleton 9,572 F3
Smithtown-Gladstone 953 G2
South Sydney 30,776 J3
South West Rocks 1,314 G2
Stephen's Creek A2
Strathfield 25,882 J3
Stroud 522 G3
Sussex Inlet 1,293 F4
Sutherland 165,336 J4
Sydney (cap.) 2,876,508 J3
Sydney †3,204,696 J3
Talbingo 481 E4
Tamworth 29,657 F2
Taralga 272 E4
Tarcutta 263 D4
Taree 14,697 G2
Tathra 1,077 F5
Temora 4,350 D4
Tenterfield 3,402 G1
Terrigal-The Entrance 37,891 F3
The Rock 693 D4
Thurloo Downs B1
Tibbita C4
Tibooburra B1
Tiltagara C2
Tingha 886 F1
Tocumwal 1,174 C4
Tongo B2
Torrowangee A2
Tottenham 366 D3
Trangie 977 D3
Trundle 515 D3
Tullamore 324 D3
Tumbarumba 1,536 D4
Tumut 5,816 E4
Tweed Heads G1
Ulladulla 6,018 F4
Ulmarra 395 G1
Ungarie 428 D3
Uralla 2,090 F2
Urana 419 D4
Urbenville 282 G1
Urunga 2,045 G2
Villawood H3
Wagga Wagga 36,837 D4
Wakool 278 C4
Walcha 1,674 F2
Walgett 2,157 E2
Walla Walla 593 D4
Wallerawang 1,855 F3
Wangi-Rathmines 5,106 F3
Warialda 1,340 F1
Warragamba 1,406 F3
Warren 2,153 D2
Warringah ○172,653 K3
Wauchope 3,645 G2
Waverley 61,575 K3
Waverley Downs B1
Wee Waa 1,904 E2
Wellington 5,280 E3
Wentworth 1,180 B4
Werris Creek 1,924 F2
West Wyalong 3,778 D3
Wetuppa B4
White Cliffs B2
Whitton 344 D4
Whyjonta B1
Wilcannia 982 B2
Willoughby 52,120 J3
Willow Tree 258 F2
Wingham 3,937 G2
Wollongong 169,381 F4
Wollongong †222,539 F4
Woodburn 647 G1
Woodenbong 409 G1
Woodstock 266 E3
Woolgoolga 2,081 G2
Wooli 457 G1
Woollahra 51,659 K3
Wyong 3,902 F3
Yallock C3
Yalpunga A1
Yamba 2,528 G1
Yancannia B2
Yanco 415 D4
Yantara B1
Yass 4,283 E4
Yenda 697 D4
Yeoval 288 E3
Young 6,906 E4

OTHER FEATURES

Ana Branch, Darling (riv.) A3
Australian Alps (mts.) D5
Barrington Tops (mt.) F2
Barwon (riv.) D2
Blue (mts.) F3
Bogan (riv.) D2
Bondi (beach) K3
Botany (bay) J4
Broken (bay) F3
Burrinjuck (res.) E4
Byron (cape) G1
Caryapundy (swamp) B1
Castlereagh (riv.) E2
Cawndilla (lake) A3
Clarence (riv.) G1
Colo (riv.) F3
Cowal (lake) D3
Culgoa (riv.) D1
Cuttaburra (creek) C1
Darling (riv.) B3
Dumaresq (riv.) F1
Eucumbene (lake) E5
George (lake) E4
Georges (riv.) H4
Gower (mt.) J2
Great Dividing (range) E3
Green (cape) F5
Gunderbooka (ranges) C2
Gwydir (riv.) E1
Howe (cape) F5
Hume (res.) D4
Hunter (riv.) F3
Kosciusko (mt.) E5
Kurnell (pen.) J4
Lachlan (range) C3
Lachlan (riv.) C3
Liverpool (range) F2
Lord Howe (isl.) 287 J2
Macintyre (riv.) E1
Macquarie (lake) F3
Macquarie (riv.) D2
Main Barrier (range) A2
Manning (riv.) F2
Marthaguy (creek) D2
McPherson (range) G1
Menindee (lake) B3
Monaro (range) E5
Moonie (riv.) E1
Moulamein (creek) C4
Mount Royal (range) F2
Murray (riv.) A4
Murrumbidgee (riv.) C4
Myall (lake) G3
Namoi (riv.) E2
Narran (lake) D1
New England (range) F1
Paroo (riv.) C1
Parramatta (riv.) J3
Poopelloe (lake) C2
Port Jackson (inlet) J3
Port Stephens (inlet) G3
Richmond (range) G1
Richmond (riv.) G1
Riverina (reg.) C4
Robe (mt.) A2
Round, The (mt.) G2
Salt, The (lake) B2
Shoalhaven (riv.) E4
Smoky (cape) G2
Snowy (mts.) E5
Snowy (riv.) E5
Stony (ranges) B2
Sturt (mt.) A1
Sugarloaf (pt.) G3
Talyawalka (creek) B2
Tandou (lake) A3
Tasman (sea) F5
The Round (mts.) G2
The Salt (lake) B2
Timbarra (riv.) G1
Tuggerah (lake) F3
Victoria (lake) A3
Warrego (riv.) C1
Willandra Billabong (creek) C3
Wollondilly (riv.) F4

VICTORIA

CITIES and TOWNS

Alexandra 1,756 C5
Altona 30,909 H5
Apollo Bay 921 B6
Ararat 8,336 B5
Avoca 1,032 B5
Bacchus Marsh 6,224 C5
Bairnsdale 9,459 D5
Ballarat 35,681 C5
Ballarat †71,930 C5
Balmoral 257 A5
Beaufort 1,214 B5
Beechworth 3,154 D5
Belgrave Heights J5
Belgrave South K5
Benalla 8,151 D5
Bendigo 31,841 C5
Bendigo †58,818 C5
Berwick 36,181 K6
Beulah 290 B4
Birchip 895 B4
Birregurra 416 B6
Boort 863 B5
Box Hill 47,579 J5
Bright 1,545 D5
Brighton 33,697 J5
Broadford 1,580 C5
Broadmeadows 103,540 H4
Brunswick 44,464 H5
Bruthen 449 D5
Bundoora J4
Camberwell 85,883 J5
Camperdown 3,545 C6
Cann River 345 E5
Casterton 1,945 A5
Castlemaine 7,583 C5
Caulfield 69,922 J5
Charlton 1,377 B5
Chelsea 26,034 J6
Churchill 4,796 D6
Clunes 761 B5
Cobden 1,453 B6
Cobram 3,817 C4
Coburg 55,035 H5
Cohuna 2,178 C4
Colac 10,587 B6
Coldstream 1,395 K4
Coleraine 1,232 A5
Collingwood 15,089 J5
Corryong 1,320 D5
Craigieburn 4,296 C5
Cranbourne 9,400 C6
Creswick 2,036 B5
Croydon 36,210 K5
Dandenong 54,962 K5
Darby D6
Dartmoor 349 A5
Daylesford 2,883 C5
Derrinallum 287 B5
Dimboola 1,675 B5
Donald 1,609 B5
Doncaster and Templestowe 90,660 J5
Drouin 3,492 C6
Dunkeld 402 B5
Dunolly 621 B5
Eaglehawk 7,355 C5
Echuca 7,943 C5
Edenhope 827 A5
Eildon 737 C5
Eltham 34,648 J4
Erica 236 D5
Essendon 56,380 H5
Euroa 2,640 C5
Fitzroy 19,112 H5
Footscray 49,756 H5
Geelong 14,471 C6
Geelong †137,173 C6
Geelong West 14,823 C6
Goroke 370 A5
Gunbower 259 C4
Hamilton 9,751 B5
Hawthorn 30,689 J5
Healesville 4,526 C5
Heathcote 1,213 C5
Heidelberg 64,757 J5
Heyfield 1,635 D6
Heywood 1,266 A6
Hopetoun 1,832 B4
Horsham 12,034 B5
Inglewood 674 B5
Inverloch 1,523 C6
Kaniva 956 A5
Keilor 81,762 H5
Kerang 4,049 B4
Kew 28,870 J5
Kilmore 1,728 C5
Knox 88,902 K5
Koroit 1,988 B6
Korumburra 2,798 D6
Kyabram 5,414 C5
Kyneton 3,185 C5
Lake Boga 502 B4
Lake Bolac 211 B5
Lakes Entrance 3,414 E5
Lara 4,231 C6
Leongatha 3,736 C6
Lillydale 62,077 J4
Macarthur 322 A6
Maffra 3,822 D5
Maldon 1,009 C5
Mallacoota 726 E5
Malvern 43,211 J5
Mansfield 1,920 D5
Maryborough 7,858 B5
Melbourne (cap.) 2,578,759 H5
Melbourne †2,722,817 H5
Melton 20,599 C5
Merbein 1,735 A4
Merino 298 A5
Mildura 15,763 A4
Minyip 567 B5
Moe 16,649 D6
Montmorency J4
Montrose K5
Moorabbin 97,810 J5
Mooroopna C5
Mordialloc 27,869 J6
Morea A5
Mornington 23,512 C6
Mortlake 1,056 B6
Morwell 16,491 D5
Mount Beauty 1,509 D5
Murrayville 313 A4
Murtoa 946 B5
Myrtleford 2,815 D5
Nagambie 1,102 C5
Narre Warren North 761 K5
Nathalia 1,222 C5
Natimuk 482 A5
Newtown 10,210 C6
Nhill 1,567 A5
Northcote 51,235 J5
Numurkah 2,713 C5
Nunawading 97,052 J5
Nyah 351 B4
Nyah West 535 B4
Oakleigh 55,612 J5
Omeo 272 D5
Orbost 2,586 E5
Ouyen 1,527 B4
Penshurst 558 B5
Porepunkah 268 D5
Port Albert 267 D6
Port Fairy 2,276 B6
Portland 9,353 A6
Port Melbourne 8,585 H5
Prahran 45,018 J5
Preston 84,519 J4
Quambatook 359 B4
Queenscliff 3,420 C6
Rainbow 700 A4
Red Cliffs 2,409 B4
Richmond 24,506 J5
Ringwood 38,665 K5
Robinvale 1,751 B4
Rochester 2,399 C5
Rushworth 994 C5
Rutherglen 1,454 D5
Saint Arnaud 2,721 B5
Saint Kilda 49,366 J5
Sale 12,968 D6
Sandringham 31,175 J5
Sea Lake 943 B4
Sebastopol 6,462 B5
Seymour 6,494 C5
Shepparton-Mooroopna ‡28,373 C5
South Barwon 35,307 C6
South Melbourne 19,955 J5
Springvale 80,186 J5
Stawell 6,160 B5
Sunbury 11,085 C5
Sunshine 94,419 H5
Swan Hill 8,398 B4
Swifts Creek 288 D5
Tallangatta 950 D5
Tatura 2,697 C5
Templestowe and Doncaster 90,660 J5
Terang 2,111 B6
Tongala 994 C5
Traralgon 18,057 D6
Underbool 274 A4
Wangaratta 16,202 D5
Warburton 2,009 C5
Warracknabeal 2,735 B5
Warragul 7,712 D6
Warrnambool 21,414 B6
Waverley 122,471 J5
Wedderburn 868 B5
Werrimull A4
Whittlesea 65,657 C5
Willaura 377 B5
Williamstown 25,554 H5
Winchelsea 825 B6
Wodonga 19,208 D5
Wonthaggi 4,797 C6
Woodend 1,785 C5
Wycheproof 938 B5
Yallourn 26 D6
Yarram 2,085 D6
Yarrawonga 3,442 C5
Yea 996 C5

OTHER FEATURES

Australian Alps (mts.) D5
Avoca (riv.) B5
Barry (mts.) D5
Bogong (mt.) D5
Bridgewater (cape) A6
Buller (mt.) D5
Campaspe (riv.) C5
Corangamite (lake) B6
Corner (inlet) D6
Dandenong (mt.) K5
Difficult (mt.) B5
Discovery (bay) A6
Eildon (lake) C5
French (isl.) 123 C6
Gippsland (reg.) D6
Glenelg (riv.) A5
Goulburn (riv.) C5
Hindmarsh (lake) A5
Hobsons (bay) H5
Hopkins (riv.) B5
Hume (lake) D4
Indian Ocean B6
Loddon (riv.) B5
Mitchell (riv.) D5
Mitta Mitta (riv.) D5
Mornington (pen.) C6
Mount Emu (creek) B5
Murray (riv.) A4
Nelson (cape) A6
Ninety Mile (beach) D6
Otway (cape) B6
Ovens (riv.) D5
Phillip (isl.) 2,832 C6
Portland (bay) A6
Port Phillip (bay) C6
Rocklands (res.) B5
Snowy (riv.) E5
South East (pt.) D6
Tasman (sea) F5
Tyrrell (lake) B4
Waratah (bay) C6
Wellington (lake) D6
Western Port (inlet) C6
Wilsons (prom.) D6
Wimmera (riv.) A5
Yarra (riv.) C5

*City and suburbs.
○ Population of district.
†Population of met. area.
‡Population of urban area.

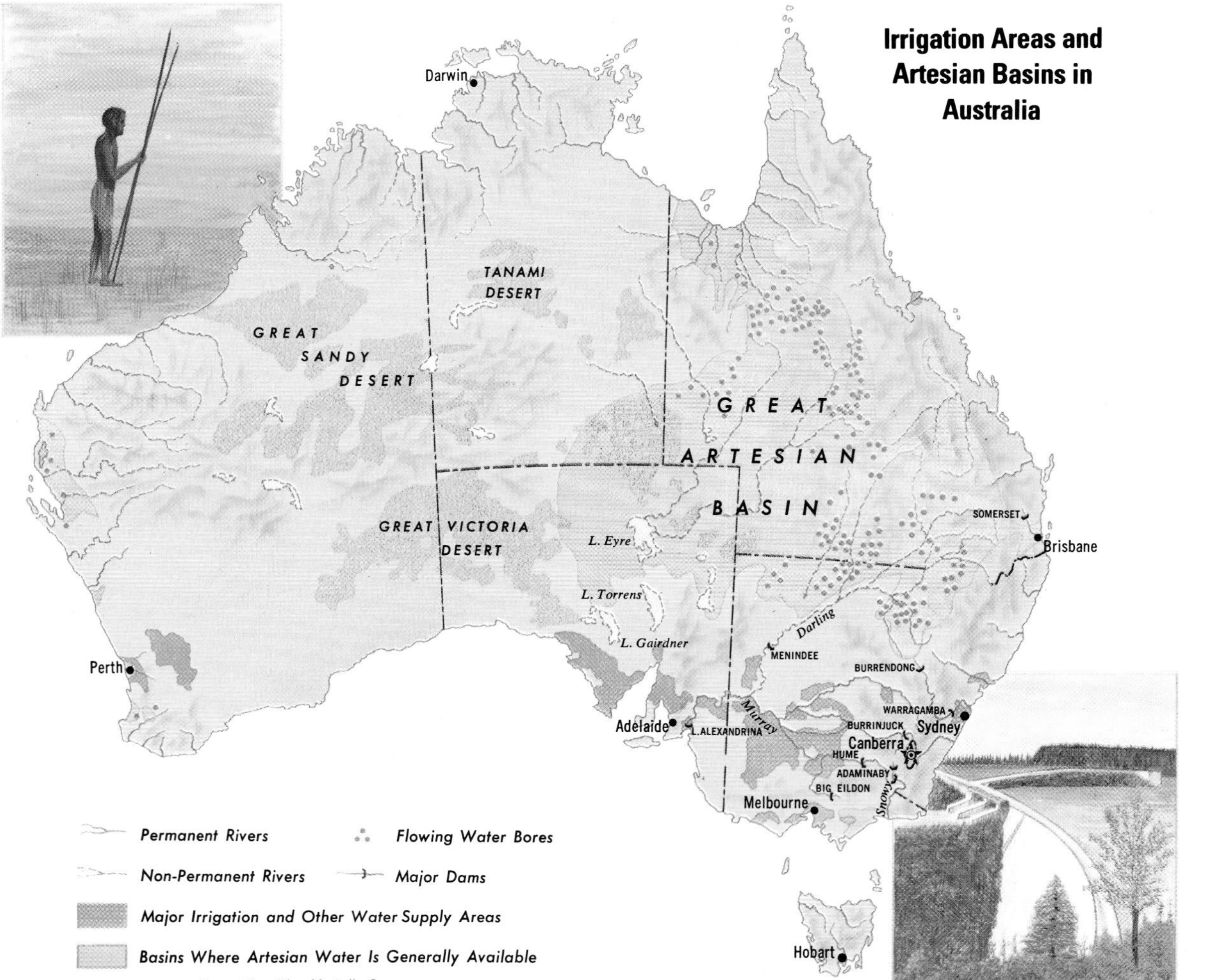

Irrigation Areas and Artesian Basins in Australia
Darwin
TANAMI DESERT
GREAT SANDY DESERT
GREAT ARTESIAN BASIN
GREAT VICTORIA DESERT
L. Eyre
L. Torrens
L. Gairdner
Darling
MENINDEE
BURRENDONG
SOMERSET
Brisbane
Perth
Adelaide
L. ALEXANDRINA
Murray
WARRAGAMBA
BURRINJUCK
Sydney
Canberra
HUME
ADAMINABY
BIG EILDON
Snowy
Melbourne
Hobart
Permanent Rivers
Non-Permanent Rivers
Flowing Water Bores
Major Dams
Major Irrigation and Other Water Supply Areas
Basins Where Artesian Water Is Generally Available

Prepared from Atlas of Australian Resources.

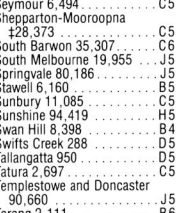

Topography

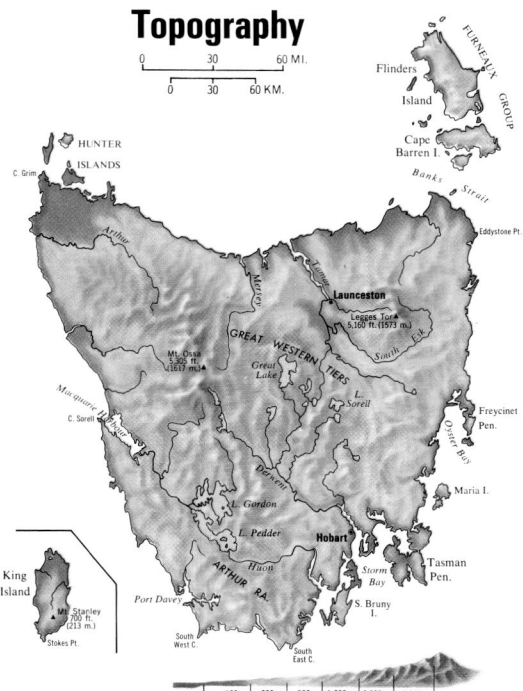

Below Sea Level — 100 m. 328 ft. — 200 m. 656 ft. — 500 m. 1,640 ft. — 1,000 m. 3,281 ft. — 2,000 m. 6,562 ft. — 5,000 m. 16,404 ft.

TASMANIA

AREA 26,178 sq. mi. (67,800 sq. km.)
POPULATION 436,353
CAPITAL Hobart
LARGEST CITY Hobart
HIGHEST POINT Mt. Ossa 5,305 ft.
(1,617 m.)

Forth (riv.)	C3	King (riv.)	B4	Ouse (riv.)	C4	South Bruny (isl.)	D5
Frankland (cape)	D1	King William (lake)	C4	Oyster (bay)	E4	South East (cape)	C5
Frankland (range)	B4	Lake (riv.)	D3	Pedder (lake)	B4	South Esk (riv.)	D3
Franklin (riv.)	B4	Legges Tor (mt.)	D3	Picton (mt.)	C5	South West (cape)	B5
Frenchmans Cap (mt.)	B4	Leven (riv.)	B3	Phoques (bay)	A1	Stanley (mt.)	A1
Freycinet (pen.)	E4	Lofty (range)	B3	Pieman (riv.)	B3	Stokes (pt.)	A1
Furneaux Group (isls.) 1,039	E1	Low Rocky (pt.)	B4	Pillar (cape)	E5	Storm (bay)	D5
Gordon (lake)	C4	Lyell (mt.)	B4	Port Davey (inlet)	B5	Strzelecki (mt.)	D2
Gordon (riv.)	B4	Maatsuyker (isls.)	C5	Portland (cape)	D2	Tamar (riv.)	D3
Great (lake)	C3	Macquarie (harb.)	B4	Ramsay (mt.)	B3	Tasman (head)	D5
Great Western Tiers (mts.)	C3	Macquarie (riv.)	D3	Raoul (cape)	D5	Tasman (pen.)	E5
Grim (cape)	A2	Maria (isl.)	E4	Reid (rapid)	B1	Tasman (sea)	E4
Hartz (mt.)	C5	Marion (bay)	E4	Ringarooma (bay)	D2	Three Hummock (isl.)	B2
Hibbs (pt.)	B4	Mersey (riv.)	C3	Robbins (isl.)	B2	Vansittart (isl.)	E2
Hogan Group (isl.)	D1	Munro (mt.)	E2	Saint Clair (lake)	C4	West (pt.)	A2
Hummock (isl.)	D2	Naturaliste (cape)	E2	Saint Helens (pt.)	E3	West Sister (isl.)	D1
Hunter (isl.)	A2	Nive (riv.)	C4	Saint Vincent (cape)	B5	Wickham (cape)	A1
Hunter (isls.)	B2	Norfolk (bay)	E4	Savage (riv.)	B3		
Huon (riv.)	C5	North (pt.)	E1	Schouten (isl.)	E4		
Indian Ocean	A4	North Bruny (isl.)	D5	Sorell (cape)	B4		
Kent Group (isls.)	D1	North Esk (riv.)	D3	Sorell (lake)	D4		
King (isl.) 2,592	A1	Ossa (mt.)	C3	South (cape)	C5		

○ Population of district.
*Population of met. area.

CITIES and TOWNS

Adventure Bay	D5	Ringarooma 223	D3
Avoca	D3	Rosebery 2,675	B3
Bagdad	D4	Ross 289	D4
Beaconsfield 898	C3	Rossarden 365	D3
Beauty Point 998	C3	Saint Helens 1,005	E3
Bell Bay	C3	Saint Marys 653	E3
Bicheno 674	E3	Sassafras	C3
Boat Harbour	B2	Savage River 1,141	B3
Bothwell 356	C4	Scottsdale 2,002	D3
Bracknell 347	C3	Sheffield 945	C3
Branxholm 273	D3	Smithton 3,378	A2
Bridgewater 6,880	D4	Snug 684	D5
Bridport 885	D3	Sorell-Midway Point 2,544	D4
Brighton 9,441	D4	Stanley 603	B2
Burnie 19,994	B3	Storeys Creek	D3
Campbell Town 879	D3	Strahan 402	B4
Chudleigh	C3	Strathgordon	C4
Colebrook	C4	Sulphur Creek 367	C3
Cressy 640	C3	Swansea 428	E4
Currie 859	A1	Tarraleah 498	C4
Cygnet 715	C5	Temma	A3
Deloraine 1,923	C3	Triabunna 924	D4
Derwent Bridge	C4	Tullah 1,894	B3
Devonport 21,424	C3	Ulverstone 9,413	C3
Dover 570	C5	Waratah 342	B3
Dunalley 203	D4	Wesley Vale	C3
Evandale 614	D3	Westbury 1,161	C3
Exeter 353	C3	Whitemark	D2
Fingal 424	E3	Woodbridge 259	D5
Forth 273	C3	Wynyard 4,582	B3
Franklin 479	C5	Zeehan 1,750	B3
Geeveston 860	C5		
George Town 5,592	C3	OTHER FEATURES	
Glenorchy 41,019	D4	Anderson (bay)	D2
Gormanston 126	B4	Anne (mt.)	C4
Gowrie Park	C3	Anser Group (isls.)	C1
Grassy 780	B1	Arthur (lake)	D4
Gravelly Beach 535	C3	Arthur (range)	C5
Hadspen 908	D3	Arthur (riv.)	B3
Hagley 232	C3	Babel (isl.)	E1
Hamilton 2,488	C4	Banks (str.)	D2
Heybridge 395	C3	Barn Bluff (mt.)	B3
Hobart (cap.) 128,603	D4	Bass (str.)	C1
Hobart *168,359	D4	Bathurst (gulf)	C5
Huonville-Ranelagh 1,347	C5	Cape Barren (isl.)	E2
Kettering	D5	Chappell (isls.)	D2
Kingston 8,556	D4	Circular (gulf)	B2
Latrobe 2,401	C3	Clarke (isl.)	E2
Lauderdale 2,117	D4	Clyde (riv.)	D4
Launceston 31,273	C3	Cox (bight)	C5
Launceston *64,555	C3	Cradle (mt.)	B3
Legana 964	D3	Cradle Mt. Lake St.	
Lilydale 308	D3	Clair Nat'l Park	B3
Longford 2,027	C3	Crescent (lake)	D4
Luina 522	B3	Curtis Group (isls.)	C1
Margate 476	C4	D'Aguilar (range)	B4
Maydena 461	C4	Davey (riv.)	B4
Meander	C3	Deal (isl.)	D1
Mole Creek 303	C3	Dee (riv.)	C4
New Norfolk 6,243	C4	Denison (range)	C4
Nubeena 225	D5	D'Entrecasteaux (chan.)	D5
Oatlands 545	D4	Derwent (riv.)	C4
Orford 378	D4	East Sister (isl.)	E1
Penguin 2,616	C3	Echo (lake)	C4
Perth 1,229	C3	Eddystone (pt.)	E2
Poatina	C3	Elliott (bay)	B5
Port Sorell 859	C3	Fires (bay)	C4
Queenstown 3,714	B4	Flinders (isl.) 2,150	D1
Railton 857	C3	Florence (riv.)	C4
Richmond 587	D4	Forestier (chan.)	E4
Ridgley 452	B3	Forestier (pen.)	E4

New Zealand

CONIC PROJECTION

SCALE OF MILES

0   50   100   150

SCALE OF KILOMETERS

0   50   100   150

Capital of Country .... ☆

© Copyright HAMMOND INCORPORATED, Maplewood, N. J.

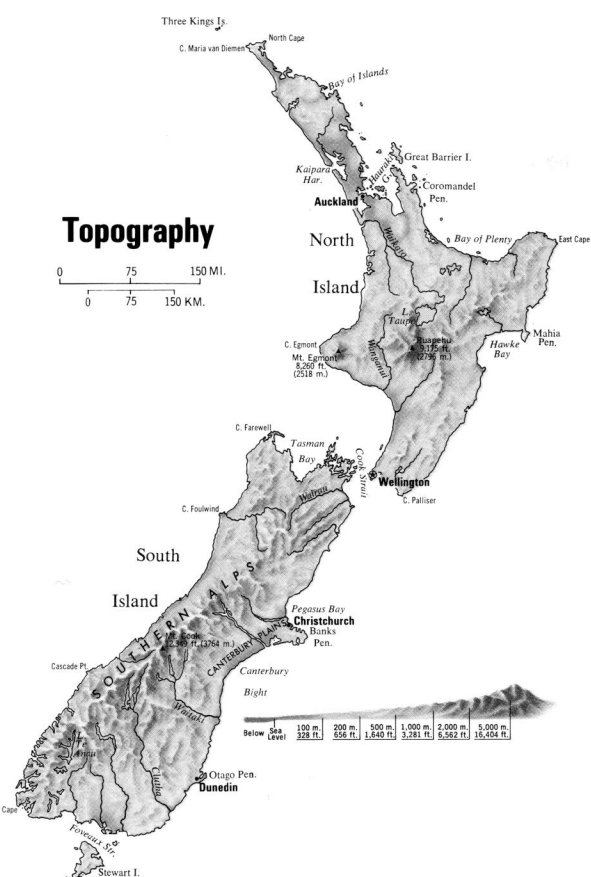

Topography

North Island

South Island

0 75 150 MI.
0 75 150 KM.

Below Sea Level | 100 m. 328 ft. | 200 m. 656 ft. | 500 m. 1,640 ft. | 1,000 m. 3,281 ft. | 2,000 m. 6,562 ft. | 5,000 m. 16,404 ft.

AREA 103,736 sq. mi. (268,676 sq. km.)
POPULATION 3,389,000
CAPITAL Wellington
LARGEST CITY Auckland
HIGHEST POINT Mt. Cook 12,349 ft. (3,764 m.)
MONETARY UNIT New Zealand dollar
MAJOR LANGUAGES English, Maori
MAJOR RELIGIONS Protestantism, Roman Catholicism

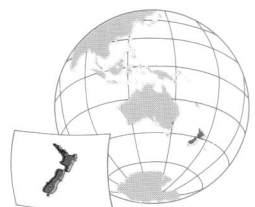

Wellington †321,004	A3	Te Anau (lake)	A6
Wellsford 1,621	E2	Tekapo (lake)	C5
Westport 4,686	C4	Terawhiti (cape)	A3
Whakatane 12,286	F2	Thames (firth)	E2
Whangamata 1,566	F2	Three Kings (isls.)	D1
Whangarei 36,550	E1	Turakirae (head)	B3
Whangarei †40,212	E1	Una (mt.)	D5
Whitianga 1,960	E2	Waiheke (isl.) 3,223	E2
Winton 2,035	B7	Waikato (riv.)	E2
Woodville 1,647	F4	Waimakariri (riv.)	D5
		Waipa (riv.)	E2
		Wairau (riv.)	D4
		Waitaki (riv.)	C6

OTHER FEATURES

Arthur's (pass)	C5	Waitemata (harb.)	B1
Aspiring (mt.)	B6	Wakatipu (lake)	B6
Banks (pen.)	D5	Wanaka (lake)	B6
Bream (bay)	E1	Wanganui (riv.)	E3
Brett (cape)	E1	West (cape)	A6
Buller (riv.)	D4	Whitcombe (mt.)	C5
Campbell (cape)	E4	
Canterbury (bight)	D6	†Population of urban area.
Cascade (pt.)	B6
Chatham (isls.) 751	D7
Cloudy (bay)	E4
Clutha (riv.)	B6
Coleridge (lake)	C5
Colville (cape)	E2
Cook (mt.)	C5
Cook (str.)	E4
Coromandel (pen.)	F2
Devil River (peak)	D4
D'Urville (isl.)	D4
Dusky (sound)	A6
East (cape)	G2
Egmont (cape)	D3
Egmont (mt.)	D3
Ellesmere (lake)	D5
Farewell (cape)	D4
Foulwind (cape)	C4
Fournier (cape)	E7
Foveaux (str.)	A7
Golden (bay)	D4
Great Barrier (isl.) 572	E2
Haast (pass)	B6
Hauraki (gulf)	C1
Hawke (bay)	F3
Hikurangi (mt.)	G2
Hokianga (harb.)	D1
Huiarau (range)	F3
Hutt (riv.)	C2
Islands (bay)	E1
Jackson (bay)	B5
Kaikoura (range)	D5
Kaimanawa (range)	F3
Kaipara (harb.)	D2
Karamea (bight)	C4
Kawhia (harb.)	E3
Kidnappers (cape)	F3
Mahia (pen.)	G3
Manapouri (lake)	A6
Manukau (harb.)	B1
Maria van Diemen (cape)	D1
Mataura (riv.)	B6
Mercury (isls.)	F2
Milford (sound)	A6
Needles (pt.)	E2
Nicholson, Port (inlet)	B3
Ninety Mile (beach)	D1
North (cape)	D1
North (isl.) 2,322,989	F1
North Taranaki (bight)	D3
Otago (pen.)	C6
Owen (mt.)	D4
Palliser (cape)	E4
Pegasus (bay)	D5
Pitt (isl.)	E7
Plenty (bay)	F2
Port Nicholson (inlet)	B3
Port Pegasus (inlet)	B7
Pukaki (lake)	B6
Puysegur (pt.)	A7
Rakaia (riv.)	C5
Rangitata (riv.)	C5
Rangitikei (riv.)	E3
Raukumara (range)	F3
Reinga (cape)	D1
Resolution (isl.)	A6
Richmond (range)	D4
Rocks (pt.)	C4
Rotorua (lake)	F3
Ruahine (range)	F4
Ruapehu (mt.)	E3
Ruapuke (isl.)	B7
South (cape)	A7
South (isl.) 852,748	B5
Southern Alps (range)	C5
South Taranaki (bight)	D3
Spenser (mts.)	D5
Stewart (isl.) 600	A7
Tararua (range)	E4
Tasman (bay)	D4
Tasman (mt.)	C5
Tasman (mts.)	D4
Tasman (sea)	B4
Taupo (lake)	F3
Tauroa (pt.)	D1

CITIES and TOWNS

Albany 2,001	B1	Invercargill 49,446	B7
Alexandra 4,348	B6	Invercargill †53,868	B7
Ashburton 14,151	C5	Kaiapoi 4,894	D5
Ashhurst 1,906	E4	Kaikohe 3,663	D1
Auckland 144,963	B1	Kaikoura 2,180	D5
Auckland †769,558	B1	Kaitaia 4,737	D1
Balclutha 4,495	B7	Kawerau 8,593	F3
Belmont 2,402	B2	Kumeu 3,414	B1
Birkenhead 21,324	B1	Levin 14,652	E4
Blenheim 17,849	D4	Lower Hutt 63,245	B2
Bluff 2,720	B7	Lyttelton 3,184	D5
Bulls 1,839	E4	Manukau 159,362	C1
Cambridge 8,514	E2	Marton 4,858	E4
Carterton 3,971	E4	Masterton 18,785	E4
Christchurch 164,680	D5	Mataura 2,345	B7
Christchurch †289,959	D5	Milton 2,193	B7
Cromwell 2,364	B6	Morrinsville 5,080	E2
Dannevirke 5,663	F4	Mosgiel 9,264	C6
Dargaville 4,747	D1	Motueka 4,693	D4
Devonport 10,410	C1	Mount Albert 26,462	B1
Dunedin 77,176	C6	Mount Eden 18,305	B1
Dunedin †107,445	C6	Mount Maunganui 11,391	F2
Eastbourne 4,561	B3	Mount Roskill 33,577	B1
East Coast Bays 28,866	B1	Mount Wellington 19,528	C1
Edgecumbe 1,929	F2	Murupara 2,964	F3
Ellerslie 5,404	C1	Napier 48,314	F3
Eltham 2,411	E3	Napier †51,330	F3
Fairfield 1,849	C6	Nelson 33,304	D4
Featherston 2,458	E4	Nelson †43,121	D4
Feilding 11,522	E4	New Lynn 10,445	B1
Foxton 2,719	E4	New Plymouth 36,048	D3
Geraldine 2,128	C6	New Plymouth †44,095	D3
Gisborne 29,986	G3	Ngaruawahia 4,435	E2
Gisborne †32,062	G3	Northcote 10,061	B1
Glen Eden 9,406	B1	Oban (Half Moon Bay) 2,448	B7
Glenfield 3,691	B1	Onehunga 15,386	B1
Gore 9,185	B7	One Tree Hill 11,078	B1
Green Bay 3,035	B1	Opotiki 3,388	F3
Green Island 6,899	C7	Orewa 5,552	E2
Greymouth 8,103	C5	Otahuhu 10,298	C1
Greytown 1,797	E4	Otaki 4,301	E4
Half Moon Bay (Oban) 2,448	B7	Otorohanga 2,574	E3
Hamilton 91,109	E2	Paeroa 3,702	E2
Hamilton †97,907	E2	Pahiatua 2,599	F4
Hastings 36,083	F3	Paihia 1,740	D1
Hastings †52,563	F3	Palmerston North 60,105	E4
Havelock North 8,507	F3	Palmerston North †66,691	E4
Hawera 8,400	E3	Papakura 22,473	E2
Helensville 1,360	B1	Papatoetoe 21,700	C1
Henderson 6,645	B1	Patea 1,938	E3
Heretaunga-Pinehaven 6,171	C2	Petone 8,113	B2
Hokitika 3,414	C5	Picton 3,220	D4
Hornby 8,215	D5	Pinehaven (Heretaunga-Pinehaven) 6,171	C2
Howick 13,866	C1	Porirua 41,104	B2
Huntly 6,534	E2	Port Chalmers 2,917	C6
Hutt (Upper and Lower) †131,257	B2	Pukekohe 9,070	E2
Inglewood 2,839	E3	Putaruru 4,222	E3
		Queenstown 3,367	B6

Raetihi 1,247	E3	Te Anau 2,610	A6
Raglan 1,414	E2	Te Aroha 3,331	E2
Rangiora 6,385	D5	Te Atatu 14,713	B1
Reefton 1,200	C5	Te Awamutu 7,922	E2
Riccarton 6,709	D5	Te Kauwhata 842	E2
Richmond 6,847	D4	Te Kuiti 4,795	E3
Riverton 1,479	B7	Temuka 3,771	C6
Rotorua 38,157	F3	Thames 6,456	E2
Rotorua †48,314	F3	The Hermitage	C5
Runanga 1,264	C5	Timaru 28,412	C6
Russell 932	E1	Timaru †29,225	C6
Saint Kilda 6,147	C7	Titirangi 8,426	B1
Shannon 1,465	E4	Tokoroa 18,713	E3
Stratford 5,518	E3	Tuakau 1,982	E2
Taihape 2,586	E3	Tuatapere 884	A7
Takapuna 64,844	B1	Turangi 5,517	E3
Tapanui 1,042	B6	Upper Hutt 31,405	B2
Taradale 4,681	F3	Waihi 3,538	E2
Taumarunui 6,541	E3	Waikanae 4,818	E4
Taupo 13,651	F3	Waikouaiti 858	C6
Tauranga 37,099	F2	Waimate 3,393	C6
Tauranga †53,097	F2	Wainuiomata 19,192	B3
Tawa 12,216	B2	Waipawa 1,732	F4
		Waipukurau 3,648	F4
		Wairoa 5,439	F3
		Waitangi	D7
		Waitara 6,012	E3
		Waitemata 87,452	B1
		Waiuku 3,654	E2
		Wanaka 1,155	B6
		Wanganui 37,012	E3
		Wanganui †39,595	E3
		Warkworth 1,734	E2
		Washdyke 949	C6
		Waverley 1,239	E3
		Wellington (cap.) 135,688	A3

Agriculture, Industry and Resources

Snapper
Fruit
Auckland
Sheep
Dairy
Wellington
Wheat
Christchurch
Sheep
Crayfish
Oysters
Soles
Dunedin
Crayfish
Lg
Lg

DOMINANT LAND USE

- Mixed Farming, Livestock
- Dairy
- Truck Farming, Horticulture
- Pasture Livestock (chiefly sheep)
- Livestock Herding
- Forests
- Nonagricultural Land

MAJOR MINERAL OCCURRENCES

- C Coal
- G Natural Gas
- J Jade
- Ka Kaolin
- Lg Lignite
- O Petroleum
- U Uranium

4 Water Power
Major Industrial Areas

AREA 11,707,000 sq. mi. (30,321,130 sq. km.)
POPULATION 648,000,000
LARGEST CITY Cairo
HIGHEST POINT Kilimanjaro 19,340 ft.
 (5,895 m.)
LOWEST POINT Lake Assal, Djibouti -512 ft.
 (-156 m.)

Population Distribution

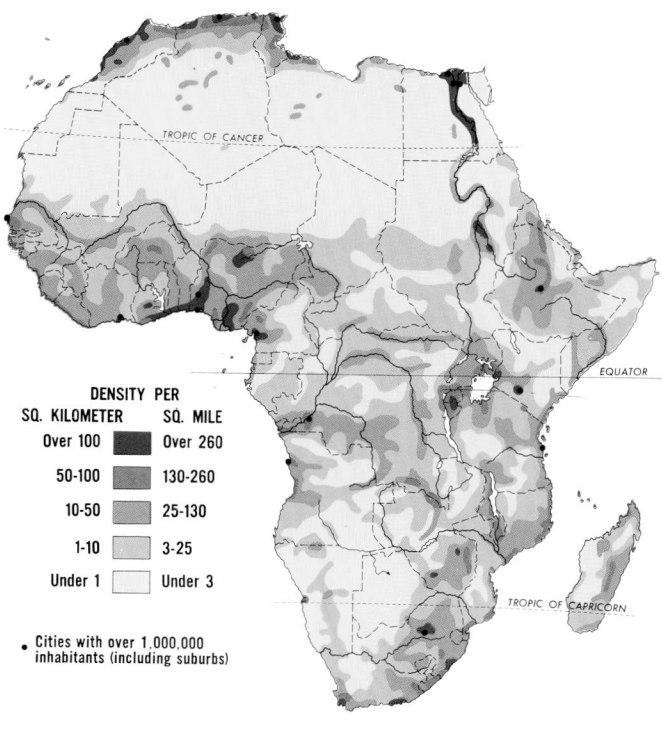

DENSITY PER

SQ. KILOMETER	SQ. MILE
Over 100	Over 260
50-100	130-260
10-50	25-130
1-10	3-25
Under 1	Under 3

• Cities with over 1,000,000
  inhabitants (including suburbs)

Vegetation

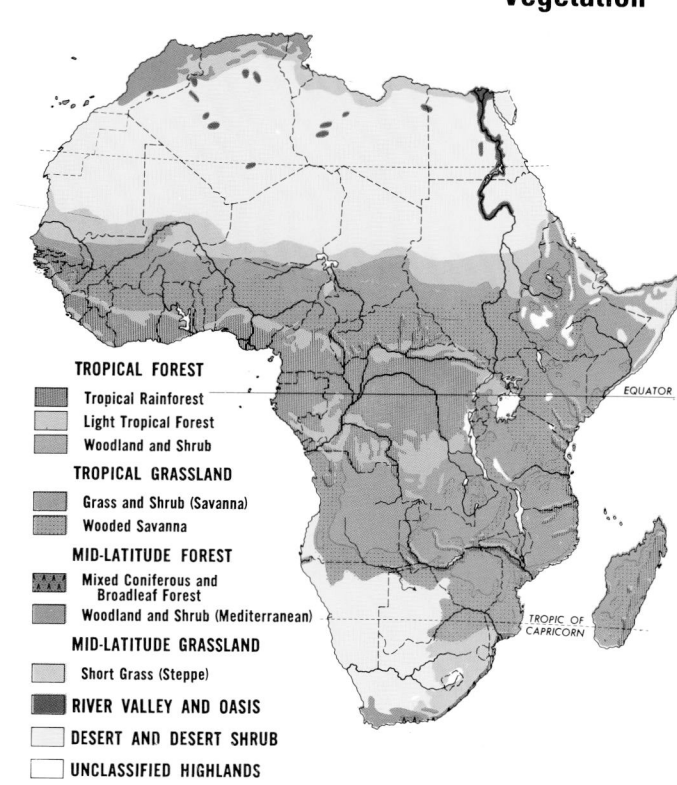

TROPICAL FOREST
- Tropical Rainforest
- Light Tropical Forest
- Woodland and Shrub

TROPICAL GRASSLAND
- Grass and Shrub (Savanna)
- Wooded Savanna

MID-LATITUDE FOREST
- Mixed Coniferous and Broadleaf Forest
- Woodland and Shrub (Mediterranean)

MID-LATITUDE GRASSLAND
- Short Grass (Steppe)

RIVER VALLEY AND OASIS

DESERT AND DESERT SHRUB

UNCLASSIFIED HIGHLANDS

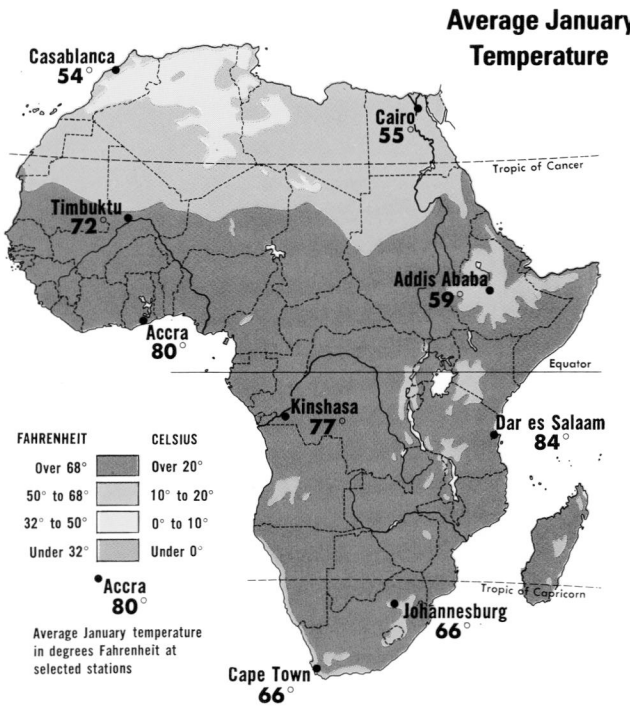

Average January Temperature

Casablanca
54°

Cairo
55°

Timbuktu
72°

Addis Ababa
59°

Accra
80°

Kinshasa
77°

Dar es Salaam
84°

Johannesburg
66°

Cape Town
66°

FAHRENHEIT	CELSIUS
Over 68°	Over 20°
50° to 68°	10° to 20°
32° to 50°	0° to 10°
Under 32°	Under 0°

•Accra
80°

Average January temperature in degrees Fahrenheit at selected stations

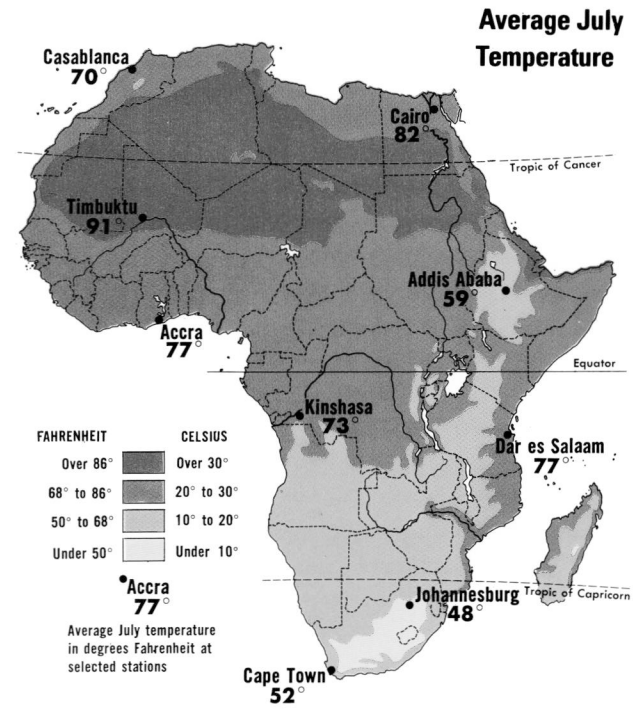

Average July Temperature

Casablanca
70°

Cairo
82°

Timbuktu
91°

Addis Ababa
59°

Accra
77°

Kinshasa
73°

Dar es Salaam
77°

Johannesburg
48°

Cape Town
52°

FAHRENHEIT	CELSIUS
Over 86°	Over 30°
68° to 86°	20° to 30°
50° to 68°	10° to 20°
Under 50°	Under 10°

•Accra
77°

Average July temperature in degrees Fahrenheit at selected stations

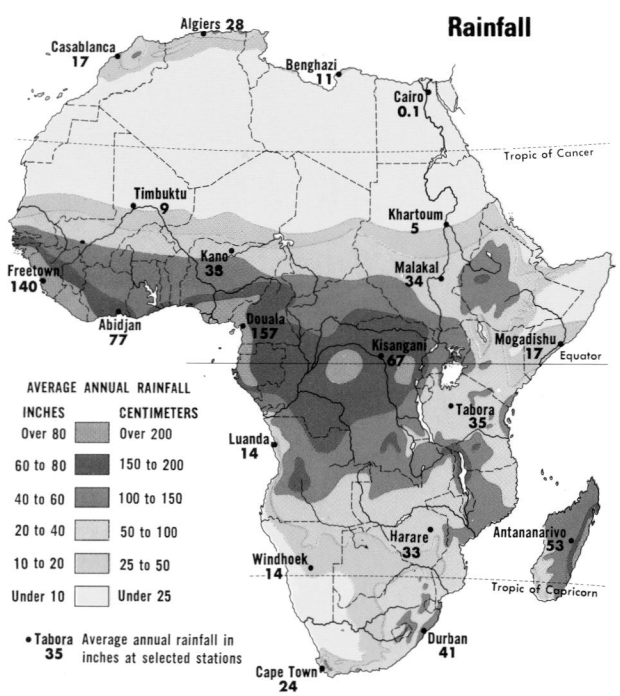

Rainfall

Algiers 28

Casablanca
17

Benghazi
11

Cairo
0.1

Timbuktu
9

Khartoum
5

Freetown
140

Kano
38

Malakal
34

Abidjan
77

Douala
157

Kisangani
67

Mogadishu
17

Tabora
35

Luanda
14

Harare
33

Antananarivo
53

Windhoek
14

Durban
41

Cape Town
24

AVERAGE ANNUAL RAINFALL

INCHES	CENTIMETERS
Over 80	Over 200
60 to 80	150 to 200
40 to 60	100 to 150
20 to 40	50 to 100
10 to 20	25 to 50
Under 10	Under 25

•Tabora Average annual rainfall in
35   inches at selected stations

Vegetation/Relief

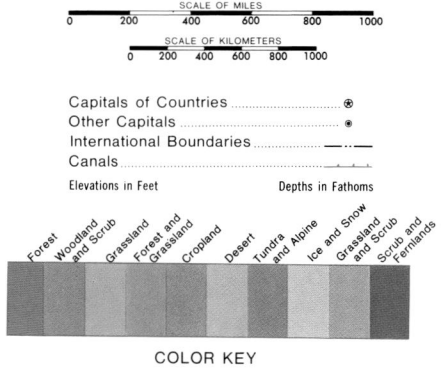

SCALE OF MILES
0 200 400 600 800 1000

SCALE OF KILOMETERS
0 200 400 600 800 1000

Capitals of Countries	⊛
Other Capitals	⊛
International Boundaries	———
Canals	

Elevations in Feet   Depths in Fathoms

Forest
Woodland and Scrub
Grassland
Forest and Grassland
Cropland
Desert
Tundra and Alpine
Ice and Snow
Grassland and Scrub
Scrub and Fernlands

COLOR KEY

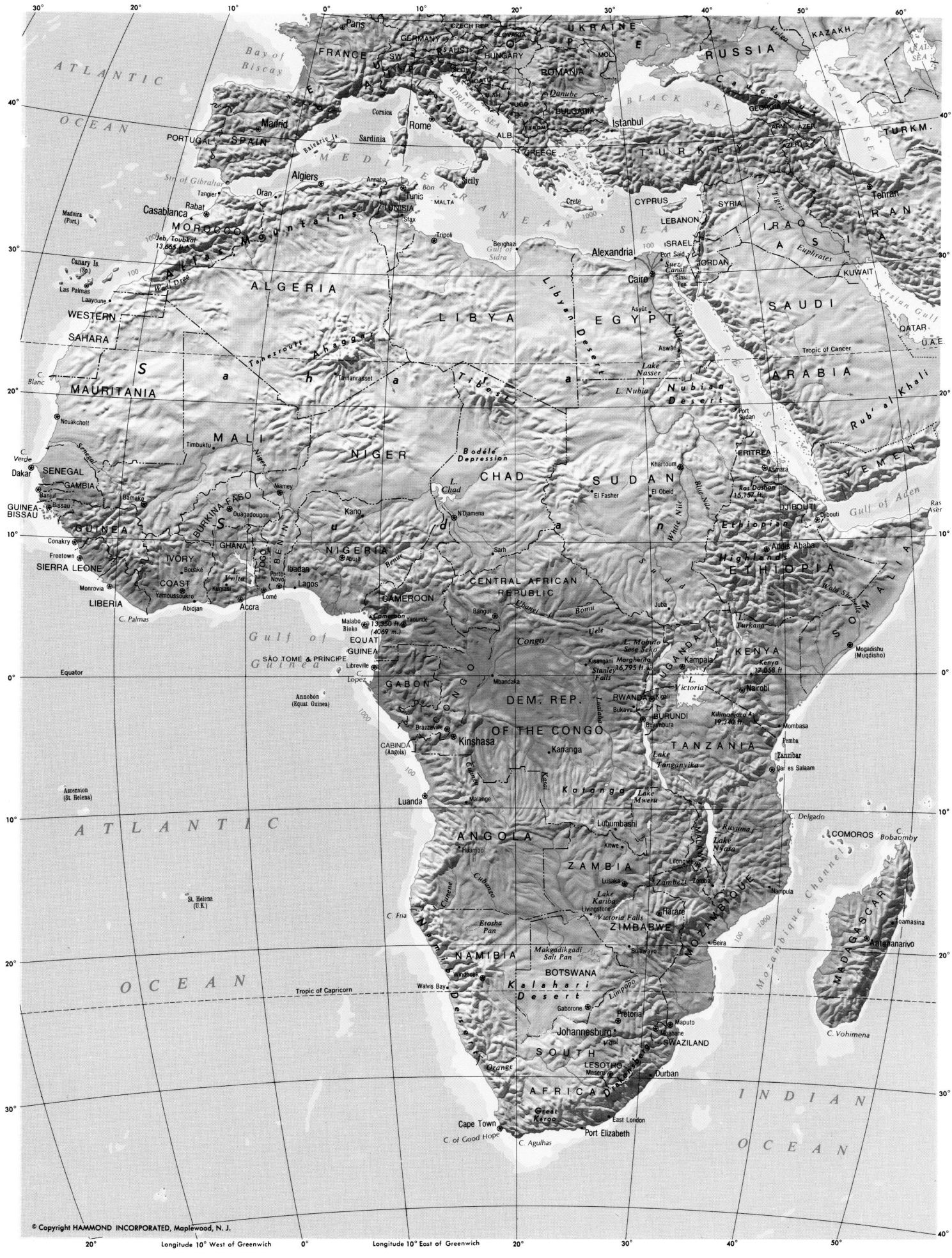

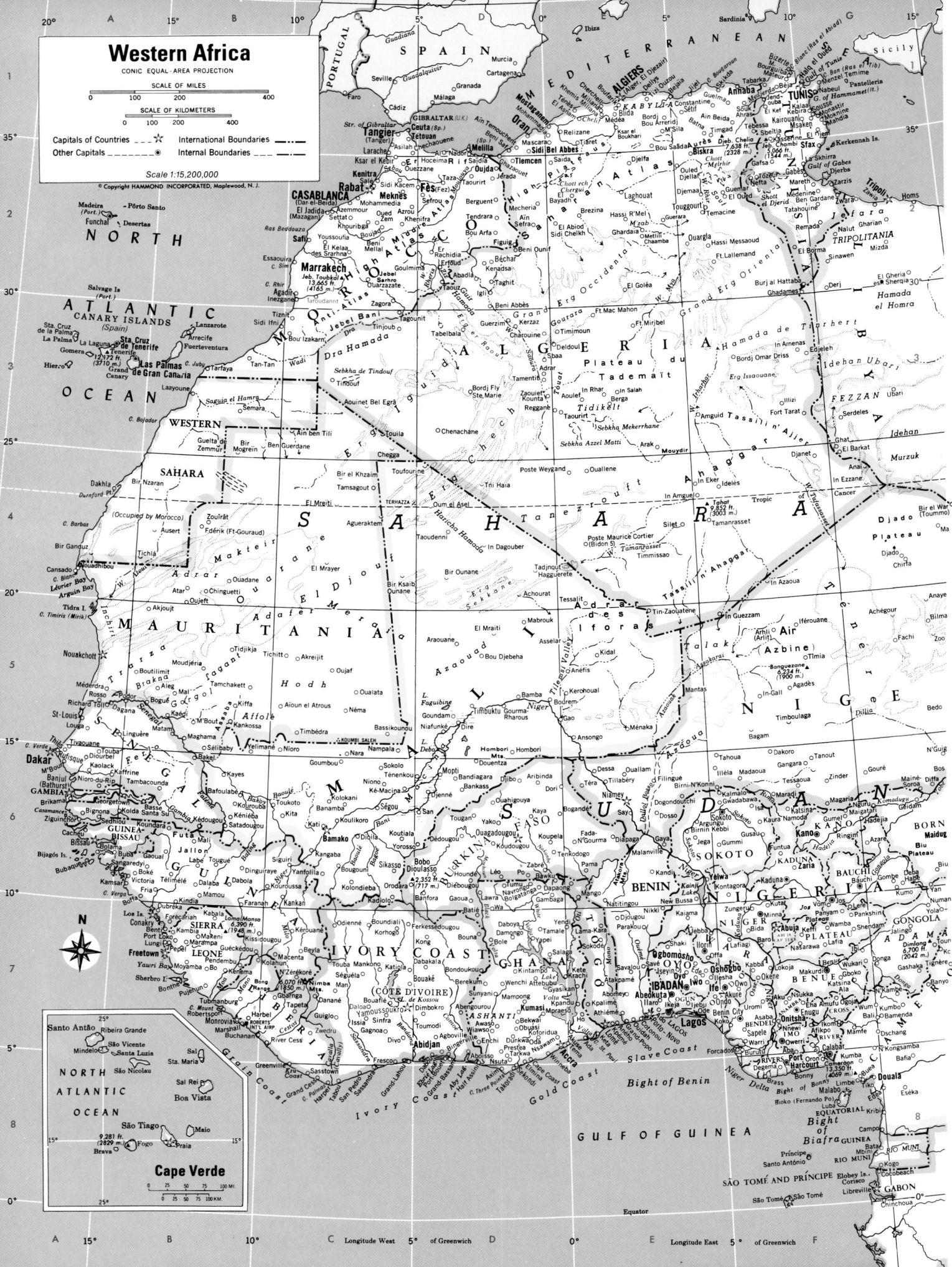

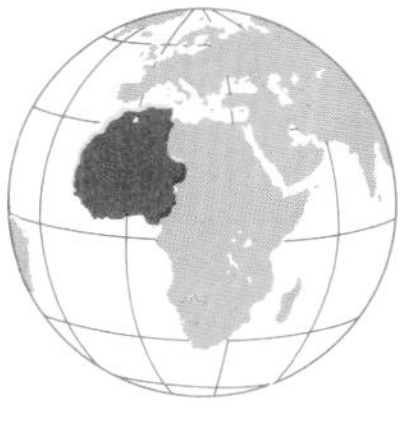

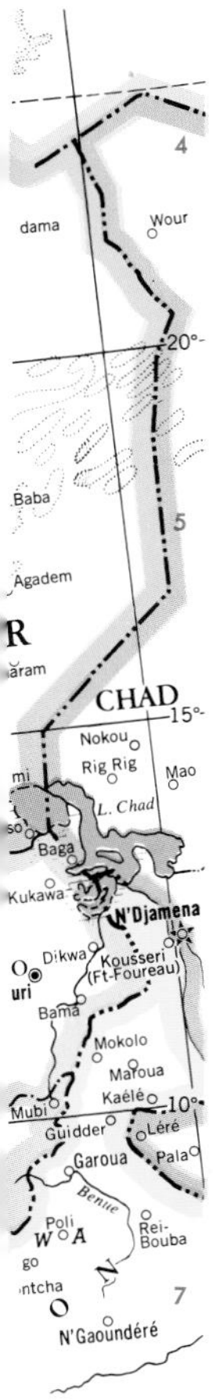

ALGERIA

AREA 919,591 sq. mi. (2,381,740 sq. km.)
POPULATION 22,971,000
CAPITAL Algiers
LARGEST CITY Algiers
HIGHEST POINT Tahat 9,852 ft. (3,003 m.)
MONETARY UNIT Algerian dinar
MAJOR LANGUAGES Arabic, Berber, French
MAJOR RELIGION Islam

BENIN

AREA 43,483 sq. mi. (112,620 sq. km.)
POPULATION 4,591,000
CAPITAL Porto-Novo
LARGEST CITY Cotonou
HIGHEST POINT Atakora Mts. 2,083 ft. (635 m.)
MONETARY UNIT CFA franc
MAJOR LANGUAGES Fon, Somba, Yoruba, Bariba, French, Mina, Dendi
MAJOR RELIGIONS Tribal religions, Islam, Roman Catholicism

CAPE VERDE

AREA 1,557 sq. mi. (4,033 sq. km.)
POPULATION 347,000
CAPITAL Praia
LARGEST CITY Praia
HIGHEST POINT 9,281 ft. (2,829 m.)
MONETARY UNIT Cape Verde escudo
MAJOR LANGUAGE Portuguese
MAJOR RELIGION Roman Catholicism

GAMBIA

AREA 4,127 sq. mi. (10,689 sq. km.)
POPULATION 688,000
CAPITAL Banjul
LARGEST CITY Banjul
HIGHEST POINT 100 ft. (30 m.)
MONETARY UNIT dalasi
MAJOR LANGUAGES Mandingo, Fulani, Wolof, English, Malinke
MAJOR RELIGIONS Islam, tribal religions, Christianity

GHANA

AREA 92,099 sq. mi. (238,536 sq. km.)
POPULATION 13,391,000
CAPITAL Accra
LARGEST CITY Accra
HIGHEST POINT Togo Hills 2,900 ft. (884 m.)
MONETARY UNIT cedi
MAJOR LANGUAGES Twi, Fante, Dagbani, Ewe, Ga, English, Hausa, Akan
MAJOR RELIGIONS Tribal religions, Christianity, Islam

GUINEA

AREA 94,925 sq. mi. (245,856 sq. km.)
POPULATION 6,706,000
CAPITAL Conakry
LARGEST CITY Conakry
HIGHEST POINT Nimba Mts. 6,070 ft. (1,850 m.)
MONETARY UNIT syli
MAJOR LANGUAGES Fulani, Mandingo, Susu, French
MAJOR RELIGIONS Islam, tribal religions

GUINEA-BISSAU

AREA 13,948 sq. mi. (36,125 sq. km.)
POPULATION 943,000
CAPITAL Bissau
LARGEST CITY Bissau
HIGHEST POINT 689 ft. (210 m.)
MONETARY UNIT Guinea-Bissau peso
MAJOR LANGUAGES Balante, Fulani, Crioulo, Mandingo, Portuguese
MAJOR RELIGIONS Islam, tribal religions, Roman Catholicism

IVORY COAST (CÔTE-D'IVOIRE)

AREA 124,504 sq. mi. (322,465 sq. km.)
POPULATION 9,300,000
CAPITAL Yamoussoukro
LARGEST CITY Abidjan
HIGHEST POINT 5,745 ft. (1,751 m.)
MONETARY UNIT CFA franc
MAJOR LANGUAGES Bale, Bete, Senufu, French, Dioula
MAJOR RELIGIONS Tribal religions, Islam

LIBERIA

AREA 43,000 sq. mi. (111,370 sq. km.)
POPULATION 2,508,000
CAPITAL Monrovia
LARGEST CITY Monrovia
HIGHEST POINT Wutivi 5,584 ft. (1,702 m.)
MONETARY UNIT Liberian dollar
MAJOR LANGUAGES Kru, Kpelle, Bassa, Vai, English
MAJOR RELIGIONS Christianity, tribal religions, Islam

MALI

AREA 464,873 sq. mi. (1,204,021 sq. km.)
POPULATION 7,960,000
CAPITAL Bamako
LARGEST CITY Bamako
HIGHEST POINT Hombori Mts. 3,789 ft. (1,155 m.)
MONETARY UNIT CFA franc
MAJOR LANGUAGES Bambara, Senufu, Fulani, Soninke, French
MAJOR RELIGIONS Islam, tribal religions

MAURITANIA

AREA 419,229 sq. mi. (1,085,803 sq. km.)
POPULATION 1,970,000
CAPITAL Nouakchott
LARGEST CITY Nouakchott
HIGHEST POINT 2,972 ft. (906 m.)
MONETARY UNIT ouguiya
MAJOR LANGUAGES Arabic, Wolof, Tukolor, French
MAJOR RELIGION Islam

MOROCCO

AREA 172,414 sq. mi. (446,550 sq. km.)
POPULATION 24,522,000
CAPITAL Rabat
LARGEST CITY Casablanca
HIGHEST POINT Jeb. Toubkal 13,665 ft. (4,165 m.)
MONETARY UNIT dirham
MAJOR LANGUAGES Arabic, Berber, French
MAJOR RELIGIONS Islam, Judaism, Christianity

NIGER

AREA 489,189 sq. mi. (1,267,000 sq. km.)
POPULATION 7,250,000
CAPITAL Niamey
LARGEST CITY Niamey
HIGHEST POINT Banguezane 6,234 ft. (1,900 m.)
MONETARY UNIT CFA franc
MAJOR LANGUAGES Hausa, Songhai, Fulani, French, Tamashek, Djerma
MAJOR RELIGIONS Islam, tribal religions

NIGERIA

AREA 357,000 sq. mi. (924,630 sq. km.)
POPULATION 104,957,000
CAPITAL Abuja
LARGEST CITY Lagos
HIGHEST POINT Dimlang 6,700 ft. (2,042 m.)
MONETARY UNIT naira
MAJOR LANGUAGES Hausa, Yoruba, Ibo, Ijaw, Fulani, Tiv, Kanuri, Ibibio, English, Edo
MAJOR RELIGIONS Islam, Christianity, tribal religions

SÃO TOMÉ AND PRÍNCIPE

AREA 372 sq. mi. (963 sq. km.)
POPULATION 116,000
CAPITAL São Tomé
LARGEST CITY São Tomé
HIGHEST POINT Pico 6,640 ft. (2,024 m.)
MONETARY UNIT dobra
MAJOR LANGUAGES Bantu languages, Portuguese
MAJOR RELIGIONS Tribal religions, Roman Catholicism

SENEGAL

AREA 75,954 sq. mi. (196,720 sq. km.)
POPULATION 7,113,000
CAPITAL Dakar
LARGEST CITY Dakar
HIGHEST POINT Futa Jallon 1,640 ft. (500 m.)
MONETARY UNIT CFA franc
MAJOR LANGUAGES Wolof, Peul (Fulani), French, Mende, Mandingo, Dida
MAJOR RELIGIONS Islam, tribal religions, Roman Catholicism

SIERRA LEONE

AREA 27,925 sq. mi. (72,325 sq. km.)
POPULATION 4,047,000
CAPITAL Freetown
LARGEST CITY Freetown
HIGHEST POINT Loma Mts. 6,390 ft. (1,947 m.)
MONETARY UNIT leone
MAJOR LANGUAGES Mende, Temne, Vai, English, Krio (pidgin)
MAJOR RELIGIONS Tribal religions, Islam, Christianity

TOGO

AREA 21,622 sq. mi. (56,000 sq. km.)
POPULATION 3,296,000
CAPITAL Lomé
LARGEST CITY Lomé
HIGHEST POINT Agou 3,445 ft. (1,050 m.)
MONETARY UNIT CFA franc
MAJOR LANGUAGES Ewe, French, Twi, Hausa
MAJOR RELIGIONS Tribal religions, Roman Catholicism, Islam

TUNISIA

AREA 63,378 sq. mi. (164,149 sq. km.)
POPULATION 7,465,000
CAPITAL Tunis
LARGEST CITY Tunis
HIGHEST POINT Jeb. Chambi 5,066 ft. (1,544 m.)
MONETARY UNIT Tunisian dinar
MAJOR LANGUAGES Arabic, French
MAJOR RELIGION Islam

BURKINA FASO

AREA 105,869 sq. mi. (274,200 sq. km.)
POPULATION 9,001,000
CAPITAL Ouagadougou
LARGEST CITY Ouagadougou
HIGHEST POINT 2,352 ft. (717 m.)
MONETARY UNIT CFA franc
MAJOR LANGUAGES Mossi, Lobi, French, Samo, Gourounsi
MAJOR RELIGIONS Islam, tribal religions, Roman Catholicism

WESTERN SAHARA

AREA 102,703 sq. mi. (266,000 sq. km.)
POPULATION 174,000
HIGHEST POINT 2,700 ft. (823 m.)
MAJOR LANGUAGE Arabic
MAJOR RELIGION Islam

Topography

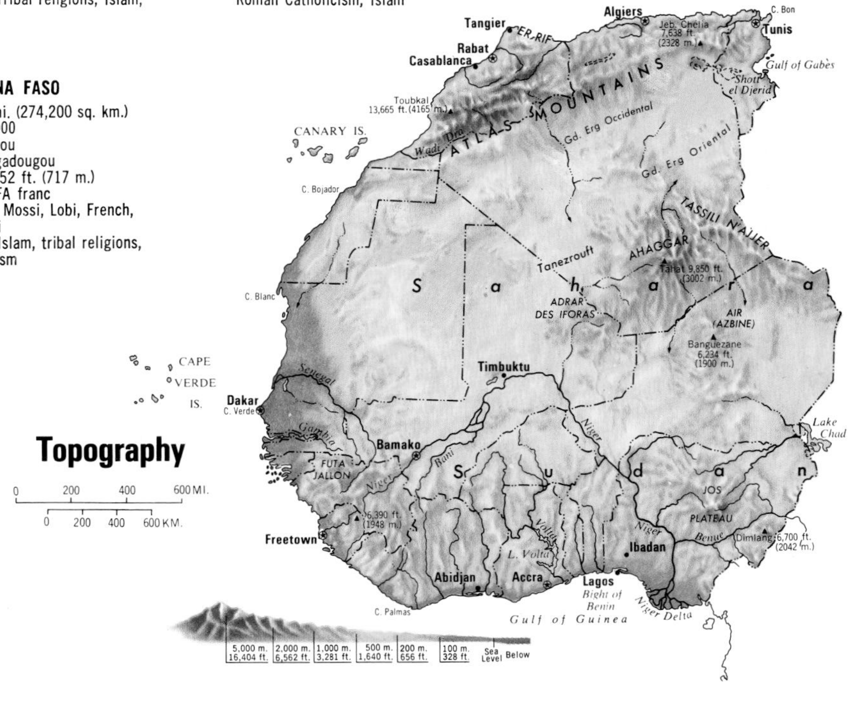

0 200 400 600 MI.
0 200 400 600 KM.

| 5,000 m. | 2,000 m. | 1,000 m. | 500 m. | 200 m. | 100 m. | Sea | Below |
| 16,404 ft. | 6,562 ft. | 3,281 ft. | 1,640 ft. | 656 ft. | 328 ft. | Level | |

(continued)

ALGERIA

CITIES and TOWNS

Adrar 28,495 ... D3
Aïn Beïda 67,281 ... F1
Aïn Sefra 22,400 ... D2
Aïn Temouchent 48,935 ... D1
Algiers (cap.) 1,687,579 ... E1
Annaba 227,795 ... F1
Aoulef 10,259 ... E3
Batna 184,833 ... F1
Béchar 107,042 ... D2
Bejaïa 118,233 ... F1
Beni Abbès 7,370 ... D2
Beni Saf 30,700 ... D1
Biskra 129,611 ... F2
Blida 131,615 ... E1
Bone (Annaba) 227,795 ... F1
Bordj Bou Arreridj 86,997 ... E1
Bordj Omar Driss 1,900 ... F3
Boufarik 54,023 ... E1
Bougie (Bejaïa) 118,233 ... F1
Bou Saâda 50,000 ... E1
Brezina 10,000 ... E2
Cherchell 32,572 ... E1
Constantine 449,602 ... F1
Dellys 29,700 ... E1
Djelfa 88,929 ... E2
Djemaa 34,600 ... F2
El Abiod Sidi Cheikh 15,300 ... E2
El Asnam 103,998 ... E1
El Bayadh 44,925 ... E2
El Djezair (Algiers) (cap.) 1,687,579 ... E1
El Goléa 24,400 ... E2
El Oued 73,093 ... F2
Ghardaïa 62,518 ... E2
Ghazaouet 29,795 ... D2
Guelma 84,826 ... F1
Guerara 22,300 ... E2
Hassi Messaoud ... F2
Hassi R'Mel 10,545 ... E2
In Guezzam 10,304 ... F5
In Salah 20,733 ... E3
Jijel 69,274 ... F1
Khemis Miliana 57,101 ... E1
Ksar el Boukhari 41,200 ... E1
Laghouat 71,808 ... E2
Mascara 70,885 ... D1
Mecheria 40,251 ... E2
Médéa 84,062 ... E1
Metlili Chaamba 21,300 ... E2
Miliana 36,400 ... E1
Mohammadia 58,967 ... D1
Mostaganem 115,302 ... D1
M'Sila 82,877 ... E1
Oran 598,525 ... D1
Orléansville (El Asnam) 103,998 ... E1
Ouargla 76,270 ... F2
Ouled Djellal 33,278 ... F2
Philippeville (Skikda) 128,503 ... F1
Reggane 10,061 ... D3
Relizane 83,864 ... E1
Saïda 84,371 ... E2
Sétif 185,786 ... F1
Sidi Bel-Abbes 154,745 ... D1
Skikda 128,503 ... F1
Souk Ahras 85,873 ... F1
Tamanrasset 38,146 ... F4
Tébessa 111,688 ... F1
Ténès 26,510 ... E1
Tiaret 105,562 ... E1
Timimoun 21,556 ... E3
Tindouf 6,500 ... C3
Tizi Ouzou 93,025 ... E1
Tlemcen 108,145 ... D2
Touggourt 75,600 ... F2
Zaouiet Kounta 10,707 ... D3

OTHER FEATURES

Adrar des Iforas (plat.) ... E5
Ahaggar (range) ... F4
Anaï (well) ... G4
Aouinet Bel Egrâ (well) ... C3
Atlas (mts.) ... E2
Aurès (lag.) ... F1
Azzel Mati, Sebkha (lake) ... E3
Bougaroun (cape) ... F1
Chech, Erg (des.) ... D3
Chelia (mt.) ... F1
Chelif (riv.) ... E1
Chergui, Chott Ech (salt lake) ... E2
Dra, Wadi (dry river) ... C3
Dra Hamada (plat.) ... C3
Gourara (oasis) ... E3
Grand Erg Occidental (des.) ... E2
Grand Erg Oriental (des.) ... F2
Guir Hamada (des.) ... D2
High Plateaus (ranges) ... D2
Iguidi, Erg (des.) ... C3
In Ezzane (well) ... G4
Irharhar, Wadi (dry river) ... F3
Kabylia (reg.) ... E1
Mediterranean (sea) ... E1
Medjerda (riv.) ... F1
Mekerrhane, Sebkha (salt lake) ... E3
Melrhir, Chott (salt lake) ... F2
Mouydir (mts.) ... E3
Mya, Wadi (dry river) ... E2
M'zab (oasis) ... E2
Raoui, Erg er (des.) ... D3
Rhir, Wadi (dry river) ... F2
Sahara (des.) ... E4
Saharan Atlas (ranges) ... E2
Saoura, Wadi (dry river) ... D3
Souf (oasis) ... F2
Tademaït, Plateau du (plat.) ... E3
Tafassasset, Wadi (dry river) ... F4
Tahat (mt.) ... F4
Tamanrasset, Wadi (dry river) ... E4
Tanezrouft (des.) ... E4
Tassili N'Ahagger (plat.) ... E4
Tassili N'Ajjer (plat.) ... F3
Tidikelt (oasis) ... E3
Timmissao (well) ... E4
Tindouf, Sebkha de (salt lake) ... C3
Tinrhert, Hamada de (des.) ... F3
Tni Haïa (well) ... D4
Touat (oasis) ... E3
Touila (well) ... C3

BENIN

CITIES and TOWNS

Abomey 38,000 ... E7
Cotonou 178,000 ... E7
Grand-Popo ... E7
Kandi ... E6
Natitingou 49,000 ... E6
Ouidah ... E7
Parakou 21,000 ... E7
Porto-Novo (cap.) 104,000 ... E7

OTHER FEATURES

Atakora (mts.) ... E6
Benin (bight) ... E8
Guinea (gulf) ... E8
Mono (riv.) ... E7
Niger (riv.) ... E6
Ouémé (riv.) ... E7
Slave Coast (reg.) ... E7
Sudan (reg.) ... E6

BURKINA FASO

CITIES and TOWNS

Banfora 12,358 ... D6
Bobo Dioulasso 115,063 ... D6
Bogandé ... E6
Dédougou ... D6
Diébougou ... D6
Djibo ... D6
Dori ... D6
Fada-N'Gourma 12,000 ... E6
Gaoua ... D6
Kaya 18,000 ... D6
Koudougou 36,838 ... D6
Koupela ... D6
Léo ... D6
Ouagadougou (cap.) 172,661 ... D6
Ouahigouya 25,690 ... D6
Po ... D6
Tenkodogo ... E6
Tougan ... D6
Yako ... D6
Zabré ... D6

OTHER FEATURES

Black Volta (Mouhoun) (riv.) ... D6
Comoé (riv.) ... D7
Mouhoun (riv.) ... D6
Nakanbe (riv.) ... D6
Nazinan (riv.) ... D6
Oti (riv.) ... E7
Red Volta (Nazinan) (riv.) ... D6
Sudan (reg.) ... D6
White Volta (Nakanbe) (riv.) ... D6

CAPE VERDE

CITIES and TOWNS

Mindelo 28,797 ... A7
Praia (cap.) 21,494 ... B8
Ribeira Grande 1,892 ... B7
Sal Rei 1,296 ... B8

OTHER FEATURES

Boa Vista (isl.) ... B8
Brava (isl.) ... B8
Fogo (isl.) ... B8
Maio (isl.) ... B8
Sal (isl.) ... B7
Santa Luzia (isl.) ... B8
Santo Antão (isl.) ... A7
São Nicolau (isl.) ... B8
São Tiago (isl.) ... B8
São Vicente (isl.) ... B7

GAMBIA

CITIES and TOWNS

Banjul (cap.) 39,476 ... A6
Basse Santa Su 2,899 ... B6
Brikama 9,483 ... A6
Georgetown 2,510 ... A6

GHANA

CITIES and TOWNS

Accra (cap.) 859,600 ... D7
Attebubu 9,800 ... D7
Axim 13,100 ... D8
Bawku 33,900 ... D6
Bekwai 11,800 ... D7
Berekum 21,900 ... D7
Bolgatanga 31,500 ... D6
Cape Coast 57,700 ... D7
Damongo 12,600 ... D7
Dunkwa 16,900 ... D7
Elmina 15,600 ... D8
Ho 37,200 ... E7
Keta 12,700 ... E7
Kintampo 14,100 ... D7
Koforidua 54,400 ... D7
Kpandu 15,800 ... D7
Kumasi 348,900 ... D7
Mampong 19,800 ... D7
Nsawam 31,900 ... D7
Obuasi 60,100 ... D7
Oda 20,957 ... D7
Prestea 16,300 ... D7

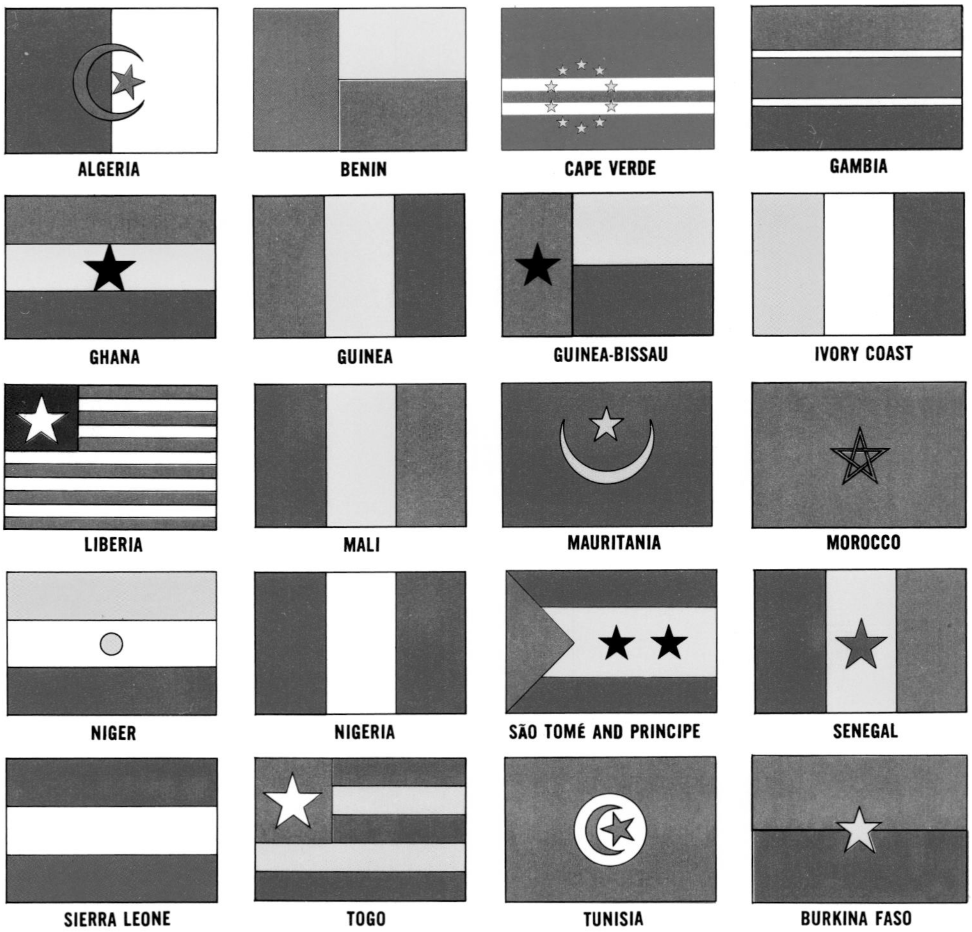

ALGERIA
BENIN
CAPE VERDE
GAMBIA
GHANA
GUINEA
GUINEA-BISSAU
IVORY COAST
LIBERIA
MALI
MAURITANIA
MOROCCO
NIGER
NIGERIA
SÃO TOMÉ AND PRINCIPE
SENEGAL
SIERRA LEONE
TOGO
TUNISIA
BURKINA FASO

Salaga 10,600 ... D7
Sekondi 32,400 ... D8
Sunyani 36,100 ... D7
Takoradi 61,500 ... D8
Tamale 136,800 ... D7
Tarkwa 22,000 ... D7
Tema 99,600 ... E7
Wa 36,000 ... D6
Wenchi 18,400 ... D7
Winneba 26,200 ... D7
Yendi 30,700 ... D7

OTHER FEATURES

Ashanti (reg.) ... D7
Benin (bight) ... E8
Black Volta (riv.) ... D7
Gold Coast (reg.) ... D8
Guinea (gulf) ... E8
Oti (riv.) ... E7
Red Volta (riv.) ... D6
Saint Paul (cape) ... E7
Three Points (cape) ... D8
Volta (lake) ... D7
Volta (riv.) ... E7
White Volta (riv.) ... D7

GUINEA

CITIES and TOWNS

Boffa ... B6
Conakry (cap.) 525,671 ... B7
Dabola ... B6
Dubréka ... B7
Fria ... B6
Kankan 85,310 ... C6
Kérouané ... C7
Kindia 79,861 ... B6
Kissidougou ... B7
Koundara 6,000 ... B6
Kouroussa ... C6
Labé 79,670 ... B6
Mali ... B6
N'Zérékoré 23,000 ... C7
Siguiri ... C6
Télimélé 12,000 ... B6
Tougué ... B6

OTHER FEATURES

Bafing (riv.) ... B6
Bakoy (riv.) ... B6
Futa Jallon (lag.) ... B6
Los (isls.) ... B7
Milo (riv.) ... C7
Moa (riv.) ... B7
Niger (riv.) ... C6
Nimba (lag.) ... C7
Verga (cape) ... B6

GUINEA-BISSAU

CITIES and TOWNS

Bissau (cap.) 109,486 ... A6
Bolama ○ 9,133 ... A6
Bubaque ○ 8,441 ... A6
Cacheu ○ 15,194 ... A6

OTHER FEATURES

Bijagós (isls.) ... A6

IVORY COAST

CITIES and TOWNS

Abengourou 31,239 ... D7
Abidjan 6 85,828 ... D7
Aboisso 14,272 ... D7
Agboville 27,192 ... D7
Bingerville 18,218 ... D7
Bondoukou 19,111 ... D7
Bouaflé 15,917 ... C7
Bouaké 1 73,248 ... D7
Dabou 23,870 ... D7
Daloa 60,958 ... C7
Danané 19,872 ... C7
Dimbokro 30,986 ... D7
Divo 37,896 ... C7
Ferkessédougou 25,307 ... C7
Gagnoa 42,362 ... C7
Grand-Bassam 25,808 ... D7
Grand-Lahou 4,070 ... C8
Guiglo 10,441 ... C7
Issia 11,143 ... C7
Katiola 21,559 ... C7
Korhogo 47,657 ... C7
Man 50,315 ... C7
Odienné 13,864 ... C7
Port-Bouet 72,616 ... D7
San Pedro 27,616 ... C8
Séguéla 12,587 ... C7
Sinfra 16,399 ... C7
Tabou 7,255 ... C8
Toumodi 12,983 ... C7
Yamoussoukro (cap.) 35,585 ... C7

OTHER FEATURES

Aby (lag.) ... D8
Bagoé (riv.) ... C6
Bandama (riv.) ... C7
Baoulé (riv.) ... C6
Black Volta (riv.) ... D6
Cavally (riv.) ... C7
Comoé (riv.) ... D7
Ebrié (lag.) ... D8
Guinea (gulf) ... E8
Ivory Coast (reg.) ... C8
Kossou, Lac de (lake) ... C7
Nimba (lag.) ... C7
Sassandra (riv.) ... C7

LIBERIA

CITIES and TOWNS

Buchanan 23,999 ... B7
Gbarnga 6,896 ... C7
Greenville 8,462 ... C8
Harbel 11,445 ... B7
Harper 10,627 ... C8
Monrovia (cap.) 166,507 ... B7
River Cess 2,041 ... C8
Robertsport 2,562 ... B7
Tapeta 3,927 ... C7
Tubmanburg 14,089 ... B7
Zwedru 6,094 ... C7

OTHER FEATURES

Bong (range) ... B7
Cavalla (riv.) ... C7
Cestos (riv.) ... C7
Grain Coast (reg.) ... B8
Kru Coast (reg.) ... C8
Mano (riv.) ... B7
Mount (cape) ... B7
Nimba (lag.) ... C7
Palmas (cape) ... C8
Roberts Field Int'l Airport ... C7

MALI

CITIES and TOWNS

Ansongo 3,485 ... E5
Bafoulabé 2,163 ... B6
Bamako (cap.) 404,022 ... C6
Banamba 6,776 ... C6
Bandiagara 8,920 ... D6
Bankass 3,229 ... D6
Bougouni 17,246 ... C6
Bourem 4,538 ... E5
Dioila 4,953 ... C6
Dire 8,941 ... D5
Djenné 10,251 ... D6
Douentza 6,746 ... D6
Gao 30,714 ... E5
Goundam 10,262 ... D5
Gourma-Rharous 4,671 ... D5
Kéniéba 4,510 ... B6
Kadiolo 3,991 ... C6
Kangaba 3,184 ... C6
Kati 24,991 ... C6
Kayes 44,736 ... B6
Ké-Macina 5,426 ... C6
Kidal 3,308 ... E5
Kita 17,538 ... C6
Kolokani 8,923 ... C6
Kolondiéba 5,882 ... C6
Koulikoro 16,376 ... C6
Koutiala 27,497 ... C6
Ménaka 3,693 ... E5
Mopti 53,885 ... D6
Nara 6,091 ... C5
Niafunké 6,399 ... D5
Niono 12,290 ... C6
Nioro 11,617 ... C5
San 22,962 ... D6
Ségou 64,890 ... C6
Sikasso 47,030 ... C6
Ténenkou 4,708 ... C6
Timbuktu (Tombouctou) 20,483 ... D5
Yanfolila 3,809 ... C6
Yelimané 1,481 ... B5
Yorosso 2,390 ... C6

OTHER FEATURES

Achourat (well) ... D4
Adrar des Iforas (plat.) ... E5
Agueraktem (well) ... C4
Asselar (well) ... D5
Azaouad (reg.) ... D5
Azaouak (dry riv.) ... E5
Bafing (riv.) ... B6
Bagoé (riv.) ... C6
Bakoy (riv.) ... B6
Bani (riv.) ... C6
Baoulé (dry riv.) ... C6
Baoulé (riv.) ... C6
Bir Ksaib Ounane (well) ... C4
Bir Ounane (well) ... D4
Chech, Erg (des.) ... D4
Debo (lake) ... D5
El Mraiti (well) ... D5
Faguibine (lake) ... D5
Falémé (riv.) ... B6
Haricha Hamada (des.) ... D4
Hombori (mts.) ... D5
In Dagouber (well) ... D4
Macina (depr.) ... D6
Niger (riv.) ... D5
Oum el Asel (well) ... D4
Sahara (des.) ... D4
Sekkane, Erg (des.) ... D4
Senegal (riv.) ... B5
Sudan (reg.) ... D6
Tadjnout Hagguerete (well) ... D4
Terhazza (ruins) ... C4
Tilemsi (valley) ... E5
Toufourine (well) ... C4

MAURITANIA

CITIES and TOWNS

Aïoun el Atrous ... C5
Akjoujt 8,044 ... B5
Aleg 6,415 ... B5
Atar 16,326 ... B4
Bassikounou ... C5
Boutilimit 7,261 ... B5
Fdérik (Fort Gouraud) 2,160 ... B4
Káedi 20,248 ... B5
Kiffa 10,629 ... B5
M'Bout ... B5
Néma 8,232 ... C5
Nouadhibou 21,961 ... A4
Nouakchott (cap.) 134,986 ... A5
Oualata ... C5
Rosso 16,466 ... A5
Sélibaby 5,994 ... B5
Tidjikja 7,870 ... B5
Timbédra 5,317 ... C5
Zouîrât 17,474 ... B4

OTHER FEATURES

Adafer (reg.) ... B5
Adrar (reg.) ... B4
Affolé (reg.) ... B5
Agueraktem (well) ... C4
Aïn ben Tili (well) ... C3
Arguin (bay) ... A4
Assaba (reg.) ... B5
Atoui, Wadi (dry riv.) ... B4
Ben Guerdane (well) ... B3
Bir el Khzaim (well) ... C4
Blanc (cape) ... A4
Brakna (reg.) ... B5
Chegga (well) ... C3
Djouf, El (des.) ... C4
El Mrayer (well) ... C4
El Mreïti (well) ... C4
Gorgol (reg.) ... B5
Hodh (reg.) ... C5
Iguidi, Erg (des.) ... C3
Inchiri (reg.) ... A5
Koumbi Saleh (ruins) ... C5

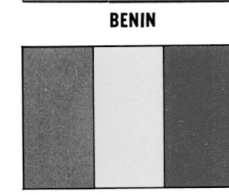

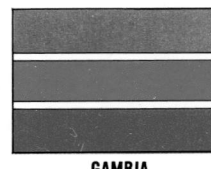

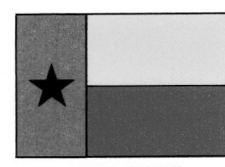

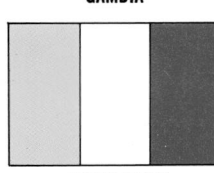

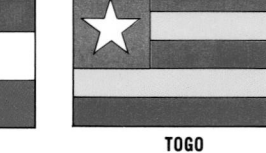

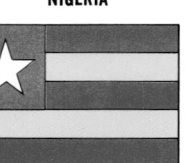

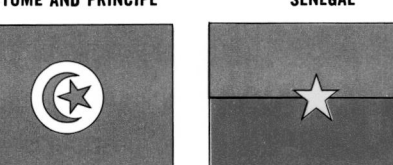

Agriculture, Industry and Resources

DOMINANT LAND USE

Cereals, Horticulture, Livestock

Market Gardening, Diversified Tropical Crops

Plantation Agriculture

Oases

Pasture Livestock

Nomadic Livestock Herding

Forests

Nonagricultural Land

MAJOR MINERAL OCCURRENCES

Al   Bauxite   Hg   Mercury
Au   Gold   Mn   Manganese
C   Coal   Na   Salt
Co   Cobalt   O   Petroleum
Cr   Chromium   P   Phosphates
Cu   Copper   Pb   Lead
D   Diamonds   Sb   Antimony
Fe   Iron Ore   Sn   Tin
G   Natural Gas   Ti   Titanium
Gn   Granite   U   Uranium
Gp   Gypsum   Zn   Zinc

⚡ Water Power

Major Industrial Areas

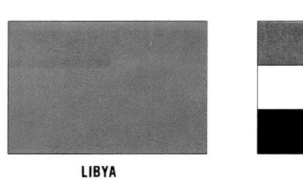 LIBYA
 EGYPT
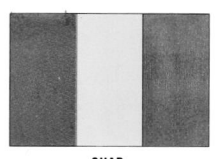 CHAD
 SUDAN
 ETHIOPIA
 ERITREA

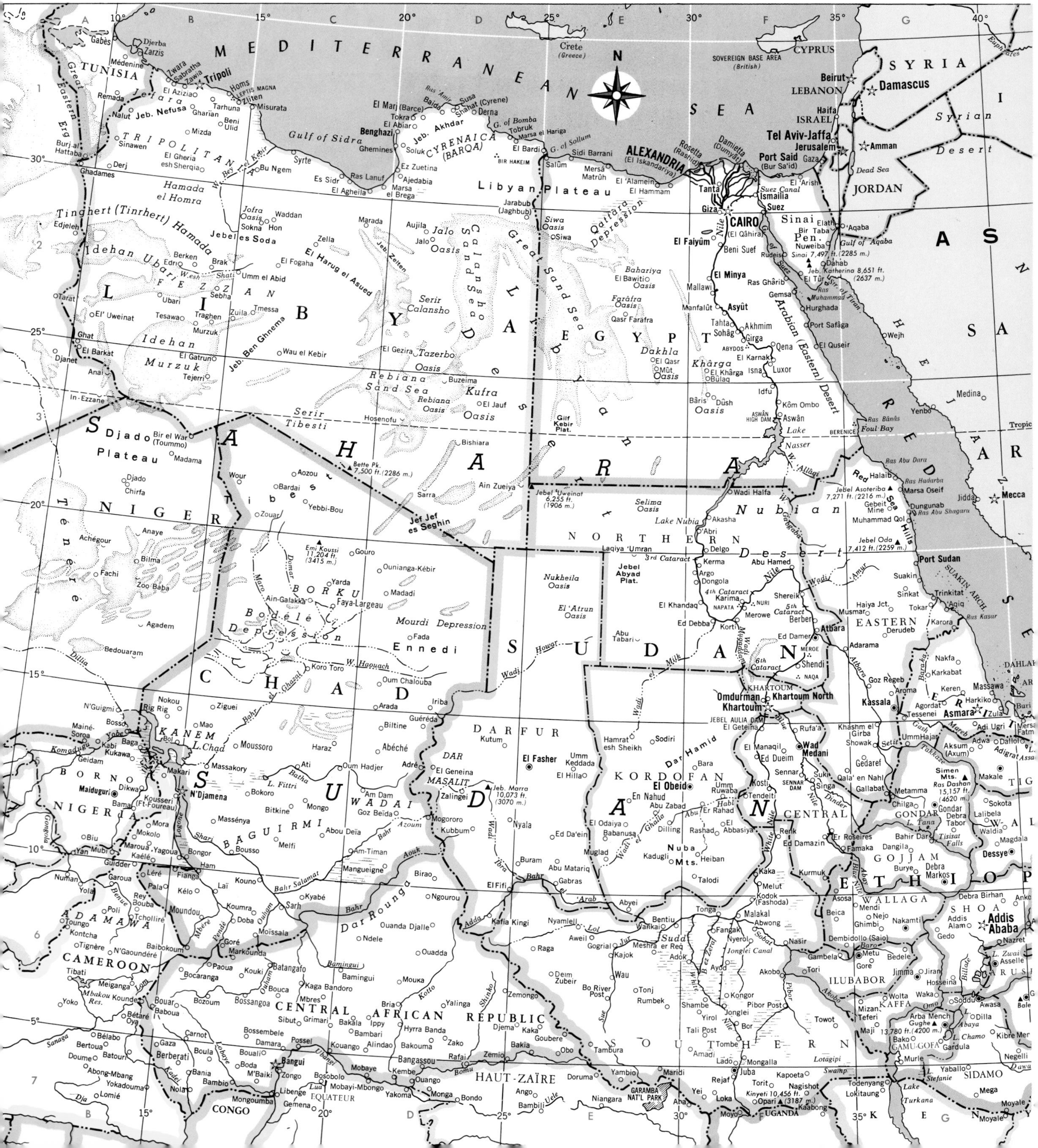

MEDITERRANEAN SEA

Crete (Greece)

SOVEREIGN BASE AREA (British) — CYPRUS

SYRIA — Beirut, LEBANON, Damascus

Haifa, ISRAEL — Tel Aviv-Jaffa, Jerusalem, Amman, JORDAN, Dead Sea

Syrian Desert

TUNISIA — Gabès, Djerba, Zarzis, Zwara, Sabratha, Tripoli, Médenine, Remada, Zawia, El Aziziya, Homs, Leptis Magna, Zliten, Nalut, Jeb. Nefusa, Gharian, Beni Ulid, Misurata, Mizda

LIBYA — Burj al Hattab, Ghadames, Derj, esh Sherqia, El Gheria, Sinawen, El Marj (Barce), Bajda, Susa, Shahat (Cyrene), Derna, Benghazi, Jeb. Akhdar, Tobruk, Marsa el Hariga, Soluk, CYRENAICA (BARQA), Gheminès, Tokra, Ez Zuetina, Ajedabia, El Agheila, Marsa el Brega, Es Sidr, Ras Lanuf, Syrte, El Abiar

ALEXANDRIA (El Iskandariya), Rosetta (Rashid), Damietta (Dumyât), Port Said (Bur Sa'id), Tanta, Giza, CAIRO (El Qâhira), Suez Canal, Ismailia, Suez

Libyan Plateau, El Hammam, El Alamein, Mersâ Matrûh, Sidi Barrani, Salûm, BIR HAKEIM

Sinai Pen., Bir Taba, Aqaba, Nuweiba, Gulf of Aqaba, Jeb. Katherina 8,651 ft. (2637 m.), Dâhab, Sinai 7,497 ft. (2285 m.)

EGYPT — El Faiyûm, Beni Suef, El Minya, Mallawi, Asyût, Manfalût, Akhmim, Sohâg, Tahta, Girga, El Kharga, Khârga Oasis, Qena, El Karnak, Luxor, Isna, Edfu, Kôm Ombo, Aswân, ASWAN HIGH DAM, Lake Nasser, Lake Nubia, Wadi Halfa

Siwa Oasis, Qattâra Depression, Great Sand Sea, Bahariya Oasis, El Bawiti, Farâfra Oasis, Qasr Farâfra, Dakhla, El Qasr, Mût, Baris, Dûsh, Kharga Oasis, Gilf Kebir Plat.

Hurghada, Port Safâga, Ras Ghârib, Gemsa, Arabian (Eastern) Desert, El Quseir, Wejh, Medina, Yenbo, HEJAZ, RED SEA, Jidda, Mecca

Jalo Oasis, Jalo, Aujila, Marada, Zella, El Fogaha, Umm el Abid, El Gezira, Tazerbo, Calansho Sand Sea, Serir Calansho, Jeb. Zeiten, El Harug el Asued, Kufra, Rebiana Oasis, El Jauf, Rebiana Sand Sea, Buzeima, Hosenofu, Bishiara

Jofra Oasis, Waddan, Sokna, Hon, Berken, Edri, Brak, Sebha, Tmessa, Zuila, Murzuk, Traghen, Ubari, Tesawa, Umm el Abid, El Gatrun, Tejerri, FEZZAN, Jebel es Soda, Hamada el Homra, Tinghert (Tinrhert) Hamada, Idehan Ubari, Idehan Murzuk, Ghat, El Barkat, Anai, In-Ezzane, Djanet, Tarat, El' Uweinat

Tibesti, Serir Tibesti, Djado Plateau, Bir el War (Toummo), Madama, Chirfa, Djado, Wour, Aozou, Bardai, Yebbi-Bou, Zouar, Ounianga-Kébir, Gouro, Madadi, Yarda, Faya-Largeau, Fada, Ennedi, Mourdi Depression, Oum Chalouba, Koro Toro, Emi Koussi 11,204 ft. (3415 m.), Bette Pk. 7,500 ft. (2286 m.), Jef Jef es Seghin, Ain Zueila, Sarra

SAHARA

NIGER — Ténéré, Achégour, Anaye, Bilma, Fachi, Zoo Baba, Agadem, Bedouaram, Dilia, N'Guigmi

CHAD — BORKU, Ain-Galakka, Bodélé Depression, Nokou, Rig Rig, Ziguei, Mao, Moussoro, Arada, Biltine, Guéréda, Iriba, BORNO, KANEM, L. Chad, Massakory, Ati, L. Fittri, Batha, Oum Hadjer, Adré, El Geneina, El Hilla, MASALIT, Abéché, Haraz, Bokoro, Mongo, Bitkine, Am Dam, WADAI, Goz Beïda, Mogororo, Kubbum, Zalingei, Jeb. Marra 10,073 ft. (3070 m.), N'Djamena, Massaguet, Masségnu, BAGUIRMI, Melfi, Am-Timan, Mangueigne, Birao, ElFifi, Haut, Abou Deïa, Kyabé, Sarh, Koumra, Moundou, Doba, Bongor, Kélo, Laï, Fianga, Pala, Kouno

NIGERIA — Maiduguri, Dikwa, Bama, Kousséri (Ft.-Foureau), Maine-Soroa, Bosso, Geidam, Kukawa, Baga, Kabi, Yobe, Komadugu, Gashua, Biu, Mubi, Yola, Numan, Garoua

CAMEROON — ADAMAWA, Rey Bouba, Tcholliré, Poli, Tignère, N'Gaoundéré, Garoua Boulaï, Meiganga, Bertoua, Batouri, Bétaré Oya, Tibati, Mbakou Kounde Res., Bocaranga, Paoua, Bouar, Baboua, Bossangoa

CENTRAL AFRICAN REPUBLIC — Bozoum, Bossembele, Bouca, Kaga Bandoro, Bamingui, Sibut, Grimari, Bambari, Ippy, Bakala, Yalinga, Bria, Ouadda, Ndélé, Birao, Ouanda Djallé, Kafia Kingi, Raga, Deim Zubeir, Bo River Post, BANGUI, M'Baiki, Mobaye, Bangassou, Rafai, Zemio, Obo, Djema

SUDAN — NORTHERN, Laqiya 'Umran, Selima Oasis, Lake Nubia, Akasha, Abri, Kerma, 3rd Cataract, Delgo, Argo, Dongola, 4th Cataract, Karima, NAPATA, Merowe, NURI, El Khandaq, Ed Debba, Korti, Shereik, Berber, 5th Cataract, Abu Hamed, Jebel Abyad Plat., Nukheila Oasis, El 'Atrun Oasis, El Atrun, Abu Tabari, Dongola, Nubian Desert, Jebel Oda 7,412 ft. (2259 m.), Jebel Asoteriba 7,271 ft. (2216 m.), Gebeit Mine, Muhammad Qol, Halaib, Marsa Oseif, Dungunab, Ras Abu Shagara, Port Sudan, Suakin, Sinkat, SUAKIN ARCH., Trinkitat, Tokar, Aqiq, Ras Kasar, Karora

Ed Damer, Atbara, Adarama, MEROE, 6th Cataract, Shendi, NAGA, Omdurman, Khartoum North, KHARTOUM, JEBEL AULIA DAM, El Geteina, Rufa'a, Wad Medani, SENNAR DAM, Sennar, Singa, Dinder, Musmar, Haiya Jct., Derudeb, EASTERN, Aroma, Kassala, Keren, Agordat, Tessenei, Goz Regeb, Gallabat, Gedaref, Showak, Khashm el Girba, UmmHajar

DARFUR, Kutum, El Fasher, Umm Keddada, DAR HAMID, Sodiri, Hamrat esh Sheikh, Bara, En Nahud, Abu Zabad, El Odaiya, Babanusa, Ed Da'ein, Muglad, Abu Matariq, Buram, Nyala, KORDOFAN, El Obeid, Umm Ruwaba, Kosti, Ed Dueim, El Manaqil, Dilling, Rashad, El Abbasiya, Nuba Mts., Kadugli, Talodi, Heiban, Kaka, Kurmuk, Renk, Melut

ETHIOPIA — Asmara, Massawa, Harkiko, Zula, Adi Ugri, Adi Quala, Om Ager, Adua, Aksum, Adigrat, TIGRE, Makale, Simen Mts., Ras Dashan 15,157 ft. (4620 m.), GONDAR, L. Tana, Chilga, Debra Tabor, Sokota, Lalibela, WALDIA, Magdala, Bahir Dar, Tissat Falls, GOJJAM, Dangila, Debra Markos, Burye, Dessye, Metamma, WALLAGA, Mendi, Nekamte, Ghimbi, Asosa, Gambela, Gore, Metu, ILUBABOR, Jimma, KAFFA, Gughe 13,780 ft. (4200 m.), SHOA, ADDIS ABABA, Addis Alam, Debra Birhan, Debra Sina, Ambo, Jiran, Hosseina, Wolta, Mizan Teferi, Maji, Bako, Arba Mench, GAMU-GOFA, L. Abaya, L. Chamo, Gardula, SIDAMO, Awasa, Bale, Soddu, L. Zwai, L. Asselle, ARUSI

SOUTHERN SUDAN — Bentiu, Wankai, Malakal, Tonga, Abwong, Fangak, Nasir, Akobo, Nyamleli, Lol, Aweil, Gogrial, Meshra er Req, Wau, Tonj, Rumbek, Shambe, Yirol, Jonglei, Bor, Kongor, Pibor Post, Abyei, Tori, Gilo, Pibor

Ngouri, Bousso, Bongor, HAUT-ZAÏRE, Juba, Torit, Kapoeta, Todenyang, Lake Turkana, Lake Stefanie, Mega, Moyale, Negelli, Dawa

GARAMBA NAT'L PARK, Kinyeti 10,456 ft. (3187 m.), Maridi, Yambio, Rejaf, Yei, Loka, Moyo, UGANDA, Lotagipi Swamp, Namuruputh

CONGO, EQUATEUR

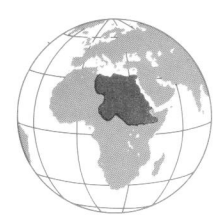

ERITREA
AREA 45,410 sq. mi. (117,600 sq. km.)
POPULATION 2,614,700
CAPITAL Asmara
LARGEST CITY Asmara
HIGHEST POINT Mount Soira 9,885 ft. (3,013 m.)
MONETARY UNIT birr
MAJOR LANGUAGES Arabic, English, Tigre, Afar
MAJOR RELIGIONS Coptic Christianity, Islam

DJIBOUTI

LIBYA
AREA 679,358 sq. mi. (1,759,537 sq. km.)
POPULATION 3,773,000
CAPITAL Tripoli
LARGEST CITY Tripoli
HIGHEST POINT Bette Pk. 7,500 ft. (2,286 m.)
MONETARY UNIT Libyan dinar
MAJOR LANGUAGES Arabic, Berber
MAJOR RELIGION Islam

EGYPT
AREA 386,659 sq. mi. (1,001,447 sq. km.)
POPULATION 53,080,000
CAPITAL Cairo
LARGEST CITY Cairo
HIGHEST POINT Jeb. Katherina 8,651 ft. (2,637 m.)
MONETARY UNIT Egyptian pound
MAJOR LANGUAGE Arabic
MAJOR RELIGIONS Islam, Coptic Christianity

CHAD
AREA 495,752 sq. mi. (1,283,998 sq. km.)
POPULATION 5,538,000
CAPITAL N'Djamena
LARGEST CITY N'Djamena
HIGHEST POINT Emi Koussi 11,204 ft. (3,415 m.)
MONETARY UNIT CFA franc
MAJOR LANGUAGES Arabic, Bagirmi, French, Sara, Massa, Moudang
MAJOR RELIGIONS Islam, tribal religions

SUDAN
AREA 967,494 sq. mi. (2,505,809 sq. km.)
POPULATION 24,485,000
CAPITAL Khartoum
LARGEST CITY Khartoum
HIGHEST POINT Jeb. Marra 10,073 ft. (3,070 m.)
MONETARY UNIT Sudanese pound
MAJOR LANGUAGES Arabic, Dinka, Nubian, Beja, Nuer
MAJOR RELIGIONS Islam, tribal religions

ETHIOPIA
AREA 426,366 sq. mi. (1,104,300 sq. km.)
POPULATION 50,576,300
CAPITAL Addis Ababa
LARGEST CITY Addis Ababa
HIGHEST POINT Ras Dashan 15,157 ft. (4,620 m.)
MONETARY UNIT birr
MAJOR LANGUAGES Amharic, Gallinya, Tigrinya, Somali, Sidamo, Arabic, Ge'ez
MAJOR RELIGIONS Coptic Christianity, Islam

DJIBOUTI
AREA 8,880 sq. mi. (23,000 sq. km.)
POPULATION 456,000
CAPITAL Djibouti
LARGEST CITY Djibouti
HIGHEST POINT Moussa Ali 6,768 ft. (2,063 m.)
MONETARY UNIT Djibouti franc
MAJOR LANGUAGES Arabic, Somali, Afar, French
MAJOR RELIGIONS Islam, Roman Catholicism

Northeastern Africa

CONIC EQUAL-AREA PROJECTION

SCALE OF MILES
0 50 100   200   300

SCALE OF KILOMETERS
0 50 100   200   300

Capitals of Countries _ _ _ _ _ _ _ _ ☆
Other Capitals _ _ _ _ _ _ _ _ _ _ _ ◉
International Boundaries _ _ _ _ _ _ ▬▬
Internal Boundaries _ _ _ _ _ _ _ _ ▬▬

Scale 1:14,300,000

© Copyright HAMMOND INCORPORATED, Maplewood, N.J.

CHAD

CITIES and TOWNS

Abéché 28,100	D5
Abou Deïa	D5
Adré	D5
Am-Timan 4,200	D4
Arada	D4
Baibokoum 5,500	C6
Biltine 3,900	D5
Bitkine 5,000	C5
Bokoro 6,500	C5
Bol 2,500	B5
Bongor 14,300	C5
Bousso 4,500	C5
Doba 13,300	C6
Fada	D4
Faya-Largeau 6,800	C4
Fianga 10,000	C6
Goré	C6
Goz Beïda	D5
Guéréda	D5
Iriba	D4
Kélo 16,800	C6
Koumra 17,000	C6
Kouno	C6

Kyabé 5,000	C6
Laï 10,400	C6
Léré	B6
Mangueigne	D5
Mao 4,900	C5
Massakory	C5
Massénya	C5
Melfi	C5
Mogororo	D5
Moïssala 5,100	C6
Mongo 8,300	C5
Moundou 39,600	C6
Moussoro 7,700	C5
N'Djamena (cap.) 179,000	C5
Oum Hadjer 5,600	C5
Ounianga-Kébir	D4
Pala 13,200	B6
Sarh 43,700	C6
Wour	C3
Zouar	C3

OTHER FEATURES

Aouk, Bahr (riv.)	D5
Azoum, Bahr (riv.)	D5
Baguirmi (reg.)	C5
Bahr el Ghazal (dry riv.)	C4
Batha (riv.)	C5

Bodélé (depr.)	C4
Borku (reg.)	C4
Chad (lake)	C5
Emi Koussi (mt.)	C4
Ennedi (plat.)	D4
Fittri (lake)	C5
Kanem (reg.)	C5
Logone (reg.)	C5
Maro (riv.)	C4
Mbéré (riv.)	C6
Mourdi (riv.)	D4
Ouham (depr.)	C6
Pendé (riv.)	C6
Sahara (riv.)	C3
Salamat, Bahr (des.)	C6
Shari (riv.)	C5
Sudan (riv.)	C5
Tibesti (mts.)	C3
Wadai (reg.)	D5

DJIBOUTI

CITIES and TOWNS

Ali Sabieh	H5
Dikhil	H5
Djibouti (cap.) 96,000	H5
Obock	H5

Tadjoura	H5

OTHER FEATURES

Abbe (lake)	H5
Aden (gulf)	J5
Bab el Mandeb (strait)	H5

EGYPT

CITIES and TOWNS

Abnûb 39,343	J4
Akhmim 53,234	F2
Alexandria 2,318,655	J2
Aswân 144,377	F3
Asyût 213,983	J4
Benha 88,992	J3
Beni Mazar 39,373	J4
Beni Suef 118,148	J3
Biba 33,074	J4
Bur Sa'id (Port Said) 262,620	K2
Cairo (cap.) 5,084,463	J3
Dairût 31,624	J4
Damanhur 188,927	J3
Damietta 93,546	J3
Disûq 58,650	J3

(continued on following page)

Topography

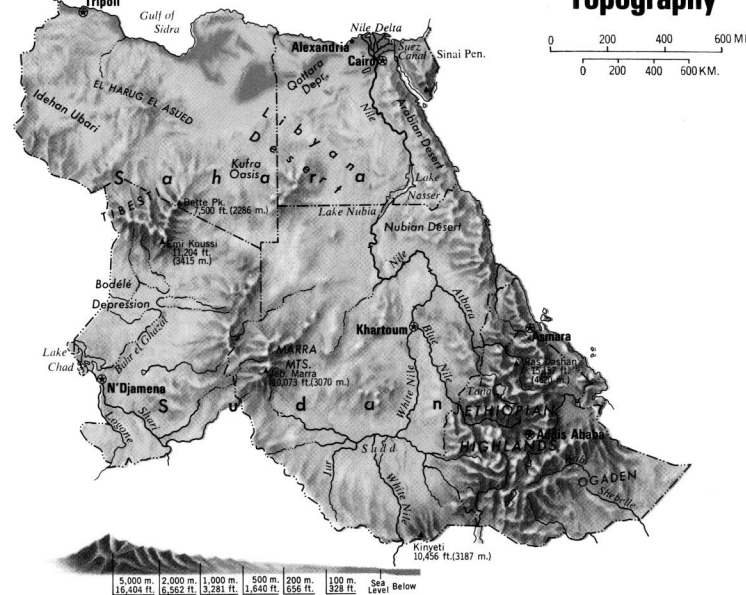

(continued on following page)

112 Northeastern Africa
(continued)

Dumyât (Damietta) 93,546 J3
El 'Alamein E1
El 'Arish F1
El Faiyûm 167,081 J3
El Fashn 33,506 J4
El Iskendariya (Alexandria) 2,318,655 J2
El Karnak F2
El Khârga 26,375 F2
El Mahalla el Kubra 292,853 J3
El Mansûra 257,866 K3
El Minya 146,423 J4
El Qahira (Cairo) (cap.) 5,084,463 J3
El Qantara 919 K3
El Quseir 12,297 F2
El Wasta 17,659 J3
Girga 51,110 F2
Giza 1,246,713 J3
Heliopolis J3
Helwân J3
Idfu 34,858 F3
Ismailia 145,978 K3
Isna 34,186 F2
Karnak (El Karnak) F2
Kôm Ombo 44,531 F3
Luxor 92,748 F2
Maghâgha 40,802 J4
Mallawi 74,256 J4
Manfalût 41,126 J4
Mersá Matrûh 27,857 E1
Minûf 55,131 J3
Mût 8,032 E2
Port Fuad K3
Port Safâga F2
Port Said 262,620 K2
Port Taufiq K3
Qalyub 62,739 J3
Qena 94,013 F2
Rashid (Rosetta) 42,962 J2
Rudeis F2
Salûm 4,161 E1
Samalût 48,146 J4
Shibin el Kom 102,844 J3
Sidi Barrani 1,574 E1
Sinnûris 42,022 J3
Sohâg 101,758 F2
Suez 194,001 K3
Tahta 45,242 F2
Tanta 284,636 J3
Zagazig 202,637 J3
Zifta 50,410 J3

OTHER FEATURES

Abu Qir (bay) J2
Abydos (ruins) F2
Aqaba (gulf) G2
Arabian (des.) F2
Aswân (dam) F3
Aswân High (dam) F3
Bahariya (oasis) E2
Bânâs, Ras (cape) G3
Berenice (ruins) F3
Birket Qârûn (lake) J3
Bitter (lakes) K3
Dakhla (oasis) E2
Eastern (Arabian) (des.) F2
Farâfra (oasis) E2
Foul (bay) G3
Gilf Kebir (plat.) E3
Great Sand Sea (des.) D2
Katherina, Jebel (mt.) F2
Khârga (oasis) F2
Libyan (des.) E2
Libyan (plat.) E1
Mediterranean (sea) E1
Memphis (ruins) J3
Muhammad, Ras (cape) F2
Nasser (lake) F3
Nile (riv.) F2
Pyramids (ruins) J3
Qattara (depr.) E2
Sahara (des.) E3
Sinai (mt.) F2
Sinai (pen.) F2
Siwa (oasis) E2
Suez (canal) K3
Suez (gulf) F2
Tiran (strait) F2
'Uweinat, Jebel (mt.) E3

ERITREA

CITIES and TOWNS

Adi Ugri 12,800 G5
Asmara (cap.) 393,800 G4
Assab 16,000 H5
Karkabat G4
Keren G4
Massawa 19,800 G4
Mersa Fatma H5
Nakfa G4
Tessenei G4
Thio H5
Umm Hajar G5
Zula G4

OTHER FEATURES

Baraka (riv.) G4
Buri (pen.) H4
Dahlak (arch.) H4
Dahlak (isl.) H4
Kasar, Ras (cape) G4
Takkaze (riv.) G5

ETHIOPIA

PROVINCES

Arusi 852,900 G6
Bale 707,800 H6
Gamu-Gofa 698,800 G6
Gojjam 1,750,100 G5
Gondar 1,355,800 G5
Harar 3,359,200 H6
Ilubabor 688,800 F6
Kaffa 1,693,000 G6
Shoa 5,369,500 G6
Sidamo 2,479,800 G7
Tigre 1,828,900 H5
Wallaga 1,269,100 G6
Wallo 2,459,900 H5

CITIES and TOWNS

Addis Ababa (cap.) 1,196,300 G6
Addis Alam 5,500 G6
Adigrat 9,400 G5
Adwa 16,400 G5
Aksum 12,800 G5
Ankober H6
Arba Mench 7,660 G6
Asselle 19,390 G6
Awareh H6
Axum (Aksum) 12,800 G5
Bahir Dar 25,100 G5
Dagabur H6
Dangila G5
Debra Birhan 16,700 G6
Debra Markos 30,260 G5
Debra Tabor 8,700 G5
Dembidollo 7,600 F6
Dessye 49,750 G5
Dilla 13,800 G6
Dire Dawa 63,700 H6
El Carre H6
Gabredarre H6
Galadi J6
Gambela F6
Gardula 5,800 G6
Gerlogubi H6
Ghimbi 8,300 G6
Ginir H6
Goba 13,500 H6
Gondar 38,600 G5
Gore 8,500 G6
Gorrahei H6
Harar 48,440 H6
Hosseina 8,500 G6
Jijiga 8,000 H6
Jimma 47,360 G6
Jiran G6
Kibre Mengist 8,300 G6
Lalibela G5
Magdala G5
Maji G6
Makale 30,780 G5
Metamma G5
Metu 6,860 G6
Miesso H6
Mizan Teferi G6
Moyale G7
Murle G6
Mustahil H6
Nakamti 18,310 G6
Nazret 42,900 G6
Negelli 8,800 G6
Nejo G6
Saio (Dembidollo) 7,600 F6
Soddu 11,900 G6
Sokota G5
Tori F6
Waka G6
Waldia 9,600 G5
Wardere J6
Wolta G6
Yaballo G7

OTHER FEATURES

Abay (riv.) G5
Abaya (lake) G6
Abbe (lake) H5
Akobo (riv.) F6
Assal (lake) H5
Assale (lake) H5
Atbara (riv.) G4
Awash (riv.) H5
Bale (mt.) G6
Blue Nile (Abay) (riv.) G5
Chamo (lake) G6
Danakil (reg.) H5
Dawa (riv.) G7
Dinder (riv.) F5
Fafan (riv.) H6
Ganale Dorya (riv.) H6
Gughe (mt.) G6
Haud (reg.) J6
Ogaden (reg.) H6
Omo (riv.) G6
Ras Dashan (mt.) G5
Red (sea) H4
Rudolf (Turkana) (lake) G7
Simen (mts.) G5
Takkaze (riv.) G5
Tana (lake) G5
Tisisat (fall) G5
Turkana (lake) G7
Zwai (lake) G6

LIBYA

CITIES and TOWNS

Ajedabia◦ 53,170 D1
Aujila◦ 6,695 D2
Baida◦ 59,765 D1
Barce (El Marj)◦ 55,444 D1
Benghazi 286,943 C1
Beni Ulid◦ 19,113 B1
Brak◦ 16,307 B2
Cyrene (Shahat)◦ 17,157 D1
Derj◦ 2,152 B1
Derna◦ 44,145 D1
El Abiar◦ 17,685 D1
El Agheila 3 C1
El Azizia◦ 34,077 B1
El Bardi◦ 4,330 D1
El Barkat◦ 2,139 B3
El Gatrun B3
El Jauf◦ 6,481 D3
El Marj◦ 55,444 D1
Es Sidr◦ 706 C1
Ez Zuetina◦ 7,256 D1
Ghadames◦ 6,172 A2
Gharian◦ 65,224 B1
Ghat◦ 6,924 B3
Ghemines◦ 4,313 C1
Homs◦ 66,890 B1
Hon◦ 2,766 C2
Jaghbub (Jarabub)◦ 1,436 D2
Jarabub◦ 1,436 D2
Marada◦ 3,201 C2
Marsa el Brega◦ 2,618 D1
Marsa el Hariga◦ 5,043 D1
Misurata◦ 102,439 C1
Mizda◦ 11,472 B1
Murzuk◦ 22,185 B2
Nalut◦ 23,535 B1
Ras Lanuf◦ 1,990 C1
Sabrathaa◦ 30,836 B1
Sebha◦ 35,879 B2
Shahat◦ 17,157 D1
Sinawen◦ 1,549 B1
Sokna◦ 3,757 C2
Soluk◦ 6,501 D1
Syrte◦ 22,797 C1
Tarhuna◦ 52,657 B1
Tobruk◦ 58,384 D1
Tokra◦ 10,714 D1
Tripoli (cap.)◦ 550,438 B1
Ubari◦ 19,132 B2
Waddan◦ 5,347 C2
Wau el Kebir C2
Zawia◦ 72,092 B1
Zella◦ 72,092 C2
Zliten◦ 58,981 B1
Zwara◦ 15,078 B1

OTHER FEATURES

Akhdar, Jebel (mts.) D1
Barqa (Cyrenaica) (reg.) D1
Ben Ghnema, Jebel (mts.) C2
Bette (peak) C3
Bey el Kebir, Wadi (dry riv.) B1
Bir Hakeim (ruins) D1
Bomba (gulf) D1
Buzeima (well) D3
Calansho Sand Sea (des.) D2
Calansho, Serir (des.) D2
Cyrenaica (reg.) D1
Fezzan (reg.) B2
Great Sand Sea (des.) D2
Harug el Asued, El (mts.) C2
Homra, Hamada el (des.) B2
Idehan Murzuk (des.) B2
Idehan Ubari (des.) B2
Jalo (oasis) D2
Jefara (reg.) B1
Jef Jef es Seghin (plat.) D3
Jofra (oasis) C2
Kufra (oasis) D3
Leptis Magna (ruins) B1
Libyan (des.) D2
Libyan (plat.) D1
Mediterranean (sea) C1
Nefusa, Jebel (mts.) B1
Rebiana (oasis) D3
Rebiana Sand Sea (des.) D3
Sahara (des.) C3
Shati, Wadi esh (dry riv.) B2
Sidra (gulf) C1
Soda, Jebel es (mts.) C2
Tazerbo (oasis) D2
Tibesti, Serir (des.) C3
Tinghert Hamada (Tinrhert) (des.) B2
Tripolitania (reg.) B1
'Uweinat, Jebel (mt.) E3
Zelten, Jebel (mts.) D2

SUDAN

PROVINCES

Central F5
Darfur D5
Eastern G4
Khartoum F4
Kordofan E5
Northern E3
Southern E6

CITIES and TOWNS

'Abri F3
Abu Hamed F4
Adok F6
Akobo F6
Amadi F6
Argo F4
Aroma G4
Atbara 66,000 F4
Babanusa E5
Bara F5
Bentiu E6
Berber F4
Bor F6
Buram E5
Damazin (Ed Damazin) 12,000 F5
Deim Zubeir E6
Dongola 6,000 F4
Dungunab G3
Ed Damazin 12,000 F5
Ed Damer 17,000 F4
Ed Dueim 27,000 F5
El Fasher 52,000 E5
El Geneina 33,000 D5
El Obeid 90,000 E5
El Odaiya E5
En Nahud 23,000 E5
Er Roseires F5
Fashoda (Kodok) F6
Gedaref 92,000 G5
Gogrial E6
Goz Regeb G4
Haiya Junction G4
Halaib G3
Jonglei F6
Juba 57,000 F7
Kadugli 18,000 E5
Kaka F5
Karima F4
Kassala 99,000 G4
Kerma F4
Khartoum (cap.) 334,000 F4
Khartoum North 151,000 F4
Khashm el Girba G5
Kodok F6
Kosti 57,000 F5
Kurmuk F5
Kutum D5
Malakal 35,000 F6
Maridi E7
Melut F5
Merowe F4
Meshra er Req E6
Mongalla F6
Muglad E5
Muhammad Qol G3
Nagishot F7
Nasir F6
Nyala 60,000 D5
Nyamlell E6
Nyerol F6
Omdurman 299,000 F4
Opari F7
Pibor Post F6
Port Sudan 133,000 G4
Qala'en Nahl F5
Raga E6
Rashad F5
Rejaf F7
Renk F5
Rufa'a F5
Rumbek 17,000 E6
Sennar F5
Shambe F6
Shendi F4
Shereik F4
Showak G5
Singa F5
Sinkat G4
Sodiri E5
Suakin G4
Suki F5
Tali Post F6
Talodi F5
Tambura E6
Tendelti F5
Tokar G4
Tombe F6
Tonga F6
Tonj E6
Torit F7
Trinkitat G4
Umm Keddada E5
Umm Ruwaba F5
Wadi Halfa F3
Wad Medani 107,000 F5
Wankai E6
Wau 53,000 E6
Yambio 7,000 E7
Yei F7
Yirol F6
Zalingei D5

OTHER FEATURES

Abu Habl, Wadi (dry riv.) F5
Abu Shagara, Ras (cape) G3
Adda (riv.) D6
Atbara (riv.) G4
Bahr Azoum (riv.) D5
Bahr el 'Arab (riv.) E6
Bahr ez Zeraf (riv.) F6
Blue Nile (riv.) F5
Dar Hamid (reg.) F5
Dar Masalit (reg.) D5
Dinder (riv.) F5
El 'Atrun (oasis) E4
Fifth Cataract (falls) F4
Fourth Cataract (falls) F4
Gabgaba, Wadi (dry riv.) F3
Ghalla, Wadi el (dry riv.) E5
Hadarba, Ras (cape) G3
Howar, Wadi (dry riv.) E4
Ibra, Wadi (dry riv.) D5
Jebel Aulia (dam) F4
Jonglei (canal) F6
Jur (riv.) E6
Kinyeti (mt.) F7
Libyan (des.) E3
Lol (dry riv.) E6
Lotagipi Swamp (plain) F6
Marra, Jebel (mt.) D5
Meroe (ruins) F4
Milk, Wadi el (dry riv.) E4
Muqaddam, Wadi (dry riv.) F4
Napata (ruins) F4
Naqa (ruins) F4
Nile (riv.) F4
Nuba (mts.) E5
Nubia (lake) F3
Nubian (des.) F3
Nukheila (oasis) E4
Nuri (ruins) F4
Oda, Jebel (mt.) G3
Pibor (riv.) F6
Red (sea) G3
Red Sea (hills) G3
Sahara (des.) E3
Selima (oasis) E3
Sennar (dam) F5
Setit (riv.) G5
Sixth Cataract F4
Sobat (riv.) F6
Suakin (arch.) G4
Sudan (reg.) E5
Sudd (swamp) F6
Sue (riv.) E6
Third Cataract (falls) E4
'Uweinat, Jebel (mt.) E3
White Nile (riv.) F5

◦ Population of sub-district or division

Agriculture, Industry and Resources

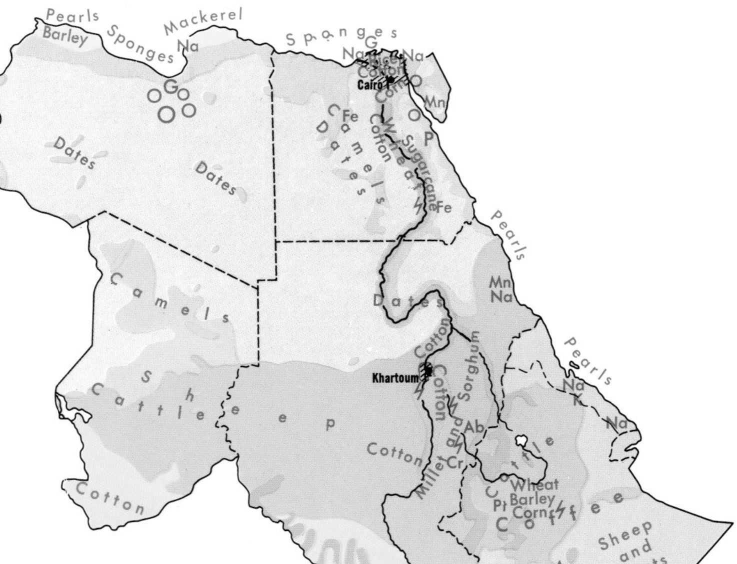
Pearls
Mackerel
Barley
Sponges
Na
Cairo
Dates
Camels
Cotton
Sugarcane
Fe
Mn
Khartoum
Sorghum
Sheep
Cattle
Millet
Ab
Cr
Wheat
Barley
Corn
Coffee
Pt
Sheep and Goats
Au

DOMINANT LAND USE

- Cereals, Horticulture, Livestock
- Cash Crops, Mixed Cereals
- Cotton, Cereals
- Market Gardening, Diversified Tropical Crops
- Plantation Agriculture
- Oases
- Pasture Livestock
- Nomadic Livestock Herding
- Forests
- Nonagricultural Land

MAJOR MINERAL OCCURRENCES

Ab	Asbestos	Mn	Manganese
Au	Gold	Na	Salt
Cr	Chromium	O	Petroleum
Fe	Iron Ore	P	Phosphates
G	Natural Gas	Pt	Platinum
K	Potash		

Water Power

Major Industrial Areas

ANGOLA
AREA 481,351 sq. mi. (1,246,700 sq. km.)
POPULATION 9,747,000
CAPITAL Luanda
LARGEST CITY Luanda
HIGHEST POINT Mt. Moco 8,593 ft. (2,620 m.)
MONETARY UNIT kwanza
MAJOR LANGUAGES Mbundu, Kongo, Lunda, Portuguese
MAJOR RELIGIONS Tribal religions, Roman Catholicism

BURUNDI
AREA 10,747 sq. mi. (27,835 sq. km.)
POPULATION 5,302,000
CAPITAL Bujumbura
LARGEST CITY Bujumbura
HIGHEST POINT 8,858 ft. (2,700 m.)
MONETARY UNIT Burundi franc
MAJOR LANGUAGES Kirundi, French, Swahili
MAJOR RELIGIONS Tribal religions, Roman Catholicism, Islam

CAMEROON
AREA 183,568 sq. mi. (475,441 sq. km.)
POPULATION 11,540,000
CAPITAL Yaoundé
LARGEST CITY Douala
HIGHEST POINT Cameroon 13,350 ft. (4,069 m.)
MONETARY UNIT CFA franc
MAJOR LANGUAGFS Fang, Bamileke, Fulani, Duala, French, English
MAJOR RELIGIONS Tribal religions, Christianity, Islam

CENTRAL AFRICAN REP.
AREA 242,000 sq. mi. (626,780 sq. km.)
POPULATION 2,740,000
CAPITAL Bangui
LARGEST CITY Bangui
HIGHEST POINT Gao 4,659 ft. (1,420 m.)
MONETARY UNIT CFA franc
MAJOR LANGUAGES Banda, Gbaya, Sangho, French
MAJOR RELIGIONS Tribal religions, Christianity, Islam

CONGO, REP. OF THE
AREA 132,046 sq. mi. (342,000 sq. km.)
POPULATION 1,843,000
CAPITAL Brazzaville
LARGEST CITY Brazzaville
HIGHEST POINT Leketi Mts. 3,412 ft. (1,040 m.)
MONETARY UNIT CFA franc
MAJOR LANGUAGES Kikongo, Bateke, Lingala, French
MAJOR RELIGIONS Christianity, tribal religions, Islam

EQUATORIAL GUINEA
AREA 10,831 sq. mi. (28,052 sq. km.)
POPULATION 341,000
CAPITAL Malabo
LARGEST CITY Malabo
HIGHEST POINT 9,868 ft. (3,008 m.)
MONETARY UNIT CFA franc
MAJOR LANGUAGES Fang, Bubi, Spanish
MAJOR RELIGIONS Tribal religions, Christianity

GABON
AREA 103,346 sq. mi. (267,666 sq. km.)
POPULATION 1,206,000
CAPITAL Libreville
LARGEST CITY Libreville
HIGHEST POINT Ibounzi 5,165 ft. (1,574 m.)
MONETARY UNIT CFA franc
MAJOR LANGUAGES Fang and other Bantu languages, French
MAJOR RELIGIONS Tribal religions, Christianity, Islam

KENYA
AREA 224,960 sq. mi. (582,646 sq. km.)
POPULATION 24,872,000
CAPITAL Nairobi
LARGEST CITY Nairobi
HIGHEST POINT Kenya 17,058 ft. (5,199 m.)
MONETARY UNIT Kenya shilling
MAJOR LANGUAGES Kikuyu, Luo, Kavirondo, Kamba, Swahili, English
MAJOR RELIGIONS Tribal religions, Christianity, Hinduism, Islam

MALAWI
AREA 45,747 sq. mi. (118,485 sq. km.)
POPULATION 8,022,000
CAPITAL Lilongwe
LARGEST CITY Blantyre
HIGHEST POINT Mulanje 9,843 ft. (3,000 m.)
MONETARY UNIT Malawi kwacha
MAJOR LANGUAGES Chichewa, Yao, English, Nyanja, Tumbuka, Tonga, Ngoni
MAJOR RELIGIONS Tribal religions, Islam, Christianity

RWANDA
AREA 10,169 sq. mi. (26,337 sq. km.)
POPULATION 6,274,000
CAPITAL Kigali
LARGEST CITY Kigali
HIGHEST POINT Karisimbi 14,780 ft. (4,505 m.)
MONETARY UNIT Rwanda franc
MAJOR LANGUAGES Kinyarwanda, French, Swahili
MAJOR RELIGIONS Tribal religions, Roman Catholicism, Islam

SOMALIA
AREA 246,200 sq. mi. (637,658 sq. km.)
POPULATION 7,339,000
CAPITAL Mogadishu
LARGEST CITY Mogadishu
HIGHEST POINT Surud Ad 7,900 ft. (2,408 m.)
MONETARY UNIT Somali shilling
MAJOR LANGUAGES Somali, Arabic, Italian, English
MAJOR RELIGION Islam

TANZANIA
AREA 363,708 sq. mi. (942,003 sq. km.)
POPULATION 24,802,000
CAPITAL Dar es Salaam
LARGEST CITY Dar es Salaam
HIGHEST POINT Kilimanjaro 19,340 ft. (5,895 m.)
MONETARY UNIT Tanzanian shilling
MAJOR LANGUAGES Nyamwezi-Sukuma, Swahili, English
MAJOR RELIGIONS Tribal religions, Christianity, Islam

UGANDA
AREA 91,076 sq. mi. (235,887 sq. km.)
POPULATION 17,804,000
CAPITAL Kampala
LARGEST CITY Kampala
HIGHEST POINT Margherita 16,795 ft. (5,119 m.)
MONETARY UNIT Ugandan shilling
MAJOR LANGUAGES Luganda, Acholi, Teso, Nyoro, Soga, Nkole, English, Swahili
MAJOR RELIGIONS Tribal religions, Christianity, Islam

CONGO, DEM. REP. OF THE
AREA 905,063 sq. mi. (2,344,113 sq. km.)
POPULATION 34,491,000
CAPITAL Kinshasa
LARGEST CITY Kinshasa
HIGHEST POINT Margherita 16,795 ft. (5,119 m.)
MONETARY UNIT zaire
MAJOR LANGUAGES Tshiluba, Mongo, Kikongo, Kingwana, Zande, Lingala, Swahili, French
MAJOR RELIGIONS Tribal religions, Christianity

ZAMBIA
AREA 290,586 sq. mi. (752,618 sq. km.)
POPULATION 8,073,000
CAPITAL Lusaka
LARGEST CITY Lusaka
HIGHEST POINT Sunzu 6,782 ft. (2,067 m.)
MONETARY UNIT Zambian kwacha
MAJOR LANGUAGES Bemba, Tonga, Lozi, Luvale, Nyanja, English
MAJOR RELIGIONS Tribal religions

ANGOLA

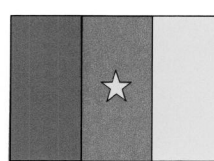
BURUNDI

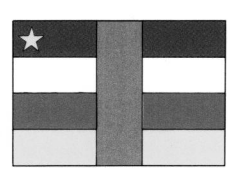
CAMEROON

CENTRAL AFRICAN REP.

CONGO, REP. OF THE

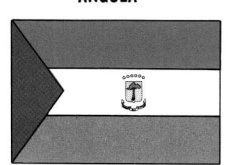
EQUATORIAL GUINEA

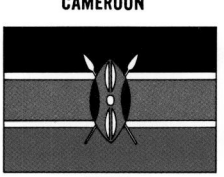
GABON

KENYA

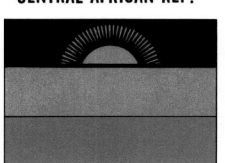
MALAWI

RWANDA

SOMALIA

TANZANIA

UGANDA

CONGO, DEM. REP. OF THE

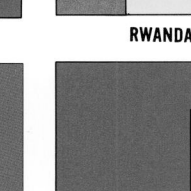
ZAMBIA

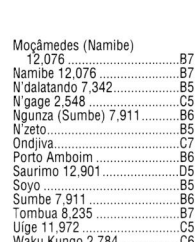

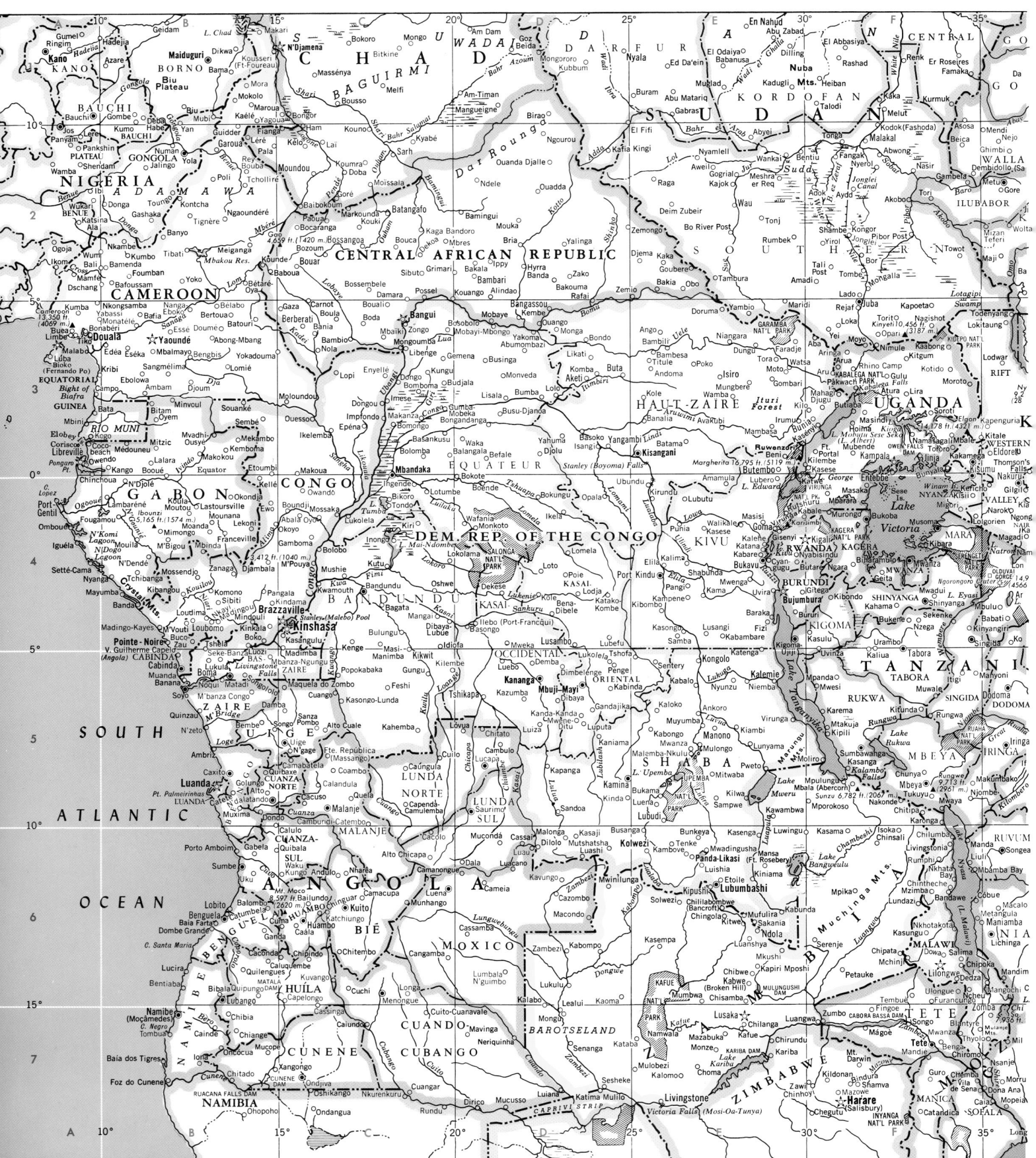

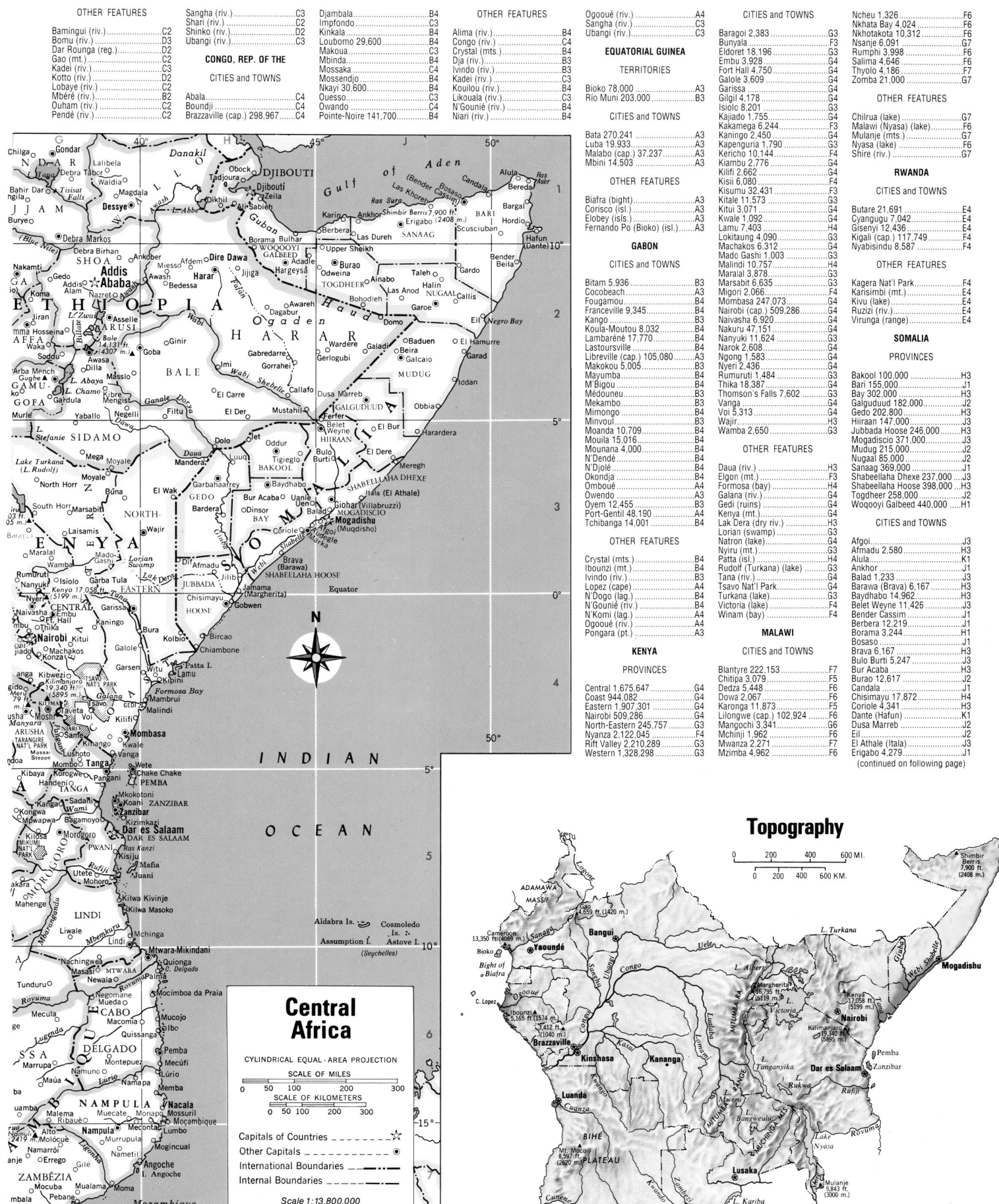

OTHER FEATURES

Bamingui (riv.) C2
Bomu (riv.) D3
Dar Rounga (reg.) D2
Gao (mt.) C2
Kadei (riv.) C3
Kotto (riv.) D2
Lobaye (riv.) C2
Mbéré (riv.) B2
Ouham (riv.) C2
Pendé (riv.) C2

Sangha (riv.) C3
Shari (riv.) C2
Shinko (riv.) D2
Ubangi (riv.) C3

CONGO, REP. OF THE

CITIES and TOWNS

Abala C4
Boundji C4
Brazzaville (cap.) 298,967 ....C4

Djambala B4
Impfondo C3
Kinkala C4
Loubomo 29,600 ....C4
Makoua C4
Mbinda B3
Mossaka C4
Mossendjo B4
Nkayi 30,600 ....B4
Ouesso C3
Owando C4
Pointe-Noire 141,700 ....B4

OTHER FEATURES

Alima (riv.) B4
Congo (riv.) C4
Crystal (mts.) B4
Dja (riv.) B3
Ivindo (riv.) B3
Kadei (riv.) C3
Kouilou (riv.) B4
Likouala (riv.) C3
N'Gounié (riv.) B4
Niari (riv.) B4

Ogooué (riv.) A4
Sangha (riv.) C3
Ubangi (riv.) C3

EQUATORIAL GUINEA

TERRITORIES

Bioko 78,000 ....A3
Río Muni 203,000 ....B3

CITIES and TOWNS

Bata 270,241 ....A3
Luba 19,933 ....A3
Malabo (cap.) 37,237 ....A3
Mbini 14,503 ....A3

OTHER FEATURES

Biafra (bight) A3
Corisco (isl.) A3
Elobey (isls.) A3
Fernando Po (Bioko) (isl.) A3

GABON

CITIES and TOWNS

Bitam 5,936 ....B3
Cocobeach A3
Fougamou B4
Franceville 9,345 ....B4
Kango B3
Koula-Moutou 8,032 ....B4
Lambaréné 17,770 ....B4
Lastoursville B4
Libreville (cap.) 105,080 .....A3
Makokou 5,005 ....B3
Mayumba B4
M'Bigou B4
Médouneu B3
Mekambo B4
Mimongo B4
Minvoul B3
Moanda 10,709 ....B4
Mouila 15,016 ....B4
Mounana 4,000 ....B4
N'Dendé B4
N'Djolé B4
Okondja B4
Omboué A4
Owendo A3
Oyem 12,455 ....B3
Port-Gentil 48,190 ....A4
Tchibanga 14,001 ....B4

OTHER FEATURES

Crystal (mts.) B4
Ibounzi (mt.) B4
Ivindo (riv.) B3
Lopez (cape) A4
N'Dogo (lag.) B4
N'Gounié (riv.) B4
N'Komi (lag.) A4
Ogooué (riv.) A4
Pongara (pt.) A3

KENYA

PROVINCES

Central 1,675,647 ....G4
Coast 944,082 ....G4
Eastern 1,907,301 ....G4
Nairobi 509,286 ....G4
North-Eastern 245,757 ....G3
Nyanza 2,122,045 ....F4
Rift Valley 2,210,289 ....G3
Western 1,328,298 ....G3

CITIES and TOWNS

Baragoi 2,383 ....G3
Bunyala F3
Eldoret 18,196 ....G3
Embu 3,928 ....G4
Fort Hall 4,750 ....G4
Galole 3,609 ....G4
Garissa G4
Gilgil 4,178 ....G3
Isiolo 8,201 ....G3
Kajiado 1,755 ....G4
Kakamega 6,244 ....F3
Kaningo 2,450 ....G4
Kapenguria 1,790 ....G3
Kericho 10,144 ....G4
Kiambu 2,776 ....G4
Kilifi 2,662 ....G4
Kisii 6,080 ....F4
Kisumu 32,431 ....F3
Kitale 11,573 ....G3
Kitui 3,071 ....G4
Kwale 1,092 ....G4
Lamu 7,403 ....H4
Lokitaung 4,090 ....G3
Machakos 6,312 ....G4
Mado Gashi 1,003 ....G3
Malindi 10,757 ....H4
Maralal 3,878 ....G3
Marsabit 6,635 ....G3
Migori 2,066 ....F4
Mombasa 247,073 ....G4
Nairobi (cap.) 509,286 ....G4
Naivasha 6,920 ....G4
Nakuru 47,151 ....G4
Nanyuki 11,624 ....G3
Narok 2,608 ....G4
Nyeri 2,436 ....G4
Rumuruti 1,484 ....G3
Thika 18,387 ....G4
Thomson's Falls 7,602 ....G4
Vanga G4
Voi 5,313 ....G4
Wajir H3
Wamba 2,650 ....G3

OTHER FEATURES

Daua (riv.) G3
Elgon (mt.) F3
Formosa (bay) H4
Galana (riv.) G4
Gedi (ruins) G4
Kenya (mt.) G4
Lak Dera (dry riv.) H3
Lorian (swamp) G3
Natron (lake) G4
Nyiru (mt.) G3
Rudolf (Turkana) (lake) G3
Tana (riv.) G4
Tsavo Nat'l Park G4
Turkana (lake) G3
Victoria (lake) F4
Winam (bay) F4

MALAWI

CITIES and TOWNS

Blantyre 222,153 ....F7
Chitipa 3,079 ....F5
Dedza 5,448 ....F6
Dowa 2,067 ....F6
Karonga 11,873 ....F5
Lilongwe (cap.) 102,924 ....F6
Mangochi 3,341 ....G6
Mchinji 1,962 ....F6
Mwanza 2,271 ....F7
Mzimba 4,962 ....F6

Ncheu 1,326 ....F6
Nkhata Bay 4,024 ....F6
Nkhotakota 10,312 ....F6
Nsanje 6,091 ....G7
Rumphi 3,998 ....F5
Salima 4,646 ....F6
Thyolo 4,186 ....F6
Zomba 21,000 ....G7

OTHER FEATURES

Chilrua (lake) G7
Malawi (Nyasa) (lake) F6
Mulanje (mts.) G7
Nyasa (lake) F6
Shire (riv.) G7

RWANDA

CITIES and TOWNS

Butare 21,691 ....E4
Cyangugu 7,042 ....E4
Gisenyi 12,436 ....E4
Kigali (cap.) 117,749 ....F4
Nyabisindu 8,587 ....F4

OTHER FEATURES

Kagera Nat'l Park F4
Karisimbi (mt.) E4
Kivu (lake) E4
Ruzizi (riv.) E4
Virunga (range) E4

SOMALIA

PROVINCES

Bakool 100,000 ....H3
Bari 155,000 ....J1
Bay 302,000 ....H3
Galguduud 182,000 ....J2
Gedo 202,800 ....H3
Hiiraan 147,000 ....H3
Jubbada Hoose 246,000 ....H3
Mogadiscio 371,000 ....J3
Mudug 215,000 ....J2
Nugaal 85,000 ....J2
Sanaag 369,000 ....J1
Shabeellaha Dhexe 237,000 .....H3
Shabeellaha Hoose 398,000 .....H3
Togdheer 258,000 ....J2
Woqooyi Galbeed 440,000 ....H1

CITIES and TOWNS

Afgoi H3
Afmadu 2,580 ....H3
Alula K1
Ankhor J3
Balad 1,233 ....J3
Barawa (Brava) 6,167 ....H3
Baydhabo 14,962 ....H3
Belet Weyne 11,426 ....H3
Bender Cassim J1
Berbera 12,219 ....H1
Borama 3,244 ....H1
Bosaso J1
Brava 6,167 ....H3
Bulo Burti 5,247 ....H3
Bur Acaba H3
Burao 12,617 ....H2
Candala J1
Chisimayu 17,872 ....H4
Coriole 4,341 ....H3
Dante (Hafun) K1
Dusa Marreb J2
Eil J2
El Athale (Itala) J3
Erigabo 4,279 ....J1

(continued on following page)

Topography

© Copyright HAMMOND INCORPORATED, Maplewood, N.J.

Central Africa

CYLINDRICAL EQUAL-AREA PROJECTION

SCALE OF MILES

SCALE OF KILOMETERS

Capitals of Countries .... ☆
Other Capitals .... ⊛
International Boundaries ....
Internal Boundaries ....

Scale 1:13,800,000

(continued)

Galcaio J2
Garbahaarrey H3
Garoe J2
Giohar 13,156 .... J3
Hafun K1
Hargeysa 40,254 .... H2
Itala J3
Jamama 5,408 .... H3
Jilib 3,232 .... H3
Kismayu (Chisimayu) 17,872 .... H4
Las Anod 2,441 .... J2
Las Khoreh 2,245 .... J1
Luuq H3
Margherita (Jamama) H3
Marka (Merka) 17,708 .... H3
Mogadishu (cap.) 371,000 .... J3
Muqdisho (Mogadishu) (cap.) 371,000 .... J3
Oddur H3
Odweina 1,422 .... J2
Uanle Uen H3
Villabruzzi (Johar) J3
Zeila 1,226 .... H1

OTHER FEATURES

Aden (gulf) J1
Asèr, Ras (cape) K1
Giuba (riv.) H3
Guban (reg.) H1
Haud (plat.) J2
Lak Dera (dry riv.) H3
Negro (bay) J2
Shimbir Berris (mt.) J1
Sura, Ras (cape) J1
Webi Shabelle (riv.) H3

TANZANIA

PROVINCES

Arusha 928,478 .... G4
Dar es Salaam G5
Dodoma 971,921 .... G5
Iringa 922,801 .... G5
Kagera 1,009,379 .... F4
Kigoma 648,950 .... F4
Kilimanjaro 902,394 .... G4
Lindi 527,902 .... G5
Mara 723,295 .... F4
Mbeya 1,080,241 .... F5
Morogoro 939,190 .... G5
Mtwara 771,726 .... G5
Mwanza 1,443,418 .... F4
Pemba 205,870 .... H5
Pwani (Coast) 516,949 .... G5
Rukwa 451,897 .... F5
Ruvuma 564,113 .... G6
Shinyanga 1,323,482 .... F4
Singida 614,030 .... F5
Tabora 818,049 .... F5
Tanga 1,088,592 .... G5
Zanzibar Mjini 143,616 .... G5
Zanzibar Shambani North 77,424 .... G5
Zanzibar Shambani South 52,325 .... G5

CITIES and TOWNS

Arusha 55,281 .... G4
Bagamoyo 5,112 .... G5
Biharamulo 1,011 .... F4
Bukene 2,288 .... F4
Bukoba 20,430 .... F4
Chake Chake 4,862 .... H5
Chunya 2,398 .... F5
Dar es Salaam (cap.) 757,346 .... G5
Dodoma 45,703 .... G5
Geita 3,066 .... F4
Iringa 57,182 .... G5
Kahama 3,211 .... F4
Kasulu F4
Kibondo F4
Kigoma-Ujiji 50,044 .... E4
Kilosa 4,458 .... G5
Kilwa Kivinje 2,790 .... G5
Kilwa Masoko G5
Koani 1,102 .... G5
Kondoa 4,514 .... G4
Kongwa G5
Korogwe 6,675 .... G5
Lindi 27,308 .... G5
Lushoto 1,803 .... G4
Mahenge G5
Manyoni G5
Masasi G6
Mbeya 76,606 .... F5
Mbulu G4
Mchinga G5
Mkokotoni 2,220 .... G5
Morogoro 61,890 .... G5
Moshi 52,223 .... G4
Mpanda F5
Mpwapwa 2,429 .... G5
Mtwara-Mikindani 48,510 .... H6
Musoma 32,658 .... F4
Mwadui 7,383 .... F4
Mwanza 110,611 .... F4
Nachingwea 3,751 .... G6
Newala G6
Ngara F4
Njombe F5
Nzega 2,386 .... F4
Pangani 2,955 .... G5
Shinyanga 21,703 .... F4
Singida 29,252 .... F4
Songea 17,954 .... G6
Sumbawanga 28,586 .... F5
Tabora 67,392 .... F5
Tanga 103,409 .... G4
Tukuyu 4,089 .... F5
Tunduru G6
Utete G5
Wete 8,469 .... G4
Zanzibar 110,669 .... G5

OTHER FEATURES

Eyasi (lake) F4
Great Ruaha (riv.) G5
Juani (isl.) G5
Kagera Nat'l Park F4
Kalambo (falls) F5
Kilimanjaro (mt.) G4
Kilombero (riv.) G5
Mafia (isl.) H5
Manyara (lake) G4
Masai (steppe) G4
Mbarangandu (riv.) G5
Mbemkuru (riv.) G5
Meru (mt.) G4
Mikumi Nat'l Park G5
Natron (lake) G4
Ngorongoro (crater) F4
Njombe (riv.) F5
Nyasa (lake) F6
Olduvai Gorge (canyon) G4
Pangani (riv.) G4
Pemba (isl.) H5
Ras Kanzi (cape) G5
Rovuma (riv.) G6
Ruaha Nat'l Park F5
Rufiji (riv.) G5
Rukwa (lake) F5
Rungwa (riv.) F5
Rungwe (mt.) F5
Serengeti Nat'l Park F4
Tanganyika (lake) E5
Tarangire Nat'l Park G4
Victoria (lake) F4
Wami (riv.) G5
Wembere (riv.) F4
Zanzibar (isl.) G5

UGANDA

CITIES and TOWNS

Arua 10,837 .... F3
Butiaba 261 .... F3
Entebbe 21,096 .... F4
Fort Portal 7,947 .... F3
Gulu 18,170 .... F3
Hoima 2,339 .... F3
Jinja 52,509 .... F3
Kabale 8,234 .... E4
Kampala (cap.) 478,895 .... F3
Kasese 7,213 .... F3
Kitgum 3,242 .... F3
Lira 7,340 .... F3
Masaka 12,987 .... F4
Masindi 2,100 .... F3
Mbale 23,544 .... F3
Mbarara 16,078 .... F4
Moroto 5,488 .... F3
Moyo 266 .... F3
Mubende 6,004 .... F3
Rhino Camp 198 .... F3
Soroti 8,130 .... F3
Tororo 15,977 .... F3

OTHER FEATURES

Albert (Mobuto Sese Seko) (lake) F3
Edward (lake) E4
Elgon (mt.) F3
George (lake) F3
Kabalega (falls) F3
Kagalega Nat'l Park F3
Kidepo Nat'l Park F3
Kioga (lake) F3
Margherita (mt.) E3
Mobuto Sese Seko (lake) F3
Owen Falls (dam) F3
Ruwenzori (range) E3
Sese (isls.) F4
Victoria (lake) F4
Virunga (range) E4
Virunga Nat'l Park F4

CONGO, DEM. REP. OF THE

PROVINCES

Bandundu 2,600,556 .... C4
Bas-Zaïre 1,504,361 .... B5
Equateur 2,431,812 .... D3
Haut-Zaïre 3,356,419 .... E3
Kasai-Occidental 2,433,861 .... D4
Kasai-Oriental 1,872,231 .... D5
Kinshasa 1,323,039 .... C4
Kivu 3,361,883 .... E4
Shaba 2,753,714 .... E5

CITIES and TOWNS

Aba 7,600 .... F3
Aketi 17,200 .... D3
Bambesa E3
Banalia E3
Banana B5
Bandundu 74,467 .... C4
Basankusu C3
Basoko 9,100 .... D3
Befale D3
Beni 22,800 .... E3
Bikoro C4
Boende 12,800 .... D4
Bokungu D4
Bolobo 10,300 .... C4
Bolomba 7,200 .... C3
Boma 61,100 .... B5
Bomongo C3
Bondo 10,000 .... D3
Bongandanga 12,900 .... D3
Bosobolo 11,100 .... D3
Budjala C3
Bukama E5
Bukavu 134,861 .... E4
Bulungu 16,300 .... C4
Bumba 34,700 .... D3
Bunia 28,800 .... E3
Bunkeya 5,100 .... E6
Businga 11,000 .... D3
Busu-Djanoa D3
Buta 19,800 .... D3
Butembo 27,800 .... E3
Dekese D4
Demba 22,000 .... D5
Dibaya 11,400 .... D5
Dibaya-Lubue 7,900 .... C4
Dilolo 14,000 .... D6
Dimbelenge D5
Djolu D3
Djugu F3
Dungu 9,100 .... E3
Faradje 10,400 .... E3
Feshi C5
Fizi E4
Gandâjika 60,100 .... D5
Gemena 37,300 .... D3
Goma 48,600 .... E4
Gungu C5
Idiofa C4
Ikela D4
Ilebo 32,200 .... D4
Ingende C4
Inongo 14,800 .... C4
Irumu 9,300 .... E3
Isangi D3
Isiro 49,300 .... E3
Kabalo 22,600 .... E5
Kabambare E4
Kabare 12,600 .... E4
Kabinda 60,500 .... D5
Kabongo 6,500 .... E5
Kahemba C5
Kalehe E4
Kalemie 62,300 .... E5
Kalima 27,500 .... E4
Kama 69,100 .... E4
Kambove 18,900 .... E6
Kamina 56,300 .... D5
Kampene 14,600 .... E4
Kananga 428,960 .... D5
Kaniama D5
Kapanga D5
Kasangulu 11,900 .... C4
Kasenga E6
Kasese E4
Kasongo 37,800 .... E4
Kasongo-Lunda C5
Katako-Kombe D4
Kazumba D5
Kenge 17,500 .... C4
Kibombo E4
Kikwit 111,960 .... C5
Kinshasa (cap.) 1,323,039 .... C4
Kipushi 32,900 .... E6
Kisangani 229,596 .... E3
Kolwezi 81,600 .... E6
Kongolo 14,800 .... E5
Kutu 10,000 .... C4
Libenge 13,000 .... C3
Lisala D3
Lodja 20,300 .... D4
Lomela D4
Lubefu D4
Lubudi 6,000 .... E5
Lubumbashi 318,000 .... E6
Lubutu E4
Luebo 21,800 .... D5
Lukula 9,400 .... B5
Luozi 7,000 .... B5
Lusambo 13,100 .... D4
Mambasa 7,400 .... E3
Mangai 41,200 .... C4
Manono 44,500 .... E5
Masi-Manimba 6,300 .... C4
Masisi E4
Matadi 110,436 .... B5
Mbandaka 107,910 .... C3
Mbanza-Ngungu 55,800 .... C5
Mbuji-Mayi 256,154 .... D5
Mitwaba E5
Monkoto D4
Muanda 6,400 .... B5
Mungbere E3
Mushie 13,700 .... C4
Muyumba E5
Mweka 24,900 .... D4
Mwene-Ditu 71,200 .... D5
Mwenga E4
Niangara 9,200 .... E3
Nyunzu 11,300 .... E5
Opala D4
Oshwe C4
Panda-Likasi 146,394 .... E6
Pangi E4
Poko E3
Popokabaka C5
Port Kindu 42,800 .... E4
Punia E4
Rutshuru E4
Sakania E6
Sandoa D5
Sentery 24,300 .... E5
Shabunda 6,900 .... E4
Songololo 4,600 .... B5
Tshela 10,700 .... B4
Tshikapa 38,900 .... D5
Ubundu 630 .... E4
Uvira 15,900 .... E4
Virunga 21,900 .... E5
Wamba 11,500 .... E3
Watsa 21,250 .... E3
Yangambi 22,600 .... D3

OTHER FEATURES

Albert (Mobutu Sese Seko) (lake) F3
Aruwimi (riv.) E3
Bomu (riv.) D3
Boyoma (Stanley) (falls) D3
Chicapa (riv.) D5
Congo (riv.) C4
Edward (lake) E4
Elila (riv.) E4
Fimi (riv.) C4
Garamba Nat'l Park E3
Giri (riv.) C3
Itimbiri (riv.) D3
Ituri (for.) E3
Karisimbi (mt.) E4
Kasai (riv.) C4
Kivu (lake) E4
Kwa (riv.) C4
Kwango (riv.) C5
Kwilu (riv.) C5
Lindi (riv.) E3
Livingstone (falls) B5
Loange (riv.) C5
Lokoro (riv.) C4
Lomami (riv.) D4
Lomela (riv.) D4
Lowa (riv.) E4
Lua (riv.) C3
Lualaba (riv.) E4
Luapula (riv.) E6
Lubilash (riv.) D5
Lufira (riv.) E5
Luilaka (riv.) C4
Lukenie (riv.) D4
Lukuga (riv.) E5
Lulua (riv.) D5
Luvua (riv.) E5
Mai-Ndombe (lake) C4
Malebo (Stanley Pool) (lake) C4
Margherita (mt.) E3
Marungu (mts.) E5
Mobuto Sese Seko (lake) F3
Mweru (lake) E5
Ruwenzori (range) E3
Ruzizi (riv.) E4
Salonga Nat'l Park D4
Sankuru (riv.) D4
Stanley (falls) D3
Stanley Pool (lake) C4
Tanganyika (lake) E5
Tshuapa (riv.) D4
Tumba (lake) C4
Ubangi (riv.) C3
Uele (riv.) E3
Ulindi (riv.) E4
Upemba (lake) E5
Upemba Nat'l Park E5
Virunga (range) E4
Virunga Nat'l Park E4

ZAMBIA

CITIES and TOWNS

Abercorn (Mbala) 11,179 .... F5
Bancroft (Chililabombwe) 61,928 .... E6
Broken Hill (Kabwe) 143,635 .... E6
Chilanga 12,503 .... E7
Chililabombwe 61,928 .... E6
Chingola 145,869 .... E6
Chipata 32,291 .... F6
Choma 17,943 .... E7
Fort Rosebery (Mansa) 34,801 .... E6
Isoka 6,832 .... F6
Kabompo 5,357 .... D6
Kabwe 143,635 .... E6
Kafue 29,794 .... E7
Kalabo 7,398 .... D6
Kalomo 5,878 .... E7
Kaoma 6,731 .... D6
Kapiri Mposhi 13,677 .... E6
Kasama 38,093 .... F6
Kawambwa 7,235 .... E5
Kitwe 314,794 .... E6
Livingstone 71,987 .... E7
Luanshya 132,164 .... E6
Lusaka (cap.) 538,469 .... E7
Mansa 34,801 .... E6
Mazabuka 29,602 .... E7
Mbala 11,179 .... F5
Mongu 24,919 .... D7
Monze 13,141 .... E7
Mpika 25,880 .... F6
Mporokoso 6,008 .... F5
Mpulungu 6,354 .... F5
Mufulira 149,778 .... E6
Mumbwa 7,570 .... E6
Ndola 282,439 .... E6
Petauke 7,531 .... F6
Senanga 7,204 .... D7
Serenje 6,008 .... F6
Solwezi 15,032 .... E6
Zambezi 8,166 .... D6

OTHER FEATURES

Bangweulu (lake) F6
Barotseland (reg.) D7
Chambeshi (riv.) F6
Cuando (riv.) D7
Dongwe (riv.) D6
Kabompo (riv.) D6
Kafue (riv.) E7
Kafue Nat'l Park E6
Kalambo (falls) F5
Kariba (dam) E7
Kariba (lake) E7
Luangwa (riv.) F6
Luapula (riv.) E6
Lungwebungu (riv.) D6
Mosi-Oa-Tunya (Victoria) (falls) E7
Mulungushi (dam) E6
Mweru (lake) E5
Sunzu (mt.) F5
Tanganyika (lake) E5
Victoria (falls) E7
Zambezi (riv.) D7

Agriculture, Industry and Resources

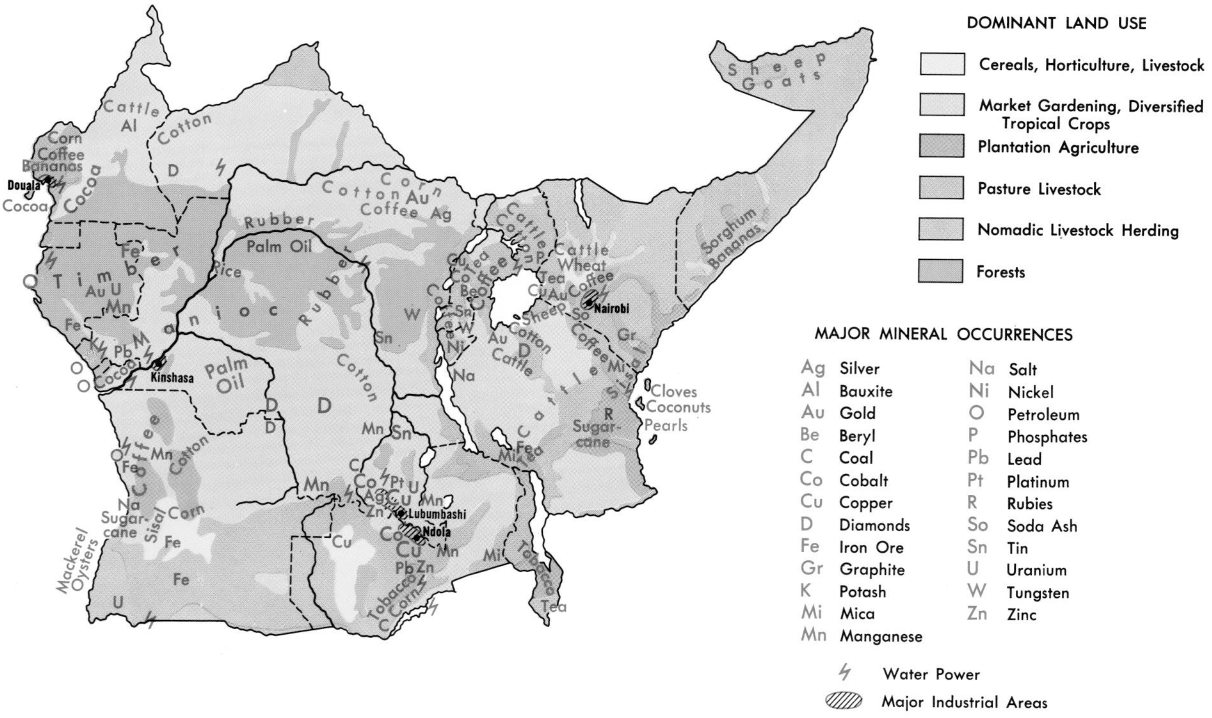

DOMINANT LAND USE
Cereals, Horticulture, Livestock
Market Gardening, Diversified Tropical Crops
Plantation Agriculture
Pasture Livestock
Nomadic Livestock Herding
Forests
MAJOR MINERAL OCCURRENCES
Ag Silver
Al Bauxite
Au Gold
Be Beryl
C Coal
Co Cobalt
Cu Copper
D Diamonds
Fe Iron Ore
Gr Graphite
K Potash
Mi Mica
Mn Manganese
Na Salt
Ni Nickel
O Petroleum
P Phosphates
Pb Lead
Pt Platinum
R Rubies
So Soda Ash
Sn Tin
U Uranium
W Tungsten
Zn Zinc
Water Power
Major Industrial Areas
Douala
Kinshasa
Nairobi
Lubumbashi
Ndola
Cattle
Cotton
Corn
Coffee
Bananas
Cocoa
Timber
Rubber
Palm Oil
Manioc
Sheep
Goats
Sorghum
Bananas
Wheat
Tea
Cloves
Coconuts
Pearls
Sugar-cane
Sisal
Tobacco
Mackerel
Oysters

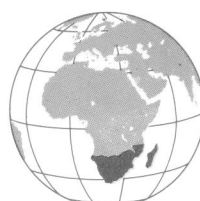

NAMIBIA

AREA 317,827 sq. mi. (823,172 sq. km.)
POPULATION 1,818,000
CAPITAL Windhoek
LARGEST CITY Windhoek
HIGHEST POINT Brandberg 8,550 ft. (2,606 m.)
MONETARY UNIT rand
MAJOR LANGUAGES Ovambo, Hottentot, Herero, Afrikaans, English
MAJOR RELIGIONS Tribal religions, Protestantism

BOTSWANA

AREA 224,764 sq. mi. (582,139 sq. km.)
POPULATION 1,256,000
CAPITAL Gaborone
LARGEST CITY Francistown
HIGHEST POINT Tsodilo Hill 5,922 ft. (1,805 m.)
MONETARY UNIT pula
MAJOR LANGUAGES Setswana, Shona, Bushman, English, Afrikaans
MAJOR RELIGIONS Tribal religions, Protestantism

ZIMBABWE

AREA 150,803 sq. mi. (390,580 sq. km.)
POPULATION 9,122,000
CAPITAL Harare
LARGEST CITY Harare
HIGHEST POINT Mt. Inyangani 8,517 ft. (2,596 m.)
MONETARY UNIT Zimbabwe dollar
MAJOR LANGUAGES English, Shona, Ndebele
MAJOR RELIGIONS Tribal religions, Protestantism

SOUTH AFRICA

AREA 455,318 sq. mi. (1,179,274 sq. km.)
POPULATION 34,492,000
CAPITALS Cape Town, Pretoria
LARGEST CITY Johannesburg
HIGHEST POINT Injasuti 11,182 ft. (3,408 m.)
MONETARY UNIT rand
MAJOR LANGUAGES Afrikaans, English, Xhosa, Zulu, Sesotho
MAJOR RELIGIONS Protestantism, Roman Catholicism, Islam, Hinduism, tribal religions

MOZAMBIQUE

AREA 303,769 sq. mi. (786,762 sq. km.)
POPULATION 15,326,000
CAPITAL Maputo
LARGEST CITY Maputo
HIGHEST POINT Mt. Binga 7,992 ft. (2,436 m.)
MONETARY UNIT metical
MAJOR LANGUAGES Makua, Thonga, Shona, Portuguese
MAJOR RELIGIONS Tribal religions, Roman Catholicism, Islam

MADAGASCAR

AREA 226,657 sq. mi. (587,041 sq. km.)
POPULATION 9,985,000
CAPITAL Antananarivo
LARGEST CITY Antananarivo
HIGHEST POINT Maromokotro 9,436 ft. (2,876 m.)
MONETARY UNIT Madagascar franc
MAJOR LANGUAGES Malagasy, French
MAJOR RELIGIONS Tribal religions, Roman Catholicism, Protestantism

MAURITIUS

AREA 790 sq. mi. (2,046 sq. km.)
POPULATION 1,068,000
CAPITAL Port Louis
LARGEST CITY Port Louis
HIGHEST POINT 2,711 ft. (826 m.)
MONETARY UNIT Mauritian rupee
MAJOR LANGUAGES English, French, French Creole, Hindi, Urdu
MAJOR RELIGIONS Hinduism, Christianity, Islam

LESOTHO

AREA 11,720 sq. mi. (30,355 sq. km.)
POPULATION 1,700,000
CAPITAL Maseru
LARGEST CITY Maseru
HIGHEST POINT 11,425 ft. (3,482 m.)
MONETARY UNIT loti
MAJOR LANGUAGES Sesotho, English
MAJOR RELIGIONS Tribal religions, Christianity

SWAZILAND

AREA 6,705 sq. mi. (17,366 sq. km.)
POPULATION 681,000
CAPITAL Mbabane
LARGEST CITY Manzini
HIGHEST POINT Emlembe 6,109 ft. (1,862 m.)
MONETARY UNIT lilangeni
MAJOR LANGUAGES siSwati, English
MAJOR RELIGIONS Tribal religions, Christianity

COMOROS

AREA 719 sq. mi. (1,862 sq. km.)
POPULATION 484,000
CAPITAL Moroni
LARGEST CITY Moroni
HIGHEST POINT Karthala 7,746 ft. (2,361 m.)
MONETARY UNIT CFA franc
MAJOR LANGUAGES Arabic, French, Swahili
MAJOR RELIGION Islam

SEYCHELLES

AREA 145 sq. mi. (375 sq. km.)
POPULATION 67,000
CAPITAL Victoria
LARGEST CITY Victoria
HIGHEST POINT Morne Seychellois 2,993 ft. (912 m.)
MONETARY UNIT Seychellois rupee
MAJOR LANGUAGES English, French, Creole
MAJOR RELIGION Roman Catholicism

REUNION

AREA 969 sq. mi. (2,510 sq. km.)
POPULATION 570,000
CAPITAL St-Denis

MAYOTTE

AREA 144 sq. mi. (373 sq. km.)
POPULATION 47,300
CAPITAL Mamoutzou

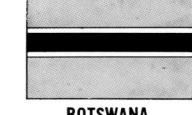
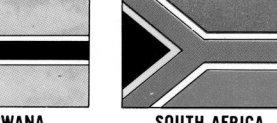

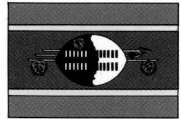

ZIMBABWE **BOTSWANA** **SOUTH AFRICA** **LESOTHO** **SWAZILAND**

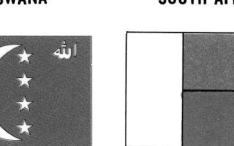
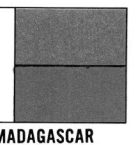
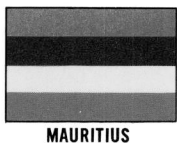

MOZAMBIQUE **COMOROS** **MADAGASCAR** **MAURITIUS** **SEYCHELLES**

NAMIBIA

Agriculture, Industry and Resources

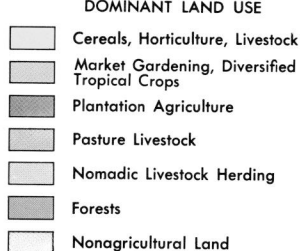

MAJOR MINERAL OCCURRENCES

Ab	Asbestos	Cu	Copper	Mn	Manganese	Sb	Antimony
Ag	Silver	D	Diamonds	Na	Salt	Sn	Tin
Al	Bauxite	Fe	Iron Ore	Ni	Nickel	U	Uranium
Au	Gold	Gr	Graphite	P	Phosphates	V	Vanadium
Be	Beryl	Lt	Lithium	Pb	Lead	W	Tungsten
C	Coal	Mg	Magnesium	Pt	Platinum	Zn	Zinc
Cr	Chromium	Mi	Mica				

DOMINANT LAND USE

Cereals, Horticulture, Livestock

Market Gardening, Diversified Tropical Crops

Plantation Agriculture

Pasture Livestock

Nomadic Livestock Herding

Forests

Nonagricultural Land

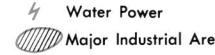 Water Power

Major Industrial Areas

BOTSWANA

CITIES and TOWNS

Dinokwe 560.....................D4
Francistown 22,000............D4
Gaborone (cap.) 21,000.....D4
Ghanzi 1,198......................C4
Kanye 10,664....................D5
Kasane 1,476....................D3
Lobatse 11,936.................D5
Mahalapye 12,056.............C4
Maun 9,614........................C4
Mochudi 6,945..................D4
Molepolole 9,448...............D4
Palapye 5,217...................D4
Ramotswa 7,991................D4
Selebi-Pikwe 20,572..........D4
Serowe 15,723..................D4

OTHER FEATURES

Chobe (riv.).......................C3
Chobe Nat'l Park...............D3
Dau (lake).........................C4
Kalahari (des.)..................C4
Kaukauveld (mts.)............C4
Limpopo (riv.)...................D4
Mababe (depr.)..................C4
Makgadikgadi (salt pan)....D3
Molopo (riv.)......................C5
Ngami (lake)......................C4
Ngamiland (reg.)...............C3
Nossob (riv.).....................B4
Okovango (riv.)..................C3
Okovango (swamps)..........C4
Orange (riv.)......................B5
Shashe (riv.).....................D4
Tati (riv.)...........................D4
Tsodilo Hill (mt.)...............C3
Xau (Dau) (lake)...............C4

COMOROS

CITIES and TOWNS

Fomboni 3,229..................G2
Mitsamiouli 3,196..............G2
Moroni (cap.) 12,000.........G2
Mutsamudu 7,652.............G2

OTHER FEATURES

Mwali (Mohéli) (isl.)..........G2
Njazidja (Grand Comoro) (isl.) G2
Nzwani (Anjouan) (isl.)......G2

LESOTHO

CITIES and TOWNS

Leribe 5,200......................D5
Mafeteng 4,600.................D5
Maseru (cap.) 71,500.........D5
Mohaleshoek 3,600............D6

MADAGASCAR

PROVINCES

Antananarivo 2,167,973.......H3
Antsiranana 597,982..........H2
Fianarantsoa 1,804,365......H4
Mahajanga 819,750............H3
Toamasina 1,179,660..........H3
Toliara 1,034,114...............G4

CITIES and TOWNS

Ambalavao 6,988...............H4
Ambanja 12,258.................H2
Ambatolampy 11,539..........H3
Ambatondrazaka 18,044......H3
Ambilobe 9,415.................H2
Ambodifototra 1,112..........J3
Ambositra 16,780..............H4
Andapa 6,275....................H2
Antalaha 17,541................J2
Antananarivo (cap.)
  451,808..........................H3
Antsirabe 32,979...............H3
Antsiranana 40,443............H2
Antsohihy 8,721................H2
Arivonimamo 8,497............H3
Belo-Tsiribihina 4,403........G3
Brickaville (Vohibinany)
  1,741..............................H3
Diégo-Suarez (Antsiranana)
  40,443............................H2
Faradofay 19,605...............H5
Farafangana 10,817...........H4
Fenoarivo, Toamasina 7,696...H3
Fianarantsoa 68,054..........H4
Fort-Dauphin (Faradofay)
  19,605............................H5
Foulpointe........................H3
Hell-Ville 6,183.................H2
Ihosy 4,521......................H4
Maevatanana 7,197...........H3
Maintirano 6,375...............H3
Majunga 65,864................H3
Manakara 19,768..............H4
Mananjary 14,638.............H4
Mandritsara 6,826.............H3
Maroantsetra 6,645...........J3
Marovoay 20,253...............H3
Moramanga 10,806............H3
Morombe 6,967.................G4
Morondava 19,061............G4
Port-Bergé 4,734..............H3
Sambava 6,215.................J2
Sosumav 10,946...............H2
Tamatave (Toamasina)
  77,395............................H3

(continued on following page)

Topography

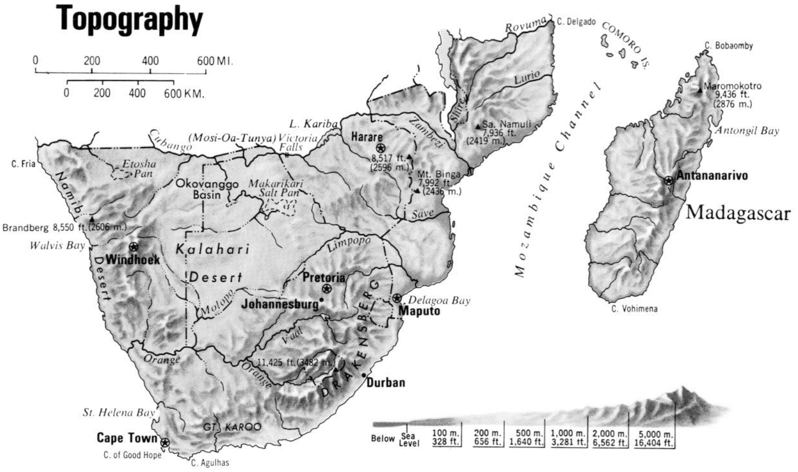
0 200 400 600 MI.
0 200 400 600 KM.
Rovuma
C. Delgado
COMORO IS.
C. Bobaomby
Lurio
Maromokotro 9,436 ft. (2876 m.)
Antongil Bay
Sa. Namuli 7,936 ft. (2419 m.)
L. Kariba
Harare
Cubango
(Mosi-Oa-Tunya) Victoria Falls
8,517 ft. (2596 m.)
Mt. Binga 7,992 ft. (2436 m.)
Antananarivo
Madagascar
C. Fria
Etosha Pan
Okavango Basin
Makarikari Salt Pan
Save
Brandberg 8,550 ft. (2606 m.)
Walvis Bay
Windhoek
Kalahari
Desert
Limpopo
Mozambique Channel
Pretoria
Johannesburg
Molopo
Delagoa Bay
Maputo
C. Vohimena
Vaal
Orange
11,425 ft. (3482 m.)
DRAKENSBERG
Durban
St. Helena Bay
Cape Town
C. of Good Hope
C. Agulhas
KAROO
Below Sea Level
100 m. 328 ft.
200 m. 656 ft.
500 m. 1,640 ft.
1,000 m. 3,281 ft.
2,000 m. 6,562 ft.
5,000 m. 16,404 ft.

Tananarive (Antananarivo) 451,808 ....H3
Tangainony 6,952 ....H4
Toamasina 77,395 ....H3
Toliara (Tuléar) 45,676 ....G4
Tsiroanomandidy 11,444 ....H3
Vangaindrano 3,249 ....H4
Vohibinany 1,741 ....H3
Vohimarina (Vohémar) 4,289 ..J2
Vohipeno 2,736 ....H4

OTHER FEATURES

Alaotra (lake) H3
Amber (Bobaomby) (cape) H2
Antongil (bay) J3
Barren (isls.) G3
Betsiboka (riv.) H3
Bobaomby (Amber) (cape) H2
Boby, Pic (mt.) H4
Chesterfield (isl.) G3
Ikopa (riv.) H3
Itasy (lake) H3
Mahajamba (bay) H2
Mananara (riv.) H4
Mananbao (riv.) G3
Mangoky (riv.) G4
Mangoro (riv.) H3
Maromokotro (mt.) H2
Masoala (pen.) J3
Menarandra (riv.) H4
Mozambique (chan.) G3
Nosy Be (isl.) H2
Nosy Boraha (isl.) J3
Onilahy (riv.) G4
Pangalanes (canal) H4
Radama (isls.) H2
Saint-André (cape) G3
Saint-Marie (Nosy Boraha) (isl.) J3
Saint-Marie (Vohimena) (cape) G5
Saint-Sébastien (cape) H2
Sofia (riv.) H3
Tsiafajavona (mt.) H3
Tsiribihina (riv.) G3
Vohimena (cape) G5

MAURITIUS

CITIES and TOWNS

Curepipe 52,709 ....G5
Mahébourg 15,463 ....G5
Port Louis (cap.) 141,022 ....G5
Quatre Bornes 51,638 ....G5
Souillac 3,361 ....G5

OTHER FEATURES

Mascarene (isls.) F5

MAYOTTE

CITIES and TOWNS

Mamoutzou (cap.) 196 ....H2

MOZAMBIQUE

PROVINCES

Cabo Delgado 940,000 ....F2
Gaza 999,900 ....E4
Inhambane 977,000 ....E4
Manica 541,200 ....E4
Maputo 491,800 ....E5
Maputo (city) 755,300 ....E5
Nampula 2,402,700 ....F2
Niassa 514,100 ....F2
Sofala 1,055,200 ....E3
Tete 831,000 ....E3
Zambézia 2,500,000 ....F3

CITIES and TOWNS

Angoche 1,714 ....G3
Bartolomeu Dias◦ 6,102 ....F4
Beira 46,293 ....F3
Beira* 130,398 ....F3
Chibuto 23,763 ....E4
Chicualacuala 2,050 ....E4
Chimoio 4,507 ....E3
Chinde 742 ....F3
Dona Ana (Mutarara) 686 ....F3
Dondo 2,112 ....F3
Funhalouro◦ 42,366 ....E4
Ibo 1,015 ....G2
Inhambane 4,975 ....F4
Inhaminga 1,607 ....F3
Inharrime 856 ....F4
Lichinga 3,011 ....F2
Lumbo◦ 11,080 ....G3
Lúrio◦ 13,417 ....G2
Mabalane◦ 13,158 ....E4
Mabote◦ 28,970 ....E4
Machanga◦ 15,754 ....F4
Machaze◦ 42,255 ....E4
Mandié◦ [illegible]1,382 ....E3
Mandimba◦ 7,634 ....F2
Manhiça◦ 1,680 ....E5
Maniamba◦ 2,045 ....F2
Maputo (cap.) 755,300 ....E5
Massangena◦ 3,301 ....E4
Massinga 517 ....F4
Moçambique 1,730 ....G3
Mocímboa da Praia 935 ....G2
Mocuba 2,293 ....F3
Montepuez 2,837 ....F2
Mualama◦ 34,992 ....F3
Mucojo◦ 15,867 ....G2
Mueda 1,583 ....F2
Mutarara (Dona Ana) 686 ....F3
Nacala 4,601 ....G2
Nampula 23,072 ....F3
Pafúri◦ 2,599 ....E4
Pemba 3,629 ....G2
Quelimane 10,522 ....F3
Quionga◦ 3,181 ....G2
Quissico 2,615 ....E4
Songo 2,230 ....E3
Tete 4,549 ....E3
Vila de Sena◦ 21,074 ....E3
Xai-Xai 5,234 ....E5

OTHER FEATURES

Angoche (isl.) G3
Bazaruto, Ilha do (isl.) F4
Binga (mt.) E3
Cabora Bassa (dam) E3
Changane (riv.) E4
Chirua (lake) F3
Delagoa (bay) E5
Delgado (cape) G2
Gorongosa Nat'l Park E3
Ligonha (riv.) F3
Limpopo (riv.) E4
Lugenda (riv.) F2
Lúrio (riv.) F2
Mazoe (riv.) E3
Mozambique (chan.) G3
Namuli, Serra (mt.) F3
Nyasa (lake) E2
Olifants (riv.) D4
Rovuma (riv.) F2
São Sebastião (pt.) F4
Save (riv.) E4
Shire (riv.) E3
Zambezi (riv.) E3

NAMIBIA

CITIES and TOWNS

Aroab 783 ....B5
Aus 767 ....B5
Berseba B5
Bethanie 1,207 ....B5
Gibeon B5
Gobabis 4,428 ....B4
Grootfontein 4,627 ....B3
Kalkfeld 587 ....B4
Kamanjab 713 ....A3
Karasburg 2,693 ....B5
Karibib 1,653 ....B4
Keetmanshoop 10,297 ....B5
Koes 514 ....B5
Lüderitz 6,642 ....A5
Maltahöhe 1,313 ....B4
Mariental 4,629 ....B4
Okahandja 1,688 ....B4
Omaruru 2,783 ....B4
Ondangwa B3
Opuno A3
Oranjemund 2,594 ....B5
Oshakati B3
Otavi 1,814 ....B3
Otjiwarongo 8,018 ....B4
Outjo 2,545 ....B4
Rehoboth 5,363 ....B4
Rundu 521 ....B3
Stampriet 271 ....B4
Swakopmund 5,681 ....A4
Tsumeb 12,338 ....B3
Usakos 2,334 ....B4
Walvis Bay 21,725 ....A4
Warmbad 810 ....B5
Windhoek (cap.) 61,369 ....B4
Witvlei 303 ....B4

OTHER FEATURES

Brandberg (mt.) A4
Caprivi Strip (reg.) C3
Chobe (riv.) C3
Cross (cape) A4
Cubango (riv.) B3
Cunene (riv.) A3
Damaraland (reg.) B4
Diamond Coast (reg.) A5
Elephant (riv.) B5
Etosha Pan (salt pan) B3
Fish (riv.) B5
Fria (cape) A3
Great Namaland (reg.) B4
Hollam's Bird (isl.) A4
Hottentot (bay) A5
Kalahari (des.) C4
Kaokoveld (reg.) A3
Kaukauveld (mts.) C3
Kuiseb (riv.) B4
Lüderitz (bay) A5
Namib (des.) A3
Nossob (riv.) B4
Okovango (riv.) C3
Omatako (riv.) B3
Ovamboland (reg.) B3
Ruacana Falls (falls) A3
Skeleton Coast (reg.) A3
Swakop (riv.) B4
Ugab (riv.) A4
Zambezi (riv.) C3

RÉUNION

CITIES and TOWNS

Le Port 21,564 ....F5
Saint-Benoît 7,778 ....G5
Saint-Denis (cap.) 80,075 ....F5
Saint-Denis* 104,603 ....F5
Saint-Joseph 8,928 ....G6
Saint-Louis 10,252 ....F5
Saint-Pierre 21,817 ....F6

OTHER FEATURES

Bassas da India (isl.) F4
Europa (isl.) G4
Glorioso (isls.) H2
Juan de Nova (isl.) G3
Mascarene (isls.) F5
Piton des Neiges (mt.) G5

SEYCHELLES

CITIES and TOWNS

Anse Boileau† 3,420 ....H5
Anse Royale† 3,182 ....H5
Cascade† 2,600 ....H5
Victoria (cap.) 15,559 ....H5
Victoria* 23,012 ....H5

OTHER FEATURES

Aldabra (isls.) H1
Assumption (isl.) H1
Astove (isl.) H2
Cerf (isl.) H5
Cosmoledo (isls.) H1
Curieuse (isl.) H5
Felicité (isl.) J5
Frigate (isl.) J5
La Digue (isl.) J5
Mahé (isl.) H5
Morne Seychellois (mt.) H5
North (isl.) H5
Praslin (isl.) H5
Sainte Anne (isl.) H5
Silhouette (isl.) H5

SOUTH AFRICA

PROVINCES

Eastern Cape 6,665,400 ....D6
Free State 2,804,600 ....D5
Gauteng 6,847,000 ....D5
KwaZulu Natal 8,549,000 ....E5
Mpumalanga 2,838,500 ....D5
Northern Cape 763,900 ....B5
Northern Province 5,120,600 ....D4
North-West 3,506,800 ....C5
Western Cape 3,620,200 ....C6

CITIES and TOWNS

Alberton 23,988 ....H6
Alexandra 57,040 ....H6
Aliwal North 12,311 ....D6
Barberton 12,382 ....E5
Beaufort West 17,862 ....C6
Bellville 49,026 ....F6
Benoni 151,294 ....J6
Bethlehem 29,918 ....D5
Bisho D6
Bloemfontein 149,836 ....D5
Boksburg 106,126 ....J6
Brakpan 73,210 ....J6
Brits 12,182 ....D5
Butterworth (Gcuwa) 2,769 ....D6
Cape Town (cap.) 854,616 ....E6
Carltonville 40,641 ....G7
Cradock 20,822 ....D6
De Aar 18,057 ....C6
Dundee 17,162 ....E5
Durban 736,852 ....E5
East London 119,727 ....D6
Edendale 41,194 ....D5
Edenvale 25,126 ....H6
Elsiesrivier 63,706 ....F6
Ermelo 19,036 ....E5
Eshowe 4,552 ....E5
Estcourt 10,922 ....D5
Fort Beaufort 11,640 ....D6
Gcuwa 2,769 ....D6
George 24,625 ....C6
Germiston 221,972 ....H6
Glencoe 10,513 ....E5
Goodwood 31,592 ....F6
Graaff-Reinet 22,392 ....C6
Grahamstown 41,302 ....D6
Grassy Park 32,709 ....E6
Griquatown 2,996 ....C5
Harrismith 16,082 ....D5
Heidelberg 12,521 ....J7
Hermanus 4,956 ....G7
Howick 12,429 ....E5
Johannesburg 654,232 ....H6
Kempton Park 37,205 ....J6
Kimberley 105,258 ....C5
King William's Town 15,798 ....D6
Klerksdorp 63,558 ....D5
Knysna 13,479 ....C6
Kokstad 10,227 ....D6
Kraaifontein 10,286 ....F6
Kroonstad 51,988 ....D5
Krugersdorp 92,725 ....H6
Ladybrand 8,757 ....D5
Ladysmith 28,920 ....D5
Lambert's Bay 3,247 ....B6
Mafikeng (Mafeking) 6,515 ....D5
Matatiele 3,853 ....D6
Messina 21,121 ....D4
Middelburg, Cape Prov. 12,121 ....D6
Middelburg, Transvaal 26,942 ....D5
Milnerton 10,893 ....F6
Mmabatho D5
Mossel Bay 17,574 ....C6
Nelspruit 25,092 ....E5
Newcastle 14,407 ....E5
Nigel 41,179 ....J7
Nyanga 15,655 ....F6
Odendaalsrus 15,603 ....D5
Oudtshoorn 26,907 ....C6
Paarl 49,244 ....F6
Parow 60,768 ....F6
Parys 17,447 ....D5
Pietermaritzburg 114,822 ....E5
Pietersburg 27,174 ....D4
Pinelands 11,769 ....F6
Pinetown 22,721 ....E6
Port Elizabeth 392,231 ....D6
Port Nolloth 2,893 ....B5
Port Saint Johns (Umzimbuvu) 1,817 ....D6
Potchefstroom 57,443 ....D5
Pretoria (cap.) 525,583 ....D5
Queenstown 39,304 ....D6
Randburg 43,257 ....H6
Randfontein 50,481 ....G6
Richards Bay 598 ....E5
Robertson 10,237 ....C6
Roodeport 115,366 ....H6
Rustenburg 22,303 ....D5
Saldanha 4,994 ....B6
Simonstown 12,137 ....E7
Sishen 2,692 ....C5
Somerset East 10,383 ....D6
Somerset West 11,828 ....F6
Soweto 602,043 ....H6
Springs 142,812 ....J6
Standerton 21,038 ....D5
Stanger 11,064 ....E5

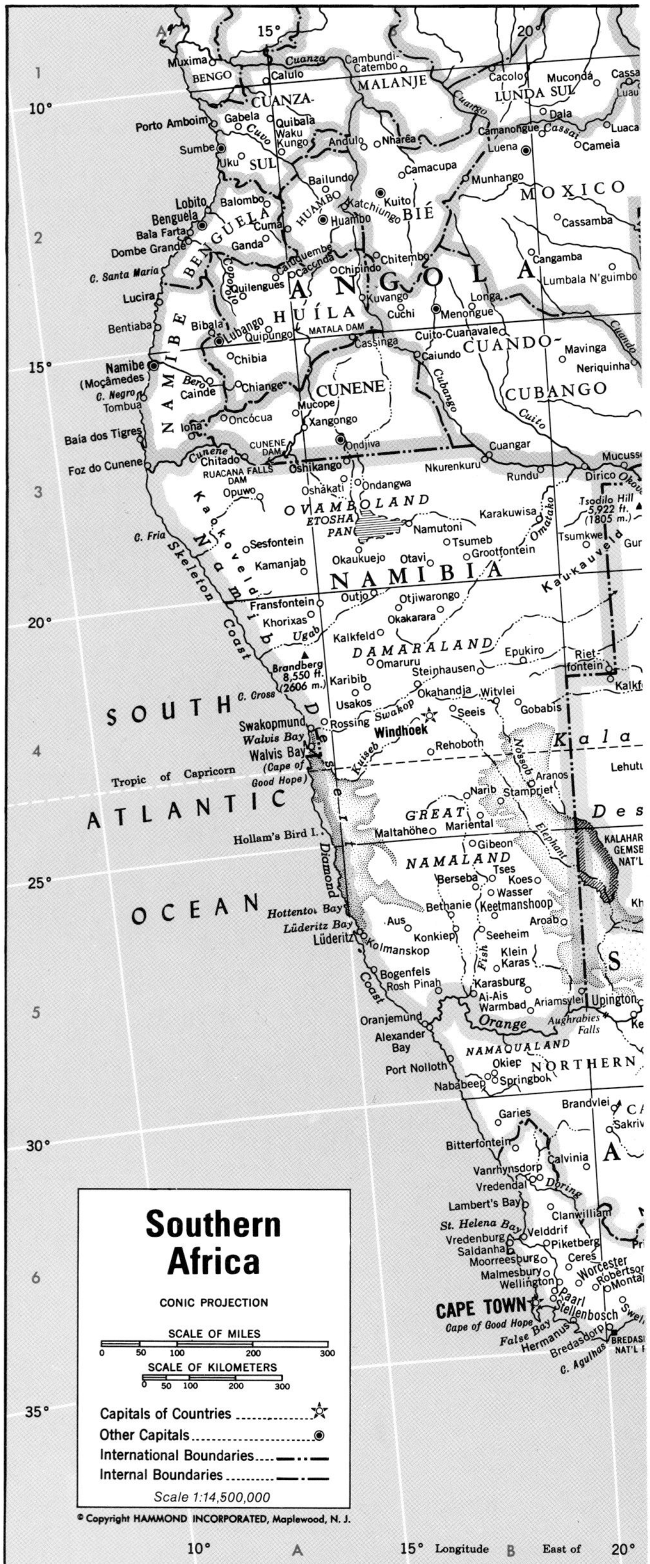
Southern Africa
CONIC PROJECTION
SCALE OF MILES
0 50 100 200 300
SCALE OF KILOMETERS
0 50 100 200 300
Capitals of Countries
Other Capitals
International Boundaries
Internal Boundaries
Scale 1:14,500,000
10° A 15° Longitude B East of 20°
ANGOLA
NAMIBIA
SOUTH ATLANTIC OCEAN
Tropic of Capricorn
Luanda
CUANZA SUL
BENGUELA
HUAMBO
BIÉ
MOXICO
HUÍLA
NAMIBE
CUNENE
CUANDO CUBANGO
Lobito
Benguela
Namibe (Moçâmedes)
OVAMBOLAND
ETOSHA PAN
Skeleton Coast
DAMARALAND
Windhoek
Walvis Bay
Swakopmund
GREAT NAMALAND
Lüderitz
Diamond Coast
Orange
NAMAQUALAND
CAPE TOWN
Cape of Good Hope
C. Agulhas

© Copyright HAMMOND INCORPORATED, Maplewood, N. J.

Population Distribution

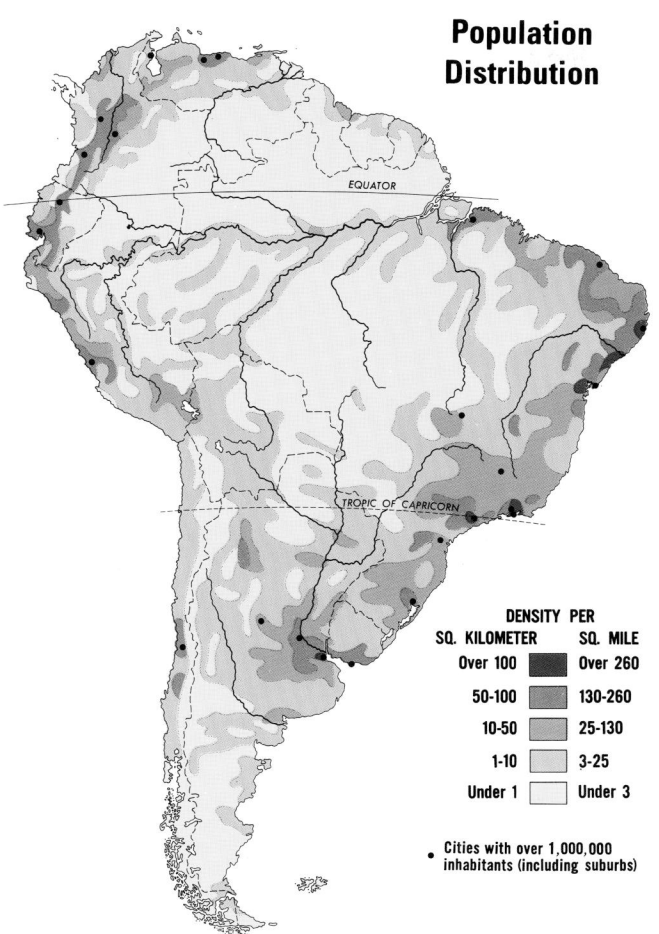

Vegetation

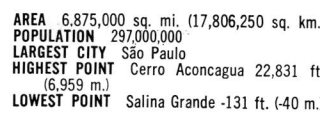

AREA 6,875,000 sq. mi. (17,806,250 sq. km.)
POPULATION 297,000,000
LARGEST CITY São Paulo
HIGHEST POINT Cerro Aconcagua 22,831 ft.
(6,959 m.)
LOWEST POINT Salina Grande -131 ft. (-40 m.)

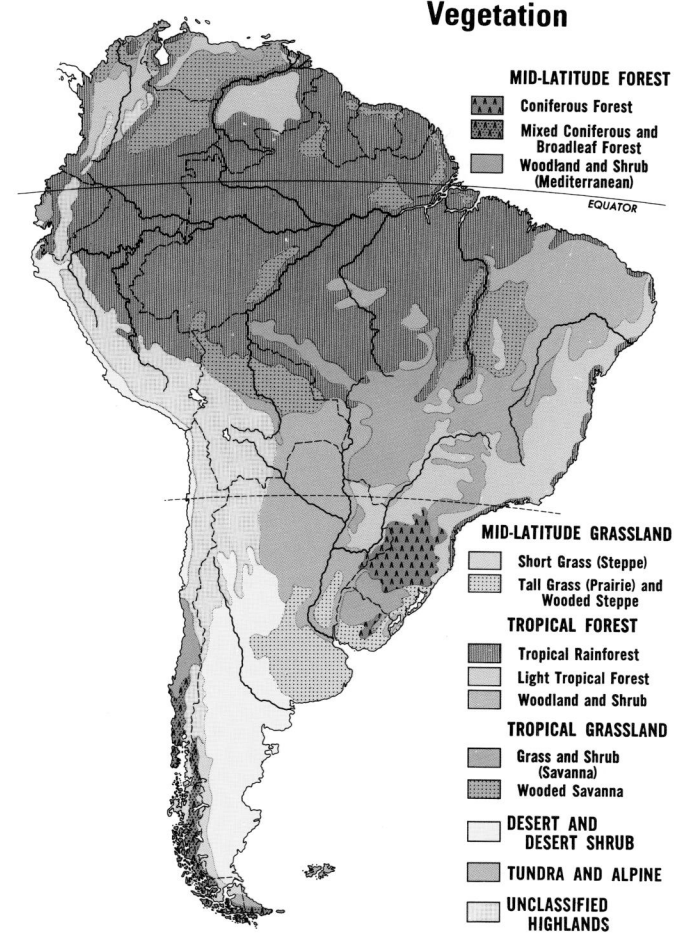

MID-LATITUDE FOREST
Coniferous Forest
Mixed Coniferous and Broadleaf Forest
Woodland and Shrub (Mediterranean)

MID-LATITUDE GRASSLAND
Short Grass (Steppe)
Tall Grass (Prairie) and Wooded Steppe

TROPICAL FOREST
Tropical Rainforest
Light Tropical Forest
Woodland and Shrub

TROPICAL GRASSLAND
Grass and Shrub (Savanna)
Wooded Savanna

DESERT AND DESERT SHRUB

TUNDRA AND ALPINE

UNCLASSIFIED HIGHLANDS

DENSITY PER

SQ. KILOMETER	SQ. MILE
Over 100	Over 260
50-100	130-260
10-50	25-130
1-10	3-25
Under 1	Under 3

• Cities with over 1,000,000 inhabitants (including suburbs)

Average January Temperature

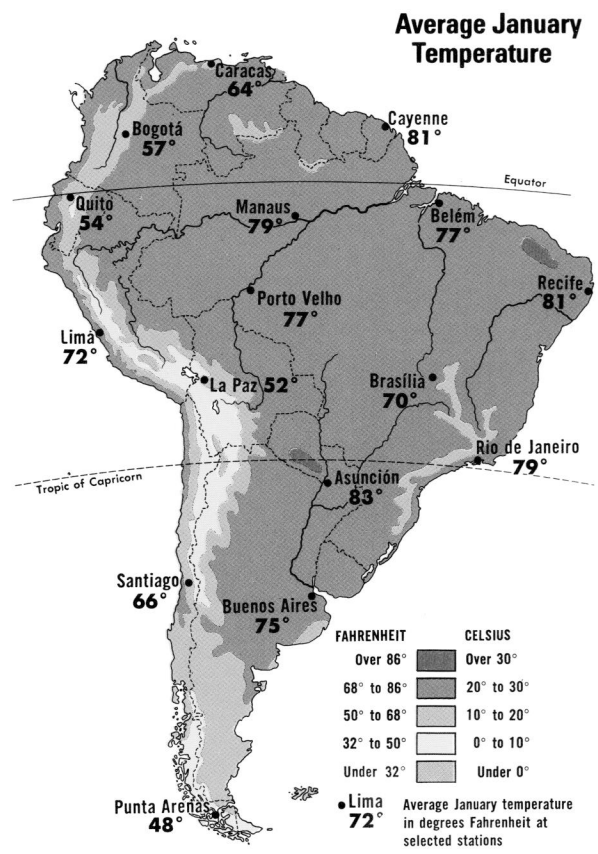

Caracas 64°
Bogotá 57°
Cayenne 81°
Quito 54°
Manaus 79°
Belém 77°
Porto Velho 77°
Recife 81°
Lima 72°
La Paz 52°
Brasília 70°
Rio de Janeiro 79°
Asunción 83°
Santiago 66°
Buenos Aires 75°
Punta Arenas 48°

Equator
Tropic of Capricorn

FAHRENHEIT	CELSIUS
Over 86°	Over 30°
68° to 86°	20° to 30°
50° to 68°	10° to 20°
32° to 50°	0° to 10°
Under 32°	Under 0°

● Lima 72°   Average January temperature in degrees Fahrenheit at selected stations

Average July Temperature

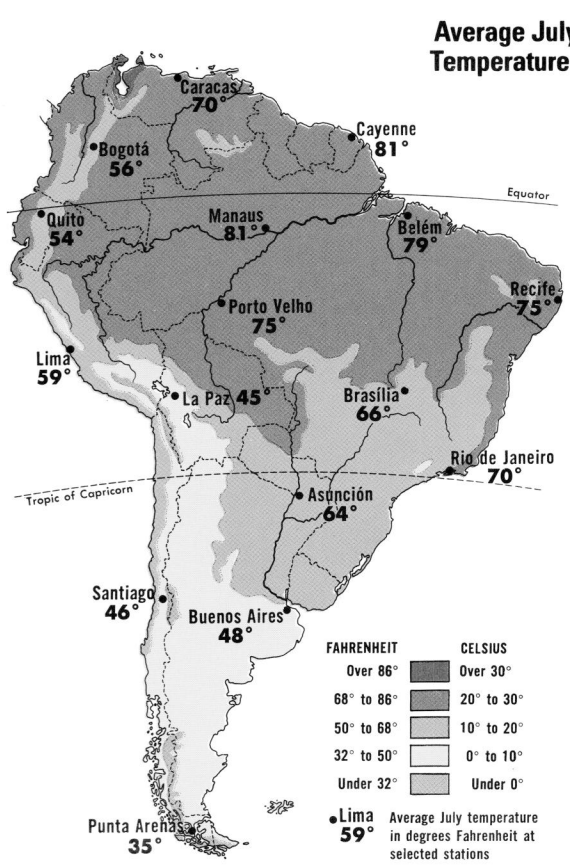

Caracas 70°
Bogotá 56°
Cayenne 81°
Quito 54°
Manaus 81°
Belém 79°
Porto Velho 75°
Recife 75°
Lima 59°
La Paz 45°
Brasília 66°
Rio de Janeiro 70°
Asunción 64°
Santiago 46°
Buenos Aires 48°
Punta Arenas 35°

Equator
Tropic of Capricorn

FAHRENHEIT	CELSIUS
Over 86°	Over 30°
68° to 86°	20° to 30°
50° to 68°	10° to 20°
32° to 50°	0° to 10°
Under 32°	Under 0°

● Lima 59°   Average July temperature in degrees Fahrenheit at selected stations

Rainfall

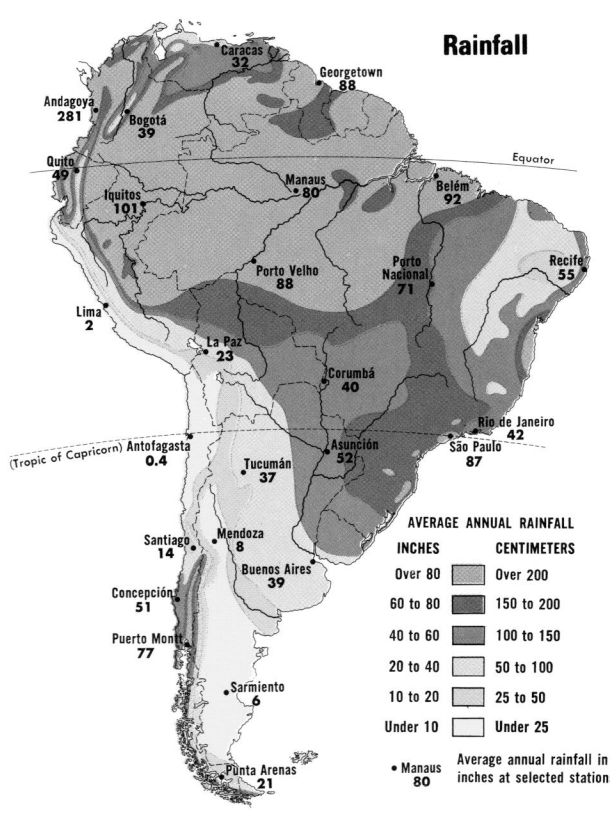

Caracas 32
Georgetown 88
Andagoya 281
Bogotá 39
Quito 49
Iquitos 101
Manaus 80
Belém 92
Porto Velho 88
Porto Nacional 71
Recife 55
Lima 2
La Paz 23
Corumbá 40
Rio de Janeiro 42
Antofagasta 0.4
(Tropic of Capricorn)
Tucumán 37
Asunción 52
São Paulo 87
Santiago 14
Mendoza 8
Buenos Aires 39
Concepción 51
Puerto Montt 77
Sarmiento 6
Punta Arenas 21

Equator

AVERAGE ANNUAL RAINFALL

INCHES	CENTIMETERS
Over 80	Over 200
60 to 80	150 to 200
40 to 60	100 to 150
20 to 40	50 to 100
10 to 20	25 to 50
Under 10	Under 25

● Manaus 80   Average annual rainfall in inches at selected stations

Vegetation/Relief

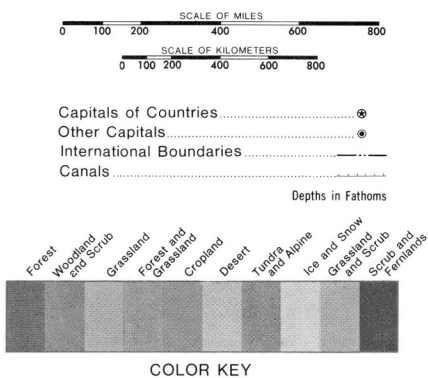

SCALE OF MILES
0 100 200 400 600 800

SCALE OF KILOMETERS
0 100 200 400 600 800

Capitals of Countries .... ⊛
Other Capitals .... ⊛
International Boundaries .... —·—·—
Canals ....

Depths in Fathoms

Forest | Woodland and Scrub | Grassland | Forest and Grassland | Cropland | Desert | Tundra and Alpine | Ice and Snow | Grassland and Scrub | Scrub and Farmlands

COLOR KEY

CARIBBEAN SEA

G. of Venezuela
Pta. Gallinas
ARUBA (Neth.)
Curaçao
Bonaire
Willemstad
NETH. ANTILLES
GRENADA
West Indies
Tobago
TRINIDAD & TOBAGO
Port of Spain
Trinidad
BARBADOS

Barranquilla
Maracaibo
L. Maracaibo
Caracas
Pico Bolívar
16,427 ft. (5007 m.)
Ciudad
Guayana
Orinoco
Guri Res.
Georgetown
Paramaribo
GUYANA
SURINAME
FRENCH
GUIANA
Cayenne

ATLANTIC

OCEAN

PANAMÁ
Panamá Canal
Panamá
G. of Panamá
Pta. Garachine

Medellín
COLOMBIA
Bucaramanga
Bogotá
Cali
Arauca
Meta
VENEZUELA
Orinoco
Guiana
Highlands
Angel Fall
Mt. Roraima
9,094 ft. 2772 m.
Cuyuni
Essequibo
Maroni
Jan

Quito
Chimborazo
20,561 ft.
(6267 m.)
ECUADOR
Gulf of Guayaquil
Pico Phelps
(Pico da Neblina)
9,889 ft.
(3014 m.)
Putumayo
Caquetá
Negro
Vaupés
Japurá
Içá
Amazon
Iquitos
Yavari
Amazon
Manaus
Tapajós
Xingu
Tocantins
B. de Marajó
I. de Marajó
Belém
Equator
São Luís
Fortaleza

Pta. Aguja
Trujillo
Huascarán
22,205 ft.
(6768 m.)
Marañón
Amazon
Juruá
Purus
Madeira
Aripuanã
Teles Pires
SELVAS
BRAZIL
Araguaia
Teresina
Cabo de
São Roque
Natal
Recife
Maceió

Callao
Lima
A N D E S
Ucayali
Mamoré
Guaporé
Planalto de
Mato Grosso
Brazilian
Caatingas
Parnaíba
Paraguaçu

Cuzco
BOLIVIA
Nev. Ancohuma
21,489 ft.
(6550 m.)
La Paz
Lake
Titicaca
Cochabamba
Grande
Brasília
Goiânia
São Francisco
Highlands
Salvador
Jequitinhonha

Arica
Lake
Poopó
Sucre
Campo Grande
Paranã
Tietê
Belo Horizonte
Pico da Bandeira
9,482 ft. (2890 m.)
C. de São Tomé

Tropic of Capricorn
Antofagasta
Volcán Llullaillaco
22,057 ft.
(6723 m.)
San Miguel de
Tucumán
Nev. Ojos del Salado
22,572 ft.
(6880 m.)
Pilcomayo
Bermejo
PARAGUAY
Asunción
ITAIPU
DAM
Iguazú
Falls
Paraguay
Iguaçu
Paraná
Curitiba
São Paulo
Santos
Rio de Janeiro
C. Frio
I. de Santa Catarina

I. de San Félix
(Chile)
I. San Ambrosio
(Chile)
CHILE
Córdoba
Santa Fe
Rosario
Salado del Norte
Gran Chaco
Pampas
Uruguay
Porto Alegre
Lagoa dos Patos

Cerro Aconcagua
22,831 ft. (6959 m.)
Valparaíso
Santiago
Mendoza
Salado
Paraná
Negro
Lagoa Mirim
URUGUAY
Montevideo

I. Alejandro Selkirk
I. Robinson Crusoe
Juan Fernández Is.
(Chile)
Buenos Aires
La Plata
Río de la Plata
C.
San Antonio

PACIFIC
OCEAN
ATLANTIC
OCEAN

Concepción
Colorado
Negro
Bahía Blanca

Puerto Montt
Golfo San Matías
Pen. Valdés
Chubut
PATAGONIA
ARGENTINA

Isla de Chiloé
Archipiélago
de los
Chonos
Golfo San Jorge
C. Tres Puntas

Pen. Taitao
Deseado
G. de Penas

Bahía
Grande
Falkland Islands
(U.K.)
Stanley

Archipiélago
Reina Adelaida
Str. of Magellan
Punta Arenas
Tierra del Fuego

Cape Horn

© Copyright HAMMOND INCORPORATED, Maplewood, N.J.

Longitude West of Greenwich

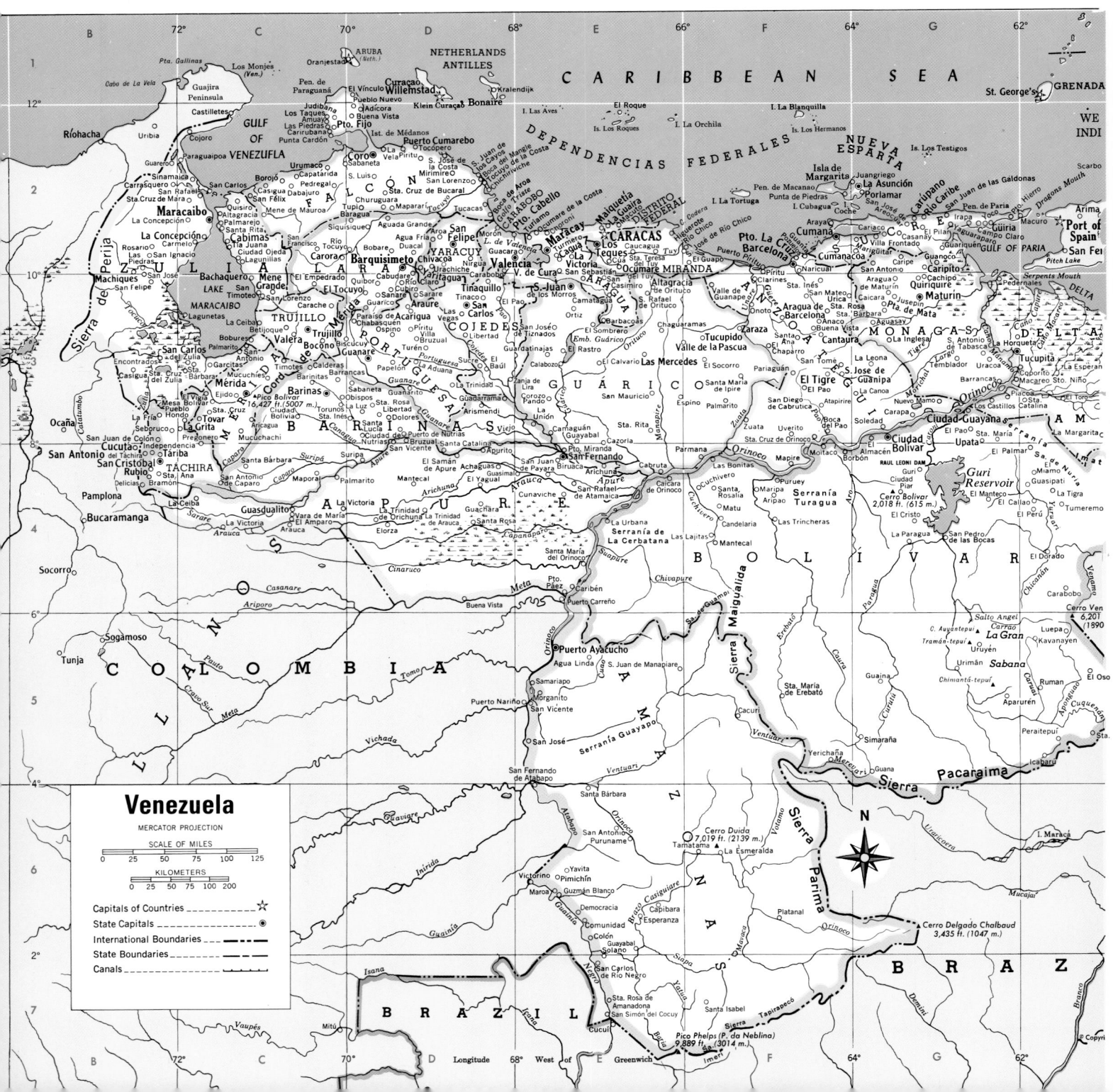

Venezuela

MERCATOR PROJECTION

SCALE OF MILES

0   25   50   75   100   125

KILOMETERS

0   25   50   75 100   200

Capitals of Countries _____ ☆
State Capitals _____ ◉
International Boundaries _ _ ___ _
State Boundaries _____
Canals _____

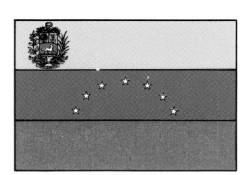

AREA 352,143 sq. mi. (912,050 sq. km.)
POPULATION 19,246,000
CAPITAL Caracas
LARGEST CITY Caracas
HIGHEST POINT Pico Bolívar 16,427 ft. (5,007 m.)
MONETARY UNIT Bolívar
MAJOR LANGUAGE Spanish
MAJOR RELIGION Roman Catholicism

Topography

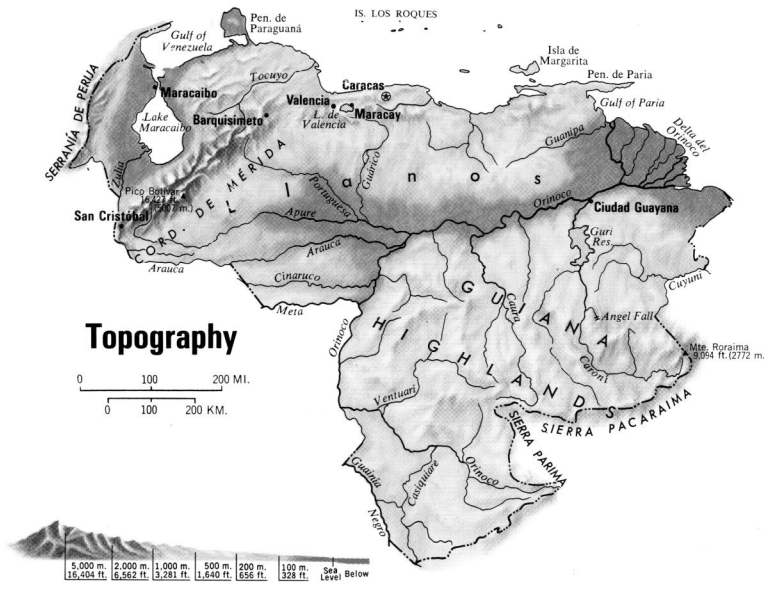

| 5,000 m. 16,404 ft. | 2,000 m. 6,562 ft. | 1,000 m. 3,281 ft. | 500 m. 1,640 ft. | 200 m. 656 ft. | 100 m. 328 ft. | Sea Level | Below |

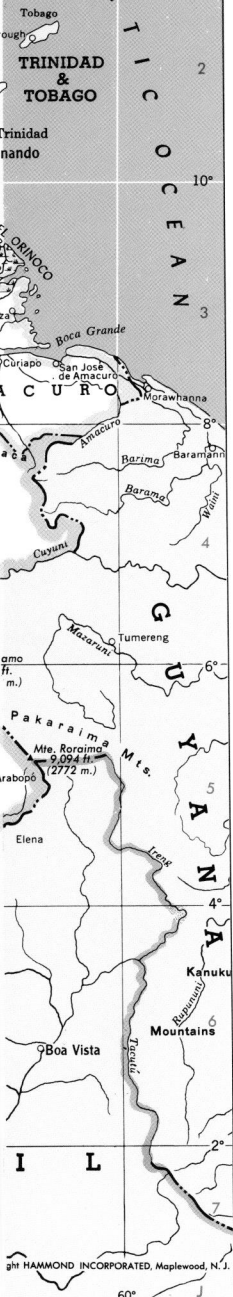

Agriculture, Industry and Resources

MAJOR MINERAL OCCURRENCES

Al Bauxite
Au Gold
C Coal
D Diamonds
Fe Iron Ore
G Natural Gas
Mn Manganese
Na Salt
O Petroleum

⚡ Water Power
▨ Major Industrial Areas

DOMINANT LAND USE

Diversified Tropical Crops (chiefly plantation agriculture)

Upland Cultivated Areas

Upland Livestock Grazing, Limited Agriculture

Extensive Livestock Ranching

Forests

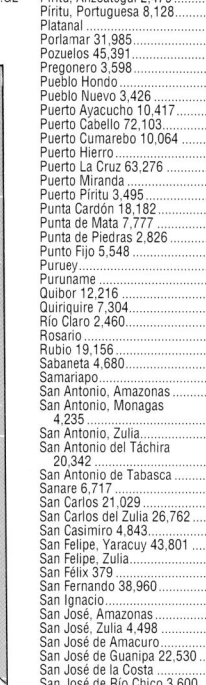

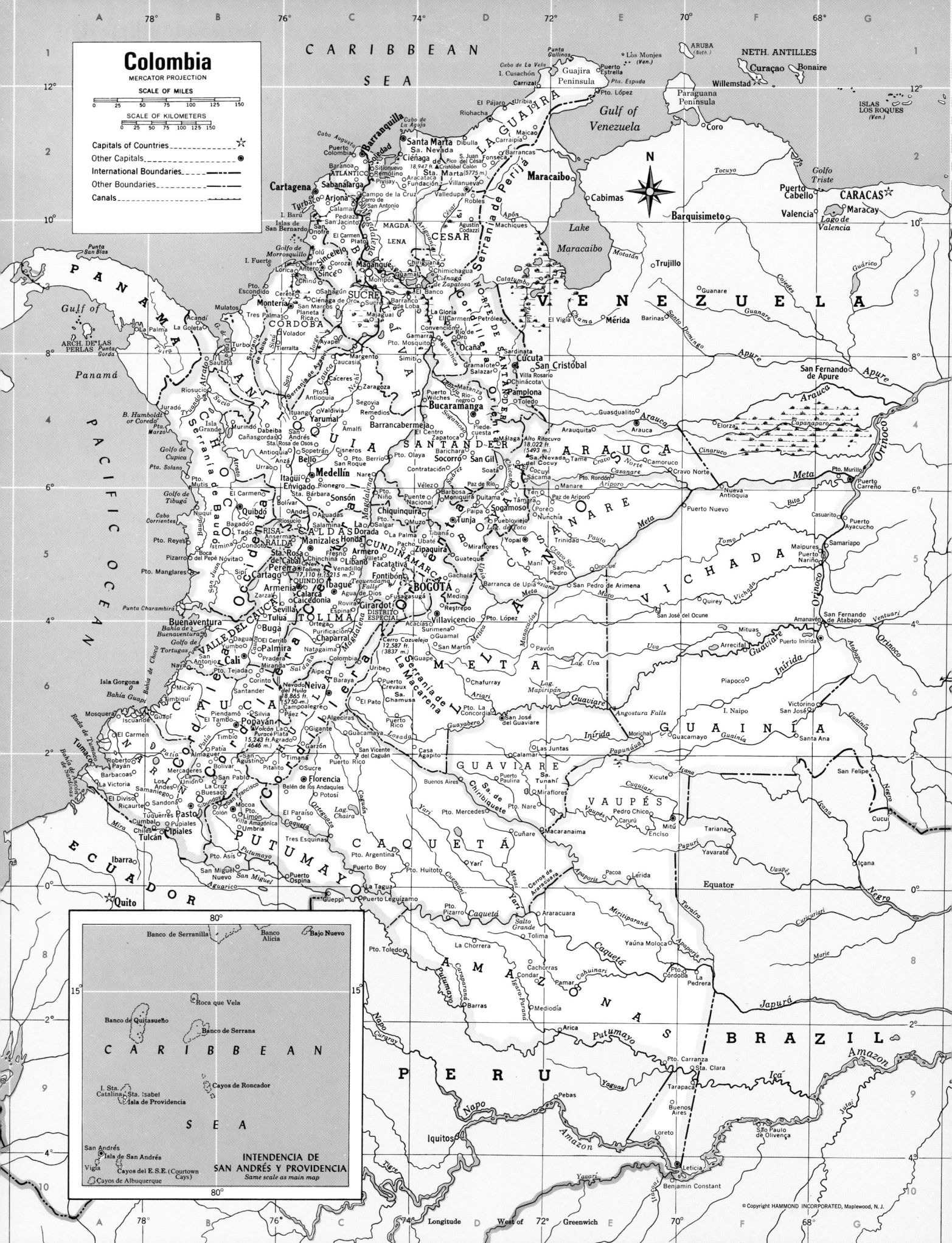

Colombia
MERCATOR PROJECTION

SCALE OF MILES

0 25 50 75 100 125 150

SCALE OF KILOMETERS

25 50 75 100 125 150

Capitals of Countries _____ ☆
Other Capitals _____ ◉
International Boundaries _____
Other Boundaries _____
Canals _____

INTENDENCIA DE
SAN ANDRÉS Y PROVIDENCIA
Same scale as main map

© Copyright HAMMOND INCORPORATED, Maplewood, N.J.

INTERNAL DIVISIONS

Amazonas (comm.) 6,825......D8
Antioquia (dept.) 3,888,067...B4
Arauca (inten.) 19,884...........E4
Atlántico (dept.) 958,560.......C2
Bolívar (dept.) 802,407..........C3
Boyacá (dept.) 1,084,766.......D5
Caldas (dept.) 700,954...........C5
Caquetá (inten.) 57,103..........C7
Casanare (inten.)....................E5
Cauca (dept.) 603,894............B6
Cesar (dept.) 339,843............D3
Chocó (dept.) 201,915............B4
Córdoba (dept.) 645,478.........C3
Cundinamarca
(dept.) 1,106,626...............C5
Distrito Especial 2,855,065....C5
Guainía (comm.) 1,792...........F6
Huila (dept.) 469,834.............C6
La Guajira (dept.) 180,520.....D2
Magdalena (dept.) 536,122.....C3
Meta (dept.) 245,176..............D6
Nariño (dept.) 807,112...........B7
Norte de Santander
(dept.) 693,298..................D3
Putumayo (inten.) 22,916.......C7
Quindío (dept.) 321,677.........C5
Risaralda (dept.) 452,626.......B5
San Andrés y Providencia
(inten.) 22,719...................B1
Santander (dept.) 1,130,977..D4
Sucre (dept.) 354,412............C3
Tolima (dept.) 903,520..........C5
Valle del Cauca
(dept.) 2,204,722...............B6
Vaupés (comm.) 6,923...........E7
Vichada (comm.) 2,172...........F5

CITIES and TOWNS

Acacías 9,238........................D6
Acandí 2,358..........................B3
Agrado 2,771..........................C6
Aguachica 16,771....................D3
Aguadas 9,995........................C5
Agua de Dios 9,689.................C5
Agustín Codazzi 28,194..........D3
Aipe 3,794..............................C6
Algeciras 5,022......................C6
Amalfi 6,494...........................C4
Andes 14,957..........................C5
Anserma 15,559......................B5
Antioquia 6,841......................B4
Aracataca 7,511......................D2
Arauca 7,613...........................E4
Arjona 29,465.........................C2
Armenia 180,221.....................B5
Armero 19,567........................C5
Ayapel 7,475...........................C3
Baranoa 27,394.......................C2
Baraya 2,581...........................C6
Barbacoas 4,653......................A7
Barbosa 7,960.........................D5
Barichara 2,548.......................D4
Barrancabermeja 137,406......C4
Barrancas 2,979......................D2
Barranco de Loba 2,215.........C3
Barranquilla 896,649..............C2
Belén de los Andaquíes 2,190 C7
Bello 206,297..........................C4
Bogotá (cap.) 3,974,813.........D5
Bolívar 13,259........................C5
Bucaramanga 341,513............D4
Buenaventura 160,342............B6
Buesaco 2,763........................B7
Buga 82,992...........................B6
Cáceres 7,154.........................C4
Caicedonia 21,959..................C5
Calamar 5,867........................C2
Calarcá 29,349........................C5
Cali 1,323,944........................B6
Campo de la Cruz 13,137.......C2
Campoalegre 11,799...............C6
Cañasgordas 3,900.................B4
Cartagena 491,368..................C2
Cartago 92,524.......................B5
Caucasia 24,138.....................C4
Cereté 25,890.........................C3
Cerro de San Antonio 3,394...C2
Chaparral 14,546....................C6
Chimichagua 6,382.................D3
Chinácota 4,478......................D4
Chinchiná 33,441....................C5
Chinú 10,023...........................C3
Chiquinquirá 21,727...............C5
Chiriguaná 6,611.....................D3
Ciénaga 56,860.......................C2
Ciénaga de Oro 10,607...........C3
Cisneros 7,226........................C4
Colombia 2,903.......................C6
Condoto 4,798.........................B5
Contratación 3,057.................D4
Convención 7,545...................D3
Corinto 6,933..........................B6
Corozal 29,471........................C3
Cúcuta 357,026.......................D4
Cumbal 2,891..........................B7
Dabeiba 7,600.........................B4
Dagua 5,392...........................B6
Duitama 56,390......................D5
El Banco 20,756......................D3
El Carmen 2,362.....................D3
El Carmen de Bolívar 30,778..C3
El Cerrito 23,575....................B6
El Cocuy 2,740.......................D4
El Tambo 2,179.......................B6
Envigado 85,539.....................C4
Espinal 37,563........................C5
Facatativá 44,331....................C5
Florencia 66,430.....................C7
Fonseca 9,988.........................D2
Fontibón..................................C5
Fresno 8,141...........................C5
Fundación 29,002....................C2
Fusagasugá 41,033..................C5
Gamarra 5,071........................D3
Garzón 13,783........................C6
Gigante 4,880.........................C6
Girardot 66,385......................C5
Gramalote 2,880.....................D4
Guamal, Meta 2,854...............D6
Guamal, Magdalena 4,986......C3
Guapí 5,005............................B6
Guateque 6,032......................D5
Honda 25,040.........................C5
Ibagué 269,495.......................C5
Ipiales 45,419.........................B7
Istmina 5,575.........................B5
Itagüí 135,797........................C4
Ituango 5,561.........................C4
La Cruz 4,353.........................B7
La Dorada 48,572...................C5
La Gloria 2,632......................D3
La Palma 5,430.......................C5
La Plata 8,047........................C6
La Unión 5,392.......................B7
Líbano 23,703.........................C5
Lorica 24,264.........................C3
Magangué 49,160...................C3
Maicao 46,033........................D2
Majagual 2,329.......................C3
Málaga 10,645........................D4
Manizales 275,067..................C5
Medellín 1,418,554.................C4
Mercaderes 3,877...................B7
Miraflores 3,584.....................D5
Miranda 6,439........................B6
Mitú 1,637..............................E7
Mocoa 6,221...........................B7
Mompós 14,076......................C3
Moniquirá 5,711.....................D5
Montería 157,466....................B3
Natagaima 7,772....................C6
Neiva 178,130.........................C6
Ocaña 51,443..........................D3
Ortega 5,150...........................C6
Pacho 6,786............................C5
Páez 2,098..............................C6
Paipa 4,260............................D5
Palmira 175,186......................B6
Pamplona 34,213....................D4
Pasto 197,407.........................B7
Patía 5,306.............................B6
Paz de Ariporo 2,584..............E5
Paz de Río 3,464.....................D4
Pereira 233,271......................C5
Piedecuesta 34,646.................D4
Piendamó 5,046......................B6
Pitalito 27,104.........................B7
Pivijay 10,172.........................C2
Planeta Rica 24,238................C3

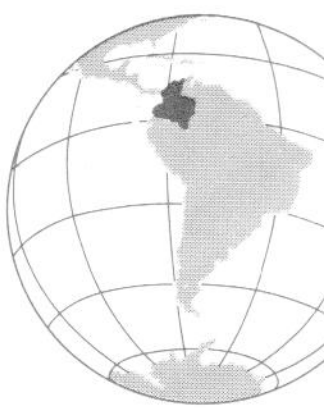

AREA 439,513 sq. mi. (1,138,339 sq. km.)
POPULATION 30,241,000
CAPITAL Bogotá
LARGEST CITY Bogotá
HIGHEST POINT Pico Cristóbal Colón 18,947 ft. (5,775 m.)
MONETARY UNIT Colombian peso
MAJOR LANGUAGE Spanish
MAJOR RELIGION Roman Catholicism

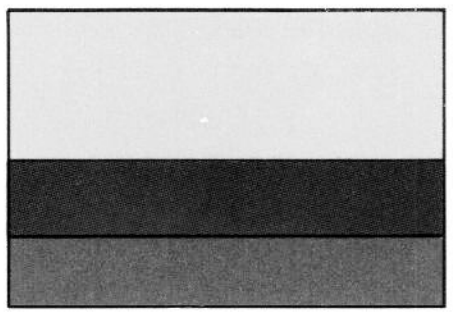

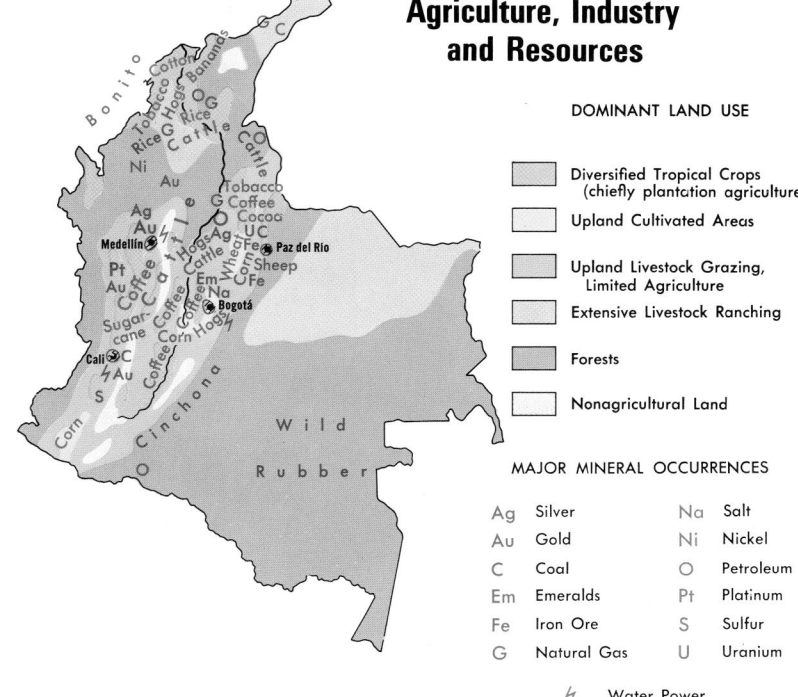

Agriculture, Industry and Resources

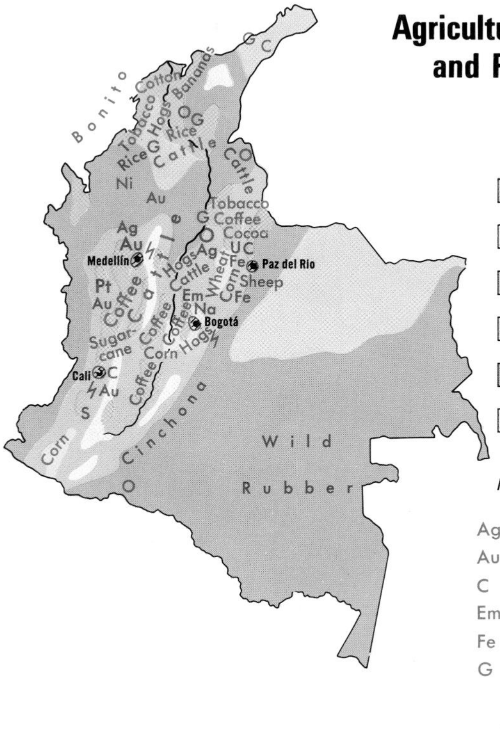

Bonito
Cotton
Tobacco
Hogs
Bananas
Rice
Cattle
Ni
Au
Ag
Medellín
Pt
Coffee
Cocoa
UC
Paz del Rio
Sheep
Em
Na
Bogotá
Sugar-cane
Corn
Hogs
Wheat
Cali
Cinchona
Wild
Rubber

DOMINANT LAND USE

- Diversified Tropical Crops (chiefly plantation agriculture)
- Upland Cultivated Areas
- Upland Livestock Grazing, Limited Agriculture
- Extensive Livestock Ranching
- Forests
- Nonagricultural Land

MAJOR MINERAL OCCURRENCES

Ag	Silver	Na	Salt
Au	Gold	Ni	Nickel
C	Coal	O	Petroleum
Em	Emeralds	Pt	Platinum
Fe	Iron Ore	S	Sulfur
G	Natural Gas	U	Uranium

Water Power

Major Industrial Areas

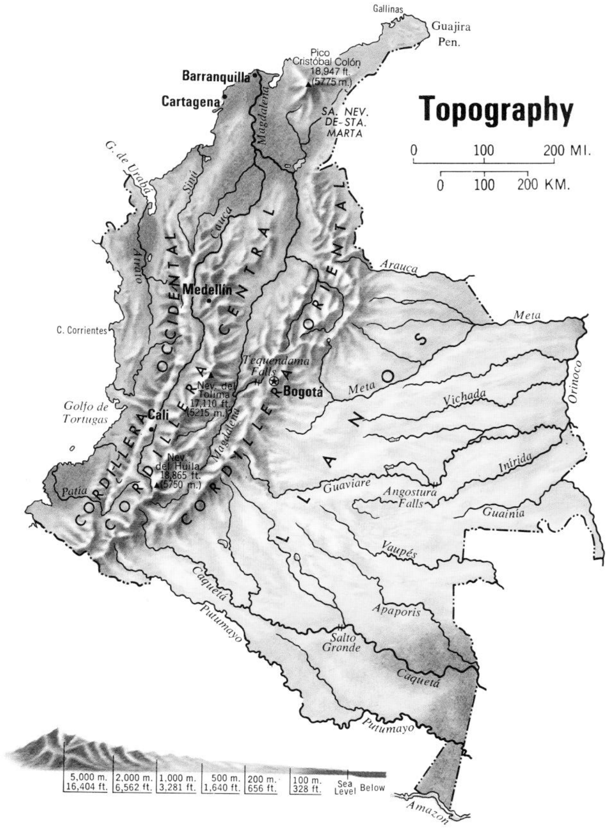

Topography
0 100 200 MI.
0 100 200 KM.
Gallinas
Guajira Pen.
Pico Cristóbal Colón 18,947 ft. (5,775 m.)
Barranquilla
Cartagena
SA. NEV. DE STA. MARTA
Magdalena
G. de Urabá
Sinú
Cauca
Atrato
Arauca
Medellín
C. Corrientes
Meta
Tequendama Falls
Nev. del Tolima 17,110 ft. (5,215 m.)
Bogotá
Golfo de Tortugas
Cali
Vichada
Orinoco
Nev. del Huila 18,865 ft. (5,750 m.)
Guaviare
Angostura Falls
Inírida
Guainía
Patía
Vaupés
Caquetá
Putumayo
Apaporis
Salto Grande
Amazon
5,000 m. 16,404 ft.
2,000 m. 6,562 ft.
1,000 m. 3,281 ft.
500 m. 1,640 ft.
200 m. 656 ft.
100 m. 328 ft.
Sea Level
Below

Plato 24,895..........................C3
Popayán 141,964....................B6
Pradera 27,152.......................B6
Puente Nacional 4,317...........D5
Puerto Asís 6,364...................B7
Puerto Berrío 21,414..............C4
Puerto Carreño 2,172.............G4
Puerto Colombia 9,255...........C2
Puerto Escondido 1,368..........B3
Puerto Inírida 1,792................F6
Puerto Leguízamo 3,179.........C8
Puerto López 4,948.................D5
Puerto Rico 4,853...................C7
Puerto Rondón 1,010...............E4
Puerto Salgar 6,396................C5
Puerto Tejada 26,573..............B6
Puerto Wilches 5,282..............D4
Pupiales 2,723........................B7
Purificación 8,164...................C6
Quibdó 47,950........................B5
Remedios 4,681......................C4
Remolino 3,408......................C2
Restrepo 2,704.......................D5
Río de Oro 2,985....................D3
Riohacha 46,667.....................D2
Rionegro, Antioquia 22,654....C4
Rionegro, Santander 3,491.....D4
Riosucio, Caldas 11,619..........C5
Riosucio, Chocó 2,184............B4
Robles 5,422...........................D2
Rovira 5,105...........................C5
Sabanalarga 35,786................C2
Sahagún 28,686......................C3
Salamina 12,136.....................C5
Salazar 2,791..........................D4
Samaniego 4,790....................B7
San Agustín 4,532...................B7
San Andrés, Antioquia 2,003..C4
San Andrés, San Andrés y
Providencia 23,325...............A9
San Antero 7,129....................C3
Sandoná 7,222........................B7
San Gil 24,599........................D4
San Jacinto 13,459..................C3
San José del Guaviare 4,138..D6
San Juan del César 9,468.......D2
San Marcos 26,542.................C3
San Martín 8,281....................D6
San Onofre 7,899....................C3
San Pablo 3,662......................B7
San Roque 4,972.....................C4
San Vicente del Caguán 3,182 C6
Santa Bárbara 11,848..............C5
Santa Marta 177,922...............D2
Santa Rosa de Cabal 37,112...C5
Santa Rosa de Osos 8,593......C4
Santander 22,644....................B6
Sardinata 3,726......................D3
Segovia 10,000.......................C4
Sevilla 31,309.........................C5
Sibundoy-Las Casas 2,853.....B7
Silvia 3,045............................B6
Simití 3,062............................C4
Sincé 11,909...........................C3
Sincelejo 120,537...................C3
Sitionuevo 5,919.....................C2
Soatá 4,294.............................D4
Socorro 15,596.......................D4
Sogamoso 64,437...................D5
Soledad 164,494.....................C2
Sonsón 15,990........................C5
Sopetrán 5,223.......................C4
Tadó 3,102..............................B5
Tame 4,811.............................E4
Tibaná 1,100...........................D5
Tierralta 7,950........................C3
Timaná 4,262..........................C7
Timbío 4,755...........................B6
Timbiquí 1,048........................B6
Toledo 2,942...........................D4
Tolú 9,118...............................C3
Trinidad 729............................E5
Tuluá 99,721...........................B5
Tumaco 45,456.......................A7
Tunjá 87,851...........................D5
Túquerres 12,058....................B7
Turbaco 28,161.......................C2
Turbo 25,992...........................B3
Ubaté 7,716............................D5
Uribia 2,193............................D2
Urrao 8,577............................B4
Valdivia 4,318.........................C4
Valledupar 142,771.................D2
Vélez 8,241.............................D4
Venadillo 8,383.......................C5
Villa Rosario 8,668..................D4
Villanueva 9,836.....................D2
Villavicencio 82,869................D5
Villeta 6,507............................C5
Yarumal 21,333.......................C4
Yopal 5,851.............................D5
Yumbo 43,508.........................B6
Zapatoca 6,258.......................D4
Zaragoza 9,660.......................C4
Zarzal 22,014..........................B5
Zipaquirá 45,676.....................D5

OTHER FEATURES

Aguarico (riv.)........................B7
Aguja, La (cape).....................C2
Alto Ritacuva (mt.).................D4
Amazon (riv.)..........................E9
Ancón de Sardinas (bay)........A7
Angostura (falls).....................E6
Apaporis (riv.).........................F8
Arauca (riv.)............................E4
Ariari (riv.)..............................D6
Ariguaní (riv.).........................D3
Ariporo (riv.)...........................E4
Atabapo (riv.)..........................G6
Atrato (riv.).............................B4
Baudó, Serranía de (mts.).......B5
Caguán (riv.)...........................C7
Cahuinarí (riv.)........................E8
Caquetá (riv.)..........................E8
Caraparaná (riv.).....................D8
Casanare (riv.).........................E4
Catatumbo (riv.)......................D3
Cauca (riv.).............................B3
Cazueleja, Cerro (mt.).............C6
César (riv.)..............................D2
Central, Cordillera (range)......C5
Charambirá (pt.)......................B5
Chicamocha (riv.)....................D4
Chocó (bay).............................B6
Corrientes (cape).....................B5
Cristóbal Colón, Pico (peak)...D2
Cuemaní (riv.).........................D7
Cupica (gulf)...........................B4
Cusachón (isl.)........................D1
Espada (pt.).............................E1
Gallinas (pt.)...........................E1
Grande (isl.)............................B4
Grande, Salto (falls)................D8
Guainía (riv.)...........................F6
Guajira (pen.)..........................E1
Guaviare (riv.).........................F6
Guayabero (riv.).......................D6
Huila, Nevado del (mt.)...........C6
Igara-Paraná (riv.)...................D8
Inírida (riv.)............................F6
Isana (riv.)..............................F7
La Aguja (cape).......................C2
La Macarena, Serranía de
(mts.)...................................D6
Llanos (plain)..........................D5
Macarena, Serranía de La
(mts.)...................................D6
Magdalena (riv.)......................C3
Manacacías (riv.).....................D6
Meta (riv.)...............................E5
Metica (riv.)............................D6
Mira (riv.)................................A7
Miritiparaná (riv.)....................E8
Morrosquillo (gulf)..................C3
Nechí (riv.)..............................C4
Negro (riv.)..............................G7
Occidental, Cordillera (range) B5
Oriental, Cordillera (range).....D5
Orinoco (riv.)..........................G5
Orteguaza (riv.).......................C7
Papurí (riv.).............................F7
Patía (riv.)...............................B6
Pauto (riv.)..............................E5
Perijá, Serranía de (mts.)........D2
Providencia (isl.).....................B9
Puracé (vol.)...........................B6
Putumayo (riv.).......................E9
Quitasueño (bank)...................A8
Roncador (cays).......................B9
Saldaña (riv.)..........................C6
Salto Grande (falls)..................D8
San Andrés (isl.)......................A10
San Jorge (riv.)........................C3
San Juan (riv.).........................B5
Santa Marta, Sierra Nevada de
(range).................................D2
Serrana (bank).........................B9
Serranilla (bank)......................B8
Sinú (riv.)................................B3
Sogamoso (riv.).......................D4
Suárez (riv.)............................D4
Taraira (riv.)...........................F8
Tequendama (falls)..................C5
Tibugá (gulf)...........................B5
Tolima (vol.)...........................C5
Tomo (riv.)..............................F5
Tortugas (gulf).........................B6
Tunahí, Sierra (mts.)................E7
Urabá (gulf).............................B3
Uva (riv.)................................E6
Vaupés (riv.)............................E7
Vela, La (cape)........................D1
Vichada (riv.)..........................F5
Yarí (riv.)................................D8
Zapatosa, Ciénaga de
(swamp)...............................D3

Peru and Ecuador

BIPOLAR OBLIQUE CONIC CONFORMAL PROJECTION

SCALE OF MILES
0   50   100   150   200

SCALE OF KILOMETERS
0   50   100   150   200

Capitals of Countries .... ☆
Other Capitals .... ◉
International Boundaries ___ ___ ___
Other Boundaries ___ . . ___ . . ___

Galápagos Islands
(Archipiélago de Colón)
(Ecuador)
Same scale as main map

PACIFIC OCEAN

I. Culpepper (Darwin)
I. Wenman (Wolf)
I. Pinta
I. Marchena
I. Genovesa
Pta. Albemarle
I. San Salvador (Santiago)
B. de Banks
I. Fernandina
Isla Isabela
Pinzón
I. Baltra
I. Sta. Cruz (Chaves)
Pto. Ayora
I. Sta. Fe
I.S. Cristobal
Baquerizo Moreno
El Progreso
Floreana (Sta. María)
I. Sta. María (Floreana)
I. Española

PROVINCES OF ECUADOR
INDICATED BY NUMBERS

1	Imbabura	C-2	5	Bolívar	C-3
2	Cotopaxi	C-3	6	Chimborazo	C-3
3	Tungurahua	C-3	7	Cañar	C-4
4	Los Ríos	C-3	8	El Oro	C-4

Copyright HAMMOND INCORPORATED, Maplewood, N.J.

Longitude 76° West of Greenwich

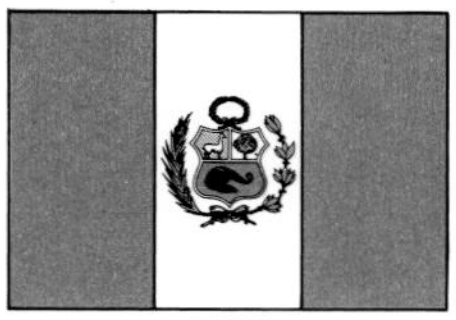
PERU

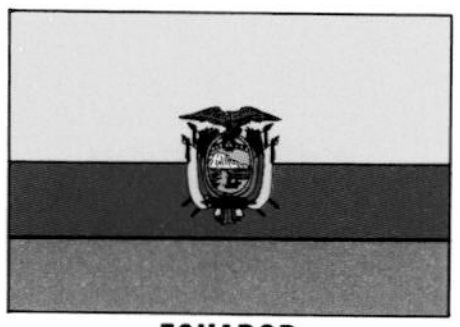
ECUADOR

PERU

AREA 496,222 sq. mi. (1,285,215 sq. km.)
POPULATION 22,332,000
CAPITAL Lima
LARGEST CITY Lima
HIGHEST POINT Huascarán 22,205 ft. (6,768 m.)
MONETARY UNIT inti
MAJOR LANGUAGES Spanish, Quechua, Aymara
MAJOR RELIGION Roman Catholicism

ECUADOR

AREA 109,483 sq. mi. (283,561 sq. km.)
POPULATION 10,490,000
CAPITAL Quito
LARGEST CITY Guayaquil
HIGHEST POINT Chimborazo 20,561 ft. (6,267 m.)
MONETARY UNIT sucre
MAJOR LANGUAGES Spanish, Quechua
MAJOR RELIGION Roman Catholicism

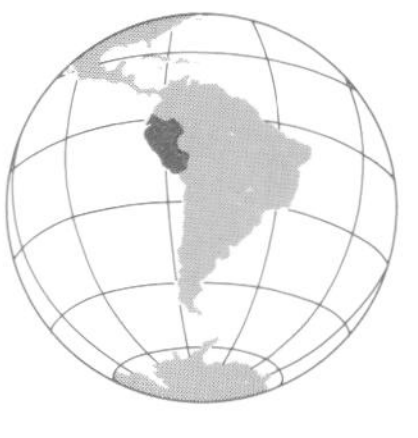

PERU

DEPARTMENTS

Amazonas 256,460 .... C5
Ancash 815,646 .... D7
Apurímac 321,936 .... F10
Arequipa 702,308 .... F10
Ayacucho 500,732 .... E9
Cajamarca 1,044,689 .... C6
Callao (prov.) 446,730 .... D9
Cusco 829,294 .... F9
Huánuco 481,924 .... D7
Huancavelica 346,460 .... E9
Ica 431,442 .... E10
Junín 848,993 .... E8
La Libertad 960,537 .... C6
Lambayeque 683,425 .... B6
Lima 4,738,266 .... D8
Loreto 446,316 .... E5
Madre de Dios 36,555 .... G8
Moquegua 99,287 .... G11
Pasco 221,219 .... E8
Piura 1,168,442 .... B5
Puno 893,586 .... G10
San Martín 319,670 .... D6
Tacna 133,240 .... G11
Tumbes 103,979 .... B4
Ucayali 200,085 .... E6

CITIES and TOWNS

Abancay 19,807 .... F9
Acarí 4,907 .... E10
Acobamba 2,156 .... E9
Acolla 5,717 .... E8
Acomayo, Cusco 1,419 .... G9
Acomayo, Huánuco 2,883 .... E7
Acora 1,910 .... H11
Acuracay 1,282 .... F5
Aija 1,843 .... D7
Alca 755 .... F10
Ambo 3,060 .... D8
Ananea 668 .... H10
Ancón 8,610 .... D8
Andahuaylas 7,654 .... F9
Anta 3,703 .... F9
Antabamba 2,223 .... F10
Aplao 1,941 .... F11
Aquia 970 .... D8
Arequipa 107,858 .... G11
Arequipa* 447,431 .... G11
Ascope 12,070 .... C6
Atalaya 2,132 .... E8
Atico 2,316 .... F11
Ayabaca 4,543 .... C5
Ayacucho 68,535 .... F9
Ayaviri 11,067 .... G10
Azángaro 7,658 .... H10
Bagua 9,735 .... C5
Bambamarca 6,867 .... C6
Barranca, Lima 21,312 .... C8
Barranca, Loreto 1,351 .... D5
Bellavista 4,906 .... C5
Bolívar 1,106 .... D6
Bretaña 1,035 .... E5
Buldibuyo 582 .... D7
Cabana 1,804 .... C7
Cailloma 1,187 .... G10
Cajabamba 7,282 .... C6
Cajamarca 60,280 .... C6
Cajatambo 1,721 .... D8
Calca 6,112 .... G9
Callalli 819 .... G10
Callao 260,581 .... D9
Callao* 441,374 .... D9
Camaná 11,386 .... F11
Candarave 1,207 .... G11
Cangallo 1,584 .... E9
Canta 3,431 .... D8
Capachica 307 .... H10
Caravelí 1,827 .... F10
Caraz 6,376 .... D7
Carhuás 3,147 .... D7
Carumás 1,031 .... G11
Cascas 2,638 .... C6
Casma 12,725 .... C7
Castrovirreyna 1,749 .... E9
Catacaos 30,927 .... B5
Celendín 8,538 .... D6
Cerro Azul 2,314 .... D9
Cerro de Pasco 71,558 .... D8
Chachapoyas 11,919 .... D6
Chala 1,646 .... E10
Chalhuanca 3,071 .... F10
Chancay 18,993 .... D8
Chepén 29,919 .... C6
Chicama 11,160 .... C6
Chiclayo 280,244 .... C6
Chilca (Pucusana) 3,329 .... D9
Chilete 2,537 .... C6
Chimbote 216,406 .... C7
Chincha Alta 37,475 .... D9
Chiquián 3,521 .... D8
Chirinos 1,061 .... C5
Chivay 3,296 .... G10
Chota 8,299 .... C6
Chulucanas 34,977 .... B5
Chupaca 5,422 .... E9
Chuquibamba 2,630 .... F10
Chuquibambilla 2,147 .... F9
Churín 1,801 .... D8
Cocachacra 5,985 .... G11
Cojata 888 .... H10
Colasay 721 .... C5
Colcamar 1,216 .... D6
Conaica 1,154 .... E9
Concepción 7,129 .... E8
Concordia 1,372 .... E5
Contamana 5,718 .... E6
Contumazá 2,491 .... C6
Coracora 4,598 .... F10
Córdova 453 .... E10
Corongo 1,762 .... D7
Cotahuasi 1,301 .... F10
Cusco (Cuzco) 85,044 .... F9
Cusco* 171,604 .... F9
Cutervo 6,890 .... C6
Cuyocuyo 1,101 .... H10
Desaguadero 2,682 .... H11
Deustua 544 .... G10
Dos de Mayo 574 .... E6
Echarate 1,071 .... F9
El Portugués C7
Esperanza 375 .... G7
Espinar 6,381 .... G10
Ferreñafe 22,200 .... C6
Fitzcarrald G8
Francisco de Orellana 445 .... F4
Guadalupe 7,613 .... E9
Huacho 43,402 .... D8
Huacrachuco 1,210 .... D7
Hualgayoc 1,691 .... C6
Hualla 4,042 .... F9
Huallanca, Ancash 930 .... D7
Huallanca, Huánuco 4,806 .... D7
Huamachuco 8,273 .... D6
Huancabamba 4,393 .... C5
Huancané 5,227 .... H10
Huancapi 2,539 .... E9
Huancavelica 20,889 .... E9
Huancayo 165,132 .... E9
Huanchaco 6,005 .... C7
Huanta 11,213 .... E9
Huánuco 52,628 .... E7
Huaral 34,235 .... D8
Huaráz 45,116 .... D7
Huari 2,344 .... D7
Huariaca 2,671 .... E8
Huarmey 11,094 .... C8
Huarochirí 1,828 .... D9
Huarocondo 2,498 .... F9
Huaura 9,338 .... D8
Huaylas 1,344 .... C7
Iberia 2,307 .... F5
Ica 111,087 .... E10
Ichuña 277 .... G11
Ilave 9,891 .... H11
Ilo 31,549 .... G11
Imperial 20,894 .... D9
Inambari H9
Iñapari 188 .... H8
Intutu 746 .... E4
Iparia 278 .... E7
Iquitos 173,629 .... F4
Jaén 24,356 .... C5
Jauja 14,630 .... E8
Jayanca 6,401 .... B6
Jeberos 1,493 .... D5
Juanjuí 9,324 .... D6
Juli 5,575 .... H11
Juliaca 77,976 .... G10
Jumbilla 1,035 .... C5
Junín 8,988 .... E8
Lagunas 4,601 .... E5
La Huaca 5,161 .... B5
La Jalca 1,769 .... D6
La Joya 5,000 .... G11
Lamas 8,937 .... D6
Lambayeque 23,746 .... B6
Lampa 4,319 .... G10
Lamud 2,405 .... C6
Lanlacuni Bajo 405 .... G9
La Oroya 33,305 .... D8
Las Piedras H9
Las Yaras 759 .... G11
La Tina B5
La Unión 2,828 .... D7
Leimebamba 1,957 .... D6
Lima (cap.) 375,957 .... D8
Lima* 3,968,972 .... D8
Limbani 728 .... H10
Lircay 5,213 .... E9
Llata 2,922 .... D7
Lobitos 2,975 .... B5
Lurín 14,405 .... D9
Machupicchu 544 .... F9
Macusani 3,389 .... G10
Madre de Dios 660 .... G9
Manú 234 .... G9
Máncora 5,358 .... B5
Marcapata 369 .... G9
Marcona 25,962 .... E10
Margos 1,622 .... D8
Masisea 1,586 .... E7
Matarani F11
Matucana 4,196 .... D8
Mazocruz 1,580 .... H11
Mendoza 1,902 .... D6
Moho 2,560 .... H10
Mollendo 21,206 .... F11
Monsefú 17,186 .... C6
Moquegua 21,488 .... G11
Morales 4,370 .... D6
Morococha 11,234 .... D8
Morropón 7,611 .... C5
Motupe 3,411 .... C6
Moyobamba 14,319 .... D6
Nauta 4,083 .... F5
Nazca 22,756 .... E10
Negritos 12,476 .... B5
Nuñoa 3,613 .... G10
Ocoña 1,062 .... F11
Ocros 1,037 .... D8
Ollachea 1,308 .... G9
Ollantaytambo 1,500 .... F9
Olmos 7,946 .... C5
Omate 1,131 .... G11
Orcotuna 3,359 .... E8
Orellana 1,550 .... E6
Otuzco 5,765 .... C6
Oxapampa 5,233 .... E8
Oyón 6,279 .... D8
Pacasmayo 17,588 .... C6
Pachiza 889 .... D6
Paiján 12,699 .... C6
Paita 18,749 .... B5
Palpa 3,393 .... E10
Pampacolca 2,010 .... F10
Pampas 3,850 .... E9
Panao 1,363 .... E7
Paruro 1,727 .... F9
Paucarbama 534 .... E9
Paucartambo, Cusco 1,620 .... G9
Paucartambo, Pasco 3,497 .... E8
Pevas 1,347 .... G4
Picota 2,288 .... D6
Pimentel 9,129 .... B6
Pisac 1,566 .... G9
Pisco 53,414 .... D9
Piura 186,354 .... B5
Pomabamba 2,489 .... D7
Pucallpa 91,953 .... E7
Pucará 2,268 .... G10
Pucaurco 628 .... G4
Pucusana 3,329 .... D9
Puerto Bermúdez 1,133 .... E8
Puerto Chicama 3,630 .... C6
Puerto Eten 2,575 .... B6
Puerto Inca 1,286 .... E7
Puerto Maldonado 12,609 .... H9
Puerto Ocopa 1,088 .... E8
Puerto Samanco 1,435 .... C7
Puno 66,477 .... G10
Punta de Bombón 4,647 .... F11
Puquina 1,026 .... G11
Putina 5,414 .... H10
Querecotillo 10,637 .... B5
Quillabamba 16,837 .... F9
Ramón Castilla 1,811 .... G5
Recuay 2,764 .... D7
Requena 8,270 .... F5
Rioja 9,876 .... D6
Salaverry 5,539 .... C7
Saña 40,144 .... C6
Sandia 1,682 .... H10
San José 4,070 .... B6
San José de Sisa 3,782 .... D6
San Miguel, Ayacucho 1,440 .... F9
San Miguel, Cajamarca 1,798 .... C6
San Pedro de Lloc 11,463 .... C6
San Ramón 7,145 .... E8
Santa 20,490 .... C7
Santa Clotilde 1,068 .... E4
Santa Cruz, Cajamarca 2,739 .... C6
Santa Cruz, Loreto 449 .... E5
Santiago 5,092 .... E10
Santiago de Cao 22,119 .... C6
Santiago de Chuco 5,189 .... C7
Santo Tomás, Amazonas 1,093 .... C6
Santo Tomás, Cusco 2,755 .... G10
San Vicente de Cañete 15,277 .... D9
Saposoa 4,541 .... D6
Saquena 2,755 .... F5
Satipo 9,208 .... E8
Sauce 2,263 .... D6
Sayán 5,129 .... D8
Sechura 11,724 .... B5
Sicuani 21,176 .... G10
Sihuas 2,178 .... D7
Sullana 80,947 .... B5
Sumbilca 1,155 .... D8
Supe 10,061 .... D8
Tacna 92,862 .... G11
Tahuamanu 2,619 .... H8
Talara 55,722 .... B5
Tambo de Mora 2,790 .... D9
Tambo Grande 10,087 .... B5
Tamshiyacu 2,040 .... F5
Tarapoto 33,429 .... D6
Tarata 2,624 .... H11
Tarma 34,369 .... E8

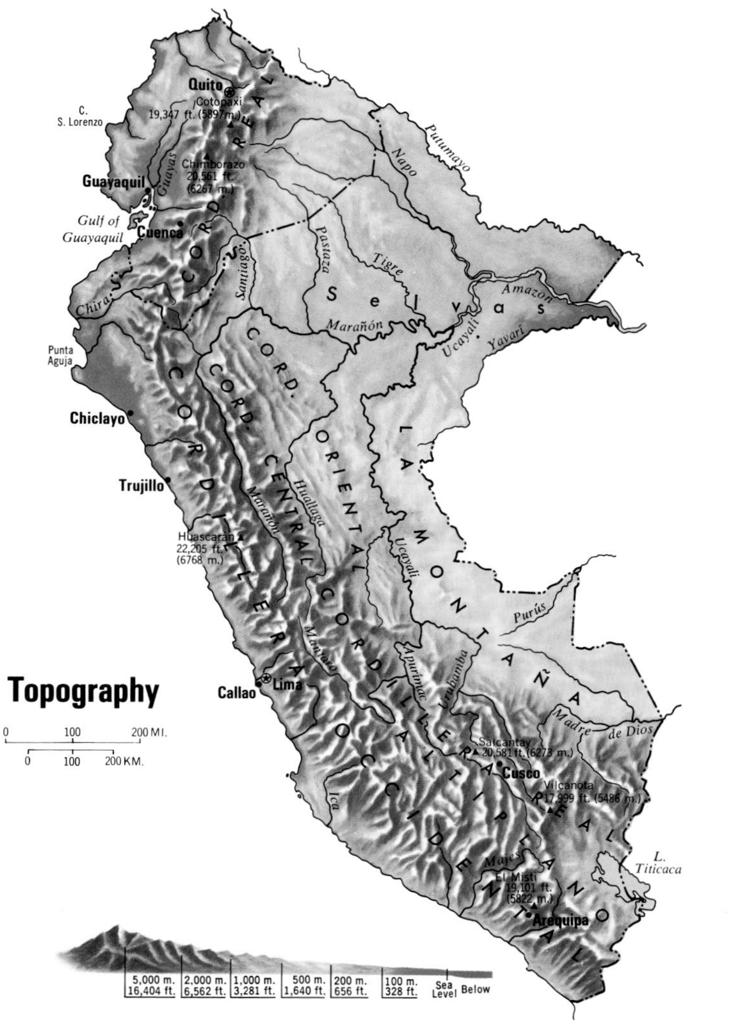
Quito
Cotopaxi
19,347 ft. (5897 m.)
C. S. Lorenzo
Guayaquil
Chimborazo
20,561 ft.
(6267 m.)
Gulf of Guayaquil
Cuenca
Putumayo
Napo
Pastaza
Tigre
Amazon
Marañón
Ucayali
Yavari
Selvas
Santiago
Chira
Punta Aguja
Chiclayo
Trujillo
CORD. OCCIDENTAL
CORD. CENTRAL
CORD. ORIENTAL
Huallaga
Huascarán
22,205 ft.
(6768 m.)
LA MONTAÑA
Purús
Callao
Lima
Mantaro
Apurímac
Urubamba
Madre de Dios
Salcantay
20,581 ft. (6273 m.)
Cusco
Vilcanota
17,999 ft. (5486 m.)
L. Titicaca
Ica
Majes
El Misti
19,101 ft.
(5822 m.)
Arequipa
Topography
0 100 200 MI.
0 100 200 KM.
5,000 m. 16,404 ft.
2,000 m. 6,562 ft.
1,000 m. 3,281 ft.
500 m. 1,640 ft.
200 m. 656 ft.
100 m. 328 ft.
Sea Level
Below

Tayabamba 1,649 .... D7
Tingo María 25,030 .... D7
Tocache 5,940 .... D7
Torata 6,320 .... G11
Trujillo 354,557 .... C7
Tumbes 48,187 .... B4
Uchiza 2,471 .... D7
Urcos 4,155 .... G9
Urubamba 4,686 .... F9
Virú 6,587 .... C7
Yambrasbamba 277 .... D5
Yanahuanca 5,109 .... D8
Yanaoca 1,152 .... G10
Yauca 1,805 .... E10
Yauli 1,020 .... D8
Yauri (Espinar) 6,381 .... G10
Yauyos 1,296 .... E9
Yunguyo 7,253 .... H11
Yurimaguas 22,858 .... E5
Zarumilla 9,713 .... B4
Zorritos 4,497 .... B4

OTHER FEATURES

Acarí (riv.) E10
Aguaytía (riv.) E7
Aguja (pt.) B5
Amazon (riv.) F4
Andes, Cordillera de los (mts.) F10
Apurímac (riv.) F9
Azángaro (riv.) G10
Azul, Cordillera (range) E7
Blanca, Cordillera (range) D7
Blanco (cape) B5
Boquerón, El (pass) E7
Cañete (riv.) D9
Chimbote (bay) C7
Chincha (isls.) D9
Chira (riv.) B5
Cóndor, Cordillera del (range) C5
Coropuna, Nudo (mt.) F10
Corrientes (riv.) E4
Ene (riv.) E8
Ferrol (pt.) C7
Grande (riv.) E10
Guañape (isls.) C7
Heath (riv.) H9
Huallaga (riv.) D5
Huasaga (riv.) D4
Huascarán (mt.) D7
Huayabamba (riv.) D6
Ica (riv.) E10
Inambari (riv.) H9
Independencia (bay) D10
Independencia (isl.) D10
Junín (lake) E8
Juruá (riv.) F7
Lobos de Afuera (isls.) B6
Lobos de Tierra (isl.) B6
Locumba (riv.) G11
Madre de Dios (riv.) G9
Majes (riv.) F1
Mantaro (riv.) E8
Manú (riv.) G8
Marañón (riv.) E5
Mayo (riv.) D6
Misti, El (mt.) G11
Montaña, La (reg.) F8
Morona (riv.) D5
Nanay (riv.) E4
Napo (riv.) F4
Negra, Cordillera (range) D7
Negra (pt.) B6
Nermete (pt.) B5
Occidental, Cordillera (range) F10
Ocoña (riv.) F11
Oriental, Cordillera (range) H10
Pachitea (riv.) E7
Paita (bay) B5
Pampas (riv.) E9
Paracas (pt.) D9
Pariñas (pt.) B5
Parinacochas (lake) F10
Pastaza (riv.) D5
Pativilca (riv.) D8
Perené (riv.) E8
Piedras, Las (riv.) G8
Pisco (bay) D9
Pisco (riv.) D9
Piura (riv.) B5
Purús (riv.) G8
Putumayo (riv.) G4
Rímac (riv.) D9
Salcantay (mt.) F9
Sama (riv.) G11
San Gallán (isl.) D9
San Lorenzo (isl.) D9
San Nicolás (bay) E10
Santa (riv.) C7
Santiago (riv.) D4
Sechura (bay) B5
Tambo (riv.) G11
Tapiche (riv.) E6
Tigre (riv.) E4
Titicaca (lake) H10
Tumbes (riv.) B4
Ucayali (riv.) F5
Urubamba (riv.) F8

(continued on following page)

Agriculture, Industry and Resources

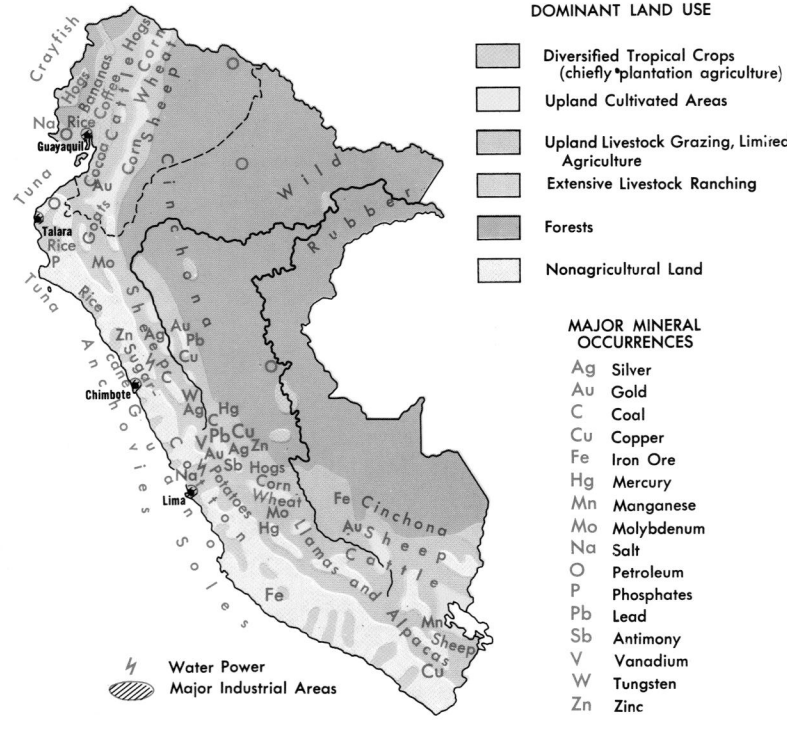

DOMINANT LAND USE

Diversified Tropical Crops
(chiefly plantation agriculture)

Upland Cultivated Areas

Upland Livestock Grazing, Limited
Agriculture

Extensive Livestock Ranching

Forests

Nonagricultural Land

MAJOR MINERAL
OCCURRENCES

Ag Silver
Au Gold
C   Coal
Cu Copper
Fe Iron Ore
Hg Mercury
Mn Manganese
Mo Molybdenum
Na Salt
O   Petroleum
P   Phosphates
Pb Lead
Sb Antimony
V   Vanadium
W   Tungsten
Zn Zinc

⚡ Water Power
▨ Major Industrial Areas

Agriculture, Industry and Resources

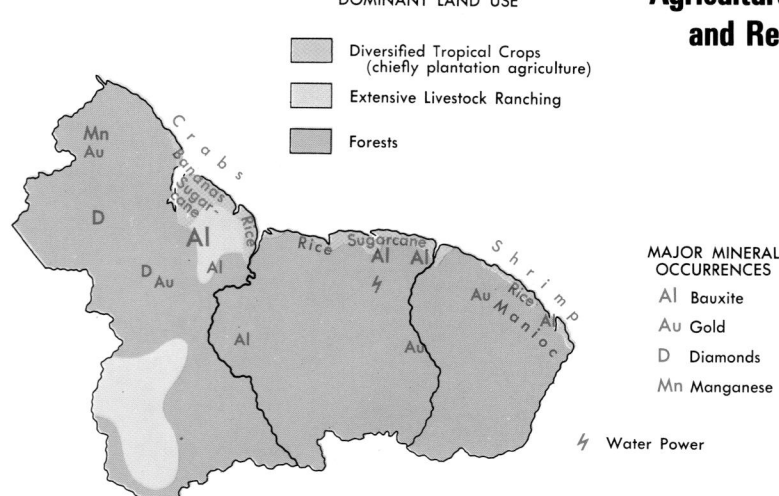

DOMINANT LAND USE

Diversified Tropical Crops
(chiefly plantation agriculture)

Extensive Livestock Ranching

Forests

MAJOR MINERAL
OCCURRENCES

Al Bauxite
Au Gold
D   Diamonds
Mn Manganese

⚡ Water Power

GUYANA

AREA 83,000 sq. mi. (214,970 sq. km.)
POPULATION 1,024,000
CAPITAL Georgetown
LARGEST CITY Georgetown
HIGHEST POINT Mt. Roraima 9,094 ft. (2,772 m.)
MONETARY UNIT Guyana dollar
MAJOR LANGUAGES English, Hindi
MAJOR RELIGIONS Christianity, Hinduism, Islam

SURINAME

AREA 55,144 sq. mi. (142,823 sq. km.)
POPULATION 400,000
CAPITAL Paramaribo
LARGEST CITY Paramaribo
HIGHEST POINT Julianatop 4,200 ft. (1,280 m.)
MONETARY UNIT Suriname guilder
MAJOR LANGUAGES Dutch, Hindi, Indonesian
MAJOR RELIGIONS Christianity, Islam, Hinduism

FRENCH GUIANA

AREA 35,135 sq. mi. (91,000 sq. km.)
POPULATION 90,000
CAPITAL Cayenne
LARGEST CITY Cayenne
HIGHEST POINT 2,723 ft. (830 m.)
MONETARY UNIT French franc
MAJOR LANGUAGE French
MAJOR RELIGIONS Roman Catholicism, Protestantism

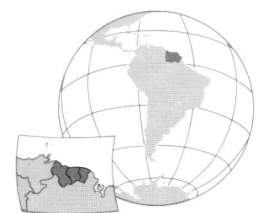

Kamaria (falls)	B2
Kuyuwini (riv.)	B4
Kwitaro (riv.)	B4
Leguan (isl.)	B2
Mazaruni (riv.)	A2
Moruka (riv.)	B2
New (riv.)	C4
Pakaraima (mts.)	A3
Pomeroon (riv.)	B2
Potaro (riv.)	B3
Puruni (riv.)	B2
Roraima (mt.)	A3
Rupununi (riv.)	B4
Takutu (riv.)	B4
Venamo (mt.)	A3
Waini (riv.)	B2
Wenamu (riv.)	A2

Lelydorp 300	D3
Mariënburg 3,500	D2
Moengo 2,100	D3
Nieuw-Amsterdam 1,400	D2
Nieuw-Nickerie 34,480	C2
Paramaribo (cap.) ⊙67,905	D2
Paranam	D3
Totness 1,300	C3
Uitkijk	D3
Wageningen 800	C3
Zanderij	D3

OTHER FEATURES

Bakhuys (mts.)	C3
Coeroeni (riv.)	C4
Commewijne (riv.)	D3
Coppename (riv.)	C3
Corantijn (riv.)	C3
Cottica (riv.)	D3
Eilerts de Haan (mts.)	C4
Frederik Willem IV (falls)	C4
Julianatop	C4
Kutari (riv.)	C4
Lely (mts.)	D3
Litani (riv.)	D3
Marowijne (riv.)	D3
Nickerie (riv.)	C3
Orange (mts.)	D4
Saramacca (riv.)	C3
Sipaliwini (riv.)	C4
Suriname (riv.)	D3
Tapanahoni (riv.)	D4

SURINAME

DISTRICTS

Brokopondo 17,763	D4
Commewijne 18,740	D3
Coronie 3,251	C3
Marowijne 25,911	D4
Nickerie 35,178	C3
Para 16,635	D3
Paramaribo 102,297	D2
Saramacca 13,554	C3
Suriname 151,585	D3

CITIES and TOWNS

Albina 1,000	D3
Brokopondo	D3
Calcutta 1,100	C3
Domburg 1,200	D3
Groningen 600	D2

*City and suburbs
⊙ Population of sub-district or division.
⊛ Population of district

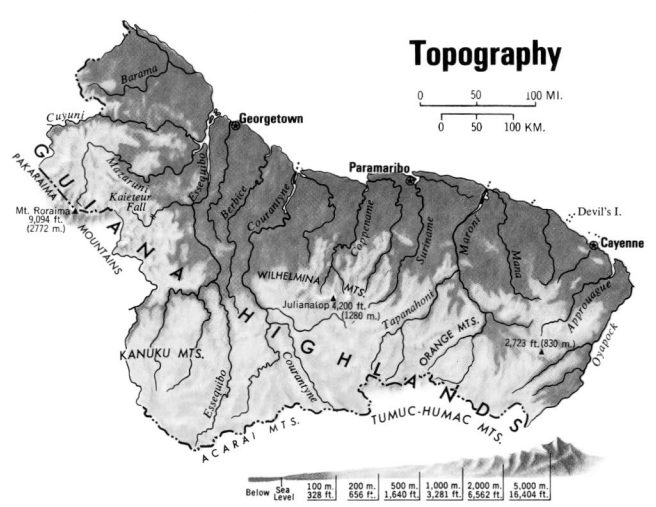

Topography

0 50 100 MI.
0 50 100 KM.

Below Sea Level / 100 m. 328 ft. / 200 m. 656 ft. / 500 m. 1,640 ft. / 1,000 m. 3,281 ft. / 2,000 m. 6,562 ft. / 5,000 m. 16,404 ft.

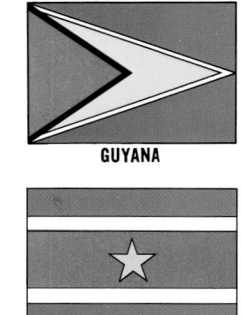

GUYANA

SURINAME

FRENCH GUIANA

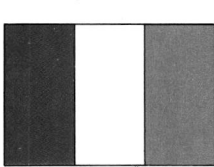

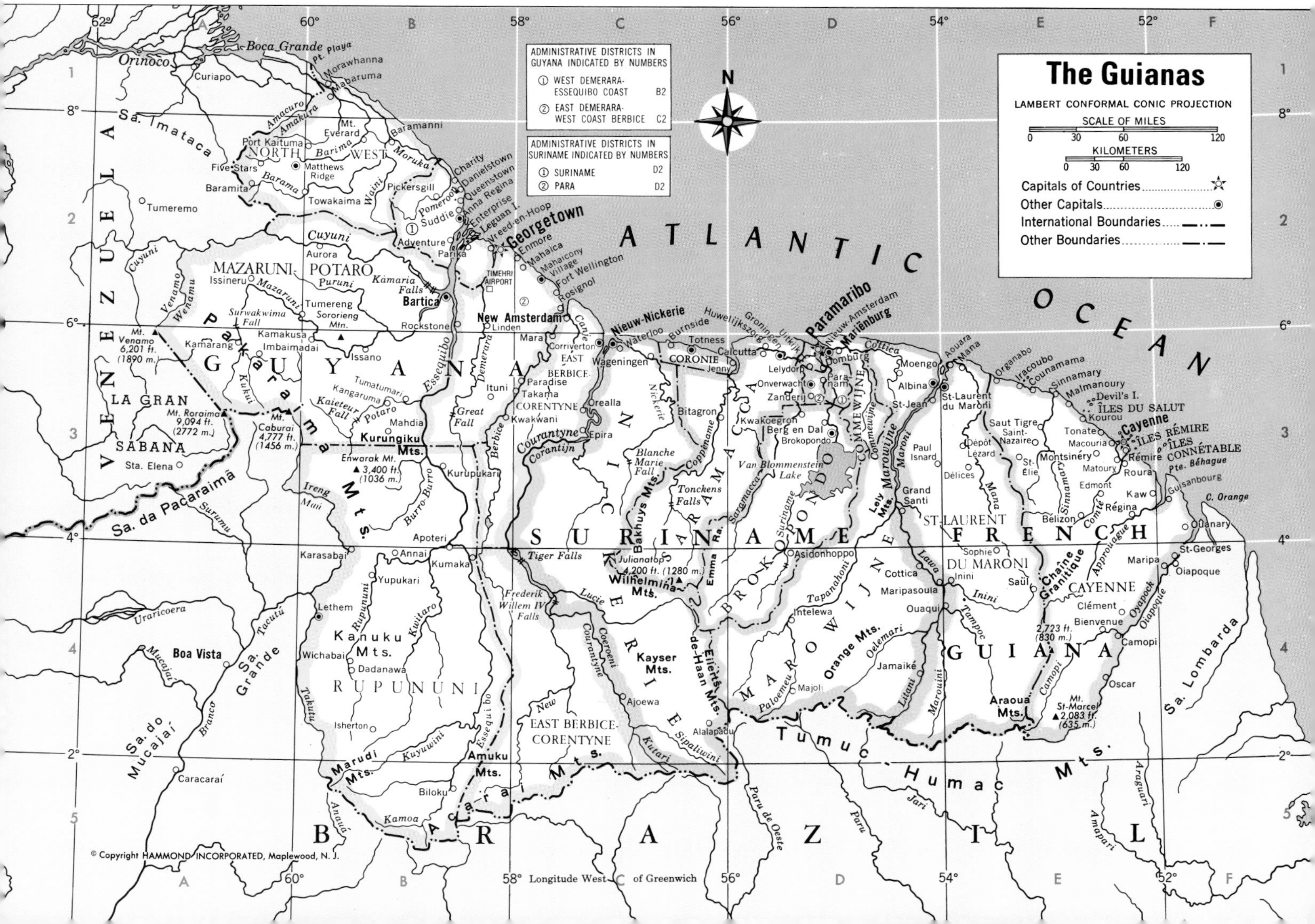

The Guianas

LAMBERT CONFORMAL CONIC PROJECTION

SCALE OF MILES
0 30 60 120

KILOMETERS
0 30 60 120

Capitals of Countries .... ☆
Other Capitals .... ⊙
International Boundaries .... —··—
Other Boundaries .... —·—

ADMINISTRATIVE DISTRICTS IN GUYANA INDICATED BY NUMBERS
① WEST DEMERARA-ESSEQUIBO COAST B2
② EAST DEMERARA-WEST COAST BERBICE C2

ADMINISTRATIVE DISTRICTS IN SURINAME INDICATED BY NUMBERS
① SURINAME D2
② PARA D2

© Copyright HAMMOND INCORPORATED, Maplewood, N.J.

STATES and TERRITORIES

Acre 301,605 . . . . . G10
Alagoas 1,987,581 . . . . . G5
Amapá (terr.) 175,634 . . . . . D2
Amazonas 1,432,066 . . . . . G9
Bahia 9,474,263 . . . . . F6
Ceará 5,294,876 . . . . . G4
Espirito Santo 2,023,821 . . . F7
Federal District 1,177,393 . . . E6
Goiás 3,865,482 . . . . . D6
Maranhão 4,002,599 . . . . . E4
Mato Grosso 1,141,661 . . . . . B6
Mato Grosso do Sul
1,370,333 . . . . . C7
Minas Gerais 13,390,805 . . . E7
Pará 3,411,868 . . . . . C4
Paralba 2,772,600 . . . . . G4
Paraná 7,630,466 . . . . . D9
Pernambuco 6,147,102 . . . . . G5
Piaul 2,140,066 . . . . . F4
Rio de Janeiro 11,297,327 . . . F8
Rio Grande do Norte
1,899,720 . . . . . G4
Rio Grande do Sul
7,777,212 . . . . . C10
Rondônia 492,810 . . . . . H10
Roraima (terr.) 79,153 . . . . . H8
Santa Catarina 3,628,751 . . . D9
São Paulo 25,040,698 . . . . . D8
Sergipe 1,141,834 . . . . . G5
Tocantins D5

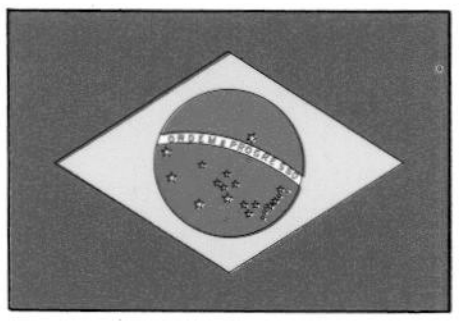

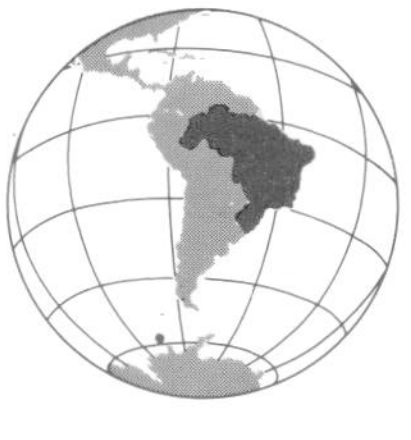

AREA 3,284,426 sq. mi. (8,506,663 sq. km.)
POPULATION 150,368,000
CAPITAL Brasília
LARGEST CITY São Paulo (greater)
HIGHEST POINT Pico da Neblina 9,889 ft. (3,014 m.)
MONETARY UNIT cruzado
MAJOR LANGUAGE Portuguese
MAJOR RELIGION Roman Catholicism

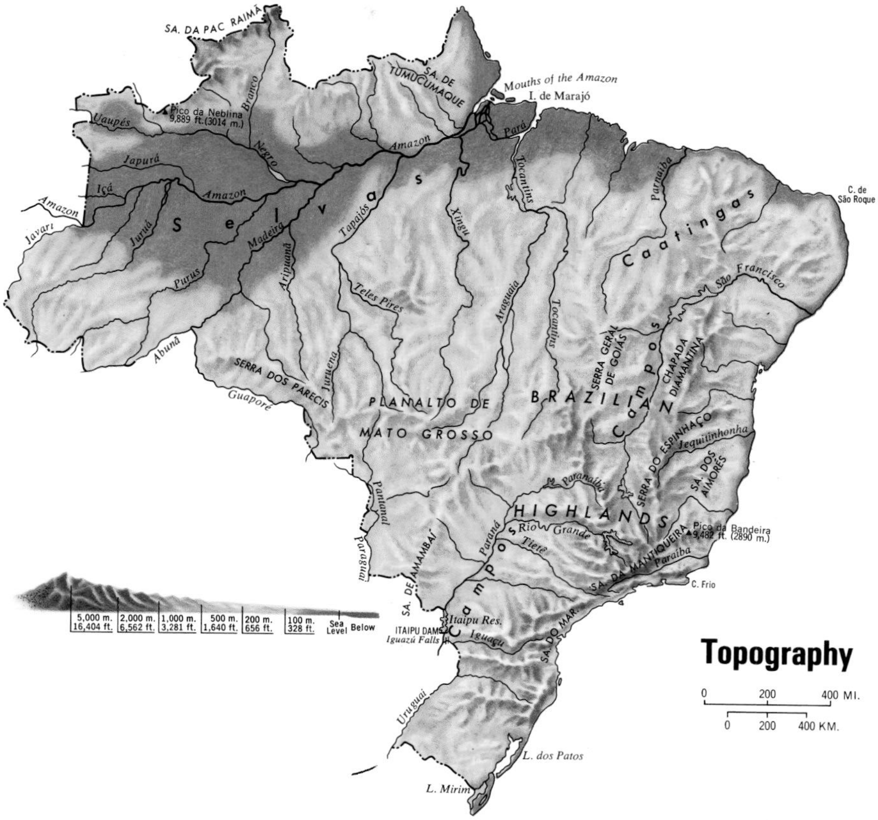

Topography

CITIES and TOWNS

Abaeté 12,861 . . . . . E7
Abaetetuba 33,031 . . . . . D3
Acaraú 7,144 . . . . . F3
Acopiara 10,747 . . . . . G4
Açu 20,544 . . . . . G4
Agudos 18,790 . . . . . *B3
Alagoa Grande 14,204 . . . . . H4
Alagoinhas 76,377 . . . . . G6
Alcobaça 3,430 . . . . . G7
Alegre 9,441 . . . . . *F2
Alegrete 54,786 . . . . . B10
Além Paralba 23,028 . . . . . *E2
Alenquer 16,477 . . . . . C3
Alfenas 31,815 . . . . . *D2
Altamira 24,846 . . . . . C3
Altos 13,621 . . . . . F4
Amambal 12,507 . . . . . C8
Amapá 2,676 . . . . . D2
Amarante 6,848 . . . . . F4
Amargosa 11,118 . . . . . F6
Americana 121,794 . . . . . *C3
Amparo 26,970 . . . . . *C3
Anápolis 160,520 . . . . . D7
Anchieta 5,741 . . . . . F8
Andaral 2,476 . . . . . F6
Andradina 42,036 . . . . . D8
Andrelândia 8,737 . . . . . *D2
Angra dos Reis 24,894 . . . . *D3
Antonina 11,950 . . . . . *B4
Aparecida 27,265 . . . . . *D3
Apiai 7,809 . . . . . *B4
Aquidauana 21,514 . . . . . C8
Aracaju 288,106 . . . . . G5
Aracati 20,282 . . . . . G4
Araçatuba 113,486 . . . . . *A2
Araçuaí 12,292 . . . . . F7
Araquari 73,302 . . . . . D7
Araranquá 22,468 . . . . . D10
Araraquara 77,202 . . . . . *B2
Araras 54,323 . . . . . *C3
Araxá 51,339 . . . . . E7
Arcoverde 40,646 . . . . . G5
Areia Branca 12,979 . . . . . G4
Assis 57,217 . . . . . *A3
Avaré 40,716 . . . . . *B3
Bacabai 43,229 . . . . . E4
Bagé 66,743 . . . . . C10
Bahia (Salvador) 1,496,276 . . G6
Baixo Guandu 13,714 . . . . . F7
Balsas 13,566 . . . . . E4
Bambui 14,172 . . . . . *C2
Barão de Cocais 11,950 . . . . *E1
Barbacena 69,675 . . . . . *E2
Bariri 15,372 . . . . . *B3
Barra 10,809 . . . . . F5
Barra do Corda 19,280 . . . . . E4
Barra do Piraí 51,214 . . . . . *E3
Barra Mansa 123,421 . . . . . *D3
Barras 8,904 . . . . . F4
Barreiras 30,355 . . . . . E6
Barreiros 19,419 . . . . . H5
Barretos 65,294 . . . . . *B2
Batatais 30,478 . . . . . *C2
Baturité 12,388 . . . . . G4
Bauru 178,861 . . . . . *B3
Bebedouro 39,070 . . . . . *B2
Bela Vista 11,936 . . . . . C8
Belém 758,117 . . . . . E3
Belém †1,000,349 . . . . . E3
Belo Horizonte 1,442,483 . . *D1
Belo Horizonte †2,541,788 *D1
Benjamin Constant 6,563 . . . G9
Bento Gonçalves 40,323 . . . C10
Betim 71,599 . . . . . *D2
Bicas 8,611 . . . . . *E2
Birigui 45,348 . . . . . *A2
Blumenau 144,819 . . . . . D9
Boa Esperanca 17,394 . . . . *D2
Boa Vista 43,131 . . . . . H8
Bocaiúva 16,616 . . . . . E7
Bom Conselho 13,196 . . . . . G5
Bom Despacho 22,941 . . . . *D1
Bom Jesus da Lapa 19,978 . . F6
Bom Sucesso 10,331 . . . . . *D2
Borba 5,366 . . . . . H9
Bragança Paulista 61,021 . . *C3
Brasiléia 4,835 . . . . . G10
Brasília (cap.) 411,305 . . . . . E6
Brasilia de Minas 10,171 . . . . . F7
Brejo 5,859 . . . . . F3
Breves 31,452 . . . . . D3
Brumado 24,663 . . . . . F6
Brusque 37,898 . . . . . D9
Cabedolo 18,581 . . . . . H4
Cabo Frio 40,668 . . . . . *F3
Caçador 25,287 . . . . . D9
Caçapava 45,258 . . . . . *D3
Caçapava do Sul 15,180 . . . . C10
Cáceres 33,472 . . . . . B7
Cachoeira 11,520 . . . . . G6
Cachoeira do Sul 59,967 . . . C10
Cachoeiro de Itapemirim
84,994 . . . . . G8
Caeté 23,331 . . . . . *E1
Caetité 8,823 . . . . . F6
Caiaponia 9,358 . . . . . C7
Caicó 30,777 . . . . . G4
Cajazeiras 30,834 . . . . . G4
Cajuru 9,670 . . . . . *C2
Camaquá 28,078 . . . . . C10
Cambará 13,218 . . . . . *A3
Cambui 8,552 . . . . . *C3
Cametá 15,539 . . . . . D3
Camocim 19,921 . . . . . F3
Campina Grande 222,229 . . . G4
Campinas 566,517 . . . . . *C3
Campo Belo 30,392 . . . . . *D2
Campo Formoso 10,324 . . . . . F5
Campo Grand 282,844 . . . . . C8
Campo Largo 34,506 . . . . . *B4
Campo Major 24,009 . . . . . F4
Campos 174,218 . . . . . *F2
Cananéia 5,581 . . . . . *C4
Canavieiras 14,076 . . . . . G6
Canindé 18,573 . . . . . G4
Canoas 214,115 . . . . . D10
Canoinhas 25,880 . . . . . D9
Capanema 28,272 . . . . . E3
Capão Bonito 24,081 . . . . . *B4
Caraguatatuba 22,932 . . . . . *D3
Carangola 15,621 . . . . . *E2
Caratinga 39,621 . . . . . *E1
Caravelas 3,704 . . . . . G7
Carazinho 41,913 . . . . . C10
Carolina 10,136 . . . . . E4
Caruaru 137,636 . . . . . G5
Casa Banca 13,739 . . . . . *C2
Cascavel 16,238 . . . . . G4
Cássia 10,701 . . . . . *C2
Castanhal 51,797 . . . . . E3
Castelo 9,162 . . . . . F8
Castro 21,079 . . . . . *B4
Castro Alves 11,286 . . . . . G6
Cataguases 40,659 . . . . . *E2
Catalão 30,516 . . . . . E7
Catanduva 64,813 . . . . . *B2
Catolé do Rocha 12,165 . . . . . G4
Caxambu 16,221 . . . . . *D2
Caxias 56,755 . . . . . F4
Caxias do Sul 198,824 . . . . . D10
Ceará (Fortaleza) 648,815 . . . G3
Ceará-Mirim 17,097 . . . . . H4
Ceres 13,671 . . . . . D6
Chapecó 53,198 . . . . . C9
Coari 14,841 . . . . . H9
Codajás 4,923 . . . . . H9
Codó 11,593 . . . . . E4
Colatina 61,057 . . . . . F7
Conceição do Araguaia
18,143 . . . . . D5
Concórdia 17,973 . . . . . D9
Conselheiro Lafaiete 66,262 E2
Corinto 17,056 . . . . . E7
Cornélio Procópio 31,201 . . . D8
Coroatá 16,070 . . . . . F3
coromandel 11,604 . . . . . F7
Corumbá 66,014 . . . . . B7
Coxim 14,876 . . . . . C7
Crateús 29,905 . . . . . F4
Crato 49,244 . . . . . G4
Criciúma 74,003 . . . . . D10
Cristalina 10,521 . . . . . E7
Cruz Alta 53,315 . . . . . C10
Cruzeiro 55,175 . . . . . *D3
Cruzeiro do Sul 11,189 . . . . . G10
Cubatão 78,327 . . . . . *C3
Cuiabá 167,894 . . . . . C6
Curitiba 843,733 . . . . . *B4
Curitiba †1,441,743 . . . . . *B4
Currais Novos 25,663 . . . . . G4
Cururupu 10,358 . . . . . E3
Curvelo 37,734 . . . . . E7
Diamantina 20,197 . . . . . F7
Divinópolis 108,344 . . . . . *D2
Dois Córregos 11,811 . . . . . *B3
Dom Pedrito 25,773 . . . . . C10
Dores do Indaiá 13,058 . . . . . E7
Dourados 76,838 . . . . . C8
Duque de Caxias 306,057 . . *E3
Erexim 46,927 . . . . . C9
Esperanca 12,964 . . . . . G4
Esplanada 9,822 . . . . . G5
Estancia 28,250 . . . . . G5
Feira de Santana 225,003 . . . G5
Fernandópolis 39,737 . . . . . *A2
Floriano 35,761 . . . . . F4
Florianópolis 153,547 . . . . . E9
Fonte Boa 3,278 . . . . . G9
Formiga 36,681 . . . . . *D2
Formosa 29,304 . . . . . E6
Fortaleza 648,815 . . . . . G3
Fortaleza †1,581,588 . . . . . G3
Foz do Iguacu 93,619 . . . . . C9
Franca 143,630 . . . . . *C2
Frutal 22,955 . . . . . *B2
Garanhuns 64,854 . . . . . G5
Garca 26,527 . . . . . *B3
Goiana 30,108 . . . . . H4
Goiânia 703,263 . . . . . D7
Goiás 15,768 . . . . . D6
Governador Valadares
173,699 . . . . . F7
Grajaú 11,147 . . . . . E4
Guacul 12,715 . . . . . *F2
Guajará-Mirim 19,992 . . . . . H10
Guarapuava 17,189 . . . . . C9
Guarantiguetá 68,370 . . . . . *D3
Guarujá 67,730 . . . . . *C4
Guarulhos 395,117 . . . . . *C3
Guaxupé 23,637 . . . . . *C2
Guirantinga 8,981 . . . . . C7
Gurupi 27,39 . . . . . D5
Humaitá 10,004 . . . . . H10
Ibaiti 11,352 . . . . . *A3
Ibiá 11,161 . . . . . E7
Ibicaraí 18,202 . . . . . G6
Ibitinga 23,359 . . . . . *B2
Icó 13,007 . . . . . G4
Igarapava 15,342 . . . . . C2
Igarapé-Miri 12,172 . . . . . D3
Iguape 16,827 . . . . . *C4
Iguatu 39,611 . . . . . G4
Ijui 51,925 . . . . . C10
Ilhéus 71,240 . . . . . G6
Imbituba 9,998 . . . . . D10
Imperatiz 111,818 . . . . . E4
Inhumas 23,455 . . . . . D7
Ipameri 14,163 . . . . . E7
Ipu 12,787 . . . . . F4
Irati 21,956 . . . . . *A4
Itabaiana, Paralba 17,843 . . . H4
Itabaiana, Sergipe 26,055 . . . G5
Itaberaba 27,590 . . . . . F6
Itabira 57,691 . . . . . F7
Itabirito 22,978 . . . . . *E2
Itabuna 129,938 . . . . . G6
Itacoatiara 26,737 . . . . . B3
Itaituba 19,644 . . . . . C4
Itajai 78,867 . . . . . D9
Itajubá 53,506 . . . . . *D3
Itanhaem 26,181 . . . . . C4
Itapecerica 10,234 . . . . . *D2
Itapecuru-Mirim 12,216 . . . . . F3
Itapemirim 16,829 . . . . . F8
Itaperuna 34,644 . . . . . *F2
Itapetinga 36,897 . . . . . G6
Itapetininga 61,344 . . . . . *B3
Itapeva 36,551 . . . . . *B3
Itapipoca 19,463 . . . . . G3
Itapira 36,308 . . . . . *C3
Itápolis 13,750 . . . . . *B2
Itaporanga 8,988 . . . . . G4
Itaqui 23,136 . . . . . B10
Itararé 24,368 . . . . . *B4
Itatiba 35,537 . . . . . *C3
Itaúna 49,372 . . . . . *D2
Itu 62,211 . . . . . *C3
Ituacu 1,749 . . . . . F6
Ituiutaba 65,178 . . . . . D7
Itumbiara 56,602 . . . . . D7
Iturama 12,363 . . . . . *A1
Ituverava 21,323 . . . . . *C2
Jaboatao 67,129 . . . . . H5
Jaboticabal 40,276 . . . . . *B2
Jacarel 103,652 . . . . . *D3
Jacarezinho 23,684 . . . . . *A3
Jacobina 26,723 . . . . . F5
Jacupiranga 7,044 . . . . . B4
Jaguaquara 11,336 . . . . . F6
Jaguarao 18,165 . . . . . C11
Jaguariaiva 8,566 . . . . . *B4
Januária 20,484 . . . . . E6
Jatai 40,957 . . . . . D7
Jaú 59,522 . . . . . *B3
Jequié 84,792 . . . . . F6
Jequitinhonha 10,900 . . . . . F7
Ji-Paraná 31,724 . . . . . H10
Joacaba 16,195 . . . . . D9
Joao Pessoa 290,424 . . . . . H4
Joao Pinheiro 17,013 . . . . . E7
Joinvile 217,074 . . . . . D9
Juazeiro 60,940 . . . . . G5
Juazeiro do Norte 125,248 . . F4
Juiz de Fora 299,728 . . . . . *E2
Jundiai 210,015 . . . . . *C3
Lages 108,768 . . . . . D9
Laguna 27,743 . . . . . D10
Lambari 9,722 . . . . . *D2
Lapa 13,314 . . . . . D9
Laranjeiras do Sul 19,329 . . . C9
Lavras 35,345 . . . . . *D2
Leme 40,155 . . . . . *C3
Leopoldina 28,554 . . . . . *E2
Limeira 137,812 . . . . . *C3
Limoeiro 36,088 . . . . . H4
Limoeiro do Norte 13,112 . . . G4
Linhares 51,575 . . . . . F7
Lins 44,633 . . . . . *B2
Londrina 258,054 . . . . . D8
Lorena 51,276 . . . . . *D3
Luis Correia 3,576 . . . . . F3
Luz 10,068 . . . . . *D1
Luziania 67,284 . . . . . E7
Macaé 39,644 . . . . . *F3
Macalba 17,036 . . . . . H4
Macapá 89,081 . . . . . D2
Macau 17,543 . . . . . G4
Maceio 376,479 . . . . . H5
Machado 16,164 . . . . . *C2
Mafra 26,226 . . . . . D9
Magé 37,597 . . . . . *E3
Mamanguape 16,321 . . . . . H4
Manacapuru 17,016 . . . . . H9
Manaus 613,068 . . . . . H9
Manhuacu 22,678 . . . . . *E2
Manhumirim 11,085 . . . . . *E2
Manicoré 9,532 . . . . . H9
Marabá 41,564 . . . . . D4
Maracaju 9,699 . . . . . C8
Maragogipe 13,512 . . . . . G6
Maranguape 20,098 . . . . . G3
Marechal Deodoro 9,400 . . . . H5
Mariana 11,785 . . . . . *E2
Marilla 103,904 . . . . . *A3
Maringá 158,047 . . . . . D8
Mata de São João 23,741 . . . G6
Mato Grosso (Vila Bela da
Santissima Trindade)
1,401 . . . . . B6
Maués 10,846 . . . . . B3
Mineiros 16,844 . . . . . C7
Miracema 15,545 . . . . . *E2
Miracema do Norte D5
Mirassol 25,173 . . . . . *B2
Mococa 33,682 . . . . . *C2
Mogi das Cruzes 122,265 . . *C3
Mogi-Mirim 41,827 . . . . . *C3
Monte Alegre 10,646 . . . . . C3
Monte Aprazivel 9,767 . . . . . *A2
Monteiro 11,051 . . . . . G4
Montenegro 27,246 . . . . . D10
Montes Claros 151,881 . . . . . E7
Morrinhos 20,154 . . . . . D7
Mossoró 118,007 . . . . . G4
Muriaé 50,040 . . . . . *E2
Muzambinho 8,803 . . . . . *C2
Nanuque 34,445 . . . . . F7
Natal 376,552 . . . . . H4
Nazaré 18,068 . . . . . G6
Niquelandia 8,828 . . . . . D6
Niterói 386,185 . . . . . *E3
Nova Cruz 12,824 . . . . . H4
Nova Era 11,126 . . . . . *E1
Nova Friburgo 88,943 . . . . . *E3
Nova Iguaçu 491,802 . . . . . *E3
Nova Lima 35,035 . . . . . *E2
Nova Russas 10,021 . . . . . F4
Novo Hamburgo 132,066 . . . D10
Novo Horizonte 18,439 . . . . *B2
Óbidos 17,143 . . . . . C3
Oeiras 12,406 . . . . . F4
Olimpia 24,376 . . . . . *B2
Olinda 266.392 . . . . . H4
Oliveira 22,642 . . . . . *D2
Oriximiná 12,078 . . . . . C3
Orlândia 22,924 . . . . . *C2
Osasco 376,689 . . . . . *C3
Ourinhos 52,698 . . . . . *B3
Ouro Preto 27,821 . . . . . *E2
Palmares 40,624 . . . . . H5
Palmas 15,823 . . . . . C9
Palmeira 11,521 . . . . . *B4
Palmeira das Missões
23,943 . . . . . C9
Pará (Belém) 758,117 . . . . . E3
Paracatu 29,911 . . . . . E7
Pará de Minas 37,127 . . . . *D1
Paraguaçu Paulista
17,399 . . . . . D8
Paralba do Sul 13,510 . . . . *E3
Paranalba 21,305 . . . . . D7
Paranaguá 68,366 . . . . . *B4
Parati 8,684 . . . . . *D3
Parintins 29,369 . . . . . B3
Parnalba 78,718 . . . . . F3
Passo Fundo 103,121 . . . . . D10
Passos 56,998 . . . . . *C2
Patos 58,735 . . . . . G4
Patos de Minas 59,896 . . . . . E7
Patrocínio 29,520 . . . . . E7
Pau dos Ferros 12,985 . . . . . G4
Paulo Afonso 62,066 . . . . . G5
Pederneiras 18,864 . . . . . *B3
Pedra Azul 13,615 . . . . . F6
Pedreiras 30,843 . . . . . E4
Pedro Segundo 9,693 . . . . . F4
Pelotas 197,092 . . . . . C10
Penápolis 32,168 . . . . . *A2
Penedo 27,064 . . . . . G5
Pernambuco (Recife)
1,184,215 . . . . . H5
Petrolina 73,436 . . . . . G5
Petrópolis 149,427 . . . . . *E3
Picos 33,098 . . . . . F4
Piedade 13,054 . . . . . *C3
Pilar 14,778 . . . . . H5
Pindamonhangaba 51,174 . *D3

(continued on following page)

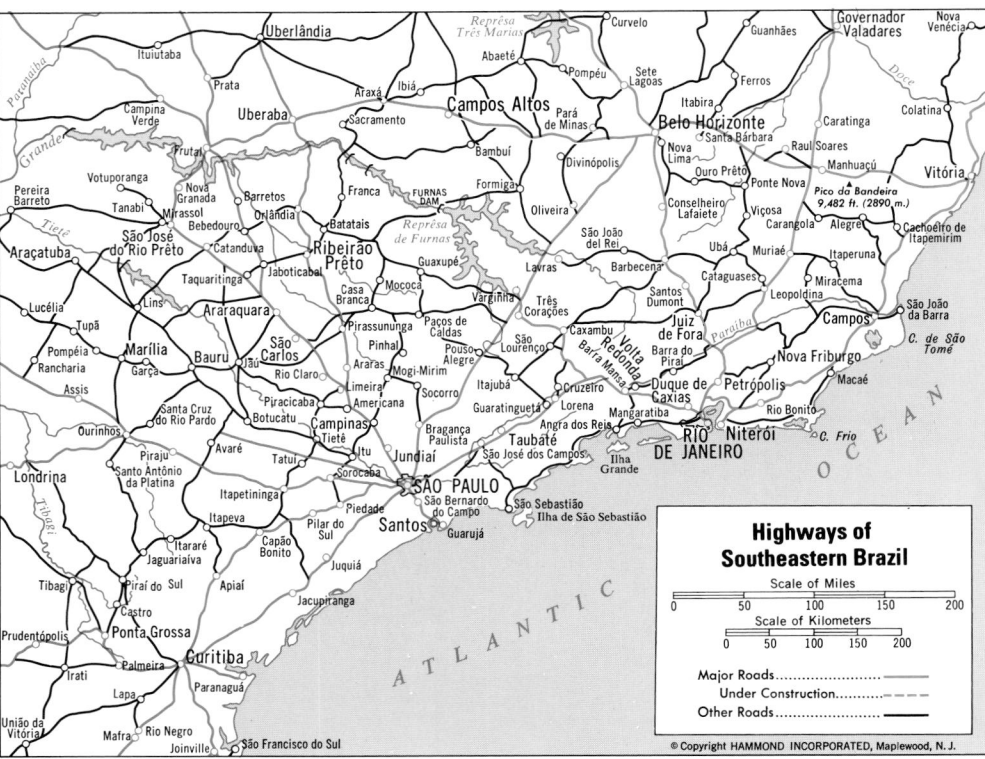

Highways of
Southeastern Brazil

Scale of Miles
0   50   100   150   200

Scale of Kilometers
0   50   100   150   200

Major Roads ....
Under Construction............
Other Roads ....

© Copyright HAMMOND INCORPORATED, Maplewood, N.J.

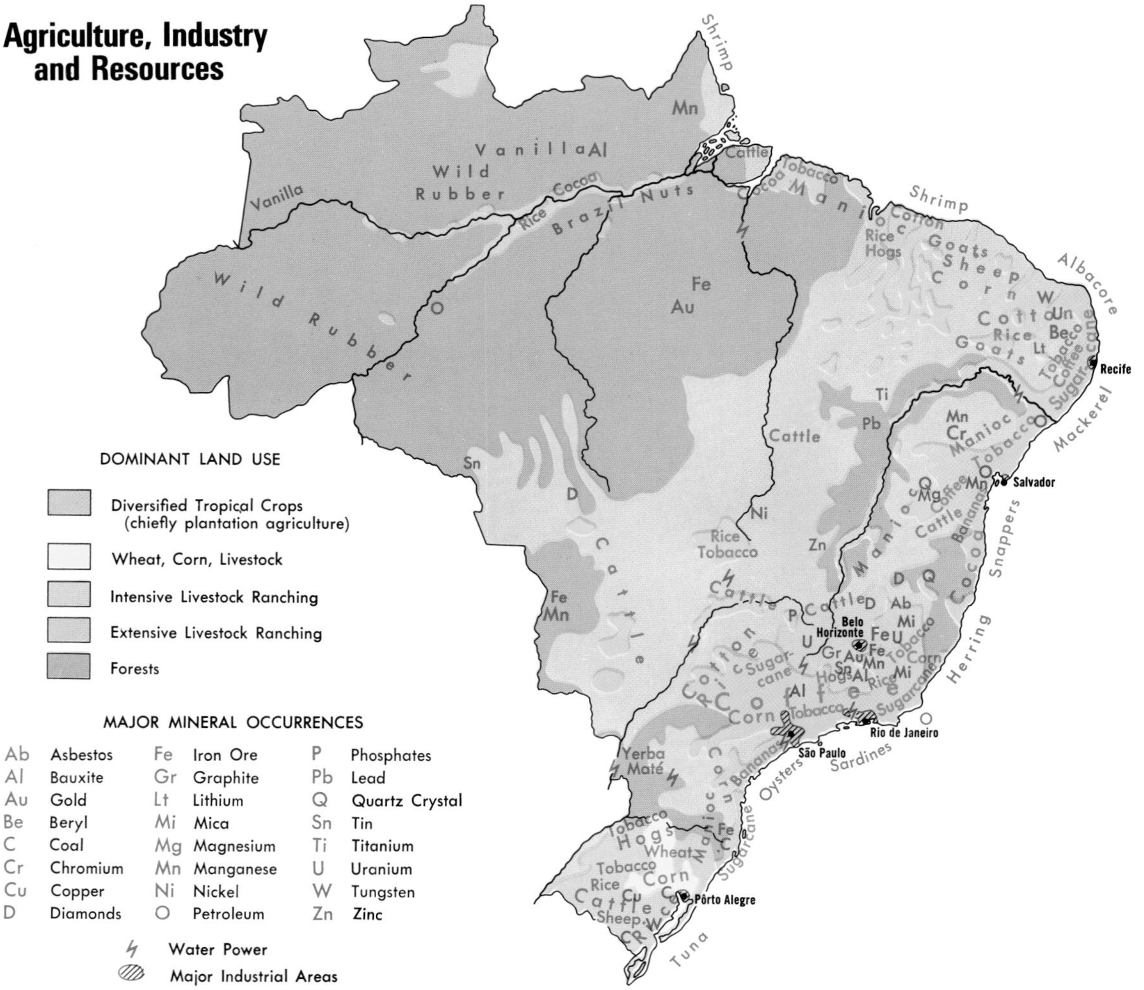

Agriculture, Industry and Resources

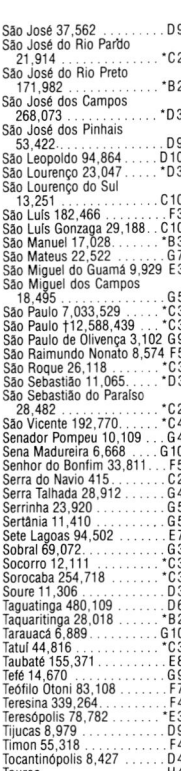

DOMINANT LAND USE

Diversified Tropical Crops
(chiefly plantation agriculture)

Wheat, Corn, Livestock

Intensive Livestock Ranching

Extensive Livestock Ranching

Forests

MAJOR MINERAL OCCURRENCES

Ab	Asbestos	Fe	Iron Ore	P	Phosphates
Al	Bauxite	Gr	Graphite	Pb	Lead
Au	Gold	Lt	Lithium	Q	Quartz Crystal
Be	Beryl	Mi	Mica	Sn	Tin
C	Coal	Mg	Magnesium	Ti	Titanium
Cr	Chromium	Mn	Manganese	U	Uranium
Cu	Copper	Ni	Nickel	W	Tungsten
D	Diamonds	O	Petroleum	Zn	Zinc

⚡ Water Power

▨ Major Industrial Areas

Três Corações 36,179 ..... *D2
Três Lagoas 45,171 ...... C8
Três Pontas 24,225 ...... *D2
Três Rios 47,497 .... *E3
Trindade 22,321 .... D7
Tubarão 64,585 .... D10
Tucuruí 27,209 .... D3
Tupã 44,450 .... *A2
Tupanciretã 13,103 ...... C10
Tutóia 4,766 .... F3
Ubá 43,080 .... *E2
Ubaitaba 9,413 .... G6
Ubatuba 23,078 .... *D3
Uberaba 180,296 .... *C1
Uberlândia 230,400 .... E7
Unaí 28,148 .... E7
União 9,396 .... F4
União da Vitória 22,682 .... D9
União dos Palmares 20,876 . H5
Uruaçu 19,607 .... D6
Uruçuí 6,047 .... E4
Uruguaiana 79,059 .... B10
Vacaria 37,370 .... D10
Valença 34,231 .... *E3
Varginha 57,448 .... *D2
Viana 9,753 .... E3
Viçosa 9,843 .... G5
Viçosa 29,198 .... *E2
Vigia 14,749 .... E3
Vila Velha Argolas 74,166 ... F8
Vilhena 12,565 .... H10
Viscondé dos Rio Branco 17,295 .... *E2
Vitória 144,143 .... G8
Vitória da Conquista 125,717 F6
Vitória de Santo Antão 62,890 .... G4
Volta Redonda 177,772 ... *D3
Votuporanga 44,169 ...... *B2
Xapuri 3,122 .... G10
Xique-Xique 17,625 .... F5

OTHER FEATURES

Abacaxis (riv.) B4
Abunã (riv.) G10
Acaraí, Serra do (range) B2
Acre (riv.) G10
Aiama (lake) H9
Amambaí, Serra de (range) . C7
Amapari (riv.) C2
Amazon (riv.) C3
Anauá (riv.) B2
Aporé (riv.) D7
Araguaia (riv.) D4
Araguari (riv.) D2
Araruama (lake) *E3
Arinos (riv.) B5
Aripuanã (riv.) A4
Armando Laydner (res.) ... *B3
Bailique (isl.) D2
Balsas (riv.) E5
Bananal (isl.) D5
Bandeira, Pico da (mt.) *E2, F8
Braço Maior do Araguaia (riv.) D5
Braço Menor do Araguaia (riv.) D6
Branco (riv.) H8
Buzios (cape) *F3
Canumã (riv.) B4
Capim (riv.) D3
Carajás, Serra dos (range) . D4
Cardoso (isl.) *C4
Cassiporé (cape) D2
Caviana (isl.) D2
Chavantes, Serra dos (range) D5
Claro (riv.) D7
Comprida (isl.) *C4
Cuiabá (riv.) B7
Culuene (riv.) C6
Curuá (riv.) C4
Doce (riv.) *E2, F7
Dois Irmãos, Serra (range) .. F5
Espigão Mestre (Geral de Goiás) (range) E6
Espinhaço, Serra do (range) . F7
Estrondo, Serra do (range) .. D4
Feia (lake) *F3
Feio (riv.) *B2
Formosa, Serra (range) C5
Frio (cape) *F3
Furnas (dam) *C2
Geral de Goiás, Serra (range) E6
Gi-Paraná (riv.) H10
Gradaús, Serra dos (range) . D4
Grajaú (riv.) E4
Grande (isl.) *D3
Grande (riv.) *B2, E8
Guanabara (bay) *E3
Guaporé (riv.) H10
Gurguéia (riv.) E5
Gurupi, Serra do (range) E4
Gurupi (riv.) E3
Ibicuí (riv.) C10
Içá (riv.) G9
Iguaçu (riv.) C9
Iguazú (falls) C9
Ilha Grande (bay) *D3
Iriri (riv.) C4
Itaipu (dam) C9
Itaipu (res.) C9
Itapecuru (riv.) F4
Itapi (riv.) B3
Itapicuru (riv.) G5
Itararé (riv.) *B3
Ival (riv.) C8
Jaculpe (riv.) F5
Jaguaribe (riv.) G4
Jamanxim (riv.) C4
Japurá (riv.) G9
Jari (riv.) C3
Jauari, Serra (mts.) C3
Javari (riv.) F9
Jequitinhonha (riv.) F7
Juruá (riv.) G10
Juruena (riv.) B5
Jutaí (riv.) G9
Lombarda, Serra (mts.) D2
Madeira (riv.) A4
Mangueira (lag.) D11
Manso (riv.) C6
Mantiqueira (range) *D3
Mapuera (riv.) B3
Mar, Serra do (range) . *C4, E9
Maracá (isl.) D2
Marajó (bay) E2
Marajó (isl.) 147,895 ..... D3
Mato Grosso, Planalto de (plat.) B6
Maués-Açu (riv.) B4
Mearim (riv.) E4
Mexiana (isl.) D2
Miranda (riv.) B8
Mirim (lag.) C11
Mogi Guaçu (riv.) *C2
Mortes (Manso) (riv.) D6
Neblina, Pico da (peak) ... G8
Negro (riv.) H9
Nhamundá (riv.) B3
Norte, Serra do (range) B5
Oiapoque (Oyapock) (riv.) ... C2
Orange (cape) D1
Órgãos (range) *E3
Oyapock (riv.) C2
Pacajá Grande (riv.) D4
Pacaraimã, Serra da (mts.) . H8
Papagaio (riv.) B6
Pará (riv.) D3
Paracatu (riv.) E7
Paraguaçu (riv.) F6
Paraguai (riv.) B8
Paraíba (riv.) *E2
Paraná (riv.) C8
Paranã (riv.) E6
Paranapanema (riv.) ... *B3, C8
Paranapiacaba (range) *B4
Paranatinga (riv.) C6
Pardo (riv.) *B2, D8
Pardo (riv.) C8
Pardo (riv.) F6
Parecis, Serra dos (range) . B6
Parnaíba (riv.) F3
Paru (riv.) C3
Patos (lag.) D10
Penitente, Serra do (range) . E5
Piauí, Serra do (range) F5
Piauí (riv.) F5
Purus (riv.) H9
Ribeira (riv.) *B4
Roncador, Serra do (range) . D5
Ronuro (riv.) C6
Roosevelt (riv.) A5
Santa Catarina (isl.) 138,556 E9
São Lourenço (riv.) C7
São Marcos (bay) F3
São Roque (cape) H4
São Francisco (riv.) ... *D2, G5
São Sebastião (isl.) 5,724. *D3, E8
São Tomé (cape) F8
Sapucaí (riv.) *D2
Sepetiba (bay) *D3
Sete Quedas (falls) C9
Sete Quedas (Grande) (isl.) . C8
Sobradinho (res.) F5
Sono (riv.) E5
Sul (chan.) D2
Tacutu (riv.) B2
Tapajós (riv.) B4
Taquari (riv.) C7
Tefé (riv.) G9
Teles Pires (riv.) B5

Tibagi (riv.) *A4
Tietê (riv.) *B2, D8
Tiracambu, Serra (range) ... E3
Tocantins (riv.) D3
Tombador, Serra do (range) . B6
Trombetas (riv.) B3
Tucuruí (res.) D4
Tumucumaque, Serra de (range) C2
Turvo (riv.) *B2
Uaupés (riv.) G9
Uraricoera (riv.) H8
Urubu (riv.) A3
Urubupungá (dam) C8
Urucún, Morro do (mt.) B7
Uruguai (riv.) C9
Vasa Barris (riv.) G5
Velhas (riv.) E7
Verde (riv.) C7
Verdinho (riv.) D7
Xavantes (res.) *B3
Xingu (riv.) C3

†Population of met. area.
*preceding reference indicates that the name will be found on S.E. Brazil map, page 135.

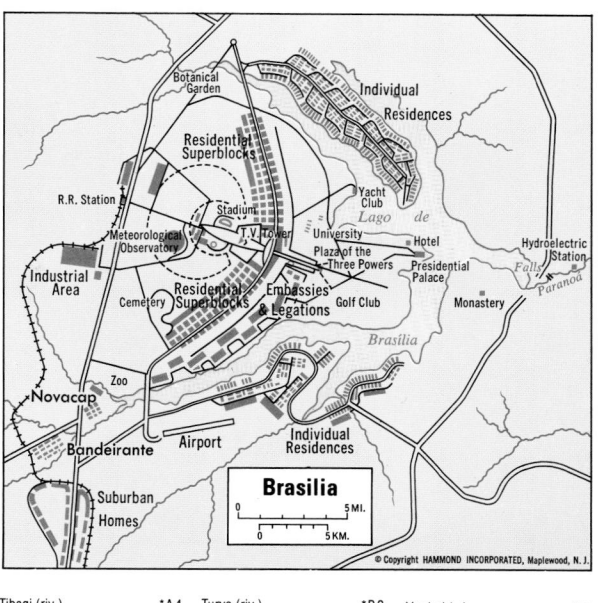

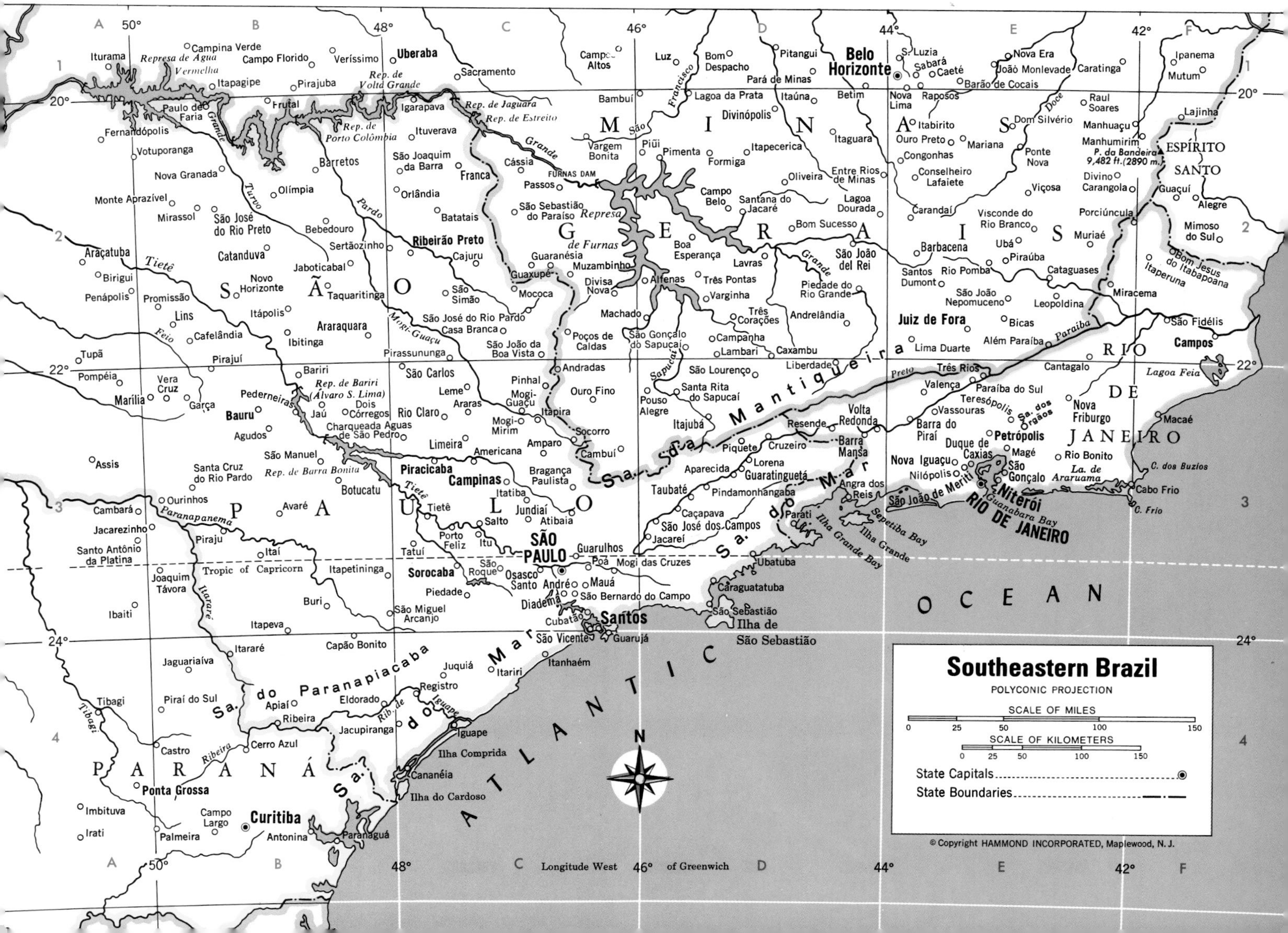

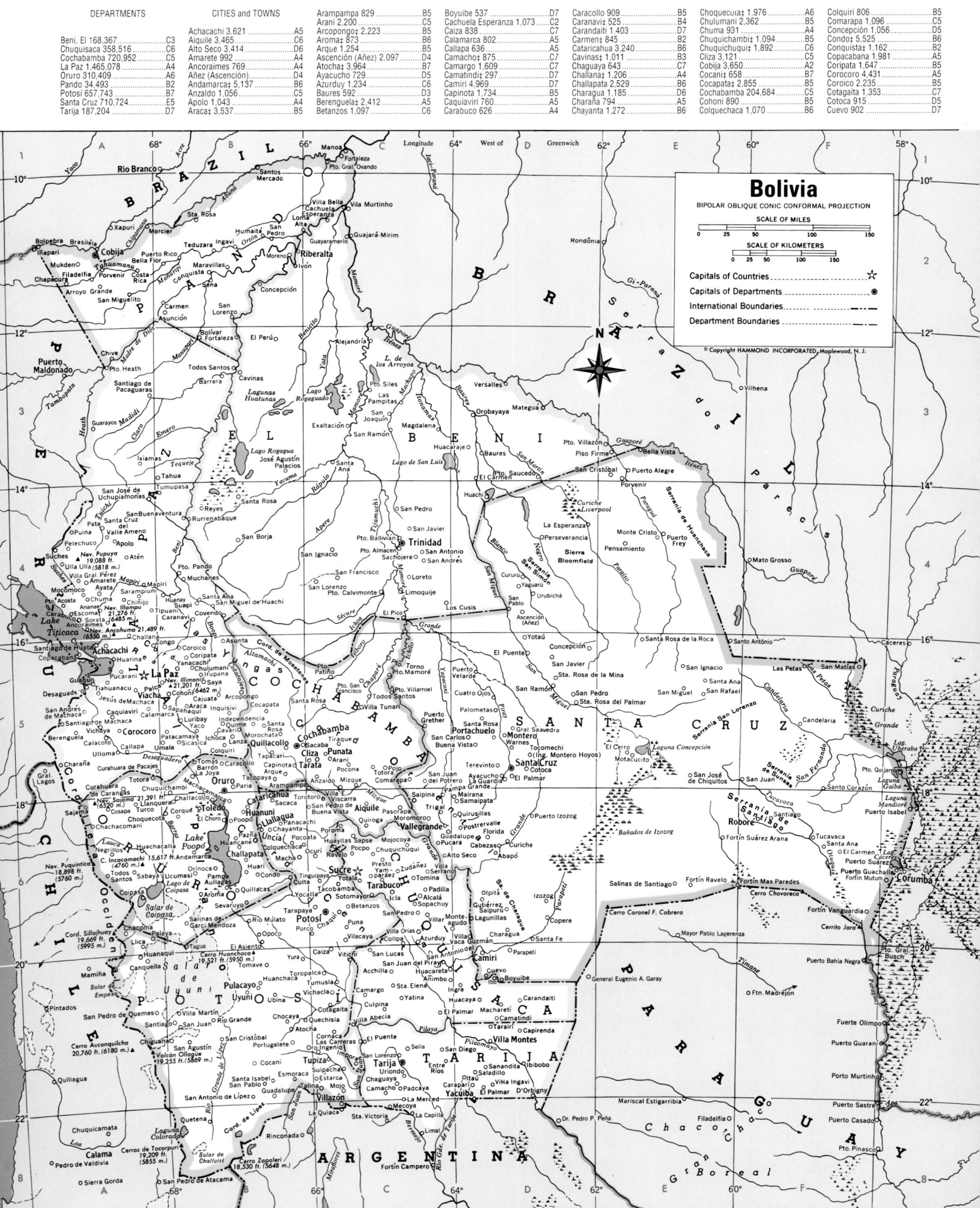

Bolivia

BIPOLAR OBLIQUE CONIC CONFORMAL PROJECTION

SCALE OF MILES

SCALE OF KILOMETERS

Capitals of Countries ....★
Capitals of Departments ....◉
International Boundaries ....
Department Boundaries ....

© Copyright HAMMOND INCORPORATED, Maplewood, N.J.

AREA 424,163 sq. mi. (1,098,582 sq. km.)
POPULATION 7,193,000
CAPITALS La Paz, Sucre
LARGEST CITY La Paz
HIGHEST POINT Nevada Ancohuma 21,489 ft.
(6,550 m.)
MONETARY UNIT Bolivian peso
MAJOR LANGUAGES Spanish, Quechua, Aymara
MAJOR RELIGION Roman Catholicism

Topography

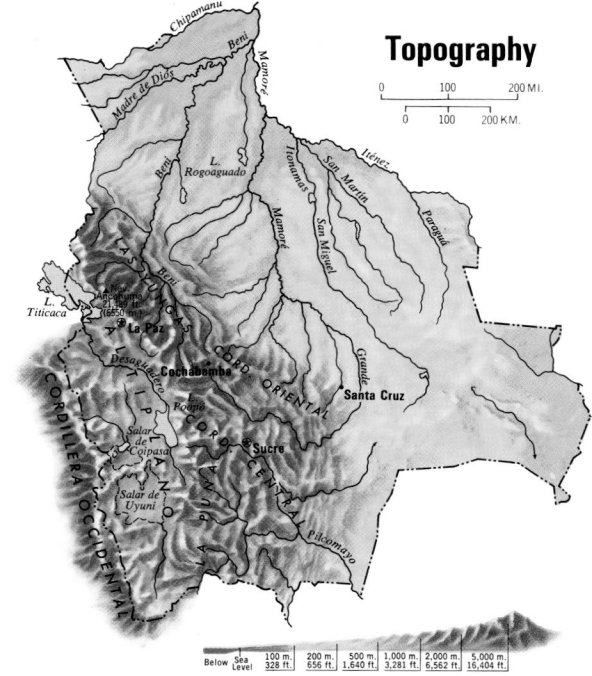

0 100 200 MI.
0 100 200 KM.

| Below Sea Level | 100 m. 328 ft. | 200 m. 656 ft. | 500 m. 1,640 ft. | 1,000 m. 3,281 ft. | 2,000 m. 6,562 ft. | 5,000 m. 16,404 ft. |

Agriculture, Industry and Resources

DOMINANT LAND USE

Diversified Tropical Crops (chiefly plantation agriculture)
Upland Cultivated Areas
Upland Livestock Grazing, Limited Agriculture
Extensive Livestock Ranching
Forests
Nonagricultural Land

MAJOR MINERAL OCCURRENCES

Ag Silver	G Natural Gas	Sb Antimony
Au Gold	O Petroleum	Sn Tin
Cu Copper	Pb Lead	W Tungsten
Fe Iron Ore	S Sulfur	Zn Zinc

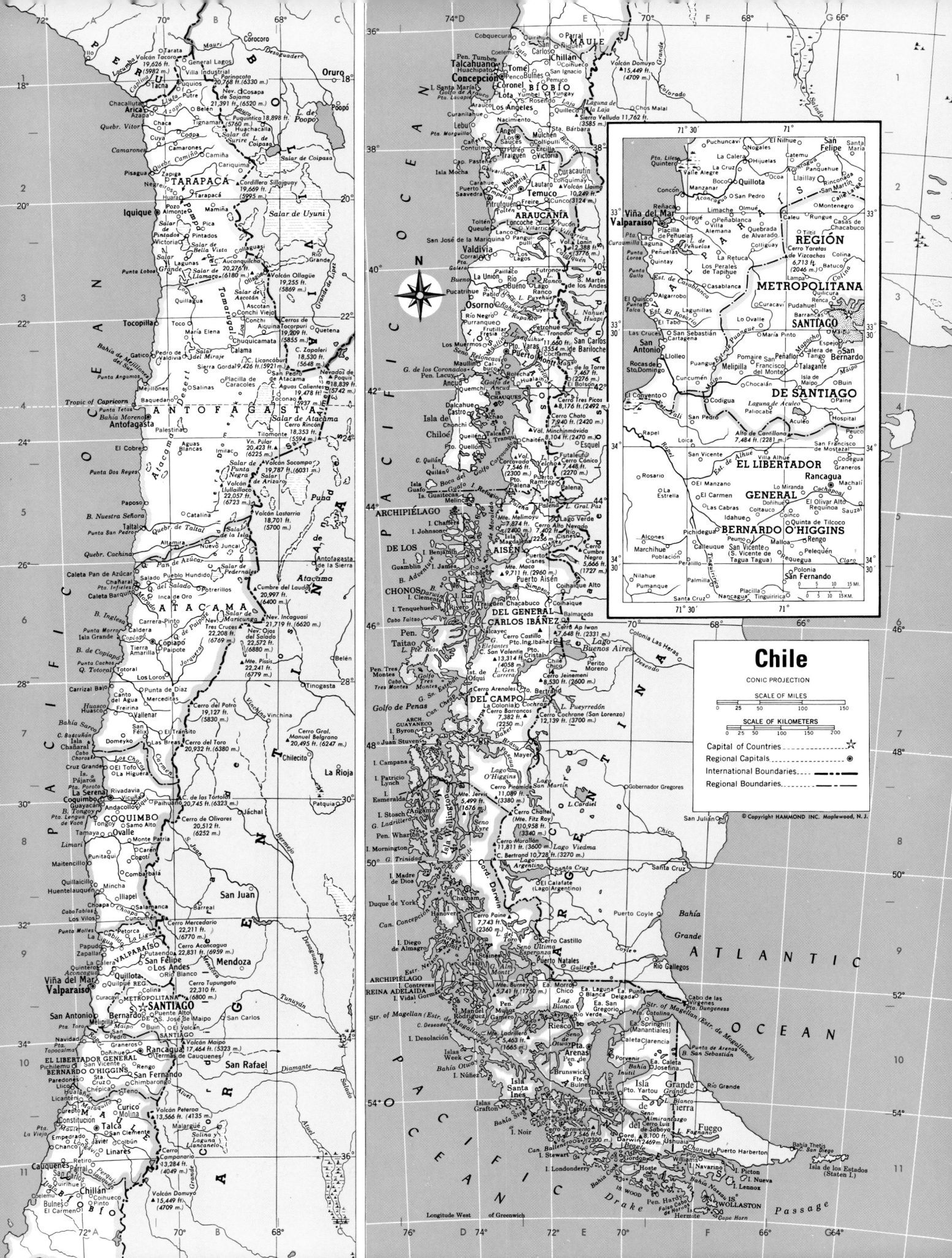

Topography

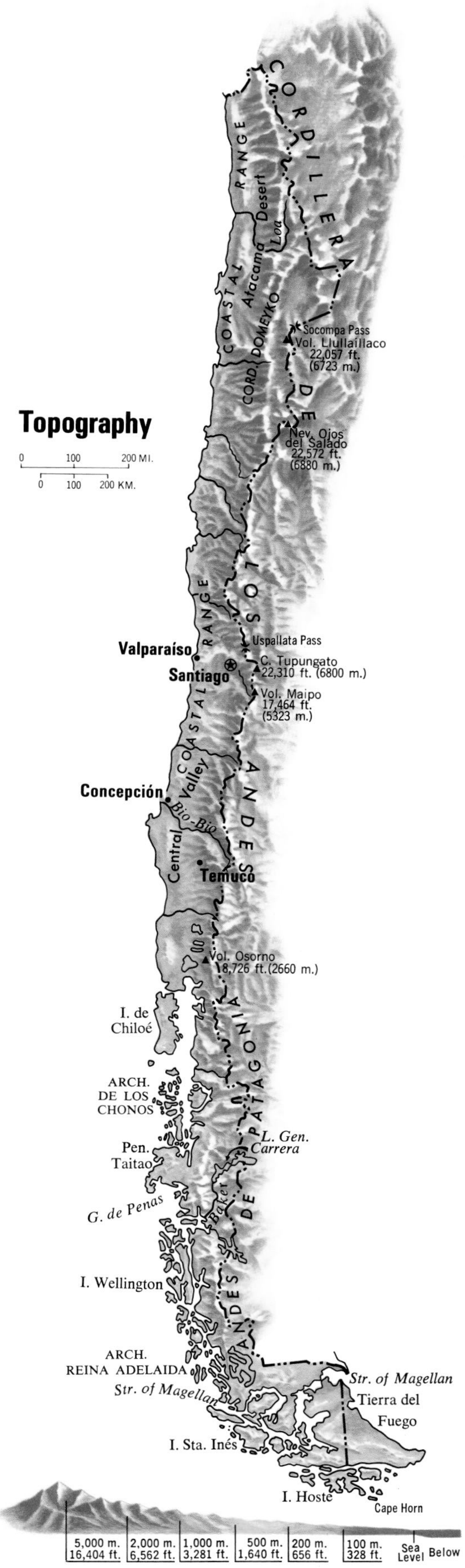
0 100 200 MI.
0 100 200 KM.
CORDILLERA DE LOS ANDES
COASTAL RANGE
Atacama Desert
Loa
CORD. DOMEYKO
Socompa Pass
Vol. Llullaillaco 22,057 ft. (6723 m.)
Nev. Ojos del Salado 22,572 ft. (6880 m.)
Valparaíso
Santiago
Uspallata Pass
C. Tupungato 22,310 ft. (6800 m.)
Vol. Maipo 17,464 ft. (5323 m.)
Central Valley
Concepción
Bío-Bío
Temuco
Vol. Osorno 8,726 ft. (2660 m.)
I. de Chiloé
ARCH. DE LOS CHONOS
Pen. Taitao
L. Gen. Carrera
PATAGONIA
Baker
G. de Penas
I. Wellington
ARCH. REINA ADELAIDA
Str. of Magellan
Tierra del Fuego
I. Sta. Inés
I. Hoste
Cape Horn
5,000 m. 16,404 ft.
2,000 m. 6,562 ft.
1,000 m. 3,281 ft.
500 m. 1,640 ft.
200 m. 656 ft.
100 m. 328 ft.
Sea Level
Below

AREA 292,257 sq. mi. (756,946 sq. km.)
POPULATION 12,961,000
CAPITAL Santiago
LARGEST CITY Santiago
HIGHEST POINT Ojos del Salado 22,572 ft. (6,880 m.)
MONETARY UNIT Chilean peso
MAJOR LANGUAGE Spanish
MAJOR RELIGION Roman Catholicism

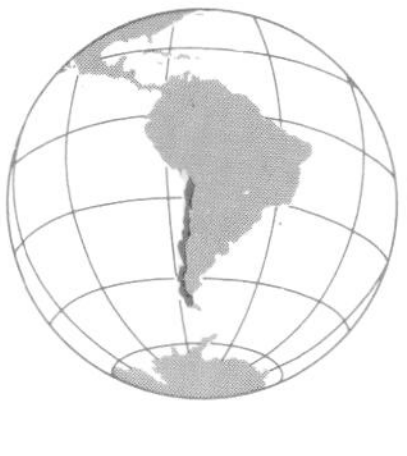

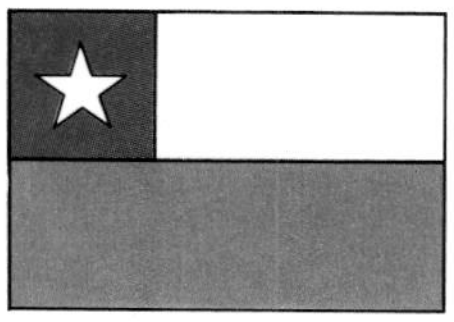

REGIONS

Aisén del General Carlos Ibáñez del Campo 65,478 .... E6
Antofagasta 341,203 .... B4
Atacama 183,071 .... B6
Bíobío 1,516,552 .... E1
Coquimbo 419,178 .... A8
El Libertador General Bernardo O'Higgins 584,989 .... A10
La Araucanía 692,924 .... E2
Los Lagos 843,430 .... D3
Magallanes 132,333 .... E10
Maule 723,224 .... A11
Santiago, Región Metropolitana de (Santiago Metropolitan Region) 4,294,938 .... A9
Tarapacá 273,427 .... B2
Valparaíso 1,204,693 .... A9

CITIES and TOWNS

Achao ○11,501 .... D4
Aguas Blancas ○203 .... B4
Algarrobo ○3,941 .... F3
Ancud 11,900 .... D4
Andacollo 6,000 .... A8
Angol 42,670 .... D1
Antofagasta 125,100 .... A4
Arauco 5,400 .... D1
Arica 87,700 .... A1
Ascotán B3
Barrancas ○184,241 .... G3
Belén ○925 .... B1
Buin 11,800 .... G4
Bulnes 6,900 .... E1
Cabildo 5,800 .... A9
Calama 45,900 .... B3
Calbuco ○21,673 .... D4
Caldera ○3,268 .... A6
Calera de Tango ○6,198 .... G4
Calle Larga ○7,172 .... G2
Cañete 7,900 .... D2
Carahue ○12,733 .... D2
Cartagena ○7,124 .... F3
Casablanca 5,500 .... F3
Casas de Chacabuco G2
Castro 11,200 .... D4
Catalina ○1,637 .... B5
Catemu ○8,728 .... G2
Cauquenes 20,200 .... A11
Cerro Castillo ○537 .... E9
Cerro Manantiales F10
Chaitén ○4,067 .... E4
Chañaral ○36,949 .... A6
Chanco ○12,433 .... A11
Chépica ○11,199 .... A10
Chillán 128,515 .... A11
Chimbarongo 5,300 .... A10
Chonchi ○8,911 .... D4
Chuquicamata 22,100 .... B3
Cobquecura ○6,298 .... D1
Cochamó ○5,042 .... E3
Codegua ○6,757 .... G4
Codpa ○950 .... B1
Coelemu 5,400 .... D1
Coihaique 32,129 .... E6
Coihueco ○17,276 .... A11
Coinco ○4,942 .... G5
Colbún ○12,924 .... A11
Colina 7,400 .... G3
Collipulli 7,200 .... E2
Coltauco ○11,857 .... F5
Combarbalá ○17,332 .... A8
Concepción 206,226 .... D1
Constitución 11,500 .... A11
Contulmo ○13,987 .... D2
Copiapó 45,200 .... B6
Coquimbo 73,953 .... A8
Coronel 37,300 .... D1
Corral ○5,533 .... D3
Cunco ○18,836 .... E2
Curacautín 9,800 .... E2
Curacaví 5,800 .... G3
Curanilahue 13,200 .... D1
Curepto ○13,020 .... A10
Curicó 41,300 .... A10
Dalcahue ○7,084 .... D4
Domeiko A7
Doñihue ○8,837 .... G5
El Carmen ○13,226 .... A11
El Monte 7,000 .... G4
El Quisco ○2,152 .... E3
El Tabo ○2,180 .... F3
El Tofo A7
Empedrado ○7,887 .... A11
Ercilla ○8,061 .... E2
Estancia Caleta Josefina ○1,042 .... F10
Estancia Morro Chico ○785 .... E9
Estancia San Gregorio ○1,156 .... E9
Estancia Springhill (Cerro Manantiales) F10
Freire ○23,313 .... E2
Freirina ○5,523 .... A7
Fresia ○15,359 .... D3
Frutillar ○12,721 .... D3
Futaleufú ○2,366 .... E4
Futrono ○7,109 .... E3
Galvarino ○9,495 .... D2
General Lagos ○810 .... B1
Graneros 8,900 .... G5
Guayacán A8
Hijuelas ○7,128 .... F2
Hualañé ○6,912 .... A10
Huara ○1,934 .... B2
Huasco ○4,971 .... A7
Illapel 12,200 .... A8
Inca de Oro 1,406 .... B6
Iquique 64,500 .... A2
Isla de Maipo ○12,903 .... G4
La Calera 24,600 .... F2
La Cruz ○8,907 .... F2
La Estrella ○3,707 .... F5
Lago Ranco ○12,767 .... E3
Lagunas ○5,653 .... B3
La Higuera ○6,991 .... A7
La Ligua 7,500 .... A9
Lampa ○10,220 .... G3
Lanco 5,200 .... D2
Las Cabras ○12,119 .... F5
La Serena 99,908 .... A8
La Unión 15,200 .... D3
Lautaro 11,900 .... E2
Lebu 12,500 .... D1
Licantén ○6,354 .... A10
Limache 15,200 .... F2
Linares 37,900 .... A11
Llay-Llay 9,700 .... G2
Loica F4
Loncoche ○17,539 .... D2
Longaví ○15,909 .... A11
Lonquimay ○9,524 .... E2
Los Andes 23,500 .... B9
Los Ángeles 49,500 .... D1
Los Lagos ○14,934 .... D3
Los Muermos ○9,296 .... D3
Los Sauces ○7,613 .... D2
Los Vilos ○10,453 .... A9
Lota 48,100 .... D1
Machall 5,800 .... G5
Maipú ○117,872 .... G3
Malloa ○9,742 .... G5
Marchigüe ○4,451 .... F5
Marla Elena 5,900 .... B3
Marla Pinto ○5,980 .... G3
Maullln ○14,544 .... D4
Mejillones ○3,333 .... A4
Melipilla 23,900 .... F4
Mincha ○11,329 .... A8
Molina 9,400 .... A10
Monte Patria ○18,927 .... A8
Mulchén 13,700 .... E1
Nacimiento ○17,651 .... D1
Nancagua ○11,076 .... F6
Navidad ○6,618 .... A10
Negreiros ○1,144 .... B2
Ñiquén ○13,640 .... E1
Nogales ○18,529 .... F2
Nueva Imperial 8,000 .... D2
Olivar Alto ○5,414 .... G5
Ollagüe B3
Olmué ○8,804 .... F2
Osorno 68,800 .... D3
Ovalle 31,700 .... A8
Paihuano ○6,048 .... B8
Paillaco 5,200 .... D3
Paine ○21,876 .... G4
Palena ○2,508 .... E5
Palmilla ○7,965 .... F6
Panguipulli 5,700 .... E2
Panquehue ○4,230 .... G2
Papudo ○2,594 .... A9
Paredones ○7,404 .... A10
Parral 17,000 .... A11
Pedro de Valdivia 6,200 .... B4
Pemuco ○7,577 .... E1
Peñaflor 15,500 .... G4
Penco ○33,962 .... D1
Peñuelas F3
Petorca ○8,343 .... A9
Petrohué E3
Peumo ○11,308 .... F5
Pica ○1,487 .... B2
Pichidegua ○13,550 .... F5
Pichilemu ○8,042 .... A10
Pinto ○8,687 .... A11
Pisagua ○1,880 .... A2
Pitrufquén 7,800 .... D2
Placilla ○6,441 .... F6
Porvenir ○4,000 .... E10
Potrerillos 5,800 .... B6
Pozo Almonte ○1,798 .... B2
Puchuncaví ○7,542 .... F2
Pucón 18,000 .... E2
Pudahuel G3
Pueblo Hundido 6,200 .... B6
Puente Alto 65,100 .... B10
Puerto Aisén 17,848 .... E6
Puerto Cisnes ○2,800 .... E5
Puerto Ingeniero Ibáñez ○1,900 .... E6
Puerto Montt 119,059 .... E4
Puerto Natales 17,280 .... E9
Puerto Quellón ○7,734 .... D4
Puerto Varas 10,900 .... E3
Puerto Williams ○949 .... F11
Pumanque ○3,137 .... F6
Punitaqui ○16,167 .... A8
Punta Arenas 2,140 .... E10
Purén ○11,604 .... D2
Purranque 5,900 .... D3
Putaendo ○12,806 .... A9
Putre ○855 .... B1
Puyehue E3
Queilén ○6,055 .... D4
Quemchi ○6,707 .... D4
Quilicura 8,100 .... G3
Quillagua B3
Quilleco ○16,043 .... E1
Quillota 36,500 .... F2
Quilpué 40,600 .... F2
Quinta de Tilcoco ○6,513 .... G5
Quintero 9,900 .... F2
Quirihue ○11,178 .... E1
Rancagua 140,589 .... G5
Renca ○67,168 .... G3
Rengo 12,400 .... G5
Requlnoa ○10,730 .... G5
Retiro ○15,146 .... A11
Rinconada San Martín ○4,118 .... G2
Río Blanco B9
Río Bueno 9,600 .... D3
Río Negro 5,100 .... D3
Río Verde ○554 .... E10
Rocas de Santo Domingo ○4,114 .... F4
Rosario ○3,383 .... F5
Salamanca ○18,741 .... A9
Samo Alto ○5,689 .... A8
San Antonio 46,700 .... F3
San Bernardo ○117,766 .... G4
San Carlos 17,000 .... E1
San Clemente ○23,273 .... A11
San Felipe 26,100 .... G2
San Fernando 23,600 .... G6
San Francisco de Mostazal ○11,439 .... G4
San Ignacio ○13,523 .... E1
San Javier 10,800 .... A11
San José de Maipo ○9,601 .... B10
San Pablo ○7,978 .... D3
San Pedro ○8,255 .... F4
San Pedro de Atacama C4
San Rosendo ○14,337 .... E1
Santa Bárbara ○14,345 .... E1
Santa Cruz 8,600 .... F6
Santa Marla ○8,162 .... G2
Santiago (cap.) 3,614,947 .... G3
Santiago *3,672,374 .... G3
San Vicente F4
San Vicente (San Vicente de Tagua Tagua) ○28,333 .... F5
Sierra Gorda ○8,805 .... B4
Talagante 16,500 .... G4
Talca 133,160 .... A11
Talcahuano 148,300 .... D1
Taltal 6,400 .... A5
Tamaya A8
Tarapacá B2
Temuco 197,232 .... E2
Teno ○17,675 .... A10
Termas de Cauquenes B10
Tierra Amarilla ○7,899 .... A6
Tiltil ○9,198 .... G2
Toco ○8,734 .... B3
Toconao C4
Tocopilla 22,000 .... A3
Toltén ○16,265 .... D2
Tomé 29,600 .... D1
Traiguén 11,400 .... D2
Valdivia 115,536 .... D3
Vallenar 26,800 .... A7
Valparaíso 271,580 .... E2
Victoria 16,500 .... D2
Vicuña 5,100 .... A8
Villa Alemana 29,600 .... F2
Villa Alhué ○5,078 .... G4
Villarrica 25,091 .... E2
Viña del Mar 281,361 .... F2
Yumbel ○21,858 .... E1
Yungay ○10,725 .... E1
Zapallar ○2,894 .... A9
Zapiga B2

OTHER FEATURES

Aconcagua (riv.) F2
Aculeo (lag.) G4
Adventure (bay) D5
Aguas Calientes, Cerro (mt.) C4
Almirantazgo (bay) F11
Almirante Montt (gulf) D9
Ancud (gulf) D4
Angamos (isl.) D8
Angamos (pt.) A4
Ap Iwan, Cerro (mt.) E6
Arauco (gulf) D1
Arenales, Cerro (mt.) D7
Atacama (des.) B4
Atacama, Salar de (salt dep.) C4
Aucanquilcha, Cerro (mt.) B3
Azapa, Quebrada (riv.) B1
Baker (riv.) D7
Ballenero (chan.) E11
Bascuñán (cape) A7
Beagle (chan.) E11
Bella Vista, Salar de (salt dep.) B3
Benjamln (isl.) D5
Blo-Blo (riv.) E2
Blanca (lag.) E10
Blanco (lake) F10
Bravo (riv.) D7
Brunswick (pen.) E10
Bueno (riv.) D3
Buenos Aires (lake) E6
Byron (isl.) D7
Cachapoal (riv.) G5
Cachina, Quebrada (riv.) A5
Cachos (pt.) A6
Calafquén (lake) E3
Camarones (riv.) A2
Camiña, Quebrada (riv.) B2
Campana (isl.) D7
Campanario, Cerro (mt.) A10
Capitán Aracena (isl.) E10
Carmen (riv.) B7
Castillo, Cerro (mt.) E6
Catalina (pt.) F10
Chaffers (isl.) D5
Chaltel, Cerro (mt.) E8
Chañaral (isl.) A7
Chatham (isl.) D9
Chauques (isls.) D4
Cheap (chan.) D7
Chiloé (isl.) 119,286 .... D4
Choapa (riv.) A9
Chonos (arch.) D6
Choros (cape) A7
Cisnes (riv.) E5
Clarence (isl.) E10
Clemente (isl.) D6
Cochrane (lake) E7
Cochrane, Cerro (mt.) E7
Cockburn (chan.) E11
Concepción (chan.) D9
Cónico, Cerro (mt.) E4
Contreras (isl.) D9
Cook (bay) E11
Copiapó (bay) A6
Copiapó (riv.) A6
Corcovado (gulf) D4
Corcovado (vol.) D5
Coronados (gulf) D4
Curaumilla (pt.) E2
Darwin (bay) D6
Darwin, Cordillera (mts.) D8
Darwin, Cordillera (mts.) E11

(continued on following page)

Agriculture, Industry and Resources

DOMINANT LAND USE

- Cereals, Livestock
- Mediterranean Agriculture (cereals, fruit, livestock)
- Pasture Livestock
- Extensive Livestock Ranching
- Limited Seasonal Grazing
- Forests
- Nonagricultural Land

MAJOR MINERAL OCCURRENCES

Ag	Silver		Hg	Mercury
Au	Gold		Id	Iodine
C	Coal		Mn	Manganese
Cu	Copper		Mo	Molybdenum
Fe	Iron Ore		N	Nitrates
G	Natural Gas		Na	Salt
Gp	Gypsum		O	Petroleum
			S	Sulfur

⚡ Water Power   ▨ Major Industrial Areas

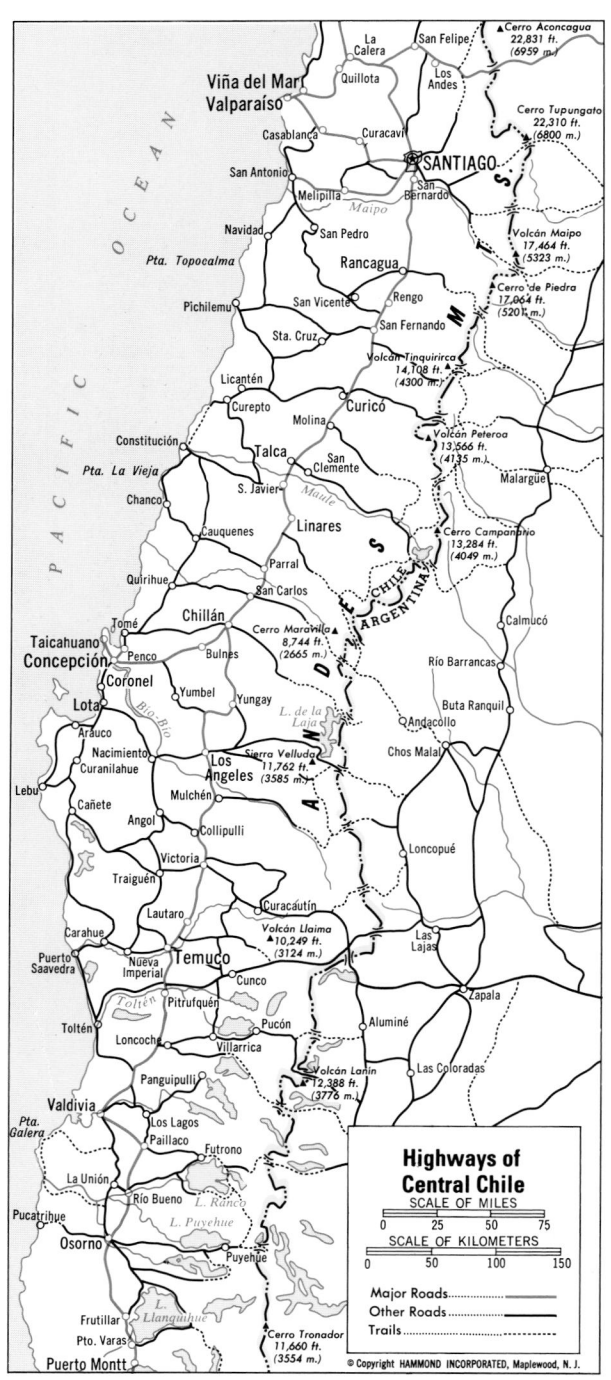

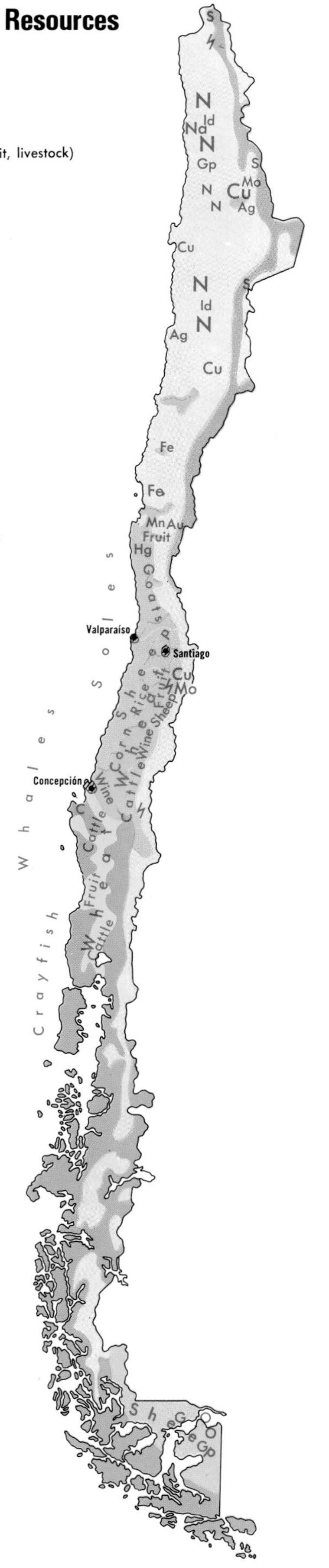

Highways of Central Chile

SCALE OF MILES

0 25 50 75

SCALE OF KILOMETERS

0 50 100 150

Major Roads ——————
Other Roads ——————
Trails ·············

© Copyright HAMMOND INCORPORATED, Maplewood, N.J.

*City and suburbs.
○ Population of commune.

PROVINCES

Buenos Aires 10,796,036 . . . . . D4
Catamarca 206,204 . . . . C2
Chaco 692,410 . . . . D2
Chubut 262,196 . . . . C5
Córdoba 2,407,135 . . . . . . D3
Corrientes 657,716 . . . . . . E2
Distrito Federal 2,908,001 . . H7
Entre Ríos 902,241 . . . . . . E3
Formosa 292,479 . . . . D1
Jujuy 408,514 . . . . C1
La Pampa 207,132 . . . . . . C4
La Rioja 163,342 . . . . C2
Mendoza 1,187,305 . . . . . . C4
Misiones 579,579 . . . . F2
Neuquén 241,904 . . . . C4
Río Negro 383,896 . . . . . . C5
Salta 662,369 . . . . D1
San Juan 469,973 . . . . . . C3
San Luis 212,837 . . . . C3
Santa Cruz 114,479 . . . . . . C6
Santa Fe 2,457,188 . . . . . . D2
Santiago del Estero 652,318 . D2
Tierra del Fuego, Antártida,
 e Islas del Atlántico
 Sur 29,451 . . . . C7
Tucumán 968,066 . . . . C2

CITIES and TOWNS

Abra Pampa 2,929 . . . . C1
Adolfo Alsina 7,707 . . . . . . D4
Aguaray 4,802 . . . . D1
Aguilares 20,286 . . . . C2
Aimogasta 4,640 . . . . C2
Alberti 6,440 . . . . G7
Alcorta 5,818 . . . . F6
Algarrobo del Águila C4
Allen 14,041 . . . . C4
Alpachiri 1,657 . . . . D4
Alta Gracia 30,628 . . . . . . D3
Aluminé 1,560 . . . . B4
Alvear 5,419 . . . . E2
Ameghino 2,775 . . . . D3
Añatuya 15,025 . . . . D2
Andalgalá 6,853 . . . . C2
Antofagasta de la Sierra . . . C2
Apóstoles 11,252 . . . . E2
Arrecifes 17,719 . . . . F7
Arroyo Seco 12,886 . . . . . . F6
Ascensión 3,031 . . . . F7
Avellaneda 330,654 . . . . . . G7
Ayacucho 12,363 . . . . E4
Azul 43,582 . . . . E4
Bahía Blanca 220,765 . . . . . D4
Bahía Bustamante C6
Bahía Thetis C7
Balcarce 28,985 . . . . E4
Balnearia 4,531 . . . . D3
Baradero 20,103 . . . . G6
Barrancas 3,602 . . . . F6
Barranqueras E2
Barreal 2,739 . . . . C3
Basavilbaso 7,657 . . . . . . G6
Belén 7,411 . . . . C2
Bella Vista, Corrientes
 14,229 . . . . E2
Bella Vista, Tucumán 9,177 . . D2
Bell Ville 26,559 . . . . D3
Bolívar 16,382 . . . . D4
Bovril 4,735 . . . . G5
Bragado 27,101 . . . . F7
Buenos Aires (cap.)
 2,908,001 . . . . H7
Buenos Aires *9,927,404 . . . . H7
Cafayate 5,048 . . . . C2
Calafate B7
Calchaquí 5,958 . . . . F5
Caleta Olivia 20,141 . . . . . . C6
Camarones C5
Campana 51,498 . . . . G6
Cañada de Gómez 24,706 . . . . F6
Canals 6,627 . . . . D3
Cañuelas 14,831 . . . . G7
Carcarañá 11,121 . . . . F6
Carlos Casares 13,286 . . . . . F7
Carlos Tejedor 4,421 . . . . . D4
Carmen de Areco 7,882 . . . . F7
Carmen de Patagones
 13,981 . . . . D5
Casilda 23,492 . . . . F6
Castelli 4,507 . . . . H7
Catamarca 88,432 . . . . C2
Caucete 14,512 . . . . C3
Ceres 10,743 . . . . D2
Chabás 5,156 . . . . F6
Chacabuco 26,492 . . . . F7
Chajarí 15,242 . . . . G5
Chamical 6,333 . . . . C3
Charadai 1,078 . . . . D2
Charata 13,070 . . . . D2
Chascomús 21,864 . . . . H7
Chepes 4,775 . . . . C3
Chicoana 1,844 . . . . C2
Chilecito 14,010 . . . . C2
Chivilcoy 43,779 . . . . F7
Choele-Choel 6,191 . . . . . . C4
Chos-Malal 4,823 . . . . C4
Cinco Saltos 15,094 . . . . . . C4
Cipolletti 40,123 . . . . C4
Clorinda 21,008 . . . . D1
Colón, Buenos Aires 16,070 . . F6
Colón, Entre Ríos 11,648 . . . G5
Colonia Las Heras 3,176 . . . . C6
Comandante Fontana 4,468 . . D2
Comandante Luis Piedrabuena
 2,492 . . . . C6
Comodoro Rivadavia 96,865 . . C6
Concepción 29,359 . . . . . . C2
Concepción de
 la Sierra 2,778 . . . . E2
Concepción del
 Uruguay 46,065 . . . . G6
Concordia 93,618 . . . . G5
Constanza 1,313 . . . . G6
Córdoba 982,018 . . . . D3
Coronda 11,554 . . . . F6
Coronel Brandsen 10,484 . . . H7
Coronel Dorrego 10,661 . . . . D4
Coronel Pringles 16,592 . . . . D4
Coronel Suárez 16,359 . . . . . D4

AREA 1,072,070 sq. mi. (2,776,661 sq. km.)
POPULATION 31,929,000
CAPITAL Buenos Aires
LARGEST CITY Buenos Aires
HIGHEST POINT Cerro Aconcagua 22,831 ft.
 (6,959 m.)
MONETARY UNIT austral
MAJOR LANGUAGE Spanish
MAJOR RELIGION Roman Catholicism

Agriculture, Industry and Resources

Coronel Vidal 4,774 . . . . . . E4
Corral de Bustos 8,613 . . . . . D3
Corrientes 179,590 . . . . . . E2
Cosquín 13,929 . . . . D3
Crespo 10,668 . . . . F6
Cruz del Eje 23,473 . . . . . . C3
Curuzú Cuatiá 24,955 . . . . . G5
Cutral-Có 25,870 . . . . C4
Daireaux 8,150 . . . . D4
Deán Funes 16,306 . . . . . . C3
Diamante 13,464 . . . . F6
Dolavon 1,778 . . . . C5
Dolores 19,307 . . . . E4
Eduardo Castex 5,397 . . . . . D4
El Bolsón 5,001 . . . . B5
Eldorado 22,821 . . . . F2
El Maitén 2,350 . . . . B5
Elortondo 4,939 . . . . F6
El Quebrachal 2,202 . . . . . . D2
Embarcación 9,016 . . . . D1
Empedrado 4,732 . . . . E2
Escobar 70,829 . . . . G7
Esperanza 22,838 . . . . F5
Esquel 17,228 . . . . B5
Esquina 10,380 . . . . G5
Famatina 1,237 . . . . C2
Federación 7,259 . . . . G5
Felipe Yofré 1,140 . . . . . . G4
Fernández 6,062 . . . . D2
Fiambalá 1,201 . . . . C2
Firmat 13,588 . . . . F6
Formosa 95,067 . . . . E2
Fortín Olmos 1,101 . . . . . . F4
Frías 20,901 . . . . D2
Gaiman 2,651 . . . . C5
Gálvez 14,711 . . . . F6
General Acha 7,647 . . . . . . C4
General Alvear, Buenos Aires
 5,481 . . . . F7
General Alvear,
 Mendoza 21,250 . . . . C3
General Arenales 3,332 . . . . F6
General Belgrano 10,909 . . . . G7
General Conesa 3,566 . . . . . C5
General Galarza 3,057 . . . . . G6
General Güemes 15,534 . . . . D1
General José de
 San Martín 16,296 . . . . . . E2
General Juan Madariaga
 13,409 . . . . E4
General La Madrid 5,154 . . . . D4
General Las Heras 6,005 . . . . G7
General Paz 5,127 . . . . H7
General Pico 30,180 . . . . . . D4
General Ramírez 5,393 . . . . . F6
General Roca 38,296 . . . . . . C4
General San Martín, Buenos
 Aires 384,306 . . . . G7
General San Martín,
 La Pampa 2,168 . . . . D4
General Viamonte 10,112 . . . . F7
General Villegas 11,307 . . . . D4
Gobernador Crespo 2,972 . . . F5
Godoy Cruz 141,553 . . . . . . C3
Goya 47,357 . . . . E2
Gualeguay 24,883 . . . . G6
Gualeguaychú 51,057 . . . . . . G6
Guandacol 1,351 . . . . C2
Hasenkamp 2,804 . . . . F5
Helvecia 3,927 . . . . F5
Hernandarias 3,002 . . . . . . F5
Hernando 8,619 . . . . D3
Huinca Renancó 7,187 . . . . . D3
Humahuaca 3,963 . . . . C1
Humberto (Humberto
 Primo) 4,163 . . . . F5
Ibarreta 5,262 . . . . D2
Ibicuy 3,082 . . . . G6
Ingeniero Huergo 3,385 . . . . C4
Ingeniero Jacobacci 4,045 . . . C5
Ingeniero Luiggi 3,002 . . . . . D4
Intendente Alvear 3,640 . . . . D4
Itatí 3,269 . . . . E2
Ituzaingó 8,687 . . . . E2
Jáchal 8,832 . . . . C3
Jesús María 17,594 . . . . . . D3
Joaquín V. González 6,054 . . . D2
Juárez 11,798 . . . . E4
Jujuy 124,487 . . . . C1
Junín 62,080 . . . . F7
Junín de los Andes 5,638 . . . B4
La Banda 46,994 . . . . D2
Laboulaye 16,883 . . . . D3
La Carlota 8,614 . . . . D3
La Cruz 4,132 . . . . E2
La Cumbre 6,110 . . . . C3
La Falda 12,502 . . . . D3
Laguna Paiva 11,129 . . . . . . F5
Lanús 465,891 . . . . H7
La Paz, Entre Ríos 14,920 . . . G5
La Paz, Mendoza 4,604 . . . . C3
La Plata 560,341 . . . . H7
Laprida 6,495 . . . . D4
La Quiaca 8,289 . . . . C1
La Rioja 66,826 . . . . C2
Larroque 3,147 . . . . F5
Las Flores 18,287 . . . . E4
Las Lomitas 4,047 . . . . D1
Las Palmas 5,061 . . . . E2
Las Parejas 7,430 . . . . F6
Las Rosas 9,725 . . . . F6
Las Varillas 10,605 . . . . . . D3
La Toma 4,325 . . . . C3
Lincoln 19,009 . . . . F7
Lobería 8,898 . . . . E4
Lobos 20,798 . . . . G7
Lomas de Zamora 508,620 . . . G7
Lucas González 3,015 . . . . . G6
Luján 38,919 . . . . G7
Lules 11,391 . . . . C2
Maciel 4,066 . . . . F6
Magdalena 7,135 . . . . H7
Maipú 7,289 . . . . E4
Malabrigo 7,135 . . . . F4
Malargüe 9,496 . . . . C4
Maquinchao 1,299 . . . . C5
Marcos Juárez 19,827 . . . . . D3
Mar del Plata 407,024 . . . . . E4
Máximo Paz 3,216 . . . . F6
Mburucuyá 3,044 . . . . E2
Médanos 4,511 . . . . D4
Mendoza 596,796 . . . . C3
Mercedes, Buenos Aires
 46,581 . . . . G7
Mercedes, Corrientes
 20,603 . . . . E2
Mercedes, San Luis 50,856 . . . C3
Merlo 293,059 . . . . G7
Metán 18,928 . . . . D2
Miramar 15,473 . . . . E4
Monte Caseros 18,247 . . . . . G5
Monte Quemado 4,707 . . . . . D2
Monteros 15,832 . . . . C2
Morón 596,769 . . . . G7
Morteros 11,456 . . . . D3
Navarro 7,176 . . . . G7
Necochea 50,939 . . . . E4
Neuquén 90,037 . . . . C4
Nogoyá 15,862 . . . . F6
Norquincó B5
Nueve de Julio 26,608 . . . . . F7
Oberá 27,311 . . . . F2
Olavarría 63,686 . . . . D4
Oliva 9,231 . . . . D3
Palo Santo 3,088 . . . . E2
Paraná 159,581 . . . . F5
Paso de Los Libres 24,112 . . . E2
Pedro Luro 3,142 . . . . D4
Pehuajó 25,613 . . . . D4
Pellegrini 3,940 . . . . D4
Pergamino 68,989 . . . . F6
Pico Truncado 9,626 . . . . . . C6
Pigüé 10,793 . . . . D4
Pilar 3,805 . . . . F5
Pirané 9,039 . . . . E2
Plaza Huincul 7,988 . . . . . . B4

DOMINANT LAND USE

- Wheat, Livestock
- Wheat, Corn, Livestock
- Diversified Tropical Crops (chiefly plantation agriculture)
- Truck Farming, Horticulture, Special Crops
- Intensive Livestock Ranching
- Upland Livestock Grazing, Limited Agriculture
- Extensive Livestock Ranching
- Forests
- Nonagricultural Land

MAJOR MINERAL OCCURRENCES

Ag	Silver	O	Petroleum
Be	Beryl	Pb	Lead
C	Coal	S	Sulfur
Cu	Copper	Sn	Tin
Fe	Iron Ore	U	Uranium
G	Natural Gas	W	Tungsten
Mn	Manganese	Zn	Zinc
Na	Salt		

⚡ Water Power

▨ Major Industrial Areas

(continued on following page)

Topography

0 150 300 MI.

0 150 300 KM.

| 5,000 m. 16,404 ft. | 2,000 m. 6,562 ft. | 1,000 m. 3,281 ft. | 500 m. 1,640 ft. | 200 m. 656 ft. | 100 m. 328 ft. | Sea Level | Below |

Highways of Central Argentina

MILES
0 25 50 75

KILOMETRES
0 50 100 150

Major Roads
Other Roads

© HAMMOND INCORPORATED, Maplewood, N.J.

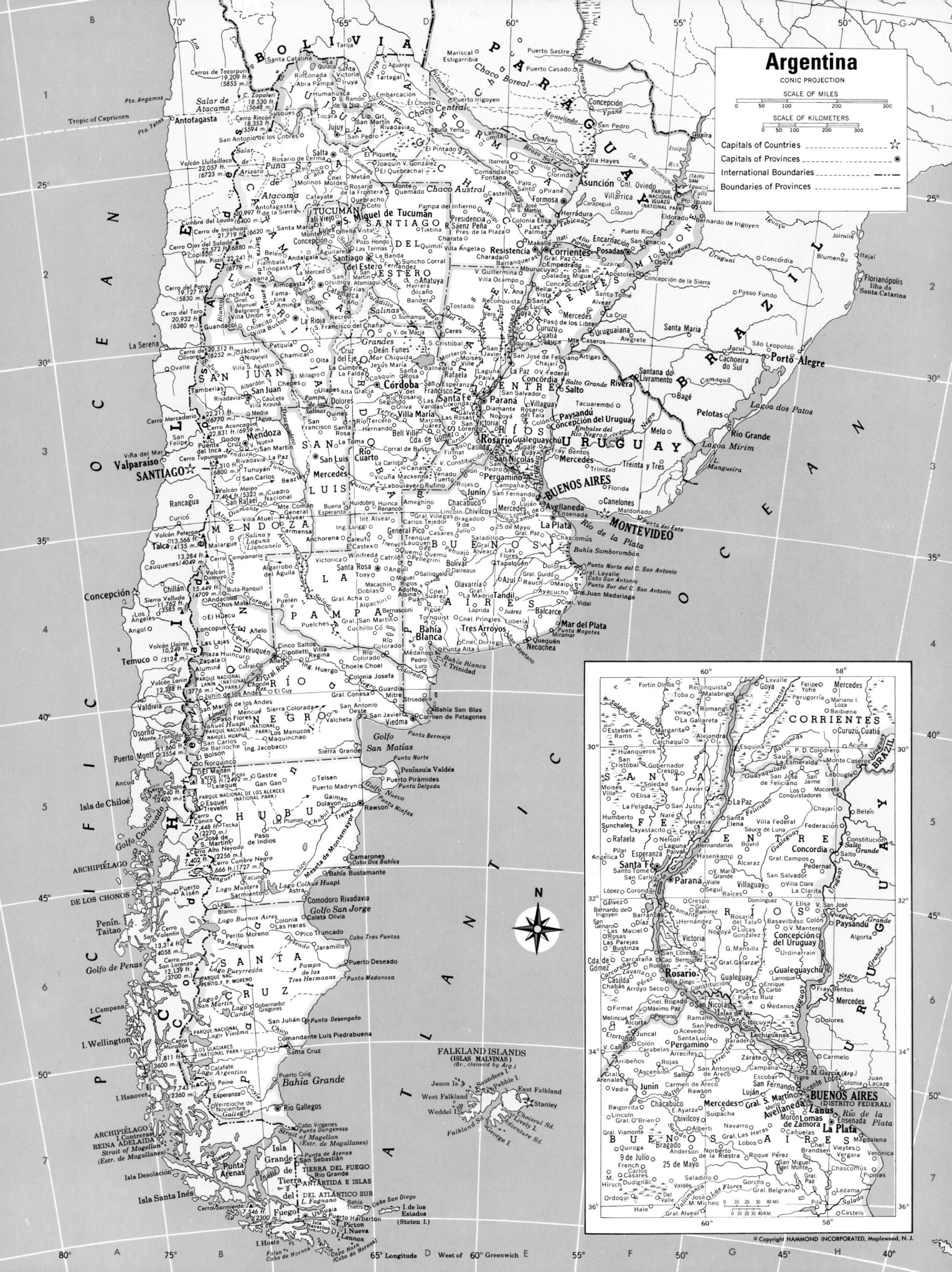

Paraguay

CONIC PROJECTION

SCALE OF MILES
0 20 40 60 80 100 120 140

SCALE OF KILOMETERS
0 20 40 60 80 100 120 140

Capitals of Countries .... ★
Capitals of Departments .... ◉
International Boundaries .... —··—
Department Boundaries .... —·—

© Copyright HAMMOND INCORPORATED, Maplewood, N.J.

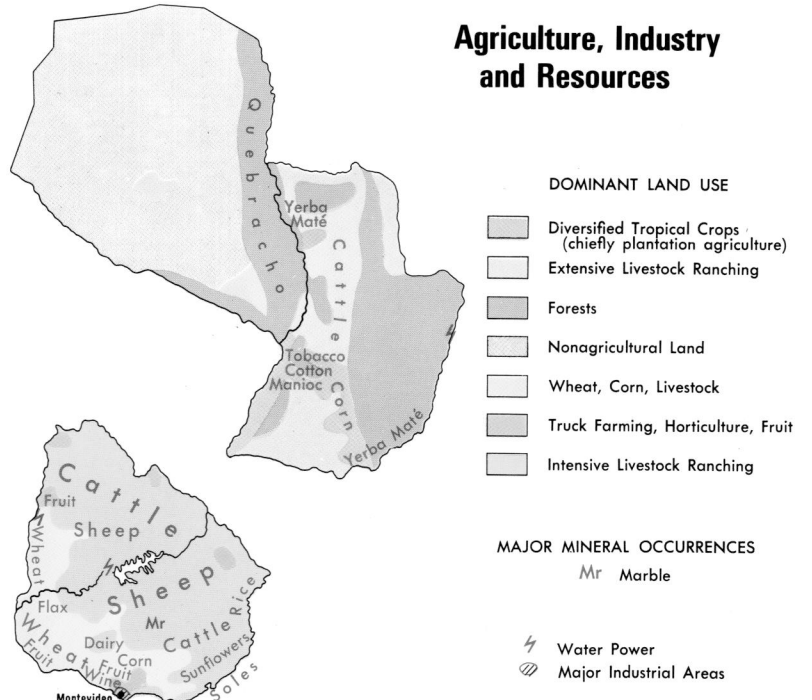

Agriculture, Industry and Resources

DOMINANT LAND USE

- Diversified Tropical Crops (chiefly plantation agriculture)
- Extensive Livestock Ranching
- Forests
- Nonagricultural Land
- Wheat, Corn, Livestock
- Truck Farming, Horticulture, Fruit
- Intensive Livestock Ranching

MAJOR MINERAL OCCURRENCES

Mr Marble

⚡ Water Power
▨ Major Industrial Areas

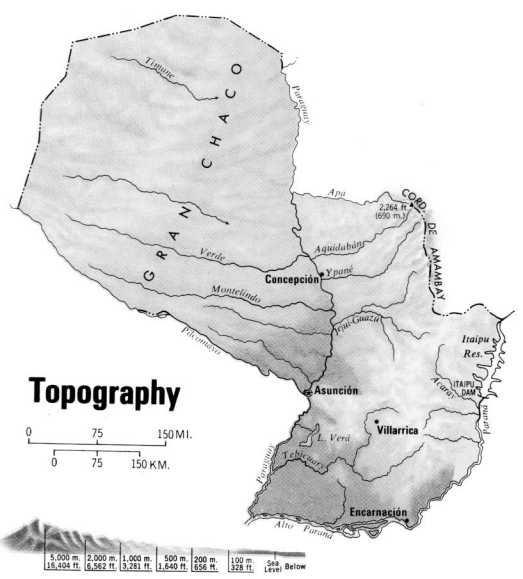

Topography

0 75 150 MI.
0 75 150 KM.

5,000 m. 2,000 m. 1,000 m. 500 m. 200 m. 100 m. Sea Level Below
16,404 ft. 6,562 ft. 3,281 ft. 1,640 ft. 656 ft. 328 ft.

URUGUAY

DEPARTMENTS

PARAGUAY

AREA 157,047 sq. mi. (406,752 sq. km.)
POPULATION 4,157,000
CAPITAL Asunción
LARGEST CITY Asunción
HIGHEST POINT Amambay Range
   2,264 ft. (690 m.)
MONETARY UNIT guaraní
MAJOR LANGUAGES Spanish, Guaraní
MAJOR RELIGION Roman Catholicism

URUGUAY

AREA 72,172 sq. mi. (186,925 sq. km.)
POPULATION 3,077,000
CAPITAL Montevideo
LARGEST CITY Montevideo
HIGHEST POINT Mirador Nacional 1,644 ft.
   (501 m.)
MONETARY UNIT Uruguayan peso
MAJOR LANGUAGE Spanish
MAJOR RELIGION Roman Catholicism

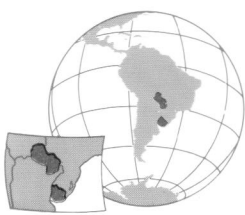

PARAGUAY

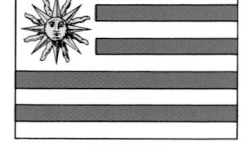

URUGUAY

Topography

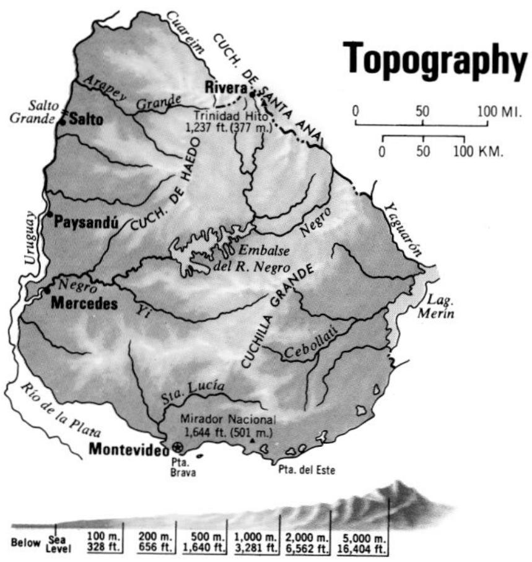

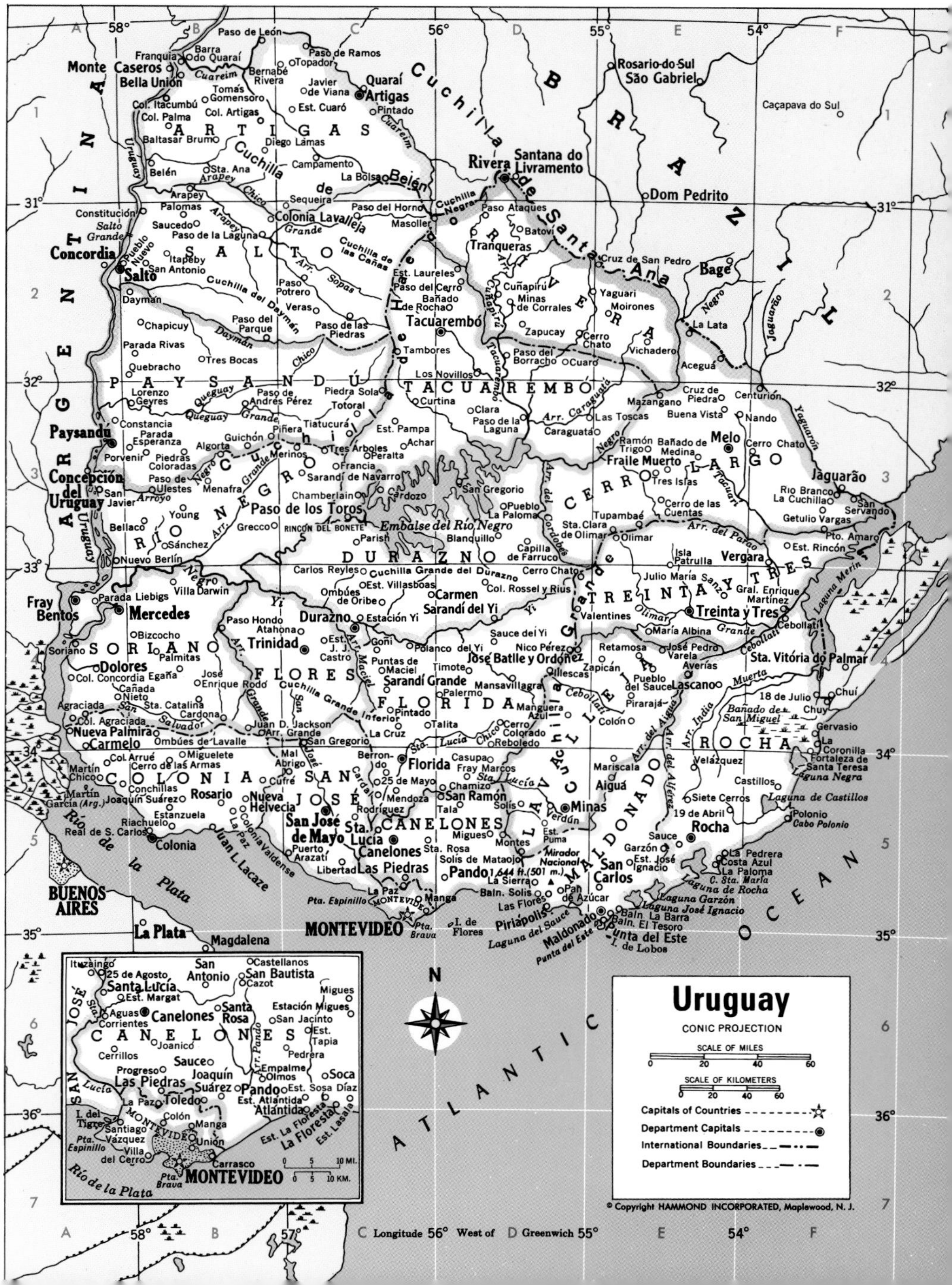

Uruguay

CONIC PROJECTION

SCALE OF MILES

SCALE OF KILOMETERS

Capitals of Countries ....★
Department Capitals ....◉
International Boundaries ......
Department Boundaries ....

North America

LAMBERT AZIMUTHAL EQUAL-AREA PROJECTION

MILES
0 100 200   400   600   800

KILOMETERS
0 100 200   400   600   800

Capitals of Countries .... ⊛
Other Capitals .... ⊚
International Boundaries .... — · —
Other Boundaries .... — · · —

© Copyright HAMMOND INCORPORATED, Maplewood, N.J.

Population Distribution

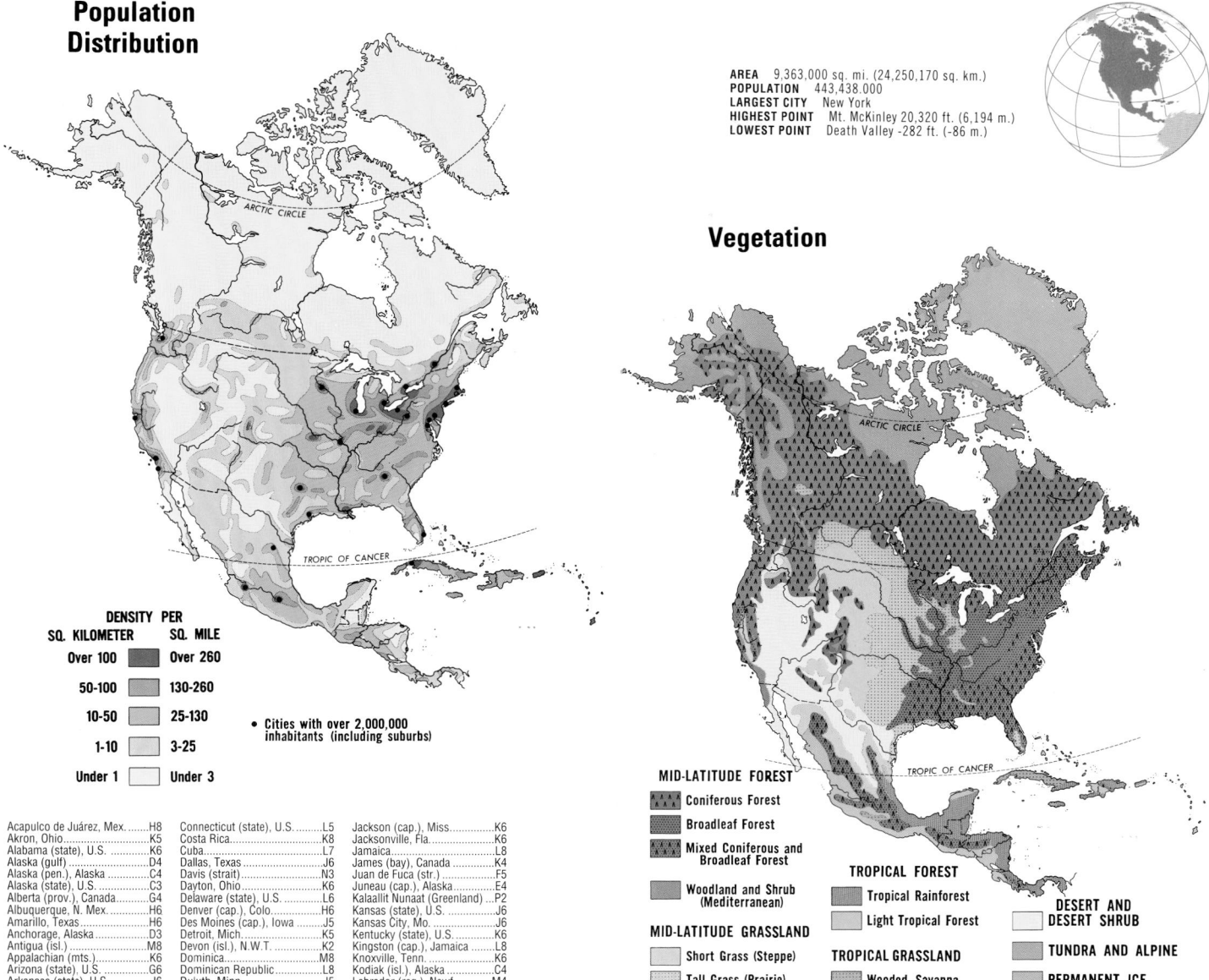

AREA	9,363,000 sq. mi. (24,250,170 sq. km.)
POPULATION	443,438,000
LARGEST CITY	New York
HIGHEST POINT	Mt. McKinley 20,320 ft. (6,194 m.)
LOWEST POINT	Death Valley -282 ft. (-86 m.)

Vegetation

DENSITY PER

SQ. KILOMETER	SQ. MILE
Over 100	Over 260
50-100	130-260
10-50	25-130
1-10	3-25
Under 1	Under 3

• Cities with over 2,000,000 inhabitants (including suburbs)

MID-LATITUDE FOREST
- Coniferous Forest
- Broadleaf Forest
- Mixed Coniferous and Broadleaf Forest
- Woodland and Shrub (Mediterranean)

MID-LATITUDE GRASSLAND
- Short Grass (Steppe)
- Tall Grass (Prairie)

TROPICAL FOREST
- Tropical Rainforest
- Light Tropical Forest

TROPICAL GRASSLAND
- Wooded Savanna

DESERT AND DESERT SHRUB

TUNDRA AND ALPINE

PERMANENT ICE

Average January Temperature

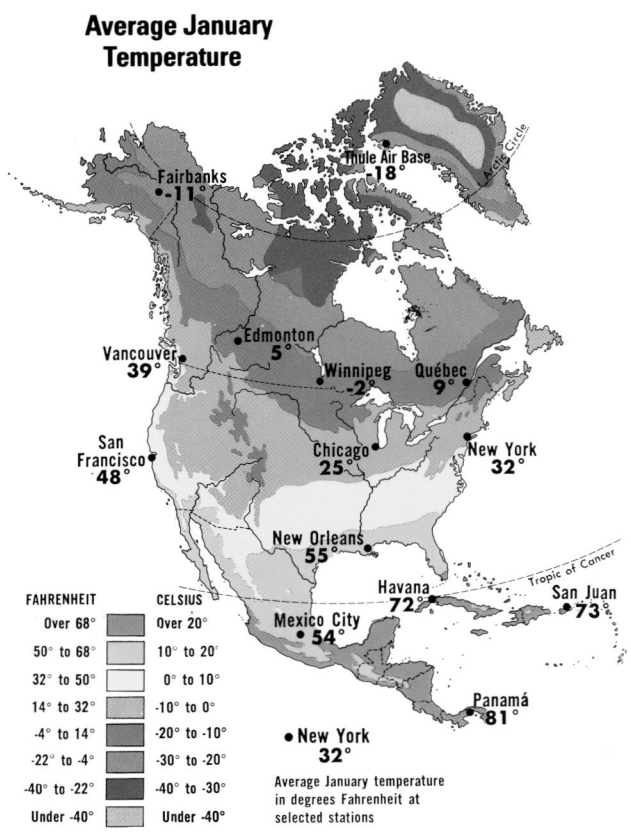

Thule Air Base
-18°

Fairbanks
-11°

Edmonton
5°

Vancouver
39°

Winnipeg
-2°

Québec
9°

San Francisco
48°

Chicago
25°

New York
32°

New Orleans
55°

Havana
72°

San Juan
73°

Mexico City
54°

Panamá
81°

FAHRENHEIT		CELSIUS	
Over 68°		Over 20°	
50° to 68°		10° to 20°	
32° to 50°		0° to 10°	
14° to 32°		-10° to 0°	
-4° to 14°		-20° to -10°	
-22° to -4°		-30° to -20°	
-40° to -22°		-40° to -30°	
Under -40°		Under -40°	

• New York
32°

Average January temperature in degrees Fahrenheit at selected stations

Average July Temperature

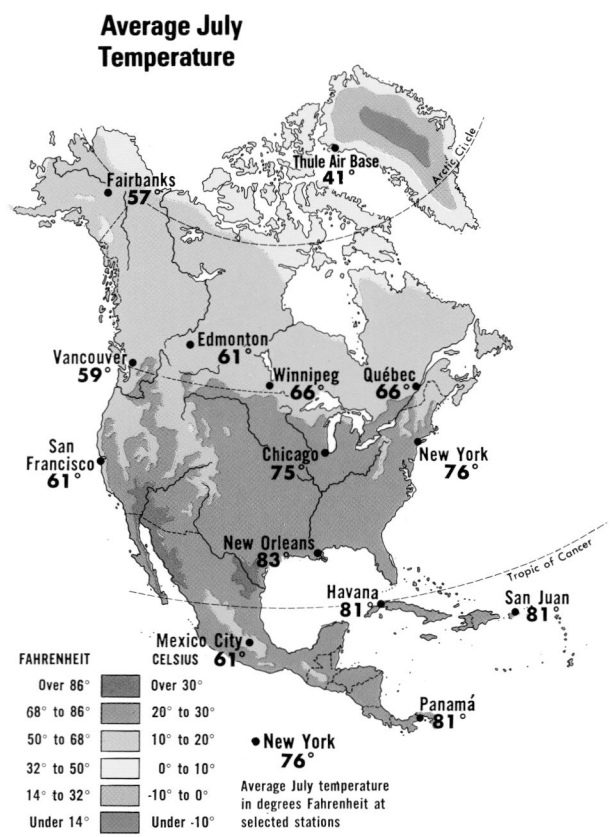

Thule Air Base
41°

Fairbanks
57°

Edmonton
61°

Vancouver
59°

Winnipeg
66°

Québec
66°

San Francisco
61°

Chicago
75°

New York
76°

New Orleans
83°

Havana
81°

San Juan
81°

Mexico City
61°

Panamá
81°

FAHRENHEIT		CELSIUS	
Over 86°		Over 30°	
68° to 86°		20° to 30°	
50° to 68°		10° to 20°	
32° to 50°		0° to 10°	
14° to 32°		-10° to 0°	
Under 14°		Under -10°	

• New York
76°

Average July temperature in degrees Fahrenheit at selected stations

Rainfall

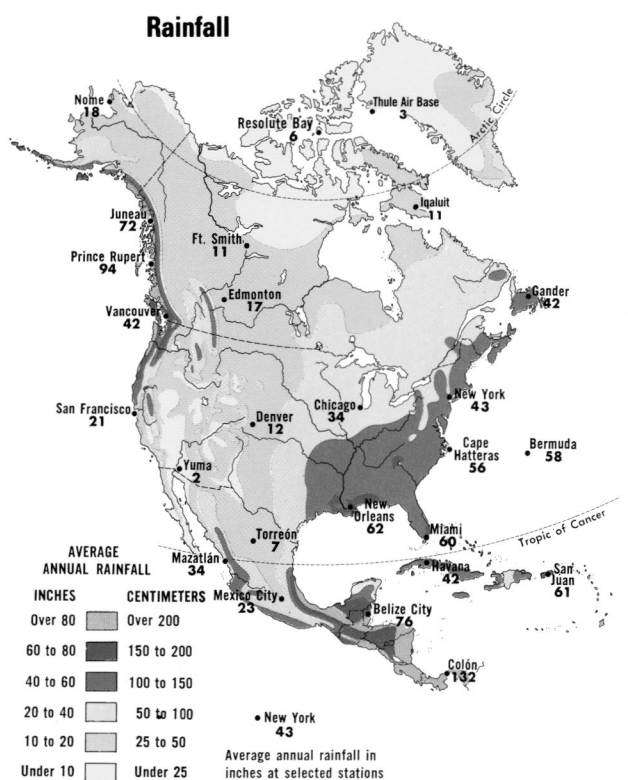

Nome
18

Thule Air Base
3

Resolute Bay
6

Juneau
72

Iqaluit
11

Prince Rupert
94

Ft. Smith
11

Vancouver
42

Edmonton
17

Gander
42

San Francisco
21

Denver
12

Chicago
34

New York
43

Yuma
2

Cape Hatteras
56

Bermuda
58

Torreón
7

New Orleans
62

Miami
60

Mazatlán
34

Havana
42

San Juan
61

Mexico City
23

Belize City
76

Colón
132

AVERAGE ANNUAL RAINFALL

INCHES		CENTIMETERS	
Over 80		Over 200	
60 to 80		150 to 200	
40 to 60		100 to 150	
20 to 40		50 to 100	
10 to 20		25 to 50	
Under 10		Under 25	

• New York
43

Average annual rainfall in inches at selected stations

Vegetation/Relief

SCALE OF MILES
0 200 400 600 800 1000

SCALE OF KILOMETERS
0 200 400 600 800 1000

Capitals of Countries .... ⊛
Other Capitals .... ◉
International Boundaries .... —·—·—
Canals ....

Depths in Fathoms

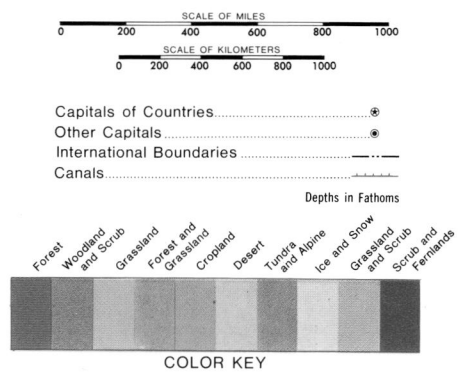

Forest
Woodland and Scrub
Grassland
Forest and Grassland
Cropland
Desert
Tundra and Alpine
Ice and Snow
Grassland and Scrub
Scrub and Fernlands

COLOR KEY

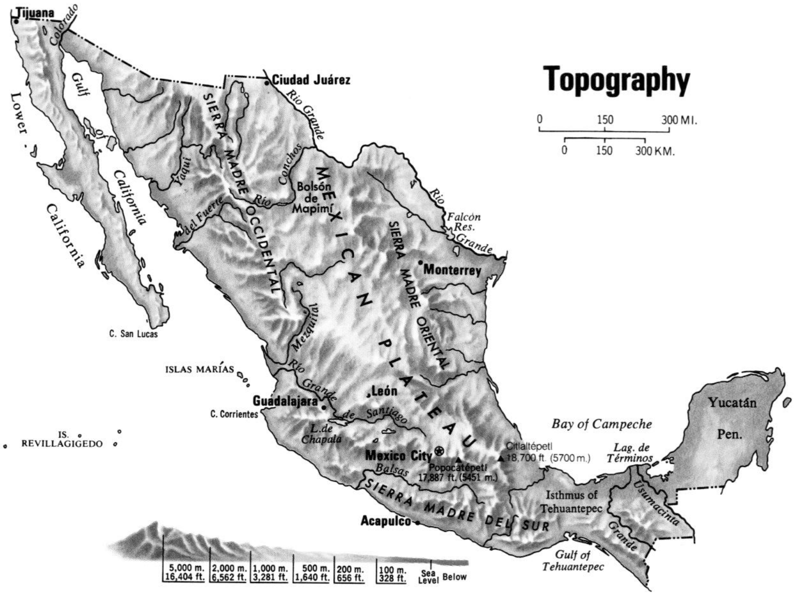
Topography
0 150 300 MI.
0 150 300 KM.
Tijuana
Colorado
Ciudad Juárez
Río Grande
Gulf
Lower
California
SIERRA MADRE OCCIDENTAL
Conchos
Bolsón de Mapimí
MEXICAN PLATEAU
Falcón Res.
Monterrey
SIERRA MADRE ORIENTAL
C. San Lucas
ISLAS MARÍAS
Guadalajara
León
C. Corrientes
IS. REVILLAGIGEDO
L. de Chapala
Mexico City
Citlaltépetl 18,700 ft. (5700 m.)
Popocatépetl 17,887 ft. (5451 m.)
Balsas
Bay of Campeche
Lag. de Términos
Yucatán Pen.
Isthmus of Tehuantepec
SIERRA MADRE DEL SUR
Acapulco
Gulf of Tehuantepec
5,000 m. 16,404 ft.
2,000 m. 6,562 ft.
1,000 m. 3,281 ft.
500 m. 1,640 ft.
200 m. 656 ft.
100 m. 328 ft.
Sea Level
Below

STATES

Aguascalientes 504,300..........H6
Baja California 1,227,400........B1
Baja California Sur 221,000....C3
Campeche 371,800..................O7
Chiapas 2,097,500..................N8
Chihuahua 1,935,100..............F2
Coahuila 1,561,000..................H3
Colima 339,400......................G7
Distrito Federal 9,377,300.......L1
Durango 1,160,300..................G4
Guanajuato 3,045,600..............J6
Guerrero 2,174,200..................J8
Hidalgo 1,518,200..................K6
Jalisco 4,296,500....................H6
México 7,542,300....................K7
Michoacán 3,049,400..............H7
Morelos 931,400......................K7
Nayarit 729,500......................G6
Nuevo León 2,463,500............K4
Oaxaca 2,517,500....................L8
Puebla 3,285,300....................L7
Querétaro 730,900..................J6
Quintana Roo 209,900............P7
San Luis Potosí 1,669,900......J5
Sinaloa 1,882,200....................F4
Sonora 1,498,100....................D2
Tabasco 1,150,000..................N7
Tamaulipas 1,924,900............K4
Tlaxcala 548,500......................N1
Veracruz 5,263,800..................L7
Yucatán 1,034,300..................P6
Zacatecas 1,144,700...............H5

CITIES and TOWNS

Acala 11,483............................N8
Acámbaro 32,257.....................J7
Acaponeta 11,844...................G5
Acapulco de Juárez 309,254...K8
Acatlán de Osorio 7,624..........K7
Acatzingo de Hidalgo 6,905....N2
Acayucan 21,173....................M8
Actopan 11,037 ....K6
Agua Dulce 21,060..................M7
Agua Prieta 20,754.................E1
Aguascalientes 181,277..........H6
Aguililla 5,715.........................H7
Ahuacatitlán 6,436..................L1
Ahuacatlán 5,350....................G6
Ahumada 6,466........................F1
Ajalpan 8,238..........................L7
Alamo 9,954.............................L6
Aldama 6,047..........................G2
Allende, Coahuila 11,076.........J2
Allende, Nuevo León 9,914.....J4
Altamira 6,053........................L5
Altepexi 6,661.........................L7
Altotonga 6,754......................P1
Alvarado 15,792.....................M7
Ameca 21,018.........................H6
Amecameca de Juárez
16,276.................................L1
Amozoc de Mota 9,203...........N2
Anáhuac, Chihuahua 10,886....F2
Anáhuac, Nuevo León 8,168...J3
Apan 13,705..........................M1
Apatzingán de la Constitución
44,849.................................H7
Apizaco 21,189.......................N1
Arandas 18,934.......................H6
Arcelia 10,024.........................J7
Ario de Rosales 8,774.............J7
Armería 10,616.......................G7
Arriaga 13,193........................N8
Arteaga 5,324.........................H7
Atlixco 41,967.......................M2
Atotonilco el Alto 16,271........H6
Atoyac de Álvarez 8,874..........J8
Autlán de Navarro 20,308.......G7
Axochiapan 8,283..................M2
Azcapotzalco 534,554.............L1
Bamoa 5,866...........................E4
Benjamín Hill 5,366................D1
Bernardino de Sahagún
12,327.................................M1
Cabo San Lucas 1,534............E5
Cacahoatán 5,079...................N9
Cadereyta Jiménez 13,586......K4
Calkiní 6,870..........................O6
Calpulálpan 8,659..................M1
Calvillo 6,453.........................H6
Campeche 69,506...................O7
Cananea 17,518......................D1
Canatlán 5,983.......................G4
Cancún 326............................Q6
Cañitas de Felipe Pescador
4,885...................................H5
Capulhuac de Mirafuentes
8,289...................................K1
Cárdenas, San Luis Potosí
12,020.................................K6
Cárdenas, Tabasco 15,643.....N8
Castaños 8,996.......................J3
Catemaco 11,786...................M7
Celaya 79,977.........................J6
Cerritos 10,421.......................J5
Cerro Azul 20,259...................L6
Chahuites 5,218.....................M8
Chalco de Díaz Covarrubias
12,172.................................M1
Champotón 6,606...................O7
Charcas 10,491.......................J5
Chetumal 23,685....................Q7
Chiapa de Corzo 8,571...........N8
Chiautempan 12,327..............N1
Chietla 4,602.........................M2
Chihuahua 327,313.................F2
Chilapa de Álvarez 9,204.........K8
Chilpancingo de los Bravos
36,193.................................K8
China, Nuevo León 4,958........K4
Chocomán 5,114....................P2
Cholula de Rivadavia 15,399..M1
Cihuatlán 9,451......................G7
Cintalapa de Figueroa 12,036 .N8
Ciudad Acuña (Villa Acuña)
30,276.................................J2
Ciudad Altamirano 8,694.........J7
Ciudad Camargo, Chihuahua
24,030.................................G3
Ciudad Camargo, Tamaulipas
5,953...................................K3
Ciudad de Río Grande 11,651.H5
Ciudad del Carmen 34,656.....N7
Ciudad del Maíz 5,241............K5
Ciudad Delicias 52,446...........G2
Ciudad Guzmán 48,166..........H7
Ciudad Hidalgo, Chiapas
4,105...................................N9
Ciudad Hidalgo, Michoacán
24,692.................................J7
Ciudad Juárez 424,135............F1
Ciudad Lerdo 19,803..............H4
Ciudad Madero 115,302..........L5
Ciudad Mante 51,247..............K5
Ciudad Mendoza 18,696.........O2
Ciudad Miguel Alemán
11,259.................................K3
Ciudad Obregón 144,795........E3
Ciudad Río Bravo 39,018........K4
Ciudad Satélite 35,083............L1
Ciudad Serdán 9,581..............O2
Ciudad Valles 47,587..............K5
Ciudad Victoria 83,897...........K5
Coalcomán de Matamoros
4,875...................................H7
Coatepec 21,542.....................P1
Coatetelco 5,268.....................L2
Coatzacoalcos 69,753............M7
Cocorit 4,478..........................E3
Colima 58,450........................H7
Colotlán 6,135........................H5
Comala 5,592.........................H7
Comalcalco 14,963.................N7
Comitán de Domínguez
21,249.................................O8
Compostela 9,801...................G6
Concepción del Oro 8,144.......J4
Contla 7,517...........................N1
Coquimatlán 6,212..................G7
Córdoba 78,495......................P2
Cosamaloapan de Carpio
19,766.................................M7
Coscomatepec de Bravo
6,023...................................P2
Costa Rica 11,795...................F4
Cotija de la Paz 9,178.............H7
Coyoacán 339,446..................L1
Coyotepec 8,888.....................L1
Coyuca de Benítez 6,328.........J8
Cozumel 5,858.......................Q6
Cuatrociénagas de Carranza
5,523...................................H3
Cuauhtémoc 26,598...............F2
Cuautepec de Hinojosa 5,501 .K6
Cuautitlán de Romero Rubio
11,439.................................L1
Cuautla Morelos 13,946..........L2
Cuernavaca 239,813...............L2
Cuitlahuac 4,813.....................P2
Culiacán 228,001....................F4
Dolores Hidalgo de la
Independencia Naci 16,849.J6
Durango 182,633....................G4
Dzidzantún 7,064....................P6
Dzitbalché 4,393.....................P6
Ebano 17,489.........................K5
Ecatepec de Morelos 11,899...L1
Ejutla de Crespo 5,263............L8
Eldorado 8,115.......................E4
El Fuerte 7,179.......................E3
El Salto 7,818.........................G5
Empalme 24,927....................D2
Encarnación de Díaz 10,474...H6
Ensenada 77,687....................A1
Escarcega 7,248.....................O7
Escuinapa de Hidalgo 16,442 .G5
Escuintla 4,111.......................N9
Esperanza, Sonora 11,762......E3
Espita 5,394...........................Q6
Fortín de las Flores 9,358.......P2
Francisco I. Madero 12,613....H4
Fresnillo de González
Echeverría 44,475..............H5
Frontera 10,066......................N7
General Terán 5,354...............K4
Gómez Palacio 79,650............G4
González 6,440......................K5
Guadalajara 1,478,383...........H6
Guadalupe, Nuevo León
51,899.................................K4
Guadalupe, Zacatecas 13,246.H5
Guadalupe Victoria, Durango
7,931...................................H4
Guamúchil 17,151...................E4
Guanajuato 36,809..................J6
Guasave 26,080......................E4
Guaymas 57,492....................D3
Gustavo Díaz Ordaz 10,154.....K3
Gutiérrez Zamora 9,099..........L6
Halachó 4,804........................O6
Hermosillo 232,691................D2
Heroica Caborca 20,771.........C1
Heroica Nogales 52,108.........D1
Hidalgo del Parral (Parral)
57,619.................................G3
Huachinango 16,826..............K7
Huajuapan de León 13,822.....L8
Huamantla 15,565..................N1
Huatabampo 18,506..............D3
Huatusco de Chicuellar 9,501 .P2
Huauchinango 16,826............L6
Huautla de Jiménez 6,132.......L7
Huejotzingo 8,552.................M1
Huejutla 6,854........................K6
Huetamo 9,333.......................J7
Huimanguillo 7,075................N8
Huitzuco de los Figueroa
9,406...................................K7
Huixtepec 5,927......................L8
Huixtla 15,737.......................N9
Hunucmá 8,020.....................O6
Iguala de la Independencia
45,355.................................K7
Irapuato 135,596....................J6
Isla, Veracruz 8,075...............M7
Isla Mujeres 2,663..................Q6
Ixmiquilpan 6,048..................K6
Ixtapa.....................................J8
Ixtapalapa 522,095.................L1
Ixtenco 5,035.........................N1
Ixtepec 14,025......................M8
Ixtlán del Río 10,986..............G6
Izamal 9,749...........................P6
Izúcar de Matamoros 21,164 .M2
Jala 4,535...............................G6
Jalapa Enríquez 161,352.........P1
Jalpa 9,904............................H6
Jalpa de Méndez 4,785...........N7
Jáltipan de Morelos 15,170....M8
Jerez de García Salinas
20,325.................................H5
Jico 7,269...............................P1
Jiménez, Chihuahua 18,095....G3
Jojutla de Juárez 14,438..........L2
José Cardel 5,396...................Q1
Juan Aldama 9,667.................H4
Juchipila 6,328.......................H6
Juchitán de Zaragoza 30,218 .M8
La Barca 18,055.....................H6
Lagos de Moreno 33,782.........J6
La Paz 46,011.........................D5
La Piedad Cavadas 34,963......H6
Las Choapas 20,166..............M7
Las Rosas 7,658.....................N8
León 468,887..........................J6
Lerdo de Tejada 11,628.........M8
Libres 4,830...........................O1
Linares 24,456........................K4
Loma Bonita 15,804...............M7
Loreto 7,132...........................J5
Los Mochis 67,953..................E4
Los Reyes de Salgado
19,452.................................H7
Macuspana 12,293.................N8
Madera 9,759.........................F2
Magdalena de Kino 10,281.....D1
Maltrata 5,457........................O2
Manzanillo 20,777..................G7
Mapastepec 5,907..................N9
Martínez de la Torre 17,203....L6
Mascota 5,674........................G6
Matamoros, Coahuila 15,125..H4
Matamoros, Tamaulipas
165,124...............................L4
Matehuala 28,799...................J5
Matías Romero 13,200..........M8
Maxcanú 6,505......................O6
Mazatlán 147,010....................F5
Melchor Múzquiz 18,868........H3
Melchor Ocampo del Balsas
4,766...................................H8
Meoquí 12,308......................G2
Mérida 233,912......................P6
Metepec 4,625......................M2
Mexicali 317,228....................B1
Mexico City (cap.) 9,377,300..L1
Miahuatlán de Porfirio Díaz
5,714...................................L8
Mier 5,636.............................K3
Miguel Auza 9,303.................H4
Minatitlán 68,397..................M8
Mineral del Monte 8,887.........K6
Misantla 8,799........................P1
Monclova 78,134....................J3
Montemorelos 18,642............K4
Monterrey 1,006,221..............J4
Morelia 199,099......................J7
Moroleón 25,620.....................J6
Motozintla de Mendoza 4,682.N9
Motul de Felipe Carrillo
Puerto 12,949.....................P6
Muna 5,491............................P6
Naica 7,190...........................G2
Namiquipa 4,875.....................F2
Nanacamilpa 6,356................M1
Naranjos 14,732.....................L6
Naucalpan de Juárez 9,425.....L1
Navojoa 43,817......................E3
Navolato 12,799.....................E4
Netzahualcóyotl 580,436.........L1
Nochistlán 8,780....................H6
Nogales 14,254......................P2
Nueva Casas Grandes 20,023..F1
Nueva Italia de Ruiz 14,718.....J7
Nueva Rosita 34,706..............J2
Nuevo Ideal 5,252..................G4
Nuevo Laredo 184,622............J3
Oaxaca de Juárez 114,948......L8
Ocampo 4,801........................K5
Ocotlán 35,361.......................H6
Ocotlán de Morelos 5,882.......L8
Ojinaga 12,757.......................G2
Ojocaliente 7,582...................H5
Ometepec 7,342.....................K8
Oriental 6,009.........................O1
Orizaba 105,150.....................P2
Oxkutzcab 8,182.....................P6
Ozumba de Alzate 6,876........M1
Pachuca de Soto 83,892.........K6
Padilla 4,581..........................K5
Palenque 2,595.......................O8
Pánuco 14,277.......................K6
Papantla de Olarte 26,773.......L6
Paraíso 7,561.........................N7
Parral 57,619.........................G3
Parras de la Fuente 18,207.....H4
Paso de Ovejas 4,371.............Q2
Pátzcuaro 17,299....................J7
Pedro Montoya 4,563.............K6
Pénjamo 9,245........................J6
Pericos 4,445..........................F4
Perote 12,742.........................O1
Petatlán 9,419.........................J8
Peto 8,362.............................P6
Pichucalco 4,615....................N8
Piedras Negras, Coahuila
41,033.................................J2
Piedras Negras, Veracruz
4,099...................................Q2
Pijijiapan 5,053......................N9
Poza Rica de Hidalgo
152,276...............................L6
Profesor Rafael Ramírez
5,338...................................O1
Progreso 17,518.....................P6

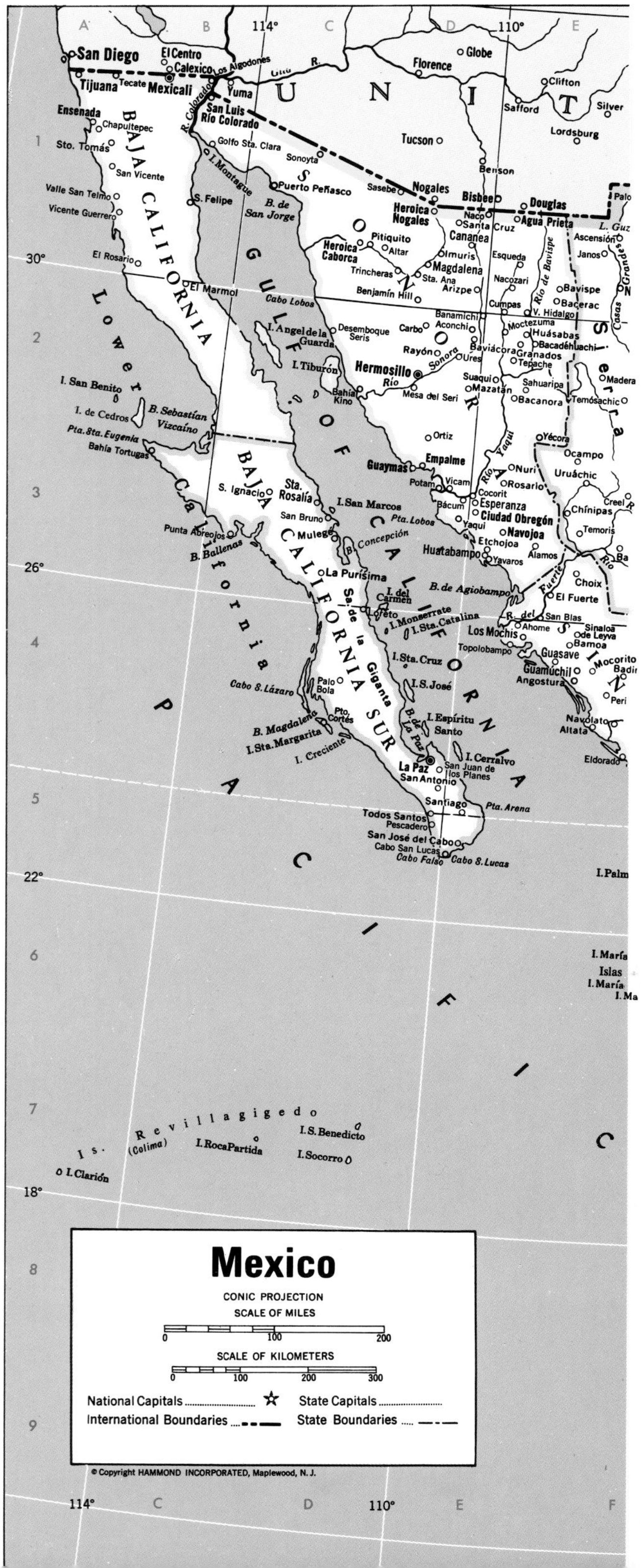
A
B
C
D
E
114°
110°
San Diego
El Centro
Calexico
Tijuana
Tecate
Mexicali
Yuma
Ensenada
San Luis Río Colorado
Globe
Florence
Clifton
Safford
Silver
Lordsburg
Tucson
Benson
UNITED
Golfo Sta. Clara
Sonoyta
Puerto Peñasco
Nogales
Bisbee
Douglas
Agua Prieta
Heroica Nogales
Cananea
Heroica Caborca
Magdalena
Arizpe
Benjamín Hill
El Rosario
San Felipe
B. de San Jorge
Desemboque
Hermosillo
Río Sonora
Mesa del Seri
I. Tiburón
I. Ángel de la Guarda
I. San Benito
I. de Cedros
B. Sebastián Vizcaíno
Pta. Sta. Eugenia
Bahía Tortugas
Guaymas
Empalme
Ciudad Obregón
Navojoa
Huatabampo
Sta. Rosalía
S. Ignacio
Mulegé
B. Concepción
Punta Abreojos
B. Ballenas
La Purísima
Loreto
I. del Carmen
I. Monserrate
I. Sta. Catalina
I. Sta. Cruz
I. San José
Los Mochis
Topolobampo
Guasave
Guamúchil
El Fuerte
Cabo S. Lázaro
B. Magdalena
I. Sta. Margarita
I. Espíritu Santo
I. Cerralvo
La Paz
San Antonio
Santiago
Pta. Arena
Todos Santos
Pescadero
San José del Cabo
Cabo San Lucas
Cabo Falso
Cabo S. Lucas
BAJA CALIFORNIA
BAJA CALIFORNIA SUR
GULF OF CALIFORNIA
Lower California
PACIFIC
SONORA
SINALOA
Sierra
Revillagigedo Is.
I. S. Benedicto
I. Roca Partida
I. Socorro
I. Clarión
(Colima)
I. Palm
I. María
Islas
I. María
30°
26°
22°
18°
1
2
3
4
5
6
7
8
9
Mexico
CONIC PROJECTION
SCALE OF MILES
0 100 200
SCALE OF KILOMETERS
0 100 200 300
National Capitals
State Capitals
International Boundaries
State Boundaries
© Copyright HAMMOND INCORPORATED, Maplewood, N. J.

AREA 761,601 sq. mi. (1,972,546 sq. km.)
POPULATION 86,154,000
CAPITAL Mexico City
LARGEST CITY Mexico City
HIGHEST POINT Citlaltépetl 18,700 ft.
  (5,700 m.)
MONETARY UNIT Mexican peso
MAJOR LANGUAGE Spanish
MAJOR RELIGION Roman Catholicism

States Indicated by Numbers

1 Tlaxcala
2 Morelos
3 Distrito Federal
4 México
5 Hidalgo
6 Querétaro
7 Guanajuato
8 Aguascalientes
9 Nayarit
10 Colima

OTHER FEATURES

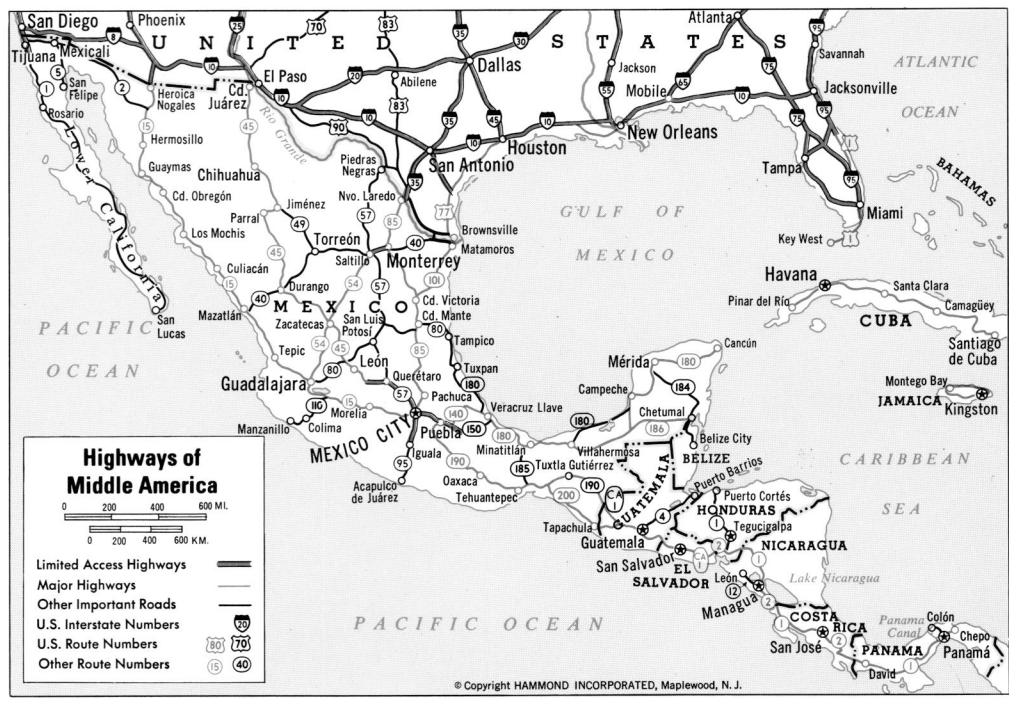

Highways of Middle America

Limited Access Highways
Major Highways
Other Important Roads
U.S. Interstate Numbers
U.S. Route Numbers
Other Route Numbers

© Copyright HAMMOND INCORPORATED, Maplewood, N.J.

Agriculture, Industry and Resources

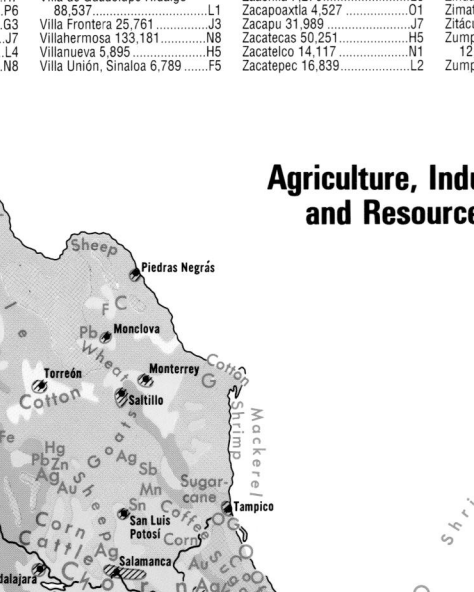

DOMINANT LAND USE

Wheat, Livestock
Cereals (chiefly corn), Livestock
Diversified Tropical Cash Crops
Cotton, Mixed Cereals
Livestock, Limited Agriculture
Range Livestock
Forests
Nonagricultural Land

MAJOR MINERAL OCCURRENCES

Ag	Silver	G	Natural Gas	O	Petroleum
Au	Gold	Gr	Graphite	Pb	Lead
C	Coal	Hg	Mercury	S	Sulfur
Cu	Copper	Mn	Manganese	Sb	Antimony
F	Fluorspar	Mo	Molybdenum	Sn	Tin
Fe	Iron Ore	Na	Salt	W	Tungsten
				Zn	Zinc

⚡ Water Power
▨ Major Industrial Areas

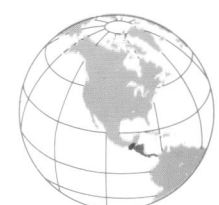

GUATEMALA

AREA 42,042 sq. mi. (108,889 sq. km.)
POPULATION 9,197,000
CAPITAL Guatemala
LARGEST CITY Guatemala
HIGHEST POINT Tajumulco 13,845 ft. (4,220 m.)
MONETARY UNIT quetzal
MAJOR LANGUAGES Spanish, Quiché
MAJOR RELIGION Roman Catholicism

BELIZE

AREA 8,867 sq. mi. (22,966 sq. km.)
POPULATION 180,000
CAPITAL Belmopan
LARGEST CITY Belize City
HIGHEST POINT Victoria Peak 3,681 ft. (1,122 m.)
MONETARY UNIT Belize dollar
MAJOR LANGUAGES English, Spanish, Mayan
MAJOR RELIGIONS Roman Catholicism, Protestantism

EL SALVADOR

AREA 8,260 sq. mi. (21,393 sq. km.)
POPULATION 5,207,000
CAPITAL San Salvador
LARGEST CITY San Salvador
HIGHEST POINT Santa Ana 7,825 ft. (2,385 m.)
MONETARY UNIT colón
MAJOR LANGUAGE Spanish
MAJOR RELIGION Roman Catholicism

HONDURAS

AREA 43,277 sq. mi. (112,087 sq. km.)
POPULATION 4,951,000
CAPITAL Tegucigalpa
LARGEST CITY Tegucigalpa
HIGHEST POINT Las Minas 9,347 ft. (2,849 m.)
MONETARY UNIT lempira
MAJOR LANGUAGE Spanish
MAJOR RELIGION Roman Catholicism

NICARAGUA

AREA 45,698 sq. mi. (118,358 sq. km.)
POPULATION 3,384,000
CAPITAL Managua
LARGEST CITY Managua
HIGHEST POINT Cerro Mocotón 6,913 ft. (2,107 m.)
MONETARY UNIT córdoba
MAJOR LANGUAGE Spanish
MAJOR RELIGION Roman Catholicism

COSTA RICA

AREA 19,575 sq. mi. (50,700 sq. km.)
POPULATION 2,959,000
CAPITAL San José
LARGEST CITY San José
HIGHEST POINT Chirripó Grande 12,530 ft. (3,819 m.)
MONETARY UNIT colón
MAJOR LANGUAGE Spanish
MAJOR RELIGION Roman Catholicism

PANAMA

AREA 29,761 sq. mi. (77,082 sq. km.)
POPULATION 2,418,000
CAPITAL Panamá
LARGEST CITY Panamá
HIGHEST POINT Vol. Baru 11,401 ft. (3,475 m.)
MONETARY UNIT balboa
MAJOR LANGUAGE Spanish
MAJOR RELIGION Roman Catholicism

Agriculture, Industry and Resources

DOMINANT LAND USE

- Cereals (chiefly corn) Livestock
- Diversified Tropical Cash Crops
- Livestock, Limited Agriculture
- Forests
- Nonagricultural Land

MAJOR MINERAL OCCURRENCES

Ag Silver
Au Gold
Cu Copper
O Petroleum
Pb Lead
Zn Zinc

⚡ Water Power
▨ Major Industrial Areas

GUATEMALA

HONDURAS

BELIZE

NICARAGUA

EL SALVADOR

COSTA RICA

PANAMA

BELIZE

CITIES and TOWNS

Belize City 39,887 ....C2
Belmopan (cap.) 2,932 ....C2
Corozal Town 6,862 ....C1
Hattieville 904 ....C2
Independence 225 ....C2
Libertad 856 ....C1

Orange Walk Town 8,441.......C1
Punta Gorda 2,219.................C2
San Ignacio 5,606.................C2
San José 420.........................C2
San Pedro 213.......................D2
Stann Creek Town 6,627........C2

OTHER FEATURES

Ambergris (cay).....................D1

Belize (riv.).............................C2
Glover (reef)...........................D2
Half Moon (cay)......................D2
Hondo (riv.).............................C1
Honduras (gulf)......................D2
Mauger (cay)..........................D2
Mya (mts.)..............................D2
New (riv.)................................C2
Saint Georges (cay)...............D2
Sarstún (riv.)..........................C3

Turneffe (isls.) D2

COSTA RICA

CITIES and TOWNS

Alajuela 33,122 ....E6
Atenas 1,728 ....E6
Bagaces 2,129 ....E5
Boruca 1,892 ....F6

Buenos Aires 302 ....F6
Cañas 6,053...........................E5
Cartago 21,753.......................F6
Ciudad Quesada 9,754 ....E5
Esparta 4,699.........................E5
Filadelfia 2,958.......................E5
Golfito 6,962...........................F6
Grecia 8,355...........................E6
Guácimo 1,168........................F6
Guápiles 3,524 ....F5

Heredia 22,700.......................E5
Las Juntas 1,129....................E5
Liberia 10,802 ....E5
Limón 29,621..........................F5
Miramar 1,673 ....E5
Nicoya 7,474...........................E5
Orotina 3,170..........................E6
Palmares 3,083.......................F6
Paquera E6
Paraíso 8,446 ....F6

Playa Bonita E6
Puerto Cortés 2,070................F6
Puntarenas 26,331..................E6
Quepos 2,155..........................E6
San Ignacio 446......................E6
San José (cap.) 215,441..........F5
San Marcos 917.......................E6
San Ramón 9,245....................E5
Santa Cruz 5,777....................E5
Santa Rosa..............................E5

(continued on following page)

Santo Domingo 5,148 ....F6
Sibube F5
Siquirres 4,361 ....F5
Suretka F6
Turrialba 12,151 ....F6
Vesta F6

OTHER FEATURES

Blanco (cape) E6
Blanco (peak) F6
Burica (pt.) F6
Cahuita (pt.) F6
Caño (isl.) F6
Chirripó Grande (mt.) F6
Coronada (bay) F6
Cuilapa Miravalles (vol.) ...E5
Dulce (gulf) F6
Góngora (mt.) E5
Guardian (bank) D6

Guionos (pt.) E6
Irazú (mt.) F6
Llerena (pt.) F6
Matapalo (cape) F6
Nicoya (gulf) F6
Nicoya (pen.) E5
Papagayo (gulf) E5
Salinas (bay) D5
San Juan (riv.) E5
Santa Elena (cape) D5
Talamanca (range) F6

EL SALVADOR

CITIES and TOWNS

Acajutla 8,598 ....B4
Ahuachapán 17,242 ....B4
Atiquizaya 7,035 ....C4
Chalatenango 7,633 ....C3

Chinameca 6,303 ....C4
Cojutepeque 20,615 ....C4
Ilobasco 6,572 ....C4
Intipucá 3,469 ....D4
La Libertad C4
La Unión 17,207 ....D4
Metapán 7,704 ....C3
Nueva San Salvador 35,106 ...C4
San Francisco Gotera 4,725 ...C4
San Miguel 59,304 ....C4
San Salvador (cap.)
   337,171 ....C4
Santa Ana 96,306 ....C4
Santa Rosa de Lima 5,707 ...D4
San Vicente 18,872 ....C4
Sensuntepeque 7,226 ......C4
Sonsonate 33,562 ....C4
Suchitoto 5,540 ....C4
Usulután 19,616 ....C4
Zacatecoluca 15,718 ....C4

OTHER FEATURES

Fonseca (gulf) D4
Güija (lake) C3
Lempa (riv.) C4
Remedios (mt.) B4
Santa Ana (mt.) C4

GUATEMALA

CITIES and TOWNS

Amatitlán 15,251 ....B3
Antigua 17,994 ....B3
Asunción Mita 7,477 ....C3
Chajul 4,329 ....B3
Champerico 5,722 ....A3
Chichicastenango 2,635 ...B3
Chimaltenango 12,860 ....B3
Chiquimula 16,126 ....C3

Coatepeque 15,979 ....A3
Cobán 11,418 ....B3
Comalapa 10,980 ....B3
Cubulco 2,021 ....B3
Cuilapa 4,287 ....B3
El Estor 2,324 ....B3
El Progreso 4,009 ....B3
Escuintla 33,205 ....B3
Flores 1,477 ....C2
Gualán 5,169 ....C3
Guatemala (cap.) 700,538 ...B3
Huehuetenango 12,570 ....B3
Ipala 3,386 ....C3
Iztapa 1,237 ....B4
Jacaltenango 4,517 ....B3
Jalapa 13,788 ....B3
Jutiapa 8,210 ....B3
La Gomera 2,394 ....B3
Livingston 2,898 ....C3
Los Amates 1,383 ....C3

Mazatenango 23,285 ....B3
Momostenango 5,210 ......B3
Morales 2,113 ....C3
Panzós 1,643 ....B3
Puerto Barrios 22,598 ......C3
Quezaltenango 53,021 .....B3
Quezaltepeque 2,222 ......B3
Rabinal 4,625 ....B3
Retalhuleu 19,060 ....B3
Río Hondo 1,416 ....B3
Sacapulas 1,439 ....B3
Salamá 5,529 ....B3
San Felipe 3,210 ....B3
San José 9,402 ....B4
San Luis Jilotepeque 6,055 ...C3
San Marcos 5,700 ....B3
San Martín Jilotepeque 3,770 ...B3
San Mateo Ixtatán 1,834 ...B3
San Pedro Carchá 4,465 ...B3
Santa Cruz del Quiché 7,651 ...B3

Santa Rosa de Lima 1,161 ...B3
Solalá 3,960 ....B3
Tacaná 1,280 ....A3
Tejutla 1,205 ....B3
Tikal C2
Totonicapán 8,568 ....B3
Zacapa 12,688 ....C3

OTHER FEATURES

Atitlán (lake) B3
Atitlán (vol.) B3
Azul (riv.) C2
Izabal (lake) C3
Minas (mts.) C3
Motagua (riv.) C3
Pasión (riv.) B2
Petén-Itzá (lake) B2
Salinas (riv.) B2

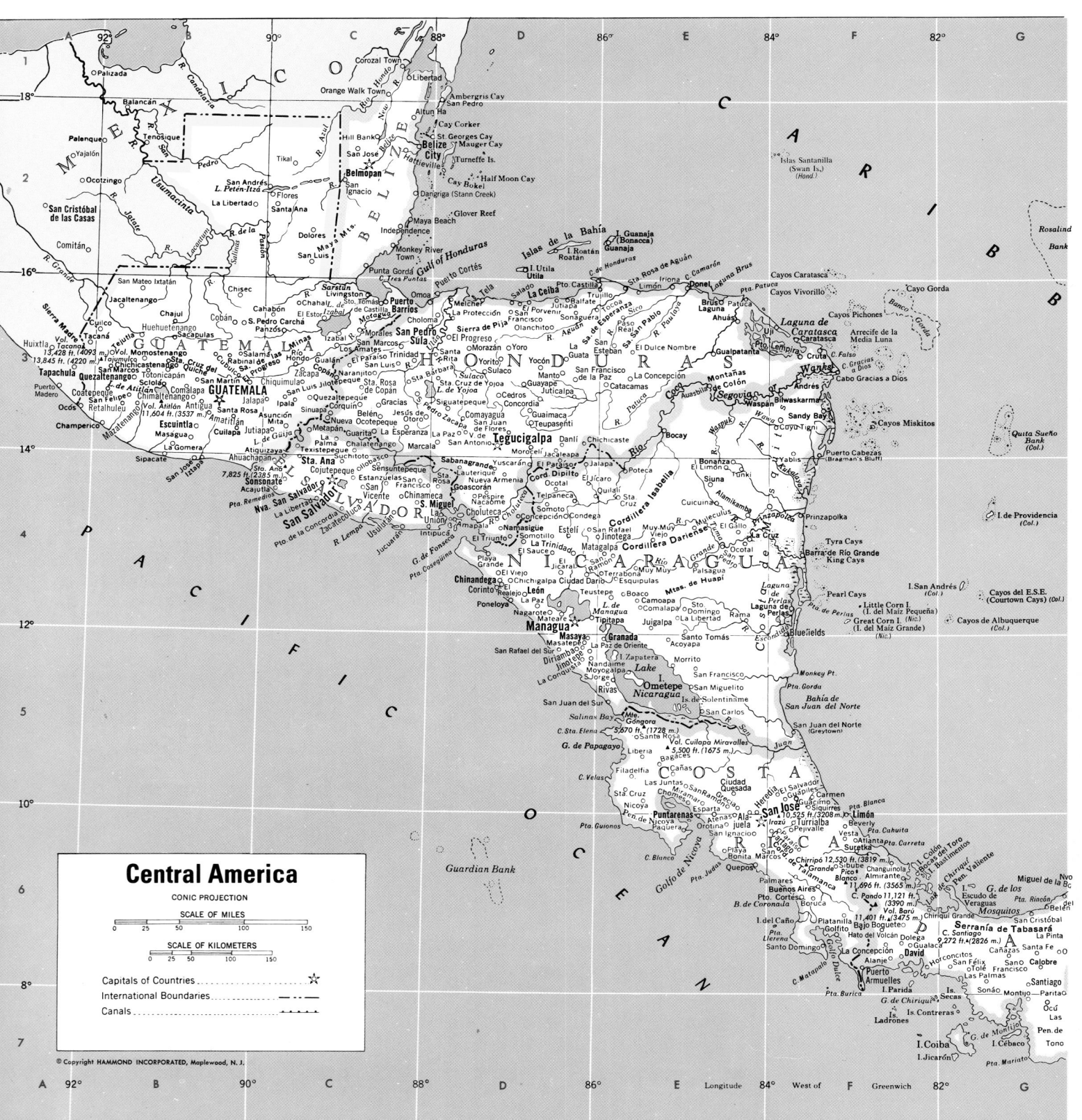

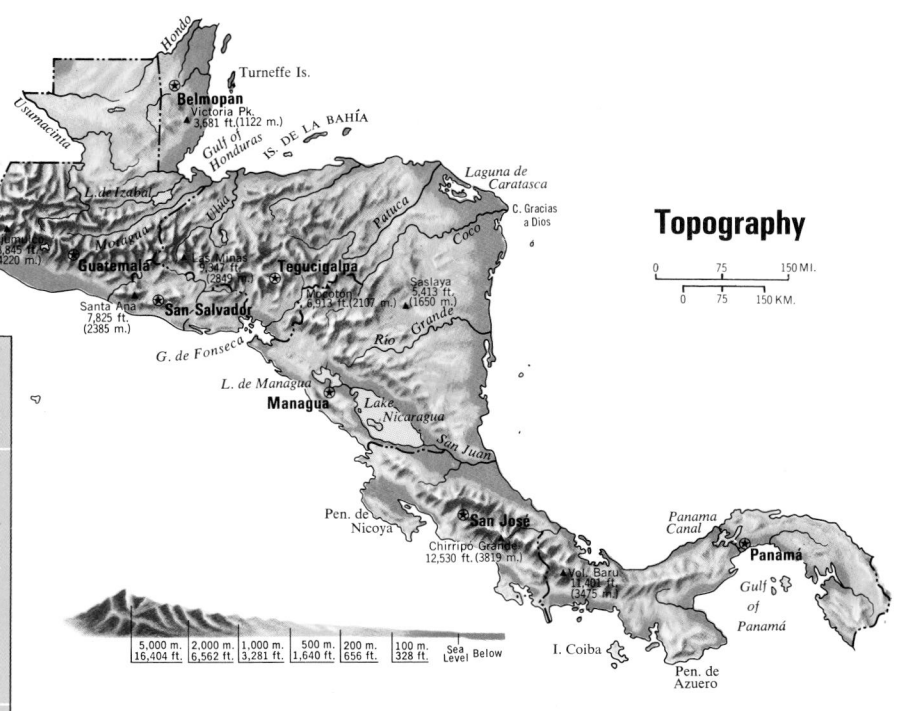

Topography

0 75 150 MI.
0 75 150 KM.

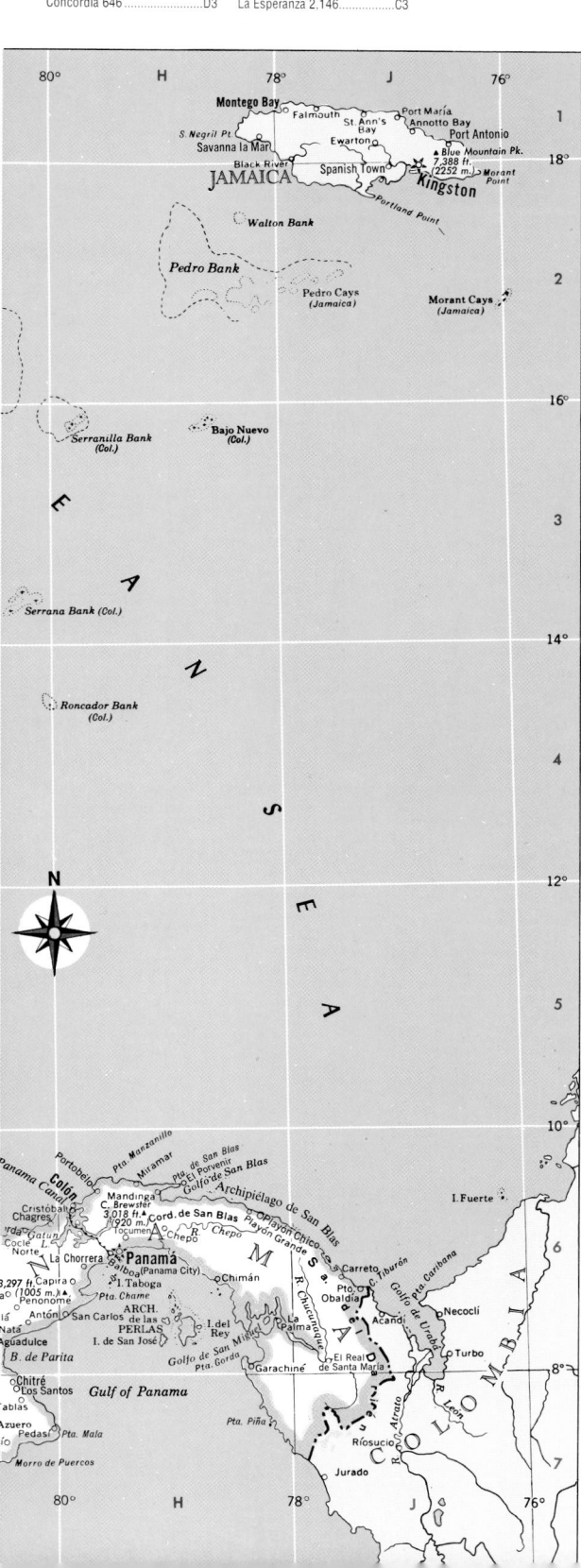

CUBA

HAITI

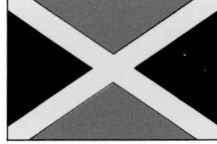
DOMINICAN REPUBLIC

JAMAICA

TRINIDAD AND TOBAGO

BARBADOS

GRENADA

BAHAMAS

DOMINICA

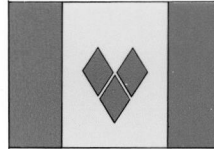
ST. LUCIA

ST. VINC. & GRENS.

ANTIGUA AND BARBUDA

CUBA
AREA 44,206 sq. mi. (114,494 sq. km.)
POPULATION 10,617,000
CAPITAL Havana
LARGEST CITY Havana
HIGHEST POINT Pico Turquino
6,561 ft. (2,000 m.)
MONETARY UNIT Cuban peso
MAJOR LANGUAGE Spanish
MAJOR RELIGION Roman Catholicism

HAITI
AREA 10,694 sq. mi. (27,697 sq. km.)
POPULATION 5,609,000
CAPITAL Port-au-Prince
LARGEST CITY Port-au-Prince
HIGHEST POINT Pic La Selle 8,793 ft. (2,680 m.)
MONETARY UNIT gourde
MAJOR LANGUAGES Creole French, French
MAJOR RELIGION Roman Catholicism

DOMINICAN REPUBLIC
AREA 18,704 sq. mi. (48,443 sq. km.)
POPULATION 6,867,000
CAPITAL Santo Domingo
LARGEST CITY Santo Domingo
HIGHEST POINT Pico Duarte
10,417 ft. (3,175 m.)
MONETARY UNIT Dominican peso
MAJOR LANGUAGE Spanish
MAJOR RELIGION Roman Catholicism

JAMAICA
AREA 4,411 sq. mi. (11,424 sq. km.)
POPULATION 2,392,000
CAPITAL Kingston
LARGEST CITY Kingston
HIGHEST POINT Blue Mountain Peak
7,402 ft. (2,256 m.)
MONETARY UNIT Jamaican dollar
MAJOR LANGUAGE English
MAJOR RELIGIONS Protestantism,
Roman Catholicism

PUERTO RICO
AREA 3,515 sq. mi. (9,104 sq. km.)
POPULATION 3,522,037
CAPITAL San Juan
MONETARY UNIT U.S. dollar
MAJOR LANGUAGES Spanish, English
MAJOR RELIGION Roman Catholicism

NETHERLANDS ANTILLES
AREA 390 sq. mi. (1,010 sq. km.)
POPULATION 246,000
CAPITAL Willemstad
MONETARY UNIT Antilles guilder
MAJOR LANGUAGES Dutch, Papiamento, English
MAJOR RELIGIONS Roman Catholicism,
Protestantism

BERMUDA
AREA 21 sq. mi. (54 sq. km.)
POPULATION 67,761
CAPITAL Hamilton
MONETARY UNIT Bermuda dollar
MAJOR LANGUAGE English
MAJOR RELIGION Protestantism

ARUBA
AREA 75 sq. mi (193 sq. km.)
POPULATION 66,790
CAPITAL Oranjestad
MONETARY UNIT Aruba guilder
MAJOR LANGUAGES Dutch, Papiamento
MAJOR RELIGION Roman Catholic

ANGUILLA
Anguilla (isl.) 6,519 . . . . F3

ANTIGUA and BARBUDA
Antigua (isl.) 76,213 . . . . G3
Barbuda (isl.) 1,071 . . . . G3
Caribbean (sea) B4
Codrington 1,071 . . . . G3
Falmouth 1,134 . . . . F3
Redonda F3
Saint John's (cap.) 21,814 . . . . G3

ARUBA
Aruba (isl.) 66,790 . . . . E4

BAHAMAS
Acklins (isl.) 616 . . . . C2
Andros (isl.) 8,397 . . . . B1
Atwood (Samana) (cay) D2
Berry (isls.) 509 . . . . B1
Biminis, The (isls.) 1,432 . . . . . B1
Caicos (passg.) D2
Cat (isl.) 2,143 . . . . C2
Cay Sal (bank) B2
Crooked (isl.) 517 . . . . D2
Eleuthera (isl.) 8,326 . . . . C1
Exuma (cays) C2
Flamingo (cay) C2
Freeport 22,301 . . . . B1
Grand Bahama (isl.) 33,102 . . . B1
Great Abaco (isl.) 7,324 . . . . . . C1
Great Bahama (bank) B1
Great Exuma (isl.) C2
Great Inagua (isl.) 939 . . . . D2
Great Isaac (isl.) B1

Gun (cay) B1
Harbour (isl.) C1
Little Inagua (isl.) D2
Long (cay) 33 . . . . C2
Long (isl.) 3,353 . . . . C2
Mayaguana (isl.) 476 . . . . D2
Mira Por Vos (cays) C2
Nassau (cap.) 135,437 . . . . C1
New Providence (isl.) 135,437 . . C1
Old Bahama (chan.) D2
Plana (cays) D2
Ragged (isl.) 146 . . . . C2
Rum (cay) C2
Samana (cay) D2
San Salvador (isl.) D1
Santaren (chan.) B1
Tongue of the Ocean (chan.) . . . C2
Verde (cay) C2
Watling (San Salvador) (isl.) . . . C1

BARBADOS
Bridgetown (cap.) 7,552 . . . . . . G4
Speightstown G4

BERMUDA
Bermuda (isls.) H3
Castle (harb.) H2
Great (sound) G3
Hamilton (cap.) 1,617 . . . . G3
Harrington (sound) H3
Ireland (isl.) G3
North (rapid) H2
Saint Davids (isl.) H3
Saint George 1,647 . . . . H2
Saint George's (isl.) H2
Somerset (isl.) G3

CAYMAN ISLANDS
Bartlett Deep B3
Cayman Brac (isl.) 1,603 . . . . . B3
George Town (cap.) 7,617 . . . . . B3
Grand Cayman (isl.) 15,000 . . . B3
Little Cayman (isl.) 74 . . . . B3
Misteriosa (bank) A3

CUBA
Bayamo 109,201 . . . . C2
Camagüey 245,235 . . . . B2
Cienfuegos 107,396 . . . . B2
Florida (str.) B1
Guanabacoa 89,741 . . . . A2
Guantánamo 178,129 . . . . C2
Havana (cap.) 1,924,886 . . . . . . A2
Holguín 190,155 . . . . C2
Juventud (Pines) (isl.) 57,879 . . A2
Manzanillo 95,420 . . . . C2
Mariano ○127,563 . . . . A2
Matanzas 103,302 . . . . B2
Pinar del Río 104,598 . . . . A2
San Felipe (cays) A2
Santa Clara 175,113 . . . . B2
Santiago de Cuba 362,432 . . . . C3
Windward (passg.) C3

DOMINICA
Portsmouth 2,329 . . . . G4
Roseau (cap.) 9,968 . . . . G4

DOMINICAN REPUBLIC
La Romana 91,571 . . . . E3
San Francisco de Macorís 64,906 . E3
San Pedro de Macorís 78,562 . . E3
Santiago 278,638 . . . . D3
Santo Domingo (cap.) 1,313,172 . . E3

GRENADA
Carriacou (isl.) 6,052 . . . . G4
Gouyave 2,498 . . . . F4
Grenadines (isls.) G4
Saint George's (cap.) 6,463 . . . . F5

GUADELOUPE
Basse-Terre (cap.) 13,397 . . . . . F4
Saint-Barthélemy (isl.) 3,059 . . . F3
Saint Martin (isl.) 8,072 . . . . . . F3

HAITI
Cap-Haïtien 64,406 . . . . D3
Gonaïves 34,209 . . . . D3
Port-au-Prince (cap.) 449,831 . . D3
Gonâve (isl.) D3
Jamaica (chan.) C3
Tortuga (isl.) D2

JAMAICA
Blue Mountain (peak) C3
Jamaica (chan.) C3
Kingston (cap.) 106,791 . . . . . . C3
Montego Bay 43,521 . . . . B3
Pedro (cays) C3
Savanna-la-Mar 11,759 . . . . . . B3

MARTINIQUE
Fort-de-France (cap.) 96,649 . . . G4
Saint-Pierre 4,923 . . . . G4
Pelée (vol.) G4

MONTSERRAT
Plymouth (cap.) 1,623 . . . . F3

NETHERLANDS ANTILLES
Bonaire (isl.) E4
Curaçao (isl.) E4
Oranjestad 10,100 . . . . D4
Saba (isl.) F3
Saint Eustatius (isl.) F3
Saint Martin (Sint Maarten) (isl.) . . F3
Willemstad (cap.) 95,000 . . . . . E4

PUERTO RICO
Bayamón 185,087 . . . . G1
Caguas 87,214 . . . . G1
Culebra (isl.) 1,265 . . . . G1
Mayagüez 82,968 . . . . F1
Mona (passg.) E3
Ponce 161,739 . . . . F1

San Juan (cap.) 424,600 . . . . . . G1
Vieques (isl.) 7,662 . . . . G1

SAINT KITTS and NEVIS
Basseterre (cap.) 14,725 . . . . . F3
Nevis (isl.) 9,300 . . . . F3
Saint Christopher (isl.) 35,104 . . F3

SAINT LUCIA
Castries (cap.) ●42,770 . . . . G4
Vieux Fort ●10,675 . . . . G4

SAINT VINCENT and THE GRENADINES
Bequia (isl.) G4
Georgetown 1,100 . . . . G4
Grenadines (isls.) 8,371 . . . . . . G4
Kingstown (cap.) 17,117 . . . . . . G4

TRINIDAD and TOBAGO
Port-of-Spain (cap.) 67,978 . . . . G5
Scarborough 6,057 . . . . G5
Tobago (isl.) 39,695 . . . . G5
Trinidad (isl.) 1,020,130 . . . . . . G5

TURKS and CAICOS ISLANDS
Caicos (isls.) 4,008 . . . . D2
Cockburn Harbour D2
Grand Caicos (isl.) 371 . . . . D2
Grand Turk (isl.) 3,146 . . . . D2
Providenciales (isl.) 979 . . . . . . D2
Turks (isls.) 3,348 . . . . D2

VIRGIN ISLANDS (British)
Anegada (isl.) 89 . . . . H1
Jost Van Dyke (isl.) 135 . . . . . . G1
Road Town (cap.) 2,200 . . . . . . H1
Tortola (isl.) 9,257 . . . . H1
Virgin Gorda (isl.) 1,443 . . . . . . H1

VIRGIN ISLANDS (U.S.)
Charlotte Amalie (cap.) 11,842 . . H1
Christiansted 2,914 . . . . H2
Fredriksted 1,046 . . . . H2
Saint Croix (isl.) 49,725 . . . . . . H2
Saint John (isl.) 2,472 . . . . H1
Saint Thomas (isl.) 44,372 . . . . G1

WEST INDIES
Antilles, Greater (isls.) B2
Antilles, Lesser (isls.) E4
Aves (Bird) (isl.) F4
Hispaniola (isl.) D2
Leeward (isls.) F3
Navassa (isl.) C3
Windward (isls.) G4

● Population of district.
○ Population of municipality.

Topography

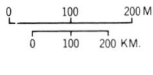
0   100   200 MI.
0   100   200 KM.

Below Sea Level | 100 m. 328 ft. | 200 m. 656 ft. | 500 m. 1,640 ft. | 1,000 m. 3,281 ft. | 2,000 m. 6,562 ft. | 5,000 m. 16,404 ft.

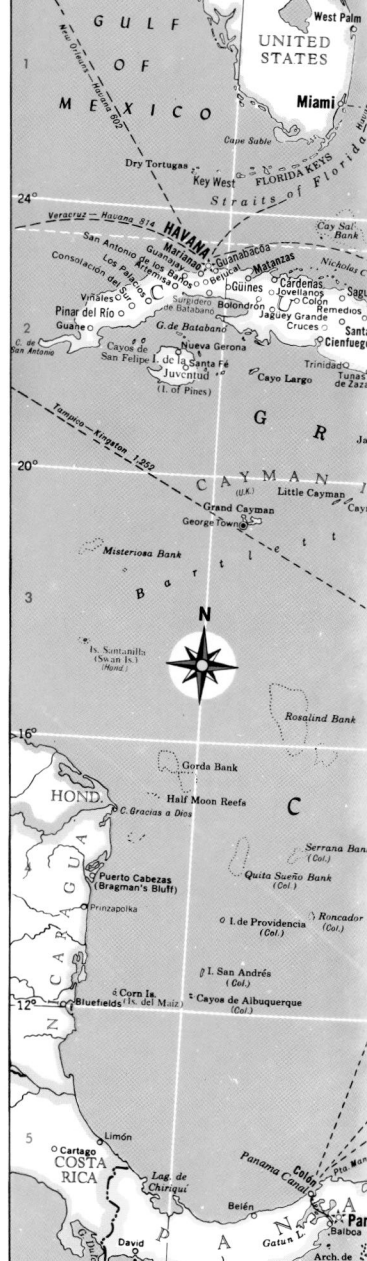

TRINIDAD AND TOBAGO
AREA 1,980 sq. mi. (5,128 sq. km.)
POPULATION 1,212,000
CAPITAL Port of Spain
LARGEST CITY Port of Spain
HIGHEST POINT Mt. Aripo 3,084 ft. (940 m.)
MONETARY UNIT Trinidad and Tobago dollar
MAJOR LANGUAGES English, Hindi
MAJOR RELIGIONS Roman Catholicism, Protestantism, Hinduism, Islam

BARBADOS
AREA 166 sq. mi. (430 sq. km.)
POPULATION 256,000
CAPITAL Bridgetown
LARGEST CITY Bridgetown
HIGHEST POINT Mt. Hillaby 1,104 ft. (336 m.)
MONETARY UNIT Barbadian dollar
MAJOR LANGUAGE English
MAJOR RELIGION Protestantism

GRENADA
AREA 133 sq. mi. (344 sq. km.)
POPULATION 103,103
CAPITAL St. George's
LARGEST CITY St. George's
HIGHEST POINT Mt. St. Catherine 2,757 ft. (840 m.)
MONETARY UNIT East Caribbean dollar
MAJOR LANGUAGES English, French patois
MAJOR RELIGIONS Roman Catholicism, Protestantism

BAHAMAS
AREA 5,382 sq. mi. (13,939 sq. km.)
POPULATION 253,000
CAPITAL Nassau
LARGEST CITY Nassau
HIGHEST POINT Mt. Alvernia 206 ft. (63 m.)
MONETARY UNIT Bahamian dollar
MAJOR LANGUAGE English
MAJOR RELIGIONS Roman Catholicism, Protestantism

DOMINICA
AREA 290 sq. mi. (751 sq. km.)
POPULATION 81,000
CAPITAL Roseau
HIGHEST POINT Morne Diablotin 4,747 ft. (1,447 m.)
MONETARY UNIT Dominican dollar
MAJOR LANGUAGES English, French patois
MAJOR RELIGIONS Roman Catholicism, Protestantism

SAINT KITTS AND NEVIS

SAINT LUCIA
AREA 238 sq. mi. (616 sq. km.)
POPULATION 148,000
CAPITAL Castries
HIGHEST POINT Mt. Gimie 3,117 ft. (950 m.)
MONETARY UNIT East Caribbean dollar
MAJOR LANGUAGES English, French patois
MAJOR RELIGIONS Roman Catholicism, Protestantism

SAINT VINCENT AND THE GRENADINES
AREA 150 sq. mi. (388 sq. km.)
POPULATION 124,000
CAPITAL Kingstown
HIGHEST POINT Soufrière 4,000 ft. (1,219 m.)
MONETARY UNIT East Caribbean dollar
MAJOR LANGUAGE English
MAJOR RELIGIONS Protestantism, Roman Catholicism

ANTIGUA AND BARBUDA
AREA 171 sq. mi. (443 sq. km.)
POPULATION 76,000
CAPITAL St. John's
HIGHEST POINT Boggy Peak 1,319 ft. (402 m.)
MONETARY UNIT East Caribbean dollar
MAJOR LANGUAGE English
MAJOR RELIGION Protestantism

SAINT KITTS & NEVIS
AREA 104 sq. mi. (269 sq. km.)
POPULATION 44,404
CAPITAL Basseterre
HIGHEST POINT Mt. Misery 4,314 ft. (1,315 m.)
MONETARY UNIT East Caribbean dollar
MAJOR LANGUAGE English
MAJOR RELIGIONS Protestantism, Roman Catholicism

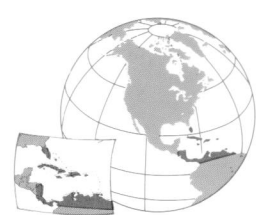

The West Indies
CONIC PROJECTION

SCALE OF MILES
0 50 100 150 200

SCALE OF KILOMETERS
0 50 100 200 300

Capitals of Countries _____ ☆
Other Capitals _____ ⊛

Puerto Rico

Bermuda Islands

© Copyright HAMMOND INCORPORATED, Maplewood, N.J.

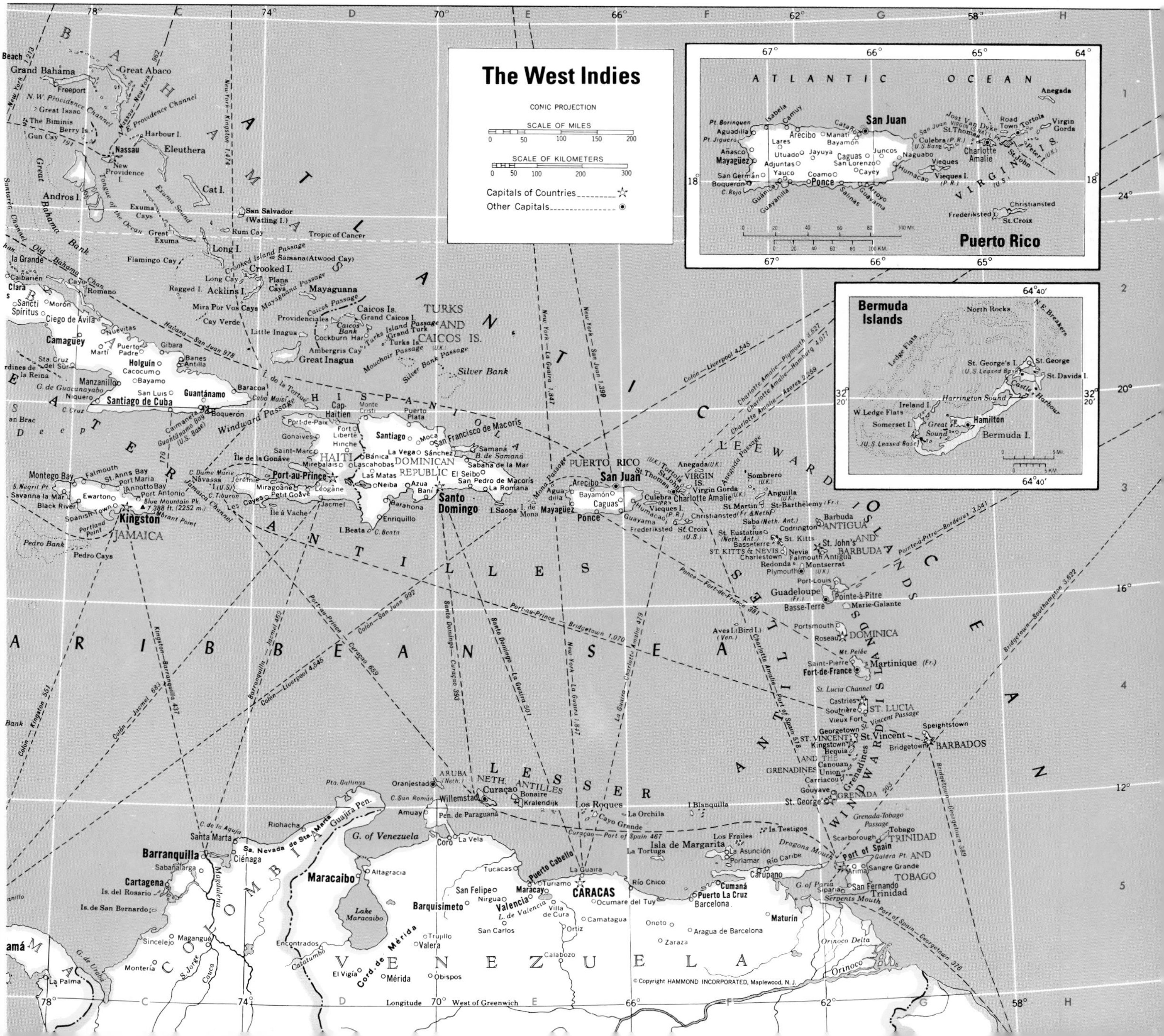

CUBA

PROVINCES

Camagüey, 664,566 . . . . . . G2
Ciego de Ávila 320,961 . . . . F2
Cienfuegos 326,412 . . . . . . E2
Granma 739,335 . . . . H4
Guantánamo 466,609 . . . . . K4
Holguín 911,034 . . . . J3
Juventud (municipio
especial) 57,879 . . . . C2
La Habana, Ciudad de
Habana 1,924,886 . . . . . C1
La Habana (Havana) 586,029 C1
Las Tunas 436,341 . . . . . . H3
Matanzas 557,628 . . . . . . D1
Pinar del Río 640,740 . . . . A2
Sancti Spíritus 399,700 . . . F2
Santiago de Cuba 909,506 . . H4
Villa Clara 764,743 . . . . . E1

CITIES and TOWNS

Abreus 14,267 . . . . D2
Agramonte 4,603 . . . . D1
Aguada de Pasajeros 20,219 D2
Alacranes 4,959 . . . . D1
Alonso Rojas 1,427 . . . . . B2
Alquízar 12,691 . . . . C1
Altagracia 1,722 . . . . G3
Alto Songo-La Maya 25,188 . J4

Amarillas 2,767 . . . . D2
Amazonas 1,066 . . . . F2
Antilla 10,052 . . . . J3
Arroyo Blanco 1,431 . . . . . F2
Artemisa 45,689 . . . . B1
Báez 4,178 . . . . E2
Báguanos 12,678 . . . . J3
Bahía Honda 16,901 . . . . . B1
Baire 4,879 . . . . H4
Banao 803 . . . . F2
Banes 38,905 . . . . J3
Baracoa 36,702 . . . . K4
Baraguá 12,633 . . . . F2
Bauta 26,826 . . . . C1
Bayamo 109,201 . . . . H4
Bejucal 15,649 . . . . C1
Bolondrón 5,840 . . . . D1
Buenaventura 4,711 . . . . . H3
Buenavista 1,303 . . . . . . F2
Buey Arriba 8,017 . . . . . . H4
Cabaiguán 36,544 . . . . . . E2
Cabañas 4,897 . . . . B1
Cabezas 5,262 . . . . D1
Cacocum 14,145 . . . . H3
Caibarién 32,094 . . . . . . F2
Caimanera 6,664 . . . . J4
Calabazar de Sagua 9,023 . . E1
Calimete 19,925 . . . . D1
Camagüey 245,235 . . . . . . G3
Camajuaní 26,653 . . . . . . E2
Campechuela 20,743 . . . . . G4
Canasí 1,637 . . . . C1

Candelaria 10,810 . . . . . . B1
Cárdenas 65,585 . . . . D1
Cartagena 2,166 . . . . D2
Cascajal 3,530 . . . . E1
Cauto del Embarcadero 949 . H4
Cauto el Cristo 1,626 . . . . J3
Central Amancio Rodríguez
22,506 . . . . G3
Central Bolivia 6,301 . . . . G2
Central Brasil 4,904 . . . . . G2
Central Cándido González
3,414 . . . . G3
Central Colombia 16,799 . . . G3
Central Frank País 9,066 . . K3
Central Guatemala 5,584 . . . J3
Central Haití 3,609 . . . . . J3
Central Los Reynaldos 3,997 J4
Central Loynaz Echevarría
3,245 . . . . G3
Central Manuel Tames 7,864 K4
Céspedes 6,634 . . . . G2
Chambas 19,877 . . . . F2
Chaparra 8,428 . . . . H3
Cidra 3,567 . . . . D1
Ciego de Ávila 80,010 . . . . F2
Cienfuegos 107,396 . . . . . E2
Colón 47,010 . . . . D1
Condado 33,115 . . . . E2
Consolación del Norte 4,681 B1
Consolación del Sur 34,334 . B2
Contramaestre 44,991 . . . . G3
Corralillo 15,822 . . . . D1

Cruces 20,324 . . . . E2
Cueto 23,183 . . . . J3
Cumanayagua 25,338 . . . . . E2
Daiquirí J4
Delicias 10,562 . . . . H3
Dos Caminos 3,772 . . . . . J4
Dos Ríos 1,786 . . . . J4
El Caney 3,921 . . . . J4
El Cobre 3,952 . . . . J4
El Santo 2,473 . . . . E1
Encrucijada 23,029 . . . . . E1
Esmeralda 17,205 . . . . . . G1
Esperanza 9,241 . . . . E2
Florencia 6,979 . . . . F2
Florida 43,881 . . . . G3
Fomento 17,310 . . . . F2
Gaspar 2,682 . . . . F2
Gibara 23,137 . . . . J3
Guáimaro 29,712 . . . . G3
Guanabacoa 89,741 . . . . . C1
Guanajay 21,042 . . . . B1
Guane 14,126 . . . . A2
Guantánamo 178,129 . . . . . K4
Guaro 3,086 . . . . J4
Guasimal 3,057 . . . . E2
Guayabal 3,703 . . . . G3
Guayos 6,753 . . . . F2
Güines 51,691 . . . . C1
Güira de Melena 19,851 . . . C1
Guisa 15,182 . . . . H4
Havana (cap.) 1,924,886 . . . C1
Herradura 3,762 . . . . B1

Holguín 190,155 . . . . J3
Ignacio Agramonte 1,487 . . G3
Imías 4,491 . . . . K4
Isabela de Sagua 3,721 . . . E1
Jagüey Grande 30,205 . . . . D2
Jamaica 5,128 . . . . K4
Jaruco 16,844 . . . . C1
Jatibonico 17,047 . . . . . . F2
Jíbaro 1,263 . . . . F2
Jiguaní 25,069 . . . . H4
Jobabo 14,899 . . . . H3
Jovellanos 35,043 . . . . . . D1
La Coloma 3,462 . . . . A2
La Maya-Alto Songo 25,188 . J4
Las Martinas 4,511 . . . . . A2
Limonar 9,629 . . . . D1
Los Arabos 10,664 . . . . . . E1
Los Palacios 21,884 . . . . . B1
Lugareño 4,396 . . . . G3
Mabay 6,176 . . . . H4
Maceo 2,652 . . . . H3
Majagua 9,110 . . . . F2
Manacas 5,914 . . . . E1
Manatí 11,054 . . . . H3
Manguito 2,739 . . . . D1
Manicaragua 33,900 . . . . . E2
Mantua 9,165 . . . . A2
Mapos (Amazonas) 1,066 . . F2
Manzanillo 95,420 . . . . . . H4
Marianao ○127,563 . . . . . . C1
Mariel 24,115 . . . . B1
Martí 11,474 . . . . D1

Matanzas 103,302 . . . . . . C1
Máximo Gómez, Ciego
de Ávila 5,116 . . . . F2
Máximo Gómez, Matanzas
4,970 . . . . D1
Mayajigua 4,425 . . . . F2
Mayarí 54,699 . . . . J3
Mayarí Arriba 2,302 . . . . . J4
Media Luna 13,794 . . . . . G4
Mendoza 2,914 . . . . A2
Meneses 4,768 . . . . F2
Minas 17,675 . . . . G2
Minas de Matahambre
14,976 . . . . A1
Moa 28,696 . . . . K3
Morón 40,396 . . . . F2
Nicaro 9,506 . . . . J3
Niquero 15,544 . . . . G4
Nueva Gerona 17,175 . . . . C2
Nuevitas 35,103 . . . . G2
Orozco 4,256 . . . . B1
Palma Soriano 66,222 . . . . J4
Palmira 19,680 . . . . E2
Pedro Betancourt 22,915 . . D1
Perico 20,633 . . . . D1
Pilón 10,194 . . . . H4
Pinar del Río 104,598 . . . . B2
Placetas 46,038 . . . . E2
Primero Enero 14,807 . . . . F2
Puerto Esperanza 3,499 . . . A1
Puerto Padre 46,806 . . . . . H3
Quemado de Güines 11,208 . E1

Rancho Veloz 3,966 . . . . . D1
Ranchuelo 34,255 . . . . . . E2
Regla 38,491 . . . . C1
Remedios 27,722 . . . . E1
República Dominicana
2,540 . . . . F2
Río Cauto 19,550 . . . . . . H4
Rodas 16,350 . . . . E2
Sagua de Tánamo 15,327 . . K3
Sagua la Grande 52,315 . . . E1
San Andrés 2,127 . . . . . . H3
San Antonio de los Baños
28,137 . . . . C1
San Cristóbal 30,769 . . . . . B1
Sancti Spíritus 79,542 . . . . E2
San Diego de los Baños
1,430 . . . . B1
San Germán 12,362 . . . . . J3
San José de las Lajas
37,149 . . . . C1
San José de los Ramos
1,726 . . . . D1
San Juan y Martínez 13,227 . B2
San Luis, Pinar del Río
5,677 . . . . B2
San Luis, Santiago de Cuba
32,826 . . . . J4
San Nicolás 12,368 . . . . . C1
San Ramón 2,676 . . . . . . H4
Santa Clara 175,113 . . . . . E1
Santa Cruz del Norte
15,239 . . . . C1

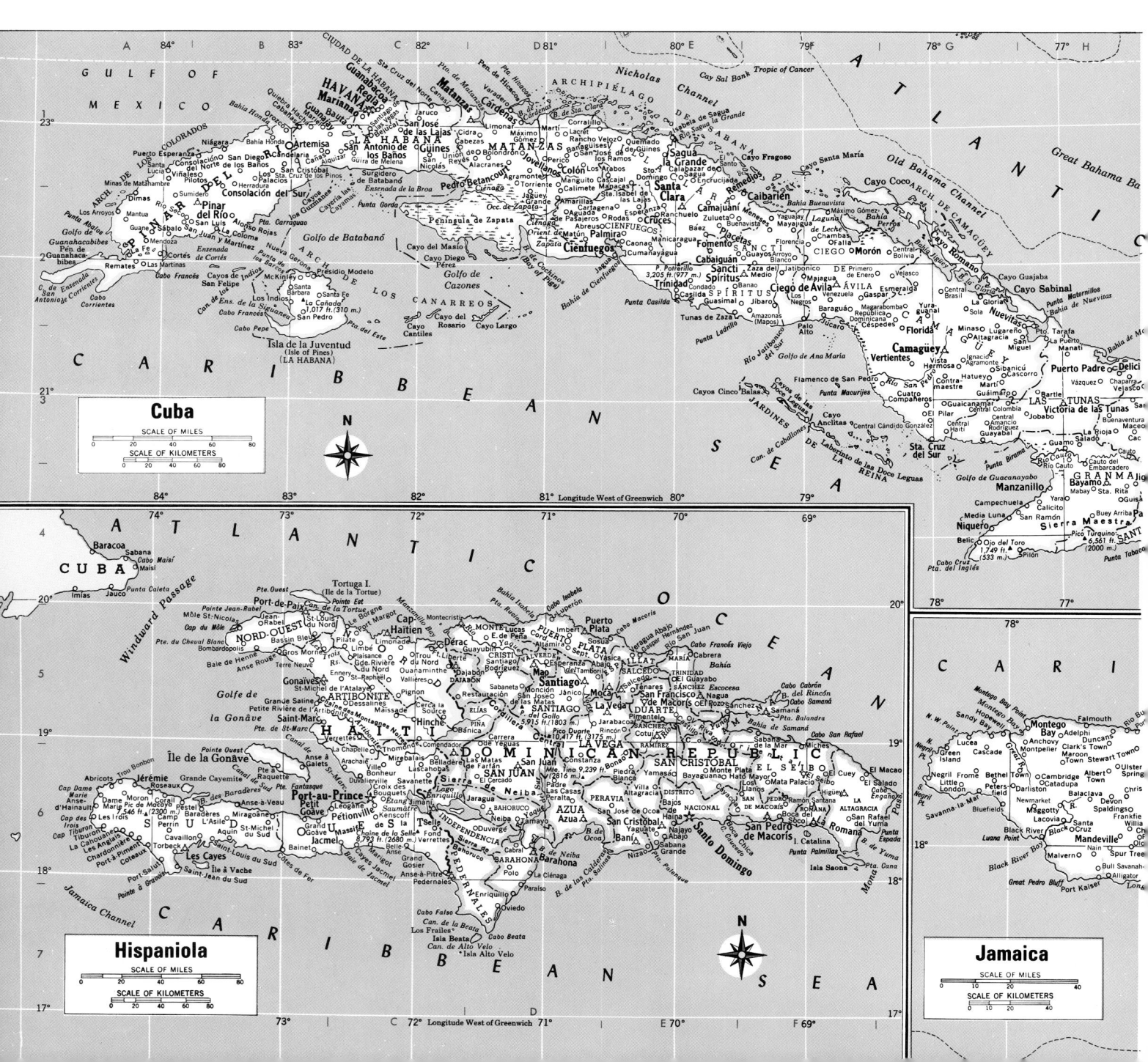

Santa Cruz de los Pinos 3,545 ... B1
Santa Cruz del Sur 27,142 .. G3
Santa Fe 3,925 ... B2
Santa Isabel de las Lajas 7,279 ... E2
Santa Lucía 3,734 ... J3
Santa Rita 6,358 ... H4
Santiago de Cuba 362,432 ..J4
Santiago de las Vegas 29,325 ... C1
Santo Domingo 32,950 .. E1
Sibanicú 14,252 ... G3
Sola 2,436 ... G2
Sumidero 980 ... C1
Surgidero de Batabanó 11,533 ... C1
Tacajó 4,469 ... J3
Torriente 1,759 ... D11
Trinidad 42,080 ... E2
Unión de Reyes 28,422 ... C1
Varadero 14,737 ... D1
Vázquez 3,851 ... H3
Velasco 5,618 ... H3
Venezuela 13,744 ... F2
Vertientes 25,178 ... G3
Victoria de las Tunas 87,522 H3
Viñales 2,049 ... A1
Yaguajay 30,720 ... F2
Yara 238,879 ... H4
Zaza del Medio 7,495 ... F2
Zulueta 5,425 ... E2

OTHER FEATURES
Abalos (pt.) ... A2
Ana María (gulf) ... F3
Anclitas (cay) ... F3
Batabanó (gulf) ... C1
Birama (pt.) ... G4
Broa (inlet) ... C1
Buenavista (bay) ... F2
Caballones (chan.) ... F3
Camagüey (arch.) ... G2
Cantiles (cay) ... G2
Cárdenas (bay) ... D1
Carraguao (pt.) ... B2
Casilda (pt.) ... E2
Cauto (riv.) ... H3
Cayamas (cays) ... C2
Cazones (gulf) ... D2
Cienfuegos (bay) ... D2
Cinco Balas (cays) ... F3
Cochinos (bay) ... D2
Coco (cay) ... G1
Corrientes (cape) ... A2
Corrientes (inlet) ... A2
Cortés (inlet) ... B2
Cristal, Sierra del (mts.) .. J3
Cruz (cape) ... G4
Diego Pérez (cay) ... C3
Doce Leguas (cays) ... F3
Este (pt.) ... C3
Fragoso (cay) ... F1
Francés (cape) ... A2

Gorda (pt.) ... C2
Gran Piedra (mt.) ... J4
Guacanayabo (gulf) ... G4
Guajaba (cay) ... G2
Guanahacabibes (gulf) ... A2
Guanahacabibes (pen.) ... A2
Guantánamo (bay) ... J4
Guantánamo Bay U.S. Nav. Reserve ... K4
Guarico (pt.) ... K3
Guzmanes (cays) ... B2
Hicacos (pen.) ... D1
Hicacos (pt.) ... D1
Honda (bay) ... B1
Indios (chan.) ... B2
Inglés (cay) ... G4
Jardines de la Reina (arch.) . F3
Jatibonico del Sur (riv.) ... F3
Jigüey (bay) ... G2
Juventud, Isla de la (Pines) (isl.) 57,879 ... B3
Laberinto de las Doce Leguas (cays) ... F3
Ladrillo (pt.) ... E3
Largo (cay) ... D2
Leche (lag.) ... F2
Los Barcos (cay) ... B2
Los Canarreos (arch.) ... C2
Los Colorados (arch.) ... A1
Lucrecia (cape) ... J3
Macurijes (pt.) ... F3
Maestra, Sierra (mts.) ... H4
Maisí (cape) ... K4
Mangle (cay) ... J3
Maslo (cay) ... C2
Matanzas (bay) ... D1
Nicholas (chan.) ... E1
Nipe (bay) ... J3
Nuevitas (bay) ... H2
Ojo del Toro (mt.) ... G4
Old Bahama (chan.) ... G1
Pepe (cape) ... B3
Perros (bay) ... G2
Pigs (Cochinos) (bay) ... D2
Pines (Isla de la Juventud) (isl.) 7,879 ... B3
Potrerillo (peak) ... E2
Quemado (pt.) ... K4
Romano (cay) ... G2
Rosario (cay) ... C2
Sabana (arch.) ... E1
Sabinal (cay) ... H2
Sagua la Grande (riv.) ... E1
San Antonio (cape) ... A2
San Felipe (cays) ... B2
San Pedro (riv.) ... G3
Santa Clara (bay) ... D1
Santa María (cay) ... F1
Siguanea (bay) ... B2
Tabacal (pt.) ... H4
Toa, Cuchillas de (mts.) .. K4
Tortuguilla (pt.) ... K4
Turquino (peak) ... H4
Zapata (pen.) ... C2
Zapata Occidental (swamp). D2
Zapata Oriental (swamp) ... D2

DOMINICAN REPUBLIC
PROVINCES
Azua 142,770 ... D6

Bahoruco 78,636 ... D6
Barahona 137,160 ... D6
Dajabón 57,709 ... D5
Distrito Nacional 1,550,739 . E6
Duarte 235,544 ... E5
Elías Piña 65,384 ... C5
El Seibo 157,866 ... F6
Espaillat 164,017 ... E5
Independencia 38,768 ... D6
La Altagracia 100,112 ... F6
La Romana 109,769 ... F6
La Vega 385,043 ... D6
María Trinidad Sánchez 112,629 ... E5
Monte Cristi 83,407 ... D5
Pedernales 17,006 ... D7
Peravia 168,123 ... E6
Puerto Plata 206,757 ... D5
Salcedo 99,191 ... E5
Samaná 65,699 ... E5
Sánchez Ramírez 126,567 .. E5
San Cristóbal 446,132 ... E6
San Juan 239,957 ... D6
San Pedro de Macorís 152,890 ... F6
Santiago 550,372 ... D5
Santiago Rodríguez 55,411 . D5
Valverde 100,319 ... D5

CITIES and TOWNS
Altamira 2,759 ... D5
Azua 31,481 ... D5
Bajos de Haina 33,135 ... E6
Baní 36,705 ... E6
Barahona 49,334 ... D6
Bonao 44,486 ... E5
Cabrera 2,542 ... E5
Comendador 5,962 ... C6
Constanza 15,141 ... D6
Cotuí 16,688 ... E5
Dajabón 8,808 ... D5
El Seibo 13,511 ... F6
Hato Mayor 17,859 ... F6
Higüey 33,501 ... F6
Imbert 5,315 ... D5
Jarabacoa 13,416 ... E5
Jimaní 3,327 ... C6
La Romana 91,571 ... F6
La Vega 52,432 ... E5
Luperón 2,500 ... D5
Mao 33,527 ... D5
Moca 31,176 ... E5
Monción 3,344 ... D5
Nagua 20,912 ... E5
Puerto Plata 45,348 ... D5
Sabana de la Mar 9,983 ... F5
Sabaneta 9,170 ... D5
Samaná 5,023 ... E5
San Cristóbal 58,520 ... E6
San Francisco de Macorís 64,906 ... E5
San Juan 49,764 ... D6
San Pedro de Macorís 78,562 ... F6
Santiago 278,638 ... D5
Santo Domingo (cap.) 1,313,172 ... E6
Tenares 4,065 ... E5
Villa Altagracia 20,890 ... E6

OTHER FEATURES
Alto Velo (chan.) ... C7
Alto Velo (isl.) ... D7
Balandra (pt.) ... F5
Beata (cape) ... D7
Beata (chan.) ... C7
Beata (isl.) ... C7
Cabrón (cape) ... F5
Calderas (bay) ... E6
Cana (pt.) ... F6
Catalina (isl.) ... F6
Caucedo (capee) ... E6
Central, Cordillera (range) . D5
Duarte (peak) ... E5
Engaño (cape) ... F6
Enriquillo (lake) ... D6
Escocesa (bay) ... F5
Espada (pt.) ... F5
Falso (cape) ... C7
Francés Viejo (cape) ... F5
Gallo (mt.) ... D5
Isabela (bay) ... D5
Isabela (cape) ... D5
Los Frailes (isl.) ... C7
Macorís (cape) ... E5
Manzanillo (bay) ... D5
Mona (passg.) ... F6
Neiba (bay) ... D6
Neiba, Sierra de (mts.) ... D6
Ocoa (bay) ... E6
Oriental, Cordillera (range) . F6
Palenque (pt.) ... E6
Palmillas (pt.) ... F6
Rincón (bay) ... F5
Rucia (pt.) ... D5
Salinas (pt.) ... E6
Samaná (bay) ... F5
Samaná (cape) ... F5
San Rafael (cape) ... C6
Saona (isl.) ... F6
Septentrional, Cordillera (range) ... D5
Tina (mt.) ... D6
Yaque del Norte (riv.) ... D5
Yaque del Sur (riv.) ... D6
Yuma (bay) ... F6
Yuna (riv.) ... E5

HAITI
DEPARTMENTS
Artibonite ... C5
Nord ... C5
Nord-Ouest ... B5
Ouest ... C6
Sud ... A6

CITIES and TOWNS
Anse-à-Galets 3,623 ... B6
Anse-d'Hainault 5,220 ... A6
Aquin 3,820 ... B6
Cap-Haïtien 64,406 ... C5
Croix des Bouquets 4,365 .. C6
Dame Marie 4,320 ... A6
Dérac 1,300 ... C5

Dessalines 7,984 ... C5
Fort Liberté 5,012 ... C5
Gonaïves 34,209 ... B5
Grande Rivière du Nord 6,007 ... C5
Gros Morne 4,739 ... B5
Hinche 10,070 ... C5
Jacmel 13,730 ... C6
Jérémie 18,493 ... A6
Kenscoff 2,605 ... C6
Lascahobas 3,805 ... C6
Léogâne 5,782 ... C6
Les Cayes 34,090 ... B6
Limbé 10,476 ... C5
Miragoâne 4,327 ... B6
Mirebalais 6,069 ... C6
Ouanaminthe 7,276 ... C5
Pétionville 35,333 ... C6
Petite Rivière de l'Artibonite 10,099 ... B5
Petit Goâve 7,310 ... B6
Pignon 4,576 ... C5
Port-au-Prince (cap.) 449,831 ... C6
Port-de-Paix 15,540 ... B5
Saint-Louis du Nord 7,203 .. B5
Saint-Marc 24,165 ... B5
Saint-Michel de l'Atalaye 7,559 ... C5
Saint-Raaphaël 3,889 ... C5
Trou du Nord 7,637 ... C5
Verrettes 3,670 ... C5

OTHER FEATURES
Artibonite (riv.) ... C5
Baradères (bay) ... B6
Cheval Blanc (pt.) ... B5
Dame Marie (cape) ... A6
Est (pt.) ... C4
Fantasque (pt.) ... B6
Gonâve (gulf) ... B6
Gonâve (isl.) ... B6
Grande Cayemite (isl.) ... B6
Gravois (pt.) ... A7
Irois (cape) ... A6
Jean-Rabel (pt.) ... B5
Macaya (mt.) ... A6
Manzanillo (bay) ... C5
Môle (pt.) ... B5
Noires (mts.) ... C5
Ouest (pt.) ... B4
Ouest (pt.) ... B6
Saint-Marc (chan.) ... B5
Saint-Marc (pt.) ... B5
Saumâtre (lake) ... C6
Selle (peak) ... C6
Sud (chan.) ... A6
Tortue (chan.) ... B5
Tortue (Tortuga) (isl.) ... C4
Tortuga (isl.) ... C4
Trois-Rivières (riv.) ... B5
Vache (isl.) ... B6
Windward (passg.) ... A5

JAMAICA
CITIES and TOWNS
Alley ... J7

Alligator Pond ... H6
Anchovy 2,558 ... H5
Annotto Bay ... K6
Bamboo 2,971 ... J6
Bath ... K6
Black River 2,701 ... H6
Bog Walk ... K6
Bowden ... K6
Browns Town 5,479 ... J6
Bull Savanna-Junction 5,110 ... H6
Cambridge 2,449 ... H6
Catadupa ... H6
Christiana ... H6
Discovery Bay 1,814 ... J5
Falmouth 3,937 ... H6
Green Island ... G6
Hope Bay ... K6
Kingston (cap.) 106,791 ..K6
Kingston *516,865 ... J7
Linstead ... J6
Lucea 3,635 ... G5
Mandeville 14,421 ... H6
Maroon Town 2,717 ... H6
May Pen 26,074 ... H6
Montego Bay 43,521 ... H5
Montpelier ... H6
Morant Bay 7,465 ... K7
Negril ... G6
Ocho Rios 5,851 ... J6
Oracabessa ... J5
Port Antonio 10,538 ... K6
Port Kaiser ... H7
Port Maria 5,259 ... J6
Port Morant ... K7
Saint Ann's Bay 7,101 ... J5
Saint Margaret's Bay ... K6
Savanna-la-Mar 11,759 ... G6
Spanish Town 40,731 ... J6
Williamsfield ... H6

OTHER FEATURES
Black (mts.) ... H6
Black River (bay) ... G6
Blue (mts.) ... J6
Blue Mountain (peak) ... K6
Galina (pt.) ... J6
Grande (riv.) ... K6
Great (riv.) ... H6
Great Pedro Bluff (prom.) .. H7
Long (bay) ... K6
Luana (pt.) ... G6
Minho (riv.) ... J6
Montego (bay) ... G5
Montego Bay (pt.) ... G5
North East (pt.) ... K6
North Negril (pt.) ... G6
North West (pt.) ... G5
Old Harbour (bay) ... J7
Portland (pt.) ... J7
Sir John's (peak) ... K6
South East (pt.) ... K6
South Negril (pt.) ... G6

*City and Suburbs.
○ Population of municipality.

LEGEND
Capitals of Countries _____ ☆
Provincial Capitals _____ △
International Boundaries _____
Provincial Boundaries _____
© Copyright HAMMOND INCORPORATED, Maplewood, N.J.

BAHAMAS

Agriculture, Industry and Resources

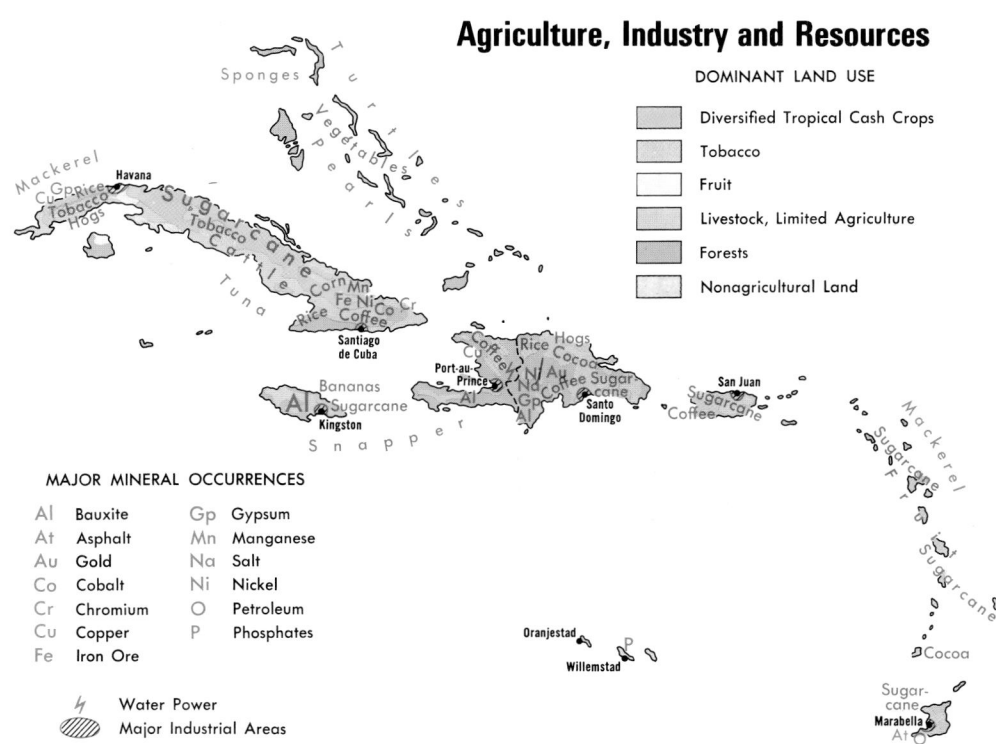

DOMINANT LAND USE

Diversified Tropical Cash Crops
Tobacco
Fruit
Livestock, Limited Agriculture
Forests
Nonagricultural Land

MAJOR MINERAL OCCURRENCES

Al Bauxite
At Asphalt
Au Gold
Co Cobalt
Cr Chromium
Cu Copper
Fe Iron Ore
Gp Gypsum
Mn Manganese
Na Salt
Ni Nickel
O Petroleum
P Phosphates

⚡ Water Power
▨ Major Industrial Areas

PUERTO RICO

DISTRICTS

Aguadilla A1
Arecibo C1
Bayamón D1
Guayama D2
Humacao E2
Mayagüez B2
Ponce C2
San Juan D1

CITIES and TOWNS

Adjuntas 5,239 B2
Aguada 5,025 A1
Aguadilla 22,039 A1
Aguas Buenas 3,766 E2
Aibonito 9,331 D2
Añasco 5,646 A1
Angeles ○2,817 B2
Arecibo 48,779 B1
Arroyo 8,435 E3
Bahomamey A1
Bajadero 3,678 C1
Barceloneta 4,502 C1
Barranquitas 3,618 D2
Bayamón 185,087 D1
Boquerón ○3,675 A3
Cabo Rojo 10,292 A2
Caguas 87,214 E2
Caguas †156,819 E2
Camuy 3,834 B1
Carolina 147,835 E1
Cataño 26,243 D1
Cayey 23,305 D2
Ceiba 4,973 F2
Central Aguirre 1,049 D3
Ciales 3,582 C1
Cidra 6,069 D2
Coamo 12,851 D2
Comerío 5,736 D2
Coquí 3,018 D3
Corozal 5,889 D1
Coto Laurel ○5,192 C2
Culebra (Dewey) 938 G1
Dorado 10,203 D1
Ensenada B3
Esperanza 1,130 G2
Fajardo 26,928 F1
Florida 3,641 C1
Guánica 9,628 B3
Guayama 21,097 E3
Guayanilla 6,163 B3
Guaynabo 65,075 D1
Gurabo 7,645 E2
Hatillo 5,019 B1
Hato Rey E1
Hormigüeros 12,031 A2
Humacao 19,147 E2
Isabela 12,087 A1
Isabel Segunda 2,330 G2
Jayuya 3,588 C2
Jobos 4,194 D3
Juana Díaz 10,469 C2
Juncos 7,851 E2
Lajas 4,275 A2
Lares 5,224 B2
Las Piedras 4,857 E2
Levittown 31,613 D1
Loíza 3,932 E1
Loíza Aldea E1
Luquillo 4,531 F1
Manatí 17,347 C1
Maricao 1,390 B2
Mayagüez 82,968 A2
Mayagüez †98,155 A2
Moca 3,960 A1
Naguabo 4,135 F2
Naranjito 2,849 D1
Palmer 1,566 F1
Palo Seco 1,172 D1
Parguera A3
Patillas 3,172 E2
Peñuelas 4,235 B2
Playa de Fajardo F1
Playa de Humacao ○5,573 F2
Ponce 161,739 C3
Ponce †168,272 C3
Puerto Nuevo D1
Puerto Real 2,390 A2
Puerto Real (Playa de Fajardo) F1
Punta Santiago (Playa de Humacao) ○5,573 F2
Quebradillas 3,770 B1
Río Blanco 1,433 F2
Río Grande 12,047 E1
Río Piedras E1
Rosario A2
Sabana Grande 7,435 B2
Sabana Seca 11,431 D1
Salinas 6,220 D3
San Antonio 2,681 A1
San Germán 13,054 A2
San Juan (cap.) 424,600 E1
San Juan †1,081,193 E1
San Lorenzo 8,880 E2
San Sebastián 10,619 B1
Santa Isabel 6,948 C3
Santurce E1
Tallaboa 1,059 B3
Toa Alta 4,427 D1
Toa Baja 1,992 D1
Trujillo Alto 41,141 E1
Utuado 11,113 B2
Vega Alta 10,582 D1
Vega Baja 18,233 D1
Vieques (Isabel Segunda) 2,330 G2
Villalba 3,469 C2
Yabucoa 6,797 E2
Yauco 14,594 B2

OTHER FEATURES

Aguadilla (bay) A1
Algarrobo (pt.) A2
Añasco (bay) A1
Arenas (pt.) F2
Bauta (riv.) C2
Bayamón (riv.) D1
Boquerón (bay) A3
Borinquen (pt.) A1
Cabullones (pt.) C3
Caja de Muertos (isl.) C3
Camuy (riv.) B1
Canovanas (riv.) E1
Caonillas (lake) C2
Carite (lake) E2
Carraízo (lake) E1
Cayey, Sierra de (mts.) D2
Central, Cordillera (range) C2
Cerro Gordo (pt.) D1
Coamo (res.) D3
Coamo (riv.) D3
Culebra (isl.) 1,265 G1
Culebrinas (riv.) A1
Culebrita (isl.) G2
El Toro (mt.) F2
El Yunque (mt.) F1
Este (pt.) G2
Fajardo (riv.) F1
Figuras (pt.) E2
Fosforescente (bay) A3
Grande de Añasco (riv.) B2
Grande de Arecibo (riv.) C1
Grande de Loíza (riv.) E1
Grande de Manatí (riv.) C1
Guajataca (lake) B1
Guanajibo (pt.) A2
Guanajibo (riv.) A2
Guánica (lake) B3
Guaniquilla (pt.) A2
Guayabal (lake) C2
Guayanés (pt.) F2
Guayanés (riv.) E2
Guayanilla (bay) B3
Guayo (lake) B2
Guilarte (mt.) B2
Honda (bay) F2
Jacaguas (riv.) C2
Jaicoa, Cordillera (mts.) B1
Jiguero (pt.) A1
Jobos (bay) D3
Lima (pt.) F2
Luquillo, Sierra de (mts.) E2
Manglillo (pt.) B3
Mayagüez (bay) A2
Miquillo (pt.) F1
Molinos (pt.) D1
Mona (passg.) A2
Negra (pt.) G2
Nigua (riv.) D2
Ola Grande (pt.) D3
Palmas Altas (pt.) C1
Patillas (lake) E2
Petrona (pt.) D3
Pirata (mt.) F2
Plata (riv.) D2
Puerca (pt.) F2
Puerto Medio Mundo (bay) F2
Punta, Cerro de (mt.) C2
Ramey A.F.B. A1
Rincón (bay) D3
Rojo (cape) A3
Roosevelt Road Naval Res. F2
Salinas (pt.) D1
San José (lag.) E1
San Juan, Cabezas de (prom.) F1
San Juan Nat'l Hist. Site D1
Soldado (pt.) G2
Sucia (bay) A3
Tanamá (riv.) B1
Toro, El (mt.) F2
Torrecilla (lag.) E1
Tortuguero (lag.) D1
Tuna (pt.) E3
Vacía Talega (pt.) E1
Vieques (isl.) 7,662 G2
Vieques (passg.) F2
Vieques (sound) G2
Yagüez (riv.) A2
Yauco (lake) B2
Yeguas ((pt.) F3

ANTIGUA

CITIES and TOWNS

All Saints 1,796 E11
Cedar Grove 1,460 E11
Falmouth 1,134 E11
Freetown 1,250 E11
Jennings 1,370 D11
Liberta 2,394 E11
Old Road 1,244 D11
Parham 1,570 E11
Saint John's (cap.) 21,814 E11
Willikies 1,843 E11

OTHER FEATURES

Antigua (isl.) 76,213 E11
Boggy (peak) D11
Boon (pt.) E11
Green (isl.) E11
Guiana (isl.) E11
Long (isl.) E11
Saint John's (harb.) C11
Standfast (pt.) E11
Willoughby (bay) E11

ARUBA

CITIES and TOWNS

Aresji D9
Balashi E10
Bubali D10
Bushiribana E10
Druif D1
Oranjestad (cap.) Aruba 10,100 D10
Sint Nicolaas E10
Westpunt D10

OTHER FEATURES

Aruba (isl.) 66,790 E9
Basora (pt.) E10
Jamanota (mt.) E10
Paarden (bay) D10
Palm (beach) D10

BARBADOS

CITIES and TOWNS

Bathsheba B8
Belleplaine B8
Bridgetown (cap.) 7,552 B9
Carlton B8
Cave Hill B9
Checker Hall B8
Codrington B8
Crab Hill B8
Crane C9
Drax Hall B8
Ellerton B9
Greenland B8
Holetown B8
Kendal B8
Lodge Hill B8
Marchfield B9
Mount Standfast B8
Oistins B9
Rose Hill B8
Rouen B9
Saint Lawrence B9
Saint Martins C9
Scarboro B9
Seawell B9
Six Mens B8
Speightstown B8
Spring Hall B8
Welchman Hall B8

OTHER FEATURES

Carlisle (bay) B9
Hillaby (mt.) B8
Long (bay) B9
North (pt.) B8
Oistins (bay) B9
Pelican (isl.) B9
Ragged (pt.) C8
Sam Lord's Castle C9
South (pt.) B9

DOMINICA

CITIES and TOWNS

Barroui 1,480 E6
Castle Bruce 1,975 F6
Coulihaut 1,735 E6
Delice F7
Grand Bay 3,152 F7
Hampstead E5
La Plaine F6
Mahout 2,095 F6
Marigot 3,183 F6
Petit Soufrière F6
Portsmouth 2,329 E5
Rosalie F6
Roseau (cap.) 9,968 E7
Roseau *16,035 E7
Saint Joseph 2,643 E6
Salybia F6
Soufrière E7
Vieille Case E5
Wesley 2,002 F5

OTHER FEATURES

Capuchin (cape) E5
Carib Reserve F6
Clyde (riv.) F6
Crumpton (pt.) F5
Diablotin, Morne (mt.) E6
Dominica (passg.) E5
Douglas (bay) E5
Grand (bay) F7
Jaquet (pt.) E5
Layou (riv.) E6
Martinique (passg.) E7
Micotrin (mt.) F6
Pagoua (bay) F6
Prince Rupert (bay) E5
Scotts (head) E7
Soufrière (bay) E7
Trois Pitons, Morne (mt.) E6

GRENADA

CITIES and TOWNS

Gouyave 2,498 C8
Grand Roy C8
Grenville 1,723 D8
Hermitage D8
La Taste D8
Marquis D8
Mount Tivoli D8
Saint George's (cap.) 6,463 C9
Saint George's *34,624 C9
Sauteurs 605 D8
Victoria 1,673 D8
Woodford C8

OTHER FEATURES

Bedford (pt.) D8
David (pt.) D8
Great Bacolet (pt.) D8
Green (isl.) D8
Grenville (bay) D8
Gros (pt.) C8
Halifax (harb.) C8
Irvin's (bay) D8
Les Tantes (isls.) D7
Molinière (pt.) C8
Prickly (pt.) C9
Ronde (isl.) D7
Saint Catherine (mt.) D8
Saline (pt.) C9
Sinai (mt.) D8
Telescope (pt.) D8

GUADELOUPE

Total Population 329,017

CITIES and TOWNS

Anse-Bertrand 1,921 A5
Baie-Mahault 5,874 A7
Baillif 3,844 A7
Bananier A7
Basse-Terre (cap.) 13,397 A7
Bouillante 1,821 A6
Bourg-des-Saintes 907 A7
Capesterre 7,541 A7
Ferry A6
Gosier 13,741 B6
Gourbeyre 5,637 A7
Goyave 1,709 A6
Grand-Bourg 3,249 B7
Lamentin 2,319 A6
Les Abymes 51,837 B6
Morne-à-l'Eau 9,457 B6
Moule 9,800 B6
Petit-Bourg 5,097 A6
Petit-Canal 1,581 A6
Pigeon A6
Pointe-à-Pitre 25,151 B6
Pointe-Noire 2,180 A6
Port-Louis 4,517 B5
Saint-Claude 6,755 A7
Sainte-Anne 11,527 B6
Sainte-Marguerite B6
Sainte-Marie A6
Sainte-Rose 4,805 A6
Saint-François 3,141 B6
Trois-Rivières 7,881 A7
Vieux-Fort 1,073 B7
Vieux-Habitants 4,065 A7

OTHER FEATURES

Allègre (pt.) A6
Antigues (pt.) A5
Basse-Terre (isl.) 138,777 A6
Châteaux (pt.) B6
Constant, Morne (hill) B7
Désirade, La (isl.) 1,602 B6
Fajou (isl.) A6
Grand Cul-de-Sac Marin (bay) A6
Grande-Terre (isl.) B6
Grande Vigie (pt.) B5
Grand-Îlet (isl.) A7
Guadeloupe (isl.) 167,896 B6
Guadeloupe (passg.) A5
Guadeloupe Nat'l Park A6
Kahouanne (isl.) A6
Marie-Galante (isl.) 13,757 B7
Nord (pt.) B7
Nord-Est (bay) B6
Petit Cul-de-Sac Marin (bay) A6
Petite-Terre (isls.) B6
Saintes (chan.) A7
Saintes (isls.) 2,901 A7
Salée (riv.) A6
Sans Toucher (mt.) A6
Soufrière (mt.) A7
Terre-de-Bas (isl.) 1,427 A7
Terre-de-Haut (isl.) 1,453 A7
Vieux-Fort (pt.) A7

MARTINIQUE

Total Population 330,220

CITIES and TOWNS

Ajoupa-Bouillon 1,569 C5
Basse-Pointe 2,163 C5
Bellefontaine 818 C6
Case-Pilote 1,776 C6
Ducos 4,429 D6
Fond-Saint-Denis 962 C6
Fort-de-France (cap.) 96,649 C6
Grand' Rivière 1,053 C5
Gros-Morne 1,976 D6
La Trinité 3,380 D6
Le Carbet 2,321 C6
Le François 2,940 D6
Le Lamentin 6,872 D6
Le Lorrain 2,024 D6
Le Marin 2,651 D7
Le Morne-Rouge 2,650 C5
Le Prêcheur 1,350 C5
Le Robert 3,610 D6
Le Saint-Esprit 3,947 D6
Les Trois-Ilets 1,484 D6
Le Vauclin 3,054 D6
Macouba 1,142 C5
Marigot 1,765 D5
Rivière-Pilote 1,587 D7
Rivière-Salée 1,859 D7
Sainte-Luce 1,502 D7
Sainte-Marie 3,966 D5
Saint-Joseph 2,052 D6
Saint-Pierre 4,923 C6
Schoelcher 16,412 C6

OTHER FEATURES

Cabet, Pitons du (mt.) C6
Cabrits (isl.) D7
Caravelle (pen.) D6
Cul-de-Sac du Marin (bay) D7
Diable (pt.) D5
Ferré (cape) E7
Fort-de-France (bay) C6
Galion (bay) D6
Lézarde (riv.) D6
Long (isl.) D6
Lorrain (riv.) D5
Martinique (passg.) C5
Pelée (vol.) C5
Pilote (riv.) D7
Ramiers (isl.) C6
Ramville (isl.) D6
Robert (harb.) D6
Rose (pt.) D6
Saint-Martin (cape) C5
Saint-Pierre (bay) C6
Salines (pt.) D7
Salomon (pt.) C7
Vauclin (mt.) D6

NETHERLANDS ANTILLES

CITIES and TOWNS

Ascension F8
Bacuna E8
Boven Bolivia E8
Dokterstuin F8
Emmastad F9
Entrejo E8
Fontein E8
Groot Sint Joris G9
Hato G8
Kralendijk (cap.), Bonaire 2,500 E8
Lagoen F8
Montanja di Reij G9
New Port G9
Noord di Salinja E8
Onima E8
Otrabanda F9
Patrick F8
Rincon E8
Rooi E8
Santa Barbara G9
Santa Catharina G9
Savonet F8
Sint Kruis F8
Sint Martha F8
Sint Michiel F9
Sint Willebrordus F8
Terra Corra E8
Westpunt F8
Willemstad (cap.) 95,000 F9
Willemstad *130,000 F9

OTHER FEATURES

Bonaire (isl.) 8,087 E9
Bullen (bay) F8
Caracas (bay) G9
Curaçao (isl.) 145,430 G7
Goto (lake) D8
Kanon (pt.) G9
Klein Bonaire (isl.) E8
Kudarebe (pt.) D9
Lac (bay) D9
Lacre (pt.) E9
Malmok (mt.) E8
Noord (pt.) D8
Noord (pt.) F8
Pekelmeer (lake) E9
Piscadera (bay) F9
Schottegat (bay) G9
Sint Anna (bay) F9
Sint Christoffel (mt.) F8
Sint Joris (bay) G9
Slag (bay) D8
Vierkant (pt.) E8

SAINT KITTS and NEVIS

CITIES and TOWNS

Basseterre (cap.) 14,725 C10
Cayon C10
Charlestown 1,326 C11
Cotton Ground 471 C11
Dieppe Bay C10
Frigate Bay C10
Gingerland D11
Golden Rock C10
Newcastle D11
Old Road Town C10
Sadlers Village C10
Sandy Point 862 C10
Tabernacle C10
Zion Hill D11

OTHER FEATURES

Brimstone (hill) C10
Dogwood (pt.) D11
Fort (pt.) C11
Great Salt (pond) D10
Heldens (pt.) C10
Horse Shoe (pt.) C11
Misery (mt.) C10
Monkey (hill) C10
Narrows, The (str.) D11
Nevis (isl.) 9,300 D11
Nevis (peak) D11
North Friars (bay) D10
Pinney's (beach) D11
Saint Christopher (Saint Kitts) (isl.) 35,104 D10
South Friars (bay) C10

SAINT LUCIA

CITIES and TOWNS

Anse la Raye •5,007 F6
Canaries •2,075 G6
Castries (cap.) •42,770 G6
Choc G5
Choiseul •6,382 F7
Dauphin G5
Dennery •9,654 G6
Gros Islet •10,329 G5
Laborie •6,944 G7
Marigot G6
Marquis G6
Micoud •12,264 G6
Preslin G6
Soufrière •7,456 F6
Vieux Fort •10,675 G7

OTHER FEATURES

Beaumont (pt.) F6
Canaries, Piton (mt.) G6
Cannelles (pt.) G7
Cannelles (riv.) G6
Cap (pt.) G5
Choc (bay) G5
Fond d'Or (bay) G6
Gimie (mt.) G6
Grand Caille (pt.) F6
Grand Cul de Sac (riv.) G6
Gros Islet (bay) G5
Gros Piton (mt.) G6
La Sorcière (mt.) G6
Maria (isls.) G7
Ministre (pt.) G7
Moule-à-Chique (cape) G7
Petit Piton (mt.) G6
Pigeon (isl.) G5
Port Castries (harb.) G6
Port Praslin (bay) G6
Roseau (riv.) G6
Saint Lucia (chan.) G5
Saint Vincent (chan.) G7
Savannes (bay) G7
Sorcière, La (mt.) G6
Soufrière (bay) F6
Vierge (pt.) G6

SAINT VINCENT and THE GRENADINES

CITIES and TOWNS

Barrouallie 1,298 A9
Calliaqua 627 A9
Camden Park A9
Colonarie A9
Georgetown 1,100 A8
Kingstown (cap.) 17,117 A9
Kingstown *23,330 A9
Layou 1,147 A9
Wallibu A8

OTHER FEATURES

Colonarie (pt.) A9
Cumberland (bay) A8
Dark (head) A8
De Volet (pt.) A8
Espagnol (pt.) A8
Greathead (bay) A9
Kingstown (bay) A9
Owia (bay) A8
Porter (pt.) A8
Richmond (peak) A8
Saint Andrew (mt.) A9
Saint Vincent (passg.) A8
Soufrière (mt.) A8
Yambou (head) A9

TRINIDAD and TOBAGO

CITIES and TOWNS

Arima 11,390 B10
Arouca B10
Basse Terre B11
Biche B10
Blanchisseuse B10
California A11
Carapichaima B10
Caroni B10
Cedros A11
Chaguanas 6,122 B10
Chaguaramas A10
Couva 3,635 B10
Cunapo B10
Flanagin Town B10
Fullarton A11
Fyzabad 1,564 A11
Grande Rivière B10
Guaico B10
Guayaguayare B11
La Brea 1,487 A11
Marabella 18,158 A11
Matelot B10
Matura B10
Mayaro 2,638 B11
Moruga B11
Mucurapo B10
Palo Seco A11
Peñal 3,606 B11
Point Fortin 6,538 A11
Port-of-Spain (cap.) 67,978 A10
Princes Town 8,288 B11
Redhead B10
Rio Claro 2,423 B11
Saint Joseph 4,132 B10
Saint Joseph B10
San Fernando 33,490 A11
San Francique A11
Sangre Grande 8,948 B1
San Juan A10
Sans Souci B10
Siparia 5,773 B11
Tabaquite 2,309 B11
Talparo B10
Toco 1,287 B10
Tunapuna 10,251 B10
Upper Manzanilla B1
Valencia B10
Waterloo A10

OTHER FEATURES

Aripo, El Cerro del (mt.) B10
Boca Grande (passg.) A10
Chacachacare (isl.) A10
Chupara (pt.) B10
Cocos (bay) B10
Dragons Mouth (str.) A10
El Tucuche (mt.) B10
Erin (bay) A11
Galeota (pt.) B11
Galera (pt.) C10
Guapo (bay) A11
Guatuaro (pt.) B11
Icacos (pt.) A11
Maracas (bay) C10
Pitch (lake) A11

VIRGIN ISLANDS (Br.)

CITIES and TOWNS

Road Town (cap.) 2,200 D3
West End C4

OTHER FEATURES

Flanagan (passg.) D4
Frenchman (cay) C4
Great Thatch (isl.) C4
Great Tobago (isl.) B3
Jost Van Dyke (isl.) 135 C3
Little Tobago (isl.) B3
Narrows, The (str.) C4
Norman (isl.) D4
Peter (isl.) D4
Road (bay) D3
Sage (mt.) D4
Sir Francis Drake (chan.) D4
Tortola (isl.) 9,257 D3

VIRGIN ISLANDS (U.S.)

CITIES and TOWNS

Bethlehem E4
Canebay E3
Charlotte Amalie (cap.) 11,842 B4
Christiansted 2,914 F4
Cruz Bay 1,928 C4
Diamond F4
Eastend D4
Emmaus C4
Fredensdal F4
Frederiksted 1,046 E4
Grove Place 3,599 E4
Kingshill F4
Longford F4
Negro Bay E4

OTHER FEATURES

Altona (lag.) F4
Annaly (bay) E3
Baron Bluff (prom.) E3
Bordeaux (mt.) C4
Brass (isls.) A4
Buck (isl.) G3
Buck Island (chan.) F3
Buck Island Reef Nat'l Mon. G3
Butler (bay) E4
Caneel (bay) B4
Capella (isls.) B5
Christiansted Nat'l Hist. Site F4
Coral (bay) C4
Crown (mt.) A4
Dutch Cap (cay) A4
Eagle (mt.) E4
East (pt.) G4
Flanagan (passg.) D4
Flat (cays) A4
Grass (pt.) G4
Great (pond) F4
Great Pond (bay) F4
Green (cay) F4
Hams Bluff (prom.) E3
Hans Lollik (isls.) B4
Hassel (isl.) B4
Jersey (bay) B4
Krause Lagoon (chan.) F4
Leeward (passg.) B4
Long (pt.) B4
Long (pt.) E4
Lovango (cay) C4
Magens (bay) B4
Maho (bay) C4
Narrows, The (str.) C4
Nulliberg (mt.) B4
Perseverance (bay) A4
Picara (pt.) B4
Pillsbury (sound) B4
Privateer (pt.) D4
Pull (pt.) F3
Ram (head) C5
Red (pt.) D4
Reef (bay) C4
Saba (isl.) A4
Saint Croix (isl.) 49,725 G4
Saint James (isls.) B4
Saint John (isl.) 2,472 C4
Saint Thomas (harb.) B4
Saint Thomas (isl.) 44,372 A4
Salt (cay) A4
Salt (riv.) F4
Salt River (bay) F3
Sandy (pt.) D4
Savana (isl.) A4
Southwest (cape) E4
Tague (bay) G4
Thatch (cay) B4
Turner Hole (bay) G4
U.S. Nav. Air Sta. A4
Virgin (str.) A4
Virgin Isls. Nat'l Park C4
Water (isl.) A4
Westend Saltpond (lag.) E4

*City and suburbs.
• Population of district.
†Population of met. area.
○ Population of municipality.

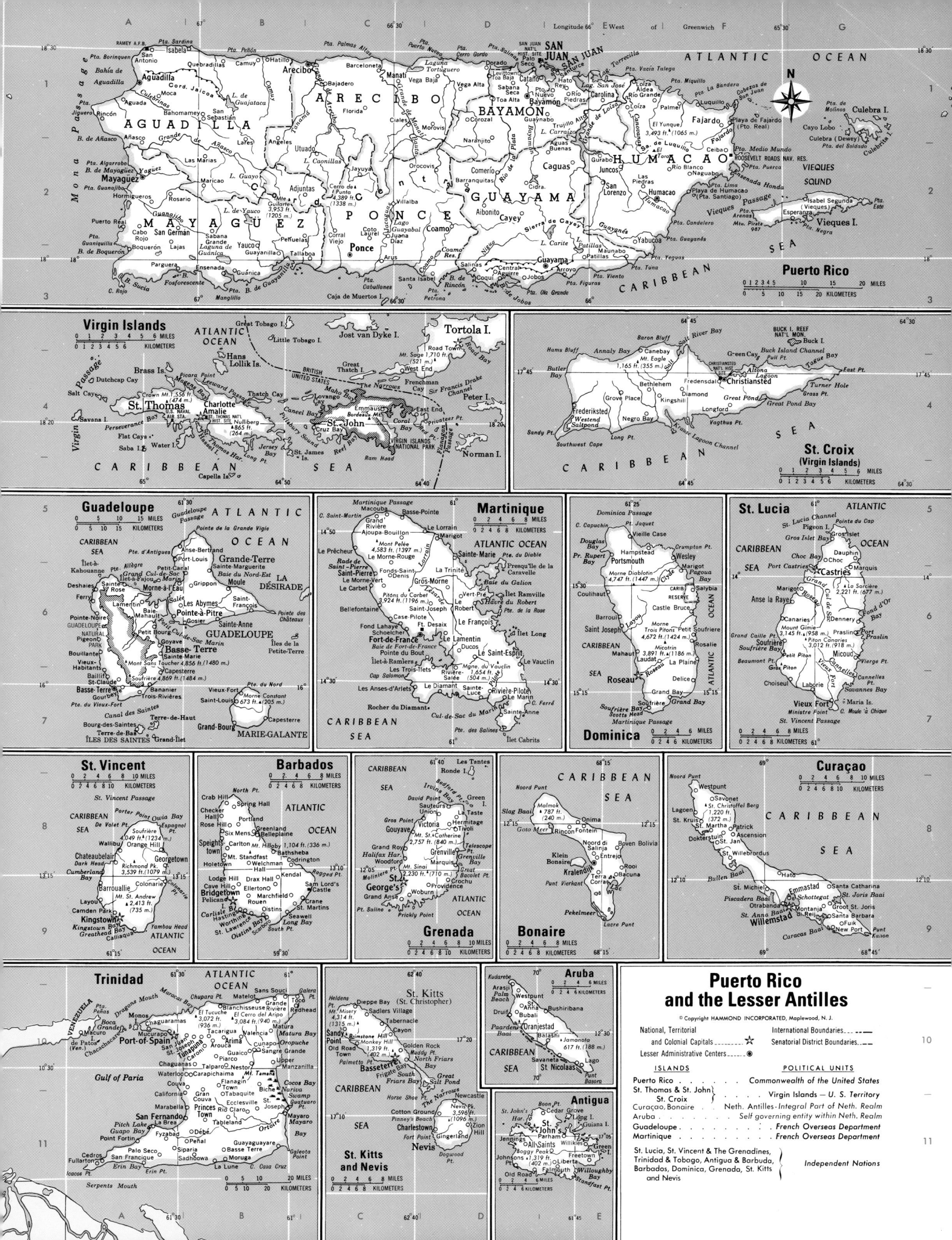

Puerto Rico and the Lesser Antilles

© Copyright HAMMOND INCORPORATED, Maplewood, N. J.

National, Territorial and Colonial Capitals .... ☆
Lesser Administrative Centers .... ◉

International Boundaries ......
Senatorial District Boundaries ......

ISLANDS   **POLITICAL UNITS**

ISLANDS	POLITICAL UNITS
Puerto Rico	Commonwealth of the United States
St. Thomas & St. John } St. Croix	Virgin Islands — U. S. Territory
Curaçao, Bonaire . . .	Neth. Antilles-Integral Part of Neth. Realm
Aruba	Self governing entity within Neth. Realm
Guadeloupe	French Overseas Department
Martinique	French Overseas Department
St. Lucia, St. Vincent & The Grenadines, Trinidad & Tobago, Antigua & Barbuda, Barbados, Dominica, Grenada, St. Kitts and Nevis	Independent Nations

Canada

CONIC PROJECTION

SCALE OF MILES
0 50 100 200 300

SCALE OF KILOMETERS
0 50 100 200 300 400 500

Capitals of Countries .... ⭐
Provincial & Territorial Capitals .... △
Administrative Centers .... ●
International Boundaries .... ▬ ▬ ▬
Provincial Boundaries .... ▬ ▬ ▬
Regional Boundaries .... ▬ ▬ ▬

© Copyright HAMMOND INCORPORATED, Maplewood, N.J.

Abitibi (lake), Ont. ....H 6	Chatham, N.Br. 6,779 ....K 6	Fort Saskatchewan, Alta.	Jonquière, Que. 60,354 .....J 6	Manitoulin (isl.), Ont. .....H 6

Abitibi (lake), Ont. ....H 6
Aklavik, N.W.T. 721 ....C 2
Albany (riv.), Ont. ....H 5
Alberta (prov.), 2,365,825 ...E 5
Amherst, N.S. 9,684 ....K 6
Amos, Que. 9,421 ....J 6
Anticosti (isl.), Que. ....K 6
Arviat, Nun. 1,022 ....G 3
Athabasca (lake) ....F 4
Athabasca (riv.), Alta. ....E 4
Axel Heiburg (isl.) Nun. .....N 3
Baffin (reg.), Nun. 8,300 ....G 1
Baffin (bay), Nun. ....J 1
Baffin (isl.), Nun. ....J 1
Baker Lake, Nun. 954 ....G 3
Banff Nat'l Park, Alta. ....E 5
Banks (isl.), N.W.T. ....D 1
Bathurst, N. Br. 15,705 ......L 5
Belle Isle (str.), Newf. ....L 5
Bonavista, Newf. 4,460 ......L 6
Boothia (pen.), Nun. ....G 1
Brandon, Man. 36,242 ....F 6
British Columbia (prov.)
  2,883,367 ....D 4
Cabot (str.) ....K 6
Calgary, Alta. 592,743 ......E5
Cambridge Bay, Nun. 815 ....F 2
Campbellton, N.Br. 9,818 ....K 6
Camrose, Alta. 12,570 ....E 5
Cape Breton (isl.), N.S. ......K 6
Cartwright, Newf. 658 ....L 5
Channel-Port aux Basques,
  Newf. 5,988 ....K 6
Charlottetown (cap.), P.E.I.
  15,282 ....K 6

Chatham, N.Br. 6,779 ....K 6
Chesterfield Inlet, Nun. 249 ..G 3
Chibougamau, Que. 10,732 ...J 6
Chicoutimi, Que. 60,064 .....J 6
Chidley (cape), Newf. ....K 3
Chilliwack, Br.Col. 40,642 ...D 6
Chisasibi, Que. 2,222 ....J 5
Churchill, Man. 1,186 ....G 4
Coast (mts.) ....C 4
Corner Brook, Newf. 24,339 ..K 6
Cornwall, Ont. 46,144 ....J 7
Cranbrook, Br.Col. 15,915 ...E 6
Cree (lake), Sask. ....F 4
Dartmouth, N.S. 62,277 .....K 7
Dauphin, Man. 8,971 ....F 5
Davis (str.), Nun. ....K 1
Dawson, Yukon 697 ....C 3
Déline, N.W.T. 521 ....D 3
Devon (isl.), Nun. ....M 3
Drumheller, Alta. 6,508 .....E 5
Echo Bay (Port Radium),
  N.W.T. 56 ....E 2
Edmonton (cap.), Alta.
  532,246 ....E 5
Ellesmere (isl.), Nun. ....N 3
Estevan, Sask. 9,174 ....F 6
Finlay (riv.), Br.Col. ....D 4
Flin Flon, Man.-Sask. 8,261 ..F 4
Fogo (isl.), Newf. ....L 6
Fort Frances, Ont. 8,906 ....G 6
Fort McMurray, Alta. 31,000 ..E 4
Fort McPherson, N.W.T. 632 ..C 2
Fort Nelson, Br.Col. 3,724 ...D 4
Fort Providence, N.W.T. 605 ..E 3

Fort Saskatchewan, Alta.
  12,169 ....E 5
Fort Simpson, N.W.T. 980 ....D 3
Fort Smith, (reg.), N.W.T.
  22,384 ....E 3
Fort Smith, N.W.T. 2,298 ....E 3
Foxe (basin), Nun. ....J 2
Fraser (riv.), Br.Col. ....C 5
Fredericton, N.Br. 43,723 ....K 6
Fundy (bay) ....K 7
Gander, Newf. 10,404 ....L 6
Gaspé, Que. 17,261 ....K 6
Georgian (bay), Ont. ....H 6
Geraldton, Ont. 2,956 ......H 6
Glace Bay, N.S. 21,466 .....L 6
Goose Bay, Newf. 7,103 .....K 5
Gouin (res.), Que. ....J 6
Grand Falls, Newf. 8,765 ....L 6
Grand Prairie, Alta. 24,263 ..E 4
Great Bear (lake), N.W.T. ....D 2
Great Slave (lake), N.W.T. ...E 3
Guelph, Ont. 71,207 ....H 7
Halifax (cap.), N.S. 114,594 ..K 7
Hamilton, Ont. 306,434 .....H 7
Harbour Grace, Newf. 2,998 ..L 6
Havre-St. Pierre, Que. 3,200 ..K 5
Hay River, N.W.T. 2,863 .....E 3
Hearst, Ont. 5,533 ....H 6
Hecate (str.), Br.Col. ....C 5
Hull, Que. 56,225 ....J 6
Inuvik (reg.), N.W.T. 7,485 ..D 2
Inuvik, N.W.T. 3,147 ....C 2
Iqaluit (cap.), Nun. 2,333 ...K 3
Iroquois Falls, Ont. 6,339 ....H 6
Jasper Nat'l Park, Alta. .....E 5

Jonquière, Que. 60,354 .....J 6
Juan de Fuca (str.), Br.Col. ..D 6
Kamloops, Br.Col. 64,048 ...D 5
Kane (basin), Nun. ....N 3
Kapuskasing, Ont. 12,014 ...H 6
Keewatin (reg.), Nun. 4,327 ..G 2
Kelowna, Br.Col. 59,196 ....D 6
Kenora, Ont. 9,817 ....G 5
Kimmirut, Nun. 252 ....J 3
Kingston, Ont. 52,616 ......J 7
Kirkland Lake, Ont. 12,219 ..H 6
Kitikmeot (reg.), Nun. 3,245 ..F 1
Kitimat, Br.Col. 12,462 .....D 5
Kluane (lake), Yukon ......C 3
Kootenay (lake), Br.Col. .....E 6
Kugluktuk, Nun. 809 ....E 2
Kuujjuac, Que. ....K 4
Kuujjuarapik, Que. 435 .....J 4
Labrador (reg.), Newf. ......K 4
Lacombe, Alta. 5,591 ....E 5
Lancaster (sound), Nun. .....H 1
Leduc, Alta. 12,471 ....E 5
Lesser Slave (lake), Alta. ....E 4
Lethbridge, Alta. 54,072 ....E 6
Liard (riv.) ....D 3
Lloydminster, Alta.-Sask.
  15,031 ....E 5
Logan (mt.), Yukon ....B 3
London, Ont. 254,280 ......H 7
Mackenzie (dist.), Nun. .....E 3
Mackenzie (riv.), Nun. ......C 2
Magdalen (isls.), Que. ......K 6
Manicouagan (riv.), Que. ....K 5
Manitoba (lake), Man. ......G 5
Manitoba (prov.), 1,063,016 ..F 5

Manitoulin (isl.), Ont. ......H 6
Maple Creek, Sask. 2,470 ...F 6
Marathon, Ont. 2,271 ......H 6
Mayo, Yukon 398 ....C 3
M'Clintock (chan.), Nun. ....F 1
Medicine Hat. Alta. 40,380 ..E 5
Melville, Sask. 5,092 ....F 5
Melville (isl.), N.W.T., Nun. ..E 1
Merritt, Br.Col. 6,110 ......D 5
Minto (lake), Que. ....J 4
Mistassibi (riv.), Que. ......J 5
Mistassini (lake), Que. ......J 5
Moncton, N.Br. 54,743 .....K 6
Mont-Joli, Que. 6,359 ......K 6
Mont-Laurier, Que. 8,405 ...J 6
Montréal, Que. 980,354 .....J 7
Moose Jaw, Sask. 33,941 ....F 6
Moosonee, Sask. 2,579 .....F 5
Moosonee, Ont. 1,433 ......H 5
Morden, Man. 4,579 ....G 6
Nain, Newf. 938 ....K 4
Nanaimo, Br.Col. 47,069 ....D 6
Nares (str.), Nun. ....N 3
Nelson, Br.Col. 9,143 ......E 6
Nelson (riv.), Man. ....G 4
New Brunswick (prov.)
  709,442 ....K 6
Newfoundland (isl.) ....L 6
Newfoundland (prov.) 568,349 .L 5
New Westminster, Br.Col.
  38,550 ....D 6
Niagara Falls, Ont. 70,960 ...J 7
Norman Wells, N.W.T. 420 ...D 2
North Battleford, Sask.
  14,030 ....F 5

North Bay, Ont. 51,268 .....J 6
North Magnetic Pole ....F 1
North Saskatchewan (riv.) ...E 5
N. Vancouver, Br.Col. 33,952 ..D 6
Northwest Territories
  39,672 ....D 3
Nove Scotia (prov.)
  873,176 ....K 7
Nunavut (terr.) 24,730 ......G 3
Okanagan (lake), Br.Col. ....D 6
Ontario (prov.) 9,101,694 ...H 5
Ottawa (cap.), Canada
  295,163 ....J 6
Ottawa (riv.) ....J 6
Owen Sound, Ont. 19,883 ...H 7
Pangnirtung, Nun. 839 .....K 2
Parry (chan.), Nun. ....E-H 1
Parry sound, Ont. 6,124 ....J 6
Peace (riv.), Alta. ....E 4
Peel (riv.) ....C 2
Pelly (riv.), Yukon ....C 3
Pembroke, Ont. 14,026 .....J 6
Péribonca (riv.), Que. ......J 5
Peterborough, Ont. 60,620 ..J 7
Pincher Creek, Alta. 3,757 ..E 6
Pond Inlet, Nun. 705 ....J 1
Portage la Prairie, Man.
  13,086 ....G 5
Povungnituk, Que. ....J 3
Prince Albert, Sask. 31,380 ..F 5
Prince Albert Nat'l Park, Sask.
  ....F 5
Prince Edward Island (prov.)
  126,646 ....K 6
Prince George, Br.Col.
  67,559 ....D 5

Prince Patrick (isl.), Nun. ...M 3
Prince Rupert, Br.Col. 16,197 ..C 5
Québec (prov.) 6,532,461 ...J 5
Québec (cap.), Que. 166,474 ..J 6
Queen Charlotte (isls.),
  Br.Col. ....C 5
Queen Elizabeth (isls.),
  Nun. ....M 3
Quesnel, Br.Col. 8,240 .....D 5
Race (cape), Newf. ....L 6
Rainy (lake), Ont. ....G 6
Rainy River, Ont. 1,061 .....G 6
Rankin Inlet, Nun. 1,109 ....G 3
Ray (cape), Newf. ....K 6
Red Deer, Alta. 46,393 .....E 5
Reindeer (lake) ....F 4
Regina (cap.), Sask. 162,613 ..F 5
Revelstoke, Br.Col. 5,544 ...E 5
Riding Mountain Nat'l Park,
  Man. ....F 5
Rimouski, Que. 29,120 .....K 6
Rivière-du-Loup, Que. 13,459 ..K 6
Roberval, Que. 11,429 ......J 6
Robson (mt.), Br.Col. ....D 5
Rocky (mts.) ....D 4
Rocky Mountain House, Alta.
  4,698 ....E 5
Rouyn, Que. 17,224 ....J 6
Sable (cape), N.S. ....K 7
Sable (isl.), N.S. ....L 7
Saint Elias (mt.), Yukon .....B 3
Saint John, N.Br. 80,521 ....K 6
St. John's (cap.), Newf.
  83,770 ....L 6
Saint Lawrence (riv.) ....K 6

Queen Elizabeth Islands

AREA 3,851,787 sq. mi. (9,976,139 sq. km.)
POPULATION 28,846,761
CAPITAL Ottawa
LARGEST CITY Montréal
HIGHEST POINT Mt. Logan 19,524 ft. (5,951 m.)
MONETARY UNIT Canadian dollar
MAJOR LANGUAGE English, French
MAJOR RELIGIONS Protestantism,
Roman Catholicism

Population Distribution

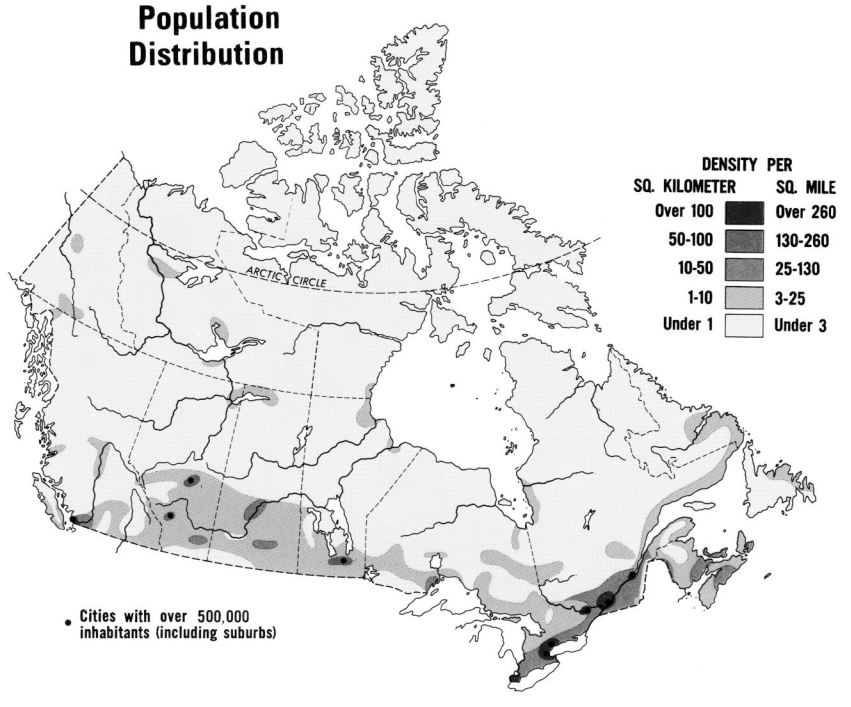

DENSITY PER	
SQ. KILOMETER	**SQ. MILE**
Over 100	Over 260
50-100	130-260
10-50	25-130
1-10	3-25
Under 1	Under 3

• Cities with over 500,000
inhabitants (including suburbs)

Vegetation

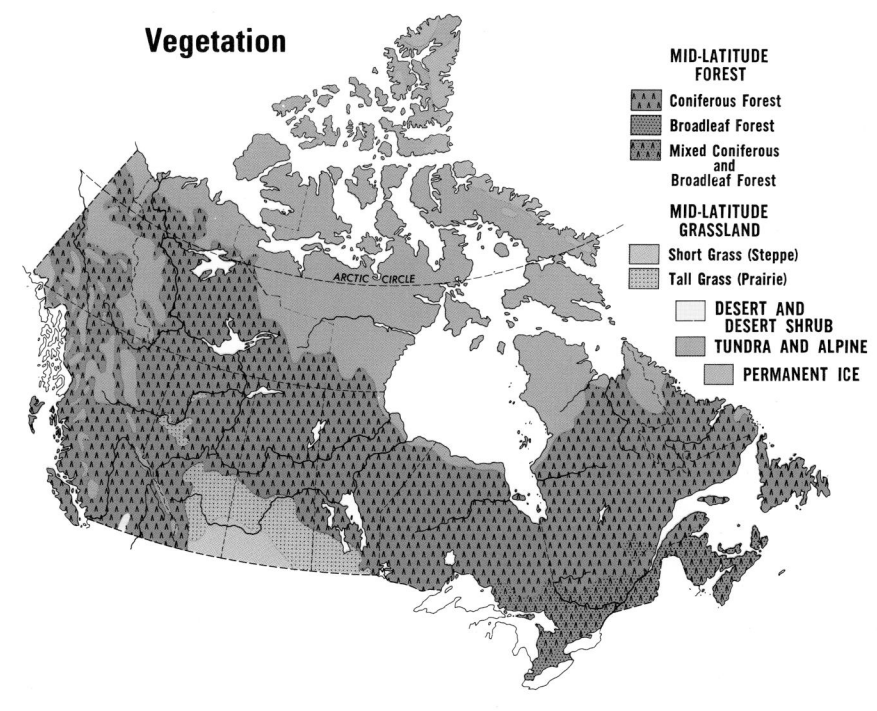

MID-LATITUDE FOREST
Coniferous Forest
Broadleaf Forest
Mixed Coniferous
and
Broadleaf Forest

MID-LATITUDE GRASSLAND
Short Grass (Steppe)
Tall Grass (Prairie)

DESERT AND DESERT SHRUB
TUNDRA AND ALPINE
PERMANENT ICE

Average January Temperature

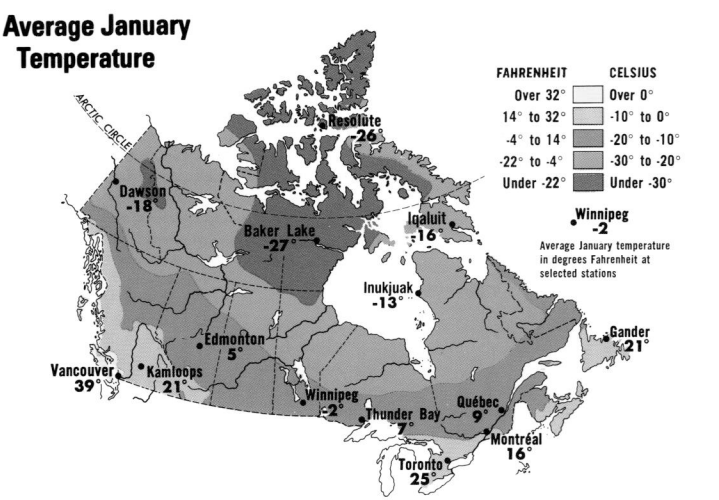

FAHRENHEIT
CELSIUS
Over 32°
Over 0°
14° to 32°
-10° to 0°
-4° to 14°
-20° to -10°
-22° to -4°
-30° to -20°
Under -22°
Under -30°
Winnipeg -2
Average January temperature in degrees Fahrenheit at selected stations
ARCTIC CIRCLE
Resolute -26°
Dawson -18°
Baker Lake -27°
Iqaluit -16°
Inukjuak -13°
Gander 21°
Edmonton 5°
Vancouver 39°
Kamloops 21°
Winnipeg -2°
Thunder Bay 7°
Québec 9°
Montréal 16°
Toronto 25°

Average July Temperature

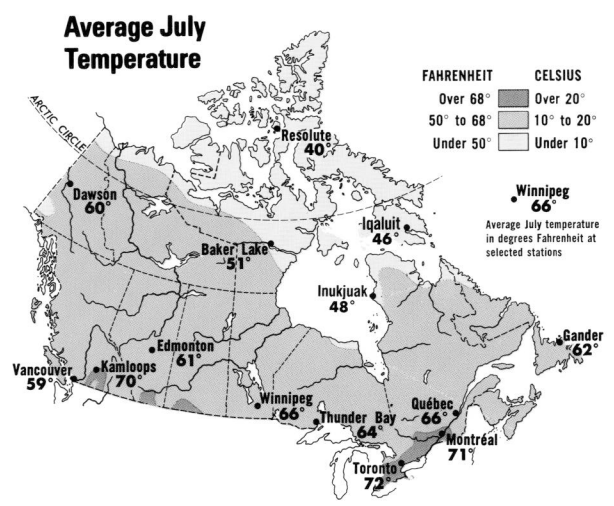

FAHRENHEIT
CELSIUS
Over 68°
Over 20°
50° to 68°
10° to 20°
Under 50°
Under 10°
Winnipeg 66°
Average July temperature in degrees Fahrenheit at selected stations
ARCTIC CIRCLE
Resolute 40°
Dawson 60°
Baker Lake 51°
Iqaluit 46°
Inukjuak 48°
Gander 62°
Edmonton 61°
Vancouver 59°
Kamloops 70°
Winnipeg 66°
Thunder Bay 64°
Québec 66°
Montréal 71°
Toronto 72°

Agriculture, Industry and Resources

Edmonton
Calgary
Vancouver
Winnipeg
Québec
Montréal
Toronto
Windsor
Salmon
Herring
Whales
Seals
Cod
Sheep
Furs
Timber
Oats
Barley
Wheat
Rye
Flax
Dairy

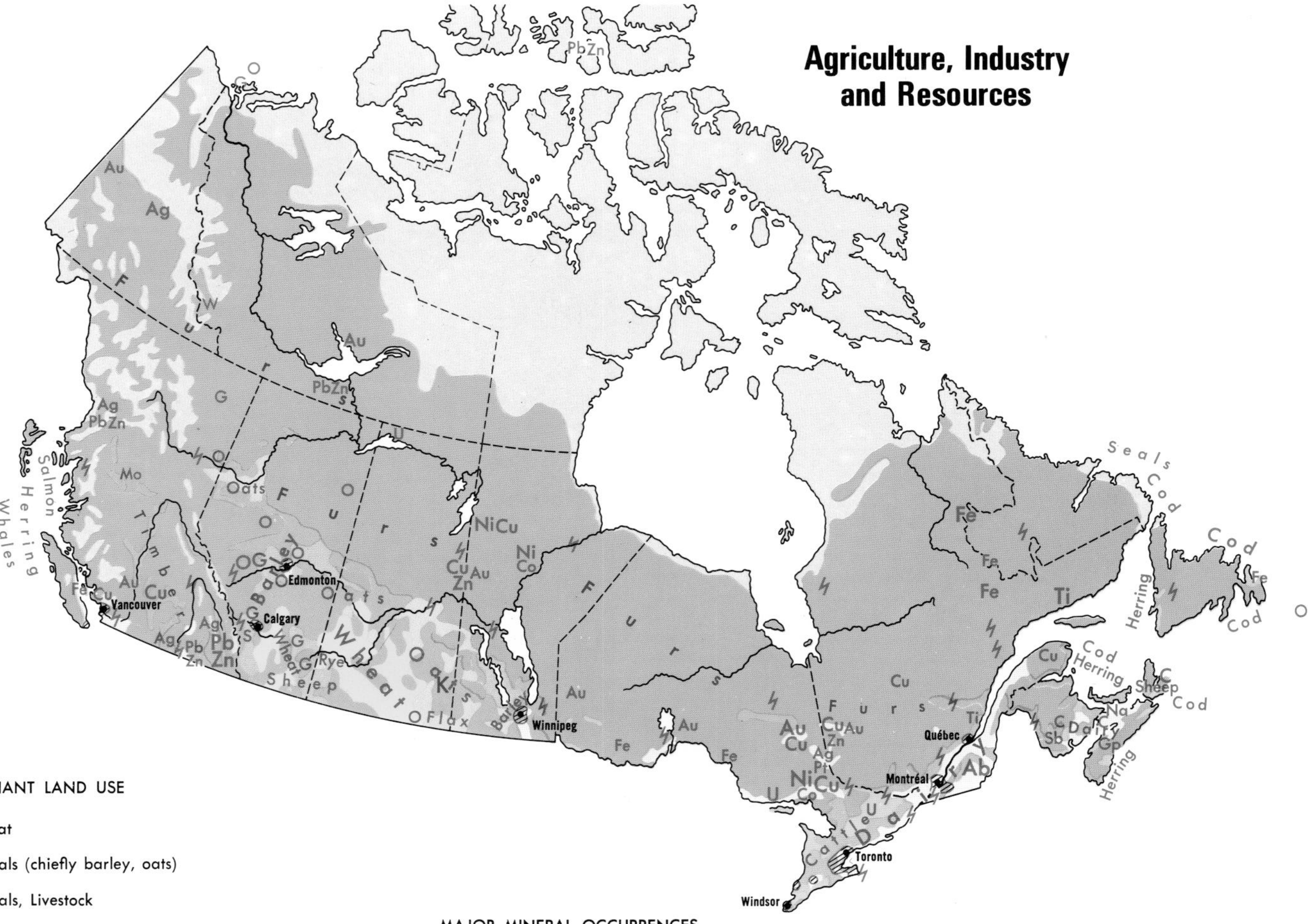

DOMINANT LAND USE

- Wheat
- Cereals (chiefly barley, oats)
- Cereals, Livestock
- General Farming, Livestock
- Dairy
- Fruit, Vegetables
- Pasture Livestock
- Range Livestock
- Forests
- Nonagricultural Land

MAJOR MINERAL OCCURRENCES

Ab	Asbestos	Fe	Iron Ore	Ni	Nickel	Sb	Antimony
Ag	Silver	G	Natural Gas	O	Petroleum	Ti	Titanium
Au	Gold	Gp	Gypsum	Pb	Lead	U	Uranium
C	Coal	K	Potash	Pt	Platinum	W	Tungsten
Co	Cobalt	Mo	Molybdenum	S	Sulfur	Zn	Zinc
Cu	Copper	Na	Salt				

- Water Power
- Major Industrial Areas

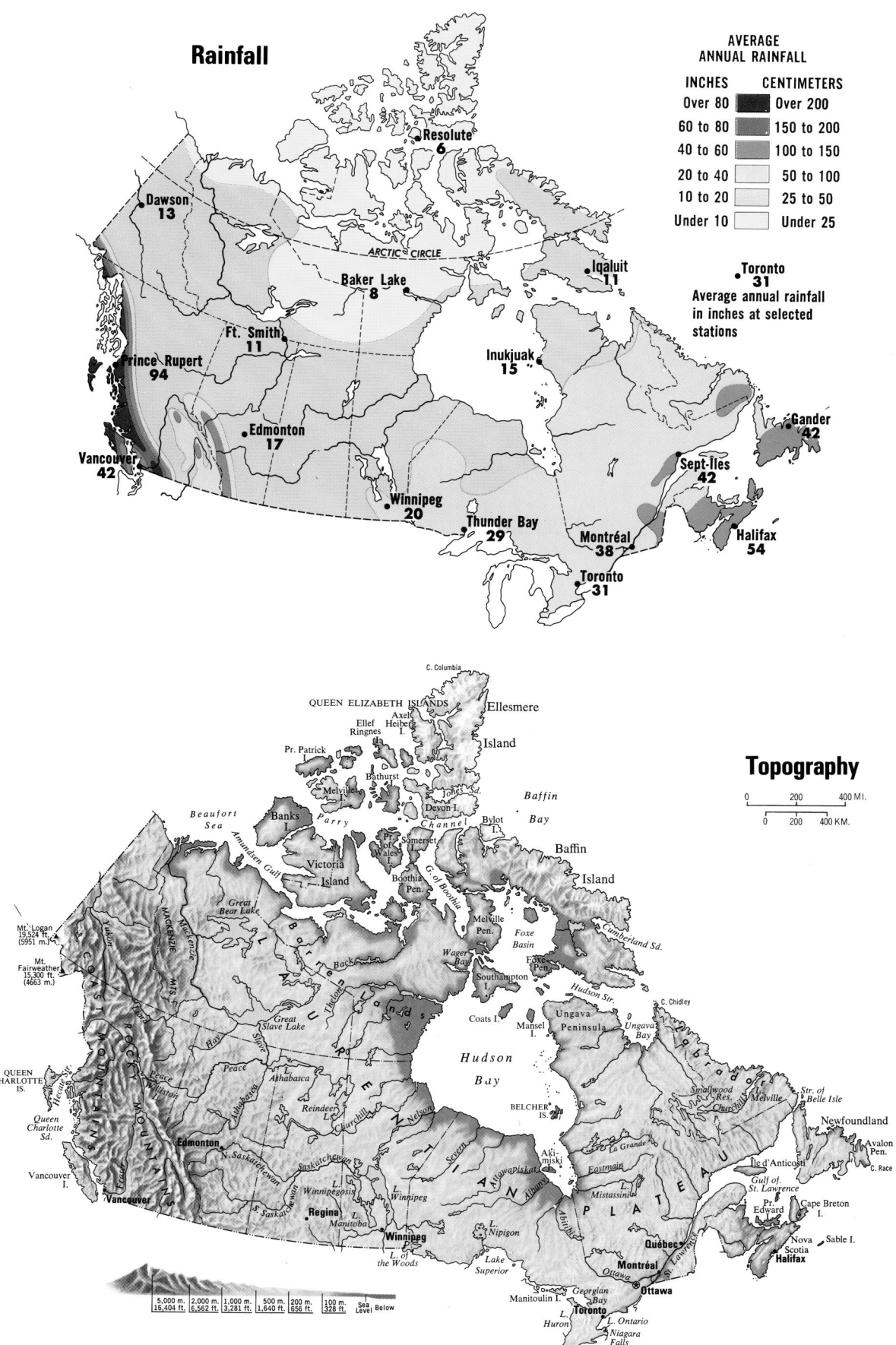

Rainfall

AVERAGE
ANNUAL RAINFALL

INCHES	CENTIMETERS
Over 80	Over 200
60 to 80	150 to 200
40 to 60	100 to 150
20 to 40	50 to 100
10 to 20	25 to 50
Under 10	Under 25

•Toronto
31
Average annual rainfall
in inches at selected
stations

•Resolute
6

Dawson
13

ARCTIC CIRCLE

•Iqaluit
11

Baker Lake
8

Ft. Smith
11

Inukjuak
15

Prince Rupert
94

Gander
42

Edmonton
17

Sept-Îles
42

Vancouver
42

Winnipeg
20

Thunder Bay
29

Montréal
38

Halifax
54

Toronto
31

Topography

QUEEN ELIZABETH ISLANDS

C. Columbia

Ellesmere

Axel
Heiberg
I.

Ellef
Ringnes
I.

Island

Pr. Patrick

Bathurst

Baffin

Melville

Bay

Beaufort
Sea

Banks
I.

Parry
Channel

Bylot
I.

Amundsen Gulf

Pr.
of
Wales
I.

Somerset

Baffin

Victoria

Boothia
Pen.

Island

Island

Great
Bear Lake

G. of Boothia

Mt. Logan
19,524 ft.
(5951 m.)

Mackenzie

Back

Melville
Pen.

Cumberland Sd.

Mt.
Fairweather
15,300 ft.
(4663 m.)

Foxe
Basin

Wager
Bay

Foxe
Pen.

Great
Slave Lake

Hudson Str.

Hay

Southampton
I.

C. Chidley

Coats I.

Mansel
I.

Ungava
Peninsula

Ungava
Bay

QUEEN
CHARLOTTE
IS.

Peace

Liard

Peace

L.
Athabasca

Reindeer

Churchill

Nelson

Hudson

Bay

BELCHER
IS.

Smallwood
Res.

Churchill

L.
Melville

Str. of
Belle Isle

Queen
Charlotte
Sd.

Edmonton

N. Saskatchewan

Saskatchewan

Seven

Akimiski

Newfoundland

Avalon
Pen.

C. Race

Vancouver
I.

S. Saskatchewan

Winnipegosis

Attawapiskat

La Grande

Eastmain

Île d'Anticosti

Vancouver

Regina

L.
Manitoba

L.
Winnipeg

Albany

Mistassini

Gulf of
St. Lawrence

Pr.
Edward

Cape Breton
I.

Winnipeg

L.
Nipigon

PLATEAU

Québec

Nova
Scotia

Sable I.

L. of
the Woods

Lake
Superior

Abitibi

Montréal

Halifax

5,000 m.
16,404 ft. 2,000 m.
6,562 ft. 1,000 m.
3,281 ft. 500 m.
1,640 ft. 200 m.
656 ft. 100 m.
328 ft. Sea
Level Below

Georgian
Bay

Ottawa

Manitoulin I.

Toronto

L. Ontario

Huron

Niagara
Falls

0   200   400 MI.

0   200   400 KM.

Newfoundland
including Labrador

SCALE

Capitals of Provinces	⊛
Provincial Boundaries	
Provincial Boundary according to Imperial Privy Council decision, 1927	

AREA 404,517 sq. km. (156,184 sq. mi.)
POPULATION 551,792
CAPITAL St. John's
LARGEST CITY St. John's
HIGHEST POINT Mt. Caubvick 1622 m. (5,321 ft.)
SETTLED IN 1610
ADMITTED TO CONFEDERATION March 23, 1949
PROV. FLOWER Pitcher Plant
PROV. BIRD Atlantic Puffin

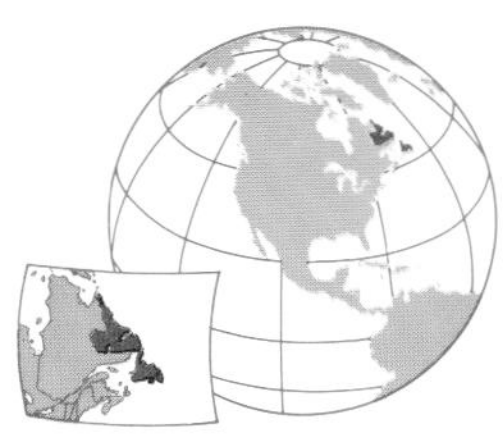

Agriculture, Industry and Resources

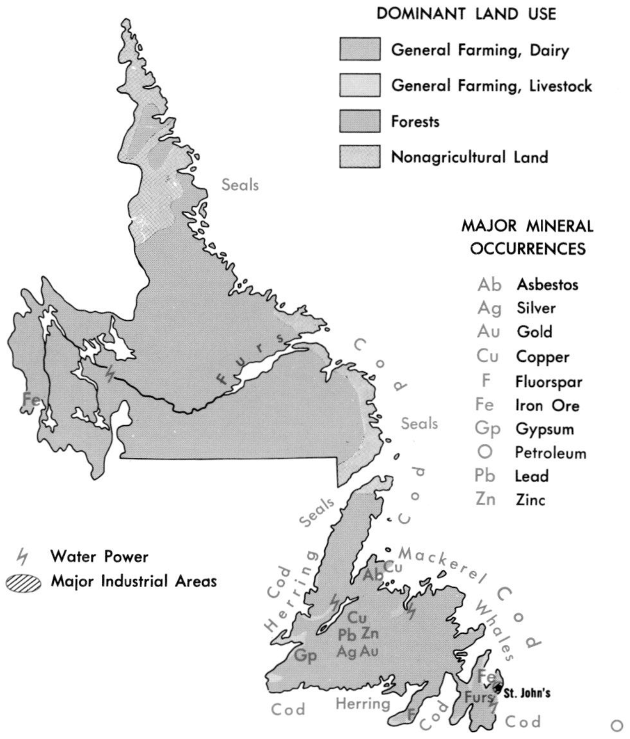
DOMINANT LAND USE
General Farming, Dairy
General Farming, Livestock
Forests
Nonagricultural Land
MAJOR MINERAL OCCURRENCES
Ab Asbestos
Ag Silver
Au Gold
Cu Copper
F Fluorspar
Fe Iron Ore
Gp Gypsum
O Petroleum
Pb Lead
Zn Zinc
Water Power
Major Industrial Areas
Seals
Furs
Cod
Mackerel
Whales
Herring
St. John's

NEWFOUNDLAND

CITIES and TOWNS

Admiral's Beach 362 . . . . D2
Admiral's Cove 99 . . . . D2
Anchor Point 368 . . . . C3
Aquaforte 200 . . . . D2
Argentia 93 . . . . C2
Arnold's Cove 1,124 . . . . C2
Avondale 890 . . . . D2
Badger 1,090 . . . . C4
Badger's Quay-Valleyfield-Pool's Island 1,566 . . . . D4
Baie Verte 2,491 . . . . C4
Battle Harbour C3
Bauline 423 . . . . D2
Bay Bulls 1,081 . . . . D2
Bay de Verde 786 . . . . D2
Bay L'Argent 483 . . . . D4
Bay Roberts 4,512 . . . . D2
Bellburns 147 . . . . C3
Belleoram 565 . . . . C4
Bellevue 286 . . . . D2
Bide Arm 339 . . . . C3
Big Pond 167 . . . . C3
Birchy Bay 707 . . . . D4
Bird Cove 400 . . . . C3
Bishop's Falls 4,395 . . . . C4
Black Tickle 194 . . . . C3
Blackhead Road 1,855 . . . . . D2
Blaketown 617 . . . . D2
Bloomfield 715 . . . . D2
Bonavista 4,460 . . . . D2
Botwood 4,074 . . . . C4
Branch 462 . . . . D2
Brigus 898 . . . . D2
Broad Cove 198 . . . . D2
Brooklyn 197 . . . . D2
Brownsdale 199 . . . . D2
Buchans 1,655 . . . . C4
Bunyan's Cove 590 . . . . C2
Burgeo 2,504 . . . . C4
Burin 2,904 . . . . C4
Burnt Islands 991 . . . . C4
Burnt Point 260 . . . . D2
Calvert 482 . . . . D2
Campbellton 703 . . . . D4
Cape Broyle 698 . . . . D2
Cape Ray 484 . . . . C4
Caplin Cove 150 . . . . D2
Carbonear 5,335 . . . . D2
Carmanville 966 . . . . D4
Cartwright 658 . . . . C3
Catalina 1,162 . . . . D2
Cavendish 343 . . . . D2
Champney's West 141 . . . . . . D2
Chance Cove 498 . . . . D2
Change Islands 580 . . . . D4
Channel-Port aux Basques 5,988 . . . . C4
Chapel Arm 689 . . . . D2
Charlottetown 330 . . . . D2
Charlottetown 250 . . . . C3
Churchill Falls 936 . . . . B3
Clarenville 2,878 . . . . C2
Clarke's Beach 1,009 . . . . . . D2
Codroy 346 . . . . C4
Colinet 318 . . . . D2
Colliers 819 . . . . D2
Come By Chance 337 . . . . . . C2
Conception Harbour 917 . . . . D2
Conche 464 . . . . C3
Cook's Harbour 388 . . . . C3
Corner Brook 24,339 . . . . . . C4
Cow Head 695 . . . . C4
Cox's Cove 980 . . . . C4
Cupids 706 . . . . D2
Daniell's Harbour 614 . . . . . . C3
Dark Cove 1,344 . . . . D4
Davis Inlet 240 . . . . B2
Deep Bight 243 . . . . C2
Deer Lake 4,348 . . . . C4
Dildo 877 . . . . D2
Dunville 1,817 . . . . D2
Durrell 1,145 . . . . D4
Eastport 597 . . . . D1
Elliston 527 . . . . D2
Embree 846 . . . . C4
Englee 998 . . . . C3
English Harbour 118 . . . . D2
English Harbour West 327 . . C4
Fermeuse 584 . . . . D2
Ferryland 795 . . . . D2
Flat Bay 322 . . . . C4
Flat Rock 808 . . . . D2
Fleur de Lys 616 . . . . C3
Flowers Cove 459 . . . . C3
Fogo 1,105 . . . . D4
Forteau 520 . . . . C3
Fortune 2,473 . . . . C4
Fox Harbour 280 . . . . C3
Fox Harbour 538 . . . . D2
François 219 . . . . C4
Freshwater 1,276 . . . . C2
Freshwater 209 . . . . D2
Gambo 2,932 . . . . D4
Gander 10,404 . . . . D4
Garnish 761 . . . . C4
Gaskiers-Point la Haye 505 . . D2
Gaultois 558 . . . . C4
Georges Brook 356 . . . . D2
Glenwood 1,129 . . . . D4
Glovertown 2,165 . . . . C1
Goobies 185 . . . . D2
Goose Bay-Happy Valley 7,103 . . . . B3
Gooseberry Cove 195 . . . . . . C2
Goose Cove 134 . . . . C2
Goose Cove 368 . . . . C3
Goulds 4,242 . . . . D2
Grand Bank 3,901 . . . . C4
Grand Falls 8,765 . . . . C4
Grates Cove 275 . . . . D2
Green Island Cove 222 . . . . . C3
Green's Harbour 785 . . . . . . D2
Greenspond 423 . . . . D4
Grey River 234 . . . . C4
Gull Island 362 . . . . D2
Hampden 838 . . . . C4
Hant's Harbour 542 . . . . D2
Happy Adventure 352 . . . . . . D2
Happy Valley-Goose Bay 7,103 . . . . B3
Harbour Breton 2,464 . . . . . . C4
Harbour Deep 278 . . . . C3
Harbour Grace 2,988 . . . . . . D2
Harbour Main-Chapel Cove-Lakeview 1,303 . . . . D2
Hare Bay 1,520 . . . . D4
Hawke's Bay 553 . . . . C3
Head of Bay d'Espoir 586 . . . C4
Heart's Content 625 . . . . D2
Heart's Delight-Islington 899 D2
Heart's Desire 416 . . . . D2
Heatherton 328 . . . . C4
Hermitage 863 . . . . C4
Hickman's Harbour 479 . . . . D2
Hillview 295 . . . . D2
Hodge's Cove 438 . . . . D2
Holyrood 1,789 . . . . D2
Hopedale 425 . . . . B2
Howley 456 . . . . C4
Isle aux Morts 1,238 . . . . . . C4
Jackson's Arm 623 . . . . C4
Jeffrey's 276 . . . . C4
Jerseyside 641 . . . . B3
Job's Cove 201 . . . . D2
Joe Batt's Arm-Barr'd Islands 1,155 . . . . D4
Keels 129 . . . . D1
Kelligrews (Foxtrap-Greeleytown-Peachtown-Kelligrews) 2,292 . . . . . . D2
Kilbride 5,014 . . . . D2
King's Cove 253 . . . . D1
King's Point 825 . . . . C4
Kippens 1,219 . . . . C4
Labrador City 11,538 . . . . . . A3
Lamaline 548 . . . . C4
L'Anse-au-Clair 267 . . . . . . C3
L'Anse-au-Loup 589 . . . . . . C3
L'Anse au Meadow 66 . . . . . C3
La Poile 186 . . . . C4
Lark Harbour 783 . . . . C4
La Scie 1,422 . . . . C4
Lawn 939 . . . . C4
Lethbridge 686 . . . . D2
Lewisporte 3,963 . . . . C4
Little Bay Islands 407 . . . . . C4
Little Catalina 750 . . . . D2
Little Heart's Ease 467 . . . . . D2
Lodge Bay 124 . . . . C3
Long Harbour-Mount Arlington Heights 660 . . . . D2
Lourdes 932 . . . . C4
Lower Island Cove 415 . . . . . D2
Lumsden 645 . . . . D4
Main Brook 514 . . . . C3
Makkovik 347 . . . . C2
Markland 344 . . . . D2
Mary's Harbour 408 . . . . C3
Marystown 6,299 . . . . D4
McCallum 243 . . . . C4
Melrose 416 . . . . D2
Middle Arm, Green Bay 575 . C4
Millertown 228 . . . . C4
Milltown-Head of Bay d'Espoir 1,376 . . . . C4
Milton 258 . . . . C2
Mobile 171 . . . . D2
Mount Carmel-Mitchell's Brook-St. Catherine's 699 . . . . . D2
Mount Pearl 11,543 . . . . D2
Musgrave Harbour 1,554 . . . D4
Musgravetown 635 . . . . C2
Nain 938 . . . . B2
New Bonaventure 106 . . . . . . D2
New Chelsea 144 . . . . D2
New Harbour 777 . . . . D2
Newmans Cove 231 . . . . D2
New Perlican 350 . . . . D2
Newtown 511 . . . . D4
Nippers Harbour 259 . . . . . . C4
Norman's Cove-Long Cove 1,152 . . . . . . D2
Norris Arm 1,216 . . . . C4
Norris Point 1,033 . . . . C4
North Harbour 151 . . . . D2
North River 245 . . . . D2
North West Brook 279 . . . . . C2
North West River 515 . . . . . B3
O'Donnells 280 . . . . D2
Old Bonaventure 111 . . . . . . D2
Old Perlican 709 . . . . D2
Paradise 2,861 . . . . D2
Parkers Cove 424 . . . . D4
Parson's Pond 605 . . . . C3
Pasadena 2,685 . . . . C4
Patrick's Cove 155 . . . . C2
Perry's Cove 141 . . . . D2
Peterview 1,119 . . . . C4
Petites 108 . . . . C4
Petley 147 . . . . D2
Petty Harbour-Maddox Cove 853 . . . . D2
Picadilly 524 . . . . C4
Pinware River 201 . . . . C3
Placentia 2,204 . . . . C2
Plate Cove 474 . . . . D2
Point La Haye 195 . . . . D2
Point Lance 141 . . . . C2
Point Leamington 848 . . . . . C4
Point Verde 296 . . . . C2
Pollards Point 502 . . . . C4
Port au Bras 366 . . . . D4
Port au Choix 1,311 . . . . . . C3
Port au Port 603 . . . . C4
Port Blandford 702 . . . . C2
Port Hope Simpson 581 . . . . C3
Port Kirwan 164 . . . . D2
Port Rexton 489 . . . . D2
Port Saunders 769 . . . . C3
Portugal Cove 2,361 . . . . . . D2
Portugal Cove South 371 . . . D2
Port Union 671 . . . . D2
Postville 223 . . . . B3
Pouch Cove 1,522 . . . . D2
Princeton 204 . . . . D2
Raleigh 373 . . . . C3
Ramea 1,386 . . . . C4
Red Bay 316 . . . . C3
Red Head Cove 225 . . . . D2
Rencontre East 230 . . . . C4
Renews-Cappahayden 578 . . D2
Rigolet 271 . . . . C3
Riverhead 431 . . . . D2
River of Ponds 304 . . . . C3
Robert's Arm 1,005 . . . . C4
Rocky Harbour 1,273 . . . . . . C4
Roddickton 1,142 . . . . C3
Rose Blanche-Harbour le Cou 975 . . . . C4
Rushoon 520 . . . . D4
Saint Alban's 1,968 . . . . C4
Saint Andrew's 262 . . . . C4
Saint Anthony 3,107 . . . . . . C3
Saint Brendan's 468 . . . . . . D4
Saint Bride's 599 . . . . C2
Saint George's 1,756 . . . . . . C4
St. John's (cap.) 83,770 . . . D2
Saint Joseph's 262 . . . . D2
Saint Lawrence 2,012 . . . . . . C4
Saint Lunaire-Griquet 1,010 . . C3
Saint Mary's 701 . . . . D2
Saint Paul's 454 . . . . C4
Saint Phillips 1,365 . . . . D2
Saint Shotts 239 . . . . D2
Saint Vincent's-Saint Stephens-Peter's River 796 . . . . D2
Sally's Cove 100 . . . . C4
Salmon Cove 786 . . . . D2
Seal Cove 751 . . . . C3
Seal Cove-White Bay 498 . . . C4
Seldom-Little Seldom 560 . . D4
Ship Harbour 265 . . . . D2
Shoal Cove 223 . . . . C3
Shoal Harbour 1,000 . . . . . . C2
South Branch 264 . . . . C4
South Brook, Hall's Bay Dist. 786 . . . . C4
South Brook, Humber Dist. 477 . . . . C4
Southern Harbour 772 . . . . . C2
South River 645 . . . . D2
Spaniard's Bay 2,125 . . . . . . D2
Springdale 3,501 . . . . C4
Stephenville 8,876 . . . . C4
Stephenville Crossing 2,172 . C4
Summerford 1,198 . . . . C4
Summerville 346 . . . . D2
Sunnyside 703 . . . . D2
Sweet Bay 204 . . . . D2
Swift Current 329 . . . . C2
Terrenceville 796 . . . . D4
Tilting 427 . . . . D4
Torbay 3,394 . . . . D2
Tors Cove 355 . . . . D2
Traytown 383 . . . . D1
Trepassey 1,473 . . . . D2
Trinity 522 . . . . D2
Trinity 375 . . . . D4
Trout River 759 . . . . C4
Twillingate 1,506 . . . . C4
Upper Island Cove 2,025 . . . D2
Victoria 1,870 . . . . D2
Wabana 4,254 . . . . D2
Wabush 3,155 . . . . A3
Wesleyville 1,125 . . . . D4
Western Bay 463 . . . . D2
West Saint Modeste 273 . . . . C3
Whitbourne 1,233 . . . . D2
Wild Cove 152 . . . . C3
Windsor 5,747 . . . . C4
Winterton 753 . . . . D2
Witless Bay 907 . . . . D2

OTHER FEATURES

Alexis (riv.) C3
Anguille (cape) C4
Annieopscotch (mts.) C4
Ashuanipi (lake) A3
Ashuanipi (riv.) A3
Atikonak (lake) B3
Attikamagen (lake) A3
Avalon (pen.) D2
Barachois Pond Prov. Park . . C4
Bauld (cape) C3
Bell (isl.) C3
Bell (isl.) D2
Belle Isle (isl.) C3
Belle Isle (str.) C3
Blackhead (bay) D2
Bonavista (bay) D1
Bonavista (cape) D1
Bonne (bay) C4
Branch (riv.) C2
Broyle (cape) D2
Bull Arm (inlet) D2
Burin (pen.) C4
Butter Pot Prov. Park D2
Cabot (str.) B4
Canada (bay) C3
Chidley (cape) B1
Churchill (falls) B3
Churchill (riv.) B3
Cirque (mt.) B2
Clode (sound) D2
Conception (bay) D2
Deep (inlet) B2
Double Mer (lake) C3
Dyke (lake) A3
Eagle (riv.) C3
Espoir (bay) C4
Exploits (riv.) C4
Fogo (isl.) D4
Fortune (bay) C4
Freels (cape) D3
Gander (bay) D4
Gander (lake) D4
Gander (riv.) D4
Glover (isl.) C4
Goose (riv.) B3
Grand (lake) B3
Grand (lake) C4
Grates (pt.) D2
Great Colinet (isl.) D2
Grey (isls.) C3
Groais (isl.) C3
Gros Morne (mt.) C4
Gros Morne Nat'l Park C4
Groswater (bay) C3
Hamilton (inlet) C3
Hamilton (sound) D4
Hare (bay) C3
Hawke (hills) D2
Hebron (fjord) B2
Hermitage (bay) C4
Holyrood (bay) D2
Horse (isls.) C3
Horse Chops (head) D2
Humber (riv.) C4
Ingornachoix (bay) C3
Ireland's Eye (isl.) D2
Islands (bay) C4
Kaipokok (bay) B2
Kanairiktok (riv.) B3
Kaumajet (mts.) B2
Kingurutik (mesa) B2
Labrador (reg.) B2
Labrador (sea) C2
La Manche Valley Prov. Park D2
La Poile (bay) C4
Little Mecatina (riv.) B3
Long (isl.) C2
Long (lake) A3
Long (pt.) C4
Long Range (mts.) C4
Main Topsail (mt.) C4
Makkovik (cape) C2
McLelan (str.) B1
Mealy (lake) C3
Meelpaeg (lake) C4
Melville (lake) C3
Menihek (lakes) A3
Merasheen (isl.) C2
Mistaken (pt.) D2
Mistastin (lake) B2
Nachvak (fjord) B2
Naskaupi (riv.) B3
Newfoundland (isl.) C4
Newman (sound) D2
New World (isl.) C4
Norman (cape) C3
North Aulatsivik (isl.) B2
Notre Dame (bay) C4
Okak (bay) B2
Ossokmanuan (res.) B3
Petitsikapau (lake) A3
Pine (cape) D2
Pinware (riv.) C3
Pistolet (bay) C3
Placentia (bay) C2
Ponds (isl.) C3
Port au Port (bay) C4
Port au Port (pen.) C4
Port Manvers (harb.) B2
Race (cape) D2
Ramah (bay) B2
Ramea (isls.) C4
Random (isl.) D2
Random (sound) D2
Ray (cape) C4
Red (isl.) C2
Red Indian (lake) C4
Red Wine (riv.) B3
Rocky (riv.) D2
Round (pond) C4
Saglek (bay) B2
Saint Francis (cape) D2
Saint George (cape) C4
Saint George's (bay) C4
Saint John (bay) C3
Saint John (cape) C3
Saint Lawrence (gulf) B4
Saint Lewis (cape) C3
Saint Mary's (bay) C2
Saint Mary's (cape) C2
Saint Michaels (bay) C3
Salmonier (riv.) D2
Sandwich (bay) C3
Shabogamo (lake) A3
Shoal (bay) D2
Smallwood (res.) B3
Smith (sound) D2
South Aulatsivik (isl.) B2
Spear (cape) D2
Squires Mem. Park C4
Swale (isl.) D1
Terra Nova (riv.) C2
Terra Nova Nat'l Park D2
Territok (cape) B2
Thoresby (mt.) B2
Torbay (pt.) D2
Torngat (mts.) B2
Trespassey (bay) D2
Trinity (bay) D2
Tunungayualok (isl.) B2
Ukasiksalik (isl.) B2
Victoria (lake) C4
White (bay) C3
White Bear (lake) C4
White Handkerchief (cape) . . B2

SAINT PIERRE and MIQUELON

CITIES and TOWNS

Saint-Pierre (cap.) 5,415 . . . C4

OTHER FEATURES

Miquelon (isl.) 626 . . . . C4
Saint Pierre (isl.) 5,415 . . . . C4

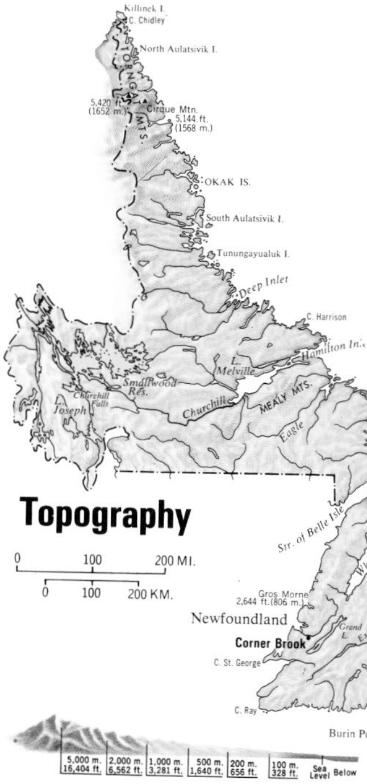
Topography
0 100 200 MI.
0 100 200 KM.
Killinek I.
C. Chidley
North Aulatsivik I.
TORNGAT MTS.
5,420 ft. (1652 m.)
Cirque Mtn. 5,144 ft. (1568 m.)
OKAK IS.
South Aulatsivik I.
Tunungayualuk I.
Deep Inlet
C. Harrison
Hamilton Inlet
Melville
Smallwood Res.
Churchill Falls
Joseph
Churchill
MEALY MTS.
Eagle
Belle Isle
C. Bauld
Str. of Belle Isle
White Bay
Notre Dame Bay
Fogo I.
Gros Morne 2,644 ft. (806 m.)
Newfoundland
Corner Brook
C. St. George
Bonavista Bay
Trinity Bay
St. John's
Avalon Pen.
C. Ray
Burin Pen.
Placentia Bay
C. Race
5,000 m. 16,404 ft.
2,000 m. 6,562 ft.
1,000 m. 3,281 ft.
500 m. 1,640 ft.
200 m. 656 ft.
100 m. 328 ft.
Sea Level
Below

PRINCE EDWARD ISLAND

AREA 5,657 sq. km. (2,184 sq. mi.)
POPULATION 134,557
CAPITAL Charlottetown
LARGEST CITY Charlottetown
HIGHEST POINT 142 m. (466 ft.)
SETTLED IN 1720
ADMITTED TO CONFEDERATION July 1, 1873
PROV. FLOWER Lady's-Slipper
PROV. BIRD Blue Jay

NOVA SCOTIA

AREA 55,491 sq. km. (21,425 sq. mi.)
POPULATION 909,282
CAPITAL Halifax
LARGEST CITY Halifax
HIGHEST POINT Cape Breton Highlands 532 m. (1,745 ft.)
SETTLED IN 1605
ADMITTED TO CONFEDERATION July 1, 1867
PROV. FLOWER Trailing Arbutus or Mayflower
PROV. BIRD Osprey

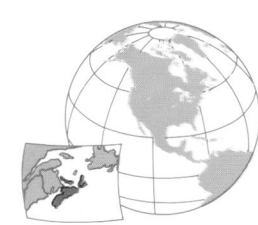

Agriculture, Industry and Resources

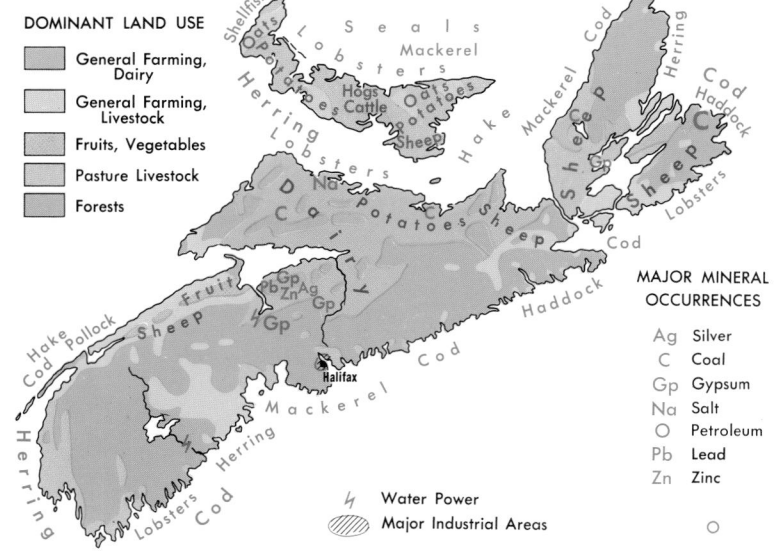

DOMINANT LAND USE

- General Farming, Dairy
- General Farming, Livestock
- Fruits, Vegetables
- Pasture Livestock
- Forests

MAJOR MINERAL OCCURRENCES

Ag	Silver
C	Coal
Gp	Gypsum
Na	Salt
O	Petroleum
Pb	Lead
Zn	Zinc

⚡ Water Power
▨ Major Industrial Areas

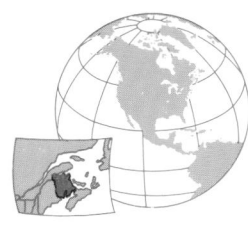

AREA 73,437 sq. km. (28,354 sq. mi.)
POPULATION 738,133
CAPITAL Fredericton
LARGEST CITY St. John
HIGHEST POINT Mt. Carleton 820 m.
(2,690 ft.)
SETTLED IN 1611
ADMITTED TO CONFEDERATION
July 1, 1867
PROV. FLOWER Purple Violet
PROV. BIRD Chickadee

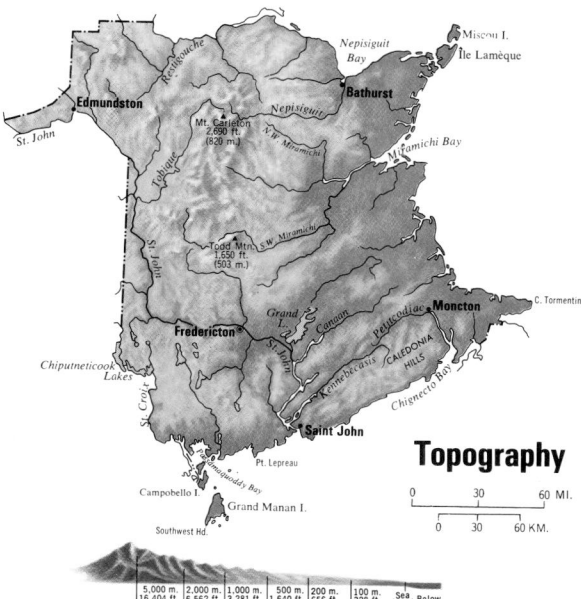

Topography

0   30   60 MI.

0   30   60 KM.

5,000 m. 2,000 m. 1,000 m. 500 m. 200 m. 100 m. Sea Below
16,404 ft. 6,562 ft. 3,281 ft. 1,640 ft. 656 ft. 328 ft. Level

Agriculture, Industry and Resources

DOMINANT LAND USE

Cereals, Livestock
Dairy
Potatoes
General Farming, Livestock
Pasture Livestock
Forests

MAJOR MINERAL OCCURRENCES

Ag Silver   Pb Lead
C Coal   Sb Antimony
Cu Copper   Zn Zinc

⚡ Water Power
Major Industrial Areas

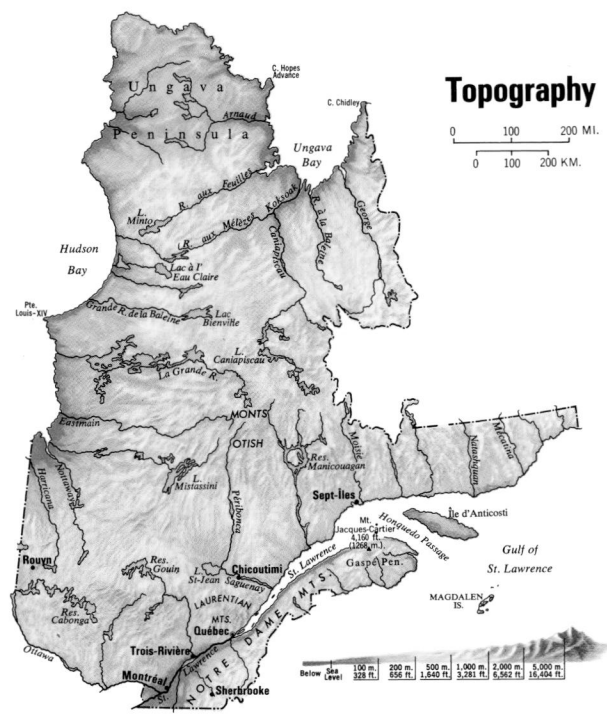

Topography

0   100   200 MI.

0   100   200 KM.

Soulanges 15,429	C 4
Stanstead 38,186	F 4
Témiscouata 52,570	J 2
Terrebonne 193,865	H 4
Vaudreuil 50,043	H 4
Verchères 63,353	J 4
Wolfe 15,635	F 4
Yamaska 14,797	E 3

CITIES and TOWNS

Acton Vale 4,371	E 4	Chandler 3,946	D 2	La Prairie⊙ 10,627	J 4	Maniwaki⊙ 5,424	B 3
Albanel 992	E 1	Charlemagne 4,827	H 4	La Providence	E 4	Manseau 626	E 3
Alma⊙ 26,322	F 1	Charlesbourg 68,326	J 3	Larouche 662	F 1	Maple Grove 2,009	H 4
Amqui⊙ 4,048	B 2	Charlesbourg 68,326	J 3	Maria 1,178	C 2
Ancienne-Lorette 12,935	J 3	Charny 8,240	J 3	La Salle 76,299	H 4	Marieville⊙ 4,048	D 4
Angers	B 4	Châteauguay 36,928	H 4	L'Ascension⊙ 1,287	F 1	Mascouche 20,345	H 4
Anjou 37,346	H 4	Château-Richer⊙ 3,628	F 3	L'Assomption⊙ 4,844	D 4	Maskinongé 1,005	D 3
Annaville 712	E 3	Chénéville 633	B 4	La Station-du-Coteau 892	C 4	Masson 4,264	B 4
Armagh 878	G 3	Chicoutimi⊙ 60,064	G 1	Laterrière 788	F 1	Massueville 671	D 4
Arthabaska⊙ 6,827	F 3	Chicoutimi-Jonquière		La Tuque 11,556	E 2	Matane⊙ 13,612	B 1
Arvida	F 1	*135,172	G 1	Laurentides 1,947	D 4	Matapédia 586	B 2
Asbestos 7,967	F 4	Chute-aux-Outardes 2,280	A 1	Laurier-Station 1,123	F 3	Melocheville 1,892	C 4
Ascot Corner 847	F 4	Clermont 3,621	G 2	Laurierville 939	F 3	Mercier 6,352	H 4
Audet 760	G 4	Coaticook 6,271	F 4	Lauzon 13,362	J 3	Metabetchouan 3,406	F 1
Ayer's Cliff⊙ 810	E 4	Coleraine 1,660	F 4	Laval 268,335	H 4	Mirabel⊙ 14,080	H 4
Aylmer 26,695	B 4	Compton 728	F 4	Lavaltrie 2,053	D 4	Mistassini 6,682	E 1
Baie-Comeau 12,866	A 1	Contrecoeur 5,449	D 4	L'Avenir 1,116	E 4	Montauban 557	E 3
Baie-d'Urfé 3,674	G 4	Cookshire⊙ 1,480	F 4	Lawrenceville 562	E 4	Mont-Carmel 807	H 2
Baie-Saint-Paul⊙ 3,961	G 2	Coteau-du-Lac 1,247	C 4	Le Moyne 6,137	J 4	Montcerf 570	A 3
Baie-Trinité 749	B 1	Coteau-Landing⊙ 1,386	C 4	Léry 2,239	H 4	Montebello 1,229	B 4
Beaconsfield 19,613	H 4	Côte-Saint-Luc 27,531	H 4	Lévis⊙ 17,895	J 3	Mont-Joli 6,359	J 1
Beauceville⊙ 4,302	G 3	Courcelles 608	G 4	Lennoxville 3,922	F 4	Mont-Laurier⊙ 8,405	B 3
Beauharnois⊙ 7,025	D 4	Courville	J 3	Les Méchins 803	B 1	Mont-Louis 756	C 1
Beaumont 791	F 3	Cowansville 12,240	E 4	Linière 1,168	G 3	Montmagny⊙ 12,405	G 3
Beauport 60,447	J 3	Crabtree 1,950	D 4	L'Islet 1,070	G 2	Montréal⊙ 980,354	H 4
Beaupré 2,740	G 2	Danville 2,200	E 4	L'Islet-sur-Mer 774	G 2	Montréal *2,828,349	H 4
Bécancour⊙ 10,247	E 3	Daveluyville 1,257	E 3	L'Isle-Verte 1,142	G 1	Montréal-Est 3,778	J 4
Bedford⊙ 2,832	E 4	Deauville 942	E 4	Longueuil⊙ 124,320	J 4	Montréal-Nord 94,914	H 4
Beebe Plain 1,072	E 4	Dégelis 3,477	J 2	Lorette⊙ville 15,060	H 3	Mont-Rolland 1,517	C 4
Bélair (Val-Bélair) 12,695	H 3	Delisle 4,011	F 1	Lorraine 6,881	H 4	Mont-Royal 19,247	H 4
Beloeil 17,540	D 4	Delson 4,935	H 4	Louiseville⊙ 3,735	E 3	Mont-Saint-Hilaire 10,066	J 4
Bernierville 2,120	F 3	Desbiens 1,541	F 1	Luceville 1,524	J 1	Morin Heights 592	C 4
Berthierville⊙ 4,049	D 3	Deschaillons-sur-Saint-		Lyster 830	F 3	Murdochville 3,396	C 1
Bic 2,994	J 1	Laurent 950	E 3	Magog 13,604	E 4	Nantes 1,167	F 4
Biencourt 824	J 2	Deschambault 977	E 3				
Black Lake 5,148	F 3	Deschênes	B 4				
Blainville 14,682	H 4	Deux-Montagnes 9,944	H 4				
Boischatel 3,345	J 3	Didyme 667	E 1				
Bois-des-Filion 4,943	H 4	Disraëli 3,181	F 4				
Bolduc 1,565	G 4	Dolbeau 8,766	E 1				
Bonaventure 1,371	C 2	Dollard-des-Ormeaux 39,940	H 4				
Boucherville 29,704	J 4	Donnacona 5,731	F 3				
Bromont 2,731	E 4	Dorion 5,749	C 4				
Bromptonville 3,035	F 4	Dorval 17,727	H 4				
Brossard 52,232	H 4	Dosquet 703	F 3				
Brownsburg 2,875	C 4	Douville	D 4				
Buckingham 7,992	B 4	Drummondville 27,347	E 4				
Cabano 3,291	J 2	Drummondville-Sud 9,220	E 4				
Cacouna 1,281	H 2	Dunham 2,887	E 4				
Calumet 729	C 4	Durham-Sud 1,045	E 4				
Candiac 8,502	H 4	East Angus 4,016	F 4				
Cap-à-l'Aigle 819	G 2	East Broughton 1,397	F 3				
Cap-Chat 3,464	C 1	East Broughton Station 1,302	F 3				
Cap-de-la-Madeleine 32,626	E 3	Eastman 612	E 4				
Caplan-Rivière Caplan 1,139	C 2	Entrelacs 1,735	C 3				
Cap-Saint-Ignace 1,485	G 2	Farnham 6,498	E 4				
Cap-Santé⊙ 671	F 3	Ferme-Neuve 2,266	B 3				
Carignan 4,544	J 4	Forestville 4,271	H 1				
Carleton 2,710	C 2	Frampton 684	G 3				
Causapscal 2,501	B 2	Francoeur 1,422	F 3				
Chambly 12,190	J 4	Gaspé 17,261	D 1				
Chambord 961	E 1	Gatineau 74,988	B 4				
		Giffard	J 3				
		Girardville 1,128	E 1				
		Gracefield 869	A 3				
		Granby 38,069	E 4				
		Grand'Mère 15,442	E 3				
		Grande-Rivière 4,420	D 2				
		Grandes-Bergeronnes 748	H 1				
		Grande-Vallée 700	D 1				
		Greenfield Park 18,527	J 4				
		Grenville 1,417	C 4				
		Gros-Morne 672	C 1				
		Hampstead 7,598	H 4				
		Ham-Sud⊙ 62	F 4				
		Hauterive 13,995	A 1				
		Hébertville 2,515	F 1				
		Hébertville-Station 1,442	F 1				
		Hemmingford 737	D 4				
		Henryville 595	D 4				
		Howick 639	D 4				
		Hudson 4,414	C 4				
		Hull⊙ 56,225	B 4				
		Huntingdon⊙ 3,018	D 4				
		Île-Perrot 5,945	G 4				
		Iberville⊙ 8,587	D 4				
		Inverness⊙ 329	F 3				
		Joliette⊙ 16,987	D 3				
		Jonquière 60,354	F 1				
		Jonquière-Chicoutimi					
		*135,172	F 1				
		Kingsey Falls 849	E 4				
		Kirkland 10,476	H 4				
		Knowlton (Lac-Brome)⊙					
		4,316	E 4				
		La Baie 20,935	G 1				
		Labelle 1,534	C 3				
		Lac-à-la-Croix 1,017	F 1				
		Lac-Alouette-Lac-Brière 1,356	D 4				
		Lac-au-Saumon 1,332	B 2				
		Lac-aux-Sables 838	E 3				
		Lac-Beaufort	F 3				
		Lac-Bouchette 1,703	E 1				
		Lac-Carré 717	C 3				
		Lac-des-Écorces 766	B 3				
		Lac-Drolet 1,120	G 4				
		Lac-Etchemin 2,729	G 3				
		Lachenaie 8,631	D 4				
		Lachine 37,521	H 4				
		Lachute⊙ 11,729	C 4				
		Lacolle 1,319	D 4				
		Lac-Mégantic⊙ 6,119	G 4				
		Lac-Saint-Charles 5,837	H 3				
		Lafontaine 4,799	C 4				
		La Guadeloupe 1,692	F 4				
		La Malbaie⊙ 4,030	G 2				
		Lambton 1,559	F 4				
		L'Annonciation 2,384	C 3				
		Lanoraie (Lanoraie-d'Autry)					
		1,613	D 4				
		La Pêche 4,977	B 4				
		La Pérade 1,039	E 3				
		La Pocatière 4,560	H 2				

COUNTIES

Argenteuil 32,454	C 4	Gaspé-Est 41,173	D 1	Missisquoi 36,161	D 4
Arthabaska 59,277	E 4	Gaspé-Ouest 18,943	C 1	Montcalm 27,557	C 3
Bagot 26,840	E 4	Gatineau 54,229	B 3	Montmagny 25,622	G 3
Beauce 73,427	G 3	Hull 131,213	B 4	Montmorency No 1 23,048	F 2
Beauharnois 54,034	C 4	Huntingdon 16,953	C 4	Montmorency No 2 6,436	G 3
Bellechasse 23,559	G 3	Iberville 23,180	D 4	Napierville 13,562	D 4
Berthier 31,096	C 2	Île-de-Montréal 1,760,122	H 4	Nicolet 33,513	E 3
Bonaventure 40,487	C 2	Île-Jésus 268,335	H 4	Papineau 37,975	B 4
Brome 17,436	E 4	Joliette 60,384	C 3	Pontiac 20,283	A 3
Chambly 307,090	J 4	Kamouraska 28,642	H 2	Portneuf 58,843	E 3
Champlain 119,595	E 2	Labelle 34,395	B 3	Québec 458,980	F 3
Charlevoix-Est 17,448	G 2	Lac-Saint-Jean-Est 47,891	F 1	Richelieu 53,058	D 4
Charlevoix-Ouest 14,172	G 2	Lac-Saint-Jean-Ouest 62,952	E 1	Richmond 40,871	E 4
Châteauguay 59,968	D 4	Laprairie 105,962	H 4	Rimouski 69,099	J 1
Chicoutimi 174,441	F 1	L'Assomption 109,705	D 4	Rivière-du-Loup 41,250	H 2
Compton 20,536	F 4	Lévis 94,104	J 3	Rouville 42,391	D 4
Deux-Montagnes 71,252	C 4	L'Islet 22,062	G 2	Saguenay 115,881	D 1
Dorchester 33,949	C 3	Lotbinière 29,653	F 3	Saint-Hyacinthe 55,888	D 4
Drummond 69,770	E 4	Maskinongé 20,763	D 3	Saint-Jean 55,576	D 4
Frontenac 26,814	G 4	Matane 29,955	B 1	Saint-Maurice 107,703	D 3
		Matapédia 23,715	B 2	Shefford 70,733	E 4
		Mégantic 57,892	F 3	Sherbrooke 115,983	E 4

Agriculture, Industry and Resources

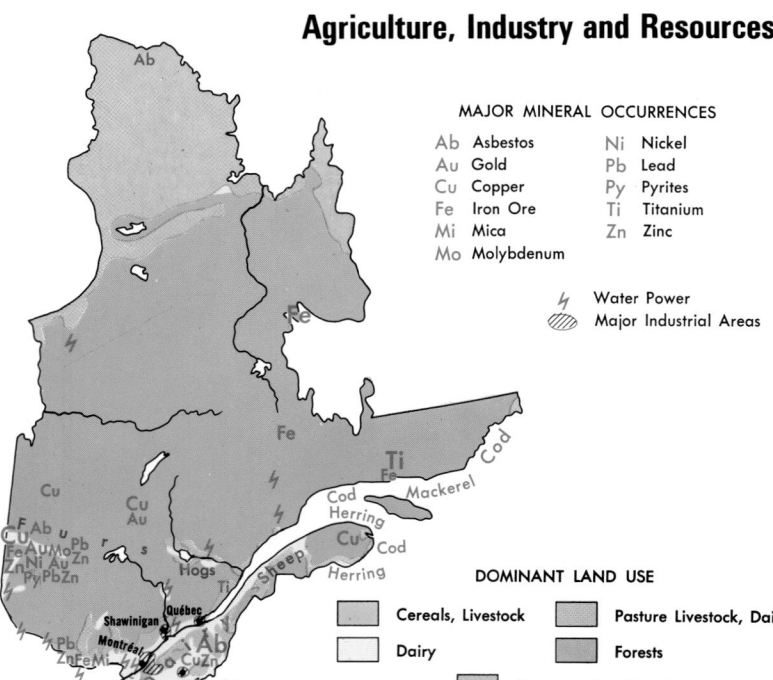

MAJOR MINERAL OCCURRENCES

Ab	Asbestos	Ni	Nickel
Au	Gold	Pb	Lead
Cu	Copper	Py	Pyrites
Fe	Iron Ore	Ti	Titanium
Mi	Mica	Zn	Zinc
Mo	Molybdenum		

⚡ Water Power

▨ Major Industrial Areas

DOMINANT LAND USE

▨ Cereals, Livestock	▨ Pasture Livestock, Dairy
▨ Dairy	▨ Forests
▨ Nonagricultural Land	

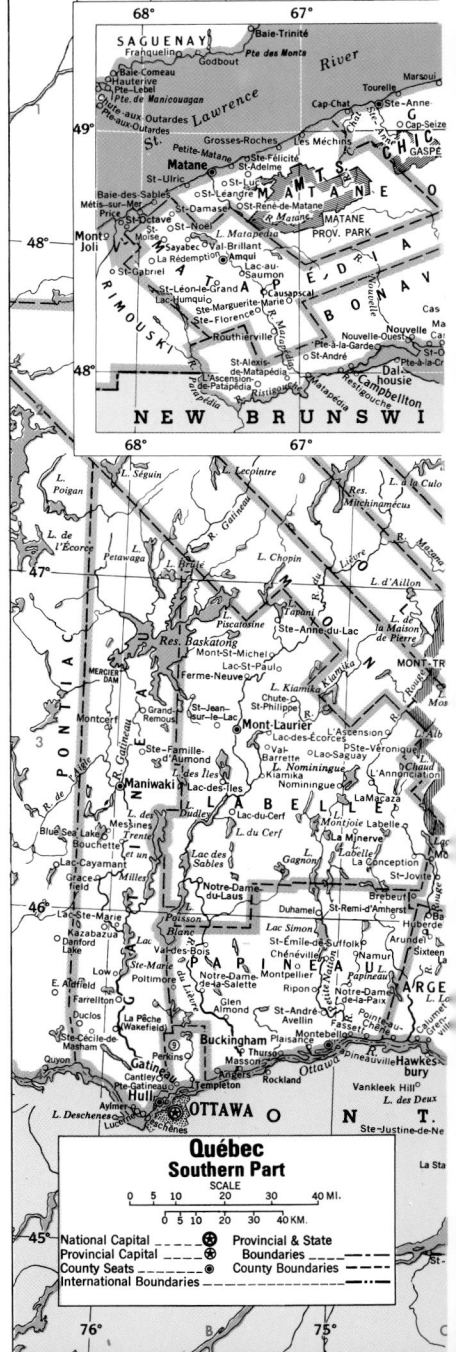

Québec
Southern Part

SCALE

0 5 10   20   30   40 MI.

AREA 1,540,680 sq. km. (594,857 sq. mi.)
POPULATION 7,138,795
CAPITAL Québec
LARGEST CITY Montréal
HIGHEST POINT Mont D'Iberville 1622 m. (5,321 ft.)
SETTLED IN 1608
ADMITTED TO CONFEDERATION July 1, 1867
PROV. FLOWER White Garden Lily
PROV. BIRD Snowy Owl

COUNTIES
indicated by numbers:
1 Iberville .... D 4
2 Napierville .... D 4
3 Rouville .... D 4
4 St-Hyacinthe .... D 4
5 Deux-Montagnes .... C 4
6 Soulanges .... C 4
7 Beauharnois .... C 4
8 Hull .... B 4
9 Laval .... H 4
10 Jésus .... B 4
11 Richelieu .... D 4
12 Vaudreuil .... C 4

Internal divisions represent Municipal Counties

© Copyright HAMMOND INCORPORATED, Maplewood, N.J.

Saint-Eustache 29,716......H 4
Saint-Fabien 1,361......J 1
Saint-Félicien 9,058......E 1
Saint-Félix-de-Valois 1,462...D 3
Saint-Ferréol-les-Neiges
1,758......G 2
Saint-Flavien 734......E 1
Saint-François-de-Sales 831...E 1
Saint-François-du-Lac© 942...E 3
Saint-Fulgence 950......G 1
Saint-Gabriel 3,161......D 3
Saint-Gabriel-de-Rimouski
779......J 1
Saint-Gédéon, Frontenac
1,569......G 4
Saint-Gédéon, Lac-St-Jean-E.
1,000......F 1
Saint-Georges, Beauce
10,342......G 4
Saint-Georges, Champlain
3,344......E 3
Saint-Georges-Ouest 6,378...G 3
Saint-Germain-de-Grantham
1,373......E 4
Saint-Gervais 973......F 3
Saint-Gilles 912......F 3
Saint-Grégoire (Mont-St-
Grégoire) 740......D 4
Saint-Henri 1,970......E 3
Saint-Honoré, Beauce 1,116...G 4
Saint-Honoré, Chicoutimi
1,790......F 1
Saint-Hubert 60,573......H 4
Saint-Hubert-de-Témiscouata
871......J 2
Saint-Hyacinthe 38,246......D 4
Saint-Isidore 811......F 3
Saint-Isidore-de-Laprairie 769...D 4
Saint-Jacques 2,152......D 4
Saint-Jacques-le-Mineur
1,203......H 4
Saint-Jean-Chrysostome
6,930......J 3
Saint-Jean-de-Dieu 1,377...J 1
Saint-Jean-de-Matha 931...D 3
Saint-Jean-Port-Joli 1,813...G 2
Saint-Jean-sur-Richelieu
35,640......D 4
Saint-Jérôme 25,123......H 4
Saint-Joachim 1,139......G 2
Saint-Joseph-de-Beauce
3,216......G 3
Saint-Jovite 3,841......C 3
Saint-Lambert 20,557......J 4
Saint-Laurent 65,900......H 4

Saint-Lazare 731......G 3
Saint-Léonard 79,429......H 4
Saint-Léonard-de-Chicoutimi 749...F 1
Saint-Léonard-d'Aston 992......E 3
Saint-Léon-de-Standon 816...G 3
Saint-Léon-le-Grand 722......B 2
Saint-Liboire© 746......E 4
Saint-Louis-de-Gonzague
615......D 4
Saint-Louis-de-Terrebonne
14,172......H 4
Saint-Louis-du-Ha! Ha! 809...H 2
Saint-Luc 6,815......D 4
Saint-Luc-de-Matane 598......B 1
Saint-Marc-des-Carrières
2,822......E 3
Saint-Méthode-de-Frontenac
925......F 3
Saint-Michel-de-Bellechasse
963......G 3
Saint-Michel-des-Saints
1,584......D 3
Saint-Nazaire-de-Chicoutimi
962......F 1
Saint-Nérée 970......G 3
Saint-Nicolas 5,074......H 3
Saint-Noël 666......B 1
Saint-Odilon 580......G 3
Saint-Omer 718......C 2
Saint-Ours 625......D 4
Saint-Pacôme 1,996......G 2
Saint-Pamphile 3,428......H 3
Saint-Pascal© 2,763......H 2
Saint-Paulin 663......D 3
Saint-Paul-de-Montminy 602...G 3
Saint-Paul-l'Ermite (Le
Gardeur) 8,312......J 4
Saint-Philippe-de-Néri 715...H 2
Saint-Pie 1,725......E 4
Saint-Pierre 5,305......H 4
Saint-Pierre-d'Orléans 880...G 3
Saint-Polycarpe 602......C 4
Saint-Prime 2,550......E 1
Saint-Prosper-de-Dorchester
2,150......G 3
Saint-Raphaël 1,346......G 3
Saint-Raymond 3,605......F 3
Saint-Rédempteur 4,463......J 3
Saint-Rémi 5,146......D 4
Saint-Roch-de-l'Achigan
1,160......D 4
Saint-Roch-de-Richelieu
1,650......D 4
Saint-Romuald-d'Etchemin©
9,849......J 3

Saint-Sauveur-des-Monts
2,348......C 4
Saint-Siméon 1,152......G 2
Saint-Simon 602......H 1
Saint-Stanislas 1,443......E 3
Saint-Sylvère 1,006......E 3
Saint-Timothée 2,113......D 4
Saint-Tite 3,031......E 3
Saint-Tite-des-Caps 626......G 2
Saint-Ubald 1,605......E 3
Saint-Ulric 792......B 1
Saint-Urbain-de-Charlevoix
1,079......G 2
Saint-Victor 1,104......G 3
Saint-Zacharie 1,284......G 3
Saint-Zotique 1,774......C 4
Sault-au-Mouton 828......H 1
Sawyerville 939......F 4
Sayabec 1,721......B 1
Scotstown 762......F 4
Senneville 1,221......H 4
Shannon 3,488......F 3
Shawbridge 942......C 4
Shawinigan 23,011......E 3
Shawinigan-Sud 11,325......E 3
Shawville 1,608......A 4
Sherbrooke© 74,075......E 4
Sherrington 614......D 4
Sillery 12,825......J 3
Sorel© 20,347......D 4
Squatec 1,000......J 2
Stanstead Plain 1,093......E 4
Sutton 1,599......E 4
Tadoussac© 900......H 1
Templeton......H 4
Terrebonne 11,769......H 4
Thetford Mines 19,965......F 3
Thurso 2,780......B 4
Tourelle (Tourelle-Grand-
Tourelle) 942......C 1
Tourville 659......H 2
Tracy 12,843......D 4
Tring-Jonction 1,315......F 3
Trois-Pistoles 4,445......H 1
Trois-Rivières 50,466......E 3
Trois-Rivières *111,453......E 3
Trois-Rivières-Ouest 13,107...E 3
Upton 926......E 4
Val-Barrette 609......C 3
Val-Brillant 687......B 1
Valcourt 2,601......E 4
Val-David 2,336......C 3
Vallée-Jonction 1,200......G 3
Valleyfield (Salaberry-de-
Valleyfield) 29,574......C 4
Vanier 10,725......J 3

Varennes 8,764......J 4
Vaudreuil 7,608......C 4
Verchères 4,473......J 4
Verdun 61,287......H 4
Victoriaville 21,838......F 3
Villeneuve......E 3
Warwick 2,847......E 3
Waterloo© 4,664......E 4
Waterville 1,397......F 4
Weedon-Centre 1,263......F 3
Westmount 20,480......H 4
Wickham 2,043......E 4
Windsor 5,233......F 4
Wottonville 673......F 4
Yamachiche© 1,258......E 3

OTHER FEATURES

Alma (isl.)......F 1
Aylmer (lake)......B 4
Baskatong (res.)......B 3
Batiscan (riv.)......E 3
Bécancour (riv.)......E 3
Bonaventure (isl.)......D 1
Bonaventure (riv.)......C 1
Brome (lake)......E 4
Brompton (lake)......E 4
Cascapédia (riv.)......C 1
Chaleur (bay)......C 2
Champlain (lake)......D 4
Chaudière (riv.)......F 3
Chic-Chocs (mts.)......C 1
Chicoutimi (riv.)......F 2
Coudres (isl.)......G 2
Deschênes (lake)......A 4
Deux Montagnes (lake)......H 4
Ditton (riv.)......F 4
Forillon Nat'l Park......D 1
Fort Chambly Nat'l Hist. Park...J 4
Gaspé (bay)......D 1
Gaspé (cape)......D 1
Gaspé (pen.)......C 1
Gaspésie Prov. Park......C 1
Gatineau (riv.)......B 3
Îles (lake)......B 3
Jacques-Cartier (mt.)......C 1
Jacques-Cartier (riv.)......F 3
Kénogami (lake)......F 1
Kiamika (lake)......C 3
La Maurice Nat'l Park......D 3
Laurentides Prov. Park......F 2
Lièvre (lake)......H 2
Lièvre (riv.)......H 2
Maskinongé (riv.)......D 3
Matane (riv.)......B 1
Matane Prov. Park......B 1

Matapédia (riv.)......B 2
Mégantic (lake)......G 4
Memphrémagog (lake)......E 4
Mercier (dam)......A 3
Métabetchouane (riv.)......F 1
Mille Îles (riv.)......H 4
Montmorency (riv.)......G 2
Mont-Tremblant Prov. Park...C 3
Nicolet (riv.)......E 3
Nominingue (lake)......C 3
Nord (riv.)......C 4
Orléans (isl.)......F 3
Ottawa (riv.)......B 4
Ouareau (riv.)......D 3
Ouelle (riv.)......H 2
Patapédia (riv.)......B 2
Péribonca (riv.)......F 1
Petite Nation (riv.)......B 4
Prairies (riv.)......H 4
Rimouski (riv.)......J 1
Ristigouche (riv.)......B 2
Saguenay (riv.)......G 1
Sainte-Anne (riv.)......E 3
Sainte-Anne (riv.)......G 2
Saint-François (riv.)......E 4
Saint-François (riv.)......E 3
Saint Lawrence (gulf)......D 2
Saint-Louis (lake)......H 4
Saint-Maurice (riv.)......E 3
Saint-Pierre (lake)......E 3
Shawinigan (riv.)......E 3
Shipshaw (riv.)......F 1
Soeurs (isl.)......H 4
Témiscouata (lake)......H 2
Tremblant (lake)......C 3
Trente et un Milles (lake)......B 3
Verte (isl.)......H 1
Yamaska (riv.)......E 4
York (riv.)......D 1

©County seat.
*Population of metropolitan area.

QUÉBEC, NORTHERN

INTERNAL DIVISIONS

Abitibi (county) 93,529......B 3
Abitibi (terr.)......B 3
Berthier (county) 31,096......D 3
Bonaventure (county) 40,487...D 3
Champlain (county) 119,595...C 3
Charlevoix-Est (co.) 17,448...C 3

Charlevoix-Ouest (county)
14,172......C 3
Chicoutimi (county) 174,441...C 3
Gaspé-Est (county) 41,173...E 3
Gaspé-Ouest (county) 18,943...D 3
Gatineau (county) 54,229......B 3
Joliette (county) 60,384......B 3
Lac-Saint-Jean-Est (county)
47,891......C 3
Lac-Saint-Jean-Ouest
(county) 62,952......C 3
Maskinongé (county) 20,763...C 3
Matane (county) 29,955......C 3
Matapédia (county) 23,715...D 3
Mistassini (terr.)......C 3
Montcalm (county) 27,557...B 3
Montmorency No. 1 (county)
23,048......C 3
Nouveau-Québec (terr.)......E 1
Pontiac (county) 20,283......A 3
Portneuf (county) 58,843......C 3
Québec (county) 458,980......C 3
Rimouski (county) 69,099......C 3
Saguenay (county) 115,881...D 2
Saint-Maurice (co.) 107,703...C 3
Témiscamingue (co.) 52,570...B 3

CITIES and TOWNS

Alma© 26,322......C 3
Amos© 9,421......B 3
Baie-Comeau 12,866......D 3
Baie-du-Poste 1,690......C 2
Chicoutimi© 60,064......C 3
Gaspé 17,261......D 3
Hauterive 13,995......D 3
Jonquière 60,354......C 3
Lévis 17,895......C 3
La Tuque 11,556......C 3
Manicouagan......D 2
Maniwaki© 5,424......B 3
Matane© 13,612......D 3
Mistassini (Baie-du-Poste)
1,690......C 2
Mont-Laurier© 8,405......C 3
Montmagny© 12,405......C 3
New Carlisle© 781......D 3
Perce© 4,839......E 3
Port-Cartier-Ouest......D 3
Port-Cartier 1,690......D 3
Port-Menier 275......D 3
Povungnituk 745......E 1
Québec (cap.)© 166,474......C 3
Rimouski© 29,120......C 3
Rivière-au-Tonnerre 480......D 2
Rivière-du-Loup 13,459......D 3
Rouyn 17,224......B 3

Sept-Îles 29,262......D 2
Shawinigan 23,011......C 3
Tadoussac 900......C 3
Val d'Or 21,371......C 3
Ville-Marie 2,651......B 3
Wemindji......B 2

OTHER FEATURES

Allard (lake)......E 2
Anticosti (isl.)......E 3
Baleine, Grand Rivière de la
(riv.)......B 1
Bell (riv.)......B 3
Betsiamites (riv.)......C 2
Bienville (lake)......C 2
Broadback (riv.)......B 3
Cabonga (res.)......B 3
Caniapiscau (res.)......D 1
Eastmain (riv.)......C 1
Eau Claire (lake)......C 1
Feuilles (riv.)......C 1
Gaspésie Prov. Park......D 3
George (riv.)......D 1
Gouin (lake)......B 3
Grande Rivière, La (riv.)......B 2
Honguedo (passage)......D 3
Hudson (bay)......A 1
Hudson (str.)......F 1
Jacques-Cartier (passage)...D 3
James (bay)......A 2
Koksoak (riv.)......E 1
Laurentides Prov. Park......C 3
Louis-XIV (pt.)......B 2
Manicouagan (res.)......D 2
Minto (lake)......C 1
Mistassini (lake)......C 2
Mistassini (riv.)......C 2
Moisie (riv.)......D 2
Natashquan (riv.)......E 2
Nottaway (riv.)......B 3
Nouveau-Québec (crater)...F 1
Otish (mts.)......C 2
Ottawa (riv.)......B 3
Péribonca (riv.)......C 2
Plétipi (lake)......D 2
Saguenay (riv.)......D 2
Saint-Jean (lake)......C 3
Saint Lawrence (gulf)......D 3
Saint Lawrence (riv.)......D 3
Ungava (bay)......E 1

©County seat.
*Population of metropolitan area.

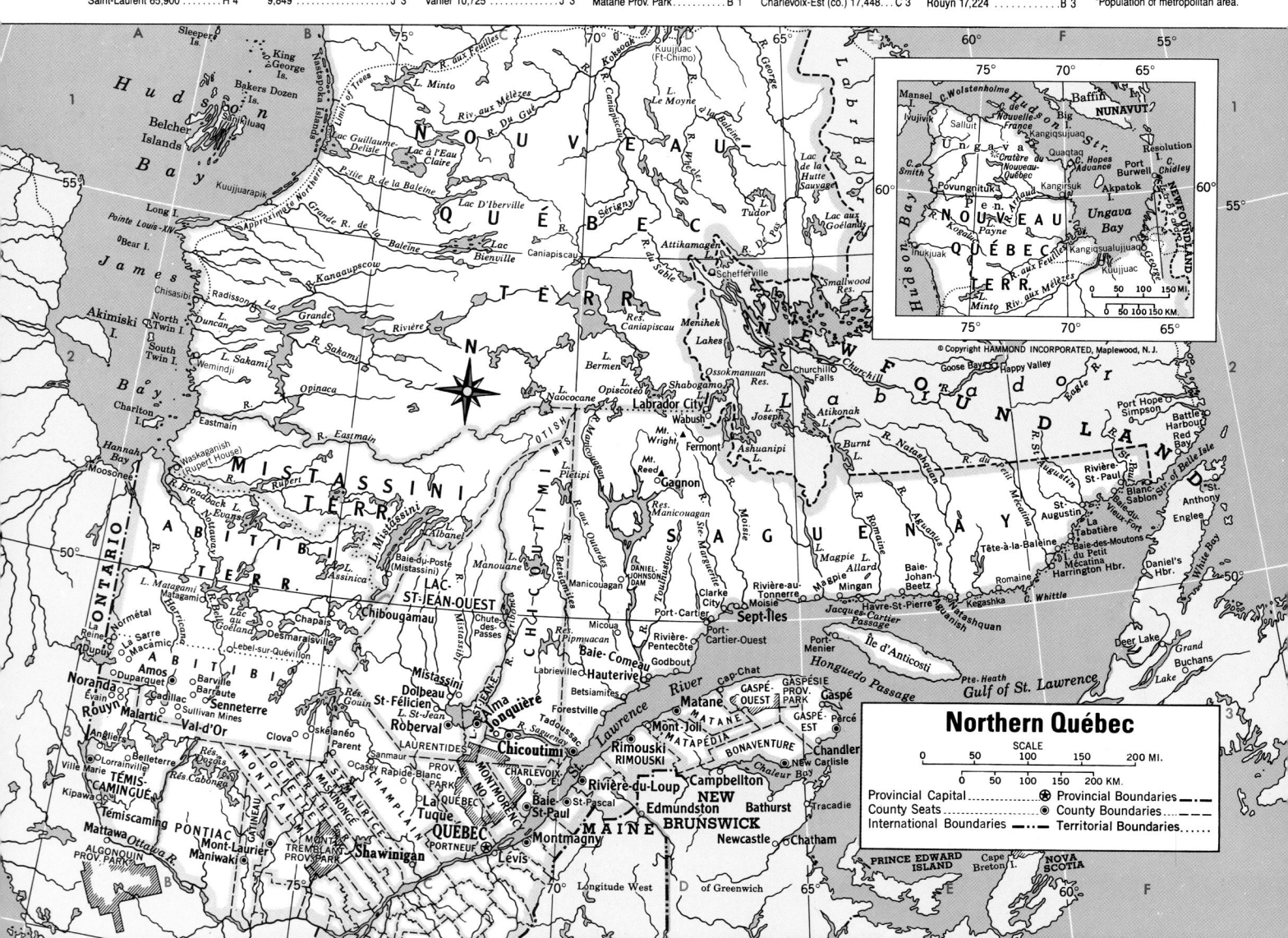

ONTARIO, NORTHERN
INTERNAL DIVISIONS

Algoma (terr. dist.) 133,553 ... D 3
Cochrane (terr. dist.) 96,875 ... D 2
Kenora (terr. dist.) 59,421 ... C 2
Manitoulin (terr. dist.) 11,001 ... D 3
Nipissing (terr. dist.) 80,268 ... E 3
Parry Sound (terr. dist.)
  33,528 ... E 3
Rainy River (terr. dist.) 22,798 B 3
Renfrew (county) 87,484 ... E 3
Sudbury (reg. munic.)
  159,779 ... D 3
Sudbury (terr. dist.) 27,068 ... D 3
Thunder Bay (terr. dist.)
  153,997 ... C 3
Timiskaming (terr. dist.)
  41,288 ... D 3

CITIES and TOWNS

Chalk River 1,010 ... E 3
Elliot Lake 16,723 ... D 3
Fort Albany 482 ... D 2
Fort Frances® 8,906 ... B 3
Kapuskasing 12,014 ... D 3
Kenora® 9,817 ... B 3
Kirkland Lake 12,219 ... D 3
Moose Factory 1,452 ... D 2
Moosonee 1,433 ... D 2
Nickel Centre 12,318 ... D 3
North Bay® 51,268 ... E 3
Pembroke® 14,026 ... E 3
Sault Sainte Marie® 82,697 ... D 3
Sudbury 91,829 ... D 3
Thunder Bay® 112,486 ... C 3
Timmins 46,114 ... D 3
Valley East 20,433 ... D 3

OTHER FEATURES

Abitibi (lake) ... E 3
Abitibi (riv.) ... D 2
Albany (riv.) ... C 2
Algonquin Prov. Park ... E 3
Asheweig (riv.) ... C 2
Attawapiskat (lake) ... C 2
Attawapiskat (riv.) ... C 2
Basswood (lake) ... B 3
Berens (riv.) ... A 2
Big Trout (lake) ... B 2
Black Duck (riv.) ... C 1
Bloodvein (riv.) ... A 2
Caribou (isl.) ... C 3

Cobham (riv.) ... A 2
Eabamet (lake) ... C 2
Ekwan (riv.) ... C 2
English (riv.) ... B 2
Fawn (riv.) ... C 2
Finger (lake) ... B 2
Georgian (bay) ... D 3
Hannah (bay) ... D 2
Henrietta Maria (cape) ... D 1
Hudson (bay) ... D 1
Huron (lake) ... D 3
James (bay) ... D 2
Kapiskau (riv.) ... D 2
Kapuskasing (riv.) ... D 3
Kenogami (riv.) ... C 2
Kesagami (riv.) ... E 2
Lake of the Woods (lake) ... B 3
Lake Superior Prov. Park ... D 3
Little Current (riv.) ... C 2
Longlac (lake) ... C 3
Manitoulin (isl.) ... D 3
Mattagami (lake) ... D 3
Michipicoten (isl.) ... C 3
Mille Lacs (lake) ... B 3
Missinaibi (lake) ... D 2
Missinaibi (riv.) ... D 2
Missisa (lake) ... D 2
Nipigon (lake) ... C 3
Nipissing (lake) ... E 3
North (chan.) ... D 3
North Caribou (lake) ... B 2
Nungesser (lake) ... B 2
Ogidaki (mt.) ... D 3
Ogoki (riv.) ... C 2
Opazatika (riv.) ... D 2
Opinnagau (riv.) ... D 2
Otoskwin (riv.) ... B 2
Ottawa (riv.) ... E 3
Pipestone (riv.) ... B 2
Polar Bear Prov. Park ... C 1
Pukaskwa Prov. Park ... C 3
Quetico Prov. Park ... B 3
Rainy (lake) ... B 3
Red (lake) ... B 2
Sachigo (riv.) ... B 2
Saganaga (lake) ... B 3
Saint Ignace (isl.) ... C 3
Saint Joseph (lake) ... B 3
Sandy (lake) ... B 2
Savant (lake) ... B 3
Seine (riv.) ... B 3
Seul (lake) ... B 3
Severn (lake) ... B 2
Severn (riv.) ... B 2
Shamattawa (riv.) ... C 2
Shibogama (lake) ... C 2

Sibley Prov. Park ... C 3
Slate (isls.) ... C 3
Stout (lake) ... B 2
Superior (lake) ... D 3
Sutton (lake) ... D 2
Sutton (riv.) ... D 1
Temigami (lake) ... D 3
Timiskaming (lake) ... D 3
Trout (lake) ... B 2
Wabuk (pt.) ... D 1
Winisk (lake) ... C 2
Winisk (riv.) ... C 2
Winnipeg (riv.) ... A 2
Woods (lake) ... B 3

ONTARIO
INTERNAL DIVISIONS

Algoma (terr. dist.) 133,553 ... J 5
Brant (county) 104,427 ... D 4
Bruce (county) 60,020 ... C 3
Cochrane (terr. dist.) 96,875 ... J 4
Dufferin (county) 31,145 ... D 3
Dundas (county) 18,946 ... J 2
Durham (reg. munic.) 283,639 F 3
Elgin (county) 69,707 ... C 5
Essex (county) 312,467 ... B 5
Frontenac (county) 108,133 ... H 3
Glengarry (county) 20,254 ... K 2
Grenville (county) 27,176 ... J 3
Grey (county) 73,824 ... D 3
Haldimand-Norfolk (reg.
  munic.) 89,456 ... E 5
Haliburton (county) 11,361 ... F 2
Halton (reg. munic.) 253,883 ... E 4
Hamilton-Wentworth (reg.
  munic.) 411,445 ... D 4
Hastings (county) 106,883 ... G 3
Huron (county) 56,127 ... C 4
Kenora (terr. dist.) 59,421 ... G 5
Kent (county) 107,022 ... H 5
Lambton (county) 123,445 ... B 5
Lanark (county) 45,676 ... H 3
Leeds (county) 53,765 ... H 3
Lennox and Addington
  (county) 33,040 ... H 3
Manitoulin (terr. dist.) 11,001 ... C 4
Middlesex (county) 318,184 ... C 4
Muskoka (dist. munic.)
  38,370 ... E 2
Niagara (reg. munic.) 368,288 ... E 4
Nipissing (terr. dist.) 80,268 ... F 2
Northumberland (county)
  64,966 ... G 3

Ottawa-Carleton (reg. munic.)
  546,849 ... J 2
Oxford (county) 85,920 ... D 4
Parry Sound (terr. dist.)
  33,528 ... D 2
Peel (reg. munic.) 490,731 ... E 4
Perth (county) 66,096 ... C 4
Peterborough (county)
  102,452 ... F 3
Prescott (county) 30,365 ... K 2
Prince Edward (county)
  22,336 ... G 3
Rainy River (terr. dist.) 22,798 G 5
Renfrew (county) 87,484 ... G 2
Russell (county) 22,412 ... J 2
Simcoe (county) 225,071 ... E 3
Stormont (county) 61,927 ... J 1
Sudbury (reg. munic.)
  159,779 ... K 6
Sudbury (terr. dist.) 27,068 ... J 5
Thunder Bay (terr. dist.)
  153,997 ... H 5
Timiskaming (terr. dist.)
  41,288 ... K 5
Toronto (metro. munic.)
  2,137,395 ... K 4
Victoria (county) 47,854 ... F 3
Waterloo (reg. munic.)
  305,496 ... D 4
Wellington (county) 129,432 ... D 4
York (reg. munic.) 252,053 ... E 4

CITIES and TOWNS

Ailsa Craig 765 ... C 4
Ajax 25,475 ... J 4
Alban 342 ... D 1
Alexandria 3,271 ... K 2
Alfred 1,057 ... K 2
Alliston 4,712 ... E 3
Almonte 3,855 ... J 2
Alvinston 736 ... C 5
Amherstburg 5,685 ... A 5
Amherst View 6,110 ... H 3
Ancaster 14,428 ... D 4
Angus 3,085 ... E 3
Apsley 264 ... F 3
Arkona 473 ... C 4
Armstrong 378 ... H 4
Aroland 291 ... H 4
Arthur 1,700 ... D 4
Astorville 340 ... E 1
Athens 948 ... J 3
Atherley 366 ... E 3
Atikokan 4,452 ... G 5

Atwood 723 ... D 4
Aurora 16,267 ... J 3
Avonmore 273 ... K 2
Aylmer 5,254 ... C 5
Ayr 1,295 ... D 4
Ayton 424 ... D 3
Baden 945 ... D 4
Bala 577 ... E 2
Bancroft 2,329 ... G 2
Barrie® 38,423 ... G 2
Barry's Bay 1,216 ... G 2
Batawa 430 ... G 3
Bath 1,071 ... H 3
Bayfield 649 ... C 4
Beachburg 682 ... H 2
Beachville 917 ... D 4
Beardmore 583 ... H 5
Beaverton 1,952 ... E 3
Beeton 1,989 ... E 3
Belle River 3,568 ... B 5
Belleville® 34,881 ... G 3
Belmont 831 ... C 5
Bethany 365 ... F 3
Bewdley 508 ... F 3
Binbrook 306 ... E 4
Blackstock 720 ... F 3
Blenheim 4,044 ... C 5
Blind River 3,444 ... J 5
Bloomfield 718 ... G 4
Blyth 926 ... C 4
Bobcaygeon 1,625 ... F 3
Bonfield 540 ... E 1
Bothwell 915 ... C 5
Bourget 1,057 ... J 2
Bracebridge® 9,063 ... E 2
Bradford 7,370 ... E 3
Braeside 492 ... H 2
Brampton® 149,030 ... J 4
Brantford® 74,315 ... D 4
Bridgenorth 1,633 ... F 3

Brigden 635 ... B 5
Brighton 3,147 ... G 3
Britt 419 ... D 2
Brockville® 19,896 ... J 3
Bruce Mines 635 ... J 5
Brussels 962 ... C 4
Burford 1,461 ... D 4
Burgessville 302 ... D 4
Burk's Falls 922 ... E 2
Burlington 114,853 ... E 4
Cache Bay 665 ... D 1
Caesarea 551 ... F 3
Calabogie 256 ... H 2
Caledon 26,645 ... E 4
Callander 1,158 ... E 1
Cambridge 77,183 ... D 4
Campbellford 3,409 ... G 3
Cannington 1,623 ... E 3
Capreol 3,845 ... K 5
Caramat 265 ... H 5
Cardinal 1,753 ... J 3
Carleton Place 5,626 ... J 2
Carlisle 781 ... D 4
Carlsbad Springs 616 ... J 2
Carp 707 ... H 2
Cartier 590 ... J 5
Casselman 1,675 ... J 2
Castleton 346 ... F 3
Chalk River 1,010 ... G 1
Chapleau 3,243 ... J 5
Charing Cross 443 ... B 5
Chatham® 40,952 ... B 5
Chatsworth 383 ... D 3
Cherry Valley 289 ... G 4
Chesley 1,840 ... C 3
Chesterville 1,430 ... J 2
Chute-à-Blondeau 365 ... K 2
City View ... J 2
Clarence Creek 796 ... J 2
Clarksburg 508 ... D 3

Clifford 645 ... D 4
Clinton 3,081 ... C 4
Cobalt 1,759 ... K 5
Cobden 997 ... H 2
Coboconk 426 ... F 3
Cobourg® 11,385 ... F 4
Cochrane® 4,848 ... K 5
Colborne 1,796 ... G 4
Colchester 711 ... B 6
Coldwater 964 ... E 3
Collingwood 12,064 ... D 3
Comber 667 ... B 5
Conseconn 295 ... G 4
Cookstown 918 ... E 3
Cornwall® 46,144 ... K 2
Cottam 404 ... B 5
Courtland 647 ... D 5
Courtright 1,024 ... B 5
Crediton 370 ... C 4
Creemore 1,182 ... D 3
Crysler 540 ... J 2
Cumberland 518 ... J 2
Cumberland Beach-Bramshot-
  Buena Vista 679 ... E 3
Dashwood 426 ... C 4
Deep River 5,095 ... G 1
Delaware 481 ... C 5
Delhi 4,043 ... D 5
Delta 360 ... H 3
Deseronto 1,740 ... G 3
Douglas 303 ... H 2
Drayton 809 ... D 4
Dresden 2,550 ... B 5
Drumbo 476 ... D 4
Dryden 6,640 ... G 5
Dublin 295 ... C 4
Dubreuilville △988 ... J 5
Dundalk 1,573 ... D 3
Dundas 19,586 ... D 4
Dungannon 284 ... C 4
Dunnville 11,353 ... E 5
Durham 2,458 ... D 3
Dutton 1,115 ... C 5
Earlton 1,028 ... K 5
East York 101,974 ... J 4
Echo Bay 786 ... J 5
Eden Mills 318 ... D 4
Eganville 1,245 ... G 2
Egmondville 465 ... C 4
Elgin 327 ... H 3
Elk Lake 526 ... K 5
Elliot Lake 16,723 ... B 1
Elmira 7,063 ... D 4
Elmvale 1,183 ... E 3
Elmwood 364 ... C 3
Elora 2,666 ... D 4
Embro 727 ... C 4
Embrun 1,883 ... J 2
Emeryville-Puce 1,611 ... B 5
Emo 762 ... G 5
Englehart 1,689 ... K 5
Enterprise 357 ... H 3
Erieau 430 ... C 5
Erin 2,313 ... D 4
Espanola 5,836 ... J 5
Essex 6,295 ... B 5
Etobicoke 298,713 ... J 4
Everett 570 ... E 3
Exeter 3,732 ... C 4
Fauquier 561 ... J 4
Fenelon Falls 1,701 ... F 3
Fergus 6,064 ... D 4
Field 462 ... E 1
Finch 353 ... J 2
Fingal 380 ... C 5
Fitzroy Harbour 446 ... H 2
Flesherton 565 ... D 3
Foleyet 484 ... J 4
Fordwich 365 ... C 4
Forest 2,671 ... C 3
Formosa 393 ... C 3
Fort Erie 24,096 ... E 4
Fort Frances® 8,906 ... F 5
Foxboro 597 ... G 3
Frankford 1,919 ... G 3
Fraserdale 303 ... J 5
Freelton 307 ... D 4
Gananoque 4,863 ... J 3
Garden Village 270 ... E 1
Geraldton 2,956 ... H 5
Glencoe 1,694 ... C 5
Glen Miller 639 ... G 3
Glen Robertson 378 ... K 2
Glen Walter 710 ... K 2
Goderich® 7,322 ... C 4
Gogama 652 ... J 5
Goodwood 335 ... E 3
Gore Bay® 777 ... B 2
Gorrie 468 ... C 4
Grafton 409 ... G 4
Grand Bend 680 ... C 4
Grand Valley 1,226 ... D 4
Granton 315 ... C 4
Gravenhurst 8,532 ... E 2
Greely 567 ... J 2
Green Valley 459 ... K 2
Grimsby 15,797 ... E 4
Guelph® 71,207 ... D 4

(continued on following page)

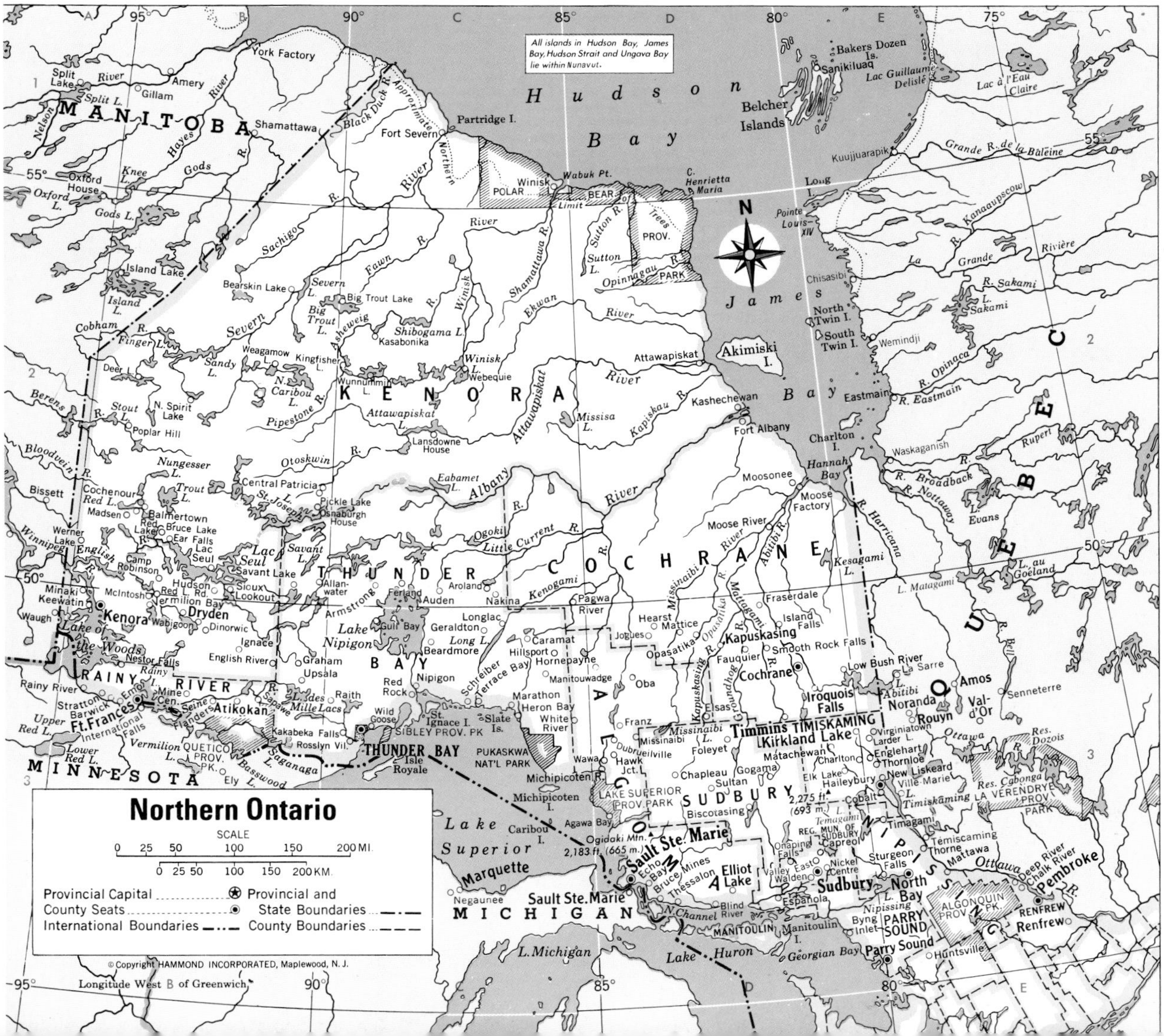

Ontario Facts

AREA 1,068,582 sq. km. (412,580 sq. mi.)
POPULATION 10,753,573
CAPITAL Toronto
LARGEST CITY Toronto
HIGHEST POINT Timiskaming District 693 m.
  (2,274 ft.)
SETTLED IN 1749
ADMITTED TO CONFEDERATION July 1, 1867
PROV. FLOWER White Trillium
PROV. BIRD Common Loon

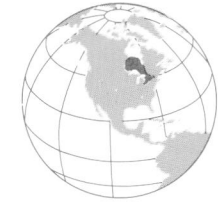

Northern Ontario

SCALE

0 25 50   100   150   200 MI.

0 25 50 100   150 200 KM.

Provincial Capital .... ⊛
County Seats .... ◉
International Boundaries ▬ ▬ ▬

Provincial and
State Boundaries ▬ ▬ ▬
County Boundaries ▬ ▬ ▬

© Copyright HAMMOND INCORPORATED, Maplewood, N.J.

Longitude West B of Greenwich

Haileybury® 4,925 .... K 5
Haldimand 16,866 .... E 5
Haliburton 1,443 .... F 2
Halton Hills 35,190 .... E 4
Hamilton 306,434 .... E 4
Hamilton *542,095 .... E 4
Hanover 6,316 .... C 3
Harriston 1,954 .... D 4
Harrow 2,274 .... B 5
Harrowsmith 599 .... H 3
Harwood 332 .... F 3
Hastings 975 .... G 3
Havelock 1,385 .... G 3
Hawkesbury 9,877 .... K 2
Hawkestone 275 .... E 3
Hawk Junction 349 .... J 5
Hearst 5,533 .... J 5
Hensall 973 .... C 4
Hepworth 393 .... C 3
Hickson 263 .... D 4
Highgate 435 .... C 5
Hillsburgh 1,065 .... D 4
Hillsdale 370 .... E 3
Holland Landing 2,771 .... E 3
Honey Harbour 505 .... E 2
Hornepayne 1,848 .... J 5
Hudson 515 .... G 4
Huntsville 11,467 .... E 2
Huron Park 1,104 .... C 4
Ignace 2,499 .... G 5
Ilderton 301 .... C 4
Ingleside 1,400 .... J 2
Ingersoll 8,494 .... C 4
Innerkip 715 .... D 4
Inverhuron 438 .... C 3
Iron Bridge 821 .... A 1

Iroquois 1,211 .... J 3
Iroquois Falls 6,339 .... J 5
Johnstown 789 .... J 3
Kakabeka Falls 300 .... G 5
Kanata 19,728 .... J 2
Kapuskasing 12,014 .... J 5
Kars 449 .... J 2
Kearney 538 .... E 2
Keene 353 .... F 3
Keewatin 1,863 .... F 5
Kemptville 2,362 .... J 2
Kenora® 9,817 .... F 4
Killaloe Station 634 .... G 2
Killarney 433 .... D 2
Kincardine 5,778 .... C 3
Kingston® 52,616 .... H 3
Kingsville 5,134 .... B 6
Kinmount 262 .... E 2
Kirkland Lake 12,219 ...... K 5
Kitchener® 139,734 .... D 4
Kitchener *287,801 .... D 4
Komoka 1,152 .... C 4
Lakefield 2,374 .... F 3
Lanark 753 .... H 2
Lancaster 637 .... K 2
Langton 348 .... D 5
Lansdowne 540 .... H 3
Larder Lake 1,084 .... K 5
Latchford 397 .... K 5
Leamington 12,528 .... B 5
Limoges 930 .... J 2
Lincoln 14,196 .... E 4
Linden Beach 579 .... B 6
Lindsay® 13,596 .... F 3
Linwood 450 .... D 4
Lion's Head 467 .... C 2

Lisle 265 .... E 3
Listowel 5,026 .... D 4
Little Britain 265 .... F 3
Little Current 1,507 .... B 2
London® 254,280 .... C 5
London *283,668 .... C 5
Longlac 2,431 .... H 5
Long Sault 1,227 .... K 2
L'Orignal 1,819 .... K 2
Lucan 1,616 .... C 4
Lucknow 1,088 .... C 4
Lyn 518 .... J 3
Lynden 451 .... D 4
Lynhurst 685 .... C 5
MacGregor's Bay 861 ..... G 2
MacTier 647 .... E 2
Madawaska 264 .... F 2
Madoc 1,249 .... G 3
Maitland 367 .... J 3
Mallorytown 368 .... J 3
Manitouwadge 3,155 ..... H 5
Manitowaning 518 .... C 2
Manotick-Hillside Gardens
  2,694 .... J 2
Marathon 2,271 .... H 5
Markdale 1,289 .... D 3
Markham 77,037 .... K 4
Markstay 444 .... D 1
Marmora 1,304 .... G 3
Martintown 388 .... K 2
Massey 1,274 .... C 1
Matachewan 444 ...... J 5
Matheson 966 .... K 5
Mattawa 2,652 .... F 1
Mattice 803 .... J 5
Maxville 836 .... K 2

Maynooth 277 .... G 2
McGregor 1,145 .... B 5
McKerrow 260 .... C 1
Meaford 4,367 .... D 3
Melbourne 346 .... C 5
Merlin 745 .... B 5
Merrickville 984 .... J 3
Metcalfe 687 .... J 2
Midhurst 1,457 .... E 3
Midland 12,132 ...... D 3
Mildmay 928 .... C 4
Milford Bay 401 .... E 2
Millbank 337 ...... D 4
Millbrook 927 .... F 3
Milton® 28,067 ... E 4
Milverton 1,463 .. D 4
Minaki 319 .... F 4
Mindemoya 376 .. B 2
Minden® 948 .... F 2
Mississauga 315,056 .. J 4
Mitchell 2,777 .... C 4
Monkton 520 .... C 4
Moonbeam 838 .... J 5
Moorefield 308 ... D 4
Mooretown 344 ... B 5
Moose Creek 393 .. K 2
Morewood 264 .... J 2
Morpeth 284 ..... C 5
Morrisburg 2,308 .. J 3
Mount Albert 1,165 .. E 3
Mount Brydges 1,557 .. C 5
Mount Forest 3,474 .. D 4
Mount Hope 557 .... E 4
Munster 1,531 ..... J 2
Nakina 936 .... H 4
Nanticoke 19,816 .. E 5

Napanee 4,803 .... G 3
Navan 419 .... J 2
Neustadt 511 .... D 3
Newboro 260 .... H 3
Newburgh 617 .... H 3
Newbury 441 .... C 5
Newcastle 32,229 ... E 3
New Hamburg 3,923 .. D 4
New Liskeard 5,551 .. K 5
Newmarket 29,753 ... E 3
Niagara Falls 70,960 .. E 4
Niagara-on-the-Lake 12,186. E 4
Nickel Centre 12,318 .. D 1
Nipigon 2,377 .... H 5
Nobel 386 .... E 2
Nobleton 1,861 ..... J 3
Noelville 702 ...... D 2
North Bay® 51,268 .. E 1
North Gower 818 ... J 2
North York 559,521 .. J 4
Norwich 2,117 ..... D 5
Norwood 1,278 .... F 3
Nottawa 960 ...... D 3
Oakville 75,773 ... E 4
Oakwood 404 .... F 3
Odessa 849 ...... H 3
Oil City 960 ..... B 5
Oil Springs 627 .. B 5
Omemee 819 .... F 3
Onaping Falls 6,198 .. J 5
Opasatika 413 ...... J 5
Orangeville® 13,740. D 4
Orillia 23,955 ...... E 3
Osgoode 1,138 .... J 2
Oshawa 117,519 .. F 4
Oshawa *154,217 .. F 4

Ottawa® (cap.), Canada
  295,163 .... J 2
Ottawa-Hull *717,978 .. J 2
Otterville 776 .... D 4
Owen Sound® 19,883 .. D 3
Paincourt 414 .... B 5
Paisley 1,039 .... C 3
Pakenham 367 ..... J 2
Palmerston 1,989 .. D 4
Paris 7,485 .... D 4
Parkhill 1,358 .... C 4
Parry Sound® 6,124 .. E 2
Pefferlaw 857 ...... E 3
Pelham 11,104 ..... E 4
Pembroke® 14,026 .. H 2
Penetanguishene 5,315.. D 3
Perth® 5,655 ...... H 3
Petawawa 5,520 .. H 2
Peterborough® 60,620.. F 3
Petrolia 4,234 .... B 5
Pickering 37,754 .. K 4
Picton® 4,361 .... G 3
Plantagenet 870 .. K 2
Plattsville 495 ... D 4
Point Edward 2,383 .. B 4
Pontypool 759 ..... F 3
Port Burwell 655 ... D 5
Port Carling 629 ... E 2
Port Colborne 19,225 .. E 4
Port Elgin 6,131 .... C 3
Port Franks 567 .... C 4
Port Hope 9,992 ... F 3
Port Lambton 921 .. B 5
Portland 271 ...... H 3
Port McNicoll 1,883 .. E 3
Port Perry 4,712 ... E 3

Port Rowan 811 .... D 5
Port Stanley 1,891 ...... C 5
Pottageville 286 .... J 3
Powassan 1,169 .... E 1
Prescott® 4,670 ...... J 2
Princeton 462 .... D 4
Puce-Emeryville 1,611 .. B 5
Rainy River 1,061 .... F 5
Ramore 382 .... K 5
Rayside-Balfour 15,017.. K 5
Red Rock 1,260 ..... H 5
Renfrew 8,283 ..... H 2
Richards Landing 405 .. J 3
Richmond 2,880 ... J 2
Richmond Hill 37,778 .. J 4
Ridgetown 3,062 .. C 5
Ripley 591 .... C 3
River Valley 275 .. D 1
Rockcliffe Park 1,869 .. J 2
Rockland 3,961 .... J 2
Rockwood 1,068 ... D 4
Rodney 1,007 ..... C 5
Rosslyn Village 362 .. G 5
Round Lake Centre 255.. G 2
Russell 1,099 ..... J 2
Ruthven 649 ...... B 6
Saint Albert 254 ... J 2
Saint Catharines® 124,018.. E 4
Saint Catharines-Niagara
  *304,353 .... E 4
Saint Charles 382 ... D 1
Saint Clair Beach 2,845 .. B 5
Saint Clements 890 .. D 4
Saint-Eugène 470 .. K 2
Saint George 865 .. D 4
Saint Isidore de Prescott 746 . K 2

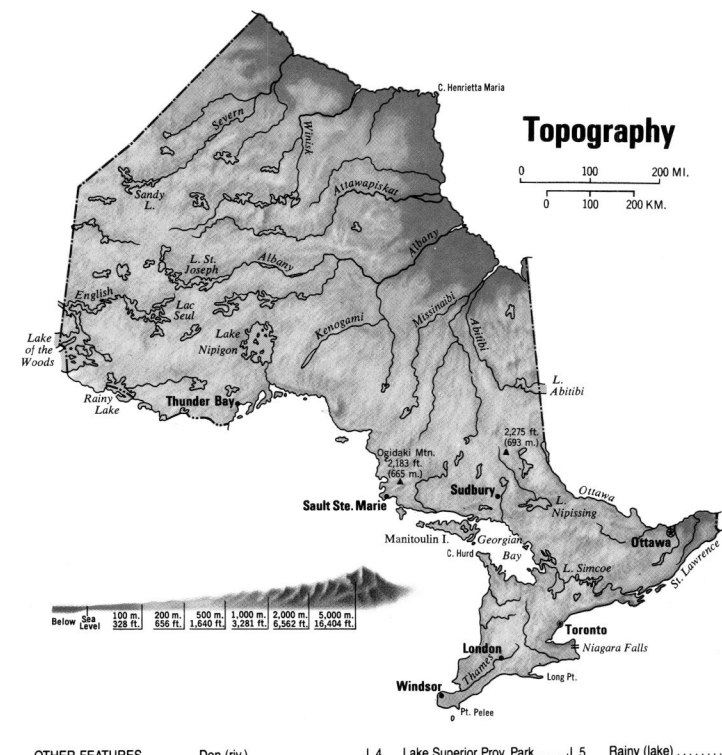

Topography

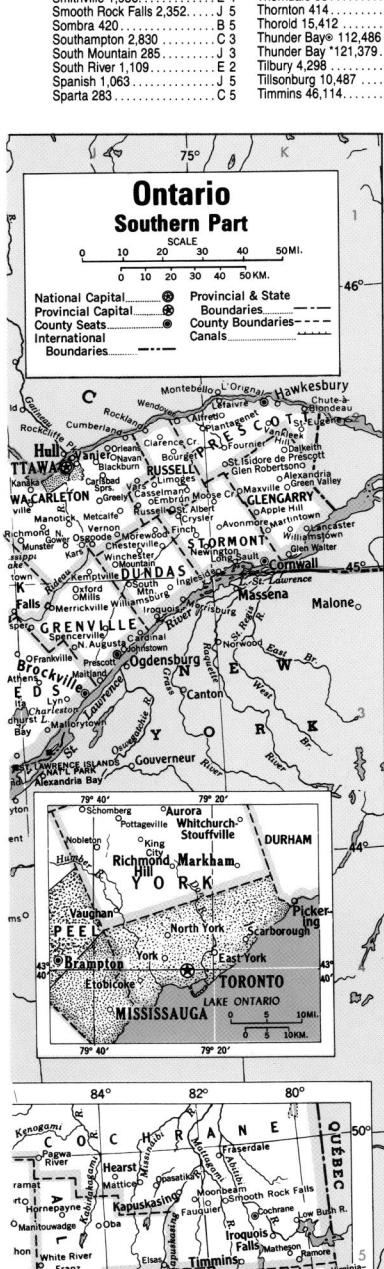

Ontario
Southern Part

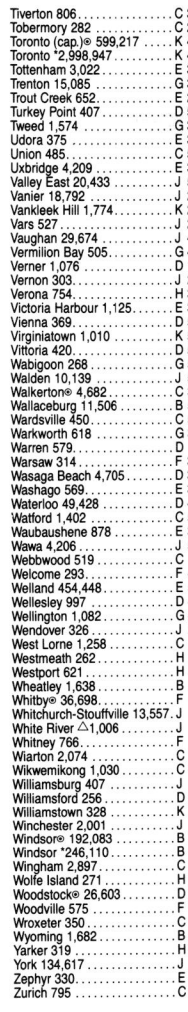

Agriculture, Industry and Resources

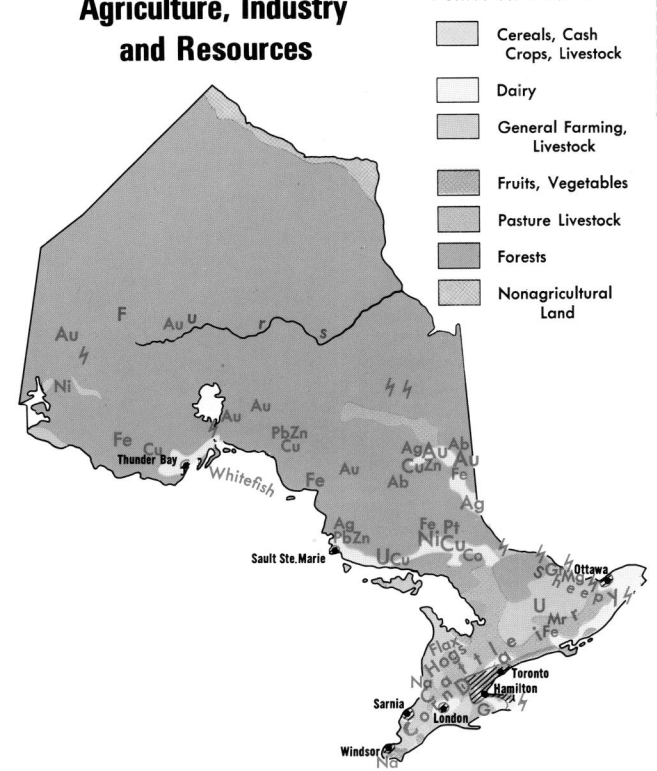

DOMINANT LAND USE

Cereals, Cash Crops, Livestock
Dairy
General Farming, Livestock
Fruits, Vegetables
Pasture Livestock
Forests
Nonagricultural Land

MAJOR MINERAL OCCURRENCES

Ab Asbestos
Ag Silver
Au Gold
Co Cobalt
Cu Copper
Fe Iron Ore
G Natural Gas
Gr Graphite
Mg Magnesium
Mr Marble
Na Salt
Ni Nickel
Pb Lead
Pt Platinum
U Uranium
Zn Zinc

Water Power
Major Industrial Areas

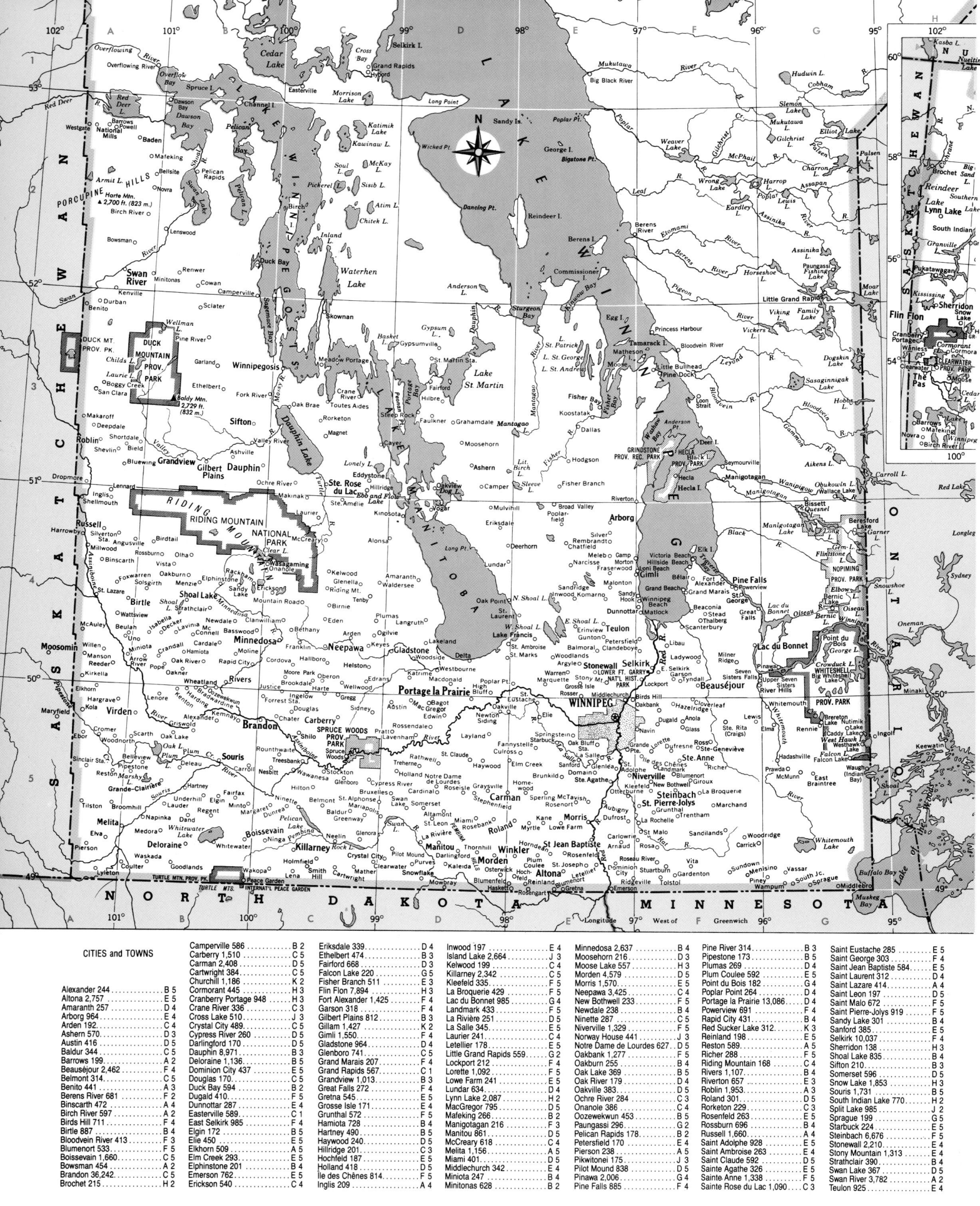

Manitoba
Northern Part

0   40   80   120 MI.
0   40   80   120 KM.

(map: Manitoba Northern Part)

Manitoba
Southern Part

SCALE

0 5 10 20   40   60 MI.

0 5 10 20   40   60 KM.

Provincial Capital .... ⊛
International Boundaries --- ⋅ ---
Provincial Boundaries --- --- ---

© Copyright HAMMOND INCORPORATED, Maplewood, N.J.

The Pas 6,390	H 3	
Thicket Portage 195	J 3	
Thompson 14,288	J 2	
Treherne 743	D 5	
Tyndall 421	F 4	
Virden 2,940	A 5	
Vita 364	F 5	
Wabowden 655	J 3	
Wallace Lake ●2,044	G 3	
Wanless 193	H 3	
Warren 459	E 4	
Waskada 239	B 5	
Wawanesa 492	C 5	
Whitemouth 320	G 5	
Whitewater ●856	E 5	
Winkler 5,046	E 5	
Winnipeg (cap.) 564,473	E 5	
Winnipeg *584,842	E 5	
Winnipeg Beach 565	F 4	
Winnipegosis 855	B 3	
Woodlands 185	E 4	
Wooodridge 170	G 5	
York Landing 229	J 2	

OTHER FEATURES

Aikens (lake)	G 3	East Shoal (lake)	E 4	Manigotagan (riv.)	G 3	Saint Andrew (lake)	E 3
Anderson (lake)	D 2	Ebb and Flow (lake)	C 3	Manitoba (lake)	D 4	Saint George (lake)	E 3
Anderson (pt.)	F 3	Egg (isl.)	E 3	Mantagao (riv.)	E 3	Saint Martin (lake)	D 3
Armit (lake)	A 2	Elbow (lake)	G 4	Marchand (lake)	B 5	Saint Patrick (lake)	E 3
Assapan (riv.)	G 2	Elk (isl.)	F 4	McKay (lake)	C 2	Sale (riv.)	E 5
Assiniboine (riv.)	C 5	Elliot (lake)	G 2	McPhail (riv.)	F 2	Sandy (lake)	D 2
Assinika (lake)	G 2	Etawney (lake)	J 2	Minnedosa (riv.)	B 4	Sasaginnigak (lake)	G 3
Assinika (riv.)	G 2	Etomami (riv.)	F 2	Moar (lake)	G 2	Seal (riv.)	J 2
Atim (lake)	C 2	Falcon (lake)	G 5	Molson (lake)	J 3	Selkirk (isl.)	C 1
Baldy (mt.)	B 3	Family (lake)	G 3	Moose (lake)	E 3	Setting (lake)	H 3
Basket (lake)	C 3	Fisher (bay)	E 3	Morrison (lake)	C 1	Shoal (lake)	G 5
Beaverhill (lake)	E 2	Fisher (riv.)	E 3	Mossy (riv.)	C 3	Shoal (lake)	B 2
Berens (isl.)	E 2	Fishing (lake)	G 2	Mukutawa (lake)	G 2	Sipiwesk (lake)	J 3
Berens (riv.)	F 2	Flintstone (lake)	G 4	Mukutawa (riv.)	E 1	Sisib (lake)	C 2
Bernic (lake)	G 4	Fox (riv.)	K 2	Muskeg (bay)	G 6	Sleeve (lake)	E 3
Big Sand (lake)	H 2	Gammon (riv.)	G 3	Nejanilini (lake)	J 1	Slemon (lake)	G 1
Bigstone (lake)	J 3	Garner (lake)	G 4	Nelson (riv.)	J 2	Snowshoe (lake)	G 4
Bigstone (pt.)	E 2	Gem (lake)	G 4	Nopiming Prov. Park	G 4	Soul (lake)	C 2
Bigstone (riv.)	J 3	George (isl.)	E 2	Northern Indian (lake)	J 2	Souris (riv.)	B 5
Birch (isl.)	C 2	George (lake)	G 4	North Knife (lake)	J 2	Southern Indian (lake)	H 2
Black (isl.)	F 3	Gilchrist (creek)	F 2	North Knife (riv.)	H 2	South Knife (riv.)	J 2
Black (riv.)	F 4	Gilchrist (riv.)	G 2	North Seal (riv.)	H 2	South Seal (riv.)	J 2
Bloodvein (riv.)	F 3	Gods (lake)	K 3	North Shoal (lake)	E 4	Split (lake)	J 2
Bonnet (lake)	G 4	Gods (riv.)	K 3	Nueltin (lake)	H 1	Spruce (isl.)	B 1
Buffalo (lake)	G 5	Granville (lake)	H 2	Oak (lake)	A 5	Spruce Woods Prov. Park	C 5
Burntwood (riv.)	J 3	Grass (riv.)	J 3	Obukowin (lake)	G 3	Stevenson (lake)	J 3
Caribou (riv.)	J 1	Grass River Prov. Park	H 3	Oiseau (lake)	G 4	Sturgeon (bay)	E 3
Carroll (lake)	G 3	Grindstone Prov. Rec. Park	F 3	Oiseau (riv.)	G 4	Swan (lake)	B 2
Cedar (lake)	B 1	Gunisao (lake)	F 3	Overflow (bay)	A 1	Swan (lake)	D 5
Channel (isl.)	B 2	Gypsum (lake)	D 3	Overflowing (riv.)	A 1	Swan (riv.)	A 3
Charron (lake)	G 2	Harrop (lake)	G 2	Owl (riv.)	K 2	Tadoule (lake)	J 2
Childs (lake)	A 3	Harte (mt.)	A 2	Oxford (lake)	J 3	Tamarack (isl.)	F 3
Chitek (lake)	C 2	Hayes (riv.)	K 3	Paint (lake)	J 2	Tatnam (cape)	K 2
Churchill (cape)	K 2	Hecla (isl.)	F 3	Palsen (riv.)	G 2	Traverse (bay)	F 4
Churchill (riv.)	J 2	Hecla Prov. Park	F 3	Pelican (bay)	B 2	Turtle (mts.)	B 5
Clear (lake)	C 4	Hobbs (lake)	G 3	Pelican (lake)	B 2	Turtle (riv.)	C 3
Clearwater Lake Prov. Park	H 3	Horseshoe (lake)	G 2	Pelican (lake)	C 5	Turtle Mountain Prov. Park	B 5
Cobham (riv.)	G 1	Hubbart (pt.)	K 2	Pembina (hills)	D 5	Valley (riv.)	B 3
Cochrane (riv.)	H 2	Hudson (bay)	K 2	Pembina (riv.)	C 5	Vickers (lake)	F 3
Commissioner (isl.)	E 2	Hudwin (lake)	G 1	Peonan (pt.)	D 3	Viking (lake)	G 3
Cormorant (lake)	H 3	Inland (lake)	C 2	Pickerel (lake)	C 2	Wanipigow (riv.)	G 3
Cross (bay)	C 1	International Peace Garden	B 5	Pigeon (riv.)	F 2	Washow (bay)	F 3
Cross (lake)	J 3	Island (lake)	K 3	Pipestone (creek)	A 5	Waterhen (lake)	C 2
Crowduck (lake)	G 4	Katimik (lake)	C 2	Plum (creek)	B 5	Weaver (lake)	F 2
Dancing (pt.)	D 2	Kawinaw (lake)	C 2	Plum (lake)	B 5	Wellman (lake)	B 3
Dauphin (lake)	C 3	Kinwow (bay)	E 2	Poplar (riv.)	E 2	West Hawk (lake)	G 4
Dauphin (riv.)	D 3	Kississing (lake)	H 2	Porcupine (hills)	A 2	West Shoal (lake)	E 4
Dawson (bay)	B 2	Knee (lake)	J 3	Portage (bay)	D 3	Whitemouth (lake)	G 5
Dog (lake)	D 3	Lake of the Woods (lake)	H 5	Punk (isl.)	F 3	Whitemouth (riv.)	G 5
Dogskin (lake)	G 3	La Salle (riv.)	E 5	Quesnel (lake)	G 4	Whiteshell Prov. Park	G 4
Duck Mountain Prov. Park	B 3	Laurie (lake)	A 3	Rat (riv.)	F 5	Whitewater (lake)	B 5
Eardley (lake)	F 2	Leaf (riv.)	F 2	Red (riv.)	F 4	Wicked (pt.)	D 2
		Lewis (lake)	G 2	Red Deer (lake)	A 2	Winnipeg (lake)	E 2
		Leyond (riv.)	F 3	Red Deer (riv.)	A 2	Winnipeg (riv.)	F 4
		Little Birch (lake)	E 3	Reindeer (isl.)	E 2	Winnipegosis (lake)	C 2
		Lonely (lake)	C 3	Reindeer (lake)	H 2	Woods (lake)	H 5
		Long (lake)	G 4	Riding (mt.)	B 4	Wrong (lake)	F 2
		Long (lake)	D 1	Riding Mountain Nat'l Park	B 4		
		Long (pt.)	D 4	Rock (lake)	C 5		
		Manigotagan (lake)	G 4	Ross (isl.)	J 3	*Population of metropolitan area.	
				Sagemace (bay)	B 3	●Population of rural municipality.	

AREA 650,087 sq. km. (250,999 sq. mi.)
POPULATION 1,113,898
CAPITAL Winnipeg
LARGEST CITY Winnipeg
HIGHEST POINT Baldy Mtn. 832 m. (2,730 ft.)
SETTLED IN 1812
ADMITTED TO CONFEDERATION July 15, 1870
PROV. FLOWER Prairie Crocus
PROV. BIRD Great Grey Owl

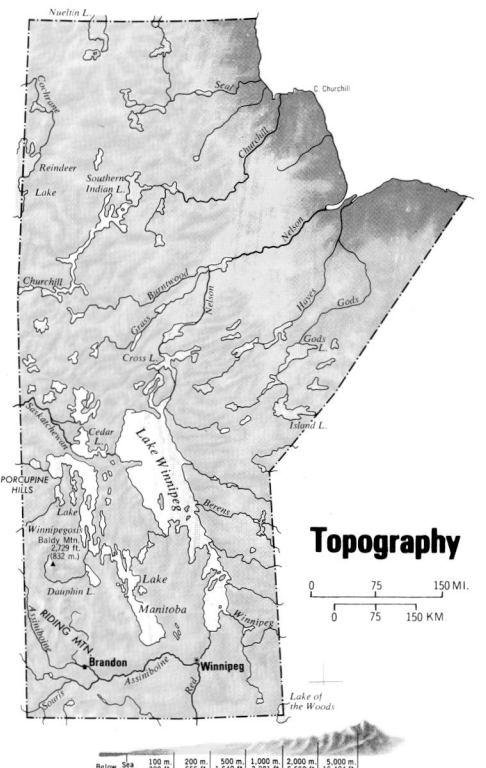

Topography

0   75   150 MI.
0   75   150 KM.

Below Sea Level | 100 m. 328 ft. | 200 m. 656 ft. | 500 m. 1,640 ft. | 1,000 m. 3,281 ft. | 2,000 m. 6,562 ft. | 5,000 m. 16,404 ft.

Agriculture, Industry and Resources

DOMINANT LAND USE

Cereals (chiefly barley, oats)
Cereals, Livestock
Dairy
Livestock
Forests
Nonagricultural Land

MAJOR MINERAL OCCURRENCES

Au Gold
Co Cobalt
Cu Copper
Na Salt

Ni Nickel
O   Petroleum
Pb Lead
Pt Platinum
Zn Zinc

⚡ Water Power
▨ Major Industrial Areas

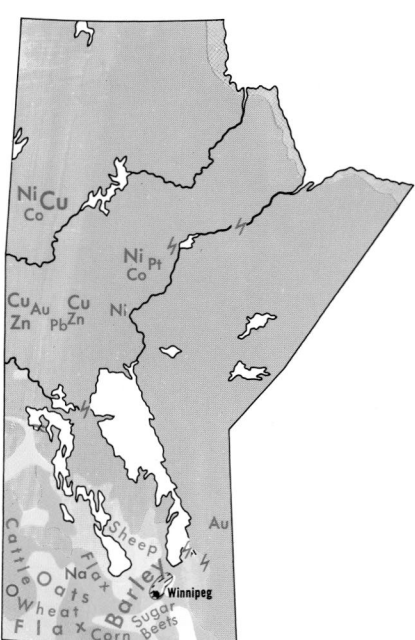

Topography

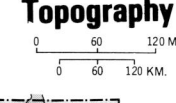

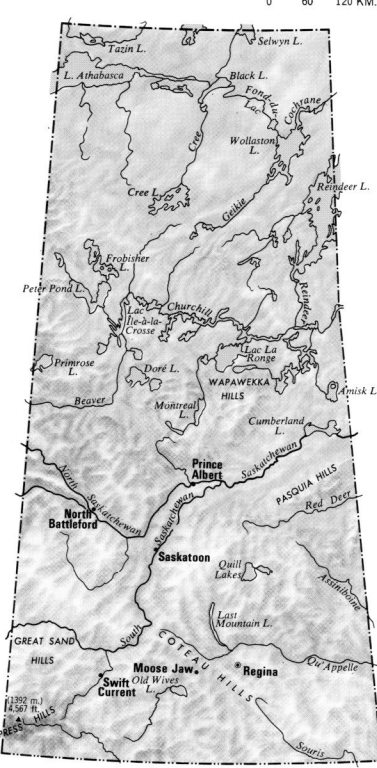

5,000 m. 2,000 m. 1,000 m. 500 m. 200 m. 100 m. Sea
16,404 ft. 6,562 ft. 3,281 ft. 1,640 ft. 656 ft. 328 ft. Level Below

Agriculture, Industry and Resources

DOMINANT LAND USE

- Wheat
- Cereals (chiefly barley, oats)
- Cereals, Livestock
- Livestock
- Forests

MAJOR MINERAL OCCURRENCES

Au Gold
Cu Copper
G Natural Gas
He Helium
K Potash
Lg Lignite
Na Salt
O Petroleum
S Sulfur
U Uranium
Zn Zinc

⚡ Water Power
Major Industrial Areas

Buffalo Pound Prov. Park	F 5	Ear (lake)	B 3	Lac La Ronge Prov. Park	M 3	Oldman (riv.)	L 2
Cabri (lake)	B 4	Echo Valley Prov. Park	G 5	Lanigan (lake)	F 4	Old Wives (lake)	E 5
Cactus (hills)	F 5	Etomami (riv.)	J 3	Last Mountain (lake)	F 4	Opuntia (lake)	C 4
Candle (lake)	F 2	Eyebrow (lake)	E 5	Leaf (lake)	J 2	Overflowing (riv.)	K 2
Cannington Manon Hist. Park	J 6	Eyehill (creek)	B 3	Leech (lake)	J 4	Pasquia (hills)	J 2
Canoe (lake)	L 3	Fife (lake)	E 6	Lenore (lake)	G 3	Pasquia (riv.)	J 2
Carrot (riv.)	L 2	File (hills)	H 5	Little Manitou (lake)	F 4	Pelican (lake)	E 5
Chaplin (lake)	E 5	Fir (riv.)	J 2	Lodge (creek)	B 6	Peter Pond (lake)	L 3
Chipman (riv.)	M 2	Fond du Lac (riv.)	M 2	Long (creek)	H 6	Pheasant (hills)	J 5
Chitek (lake)	D 3	Forrest (lake)	M 2	Loon (creek)	G 4	Pine Lake Prov. Park	E 4
Churchill (riv.)	M 3	Fort Battleford Nat'l Hist. Park	C 3	Makwa (lake)	B 1	Pinto (creek)	D 6
Clearwater (riv.)	L 3	Fort Carlton Hist. Park	E 3	Makwa (riv.)	B 1	Pipestone (creek)	K 6
Cochrane (riv.)	N 2	Fort Pitt Hist. Park	B 2	Manito (lake)	B 3	Pipestone (creek)	K 6
Coteau (hills)	D 2	Fort Walsh Nat'l Hist. Park	A 6	Maple (lake)	B 5	Ponass (lake)	H 3
Cowan (lake)	D 2	Foster (riv.)	M 3	McFarlane (riv.)	L 2	Poplar (riv.)	E 6
Crane (lake)	C 6	Frenchman (riv.)	B 6	Meadow (lake)	C 1	Porcupine (hills)	K 3
Crean (lake)	E 1	Frobisher (lake)	L 2	Meadow Lake Prov. Park	K 4	Primrose (lake)	K 3
Cree (lake)	L 2	Gap (lake)	B 6	Meeting (lake)	D 2	Primrose Lake Air Weapons	
Cree (riv.)	M 2	Gardiner (dam)	D 4	Midnight (lake)	C 2	Range	E 1
Cumberland (lake)	J 1	Geikie (riv.)	M 3	Ministikwan (lake)	B 1	Prince Albert Nat'l Park	E 1
Cypress (hills)	B 6	Good Spirit (lake)	J 4	Missouri Coteau (hills)	F 5	Qu'Appelle (riv.)	J 5
Cypress (lake)	B 6	Goodspirit Lake Prov. Park	J 4	Montreal (lake)	F 1	Quill (lake)	G 4
Cypress Hills Prov. Park	B 6	Great Sand (hills)	B 5	Moose (mt.)	J 6	Red Deer (riv.)	A 5
Danielson Prov. Park	E 4	Green (lake)	D 1	Moose Jaw (riv.)	G 5	Red Deer (riv.)	K 2
Delaronde (lake)	E 1	Greenwater Lake Prov. Park	H 3	Moose Mountain (creek)	J 6	Reindeer (lake)	N 3
Diefenbaker (lake)	E 4	Haultain (riv.)	L 3	Moose Mountain Prov. Park	J 6	Reindeer (riv.)	M 3
Doré (lake)	L 3	Île-à-la-Crosse (lake)	L 3	Mossy (riv.)	H 1	Riou (lake)	M 2
Douglas Prov. Park	E 4	Ironspring (creek)	G 3	Muddy (lake)	B 3	Rivers (lake)	F 6
Duck Lake Hist. Park	E 3	Jackfish (lake)	C 2	Mudjatik (riv.)	L 3	Ronge, La (lake)	M 2
Duck Mountain Prov. Park	K 4	Katepwa Prov. Park	H 5	Nipawin Prov. Park	G 1	Rowans Ravine Prov. Park	F 4
Eagle (lake)	C 3	Kingsmere (lake)	E 1	North Saskatchewan (riv.)	D 3	St. Victor Petroglyphs Hist.	
Eaglehill (creek)	D 4	Kiyiu (lake)	C 4	Notukeu (creek)	D 6	Park	E 6

Saskatchewan (riv.)	H 2	Thickwood (hills)	D 2	White Fox (riv.)	G 2
Saskatchewan Landing Prov.		Thunder (hills)	L 4	White Gull (creek)	G 2
Park	C 5	Tobin (lake)	H 2	Whiteshore (lake)	C 3
Saskeram (riv.)	K 2	Torch (riv.)	H 2	Whiteswan (lakes)	F 1
Scott (lake)	M 2	Touchwood (hills)	G 4	William (riv.)	L 2
Selwyn (lake)	M 2	Tramping (lake)	C 3	Willow Bunch (lake)	F 6
Souris (riv.)	H 6	Trout (riv.)	L 2	Witchekan (lake)	D 2
South Saskatchewan (riv.)	D 3	Turtle (lake)	C 2	Wollaston (lake)	N 2
Steele Narrows Hist. Park	B 2	Twelvemile (lake)	E 6	Wood (mt.)	E 6
Stripe (lake)	C 4	Vermilion (hills)	E 5	Wood (lake)	E 6
Sturgeon (riv.)	E 2	Wapawekka (hills)	M 4	Wood Mountain Hist. Park	E 6
Swan (lake)	J 3	Waskana (creek)	E 2	
Swift Current (creek)	D 5	Waskesiu (lake)	E 2	
Tazin (riv.)	M 3	Watanan (riv.)	M 3	
The Battlefords Prov. Park	C 2	Weed (hills)	J 5	

AREA 651,900 sq. km. (251,699 sq. mi.)
POPULATION 990,237
CAPITAL Regina
LARGEST CITY Regina
HIGHEST POINT Cypress Hills 1468 m. (4,816 ft.)
SETTLED IN 1774
ADMITTED TO CONFEDERATION September 1, 1905
PROV. FLOWER Prairie Lily
PROV. BIRD Sharp-tailed Grouse

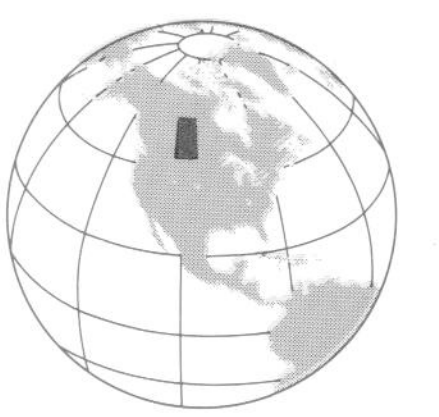

*Population of metropolitan area.
•Population of rural municipality.

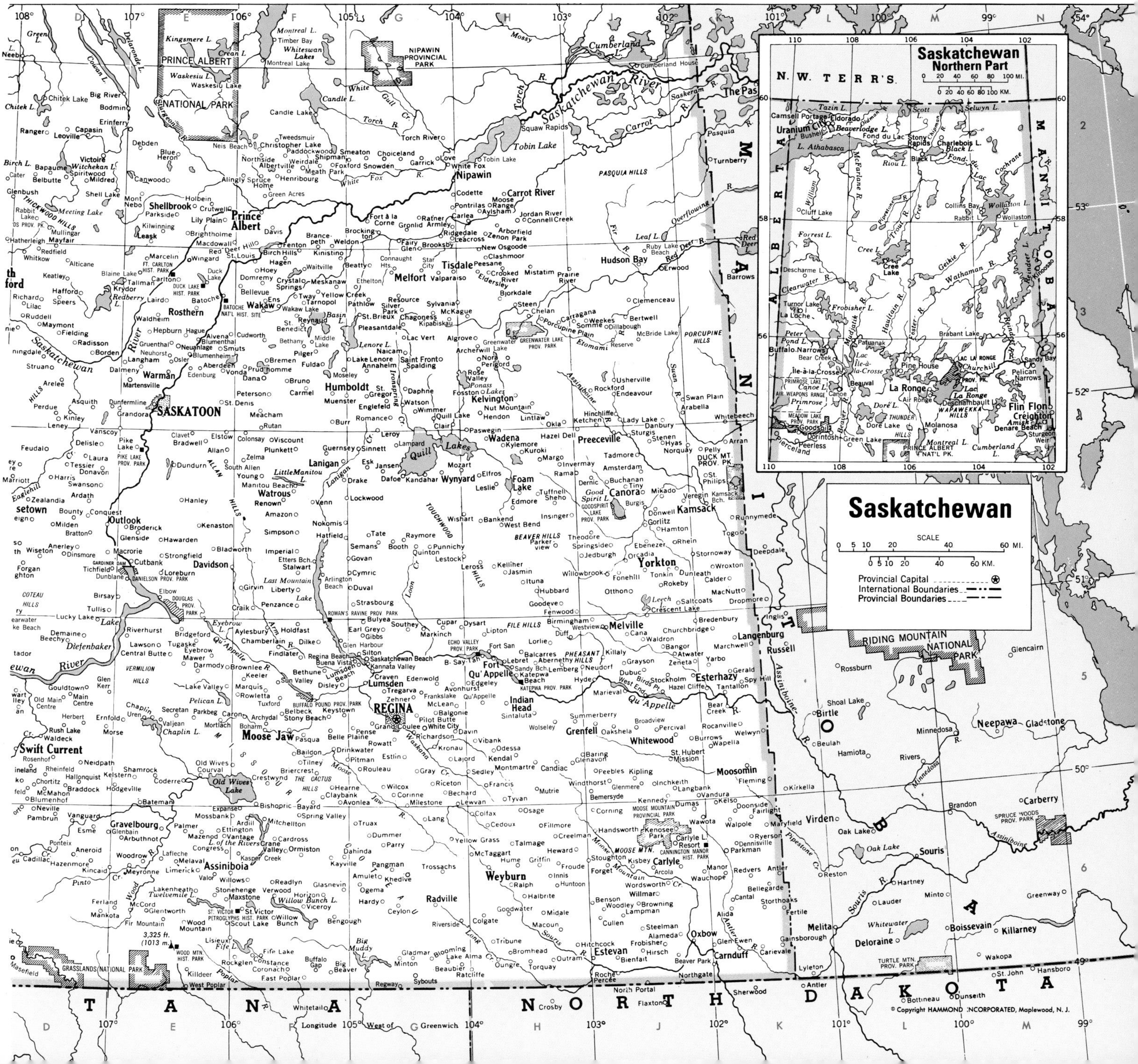

Saskatchewan
Northern Part

Saskatchewan

Provincial Capital
International Boundaries
Provincial Boundaries

© Copyright HAMMOND INCORPORATED, Maplewood, N.J.

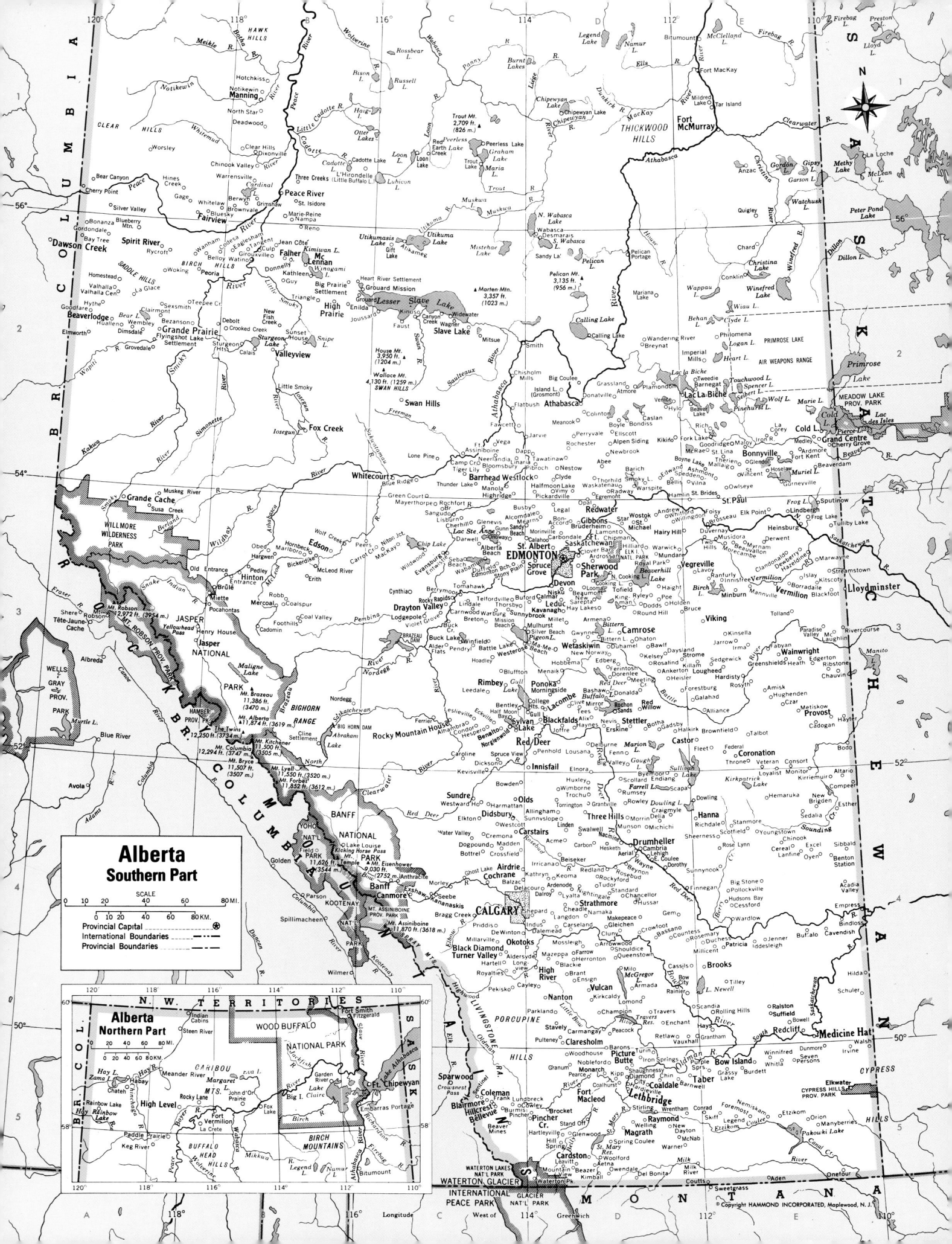

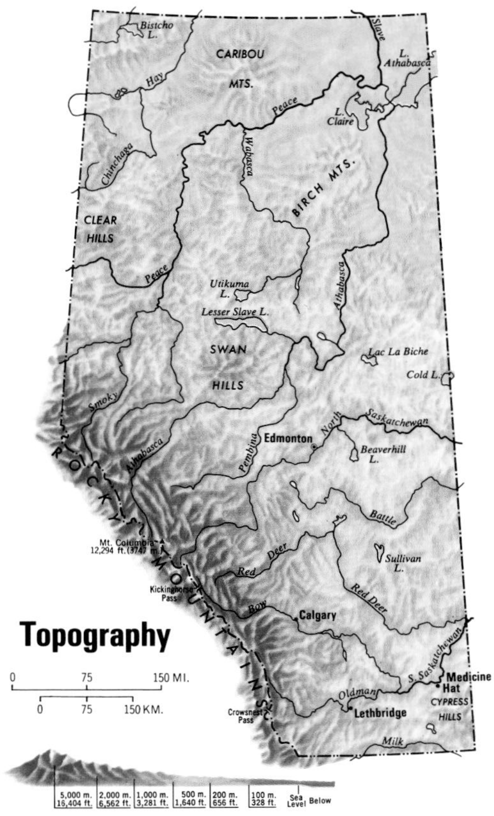
Bistcho L.
CARIBOU MTS.
L. Athabasca
Slave
Hay
Peace
L. Claire
Chinchaga
Wabasca
BIRCH MTS.
CLEAR HILLS
Utikuma L.
Lesser Slave L.
Athabasca
SWAN HILLS
Lac La Biche
Cold L.
Smoky
North Saskatchewan
Edmonton
Pembina
Beaverhill L.
ROCKY MOUNTAINS
Battle
Mt. Columbia 12,294 ft. (3747 m.)
Sullivan L.
Red Deer
Kickinghorse Pass
Bow
Calgary
South Saskatchewan
Medicine Hat
Oldman
Crowsnest Pass
Lethbridge
CYPRESS HILLS
Milk
0 75 150 MI.
0 75 150 KM.
5,000 m. 16,404 ft.
2,000 m. 6,562 ft.
1,000 m. 3,281 ft.
500 m. 1,640 ft.
200 m. 656 ft.
100 m. 328 ft.
Sea Level
Below

Topography

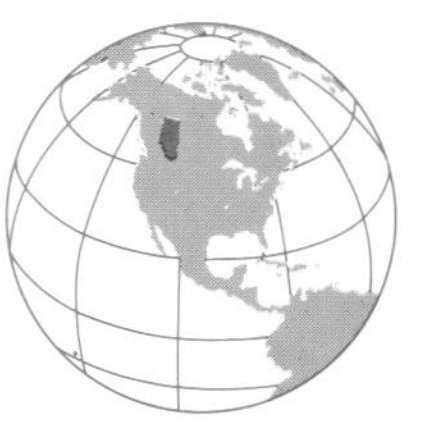

AREA 611,185 sq. km. (255,285 sq. mi.)
POPULATION 2,696,826
CAPITAL Edmonton
LARGEST CITY Edmonton
HIGHEST POINT Mt. Columbia 3747 m. (12,293 ft.)
SETTLED IN 1861
ADMITTED TO CONFEDERATION September 1, 1905
PROV. FLOWER Wild Rose
PROV. BIRD Great Horned Owl

CITIES and TOWNS

Acme 457 D 4
Airdrie 8,414 C 4
Alberta Beach 485 C 3
Alix 837 D 3
Andrew 548 D 3
Antler Lake 334 D 3
Ardmore 224 E 2
Arrowwood 156 D 4
Athabasca 1,731 D 2
Banff 4,208 C 4
Barnwell 359 D 5
Barons 315 D 4
Barrhead 3,736 C 2
Bashaw 875 D 3
Bassano 1,200 D 4
Bawlf 350 D 3
Beaumont 2,638 D 3
Beaverlodge 1,937 A 2
Beiseker 580 D 4
Bentley 823 C 3
Berwyn 557 B 1
Big Valley 360 D 3
Black Diamond 1,444 C 4
Blackfalds 1,488 D 3
Blackfoot 220 E 3
Blackie 298 D 4
Bon Accord 1,376 D 3
Bonnyville 4,454 E 2
Bowden 989 C 4
Bow Island 1,491 E 5
Boyle 638 D 2
Bragg Creek 505 C 4
Breton 552 C 3
Brooks 9,421 E 4
Bruce 88 E 3
Bruderheim 1,136 D 3
Burdett 220 E 5
Calgary 592,743 C 4
Calgary *592,743 C 4
Calmar 1,003 D 3
Camrose 12,570 D 3
Canmore 3,484 C 4
Carbon 434 D 4
Cardston 3,267 D 5
Carmangay 266 D 4
Caroline 436 C 3
Carseland 484 D 4
Carstairs 1,587 D 4
Castor 1,123 D 3
Cereal 249 E 4
Champion 339 D 4
Chauvin 298 E 3
Chipman 266 D 3
Clairmont 469 A 2
Claresholm 3,493 D 4
Clive 364 D 3
Clyde 364 D 2
Coaldale 4,579 D 5
Coalhurst 882 D 5
Cochrane 3,544 C 4
Cold Lake 2,110 E 2
College Heights 267 D 3
Consort 632 E 3
Cooking Lake 218 D 3
Coronation 1,309 E 3
Coutts 400 D 5
Cowley 304 D 5
Cremona 382 C 4
Crossfield 1,217 C 4
Daysland 679 D 3
Delburne 574 D 3
Desmarais 260 D 2
Devon 3,885 D 3
Didsbury 3,095 C 4
Donalda 280 D 3
Donnelly 336 B 2
Drayton Valley 5,042 C 3
Drumheller 6,508 D 4
Duchess 429 E 4
East Coulee 218 D 4
Eckville 870 C 3
Edgerton 387 E 3
Edmonton (cap.) 532,246 D 3
Edmonton *657,057 D 3
Edmonton Beach 280 C 3
Edson 5,835 B 3
Elk Point 1,022 E 3
Elnora 249 D 3
Entwistle 462 C 3
Erskine 259 D 3
Evansburg 779 C 3
Exshaw 353 C 4
Fairview 2,869 A 1
Falher 1,102 B 2
Faust 399 C 2
Foremost 568 E 5
Forestburg 924 E 3
Fort Assiniboine 207 C 2
Fort Chipewyan 944 C 5
Fort Macleod 3,139 D 5
Fort McKay 267 E 1
Fort McMurray 31,000 E 1
Fort Saskatchewan 12,169 D 3
Fort Vermilion 752 B 5
Fox Creek 1,978 B 2
Fox Lake 634 B 5
Gibbons 2,276 D 3
Gift Lake 428 C 2
Girouxville 325 B 2
Gleichen 381 D 4
Glendon 430 E 2
Glenwood 259 D 5
Grand Centre 3,146 E 2
Grande Cache 4,523 A 3
Grande Prairie 24,263 A 2
Granum 399 D 5
Grimshaw 2,316 B 1
Grouard Mission 221 C 2
Hanna 2,806 E 4
Hardisty 641 E 3
Hay Lakes 302 D 3
Heisler 212 D 3
High Level 2,194 A 5
High Prairie 2,506 B 2
High River 4,792 D 4
Hines Creek 575 A 1
Hinton 8,342 B 3
Holden 430 D 3
Hughenden 267 E 3
Hythe 639 A 2
Innisfail 5,247 D 3
Innisfree 255 E 3
Irma 474 E 3
Irricana 558 D 4
Irvine 360 E 5
Jasper 3,269 B 3
John d'Or Prairie 437 B 5
Joussard 330 B 2
Killam 1,005 E 3
Kinuso 285 C 2
Kitscoty 497 E 3
Lac La Biche 2,007 E 2
Lacombe 5,591 D 3
La Crete 479 B 5
Lake Louise 355 B 4
Lamont 1,563 D 3
Leduc 12,471 D 3
Legal 1,022 D 3
Lethbridge 54,072 D 5
Linden 407 D 4
Little Buffalo Lake 253 B 1
Lloydminster 8,997 E 3
Longview 301 C 4
Lougheed 226 E 3
Lundbreck 244 C 5
Magrath 1,576 D 5
Manning 1,173 B 1
Mannville 788 E 3
Marlboro 211 B 3
Marwayne 500 E 3
Mayerthorpe 1,475 C 3
McLennan 1,125 B 2
Medicine Hat 40,380 E 4
Milk River 894 D 5
Millet 1,120 D 3
Mirror 507 D 3
Monarch 212 D 5
Morinville 4,657 D 3
Morrin 244 D 4
Mundare 604 D 3
Myrnam 397 E 3
Nacmine 369 D 4
Nampa 334 B 1
Nanton 1,641 D 4
New Norway 291 D 3
New Sarepta 417 D 3
Nobleford 534 D 5
North Calling Lake 234 D 2
Okotoks 3,847 C 4
Olds 4,813 D 4
Onoway 621 C 3
Oyen 975 E 4
Peace River 5,907 B 1
Penhold 1,531 D 3
Picture Butte 1,404 D 5
Pincher Creek 3,757 D 5
Plamondon 259 D 2
Pollockville 19 E 4
Ponoka 5,221 D 3
Provost 1,645 E 3
Rainbow Lake 504 A 5
Ralston 357 E 4
Raymond 2,837 D 5
Redcliff 3,876 E 4
Red Deer 46,393 D 3
Redwater 1,932 D 3
Rimbey 1,685 C 3
Robb 230 B 3
Rockyford 329 D 4
Rocky Mountain House 4,698 C 3
Rosemary 328 E 4
Rycroft 649 A 2
Ryley 483 D 3
Saint Albert 31,996 D 3
Saint Paul 4,884 E 3
Sangudo 398 C 3
Sedgewick 879 E 3
Sexsmith 1,180 A 2
Shaughnessy 270 D 5
Sherwood Park 29,285 D 3
Slave Lake 4,506 C 2
Smith 216 D 2
Smoky Lake 1,074 D 2
Spirit River 1,104 A 2
Spruce Grove 10,326 D 3
Standard 379 D 4
Stavely 504 D 4
Stettler 5,136 D 3
Stirling 688 D 5
Stony Plain 4,839 C 3
Strathmore 2,986 D 4
Strome 281 E 3
Sundre 1,742 C 4
Swan Hills 2,497 C 2
Sylvan Lake 3,779 C 3
Taber 5,988 E 5
Thorhild 576 D 2
Thorsby 737 C 3
Three Hills 1,787 D 4
Tilley 345 E 4
Tofield 1,504 D 3
Trochu 880 D 4
Turner Valley 1,311 C 4
Two Hills 1,193 E 3
Valleyview 2,061 B 2
Vauxhall 1,049 D 4
Vegreville 5,251 E 3
Vermilion 3,766 E 3
Veteran 314 E 3
Viking 1,232 E 3
Vilna 345 E 2
Vulcan 1,489 D 4
Wabamun 662 C 3
Wabasca 701 D 2
Wainwright 4,266 E 3
Warburg 501 C 3
Warner 477 D 5
Waskatenau 290 D 2
Wembley 1,169 A 2
Westlock 4,424 C 2
Wetaskiwin 9,597 D 3
Whitecourt 5,585 C 2
Wildwood 441 C 3
Willingdon 366 E 3
Youngstown 297 E 4

OTHER FEATURES

Abraham (lake) B 3
Alberta (mt.) B 3
Assiniboine (mt.) C 4
Athabasca (lake) C 5
Athabasca (riv.) D 1
Banff Nat'l Park B 4
Battle (riv.) D 3
Bear (lake) A 2
Beaver (riv.) E 2
Beaverhill (lake) D 3
Behan (lake) E 2
Belly (riv.) D 5
Berland (riv.) A 3
Berry (creek) E 4
Biche (lake) E 2
Big (isl.) B 5
Big Horn (dam) B 3
Bighorn (range) B 3
Birch (hills) A 2
Birch (lake) E 3
Birch (mts.) B 5
Birch (riv.) B 5
Bison (lake) B 1
Bittern (lake) D 3
Botha (riv.) B 1
Bow (riv.) D 4
Boyer (riv.) A 5
Brazeau (mt.) B 3
Brazeau (riv.) B 3
Buffalo (lake) D 3
Buffalo Head (hills) B 5
Burnt (lakes) C 1
Cadotte (lake) B 1
Cadotte (riv.) B 1
Calling (lake) D 2
Canal (creek) E 5
Cardinal (lake) B 1
Caribou (mts.) B 5
Chinchaga (riv.) A 5
Chip (lake) C 3
Chipewyan (lake) D 1
Chipewyan (riv.) D 1
Christina (lake) E 2
Christina (riv.) E 1
Claire (lake) B 5
Clear (hills) A 1
Clearwater (riv.) C 4
Clearwater (riv.) E 1
Clyde (lake) E 2
Cold (lake) E 2
Columbia (mt.) B 3
Crowsnest (pass) C 5
Cypress (hills) E 5
Cypress Hills Prov. Park E 5
Dillon (riv.) E 2
Dowling (lake) D 4
Dunkirk (riv.) D 1
Eisenhower (mt.) C 4
Elbow (riv.) C 4
Elk Island Nat'l Park D 3
Ells (riv.) D 1
Etzikom Coulee (riv.) E 5
Eva (lake) B 5
Farrell (lake) D 4
Firebag (riv.) E 1
Forbes (mt.) B 4
Freeman (riv.) C 2
Frog (lake) E 3
Garson (lake) E 1
Gipsy (lake) E 1
Gordon (lake) E 1
Gough (lake) D 3
Graham (lake) C 1
Gull (lake) C 3
Haig (lake) B 1
Hawk (hills) B 1
Hay (lake) A 5
Hay (riv.) A 5
Heart (lake) E 2
Highwood (riv.) C 4
House (mt.) C 2
House (riv.) D 2
Iosegun (lake) B 2
Iosegun (riv.) B 2
Jackfish (riv.) B 5
Jasper Nat'l Park A 3
Kakwa (riv.) A 2
Kickinghorse (pass) B 4
Kimiwan (lake) B 2
Kirkpatrick (lake) E 4
Kitchener (mt.) B 3
Legend (lake) D 1
Lesser Slave (lake) C 2
Liége (riv.) D 1
Little Bow (riv.) D 4
Little Cadotte (riv.) B 1
Little Smoky (riv.) B 2
Livingstone (range) C 4
Logan (lake) E 2
Loon (lake) C 1
Loon (riv.) C 1
Lubicon (lake) C 1
Lyell (mt.) B 4
MacKay (riv.) D 1
Maligne (lake) B 3
Margaret (lake) B 5
Marie (lake) E 2
Marion (lake) D 3
Marten (mt.) C 2
McClelland (lake) E 1
McGregor (lake) D 4
McLeod (riv.) B 3
Meikle (riv.) A 1
Mikkwa (riv.) B 5
Milk (riv.) D 5
Mistehae (lake) C 2
Muriel (lake) E 2
Muskwa (lake) C 1
Muskwa (riv.) C 1
Namur (lake) D 1
Newell (lake) E 4
Nordegg (riv.) C 3
North Saskatchewan (riv.) E 3
North Wabasca (lake) D 1
Notikewin (riv.) A 1
Oldman (riv.) D 5
Otter (lakes) B 1
Pakowki (lake) E 5
Panny (riv.) C 1
Peace (riv.) B 1
Peerless (lake) C 1
Pelican (lake) D 2
Pelican (mts.) D 2
Pembina (riv.) C 3
Pigeon (lake) D 3
Pinehurst (lake) E 2
Porcupine (hills) C 4
Primrose (lake) E 2
Rainbow (lake) A 5
Red Deer (lake) D 3
Red Deer (riv.) D 4
Richardson (riv.) C 5
Rocky (mts.) B-C 4
Rosebud (riv.) D 4
Russell (lake) C 1
Saddle (hills) A 2
Sainte Anne (lake) C 3
Saint Mary (res.) D 5
Saint Mary (riv.) D 5
Saulteaux (riv.) C 2
Seibert (lake) E 2
Simonette (riv.) A 2
Slave (riv.) C 5
Smoky (riv.) A 2
Snake Indian (riv.) A 3
Snipe (lake) B 2
Sounding (creek) E 4
South Saskatchewan (riv.) E 4
South Wabasca (lake) D 2
Spencer (lake) E 2
Spray (mts.) C 4
Sturgeon (lake) B 2
Sullivan (lake) D 3
Swan (hills) C 2
Swan (riv.) C 2
Temple (mt.) B 4
The Twins (mt.) B 3
Thickwood (hills) D 1
Touchwood (lake) E 2
Travers (res.) D 4
Trout (mt.) C 1
Trout (riv.) C 1
Utikuma (lake) C 2
Utikuma (riv.) C 1
Utikumasis (lake) C 2
Vermilion (riv.) E 3
Wabasca (riv.) C 1
Wallace (mt.) C 2
Wapiti (riv.) A 2
Wappau (lake) E 2
Watchusk (lake) E 1
Waterton-Glacier Int'l Peace Park C 5
Waterton Lakes Nat'l Park C 5
Whitemud (riv.) A 1
Wildhay (riv.) B 3
Willmore Wilderness Prov. Park A 3
Winagami (lake) B 2
Winefred (lake) E 2
Winefred (riv.) E 2
Wolf (lake) E 2
Wolverine (riv.) B 1
Wood Buffalo Nat'l Park B 5
Yellowhead (pass) A 3
Zama (lake) A 5

*Population of metropolitan area.

Agriculture, Industry and Resources

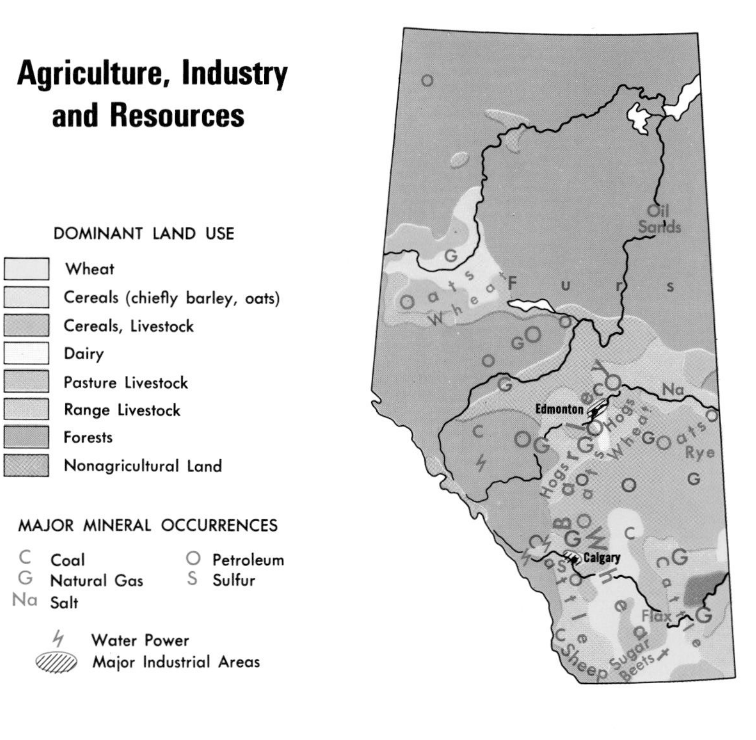
DOMINANT LAND USE
Wheat
Cereals (chiefly barley, oats)
Cereals, Livestock
Dairy
Pasture Livestock
Range Livestock
Forests
Nonagricultural Land
MAJOR MINERAL OCCURRENCES
C Coal
O Petroleum
G Natural Gas
S Sulfur
Na Salt
Water Power
Major Industrial Areas
Oil Sands
Furs
Oats
Wheat
Edmonton
Na
Hogs
Rye
Calgary
Flax
Sheep
Sugar Beets

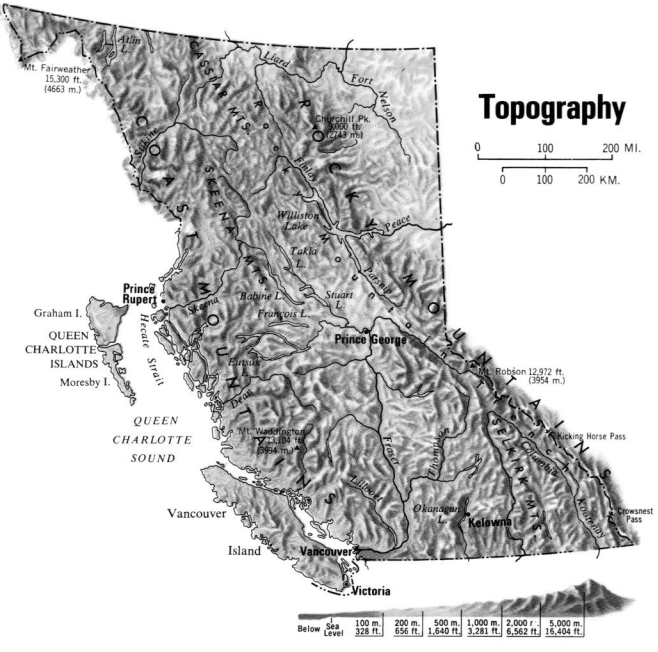

Topography

0 — 100 — 200 MI.
0 — 100 — 200 KM.

Below Sea Level	100 m. 328 ft.	200 m. 656 ft.	500 m. 1,640 ft.	1,000 m. 3,281 ft.	2,000 r. 6,562 ft.	5,000 m. 16,404 ft.

Agriculture, Industry and Resources

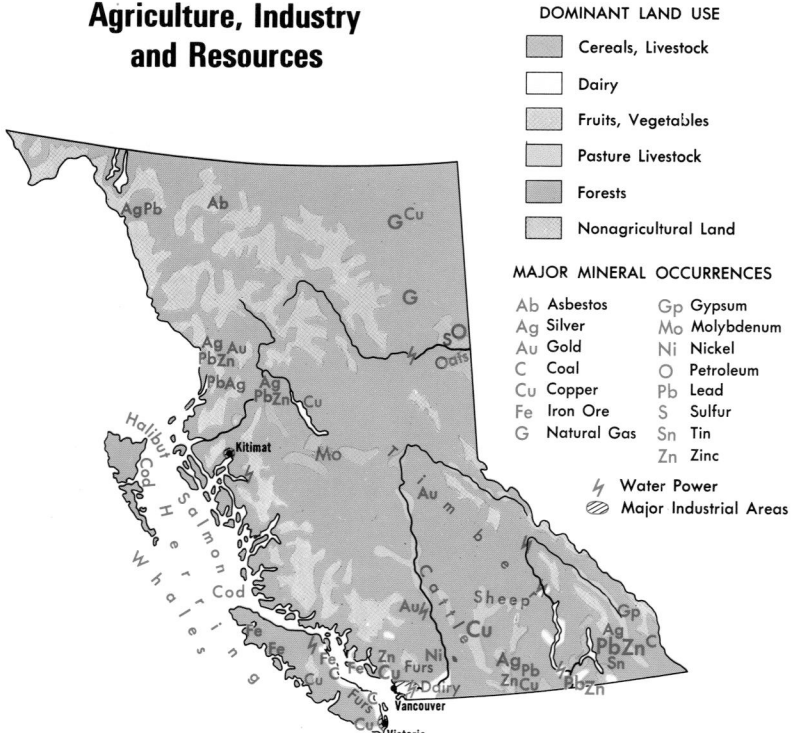

DOMINANT LAND USE

- Cereals, Livestock
- Dairy
- Fruits, Vegetables
- Pasture Livestock
- Forests
- Nonagricultural Land

MAJOR MINERAL OCCURRENCES

Ab	Asbestos	Gp	Gypsum
Ag	Silver	Mo	Molybdenum
Au	Gold	Ni	Nickel
C	Coal	O	Petroleum
Cu	Copper	Pb	Lead
Fe	Iron Ore	S	Sulfur
G	Natural Gas	Sn	Tin
		Zn	Zinc

⚡ Water Power
▨ Major Industrial Areas

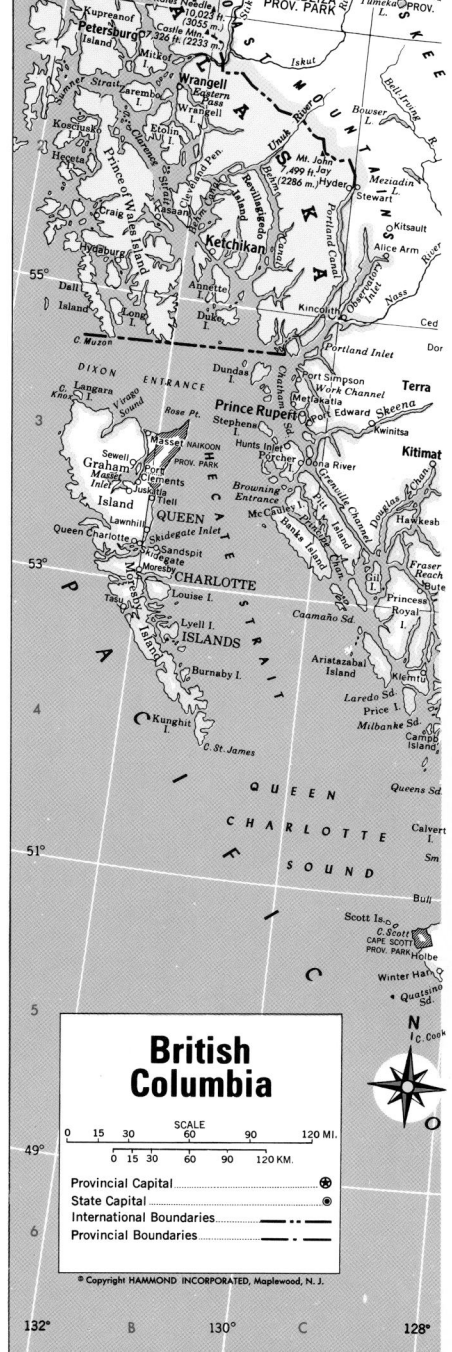

British Columbia

SCALE
0 — 15 — 30 — 60 — 90 — 120 MI.
0 — 15 — 30 — 60 — 90 — 120 KM.

Provincial Capital .... ⊛
State Capital .... ◉
International Boundaries .... — ∙ — ∙ —
Provincial Boundaries .... — — —

AREA 948,596 sq. km. (366,253 sq. mi.)
POPULATION 3,724,500
CAPITAL Victoria
LARGEST CITY Vancouver
HIGHEST POINT Mt. Fairweather 4663 m. (15,298 ft.)
SETTLED IN 1806
ADMITTED TO CONFEDERATION July 20, 1871
PROV. FLOWER Dogwood
PROV. BIRD Stellar's Jay

*Population of metropolitan area.
○Population of municipality.

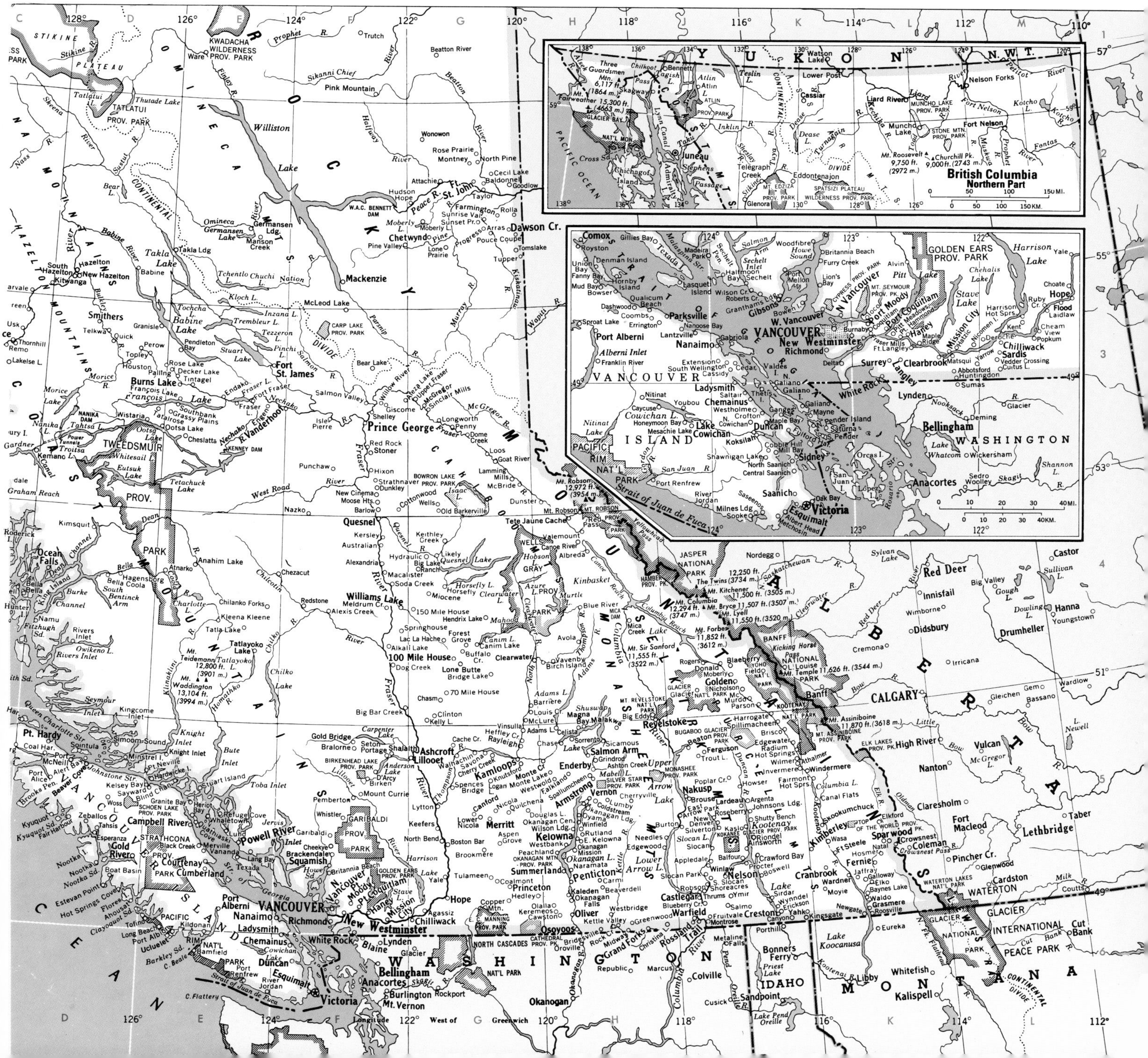

NORTHWEST TERRITORIES

CITIES and TOWNS

Topography

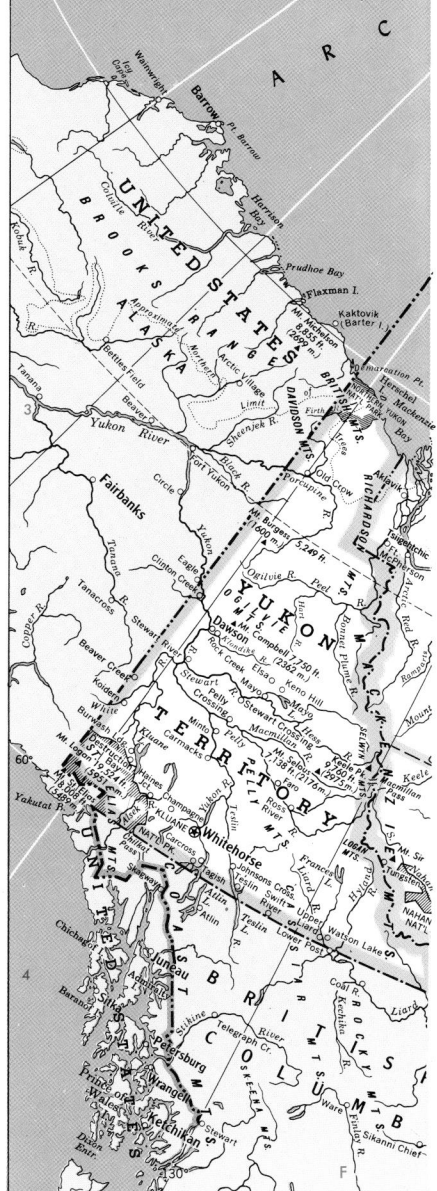

Scale bars:
0 200 400 MI.
0 200 400 KM.

Mt. Logan 19,524 ft. (5951 m.)

Mt. Sir James MacBrien 9,062 ft. (2762 m.)

5,000 m. 16,404 ft. | 2,000 m. 6,562 ft. | 1,000 m. 3,281 ft. | 500 m. 1,640 ft. | 200 m. 656 ft. | 100 m. 328 ft. | Sea Level | Below

Agriculture, Industry and Resources

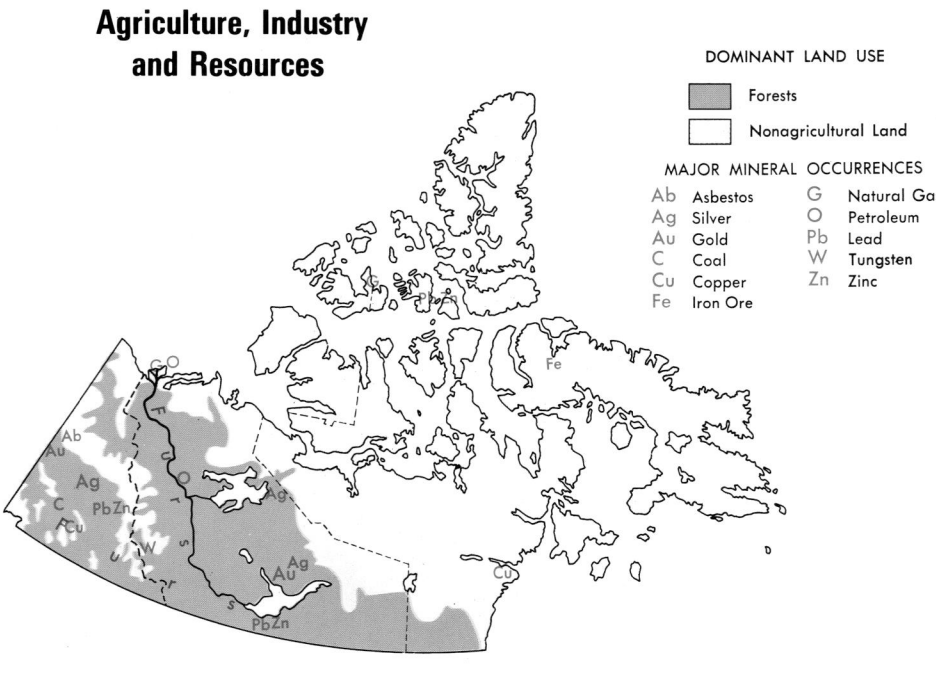

DOMINANT LAND USE

Forests

Nonagricultural Land

MAJOR MINERAL OCCURRENCES

Ab — Asbestos
Ag — Silver
Au — Gold
C — Coal
Cu — Copper
Fe — Iron Ore
G — Natural Gas
O — Petroleum
Pb — Lead
W — Tungsten
Zn — Zinc

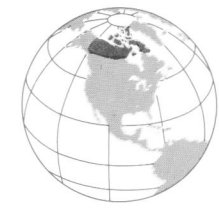

Prince Charles (isl.)	L3
Prince Gustav Adolf (sea)	H2
Prince of Wales (isl.)	J2
Prince Regent (inlet)	J2
Queen Elizabeth (isls.)	H1
Queen Maud (gulf)	H3
Queens (chan.)	J2
Raanes (pen.)	K2
Rae (isth.)	K3
Rae (riv.)	G3
Rae (str.)	J3
Resolution (isl.)	M3
Robeson (chan.)	M1
Ross Welcome (sound)	K3
Rowley (isl.)	K3
Royal Geographic Society (isls.)	J3
Russell (isl.)	J2
Sabine (pen.)	K3
Salisbury (isl.)	L3
Seahorse (pt.)	L3
Simpson (pen.)	K3
Smith (bay)	L2
Smith (cape)	L3
Smith (sound)	L2
Somerset (isl.)	J2
South (bay)	K3
Southampton (isl.)	K3
Stallworthy (cape)	J1
Steensby (inlet)	L2
Stefansson (isl.)	H2
Sverdrup (chan.)	J1
Takijug (lake)	G3
Talbot (inlet)	L2

NORTHWEST TERRITORIES

AREA 519,731 sq. mi.
(1,346,106 sq. km.)
POPULATION 39,672
CAPITAL Yellowknife
LARGEST CITY Yellowknife
HIGHEST POINT Mt. Sir James McBrien
9,062 ft. (2,762 m.)
SETTLED IN 1800
ADMITTED TO CONFEDERATION 1870
PROV. FLOWER Mountain Avens
PROV. BIRD Gyrfalcon

Tha'ane (riv.)	J3
Thelon (riv.)	H3
Thlewiasa (riv.)	J3
Ungava (bay)	M4
Vansittart (isl.)	K3
Victoria (isl.)	G2
Victoria (str.)	G2
Viscount Melville (sound)	G2
Wager (bay)	K3
Wales (isl.)	K3
Walsingham (cape)	M3
Wellington (chan.)	J2
Winter (harb.)	H3
Wollaston (pen.)	G3
Yathkyed (lake)	J3

YUKON TERRITORY
CITIES and TOWNS

Beaver Creek 113	D3
Burwash Landing 64	D3
Carcross 209	E3
Carmacks 280	E3
Champagne 57	E3
Clinton Creek	D3
Cowley	E3
Dawson 1,287	E3
Destruction Bay 48	E3
Elsa 294	E3
Faro 1251	E3

NUNAVUT

AREA 808,180 sq. mi.
(2,093,190 sq. km.)
POPULATION 24,730
CAPITAL Iqaluit
LARGEST CITY Iqaluit
HIGHEST POINT Barbeau Pk. 8,583 ft.
(2,616 m.)
SETTLED IN 1850
ADMITTED TO CONFEDERATION 1999
PROV. FLOWER Purple Saxifrage
PROV. BIRD Ptarmigan

Haines Junction 340	E3
Johnson's Crossing 18	E3
Keno Hill 47	D3
Koidern	D3
Mayo 324	E3
Minto	E3
Old Crow 232	E3
Pelly Crossing 177	E3
Rock Creek 75	E3
Ross River 352	E3
Stewart Crossing 40	E3
Stewart River	D3
Swift River 5	E3
Tagish 103	E3
Teslin 181	E3

YUKON TERRITORY

AREA 186,660 sq. mi.
(483,450 sq. km.)
POPULATION 30,766
CAPITAL Whitehorse
LARGEST CITY Whitehorse
HIGHEST POINT Mt. Logan 19,524 ft.
(5,951 m.)
SETTLED IN 1897
ADMITTED TO CONFEDERATION 1898
PROV. FLOWER Fireweed
PROV. BIRD Raven

Upper Liard 130	E3
Watson Lake 993	F3
Whitehorse (cap.) 19,157	E3

OTHER FEATURES

Alsek (riv.)	D3
Bonnet Plume (riv.)	D3
British (mts.)	D3
Campbell (mt.)	E3
Cassiar (mts.)	E3
Frances (lake)	E3
Herschel (isl.)	D3
Hess (riv.)	E3
Hyland (riv.)	F3

Keele (peak)	E3
Klondike (riv.)	E3
Kluane (lake)	E3
Kluane Nat'l Park	E3
Liard (riv.)	E3
Logan (mt.)	D3
Logan (mts.)	F3
Mackenzie (mts.)	E3
Macmillan (riv.)	E3
Mayo (lake)	E3
Northern Yukon Nat'l Pk.	E3
Ogilvie (lake)	E3
Ogilvie (riv.)	E3
Peel (riv.)	E3
Pelly (mts.)	E3

Pelly (riv.)	E3
Porcupine (riv.)	E3
Richardson (mts.)	F4
Rocky (mts.)	E3
Saint Elias (mt.)	D3
Saint Elias (mts.)	E3
Selous (mt.)	E3
Selwyn (mts.)	E3
Stewart (riv.)	E3
Teslin (lake)	E4
Teslin (riv.)	E3
White (riv.)	D3
Yukon (riv.)	E3

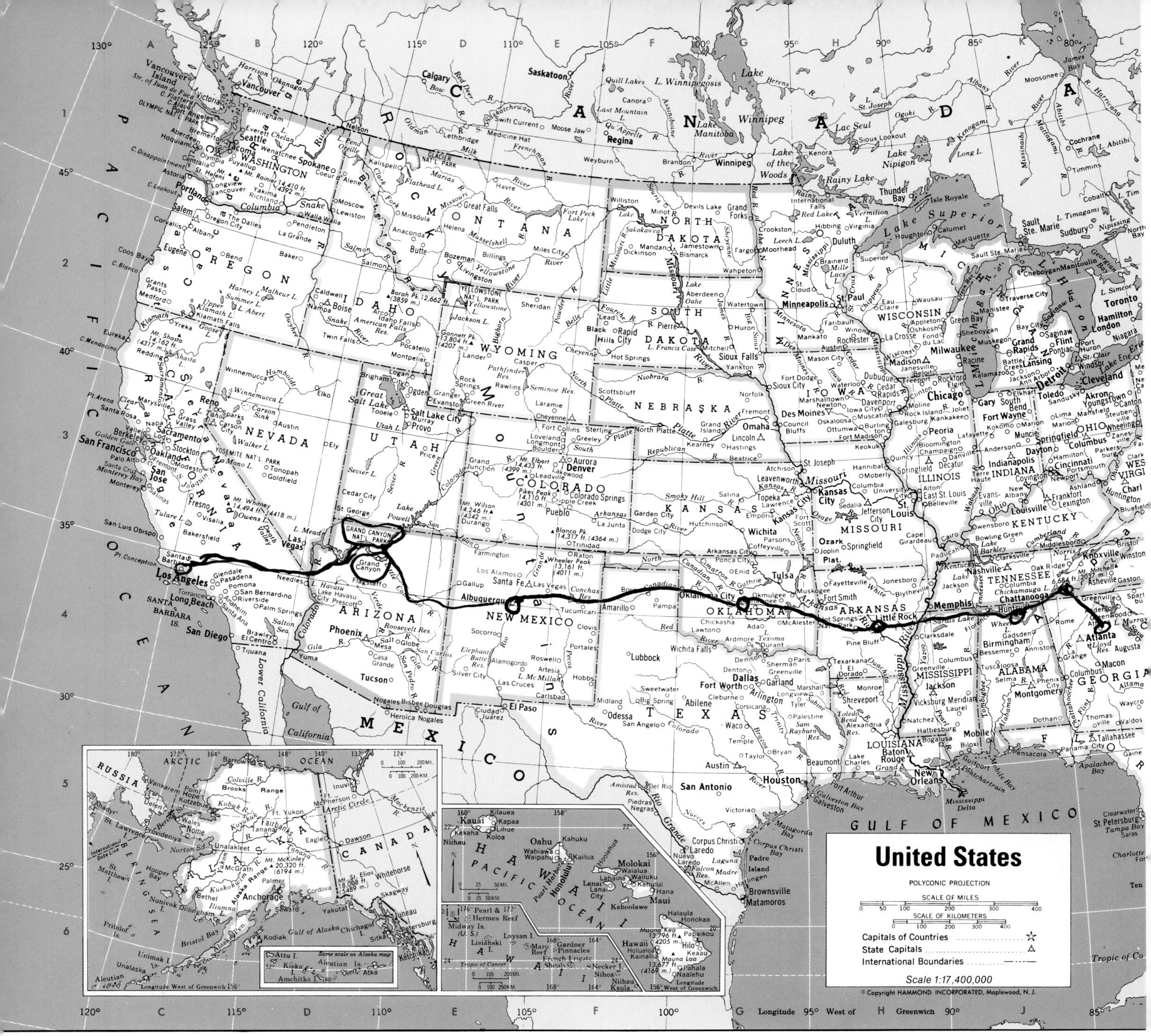

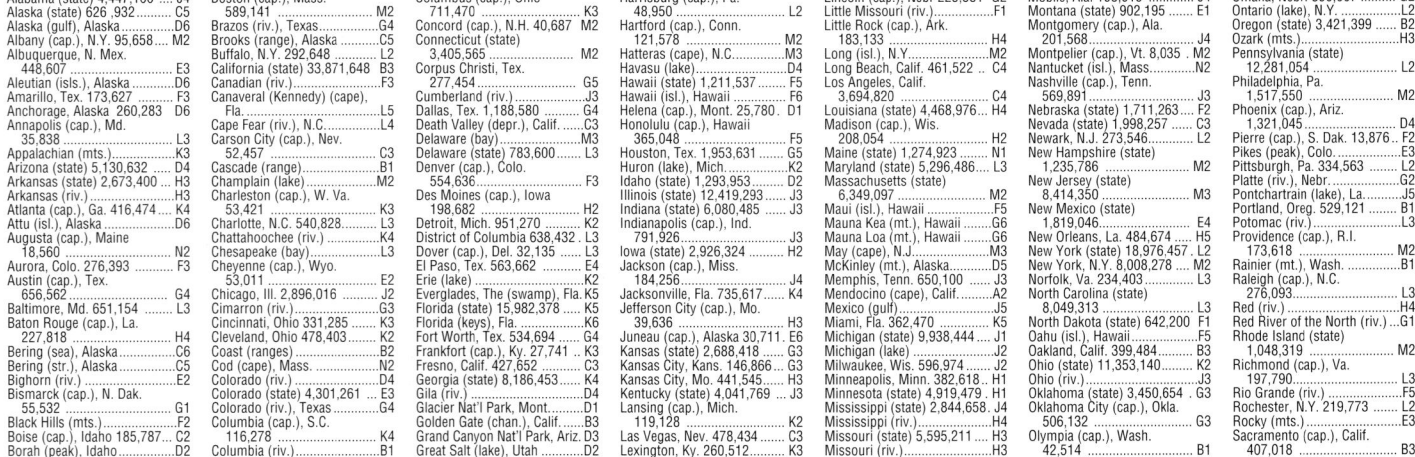

Alabama (state) 4,447,100 .... J4	Boston (cap.), Mass.	Columbus (cap.), Ohio	Harrisburg (cap.), Pa.	Lincoln (cap.), Nebr 225,581 . G2	Mobile, Ala. 198,915 .... J4	Omaha, Nebr. 390,007 .... G2
Alaska (state) 626,932 ..... C5	589,141 ....M2	711,470 ....K3	48,950 ....L2	Little Missouri (riv.)........F1	Montana (state) 902,195 ...... E1	Ontario (lake), N.Y......... L2
Alaska (gulf), Alaska D6	Brazos (riv.), Texas.........G4	Concord (cap.), N.H. 40,687 M2	Hartford (cap.), Conn.	Little Rock (cap.), Ark.	Montgomery (cap.), Ala.	Oregon (state) 3,421,399 ..... B2
Albany (cap.), N.Y. 95,658.... M2	Brooks (range), Alaska C5	Connecticut (state)	121,578 ....M2	183,133 ....H4	201,568 ....J4	Ozark (mts.).................H3
Albuquerque, N. Mex.	Buffalo, N.Y. 292,648 ...... L2	3,405,565 ....M2	Hatteras (cape), N.C.........M4	Long (isl.), N.Y..............M2	Montpelier (cap.), Vt. 8,035 . M2	Pennsylvania (state)
448,607 ....E3	California (state) 33,871,648 B3	Corpus Christi, Tex.	Havasu (lake)................D3	Long Beach, Calif. 461,522 .. C4	Nantucket (isl.), Mass......N2	12,281,054 ....L2
Aleutian (isls.), Alaska D6	Canadian (riv.)..............F3	277,454 ....G5	Hawaii (state) 1,211,537 .... F5	Los Angeles, Calif.	Nashville (cap.), Tenn.	Philadelphia, Pa.
Amarillo, Tex. 173,627 ......F3	Canaveral (Kennedy) (cape),	Cumberland (riv.)............J3	Hawaii (isl.), Hawaii........F6	3,694,820 ....C4	569,891 ....J3	1,517,550 ....M2
Anchorage, Alaska 260,283 .. D6	Fla....................L5	Dallas, Tex. 1,188,580 ..... G4	Helena (cap.), Mont. 25,780 . D1	Louisiana (state) 4,468,976.. H4	Nebraska (state) 1,711,263.. F2	Phoenix (cap.), Ariz.
Annapolis (cap.), Md.	Cape Fear (riv.), N.C........L4	Death Valley (depr.), Calif... C4	Honolulu (cap.), Hawaii	Madison (cap.), Wis.	Nevada (state) 1,998,257 ... C3	1,321,045 ....D4
35,838 ....L3	Carson City (cap.), Nev.	Delaware (bay)...............M3	365,048 ....F5	208,054 ....H2	Newark, N.J. 273,546........ L2	Pierre (cap.), S. Dak. 13,876.. F2
Appalachian (mts.)..........K3	52,457 ....C3	Delaware (state) 783,600 .... L3	Houston, Tex. 1,953,631 .... G5	Maine (state) 1,274,923 .... N1	New Hampshire (state)	Pikes (peak), Colo...........F3
Arizona (state) 5,130,632 .. D4	Cascade (range)..............B1	Denver (cap.), Colo.	Huron (lake).................K2	Maryland (state) 5,296,486... L3	1,235,786 ....M2	Pittsburgh, Pa. 334,563 .... L2
Arkansas (state) 2,673,400 .. H3	Champlain (lake).............M2	554,636 ....F3	Idaho (state) 1,293,953 ..... D2	Massachusetts (state)	New Jersey (state)	Platte (riv.), Nebr..........G2
Arkansas (riv.)..............H3	Charleston (cap.), W. Va.	Des Moines (cap.), Iowa	Illinois (state) 12,419,293 .. J3	6,349,097 ....M2	8,414,350 ....M3	Pontchartrain (lake), La.....J5
Atlanta (cap.), Ga. 416,474 .. K4	53,421 ....K3	198,682 ....H2	Indiana (state) 6,080,485 ... J3	Maui (isl.), Hawaii..........F5	New Mexico (state)	Portland, Oreg. 529,121 .... B1
Attu (isl.), Alaska..........D6	Charlotte, N.C. 540,828...... L3	Detroit, Mich. 951,270 ...... K2	Indianapolis (cap.), Ind.	Mauna Kea (mt.), Hawaii G6	1,819,046 ....E4	Potomac (riv.)...............L3
Augusta (cap.), Maine	Chattahoochee (riv.).........K4	District of Columbia 638,432 . L3	791,926 ....J3	Mauna Loa (mt.), Hawaii G6	New Orleans, La. 484,674 ... H5	Providence (cap.), R.I.
18,560 ....N2	Chesapeake (bay).............L3	Dover (cap.), Del. 32,135 .... L3	Iowa (state) 2,926,324 ..... H2	May (cape), N.J..............M3	New York (state) 18,976,457 . L2	173,618 ....M2
Aurora, Colo. 276,393 ...... F3	Cheyenne (cap.), Wyo.	El Paso, Tex. 563,662 ...... E4	Jackson (cap.), Miss.	McKinley (mt.), Alaska...... D5	New York, N.Y. 8,008,278 .. M2	Raleigh (cap.), N.C.
Austin (cap.), Tex.	53,011 ....E2	Erie (lake)..................K2	184,256 ....J4	Memphis, Tenn. 650,100 .... J3	Norfolk, Va. 234,403 .... L3	276,093 ....L3
656,562 ....G4	Chicago, Ill. 2,896,016 ..... J2	Everglades, The (swamp), Fla. K5	Jacksonville, Fla. 735,617... K4	Mexico (gulf)................J5	North Carolina (state)	Red (riv.)...................H4
Baltimore, Md. 651,154 .... L3	Cimarron (riv.)..............G3	Florida (state) 15,982,378 ... K5	Jefferson City (cap.), Mo.	Miami, Fla. 362,470 .... K5	8,049,313 ....L3	Red River of the North (riv.).. G1
Baton Rouge (cap.), La.	Cincinnati, Ohio 331,285 .... K3	Florida (keys)...............K6	39,636 ....H3	Michigan (state) 9,938,444 .. J1	North Dakota (state) 642,200 . F1	Rhode Island (state)
227,818 ....H4	Cleveland, Ohio 478,403 .... K2	Fort Worth, Tex. 534,694 ... G4	Juneau (cap.), Alaska 30,711 . E6	Michigan (lake)..............J2	Ohio (state) 11,353,140 .... K2	1,048,319 ....M2
Bering (sea), Alaska........C6	Coast (ranges)...............B2	Frankfort (cap.), Ky. 27,741 . K3	Kansas (state) 2,688,418 .... G3	Milwaukee, Wis. 596,974.... J2	Ohio (riv.)..................J3	Richmond (cap.), Va.
Bering (str.), Alaska........C5	Cod (cape), Mass.............N2	Fresno, Calif. 427,652 ...... C3	Kansas City, Kans. 146,866 .. G3	Minneapolis, Minn. 382,618.. H1	Oklahoma (state) 3,450,654 . G3	197,790 ....L3
Bighorn (riv.)...............E2	Colorado (riv.)..............D4	Georgia (state) 8,186,453 ... K4	Kansas City, Mo. 441,545.... H3	Minnesota (state) 4,919,479 . H1	Oklahoma City (cap.), Okla.	Rio Grande (riv.)............F5
Bismarck (cap.), N. Dak.	Colorado (state) 4,301,261 .. E3	Gila (riv.)..................D4	Kentucky (state) 4,041,769 .. J3	Mississippi (state) 2,844,658. J4	506,132 ....G3	Rochester, N.Y. 219,773 .... L3
55,532 ....G1	Colorado (riv.), Texas.......G4	Glacier Nat'l Park, Mont......D1	Lansing (cap.), Mich.	Mississippi (riv.)...........H4	Olympia (cap.), Wash.	Rocky (mts.).................E3
Black Hills (mts.)...........F2	Columbia (cap.), S.C.	Golden Gate (chan.), Calif... B3	119,128 ....K2	Missouri (state) 5,595,211 .. H3	42,514 ....B1	Sacramento (cap.), Calif.
Boise (cap.), Idaho 185,787... C2	116,278 ....K4	Grand Canyon Nat'l Park, Ariz. D3	Las Vegas, Nev. 478,434 .... C3	Missouri (riv.)..............H3		407,018 ....B3
Borah (peak), Idaho D2	Columbia (riv.)..............B1	Great Salt (lake), Utah D2	Lexington, Ky. 260,512...... K3			

AREA 3,623,420 sq. mi.
(9,384,658 sq. km.)
POPULATION 281,421,906
CAPITAL Washington
LARGEST CITY New York
HIGHEST POINT Mt. McKinley 20,320 ft.
(6,194 m.)
MONETARY UNIT U.S. dollar
MAJOR LANGUAGE English
MAJOR RELIGIONS Protestantism,
Roman Catholicism, Judaism

Population Distribution

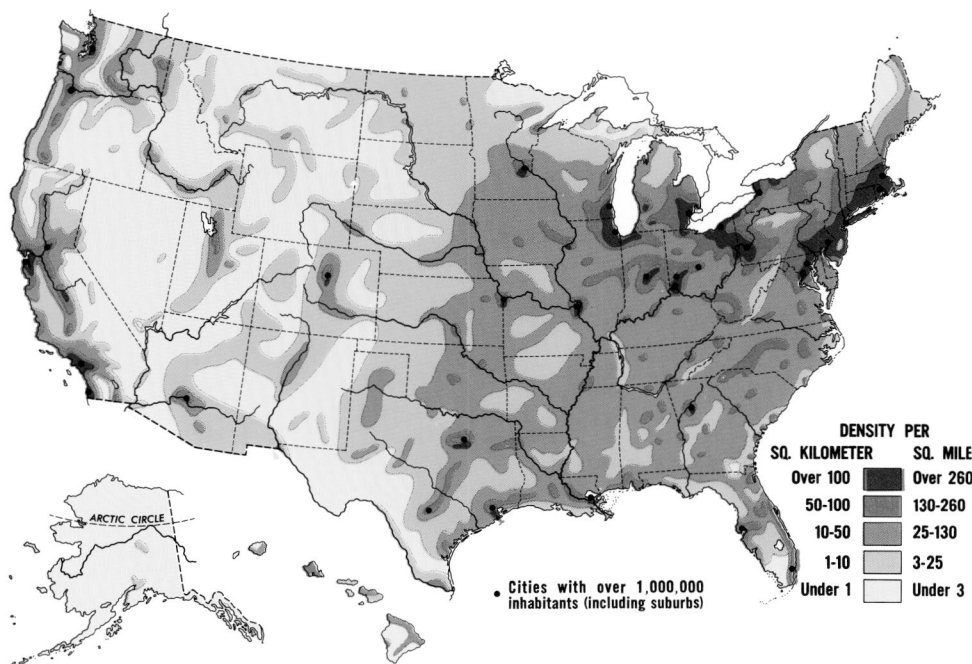

DENSITY PER

SQ. KILOMETER	SQ. MILE
Over 100	Over 260
50-100	130-260
10-50	25-130
1-10	3-25
Under 1	Under 3

• Cities with over 1,000,000 inhabitants (including suburbs)

Vegetation

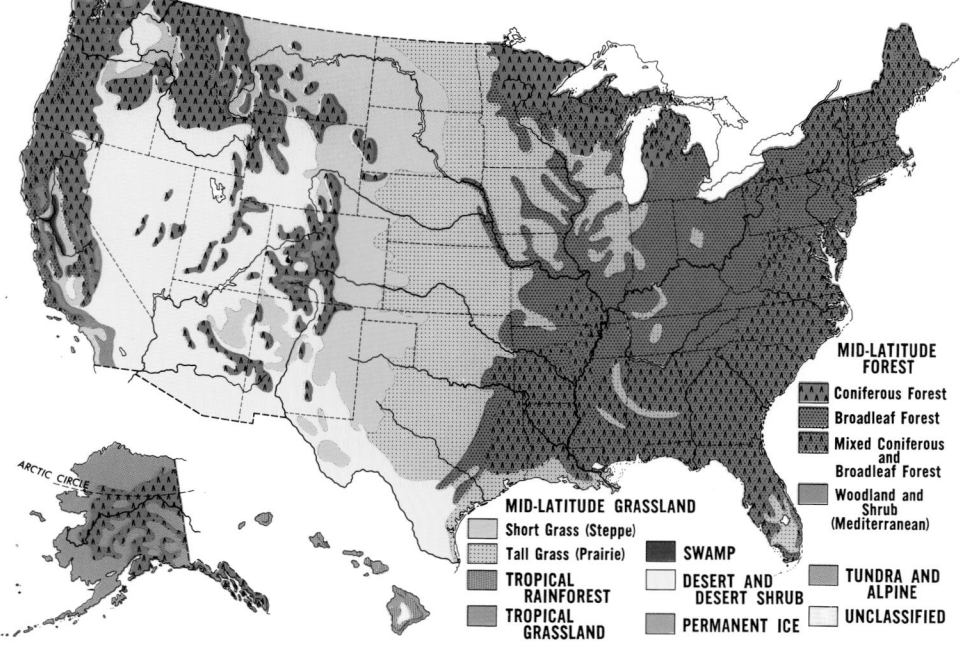

MID-LATITUDE FOREST
Coniferous Forest
Broadleaf Forest
Mixed Coniferous and Broadleaf Forest
Woodland and Shrub (Mediterranean)

MID-LATITUDE GRASSLAND
Short Grass (Steppe)
Tall Grass (Prairie)

TROPICAL RAINFOREST
TROPICAL GRASSLAND

SWAMP
DESERT AND DESERT SHRUB
PERMANENT ICE

TUNDRA AND ALPINE
UNCLASSIFIED

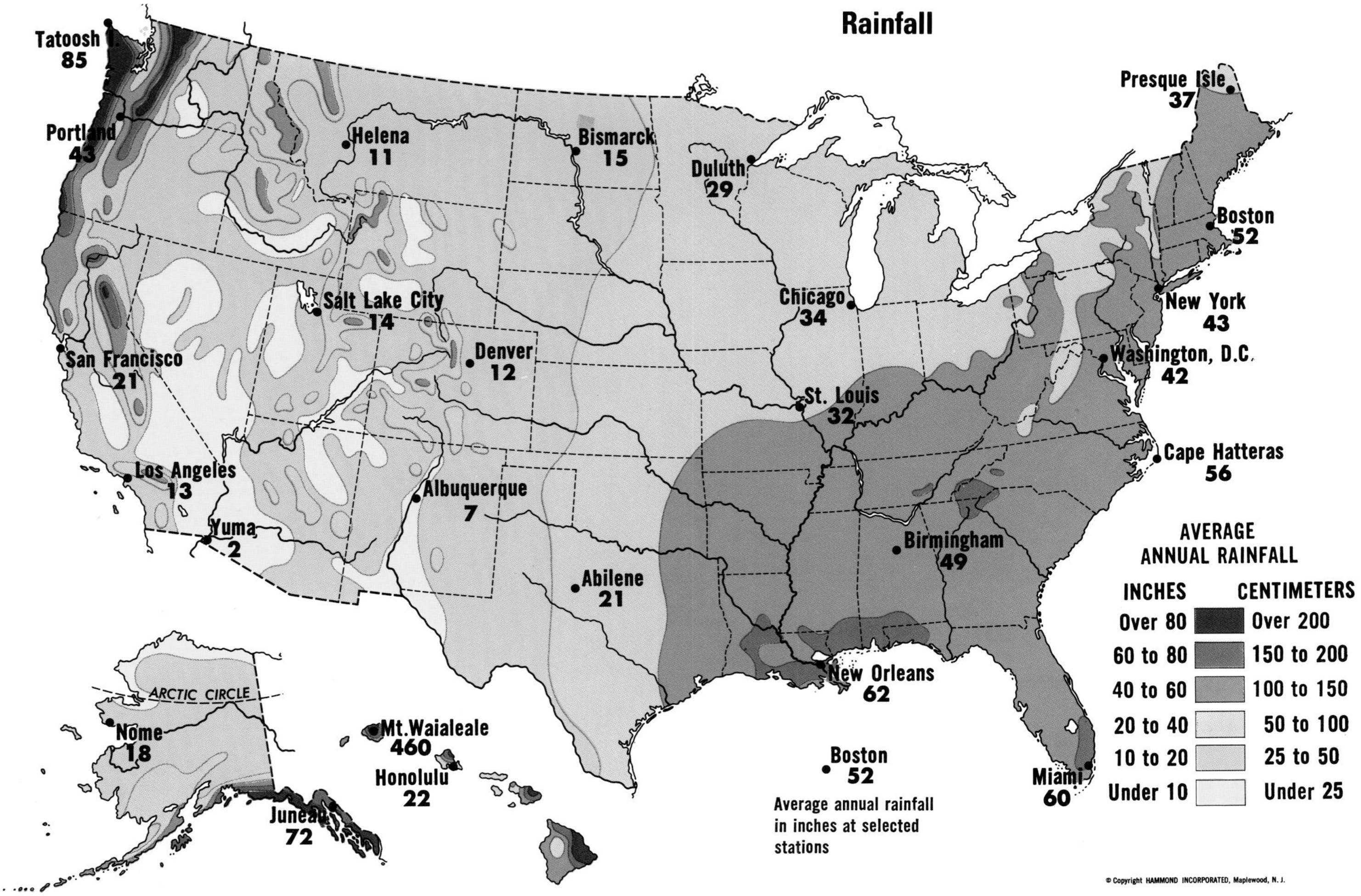

Rainfall

Tatoosh I. 85

Portland 43

Helena 11

Bismarck 15

Duluth 29

Presque Isle 37

Boston 52

New York 43

Chicago 34

Washington, D.C. 42

Salt Lake City 14

San Francisco 21

Denver 12

St. Louis 32

Cape Hatteras 56

Los Angeles 13

Albuquerque 7

Birmingham 49

Yuma 2

Abilene 21

New Orleans 62

ARCTIC CIRCLE

Nome 18

Mt.Waialeale 460

Honolulu 22

Boston 52

Miami 60

Juneau 72

Average annual rainfall in inches at selected stations

AVERAGE ANNUAL RAINFALL

INCHES	CENTIMETERS
Over 80	Over 200
60 to 80	150 to 200
40 to 60	100 to 150
20 to 40	50 to 100
10 to 20	25 to 50
Under 10	Under 25

© Copyright HAMMOND INCORPORATED, Maplewood, N. J.

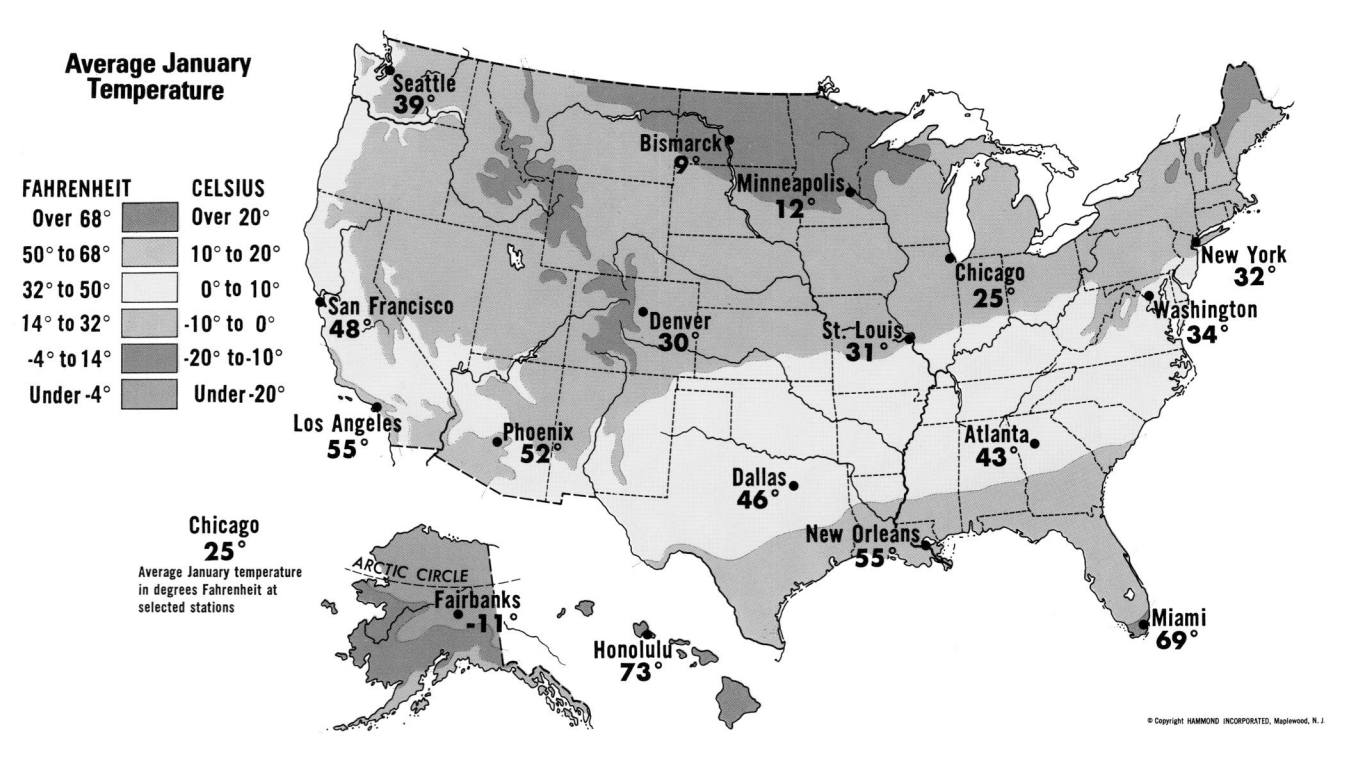

Average January Temperature

FAHRENHEIT	CELSIUS
Over 68°	Over 20°
50° to 68°	10° to 20°
32° to 50°	0° to 10°
14° to 32°	-10° to 0°
-4° to 14°	-20° to -10°
Under -4°	Under -20°

Seattle 39°

Bismarck 9°

Minneapolis 12°

New York 32°

Chicago 25°

Washington 34°

San Francisco 48°

Denver 30°

St. Louis 31°

Los Angeles 55°

Phoenix 52°

Dallas 46°

Atlanta 43°

New Orleans 55°

Chicago 25°

Average January temperature in degrees Fahrenheit at selected stations

ARCTIC CIRCLE

Fairbanks -11°

Honolulu 73°

Miami 69°

© Copyright HAMMOND INCORPORATED, Maplewood, N. J.

Topography

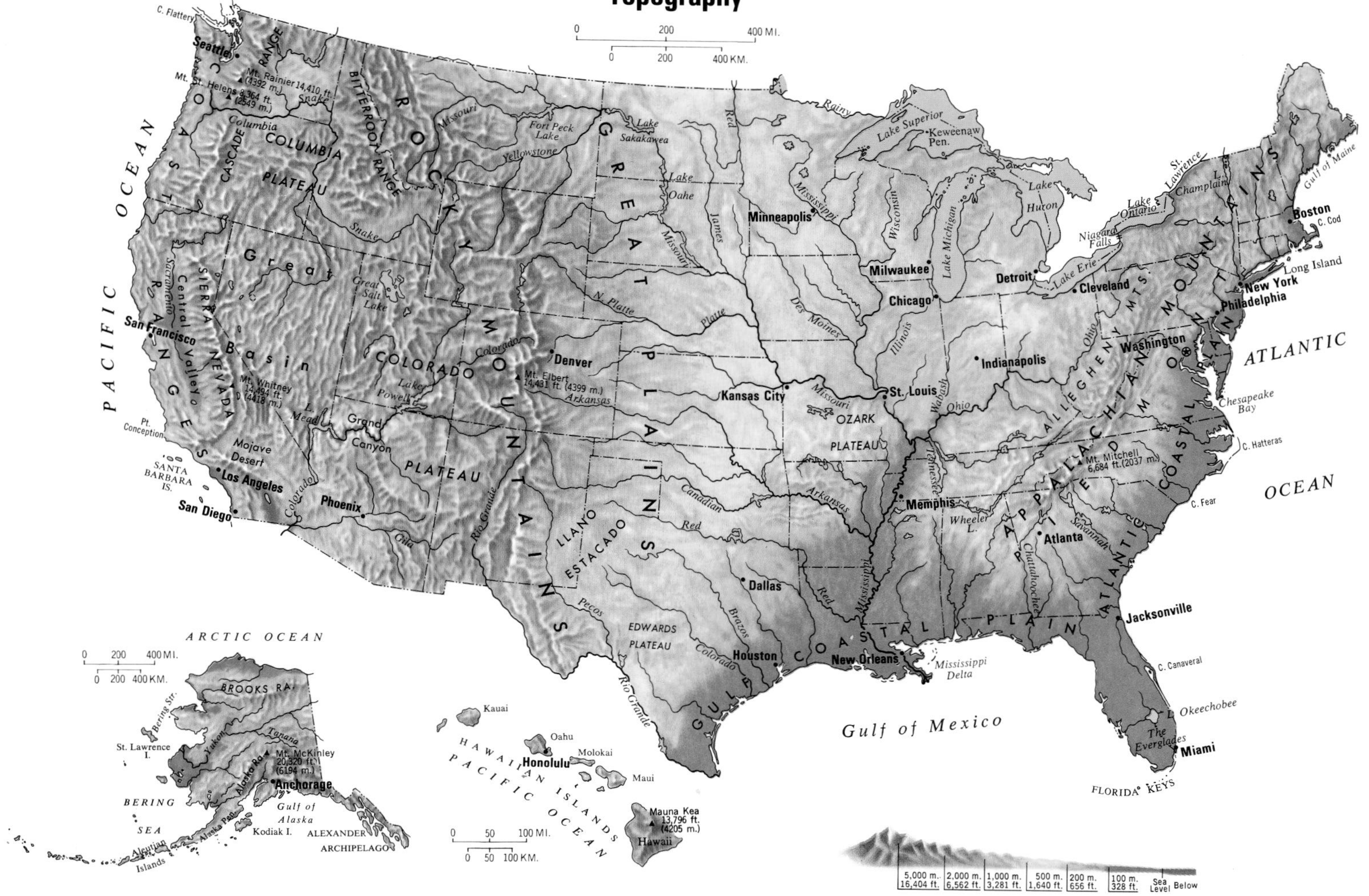

0 200 400 MI.
0 200 400 KM.
C. Flattery
Seattle
COAST RANGE
Mt. Rainier 14,410 ft. (4392 m.)
Mt. St. Helens 8,364 ft. (2549 m.)
Snake
BITTERROOT RANGE
ROCKY MOUNTAINS
Missouri
Fort Peck Lake
Yellowstone
Lake Sakakawea
GREAT PLAINS
Red
Rainy
Lake Superior
Keweenaw Pen.
Columbia
COLUMBIA PLATEAU
CASCADE RANGE
PACIFIC OCEAN
Lake Oahe
James
Minneapolis
Mississippi
Wisconsin
Lake Michigan
Lake Huron
Lake Ontario
St. Lawrence
L. Champlain
Gulf of Maine
Boston
C. Cod
Niagara Falls
Snake
Great Basin
Great Salt Lake
Sacramento
Central Valley
SIERRA NEVADA
COAST RANGES
San Francisco
Milwaukee
Detroit
Lake Erie
Cleveland
Long Island
New York
Philadelphia
Chicago
N. Platte
Platte
Des Moines
Illinois
Colorado
COLORADO PLATEAU
Denver
Mt. Elbert 14,431 ft. (4399 m.)
Arkansas
Mt. Whitney 14,494 ft. (4418 m.)
Lake Powell
L. Mead
Grand Canyon
Kansas City
Missouri
St. Louis
OZARK PLATEAU
Indianapolis
Wabash
Ohio
ALLEGHENY MTS.
APPALACHIAN MOUNTAINS
Washington
ATLANTIC
Chesapeake Bay
Pt. Conception
Mojave Desert
SANTA BARBARA IS.
Los Angeles
San Diego
Phoenix
Gila
Rio Grande
LLANO ESTACADO
Canadian
Red
Arkansas
Memphis
Tennessee
Mt. Mitchell 6,684 ft. (2037 m.)
C. Hatteras
OCEAN
C. Fear
PIEDMONT
Wheeler L.
Atlanta
Savannah
Chattahoochee
ATLANTIC COASTAL PLAIN
Dallas
Brazos
Red
Pecos
EDWARDS PLATEAU
Colorado
Houston
GULF COASTAL PLAIN
New Orleans
Mississippi Delta
Jacksonville
C. Canaveral
L. Okeechobee
The Everglades
Miami
FLORIDA KEYS
Gulf of Mexico
ARCTIC OCEAN
0 200 400 MI.
0 200 400 KM.
BROOKS RA.
Bering Str.
Yukon
Tanana
St. Lawrence I.
Mt. McKinley 20,320 ft. (6194 m.)
Alaska Ra.
Anchorage
Gulf of Alaska
BERING SEA
Alaska Pen.
Kodiak I.
ALEXANDER ARCHIPELAGO
Aleutian Islands
Kauai
Oahu
Honolulu
Molokai
Maui
HAWAIIAN ISLANDS
PACIFIC OCEAN
Mauna Kea 13,796 ft. (4205 m.)
Hawaii
0 50 100 MI.
0 50 100 KM.
5,000 m. 16,404 ft.
2,000 m. 6,562 ft.
1,000 m. 3,281 ft.
500 m. 1,640 ft.
200 m. 656 ft.
100 m. 328 ft.
Sea Level
Below

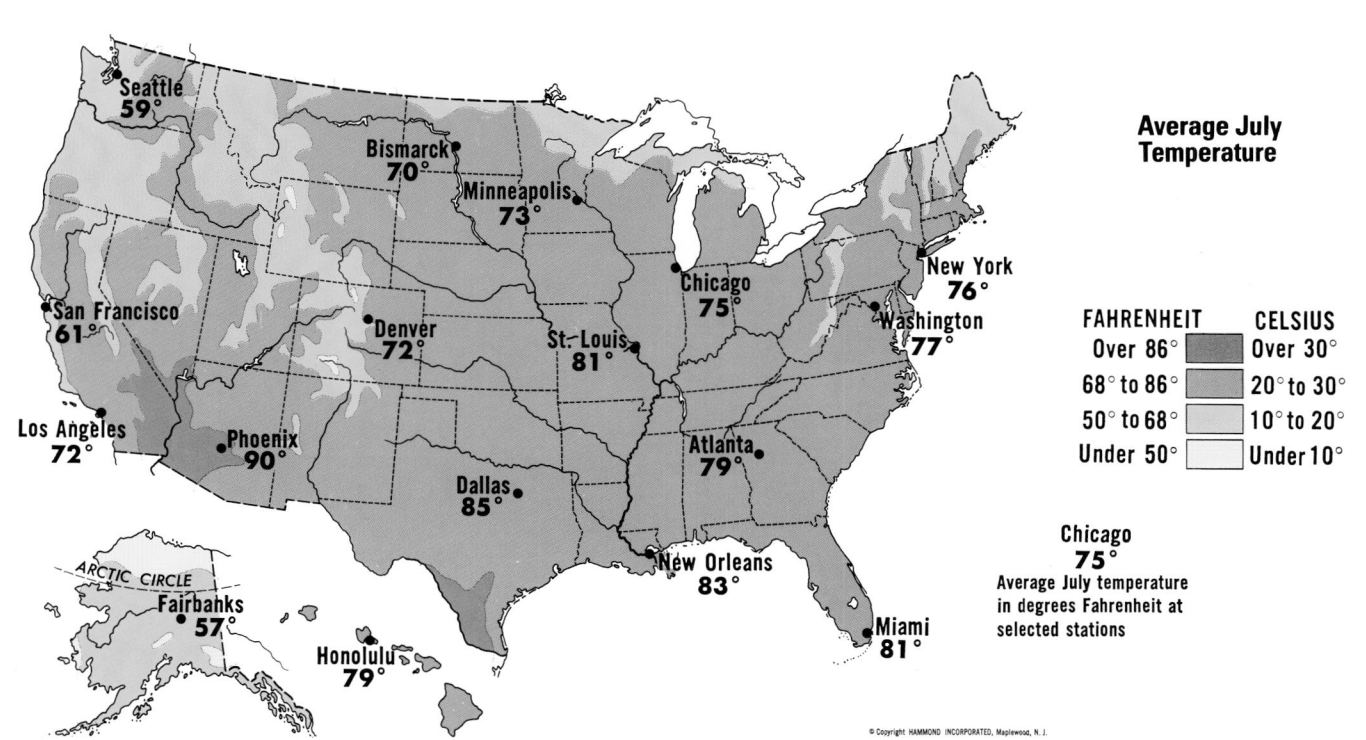

Average July Temperature

Seattle 59°
Bismarck 70°
Minneapolis 73°
Chicago 75°
New York 76°
San Francisco 61°
Denver 72°
St. Louis 81°
Washington 77°
Los Angeles 72°
Phoenix 90°
Atlanta 79°
Dallas 85°
New Orleans 83°
Miami 81°
ARCTIC CIRCLE
Fairbanks 57°
Honolulu 79°
FAHRENHEIT CELSIUS
Over 86° Over 30°
68° to 86° 20° to 30°
50° to 68° 10° to 20°
Under 50° Under 10°
Chicago 75°
Average July temperature in degrees Fahrenheit at selected stations

United States Standard Time Zones

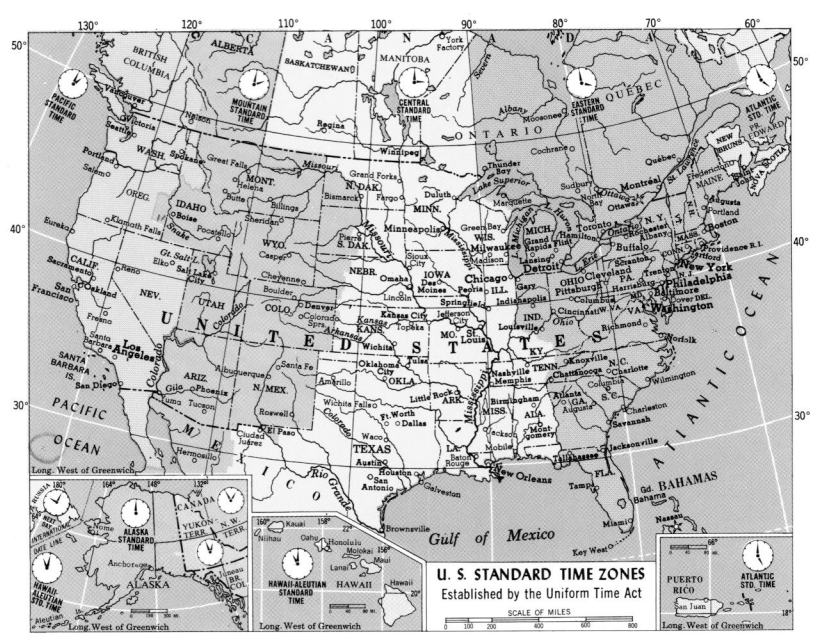

Agriculture, Industry and Resources

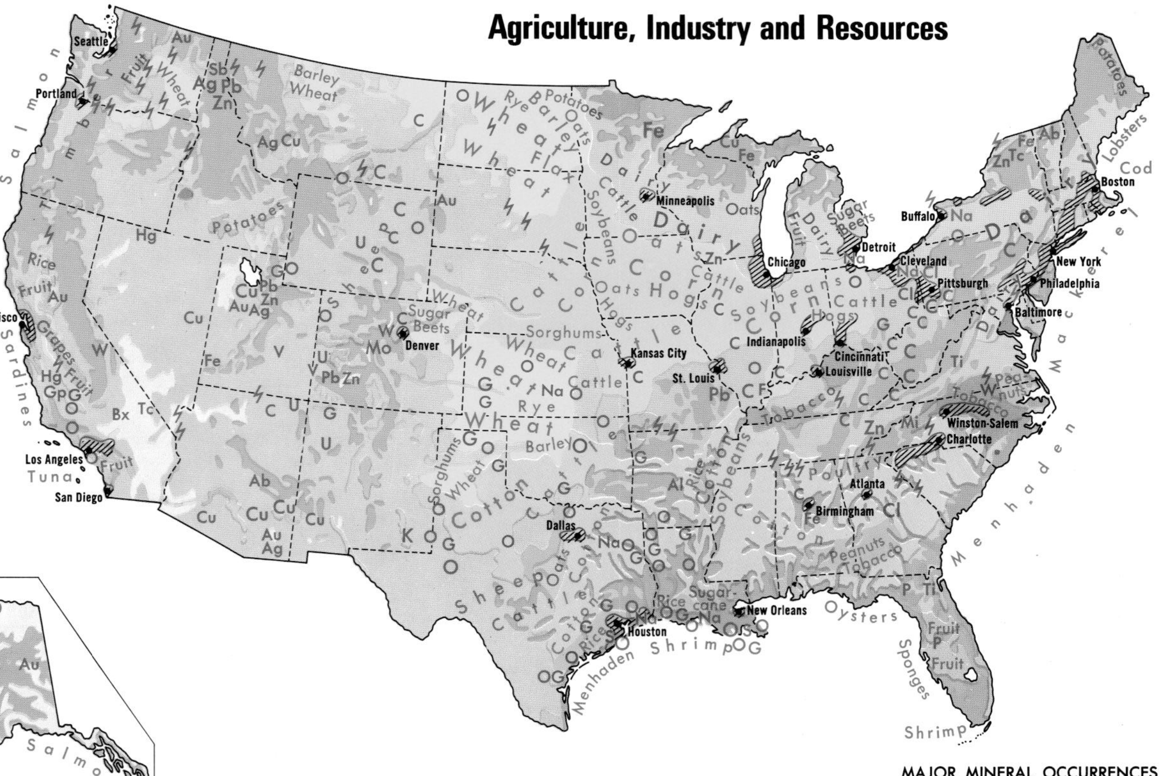

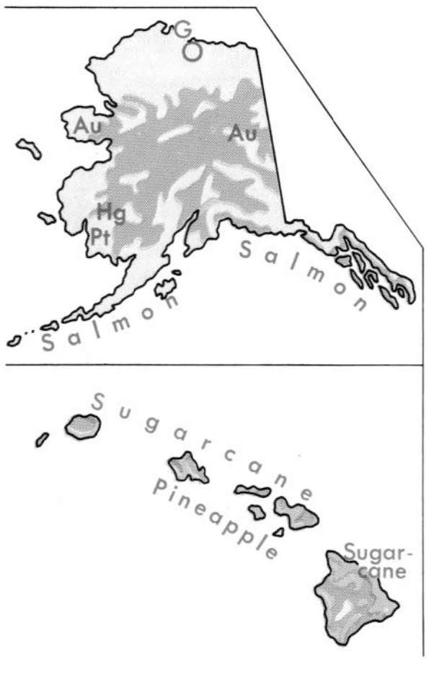

MAJOR MINERAL OCCURRENCES

Ab	Asbestos	Gp	Gypsum	Sb	Antimony
Ag	Silver	Hg	Mercury	Tc	Talc
Al	Bauxite	K	Potash	Ti	Titanium
Au	Gold	Mi	Mica	U	Uranium
Bx	Borax	Mo	Molybdenum	V	Vanadium
C	Coal	Na	Salt	W	Tungsten
Cl	Clay	O	Petroleum	Zn	Zinc
Cu	Copper	P	Phosphates		
F	Fluorspar	Pb	Lead	⚡	Water Power
Fe	Iron Ore	Pt	Platinum	▨	Major Industrial
G	Natural Gas	S	Sulfur		Areas

DOMINANT LAND USE

- Wheat and Small Grains
- Feed Grains and Livestock
- Dairy
- General Farming
- Cotton
- Fruit, Truck and Mixed Farming
- Tobacco and General Farming
- Special Crops and General Farming
- Range Livestock
- Forests
- Swampland
- Nonagricultural Land

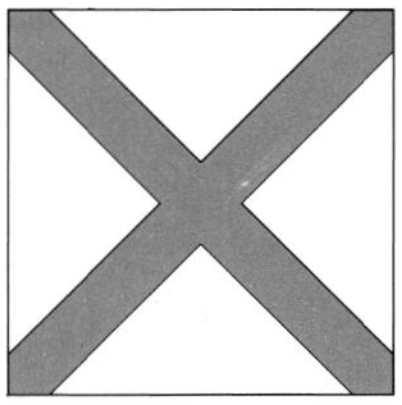

AREA 51,705 sq. mi. (133,916 sq. km.)
POPULATION 4,447,100
CAPITAL Montgomery
LARGEST CITY Birmingham
HIGHEST POINT Cheaha Mtn. 2,405 ft. (733 m.)
SETTLED IN 1702
ADMITTED TO UNION December 14, 1819
POPULAR NAME Heart of Dixie; Cotton State
STATE FLOWER Camellia
STATE BIRD Yellowhammer

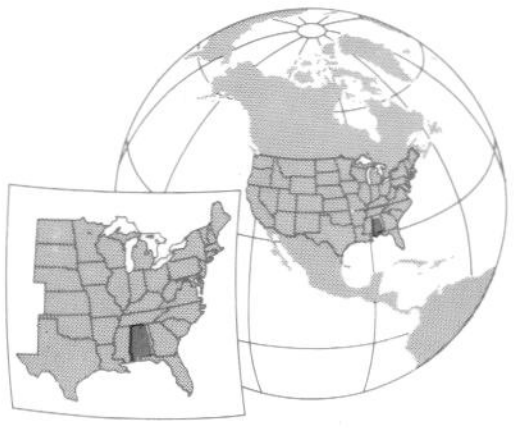

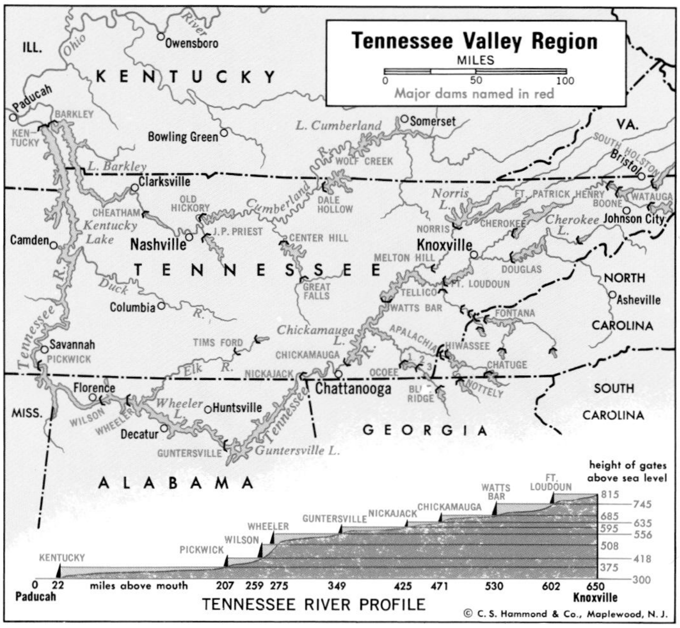

Tennessee Valley Region
MILES
0 50 100
Major dams named in red
ILL.
Owensboro
KENTUCKY
Paducah
BARKLEY
KENTUCKY
Bowling Green
L. Cumberland
Somerset
WOLF CREEK
VA.
Bristol
SOUTH HOLSTON
L. Barkley
Clarksville
OLD HICKORY
DALE HOLLOW
Norris L.
FT. PATRICK HENRY
BOONE
WATAUGA
CHEATHAM
Kentucky Lake
Camden
Nashville
J. P. PRIEST
CENTER HILL
NORRIS
CHEROKEE
Cherokee L.
Johnson City
Knoxville
MELTON HILL
TENNESSEE
DOUGLAS
NORTH
Asheville
GREAT FALLS
TELLICO
FT. LOUDOUN
Columbia
Duck R.
WATTS BAR
FONTANA
CAROLINA
Chickamauga L.
Savannah
PICKWICK
TIMS FORD
CHICKAMAUGA
APALACHIA
HIWASSEE
CHATUGE
NICKAJACK
Chattanooga
OCOEE
NOTTELY
BLUE RIDGE
SOUTH CAROLINA
Elk R.
Florence
Wheeler L.
Huntsville
MISS.
WILSON
WHEELER
Decatur
GEORGIA
GUNTERSVILLE
Guntersville L.
ALABAMA
height of gates above sea level
KENTUCKY
PICKWICK
WILSON
WHEELER
GUNTERSVILLE
NICKAJACK
CHICKAMAUGA
WATTS BAR
FT. LOUDOUN
815
745
685
635
595
556
508
418
375
300
0 22 miles above mouth 207 259 275 349 425 471 530 602 650
Paducah
Knoxville
TENNESSEE RIVER PROFILE
© C. S. Hammond & Co., Maplewood, N. J.

COUNTIES

Autauga 43,671 E5
Baldwin 140,415 C9
Barbour 29,038 H7
Bibb 20,826 D5
Blount 51,024 E2
Bullock 11,714 G6
Butler 21,399 E7
Calhoun 112,249 G3
Chambers 36,583 H5
Cherokee 23,988 G2
Chilton 39,593 E5
Choctaw 15,922 B6
Clarke 27,867 C7
Clay 14,254 G4
Cleburne 14,123 G3
Coffee 43,615 G8
Colbert 54,984 C1
Conecuh 14,089 E8
Coosa 12,202 F5
Covington 37,631 F8
Crenshaw 13,665 F7
Cullman 77,483 E2
Dale 49,129 G8
Dallas 46,365 D6
De Kalb 64,452 G2
Elmore 65,874 F5
Escambia 38,440 D8
Etowah 103,459 F2
Fayette 18,495 C3
Franklin 31,223 C2
Geneva 25,764 G8
Greene 9,974 C5
Hale 17,185 C5
Henry 16,310 H7
Houston 88,787 H8
Jackson 53,926 F1
Jefferson 662,047 E3
Lamar 15,904 B3
Lauderdale 87,966 C1
Lawrence 34,803 D1
Lee 115,092 H5
Limestone 65,676 E1
Lowndes 13,473 E6
Macon 24,105 G6
Madison 276,700 E1
Marengo 22,539 C6
Marion 31,214 C2
Marshall 82,231 F2
Mobile 399,843 B9
Monroe 24,324 D7
Montgomery 223,510 F6
Morgan 111,064 E2
Perry 11,861 D5
Pickens 20,949 B4
Pike 29,605 G7
Randolph 22,380 H4
Russell 49,756 H6
Saint Clair 64,742 F3
Shelby 143,293 E4
Sumter 14,798 B5
Talladega 80,321 F4
Tallapoosa 41,475 G5
Tuscaloosa 164,875 C4
Walker 70,713 D3
Washington 18,097 B8
Wilcox 13,183 D7
Winston 24,843 D2

CITIES and TOWNS

Abbeville▲ 2,987 H7
Abernant 405 D4
Ackerville 200 D6
Adamsville 4,965 D3
Addison 723 D2
Adger 400 D4
Akron 521 C5
Alabaster 22,619 E4
Alberta 100 D6
Albertville 17,247 F2
Aldrich 500 E4
Alexander City 15,008 G5
Alexandria 3,692 G3
Aliceville 2,567 B4
Allen 149 C7
Allgood 629 F3
Allsboro 300 B1
Alma 500 C8
Alpine 150 F4
Alton 150 E3
Altoona 984 F2
Andalusia▲ 8,794 E8
Anderson 354 D1
Annemanie 100 D6
Anniston▲ 24,276 G3
Arab 7,174 E2
Ardmore 1,034 E1
Argo 1,780 E3
Ariton 772 G7
Arkadelphia 150 E3
Arley 290 D2
Arlington 200 C6
Ashby 500 E4
Ashford 1,853 H8
Ashland▲ 1,965 G4
Ashville▲ 2,260 F3
Athens▲ 18,967 E1
Atmore 7,676 C8
Attalla 6,592 F2
Auburn 42,987 H5
Autaugaville 820 E6
Avon 466 H8
Axis 500 B9
Babbie 627 F8
Baileyton 684 E2
Baker Hill 300 H7
Banks 224 G7
Bankston 125 C3
Barlow Bend 300 C8
Barnwell 700 C10
Barton 150 C1
Bashi 225 C7
Batesville 100 H6
Battles Wharf 300 C10
Bay Minette▲ 224 C9
Bayou La Batre 2,313 B10
Bear Creek 1,053 C2
Beatrice 412 D7
Beaverton 226 B3
Belgreen 500 C2
Belk 214 C3
Bellamy 700 B6
Belle Mina 675 E1
Bellview 200 D7
Bellwood 400 G8
Beloit 100 D6
Bermuda 120 D8
Berry 1,238 C3
Bessemer 29,672 D4
Beulah 500 H5
Billingsley 116 E5
Birmingham▲ 242,820 D3
Black 202 G8
Blacksher 200 C8
Bladon Springs 125 B7
Blanton 100 H5
Bleecker 250 H5
Blountsville 1,768 E2
Blue Mountain 233 G3
Blue Springs 121 G7
Boaz 7,411 F2
Boligee 369 C5
Bolinger 175 B7
Bolling 100 E7
Bon Air 96 F4
Bon Secour 850 C10
Booth 200 E6
Boyd 100 B5
Braggs 180 E6
Branchville 825 F3
Brantley 920 F7
Bremen 125 E3
Brent 4,024 D5
Brewton▲ 5,498 D8
Bridgeport 2,728 G1
Brierfield 250 E4
Brighton 3,640 D4
Brilliant 762 C2
Brooklyn 300 E8
Brookside 1,393 E3
Brooksville 120 F2
Brookwood 1,483 D4
Browns 375 D6
Brownsboro 150 F1
Brownville 2,386 C4
Brundidge 2,341 G7
Bryant 300 G1
Bucks 201 B8
Buhl 100 C4
Burkville 250 E6
Burnsville 100 E6
Butler▲ 1,952 B6
Cahaba D6
Calcis 200 F4
Calera 3,158 E4
Calhoun 950 F6
Calvert 600 B8
Camden▲ 2,257 D7
Campbell 200 C7
Camp Hill 1,273 G5
Canoe 560 D8
Canton Bend 300 D6
Carbon Hill 2,071 D3
Cardiff 82 E3
Carlowville 100 D6
Carlton 275 C8
Carolina 248 E8
Carrollton▲ 987 B4
Carson 400 C8
Castleberry 590 D8
Catherine 250 D6
Cedar Bluff 1,467 G2
Cedar Cove 100 D4
Central 300 F5
Centre▲ 3,216 G2
Centreville▲ 2,466 D5
Chancellor 200 G8
Chandler Springs 100 F4
Chapman 300 E7
Chase 175 E1
Chastang 200 B8
Chatom▲ 1,193 B8
Chelsea 2,949 E4
Cherokee 1,237 C1
Chestnut 125 D7
Chickasaw 6,364 B9
Childersburg 4,927 F4
Choccolocco 500 G3
Choctaw 600 B6
Chrysler 400 C8
Chunchula 700 B9
Citronelle 3,659 B8
Claiborne 125 D7
Clanton▲ 7,800 E5
Clayhatchee 501 G8
Clayton▲ 1,475 G7
Cleveland 1,241 E3
Clinton 150 C5
Clio 2,206 G7
Cloverdale 100 C1
Coaling 1,115 D4
Coden 600 B10
Coffee Springs 251 G8
Coffeeville 360 B7
Coker 808 C4
Collinsville 1,644 G2
Collirene 100 E6
Columbia 804 H8
Columbiana▲ 3,316 E4
Cooper 250 E5
Coosada 1,382 F5
Copeland 160 B7
Cordova 2,423 D3
Corona 300 C3
Cottondale 500 D4
Cottonton 324 H6
Cottonwood 1,170 H8
County Line 257 E3
Courtland 769 D1
Cowarts 1,546 H8
Coy 950 D7
Crane Hill 355 D2
Creola 2,002 B9
Cromwell 650 B6
Crossville 1,431 G2
Cuba 363 B6
Cullman▲ 13,995 E2
Cullomburg 325 B7
Cusseta 650 H5
Cypress 300 C5
Dadeville▲ 3,212 G5
Daleville 4,653 G8
Dancy 116 B4
Danville 300 D2
Daphne 16,581 C9
Darlington 150 D7
Dauphin Island 1,371 B10
Daviston 267 G4
Dayton 60 C6
De Armanville 350 G3
Deatsville 340 F5
Decatur▲ 53,929 D1
Deer Park 300 B8
Delmar 200 C2
Delta 100 G4
Demopolis 7,540 C6
Detroit 247 B2
Dickinson 250 C7
Dixons Mills 100 C6
Dixonville 125 E8
Dora 2,413 D3
Dothan▲ 57,737 H8
Double Springs▲ 1,003 D2
Douglas 530 F2
Dozier 391 F7
Duke 250 G3
Duncanville 150 D4
Dutton 310 G1
Dyas 250 C9
Eastaboga 300 F3
East Brewton 2,496 E8
Echo 200 G8
Echola 300 C4
Eclectic 1,037 F5
Edwardsville 186 H3
Edwin 296 H7
Elamville 180 G7
Elba▲ 4,185 F8
Elberta 552 C10
Eldridge 184 C3
Eliska 200 C8
Elkmont 470 E1
Elmore 470 F5
Elon 125 F1
Elrod 746 C4
Emelle 31 B5
Empire 600 D3
Enterprise 21,178 G8
Epes 206 B5
Equality 125 F5
Estillfork 200 F1
Eufaula 13,908 H7
Eunola 182 G8
Eutaw▲ 1,878 C5
Eva 491 E2
Evergreen▲ 3,630 E8
Excel 582 D8
Fabius 150 G1
Fackler 250 G1
Fairfield 12,381 E4
Fairford 200 B8
Fairhope 12,480 C10
Fairview 522 E2
Falkville 1,202 E2
Farmersville 200 E6
Faunsdale 87 C6
Fayette▲ 4,922 C3
Fayetteville 200 F4
Finchburg 150 D7
Fitzpatrick 108 G6
Five Points 146 H4
Flat Rock 750 G1
Flatwood 300 C6
Flint City 1,033 D1
Flomaton 1,588 D8
Florala 1,964 F8
Florence▲ 36,264 C1
Foley 7,590 C10
Forestdale 10,509 E3
Forkland 629 C5
Forney 100 H2
Fort Davis 500 G6
Fort Deposit 1,270 E7
Fort Mitchell 900 H6
Fort Payne▲ 12,938 G2
Fosters 400 C4
Fostoria 200 E6
Frankfort 125 C1
Franklin 149 G6
Frankville 200 B7
Fredonia 300 H5
Freemanville 200 D8
Frisco City 1,460 D8
Fruitdale 500 B8
Fruithurst 270 G3
Fulton 308 C7
Fultondale 6,595 E3
Furman 200 E6
Fyffe 971 G2
Gadsden▲ 38,978 G2
Gainestown 300 C8
Gainesville 220 B5
Gallant 475 F2
Gallion 239 C6
Gantt 241 E8
Garden City 564 E2
Gardendale 11,626 E3
Garland 155 E7
Gasque 100 C10
Gateswood 200 C9
Gaylesville 140 G2
Geiger 161 B5
Geneva▲ 4,388 G8
Georgiana 1,737 E7
Geraldine 786 G2
Gilbertown 187 B7
Glen Allen 442 C3
Glencoe 5,152 G3
Glenwood 191 F7
Good Hope 1,966 E2
Goodsprings 360 D3
Goodwater 1,633 F4
Goodway 200 D8
Gordo 1,677 C4
Gordon 408 H8
Gordonsville 318 E6
Gorgas 500 D3
Goshen 300 F7
Gosport 500 C7
Graham 100 H4
Grand Bay 3,918 B10
Grant 665 F1
Graysville 2,344 D3
Greenbrier 100 E1
Green Pond 750 D4
Greensboro▲ 2,731 C5
Greenville▲ 7,228 E7
Grimes 459 H8
Grove Hill▲ 1,438 C7
Guin 2,389 C3
Gulf Crest 200 B8
Gulf Shores 5,044 C10
Guntersville▲ 7,395 F2
Gurley 876 F1
Gu-Win 204 C3
Hackleburg 1,527 C2
Haleburg 108 H8
Haleyville 4,182 C2
Halsell 250 B6
Hamilton▲ 6,786 C2
Hammondville 486 G1
Hanceville 2,951 E2
Hardaway 600 G6
Harpersville 1,620 F4
Hartford 2,369 G8
Hartselle 12,019 E2
Harvest 3,054 E1
Hatchechubbee 840 H6
Hatton 950 D1
Havana 150 C5
Hayden 470 E3
Hayneville▲ 1,177 E6
Hazel Green 3,805 E1
Headland 3,523 H8
Healing Springs 100 B7
Heath 249 F8
Heflin▲ 3,002 G3
Heiberger 310 D5
Helena 10,296 E4
Henagar 2,400 G1
Higdon 925 G1
Highland Home 150 F7
Highland Lake 408 F3
Hillsboro 608 D1
Hissop 250 F5
Hobbs Island 100 F1
Hobson City 878 G3
Hodges 261 C2
Hokes Bluff 4,149 G3
Hollins 500 F4
Holly Pond 645 E2
Hollywood 950 G1
Holt 4,103 C4
Holy Trinity 400 H6
Homewood 25,043 E4
Honoraville 200 F7
Hoover 62,742 E4
Hope Hull 975 F6
Horton 100 F2
Houston 100 D2
Hueytown 15,364 D4
Hulaco 225 E2
Huntsville▲ 158,216 E1
Hurricane 300 C9
Hurtsboro 592 H6
Huxford 141 D8
Hybart 200 D7
Hytop 315 F1
Ider 664 G1
Irondale 9,813 E3
Irvington 150 B9
Isbell 250 C2
Isney 145 B7
Jachin 150 B6
Jack F7
Jackson 5,419 C8
Jacksons Gap 761 G5
Jacksonville 8,404 G3
Jamestown 147 G2
Jasper▲ 14,052 D3
Jeff 150 E1
Jefferson 300 C6
Jemison 2,248 E5
Jenifer 300 G3
Jones 135 E5
Joppa 200 E2
Josephine 200 C10
Kansas 260 C3
Keener 125 G2
Kellerman 100 D4
Kellyton 375 F5
Kennedy 541 B3
Kent 180 G5
Key 400 G2
Killen 1,119 D1
Kimberly 1,801 E3
Kimbrough 150 C6
Kinsey 1,796 H8
Kinston 602 F8
Knoxville 200 C4
Laceys Spring 400 E1
Lafayette▲ 3,234 H5
Lakeview 163 G2
Lamison 200 C6
Landersville 150 D2
Lanett 7,897 H5
Langston 254 G1
Lapine 300 F7
Larkinsville 425 F1
Latham 133 C8
Lavaca 500 B6
Lawley 125 E5
Leeds 10,455 E3
Leesburg 799 G2
Leighton 849 D1
Lenox 150 D8
Leroy 699 B8
Lester 107 D1
Letohatchee 250 E6
Level Plains 1,544 G8
Lexington 840 D1
Libertyville 106 F8
Lillian 350 D10
Lincoln 4,577 F3
Linden▲ 2,424 C6
Lineville 2,401 G4
Lipscomb 2,458 E4
Lisman 653 B6
Little River 400 C8
Littleville 978 C1
Livingston▲ 3,297 B5
Loachapoka 165 G5
Lockhart 548 F8
Locust Fork 1,016 E3
Logan 300 E2
Lomax 300 E5
Longview 475 E4
Lottie 150 C8
Louisville 612 G7
Lower Peach Tree 926 C7
Lowery 100 F8
Lowndesboro 140 E6
Loxley 1,348 C9
Luverne▲ 2,635 F7
Lynn 597 C2
Madison 29,329 E1
Madrid 303 H8
Magnolia 100 C6
Magnolia Springs 800 C10
Malcolm 300 B8
Malvern 1,215 G8
Manchester 400 D3
Manila 100 C7
Maplesville 672 E5
Marbury 300 E5
Margaret 1,169 F3
Margerum 250 B1
Marion▲ 3,511 D5
Marion Junction 400 D6
Marvel 100 D4
Marvyn 300 H6
Maud 150 B1
Maylene 500 E4
McCalla 657 E4
McCullough 500 D8
McIntosh 244 B8
McKenzie 644 E7
McKinley 100 C6
McShan 150 B4
McWilliams 305 D7
Megargel 240 D8
Melvin 300 B7
Mentone 451 G1
Meridianville 4,117 F1
Mexia 200 D8
Midfield 5,626 E4
Midland City 1,703 H8
Midway 457 H6
Miflin 150 C10
Mignon 1,348 F4
Millbrook 10,386 F6
Millers Ferry 300 D6
Millerville 345 G4
Millport 1,160 B3
Millry 615 B7
Milltown 125 H4
Milstead 150 G6
Minter 450 D6
Mobile▲ 198,915 B9
Monroeville▲ 6,862 D7
Monrovia 500 E1
Montevallo 4,825 E4
Montgomery (cap.)▲ 201,568 F6
Montrose 750 C9
Moody 8,053 F3

(continued on following page)

Agriculture, Industry and Resources

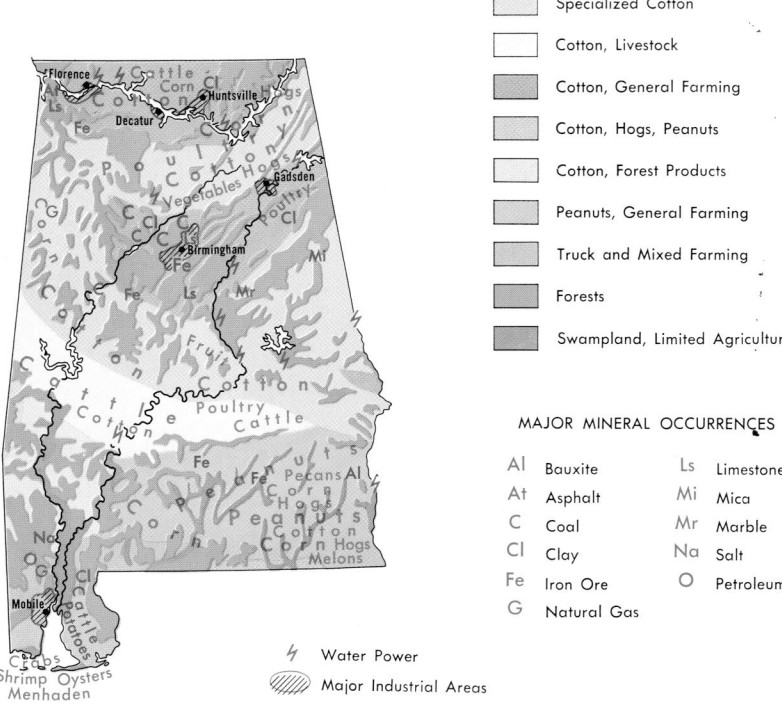

DOMINANT LAND USE

- Specialized Cotton
- Cotton, Livestock
- Cotton, General Farming
- Cotton, Hogs, Peanuts
- Cotton, Forest Products
- Peanuts, General Farming
- Truck and Mixed Farming
- Forests
- Swampland, Limited Agriculture

MAJOR MINERAL OCCURRENCES

Al	Bauxite	Ls	Limestone
At	Asphalt	Mi	Mica
C	Coal	Mr	Marble
Cl	Clay	Na	Salt
Fe	Iron Ore	O	Petroleum
G	Natural Gas		

⚡ Water Power

▨ Major Industrial Areas

Topography

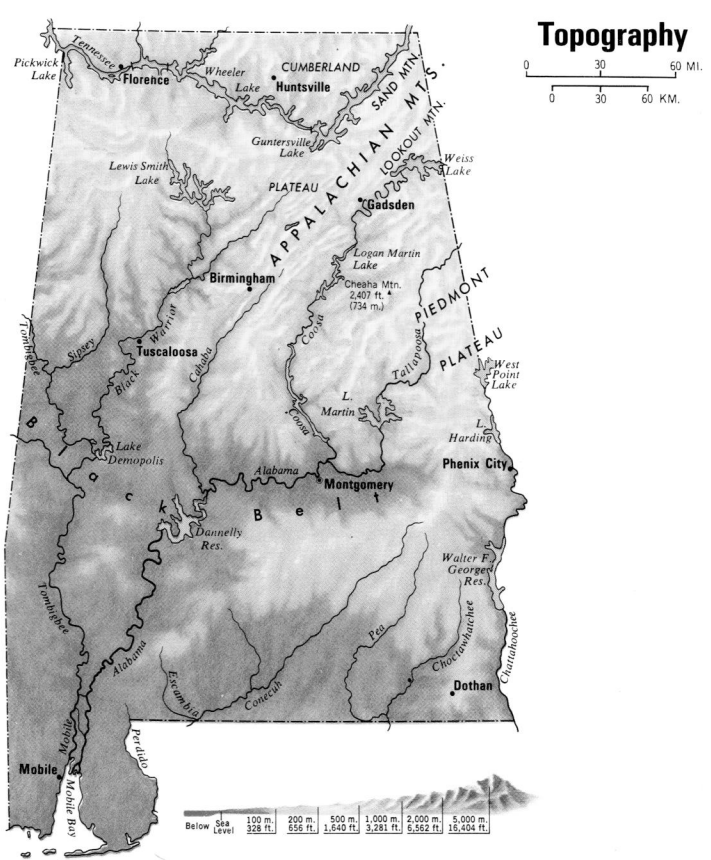

0   30   60 MI.

0   30   60 KM.

Below Sea Level — 100 m. 328 ft. — 200 m. 656 ft. — 500 m. 1,640 ft. — 1,000 m. 3,281 ft. — 2,000 m. 6,562 ft. — 5,000 m. 16,404 ft.

Alabama

SCALE
0 5 10 20 30 40 MI.
0 5 10 20 30 40 KM.
State Capitals............⊛
County Seats.............◉
Major Limited Access Hwys.

© Copyright HAMMOND INCORPORATED, Maplewood, N. J.

CITIES and TOWNS

Akiachak 585................F2
Akolmiut (Kasigluk) 425........F2
Akutan 713................E4
Alakanuk 652................E2
Anchor Point 1,845........B2
Anchorage 260,283........B1
Anderson 367................J2
Angoon 572................M1
Aniak 572................G2
Barrow 4,581................G1
Bethel 5,471................F2
Big Lake 2,635................C1
Butte 2,561................C1
Chevak 765................E2
Clear 504................J2
Clover Pass 451................N2
Cohoe 1,168................B2
College 11,402................J1
Copper Center 362................J2
Cordova 2,454................D1
Craig 1,397................M2
Delta Junction 840................J2
Dillingham 2,466................G3
Emanguk (Emmonak) 767........E2
Fairbanks 30,224................J2
Fort Yukon 595................J1
Fritz Creek 1,603................B2
Galena 675................G2
Gambell 649................D1
Glennallen 554................J2
Haines 1,811................M1
Healy 1,000................J2

Homer 3,946................B2
Hoonah 795................M1
Hooper Bay 1,014................E2
Houston 1,202................B1
Hydaburg 382................M2
Juneau (cap.) 30,711................N1
Kachemak City 431................B2
Kake 710................M1
Kasigluk 543................F2
Kasilof 471................B1
Kenai 6,942................B1
Ketchikan 7,922................N2
Kiana 388................F1
King Cove 792................F4
King Salmon 442................G3
Kipnuk 644................F2
Klawock (Klawak) 854........M2
Kodiak 6,334................H3
Kotlik 591................F2
Kotzebue 3,082................F1
Kwethluk 713................F2
Manokotak 399................G3
McGrath 401................H2
Metlakatla 1,375................N2
Mountain Point 396................N2
Mountain Village 755................E2
Naknek 678................G3
Nenana 402................J2
New Stuyahok 471................G3
Ninilchik 772................B1
Nome 3,505................E2
Noorvik 634................F1
North Pole 1,570................J2
Nulato 336................G2

Nunapitchuk 466................F2
Palmer 4,533................C1
Petersburg 3,224................N2
Pilot Station 550................F2
Point Hope 757................E1
Quinhagak 555................F3
Saint Marys (Andreafski) 500..F2
Saint Paul Island 532................D3
Sand Point 952................G3
Savoonga 643................E2
Saxman 431................N2
Selawik 772................G1
Seward 2,830................C1
Shishmaref 562................E1
Sitka 8,835................M1
Skagway 862................M1
Soldotna 3,759................B1
Stebbins 547................F2
Sterling 3,802................B1
Thorne Bay 557................M2
Togiak 809................F3
Tok 1,393................K2
Toksook Bay 532................F2
Unalakleet 747................G2
Unalaska 4,283................E4
Valdez 4,036................D1
Wainwright 546................F1
Wasilla 5,469................B1
Wrangell 2,308................N2
Yakutat 680................L3

OTHER FEATURES

Adak (isl.)................L4

Admiralty (isl.)................M1
Afognak (isl.)................H3
Agattu (isl.)................J3
Akutan (isl.)................E4
Alaska (gulf)................H2
Alaska (range)................H2
Aleutian (isls.)................J4
Aleutian (range)................G3
Alexander (arch.)................L1
Amchitka (isl.)................K4
Amlia (passage)................L4
Amukta (isl.)................D4
Andreanof (isls.)................L4
Atka (isl.)................L4
Attu (isl.)................J2
Baird (mts.)................F1
Baranof (isl.)................M1
Barrow (pt.)................G1
Bear (mt.)................K2
Beaufort (sea)................K1
Becharof (lake)................G3
Bering (glac.)................K2
Bering (sea)................D1
Bering (str.)................E1
Blackburn (mt.)................K2
Bona (mt.)................K2
Bristol (bay)................G3
British (mts.)................K1
Brooks (range)................G1
Chandalar (riv.)................J1
Chatham (str.)................M1
Chichagof (isl.)................M1
Chignik (bay)................G3
Chilkoot (pass)................M1

Chirikof (isl.)................G3
Chitina (riv.)................K2
Christian (sound)................M2
Chugach (mts.)................C1
Chukchi (sea)................E1
Clarence (str.)................N2
Clark (lake)................H2
Clear (cape)................D1
Coast (mts.)................N1
Columbia (glac.)................C1
Colville (riv.)................G1
Constantine (cape)................G3
Cook (inlet)................B1
Cook (mt.)................K2
Copper (riv.)................J2
Cordova (bay)................M2
Coronation (isl.)................M2
Cross (sound)................L1
Dease (inlet)................H1
Decision (cape)................M2
Denali Nat'l Park................H2
Devils Paw (mt.)................N1
Dixon Entrance (chan.)........M2
Douglas (mt.)................H3
Dry (bay)................L3
Eielson A.F.B. 5,400........J2
Elmendorf A.F.B................B1
Endicott (mts.)................H1
Etolin (isl.)................N2
Fairweather (cape)................L1
Fairweather (mt.)................L1
Firth (riv.)................K1
Foraker (mt.)................H2
Fort Davis................E2

Fort Greely 461................J2
Fort Richardson................C1
Fort Wainwright................J1
Four Mountains (isls.)........E4
Fox (isls.)................E1
Frederick (sound)................N1
Gates of the Arctic Nat'l Park..H1
Glacier (bay)................M1
Glacier Bay Nat'l Park........M1
Goodhope (bay)................F1
Great Sitkin (isl.)................L4
Guyot (glac.)................K2
Hagemeister (isl.)................F3
Halkett (cape)................H1
Hall (isl.)................D2
Harding Icefield................C2
Harrison (bay)................H1
Hayes (mt.)................J2
Hazen (bay)................F2
Hinchinbrook (isl.)................D1
Hoonah (sound)................M1
Hope (pt.)................E1
Howard (pass)................G1
Icy (bay)................K3
Icy (cape)................G1
Icy (pt.)................L1
Icy (str.)................M1
Iliamna (lake)................G3
Iliamna (vol.)................H3
Innoko (riv.)................G2
Kachemak (bay)................B2
Kanaga (isl.)................L4
Kates Needle (mt.)................N1
Katmai (vol.)................H3

Katmai Nat'l Park................H3
Kayak (isl.)................K3
Kenai (lake)................C1
Kenai (mts.)................C2
Kenai (pen.)................C2
Kenai Fjords Nat'l Park........C3
Kennedy Entrance (str.)........H3
King (isl.)................D1
Kiska (isl.)................J4
Kiska (vol.)................J4
Klondike Gold Rush Nat'l
  Hist. Park................N1
Knight (isl.)................D1
Knik Arm (inlet)................B1
Kobuk (riv.)................G1
Kobuk Valley Nat'l Park........F1
Kodiak (isl.)................H3
Kotzebue (sound)................F1
Koyukuk (riv.)................G1
Krusenstern (cape)................F1
Kuiu (isl.)................M1
Kuk (riv.)................G1
Kuskokwim (bay)................F3
Kuskokwim (mts.)................G2
Kuskokwim (riv.)................G2
Kvichak (bay)................G3
Lake Clark Nat'l Park........G2
Lisburne (cape)................E1
Little Diomede (isl.)................E1
Little Sitkin (isl.)................K4
Lynn Canal (inlet)................M1
Makushin (vol.)................E4
Malaspina (glac.)................K3
Marcus Baker (mt.)................C1

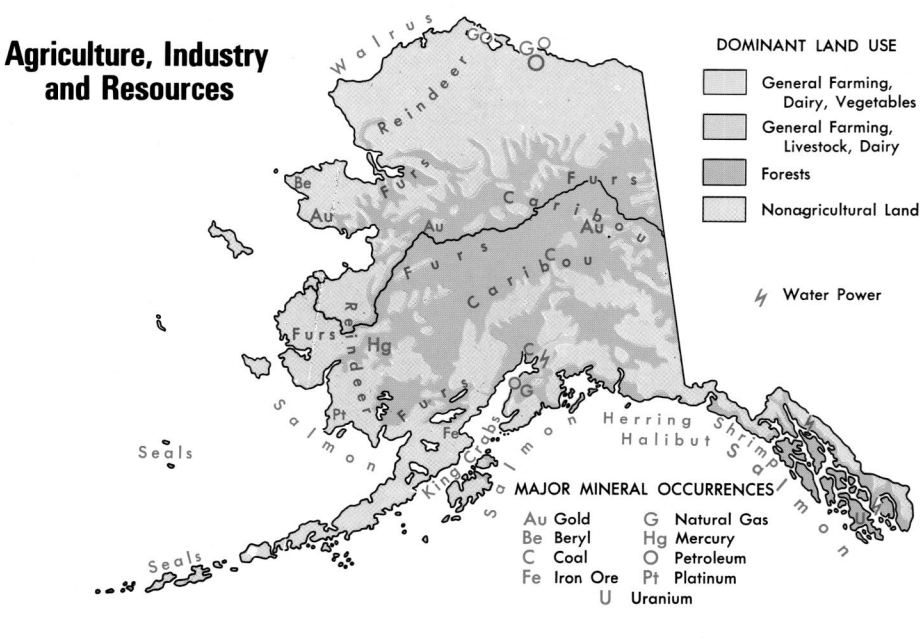

Agriculture, Industry and Resources

DOMINANT LAND USE

General Farming, Dairy, Vegetables
General Farming, Livestock, Dairy
Forests
Nonagricultural Land

Water Power

MAJOR MINERAL OCCURRENCES

Au Gold
Be Beryl
C Coal
Fe Iron Ore
U Uranium
G Natural Gas
Hg Mercury
O Petroleum
Pt Platinum

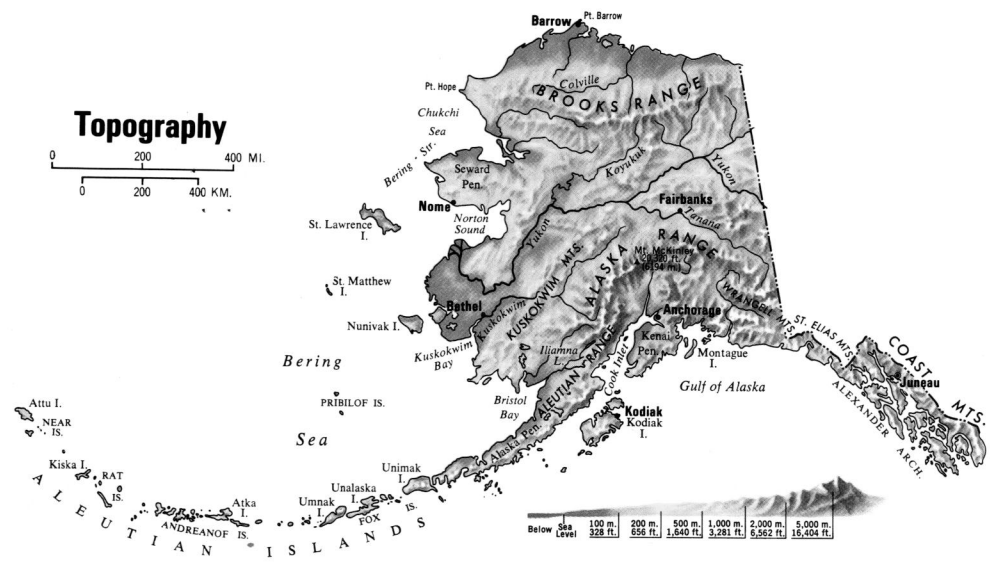

Topography

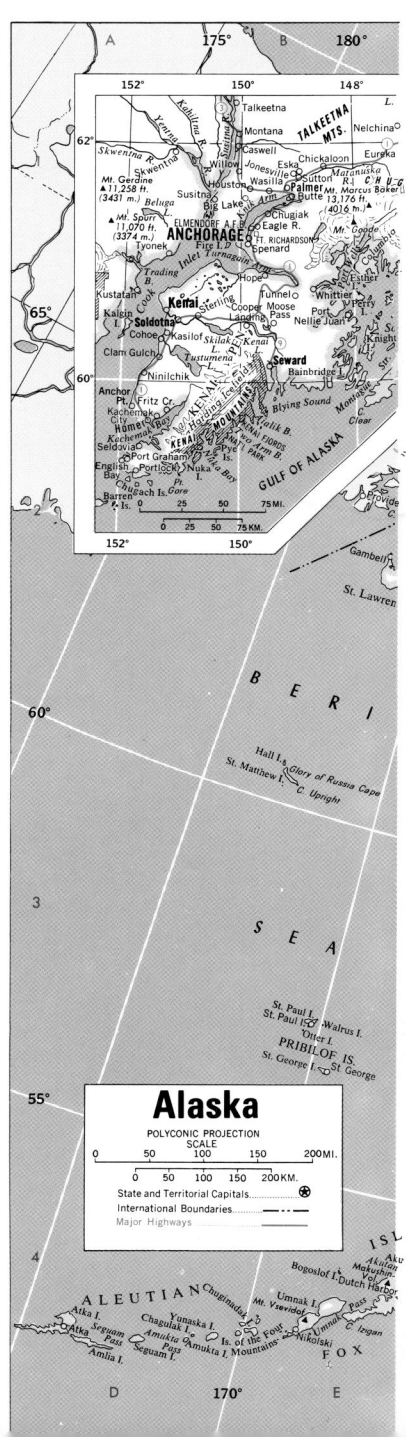

Alaska
POLYCONIC PROJECTION
SCALE

State and Territorial Capitals
International Boundaries
Major Highways

Marmot (isl.) H3
Matanuska (riv.) C1
McKinley (mt.) H2
Meade (riv.) G1
Mendenhall (cape) E3
Mentasta (pass) K2
Merrill (pass) H2
Michelson (pass) K1
Middleton (isl.) J3
Misty Fjords Nat'l Mon. ..N2
Mitkof (isl.) N2
Montague (isl.) D1
Muir (glac.) M1
Mulchatna (riv.) G2
Muzon (cape) M2
Naknek (lake) G3
Near (isls.) H3
Nelson (isl.) E2
Newenham (cape) F3
Noatak (riv.) F2
Norton (bay) F1
Norton (sound) E2
Nowitna (riv.) H2
Nuka (bay) E3
Nunivak (isl.) E3
Nushagak (riv.) G2
Nuyakuk (lake) F3
Ommaney (cape) M2
Otter (isl.) D3
Pastol (bay) F3
Pavlof (bay) F3
Pavlof (vol.) F3
Philip Smith (mts.) J1
Porcupine (riv.) K1

Port Clarence (inlet) E1
Port Heiden (inlet) G3
Portland Canal (inlet) N2
Port Moller (inlet) F3
Port Wells (inlet) C1
Pribilof (isls.) D3
Prince of Wales (cape) ...E1
Prince of Wales (isl.) N2
Prince William (sound) ...D1
Prudhoe (bay) J1
Rat (isls.) K4
Redoubt (vol.) H2
Revillagigedo (chan.) N2
Revillagigedo (isl.) N2
Romanzof (cape) E2
Sagavanirktok (riv.) J1
Saint Elias (cape) K3
Saint Elias (mt.) K2
Saint George (isl.) D3
Saint Lawrence (isl.) D2
Saint Matthew (isl.) D2
Saint Paul (isl.) D3
Salisbury (sound) M1
Sanak (isl.) F4
Sanford (mt.) K2
Schwatka (mts.) G1
Seguam (isl.) D4
Selawik (riv.) F1
Semichi (isls.) J3
Semidi (isls.) G3
Semisopochnoi (isl.) K4
Seward (pen.) E1
Sheenjek (riv.) K1
Shelikof (str.) H3

Shemya (isl.) J3
Shishaldin (vol.) E4
Shumagin (isls.) G4
Shuyak (isl.) H3
Sitka (sound) M1
Sitka Nat'l Hist. Park M1
Sitkinak (str.) H3
Skilak (lake) C1
Skwentna (riv.) A1
Smith (bay) H1
Spencer (cape) L1
Stephens (passage) N1
Stevenson Entrance (str.) H3
Stikine (riv.) N2
Stikine (str.) N2
Stony (riv.) G2
Stuart (isl.) F2
Suemez (isl.) M2
Sumner (str.) M2
Susitna (riv.) B1
Sutwik (isl.) G3
Taku (glac.) N1
Taku (riv.) N1
Talkeetna (mts.) J2
Tanaga (isl.) K4
Tanaga (vol.) K4
Tanana (riv.) J2
Taylor (mts.) G2
Tazlina (lake) D1
Tazlina (riv.) D1
Teshekpuk (lake) H1
Tigalda (isl.) F4
Tikchik (lakes) G2
Togiak (bay) F3

Tugidak (isl.) G3
Turnagain Arm (inlet) B1
Tustumena (lake) C1
Two Arm (bay) C2
Ugashik (lakes) G3
Umnak (isl.) E4
Umnak (passage) E4
Unalaska (isl.) E4
Unga (isl.) F3
Unimak (isl.) F4
Unimak (passage) F4
Utukok (riv.) F1
Valley of Ten
   Thousand Smokes G3
Vancouver (mt.) L2
Veniaminof (crater) F3
Vsevidof (mt.) E4
Walrus (isl.) E3
Walrus (isls.) E3
Waring (mts.) G1
West Point (mt.) K2
White (pass) N1
White (riv.) K2
White Mountains Nat'l
   Rec. Area J1
Witherspoon (mt.) C1
Wrangell (cape) H3
Wrangell (isl.) N2
Wrangell (mts.) N2
Wrangell-St. Elias Nat'l Park .N2
Yakobi (isl.) M1
Yakutat (bay) K3
Yentna (riv.) A1
Yukon (riv.) F2

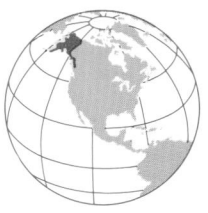

AREA 591,004 sq. mi. (1,530,700 sq. km.)
POPULATION 626,932
CAPITAL Juneau
LARGEST CITY Anchorage
HIGHEST POINT Mt. McKinley 20,320,ft.
   (6194 m.)
SETTLED IN 1801
ADMITTED TO UNION January 3, 1959
POPULAR NAME Great Land; Last Frontier
STATE FLOWER Forget-me-not
STATE BIRD Willow Ptarmigan

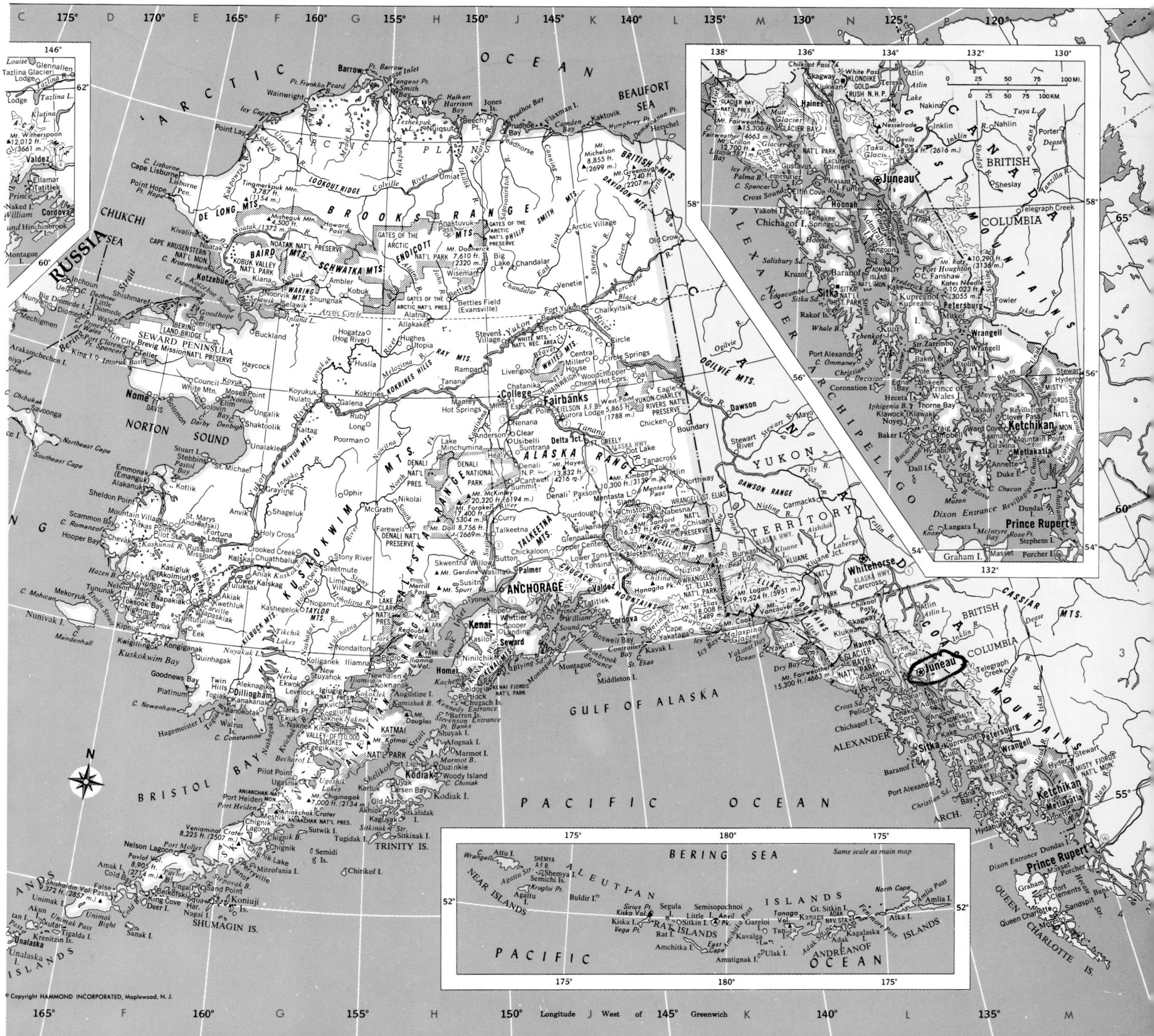

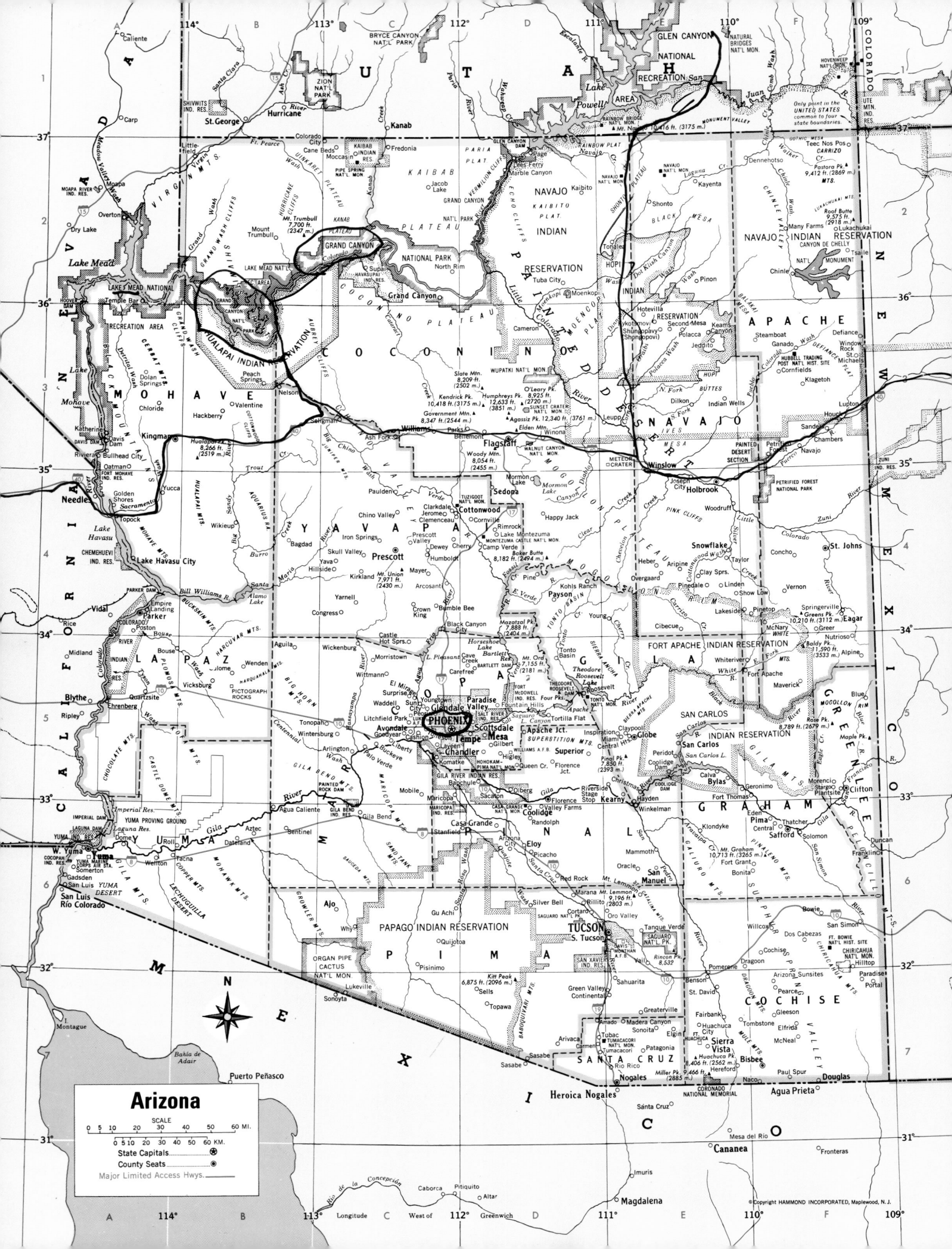

Arizona

SCALE
0 5 10 20 30 40 50 60 MI.
0 5 10 20 30 40 50 60 KM.

⊛ State Capitals
◉ County Seats
Major Limited Access Hwys.

© Copyright HAMMOND INCORPORATED, Maplewood, N.J.

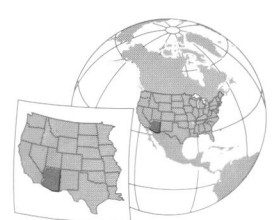

AREA 114,000 sq. mi. (295,260 sq. km.)
POPULATION 5,130,632
CAPITAL Phoenix
LARGEST CITY Phoenix
HIGHEST POINT Humphreys Pk. 12,633 ft.
    (3851 m.)
SETTLED IN 1752
ADMITTED TO UNION February 14, 1912
POPULAR NAME Grand Canyon State
STATE FLOWER Saguaro Cactus Blossom
STATE BIRD Cactus Wren

Agriculture, Industry and Resources

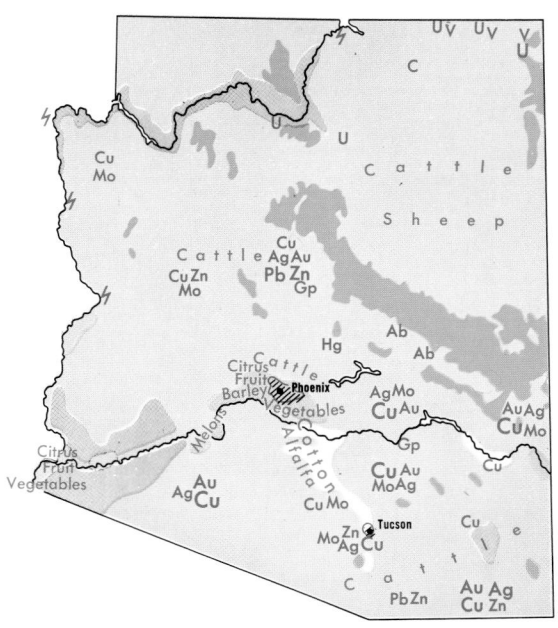

MAJOR MINERAL OCCURRENCES

Ab	Asbestos	Cu	Copper	Pb	Lead
Ag	Silver	Gp	Gypsum	U	Uranium
Au	Gold	Hg	Mercury	V	Vanadium
C	Coal	Mo	Molybdenum	Zn	Zinc

DOMINANT LAND USE

Fruit, Truck and Mixed Farming

Cotton and Alfalfa

General Farming, Livestock, Special Crops

Range Livestock

Forests

Nonagricultural Land

 Water Power

Major Industrial Areas

COUNTIES

Apache 69,423 ....F3
Cochise 117,755 ....F7
Coconino 116,320 ....C3
Gila 51,335 ....E5
Graham 33,489 ....E6
Greenlee 8,547 ....F5
La Paz 19,715 ....A5
Maricopa 3,072,149 ....C5
Mohave 155,032 ....A3
Navajo 97,470 ....E3
Pima 843,746 ....D6
Pinal 179,727 ....D6
Santa Cruz 38,381 ....E7
Yavapai 167,517 ....C4
Yuma 160,026 ....A5

CITIES and TOWNS

Aguila 900............................B5
Ajo 3,705 ....C6
Alpine 450 ....F5
Apache Junction 31,814 ....D5
Arivaca 400..........................D7
Arizona City 4,385 ....D6
Arizona Sunsites 825 ....F7
Arlington 950 ....C5
Ash Fork 457 ....C3
Avondale 35,883 ....C5
Bagdad 1,578........................B4
Bapchule 400........................D5
Bellemont 210 ....D3

Benson 4,711 ....E7
Bisbee▲ 6,090......................F7
Black Canyon City 2,697 ....C4
Bouse 615.............................A5
Bowie 600 ....F6
Buckeye 6,537 ....C5
Bullhead
  City-Riviera 33,769 ....A3
Bylas 1,219 ....E5
Cameron 978 ....D3
Camp Verde 9,451 ....D4
Carefree 2,927 ....C5
Casa Grande 25,224 ....D6
Cashion 3,014.......................C5
Cave Creek 3,728 ....D5
Central 300 ....F6
Central Heights-Midland
  City 2,694............................E5
Chambers 500.......................F3
Chandler 176,581 ....C5
Chinle 5,366 ....F2
Chino Valley 7,835 ....C4
Chloride 225 ....A3
Christmas 201.......................E5
Cibecue 1,331 ....E4
Clarkdale 3,422 ....C4
Claypool 1,794 ....E5
Clay Springs 500...................E4
Clemenceau 300 ....C4
Clifton▲ 2,596 ....F5
Cochise 150 ....F6
Colorado City 3,334 ....B2
Congress 1,717 ....C4

Continental 250 ....D7
Coolidge 7,786 ....D6
Cornfields 200......................F3
Cornville 3,335 ....D4
Cortaro 375 ....D6
Cottonwood 9,179.................D4
Davis Dam 125 ....A3
Dennehotso 734 ....F2
Dolan Springs 1,867 ....A3
Douglas 14,312 ....F7
Dragoon 150 ....F6
Duncan 662...........................F6
Eagar 4,033 ....F4
Ehrenberg 1,357 ....A5
Elfrida 700 ....F7
Elgin 309 ....E7
El Mirage 7,609.....................C5
Eloy 10,375 ....D6
Flagstaff▲ 52,894 ....D3
Florence▲ 17,054 ....D5
Fort Apache 500 ....F5
Fort Defiance 4,061 ....F3
Fort Grant 240 ....E6
Fort Thomas 450...................E5
Fountain Hills 20,235 ....D5
Franklin 300 ....F6
Fredonia 1,036......................C2
Gadsden 953.........................A6
Ganado 1,505 ....F3
Geronimo 25 ....F5
Gila Bend 1,980 ....C6
Gilbert 109,697 ....D5
Glendale 218,812..................C5

(continued on following page)

Topography

0   50   100 MI.

0   50   100 KM.

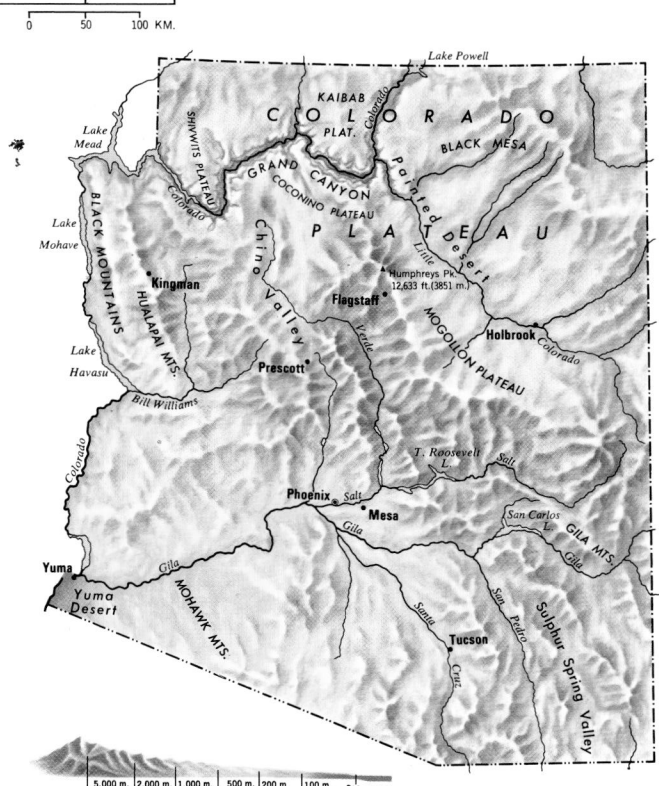

| 5,000 m. 16,404 ft. | 2,000 m. 6,562 ft. | 1,000 m. 3,281 ft. | 500 m. 1,640 ft. | 200 m. 656 ft. | 100 m. 328 ft. | Sea Level | Below |

Globe▲ 7,486 E5
Goodyear 18,911 C5
Grand Canyon 1,460 C2
Green Valley 17,283 D7
Greer 385 F4
Gu Achi 339 C6
Hackberry 250 B3
Hayden 892 E5
Heber-Overgaard 2,722 E4
Higley 500 D5
Holbrook▲ 4,917 E4
Hotevilla 767 E3
Houck 1,087 F3
Huachuca City 1,751 E7
Humboldt 787 C4
Indian Wells 150 E3
Inspiration 500 D5
Iron Springs 175 C4
Jerome 329 C4
Joseph City 650 E4
Kaibito 1,607 D2
Kayenta 4,922 E2
Keams Canyon 260 E3
Kearny 2,249 E5
Kingman▲ 20,069 A3
Klagetoh 200 F3
Kohls Ranch 100 D4
Komatke 1,116 C5
Kykotsmovi 600 E3
Lake Havasu City 41,938 A4
Lake Montezuma 3,344 D4
Lakeside 1,333 E4
Laveen 800 C5
Leupp 970 E3
Liberty 150 C5
Litchfield Park 3,810 C5
Lukachukai 1,565 F2
Lupton 250 F3
Mammoth 1,762 E6
Many Farms 1,548 F2
Marana 13,556 D6
Maricopa 1,040 C5
Mayer 1,408 C4
McNary 349 F4
Mesa 396,375 D5
Miami 1,936 E5
Moccasin 150 C2
Moenkopi 901 D2
Morenci 1,879 F5
Morristown 400 C5
Mount Lemmon 400 E6
Naco 833 E7
Navajo 100 F3
Nogales▲ 20,878 E7
Nutrioso 500 F5
Oatman 175 A3
Oracle 3,563 E6
Oro Valley 29,700 E6
Page 6,809 D2
Palo Verde 500 C5
Paradise Valley 13,664 D5
Parker 3,140 A4
Parks 1,137 C3
Patagonia 881 E7
Paulden 3,420 C4
Payson 13,620 D4
Peach Springs 600 B3
Pearce 700 F7
Peoria 108,364 C5
Peridot 1,266 E5
Petrified Forest 80 F3
Phoenix (cap.)▲ 1,321,045 C5
Picacho 850 D6
Pima 1,989 F6
Pine 1,931 D4
Pinedale 400 E4
Pinetop 3,582 F4
Pinon 1,190 E2
Pisinimo 237 C6
Polacca 1,108 E3
Pomerene 365 E6
Poston 389 A4
Prescott▲ 33,938 C4
Prescott Valley 23,535 C4
Quartzsite 3,354 A5
Queen Creek 4,316 D5
Quijotoa 200 C6
Randolph 350 D6
Red Rock 250 D6
Rillito 400 D6
Rimrock 217 D4
Riverside Stage Stop 418 D5
Riviera-Bullhead
City 21,951 A3
Roll 700 A6
Sacaton 1,584 D5
Safford▲ 9,232 F6
Sahuarita 3,242 E7
Saint David 1,744 E7
Saint Johns▲ 3,269 F4
Saint Michaels 1,295 F3
Salome 1,690 B5
San Carlos 3,716 E5
Sanders 900 F3
San Luis 15,322 A6
San Manuel 4,375 E6
San Simon 400 F6
Scottsdale 202,705 D5
Second Mesa 814 E3
Sedona 10,192 D4
Seligman 456 B3
Sells 2,799 D7
Shonto 568 E2
Show Low 7,695 F4
Shungopavy
(Shongopovi) 632 E3
Sierra Vista 37,775 E7
Silver Bell 900 D6
Skull Valley 250 C4
Snowflake 4,460 E4
Solomon 700 F6
Somerton 7,266 A6
Sonoita 826 E7
South Tucson 5,490 D6
Springerville 1,972 F4
Stanfield 651 C6
Stargo 1,038 F5
Sun City 38,309 C5
Supai 503 C2
Superior 3,254 D5
Surprise 30,848 C5
Tacna 555 B6
Tanque Verde 16,195 E6
Taylor 3,176 E4
Teec Nos Pos 799 F2
Tempe 158,625 D5
Thatcher 4,022 F6
Tolleson 4,974 C5
Tombstone 1,504 F7
Tonalea 562 E2
Tonopah 54 B5
Tonto Basin 840 D5
Topawa 500 D7
Topock 325 A4
Tortilla Flat 37 D5
Tsaile 1,078 F2
Tubac 949 E7
Tuba City 8,225 D2
Tucson▲ 486,699 D6
Tumacacori-Carmen 569 E7
Vail 2,484 E6
Valley Farms 240 D6
Wellton 1,829 A6
Wenden 556 B5
Whiteriver 5,220 E5
Wickenburg 5,082 C5
Wikieup 150 B4
Willcox 3,733 F6
Williams 2,842 C3
Window Rock 3,059 F3
Winkelman 443 E6
Winslow 9,520 E3
Wintersburg 400 B5
Wittmann 600 C5
Woodruff 280 E4
Yarnell 645 C4
Young 561 D4
Youngtown 3,010 C5
Yucca 250 A4
Yuma▲ 77,515 A6

OTHER FEATURES

Agassiz (peak) D3
Agua Fria (riv.) C5
Alamo (lake) B4
Apache (lake) D5
Aquarius (range) B4
Aravaipa (creek) E6
Aubrey (cliffs) B3
Baboquivari (mts.) D7
Baker Butte (mt.) D4
Balakai (mesa) F3
Baldy (peak) F5
Bartlett (dam) D5
Bartlett (res.) D5
Big Chino Wash (dry riv.) C3
Big Horn (mts.) B5
Big Sandy (riv.) B4
Bill Williams (riv.) B4
Black (mesa) E2
Black (mts.) A3
Black (riv.) E5
Blue (riv.) F5
Bouse Wash (dry riv.) A4
Buckskin (mts.) B4
Burro (creek) B4
Canyon (lake) D5
Canyon de Chelly Nat'l Mon. F2
Carrizo (creek) E4
Carrizo (mts.) F2
Casa Grande Ruins Nat'l Mon. D6
Castle Dome (mts.) A5
Cataract (creek) C3
Centennial Wash (dry riv.) B5
Cerbat (mts.) A3
Cherry (creek) E4
Chevelon (creek) E4
Chinle (creek) F2
Chinle (valley) F2
Chinle Wash (dry riv.) F2
Chino (valley) C3
Chiricahua (mts.) F6
Chiricahua Nat'l Mon. F6
Chocolate (mts.) A5
Clear (creek) D4
Coconino (plat.) C3
Cocopah Ind. Res. A6
Colorado (riv.) A5
Colorado River Ind. Res. A5
Coolidge (dam) E5
Copper (mts.) B6
Corn (creek) E3
Coronado Nat'l Memorial E7
Cottonwood (cliffs) B3
Cottonwood Wash (dry riv.) E4
Davis (dam) A3
Defiance (plat.) F3
Detrital Wash (dry riv.) A3
Diablo (canyon) D4
Dinnebito Wash (dry riv.) E3
Dot Klish (canyon) E2
Dragoon (mts.) F7
Eagle (creek) F5
East Verde (riv.) D4
Echo (cliffs) D2
Elden (mt.) D3
Fort Apache Ind.Res. E5
Fort Bowie Nat'l Hist. Site F6
Fort Huachuca E7
Fort McDowell Ind. Res. D5
Fort Mohave Ind. Res. A4
Fort Pearce Wash (dry riv.) B2
Fossil (creek) D4
Four Peaks (mt.) D5
Galiuro (mts.) E6
Gila (mts.) A6
Gila (mts.) F5
Gila (riv.) B6
Gila Bend (mts.) B5
Gila Bend Ind. Res. C6
Gila River Ind. Res. C5
Glen Canyon (dam) D2
Glen Canyon Nat'l Rec. Area E1
Gothic (mesa) F2
Government (mt.) C3
Graham (mt.) F6
Grand Canyon Nat'l Park C2
Grand Wash (butte) B2
Grand Wash (riv.) B2
Greens (peak) F4
Growler (mts.) B6
Harcuvar (mts.) B5
Harquahala (mts.) B5
Hassayampa (riv.) C5
Havasu (lake) A4
Havasupai Ind. Res. C2
Hohokam-Pima Nat'l Mon. D5
Hoover (dam) A2
Hopi (buttes) E3
Hopi Ind. Res. E2
Horseshoe (lake) D5
Huachuca (peak) E7
Hualapai (mts.) B4
Hualapai (peak) B3
Hualapai Ind. Res. B3
Hubbell Trading Post
Nat'l Hist. Site F3
Humphreys (peak) D3
Hurricane (cliffs) B2
Imperial (res.) A6
Ives (mesa) E3
Juniper (mts.) C3
Kaibab (plat.) C2
Kaibab Ind. Res. C2
Kaibito (plat.) D2
Kanab (creek) C2
Kanab (plat.) C2
Kendrick (peak) D3
Kitt (peak) D7
Kofa (mts.) B5
Laguna (creek) E2
Laguna (res.) A6
Lake Mead Nat'l Rec. Area A2
Lechuguilla (des.) A6
Lemmon (mt.) E6
Little Colorado (riv.) D3
Lukachukai (mts.) F2
Luke A.F.B. C5
Maple (peak) F5
Marble Canyon Nat'l Mon. D2
Maricopa (mts.) C5
Maricopa Ind. Res. C6
Mazatzal (peak) D4
Mead (lake) A2
Meteor (crater) E3
Miller (peak) E7
Moencopi (plat.) D3
Moenkopi Wash (dry riv.) D2
Mogollon (plat.) D4
Mogollon Rim (cliffs) D4
Mohave (lake) A3
Mohave (mts.) A4
Mohawk (mts.) B6
Montezuma Castle Nat'l Mon. D4
Mormon (lake) D4
Mule (mts.) E7
Navajo (creek) D2
Navajo Ind. Res. D2
Navajo Nat'l Mon. E2
Navajo Ord. Depot D3
O'Leary (peak) D3
Oraibi Wash (dry riv.) E3
Ord (mt.) D5
Organ Pipe Cactus Nat'l Mon. C6
Painted (des.) D2
Painted Desert Section
(Petrified Forest) F3
Painted Rock (dam) C5
Papago Ind. Res. C6
Paria (plat.) D2
Paria (riv.) D1
Parker (dam) A4
Pastora (peak) F2
Peloncillo (mts.) F6
Petrified Forest Nat'l Park F4
Pinal (peak) E5
Pinaleno (mts.) F6
Pink (cliffs) E4
Pipe Spring Nat'l Mon. C2
Pleasant (lake) C5
Plomosa (mts.) A5
Polacca Wash (dry riv.) E3
Powell (lake) D1
Pueblo Colorado Wash
(dry riv.) F3
Puerco (riv.) F3
Quajote Wash (dry riv.) D6
Rainbow (plat.) D2
Rincon (peak) E6
Roof Butte (mt.) F2
Rose (peak) F5
Sacramento Wash (dry riv.) A4
Saguaro (lake) D5
Saguaro Nat'l Park E6
Salt (riv.) D5
Salt River Ind. Res. D5
San Carlos (lake) E5
San Carlos (riv.) E5
San Carlos Ind. Res. E5
Sand Tank (mt.) C6
San Francisco (riv.) F5
San Pedro (riv.) E6
San Simon (riv.) F6
Santa Catalina (mts.) E6
Santa Cruz (riv.) D6
Santa Maria (riv.) B4
Santa Rosa Wash (dry riv.) D6
San Xavier Ind. Res. D6
Sauceda (mts.) C6
Shivwits (plat.) B2
Shonto (plat.) E2
Sierra Ancha (mts.) D5
Sierra Apache (mts.) E5
Silver (creek) E4
Slate (mt.) D3
Sulphur Spring (valley) F6
Sunset Crater Nat'l Mon. D3
Superstition (mts.) D5
Theodore Roosevelt (lake) D5
Tonto (creek) D4
Tonto Nat'l Mon. D5
Trout (creek) B3
Trumbull (mt.) B2
Tumacacori Nat'l Mon. E7
Tuzigoot Nat'l Mon. D4
Tyson Wash (dry riv.) A5
Uinkaret (plat.) B2
Union (mt.) C4
Verde (riv.) D5
Virgin (mts.) B2
Walker (creek) F2
Walnut Canyon Nat'l Mon. D3
White (riv.) E5
Williams A.F.B. D5
Woody (mt.) D3
Wupatki Nat'l Mon. D3
Yuma (des.) A6
Yuma Proving Ground A6
Zuni (riv.) F4

▲County seat.

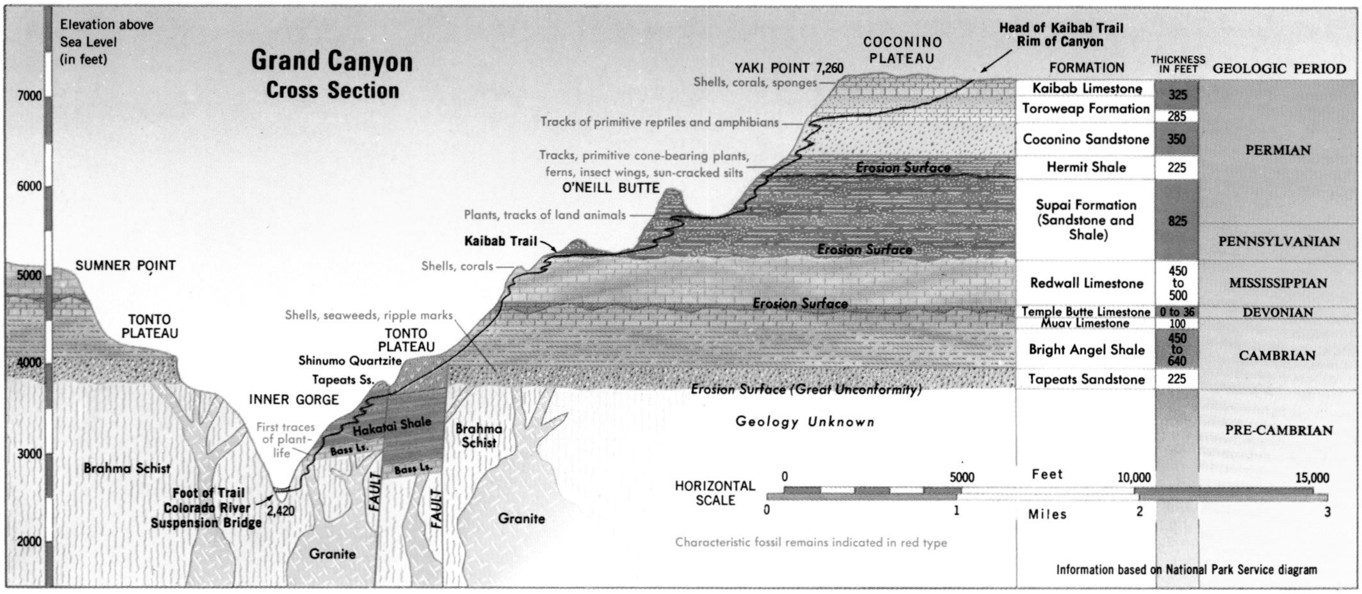

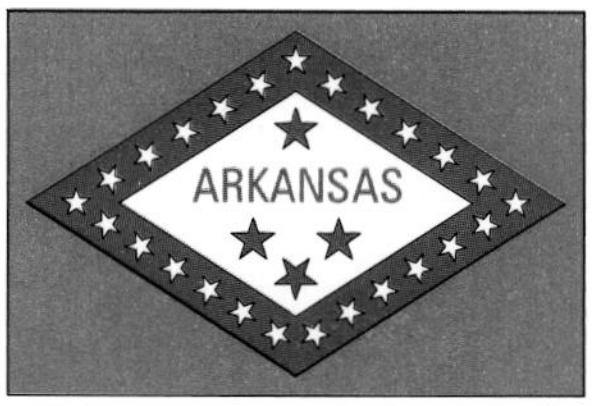

AREA 53,187 sq. mi. (137,754 sq. km.)
POPULATION 2,673,400
CAPITAL Little Rock
LARGEST CITY Little Rock
HIGHEST POINT Magazine Mtn. 2,753 ft. (839 m.)
SETTLED IN 1685
ADMITTED TO UNION June 15, 1836
POPULAR NAME Land of Opportunity
STATE FLOWER Apple Blossom
STATE BIRD Mockingbird

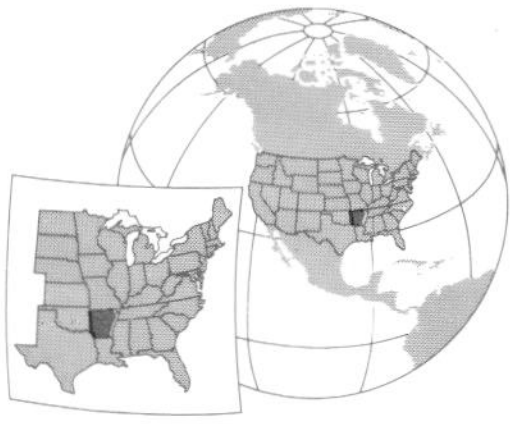

COUNTIES

Arkansas 20,749 ... H5
Ashley 24,209 ... G7
Baxter 38,386 ... F1
Benton 153,406 ... B1
Boone 33,948 ... D1
Bradley 12,600 ... F7
Calhoun 5,744 ... E6
Carroll 25,357 ... C1
Chicot 14,117 ... H7
Clark 23,546 ... D5
Clay 17,609 ... K1
Cleburne 24,046 ... F2
Cleveland 8,571 ... F6
Columbia 25,603 ... D7
Conway 20,336 ... E3
Craighead 82,148 ... J2
Crawford 53,247 ... B2
Crittenden 50,866 ... K3
Cross 19,526 ... J3
Dallas 9,210 ... E6
Desha 15,341 ... H6
Drew 18,723 ... G6
Faulkner 86,014 ... F3
Franklin 17,771 ... C2
Fulton 11,642 ... G1
Garland 88,068 ... D4
Grant 16,464 ... F5
Greene 37,331 ... J1
Hempstead 23,587 ... C6
Hot Spring 30,353 ... E5
Howard 14,300 ... C5
Independence 34,233 ... G2
Izard 13,249 ... G1
Jackson 18,418 ... H2
Jefferson 84,278 ... G5
Johnson 22,781 ... C2
Lafayette 8,559 ... C7
Lawrence 17,774 ... H1
Lee 12,580 ... J4
Lincoln 14,492 ... G6
Little River 13,628 ... B6
Logan 22,486 ... C3
Lonoke 52,828 ... G4
Madison 14,243 ... C1
Marion 16,140 ... E1
Miller 40,443 ... C7
Mississippi 51,979 ... K2
Monroe 10,254 ... H4
Montgomery 9,245 ... C4
Nevada 9,955 ... D6
Newton 8,608 ... D2
Ouachita 28,790 ... E6
Perry 10,209 ... E4
Phillips 26,445 ... J5
Pike 11,303 ... C5
Poinsett 25,614 ... J2
Polk 20,229 ... B5
Pope 54,469 ... D3
Prairie 9,539 ... G4
Pulaski 361,474 ... F4
Randolph 18,195 ... H1
Saint Francis 29,329 ... J3
Saline 83,529 ... E4
Scott 10,996 ... B4
Searcy 8,261 ... E2
Sebastian 115,071 ... B3
Sevier 15,757 ... B6
Sharp 17,119 ... G1
Stone 11,499 ... F2
Union 45,629 ... E7
Van Buren 16,192 ... E2
Washington 157,715 ... B2
White 67,165 ... G3
Woodruff 8,741 ... H3
Yell 21,139 ... D3

CITIES and TOWNS

Adona 187 ... E3
Agnos 150 ... G1
Alco 200 ... F2
Alexander 614 ... F4
Alicia 145 ... H2
Alix 225 ... C3
Alleene 200 ... B6
Allport 127 ... G4
Alma 4,160 ... B3
Almyra 319 ... H5
Alpena 371 ... D1
Alpine 100 ... D5
Altheimer 1,192 ... G5
Altus 817 ... C3
Aly 86 ... D4
Amagon 95 ... H2
Amity 762 ... D5
Antoine 156 ... D5
Appleton 150 ... E3
Arden 80 ... B6
Arkadelphia▲ 10,912 ... D5
Arkansas City▲ 589 ... H6
Armorel 500 ... L2
Ashdown▲ 4,781 ... B6
Ash Flat▲ 977 ... G1
Atkins 2,878 ... E3
Aubrey 221 ... J4
Augusta▲ 2,665 ... H3
Austin 605 ... G4
Auvergne 150 ... H2
Avoca 423 ... B1
Bald Knob 3,210 ... G3
Banks 120 ... F6
Barling 4,176 ... B3
Barton 100 ... J4
Bassett 168 ... K2
Batesville▲ 9,445 ... G2
Bauxite 432 ... F4
Bay 1,800 ... J2
Bearden 1,125 ... E6
Beebe 4,930 ... G3
Beedeville 105 ... H3
Beirne 100 ... D6
Bella Vista 16,582 ... B1
Bellefonte 400 ... D1
Belleville 371 ... D3
Ben Lomond 126 ... B6
Benton▲ 21,906 ... E4
Bentonville▲ 19,730 ... B1
Bergman 407 ... E1
Berryville▲ 4,433 ... C1
Bethel Heights 714 ... B1
Bethesda 285 ... G2
Bigelow 329 ... E3
Big Flat 104 ... F1
Biggers 355 ... J1
Biscoe 484 ... H4
Bismarck 200 ... D5
Black Oak 286 ... K2
Black Rock 717 ... H1
Black Springs 114 ... C5
Blackton 175 ... H4
Blackwell 150 ... E3
Blevins 365 ... C6
Bloomer 200 ... B3
Blue Mountain 132 ... C3
Bluff City 158 ... D6
Bluffton 105 ... C4
Blytheville▲ 18,272 ... L2
Board Camp 90 ... B4
Bodcaw 154 ... D6
Boles 192 ... B4
Bonanza 514 ... B3
Bono 1,512 ... J2
Booneville▲ 4,117 ... C3
Boswell 60 ... F1
Boydell 144 ... H7
Bradford 800 ... G3
Bradley 563 ... C7
Branch 357 ... C3
Brasfield 200 ... H4
Brentwood 75 ... B2
Briggsville 127 ... C4
Brinkley 3,940 ... H4
Brookland 1,332 ... J2
Bryant 9,764 ... F4
Buckner 396 ... D7
Buckville 65 ... D4
Bull Shoals 2,000 ... E1
Burdette 129 ... L2
Butlerville 60 ... G4
Butterfield 105 ... E5
Cabot 15,261 ... F4
Caddo Gap 125 ... C5
Caddo Valley 563 ... D5
Caldwell 465 ... J3
Cale 75 ... D6
Calico Rock 991 ... F1
Calion 516 ... E7
Calmer 55 ... F6
Camden▲ 13,154 ... E6
Cammack Village 831 ... E4
Campbell Station 228 ... H2
Canehill 100 ... B2
Canfield 365 ... C7
Caraway 1,349 ... K2
Carlisle 2,304 ... G4
Carson Lake 100 ... K2
Carthage 442 ... E5
Casa 209 ... D3
Cash 294 ... J2
Cass 225 ... C2
Casscoe 297 ... H4
Cato 75 ... F4
Caulksville 233 ... C3
Cave City 1,946 ... G2
Cave Springs 1,103 ... B1
Cecil 100 ... C3
Cedar Creek 123 ... C4
Cedarville 1,133 ... B2
Center Ridge 120 ... E3
Centerton 2,146 ... B1
Centerville 300 ... D3
Central City 531 ... B3
Chapel Hill 154 ... B5
Charleston▲ 2,965 ... B3
Chatfield 130 ... K3
Cherokee Village-Hidden Valley 4,648 ... G1
Cherry Hill 250 ... B4
Cherry Valley 704 ... J3
Chester 99 ... B2
Chidester 335 ... D6
Choctaw 97 ... F2
Cincinnati 46 ... B1
Clarendon▲ 1,960 ... H4
Clarkedale 300 ... K3
Clarksville▲ 7,719 ... D3
Clinton▲ 2,283 ... F2
Clover Bend 90 ... H2
Coal Hill 1,001 ... C3
College City 269 ... J1
Colt 368 ... J3
Columbus 265 ... C6
Combs 125 ... C2
Concord 255 ... G2
Conway▲ 43,167 ... F3
Cord 250 ... H2
Corinth 65 ... C3
Cornerstone 107 ... G5
Corning▲ 3,679 ... J1
Cotter 921 ... E1
Cotton Plant 960 ... H3
Cove 383 ... B5
Coy 116 ... G4
Crawfordsville 514 ... K3
Crocketts Bluff 70 ... H5
Crossett 6,097 ... G7
Crumrod 200 ... H5
Crystal Springs 215 ... D5
Curtis 300 ... D6
Cushman 461 ... G2
Daisy 118 ... C5
Dalton 200 ... H1
Damascus 306 ... F3
Danville▲ 2,392 ... D3
Dardanelle▲ 4,228 ... D3
Datto 97 ... J1
Decatur 1,314 ... A1
Delaplaine 127 ... J1
Delaware 200 ... D3
Delight 311 ... C5
Dell 251 ... K2
Dennard 200 ... E2
Denning 270 ... C3
De Queen▲ 5,765 ... B5
Dermott 3,292 ... H7
Des Arc▲ 1,933 ... G4
De Valls Bluff▲ 783 ... H4
De Witt▲ 3,552 ... H5
Diamond City 730 ... E1
Diaz 1,284 ... H2
Dierks 1,230 ... B5
Doddridge 75 ... C7
Dogpatch 75 ... D1
Donaldson 326 ... E5
Dover 1,329 ... D3
Drasco 200 ... G2
Dumas 5,238 ... H6
Durham 100 ... C2
Dutch Mills 22 ... B2
Dyer 585 ... B3
Dyess 515 ... K2
Eagle Mills 149 ... E6
Earle 3,036 ... K3
East Camden 902 ... E6
Edgemont 99 ... F2
Edmondson 513 ... K3
El Dorado▲ 21,530 ... E7
El Paso 133 ... F3
Elaine 865 ... J5
Elkins 1,251 ... C1
Elm Springs 1,044 ... B1
Emerson 359 ... D7
Emmet 506 ... D6
England 2,972 ... G4
Enola 188 ... F3
Ethel 200 ... H5
Etowah 366 ... K2
Eudora 2,819 ... H7
Eureka Springs▲ 2,278 ... C1
Evansville 80 ... B2
Evening Shade 465 ... G1
Everton 170 ... E1
Fair Oaks 150 ... J3
Fargo 118 ... H4
Farmington 3,605 ... B1
Fayetteville▲ 58,047 ... B1
Felsenthal 152 ... F7
Felton 97 ... J4
Fifty-Six 163 ... F2
Fisher 265 ... J2
Flippin 1,357 ... E1
Floral 150 ... G2
Florence 100 ... G6
Fordyce▲ 4,799 ... F6
Foreman 1,125 ... B6
Formosa 224 ... E3
Forrest City▲ 14,774 ... J3
Fort Smith▲ 80,268 ... B3
Forum 75 ... C1
Fouke 814 ... C7
Fountain Hill 159 ... G7
Fox 182 ... F2
Franklin 184 ... G1
Fredonia (Biscoe) 476 ... H4
Friendship 206 ... E5
Fulton 245 ... C6
Garfield 490 ... C1
Garland City 352 ... C7
Garner 284 ... G3
Gassville 1,706 ... F1
Gateway 116 ... C1
Genoa 350 ... C7
Gentry 2,165 ... A1
Georgetown 126 ... G3
Gillett 819 ... H5
Gillham 188 ... B5
Gilmore 292 ... K3
Glenwood 1,751 ... C5
Goodwin 225 ... J4
Goshen 752 ... C1
Gosnell 3,968 ... K2
Gould 1,305 ... G6
Grady 523 ... G5
Grand Glaise 50 ... G3
Grannis 575 ... B5
Grapevine 75 ... F5
Gravelly 300 ... C4
Gravette 1,810 ... B1
Greenbrier 3,042 ... F3
Green Forest 2,717 ... D1
Greenland 907 ... B1
Greenway 244 ... K1
Greenwood▲ 7,112 ... B3
Greers Ferry 930 ... F2
Gregory 145 ... H3
Griffithville 262 ... G3
Grubbs 438 ... H2
Guion 90 ... G2
Gum Springs 194 ... D5
Gurdon 2,276 ... D6
Guy 202 ... F3
Hackett 694 ... B3
Hagarville 190 ... D2
Hamburg▲ 3,039 ... G7
Hampton▲ 1,579 ... F6
Hardy 578 ... H1
Harrell 293 ... F7
Harriet 60 ... E2

(continued on following page)

Agriculture, Industry and Resources

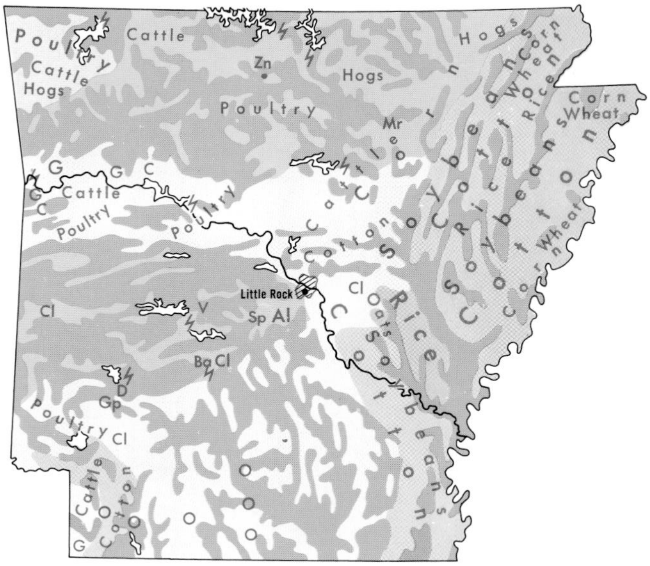

DOMINANT LAND USE

- Fruit and Mixed Farming
- Specialized Cotton
- Cotton, General Farming
- Rice, General Farming
- General Farming, Livestock, Truck Farming, Cotton
- Forests
- Swampland, Limited Agriculture

MAJOR MINERAL OCCURRENCES

Al	Bauxite	Gp	Gypsum
Ba	Barite	Mr	Marble
C	Coal	O	Petroleum
Cl	Clay	Sp	Soapstone
D	Diamonds	V	Vanadium
G	Natural Gas	Zn	Zinc

Water Power — Major Industrial Areas

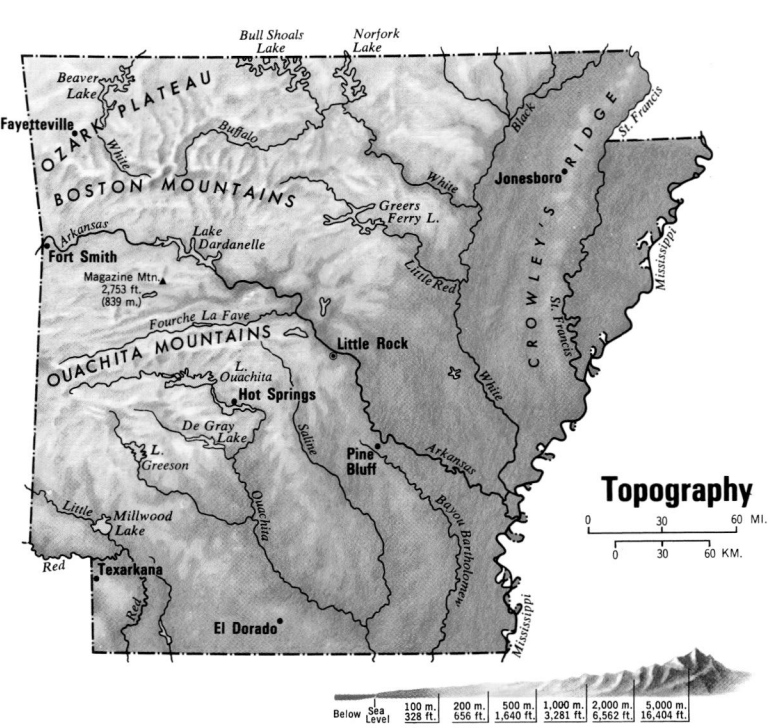

Topography

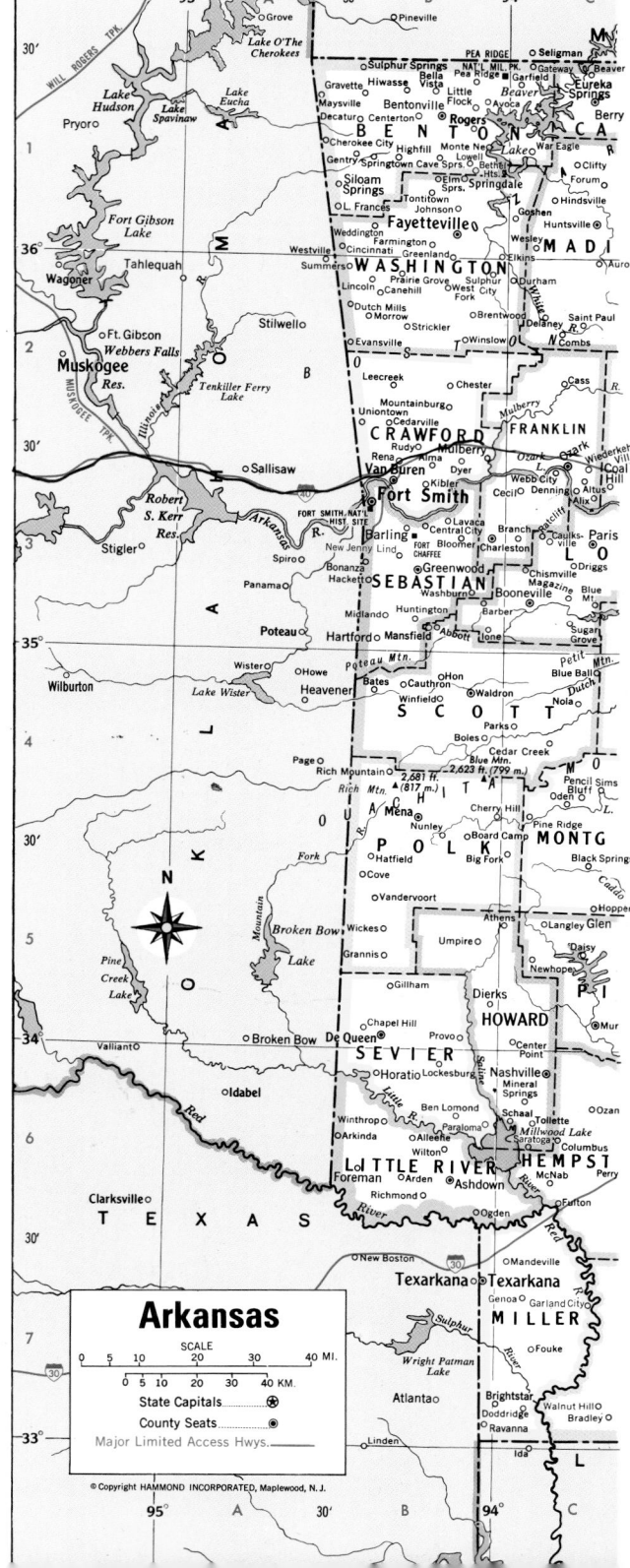

Arkansas
SCALE
0 5 10 20 30 40 MI.
0 5 10 20 30 40 KM.
State Capitals ....⊛
County Seats ....◉
Major Limited Access Hwys.

© Copyright HAMMOND INCORPORATED, Maplewood, N.J.

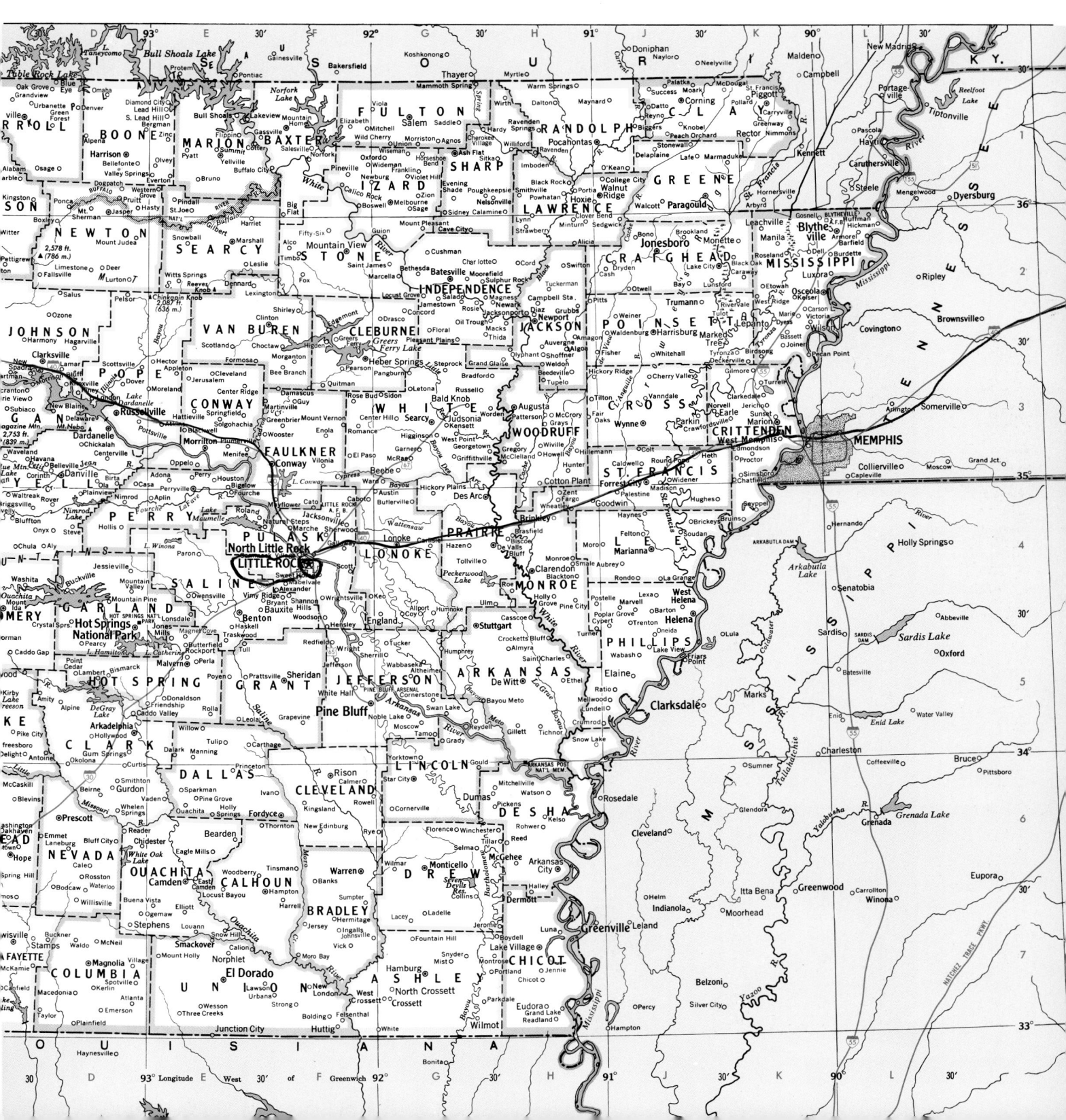

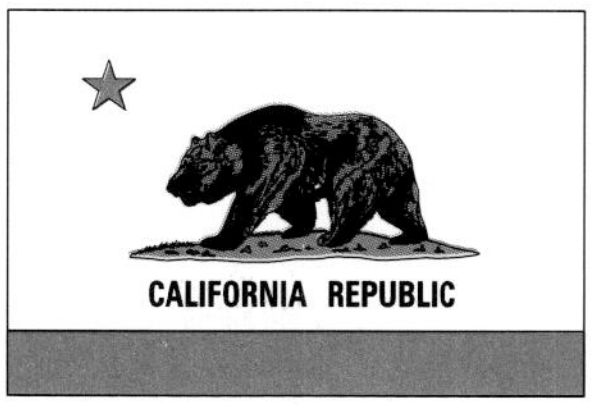
CALIFORNIA REPUBLIC

AREA 158,706 sq. mi. (411,049 sq. km.)
POPULATION 33,871,648
CAPITAL Sacramento
LARGEST CITY Los Angeles
HIGHEST POINT Mt. Whitney 14,494 ft. (4418 m.)
SETTLED IN 1769
ADMITTED TO UNION September 9, 1850
POPULAR NAME Golden State
STATE FLOWER Golden Poppy
STATE BIRD California Valley Quail

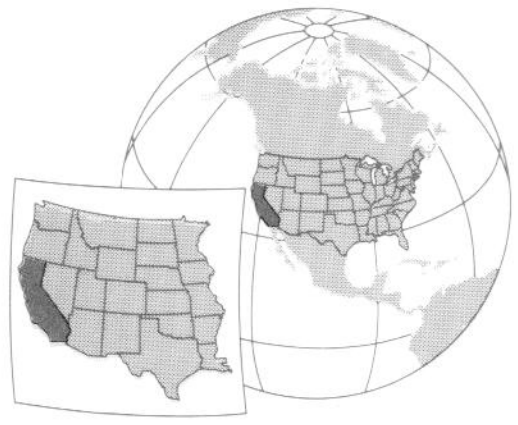

COUNTIES

Alameda 1,443,741 ....D6
Alpine 1,208 ....F5
Amador 35,100 ....E5
Butte 203,171 ....D4
Calaveras 40,554 ....E5
Colusa 18,804 ....C4
Contra Costa 948,816 ....D6
Del Norte 27,507 ....B2
El Dorado 156,299 ....E5
Fresno 799,407 ....E7
Glenn 26,453 ....C4
Humboldt 126,518 ....B3
Imperial 142,361 ....K10
Inyo 17,945 ....H7
Kern 661,645 ....G8
Kings 129,461 ....E8
Lake 58,309 ....C4
Lassen 33,828 ....E3
Los Angeles 9,519,338 ....G9
Madera 123,109 ....F6
Marin 247,289 ....C5
Mariposa 17,130 ....E6
Mendocino 86,265 ....B4
Merced 210,554 ....E6
Modoc 9,449 ....E2
Mono 12,853 ....F5
Monterey 401,762 ....D7
Napa 124,279 ....C5
Nevada 92,033 ....E4
Orange 2,846,289 ....H10
Placer 248,399 ....E4
Plumas 20,824 ....E4
Riverside 1,545,387 ....J10
Sacramento 1,223,499 ....D5
San Benito 53,234 ....D7
San Bernardino 1,709,434 ....J9
San Diego 2,813,833 ....J10
San Francisco (city county) 776,733 ....J2
San Joaquin 563,598 ....D6
San Luis Obispo 246,681 ....E8
San Mateo 707,161 ....J3
Santa Barbara 399,347 ....E9
Santa Clara 1,682,585 ....D6
Santa Cruz 255,602 ....C6
Shasta 163,256 ....C3
Sierra 3,555 ....E4
Siskiyou 44,301 ....C2
Solano 394,542 ....D5
Sonoma 458,614 ....C5
Stanislaus 446,997 ....D6
Sutter 78,930 ....D4
Tehama 56,039 ....C3
Trinity 13,022 ....B3
Tulare 368,021 ....G7
Tuolumne 54,501 ....F5
Ventura 753,197 ....F9
Yolo 168,660 ....D5
Yuba 60,219 ....D4

CITIES and TOWNS

Adelanto 18,130 ....H9
Alameda 72,259 ....J2
Alamo 15,626 ....K2
Albany 16,444 ....J2
Alhambra 85,804 ....C10
Alpine 13,143 ....J11
Altadena 42,610 ....C10
Alturas▲ 2,892 ....E2
Anaheim 328,014 ....D11
Anderson 9,022 ....C3
Angels Camp 3,004 ....E5
Angwin 3,148 ....C5
Antioch 90,532 ....L1
Apple Valley 54,239 ....H9
Aptos 9,396 ....K4
Arbuckle 2,332 ....C4
Arcadia 53,054 ....C10
Arcata 16,651 ....A3
Arden-Arcade 96,025 ....B8
Armona 3,239 ....F7
Arnold 4,218 ....E5
Aromas 2,797 ....D7
Arroyo Grande 15,851 ....E8
Artesia 15,851 ....C11
Arvin 12,956 ....G8
Ashland 20,793 ....K2
Atascadero 26,411 ....E8
Atherton 7,194 ....K3
Atwater 23,113 ....E6
Auberry 2,053 ....F6
Auburn▲ 12,462 ....C8
Avalon 3,127 ....G10
Avenal 14,674 ....E8
Azusa 44,712 ....D10
Baker 174,820 ....J8
Bakersfield▲ 247,057 ....G8
Baldwin Park 75,837 ....D10
Banning 23,562 ....J10
Barstow 21,119 ....H9
Baywood Park-Los Osos 14,351 ....E8
Beaumont 11,384 ....J10
Bell 36,664 ....C11
Bellflower 72,878 ....C11
Bell Gardens 44,054 ....C11
Belmont 25,123 ....J3
Belvedere 2,125 ....H2
Benicia 26,865 ....K1
Ben Lomond 2,364 ....K4
Berkeley 102,743 ....J2
Bethel Island 2,312 ....L1
Beverly Hills 33,784 ....B10
Big Bear City (Sugarloaf Post Office) 5,779 ....J9
Big Bear Lake 5,438 ....J9
Biggs 1,793 ....D4
Big Pine 1,350 ....G6
Bishop 3,575 ....G6
Bloomington 19,318 ....E10
Blythe 12,155 ....L10
Bodfish 1,823 ....G8
Boron 2,025 ....H8
Borrego Springs 2,535 ....J10
Boulder Creek 4,081 ....J4
Brawley 22,052 ....K11
Brea 35,410 ....D11
Brentwood 23,302 ....L2
Bridgeport▲ 525 ....F5
Brisbane 3,597 ....J2
Broderick-Bryte 10,194 ....B8
Bryte-Broderick 10,194 ....B8
Buellton 3,828 ....E9
Buena Park 78,282 ....D11
Burbank 100,316 ....C10
Burlingame 28,158 ....J2
Burney 3,217 ....D3
Buttonwillow 1,266 ....F8
Calexico 27,109 ....K11
California City 8,385 ....H8
Calipatria 7,289 ....K10
Calistoga 5,190 ....C5
Calwa 762 ....F7
Camarillo 57,077 ....G9
Cambria 6,232 ....D8
Campbell 38,138 ....K3
Canyon 7,938 ....K2
Capistrano Beach 6,168 ....H10
Capitola 10,033 ....K4
Cardiff-by-the-Sea 10,054 ....H10
Carlsbad 10,033 ....H10
Carmel 4,081 ....D7
Carmel Valley 4,700 ....D7
Carmichael 49,742 ....C8
Carpinteria 14,194 ....F9
Carson 89,730 ....C11
Caruthers 2,103 ....E7
Castro Valley 57,292 ....K2
Castroville 6,724 ....D7
Cathedral City 42,647 ....J10
Cayucos 2,943 ....E8
Central Valley 4,340 ....C3
Ceres 34,609 ....D6
Cerritos 51,488 ....C11
Chemeketa Park-Redwood Estates 1,847 ....K4
Cherryland 13,837 ....K2
Chester 2,316 ....D3
Chico 59,954 ....D4
China Lake 1,761 ....H8
Chinese Camp 146 ....E6
Chino 67,168 ....D10
Chowchilla 11,127 ....E6
Chula Vista 173,556 ....J11
Citrus Heights 85,071 ....C8
Claremont 33,998 ....D10
Clay 7,317 ....C9
Clayton 10,762 ....K2
Clearlake 10,762 ....C5
Clearlake Oaks 2,402 ....C4
Cloverdale 6,831 ....B5
Clovis 68,468 ....F7
Coachella 22,724 ....J10
Coalinga 11,668 ....E7
Colton 47,662 ....E10
Colusa▲ 5,402 ....C4
Commerce 12,568 ....C10
Compton 93,493 ....C11
Concord 121,780 ....K1
Corcoran 14,458 ....F7
Corning 6,741 ....C4
Corona 124,966 ....E11
Coronado 24,100 ....H11
Corralitos 2,431 ....L4
Corte Madera 9,100 ....J2
Costa Mesa 108,724 ....D11
Cotati 6,471 ....C5
Cottonwood 2,960 ....C3
Covina 46,837 ....D10
Crescent City▲ 4,006 ....A2
Crestline 10,218 ....H9
Crockett 3,194 ....J1
Cudahy 24,208 ....C11
Culver City 38,816 ....B10
Cupertino 50,546 ....K3
Cutler 4,491 ....F7
Cutten 2,933 ....A3
Cypress 46,229 ....D11
Daly City 103,621 ....H2
Dana Point 35,110 ....H10
Danville 41,715 ....K2
Davis 60,308 ....B8
Deer Park 1,433 ....C5
Delano 38,824 ....F8
Delhi 8,022 ....E6
Del Mar 4,389 ....H11
Del Rey Oaks 1,650 ....D7
Del Rosa F10
Desert Hot Springs 16,582 ....J9
Desert View Highlands 2,337 ....G9
Diamond Springs 4,888 ....D8
Dinuba 16,844 ....F7
Dixon 16,103 ....B9
Dos Palos 4,581 ....E6
Downey 107,323 ....C11
Downieville▲ 500 ....E4
Duarte 21,486 ....D10
Dublin 29,973 ....K2
Dunsmuir 1,923 ....C2
Durham 5,220 ....D4
Earlimart 6,583 ....F8
East Blythe 3 ....L10
East Los Angeles 124,283 ....C10
Easton 1,966 ....F7
El Cajon 94,869 ....J11
El Centro▲ 37,835 ....K11
El Cerrito 23,171 ....J2
El Dorado 6,395 ....C8
El Dorado Hills 18,016 ....C8
El Granada 5,724 ....H3
Elk 17,483 ....B4
Elk Grove 59,984 ....B9
El Monte 115,965 ....D10
El Rio 6,193 ....F9
El Segundo 16,033 ....B11
El Toro 62,685 ....E11
Emeryville 6,882 ....J2
Encinitas 58,014 ....H10
Encino B10
Escalon 5,963 ....E6
Escondido 133,559 ....J10
Esparto 1,858 ....C5
Eureka▲ 26,128 ....A3
Exeter 9,168 ....F7
Fairfax 7,319 ....H1
Fairfield▲ 96,178 ....K1
Fair Oaks 28,008 ....C8
Fallbrook 29,100 ....H10
Farmersville 8,737 ....F7
Felton 1,051 ....K4
Ferndale 1,382 ....A3
Fillmore 13,643 ....G9
Firebaugh 5,743 ....E7
Florin 27,653 ....B8
Folsom 51,884 ....C8
Fontana 128,929 ....E10
Ford City 3,512 ....F8
Foresthill 1,791 ....E4
Forest Knolls-Lagunitas H1
Fort Bragg 7,026 ....B4
Fortuna 10,497 ....A3
Foster City 28,803 ....J2
Fountain Valley 54,978 ....D11
Fowler 3,979 ....F7
Frazier Park 2,348 ....F9
Freedom 6,000 ....L4
Fremont 203,413 ....K3
Fresno▲ 427,652 ....F7
Fullerton 126,003 ....D11
Galt 19,472 ....C9
Gardena 57,746 ....C11
Garden Grove 165,196 ....D11
Gilroy 41,464 ....D6
Glen Avon Heights 14,853 ....E10
Glendale 194,973 ....C10
Glendora 49,415 ....D10
Gonzales 7,525 ....D7
Goshen 2,394 ....F7
Grand Terrace 11,626 ....E10
Grass Valley 10,922 ....D4
Greenacres 7,379 ....F8
Greenfield 12,583 ....D7
Greenville 1,160 ....E3
Gridley 5,382 ....D4
Groveland 3,388 ....E6
Grover Beach 13,067 ....E8
Guadalupe 5,659 ....E9
Guerneville 2,441 ....B5
Gustine 4,698 ....D6
Half Moon Bay 11,842 ....H3
Hamilton City 1,903 ....C4
Hanford▲ 41,686 ....F7
Hawthorne 84,112 ....C11
Hayfork 2,315 ....B3
Hayward 140,030 ....K2
Healdsburg 10,722 ....B5
Heber 2,988 ....K11
Hemet 58,812 ....H10
Hercules 19,488 ....J1
Hermosa Beach 18,566 ....B11
Hesperia 62,582 ....H9
Hidden Hills 1,875 ....B10
Highgrove 3,445 ....E10
Highland 44,605 ....H9
Hillsborough 10,825 ....J2
Hilmar-Irwin 4,807 ....E6
Hollister▲ 34,413 ....D7
Hollywood C10
Holt 4,820 ....D6
Holtville 5,612 ....K11
Home Gardens 9,461 ....E11
Homeland 3,710 ....H10
Hughson 3,980 ....E6
Huntington Beach 189,594 ....C11
Huntington Park 61,348 ....C11
Huron 6,306 ....E7
Idyllwild-Pine Cove 3,504 ....J10
Imperial 7,560 ....K11
Imperial Beach 26,992 ....H11
Independence▲ 574 ....H7
Indian Wells 3,816 ....J10
Indio 49,116 ....J10
Inglewood 112,580 ....B11
Ione 7,129 ....C9
Irvine 143,072 ....D11
Isla Vista 18,344 ....E9
Ivanhoe 4,474 ....F7
Jackson▲ 3,989 ....C9
Jamestown 3,017 ....E6
Joshua Tree 4,207 ....J9
Julian 1,621 ....J10
Kelseyville 2,928 ....C5
Kensington 4,936 ....J2
Kerman 8,551 ....E7
Kernville 1,736 ....G8
Keyes 4,575 ....E6
King City 11,094 ....D7
Kings Beach 4,037 ....F4
Kingsburg 9,199 ....F7
La Canada Flintridge 20,318 ....C10
La Crescenta-Montrose 18,532 ....C10
Lafayette 23,908 ....K2
Laguna Beach 23,727 ....G10
Laguna Hills 31,178 ....D11
Laguna Niguel 61,891 ....H10
Lagunitas-Forest Knolls 1,835 ....H10
La Habra 58,974 ....D11
Lake Arrowhead 8,934 ....H9
Lake Elsinore 28,928 ....H10
Lake Isabella 3,315 ....G8
Lakeland Village 5,626 ....E11
Lakeport▲ 4,820 ....C4
Lakewood 79,345 ....C11
La Mesa 54,749 ....H11
La Mirada 46,783 ....D11
Lamont 13,296 ....G8
Lancaster 118,718 ....G9
La Puente 41,063 ....D10
Larkspur 12,014 ....H1
La Selva Beach 1,603 ....K4
Lathrop 10,445 ....D6
La Verne 31,638 ....D10
Lawndale 31,711 ....B11
Lemon Grove 24,918 ....J11
Lemoore 19,712 ....F7
Lenwood 3,222 ....H9
Leucadia 9,478 ....H10
Lincoln 11,205 ....B8
Linda 13,474 ....D4
Lindsay 10,297 ....F7
Live Oak 16,628 ....K4
Live Oak 6,229 ....D4
Livermore 73,345 ....L2
Livingston 10,473 ....E6
Lockeford 3,179 ....C9
Lodi 56,999 ....C9
Loma Linda 18,681 ....F10
Lomita 20,046 ....C11
Lompoc 41,103 ....E9
Lone Pine 1,655 ....H7
Long Beach 461,522 ....C11
Loomis 6,260 ....C8
Los Alamitos 11,536 ....D11
Los Altos 27,693 ....K3
Los Altos Hills 7,902 ....J3
Los Angeles▲ 3,694,820 ....C10
Los Banos 25,869 ....E6
Los Gatos 28,592 ....K4
Los Osos-Baywood Park 14,351 ....E8
Lucerne 2,870 ....C4
Lynwood 69,845 ....C11
Madera▲ 43,207 ....E7
Magalia 10,569 ....D4
Mammoth Lakes 7,093 ....G6
Manhattan Beach 33,852 ....B11
Manteca 49,258 ....D6
Maricopa 1,111 ....F8
Marina 25,101 ....D7
Mariposa▲ 1,373 ....F6
Markleeville▲ 197 ....F5
Martinez▲ 35,866 ....K1
Marysville▲ 12,268 ....D4
Maywood 28,083 ....C10
McCloud 1,343 ....C2
McFarland 9,618 ....F8
McKinleyville 13,599 ....A3
Mecca 5,402 ....K10
Meiners Oaks-Mira Monte 10,927 ....F9
Mendota 7,890 ....E7
Menlo Park 30,785 ....J3
Mentone 7,803 ....H9
Merced▲ 63,893 ....E6
Millbrae 20,718 ....J2
Mill Valley 13,600 ....H2
Milpitas 62,698 ....L3
Mira Loma 17,617 ....E10
Mission Viejo 93,102 ....D11
Modesto▲ 188,856 ....D6
Mojave 3,836 ....G8
Monrovia 36,929 ....D10
Montague 1,456 ....C2
Montara 2,950 ....H3
Montclair 33,049 ....D10
Montebello 62,150 ....C10
Monterey 29,674 ....D7
Monterey Park 60,051 ....C10
Montrose-La Crescenta 18,532 ....C10
Monte Sereno 3,483 ....K4
Moorpark 31,415 ....G9
Moraga 16,290 ....K2
Moreno Valley 142,381 ....F11
Morgan Hill 33,556 ....L4
Morro Bay 10,350 ....D8
Moss Beach 1,953 ....H3
Mountain View 70,708 ....K3
Mount Shasta 3,621 ....C2
Mulberry 1,946 ....D4
Murrieta 44,282 ....H10
Muscoy 8,919 ....E10
Napa▲ 72,585 ....C5
National City 54,260 ....J11
Needles 4,830 ....L9
Nevada City▲ 3,001 ....D4
Newark 42,471 ....K3
Newhall 12,029 ....G9
Newman 7,093 ....D6
Newport Beach 70,032 ....D11
Nipomo 12,626 ....E8
Norco 24,157 ....E11
North Highlands 44,187 ....B8
Norwalk 103,298 ....C11
Novato 47,630 ....H1
Oakdale 15,503 ....E6
Oakhurst 2,868 ....F6
Oakland▲ 399,484 ....J2
Oakley 25,619 ....L1
Oak View 4,199 ....F9
Oceano 7,260 ....E8
Oceanside 161,029 ....H10
Oildale 27,885 ....F8
Ojai 7,862 ....F9
Ontario 158,007 ....D10
Opal Cliffs 6,458 ....K4
Orange 128,821 ....D11
Orange Cove 7,722 ....F7
Orinda 17,599 ....J2
Orland 6,281 ....C4
Orosi 7,318 ....F7
Oroville▲ 13,004 ....D4
Oxnard 170,358 ....F9
Pacheco-Vine Hill 6,822 ....K1
Pacifica 38,390 ....H2
Pacific Grove 15,522 ....C7
Pajaro 3,384 ....D7
Palermo 5,720 ....D4
Palmdale 116,670 ....G9
Palm Desert 41,155 ....J10
Palm Springs 42,807 ....J10
Palo Alto 58,598 ....K3
Palos Verdes Estates 13,340 ....B11
Paradise 26,408 ....D4
Paramount 55,266 ....C11
Parlier 11,145 ....F7
Pasadena 133,936 ....C10
Paso Robles 18,583 ....E8
Patterson 11,606 ....D6
Pebble Beach C7
Pedley 11,207 ....E10
Perris 36,189 ....F11
Petaluma 54,548 ....H1
Pico Rivera 63,428 ....C10
Piedmont 10,952 ....J2
Pinole 19,039 ....J1
Piru 1,196 ....G9
Pismo Beach 8,551 ....E8
Pittsburg 56,769 ....L1
Pixley 2,586 ....F8
Placentia 46,488 ....D11
Placerville▲ 9,610 ....C8
Planada 4,369 ....E6
Pleasant Hill 32,837 ....K2
Pleasanton 63,654 ....L2
Pollock Pines 4,728 ....E5
Pomona 149,473 ....D10
Poplar-Cotton Center 1,496 ....F7
Porterville 39,615 ....G7
Port Hueneme 21,845 ....F9
Portola 2,227 ....E4
Portola Valley 4,462 ....J3
Poway 48,044 ....J11
Project City 1,657 ....C3
Quartz Hill 9,890 ....G9
Quincy-East Quincy▲ 4,277 ....E4
Ramona 15,691 ....J10
Rancho Cordova 55,060 ....C8
Rancho Cucamonga (Cucamonga) 127,743 ....E10
Rancho Mirage 13,249 ....J10
Rancho Palos Verdes 41,145 ....B11
Rancho Santa Fe 3,252 ....H10
Red Bluff▲ 13,147 ....C3
Redding▲ 80,865 ....C3
Redlands 63,591 ....H9
Redondo Beach 63,261 ....B11
Redwood City▲ 75,402 ....J3
Redwood Estates-Chemeketa Park 1,847 ....K4
Reedley 20,756 ....F7
Rialto 91,873 ....E10
Richgrove 2,723 ....F8
Richmond 99,216 ....J1
Ridgecrest 24,927 ....H8
Rio Dell 3,174 ....A3
Rio Linda 10,466 ....B8
Rio Vista 4,571 ....L1
Ripon 10,146 ....D6
Riverbank 15,826 ....E6
Riverdale 2,416 ....E7
Riverside▲ 255,166 ....E11
Rocklin 36,330 ....B8
Rodeo 8,717 ....J1
Rohnert Park 42,236 ....C5
Rolling Hills 1,871 ....B11
Rolling Hills Estates 7,676 ....B11
Rosamond 14,349 ....G9
Rosemead 53,505 ....C10
Roseville 79,921 ....B8

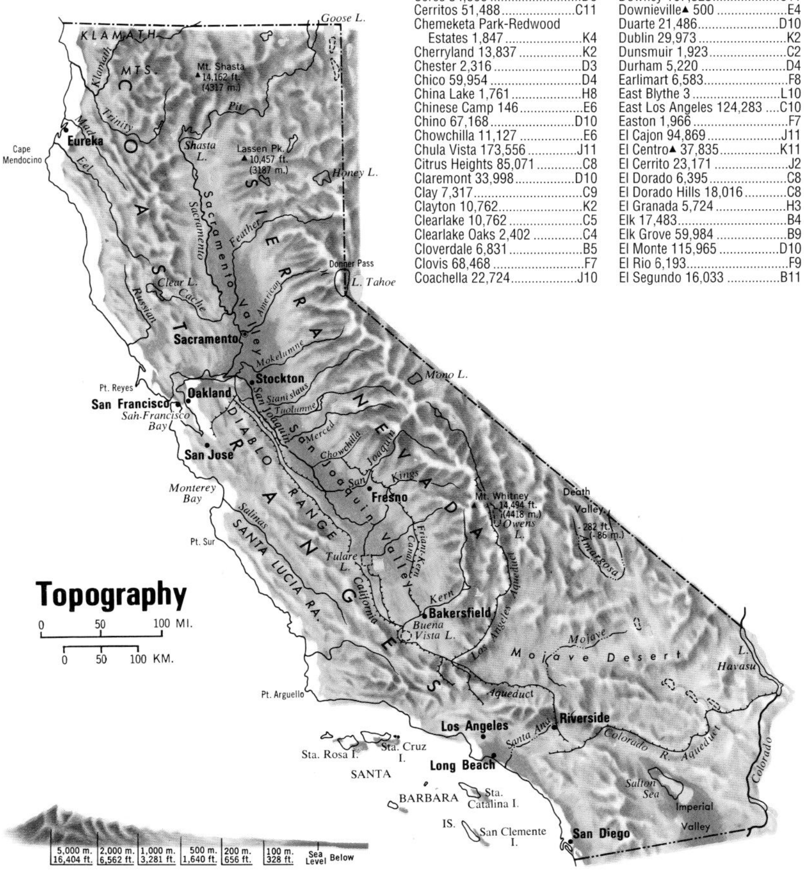
Topography
0 50 100 MI.
0 50 100 KM.
5,000 m. 16,404 ft.
2,000 m. 6,562 ft.
1,000 m. 3,281 ft.
500 m. 1,640 ft.
200 m. 656 ft.
100 m. 328 ft.
Sea Level
Below
Goose L.
KLAMATH MTS.
Mt. Shasta 14,162 ft. (4317 m.)
Pit
Trinity
Klamath
Mad
Eureka
Cape Mendocino
Eel
Shasta L.
Lassen Pk. 10,457 ft. (3187 m.)
Honey L.
COAST RANGES
SIERRA NEVADA
Sacramento
Feather
Clear L.
Cache
Russian
American
Donner Pass
L. Tahoe
Sacramento
Mokelumne
Stockton
Pt. Reyes
San Francisco
San Francisco Bay
Oakland
Stanislaus
Tuolumne
Merced
San Jose
DIABLO RANGE
San Joaquin
Chowchilla
Fresno
Kings
Monterey Bay
Salinas
Pt. Sur
SANTA LUCIA RA.
Mono L.
Mt. Whitney 14,494 ft. (4418 m.)
Owens L.
Death Valley
-282 ft. (-86 m.)
Amargosa
Tulare L.
California
Kern
Bakersfield
Buena Vista L.
Los Angeles Aqueduct
Mojave Desert
Mojave
L. Havasu
Colorado
Pt. Arguello
Aqueduct
Riverside
Los Angeles
Santa Ana
Colorado R. Aqueduct
Sta. Rosa I.
Sta. Cruz I.
SANTA BARBARA IS.
Long Beach
Sta. Catalina I.
San Clemente I.
San Diego
Salton Sea
Imperial Valley

(continued on following page)

Agriculture, Industry and Resources

DOMINANT LAND USE

Wheat, Small Grains
Specialized Dairy
Fruit and Mixed Farming
Fruit, Truck and Mixed Farming
General Farming, Livestock, Special Crops
Cotton, Alfalfa
Potatoes, General Farming
Range Livestock
Forests
Urban Areas
Nonagricultural Land

MAJOR MINERAL OCCURRENCES

Ab Asbestos
Ag Silver
Au Gold
Bx Borax
Cl Clay
Cu Copper
Fe Iron Ore
G Natural Gas
Gp Gypsum
Hg Mercury
K Potash
Lt Lithium
Mg Magnesium
Mo Molybdenum
Mr Marble
Na Salt
O Petroleum
Pb Lead
Pt Platinum
Tc Talc
W Tungsten
Zn Zinc

Water Power
Major Industrial Areas

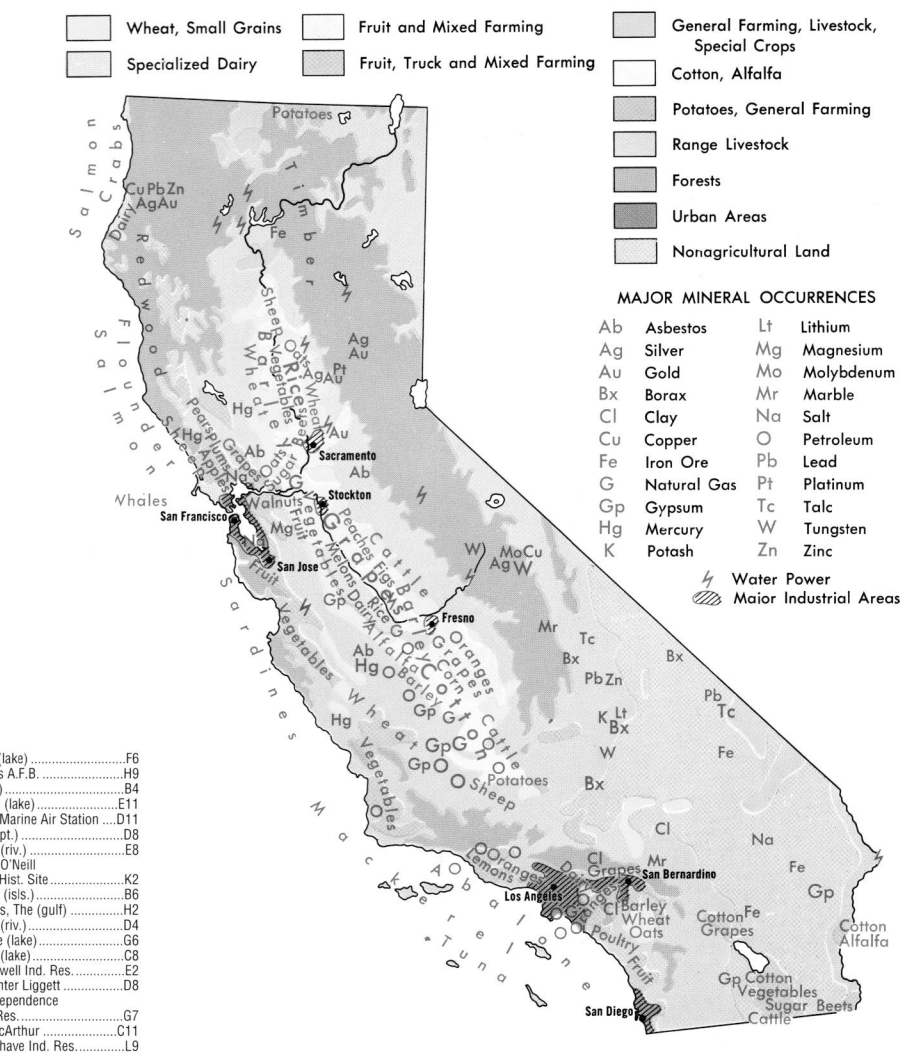

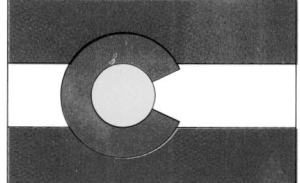

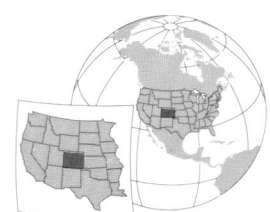

AREA 104,091 sq. mi. (269,596 sq. km.)
POPULATION 4,301,261
CAPITAL Denver
LARGEST CITY Denver
HIGHEST POINT Mt. Elbert 14,433 ft.
(4399 m.)
SETTLED IN 1858
ADMITTED TO UNION August 1, 1876
POPULAR NAME Centennial State
STATE FLOWER Rocky Mountain Columbine
STATE BIRD Lark Bunting

(continued on following page)

Agriculture, Industry and Resources

DOMINANT LAND USE

Specialized Wheat

Wheat, Range Livestock

Wheat, Grain Sorghums, Range Livestock

Dry Beans, General Farming

Sugar Beets, Dry Beans, Livestock, General Farming

Fruit, Mixed Farming

General Farming, Livestock, Special Crops

Range Livestock

Forests

Urban Areas

Nonagricultural Land

MAJOR MINERAL OCCURRENCES

Ag Silver
Au Gold
Be Beryl
C   Coal
Cl Clay
Cu Copper
F   Fluorspar
Fe Iron Ore
G   Natural Gas

Mi Mica
Mo Molybdenum
Mr Marble
O   Petroleum
Pb Lead
U   Uranium
V   Vanadium
W   Tungsten
Zn Zinc

⚡ Water Power
▨ Major Industrial Areas

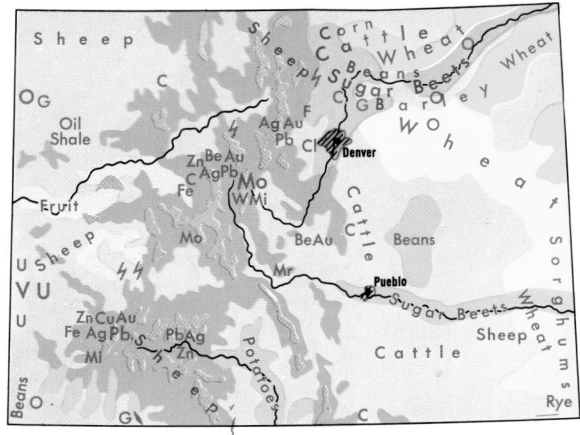

Topography

50   100 MI.

50   100 KM.

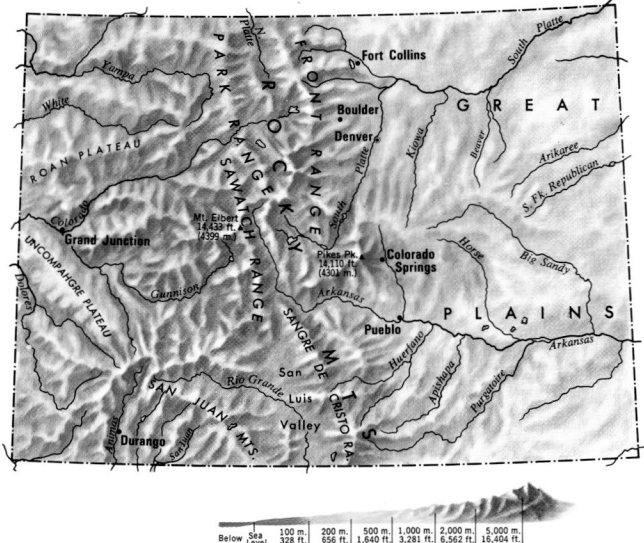

Below Sea Level | 100 m. 328 ft. | 200 m. 656 ft. | 500 m. 1,640 ft. | 1,000 m. 3,281 ft. | 2,000 m. 6,562 ft. | 5,000 m. 16,404 ft.

Colorado
SCALE

0 5 10   20   30   40 MI.

0 5 10 20   30   40 KM.

State Capitals........⊛ County Seats......⦿
Major Limited Access Hwys.

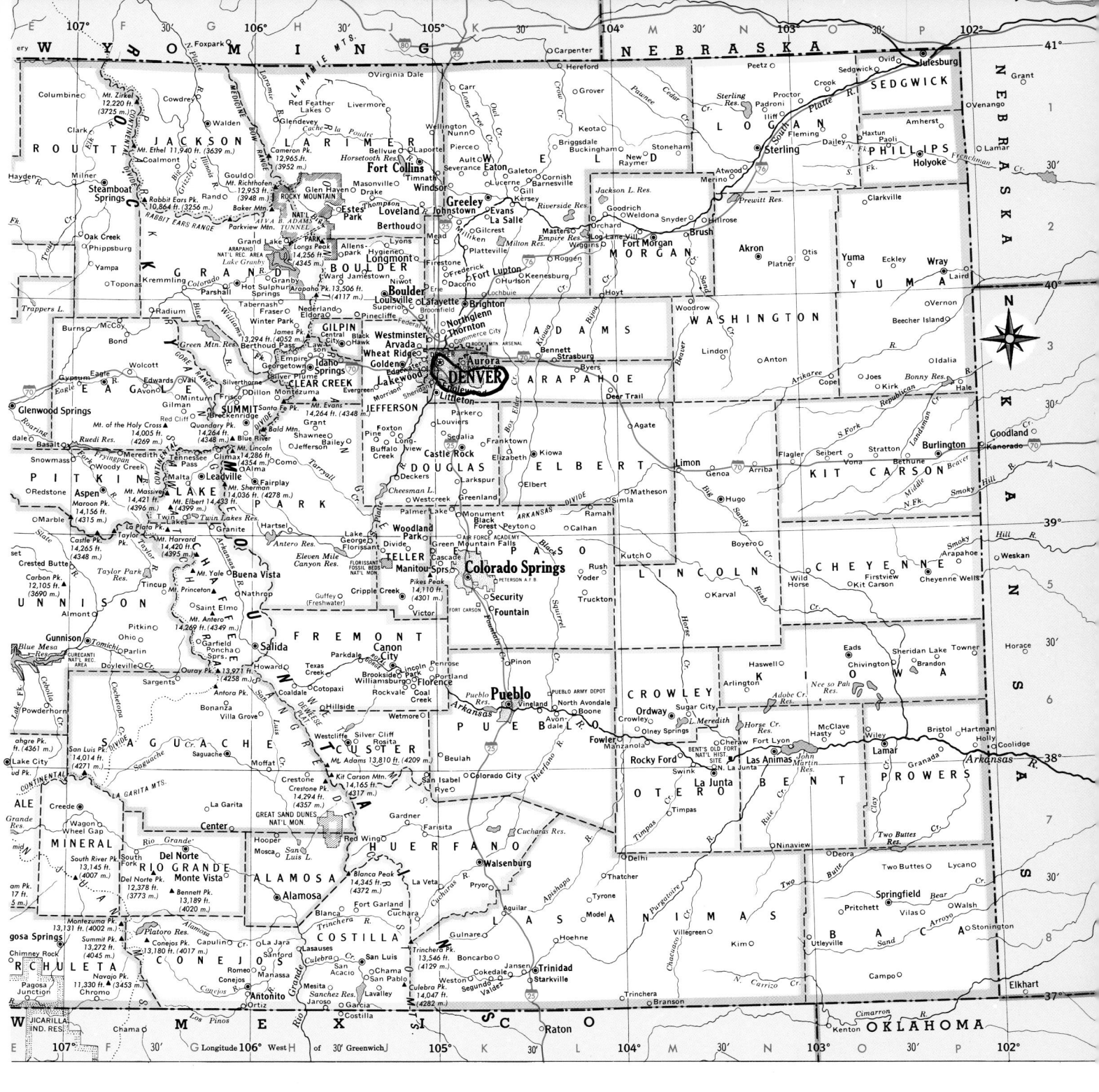

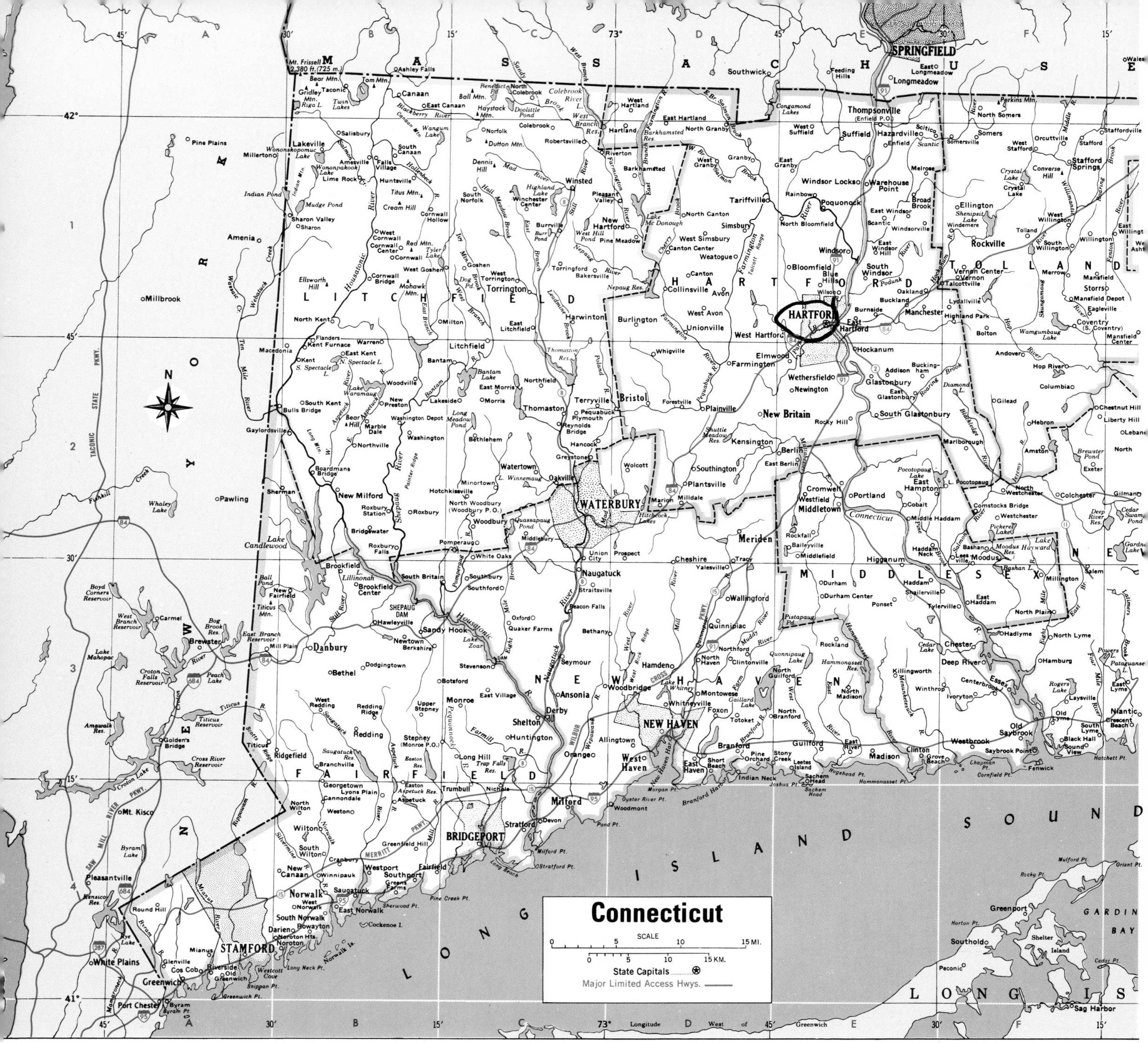

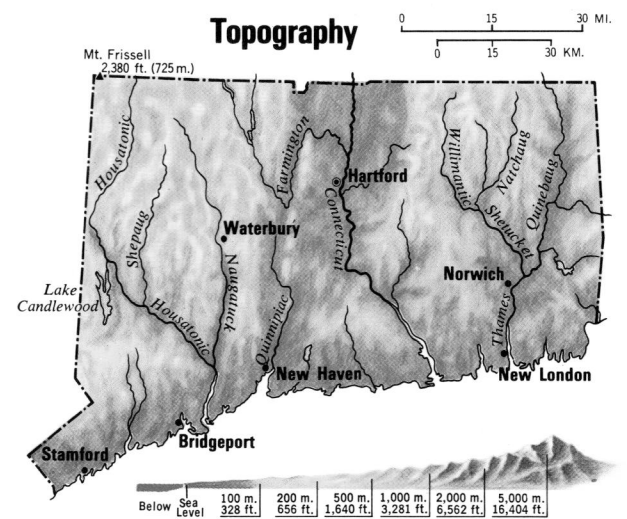

Connecticut
SCALE
0 5 10 15 MI.
0 5 10 15 KM.
State Capitals
Major Limited Access Hwys.
MASSACHUSETTS
NEW YORK
LITCHFIELD
HARTFORD
TOLLAND
FAIRFIELD
NEW HAVEN
MIDDLESEX
LONG ISLAND SOUND
HARTFORD
SPRINGFIELD
WATERBURY
NEW HAVEN
BRIDGEPORT
STAMFORD
Mt. Frissell 2,380 ft. (725 m.)
73° Longitude D West of 45′ Greenwich

Topography
0 15 30 MI.
0 15 30 KM.
Mt. Frissell 2,380 ft. (725 m.)
Housatonic
Farmington
Hartford
Connecticut
Willimantic
Natchaug
Quinebaug
Shetucket
Waterbury
Shepaug
Naugatuck
Quinnipiac
Norwich
Thames
Lake Candlewood
New Haven
New London
Stamford
Bridgeport
Below Sea Level
100 m. 328 ft.
200 m. 656 ft.
500 m. 1,640 ft.
1,000 m. 3,281 ft.
2,000 m. 6,562 ft.
5,000 m. 16,404 ft.

COUNTIES

Fairfield 882,567 .... B3
Hartford 857,183 .... D1
Litchfield 182,193 .... B1
Middlesex 155,071 .... E3
New Haven 824,008 .... D3
New London 259,088 .... G2
Tolland 136,364 .... F1
Windham 109,091 .... H1

CITIES and TOWNS

Abington 600 .... G1
Addison 700 .... E2
Allingtown D3
Amston 900 .... F2
Andover • 3,036 .... F2
Ansonia 18,554 .... C3
Ashford • 4,098 .... G1
Ashford P.O. (Warrenville) 500 .... G1
Attawaugan 400 .... H1
Avon 1,434 .... D1
Avon • 15,832 .... D1
Bakersville 750 .... C1
Ballouville 800 .... H1
Baltic G2
Bantam 802 .... B2
Barkhamsted • 3,494 .... D1
Beacon Falls • 5,246 .... C3
Berkshire 500 .... B3
Berlin • 18,215 .... E2
Bethany • 5,040 .... C3
Bethel 9,137 .... B3
Bethel • 18,067 .... B3
Bethlehem • 3,422 .... C2
Bethlehem 2,022 .... C2
Bloomfield • 19,587 .... E1
Blue Hills 3,020 .... E1
Bolton • 5,017 .... F1
Botsford 400 .... C3
Branchville 600 .... B3
Branford • 28,683 .... D3
Branford 5,735 .... D3
Bridgeport 139,529 .... C4
Bridgewater • 1,824 .... B2
Bristol 60,062 .... D2
Broad Brook 3,469 .... E1
Brookfield • 15,664 .... B3
Brookfield Center B3
Brooklyn • 7,173 .... H1
Buckingham 800 .... E2
Burlington • 8,190 .... D1
Burnside E1
Byram A4
Canaan • 1,288 .... B1
Canaan 1,081 .... B1
Cannondale 400 .... B4
Canterbury • 4,692 .... H2
Canton • 8,840 .... D1
Canton 1,565 .... D1
Canton Center 312 .... D1
Centerbrook 800 .... F3
Center Groton 600 .... G3
Central Village 950 .... H2
Chaplin • 2,250 .... G1
Cheshire • 28,543 .... D2
Cheshire 5,789 .... D2
Chester • 3,743 .... F3
Chester 1,546 .... F3
Clinton • 13,094 .... E3
Clinton 3,516 .... E3
Clintonville D3
Cobalt 700 .... E2
Colchester • 14,551 .... F2
Colchester 3,212 .... F2
Colebrook 1,471 .... C1
Collinsville 2,686 .... D1
Columbia • 4,971 .... F2
Cornwall • 1,434 .... B1
Cos Cob A4
Coventry • 11,504 .... F1
Coventry 2,914 .... F1
Cranbury 700 .... B4
Cromwell • 12,871 .... E2
Crystal Lake 1,459 .... F1
Danbury 74,848 .... B3
Danielson 4,265 .... H1
Darien • 19,607 .... B4
Dayville H1
Deep River • 4,610 .... F3
Deep River 2,470 .... F3
Derby 12,391 .... C3
Devon C4
Durham • 6,627 .... E3
Durham 2,773 .... E3
Durham Center 500 .... E3
Eagleville 310 .... F1
East Berlin 950 .... E2
East Brooklyn 1,473 .... H1
East Canaan 800 .... B1
East Glastonbury 300 .... E2
Eastford • 1,618 .... G1
East Granby • 4,745 .... E1
East Haddam • 8,333 .... F3
East Hampton • 13,352 .... E2
East Hampton 2,254 .... E2
East Hartford • 49,575 .... E1
East Hartland 900 .... D1
East Haven • 28,189 .... D3
East Killingly 900 .... H1
East Litchfield 300 .... C1
East Lyme • 18,118 .... G3
East Morris 800 .... C2
East Norwalk B4
Easton • 7,272 .... B4
East Putnam 500 .... H1
East River 500 .... E3
East Thompson 350 .... H1
East Windsor • 9,818 .... E1
East Windsor Hill 500 .... E1
East Woodstock 400 .... H1
Ellington • 12,921 .... F1
Elmwood D2
Enfield • 45,212 .... E1
Enfield 8,151 .... E1
Essex • 6,505 .... F3
Essex 2,573 .... F3
Exeter 160 .... F2
Fabyan 600 .... H1

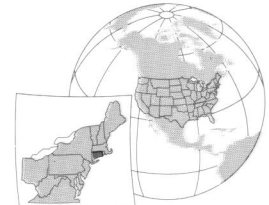

AREA 5,018 sq. mi. (12,997 sq. km.)
POPULATION 3,405,565
CAPITAL Hartford
LARGEST CITY Bridgeport
HIGHEST POINT Mt. Frissell 2,380 ft. (725 m.)
SETTLED IN 1858
ADMITTED TO UNION January 9, 1788
POPULAR NAME Constitution State; Nutmeg State
STATE FLOWER Mountain Laurel
STATE BIRD Robin

© Copyright HAMMOND INCORPORATED, Maplewood, N.J.

Mystic 4,001	H3	Quinebaug 1,122	H1
Naugatuck 30,989	C3	Quinnipiac	D3
New Britain 71,538	E2	Redding • 8,270	B3
New Canaan • 19,395	B4	Redding Ridge 550	B3
New Fairfield • 13,953	B3	Ridgefield • 23,643	B3
New Hartford 1,049	C1	Ridgefield 7,212	B3
New Hartford • 6,088	C1	Riverside	A4
New Haven 123,626	D3	Rockfall 900	E2
Newington • 29,306	E2	Rockville	F1
New London 25,671	G3	Rocky Hill • 17,966	E2
New Milford 27,121	B2	Rogers 650	H1
New Milford 6,633	B2	Round Hill 900	A4
New Preston 1,110	B2	Rowayton	B4
Newtown • 25,031	B3	Roxbury • 2,136	B2
Newtown 1,843	B3	Roxbury Station 250	B2
Niantic 3,085	G3	Sachem Head 250	E3
Nichols	C4	Salem • 3,858	F3
Noank 1,830	G3	Salisbury • 3,977	B1
Norfolk • 1,660	C1	Sandy Hook	B3
Noroton	B4	Saugatuck	B4
Noroton Heights	B4	Saybrook Point 700	F3
North Bloomfield 500	E1	Scantic 500	E1
North Branford • 13,906	E3	Scotland • 1,556	G2
Northfield 600	C2	Seymour • 15,454	C3
Northford	D3	Sharon • 2,968	B1
North Franklin 500	G2	Shelton 38,101	C3
North Granby 1,720	D1	Sherman • 3,827	B2
North Grosvenor Dale 1,424	H1	Short Beach	D3
North Guilford	E3	Simsbury • 23,234	D1
North Haven • 23,035	D3	Simsbury 5,603	D1
North Lyme	F3	Somers • 10,417	F1
North Stonington • 4,991	H3	Somers 1,643	F1
Northville 700	B2	Somersville 750	F1
North Wilton 900	B4	South Britain 390	B3
North Woodbury 900	C2	Southbury • 18,567	C3
North Woodstock 400	G1	South Coventry (Coventry) 1,381	
Norwalk 82,951	B4	South Glastonbury	E2
Norwich 36,117	G2	Southington • 39,728	D2
Norwichtown	G2	South Kent 450	B2
Oakdale 608	G3	South Killingly 500	H1
Oakville 8,618	C2	South Lyme 250	F3
Occum	G2	South Norfolk 400	C1
Old Greenwich	A4	South Norwalk	B4
Old Lyme • 7,406	F3	Southport	B4
Old Mystic 3,205	H3	South Willington 450	F1
Old Saybrook 1,962	F3	South Wilton	B4
Old Saybrook • 10,367	F3	South Windham 1,278	G2
Oneco 550	H2	South Windsor • 24,412	E1
Orange • 13,233	C3	South Woodstock 1,211	G1
Oxford • 9,821	C3	Stafford • 11,307	F1
Pachaug 400	H2	Stafford Springs 4,100	F1
Pawcatuck 5,474	H3	Staffordville 500	G1
Pequabuck 642	C2	Stamford 117,083	A4
Pine Meadow 400	D1	Stepney	B3
Plainfield • 14,619	H2	Sterling • 3,099	H2
Plainfield 2,638	H2	Stonington • 17,906	H3
Plainville • 17,328	D2	Stonington 1,032	H3
Plantsville	D2	Stony Creek	E3
Pleasant Valley 300	C1	Storrs 10,996	F1
Pleasure Beach 1,356	G3	Stratford • 49,976	C4
Plymouth • 11,634	C2	Suffield • 8,983	D1
Pomfret • 3,798	H1	Suffield 13,552	E1
Pomfret Center 175	H1	Taconic 400	B1
Poquonock	E1	Taftville	G2
Poquonock Bridge 1,592	G3	Talcottville 875	F1
Portland • 8,732	E2	Tariffville 1,371	D1
Portland 5,534	E2	Terryville 5,360	C2
Preston 4,688	H2	Thamesville	G2
Prospect • 8,707	D2	Thomaston • 7,503	C2
Putnam • 9,002	H1	Thompson • 8,878	H1
Putnam 6,746	H1	Thompsonville 8,125	E1
Putnam Heights 500	H1	Titicus 450	A3
Quaddick 400	H1	Tolland 13,146	F1
Quaker Hill 2,052	G3		

Torringford	C1	Byram (riv.)	A4
Torrington 35,202	C1	Candlewood (lake)	A2
Totoket 950	D3	Coast Guard Academy	G3
Trumbull • 34,243	C4	Colebrook River (lake)	C1
Uncasville 1,597	G3	Congamond (lakes)	E1
Union • 693	G1	Connecticut (riv.)	E2
Union City	C2	Dennis (hill)	C1
Unionville	D1	Easton (res.)	B3
Vernon • 28,063	F1	Eight Mile (riv.)	F3
Vernon Center	F1	Farmington (riv.)	D1
Versailles 540	G2	French (riv.)	H1
Voluntown • 2,528	H2	Frissell (mt.)	B1
Wallingford • 43,026	D3	Gaillard (lake)	D3
Wallingford 17,509	D3	Gardner (lake)	G2
Warehouse Point	E1	Hammonasset (pt.)	E3
Warren • 1,254	B2	Hammonasset (res.)	E3
Warrenville 500	G1	Haystack (mt.)	C1
Washington • 3,596	B2	Highland (lake)	C1
Washington Depot 900	B2	Hockanum (riv.)	E1
Waterbury 107,271	C2	Hop (riv.)	F1
Waterford • 19,152	G3	Housatonic (riv.)	C3
Waterford 2,736	G3	Lillinonah (lake)	B3
Watertown • 21,661	C2	Little (riv.)	G2
Wauregan 1,085	H2	Long Island (sound)	C4
Weatogue 2,805	D1	Mad (riv.)	C1
West Avon	D1	Mashapaug (lake)	G1
Westbrook • 6,292	F3	Mason (isl.)	H3
Westbrook 2,238	F3	Mattabesset (riv.)	E2
West Cornwall 425	B1	Mianus (riv.)	A4
Westfield	E2	Mohawk (mt.)	B1
West Granby 567	D1	Moosup (riv.)	H2
West Hartford 63,589	D1	Mount Hope (riv.)	G1
West Haven 52,360	D3	Mudge (pond)	B1
West Mystic 3,595	H3	Mystic (riv.)	H3
West Norwalk 950	B4	Natchaug (riv.)	G1
Weston • 10,037	B4	Naugatuck (riv.)	C3
Westport • 25,749	B4	Nepaug (riv.)	D1
West Reading 500	B3	Niantic (riv.)	G3
West Simsbury 2,395	D1	Norwalk (riv.)	B4
West Suffield	E1	Pachaug (pond)	H2
West Woodstock 500	G1	Pawcatuck (riv.)	H3
Wethersfield • 26,271	E2	Pequabuck (riv.)	D2
Whigville 500	D2	Pequonnock (riv.)	C3
Whitneyville	D3	Pocotopaug (lake)	E2
Willimantic 15,823	G1	Quaddick (res.)	H1
Willington • 5,959	F1	Quinebaug (riv.)	H2
Wilson • 300	E1	Quinnipiac (riv.)	D3
Wilton 15,989	B4	Rippowam (riv.)	A4
Winchester • 10,664	C1	Sachem (head)	E4
Winchester Center 350	C1	Salmon (brook)	D1
Wincham • 22,857	G2	Salmon (riv.)	F2
Windsor • 28,237	E1	Saugatuck (res.)	B3
Windsor 17,517	E1	Scantic (riv.)	E1
Windsor Locks • 12,043	E1	Shenipsit (lake)	F1
Windsorville 450	E1	Shepaug (riv.)	B2
Winnipauk 650	B4	Shetucket (riv.)	G2
Winsted 7,321	C1	Silvermine (riv.)	B4
Winthrop 750	F3	Spectacle (lakes)	F1
Wolcott • 15,215	D2	Still (riv.)	B3
Woodbridge • 8,983	D3	Still (riv.)	C1
Woodbury • 9,198	C2	Talcott (range)	D1
Woodbury 1,298	C2	Thames (riv.)	G3
Woodmont 1,711	D4	Thomaston (res.)	C2
Woodstock • 7,221	H1	Titicus (riv.)	A3
Yalesville	D3	Trap Falls (res.)	C3
Yantic	G2	Twin (lakes)	B1
		Wamgumbaug (lake)	F1
OTHER FEATURES		Waramaug (lake)	B2
		West Rock Ridge (hills)	D3
Aspetuck (res.)	B4	Willimantic (riv.)	F1
Bantam (lake)	C2	Wononskopomuc (lake)	B1
Barkhamsted (res.)	D1	Yantic (riv.)	G2
Bear (mt.)	B1		

• Population of town or township

Fairfield • 57,340	B4	Hartland • 2,012	D1	Manchester • 54,740	E1
Falls Village 600	B1	Harwinton • 5,283	C1	Manchester 31,058	E1
Farmington • 23,641	D2	Harwinton 3,293	C1	Mansfield • 20,720	F1
Fenwick 52	F3	Hawleyville 600	B3	Mansfield Center 973	G1
Fitchville 400	G2	Hazardville 4,900	E1	Mansfield Depot 120	F1
Forestville	D2	Hebron • 8,610	F2	Marble Dale 300	B2
Foxon	D3	Higganum 1,671	E2	Marion 900	D2
Franklin • 1,835	G2	Highland Park 500	F1	Marlborough • 5,709	F2
Gales Ferry 1,191	G3	Hockanum	E2	Marlborough 1,039	F2
Gaylordsville 960	A2	Huntington	C3	Massapeag 350	G3
Georgetown 1,650	B4	Indian Neck	D3	Mechanicsville 425	H1
Gilead 350	F2	Ivoryton	F3	Melrose 350	E1
Gilman 350	G2	Jewett City 3,053	H2	Meriden 58,244	D2
Glasgo 450	H2	Kensington 8,541	D2	Merrow 290	F1
Glastonbury • 31,876	E2	Kent • 2,858	B2	Mianus 300	A4
Glastonbury 7,157	E2	Killingly • 16,472	H1	Middlebury • 6,451	C2
Glenville	A4	Killingworth • 6,018	E3	Middlefield • 4,203	E2
Goshen • 2,697	C1	Lake Pocotopaug 3,169	F2	Middle Haddam 325	E2
Granby • 10,347	D1	Lakeside 350	B2	Middletown 43,167	E2
Granby 1,912	D1	Lakeville	B1	Milford 52,305	C4
Greenfield Hill	B4	Laysville 350	F3	Milldale 975	D2
Greenwich • 61,101	A4	Lebanon • 6,907	G2	Mill Plain 750	A3
Grosvenor Dale 700	H1	Ledyard • 14,687	G3	Milton 600	C1
Groton • 39,907	G3	Leetes Island 500	E3	Mohegan 700	G3
Groton 10,010	G3	Lime Rock 350	B1	Monroe • 19,247	C3
Guilford • 21,398	E3	Lisbon • 4,069	G2	Monroe P.O. (Stepney)	B3
Guilford 2,603	E3	Litchfield • 8,316	C2	Montowese	D3
Haddam • 7,157	E3	Litchfield 1,328	C2	Montville • 18,546	G3
Hadlyme •	F3	Long Hill 3,534	C3	Montville 1,711	G3
Hamden • 56,913	D3	Lords Point 500	H3	Moodus 1,263	F2
Hampton • 1,758	G1	Lyons Plain 700	B4	Moosup 3,237	H2
Hanover 500	H1	Madison • 17,858	E3	Morningside Park	G3
Hartford (cap.) 121,578	E1	Madison 2,222	E3	Morris • 2,301	C2

Agriculture, Industry and Resources

DOMINANT LAND USE

- Specialized Dairy
- Dairy, Poultry, Mixed Farming
- Forests
- Urban Areas

MAJOR MINERAL OCCURRENCES

- Cl Clay
- Mi Mica
- Major Industrial Areas

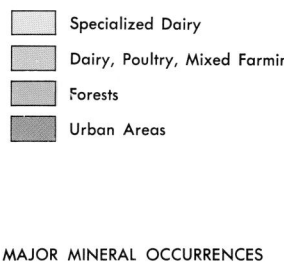

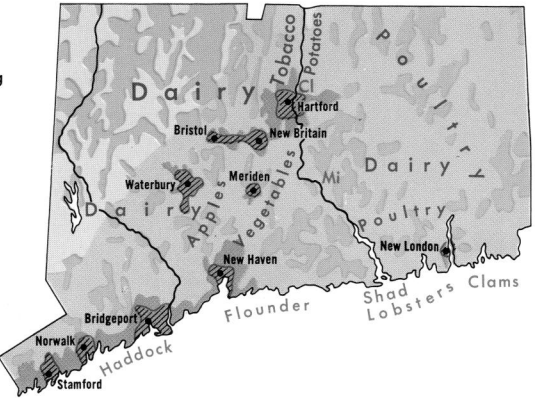

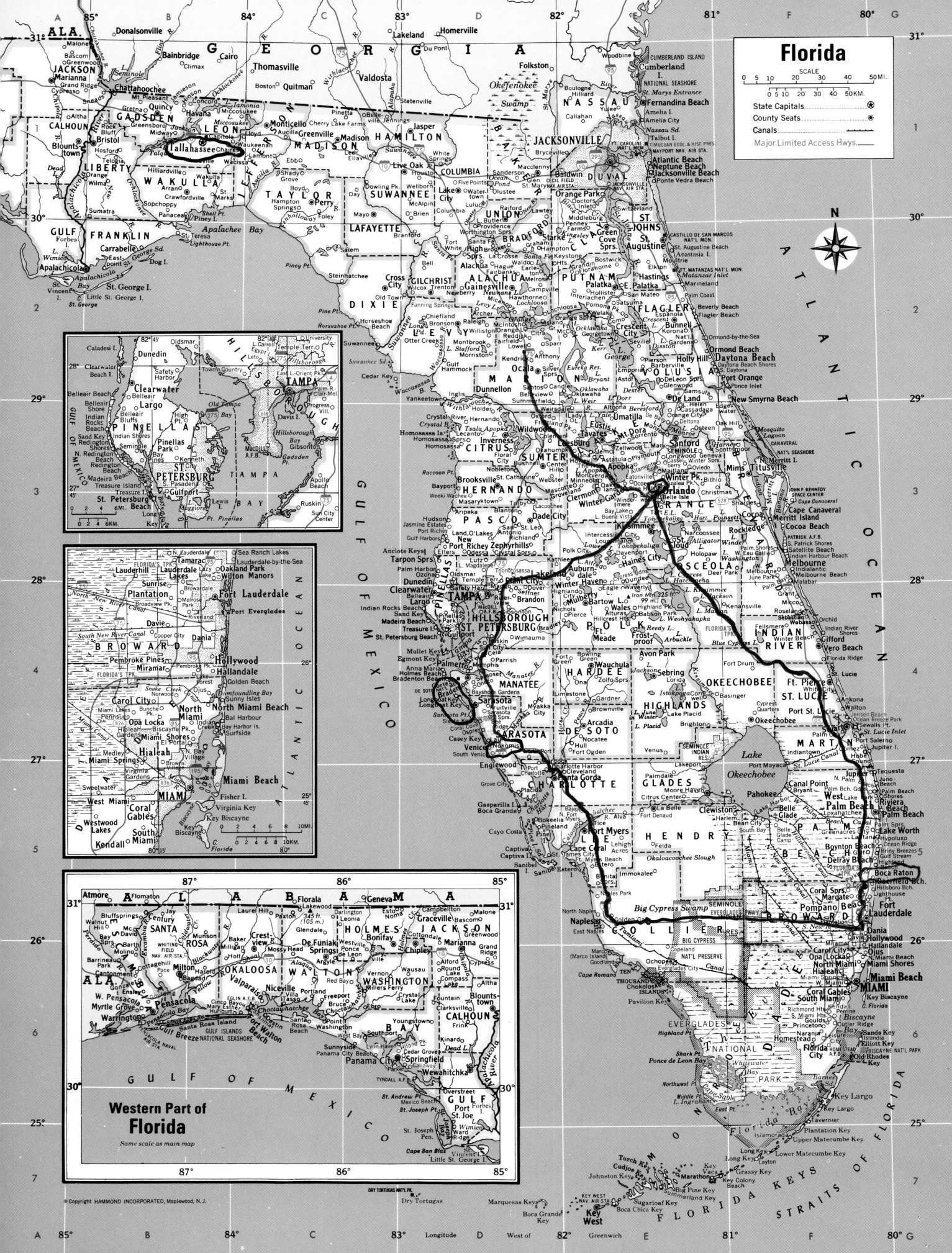

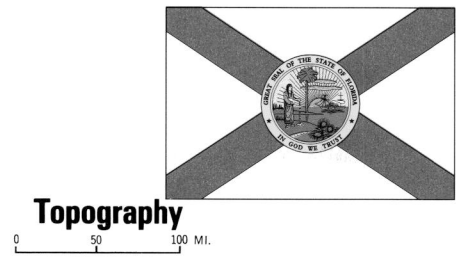

AREA 58,664 sq. mi. (151,940 sq. km.)
POPULATION 15,982,378
CAPITAL Tallahassee
LARGEST CITY Jacksonville
HIGHEST POINT (Walton County) 345 ft. (105 m.)
SETTLED IN 1565
ADMITTED TO UNION March 3, 1845
POPULAR NAME Sunshine State; Peninsula State
STATE FLOWER Orange Blossom
STATE BIRD Mockingbird

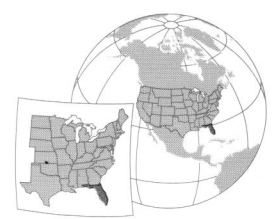

Topography

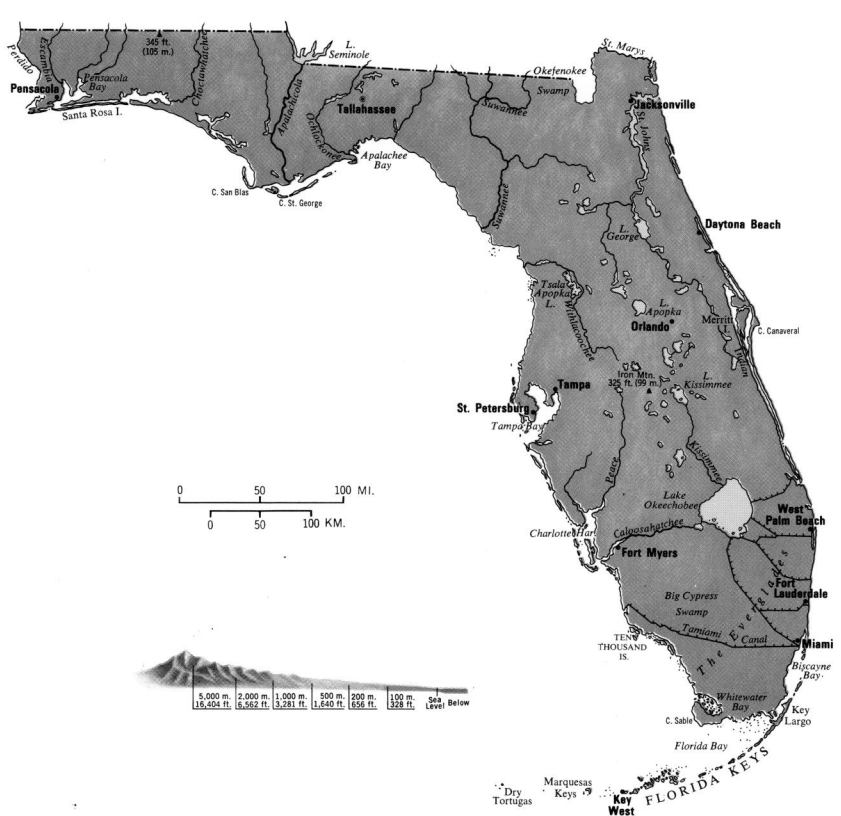

De Land▲ 20,904	E2
De Leon Springs 2,358	E2
Delray Beach 60,020	F5
Deltona 69,543	E3
Destin 11,119	C6
Doctors Inlet 800	E1
Dover 2,798	D4
Dowling Park 250	C1
Dundee 2,912	E3
Dunedin 35,691	B2
Dunnellon 1,898	D2
Eagle Lake 2,496	E4
Earleton 350	E2
East Lake-Orient Park 5,703	C2
East Naples 22,951	E5
East Palatka 1,707	E2
Eastpoint 2,158	B2
Eatonville 2,432	E3
Ebro 250	C6
Edgewater 18,668	F3
Edgewood 1,901	E3
Egypt Lake-Leto 32,782	C2
Elfers 13,161	D3
Elkton 240	E2
El Portal 2,505	B4
Englewood 16,196	D5
Ensley 18,752	B6
Espanola 300	E2
Estero 9,503	E5
Esto 356	C5
Eustis 15,106	E3
Everglades 479	E6
Fairbanks 300	D2
Fairfield 450	D2
Fanning Springs 737	D2
Felda 500	E5
Fellsmere 3,813	F4
Fernandina Beach▲ 10,549	E1
Five Points 1,362	D1
Flagler Beach 4,954	E2
Florahome 400	E2
Floral City 4,989	D3
Florida City 7,843	F6
Florida Ridge 15,217	F4
Foley 525	C1
Fort Denaud 600	E5
Fort Green 300	E4
Fort Lauderdale▲ 152,397	C4
Fort McCoy 600	E2
Fort Meade 5,691	E4
Fort Myers▲ 48,208	E5
Fort Myers Beach 6,561	E5
Fort Ogden 900	E4
Fort Pierce 37,516	F4
Fort Walton Beach 19,973	C6
Fort White 409	D2
Fountain 900	D6
Freeport 1,190	C6
Frink 275	D6
Frostproof 2,975	E4
Fruitland Park 3,186	D3
Fruitville 12,741	D4
Gainesville▲ 95,447	D2
Geneva 2,601	E3
Georgetown 687	E2
Gibsonton 8,752	C3
Gifford 7,599	F4
Glen Saint Mary 473	D1
Glenwood 400	E2
Golden Beach 774	C4
Golden Gate 20,951	E5
Golf 230	F5
Gomez 400	F4
Gonzalez 11,365	B6
Goodland 320	E6
Goulding 4,484	B6
Goulds 7,453	F6
Graceville 2,402	D5
Graham 225	D2
Grandin 250	E2
Grand Ridge 536	A1
Grant 500	F4
Greenacres City 27,569	F5
Green Cove Springs▲ 5,378	E2
Greensboro 619	B1
Greenville 837	C1
Greenwood 735	A1
Gretna 1,709	B1
Grove City 2,092	D5
Groveland 2,360	E3
Gulf Breeze 5,665	B6
Gulf Hammock 325	D2
Gulf Harbors	D3
Gulfport 12,527	B3
Gulf Stream 716	F5
Haines City 13,174	E3
Hallandale 34,282	B4
Hampton 431	D2
Harlem 2,730	F5
Harold 500	B6
Hastings 521	E2
Havana 1,713	B1
Hawthorne 1,415	D2
Hernando 8,253	D3
Hialeah 226,419	B4
Hialeah Gardens 19,297	A4
Highland Beach 3,775	F5
Highland City 2,051	E4
Highland Park 244	E4
High Point 2,288	B3

High Springs 3,863	D2
Hiland Park 999	C6
Hillcrest Heights 266	E4
Hilliard 2,702	E1
Hillsboro Beach 2,163	F5
Hinson 250	B1
Hobe Sound 11,376	F4
Holder 350	D3
Hollister 980	E2
Holly Hill 12,119	E2
Hollywood 139,357	B4
Holmes Beach 4,966	D4
Holt 850	C6
Homestead 31,909	F6
Homosassa 2,294	D3
Homosassa Springs 12,458	D3
Horseshoe Beach 206	C2
Hosford 750	B1
Howey In The Hills 956	E3
Hudson 12,765	D3
Hurlburt	B6
Hypoluxo 2,015	F5
Immokalee 19,763	E5
Indialantic 2,944	F3
Indian Creek 33	B4
Indian Harbour Beach 8,152	F3
Indian River Shores 3,448	F4
Indian Rocks Beach 5,072	B3
Indian Shores 1,705	B4
Indiantown 5,588	F4
Inglis 1,241	D2
Intercession City 600	E3
Interlachen 1,475	E2
Inverness▲ 6,789	D3
Islamorada 6,846	F7
Islandia 6	F6
Jacksonville▲ 735,617	E1
Jacksonville Beach 20,990	E1
Jasmine Estates 18,213	D3
Jasper▲ 1,780	D1
Jay 579	B5
Jennings 833	C1
Jensen Beach 11,100	F4
June Park 4,367	F3
Juno Beach 3,262	F5
Jupiter 39,328	F5
Jupiter Island 620	F4
Kathleen 3,280	D3
Kenansville 650	F4
Kendall 75,226	B5
Kenneth City 4,400	B3
Key Biscayne 10,507	B5
Key Colony Beach 788	F7
Key Largo 11,886	F6
Keystone Heights 1,349	E2
Key West▲ 25,478	C5
Kinard 295	D6
Kissimmee▲ 47,814	E3
La Belle▲ 4,210	E5
Lacoochee 1,345	D3
La Crosse 143	D2
Lady Lake 11,828	E3
Lake Alfred 3,890	E3
Lake Buena Vista 16	E3
Lake Butler▲ 1,927	D1
Lake Carroll 9,980	C2
Lake City▲ 9,980	D1
Lake Como 340	E2
Lake Forest 4,994	B4
Lake Harbor 195	F5
Lake Helen 2,743	E3
Lake Jem 314	E3
Lakeland 78,452	D4
Lake Magdalene 28,755	D3
Lake Mary 11,458	E3
Lake Monroe 500	E3
Lake Park 8,721	F5
Lake Placid 1,668	E4
Lakeport 375	E4
Lake Wales 10,194	E4
Lakewood (Lakewood Park) 10,458	C5
Lake Worth 35,133	G5
Land O'Lakes 20,971	D3
Lantana 9,437	F5
Largo 69,371	B3
Lauderdale-by-the-Sea 2,563	C3
Lauderdale Lakes 31,705	B3
Lauderhill 57,585	B3
Laurel 8,393	D4
Laurel Hill 549	C5
Lawtey 656	D1
Layton 186	F7
Lazy Lake 38	B3
Lecanto 5,161	D3
Lee 352	C1
Leesburg 15,956	E3
Lehigh Acres 33,430	E5
Leisure City 22,152	F6
Leonia 350	C5
Lighthouse Point 10,767	F5
Live Oak▲ 6,480	D1
Lloyd 500	C1
Lochloosa 400	E2
Longboat Key 7,603	D4
Longwood 13,745	E3
Lorida 950	E4
Loughman 1,385	E3
Lowell 250	D2

COUNTIES

Alachua 217,955	D2
Baker 22,259	D1
Bay 148,217	C6
Bradford 26,088	D2
Brevard 476,230	F3
Broward 1,623,018	E5
Calhoun 13,017	D6
Charlotte 141,627	E5
Citrus 118,085	D3
Clay 140,814	E2
Collier 251,377	E5
Columbia 56,513	D1
De Soto 32,209	E4
Dixie 13,827	C2
Duval 778,879	E1
Escambia 294,410	A6
Flagler 49,832	E2
Franklin 11,057	B2
Gadsden 45,087	B1
Gilchrist 14,437	D2
Glades 10,576	E5
Gulf 13,332	D7
Hamilton 13,327	C1
Hardee 26,938	E4
Hendry 36,210	E5
Hernando 130,802	D3
Highlands 87,366	E4
Hillsborough 998,948	D4
Holmes 18,564	C5
Indian River 112,947	F4
Jackson 46,755	C5
Jefferson 12,902	C1
Lafayette 7,022	C2
Lake 210,528	E3
Lee 440,888	E5
Leon 239,452	B1
Levy 34,450	D2
Liberty 7,021	B1
Madison 18,733	C1
Manatee 264,002	D4
Marion 258,916	D2

Martin 126,731	F4
Miami-Dade 2,253,362	F6
Monroe 79,589	E7
Nassau 57,663	E1
Okaloosa 170,498	C6
Okeechobee 35,910	E4
Orange 896,344	E3
Osceola 172,493	E3
Palm Beach 1,131,184	F5
Pasco 344,765	D3
Pinellas 921,482	D4
Polk 483,924	E4
Putnam 70,423	E2
Saint Johns 123,135	E2
Saint Lucie 192,695	F4
Santa Rosa 117,743	B6
Sarasota 325,957	D4
Seminole 365,196	E3
Sumter 53,345	D3
Suwannee 34,844	C1
Taylor 19,256	C1
Union 13,442	D1
Volusia 443,343	E2
Wakulla 22,863	B1
Walton 40,601	C6
Washington 20,973	C6

CITIES and TOWNS

Alachua 6,098	D2
Alford 466	D6
Altamonte Springs 41,200	E3
Altha 506	A1
Altoona 88	E3
Alturas 900	E4
Alva 2,182	E5
Anna Maria 1,814	D4
Anthony 500	D2
Apalachicola▲ 2,334	A2
Apollo Beach 7,444	D3
Apopka 26,642	E3
Arcadia▲ 6,604	E4
Archer 1,289	D2

Aripeka 450	D3
Astatula 1,298	E3
Astor 1,487	E2
Atlantic Beach 13,368	E1
Auburndale 11,032	E3
Avon Park 8,542	E4
Azalea Park 11,073	E3
Babson Park 1,182	E4
Bagdad 1,490	B6
Baker 500	C5
Baldwin 1,634	E1
Bal Harbour 3,305	C4
Barberville 500	E2
Bartow▲ 15,340	E4
Bascom 106	A1
Basinger 300	F4
Bay Harbor Islands 5,146	B4
Bay Lake 23	E3
Bay Pines 3,065	B3
Bayshore	E5
Bayshore Gardens 17,350	D4
Bee Ridge 8,744	D4
Bell 349	D2
Belleair 4,067	B2
Belleair Beach 1,751	B2
Belleair Bluffs 2,243	B3
Belleair Shore	B3
Belle Glade 14,906	F5
Belle Glade Camp 1,141	F5
Belle Isle 5,531	E3
Belleview 3,478	D2
Beverly Beach 547	E2
Biscayne Park 3,269	B4
Bithlo 4,626	E3
Blountstown▲ 2,444	A1
Boca Grande 900	D5
Boca Raton 74,764	F5
Bokeelia 1,997	D5
Bonifay▲ 4,078	C5
Bonita Springs 32,797	E5
Bostwick 500	E2
Boulogne	E1
Bowling Green 2,892	E4

Boynton Beach 60,389	F5
Bradenton▲ 49,504	D4
Bradenton Beach 1,482	D4
Bradley 1,108	D4
Brandon 77,895	D4
Branford 695	D2
Briny Breezes 411	G5
Broadview Park 5,314	B4
Bronson▲ 964	D2
Brooker 352	D2
Brooksville▲ 7,264	D3
Browardale 6,257	B4
Browns Village	B4
Bruce 221	C6
Bunche Park 3,972	B4
Bunnell▲ 2,122	E2
Bushnell▲ 2,050	D3
Callahan 962	E1
Callaway 14,233	C6
Campbellton 212	D5
Canal Point 525	F5
Candler 275	D2
Cantonment	B6
Cape Canaveral 8,829	F3
Cape Coral 102,286	E5
Carol City 59,443	B4
Carrabelle 1,303	B2
Caryville 216	C6
Cassadaga 325	E3
Casselberry 22,629	E3
Cedar Grove 5,367	D6
Cedar Key 790	C2
Center Hill 910	D3
Century 1,714	B5
Charlotte Harbor 3,647	D5
Chattahoochee 3,287	A1
Cherry Lake Farms 400	C1
Chiefland 1,993	D2
Chipley▲ 3,592	D6
Chokoloskee 404	E6
Christmas 1,162	E3
Cinco Bayou 377	B6

Citra 500	D2
Clarksville 350	D6
Clearwater▲ 108,787	B2
Clermont 9,333	E3
Cleveland 3,268	E5
Clewiston 6,460	E5
Cocoa 16,412	F3
Cocoa Beach 12,482	F3
Coconut Creek 43,566	F5
Coleman 869	D3
Compass Lake 296	D6
Concord 300	B1
Cooper City 27,939	B4
Copeland 350	E6
Coral Cove 2,042	D4
Coral Gables 42,249	B5
Coral Springs 117,549	F5
Cornwell 700	E4
Cortez 4,491	D4
Cottagehill 500	B6
Cottondale 869	D6
Crawfordville▲ 1,110	B1
Crescent City 1,776	E2
Crestview▲ 14,766	C6
Cross City▲ 1,775	C2
Crystal Lake 5,341	D6
Crystal River 3,485	D3
Crystal Springs 1,175	D4
Cutler Ridge 24,781	F6
Cypress	A1
Cypress Gardens 8,844	E3
Cypress Quarters 1,150	F4
Dade City▲ 6,188	D3
Dania 20,061	B4
Davenport 1,924	E3
Davie 75,720	B4
Daytona Beach 64,112	F2
Daytona Beach Shores 4,299	F2
De Bary 15,559	E3
Deerfield Beach 64,583	F5
Deer Park 250	F3
De Funiak Springs▲ 5,089	C6

(continued on following page)

(continued)

Loxahatchee 950 ... F5
Lutz 17,081 ... D3
Lynn Haven 12,451 ... C6
Macclenny▲ 4,459 ... D1
Madeira Beach 4,511 ... B3
Madison▲ 3,061 ... C1
Maitland 12,019 ... E3
Malabar 2,622 ... F3
Malone 2,007 ... A1
Mango 8,842 ... D4
Marathon 10,255 ... E7
Marco (Marco Island) 14,879 ... E6
Margate 53,909 ... F5
Marianna▲ 6,230 ... A1
Marineland 6 ... E2
Mary Esther 4,055 ... B6
Masaryktown 920 ... D3
Mascotte 2,687 ... E3
Mayo▲ 988 ... C1
McDavid 500 ... B5
McIntosh 453 ... D2
Medley 1,098 ... B4
Melbourne 71,382 ... F3
Melbourne Beach 3,335 ... F3
Melrose 6,477 ... D2
Melrose Park 7,114 ... B4
Memphis 7,264 ... D4
Merritt Island 36,090 ... F3
Mexico Beach 1,017 ... D6
Miami▲ 362,470 ... B5
Miami Beach 87,933 ... C5
Miami Lakes 22,676 ... B4
Miami Shores 10,380 ... B4
Miami Springs 13,712 ... B5
Micanopy 653 ... D2
Micco 9,498 ... F4
Miccosukee 300 ... B1
Middleburg 10,338 ... E1
Midway 1,446 ... B1
Milligan 950 ... C6
Milton▲ 7,045 ... B6
Mims 9,147 ... F3
Minneola 5,435 ... E3
Miramar 72,739 ... B4
Molino 1,312 ... B6
Montbrook 250 ... D2
Monticello▲ 2,533 ... C1
Montverde 882 ... E3
Moore Haven▲ 1,635 ... E5
Mossy Head 280 ... C6
Mount Dora 9,418 ... E3
Mulberry 3,230 ... E4
Murdock 272 ... D4
Myakka City 672 ... D4
Myrtle Grove 17,211 ... B6
Naples▲ 20,976 ... E5
Naples Park 6,741 ... E5
Naranja 4,034 ... F6
Neptune Beach 7,270 ... E1
Newberry 3,316 ... D2
New Port Richey 16,117 ... D3
New Smyrna Beach 20,048 ... F2
Niceville 11,684 ... C6
Nichols 300 ... E4
Nocatee 950 ... E4
Nokomis 3,334 ... D4
Noma 213 ... C5
Norland 22,995 ... B4
North Bay Village 6,733 ... B4
North Fort Myers 40,214 ... E5
North Lauderdale 32,264 ... B3
North Miami 59,880 ... B4
North Miami Beach 40,786 ... C4
North Naples 13,422 ... E5
North Palm Beach 12,064 ... F5
North Port 22,797 ... D4
North Redington Beach 1,474 ... B3
Oak Hill 1,378 ... F3
Oakland 936 ... E3
Oakland Park 30,966 ... B3
Ocala▲ 45,943 ... D2
Ocean Breeze Park 463 ... F4
Ocean Ridge 1,636 ... F5
Ochopee 750 ... E6
Ocklawaha 700 ... E2
Ocoee 24,391 ... E3
Odessa 3,173 ... D3
Ojus 16,642 ... B4
Okahumpka 251 ... D3
Okeechobee▲ 5,376 ... F4
Old Town 850 ... C2
Oldsmar 11,910 ... B2
Olustee 400 ... D1
Ona ... E4
Oneco ... D4
Opa Locka 14,951 ... B4
Orange ... B1
Orange City 6,604 ... E3
Orange Lake 900 ... D2
Orange Park 9,081 ... E1
Orange Springs 500 ... E2
Orchid 140 ... F4
Orlando▲ 185,951 ... E3
Ormond Beach 36,301 ... E2
Ormond-by-the-Sea 8,430 ... E2
Osprey 4,143 ... D4
Osteen 875 ... E3
Otter Creek 121 ... D2
Oviedo 26,316 ... E3
Oxford ... D3
Ozona 900 ... D3
Pace 7,393 ... B6
Pahokee 5,985 ... F5
Palatka▲ 10,033 ... E2
Palm Bay 79,413 ... F3
Palm Beach 10,468 ... G5
Palm Beach Gardens 35,058 ... F5
Palm Beach Shores 1,269 ... G5
Palm City 20,097 ... F4
Palm Coast 32,732 ... E2
Palmetto 12,571 ... D4
Palm Harbor 59,248 ... D3
Palm River-Clair Mel 17,589 ... C3
Palm Shores 794 ... F3
Palm Springs 11,699 ... F5
Panacea 950 ... B1
Panama City▲ 36,417 ... C6
Panama City Beach 7,671 ... C6
Parker 4,623 ... C6
Parkland 13,835 ... F5
Parrish 950 ... D4
Paxton 656 ... C5
Pembroke Park 6,299 ... B4
Pembroke Pines 137,427 ... B4
Penney Farms 580 ... E2
Pennsuco 15 ... B4
Pensacola▲ 56,255 ... B6
Perrine 15,576 ... F6
Perry▲ 6,847 ... C1
Pierce 500 ... E4
Pierson 2,596 ... E2
Pine Hills 41,764 ... E3
Pineland 444 ... D5
Pinellas Park 45,658 ... B3
Placida 250 ... D5
Plantation 82,934 ... B4
Plant City 29,915 ... D3
Plymouth 950 ... E3
Polk City 1,516 ... E3
Pomona Park 789 ... E2
Pompano Beach 78,191 ... F5
Ponce de Leon 457 ... C6
Ponce Inlet 2,513 ... F2
Ponte Vedra Beach ... E1
Port Charlotte 46,451 ... D5
Portland 300 ... C6
Port Mayaca 400 ... F5
Port Orange 45,823 ... F2
Port Richey 3,021 ... D3
Port Saint Joe 3,644 ... D6
Port Saint Lucie 88,769 ... F4
Port Salerno 10,141 ... F4
Princeton 10,090 ... F6
Progress Village 2,482 ... C3
Punta Gorda▲ 14,344 ... E5
Quincy▲ 6,982 ... B1
Raiford 187 ... D1
Raleigh 275 ... D2
Red Bay 300 ... C6
Reddick 571 ... D2
Redington Beach 1,539 ... B3
Redington Shores 2,338 ... B3
Richland 250 ... D3
Richmond Heights 8,479 ... F6
Riverland 2,108 ... B4
Riverview 12,035 ... D4
Riviera Beach 29,884 ... G5
Rockledge 20,170 ... F3
Roseland 1,775 ... F4
Round Lake 275 ... D6
Ruskin 8,321 ... C3
Safety Harbor 17,203 ... B2
Saint Augustine▲ 11,592 ... E2
Saint Augustine Beach 4,683 ... E2
Saint Catherine 486 ... D3
Saint Cloud 20,074 ... E3
Saint James City 4,105 ... D5
Saint Leo 595 ... D3
Saint Lucie 604 ... F4
Saint Marks 272 ... B1
Saint Petersburg 248,232 ... B3
Saint Petersburg Beach 9,929 ... B3
Samoset 3,440 ... D4
Samsula (Samsula-Spruce Creek) 4,877 ... E2
San Antonio 655 ... D3
Sanderson 800 ... D1
Sanford▲ 38,291 ... E3
Sanibel 6,064 ... D5
San Mateo 975 ... E2
Sarasota▲ 52,715 ... D4
Sarasota Springs 15,875 ... D4
Satellite Beach 9,577 ... F3
Satsuma 610 ... E2
Scottsmoor 900 ... F3
Sea Ranch Lakes 1,392 ... C3
Sebastian 16,181 ... F4
Sebring▲ 9,667 ... E4
Seffner 5,467 ... D4
Seminole 10,890 ... B3
Seville 500 ... E2
Sewall's Point 1,946 ... F4
Shalimar 718 ... C6
Sharpes 3,415 ... F3
Siesta Key 7,150 ... D4
Silver Springs ... D2
Sneads 1,919 ... B1
Sopchoppy 426 ... B1
Sorrento 765 ... E3
South Bay 3,859 ... F5
South Daytona 13,177 ... F2
South Miami 10,741 ... B5
South Miami Heights 33,522 ... F6
South Pasadena 5,778 ... B3
South Patrick Shores 8,913 ... F3
Southport 1,992 ... C6
South Venice 13,539 ... D4
Sparr 902 ... D2
Springfield 8,810 ... D6
Starke▲ 5,593 ... D2
Steinhatchee 800 ... C2
Stuart▲ 14,633 ... F4
Summerfield 780 ... D2
Summerland Key 350 ... E7
Sun City ... D4
Sun City Center 8,326 ... C3
Sunny Isles 15,315 ... C4
Sunnyside 1,008 ... C6
Sunrise 85,779 ... B4
Surfside 4,909 ... B4
Suwannee 365 ... C2
Sweetwater 14,226 ... B5
Switzerland ... E1
Taft 1,938 ... E3
Tallahassee (cap.)▲ 150,624 ... B1
Tamarac 55,588 ... B3
Tampa▲ 303,447 ... C2
Tarpon Springs 21,003 ... D3
Tavares▲ 9,700 ... E3
Tavernier 2,173 ... F6
Telogia 400 ... B1
Temple Terrace 20,918 ... C2
Tequesta 5,273 ... F5
Terra Ceia 450 ... D4
Thonotosassa 6,091 ... D3
Tice 4,538 ... E5
Titusville▲ 40,670 ... F3
Town'n Country 72,523 ... B2
Treasure Island 7,450 ... B3
Trenton▲ 1,617 ... D2
Trilby 930 ... D3
Umatilla 2,214 ... E3
University (University West) 30,736 ... C2
Valparaiso 6,408 ... C6
Venice 17,764 ... D4
Venus 500 ... E4
Vernon 743 ... C6
Vero Beach▲ 17,705 ... F4
Villa Tasso 365 ... C6
Virginia Gardens 2,348 ... B5
Wabasso 918 ... F4
Wacissa 350 ... B1
Wakulla 225 ... B1
Waldo 821 ... D2
Walnut Hill 500 ... B5
Ward Ridge 104 ... D7
Warrington 15,207 ... B6
Watertown 2,837 ... D1
Wauchula▲ 2,837 ... E4
Wausau 398 ... D6
Waverly 1,927 ... E4
Webster 805 ... D3
Weeki Wachee 12 ... D3
Weirsdale 995 ... D3
Welaka 533 ... E2
Wellborn 500 ... D1
West Bay 500 ... C6
West Eau Gallie ... F3
West Melbourne 9,824 ... F3
West Miami 5,863 ... B5
West Palm Beach▲ 82,103 ... F5
West Pensacola 21,939 ... B6
Westville 221 ... C6
Westwood Lakes 12,005 ... B5
Wewahitchka▲ 1,722 ... D6
White City 4,221 ... F4
White Springs 819 ... D1
Wildwood 3,924 ... D3
Williston 2,297 ... D2
Wilton Manors 12,697 ... B3
Wimauma 4,246 ... D4
Windermere 1,897 ... E3
Winter Beach 965 ... F4
Winter Garden 14,351 ... E3
Winter Haven 26,487 ... E3
Winter Park 24,090 ... E3
Winter Springs 31,666 ... E3
Woodville 3,006 ... B1
Worthington Springs 193 ... D2
Yalaha 1,175 ... E3
Yankeetown 629 ... D2
Youngstown 900 ... D6
Yulee 8,392 ... E1
Zellwood 2,540 ... E3
Zephyrhills 10,833 ... D3
Zolfo Springs 1,641 ... E4

OTHER FEATURES

Alapaha (riv.) ... C1
Alaqua (creek) ... C6
Alligator (lake) ... E3
Amelia (isl.) ... E1
Anastasia (isl.) ... E2
Anclote (keys) ... D3
Apalachee (bay) ... B2
Apalachicola (bay) ... B2
Apalachicola (riv.) ... A1
Apopka (lake) ... E3
Arbuckle (lake) ... E4
Aucilla (riv.) ... C1
Banana (riv.) ... F3
Barnes (sound) ... F6
Beresford (lake) ... E3
Big Cypress (swamp) ... E5
Big Cypress Nat'l Preserve ... E5
Big Pine (key) ... E7
Biscayne (bay) ... F6
Biscayne (key) ... B5
Biscayne Nat'l Park ... F6
Blackwater (riv.) ... B6
Blue Cypress (lake) ... F4
Boca Chica (key) ... E7
Boca Ciega (key) ... B3
Boca Grande (key) ... D7
Bryant (lake) ... E2
Caladesi (isl.) ... B2
Caloosahatchee (riv.) ... E5
Canaveral (cape) ... F3
Captiva (isl.) ... D5
Casey (key) ... D4
Castillo de San Marcos Nat'l Mon. ... E2
Cayo Costa (isl.) ... D5
Charlotte (harb.) ... D5
Chattahoochee (riv.) ... B1
Chipola (riv.) ... D6
Choctawhatchee (bay) ... C6
Choctawhatchee (riv.) ... C6
Clearwater Beach (isl.) ... B2
Crescent (lake) ... E2
Crystal (bay) ... D3
Cudjoe (key) ... E7
Cumberland Island Nat'l Seashore ... E1
Cypress (lake) ... E3
Davis (isl.) ... C3
De Soto Nat'l Mem. ... D4
Dead (lake) ... D6
Dexter (lake) ... E2
Disston (lake) ... E2
Dog (isl.) ... B2
Dorr (lake) ... E2
Dry Tortugas (keys) ... D7
Dumfoundling (bay) ... C4
East (pt.) ... E6
East Tohopekaliga (lake) ... E3
Eglin A.F.B 8,082 ... C6
Egmont (key) ... D4
Elliott (key) ... F6
Escambia (riv.) ... B6
Estero (isl.) ... E5
Eureka (res.) ... E2
Everglades, The (swamp) ... F6
Everglades Nat'l Park ... E6
Fenholloway (riv.) ... C1
Fisher (isl.) ... B5
Florida (bay) ... F6
Florida (cape) ... F6
Florida (keys) ... E7
Florida (strs.) ... F7
Forbes (isl.) ... D6
Fort Caroline Nat'l Mem. ... E1
Fort Matanzas Nat'l Mon. ... E2
Gadston (pt.) ... C3
Gasparilla (isl.) ... D5
George (lake) ... E2
Grassy (key) ... F7
Gulf Islands Nat'l Seashore ... B6
Harney (lake) ... F3
Hart (lake) ... E3
Hatchineha (lake) ... E3
Highland (pt.) ... E6
Hillsborough (bay) ... C3
Hillsborough (canal) ... F5
Hillsborough (riv.) ... C2
Holmes (creek) ... D5
Horse (creek) ... E4
Horseshoe (pt.) ... C2
Homosassa (isls.) ... D3
Iamonia (lake) ... B1
Indian (riv.) ... F3
Iron (mt.) ... E4
Istokpoga (lake) ... E4
Jackson (lake) ... B1
Jackson (lake) ... E4
Jacksonville Naval Air Sta. ... E1
John F. Kennedy Space Center ... F3
Johnston (key) ... E7
June in Winter (lake) ... E4
Kerr (lake) ... E2
Key Largo (key) ... F6
Key Vaca (key) ... E7
Key West Naval Air Sta. ... E7
Kingsley (lake) ... E2
Kissimmee (lake) ... E4
Kissimmee (riv.) ... E4
Largo (key) ... F6
Levy (lake) ... D2
Lewis (isl.) ... B3
Lighthouse (pt.) ... B2
Little Saint George (isl.) ... B2
Lochloosa (lake) ... D2
Long (key) ... B3
Long (key) ... F7
Longboat (key) ... D4
Lower Matecumbe (key) ... F7
Lowery (lake) ... E3
MacDill A.F.B. ... C3
Maggiore (lake) ... B3
Manatee (riv.) ... D4
Marco (isl.) ... E6
Marian (lake) ... E4
Marquesas (keys) ... D7
Matanzas (inlet) ... E2
Mayport Naval Air Sta. ... E1
McCoy A.F.B. ... E3
Merritt (isl.) ... F3
Mexico (gulf) ... C4
Miami (canal) ... F5
Miami (riv.) ... B5
Miccosukee (lake) ... B1
Middle (pt.) ... E6
Monroe (lake) ... E3
Mosquito (lag.) ... F3
Mullet (key) ... D4
Myakka (riv.) ... D4
Nassau (riv.) ... E1
Nassau (sound) ... E1
New (riv.) ... B1
New (riv.) ... D1
Newnans (lake) ... D2
North Merritt (isl.) ... F3
North New River (canal) ... F5
Northwest (pt.) ... E6
Ocean (pond) ... D1
Ochlockonee (riv.) ... B1
Ocklawaha (riv.) ... E2
Okaloacoochee Slough (swamp) ... E5
Okeechobee (lake) ... F5
Okefenokee (swamp) ... D1
Old Rhodes (key) ... F6
Old Tampa (bay) ... B3
Olustee (riv.) ... D1
Orange (lake) ... D2
Patrick A.F.B. ... F3
Pavilion (key) ... E6
Peace (riv.) ... E4
Pensacola (bay) ... B6
Pensacola Naval Air Sta. ... B6
Perdido (riv.) ... B6
Pine (isl.) ... D5
Pine (pt.) ... C2
Pine Island (sound) ... D5
Pinellas (pt.) ... B3
Piney (isl.) ... B1
Piney (pt.) ... C2
Placid (lake) ... E4
Plantation (key) ... F7
Poinsett (lake) ... F3
Ponce de Leon (bay) ... E6
Port Everglades (harb.) ... C4
Poet Tampa (harb.) ... B3
Raccoon (pt.) ... D3
Reedy (lake) ... E4
Romano (cape) ... E6
Sable (cape) ... E6
Saint Andrew (pt.) ... D6
Saint George (cape) ... A2
Saint George (isl.) ... B2
Saint George (sound) ... B2
Saint Johns (riv.) ... E2
Saint Joseph (bay) ... D6
Saint Joseph (pt.) ... D6
Saint Joseph (pen.) ... D7
Saint Lucie (canal) ... F4
Saint Lucie (inlet) ... F4
Saint Marys (riv.) ... D1
Saint Marys Entrance (inlet) ... E1
Saint Vincent (isl.) ... D7
San Blas (cape) ... D7
Sand (key) ... B3
Sands (key) ... F6
Sanibel (isl.) ... D5
Santa Fe (lake) ... D2
Santa Fe (riv.) ... D2
Santa Rosa (isl.) ... B6
Santa Rosa (sound) ... B6
Sarasota (pt.) ... D4
Seminole (lake) ... B1
Seminole Ind. Res. ... E4
Seminole Ind. Res. ... E5
Shark (pt.) ... E6
Shell (pt.) ... B1
Shoal (riv.) ... C6
Snake Creek (canal) ... B4
South New River (canal) ... F5
Stafford (lake) ... D2
Sugarloaf (key) ... E7
Suwannee (riv.) ... C2
Suwannee (sound) ... C2
Talbot (isl.) ... E1
Talquin (lake) ... B1
Tamiami (canal) ... E5
Tampa (bay) ... D4
Ten Thousand (isls.) ... E6
Timucuan Ecological and Historic Preserve ... E1
Torch (key) ... E7
Treasure (isl.) ... B3
Tsala Apopka (lake) ... D3
Tyndall A.F.B. ... C6
Upper Matecumbe (key) ... F7
Vaca (key) ... E7
Virginia (key) ... B5
Waccasassa (bay) ... D2
Waccasassa (riv.) ... D2
Washington (lake) ... F3
Weir (lake) ... E2
Weohyakapka (lake) ... E4
West Palm Beach (canal) ... F5
Whitewater (bay) ... F6
Whiting Field Naval Air Sta. ... B6
Wimico (lake) ... A2
Winder (lake) ... F3
Withlacoochee (riv.) ... C1
Withlacoochee (riv.) ... D2
Yale (lake) ... E3
Yellow (riv.) ... B6

▲County seat

Agriculture, Industry and Resources

Cotton Peanuts
Corn
Hogs
Cotton
Tobacco
Peanuts
Pensacola
Oysters
Red Snapper
Mullet
Groupers
Crabs
Jacksonville
Cattle
Potatoes
Menhaden
Mackerel
Hogs
Vegetables
Melons
Oranges
Grapefruit
Orlando
Sponges
Tampa
St. Petersburg
Cattle
Oranges
Citrus Fruit
Kingfish
Vegetables
W. Palm Beach
Sugar cane
Mackerel
Miami
Shrimp
Clams
Lobsters
Sponges
Shrimp
DOMINANT LAND USE
Fruit, Truck & Mixed Farming
Truck & Mixed Farming
Truck Farming
Cotton, Tobacco, Hogs, Peanuts
Peanuts, General Farming
General Farming, Forest Products, Truck Farming, Cotton
Livestock Grazing
Forests
Swampland, Limited Agriculture
Urban Areas
Nonagricultural Land
MAJOR MINERAL OCCURRENCES
Cl Clay
Ls Limestone
O Petroleum
P Phosphates
Pe Peat
Ti Titanium
Zr Zirconium
Water Power
Major Industrial Areas

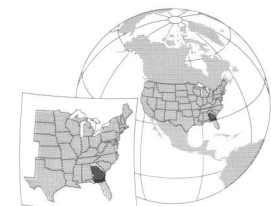

AREA 58,910 sq. mi. (152,577 sq. km.)
POPULATION 8,186,453
CAPITAL Atlanta
LARGEST CITY Atlanta
HIGHEST POINT Brasstown Bald 4,784 ft.
(1458 m.)
SETTLED IN 1733
ADMITTED TO UNION January 2, 1788
POPULAR NAME Empire of the South;
Peach State
STATE FLOWER Cherokee Rose
STATE BIRD Brown Thrasher

COUNTIES

Appling 17,419	H7
Atkinson 7,609	G8
Bacon 10,103	G7
Baker 4,074	D8
Baldwin 44,700	F4
Banks 14,422	E2
Barrow 46,144	E2
Bartow 76,019	C2
Ben Hill 17,484	F7
Berrien 16,235	F8
Bibb 153,887	E5
Bleckley 11,666	F6
Brantley 14,629	J8
Brooks 16,450	E9
Bryan 23,417	K6
Bulloch 55,983	J6
Burke 22,243	J4
Butts 19,522	E4
Calhoun 6,320	C7
Camden 43,664	J9
Candler 9,577	H6
Carroll 87,268	B3
Catoosa 53,282	B1
Charlton 10,282	H9
Chatham 232,048	K6
Chattahoochee 14,882	C6
Chattooga 25,470	B1
Cherokee 141,903	D2
Clarke 101,489	F3
Clay 3,357	C7
Clayton 236,517	D3
Clinch 6,878	G9
Cobb 607,751	C3
Coffee 37,413	G8
Colquitt 42,053	E8
Columbia 89,288	H3
Cook 15,771	F8
Coweta 89,215	C4
Crawford 12,495	E5
Crisp 21,996	E7
Dade 15,154	A1
Dawson 15,999	D2
Decatur 28,240	C9
De Kalb 665,865	D3
Dodge 19,171	F6
Dooly 11,525	E6
Dougherty 96,065	D7
Douglas 92,174	C3
Early 12,354	C8
Echols 3,754	G9
Effingham 37,535	K6
Elbert 20,511	G2
Emanuel 21,837	H5
Evans 10,495	J6
Fannin 19,798	D1
Fayette 91,263	C4
Floyd 90,565	B2
Forsyth 98,407	D2
Franklin 20,285	F2
Fulton 816,006	D3
Gilmer 23,456	D1
Glascock 2,556	G4
Glynn 67,568	J8
Gordon 44,104	C2
Grady 23,659	D9
Greene 14,406	F3
Gwinnett 588,448	D2
Habersham 35,902	E1
Hall 139,277	E2
Hancock 10,076	G4
Haralson 25,690	B3
Harris 23,695	C5
Hart 22,997	G2
Heard 11,012	B4
Henry 119,341	D4
Houston 110,765	E6
Irwin 9,931	F7
Jackson 41,589	E2
Jasper 11,426	E4
Jeff Davis 12,684	G7
Jefferson 17,266	H4
Jenkins 8,575	J5
Johnson 8,560	G5
Jones 23,639	E5
Lamar 15,912	D4
Lanier 7,241	F8
Laurens 44,874	G6
Lee 24,757	D7
Liberty 61,610	J7
Lincoln 8,348	H3
Long 10,304	J7
Lowndes 92,115	F9
Lumpkin 21,016	D1
Macon 14,074	D6
Madison 25,730	F2
Marion 7,144	C6
McDuffie 21,231	H4
McIntosh 10,847	K7
Meriwether 22,534	C4
Miller 6,383	C8
Mitchell 23,932	D8
Monroe 21,757	E4
Montgomery 8,270	G6
Morgan 15,457	F3
Murray 36,506	C1
Muscogee 186,291	C6
Newton 62,001	E3
Oconee 26,225	F3
Oglethorpe 12,635	F3
Paulding 81,678	C3
Peach 23,668	E5
Pickens 22,983	D2
Pierce 15,636	H8
Pike 13,688	D4
Polk 38,127	B3
Pulaski 9,588	E6
Putnam 18,812	F4
Quitman 2,598	B7
Rabun 15,050	F1
Randolph 7,791	C7
Richmond 199,775	H4
Rockdale 70,111	D3
Schley 3,766	D6
Screven 15,374	J5
Seminole 9,369	C9
Spalding 58,417	D4
Stephens 25,435	F1
Stewart 5,252	C6
Sumter 33,200	D6
Talbot 6,498	C5
Taliaferro 2,077	G3
Tattnall 22,305	J6
Taylor 8,815	D5
Telfair 11,794	G7
Terrell 10,970	D7
Thomas 42,737	E9
Tift 38,407	E7
Toombs 26,067	H6
Towns 9,319	E1
Treutlen 6,854	G6
Troup 58,779	B4
Turner 9,504	E7
Twiggs 10,590	F5
Union 17,289	E1
Upson 27,597	D5
Walker 61,053	B1
Walton 60,687	E3
Ware 35,483	H8
Warren 6,336	G4
Washington 21,176	G4
Wayne 26,565	J7
Webster 2,390	C6
Wheeler 6,179	G6
White 19,944	E1
Whitfield 83,525	B1
Wilcox 8,577	F7
Wilkes 10,687	G3
Wilkinson 10,220	F5
Worth 21,967	E8

CITIES and TOWNS

Abbeville▲ 2,298	F7
Acworth 13,422	C2
Adairsville 2,542	C2
Adel▲ 5,307	F8
Adrian 579	G5
Ailey 394	G6
Alamo▲ 1,943	G6
Alapaha 682	F8
Albany▲ 76,939	D7
Aldora 98	D4
Allenhurst 788	J7
Allentown 287	F5
Alma▲ 3,236	G7
Almon 400	E3
Alpharetta 34,854	D2
Alston 159	H6
Alto 876	E2
Alvaton 91	C4
Ambrose 320	G7
Americus▲ 17,013	D6
Andersonville 331	D6
Apalachee 150	E3
Appling▲ 150	H3
Arabi 456	E7
Arco 6,189	J8
Aragon 1,039	B2
Arcade 1,643	E2
Argyle 151	G8
Arlington 1,602	C8
Armuchee 600	B2
Arnoldsville 312	F3
Ashburn▲ 4,419	E7
Ashland 350	F2
Athens▲ 100,266	F3
Atlanta (cap.)▲ 416,474	K1
Attapulgus 492	D9
Auburn 6,904	E2
Augusta▲ 199,775	J4
Austell 5,359	J1
Avalon 278	F1
Avera 217	G4
Avondale Estates 2,609	L1
Axson 300	G8
Baconton 804	D8
Bainbridge▲ 11,722	C9
Baldwin 2,425	E2
Ball Ground 730	D2
Barretts 275	F8
Barnesville▲ 5,972	D4
Barney 146	E8
Bartow 223	G5
Barwick 444	E9
Baxley▲ 4,150	H7
Bellville 130	H6
Belvedere 18,945	L1
Benevolence 138	C7
Berkeley Lake 1,695	D3
Berlin 595	E8
Berryton 250	B2
Bethlehem 716	E3
Between 148	E3
Bibb City 510	B5
Bishop 146	F3
Blackshear▲ 3,283	H8
Blairsville▲ 659	E1
Blakely▲ 5,696	C8
Blitchton 130	J6
Bloomingdale 2,665	K6
Blue Ridge▲ 1,210	D1
Bluffton 118	C7
Blythe 718	H4
Bogart 1,049	E3
Bolingbroke 220	E5
Boneville 285	G4
Boston 1,417	E9
Bostwick 322	E3
Bowdon 513	B3
Bowersville 334	F2
Bowman 898	G2
Box Springs 518	C5
Braselton 1,206	E2
Braswell 80	C3
Bremen 4,579	B3
Brinson 225	C9
Bristol 225	H8
Bronwood 513	D7
Brookfield 600	F8
Brookhaven	K1
Brooklet 1,113	J6
Brooks 553	D4
Broxton 1,428	G7
Brunswick▲ 15,600	K8
Buchanan▲ 941	B3
Buckhead 205	F3
Buena Vista▲ 1,664	D6
Buford 10,668	D2
Bullard 230	F5
Butler▲ 1,907	D5
Byromville 415	E6
Byron 2,887	E5
Cadwell 329	G6
Cairo▲ 9,239	D9
Calhoun▲ 10,667	C1
Calvary 500	D9
Camak 165	G4
Camilla▲ 5,669	D8
Campton	E3
Canon 755	F2
Canton▲ 7,709	C2
Carl 205	E3
Carlton 233	F2
Carnesville▲ 541	F2
Carrollton▲ 19,843	C3
Cartecay 250	D1
Carters 12,035	C1
Cartersville▲ 15,925	C2
Cassville 350	C2
Cataula 500	C5
Cave Spring 975	B2
Cecil 265	F8
Cedar Grove	A1
Cedartown▲ 9,470	B2
Center 3,251	F2
Centerville 4,278	E5
Centralhatchee 383	B4
Chalybeate Springs 265	C5
Chamblee 9,552	K1
Charles	H6
Chatsworth▲ 3,531	C1
Chauncey 295	F6
Chester 305	G6
Chickamauga 2,245	B1
Chula 500	E7
Cisco 488	C1
Clarkesville▲ 1,248	E1
Clarkston 7,231	L1
Claxton▲ 2,276	J6
Clayton▲ 2,019	F1
Clem 350	B3
Clermont 419	E2
Cleveland▲ 1,907	E1
Climax 297	D9
Clyattville 500	F9
Clyo 300	K6
Cobb 338	E7
Cobbtown 311	H6
Cochran▲ 4,455	F6
Cohutta 582	C1
Colbert 488	F2
Coleman 149	C7
College Park 20,382	K2
Collins 528	H6
Colquitt▲ 1,939	C8
Columbus▲ 186,291	C6
Comer 1,052	F2
Commerce 5,292	E2
Concord 336	D4
Conley 6,188	K2
Constitution	K2
Conyers▲ 10,689	D3
Coolidge 552	E8
Coosa 600	B2
Cordele▲ 11,608	E7
Corinth 213	B4
Cornelia 3,674	E1
Cotton 122	D8
Covington▲ 11,547	E3
Crandall	C1
Crawford 807	F3
Crawfordville▲ 572	G3
Crosland	E8
Crystal Springs 500	B2
Culloden 223	D5
Cumming▲ 4,220	D2
Cusseta▲ 1,196	C6
Cuthbert▲ 3,731	C7
Dacula 3,848	E3
Dahlonega▲ 3,638	D1
Daisy 126	J6
Dallas▲ 5,056	C3
Dalton▲ 27,912	C1
Damascus 277	C8
Danielsville▲ 457	F2
Danville 373	F5
Darien▲ 1,719	K8
Dasher 834	F9
Davisboro 1,544	G5
Dawson▲ 5,058	D7
Dawsonville▲ 619	D2
Dearing 441	H4
Decatur▲ 18,147	K1
Deenwood 1,836	H8
Deepstep 132	G4
Demorest 1,465	F1
Denton 269	G7
De Soto 214	D7
Devereux 300	F4
Dexter 509	G6
Dickey	C7
Dillard 198	F1
Dixie 259	F9
Dock Junction (Arco) 6,951	J8
Doerun 828	E8
Donalsonville▲ 2,796	C8
Dooling 163	E6
Doraville 9,862	K1
Douglas▲ 10,639	G7
Douglasville▲ 20,065	C3
Draketown 300	B3
Dry Branch 700	F5
Dublin▲ 15,857	G5
Ducktown	D2
Dudley 447	F5
Duluth 22,122	D2
Dunwoody 32,808	K1
Du Pont 139	G9
Durand 206	C5
Eastanollee 365	F1
East Dublin 2,484	G5
East Ellijay 707	C1
East Juliette	E4
Eastman▲ 5,440	F6
East Newnan 1,305	C4
East Point 39,595	K2
Eastville	E3
Eatonton▲ 6,764	F4
Eden 990	K6
Edge Hill 22	G4
Edison 1,340	C7
Elberta 1,559	E5
Elberton▲ 4,743	G2
Elizabeth 950	J1
Ellabell 500	K6
Ellaville▲ 1,609	D6
Ellenton 336	E8
Ellenwood	L2
Ellerslie 700	C5
Ellijay▲ 1,584	C1
Emerson 1,092	C2
Empire 250	F6
Enigma 869	F8
Ephesus 388	B4
Epworth 300	D1
Eton 319	C1
Euharlee 3,208	C2
Evans 17,727	H3
Everett 250	J8
Experiment 3,233	D4
Fairburn 5,464	J2
Fairmount 745	C2

(continued on following page)

Agriculture, Industry and Resources

[Map of Georgia showing agricultural and resource distribution]

DOMINANT LAND USE

- Specialized Cotton
- Cotton, General Farming
- Cotton, Tobacco, Hogs, Peanuts
- Peanuts, General Farming
- General Farming, Livestock, Fruit, Tobacco
- General Farming, Forest Products, Cotton, Truck Farming
- Forests
- Swampland, Limited Agriculture
- Urban Areas

MAJOR MINERAL OCCURRENCES

- Al Bauxite
- Ba Barite
- C Coal
- Cl Clay
- Fe Iron Ore
- Gn Granite
- Mi Mica
- Mn Manganese
- Mr Marble
- Sl Slate
- Tc Talc
- Ti Titanium

Water Power

Major Industrial Areas

OTHER FEATURES

Georgia

Topography

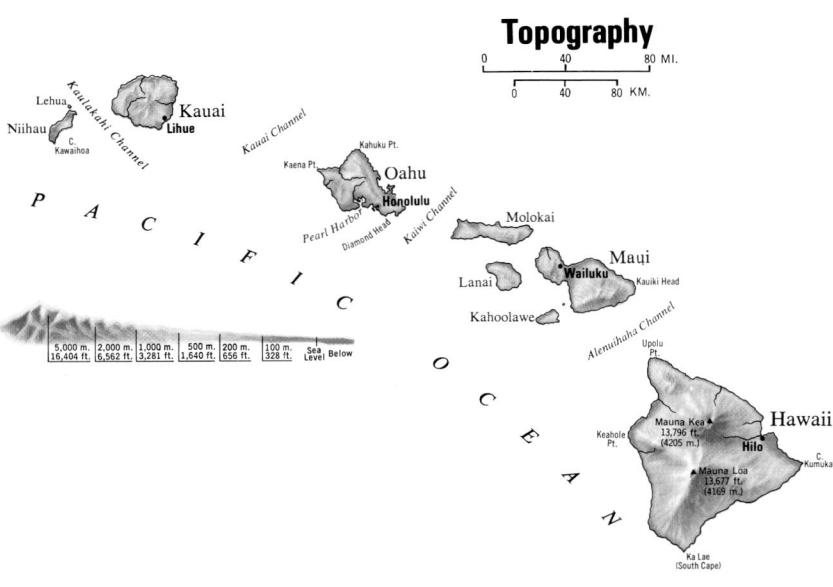

Agriculture, Industry and Resources

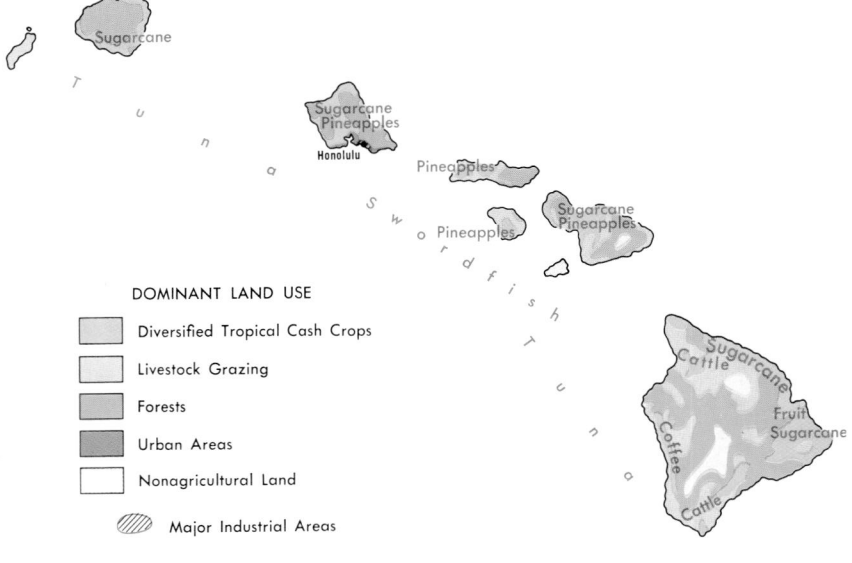

DOMINANT LAND USE

Diversified Tropical Cash Crops

Livestock Grazing

Forests

Urban Areas

Nonagricultural Land

Major Industrial Areas

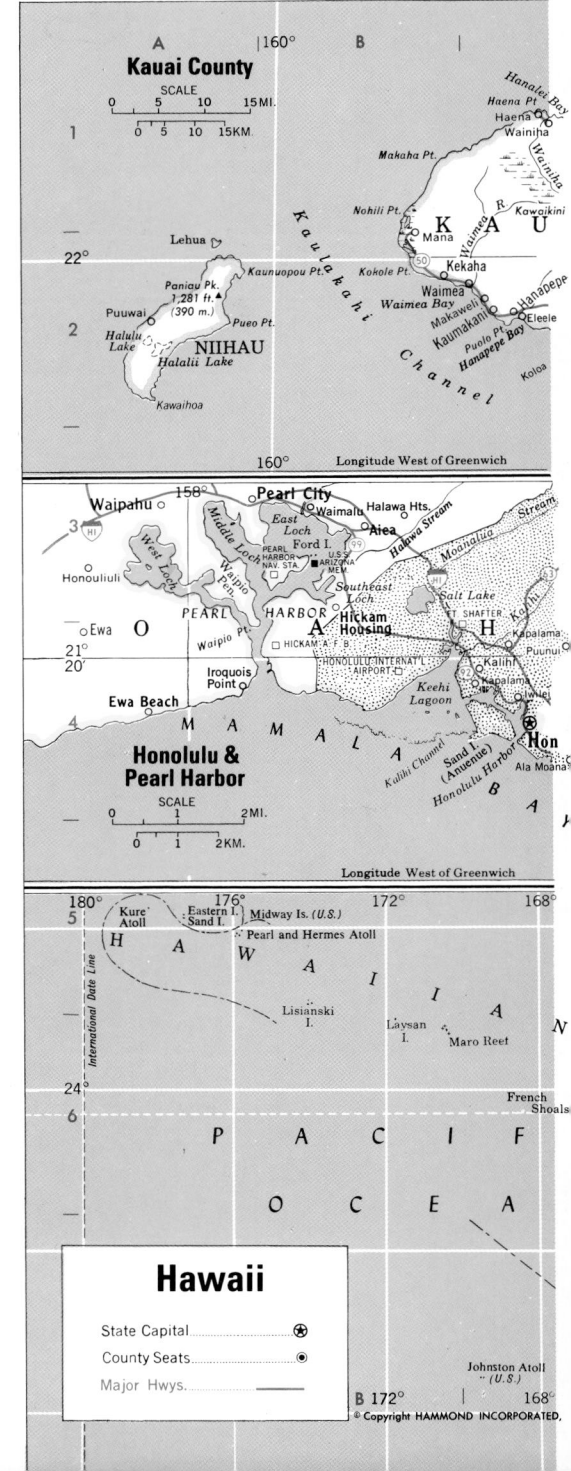

Kauai County

Honolulu & Pearl Harbor

Hawaii

State Capital ....⊛

County Seats ....◉

Major Hwys. ....

Johnston Atoll
(U.S.)

© Copyright HAMMOND INCORPORATED

HAWAII

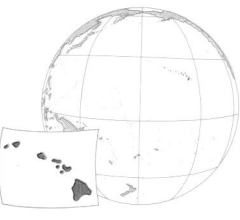

AREA 6,471 sq. mi. (16,760 sq. km.)
POPULATION 1,211,537
CAPITAL Honolulu
LARGEST CITY Honolulu
HIGHEST POINT Mauna Kea 13,796 ft. (4205 m.)
SETTLED IN —
ADMITTED TO UNION August 21, 1959
POPULAR NAME Aloha State
STATE FLOWER Hibiscus
STATE BIRD Nene (Hawaiian Goose)

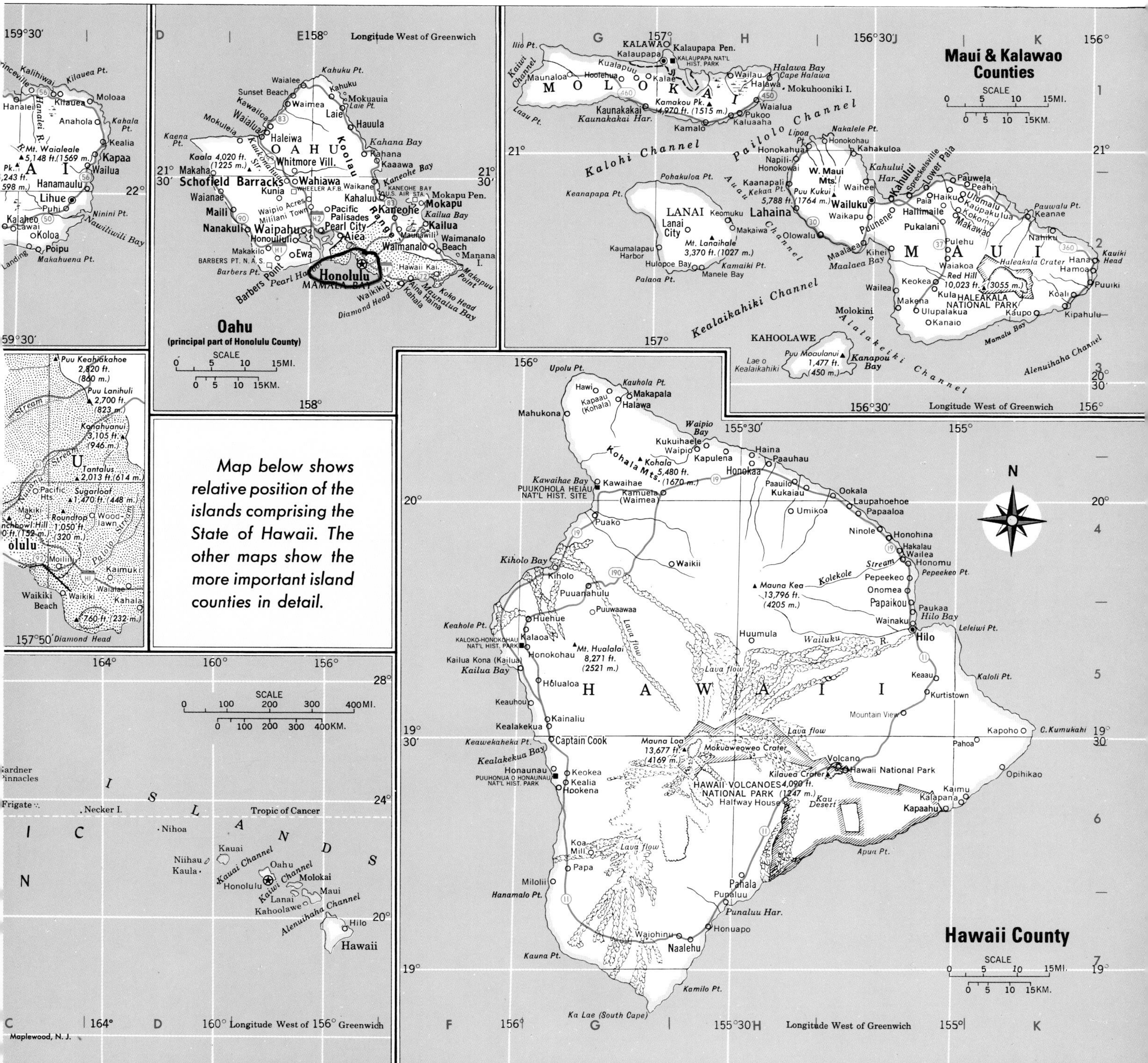

Oahu
(principal part of Honolulu County)

Maui & Kalawao Counties

Map below shows relative position of the islands comprising the State of Hawaii. The other maps show the more important island counties in detail.

Hawaii County

Maplewood, N.J.

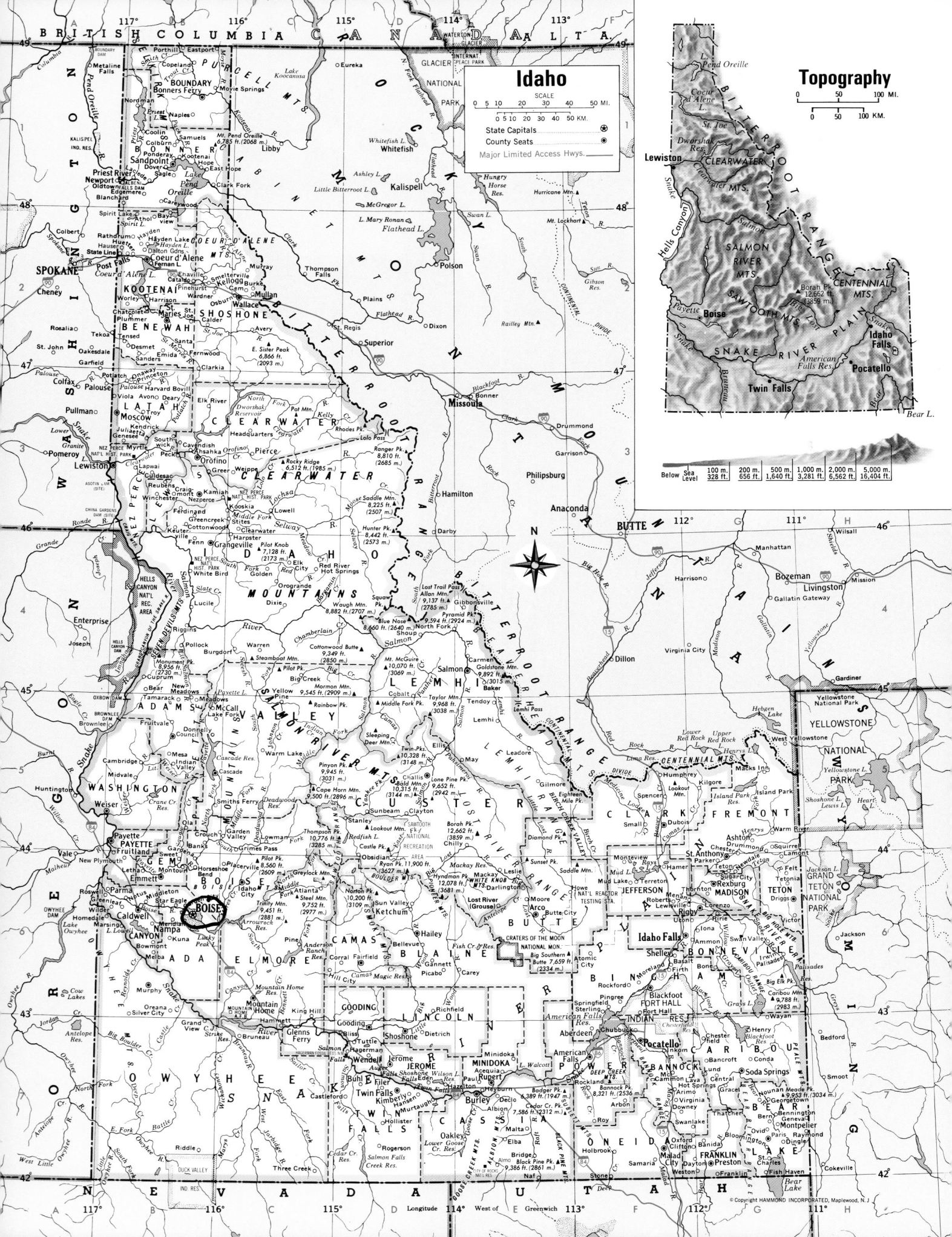

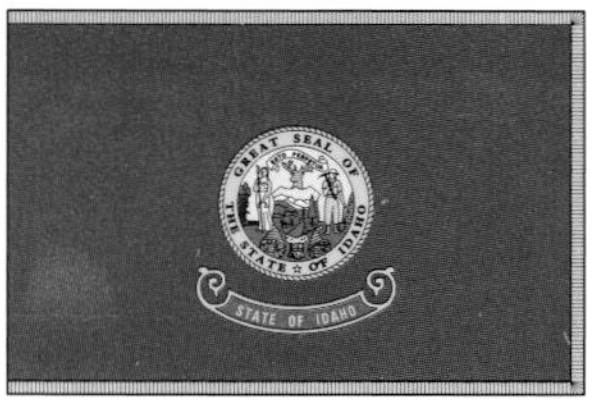
STATE OF IDAHO

AREA 83,564 sq. mi. (216,431 sq. km.)
POPULATION 1,293,953
CAPITAL Boise
LARGEST CITY Boise
HIGHEST POINT Borah Pk. 12,662 ft. (3859 m.)
SETTLED IN 1842
ADMITTED TO UNION July 3, 1890
POPULAR NAME Gem State
STATE FLOWER Syringa
STATE BIRD Mountain Bluebird

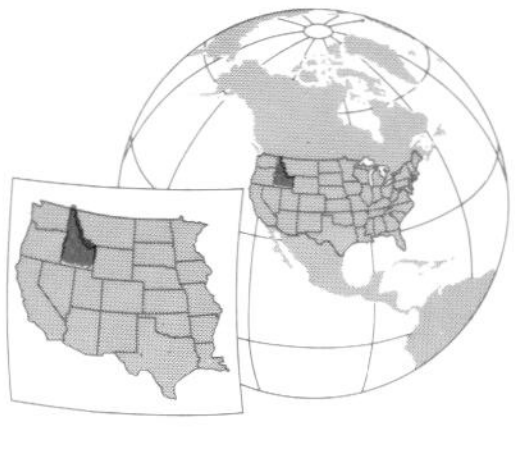

COUNTIES

Ada 300,904 B6
Adams 3,476 B5
Bannock 75,565 F7
Bear Lake 6,411 G7
Benewah 9,171 B2
Bingham 41,735 F6
Blaine 18,991 D6
Boise 6,670 C6
Bonner 36,835 B1
Bonneville 82,522 G6
Boundary 9,871 B1
Butte 2,899 E6
Camas 991 D6
Canyon 131,441 B6
Caribou 7,304 G7
Cassia 21,416 E7
Clark 1,022 F5
Clearwater 8,930 C3
Custer 4,342 D5
Elmore 29,130 C6
Franklin 11,329 G7
Fremont 11,819 G5
Gem 15,181 B6
Gooding 14,155 D6
Idaho 15,511 C4
Jefferson 19,155 F6
Jerome 18,342 D7
Kootenai 108,685 B2
Latah 34,935 B3
Lemhi 7,806 D4
Lewis 3,747 B3
Lincoln 4,044 D6
Madison 27,467 G6
Minidoka 20,174 E7
Nez Perce 37,410 B3
Oneida 4,125 F7
Owyhee 10,644 B7
Payette 20,578 B5
Power 7,538 F7
Shoshone 13,771 B2
Teton 5,999 G6
Twin Falls 64,284 D7
Valley 7,651 C5
Washington 9,977 B5

CITIES and TOWNS

Aberdeen 1,840 F7
Acequia 144 E7
Ahsahka 160 B3
Albion 262 E7
American Falls▲ 4,111 E7
Ammon 6,187 G6
Arco▲ 1,026 E6
Arimo 348 F7
Ashton 1,129 G5
Athol 676 B2
Atomic City 25 F6
Bancroft 382 G7
Basalt 419 F6
Bellevue 1,876 D6
Blackfoot▲ 10,419 F6
Bliss 275 D7
Bloomington 251 G7
Boise (cap.)▲ 185,787 B6
Bonners Ferry▲ 2,515 B1
Bovill 305 B3
Buhl 3,985 D7
Butte City 76 E6
Caldwell▲ 25,967 B6
Cambridge 360 B5
Carey 513 E6
Cascade▲ 997 C5
Castleford 277 C7
Challis▲ 909 D5
Chatcolet 72 B2
Chubbuck 9,700 F7
Clark Fork 530 B1
Clarkia 175 B2
Clayton 27 D5
Clifton 213 F7
Coeur d'Alene▲ 34,514 B2
Cottonwood 944 B3
Council▲ 816 B5
Craigmont 556 B3
Crouch 154 B5
Culdesac 378 B3
Dalton Gardens 2,278 B2
Dayton 444 F7
Deary 552 B3
Declo 338 E7
Dietrich 150 D7
Dingle 300 G7
Donnelly 138 B5
Downey 613 F7
Driggs▲ 1,100 G6
Drummond 15 G5
Dubois▲ 647 F5
Eagle 11,085 B6
East Hope 200 B1
Eden 411 D7
Elk City 500 C4
Elk River 156 B3
Emmett▲ 5,490 B6
Fairfield▲ 395 D6
Ferdinand 145 B3
Fernan Lake 186 B2
Fernwood 608 B2
Filer 1,620 D7
Firth 408 F6
Fort Hall 3,193 F6
Franklin 641 G7
Fruitland 3,805 B6
Fruitvale 200 B5
Garden City 10,624 B6
Garden Valley 250 C5
Genesee 946 B3
Georgetown 538 G7
Glenns Ferry 1,611 C7
Gooding▲ 3,384 D7
Grace 990 G7
Grand View 470 B7
Grangeville▲ 3,228 B4
Greenleaf 862 B6
Hagerman 656 D7
Hailey▲ 6,200 D6
Hamer 12 F6
Hammett 180 C7
Hansen 970 D7
Harrison 267 B2
Hauser 668 A2
Hayden 9,159 B2
Hayden Lake 494 B2
Hazelton 687 E7
Heise 84 G6
Heyburn 2,899 E7
Hollister 237 D7
Homedale 2,528 A6
Hope 79 B1
Horseshoe Bend 770 B6
Huetter 96 B2
Idaho City▲ 458 C6
Idaho Falls▲ 50,730 F6
Inkom 1,201 F7
Iona 1,201 G6
Irwin 157 G6
Island Park 215 G5
Jerome▲ 7,780 D7
Juliaetta 609 B3
Kamiah 1,160 B3
Kellogg 2,395 B2
Kendrick 369 B3
Ketchum 3,003 D6
Kimberly 2,614 D7
Kooskia 675 C3
Kootenai 441 B1
Kuna 5,382 B6
Laclede 400 B1
Lapwai 1,134 B3
Lava Hot Springs 521 F7
Leadore 90 E5
Lewiston▲ 30,904 A3
Lewisville 467 F6
Lost River (Grouse) 26 E6
Mackay 566 E6
Malad City▲ 2,158 F7
Malta 177 E7
Marsing 890 B6
McCall 2,084 C5
McCammon 805 F7
Melba 439 B6
Menan 707 F6
Meridian 34,919 B6
Middleton 2,978 B6
Midvale 176 B5
Minidoka 129 E7
Monteview 200 F6
Montpelier 2,785 G7
Moore 196 E6
Moreland 600 F6
Moscow▲ 21,291 B3
Mountain Home▲ 11,143 C6
Moyie Springs 656 B1
Mud Lake 270 F6
Mullan 840 C2
Murphy▲ 200 B6
Murtaugh 139 D7
Nampa 51,867 B6
Newdale 358 G6
New Meadows 533 B4
New Plymouth 1,400 B6
Nezperce▲ 523 B3
Notus 458 B6
Oakley 668 D7
Oldtown 190 A1
Onaway 230 B3
Orofino▲ 3,247 B3
Osburn 1,545 B2
Oxford 53 F7
Paris▲ 576 G7
Parker 319 G6
Parma 1,771 B6
Patterson 4 E5
Paul 998 E7
Payette▲ 7,054 B5
Peck 186 B3
Pearl 8 B6
Pierce 617 C3
Pinehurst 1,661 B2
Placerville 60 C6
Plummer 990 B2
Pocatello▲ 51,466 F7
Ponderay 638 B1
Post Falls 17,247 A2
Potlatch 791 A3
Preston▲ 4,682 G7
Priest River 1,754 A1
Rathdrum 4,816 A2
Reubens 72 B3
Rexburg▲ 17,257 G6
Richfield 412 D6
Rigby▲ 2,998 F6
Riggins 410 B4
Ririe 545 G6
Roberts 647 F6
Rockland 316 F7
Rupert▲ 5,645 E7
Sagle 600 B1
Saint Anthony▲ 3,342 G6
Saint Charles 156 G7
Saint Maries▲ 2,652 B2
Salmon▲ 3,122 D4
Samuels 467 B1
Sandpoint▲ 6,835 B1
Shelley 3,813 F6
Shoshone▲ 1,398 D7
Silver City 1 B6
Smelterville 651 B2
Soda Springs▲ 3,381 G7
Spencer 38 F5
Spirit Lake 1,376 A2
Stanley 100 D5
Star 1,795 B6
State Line 28 A2
Stites 226 C3
Sugar City 1,242 G6
Sun Valley 1,427 D6
Swan Valley 213 G6
Sweet 290 B6
Tensed 126 B2
Terreton 400 F6
Teton 569 G6
Tetonia 247 G6
Troy 798 B3
Twin Falls▲ 34,469 D7
Ucon 943 F6
Victor 840 G6
Wallace▲ 960 C2
Wardner 215 B2
Weippe 416 C3
Weiser▲ 5,343 B5
Wendell 2,338 D7
Weston 425 F7
White Bird 106 B4
Wilder 1,462 A6
Winchester 308 B3
Worley 223 B2

OTHER FEATURES

Albeni Falls (dam) B1
Albion (mts.) E7
Allan (mt.) D4
American Falls (res.) F6
Anderson Ranch (res.) C6
Antelope (creek) E6
Arrowrock (res.) C6
Auger (falls) D7
Badger (peak) E7
Bald (mt.) D5
Bannock (creek) F7
Bannock (peak) F7
Bannock (range) F7
Bargamin (creek) C4
Battle (creek) B7
Bear (lake) G7
Bear (riv.) G7
Bear River (range) G7
Beaver (creek) F5
Beaverhead (mts.) E4
Big (creek) C4
Big Boulder (creek) B7
Big Elk (peak) G6
Big Hole (mts.) G6
Big Lost (riv.) E6
Big Southern (butte) E6
Big Wood (riv.) D6
Birch (creek) F5
Birch Creek (valley) E5
Bitterroot (range) D3
Blackfoot (res.) G7
Black Pine (mts.) E7
Blue Nose (mt.) D4
Boise (mts.) B6
Boise (riv.) B6
Borah (peak) E5
Boulder (mts.) D6
Brownlee (dam) B5
Bruneau (riv.) C7
Camas (creek) D5
Camas (creek) D6
Camas (creek) F5
Canyon (creek) C6
Cape Horn (mt.) C5
Caribou (mt.) G6
Caribou (range) G6
Cascade (res.) C5
Castle (creek) B7
Castle (peak) D5
Cedar Creek (peak) E7
Cedar Creek (res.) D7
Centennial (mts.) F5
Chesterfield (res.) F7
City of Rocks Nat'l Reserve E7
Clearwater (mts.) C3
Clearwater (riv.) B3
Coeur d'Alene (lake) B2
Coeur d'Alene (mts.) C2
Coeur d'Alene (riv.) B2
Cottonwood (butte) C4
Craig (mts.) B4
Crane Creek (res.) B5
Craters of the Moon Nat'l Mon. E6
Deadwood (res.) C5
Deep (creek) B7
Deep (creek) F7
Deep Creek (mts.) F7
Diamond (peak) E5
Duck Valley Ind. Res. B7
Dworshak (res.) C3
East Sister (peak) C2
Eighteen Mile (peak) E5
Fish Creek (res.) E6
Fort Hall Ind. Res. F6
Goldstone (mt.) E4
Goose (creek) E7
Goose Creek (mts.) E7
Grand Canyon of the Snake River (canyon) B4
Grays (lake) G6
Grays Lake Outlet (creek) G6
Greylock (mt.) C6
Hagerman Fossil Beds Nat'l Mon. D7
Hayden (lake) B2
Hells (canyon) B4
Hells Canyon Nat'l Rec. Area B4
Henrys (lake) G5
Henrys Fork, Snake (riv.) G5
Hunter (peak) D3
Hyndman (peak) D6
Indian (creek) C5
Island Park (res.) G5
Jarbidge (riv.) C7
Johnson (creek) C5
Jordan (creek) A7
Kootenai (riv.) C1
Lemhi (pass) E5
Lemhi (range) E5
Lemhi (riv.) E5
Little Lost (riv.) E5
Little Owyhee (riv.) B7
Little Salmon (riv.) B4
Little Weiser (riv.) B5
Little Wood (riv.) D6
Lochsa (riv.) C3
Lolo (creek) C3
Lolo (pass) D3
Lone Pine (peak) D5
Lookout (mt.) D5
Lookout (mt.) F5
Lost River (range) E5
Lost Trail (pass) D4
Lowell (lake) B6
Lower Goose Creek (res.) D7
Lower Granite (lake) A3
Lucky Peak (lake) B6
Mackay (res.) E6
Magic (res.) D6
Malad (riv.) F7
Marsh (creek) F7
McGuire (mt.) D4
Meade (peak) G7
Meadow (creek) C4
Medicine Lodge (creek) F5
Middle Fork (peak) D5
Monument (peak) B4
Moose (creek) D3
Mores (creek) C6
Mormon (mt.) C4
Mountain Home (res.) C6
Mountain Home A.F.B. 8,894 C6
Moyie (riv.) B1
Mud (lake) F6
National Reactor Testing Sta. F6
Nez Perce Nat'l Hist. Park C3
Norton (peak) D6
Orofino (creek) C3
Owyhee (mts.) B6
Owyhee, East Fork (riv.) B7
Oxbow (dam) B5
Pack (riv.) B1
Pahsimeroi (riv.) E5
Palisades (res.) G6
Palouse (riv.) B3
Panther (creek) D4
Payette (lake) C4
Payette (mts.) B5
Payette (riv.) B6
Peale (mts.) G7
Pend Oreille (lake) B1
Pend Oreille (mt.) B1
Pend Oreille (riv.) A1
Pilot (peak) C4
Pilot (peak) C6
Pilot Knob (mt.) C4
Pinyon (peak) C5
Pioneer (mts.) D6
Pot (mt.) C3
Potlatch (riv.) B3
Priest (lake) B1
Priest (riv.) B1
Purcell (mts.) B1
Pyramid (peak) E4
Raft (riv.) E7
Rainbow (mt.) C5
Ranger (peak) D3
Rays (lake) F6
Red (riv.) C4
Redfish (lake) D5
Reynolds (creek) B6
Rhodes (peak) D3
Rock (creek) F7
Rocky (mts.) D1
Rocky Ridge (mt.) C3
Ryan (peak) D6
Saddle (mt.) D3
Saddle (mt.) F6
Sailor (creek) C7
Saint Joe (riv.) B2
Saint Maries (riv.) B2
Salmon (falls) C7
Salmon (riv.) B4
Salmon Falls (creek) D7
Salmon Falls Creek (res.) D7
Salmon River (mts.) C5
Sawtooth (range) C6
Sawtooth Nat'l Rec. Area D5
Secesh (riv.) C4
Selkirk (mts.) B1
Selway (riv.) C3
Seven Devils (mts.) B4
Shoshone (falls) D7
Sleeping Deer (mt.) D5
Smith (creek) B1
Smoky (mts.) D6
Snake (riv.) A3
Snake River (plain) D7
Snake River (range) G6
Spirit (lake) B2
Squaw (creek) B5
Squaw (peak) D4
Steamboat (mt.) C4
Steel (mt.) C6
Strike, C.J. (res.) C7
Sublett (mts.) E7
Sunset (peak) E6
Taylor (mt.) D5
Teton (riv.) G6
Thompson (peak) C5
Trinity (mt.) C6
Trout (creek) B1
Twin (falls) D7
Twin Peaks (mt.) D5
Walcott (lake) E7
Waugh (mt.) D4
Weiser (riv.) B5
White Knob (mts.) E6
Wickahoney (creek) C7
Willow (creek) G6
Wilson Lake (res.) D7
Yankee Fork, Salmon (riv.) D5
Yellowstone Nat'l Park H5

▲County seat

Agriculture, Industry and Resources

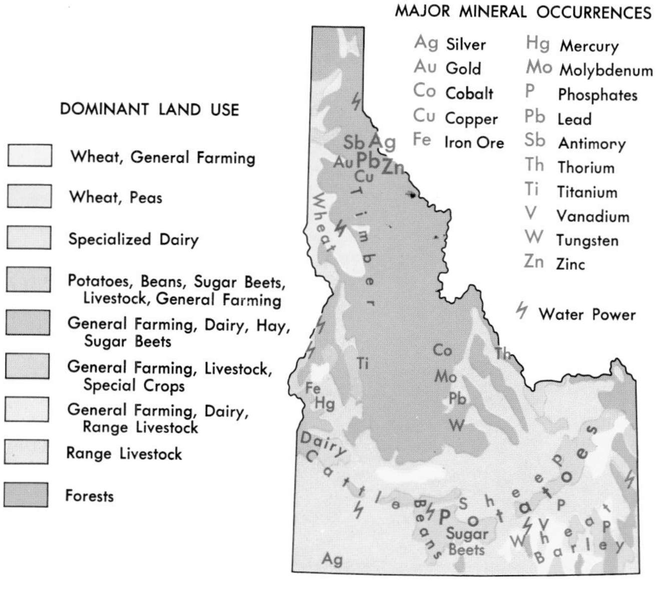
MAJOR MINERAL OCCURRENCES
Ag Silver
Au Gold
Co Cobalt
Cu Copper
Fe Iron Ore
Hg Mercury
Mo Molybdenum
P Phosphates
Pb Lead
Sb Antimory
Th Thorium
Ti Titanium
V Vanadium
W Tungsten
Zn Zinc
Water Power
DOMINANT LAND USE
Wheat, General Farming
Wheat, Peas
Specialized Dairy
Potatoes, Beans, Sugar Beets, Livestock, General Farming
General Farming, Dairy, Hay, Sugar Beets
General Farming, Livestock, Special Crops
General Farming, Dairy, Range Livestock
Range Livestock
Forests
Timber
Wheat
Dairy
Cattle
Sheep
Potatoes
Beans
Sugar Beets
Wheat
Barley

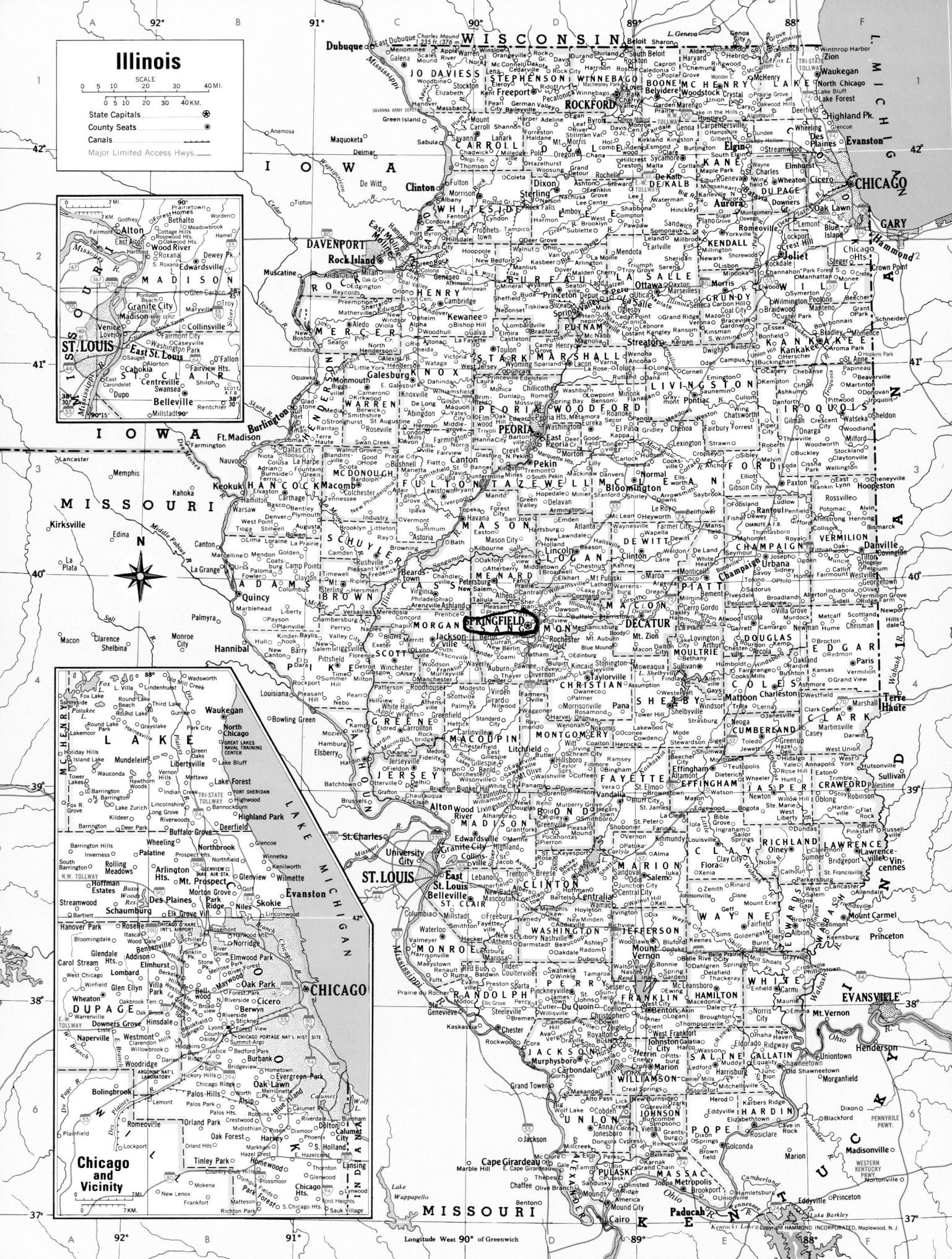

Illinois 223

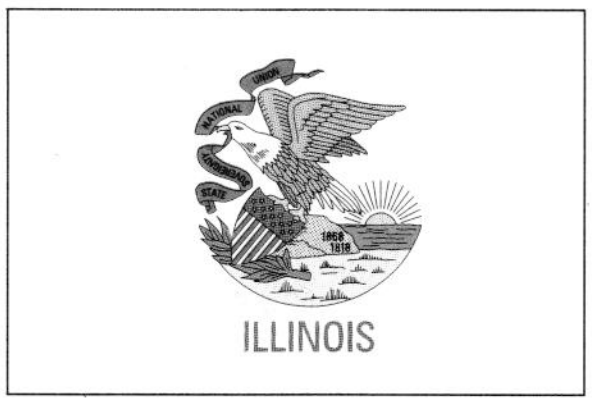
ILLINOIS

AREA 56,345 sq. mi. (145,934 sq. km.)
POPULATION 12,419,293
CAPITAL Springfield
LARGEST CITY Chicago
HIGHEST POINT Charles Mound 1,235 ft. (376 m.)
SETTLED IN 1720
ADMITTED TO UNION December 3, 1818
POPULAR NAME Prairie State; Land of Lincoln
STATE FLOWER Native Violet
STATE BIRD Cardinal

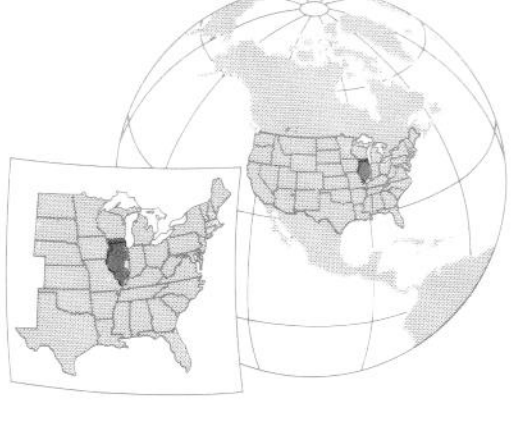

COUNTIES

Adams 68,277 B4
Alexander 9,590 D6
Bond 17,633 D5
Boone 41,786 E1
Brown 6,950 C4
Bureau 35,503 D2
Calhoun 5,084 C4
Carroll 16,674 D1
Cass 13,695 C4
Champaign 179,669 E3
Christian 35,372 D4
Clark 17,008 F4
Clay 14,560 E5
Clinton 35,535 D5
Coles 53,196 E4
Cook 5,376,741 F2
Crawford 20,452 F4
Cumberland 11,253 E4
De Kalb 88,969 E2
De Witt 16,798 E3
Douglas 19,922 E4
Du Page 904,161 E2
Edgar 19,704 F4
Edwards 6,971 E5
Effingham 34,264 E4
Fayette 21,802 D4
Ford 14,241 E3
Franklin 39,018 E5
Fulton 38,250 C3
Gallatin 6,445 E6
Greene 14,761 C4
Grundy 37,535 E2
Hamilton 8,621 E5
Hancock 20,121 B3
Hardin 4,800 E6
Henderson 8,213 C3
Henry 51,020 C2
Iroquois 31,334 F3
Jackson 59,612 D6
Jasper 10,117 E4
Jefferson 40,045 E5
Jersey 21,668 C4
Jo Daviess 22,289 C1
Johnson 12,878 E6
Kane 404,119 E2
Kankakee 103,833 F2
Kendall 54,544 E2
Knox 55,836 C3
Lake 644,356 E1
La Salle 111,509 E2
Lawrence 15,452 F5
Lee 36,062 D2
Livingston 39,678 E3
Logan 31,183 D3
Macon 114,706 E4
Macoupin 49,019 D4
Madison 258,941 D5
Marion 41,691 E5
Marshall 13,180 D2
Mason 16,038 D3
Massac 15,161 E6
McDonough 32,913 C3
McHenry 260,077 E1
McLean 150,433 E3
Menard 12,486 D3
Mercer 16,957 C2
Monroe 27,619 C5
Montgomery 30,652 D4
Morgan 36,616 C4
Moultrie 14,287 E4
Ogle 51,032 D1
Peoria 183,433 D3
Perry 23,094 D5
Piatt 16,365 E4
Pike 17,384 C4
Pope 4,413 E6
Pulaski 7,348 D6
Putnam 6,086 D2
Randolph 33,893 D5
Richland 16,149 E5
Rock Island 149,374 C2
Saint Clair 256,082 D5
Saline 26,733 E6
Sangamon 188,951 D4
Schuyler 7,189 C3
Scott 5,537 C4
Shelby 22,893 E4
Stark 6,332 D2
Stephenson 48,979 D1
Tazewell 128,485 D3
Union 18,293 D6
Vermilion 83,919 F3
Wabash 12,937 F5
Warren 18,735 C3
Washington 15,148 D5
Wayne 17,151 E5
White 15,371 E5
Whiteside 60,653 D2
Will 502,266 F2
Williamson 61,296 E6
Winnebago 278,418 D1
Woodford 35,469 D3

CITIES and TOWNS

Abingdon 3,612 C3
Addison 35,914 B5
Albany 895 C2
Albers 878 D5
Albion▲ 1,933 E5
Aledo▲ 3,613 C2
Alexis 863 C2
Algonquin 23,276 E1
Alhambra 630 D5
Alorton 2,749 B2
Alpha 726 C2
Alsip 19,725 B6
Altamont 2,283 E4
Alton 30,496 A2
Amboy 2,561 D2
Andalusia 1,050 C2
Anna 5,136 D6
Annawan 868 C2
Antioch 8,788 E1
Arcola 2,652 E4
Argenta 921 E4
Arlington Heights 76,031 B5
Aroma Park 821 F2
Arthur 2,203 E4
Ashkum 724 E3
Ashland 1,361 C4
Ashley 613 D5
Ashmore 809 F4
Ashton 1,142 D2
Assumption 1,261 E4
Astoria 1,193 C3
Athens 1,726 D4
Atkinson 1,001 C2
Atlanta 1,649 D3
Atwood 1,290 E4
Auburn 4,317 D4
Augusta 657 C3
Aurora 142,990 E2
Ava 662 D6
Aviston 1,231 D5
Avon 915 C3
Bannockburn 3,627 B5
Barrington 10,168 A5
Barrington Hills 3,915 A5
Barry 1,368 B4
Bartlett 36,706 A5
Bartonville 6,310 D3
Batavia 23,866 E2
Beardstown 5,766 C3
Beckemeyer 1,043 D5
Bedford Park 574 B6
Beecher 2,033 F2
Belleville▲ 41,410 B3
Bellwood 20,535 B5
Belvidere▲ 20,820 E1
Bement 1,784 E4
Benld 1,541 D4
Bensenville 20,703 B5
Benton▲ 6,880 E6
Berkeley 5,245 B5
Berwyn 54,016 B6
Bethalto 9,454 B2
Bethany 1,287 E4
Blandinsville 777 C3
Bloomingdale 21,675 A5
Bloomington▲ 64,808 D3
Blue Island 23,463 B6
Blue Mound 1,129 D4
Bluffs 748 C4
Bluford 785 E5
Bolingbrook 56,321 A6
Bourbonnais 15,256 F2
Braceville 792 E2
Bradford 787 D2
Bradley 12,784 F2
Braidwood 5,203 E2
Breese 4,048 D5
Bridgeport 2,168 F5
Bridgeview 15,335 B6
Brighton 2,196 C4
Brimfield 933 D3
Broadview 8,264 B6
Brookfield 19,085 B6
Brooklyn (Lovejoy) 676 A2
Brookport 1,054 E6
Brownstown 705 E5
Buda 592 D2
Buffalo Grove 42,909 B5
Bunker Hill 1,801 D4
Burbank 27,902 B6
Burnham 4,170 C6
Burr Ridge 10,408 B6
Bushnell 3,221 C3
Byron 2,917 D1
Cahokia 16,391 A3
Cairo▲ 3,632 D6
Calumet City 39,071 C6
Calumet Park 8,516 C6
Cambria 1,330 D6
Cambridge▲ 2,180 C2
Camp Point 1,244 B3
Canton 15,288 C3
Capron 961 E1
Carbon Cliff 1,689 C2
Carbondale 20,681 D6
Carlinville▲ 5,685 D4
Carlyle▲ 3,406 D5
Carmi▲ 5,422 E5
Carol Stream 40,438 A5
Carpentersville 30,586 E1
Carrier Mills 1,886 E6
Carrollton▲ 2,605 C4
Carterville 4,616 D6
Carthage▲ 2,725 B3
Cary 15,531 E1
Casey 2,942 F4
Caseyville 4,310 B2
Catlin 2,087 F3
Cedarville 719 D1
Central City 1,371 D5
Centralia 14,136 D5
Centreville 5,951 B3
Cerro Gordo 1,436 E4
Champaign 67,518 E3
Chandlerville 704 C3
Channahon 7,344 E2
Chapin 592 C4
Charleston▲ 21,039 E4
Chatham 8,583 D4
Chatsworth 1,265 E3
Chebanse 1,148 F3
Chenoa 1,845 E3
Cherry Valley 2,191 D1
Chester▲ 5,185 D6
Chicago▲ 2,896,016 C5
Chicago Heights 32,776 C6
Chicago Ridge 14,127 B6
Chillicothe 5,996 D3
Chrisman 1,318 F4
Christopher 2,836 D6
Cicero 85,616 B5
Cisne 673 E5
Cissna Park 811 F3
Clarendon Hills 7,610 B6
Clay City 1,000 E5
Clayton 904 B3
Clifton 1,317 F3
Clinton▲ 7,485 E3
Coal City 4,797 E2
Coal Valley 3,606 C2
Cobden 1,116 D6
Coffeen 709 D4
Colchester 1,493 C3
Colfax 989 E3
Collinsville 24,707 B2
Colona 5,173 C2
Columbia 7,922 C5
Cordova 633 C2
Cortland 2,066 E2
Coulterville 1,230 D5
Country Club
Hills 16,169 B6
Countryside 5,991 B6
Creal Springs 702 E6
Crescent City 631 F3
Crest Hill 13,329 E2
Crestwood 11,251 B6
Crete 7,346 F2
Creve Coeur 5,448 D3
Crossville 782 F5
Crystal Lake 38,000 E1
Cuba 1,418 C3
Dallas City 1,055 B3
Dalzell 717 D2
Danvers 1,183 D3
Danville▲ 33,904 F3
Darien 22,860 B6
Decatur▲ 81,860 E4
Deer Creek 605 D3
Deerfield 18,420 B5
Deer Park 3,102 A5
De Kalb 39,018 E2
De Land 475 E3
Delavan 1,825 D3
Depue 1,842 D2
De Soto 1,653 D6
Des Plaines 58,720 B5
Divernon 1,201 D4
Dixmoor 3,934 C6
Dixon▲ 15,941 D2
Dolton 25,614 C6
Dongola 806 D6
Downers Grove 48,724 A6
Dundee (East and West
Dundee) 8,383 E1
Dunlap 926 D3
Dupo 3,933 A3
Du Quoin 6,448 D5
Durand 1,081 D1
Dwight 4,363 E2
Earlville 1,778 E2
East Alton 6,830 A2
East Dubuque 1,995 C1
East Dundee (Dundee)
2,955 E1
East Galesburg 839 C3
East Hazelcrest 1,607 C6
East Moline 20,333 C2
East Peoria 22,638 D3
East Saint Louis 31,542 A2
Edinburg 1,135 D4
Edwards 14,579 D3
Edwardsville▲ 21,491 B2
Effingham▲ 12,384 E4
Elburn 2,756 E2
Eldorado 4,534 E6
Elgin 94,487 E1
Elizabeth 682 C1
Elizabethtown▲ 348 E6
Elk Grove Village 34,727 B5
Elkville 1,001 D6
Elmhurst 42,762 B5
Elmwood 1,945 D3
Elmwood Park 25,405 B5
El Paso 2,695 D3
Elsah 635 C5
Elwood 1,620 E2
Energy 1,175 E6
Enfield 625 E5
Equality 721 E6
Erie 1,589 C2
Eureka▲ 4,871 D3
Evanston 74,239 B5
Evansville 724 D5
Evergreen Park 20,821 B6
Fairbury 3,968 E3
Fairfield▲ 5,421 E5
Fairmont City 2,436 B2
Fairmount 640 F3
Fairview Heights 15,034 B3
Farmer City 2,055 E3
Farmersville 768 D4
Farmington 2,601 C3
Findlay 723 E4
Fisher 1,647 E3
Flanagan 1,083 E3
Flat Rock 415 F5
Flora 5,086 E5
Flossmoor 9,301 B6
Ford Heights 3,456 C6
Forest Homes 1,701 B2
Forest Park 15,688 B5
Forest View 778 B6
Forrest 1,225 E3
Forreston 1,469 D1
Forsyth 2,434 D4
Fox Lake 9,178 A4
Fox River Grove 4,862 A5
Frankfort 10,391 B6
Franklin 586 C4
Franklin Grove 1,052 D2
Franklin Park 19,434 B5
Freeburg 3,872 D5
Freeport▲ 26,443 D1
Fulton 3,881 C2
Galatia 1,013 E6
Galena▲ 3,460 C1
Galesburg▲ 33,706 C3
Galva 2,758 D2
Gardner 1,406 E2
Geneseo 6,480 C2
Geneva▲ 19,515 E2
Genoa 4,169 E1
Georgetown 3,628 F4
Germantown 1,118 D5
Gibson City 3,373 E3
Gifford 815 E3
Gillespie 3,412 D4
Gilman 1,793 E3
Glasford 1,076 D3
Glen Carbon 10,425 B2
Glencoe 8,762 B5
Glendale Heights 31,765 A5
Glen Ellyn 26,999 A5
Glenview 41,847 B5
Glenwood 9,000 C6
Golconda▲ 726 E6
Goreville 938 E6
Grafton 609 C5
Grand Ridge 546 E2
Grand Tower 624 D6
Grandview 1,537 D4
Granite City 31,301 A2
Grant Park 1,358 F2
Granville 1,414 D2
Grayslake 18,506 B4
Grayville 1,725 E5
Greenfield 1,179 C4
Green Oaks 3,572 B4
Green Rock 2,615 C2
Greenup 1,532 E4
Green Valley 728 D3
Greenview 862 D3
Greenville▲ 6,955 D5
Gridley 1,411 E3
Griggsville 1,258 C4
Gurnee 28,834 B4
Hamilton 3,029 B3
Hampshire 2,900 E1
Hampton 1,626 C2
Hanna City 1,013 D3
Hanover 836 C1
Hanover Park 38,278 A5
Hardin▲ 959 C4
Harrisburg▲ 9,860 E6
Harristown 1,338 D4
Hartford 1,545 A2
Harvard 7,996 E1
Harvey 30,000 B6
Harwood Heights 8,297 B5
Havana▲ 3,577 D3
Hawthorn Woods 6,002 B5
Hazel Crest 14,816 B6
Hebron 1,038 E1
Hegeler 1,853 F3
Hennepin▲ 707 D2
Henry 2,540 D2
Herrin 11,298 E6
Herscher 1,523 E2
Heyworth 2,431 E3
Hickory Hills 13,926 B6
Highland 8,438 D5
Highland Park 31,365 B5
Highwood 4,143 B5
Hillcrest 1,158 D2
Hillsboro▲ 4,359 D4
Hillsdale 588 C2
Hillside 8,155 B5
Hinckley 1,994 E2
Hinsdale 17,349 B6
Hodgkins 2,134 B6
Hoffman Estates 49,495 A5
Holiday Hills 831 A4
Homer 1,200 F3
Hometown 4,467 B6
Homewood 19,543 B6
Hoopeston 5,965 F3
Hopedale 929 D3
Hudson 1,510 E3
Huntley 5,730 E1
Hurst 805 D6
Hutsonville 568 F4
Illiopolis 916 D4
Inverness 6,749 A5
Ipava 506 C3
Irvington 736 D5
Island Lake 8,153 A4
Itasca 8,302 B5
Jacksonville▲ 18,940 C4
Jerome 1,414 D4
Jerseyville▲ 7,984 C4
Johnsburg 5,391 A4
Johnston City 3,557 E6
Joliet▲ 106,221 E2
Jonesboro▲ 1,853 D6
Justice 12,193 B6
Kankakee▲ 27,491 F2
Kansas 842 F4
Karnak 619 E6
Kaskaskia 9 C6
Keithsburg 714 B2
Kenilworth 2,494 B5
Kewanee 12,944 C2
Kildeer 3,460 A5
Kincaid 1,441 D4
Kinmundy 892 E5
Kirkland 1,166 E1
Kirkwood 794 C3

(continued on following page)

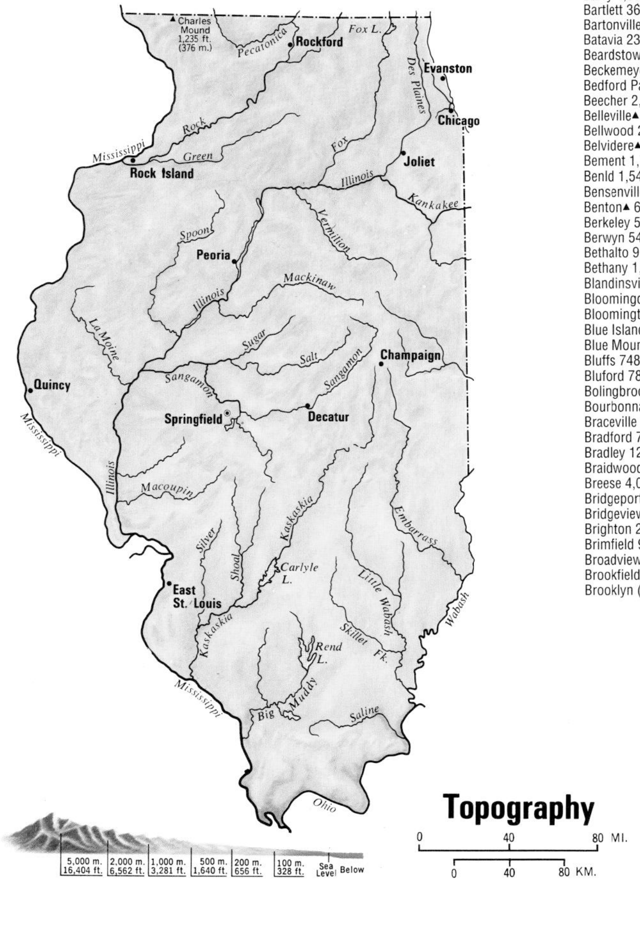
Charles Mound 1,235 ft. (376 m.)
Rockford
Pecatonica
Fox L.
Des Plaines
Evanston
Chicago
Rock
Mississippi
Green
Fox
Joliet
Rock Island
Illinois
Kankakee
Spoon
Vermilion
Peoria
Mackinaw
Illinois
La Moine
Sugar
Salt
Champaign
Sangamon
Quincy
Sangamon
Springfield
Decatur
Mississippi
Illinois
Macoupin
Kaskaskia
Embarrass
Silver
Shoal
Carlyle L.
Little Wabash
East St. Louis
Wabash
Kaskaskia
Skillet Fk.
Rend L.
Mississippi
Big Muddy
Saline
Ohio
5,000 m. 16,404 ft.
2,000 m. 6,562 ft.
1,000 m. 3,281 ft.
500 m. 1,640 ft.
200 m. 656 ft.
100 m. 328 ft.
Sea Level
Below
0 40 80 MI.
0 40 80 KM.

Topography

Agriculture, Industry and Resources

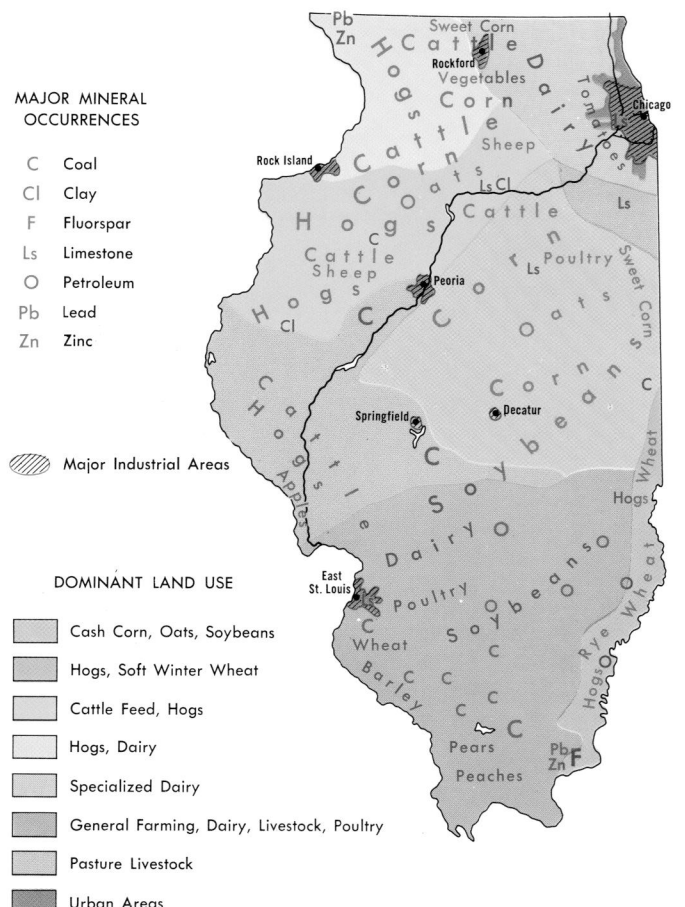

MAJOR MINERAL OCCURRENCES

C   Coal
Cl   Clay
F   Fluorspar
Ls   Limestone
O   Petroleum
Pb   Lead
Zn   Zinc

▨ Major Industrial Areas

DOMINANT LAND USE

- Cash Corn, Oats, Soybeans
- Hogs, Soft Winter Wheat
- Cattle Feed, Hogs
- Hogs, Dairy
- Specialized Dairy
- General Farming, Dairy, Livestock, Poultry
- Pasture Livestock
- Urban Areas

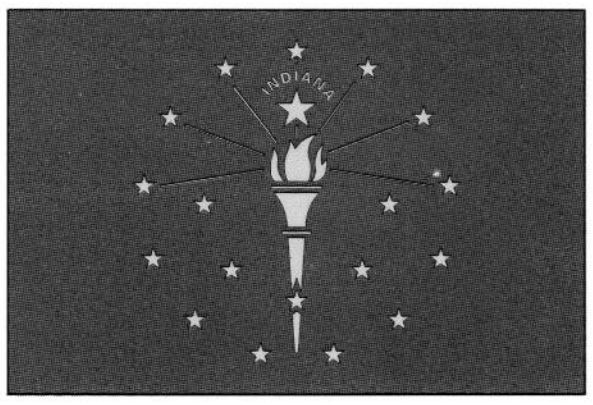
INDIANA

AREA 36,185 sq. mi. (93,719 sq. km.)
POPULATION 6,080,485
CAPITAL Indianapolis
LARGEST CITY Indianapolis
HIGHEST POINT (Wayne County) 1,257 ft. (383 m.)
SETTLED IN 1730
ADMITTED TO UNION December 11, 1816
POPULAR NAME Hoosier State
STATE FLOWER Peony
STATE BIRD Cardinal

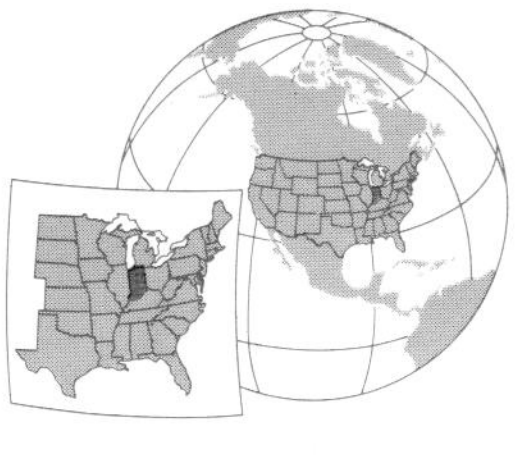

COUNTIES

Adams 33,625 H3
Allen 331,849 G2
Bartholomew 71,435 F6
Benton 9,421 C3
Blackford 14,048 G4
Boone 46,107 E4
Brown 14,957 E6
Carroll 20,165 D3
Cass 40,930 E3
Clark 96,472 F8
Clay 26,556 C6
Clinton 33,866 E4
Crawford 29,820 E8
Daviess 29,820 C7
Dearborn 46,109 H6
Decatur 24,555 G6
De Kalb 40,285 H2
Delaware 118,769 G4
Dubois 39,674 D8
Elkhart 182,791 F1
Fayette 25,588 G5
Floyd 70,823 F8
Fountain 17,954 C4
Franklin 22,151 G6
Fulton 20,511 E2
Gibson 32,500 B8
Grant 73,403 F3
Greene 33,157 D6
Hamilton 182,740 E4
Hancock 55,391 F5
Harrison 34,325 E8
Hendricks 104,093 E5
Henry 48,508 G5
Howard 84,964 E4
Huntington 38,075 G3
Jackson 41,335 E7
Jasper 30,043 C2
Jay 21,806 G4
Jefferson 31,705 G7
Jennings 27,554 F7
Johnson 115,209 E6
Knox 39,256 C7
Kosciusko 74,057 F2
Lagrange 34,909 G1
Lake 484,564 C2
LaPorte 110,106 D1
Lawrence 45,922 E7
Madison 133,358 F4
Marion 860,454 E5
Marshall 45,128 E2
Martin 10,369 D7
Miami 36,082 E3
Monroe 120,563 D6
Montgomery 37,629 D4
Morgan 66,689 E6
Newton 14,566 C3
Noble 46,275 G2
Ohio 5,623 H7
Orange 19,306 E7
Owen 21,786 D6
Parke 17,241 C5
Perry 18,899 D8
Pike 12,837 C8
Porter 146,798 C2
Posey 27,061 B8
Pulaski 13,755 D2
Putnam 36,019 D5
Randolph 27,401 G4
Ripley 26,523 G6
Rush 18,261 G5
Saint Joseph 265,559 E1
Scott 22,960 F7
Shelby 43,445 F5
Spencer 20,391 C9
Starke 23,556 D2
Steuben 33,214 G1
Sullivan 21,751 C6
Switzerland 9,065 G7
Tippecanoe 148,955 D4
Tipton 16,577 E4
Union 7,349 H5
Vanderburgh 171,922 B8
Vermillion 16,788 C5
Vigo 105,848 C6
Wabash 34,960 F3
Warren 8,419 C4
Warrick 52,383 C8
Washington 27,223 E7
Wayne 71,097 G5
Wells 27,600 G3
White 25,267 D3
Whitley 30,707 F2

CITIES and TOWNS

Abington 200 H5
Adams 250 F6
Adamsboro 325 E3
Advance 562 D5
Akron 1,076 E2
Alamo 137 C5
Albany 2,368 G4
Albion▲ 2,284 G2
Alexandria 6,260 F4
Altona 198 G2
Ambia 197 C4
Amboy 360 F3
Amity 200 E6
Amo 414 D5
Anderson▲ 59,734 F4
Andersonville 225 G5
Andrews 1,290 F3
Angola▲ 7,344 G1
Arcadia 1,747 E4
Arcola 300 G2
Ardmore 800 E1
Argos 1,613 E2
Arlington 500 F5
Ashley 1,010 G1
Atlanta 761 E4
Attica 3,491 C4
Atwood 300 F2
Auburn▲ 12,074 G2
Aurora 3,965 H6
Austin 4,724 F7
Avilla 2,049 G2
Avoca 400 D7
Azalia 194 F6
Bainbridge 743 D5
Bargersville 2,120 E5
Batesville 6,033 G6
Battle Ground 1,323 D3
Bedford▲ 13,768 E7
Beech Grove 14,880 E5
Bellmore 160 C5
Benton 220 F2
Berne 4,150 H3
Beverly Shores 708 D1
Bicknell 3,378 C7
Bippus 300 F3
Birdseye 465 D8
Blanford 500 B5
Blocher 400 F7
Bloomfield▲ 2,542 D6
Bloomingdale 319 C5
Blooming Grove 300 G5
Bloomington▲ 69,291 D6
Blountsville 166 G4
Blue Ridge 219 F5
Bluffton▲ 9,536 G3
Boggstown 200 F5
Boone Grove 220 C2
Boonville▲ 6,834 C8
Borden 818 F8
Boston 177 H5
Boswell 827 C3
Bourbon 1,691 E2
Bowling Green 200 D6
Bradford 350 E8
Brazil▲ 8,188 C5
Bremen 4,486 E2
Bridgeton 250 C5
Bright 5,405 H6
Brimfield 292 G2
Bringhurst 275 E3
Bristol 1,382 F1
Brook 1,062 C3
Brooklyn 1,545 E5
Brookston 1,717 D3
Brookville▲ 2,652 G6
Brownsburg 14,520 E5
Brownstown▲ 2,978 F7
Brownsville 250 H5
Bruceville 469 C7
Bryant 272 G3
Buck Creek 225 D4
Buckskin 200 C8
Buffalo 672 D3
Bunker Hill 987 E3
Burket 195 F2
Burlington 444 E4
Burnettsville 373 D3
Burney 300 F6
Burns Harbor 766 C1
Burrows 250 E3
Butler 2,725 H2
Butlerville 300 F6
Byron 200 C5
Cadiz 161 G5
Cambridge City 2,121 G5
Camden 582 D3
Cammack 250 G4
Campbellsburg 578 E7
Cannelton▲ 1,209 D9
Carbon 334 C5
Carlisle 2,660 C7
Carmel 37,733 E5
Cartersburg 300 E5
Carthage 928 F5
Cayuga 1,109 C5
Cedar Grove 185 H6
Cedar Lake 9,279 C2
Center 278 E4
Centerpoint 292 C6
Centerton 250 E5
Centerville 2,427 H5
Chalmers 513 D3
Chandler 3,094 C8
Chapel Hill 175 E6
Charlestown 5,993 F8
Charlottesville 300 F5
Chelsea 200 F7
Chesterfield 2,969 F4
Chesterton 10,488 D1
Chili 280 F3
Chrisney 544 C8
Churubusco 1,666 G2
Cicero 4,303 E4
Clarksburg 300 G6
Clarks Hill 680 D4
Clarksville 21,400 F8
Clay City 1,019 C6
Claypool 311 F2
Clayton 693 D5
Clear Creek 200 E6
Clear Lake 244 H1
Clifford 291 F6
Clinton 5,126 C5
Cloverdale 2,243 D5
Cloverland 175 C6
Coal City 225 D6
Coalmont 450 C6
Coatesville 516 D5
Colburn 300 D3
Colfax 768 D4
Collegeville 865 C3
Columbia City▲ 7,077 G2
Columbus▲ 39,059 E6
Connersville▲ 15,411 G5
Converse 1,137 F3
Cortland 175 F7
Corunna 254 G2
Corydon▲ 2,715 E8
Covington▲ 2,565 C4
Cowan 428 G4
Crandall 131 E8
Crane 203 D7
Crawfordsville▲ 15,243 D4
Cromwell 452 F2
Cross Plains 254 G7
Crothersville 1,570 F7
Crown Point▲ 19,806 C2
Crumstown 175 E1
Culver 1,539 E2
Cumberland 5,500 F5
Cynthiana 693 B8
Dale 1,568 D8
Daleville 1,658 F4
Dana 662 C5
Danville▲ 6,418 D5
Darlington 854 D4
Darmstadt 1,313 B8
Dayton 1,120 D4
Decatur▲ 9,528 H3
Decker 283 B7
Deer Creek 250 E3
Deerfield 300 H4
Delphi▲ 3,015 D3
DeMotte 3,234 C2
Denver 541 E3
Deputy 200 F7
Desoto 385 G4
Dillsboro 1,436 G6
Donaldson 320 E2
Doolittle Mills 200 D8
Dublin 697 G5
Dubois 550 D8
Dugger 955 C6
Dune Acres 213 C1
Dunkirk 2,646 G4
Dunlap 5,887 F1
Dunreith 184 F5
Dupont 392 G7
Dyer 13,895 C1
Eagletown 306 E4
Earl Park 485 C3
East Chicago 32,414 C1
East Enterprise 250 H7
East Germantown (Pershing) 243 G5
Eaton 1,603 G4
Economy 200 G5
Edgewood 1,988 F4
Edinburgh 4,505 E6
Edwardsport 363 C7
Edwardsville 700 F8
Elberfeld 636 C8
Elizabeth 137 F8
Elizabethtown 391 F6
Elkhart 51,874 F1
Ellettsville 5,078 D6
Elnora 721 C7
Elrod 200 G6
Elston 500 D4
Elwood 9,737 F4
Eminence 200 D5
English▲ 673 E8
Etna Green 663 E2
Eugene 400 B5
Evansville▲ 121,582 C9
Everton 500 G5
Fairbanks 165 B6
Fairland 1,276 F5
Fairmount 2,992 F4
Fair Oaks 175 C2
Fairview Park 1,496 C5
Farmersburg 1,180 C6
Farmland 1,456 G4
Fayetteville 180 D7
Ferdinand 2,277 D8
Fillmore 545 D5
Finly 400 F5
Fishers 37,835 E5
Flat Rock 323 F6
Flora 2,227 E3
Floyds Knobs 500 F8
Fontanet 325 C5
Forest 400 E4
Fort Branch 2,320 B8
Fortville 3,444 F5
Fort Wayne▲ 205,727 G2
Fountain City 735 H5
Fountaintown 225 F5
Fowler▲ 2,415 C3
Fowlerton 298 F4
Francesville 905 D3
Francisco 543 B8
Frankfort▲ 16,662 E4
Franklin▲ 19,463 E6
Frankton 1,905 F4
Fredericksburg 92 E8
Freelandville 600 C7
Freetown 600 E7
Fremont 1,696 H1
French Lick 1,941 D7
Fulton 326 E3
Galena 1,831 F8
Galveston 1,532 E3
Garrett 5,803 G2
Gary 102,746 C1
Gas City 5,940 F4
Gaston 1,010 G4
Geneva 1,368 H3
Gentryville 262 C8
Georgetown 2,227 F8
Glenwood 318 G5
Glezen 300 C8
Goldsmith 235 E4
Goodland 1,096 C3

(continued on following page)

Agriculture, Industry and Resources

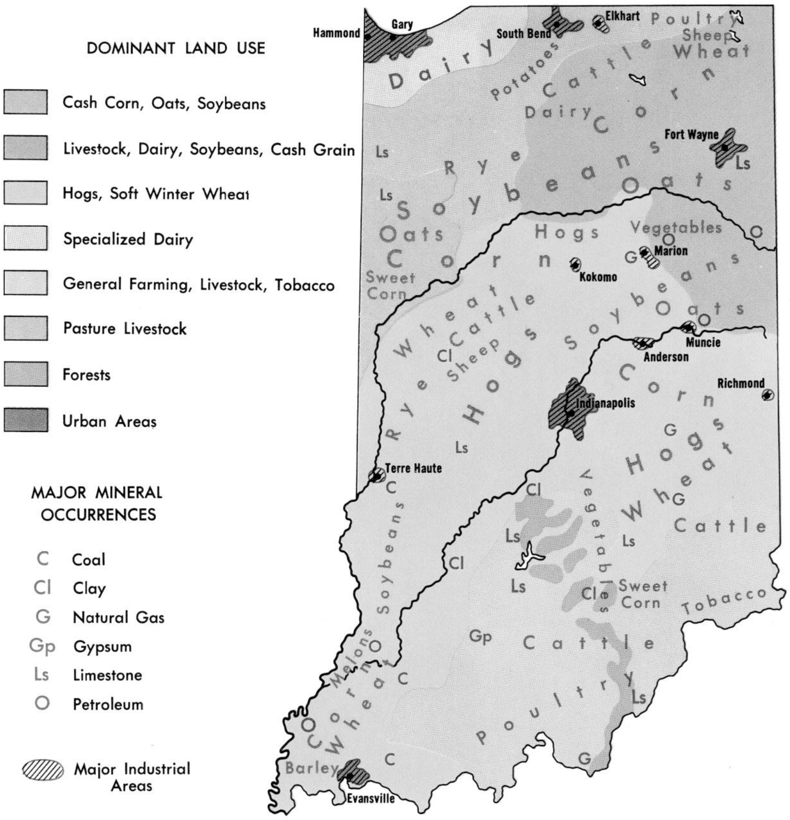
DOMINANT LAND USE
Cash Corn, Oats, Soybeans
Livestock, Dairy, Soybeans, Cash Grain
Hogs, Soft Winter Wheat
Specialized Dairy
General Farming, Livestock, Tobacco
Pasture Livestock
Forests
Urban Areas
MAJOR MINERAL OCCURRENCES
C Coal
Cl Clay
G Natural Gas
Gp Gypsum
Ls Limestone
O Petroleum
Major Industrial Areas
Hammond
Gary
South Bend
Elkhart
Fort Wayne
Marion
Kokomo
Muncie
Anderson
Richmond
Indianapolis
Terre Haute
Evansville
Dairy
Potatoes
Cattle
Poultry
Sheep
Wheat
Corn
Rye
Soybeans
Oats
Hogs
Vegetables
Sweet Corn
Tobacco
Melons
Barley

Topography

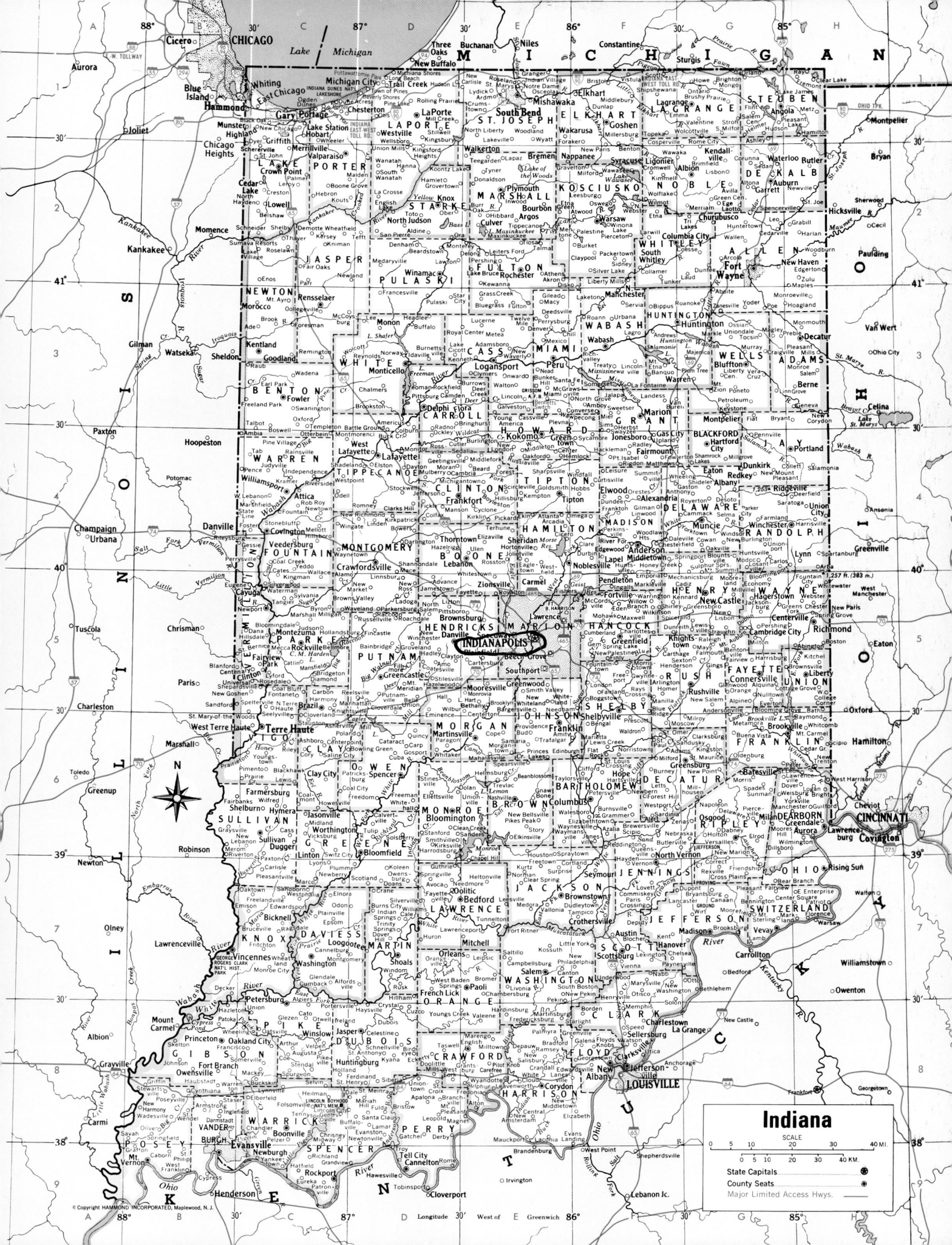

Indiana

SCALE

0 5 10   20   30   40 MI.

0 5 10 20   30   40 KM.

State Capitals .... ◉

County Seats .... ◉

Major Limited Access Hwys. _____

© Copyright HAMMOND INCORPORATED, Maplewood, N.J.

IOWA

AREA 56,275 sq. mi. (145,752 sq. km.)
POPULATION 2,926,324
CAPITAL Des Moines
LARGEST CITY Des Moines
HIGHEST POINT (Osceola Co.) 1,670 ft. (509 m.)
SETTLED IN 1788
ADMITTED TO UNION December 28, 1846
POPULAR NAME Hawkeye State
STATE FLOWER Wild Rose
STATE BIRD Eastern Goldfinch

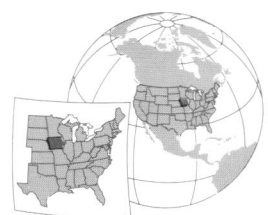

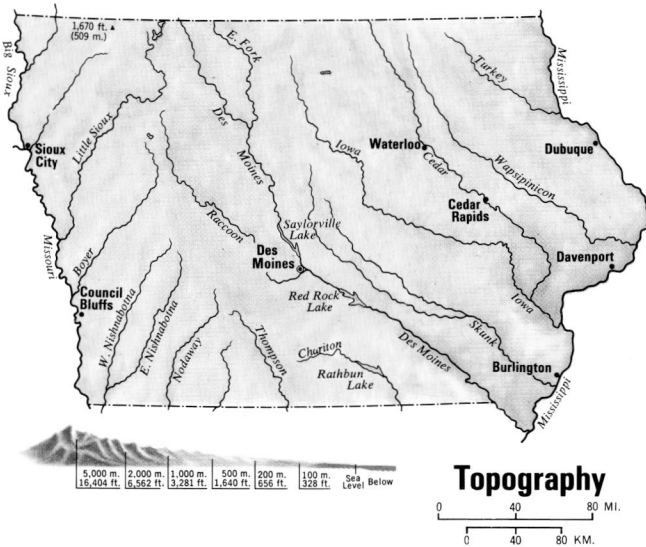

Topography

(continued on following page)

Agriculture, Industry and Resources

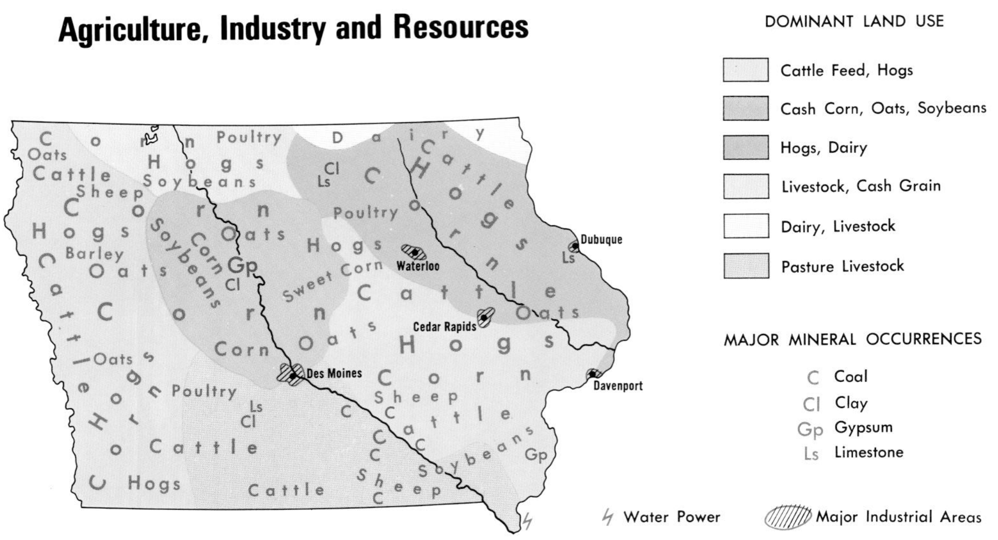
DOMINANT LAND USE
Cattle Feed, Hogs
Cash Corn, Oats, Soybeans
Hogs, Dairy
Livestock, Cash Grain
Dairy, Livestock
Pasture Livestock
MAJOR MINERAL OCCURRENCES
C Coal
Cl Clay
Gp Gypsum
Ls Limestone
Water Power
Major Industrial Areas
Dubuque
Waterloo
Cedar Rapids
Des Moines
Davenport
Poultry
Dairy
Corn
Oats
Cattle
Sheep
Hogs
Soybeans
Barley
Sweet Corn

Grand Junction 964 ... E4
Grand Mound 676 ... M5
Grand River 225 ... F7
Grandview 600 ... L6
Granger 583 ... F5
Grant 102 ... C6
Granville 325 ... B3
Gravity 218 ... D7
Greeley 276 ... L3
Greene 1,099 ... H3
Greenfield▲ 2,129 ... D6
Green Mountain 200 ... H4
Grimes 5,098 ... F5
Grinnell 9,105 ... H5
Griswold 1,039 ... C6
Grundy Center▲ 2,596 ... H4
Gruver 106 ... D2
Guthrie Center▲ 1,668 ... D5
Guttenberg 1,987 ... L3
Halbur 202 ... D4
Hamburg 1,240 ... B7
Hamilton 144 ... H6
Hampton▲ 4,218 ... G3
Hancock 207 ... C6
Hanlontown 229 ... G2
Harcourt 340 ... E4
Harlan▲ 5,282 ... C5
Harper 134 ... J6
Harpers Ferry 330 ... L2
Harris 200 ... C2
Hartford 759 ... G6
Hartley 1,733 ... C2
Harvey 277 ... H6
Hastings 214 ... C6
Havelock 177 ... D3
Haverhill 170 ... H5
Hawarden 2,478 ... A2
Hawkeye 489 ... J3
Hazleton 950 ... K3
Hedrick 837 ... J6
Henderson 171 ... B6
Hiawatha 6,480 ... K4
High 150 ... K5
Hills 679 ... K5
Hillsboro 205 ... K7
Hinton 808 ... A3
Holland 250 ... H4
Holmes 150 ... F3
Holstein 1,470 ... B4
Holy Cross 339 ... L3
Homestead 200 ... K5
Hopkinton 681 ... L4
Hornick 253 ... A4
Hospers 672 ... B2
Hubbard 885 ... G4
Hudson 2,117 ... H4
Hull 1,960 ... A2
Humboldt▲ 4,452 ... E3
Humeston 543 ... G7
Huxley 2,316 ... F5
Ida Grove▲ 2,350 ... B4
Imogene 66 ... C7
Independence▲ 6,014 ... K4
Indianola▲ 12,998 ... F6
Inwood 875 ... A2
Ionia 277 ... J2
Iowa City▲ 62,220 ... L5
Iowa Falls 5,193 ... G3
Ireton 585 ... A3
Irwin 372 ... C5
Jamaica 237 ... E5
Janesville 829 ... J3
Jefferson▲ 4,626 ... E4
Jesup 2,212 ... J4
Jewell 1,239 ... F4
Johnston 8,649 ... F5
Joice 231 ... G2
Kalona 2,293 ... K6
Kamrar 229 ... F4
Kanawha 739 ... F3
Kellerton 372 ... E7
Kelley 300 ... F5
Kellogg 606 ... H5
Kensett 280 ... G2
Keokuk▲ 11,427 ... L8
Keota 1,025 ... K6
Keswick 295 ... J6
Keystone 687 ... J5
Killduff 197 ... H5
Kimballton 342 ... D5
Kingsley 1,245 ... A3
Kirkville 214 ... H6
Kiron 273 ... C4
Klemme 593 ... F3
Knoxville▲ 7,731 ... G6
Lacona 360 ... G6
Ladora 287 ... J5
Lake City 1,787 ... D4
Lake Mills 2,140 ... F2
Lake Park 1,023 ... C2
Lakeside 484 ... C3
Lake View 1,278 ... C4
Lakota 255 ... E2
Lambs Grove 225 ... G5
La Moille 25 ... G4
Lamoni 2,444 ... E7
Lamont 503 ... K3
La Motte 272 ... M4
Lanesboro 152 ... D4
Lansing 1,012 ... L2
La Porte City 2,275 ... J4
Larchwood 788 ... A2
Larrabee 149 ... B3
Latimer 535 ... G3
Laurel 266 ... H5
Laurens 1,476 ... D3
Lawler 461 ... J2
Lawton 697 ... A4
Le Claire 2,847 ... N5
Ledyard 147 ... E2
Le Grand 883 ... H5
Lehigh 497 ... E4
Leland 258 ... F2
Le Mars▲ 9,237 ... A3
Lenox 1,401 ... D7
Leon▲ 1,983 ... F7
Lester 251 ... A2
Letts 392 ... L6
Lewis 438 ... C6
Libertyville 325 ... K7
Lidderdale 186 ... D4
Lime Springs 496 ... J2
Lincoln 182 ... H4
Linden 226 ... E5
Lineville 273 ... G7
Linn Grove 211 ... C3
Lisbon 1,898 ... L5
Liscomb 272 ... H4
Little Rock 489 ... B2
Little Sioux 217 ... B5
Littleton 200 ... K3
Livermore 431 ... E3
Lockridge 275 ... K7
Logan▲ 1,545 ... B5
Lohrville 431 ... D4
Lone Rock 157 ... E2
Lone Tree 1,151 ... L6
Long Grove 597 ... M5
Lorimor 427 ... E6
Lost Nation 497 ... M5
Lovilia 583 ... H6
Lowden 794 ... L5
Low Moor 240 ... N5
Luana 249 ... K2
Lucas 243 ... G6
Luther 158 ... F5
Lu Verne 299 ... E3
Luxemburg 246 ... L3
Lynnville 366 ... H5
Lytton 305 ... D4
Macedonia 325 ... C6
Madrid 2,264 ... F5
Magnolia 200 ... B5
Malcom 352 ... H5
Mallard 298 ... D3
Malvern 1,256 ... B7
Manchester▲ 5,257 ... L3
Manilla 839 ... C5
Manly 1,342 ... G2
Manning 1,490 ... C5
Manson 1,893 ... D3
Mapleton 1,416 ... B4
Maquoketa▲ 6,112 ... M4
Marathon 302 ... C3
Marble Rock 326 ... H3
Marcus 1,139 ... B3
Marengo▲ 2,535 ... J5
Marion 26,294 ... K4
Marne 149 ... C6
Marquette 421 ... L2
Marshalltown▲ 26,009 ... G4
Martelle 280 ... L4
Martensdale 467 ... F6
Martinsburg 126 ... J6
Mason City▲ 104 ... G2
Masonville 104 ... K4
Massena 414 ... D6
Maurice 254 ... A3
Maxwell 807 ... G5
Maynard 500 ... K3
Maysville 163 ... M5
McCallsburg 318 ... G4
McCausland 299 ... M5
McClelland 129 ... B6
McGregor 871 ... L2
McIntire 173 ... H2
Mechanicsville 1,173 ... L5
Mediapolis 1,644 ... L6
Melbourne 794 ... G5
Melcher-Dallas 1,298 ... G6
Melrose 130 ... G7
Melvin 243 ... B2
Menlo 365 ... E5
Meriden 184 ... B3
Merrill 754 ... A3
Meservey 252 ... G3
Middle 386 ... K5
Middletown 535 ... L7
Miles 462 ... N4
Milford 2,474 ... C2
Millersburg 184 ... J5
Milo 839 ... G6
Milton 550 ... J7
Minburn 391 ... E5
Minden 564 ... B5
Mineola 155 ... B6
Mingo 269 ... G5
Missouri Valley 2,992 ... B5
Mitchell 155 ... H2
Mitchellville 1,715 ... G5
Modale 303 ... B5
Mondamin 423 ... B5
Monmouth 180 ... M4
Monona 1,550 ... L2
Monroe 1,808 ... G5
Montezuma▲ 1,440 ... H5
Monticello 3,607 ... L4
Montour 285 ... H5
Montpelier 250 ... M6
Montrose 957 ... L7
Mooar 350 ... L8
Moorhead 232 ... B5
Moorland 197 ... E4
Moravia 713 ... H7
Morning Sun 872 ... L6
Morrison 97 ... H4
Moscow 350 ... L5
Moulton 658 ... H7
Mount Auburn 160 ... J4
Mount Ayr▲ 1,822 ... E7
Mount Pleasant▲ 8,751 ... L7
Mount Union 132 ... L6
Mount Vernon 3,390 ... K5
Moville 1,583 ... A4
Muscatine▲ 22,697 ... L6
Mystic 588 ... H7
Nashua 1,618 ... J3
Neola 845 ... B6
Nevada▲ 6,658 ... G5
New Albin 527 ... L2
Newell 887 ... D3
Newhall 886 ... K5
New Hampton▲ 3,692 ... J2
New Hartford 659 ... H3
New London 1,937 ... L7
New Market 456 ... D7
New Providence 227 ... G4
New Sharon 1,301 ... H6
Newton▲ 15,579 ... H5
New Vienna 400 ... L3
New Virginia 469 ... F6
Nichols 374 ... L6
Nodaway 132 ... D7
Nora Springs 1,532 ... H2
North Buena Vista 124 ... L3
North English 991 ... J5
North Liberty 5,367 ... K5
North Washington 118 ... J2
Northwood▲ 2,050 ... G2
Norwalk 6,884 ... F6
Norway 601 ... K5
Numa 109 ... G7
Oakdale 300 ... K5
Oakland 1,487 ... C6
Oakville 439 ... L6
Ocheyedan 536 ... B2
Odebolt 1,158 ... C4
Oelwein 6,692 ... K3
Ogden 2,023 ... E4
Okoboji 820 ... C2
Olds 249 ... K6
Olin 716 ... L5
Ollie 224 ... J6
Onawa▲ 3,091 ... A4
Onslow 223 ... M4
Ontario 140 ... F4
Oran 4,940 ... J3
Orange City▲ 5,582 ... A2
Orient 402 ... E6
Orleans 583 ... C2
Osage▲ 3,451 ... H2
Osceola▲ 4,659 ... F6
Oskaloosa▲ 10,938 ... H6
Ossian 853 ... K2
Otho 571 ... E4
Otley 150 ... G6
Oto 145 ... B4
Otterville 150 ... K3
Ottumwa▲ 24,998 ... J6
Oxford 705 ... K5
Oxford Junction 573 ... M4
Oyens 132 ... A3
Pacific Junction 507 ... B6
Packwood 223 ... J6
Palmer 214 ... D3
Palo 614 ... K4
Panama 212 ... B5
Panora 1,175 ... E5
Panorama Park 111 ... N5
Parkersburg 1,889 ... H3
Parnell 220 ... J5
Paton 265 ... E4
Paullina 1,124 ... B3
Pella 9,832 ... H6
Perry 7,633 ... E5
Pershing 325 ... G6
Persia 363 ... B5
Peterson 372 ... C3
Pierson 371 ... B3
Pilot Mound 214 ... F4
Pisgah 316 ... B5
Plainfield 438 ... J3
Pleasant Plain 131 ... K6
Pleasantville 1,539 ... G6
Plymouth 429 ... G2
Pocahontas▲ 1,970 ... D3
Polk City 2,344 ... F5
Pomeroy 710 ... D3
Portsmouth 225 ... C5
Postville 2,273 ... K2
Prairieburg 175 ... L4
Prairie City 1,365 ... G5
Prescott 266 ... D6
Preston 949 ... N4
Primghar▲ 891 ... B2
Princeton 946 ... N5
Promise City 105 ... G7
Protivin 317 ... J2
Pulaski 249 ... J7
Quasqueton 574 ... K4
Quimby 368 ... B4
Radcliffe 607 ... G4
Rake 227 ... F2
Randall 148 ... F4
Randolph 209 ... B7
Raymond 537 ... J4
Readlyn 786 ... J3
Reasnor 194 ... G5
Redfield 833 ... E5
Red Oak▲ 6,197 ... C6
Reinbeck 1,751 ... H4
Rembrandt 228 ... C3
Remsen 1,762 ... B3
Renwick 306 ... E3
Rhodes 294 ... G5
Riceville 840 ... H2
Richland 587 ... K6
Richmond 152 ... K6
Rickardsville 191 ... M3
Ricketts 144 ... B4
Ridgeway 293 ... K2
Ringsted 436 ... D2
Rippey 319 ... E5
Riverdale 656 ... N5
Riverside 928 ... K6
River Sioux 150 ... B5
Riverton 304 ... B7
Robins 1,806 ... K4
Rochester 146 ... L5
Rock Falls 170 ... G2
Rockford 907 ... H2
Rock Rapids▲ 2,573 ... A2
Rock Valley 2,702 ... A2
Rockwell 989 ... G3
Rockwell City▲ 2,264 ... D4
Roland 1,324 ... F4
Rolfe 675 ... D3
Rose Hill 205 ... J6
Rossie 58 ... C2
Rossville 183 ... L2
Rowan 218 ... F3
Rowley 290 ... K4
Royal 479 ... C2
Rudd 431 ... H2
Runnells 352 ... G5
Russell 559 ... G7
Ruthven 711 ... D2
Rutland 145 ... E3
Ryan 410 ... K4
Sabula 670 ... N4
Sac City▲ 2,368 ... C4
Sageville 203 ... M3
Saint Ansgar 1,031 ... H2
Saint Anthony 109 ... G4
Saint Charles 619 ... F6
Saint Donatus 140 ... M4
Saint Lucas 178 ... K2
Saint Paul 118 ... L7
Salem 464 ... K7
Salix 370 ... A4
Sanborn 1,353 ... B2
Schaller 779 ... C4
Schleswig 833 ... B4
Scranton 604 ... D4
Sergeant Bluff 3,321 ... A4
Seymour 810 ... G7
Shambaugh 188 ... D7
Sheffield 930 ... G3
Shelby 696 ... C5
Sheldahl 336 ... F5
Sheldon 4,914 ... B2
Shell Rock 1,298 ... H3
Shellsburg 938 ... K4
Shenandoah 5,546 ... C7
Sherrill 186 ... M3
Shueyville 250 ... K5
Sibley▲ 2,796 ... B2
Sidney▲ 1,300 ... B7
Sigourney▲ 2,209 ... J6
Silver City 259 ... B6
Sioux Center 6,002 ... A2
Sioux City▲ 85,013 ... A3
Sioux Rapids 720 ... C3
Slater 1,306 ... F5
Sloan 1,032 ... A4
Smithland 221 ... B4
Soldier 207 ... B5
Solon 1,177 ... L5
Somers 165 ... E4
South Amana 160 ... J5
South English 213 ... J6
Spencer▲ 11,317 ... C2
Spillville 386 ... J2
Spirit Lake▲ 4,261 ... C2
Spragueville 89 ... N4
Springbrook 182 ... N4
Springville 1,068 ... L4
Stacyville 1,091 ... H2
Stanhope 488 ... F4
Stanley 128 ... K3
Stanton 714 ... C7
Stanwood 680 ... L5
State Center 1,349 ... G5
Steamboat Rock 336 ... G4
Stockport 284 ... K7
Storm Lake▲ 10,076 ... C3
Story City 3,228 ... F4
Stout 217 ... H3
Stratford 746 ... F4
Strawberry Point 1,386 ... K3
Stuart 1,712 ... E6
Sully 904 ... H5
Sumner 2,106 ... J3
Superior 142 ... D2
Sutherland 707 ... B3
Swaledale 174 ... G3
Swea City 642 ... E2
Tabor 993 ... B7
Tama 2,731 ... H5
Templeton 334 ... D5
Terril 404 ... C2
Thompson 596 ... F2
Thor 174 ... E3
Thornton 422 ... G3
Thurman 236 ... B7
Tiffin 975 ... K5
Tingley 171 ... E7
Tipton▲ 3,155 ... L5
Titonka 584 ... E2
Toddville 225 ... K4
Toledo▲ 2,539 ... H4
Toronto 134 ... M5
Tracy 325 ... H6
Traer 1,594 ... J4
Treynor 950 ... B6
Tripoli 1,310 ... J3
Troy Mills 200 ... K4
Truro 427 ... F6
Underwood 688 ... B6
Union 427 ... G4
Unionville 127 ... H7
University Heights 987 ... K5
University Park 536 ... H6
Urbana 1,019 ... K4
Urbandale 29,072 ... F5
Ute 378 ... B4
Vail 452 ... C4
Van Horne 716 ... J4
Van Meter 866 ... E5
Van Wert 231 ... F7
Ventura 670 ... F2
Victor 952 ... J5
Villisca 1,344 ... C7
Vincent 158 ... E3
Vinton▲ 5,102 ... J4
Viola 150 ... L4
Volga 247 ... L3
Wadena 243 ... K3
Wahpeton 462 ... C2
Walcott 1,528 ... M5
Walford 1,224 ... K5
Walker 750 ... K4
Wallingford 210 ... D2
Wall Lake 841 ... C4
Walnut 778 ... C6
Wapello▲ 2,124 ... L6
Washington▲ 7,047 ... K6
Washta 282 ... B3
Waterloo▲ 68,747 ... J4
Waterville 145 ... L2
Waucoma 299 ... J2
Waukee 5,126 ... F5
Waukon▲ 4,131 ... L2
Waverly▲ 8,968 ... J3
Wayland 945 ... K6
Webb 165 ... D3
Webster City▲ 8,176 ... F4
Weldon 145 ... F7
Wellman 1,393 ... K6
Wellsburg 716 ... H4
Welton 159 ... M5
Wesley 467 ... E2
West 862 ... J5
West Bend 834 ... D3
West Branch 2,188 ... L5
West Burlington 3,161 ... L7
West Chester 159 ... K6
West Des Moines 46,403 ... F5
Westfield 189 ... A3
Westgate 234 ... K3
West Liberty 3,332 ... L5
West Okoboji 432 ... C2
Westphalia 160 ... C5
West Point 980 ... K7
Westside 327 ... C4
West Union▲ 2,549 ... K3
Wever 260 ... L7
What Cheer 678 ... J6
Wheatland 772 ... M5
Whiting 707 ... A4
Whittemore 530 ... E2
Whitten 160 ... H4
Williams 427 ... F3
Williamsburg 2,622 ... J5
Williamson 163 ... G6
Wilton 2,829 ... M5
Windsor Heights 4,805 ... F5
Winfield 1,131 ... L6
Winterset▲ 4,768 ... E6
Winthrop 772 ... K4
Wiota 149 ... D6
Woden 243 ... F2
Woodbine 1,564 ... B5
Woodburn 244 ... F7
Woodward 1,200 ... E5
Woolstock 204 ... F3
Worthington 381 ... L4
Wyoming 626 ... L4
Yale 287 ... E5
Zearing 617 ... G4

OTHER FEATURES

Big Sioux (riv.) ... A3
Boyer (riv.) ... B5
Cedar (riv.) ... K4
Chariton (riv.) ... G7
Clear (lake) ... G2
Des Moines (riv.) ... J7
Eagle (lake) ... F2
East Nishnabotna (riv.) ... C6
Effigy Mounds Nat'l Mon. ... L2
Five Island (lake) ... D2
Floyd (riv.) ... A3
Herbert Hoover Nat'l Hist. Site ... L5
Iowa (riv.) ... H4
Little Sioux (riv.) ... B3
Lost Island (lake) ... D2
Mississippi (riv.) ... L7
Missouri (riv.) ... A5
Nodaway (riv.) ... D7
Palo Alto (lake) ... D2
Platte (riv.) ... D8
Raccoon (riv.) ... D4
Rathbun (lake) ... G7
Red Rock (lake) ... G6
Rock (riv.) ... A2
Sac and Fox Ind. Res. ... H5
Skunk (riv.) ... K6
Spirit (lake) ... C2
Storm (lake) ... C3
Summit (lake) ... E6
Thompson (riv.) ... E7
Trumbull (lake) ... C2
Turkey (riv.) ... K2
Upper Iowa (riv.) ... K2
Wapsipinicon (riv.) ... J3
West Nishnabotna (riv.) ... C6

▲County seat

KANSAS

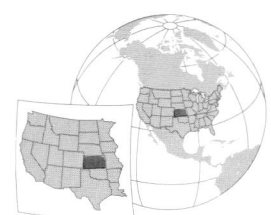

AREA 82,277 sq. mi. (213,097 sq. km.)
POPULATION 2,688,418
CAPITAL Topeka
LARGEST CITY Wichita
HIGHEST POINT Mt. Sunflower 4,039 ft.
(1231 m.)
SETTLED IN 1831
ADMITTED TO UNION January 29, 1861
POPULAR NAME Sunflower State
STATE FLOWER Sunflower
STATE BIRD Western Meadowlark

COUNTIES

Allen 14,385	G4
Anderson 8,110	G3
Atchison 16,774	G2
Barber 5,307	D4
Barton 28,205	D3
Bourbon 15,379	H4
Brown 10,724	G2
Butler 59,482	F4
Chase 3,030	F3
Chautauqua 4,359	F4
Cherokee 22,605	H4
Cheyenne 3,165	A2
Clark 2,390	C4
Clay 8,822	E2
Cloud 10,268	E2
Coffey 8,865	G3
Comanche 1,967	C4
Cowley 36,291	F4
Crawford 38,242	H4
Decatur 3,472	B2
Dickinson 19,344	E3
Doniphan 8,249	G2
Douglas 99,962	G3
Edwards 3,449	C4
Elk 3,261	F4
Ellis 27,507	C3
Ellsworth 6,525	D3
Finney 40,523	B3
Ford 32,458	C4
Franklin 24,784	G3
Geary 27,947	F3
Gove 3,068	B3
Graham 2,946	C2
Grant 7,909	A4
Gray 5,904	B4
Greeley 1,534	A3
Greenwood 1,534	F4
Hamilton 2,670	A3
Harper 6,536	D4
Harvey 32,869	E3
Haskell 4,307	B4
Hodgeman 2,085	C3
Jackson 12,657	G2
Jefferson 18,426	G2
Jewell 3,791	D2
Johnson 451,086	H3
Kearny 4,531	A3
Kingman 8,673	D4
Kiowa 3,278	C4
Labette 22,835	G4
Lane 2,155	B3
Leavenworth 68,691	G2
Lincoln 3,578	D2
Linn 9,570	H3
Logan 3,046	A3
Lyon 35,935	F3
Marion 13,361	E3
Marshall 10,965	F2
McPherson 29,554	E3
Meade 4,631	B4
Miami 28,351	H3
Mitchell 6,932	D2
Montgomery 36,252	G4
Morris 6,104	F3
Morton 3,496	A4
Nemaha 10,717	F2
Neosho 16,997	G4
Ness 3,454	C3
Norton 5,953	C2
Osage 16,712	G3
Osborne 4,452	D2
Ottawa 6,163	E2
Pawnee 7,233	C3
Phillips 6,001	C2
Pottawatomie 18,209	F2
Pratt 9,647	D4
Rawlins 2,966	A2
Reno 64,790	D4
Republic 5,835	E2
Rice 10,761	D3
Riley 62,843	F2
Rooks 5,685	C2
Rush 3,551	C3
Russell 7,370	D3
Saline 53,597	E3
Scott 5,120	B3
Sedgwick 452,869	E4
Seward 22,510	B4
Shawnee 169,871	G2
Sheridan 2,813	B2
Sherman 6,760	A2
Smith 4,536	D2
Stafford 4,789	D3
Stanton 2,406	A4
Stevens 5,463	A4
Sumner 25,946	E4
Thomas 8,180	A2
Trego 3,319	C3
Wabaunsee 6,885	F3
Wallace 1,749	A3
Washington 6,483	E2
Wichita 3,043	A3
Wilson 10,332	G4
Woodson 3,944	G4
Wyandotte 157,882	H2

CITIES and TOWNS

Abbyville 128	D4
Abilene▲ 6,543	E3
Ada 120	E2
Admire 177	F3
Agenda 81	E2
Agra 306	C2
Agricola 43	G3
Alamota 50	B3
Albert 181	D3
Alden 168	D3
Alexander 75	C3
Aliceville 60	G3
Allen 211	F3
Alma▲ 797	F2
Almena 469	C2
Altamont 1,092	G4
Alta Vista 442	F3
Alton 117	D2
Altoona 485	G4
Americus 938	F3
Ames 65	E2
Andale 766	E4
Andover 6,698	E4

Angola 55	G4
Anson 32	E4
Antelope 35	F3
Anthony▲ 2,440	D4
Antonino 38	C3
Arcadia 391	H4
Argonia 534	E4
Arkansas City 11,963	E4
Arlington 459	D4
Arma 1,529	H4
Arnold 48	B3
Arrington 45	G2
Asherville 32	D2
Ash Grove 28	D2
Ashland▲ 975	C4
Ashton 30	E4
Ash Valley 50	C3
Assaria 438	E3
Atchison▲ 10,232	G2
Athol 51	D2
Atlanta 255	F4
Attica 636	D4
Atwood▲ 1,279	B2
Auburn 1,121	G3
Augusta 8,423	F4
Aurora 79	E2
Axtell 445	F2
Baileyville 130	F2
Bala 26	E2
Baldwin City 3,400	G3
Barnard 123	D2
Barnes 152	F2
Bartlett 124	G4
Basehor 2,238	G2
Bassett 22	G4
Bavaria 100	E3
Baxter Springs 4,602	H4
Bazaar 40	F3
Bazine 311	C3
Beagle 88	G3
Beattie 277	F2
Beaumont 112	F4
Beaver 57	D3
Beeler 80	B3
Bellefont 28	C4
Belle Plaine 1,708	E4
Belleville▲ 2,239	E2
Belmont 33	D4
Beloit▲ 4,019	D2
Belpre 104	C4
Belvidere 25	C4
Belvue 228	F2
Bendena 125	G2
Benedict 103	G4
Bennington 623	E2
Bentley 368	E4
Benton 827	E4
Bern 204	F2
Berryton 150	G3
Beverly 199	D2
Big Bow 55	A4
Bird City 482	A2
Bison 235	C3
Blaine 50	F2
Blair 40	H2
Bloom 65	C4
Blue Mound 277	H3
Blue Rapids 1,088	F2
Bluff City 80	E4
Bogue 179	C2
Bonner Springs 6,768	H2
Brazilton 91	H4
Bremen 60	F2
Brewster 285	A2
Bridgeport 95	E3
Bronson 346	H4
Brookville 259	E3
Brownell 48	C3
Bucklin 725	C4
Bucyrus 135	H3
Buffalo 284	G4
Buhler 1,358	E3
Bunker Hill 101	D3
Burden 564	F4
Burdett 256	C3
Burdick 100	F3
Burlingame 1,017	G3
Burlington▲ 2,790	G3
Burns 268	E3
Burr Oak 265	D2
Burrton 932	E3
Bushong 55	F3
Bushton 314	D3
Byers 50	D4
Cairo 38	D4
Caldwell 1,284	E4
Calvert 35	C2
Cambridge 103	F4
Caney 2,092	G4
Canton 829	E3
Carbondale 1,478	G3
Carlton 38	E3
Carlyle 75	G4
Carneiro 28	D3
Cassoday 130	F3
Catharine 121	C3
Cawker City 521	D2
Cedar 26	D2
Cedar Point 53	F3
Cedar Vale 723	F4
Centerville 104	H3
Centralia 534	F2
Chanute 9,411	G4
Chapman 1,241	E3
Chase 490	D3
Chautauqua 113	F4
Cherokee 722	H4
Cherryvale 2,386	G4
Chetopa 1,281	G4
Chicopee 300	H4
Cimarron▲ 1,934	B4
Circleville 185	G2
Claflin 705	D3
Clay Center▲ 4,564	E2
Clayton 66	C2
Clearwater 2,178	E4
Clifton 557	E2
Climax 64	F4
Clyde 740	E2
Coats 112	D4
Codell 90	C2
Coffeyville 11,021	G4

Agriculture, Industry and Resources

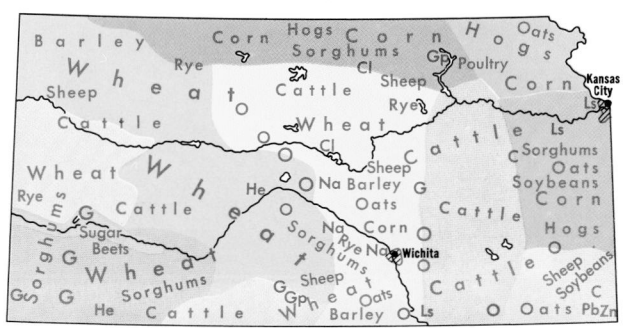

DOMINANT LAND USE

- Specialized Wheat
- Wheat, General Farming
- Wheat, Range Livestock
- Wheat, Grain Sorghums, Range Livestock
- Cattle Feed, Hogs
- Livestock, Cash Grain
- Livestock, Cash Grain, Dairy
- General Farming, Livestock, Cash Grain
- General Farming, Livestock, Special Crops
- Range Livestock

MAJOR MINERAL OCCURRENCES

C	Coal	Ls	Limestone
Cl	Clay	Na	Salt
G	Natural Gas	O	Petroleum
Gp	Gypsum	Pb	Lead
He	Helium	Zn	Zinc

///// Major Industrial Areas

Colby▲ 5,450	A2	De Graff 31	F4	Edgerton 1,440	H3	Esbon 148	D2	Galatia 61	D3
Coldwater▲ 792	C4	Delavan 40	F3	Edmond 47	C2	Eskridge 589	F3	Galena 3,287	H4
Collyer 133	B2	Delia 179	G2	Edna 423	G4	Eudora 4,307	G3	Galesburg 150	G4
Colony 397	G3	Delphos 469	E2	Edson 55	A2	Eureka▲ 2,914	F4	Galva 701	E3
Columbus▲ 3,396	H4	Denison 231	G2	Edwardsville 4,146	H2	Everest 314	G2	Garden City▲ 28,451	B4
Colwich 1,229	E4	Denmark 40	D2	Effingham 588	G2	Fairport 41	D2	Garden Plain 797	E4
Concordia▲ 5,714	E2	Dennis 96	G4	Elbing 218	E3	Fairview 271	G2	Gardner 9,396	H3
Conway 1,384	D3	Densmore 34	C2	El Dorado▲ 12,057	F4	Fairway 3,952	H3	Garfield 198	C3
Conway Springs 1,322	E4	Denton 186	G2	Elgin 82	F4	Fall River 156	F4	Garland 112	H4
Coolidge 86	A3	Derby 17,807	E4	Elk City 305	G4	Falun 89	E3	Garnett▲ 3,368	G3
Copeland 339	B4	De Soto 4,561	H3	Elk Falls 112	F4	Farlington 80	H4	Gas 556	G4
Corbin 36	E4	Detroit 90	E3	Elkhart▲ 2,233	A4	Fellsburg 41	C4	Gaylord 145	D2
Corning 170	F2	Devon 108	H4	Ellinwood 2,164	D3	Florence 671	E3	Gem 96	B2
Corwin 25	D4	Dexter 364	F4	Ellis 1,873	C3	Fontana 149	H3	Geneseo 272	D3
Cottonwood Falls▲ 966	F3	Dighton▲ 1,261	B3	Ellsworth▲ 2,965	D3	Ford 314	C4	Geuda Springs 212	E4
Council Grove▲ 2,321	F3	Dillon 25	E3	Elmdale 50	F3	Formoso 129	D2	Girard▲ 2,773	H4
Courtland 334	E2	Dodge City▲ 25,176	B4	Elmo 41	E3	Fort Dodge 400	C4	Glade 114	C2
Coyville 71	G4	Dorrance 205	D3	Elmont 112	G2	Fort Leavenworth		Glasco 556	E2
Crestline 85	H4	Douglass 1,813	E4	Elmore 73	G4	Fort Scott▲ 8,297	H4	Glen Elder 439	D2
Cuba 231	E2	Dover 192	G3	Elwood 1,145	H2	Fostoria 49	F2	Goddard 2,037	E4
Cullison 98	D4	Downs 1,038	D2	Elyria 160	E3	Fowler 567	B4	Goessel 565	E3
Culver 164	E3	Dresden 51	B2	Emmett 277	F2	Frankfort 855	F2	Goff 181	G2
Cummings 150	G2	Dunlap 81	F3	Engelvale 29	H4	Franklin 400	H4	Goodland▲ 4,948	A2
Cunningham 514	D4	Durham 114	E3	Englewood 109	C4	Fredonia▲ 2,600	G4	Gordon 30	F4
Damar 155	C2	Dwight 330	F3	Ensign 203	B4	Friend 35	B3	Gorham 360	D3
Danville 59	E4	Earlton 80	G4	Enterprise 836	E3	Frontenac 2,996	H4	Gove▲ 105	B3
Dearing 415	G4	Eastborough 826	E4	Erie▲ 1,211	G4	Fulton 184	H4	Grainfield 327	B2
Deerfield 884	A4	Easton 362	G2					Grandview Plaza 1,184	F2

(continued on following page)

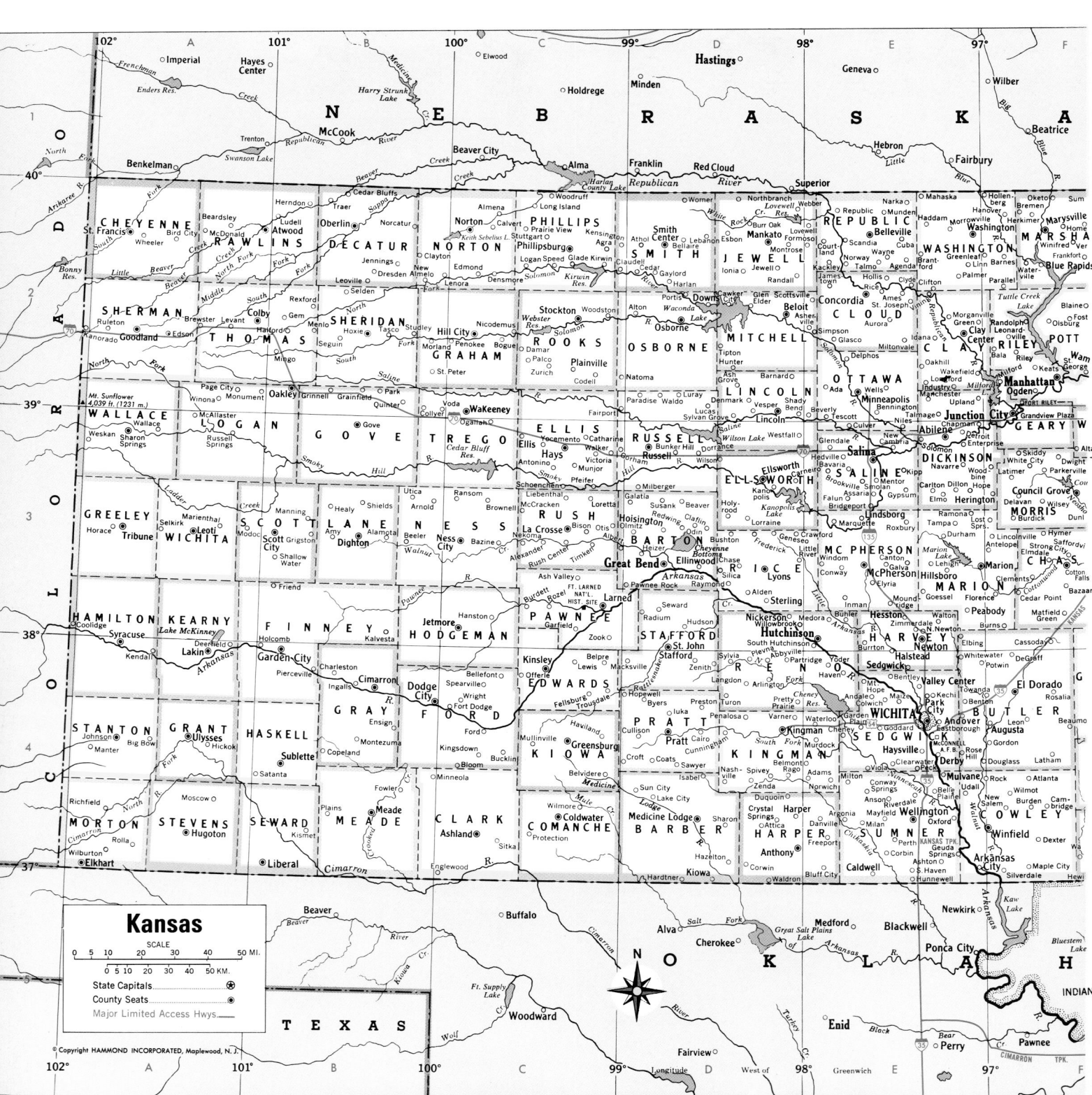

Matfield Green 60 ... F3
Mayetta 312 ... G2
Mayfield 113 ... E4
McCracken 211 ... C3
McCune 426 ... G4
McDonald 159 ... A2
McFarland 271 ... F2
McLouth 868 ... G2
McPherson▲ 13,770 ... E3
Meade▲ 1,672 ... B4
Medicine Lodge▲ 2,193 ... D4
Melvern 429 ... G3
Mentor 100 ... E3
Mercier 85 ... G2
Meriden 706 ... G2
Merriam 11,008 ... H3
Michigan Valley 65 ... G3
Milan 137 ... E4
Milford 502 ... F2
Miller 50 ... F3
Milton 484 ... E4
Miltonvale 523 ... E2
Minneapolis▲ 2,046 ... E2
Minneola 717 ... C4
Mission 9,727 ... H2
Moline 457 ... F4
Montezuma 966 ... B4
Monument 85 ... A2
Moran 562 ... G4
Morehead 65 ... G4
Morganville 198 ... E2
Morland 164 ... B2
Morrill 277 ... G2
Morrowville 168 ... E2
Moscow 247 ... A4
Mound City▲ 821 ... H3
Moundridge 1,593 ... E3
Mound Valley 418 ... G4
Mount Hope 830 ... E4
Mulberry 577 ... H4
Mullinville 279 ... C4
Mulvane 5,155 ... E4
Munden 122 ... E2
Munjor 200 ... C3
Murdock 75 ... E4
Muscotah 200 ... G2
Narka 93 ... E2
Nashville 111 ... D4
Natoma 367 ... D2
Navarre 66 ... E3

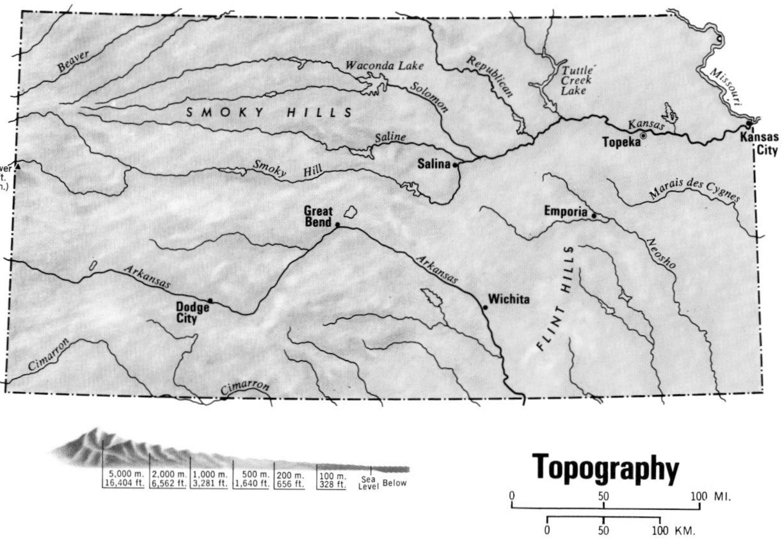
SMOKY HILLS
FLINT HILLS
Beaver
Waconda Lake
Republican
Tuttle Creek Lake
Missouri
Solomon
Kansas
Topeka
Kansas City
Saline
Salina
Smoky Hill
Mt. Sunflower 4,039 ft. (1231 m.)
Marais des Cygnes
Emporia
Great Bend
Neosho
Arkansas
Wichita
Dodge City
Cimarron
5,000 m. 16,404 ft.
2,000 m. 6,562 ft.
1,000 m. 3,281 ft.
500 m. 1,640 ft.
200 m. 656 ft.
100 m. 328 ft.
Sea Level
Below
0 50 100 MI.
0 50 100 KM.

Topography

Neal 72 ... F4
Neodesha 2,848 ... G4
Neosho Falls 179 ... G3
Neosho Rapids 274 ... F3
Ness City▲ 1,534 ... C3
Netawaka 170 ... G2
New Albany 73 ... G4
New Almelo 80 ... B2
New Cambria 150 ... E3
New Lancaster 52 ... H3
New Salem 100 ... F4
New Strawn (Strawn) 425 ... G3
Newton▲ 17,190 ... E3
Nickerson 1,194 ... D3
Niles 75 ... E2
Niotaze 122 ... F4
Norcatur 169 ... B2
North Newton 1,522 ... E3
Norton▲ 3,012 ... C2
Nortonville 620 ... G2
Norway 60 ... E2
Norwich 551 ... E4
Oakhill 35 ... E2
Oakley▲ 2,173 ... B2
Oberlin▲ 1,994 ... B2
Odin 105 ... D3
Offerle 220 ... C4
Ogallah 75 ... C3
Ogden 1,762 ... F2
Oketo 87 ... F2
Olathe▲ 92,962 ... H3
Olivet 64 ... G3
Olmitz 138 ... D3
Olpe 504 ... F3
Olsburg 192 ... F2
Onaga 704 ... F2
Oneida 70 ... G2
Opolis 165 ... H4
Osage City 3,034 ... G3
Osawatomie 4,645 ... H3
Osborne▲ 1,607 ... D2
Oskaloosa▲ 1,165 ... G2
Oswego▲ 2,046 ... G4
Otis 325 ... C3
Ottawa▲ 11,921 ... G3
Overbrook 947 ... G3
Overland Park 149,080 ... H3
Oxford 1,173 ... E4
Ozawkie 552 ... G2
Page City 60 ... A2
Palco 248 ... C2
Palmer 121 ... E2
Paola▲ 5,011 ... H3
Paradise 64 ... D2
Park 151 ... B2
Park City 5,814 ... E4
Parker 281 ... H3
Parkerville 73 ... F3
Parsons 11,514 ... G4
Partridge 259 ... D4
Pauline 250 ... G3
Pawnee Rock 356 ... D3
Paxico 211 ... F2
Peabody 1,384 ... E3
Peck 250 ... E4
Penalosa 27 ... D4
Penokee 73 ... C2
Perry 901 ... G2
Perth 46 ... E4
Peru 183 ... F4
Petrolia 69 ... G4
Pfeifer 80 ... C3
Phillipsburg▲ 2,668 ... C2
Piedmont 96 ... F4
Pierceville 135 ... B4
Piqua 103 ... G4
Pittsburg 19,243 ... H4
Plains 1,163 ... B4
Plainville 2,029 ... C2
Pleasanton 1,387 ... H3
Plevna 99 ... D4
Pomona 923 ... G3
Portis 123 ... D2
Potter 175 ... G2
Potwin 457 ... F4
Powhattan 91 ... G2
Prairie View 141 ... C2
Prairie Village 22,072 ... H2
Pratt▲ 6,570 ... D4
Prescott 280 ... H3
Preston 164 ... D4
Pretty Prairie 615 ... D4
Princeton 317 ... G3
Protection 558 ... C4
Purcell 34 ... G2
Quenemo 468 ... G3
Quincy 35 ... F4
Quinter 961 ... B2
Ramona 94 ... E3
Randall 90 ... D2
Randolph 175 ... F2
Ransom 338 ... C3
Rantoul 241 ... G3
Raymond 95 ... D3
Reading 247 ... F3
Redfield 140 ... H4
Reece 60 ... F4
Republic 161 ... E2
Reserve 100 ... G2
Rexford 157 ... B2
Richfield 48 ... A4
Richland 100 ... G3
Richmond 510 ... G3
Riley 886 ... F2
Riverdale 46 ... E4
Riverton 650 ... H4
Robinson 216 ... G2
Rock 250 ... F4
Rock Creek 93 ... G2
Roeland Park 6,817 ... H2
Rolla 482 ... A4
Rosalia 125 ... F4
Rose Hill 3,432 ... E4
Roseland 101 ... H4
Rossville 1,014 ... G2
Roxbury 115 ... E3
Rozel 182 ... C3
Rush Center 176 ... C3
Russell▲ 4,696 ... D3
Sabetha 2,589 ... G2
Saffordville 50 ... F3
Saint Benedict 55 ... F2
Saint Francis▲ 1,497 ... A2
Saint George 434 ... F2
Saint John▲ 1,318 ... D3
Saint Joseph 28 ... E2
Saint Marys 2,198 ... G2
Saint Paul 646 ... G4
Saint Peter 40 ... C2
Salina▲ 45,679 ... E3
Satanta 1,239 ... B4
Savonburg 91 ... G4
Sawyer 124 ... D4
Scammon 496 ... H4
Scandia 436 ... E2
Schoenchen 214 ... C3
Scott City▲ 3,855 ... B3
Scottsville 21 ... D2
Scranton 724 ... G3
Sedan▲ 1,342 ... F4
Sedgwick 1,537 ... E4
Selden 201 ... B2
Seneca▲ 2,122 ... F2
Severance 108 ... G2
Severy 359 ... F4
Seward 63 ... D3
Shady Bend 28 ... D2
Shallow Water 100 ... B3
Sharon 210 ... D4
Sharon Springs▲ 835 ... A3
Shawnee 47,996 ... H2
Sherman 44 ... H4
Shields 110 ... B3
Silver Lake 1,358 ... G2
Simpson 114 ... E2
Skiddy 55 ... F3
Smith Center▲ 1,931 ... D2
Smolan 218 ... E3
Soldier 122 ... G2
Solomon 1,072 ... E3
South Haven 390 ... E4
South Hutchinson 2,539 ... D3
South Mound 46 ... G4
Sparks 71 ... G2
Spearville 813 ... C4
Spivey 80 ... D4
Spring Hill 2,727 ... H3
Stafford 1,161 ... D4
Stanley 450 ... H3
Stark 106 ... G4
Sterling 2,642 ... D3
Stilwell 350 ... H3
Stockton▲ 1,558 ... C2
Strawn 428 ... G3
Strong City 584 ... F3
Studley 60 ... B2
Stuttgart 70 ... C2
Sublette▲ 1,592 ... B4
Summerfield 211 ... F2
Sun City 81 ... D4
Susank 57 ... D3
Sycamore 150 ... G4
Sylvan Grove 324 ... D2
Sylvia 297 ... D4
Syracuse▲ 1,824 ... A3
Talmage 130 ... E2
Tampa 144 ... E3
Tecumseh 300 ... G2
Tescott 339 ... E2
Thayer 500 ... G4
Timken 83 ... C3
Tipton 243 ... D2
Tonganoxie 2,728 ... G2
Topeka (cap.)▲ 122,377 ... G2
Toronto 312 ... G4
Towanda 1,338 ... E4
Trading Post 35 ... H3
Treece 149 ... H4
Tribune▲ 835 ... A3
Trousdale 54 ... C4
Troy▲ 1,054 ... G2
Turon 436 ... D4
Tyro 226 ... G4
Udall 794 ... E4
Ulysses▲ 5,960 ... A4
Uniontown 288 ... G4
Utica 223 ... B3
Valeda 45 ... G4
Valley Center 4,883 ... E4
Valley Falls 1,254 ... G2
Vassar 90 ... G3
Vermillion 107 ... F2
Vesper 48 ... D2
Victoria 1,208 ... C3
Vining 58 ... E2
Vinland 102 ... G3
Viola 211 ... E4
Virgil 113 ... F4
Vliets 56 ... F2
Wabaunsee 75 ... F2
Wakarusa 171 ... G3
WaKeeney▲ 1,924 ... C2
Wakefield 838 ... E2
Waldo 48 ... D2
Walker 104 ... C3
Wallace 67 ... A3
Walnut 221 ... G4
Walton 284 ... E3
Wamego 4,246 ... F2
Washington▲ 1,223 ... F2
Waterloo 30 ... E4
Waterville 681 ... F2
Wathena 1,348 ... H2
Waverly 589 ... G3
Wayside 85 ... G4
Webber 37 ... E2
Weir 780 ... H4
Welda 150 ... G3
Wellington▲ 8,647 ... E4
Wells 1,563 ... E2
Wellsville 1,606 ... G3
Weskan 200 ... A3
West Mineral 243 ... H4
Westmoreland▲ 631 ... F2
Westphalia 165 ... G3
West Plains (Plains) 1,163 ... B4
Wetmore 362 ... G2
Wheaton 92 ... F2
White City 518 ... F3
White Cloud 239 ... G2
Whitewater 653 ... E4
Whiting 206 ... G2
Wichita▲ 344,284 ... E4
Wilburton 40 ... A4
Willard 86 ... G2
Williamsburg 351 ... G3
Williamstown 89 ... G2
Willis 69 ... G2
Willowbrook 36 ... D3
Wilmore 57 ... C4
Wilsey 191 ... F3
Wilson 799 ... D3
Winchester 579 ... G2
Windom 137 ... E3
Winfield▲ 12,206 ... F4
Winifred 80 ... F2
Winona 228 ... A2
Woodbine 207 ... E3
Woodruff 33 ... C2
Woodston 116 ... C2
Wright 300 ... C4
Yates Center▲ 1,599 ... G4
Yoder 160 ... E4
Zenda 123 ... D4
Zurich 126 ... C2

OTHER FEATURES

Arkansas (riv.) ... D3
Beaver (creek) ... A2
Big Blue (riv.) ... F1
Cedar Bluff (res.) ... C3
Cheney (res.) ... E4
Cheyenne Bottoms (lake) ... D3
Chikaskia (riv.) ... E4
Cimarron (riv.) ... B4
Cottonwood (riv.) ... F3
Council Grove (lake) ... F3
Crooked (creek) ... B4
Elk (riv.) ... F4
Fall (riv.) ... G4
Fall River (lake) ... F4
Fort Larned Nat'l Hist. Site ... C3
Fort Riley-Camp Whiteside ... F2
Fort Scott Nat'l Hist. Site ... H4
Hulah (lake) ... F5
John Redmond (res.) ... G3
Kanopolis (lake) ... D3
Kansas (riv.) ... F2
Keith Sebelius (res.) ... C2
Kickapoo Ind. Res. ... G2
Kirwin (res.) ... C2
Little Arkansas (riv.) ... E3
Little Blue (riv.) ... E1
Lovewell (res.) ... D2
Marion (lake) ... E3
McConnell A.F.B. ... E4
McKinney (lake) ... A3
Medicine Lodge (riv.) ... D4
Milford (lake) ... E2
Missouri (riv.) ... G1
Nemaha (riv.) ... G1
Neosho (riv.) ... G4
Ninnescah (riv.) ... E4
Pawnee (riv.) ... B3
Perry (lake) ... G2
Pomona (lake) ... G3
Potawatomi Ind. Res. ... G2
Rattlesnake (creek) ... D4
Republican (riv.) ... E2
Sac-Fox-Iowa Ind. Res. ... G2
Saline (riv.) ... D3
Sappa (creek) ... B2
Smoky Hill (riv.) ... C3
Solomon (riv.) ... E2
Sunflower (mt.) ... A2
Toronto (lake) ... F4
Tuttle Creek (lake) ... F2
Verdigris (riv.) ... G5
Walnut (riv.) ... E4
Webster (res.) ... C2
White Rock (creek) ... D2
Wilson (lake) ... D3

▲County seat

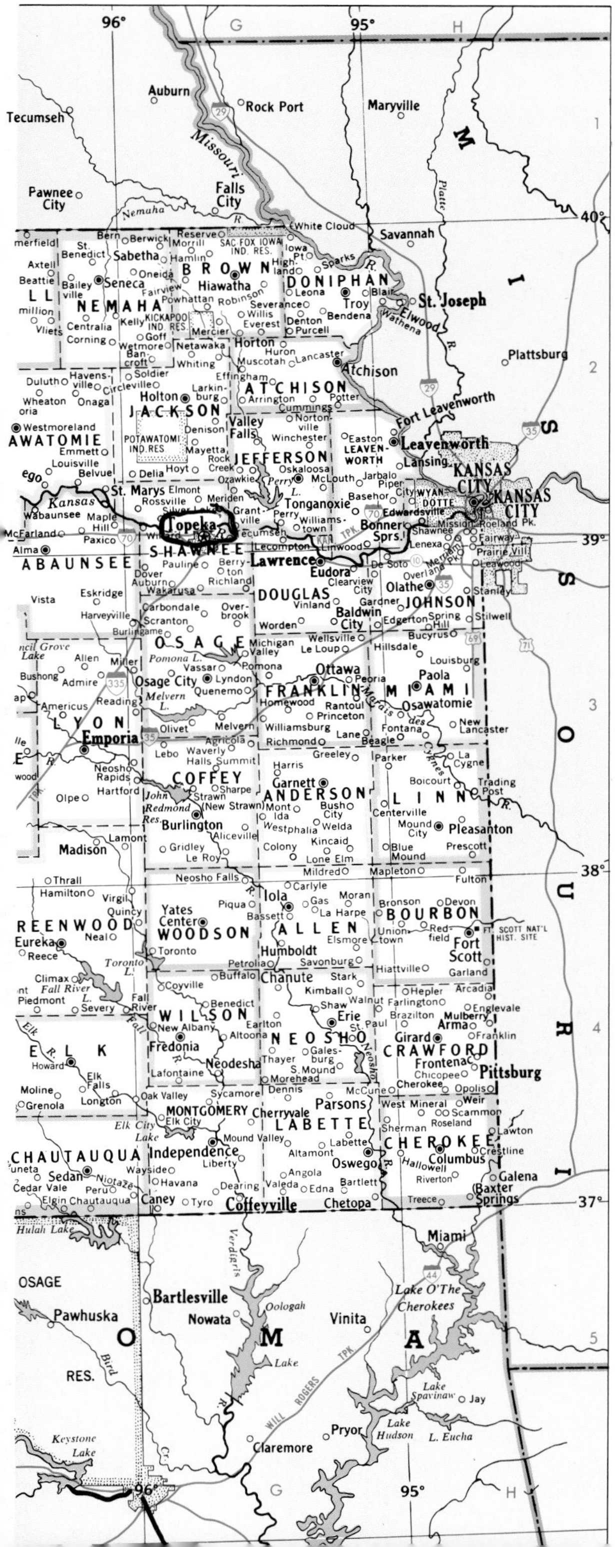
96°
G
95°
H
Auburn
Rock Port
Maryville
Tecumseh
Missouri
Falls City
Pawnee City
Nemaha
White Cloud
Savannah
BROWN
DONIPHAN
St. Joseph
NEMAHA
Hiawatha
Troy
Atchison
Plattsburg
ATCHISON
JACKSON
Holton
JEFFERSON
Leavenworth
LEAVENWORTH
WABAUNSEE
SHAWNEE
Topeka
Lawrence
KANSAS CITY
DOUGLAS
JOHNSON
Olathe
Paola
OSAGE
FRANKLIN
Ottawa
MIAMI
Osawatomie
LYON
Emporia
COFFEY
Burlington
ANDERSON
Garnett
LINN
Pleasanton
WOODSON
Yates Center
ALLEN
Iola
BOURBON
Fort Scott
Humboldt
Chanute
WILSON
Fredonia
NEOSHO
Erie
CRAWFORD
Pittsburg
Girard
Neodesha
MONTGOMERY
Independence
LABETTE
Parsons
Oswego
CHEROKEE
Columbus
Galena
Baxter Springs
CHAUTAUQUA
Sedan
Coffeyville
Caney
Chetopa
ELK
GREENWOOD
Eureka
Miami
Bartlesville
Nowata
Vinita
Pawhuska
OSAGE
Lake O'The Cherokees
Claremore
Pryor
Jay
Hulah Lake
Keystone Lake
Oologah Lake
Lake Hudson
L. Eucha
Lake Spavinaw
M
I
S
S
O
U
R
I
O
M
A
40°
39°
38°
37°
1
2
3
4
5

KENTUCKY

COUNTIES

Agriculture, Industry and Resources

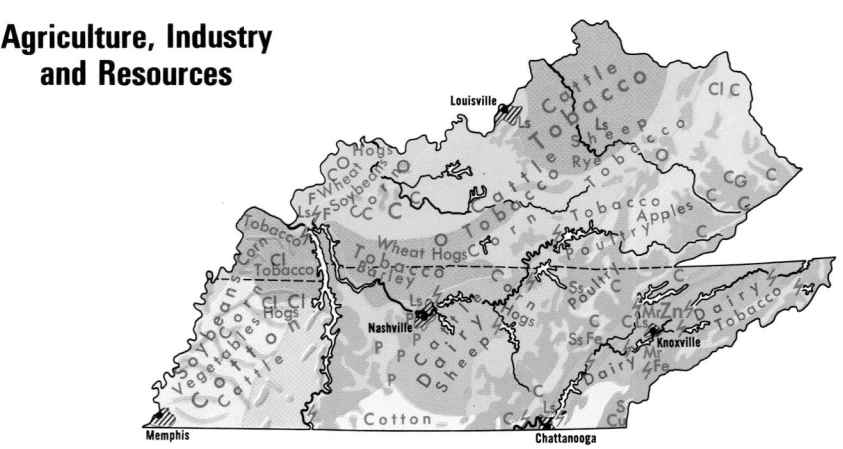

DOMINANT LAND USE

- Hogs, Soft Winter Wheat
- Tobacco, General Farming
- General Farming, Livestock, Tobacco
- General Farming, Livestock, Dairy
- General Farming, Livestock, Fruit, Tobacco
- Specialized Cotton
- Cotton, General Farming
- Cotton, Livestock
- Forests
- Swampland, Limited Agriculture

MAJOR MINERAL OCCURRENCES

C Coal
Cl Clay
Cu Copper
F Fluorspar
Fe Iron Ore
G Natural Gas
Ls Limestone
Mr Marble
O Petroleum
P Phosphates
S Pyrites
Ss Sandstone
Zn Zinc

⚡ Water Power ⟋⟋⟋ Major Industrial Areas

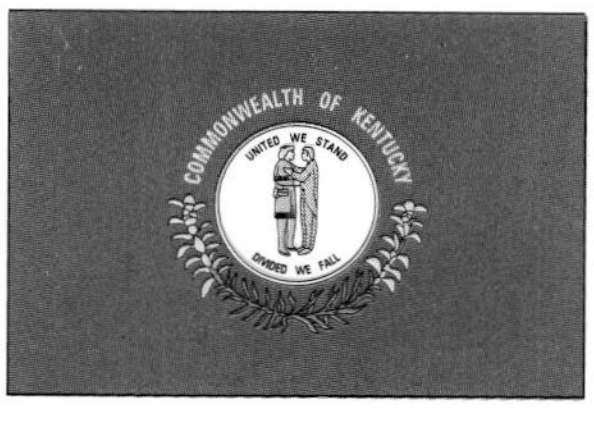

KENTUCKY

KENTUCKY

AREA 40,409 sq. mi. (104,659 sq. km.)
POPULATION 4,041,769
CAPITAL Frankfort
LARGEST CITY Louisville
HIGHEST POINT Black Mtn. 4,139 ft. (1262 m.)
SETTLED IN 1774
ADMITTED TO UNION June 1, 1792
POPULAR NAME Bluegrass State
STATE FLOWER Goldenrod
STATE BIRD Cardinal

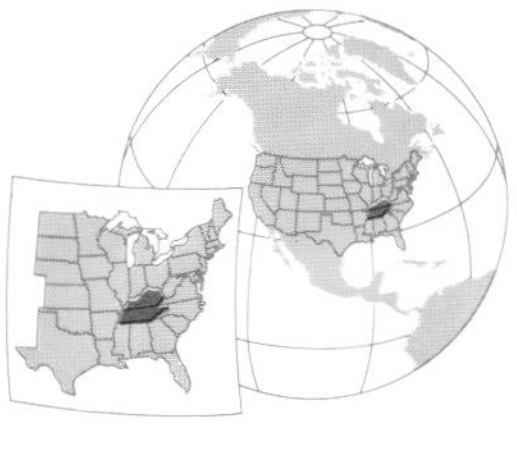

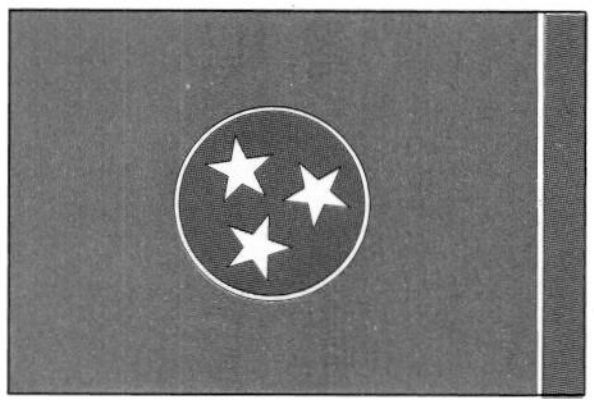

TENNESSEE

TENNESSEE

AREA 42,144 sq. mi. (109,153 sq. km.)
POPULATION 5,689,283
CAPITAL Nashville
LARGEST CITY Memphis
HIGHEST POINT Clingmans Dome 6,643 ft. (2025 m.)
SETTLED IN 1757
ADMITTED TO UNION June 1, 1796
POPULAR NAME Volunteer State
STATE FLOWER Iris, Passionflower
STATE BIRD Mockingbird

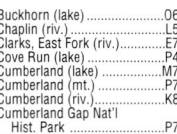

Salem 769 ... E6
Salt Lick 342 ... O4
Salyersville▲ 1,604 ... P5
Sanders 246 ... M3
Sandy Hook▲ 678 ... P4
Sardis 149 ... O3
Scalf 500 ... O7
Science Hill 634 ... M6
Scottsville▲ 4,327 ... J7
Sebree 1,558 ... F5
Sedalia ... D7
Seneca Gardens 699 ... K2
Sextons Creek 975 ... O6
Shelbiana 500 ... R6
Shelbyville▲ 10,085 ... L4
Shepherdsville▲ 8,334 ... K4
Shively 15,157 ... K4
Silver Grove 1,215 ... T2
Simpson 907 ... P5
Simpsonville 1,281 ... L4
Slaughters 238 ... F6
Slemp 500 ... P6
Smilax 987 ... P6
Smithland▲ 401 ... E6
Smiths Grove 784 ... J6
Somerset▲ 11,352 ... M6
Sonora 350 ... K5
South 202 ... J6
Southgate 3,472 ... T2
South Portsmouth 900 ... P3
South Shore 1,226 ... R3
South Williamson 1,016 ... S5
Sparta 230 ... M3
Spottsville 914 ... G5
Springfield▲ 2,634 ... L5
Springlee 426 ... K2
Staffordsville 700 ... R5
Stamping Ground 566 ... M4
Stanford▲ 3,430 ... M5
Stanley 475 ... G5
Stanton▲ 3,029 ... O5
Stearns 1,586 ... N7
Stone 900 ... S5
Strathmoor Village 625 ... K2
Sturgis 2,030 ... F5
Tateville 680 ... M7
Taylor Mill 6,913 ... S2
Taylorsville▲ 1,009 ... L4
Thealka 600 ... R5
Tollesboro 808 ... O3
Tompkinsville▲ 2,660 ... K7
Trenton 419 ... G7
Tyner 590 ... O6
Union 2,893 ... M3
Uniontown 1,064 ... F5
Upton 654 ... K6
Valley Station 22,946 ... K4
Van 1,050 ... R6
Van Lear 2,035 ... R5
Vanceburg▲ 1,731 ... P3
Verda 1,133 ... P7
Verona 500 ... M3
Versailles▲ 7,511 ... M4
Vicco 318 ... P6
Villa Hills 7,948 ... R2
Vine Grove 4,169 ... K5
Virgie 600 ... R6
Wallins Creek 257 ... O7
Walton 2,450 ... M3
Warfield 284 ... S5
Warsaw▲ 1,811 ... M3
Washington 795 ... O3
Water Valley 316 ... D7
Wayland 298 ... R6
Waynesburg 500 ... M6
Webbville 400 ... R4
Weeksbury 850 ... R6
Wellington 593 ... O5
Wellington 561 ... K2
West Buechel 1,301 ... K2
West Liberty▲ 3,277 ... P5
West Point 1,100 ... J4
West Somerset 850 ... M6
Westwood 4,888 ... R4
Westwood 612 ... L1
Wheatcroft 173 ... F5
Wheelwright 1,042 ... R6
White Mills 500 ... J5
White Plains 800 ... G6
Whitesburg▲ 1,600 ... R6
Whitesville 632 ... H5
Whitley City▲ 1,111 ... N7
Wickliffe▲ 794 ... C7
Wilders 2,624 ... S2
Williamsburg▲ 5,143 ... N7
Williamstown▲ 3,227 ... M3
Willisburg 304 ... L5
Wilmore 5,905 ... M5
Winchester▲ 16,724 ... N5
Windy Hills 2,480 ... K1
Wingo 581 ... D7
Wolf Creek 600 ... J4
Woodbine 900 ... N7
Woodburn 323 ... J7
Woodland Hills 657 ... L2
Woodlawn (Oakdale) 4,937 ... D6
Woodlawn 268 ... T2
Woodlawn Park 1,033 ... K2
Wooton 750 ... P6
Worthington 1,673 ... R3
Worthville 215 ... L3
Wurtland 1,049 ... R3
Yeaddiss 430 ... P6
Yosemite 400 ... M6
Zebulon 750 ... S5

OTHER FEATURES

Abraham Lincoln Birthplace Nat'l Hist. Site ... K5
Barkley (dam) ... E6
Barkley (lake) ... F7
Barren (riv.) ... H6
Barren River (lake) ... J7
Beech Fork (riv.) ... L5
Big Sandy (riv.) ... R4
Black (mt.) ... R7
Buckhorn (lake) ... O6
Chaplin (riv.) ... L5
Clarks, East Fork (riv.) ... E7
Cove Run (lake) ... P4
Cumberland (lake) ... M7
Cumberland (mt.) ... P7
Cumberland (riv.) ... K8
Cumberland Gap Nat'l Hist. Park ... P7
Dale Hollow (lake) ... L7
Dewey (lake) ... R5
Dix (riv.) ... M5
Drakes (creek) ... J7
Dry (creek) ... R2
Eagle (creek) ... M3
Fishtrap (lake) ... S6
Fort Campbell ... G7
Grayson (lake) ... P4
Green (riv.) ... G6
Green River (lake) ... L6
Herrington (lake) ... M5
Hinkston (creek) ... N4
Kentucky (dam) ... E7
Kentucky (lake) ... E8
Kentucky (riv.) ... M3
Land Between The Lakes Rec. Area ... E7
Laurel River (lake) ... N6
Lexington Blue Grass Army Depot ... N4
Licking (riv.) ... N3
Mammoth Cave Nat'l Park ... J6
Mayfield (creek) ... C7
Mississippi (riv.) ... 10
Mud (riv.) ... H7
Nolin (lake) ... K6
Nolin (riv.) ... J6
Obion (creek) ... C7
Ohio (riv.) ... F5
Paint Lick (riv.) ... M5
Panther (creek) ... G5
Pine (mt.) ... O7
Pond (riv.) ... G6
Red (riv.) ... G7
Red (riv.) ... O5
Rockcastle (riv.) ... N6
Rolling Fork (riv.) ... K5
Rough (riv.) ... H5
Rough River (lake) ... J5
Salt (riv.) ... K5
Tennessee (riv.) ... D6
Tradewater (riv.) ... F6
Tug Fork (riv.) ... S5

TENNESSEE

COUNTIES

Anderson 71,330 ... N8
Bedford 37,586 ... J9
Benton 16,537 ... E8
Bledsoe 12,367 ... L9
Blount 105,823 ... O9
Bradley 87,965 ... M10
Campbell 39,854 ... N8
Cannon 12,826 ... J9
Carroll 29,475 ... E9
Carter 56,742 ... S8
Cheatham 35,912 ... G8
Chester 15,540 ... D10
Claiborne 29,862 ... O8
Clay 7,976 ... K7
Cocke 33,565 ... P9
Coffee 48,014 ... J9
Crockett 14,532 ... C9
Cumberland 46,802 ... L9
Davidson 569,891 ... H8
Decatur 11,731 ... E9
De Kalb 17,423 ... K9
Dickson 43,156 ... G8
Dyer 37,279 ... C8
Fayette 28,806 ... C10
Fentress 16,625 ... M8
Franklin 39,270 ... J10
Gibson 48,152 ... D9
Giles 29,447 ... G10
Grainger 20,659 ... O8
Greene 62,909 ... R8
Grundy 14,332 ... K10
Hamblen 58,128 ... P8
Hamilton 307,896 ... L10
Hancock 6,786 ... P7
Hardeman 28,105 ... C10
Hardin 25,578 ... E10
Hawkins 53,563 ... P8
Haywood 19,797 ... C9
Henderson 25,522 ... E9
Henry 31,115 ... E8
Hickman 22,295 ... G9
Houston 8,088 ... F8
Humphreys 17,929 ... F8
Jackson 10,984 ... K8
Jefferson 44,294 ... P8
Johnson 17,499 ... T7
Knox 382,032 ... O9
Lake 7,954 ... B8
Lauderdale 27,101 ... B9
Lawrence 39,926 ... G10
Lewis 11,367 ... F9
Lincoln 31,340 ... H10
Loudon 39,086 ... N9
Macon 20,386 ... J7
Madison 91,837 ... D9
Marion 27,776 ... K10
Marshall 26,767 ... H10
Maury 69,498 ... G9
McMinn 49,015 ... M10
McNairy 24,653 ... D10
Meigs 11,086 ... M9
Monroe 38,961 ... N10
Montgomery 134,768 ... G8
Moore 5,740 ... J10
Morgan 19,757 ... M8
Obion 32,450 ... C8
Overton 20,118 ... L8
Perry 7,631 ... F9
Pickett 4,945 ... M7
Polk 16,050 ... N10
Putnam 62,315 ... K8
Rhea 28,400 ... M9
Roane 51,910 ... M9
Robertson 54,433 ... H7
Rutherford 182,023 ... J9
Scott 21,127 ... M8
Sequatchie 11,370 ... L10
Sevier 71,170 ... O9
Shelby 897,472 ... B10
Smith 17,712 ... J8
Stewart 12,370 ... F7
Sullivan 153,048 ... S7
Sumner 130,449 ... J8
Tipton 51,271 ... B9
Trousdale 7,259 ... J8
Unicoi 17,667 ... S8
Union 17,808 ... O8
Van Buren 5,508 ... L9
Warren 38,276 ... K9
Washington 107,198 ... R8
Wayne 16,842 ... F10
Weakley 34,895 ... D8
White 23,102 ... L9
Williamson 126,638 ... H9
Wilson 88,809 ... J8

CITIES and TOWNS

Adams 566 ... G7
Adamsville 1,983 ... E10
Afton 800 ... R8
Alamo▲ 2,392 ... C9
Alcoa 7,734 ... N9
Alexandria 814 ... J8
Algood 2,942 ... K8
Allardt 642 ... M8
Allons 600 ... L8
Altamont▲ 1,136 ... K10
Apison 750 ... L10
Ardmore 1,082 ... H10
Arlington 2,569 ... B10
Armathwaite 700 ... M8
Arthur 500 ... O7
Ashland City▲ 3,641 ... G8
Athens▲ 13,220 ... M10
Atoka 3,235 ... B10
Atwood 1,000 ... D9
Auburntown 252 ... J9
Baileyton 504 ... R8
Banner Hill 1,053 ... R8
Bartlett 40,543 ... B10
Bath Springs 800 ... E10
Baxter 1,279 ... K8
Bean Station 500 ... P8
Beechgrove 550 ... J9
Beersheba Springs 553 ... K10
Bell Buckle 391 ... J9
Belle Meade 2,943 ... H8
Bells 2,171 ... C9
Benton▲ 1,138 ... M10
Berry Hill 674 ... H8
Berry's Chapel 2,703 ... H9
Bethel Springs 763 ... D10
Big Sandy 518 ... E8
Birchwood 550 ... M10
Blaine 1,585 ... O8
Bloomingdale 10,350 ... R7
Bloomington Springs 800 ... K8
Blountville▲ 2,959 ... S7
Bluff City 1,559 ... S8
Bolivar▲ 5,802 ... C10
Bradford 1,113 ... D8
Braemar ... S8
Brazil 325 ... C9
Brentwood 23,445 ... H8
Briceville 850 ... N8
Brighton 1,719 ... B10
Bristol 24,821 ... S7
Brownsville▲ 10,748 ... C9
Bruceton 1,554 ... E8
Buena Vista 500 ... E9
Bulls Gap 714 ... P8
Burlison 453 ... B9
Burns 1,366 ... G8
Butler 500 ... T8
Byrdstown▲ 903 ... L7
Calhoun 496 ... M10
Camden▲ 3,828 ... E8
Carthage▲ 2,251 ... K8
Caryville 2,243 ... N8
Castalian Springs 650 ... J8
Cedar Hill 298 ... H7
Celina▲ 1,379 ... K7
Centertown 257 ... K9
Centerville▲ 3,793 ... G9
Chapel Hill 943 ... H9
Charleston 630 ... M10
Charlotte▲ 1,153 ... G8
Chattanooga▲ 155,554 ... K10
Chuckey 500 ... R8
Church Hill 5,916 ... R7
Clairfield 650 ... O7
Clarksburg 285 ... E9
Clarksville▲ 103,455 ... G7
Cleveland▲ 37,192 ... M10
Clifton 2,699 ... F10
Clinton▲ 9,409 ... N8
Coalfield 712 ... N8
Coalmont 948 ... K10
Cokercreek 500 ... N10
Collegedale 6,514 ... M10
College Grove 580 ... H9
Collierville 31,872 ... B10
Collinwood 1,024 ... F10
Colonial Heights 7,067 ... R8
Columbia▲ 33,055 ... G9
Concord 8,569 ... N9
Cookeville▲ 23,923 ... L8
Copperhill 511 ... N10
Cordova 600 ... B10
Cornersville 962 ... H10
Corryton 500 ... O8
Counce 975 ... E10
Covington▲ 8,463 ... B9
Cowan 1,770 ... J10
Crab Orchard 838 ... M9
Crockett Mills 500 ... C9
Cross Plains 1,381 ... H7
Crossville▲ 8,981 ... L9
Crump 1,521 ... E10
Cumberland City 316 ... F8
Cumberland Gap 204 ... O7
Cypress Inn 500 ... F10
Dandridge▲ 2,078 ... O8
Dayton▲ 6,180 ... L9
Decatur▲ 1,395 ... M9
Decaturville▲ 859 ... E9
Decherd 2,246 ... J10
Dickson 12,244 ... G8
Dover▲ 1,442 ... F8
Dowelltown 302 ... K8
Doyle 525 ... K9
Dresden▲ 2,855 ... D8
Drummonds 800 ... A10
Duck River 750 ... G9
Ducktown 427 ... N10
Dunlap▲ 4,173 ... L10
Dyer 2,406 ... D8
Dyersburg▲ 17,452 ... C8
Eads 550 ... B10
Eagleton Village 4,883 ... O9
Eagleville 464 ... H9
East Ridge 20,640 ... L11
Eastview 618 ... D10
Elgin 700 ... M8
Elizabethton▲ 13,372 ... S8
Elkton 510 ... H10
Elk Valley 750 ... N7
Ellendale 850 ... B10
Embreeville Junction ... R8
Emory Gap 500 ... M9
Englewood 1,590 ... M10
Enville 230 ... E10
Erin▲ 1,490 ... F8
Erwin▲ 5,610 ... S8
Estill Springs 2,152 ... J10
Ethridge 536 ... G10
Etowah 3,663 ... M10
Eva 500 ... E8
Fairfield 4,885 ... J9
Fairview 5,800 ... G9
Fall Branch 1,313 ... R8
Farner 750 ... N10
Fayetteville▲ 6,994 ... H10
Finger 350 ... D10
Finley 1,014 ... B8
Flintville 500 ... J10
Forest Hills 4,710 ... H8
Fort Pillow 700 ... B9
Fowlkes 700 ... C9
Franklin▲ 41,842 ... H9
Friendship 608 ... C9
Friendsville 890 ... N9
Gadsden 553 ... D9
Gainesboro▲ 879 ... K8
Gallatin▲ 23,230 ... H8
Gallaway 666 ... B10
Garland 309 ... B9
Gates 901 ... C9
Gatlinburg 3,382 ... O9
Germantown 37,348 ... B10
Gibson 305 ... D9
Gilt Edge 489 ... B9
Gleason 1,463 ... D8
Goodlettsville 13,780 ... H8
Gordonsville 1,066 ... K8
Grand Junction 301 ... C10
Grandview ... M9
Graysville 1,411 ... L10
Greenback 954 ... N9
Greenbrier 4,940 ... H8
Greeneville▲ 15,198 ... R8
Greenfield 2,208 ... D8
Grimsley 650 ... L8
Gruetli 1,867 ... K10
Guys 483 ... D10
Habersham 750 ... N8
Halls 2,311 ... C9
Halls Crossroads ... O8
Hampshire 788 ... G9
Hampton 2,236 ... S8
Harriman 6,744 ... M9
Harris 7,191 ... C8
Harrison 7,630 ... L10
Harrogate-Shawanee 2,865 ... O8
Hartsville▲ 2,395 ... J8
Helenwood 846 ... M8
Henderson▲ 5,670 ... D10
Hendersonville 40,620 ... H8
Henning 970 ... B9
Henry 520 ... E8
Hickory Valley 136 ... C10
Hickory Withe 2,574 ... B10
Hohenwald▲ 3,754 ... F9
Hollow Rock 963 ... E8
Hornbeak 435 ... C8
Hornsby 306 ... D10
Humboldt 9,467 ... D9
Huntingdon▲ 4,349 ... E8
Huntland 916 ... J10
Huntsville▲ 981 ... N8
Hurricane Mills 850 ... F9
Iron City 368 ... F10
Jacksboro▲ 1,887 ... N8
Jackson▲ 59,643 ... D9
Jamestown▲ 1,839 ... M8
Jasper▲ 3,214 ... K10
Jefferson City 7,760 ... P8
Jellico 2,448 ... N7
Johnson City 55,469 ... S8
Jones 3,091 ... C9
Jonesborough▲ 4,168 ... R8
Karns 1,458 ... N9
Kenton 1,306 ... C8
Kimball 1,312 ... K10
Kimberlin Heights 500 ... O9
Kingsport 44,905 ... R7
Kingston▲ 5,264 ... N9
Kingston Springs 2,773 ... G8
Knoxville▲ 173,890 ... O9
Kodak 700 ... O9
Laager 675 ... K10
Lafayette▲ 3,885 ... J7
La Follette 7,926 ... N8
La Grange 136 ... C10
Lake City 1,888 ... N8
Lakeland 6,862 ... B10
Lakesite 1,845 ... L10
Lakewood 2,341 ... H8
La Vergne 18,687 ... H9
Lawrenceburg▲ 10,796 ... G10
Lebanon▲ 20,235 ... J8
Lenoir City 6,819 ... N9
Leoma 600 ... G10
Lewisburg▲ 10,413 ... H10
Lexington▲ 7,393 ... E9
Liberty 367 ... K8
Linden▲ 1,015 ... F9
Livingston▲ 3,498 ... L8
Lobelville 915 ... F9
Lone Mountain 150 ... O8
Lookout Mountain 2,000 ... L11
Loretto 1,665 ... G10
Loudon▲ 4,476 ... N9
Louisville 2,001 ... N9
Luttrell 915 ... O8
Lutts 740 ... F10
Lyles 500 ... G9
Lynchburg▲ 5,740 ... J10
Lynnville 345 ... G10
Madisonville▲ 3,939 ... N9
Malesus 600 ... D9
Manchester▲ 8,294 ... J10
Martel 500 ... N9
Martin 10,515 ... D8
Maryville▲ 23,120 ... O9
Mascot 2,119 ... O8
Mason 1,089 ... B10
Maury City 704 ... C9
Maynardville▲ 1,782 ... O8
McDonald 500 ... M10
McEwen 1,702 ... F8
McKenzie 5,295 ... E8
McLemoresville 259 ... D9
McMinnville▲ 12,749 ... K9
Medina 969 ... D9
Medon 191 ... D10
Memphis▲ 650,100 ... B10
Michie 647 ... E10
Middleton 602 ... D10
Midway 2,491 ... P8
Milan 7,664 ... D9
Milledgeville 287 ... E10
Milligan College 600 ... S8
Millington 10,433 ... B10
Minor Hill 437 ... G10
Mitchellville 207 ... J7
Monteagle 1,238 ... K10
Monterey 2,717 ... L8
Morley 600 ... N7
Morrison 684 ... K9
Morrison City 2,032 ... R7
Morristown▲ 24,965 ... P8
Moscow 422 ... C10
Mosheim 1,749 ... R8
Mountain City▲ 2,383 ... T8
Mount Carmel 4,795 ... R8
Mount Juliet 12,366 ... H8
Mount Pleasant 4,491 ... G9
Munford 4,708 ... B10
Murfreesboro▲ 68,816 ... J9
Murray Lake Hills ... L10
Nashville (cap.)▲ 569,891 ... H8
Neubert 800 ... O9
Newbern 2,988 ... C8
New Hope 1,043 ... K11
New Johnsonville 1,905 ... E8
New Market 1,234 ... O8
Newport▲ 7,242 ... P9
New River 250 ... M8
New Tazewell 2,871 ... O8
Niota 781 ... M9
Norris 1,446 ... N8
Nunnelly 375 ... G9
Oak Hill 4,493 ... H8
Oakland 1,279 ... B10
Oak Ridge 27,387 ... N8
Obion 1,134 ... C8
Olivehill 450 ... E10
Oliver Springs 3,303 ... N8
Oneida 3,615 ... N7
Ooltewah 5,681 ... M10
Orebank 1,284 ... R7
Orlinda 594 ... H7
Orme 124 ... K10
Pall Mall 750 ... M7
Palmer 726 ... K10
Paris▲ 9,763 ... E8
Parrotsville 207 ... P8
Parsons 2,452 ... E9
Pegram 2,146 ... H8
Petersburg 580 ... H10
Petros 1,286 ... M8
Philadelphia 533 ... M9
Pickwick Dam 650 ... E10
Pigeon Forge 5,083 ... O9
Pikeville▲ 1,781 ... L9
Piperton 589 ... B10
Pittman Center 477 ... P9
Pleasant Hill 544 ... L9
Pleasant View 2,934 ... G8
Portland 8,458 ... H7
Powder Springs 600 ... O8
Powell 7,534 ... N8
Powells Crossroads 1,286 ... L10
Primm Springs 750 ... G9
Pulaski▲ 7,871 ... G10
Puryear 667 ... E8
Red Bank 12,418 ... L10
Red Boiling Springs 1,023 ... K7
Riceville 500 ... M10
Ridgely 1,667 ... B8
Ridgeside 389 ... L10
Ridgetop 1,083 ... H8
Ripley▲ 7,844 ... B9
Rives 331 ... C8
Roan Mountain 1,160 ... S8
Rockford 798 ... O9
Rockwood 5,774 ... M9
Rogersville▲ 4,240 ... P8
Rosemark 950 ... B10
Rossville 380 ... B10
Russellville 1,069 ... P8
Rutherford 1,272 ... C8
Rutledge▲ 1,187 ... P8
Saint Joseph 829 ... G10
Sale Creek 900 ... L10
Saltillo 342 ... E10
Samburg 260 ... C8
Sardis 445 ... E10
Saulsbury 99 ... C10
Savannah▲ 6,917 ... E10
Scotts Hill 894 ... E10
Selmer▲ 4,541 ... D10
Sequatchie 800 ... K10
Sevierville▲ 11,757 ... P9
Sewanee 2,361 ... K10
Seymour 8,850 ... O9
Shady Valley 475 ... T7
Sharon 988 ... D8
Shelbyville▲ 16,105 ... H10
Sherwood 900 ... K10
Signal Mountain 7,429 ... L10
Smithville▲ 3,994 ... K9
Smyrna 25,569 ... H9
Sneedville▲ 1,257 ... P7
Soddy-Daisy 11,530 ... L10
Somerville▲ 2,519 ... C10
South Carthage 1,302 ... K8
South Cleveland 6,216 ... M10
South Clinton 1,671 ... N8
South Fulton 2,517 ... D8
South Pittsburg 3,295 ... K10
Southside 800 ... G8
South Tunnel 400 ... H8
Sparta▲ 4,599 ... K9
Spencer▲ 1,713 ... L9
Spring City 2,025 ... M9
Springfield▲ 14,329 ... H8
Spring Hill 7,715 ... H9
Stanton 615 ... C10
Stantonville 312 ... E10
Strawberry Plains 680 ... O8
Sullivan Gardens 2,513 ... R8
Summertown 850 ... G10
Surgoinsville 1,484 ... R8
Sweetwater 5,586 ... N9
Talbott 975 ... P8
Tazewell▲ 2,165 ... P8
Tellico Plains 859 ... N10
Ten Mile 700 ... M9
Tennessee Ridge 1,334 ... F8
Thompsons Station 1,283 ... H9
Tipton 2,149 ... B10
Tiptonville▲ 2,439 ... B8
Toone 330 ... D10
Townsend 244 ... O9
Tracy City 1,679 ... K10
Treadway 712 ... P8
Trenton▲ 4,683 ... D9
Trezevant 901 ... D8
Trimble 728 ... C8
Troy 1,273 ... C8
Tullahoma 17,994 ... J10
Tusculum 2,004 ... R8
Unicoi 3,519 ... S8
Union City▲ 10,876 ... C8
Vanleer 310 ... G8
Victoria 800 ... K10
Vonore 1,162 ... N9
Walden 1,960 ... L10
Walterhill 1,523 ... J9
Wartburg▲ 890 ... M8
Wartrace 548 ... J9

(continued on following page)

Washburn 500 .... O8
Watertown 1,358 .... J8
Waverly▲ 4,028 .... F8
Waynesboro▲ 2,228 .... F10
Western Institute 850 .... C10
Westmoreland 2,093 .... J7
White Bluff 2,142 .... G8
White House 7,220 .... H8
White Pine 1,997 .... P8
Whiteville 3,148 .... C10
Whitwell 1,660 .... K10
Wildersville 340 .... E9
Winchester▲ 7,329 .... J10
Winfield 911 .... M7
Woodbury▲ 2,428 .... J9
Woodland Mills 296 .... C8
Woodlawn 500 .... F7

OTHER FEATURES

Andrew Johnson
Nat'l Hist. Site R8
Appalachian (mts.) N10
Bald (mts.) R9
Barkley (lake) F7
Big Sandy (riv.) E9
Boone (lake) S8
Buffalo (riv.) F9
Caney Fork (riv.) L9
Center Hill (lake) K9
Cheatham (dam) G8
Cheatham (lake) H8
Cherokee (dam) O8
Cherokee (lake) P8
Chickamauga (dam) L10
Chickamauga (lake) L10
Chilhowee (mt.) O9
Clinch (riv.) N9
Clingmans Dome (mt.) O9
Collins (riv.) K9
Conasauga (riv.) M11
Cordell Hull (res.) K8
Cumberland (plat.) L9
Cumberland (riv.) K8
Cumberland Gap
Nat'l Hist. Park O7
Dale Hollow (lake) L7
Douglas (lake) P9
Duck (riv.) F9
Elk (riv.) H10
Emory (riv.) M8
Forked Deer (riv.) C9
Fort Campbell G7
Fort Donelson Nat'l Battlefield . F8
Fort Loudoun (lake) N9
French Broad (riv.) R9
Great Falls (dam) K9
Great Smoky (mts.) P9
Great Smoky Mountains
Nat'l Park O9
Green (riv.) F10
Guyot (mt.) P9
Harpeth (riv.) G8
Hatchie (riv.) B9
Hiwassee (riv.) O10
Holston (riv.) O8
Iron (mts.) S8
Kentucky (lake) E8
Land Between The Lakes
Rec. Area E7
Lick (creek) R8
Little Tennessee (riv.) N10
Loosahatchie (riv.) B10
Melton Hill (lake) N9
Memphis Naval Air Sta. B10
Meriwether Lewis Park,
Natchez Trace Pkwy G10
Mississippi (riv.) A10
Nolichucky (riv.) R8
Norris (dam) O8
Norris (lake) O8
Obed (riv.) M8
Obion (riv.) C8
Ocoee (riv.) M10
Old Hickory (dam) H8
Old Hickory (lake) J8
Pickwick (lake) E11
Powell (riv.) P8
Priest, J. Percy (lake) J8
Red (riv.) G7
Reelfoot (lake) C8
Richland (creek) G10
Rutherford Fork, Obion (riv.) .. D8
Sequatchie (riv.) L10
Sewart A.F.B. J8
Shiloh Nat'l Mil. Park E10
Shoal (creek) F10
South Holston (lake) S7
Stone (mts.) T8
Stones (riv.) J9
Stones River Nat'l Battlefield ... J9
Sulphur Fork, Red (riv.) G8
Tellico (riv.) N10
Tennessee (riv.) E10
Tims Ford (lake) J10
Unaka (mts.) S8
Unicoi (mts.) N10
Watauga (lake) T8
Watts Bar (dam) M9
Watts Bar (lake) M9
Whiteoak (creek) F8
Wolf (riv.) B10
Woods (res.) J10
Yellow (creek) F2

▲ County seat.

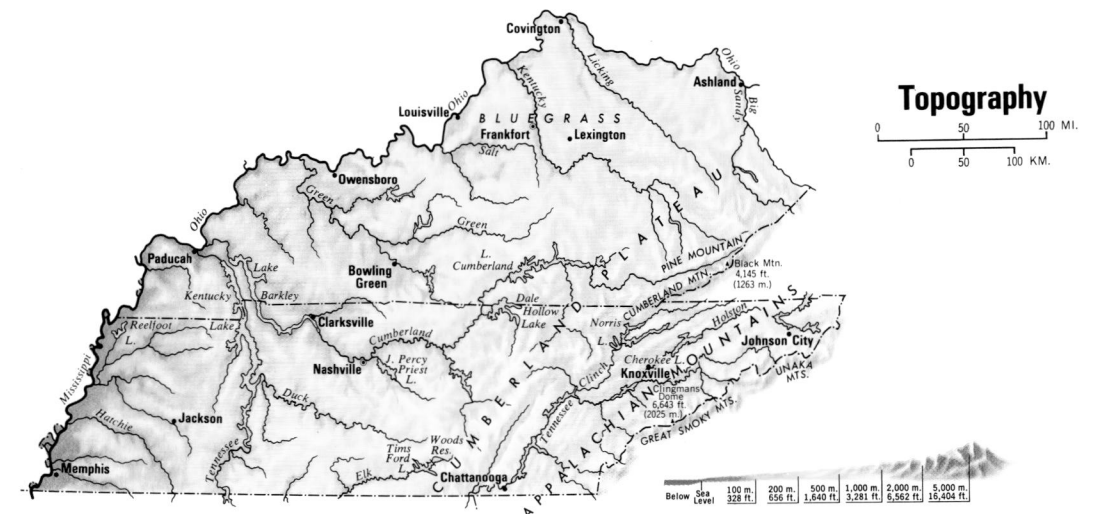

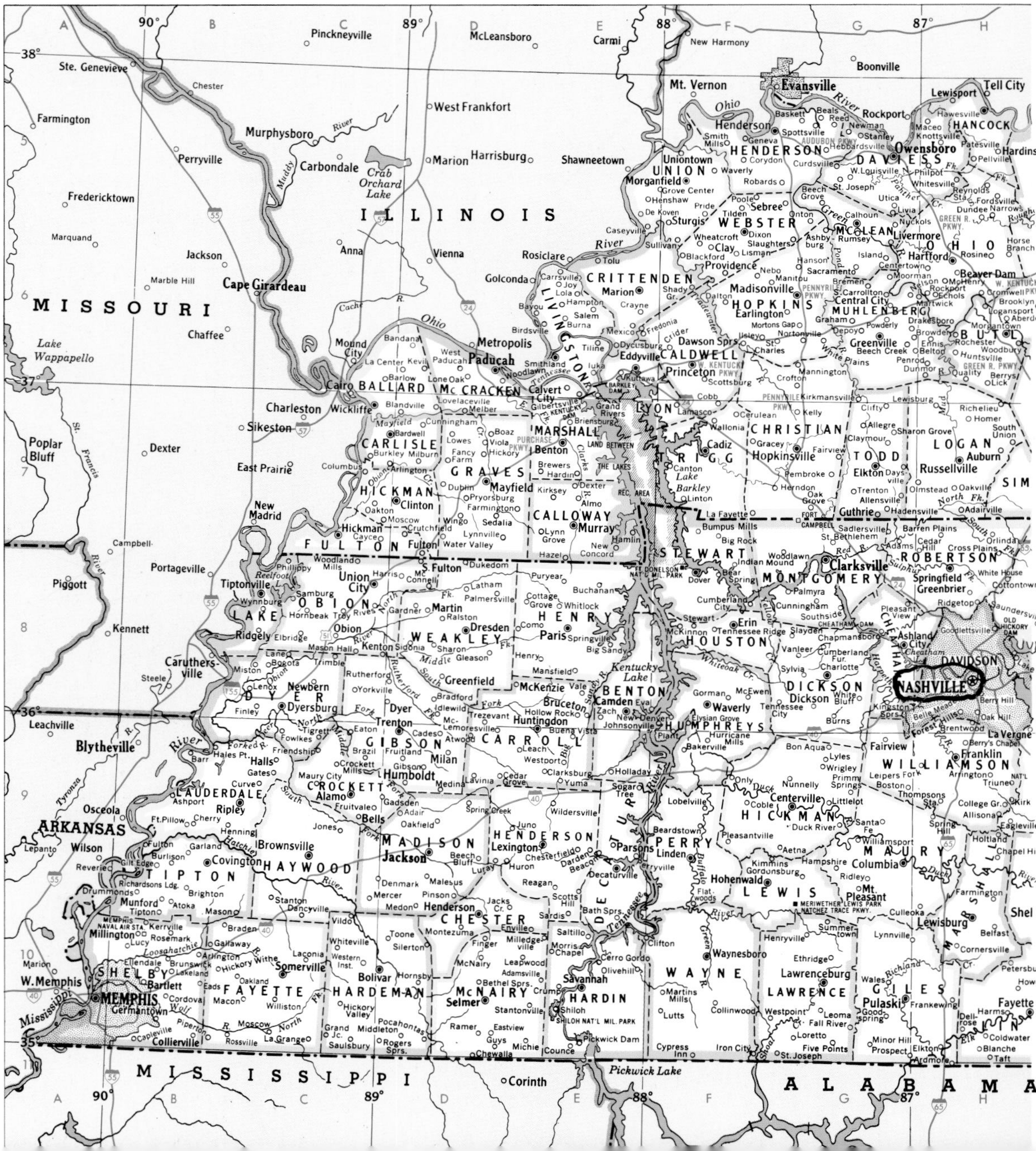

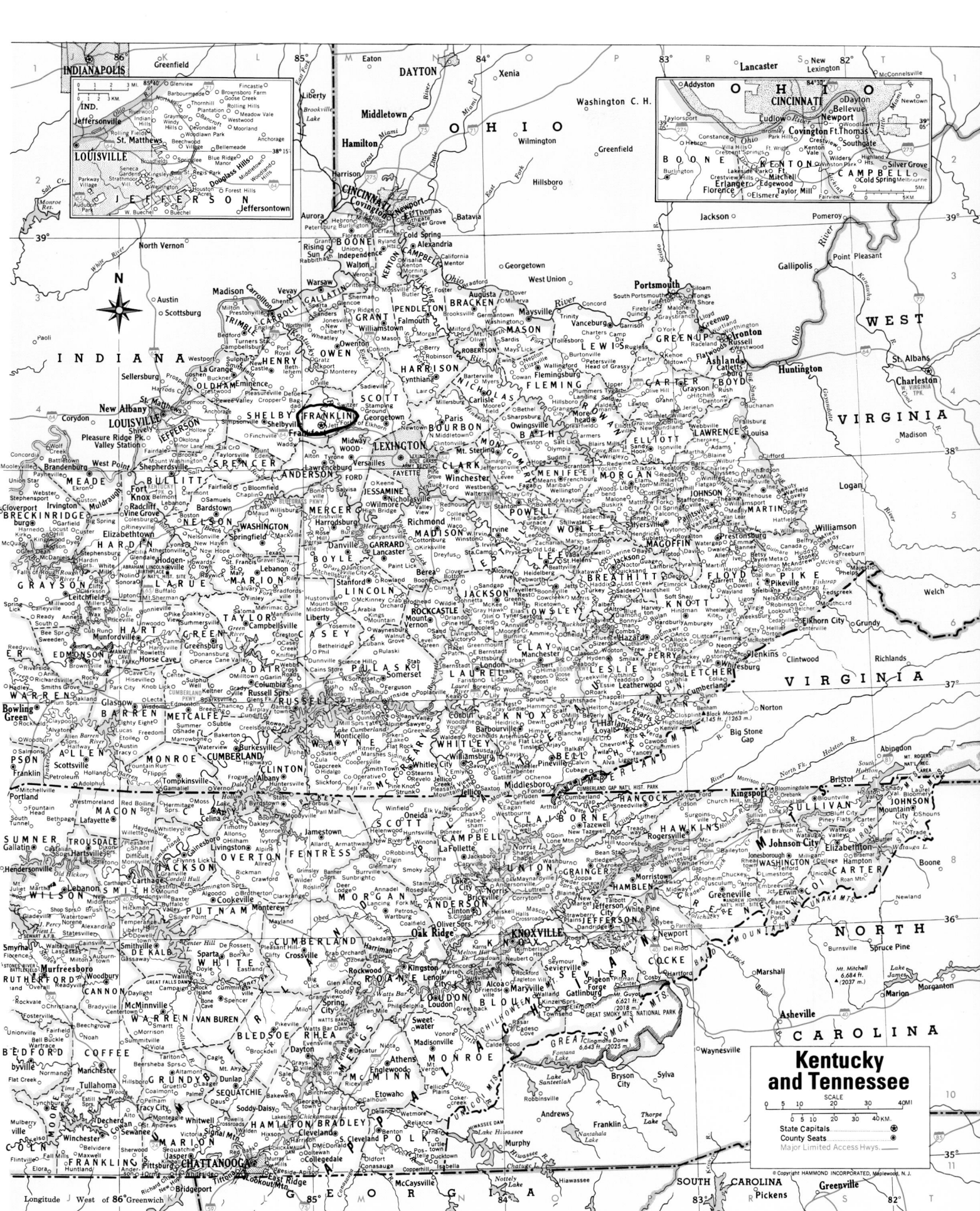

Kentucky
and Tennessee

Topography

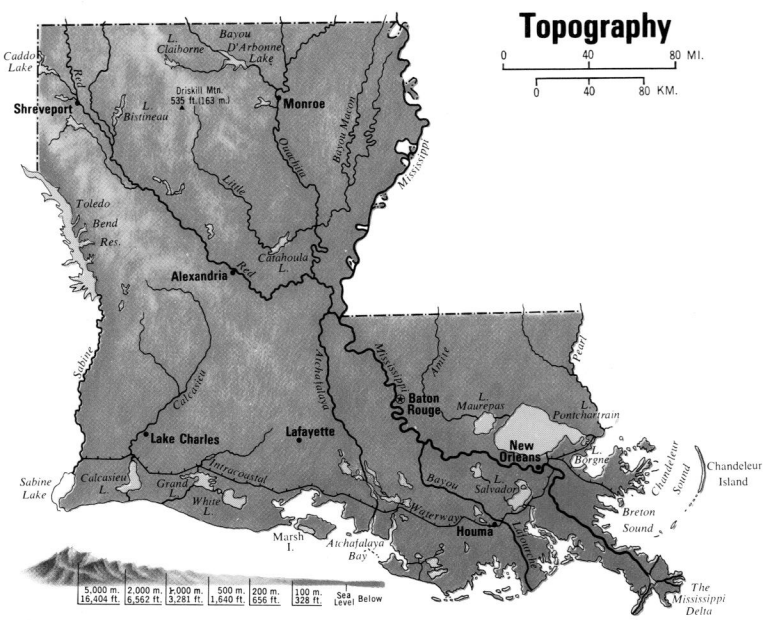

Driskill Mtn. 535 ft.(163 m.)

5,000 m. | 2,000 m. | 1,000 m. | 500 m. | 200 m. | 100 m. | Sea
16,404 ft. | 6,562 ft. | 3,281 ft. | 1,640 ft. | 656 ft. | 328 ft. | Level Below

The Mississippi Delta

Louisiana

SCALE

0 5 10   20   30   40 MI.
0 5 10   20   30   40 KM.

State Capitals...................⊛
Parish Seats.....................◉
Canals............................
Major Limited Access Hwys.

PARISHES

Acadia 58,861 ....F6
Allen 25,440 ....E5
Ascension 76,627 ....L3
Assumption 23,388 ....H7
Avoyelles 41,481 ....G4
Beauregard 32,986 ....D5
Bienville 15,752 ....C1
Bossier 98,310 ....C1
Caddo 252,161 ....C1
Calcasieu 183,577 ....D6
Caldwell 10,560 ....E2
Cameron 9,991 ....D7
Catahoula 10,920 ....E2
Claiborne 16,851 ....D1
Concordia 20,247 ....F4
De Soto 25,494 ....C2
East Baton Rouge 412,852 ...K1
East Carroll 9,421 ....H1
East Feliciana 21,360 ......H5
Evangeline 35,434 ....F5
Franklin 21,263 ....G2
Grant 18,698 ....E3
Iberia 73,266 ....G6
Iberville 33,320 ....H6
Jackson 15,397 ....E2
Jefferson 454,466 ....K7
Jefferson Davis 31,435 .....E6
Lafayette 190,503 ....F6
Lafourche 89,974 ....K7
La Salle 14,282 ....F3
Lincoln 42,509 ....E2
Livingston 91,814 ....L2
Madison 13,728 ....H2
Morehouse 31,021 ....G1
Natchitoches 39,080 ....D3
Orleans 484,674 ....L6
Ouachita 147,250 ....G2
Plaquemines 26,757 ....L8
Pointe Coupee 22,763 ....G5
Rapides 126,337 ....E4
Red River 9,622 ....D2
Richland 20,981 ....G2
Sabine 23,459 ....C3
Saint Bernard 67,229 ....L7
Saint Charles 48,072 ....K7
Saint Helena 10,525 ....J5
Saint James 21,216 ....L3
Saint John the Baptist
43,044 ....M3
Saint Landry 87,700 ....F5
Saint Martin 48,583 ....G6
Saint Mary 53,500 ....H7
Saint Tammany 191,268 ......L6
Tangipahoa 100,588 ....L5
Tensas 6,618 ....H2
Terrebonne 104,503 ....J8
Union 22,803 ....F1
Vermilion 53,807 ....F7
Vernon 52,531 ....D4
Washington 43,926 ....K5
Webster 41,831 ....D1
West Baton Rouge 21,601 ....H6
West Carroll 12,314 ....H1
West Feliciana 15,111 ......H5
Winn 16,894 ....E3

CITIES and TOWNS

Abbeville▲ 11,887 ....F7
Abita Springs 1,957 ....L6
Acme 235 ....G4
Acy 570 ....L3
Addis 2,238 ....J2
Adeline 200 ....G7

Akers 150 ....N2
Albany 865 ....M1
Alberta 150 ....D2
Alexandria▲ 46,342 ....E4
Allen 175 ....D3
Alto 132 ....G2
Alton 500 ....L6
Amelia 2,423 ....H7
Amite▲ 4,110 ....K5
Anacoco 866 ....D4
Anandale F4
Andrew 100 ....F6
Angie 240 ....L5
Angola 600 ....G5
Ansley 100 ....E2
Arabi 8,093 ....P4
Arbroth 250 ....H5
Arcadia▲ 3,041 ....E1
Archibald 425 ....G2
Archie 280 ....G3
Arcola 200 ....K5
Arnaudville 1,398 ....G6
Ashland 291 ....D2
Athens 262 ....E1
Atlanta 150 ....E3
Avery Island 500 ....G7
Bains 400 ....H5
Baker 13,793 ....K1
Baldwin 2,497 ....G7
Baptist 150 ....M1
Barataria 1,333 ....K7
Basile 1,660 ....E5
Baskin 188 ....G2
Bastrop▲ 12,988 ....G1
Batchelor 500 ....G5
Baton Rouge (cap.)▲ 227,818.K2
Bayou Barbary 200 ....M2
Bayou Cane 17,046 ....J7
Bayou Goula 850 ....J3
Bayou Vista 4,351 ....H7
Baywood 100 ....K1
Beaver 350 ....E5
Beekman 150 ....G1
Bel 150 ....D6
Belcher 272 ....C1
Bell City 400 ....D6
Belle Alliance H6
Belle Chasse 9,848 ....O4
Belle D'Eau 120 ....F4
Belle Rose 1,944 ....K3
Bellwood 150 ....D3
Belmont 350 ....C3
Benson 200 ....C3
Bentley 120 ....E3
Benton▲ 2,035 ....C1
Bernice 1,809 ....E1
Bertrandville 175 ....L7
Berwick 4,418 ....H7
Bienville 262 ....D2
Bogalusa 13,365 ....L5
Bolinger 200 ....C1
Bonita 335 ....G1
Boothville 2,220 ....M8
Bordelonville 350 ....G4
Bosco 480 ....F2
Bossier City 56,461 ....C1
Boudreaux 275 ....J8
Bourg 2,073 ....J7
Boutte 2,181 ....N4
Boyce 1,190 ....E4
Braithwaite 350 ....P4
Branch 200 ....F6

Breaux Bridge 7,281 ....G6
Brittany 475 ....L3
Broussard 5,874 ....G6
Brusly 2,020 ....J2
Bryceland 114 ....E2
Buckeye 280 ....F4
Bunkie 4,662 ....F5
Buras-Triumph 3,358 ....L8
Burnside 500 ....L3
Bush 275 ....L5
Cade 175 ....G6
Calcasieu 400 ....D6
Calhoun 350 ....F2
Calumet 280 ....H7
Calvin 236 ....E3
Cameron▲ 1,965 ....D7
Campti 1,057 ....D3
Cankton 362 ....G6
Carencro 6,120 ....G6
Carlisle 975 ....L7
Carville 1,108 ....K3
Castor 209 ....D2
Cecelia 1,505 ....G6
Center Point 850 ....F4
Centerville 600 ....H7
Central 546 ....L3
Chacahoula 150 ....J7
Chalmette▲ 32,069 ....P4
Charenton 1,944 ....H7
Chase 200 ....G2
Chataignier 383 ....F5
Chatham 623 ....F2
Chauvin 3,229 ....J8
Cheneyville 901 ....F4
Chopin 175 ....E4
Choudrant 582 ....F1
Church Point 4,756 ....F6
Clarence 516 ....E3
Clarks 1,071 ....F2
Clay 400 ....F3
Clayton 858 ....H3
Clear Lake 100 ....E3
Clinton▲ 1,998 ....J5
Clio 125 ....M2
Cloutierville 100 ....E3
Colfax▲ 1,659 ....E3
Collinston 327 ....G1
Columbia▲ 477 ....F2
Conventa 400 ....L3
Converse 400 ....C3
Corey 110 ....F2
Cottonport 2,316 ....F5
Cotton Valley 1,189 ....D1
Couchwood 150 ....D1
Coushatta▲ 2,299 ....D2
Covington▲ 8,483 ....K5
Cow Island 200 ....F7
Cravens 200 ....E5
Creole 175 ....D7
Crescent 300 ....J2
Creston 135 ....E3
Crowley▲ 14,225 ....F6
Crowville 400 ....G2
Cullen 1,296 ....D1
Curtis 110 ....C2
Cut Off 5,635 ....K7
Dalcour 275 ....P4
Danville 100 ....E2
Darrow 500 ....K3
Davant 600 ....L7
Deerford 100 ....K1
Delcambre 2,168 ....G7
Delhi 3,066 ....H2
Delta 239 ....J2
Denham Springs 8,757 ....L2
De Quincy 3,398 ....D6

De Ridder▲ 9,808 ....D5
Des Allemands 2,500 ....N4
Destrehan 11,260 ....N4
Deville 1,007 ....F4
Diamond 370 ....L7
Dixie 330 ....C1
Dixie Inn 352 ....D1
Dodson 357 ....E2
Donaldsonville▲ 7,605 ......K3
Donner 500 ....J7
Downsville 118 ....F1
Doyline 841 ....D1
Dry Creek 300 ....D5
Dry Prong 421 ....E3
Dubach 800 ....E1
Dubberly 290 ....D1
Dulac 2,458 ....J8
Dunn 225 ....G2
Duplessis 500 ....K2
Duson 1,672 ....F6
East Hodge 366 ....E2
Easton 365 ....E5
East Point 100 ....C2
Echo 525 ....F4
Edgard▲ 2,637 ....M3
Edgefield 190 ....D2
Edgerly 250 ....C6
Effie 300 ....F4
Elizabeth 574 ....E5
Elmer 200 ....E4
Elm Grove 100 ....C2
Elton 1,261 ....E6
Empire 2,211 ....L8
Enterprise 375 ....G3
Epps 1,153 ....G1
Erath 2,187 ....F7
Eros 202 ....F2
Erwinville 790 ....H5
Esther 745 ....F7
Estherwood 807 ....F6
Ethel 250 ....H5
Eunice 11,499 ....F6
Eva 100 ....G4
Evangeline 400 ....F6
Evans 500 ....D5
Evergreen 314 ....F5
Extension 950 ....G3
Fairbanks 300 ....F1
Farmerville▲ 3,808 ....F1
Fenton 380 ....E6
Ferriday 3,723 ....G3
Fields 125 ....C5
Fisher 268 ....D3
Flatwoods 360 ....E3
Flora 300 ....D3
Florien 692 ....D3
Fluker 400 ....K5
Folsom 525 ....K5
Forbing 100 ....C1
Fordoche 933 ....G5
Forest 275 ....H1
Forest Hill 456 ....E4
Fort Jesup 100 ....D3
Fort Necessity 150 ....G3
Franklin▲ 8,354 ....G7
Franklinton▲ 3,657 ....K5
French Settlement 945 ......L2
Frierson 700 ....C2
Frost 500 ....L2
Fryeburg 150 ....D2
Fullerton 120 ....D4
Galliano 7,356 ....K8
Galvez 200 ....L2
Garden City 225 ....H7
Garyville 2,775 ....M3

(continued)

AREA 47,752 sq. mi. (123,678 sq. km.)
POPULATION 4,468,976
CAPITAL Baton Rouge
LARGEST CITY New Orleans
HIGHEST POINT Driskill Mtn. 535 ft. (163 m.)
SETTLED IN 1699
ADMITTED TO UNION April 30, 1812
POPULAR NAME Pelican State
STATE FLOWER Magnolia
STATE BIRD Eastern Brown Pelican

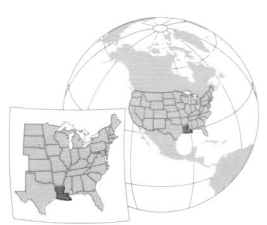

New Orleans, Baton Rouge and Vicinity

0   5   10   15   20 MI.
0   5   10   15   20 KM.

Longitude 91° West of Greenwich

© Copyright HAMMOND INCORPORATED, Maplewood, N.J.

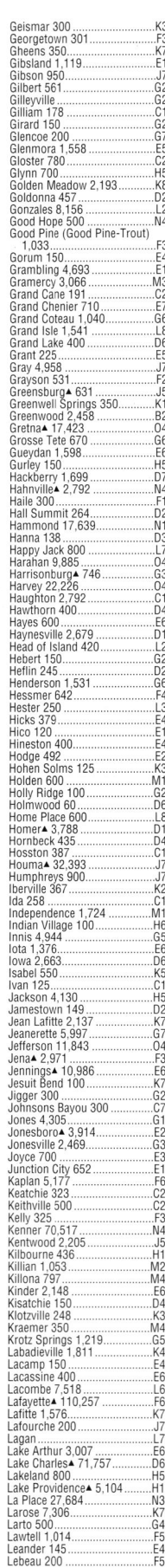

Agriculture, Industry and Resources

DOMINANT LAND USE

Specialized Cotton

Cotton, General Farming

Cotton, Livestock

Cotton, Sugarcane

Cotton, Forest Products

Truck and Mixed Farming

General Farming, Forest Products, Truck Farming, Cotton

Sugarcane, General Farming

Rice, General Farming

Forests

Swampland, Limited Agriculture

Major Industrial Areas

MAJOR MINERAL OCCURRENCES

G Natural Gas   Na Salt   S Sulfur
Gp Gypsum   O Petroleum

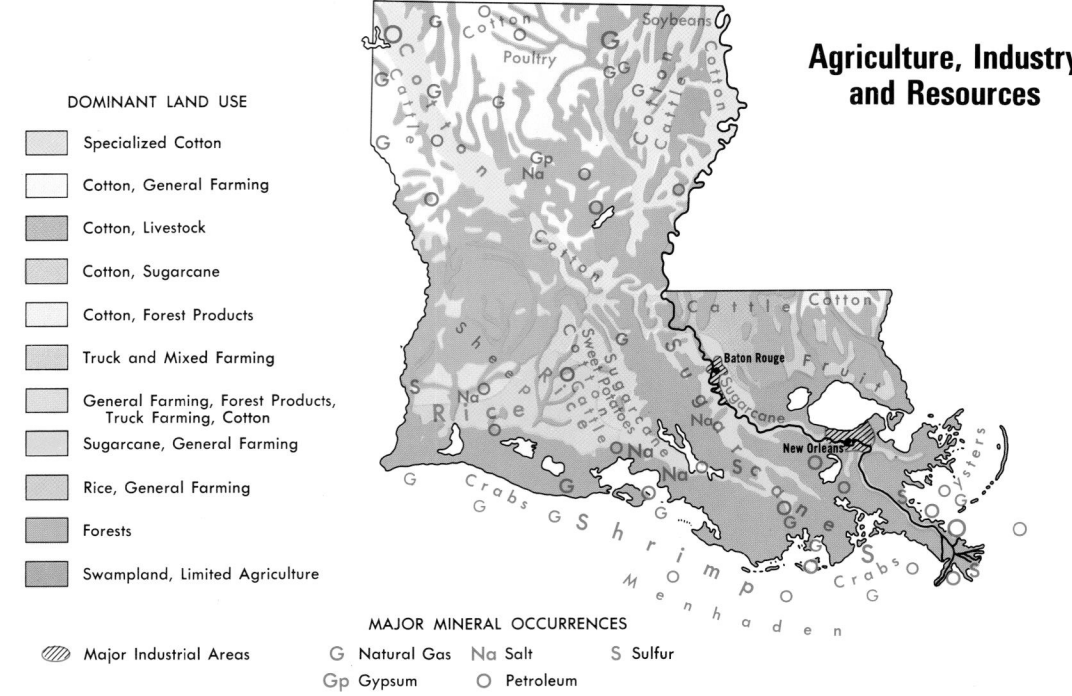

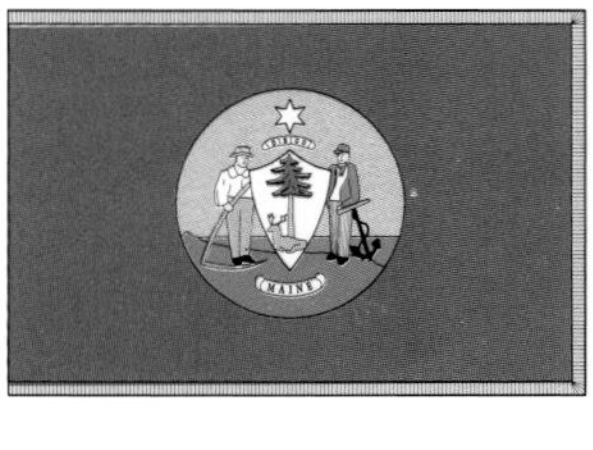

AREA 33,265 sq. mi. (86,156 sq. km.)
POPULATION 1,274,923
CAPITAL Augusta
LARGEST CITY Portland
HIGHEST POINT Katahdin 5,267 ft. (1605 m.)
SETTLED IN 1624
ADMITTED TO UNION March 15, 1820
POPULAR NAME Pine Tree State
STATE FLOWER White Pine Cone & Tassel
STATE BIRD Chickadee

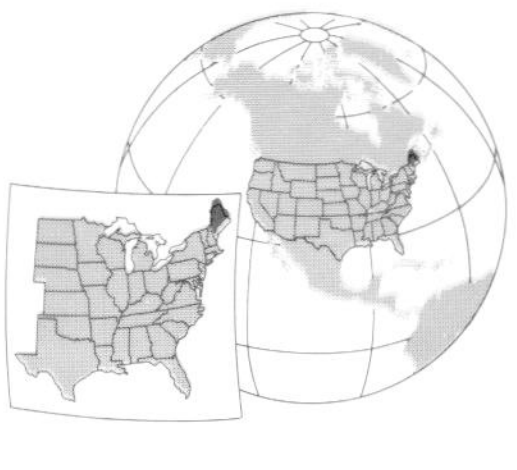

COUNTIES

Androscoggin 103,793 ... C7
Aroostook 73,938 ... F2
Cumberland 265,612 ... C8
Franklin 29,467 ... B5
Hancock 51,791 ... G6
Kennebec 117,114 ... D7
Knox 39,618 ... E7
Lincoln 33,616 ... D7
Oxford 54,755 ... B7
Penobscot 144,919 ... F5
Piscataquis 17,235 ... E4
Sagadahoc 35,214 ... D7
Somerset 50,888 ... C4
Waldo 36,280 ... E6
Washington 33,941 ... H6
York 186,742 ... B9

CITIES and TOWNS

Abbot Village• 576 ... D5
Acton 850 ... B8
Acton• 2,145 ... B8
Addison 350 ... H6
Addison• 1,209 ... H6
Albion• 1,946 ... E6
Alexander• 514 ... H5
Alfred▲ 1,890 ... B9
Alfred• 2,497 ... B9
Allagash• 277 ... F1
Alna• 675 ... D7
Alton• 816 ... F5
Amherst• 230 ... G6
Andover 350 ... B6
Andover• 864 ... B6
Anson 818 ... D6
Anson• 2,583 ... D6
Appleton• 1,271 ... E7
Argyle 253 ... F5
Ashland 750 ... G2
Ashland• 1,474 ... G2
Athens 300 ... D6
Athens• 847 ... D6
Atkinson• 323 ... E5
Auburn▲ 23,203 ... C7
Augusta (cap.)▲ 18,560 ... D7
Aurora• 121 ... G6
Bailey Island 500 ... D8
Bancroft ... H4
Bancroft• 61 ... H4
Bangor▲ 31,473 ... F6
Bar Harbor 2,680 ... G7
Bar Harbor• 4,820 ... G7
Baring 235 ... J5
Baring• 273 ... J5
Bar Mills 800 ... C8
Bass Harbor 450 ... G7
Bath▲ 9,266 ... D8
Bayside ... F7
Beals• 618 ... H7
Beddington• 29 ... H6
Belfast▲ 6,381 ... F7
Belgrade 950 ... D7
Belgrade• 2,978 ... D7
Belgrade Lakes 700 ... D6
Belmont• 821 ... E7
Benedicta• 225 ... G4
Benton• 2,557 ... D6
Berwick 1,993 ... B9
Berwick• 6,353 ... B9
Bethel 750 ... B7
Bethel• 2,411 ... B7
Biddeford 20,942 ... B9
Biddeford Pool 500 ... C9
Bingham 856 ... D5
Bingham• 989 ... D5
Birch Harbor 300 ... H7
Blaine-Mars Hill 1,717 ... H2
Blaine• 806 ... H2
Blanchard 83 ... D5
Blue Hill 850 ... F7
Blue Hill• 2,390 ... F7
Bolsters Mills 150 ... B7
Boothbay 200 ... D8
Boothbay• 2,960 ... D8
Boothbay Harbor 1,237 ... D8
Bowdoinham• 2,612 ... D7
Bowerbank• 123 ... E5
Bradford 150 ... F5
Bradford• 1,186 ... F5
Bradley• 1,242 ... F6
Brewer 8,987 ... F6
Bridgewater• 612 ... H3
Bridgton 2,359 ... B7
Bridgton• 4,883 ... B7
Brighton• 94 ... D5
Bristol 450 ... D8
Bristol• 2,644 ... D8
Brooklin• 841 ... F7
Brooks• 1,022 ... E6
Brooksville• 911 ... F7
Brookton 175 ... H4
Brownfield 300 ... B8
Brownfield• 1,251 ... B8
Brownville 600 ... E5
Brownville• 1,259 ... E5
Brownville Junction 950 ... E5
Brunswick 14,816 ... C8
Brunswick• 21,172 ... C8
Bryant Pond 600 ... B7
Buckfield• 1,723 ... C7
Bucks Harbor 300 ... J6
Bucksport 2,970 ... F6
Bucksport• 4,908 ... F6
Burlington• 351 ... G5
Burnham• 1,142 ... E6
Buxton• 7,452 ... C8
Byron• 121 ... B6
Calais 3,447 ... J5
Cambridge• 492 ... E5
Camden 3,934 ... F7
Camden• 5,254 ... F7
Canaan• 2,017 ... D6
Canton• 1,121 ... C7
Cape Neddick 2,997 ... B9
Cape Porpoise 500 ... C9
Caratunk• 108 ... C5
Cardville 223 ... F5
Caribou 8,312 ... G2
Carmel• 2,416 ... E6
Carrabassett Valley• 399 ... C5
Carroll• 144 ... G5
Carthage• 520 ... C6
Cary• 217 ... H4
Casco 400 ... B7
Casco• 3,469 ... B7
Castine• 1,343 ... F7
Centerville• 26 ... H6
Chapman• 465 ... G2
Charleston• 1,397 ... F5
Charlotte• 324 ... J5
Chebeague Island 900 ... C8
Chelsea• 2,559 ... D7
Cherryfield• 1,157 ... H6
Chester• 525 ... F5
Chesterville• 1,170 ... C6
China 2,918 ... E7
China• 4,106 ... E7
Chisholm 1,399 ... C7
Clifton• 743 ... G6
Clinton 1,305 ... D6
Clinton• 3,340 ... D6
Columbia• 459 ... H6
Columbia Falls• 599 ... H6
Cooper• 145 ... H6
Coopers Mills 200 ... E7
Corea 375 ... H7
Corinna• 2,145 ... E6
Cornish• 1,269 ... B8
Cornville• 1,208 ... D6
Costigan 200 ... F5
Cranberry Isles• 128 ... G7
Crawford• 108 ... H5
Crescent Lake 325 ... C7
Criehaven ... F8
Crouseville 450 ... G2
Crystal• 285 ... G4
Cumberland Center 2,596 ... C8
Cumberland Center• 1,890 ... C8
Cundys Harbor 150 ... D8
Cushing• 1,322 ... E7
Cutler 400 ... J6
Cutler• 623 ... J6
Damariscotta• 2,041 ... E7
Damariscotta-Newcastle 1,751 ... E7
Danforth 650 ... H4
Danforth• 629 ... H4
Deblois• 49 ... H6
Dedham• 1,422 ... F6
Deer Isle 600 ... F7
Deer Isle• 1,876 ... F7
Denmark• 1,004 ... B8
Dennysville• 319 ... J6
Derby 300 ... E5
Detroit• 816 ... E6
Dexter 2,201 ... E5
Dexter• 3,890 ... E5
Dixfield 1,137 ... C6
Dixfield• 2,514 ... C6
Dixmont• 1,065 ... E6
Dover-Foxcroft▲ 2,592 ... E5
Dover-Foxcroft• 4,211 ... E5
Dresden• 1,625 ... D7
Dry Mills 700 ... C8
Dryden 675 ... C6
Dyer Brook• 243 ... G3
Eagle Lake 675 ... F1
Eagle Lake• 815 ... F1
East Andover 250 ... B6
East Baldwin 175 ... B8
East Blue Hill 150 ... G7
East Boothbay 800 ... D8
East Corinth 525 ... F5
East Dixfield 250 ... C6
East Eddington 200 ... F6
East Hiram 198 ... B8
East Holden 600 ... F6
East Lebanon 950 ... B9
East Limington 200 ... B8
East Livermore 500 ... C7
East Machias 850 ... J6
East Machias• 1,298 ... J6
East Madison 400 ... D6
East Millinocket 1,701 ... F4
East Millinocket• 1,828 ... F4
Easton• 1,249 ... H2
East Parsonfield 400 ... B8
East Peru 200 ... C7
East Poland 200 ... C7
Eastport 1,640 ... K6
East Stoneham 300 ... B7
East Sullivan 496 ... G6
East Vassalboro 300 ... D7
East Waterboro 365 ... B8
East Wilton 650 ... C6
Eddington 250 ... F6
Eddington• 2,052 ... F6
Edgecomb• 1,090 ... D8
Edmunds 430 ... J6
Eliot• 5,954 ... B9
Ellsworth▲ 6,456 ... F6
Enfield 150 ... F5
Enfield• 1,616 ... F5
Etna• 1,012 ... E6
Eustis• 685 ... B5
Exeter• 997 ... E6
Fairbanks 400 ... C6
Fairfield 2,569 ... D6
Fairfield• 6,573 ... D6
Fairfield Center 975 ... D6
Falmouth 1,655 ... C8
Falmouth• 10,310 ... C8
Farmingdale 1,935 ... D7
Farmingdale• 2,804 ... D7
Farmington▲ 4,098 ... C6
Farmington• 7,410 ... C6
Farmington Falls 500 ... C6
Fayette• 1,040 ... C7
Five Islands 225 ... D8
Fort Fairfield 1,600 ... H2
Fort Fairfield• 3,579 ... H2
Fort Kent 1,978 ... F1
Fort Kent• 4,233 ... F1
Fort Kent Mills 200 ... F1
Foxcroft 2,974 ... E5
Frankfort• 1,041 ... F6
Franklin 350 ... G6
Franklin• 1,370 ... G6
Freedom• 645 ... E7
Freeport 1,813 ... C8
Freeport• 7,800 ... C8
Frenchboro• 38 ... G7
Frenchville 980 ... G1
Frenchville• 1,225 ... G1
Friendship 700 ... E7
Friendship• 1,204 ... E7
Fryeburg 1,549 ... A7
Fryeburg• 3,083 ... A7
Gardiner• 6,198 ... D7
Garland 300 ... E5
Garland• 990 ... E5
Georgetown 190 ... D8
Georgetown• 1,020 ... D8
Gilead• 156 ... B7
Glen Cove 250 ... E7
Glenburn• 3,964 ... F6
Goodwins Mills 340 ... B8
Goose Rocks Beach 200 ... C9
Gorham 4,164 ... C8
Gorham• 14,141 ... C8
Gouldsboro 498 ... H7
Gouldsboro• 1,941 ... H7
Grand Isle 600 ... G1
Grand Isle• 518 ... G1
Grand Lake Stream• 150 ... H5
Gray 525 ... C8
Gray• 6,820 ... C8
Great Pond• 47 ... G6
Greene• 4,076 ... C7
Greenville 1,319 ... D5
Greenville• 1,623 ... D5
Greenville Junction 650 ... D5
Guilford 945 ... E5
Guilford• 1,531 ... E5
Hallowell 2,467 ... D7
Hamlin• 257 ... H1
Hampden 4,126 ... F6
Hampden• 6,327 ... F6
Hampden Highlands 950 ... F6
Hancock• 2,147 ... G6
Hanover• 251 ... B7
Harmony 450 ... D6
Harmony• 954 ... D6
Harpswell• 5,239 ... D8
Harrington• 882 ... H6
Harrison• 2,315 ... B7
Hartford• 963 ... C7
Hartland 872 ... D6
Hartland• 1,816 ... D6
Haynesville• 122 ... G4
Hebron• 1,053 ... C7
Hermon• 4,437 ... F6
Highland Lake 600 ... C8
Hiram 175 ... B8
Hiram• 1,423 ... B8
Hodgdon• 1,240 ... H3
Hollis Center• 2,892 ... B8
Hope 175 ... E7
Hope• 1,310 ... E7
Houlton▲ 5,270 ... H3
Houlton• 6,476 ... H3
Howland 1,210 ... F5
Howland• 1,362 ... F5
Hudson• 1,048 ... F5
Hulls Cove 200 ... G7
Island Falls• 793 ... G3
Isle Au Haut• 79 ... F7
Islesboro 200 ... F7
Islesboro• 603 ... F7
Jackman 700 ... C4
Jackman• 718 ... C4
Jacksonville 200 ... J6
Jay 850 ... C7
Jay• 4,985 ... C7
Jefferson• 2,388 ... D7
Jonesboro• 594 ... J6
Jonesport 1,050 ... H6
Jonesport• 1,408 ... H6
Keegan 450 ... G1
Kenduskeag• 1,171 ... E6
Kennebunk 4,804 ... B9
Kennebunk• 10,476 ... B9
Kennebunk Beach 200 ... C9
Kennebunkport 1,376 ... C9
Kennebunkport• 3,720 ... C9
Kents Hill 300 ... D7
Kezar Falls 680 ... B8
Kingfield• 1,103 ... C6
Kingman 213 ... G4
Kingsbury• 13 ... D5
Kittery 4,884 ... B9
Kittery• 9,543 ... B9
Kittery Point 1,135 ... B9
Knox• 747 ... E6
Lagrange 250 ... F5
Lagrange• 747 ... F5
Lake View• 43 ... F5
Lamoine• 1,495 ... G7
Lee• 845 ... G5
Leeds• 2,001 ... C7
Levant• 2,171 ... F6
Lewiston 35,690 ... C7
Liberty 200 ... E7
Liberty• 927 ... E7
Lille 300 ... G1
Limerick• 2,240 ... B8
Limestone 1,453 ... H2
Limestone• 2,361 ... H2
Limington• 3,403 ... B8
Lincoln 2,933 ... G5
Lincoln• 5,221 ... G5
Lincoln Center 325 ... G5
Lincolnville 800 ... E7
Lincolnville• 2,042 ... E7
Lincolnville Center 200 ... E7
Linneus• 892 ... H3
Lisbon• 9,077 ... C7
Lisbon Falls 4,420 ... D7
Lisbon-Lisbon Center 1,865 ... C7
Litchfield• 3,110 ... D7
Little Deer Isle 475 ... F7
Little Falls-South Windham 1,792 ... C8
Littleton• 955 ... H3
Livermore 2,106 ... C7
Livermore• 1,950 ... C7
Livermore Falls 1,626 ... C7
Livermore Falls• 3,227 ... C7
Locke Mills 600 ... B7
Lovell 180 ... B7
Lovell• 974 ... B7
Lowell• 291 ... F5
Lubec 900 ... K6
Lubec• 1,652 ... K6
Ludlow• 402 ... G3
Machias▲ 1,376 ... J6
Machias• 2,353 ... J6
Machiasport 374 ... H6
Machiasport• 1,160 ... H6
Macwahoc• 98 ... G4
Madawaska 3,326 ... G1
Madawaska• 4,534 ... G1
Madison 2,733 ... D6
Madison• 4,523 ... D6
Madrid• 173 ... B6
Manchester• 2,465 ... D7
Mapleton• 1,889 ... G2
Mars Hill• 1,480 ... H2
Mars Hill-Blaine 1,428 ... H2
Masardis• 255 ... G3

(continued on following page)

Agriculture, Industry and Resources

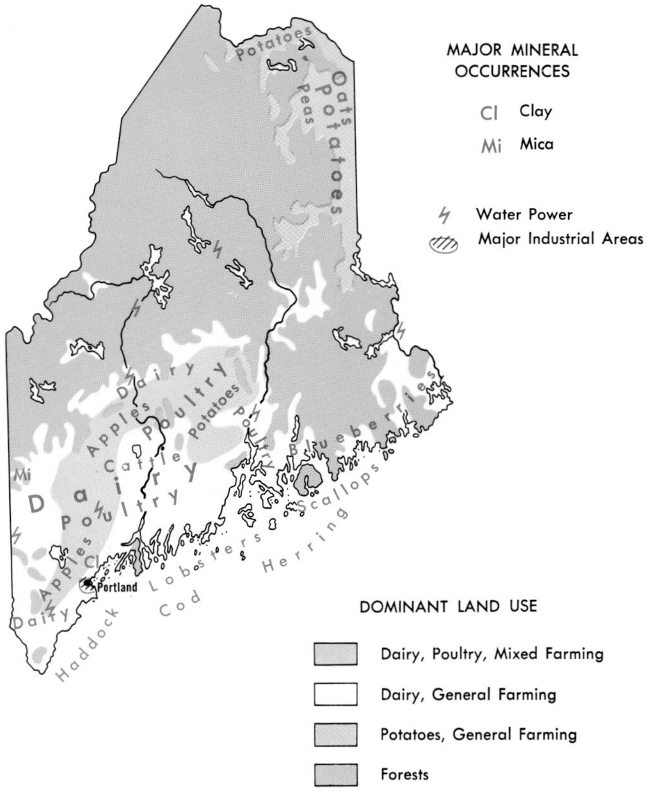

Potatoes
Oats
Peas
Potatoes
Dairy
Poultry
Potatoes
Apples
Poultry
Cattle
Dairy
Blueberries
Scallops
Dairy
Poultry
Apples
Mi
Cl
Portland
Lobsters
Herring
Cod
Haddock
Dairy
MAJOR MINERAL OCCURRENCES
Cl Clay
Mi Mica
Water Power
Major Industrial Areas
DOMINANT LAND USE
Dairy, Poultry, Mixed Farming
Dairy, General Farming
Potatoes, General Farming
Forests

(continued)

Matinicus 66 ....F8
Mattawamkeag• 825....G5
Mechanic Falls 2,450....C7
Mechanic Falls• 3,138....C7
Meddybemps• 150....J5
Medford• 231....F5
Medway• 1,489....G4
Mercer• 647....D6
Mexico 1,946....B6
Mexico• 2,959....B6
Milbridge• 1,279....H6
Milford 2,197....F6
Milford• 2,950....F6
Millinocket• 5,203....F4
Milo 1,898....F5
Milo• 2,383....F5
Minot 250....C7
Minot• 2,248....C7
Minturn 150....G7
Monhegan• 75....E8
Monmouth 500....D7
Monmouth• 3,785....C7
Monroe• 882....E6
Monson• 666....E5
Monticello• 790....H3
Montville• 1,002....E7
Moody 500....B9
Moose River• 219....C4
Morrill• 774....E7
Mount Desert 150....G7
Mount Desert• 2,109....G7
Mount Vernon• 1,524....D7
Naples• 3,274....B8
Newburgh• 1,394....F6
Newcastle• 1,748....D7
Newcastle-Damariscotta 1,567....E7
Newfield 200....B8
Newfield• 1,328....B8
New Gloucester 400....C8
New Gloucester• 4,803....C8
New Harbor 850....E8
New Limerick• 523....G3
Newport 1,754....E6
Newport• 3,017....E6
New Portland 300....C6
New Portland• 785....C6
Newry• 344....B6
New Sharon• 1,297....C6
New Sweden 175....G2
New Sweden• 621....G2
New Vineyard• 725....C6
Nobleboro• 1,626....D7
Norridgewock 1,557....D6
Norridgewock• 3,294....D6
North Anson 950....D6
North Belgrade 300....D7
North Berwick 1,580....B9
North Berwick• 4,293....B9
North Bridgton 300....B7
North Cutler 153....J6
Northeast Harbor 800....G7
Northfield• 131....H6
North Fryeburg 250....B7
North Haven 400....F7
North Haven• 381....F7
North Jay 800....C6
North Limington 400....B8
North Livermore 250....C7
North Lubec 250....J6
North New Portland 500....C6
North Penobscot 443....F7
Northport• 1,331....E7
North Raymond 225....C8
North Turner 350....C7
North Vassalboro 950....D7
North Waldoboro 250....E7
North Waterboro 200....B8
North Waterford 390....B7
North Wayne 175....C7
North Whitefield 300....D7
North Windham 4,568....C8
North Yarmouth 500....C8
North Yarmouth• 3,210....C8
Norway 2,623....B7
Norway• 4,611....B7
Oakfield• 732....G3
Oakland 2,758....D6
Oakland• 5,959....D6
Ocean Park 200....C9
Ogunquit• 1,226....B9
Olamon 150....F5
Old Orchard Beach 8,856....C9
Old Orchard Beach• 8,856....C9
Old Town 8,130....F6
Oquossoc 150....B6
Orient• 145....H4
Orland 200....F6
Orland• 2,134....F6
Orono 8,253....F6
Orono• 9,112....F6
Orrington 250....F6
Orrington• 3,526....F6
Orrs Island 600....D8
Otisfield• 1,560....B7
Otter Creek 260....G7
Owls Head• 1,601....F7
Oxbow• 56....G3
Oxford 1,300....B7
Oxford• 3,960....B7
Palermo• 1,220....E7
Palmyra• 1,953....E6
Paris• 4,793....B7
Parkman• 811....D5
Passadumkeag• 441....F5
Patten 1,057....F4
Patten• 1,111....F4
Pejepscot 200....D8
Pemaquid 200....E8
Pembroke 300....J6
Pembroke• 879....J6
Penobscot 150....F7
Penobscot• 1,344....F7
Perham• 434....G2
Perry• 847....J6
Peru• 1,515....C6
Phillips• 990....C6
Phippsburg 1,527....D8
Phippsburg• 2,106....D8
Pine Point 650....C8
Pittsfield 3,217....E6
Pittsfield• 4,214....E6
Pittston• 2,548....D7
Plymouth• 1,257....E6
Poland 500....C7
Poland• 4,866....C7
Portage• 562....G2
Port Clyde 400....E8
Porter 225....B8
Porter• 1,438....B8
Portland▲ 64,249....C8
Pownal• 1,491....C8
Prentiss• 245....G5
Presque Isle 9,511....H2
Princeton• 892....H5
Prospect• 642....F6
Prospect Harbor 445....H7
Randolph• 1,911....D7
Rangeley 123....B6
Rangeley• 1,052....B6
Raymond 550....B8
Raymond• 4,299....B8
Readfield 300....D7
Readfield• 2,360....D7
Red Beach 210....J5
Richmond 1,864....D7
Richmond• 3,298....D7
Richmond Corner 200....D7
Ripley• 452....E5
Robbinston 200....J5
Robbinston• 525....J5
Robinsons 160....H3
Rockland▲ 7,609....E7
Rockport 875....E7
Rockport• 3,209....E7
Rockville 250....E7
Rockwood 265....D4
Rome• 980....D6
Roque Bluffs• 264....H6
Round Pond 400....E8
Roxbury• 384....B6
Rumford 4,795....B6
Rumford• 6,472....B6
Rumford Center 325....B7
Rumford Point 320....B6
Sabattus 1,234....C7
Sabattus• 4,486....C7
Saco 16,822....C8
Saint Agatha• 802....G1
Saint Albans• 1,836....E6
Saint David 915....G1
Saint Francis• 577....E1
Saint George 700....E7
Saint George• 2,580....E7
Saint John• 282....F1
Sandy Point 350....F7
Sanford 10,133....B9
Sanford• 20,806....B9
Sangerville• 1,270....E5
Scarborough 3,867....C8
Scarborough• 16,970....C8
Seal Cove 215....G7
Seal Harbor 500....G7
Searsmont 400....E7
Searsmont• 1,174....E7
Searsport 1,102....F7
Searsport• 1,102....F7
Sebago Lake 800....B8
Sebec• 612....E5
Seboeis• 41....F5
Sedgwick• 1,102....F7
Shapleigh• 2,326....B8
Shawmut 500....D6
Sheepscott 150....D7
Sheridan 300....F2
Sherman 1,021....G4
Sherman• 937....G4
Sherman Mills 600....G4
Sherman Station 650....F4
Shirley Mills 242....D5
Shirley Mills• 208....D5
Sidney• 3,514....D7
Sinclair• 264....G1
Skowhegan▲ 6,696....D6
Skowhegan• 8,824....D6
Smithfield• 930....D6
Smyrna Mills• 354....G3
Soldier Pond 500....F1
Solon• 940....D6
Somerville• 509....D7
Somesville (Mount Desert) 150....D7
Sorrento• 290....G7
South Berwick 2,120....B9
South Berwick• 6,671....B9
South Bridgton 373....B8
South Bristol• 897....D8
South Casco 750....B8
South China 225....D7
South Eliot 3,445....B9
South Harpswell 650....C8
South Hiram 350....B8
South Hope 200....E7
South La Grange 150....F5
South Lebanon 200....A9
South Lincoln 150....F5
South Monmouth 400....D7
South Orrington 400....F6
South Paris▲ 2,237....C7
South Penobscot 150....F7
Southport 400....D8
Southport• 684....D8
South Portland 23,324....C8
South Sanford 4,173....B9
South Thomaston• 1,416....E7
South Waldoboro 300....E7
South Waterford 300....B7
Southwest Harbor 1,966....G7
Southwest Harbor• 1,952....G7
South Windham (Little Falls-South Windham)....C8
Springfield• 379....G5
Springvale 3,488....B9
Stacyville 155....F4
Stacyville• 405....F4
Standish 700....B8
Standish• 9,285....B8
Starks• 578....D6
Steep Falls 500....B8
Stetson• 981....E6
Steuben 190....H6
Steuben• 1,126....H6
Stillwater 700....F6
Stockholm• 271....G1
Stockton Springs 500....F7
Stockton Springs• 1,481....F7
Stonington• 1,152....F7
Stow• 288....A7
Stratton 600....B5
Strong• 1,259....C6
Sullivan• 1,185....G6
Sumner• 854....C7
Sunset 165....F7
Surry• 1,361....F7
Swans Island• 327....G7
Swanville• 1,357....E6
Sweden• 324....B7
Temple• 572....C6
Tenants Harbor 900....E8
Thomaston 2,714....E7
Thomaston• 3,748....E7
Thorndike• 712....E6
Topsfield• 225....H5
Topsham 6,271....D8
Topsham• 9,100....D8
Tremont 175....G7
Tremont• 1,529....G7
Trenton• 1,370....G7
Trevett 400....D8
Troy• 963....E6
Turner 400....C7
Turner• 4,972....C7
Union 300....E7
Union• 2,209....E7
Unity• 31....E6
Upper Frenchville 405....G1
Upton• 62....B6
Van Buren 2,369....G1
Van Buren• 2,631....G1
Vanceboro• 147....J4
Vassalboro• 4,047....D7
Veazie• 1,744....F6
Vienna• 527....D6
Vinalhaven• 1,235....F7
Waite• 105....H5
Waldo• 733....E7
Waldoboro 1,291....E7
Waldoboro• 4,916....E7
Walnut Hill 400....C8
Waltham• 306....G6
Warren 770....E7
Warren• 3,794....E7
Washburn 1,221....G2
Washburn• 1,627....G2
Washington• 1,345....E7
Waterboro 700....B8
Waterboro• 6,214....B8
Waterford• 1,455....B7
Waterville 15,605....D6
Wayne 175....D7
Wayne• 1,112....D7
Weeks Mills 235....E7
Weld• 402....C6
Wellington• 270....D5
Wells 950....B9
Wells• 9,400....B9
Wells Beach 600....B9
Wesley• 114....H6
West Baldwin 198....B8
West Bath• 1,798....D8
West Bethel 160....B7
Westbrook 16,142....C8
West Brooksville 156....F7
West Buxton 185....B8
West Enfield 609....F5
West Farmington 700....C6
Westfield• 559....G2
West Forks• 47....D5
West Franklin 350....G6
West Gouldsboro 225....G7
West Jonesport 400....H6
West Kennebunk 809....B9
West Lubec 275....J6
West Minot 400....C7
West Newfield 300....B8
Weston• 207....H4
West Paris• 1,722....B7
West Peru 700....C7
West Poland 250....C7
West Rockport 350....E7
West Scarborough 500....C8
West Tremont 250....G7
Whitefield 550....D7
Whitefield• 2,273....D7
Whiting• 430....J6
Whitneyville• 262....H6
Willimantic• 135....E5
Wilton 2,290....C6
Wilton• 4,123....C6
Windsor• 2,204....D7
Winn 250....G5
Winn• 420....G5
Winslow 7,743....D6
Winslow• 7,743....D6
Winter Harbor• 988....G7
Winterport 1,307....F6
Winterport• 3,602....F6
Winterville• 196....F2
Winthrop 2,893....C7
Winthrop• 6,232....C7
Wiscasset 1,203....D7
Wiscasset• 3,603....D7
Woodland• 1,403....H5
Woolwich• 2,810....D8
Wyman Dam 300....D5
Yarmouth 3,560....C8
Yarmouth• 8,360....C8
York 4,530....B9
York• 12,854....B9
York Beach 900....B9
York Harbor 3,321....B9

OTHER FEATURES

Abraham (mt.)....C5
Acadia Nat'l Park....G7
Allagash (lake)....D3
Allagash (riv.)....E2
Androscoggin (riv.)....C7
Aroostook (riv.)....G2
Attean (pond)....C4
Baker (lake)....D3
Baskahegan (lake)....H5
Bear (riv.)....B6
Big (brook)....E2
Big (lake)....H5
Big Black (riv.)....D2
Bigelow (bight)....C9
Big Spencer (mt.)....E4
Black (pond)....D3
Blue (mt.)....C6
Blue Hill (bay)....G7
Bog (lake)....H6
Brassua (lake)....D4
Casco (bay)....C8
Cathance (lake)....J6
Caucomgomoc (lake)....D3
Center (pond)....E5
Chamberlain (lake)....E3
Chemquasabamticook (lake)....D3
Chesuncook (lake)....E3
Chiputneticook (lakes)....H4
Clayton (lake)....D2
Clifford (lake)....H5
Cold Stream (pond)....G5
Crawford (lake)....H5
Cross (isl.)....J6
Cross (lake)....G1
Cupsuptic (riv.)....B5
Dead (riv.)....C5
Deer (isl.)....F7
Duck (isls.)....G7
Eagle (lake)....E3
Eagle (lake)....F1
East Machias (riv.)....H6
East Musquash (lake)....H5
Elizabeth (cape)....C8
Ellis (pond)....B6
Ellis (riv.)....B6
Embden (pond)....D6
Endless (lake)....F5
Englishman (bay)....J6
Eskutassis (pond)....G5
Fifth (lake)....H5
Fish (riv.)....F2
Fish River (lake)....F2
Flagstaff (lake)....C5
Fourth (lake)....H5
Frenchman (bay)....G7
Gardner (lake)....J6
Georges (isls.)....E8
Graham (lake)....G6
Grand (lake)....H4
Grand Falls (lake)....H5
Grand Lake Seboeis (lake)....F3
Grand Manan (chan.)....K6
Great Moose (lake)....D6
Great Wass (isl.)....J7
Green (isl.)....F8
Harrington (lake)....E4
Haut (isl.)....G7
Indian (pond)....D4
Islesboro (isl.)....F7
Jo-Mary (lakes)....E4
Katahdin (mt.)....F4
Kennebec (riv.)....D7
Kezar (lake)....B7
Kezar (pond)....B7
Kingsbury (pond)....D5
Little Black (riv.)....E1
Little Madawaska (riv.)....G2
Lobster (lake)....E4
Long (lake)....B7
Long (lake)....E2
Long (lake)....G1
Long (pond)....C4
Long (pond)....D6
Long (pond)....E5
Long Falls (dam)....C5
Longfellow (mts.)....B6
Loon (lake)....D3
Loring A.F.B. 225....H2
Lower Roach (pond)....E4
Lower Sysladobsis (lake)....G5
Machias (bay)....J6
Machias (riv.)....F2
Machias (riv.)....H6
Machias Seal (isl.)....J7
Madagascal (pond)....G5
Marshall (isl.)....G7
Matinicus Rock (isl.)....F8
Mattamiscontis (lake)....F4
Mattawamkeag (lake)....G4
Mattawamkeag (riv.)....G4
Meddybemps (lake)....J5
Metinic (isl.)....E8
Millinocket (lake)....E3
Millinocket (lake)....F4
Molunkus (lake)....G4
Monhegan (isl.)....E8
Moose (pond)....B7
Moose (riv.)....D4
Moosehead (lake)....D4
Mooseleuk (stream)....F2
Mooselookmeguntic (lake)....B6
Mopang (lake)....H6
Mount Desert (isl.)....G7
Mount Desert Rock (isl.)....G8
Moxie (lake)....D5
Munsungan (lake)....E3
Muscongus (bay)....E8
Musquacook (lakes)....E2
Nahmakanta (lake)....E4
Nicatous (lake)....G5
Nollesemic (lake)....F4
Old (stream)....H6
Onawa (lake)....E5
Parlin (pond)....C4
Parmachenee (lake)....B5
Passamaquoddy (bay)....J5
Passamaquoddy Ind. Res....J6
Pemadumcook (lake)....E4
Penobscot (bay)....F7
Penobscot (lake)....C4
Penobscot (riv.)....F5
Penobscot Ind. Res....F6
Pierce (pond)....C5
Piscataqua (riv.)....B9
Piscataquis (riv.)....E5
Pleasant (lake)....E3
Pleasant (lake)....G3
Pleasant (lake)....H5
Pleasant (riv.)....H6
Pocomoonshine (lake)....H5
Portage (lake)....F2
Presque Isle A.F.B....G2
Priestly (lake)....E2
Pushaw (lake)....F6
Ragged (isl.)....F8
Ragged (lake)....E4
Rainbow (lake)....E4
Rangeley (lake)....B6
Richardson (lakes)....B6
Rocky (lake)....J6
Round (pond)....E2
Rowe (lake)....F2
Saco (riv.)....B8
Saint Croix (riv.)....J5
Saint Croix Island Nat'l Mon....J5
Saint Francis (riv.)....E1
Saint Froid (lake)....F2
Saint John (pond)....D3
Saint John (riv.)....G1
Salmon Falls (riv.)....B9
Sandy (riv.)....C6
Schoodic (lake)....F5
Scraggly (lake)....F3
Scraggly (lake)....H5
Seal (isl.)....F8
Sebago (lake)....B8
Sebasticook (lake)....E6
Seboeis (lake)....F5
Seboeis (riv.)....F3
Seboomook (lake)....D4
Shallow (lake)....E3
Small (cape)....D8
Sourdnahunk (lake)....F3
Spencer (pond)....D4
Spencer (stream)....C5
Spider (lake)....E3
Squa Pan (lake)....G2
Square (lake)....G1
Sunday (riv.)....B6
Swift (riv.)....B6
Sysladobsis, Lower (lake)....G5
Third (lake)....H5
Twin (lakes)....F4
Umbagog (lake)....A6
Umcalcus (lake)....G3
Umsaskis (lake)....E2
Union, West Branch (riv.)....G6
Vinalhaven (isl.)....F7
Wassataquoik (stream)....F4
Webb (lake)....C6
Webster (brook)....E3
West Grand (lake)....H5
West Musquash (lake)....H5
West Quoddy (head)....K6
Wilson (ponds)....E5
Winnecook (lake)....E6
Wooden Ball (isl.)....F8
Wyman (lake)....C5
Wytopitlock (lake)....G4

▲County seat.
•Population of town or township.

Topography

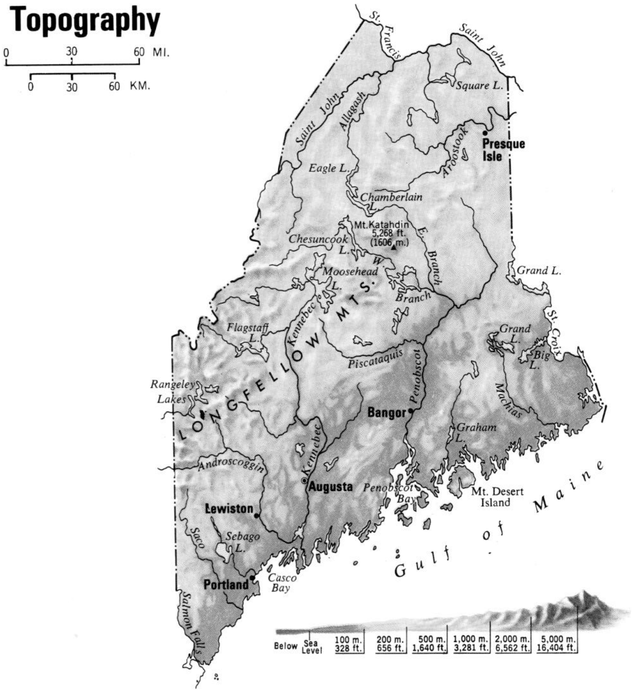

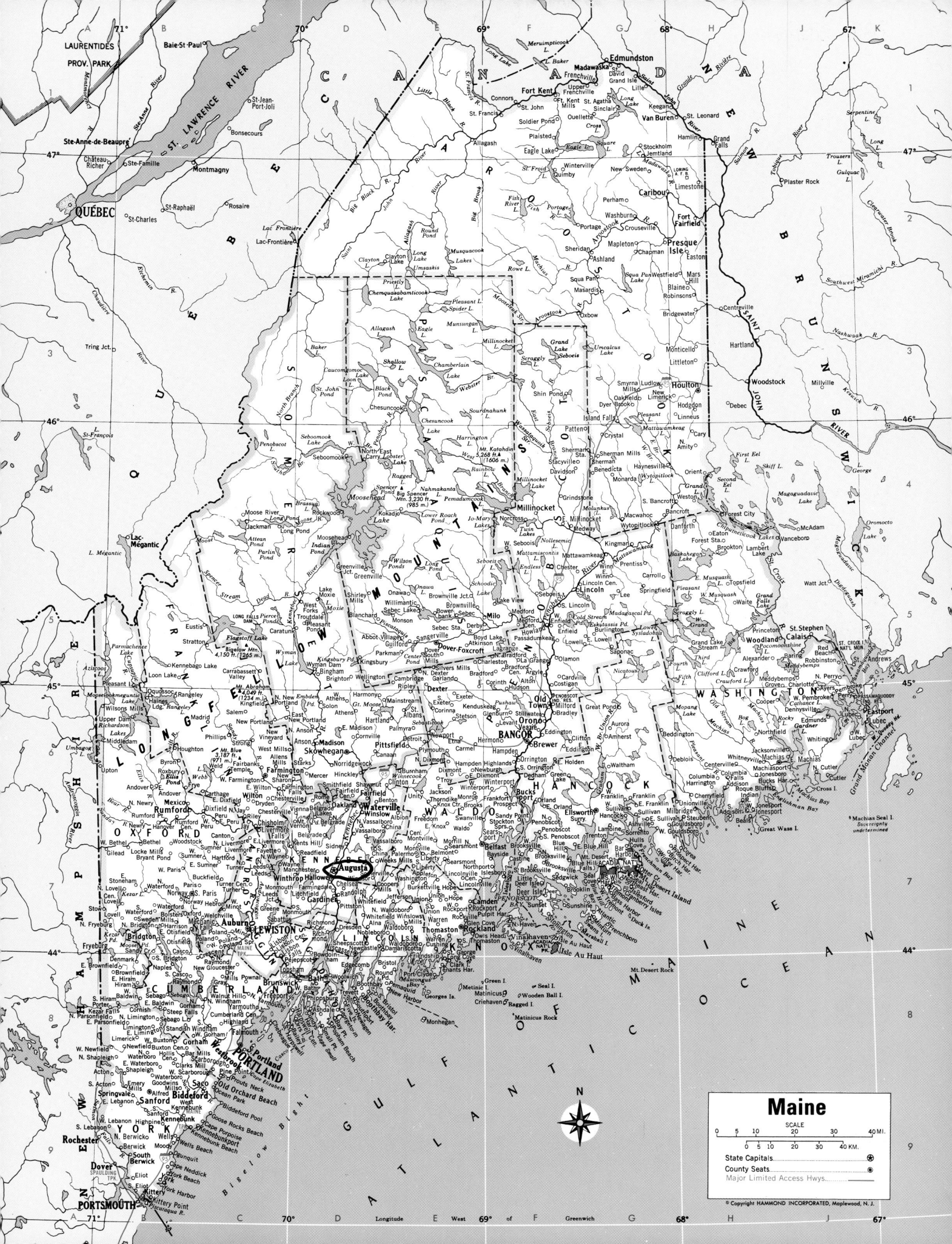

Maine

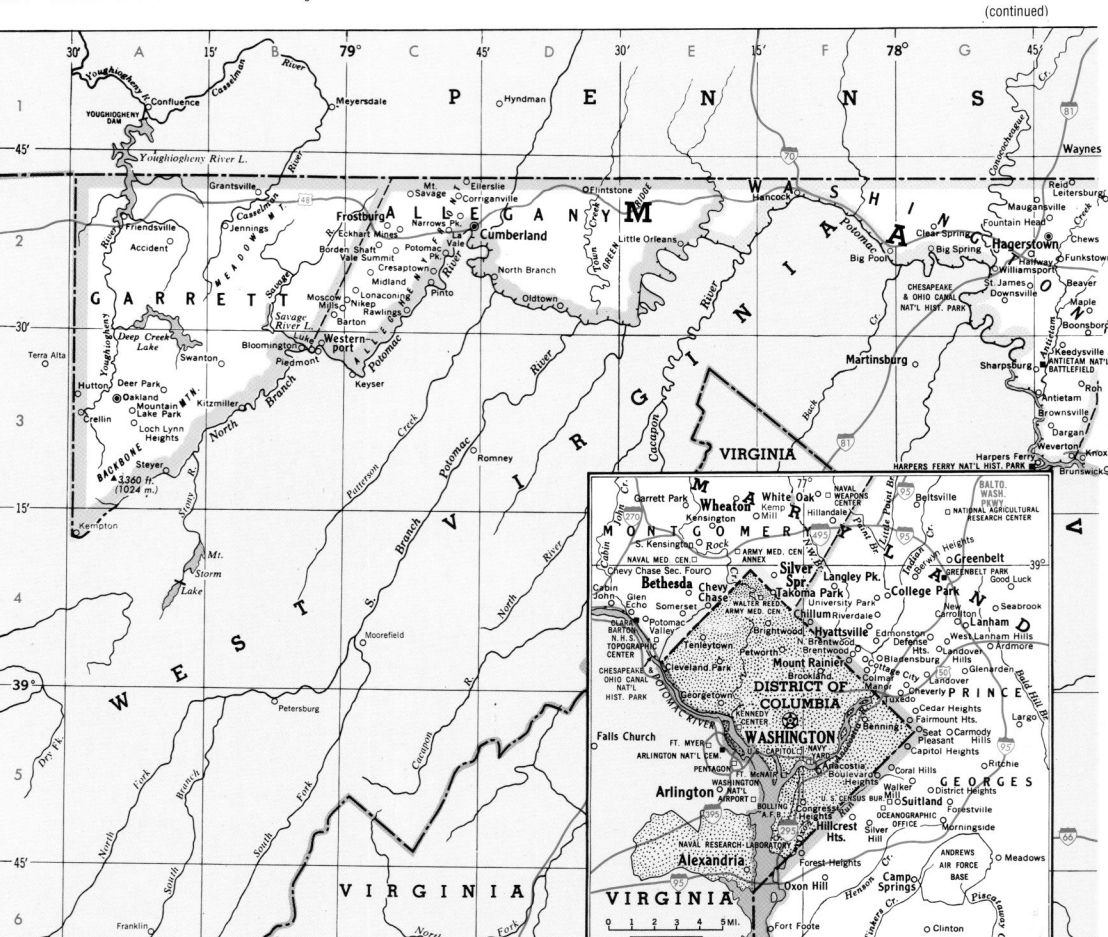

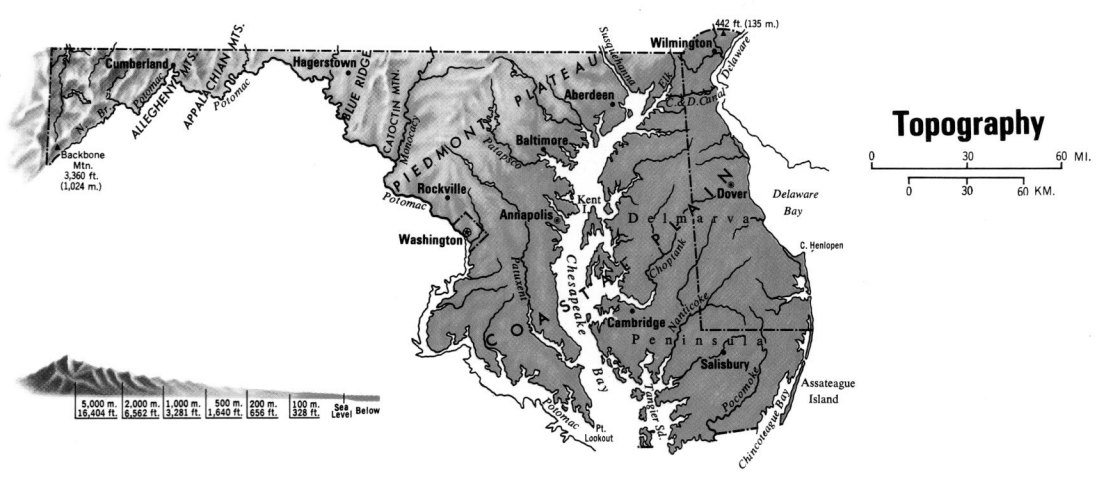

Topography

0 · · · 30 · · · 60 MI.

0 · · · 30 · · · 60 KM.

5,000 m. 2,000 m. 1,000 m. 500 m. 200 m. 100 m. Sea Below
16,404 ft. 6,562 ft. 3,281 ft. 1,640 ft. 656 ft. 328 ft. Level

MARYLAND
AREA 10,460 sq. mi. (27,091 sq. km.)
POPULATION 5,296,486
CAPITAL Annapolis
LARGEST CITY Baltimore
HIGHEST POINT Backbone Mtn. 3,360 ft. (1024 m.)
SETTLED IN 1634
ADMITTED TO UNION April 28, 1788
POPULAR NAME Old Line State; Free State
STATE FLOWER Black-eyed Susan
STATE BIRD Baltimore Oriole

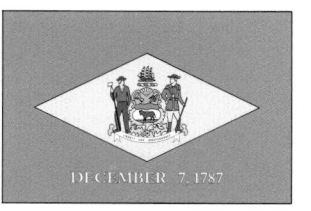

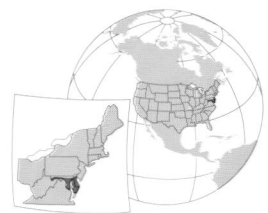

DELAWARE
AREA 2,044 sq. mi. (5294 sq. km.)
POPULATION 783,600
CAPITAL Dover
LARGEST CITY Wilmington
HIGHEST POINT Ebright Road 442 ft. (135 m.)

SETTLED IN 1627
ADMITTED TO UNION December 7, 1787
POPULAR NAME First State; Diamond State
STATE FLOWER Peach Blossom
STATE BIRD Blue Hen Chicken

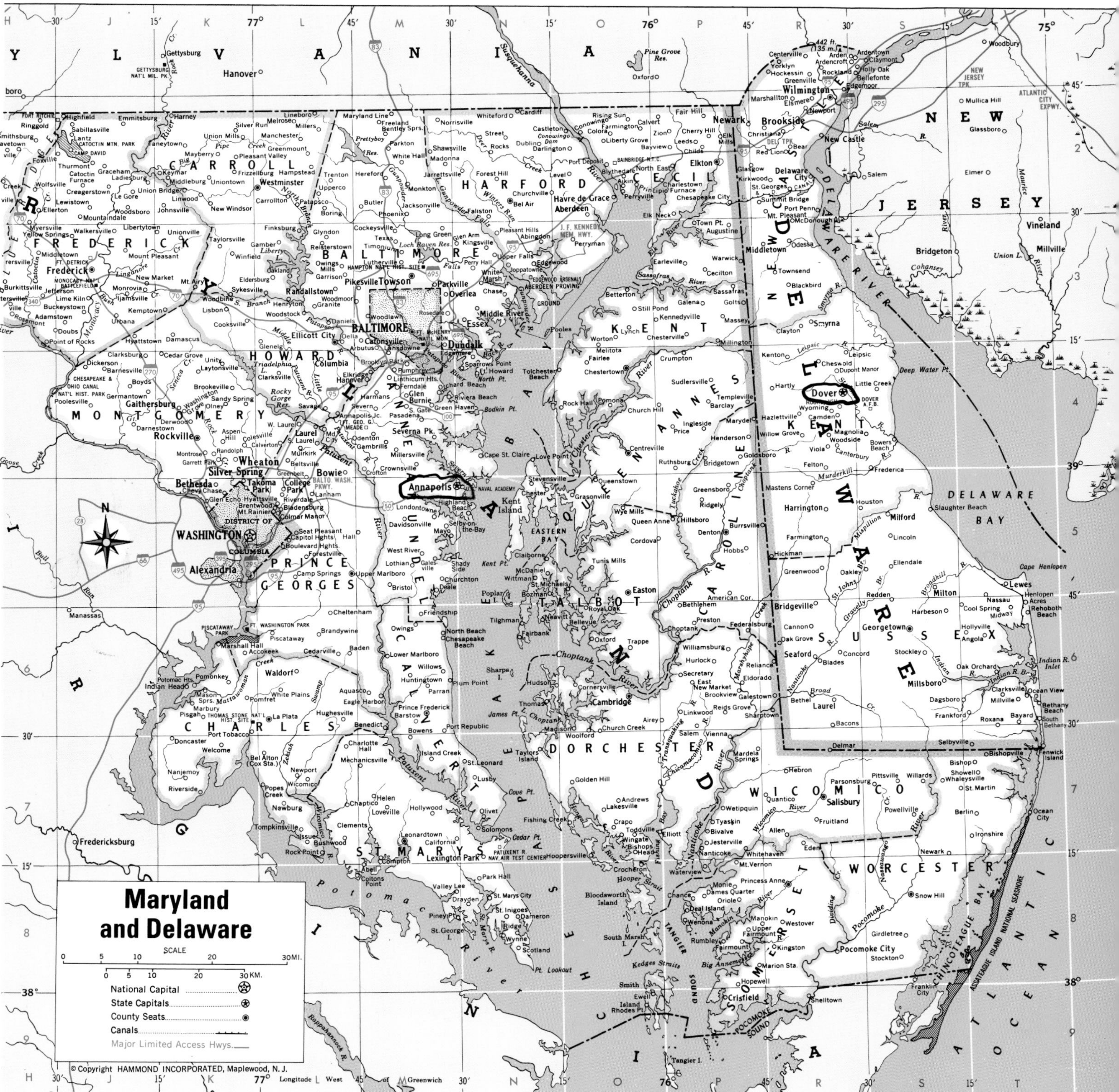

Maryland and Delaware

SCALE

0	5	10	20	30 MI.
0	5 10	20	30 KM.	

National Capital .... ★
State Capitals .... ✪
County Seats .... ◉
Canals .... ━━━
Major Limited Access Hwys. ━━━

Agriculture, Industry and Resources

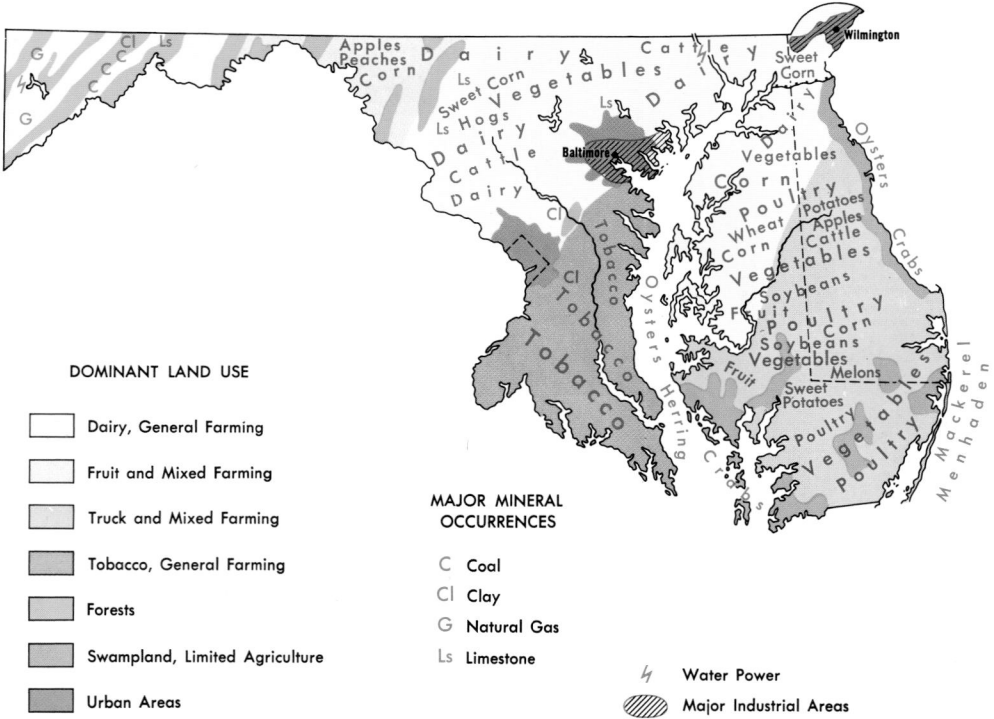

DOMINANT LAND USE

- Dairy, General Farming
- Fruit and Mixed Farming
- Truck and Mixed Farming
- Tobacco, General Farming
- Forests
- Swampland, Limited Agriculture
- Urban Areas

MAJOR MINERAL OCCURRENCES

- C   Coal
- Cl   Clay
- G   Natural Gas
- Ls   Limestone

⚡ Water Power

▨ Major Industrial Areas

MASSACHUSETTS
AREA 8,284 sq. mi. (21,456 sq. km.)
POPULATION 6,349,097
CAPITAL Boston
LARGEST CITY Boston
HIGHEST POINT Mt. Greylock 3,487 ft.
    (1063 m.)
SETTLED IN 1620
ADMITTED TO UNION February 6, 1788
POPULAR NAME Bay State; Old Colony
STATE FLOWER Mayflower
STATE BIRD Chickadee

RHODE ISLAND
AREA 1,212 sq. mi. (3139 sq. km.)
POPULATION 1,048,319
CAPITAL Providence
LARGEST CITY Providence
HIGHEST POINT Jerimoth Hill 812 ft.
    (247 m.)
SETTLED IN 1636
ADMITTED TO UNION May 29, 1790
POPULAR NAME Little Rhody;
    Ocean State
STATE FLOWER Violet
STATE BIRD Rhode Island Red

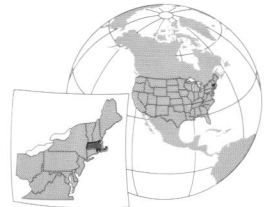

Agriculture, Industry and Resources

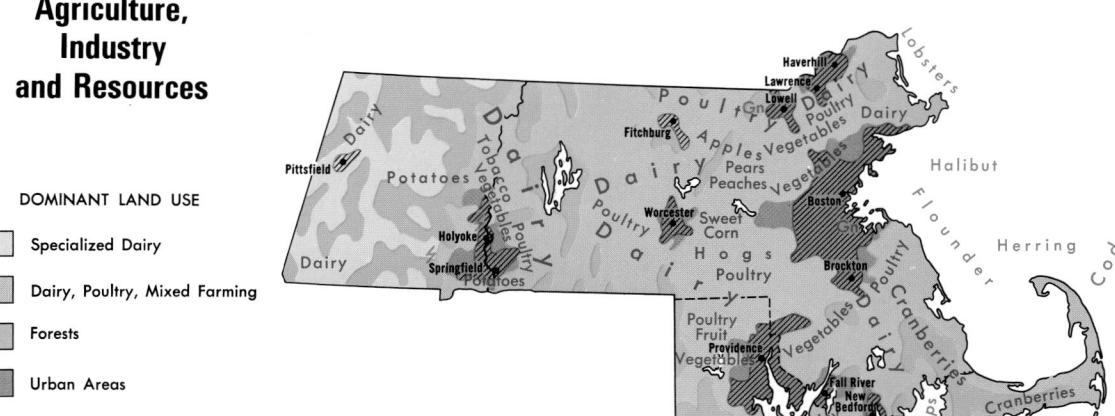

DOMINANT LAND USE

- Specialized Dairy
- Dairy, Poultry, Mixed Farming
- Forests
- Urban Areas

MAJOR MINERAL OCCURRENCES

Gn Granite

⚡ Water Power   ▨ Major Industrial Areas

(continued on following page)

Boston and Vicinity

Vineyard Haven 2,048 M7
Waban B7
Wakefield• 24,804 C5
Wales• 1,737 F4
Walpole• 22,824 B8
Walpole 5,867 B8
Waltham 59,226 B6
Waquoit 450 M6
Ware• 9,707 E3
Ware 6,174 E3
Wareham• 20,335 L5
Wareham 19,232 L5
Wareham Center 2,874 L5
Warren• 4,776 F4
Warren 1,452 F4
Warwick• 750 E2
Watertown 32,986 C6
Waverley B6
Wayland• 13,100 A7
Webster• 16,415 G4
Webster 11,600 G4
Wellesley• 26,613 B7
Wellesley Hills B7
Wellfleet• 2,749 O5
Wendell• 986 E2
Wenham• 4,440 L2
West Acton 975 H3
West Barnstable 1,508 N6
Westborough• 17,997 H3
Westborough 3,983 H3
West Boxford 950 K2
West Boylston• 7,481 G3
West Bridgewater• 6,634 K4
West Brookfield• 3,804 F4
West Brookfield 1,610 F4
West Chatham 1,446 O6
West Chelmsford J2
West Concord 5,632 A6
West Dennis 2,570 O6
West Falmouth 1,867 M6
Westfield 40,072 D4
Westford• 20,754 J2
West Groton 950 H2
Westhampton• 1,468 C3
West Hanover L4
West Harwich 883 O6
West Mansfield 950 K5
West Medway J4
Westminster• 6,907 G2
West Newbury• 4,149 L1
West Newton B7
Weston• 11,469 B6
Westport• 14,183 K6
Westport 13,852 K6
West Springfield• 27,899 D4
West Stockbridge• 1,416 A3
West Tisbury• 2,467 M7
West Townsend 950 H2
West Upton-Upton 4,677 H4
West Wareham 1,908 L5
West Warren F4
Westwood• 14,117 B8
West Yarmouth 6,460 N6
Weymouth• 53,988 D8
Whately• 1,573 D3
Whitinsville 6,340 H4
Whitman• 13,882 L4
Wilbraham• 13,473 E4
Wilbraham 3,544 E4
Williamsburg• 2,427 C3
Williamstown• 8,424 B2
Williamstown 4,754 B2
Wilmington• 21,363 C5
Winchendon 4,246 F2
Winchendon• 9,611 F2
Winchester• 20,810 C6
Windsor• 875 B2
Winthrop• 18,303 D6
Woburn 37,258 C6
Woods Hole 925 M6
Worcester▲ 172,648 H3
Worthington• 1,270 C3
Wrentham• 10,554 J4
Yarmouth• 24,807 O6
Yarmouth Port 5,395 N6

OTHER FEATURES

Adams Nat'l Hist. Site D7
Agawam (riv.) M5
Allerton (pt.) E7
Ann (cape) M2
Ashmere (lake) B3
Assabet (riv.) H3
Assawompset (pond) L5
Bachelor (brook) D3
Berkshire (hills) B4
Big (pond) B4
Bigelow (bight) M1
Blackstone (riv.) G3
Blue (hills) C8
Boston (bay) E6
Boston (harb.) D7
Boston Nat'l Hist. Park D6
Brewster (isls.) E7
Buel (lake) A4
Buzzards (bay) L7
Cambridge (res.) B6
Cape Cod (bay) N5
Cape Cod (canal) N5
Cape Cod Nat'l Seashore P5
Chappaquiddick (isl.) N7
Charles (riv.) C7
Chicopee (riv.) D4
Cobble Mountain (res.) C4
Cochituate (lake) A7
Cod (cape) O4
Concord (riv.) J2
Congamond (lakes) D4
Connecticut (riv.) D2
Cuttyhunk (isl.) L7
Deer (isl.) E7
Deerfield (riv.) C2
East (pt.) E6
East Chop (pt.) M7
Eastern (pt.) M2
Elizabeth (isls.) L7
Everett (mt.) A4
Falls (riv.) D2
Fort Devens 1,017 H2
Fresh (pond) C6
Gammon (pt.) N6
Gay Head (prom.) L7
Grace (mt.) E2
Great (pt.) O7
Green (riv.) B2
Greylock (mt.) B2
Gurnet (pt.) M4
Hingham (bay) E7
Holyoke (range) D3
Hoosac (mts.) B2
Hoosic (riv.) A1
Housatonic (riv.) A4
Ipswich (riv.) L2
John F. Kennedy Nat'l Hist. Site C7
Knightville (res.) C3
Laurence G. Hanscom Field B6
Little (riv.) C4
Logan Int'l Airport D7
Long (isl.) E7
Long (pt.) O4
Long (pond) L5
Longfellow Nat'l Hist. Site C6
Lowell Nat'l Hist. Park J2
Maine (gulf) M2
Manhan (riv.) D4
Manomet (pt.) N5
Marblehead (neck) F6
Martha's Vineyard (isl.) M7
Massachusetts (bay) M4
Merrimack (riv.) K1
Mill (riv.) C3
Mill (riv.) D3
Millers (riv.) E2
Minute Man Nat'l Hist. Park B6
Mishaum (pt.) L6
Monomonac (lake) G2
Monomoy (isl.) O6
Monomoy (pt.) O6
Mount Hope (bay) K6
Muskeget (chan.) N7
Muskeget (isl.) N7
Mystic (lake) C6
Mystic (riv.) C6
Nahant (bay) E6
Nantucket (isl.) O8
Nantucket (sound) N6
Nashawena (isl.) L7
Nashua (riv.) H3
Naushon (isl.) L7
Neponset (riv.) C8
Nomans Land (isl.) L7
Nonamesset (isl.) M6
North (riv.) D2
North (riv.) L4
Onota (lake) A3
Otis (res.) B4
Otis A.F.B. M6
Pasque (isl.) L7
Plum (isl.) L2
Plymouth (bay) M5
Poge (cape) N7
Pontoosuc (lake) A3
Quabbin (res.) E3
Quaboag (riv.) F4
Quincy (bay) D7
Quinebaug (riv.) F4
Race (pt.) N4
Salem Maritime Nat'l Hist. Site E5
Saugus Iron Works Nat'l Hist. Site D6
Shawshine (riv.) K2
Silver (lake) L4
South (riv.) D2
South Weymouth Nav. Air Sta. E8
Springfield Armory Nat'l Hist. Site D4
Squibnocket (pt.) M7
Stillwater (riv.) G3
Sudbury (res.) H3
Sudbury (riv.) A6
Swift (riv.) E4
Taconic (mts.) A2
Taunton (riv.) K5
Thompson (isl.) D7
Toby (mt.) E3
Tom (mt.) D4
Tuckernuck (isl.) N7
Vineyard (sound) L7
Wachusett (mt.) G3
Wachusett (res.) G3
Walden (pond) A6
Ware (riv.) F3
Watuppa (pond) K6
Webster (lake) G4
Wellfleet (harb.) O5
West (riv.) H4
West Chop (pt.) M7
Westfield (riv.) C3
Westover A.F.B. D4
Weweantic (riv.) L5
Whitman (riv.) G2
Winter I. Coast Guard Air Sta. E5

RHODE ISLAND

COUNTIES

Bristol 50,648 J6
Kent 167,090 H6
Newport 85,433 K6
Providence 621,602 H5
Washington 123,546 H7

CITIES and TOWNS

Adamsville 500 K6
Albion 800 J5
Allenton 975 J6
Anthony H6
Apponaug J6
Arctic H6
Arnold Mills J5
Ashaway 1,537 G7
Ashton J5
Barrington• 16,819 J5
Block Island• 836 H8
Bradford 1,497 H7
Bristol▲• 22,469 J6
Centerdale J5
Central Falls 18,928 J5
Charlestown 7,859 H7
Chepachet 875 H5
Conimicut J6
Coventry (Washington)• 33,668 H6
Coventry Center 200 H6
Cranston 79,269 J5
Davisville 500 J6
East Greenwich▲• 12,948 H6
East Providence 48,688 J5
Esmond H5
Exeter• 6,045 H6
Fiskeville 600 H6
Foster• 4,274 H5
Georgiaville H5
Glendale 750 H5
Greenville 8,626 H5
Hamilton 950 J6
Harmony 975 H5
Harrisville 1,561 H5
Hillsgrove J6
Hope 975 H6
Hope Valley 1,649 H6
Hopkinton• 7,836 H7
Island Park J6
Jamestown• 5,622 J6
Jamestown J6
Kingston 5,446 J7
La Fayette H6
Little Compton• 3,593 K6
Lonsdale J5
Manville H5
Mapleville 975 H5
Middletown• 17,334 J6
Narragansett• 16,361 J7
Narragansett 3,721 J7
Nasonville 800 H5
Natick H6
Newport▲ 26,475 J7
New Shoreham (Block Island)• 1,010 H8
North Kingstown• 26,326 J6
North Providence• 32,411 J5
North Scituate 850 H5
North Tiverton K6
Norwood J6
Oakland Beach J6
Pascoag 4,742 H5
Pawtucket 72,958 J5
Peace Dale-Wakefield 8,468 J7
Pontiac J6
Portsmouth• 17,149 J6
Providence (cap.)▲ 173,618 H5
Riverside J5
Rumford J5
Saunderstown 600 J6
Tiverton• 15,260 K6
Tiverton 7,282 K6
Valley Falls 11,599 J5
Wakefield-Peace Dale 8,468 J7
Warren• 11,360 J6
Warwick 85,808 J6
Washington (Coventry)• 33,668 H6
Westerly• 22,966 G7
Westerly▲ 17,682 G7
West Kingston▲ 950 H7
West Warwick 29,581 H6
Woonsocket▲ 43,224 J4

OTHER FEATURES

Black Rock (pt.) H8
Block (isl.) H8
Block Island (sound) H8
Brenton (pt.) J7
Conanicut (isl.) J6
Dickens (pt.) H8
Durfee (hill) G5
Grace (pt.) H8
Jerimoth (hill) G5
Judith (pt.) J7
Mount Hope (bay) K6
Narragansett (bay) J6
Noyes (pt.) H7
Pawcatuck (riv.) G7
Prudence (isl.) J6
Rhode Island (isl.) J6
Rhode (sound) J7
Roger Williams Nat'l Mem. J5
Sakonnet (pt.) K7
Sakonnet (riv.) K7
Sandy (pt.) H8
Scituate (res.) H5
Touro Synagogue Nat'l Hist. Site J7
Watch Hill (pt.) G7

▲County seat or Shire town
•Population of town or township

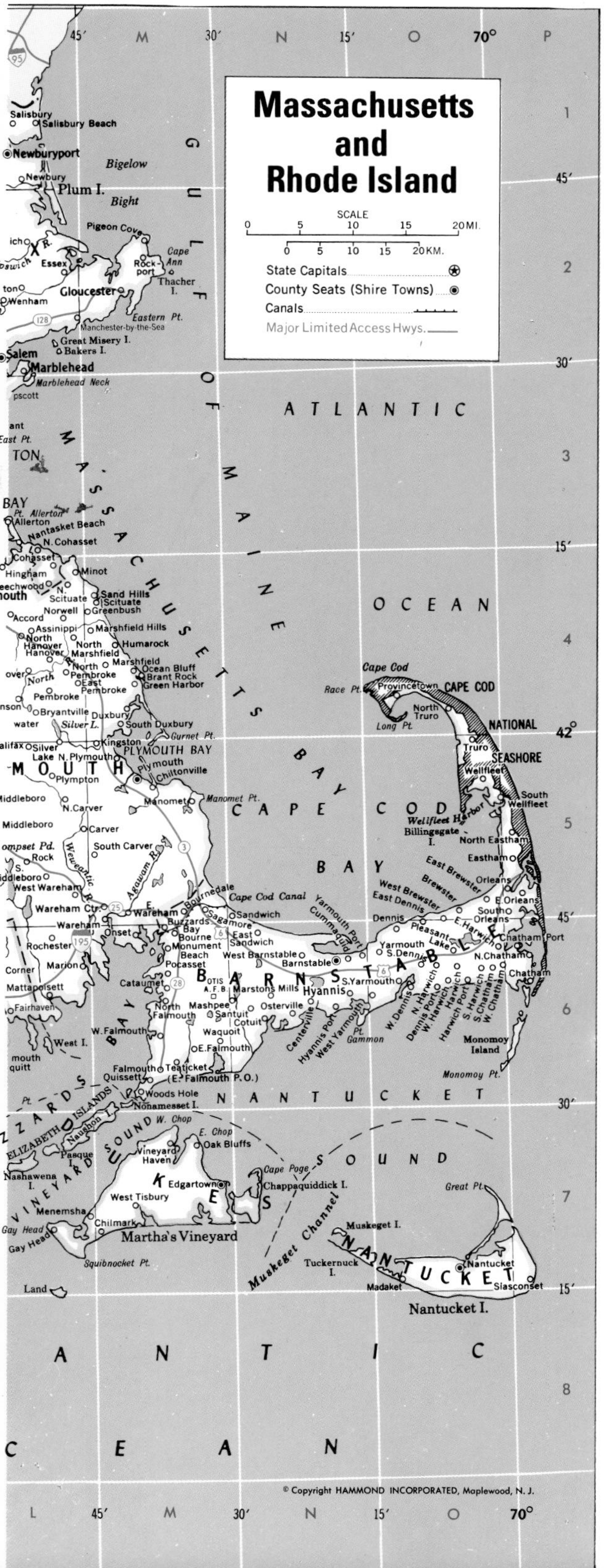

Massachusetts and Rhode Island
SCALE
0 5 10 15 20 MI.
0 5 10 15 20 KM.
State Capitals
County Seats (Shire Towns)
Canals
Major Limited Access Hwys.
ATLANTIC OCEAN
GULF OF MAINE
MASSACHUSETTS BAY
CAPE COD BAY
NANTUCKET SOUND
CAPE COD NATIONAL SEASHORE
Martha's Vineyard
Nantucket I.
Provincetown
Hyannis
Barnstable
Falmouth
Plymouth
Gloucester
Salem
Marblehead
Newburyport

© Copyright HAMMOND INCORPORATED, Maplewood, N. J.

Topography

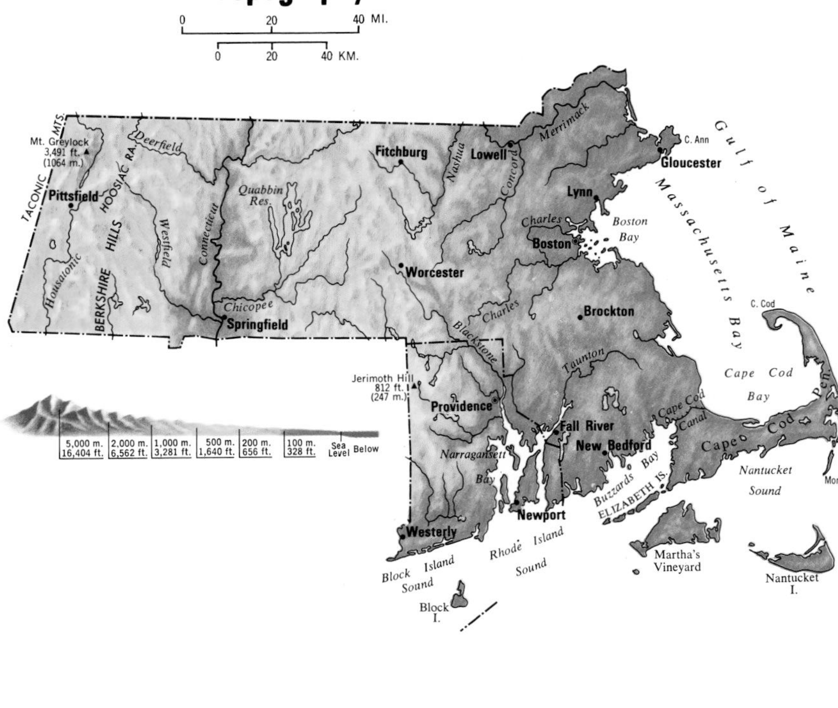

0 20 40 MI.
0 20 40 KM.
Mt. Greylock 3,491 ft. (1064 m.)
Jerimoth Hill 812 ft. (247 m.)
5,000 m. 16,404 ft.
2,000 m. 6,562 ft.
1,000 m. 3,281 ft.
500 m. 1,640 ft.
200 m. 656 ft.
100 m. 328 ft.
Sea Level
Below
Pittsfield
Springfield
Worcester
Fitchburg
Lowell
Lynn
Gloucester
Boston
Brockton
Providence
Fall River
New Bedford
Newport
Westerly
Gulf of Maine
Massachusetts Bay
Cape Cod Bay
Nantucket Sound
Martha's Vineyard
Nantucket I.
Block I.
Block Island Sound
Rhode Island Sound

Michigan

COUNTIES

Alcona 11,719 F4
Alger 9,862 C2
Allegan 105,665 D6
Alpena 31,314 F4
Antrim 23,110 D3
Arenac 17,269 F4
Baraga 8,746 A2
Barry 56,755 D6
Bay 110,157 E5
Benzie 15,998 C4
Berrien 162,453 C7
Branch 45,787 D7
Calhoun 137,985 D6
Cass 51,104 C7
Charlevoix 26,090 D3
Cheboygan 26,448 E3
Chippewa 38,543 E2
Clare 31,252 E5
Clinton 64,753 E6
Crawford 14,273 E4
Delta 38,520 C2
Dickinson 27,472 B2
Eaton 103,655 E6
Emmet 31,437 E3
Genesee 436,141 F5
Gladwin 26,023 E4
Gogebic 17,370 F2
Grand Traverse 77,654 D4
Gratiot 42,285 E5
Hillsdale 46,527 E7
Houghton 36,016 G1
Huron 36,079 F5
Ingham 279,320 E6
Ionia 61,518 D6
Iosco 27,339 F4
Iron 13,138 G2
Isabella 63,351 E5
Jackson 158,422 E6
Kalamazoo 238,603 D6
Kalkaska 16,571 D4
Kent 574,335 D5
Keweenaw 2,301 A1
Lake 11,333 D5
Lapeer 87,904 F5
Leelanau 21,119 D4
Lenawee 98,890 E7
Livingston 156,951 F6
Luce 7,024 D2
Mackinac 11,943 D2
Macomb 788,149 G6
Manistee 24,527 C4
Marquette 64,634 B2
Mason 28,274 C4
Mecosta 40,553 D5
Menominee 25,326 B3
Midland 82,874 E5
Missaukee 14,478 D4
Monroe 145,945 F7
Montcalm 61,266 D5
Montmorency 10,315 E3
Muskegon 170,200 C5
Newaygo 47,874 D5
Oakland 1,194,156 F6
Oceana 26,873 C5
Ogemaw 21,645 E4
Ontonagon 7,818 F1
Osceola 23,197 D5
Oscoda 9,418 E4
Otsego 23,301 E3
Ottawa 238,314 C6
Presque Isle 14,411 F3
Roscommon 25,469 E4
Saginaw 210,039 E5
Saint Clair 164,235 G6
Saint Joseph 62,422 D7
Sanilac 44,547 G5
Schoolcraft 8,903 C2
Shiawassee 71,687 E6
Tuscola 58,266 F5
Van Buren 76,263 C6
Washtenaw 322,895 F6
Wayne 2,061,162 F6
Wexford 30,484 D4

CITIES and TOWNS

Addison 627 E7
Adrian▲ 21,574 F7
Advance 600 D3
Afton 300 E3
Akron 461 F5
Alanson 785 E3
Alba 300 E4
Albion 9,144 E6
Alger 500 E4
Algonac 4,613 G6
Allegan▲ 4,838 D6
Allen Park 29,376 B7
Alma 9,275 E5
Almont 2,803 F6
Alpena▲ 11,304 F3
Amasa 450 G2
Anchorville 3,202 G6
Ann Arbor▲ 114,024 F6
Antrim 475 D4
Arcadia 780 C4
Armada 1,573 G6
Ashley 526 E5
Athens 1,111 D6
Atlanta▲ 757 E3
Atlantic Mine 809 G1
Auburn 2,011 F5
Auburn Heights 7,500 F6
Au Gres 1,028 F4
Augusta 899 D6
Aura 325 A2
Au Sable 1,533 F4
Averill 800 E5
Bad Axe▲ 3,462 G5
Baldwin▲ 1,107 D5
Bancroft 616 E6
Bangor 1,933 C6
Baraga 1,285 G1
Barbeau 400 E2
Bark River 800 B3
Baroda 858 C7
Barryton 381 D5
Barton Hills 335 F6
Bath 600 E6
Battle Creek 53,364 D6
Bay City▲ 36,817 F5
Bay Port 750 F5
Bay View 500 E3
Bear Lake 318 C4
Beaverton 1,106 E5
Bedford 450 D6
Beechwood 2,963 C6
Belding 5,877 D5
Bellaire▲ 1,164 D4
Belleville 3,997 F6
Bellevue 1,365 E6
Bentley 350 E5
Benton Harbor 11,182 C6
Benton Heights 5,458 C6
Benzonia 519 D4
Bergland 650 F1
Berkley 15,531 B6
Berrien Springs 1,862 C7
Bessemer▲ 2,148 F2
Beulah▲ 363 C4
Beverly Hills 10,437 B6
Big Bay 265 B2
Big Rapids▲ 10,849 D5
Birch Run 1,653 F5
Birmingham 19,291 B6
Bitely 750 D5
Blanchard 350 D5
Blissfield 3,223 F7
Bloomfield Hills 3,940 B6
Bloomingdale 528 C6
Boyne City 3,503 E3
Boyne Falls 370 E3
Brampton 400 B3
Breckenridge 1,339 E5
Bridgeport 7,849 F5
Bridgman 2,428 C7
Brighton 6,701 F6
Brimley 550 E2
Britton 699 F6
Bronson 2,421 D7
Brooklyn 1,176 E6
Brown City 1,334 G5
Bruce Crossing 400 G2
Buchanan 4,681 C7
Buckley 550 D4
Burnips 528 D6
Burr Oak 797 D7
Burt 1,122 F5
Burton 30,308 F6
Byron 595 E6
Byron Center 3,777 D6
Cadillac▲ 10,000 D4
Caledonia 1,102 D6
Calumet 879 A1
Camden 550 E7
Capac 1,775 G5
Carleton 2,562 F6
Caro▲ 4,145 F5
Carp Lake 464 E3
Carrollton 6,602 E5
Carson City 1,190 E5
Carsonville 502 G5
Caseville 888 F5
Casnovia 315 D5
Caspian 997 G2
Cass City 2,276 F5
Cassopolis▲ 1,740 C7
Castle Park 500 C6
Cedar Springs 3,112 D5
Cedarville 500 E2
Cement City 452 E6
Center Line 8,531 B6
Central Lake 990 D3
Centreville▲ 1,579 D7
Channing 456 B2
Charlevoix▲ 2,994 D3
Charlotte▲ 8,389 E6
Cheboygan▲ 5,295 E3
Chelsea 4,398 E6
Chesaning 2,548 E5
Chippewa Lake 580 D5
Clare 3,173 E5
Clarklake 500 E6
Clarkston 1,005 F6
Clarksville 317 D6
Clawson 12,732 B6
Clayton 326 E7
Clifford 324 F5
Climax 791 D6
Clinton 2,293 F6
Clio 2,483 F5
Cohoctah 300 F6
Coldwater▲ 12,697 D7
Coleman 1,296 E5
Coloma 1,595 C6
Colon 1,227 D7
Columbiaville 815 F5
Comstock• 13,851 D6
Concord 1,101 E6
Constantine 2,095 D7
Conway 560 E3
Cooks 620 C3
Coopersville 3,910 C5
Corunna▲ 3,381 E6
Covert 650 C6
Covington 380 G2
Croswell 2,467 G5
Crystal 800 E5
Crystal Falls▲ 1,791 A2
Curtis 800 D2
Custer 318 C5
Cutlerville 15,114 D6
Dalton• 8,047 C5
Dansville 429 E6
Davison 5,536 F5
Dayton 125 C7
Dearborn 97,775 B7
Dearborn Heights 58,264 B7
Decatur 1,838 C6
Deckerville 944 G5
Deerfield 1,005 F7
Delton 400 D6
De Tour Village 421 E3
Detroit▲ 951,270 B7
Detroit Beach 2,289 F7
De Witt 4,702 E6
Dexter 2,338 F6
Dimondale 1,342 E6
Dollar Bay 950 G1
Dorr• 6,579 D6
Douglas 1,214 C6
Dowagiac 6,147 C7
Drummond Island• 746 F3
Dryden 815 F6
Dundee 3,522 F7
Durand 3,933 E6
Eagle River▲ 20 A1
East Grand Rapids 10,764 D6
East Jordan 2,507 D3
Eastlake 441 C4
East Lansing 46,525 E6
Eastpointe 34,077 B6
East Tawas 2,951 F4
Eastwood 6,265 D6
Eaton Rapids 5,330 E6
Eau Claire 656 C6
Eben Junction 450 B2
Ecorse 11,229 B7
Edmore 1,126 E5
Edwardsburg 1,147 C7
Elberta 457 C4
Elk Rapids 1,700 D4
Elkton 863 F5
Ellsworth 483 D3
Elsie 1,055 E5
Empire 378 C4
Engadine 500 D2
Erie• 4,850 F7
Escanaba▲ 13,140 C3
Essexville 3,766 F5
Estey 550 E5
Estral Beach 486 F7
Evart 1,738 D5
Ewen 821 F2
Fairgrove 627 F5
Fair Haven 1,505 G6
Fair Plain 7,828 C6
Fairview 600 E4
Falmouth 350 E4
Farmington 10,423 F6
Farmington Hills 82,111 F6
Farwell 855 E5
Felch 300 B3
Fennville 1,459 C6
Fenton 10,582 F6
Ferndale 22,105 B6
Ferrysburg 3,040 C5
Fife Lake 466 D4
Flat Rock 8,488 F6
Flint▲ 124,943 F5
Flushing 8,348 F5
Fowler 1,136 E5
Fowlerville 2,972 F6
Frankenmuth 4,838 F5
Frankfort 1,513 C4
Franklin 2,937 B6
Fraser 15,297 B6
Frederic 500 E4
Freeland 5,147 E5
Freeport 444 D6
Fremont 4,224 D5
Fruitport 1,124 C5
Fulton 500 D6
Gaastra 339 G2
Gagetown 389 F5
Gaines 366 F6
Galesburg 1,988 D6
Galien 593 C7
Garden City 30,047 F6
Gaylord▲ 3,681 E3
Germfask• 491 C2
Gibraltar 4,264 F6
Gladstone 5,032 C3
Gladwin▲ 3,001 E5
Glen Arbor• 788 C4
Gobles 815 D6
Goetzville 456 E2
Good Hart 500 D3
Goodrich 1,353 F6
Gould City 450 D2
Gowen 300 D5
Grand Blanc 8,242 F6
Grand Haven▲ 11,168 C5
Grand Junction 300 C6
Grand Ledge 7,813 E6
Grand Marais 700 D2
Grand Rapids▲ 197,800 D5
Grandville 16,263 D6
Grant 881 D5
Grass Lake 1,082 E6
Grayling▲ 1,952 E4
Greenbush 650 F4
Greenland 870 G1
Greenville 7,935 D5
Gregory 550 E6
Grosse Ile 10,894 B7
Grosse Pointe 5,670 B7
Grosse Pointe Farms 9,764 B6
Grosse Pointe Park 12,443 B7
Grosse Pointe Shores 2,823 B6
Grosse Pointe Woods 17,080 B6
Gulliver 962 D2
Gwinn 1,965 B2
Hale 500 F4
Hamburg• 20,627 F6
Hamilton 950 C6
Hamtramck 22,976 B6
Hancock 4,323 G1
Hanover 424 E6
Harbor Beach 1,837 G5
Harbor Springs 1,567 D3
Harper Woods 14,254 B6
Harrison▲ 2,108 E4
Harrisville▲ 514 F4
Harsens Island 750 G6
Hart▲ 1,950 C5
Hartford 2,476 C6
Haslett 11,283 E6
Hastings▲ 7,095 D6
Hazel Park 18,963 B6
Hemlock 1,585 E5
Hermansville 950 B3
Herron 300 F3
Hersey 374 D5
Hesperia 954 D5
Hessel 640 E2
Higgins Lake 500 E4
Highland Park 16,746 B6
Hillman 685 F3
Hillsdale▲ 8,233 E7
Holland 35,048 C6
Holly 6,135 F6
Holt 11,315 E6
Holton 500 C5
Homer 1,851 E6
Hopkins 592 D6
Horton 400 E6
Houghton▲ 7,010 G1
Houghton Lake 3,749 E4
Howard City 1,585 D5
Howell▲ 9,232 E6
Hubbardston 394 E5
Hubbell 1,105 A1
Hudson 2,499 E7
Hudsonville 7,160 D6
Huntington Woods 6,151 B6
Ida 970 F7
Idlewild 500 D5
Imlay City 3,869 F5
Indian River 2,008 E3
Inkster 30,115 B7
Interlochen 600 D4
Ionia▲ 10,569 D6
Iron Mountain▲ 8,154 B3
Iron River 1,929 G2
Ironwood 6,293 F2
Ishpeming 6,686 B2
Ithaca▲ 3,098 E5
Jackson▲ 36,316 E6
Jasper 350 E7
Jenison 17,211 D6
Jones 420 D7
Jonesville 2,337 E6
Kalamazoo▲ 77,145 D6
Kaleva 509 C4
Kalkaska▲ 4,830 D4
Kawkawlin• 5,104 F5
Keego Harbor 2,769 F6
Kent City 1,061 D5
Kentwood 45,255 D6
Kinde 534 G5
Kingsford 5,549 A3
Kingsley 1,469 D4
Kingston 450 F5
Laingsburg 1,223 E6
Lake City▲ 923 D4
Lake George 950 E5
Lakeland 720 F6
Lake Leelanau 350 D4
Lake Linden 1,081 A1
Lake Michigan Beach 1,509 C6
Lake Odessa 2,272 D6
Lake Orion 2,715 F6
Lakeview 1,112 D5
Lakewood Club 1,006 C5
Lambertville 9,299 F7
L'Anse▲ 2,107 G1
Lansing (cap.) 119,128 E6
Lapeer▲ 9,072 F5
Laporte 350 E5
Laurium 2,126 A1
Lawrence 1,059 C6
Lawton 1,859 D6
Leland• 2,033 D3
Lennon 517 E5
Leonard 332 F6
Leslie 2,044 E6
Levering 967 E3
Lewiston 990 E4
Lexington 1,104 G5
Lincoln 364 F4
Lincoln Park 40,008 B7
Linden 2,861 F6
Linwood 400 F5
Litchfield 1,458 E6
Little Lake 975 B2
Livonia 100,545 F6
Long Lake 500 F4
Lowell 4,013 D6
Ludington▲ 8,357 C5
Luna Pier 1,483 F7
Luther 339 D4
Lyons 726 E6

(continued on following page)

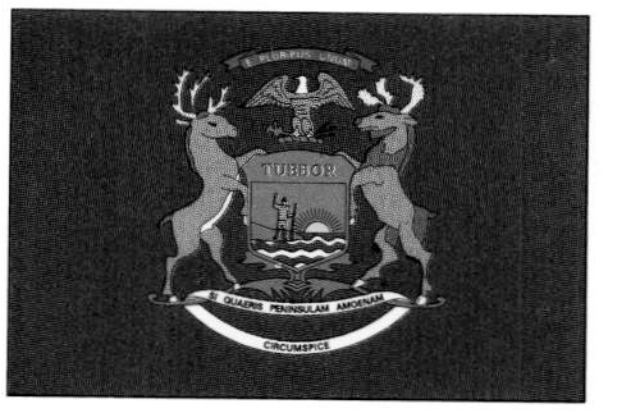

AREA 58,527 sq. mi. (151,585 sq. km.)
POPULATION 9,938,444
CAPITAL Lansing
LARGEST CITY Detroit
HIGHEST POINT Mt. Arvon 1,979 ft. (603 m.)
SETTLED IN 1831
ADMITTED TO UNION January 26, 1837
POPULAR NAME Wolverine State
STATE FLOWER Apple Blossom
STATE BIRD Robin

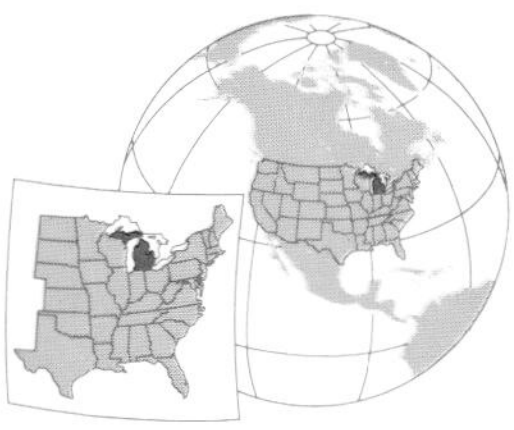

Topography

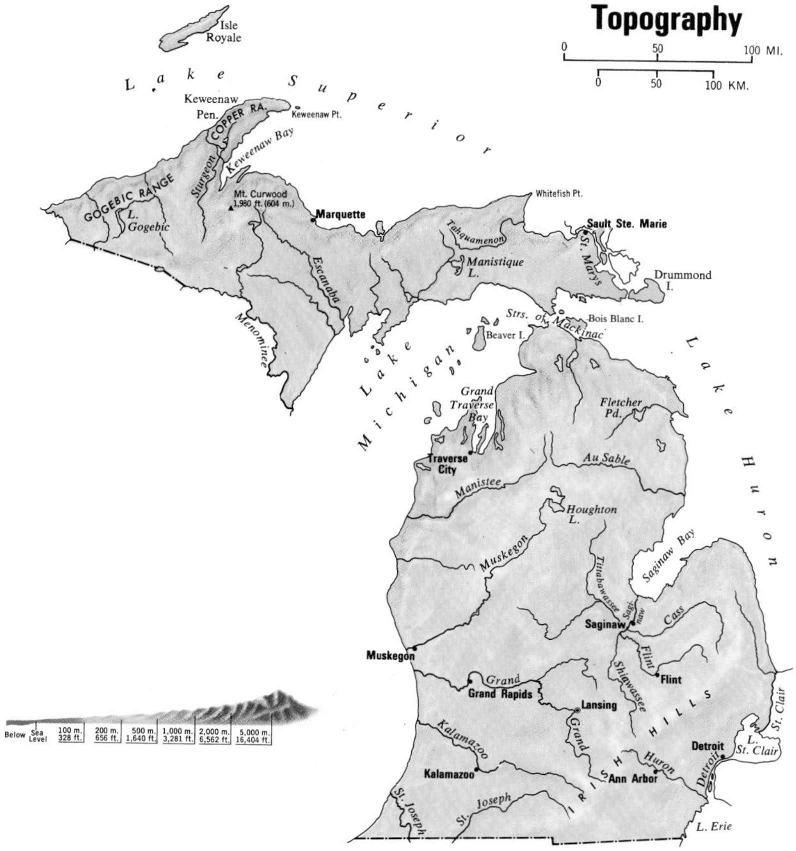
0 50 100 MI.
0 50 100 KM.
Isle Royale
Lake Superior
Keweenaw Pen.
Keweenaw Pt.
COPPER RA.
Keweenaw Bay
GOGEBIC RANGE
L. Gogebic
Sturgeon
Mt. Curwood 1,980 ft. (604 m.)
Marquette
Whitefish Pt.
Tahquamenon
Sault Ste. Marie
St. Marys
Manistique L.
Drummond I.
Escanaba
Menominee
Strs. of Mackinac
Bois Blanc I.
Beaver I.
Lake Michigan
Lake Huron
Grand Traverse Bay
Fletcher Pd.
Traverse City
Au Sable
Manistee
Houghton L.
Muskegon
Saginaw Bay
Tittabawassee
Saginaw
Cass
Muskegon
Flint
Grand
Grand Rapids
Shiawassee
Lansing
Flint
HILLS
IRISH
St. Clair
L. St. Clair
Detroit
Kalamazoo
Kalamazoo
Ann Arbor
Huron
St. Joseph
Joseph
L. Erie
Below Sea Level
100 m. 328 ft.
200 m. 656 ft.
500 m. 1,640 ft.
1,000 m. 3,281 ft.
2,000 m. 6,562 ft.
5,000 m. 16,404 ft.

Mackinac Island 523 E3
Mackinaw City 859 E3
Madison Heights 31,101 B6
Mancelona 1,408 E4
Manchester 2,160 E6
Manistee▲ 6,586 C4
Manistique▲ 3,583 C3
Manton 1,221 D4
Maple Rapids 643 E5
Marcellus 1,162 D6
Marenisco• 1,051 F2
Marine City 4,652 G6
Marion 836 D4
Marlette 2,104 G5
Marne 450 D5
Marquette▲ 19,661 B2
Marshall▲ 7,459 E6
Martin 435 D6
Marysville 9,684 G6
Mason▲ 6,714 E6
Mass City 850 G1
Mattawan 2,536 D6
Maybee 505 F6
Mayville 1,055 F5
McBain 584 D4
Mears 375 C5
Mecosta 440 D5
Melvindale 10,735 B7
Memphis 1,129 G6
Mendon 917 D7
Menominee▲ 9,131 B3
Merrill 782 E5
Mesick 447 D4
Metamora 507 F6
Metz 495 F3
Michigamme 287 B2
Michigan Center 4,641 E6
Middleton 540 E5
Middleville 2,721 D6
Midland▲ 41,685 E5
Milan 4,775 F6
Milford 6,272 F6
Millington 1,137 F5
Mio▲ 2,016 E4
Mohawk 800 A1
Moline 750 D6
Monroe▲ 22,076 F7
Montague 2,407 C5
Montgomery 386 E7
Montrose 1,619 F5
Morenci 2,398 E7
Morley 495 D5
Morrice 882 E6
Mount Clemens▲ 17,312 G6
Mount Morris 3,194 F5
Mount Pleasant▲ 25,946 E5
Muir 634 D5
Mulliken 590 E6
Munger 366 F5
Munising▲ 2,539 C2
Munith 600 E6
Muskegon▲ 40,105 C5
Muskegon Heights 12,049 C5
Napoleon 1,254 E6
Nashville 1,684 D6
National City 500 F4
National Mine 565 B2
Naubinway 850 D2
Negaunee 4,576 B2
Newaygo 1,670 D5
New Baltimore 7,405 G6
Newberry▲ 2,686 D2
New Boston 1,200 F6
New Buffalo 2,200 C7
New Era 461 C5
New Haven 3,071 G6
New Lothrop 603 F5
New Troy 350 C7
Niles 12,204 C7
North Adams 514 E7
North Bradley 300 E5
North Branch 1,027 F5
North Muskegon 4,031 C5
Northport 648 D3
Northville 6,459 F6
Norton Shores 22,527 C5
Norway 2,959 B3
Novi 47,386 F6
Oakley 339 E5
Oak Park 29,793 B6
Ocqueoc• 634 F3
Okemos 22,805 E6
Olivet 1,758 E6
Omer 337 F4
Onaway 993 E3
Onekama 647 C4
Onsted 813 E6
Ontonagon▲ 1,769 F1
Orchard Lake 2,215 F6
Ortonville 1,535 F6
Oscoda 992 F4
Osseo 400 E7
Otisville 882 F5
Otsego 3,933 D6
Otsego Lake• 2,532 E4
Ottawa Lake 350 F7
Otter Lake 437 F5
Ovid 1,514 E5
Owendale 296 F5
Owosso 3,540 E5
Oxford 3,540 F6
Painesdale 650 G1
Palmer 449 B2
Palmyra• 2,366 E7
Paradise 400 D2
Parchment 1,936 D6
Parma 907 E6
Paw Paw▲ 3,363 D6
Paw Paw Lake 3,944 C6
Pearl Beach 3,224 G6
Peck 599 G5
Pellston 771 E3
Pentwater 958 C5
Perkins 350 B3
Perrinton 439 E5
Perry 2,065 E6
Petersburg 1,157 F7
Petoskey▲ 6,080 E3
Pewamo 560 E5
Pickford• 1,584 E2
Pigeon 1,207 F5
Pinckney 2,141 F6
Pinconning 1,386 F5
Pittsford• 1,600 E7
Plainwell 3,933 D6
Pleasant Ridge 2,594 B6
Plymouth 9,022 F6
Pontiac▲ 66,337 F6
Portage 44,897 D6
Port Austin 737 F4
Port Hope 310 G5
Port Huron▲ 32,338 G6
Portland 3,789 E6
Port Sanilac 658 G5
Potterville 2,168 E6
Prescott 286 F4
Presque Isle• 1,691 F3
Prudenville 1,737 E4
Pullman 500 C6
Quincy 1,701 E7
Quinnesec 1,187 A3
Rapid City 450 D4
Rapid River 950 C3
Ravenna 1,206 D5
Reading 1,134 E7
Reed City▲ 2,430 D5
Reese 1,375 F5
Remus 425 D5
Republic 614 B2
Richland 593 D6
Richmond 4,897 G6
Richville 500 F5
Riverdale 380 E5
River Rouge 9,917 B7
Riverside 750 C6
Riverview 13,272 B7
Rives Junction 500 E6
Rochester 10,467 F6
Rock 520 B2
Rockford 4,626 D5
Rockland• 324 G1
Rockwood 3,442 F6
Rogers City▲ 3,322 F3
Romeo 3,721 F6
Romulus 22,979 F6
Roosevelt Park 3,890 C5
Roscommon▲ 1,133 E4
Rosebush 379 E5
Rose City 721 E4
Roseville 48,129 B6
Rothbury 416 C5
Royal Oak 60,062 B6
Rudyard 950 E2
Saginaw▲ 61,799 F5
Saint Charles 2,215 E5
Saint Clair 5,802 G6
Saint Clair Shores 63,096 B6
Saint Helen 2,993 E4
Saint Ignace▲ 2,678 E3
Saint Johns▲ 7,485 E6
Saint Joseph▲ 8,789 C6
Saint Louis 4,494 E5
Salem• 5,562 F6
Saline 8,034 F6
Sand Lake 492 D5
Sands• 2,127 B2
Sandusky▲ 2,745 G5
Sanford 943 E5
Saranac 1,326 D6
Saugatuck 1,065 C6
Sault Sainte Marie▲ 16,542 E2
Sawyer 500 C7
Schoolcraft 1,587 D6
Scottville 1,266 C5
Sebewaing 1,974 F5
Shelby 1,914 C5
Shepherd 1,536 E5
Sheridan 705 D5
Sherwood 324 D6
Shoreham 860 C6
Shorewood Hills 1,735 C7
Sister Lakes 700 C6
Six Lakes 350 D5
Snover 400 G5
Sodus• 2,139 C6
South Branch 400 E4
Southfield 78,296 F6
Southgate 30,136 F6
South Haven 5,021 C6
South Lyon 10,036 F6
South Monroe 6,370 F7
South Range 727 G1
South Rockwood 1,284 F7
Sparlingville 1,974 G6
Sparta 4,159 D5
Spring Arbor 2,188 E6
Springfield 5,189 D6
Spring Lake 2,514 C5
Springport 704 E6
Standish▲ 1,581 F5
Stanton▲ 1,504 D5
Stephenson 875 B3
Sterling 533 E4
Sterling Heights 124,471 B6
Stevensville 1,191 C6
Stockbridge 1,260 E6
Stonington 323 C3
Strongs 312 E2
Sturgis 11,285 D7
Sunfield 591 D6
Suttons Bay 589 D3
Swartz Creek 5,102 F6
Sylvan Lake 1,735 F6
Tawas City▲ 2,005 F4
Taylor 65,868 B7
Tecumseh 8,574 E7
Tekonsha 712 E6
Temperance 7,757 F7
Thompson• 671 C3
Thompsonville 457 C4
Three Oaks 1,829 C7
Three Rivers 7,328 D7
Tower 560 E3
Traverse City▲ 14,532 D4
Trenary 350 C2
Trenton 19,584 B7
Trout Lake• 465 E2
Troy 80,959 B6
Trufant 400 D5
Tuscola• 2,152 F5
Twin Lake 1,613 C5
Ubly 873 G5
Union City 1,804 D6
Union Pier 1,039 C7
Unionville 605 F5
Utica 4,577 F6
Vandalia 429 D7
Vanderbilt 587 E3
Vassar 2,823 F5
Vermontville 789 E6
Vernon 847 F6
Vestaburg 500 E5
Vicksburg 2,320 D6
Vulcan 400 B3
Wakefield 2,085 F2
Waldron 590 E7
Walhalla 600 C5
Walker 21,842 D6
Walled Lake 6,713 F6
Walloon Lake 325 E3
Waltz 300 F6
Warren 138,247 B6
Waters 500 E4
Watersmeet• 1,472 G2
Watervliet 1,843 C6
Wayland 3,939 D6
Wayne 19,051 F6
Webberville 1,503 E6
Weidman 879 D5
Wellston 325 D4
West Branch▲ 1,926 E4
Westland 86,602 F6
West Olive 304 C6
Weston 350 E7
Westphalia 876 E6
Wheeler• 2,785 E5
White Cloud▲ 1,420 D5
Whitehall 2,884 C5
White Pigeon 1,627 D7
White Pine 1,142 F1
Whitmore Lake 6,574 F6
Whittaker 400 F6
Whittemore 476 F4
Williamston 3,441 E6
Willis 500 F6
Winn 600 E5
Wixom 13,263 F6
Wolf Lake 4,455 D5
Wolverine 359 E3
Woodhaven 12,530 F6
Woodland 495 D6
Wyandotte 28,006 B7
Wyoming 69,368 D6
Yale 2,063 G5
Yalmer 300 B2
Ypsilanti 22,362 F6
Zeeland 5,805 D6
Zilwaukee 1,799 F5

OTHER FEATURES

Abbaye (pt.) B2
Au Sable (pt.) C2
Au Sable (pt.) F4
Au Sable (riv.) E4
Au Train (bay) C2
Bad (riv.) E5
Barques (pt.) C3
Beaver (isl.) D3
Beaver (lake) F4
Belle (riv.) G6
Bete Grise (bay) B1
Big Bay (pt.) B2
Big Bay de Noc (bay) C3
Big Iron (riv.) F1
Big Sable (pt.) C4
Big Sable (riv.) C4
Big Star (lake) C5
Black (lake) E3
Black (riv.) E3
Black (riv.) G5
Blake (pt.) E1
Boardman (riv.) D4
Bois Blanc (isl.) E3
Bond Falls (res.) G2
Brevoort (lake) D3
Brule (riv.) A3
Burt (lake) E3
Cass (riv.) F5
Cedar (lake) F4
Charlevoix (lake) D3
Chippewa (riv.) E5
Crisp (pt.) D2
Crystal (lake) C4
Curwood (mt.) A2
Dead (riv.) B2
Deer (riv.) A2
De Tour (passage) E3
Detour (pt.) C3
Detroit (riv.) B7
Drummond (isl.) F2
Duck (lake) F4
Elk (lake) D4
Erie (lake) G7
Escanaba (riv.) B2
False Detour (chan.) F3
Father Marquette Nat'l Mem. E3
Fawn (riv.) D7
Fence (riv.) A2
Firesteel (riv.) G1
Fletcher (pond) F4
Flint (riv.) F5
Ford (riv.) B2
Forty Mile (pt.) F3
Fourteen Mile (pt.) F1
Garden (isl.) D3
Garden (pen.) C3
Glen (lake) C4
Gogebic (lake) F2
Good Harbor (bay) D3
Government (peak) F1
Grand (isl.) C2
Grand (lake) F3
Grand (riv.) D6
Grand Traverse (bay) D3
Granite (pt.) B2
Green (bay) B4
Gun (lake) D6
Hamlin (lake) C4
Higgins (lake) E4
High (isl.) D3
Hog (isl.) D3
Houghton (lake) E4
Hubbard (lake) F4
Huron (bay) A2
Huron (lake) G4
Huron (riv.) F6
Huron River (pt.) B2
Independence (lake) B2
Indian (lake) C2
Isle Royale (isl.) D1
Isle Royale Nat'l Park E1
Kalamazoo (riv.) C6
Keweenaw (bay) A1
Keweenaw (pt.) B1
K.I. Sawyer A.F.B. 1,443 B2
L'Anse Ind. Res. A2
Laughing Fish (pt.) B2
Leelanau (lake) D4
Light House (pt.) D3
Little Bay de Noc (bay) B3
Little Girl (pt.) E1
Little Sable (pt.) C5
Little Summer (isl.) C3
Little Traverse (bay) D3
Long (lake) F3
Lookingglass (riv.) E6
Mackinac (isl.) E3
Mackinac (str.) E3
Manistee (riv.) C4
Manistique (riv.) D2
Manistique (riv.) C2
Manitou (isl.) B1
Maple (riv.) E5
Margrethe (lake) E4
Marquette (isl.) E3
Maumee (bay) F7
Menominee (riv.) B3
Michigamme (lake) A2
Michigamme (res.) B2
Michigamme (riv.) A2
Michigan (lake) B5
Mill (creek) G5
Millecoquins (lake) D2
Misery (bay) G1
Misery (riv.) G1
Montreal (riv.) F1
Mullett (lake) E3
Munuscong (lake) E2
Muskegon (riv.) C5
Neebish (isl.) E2
Net (riv.) G2
Ninemile (pt.) E3
North (chan.) F2
North (pt.) F3
North Fox (isl.) D3
North Manitou (isl.) C3
Oak (pt.) F5
Ontonagon (riv.) G1
Ontonagon Ind. Res. F1
Otsego (lake) E4
Paint (riv.) A2
Paradise (lake) E3
Passage (isl.) E1
Patterson (pt.) D3
Paw Paw (riv.) C6
Peninsula (pt.) C3
Perch (lake) G2
Perch (riv.) G2
Pere Marquette (riv.) D5
Pictured Rocks (cliff) C2
Pictured Rocks Nat'l Lakeshore C2
Pigeon (riv.) D7
Pigeon (riv.) E3
Pine (riv.) D4
Pine (riv.) E5
Platte (lake) C4
Porcupine (mts.) F1
Potagannissing (bay) F2
Poverty (isl.) C3
Prairie (riv.) D7
Presque Isle (riv.) F1
Rabbit (riv.) D6
Raisin (riv.) F7
Rapid (riv.) B2
Reedsburg (res.) E4
Rifle (riv.) E4
Royale (isl.) E1
Saginaw (bay) F5
Saginaw (riv.) F5
Saint Clair (lake) G6
Saint Clair (riv.) G6
Saint Joseph (riv.) C7
Saint Martin (bay) E3
Saint Martin (isl.) C3
Saint Marys (riv.) E2
Salt (pt.) E2
Sand (pt.) F5
Seul Choix (pt.) D3
Shelldrake (riv) D2
Shiawassee (riv.) E5
Siskiwit (bay) E1
Sleeping Bear Dunes Nat'l Lakeshore C4
South (bay) C2
South (pt.) F4
South (chan.) E3
South Fox (isl.) D3
South Manitou (isl.) C3
Sturgeon (riv.) C2
Sugar (isl.) E2
Summer (isl.) C3
Superior (lake) C2
Tahquamenon (falls) D2
Tahquamenon (riv.) D2
Tawas (lake) F4
Tawas (pt.) F4
Thunder (bay) F4
Thunder Bay (riv.) F3
Tittabawassee (riv.) E5
Torch (lake) D3
Traverse (isl.) A1
Traverse (pt.) A1
Turtle (lake) F4
Two Hearted (riv.) D2
Vieux Desert (lake) G2
Walloon (lake) E3
White (riv.) C5
Whitefish (bay) E2
Whitefish (pt.) E2
Whitefish (riv.) C2
Wood (isl.) C2
Wurtsmith A.F.B. F4
Yellow Dog (riv.) B2

▲County Seat
•Population of town or township

Agriculture, Industry and Resources

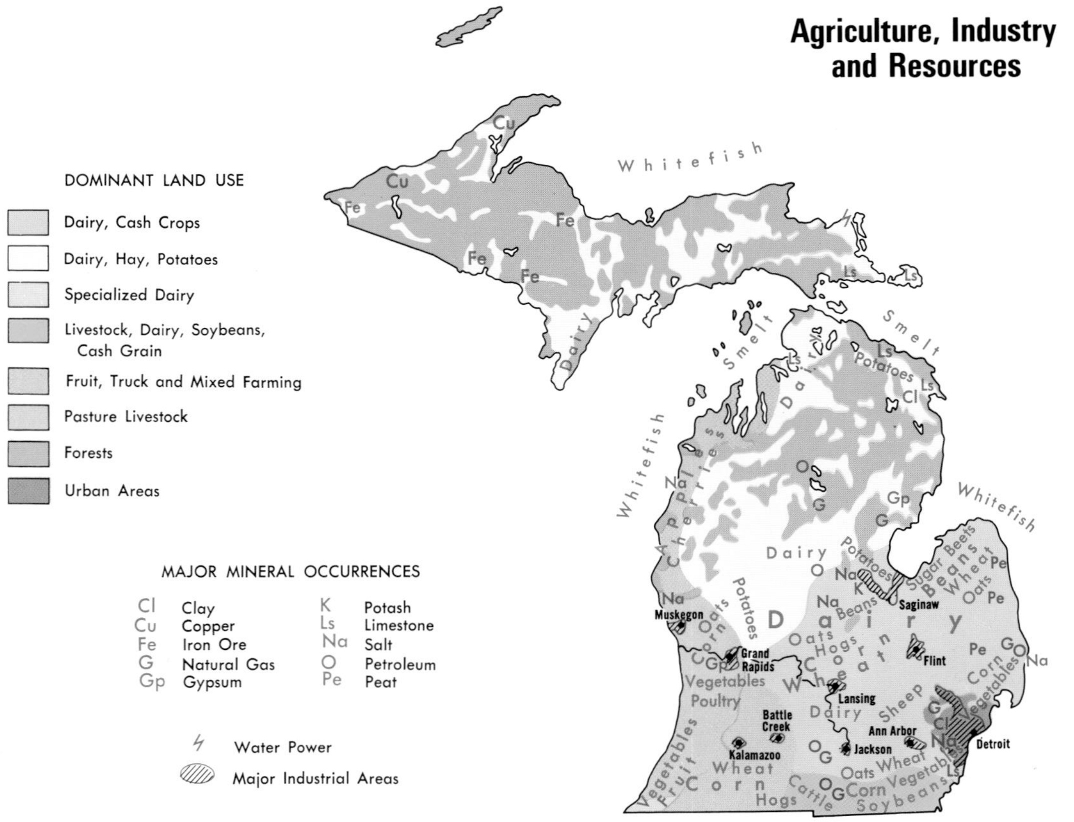
DOMINANT LAND USE
Dairy, Cash Crops
Dairy, Hay, Potatoes
Specialized Dairy
Livestock, Dairy, Soybeans, Cash Grain
Fruit, Truck and Mixed Farming
Pasture Livestock
Forests
Urban Areas
MAJOR MINERAL OCCURRENCES
Cl Clay
Cu Copper
Fe Iron Ore
G Natural Gas
Gp Gypsum
K Potash
Ls Limestone
Na Salt
O Petroleum
Pe Peat
Water Power
Major Industrial Areas
Whitefish
Smelt
Dairy
Potatoes
Sugar Beets
Beans
Wheat
Oats
Corn
Vegetables
Poultry
Fruit
Hogs
Cattle
Soybeans
Sheep
Muskegon
Grand Rapids
Saginaw
Flint
Lansing
Battle Creek
Kalamazoo
Jackson
Ann Arbor
Detroit

AREA 84,402 sq. mi. (218,601 sq. km.)
POPULATION 4,919,479
CAPITAL St. Paul
LARGEST CITY Minneapolis
SETTLED IN 1805
HIGHEST POINT Eagle Mtn. 2,301 ft. (701 m.)
ADMITTED TO UNION May 11, 1858
POPULAR NAME North Star State; Gopher State
STATE FLOWER Pink & White Lady's-Slipper
STATE BIRD Common Loon

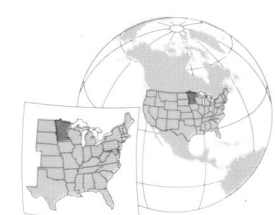

COUNTIES

Aitkin 15,301	E4
Anoka 298,084	E5
Becker 30,000	C4
Beltrami 39,650	C2
Benton 34,226	D5
Big Stone 5,820	B5
Blue Earth 55,941	D6
Brown 26,911	D6
Carlton 31,671	F4
Carver 70,205	E6
Cass 27,150	D4
Chippewa 13,088	C5
Chisago 41,101	F5
Clay 51,229	B4
Clearwater 8,423	C3
Cook 5,168	H3
Cottonwood 12,167	C6
Crow Wing 55,099	D4
Dakota 355,904	E6
Dodge 17,731	F7
Douglas 32,821	C5
Faribault 16,181	D7
Fillmore 21,122	F7
Freeborn 32,584	E7
Goodhue 44,127	F6
Grant 6,289	B5
Hennepin 1,116,200	E5
Houston 19,718	G7
Hubbard 18,376	D3
Isanti 31,287	E5
Itasca 43,992	E3
Jackson 11,268	C7
Kanabec 14,996	E5
Kandiyohi 41,203	C5
Kittson 5,285	B2
Koochiching 14,355	E2
Lac qui Parle 8,067	B6
Lake 11,058	G3
Lake of the Woods 4,522	D2
Le Sueur 25,426	E6
Lincoln 6,429	B6
Lyon 25,425	C6
Mahnomen 5,190	C3
Marshall 10,155	B2
Martin 21,802	D7
McLeod 34,898	D6
Meeker 22,644	D5
Mille Lacs 22,330	E5
Morrison 31,712	D4
Mower 38,603	F7
Murray 9,165	C6
Nicollet 29,771	D6
Nobles 20,832	C7
Norman 7,442	B3
Olmsted 124,277	F7
Otter Tail 57,159	C4
Pennington 13,584	B2
Pine 26,530	F4
Pipestone 9,895	B6
Polk 31,369	B3
Pope 11,236	C5
Ramsey 511,035	E5
Red Lake 4,299	B3
Redwood 16,815	C6
Renville 17,154	C6
Rice 56,665	E6
Rock 9,721	B7
Roseau 16,338	C2
Saint Louis 200,528	F3
Scott 89,498	E6
Sherburne 64,417	E5
Sibley 15,356	D6
Stearns 133,166	D5
Steele 33,680	E7
Stevens 10,053	B5
Swift 11,956	C5
Todd 24,426	D4
Traverse 4,134	B5
Wabasha 21,610	F6
Wadena 13,713	D4
Waseca 19,526	E6
Washington 201,130	F5
Watonwan 11,876	D7
Wilkin 7,138	B4
Winona 49,985	G6
Wright 89,986	D5
Yellow Medicine 11,080	B6

CITIES and TOWNS

Ada▲ 1,657	B3
Adams 800	F7
Adrian 1,234	C7
Afton 2,839	F6
Aitkin▲ 1,984	E4
Akeley 412	D4
Albany 1,796	D5
Alberta 142	B5
Albert Lea▲ 18,356	E7
Albertville 3,621	E5
Alborn• 399	F4
Alden 652	E7
Alexandria▲ 8,820	C5
Alpha 126	D7
Altura 417	F6
Alvarado 371	B2
Amboy 575	D7
Andover 26,588	E5
Annandale 2,684	D5
Anoka▲ 18,076	E5
Apple Valley 45,527	G6
Appleton 2,871	C5
Argyle 656	B2
Arlington 2,048	D6
Arnold 3,032	F4
Ashby 472	C4
Askov 368	F4
Atwater 1,079	D5
Audubon 445	C4
Aurora 1,850	F3
Austin▲ 23,314	E7
Avoca 146	C7
Avon 1,242	D5
Babbitt 1,670	G3
Backus 311	D4
Badger 470	B2
Bagley▲ 1,235	C3
Balaton 637	C6
Barnesville 2,173	B4
Barnum 525	F4
Barrett 355	B5
Battle Lake 686	C4
Baudette▲ 1,104	D2
Baxter 5,555	D4
Bayport 3,162	F5
Beardsley 262	B5
Beaver Creek 250	B7
Becker 2,673	E5
Belgrade 750	C5
Bellechester 172	F6
Belle Plaine 3,789	E6
Bellingham 205	B5
Belview 412	C6
Bemidji▲ 11,917	D3
Benson▲ 3,376	C5
Bertha 470	C4
Bethel 443	E5
Bigelow 231	C7
Big Falls 264	E2
Bigfork 469	E3
Big Lake 6,063	E5
Bingham Lake 167	C7
Bird Island 1,195	D6
Biwabik 954	F3
Blackduck 696	D3
Blaine 44,942	G5
Blomkest 186	D6
Blooming Prairie 1,933	E7
Bloomington 85,172	G6
Blue Earth▲ 3,621	D7
Bluffton 210	C4
Bovey 662	E3
Bowlus 260	D5
Boyd 210	C6
Braham 1,276	E5
Brainerd▲ 13,178	D4
Branch 2,400	F5
Brandon 450	C5
Breckenridge▲ 3,559	B4
Breezy Point 979	D4
Brewster 502	C7
Bricelyn 379	E7
Brooklyn Center 29,172	G5
Brooklyn Park 67,388	G5
Brook Park 156	F5
Brooks 141	B3
Brooten 649	C5
Browerville 735	D4
Brownsdale 718	F7
Browns Valley 690	B5
Brownsville 517	G7
Brownton 807	D6
Buckman 208	D5
Buffalo▲ 10,097	E5
Buffalo Lake 768	D6
Buhl 983	F3
Burnsville 60,220	E6
Burtrum 146	D5
Butterfield 564	D7
Byron 3,500	F6
Caledonia▲ 2,965	G7
Callaway 200	C3
Calumet 383	E3
Cambridge▲ 5,520	E5
Campbell 241	B4
Canby 1,903	B6
Cannon Falls 3,795	F6
Canton 343	F7
Carlos 329	C5
Carlton▲ 810	F4
Carver 1,266	E6
Cass Lake 860	D3
Cedar Mills 53	D6
Center City▲ 582	F5
Centerville 3,202	E5
Ceylon 413	D7
Champlin 22,193	G5
Chandler 276	C7
Chanhassen 20,321	F6
Chaska▲ 17,449	F6
Chatfield 2,394	F7
Chisago City 2,622	E5
Chisholm 4,960	E3
Chokio 443	B5
Circle Pines 4,663	G5
Clara City 1,393	C6
Claremont 620	E6
Clarissa 609	C4
Clarkfield 944	C6
Clarks Grove 734	E7
Clearbrook 551	C3
Clear Lake 266	E5
Clearwater 858	D5
Clements 191	D6
Cleveland 673	E6
Climax 243	B3
Clinton 453	B5
Clontarf 173	C5
Cloquet 11,201	F4
Coates 163	E6
Cokato 2,727	D5
Cold Spring 2,975	D5
Coleraine 1,110	E3
Cologne 1,012	E6
Columbia Heights 18,520	G5
Comfrey 367	D6
Cook 622	F3
Coon Rapids 61,607	G5
Corcoran 5,630	F5
Cosmos 582	D6
Cottage Grove 30,582	F6
Cotton• 506	F3
Cottonwood 1,148	C6
Courtland 538	D6
Cromwell 143	F4
Crookston▲ 8,192	B3
Crosby 2,299	D4
Crosslake 1,893	E4
Crystal 22,698	G5
Currie 225	C6
Cuyuna 231	E4
Cyrus 303	C5
Dakota 329	G7
Dalton 258	C4
Danube 529	C6
Darwin 276	D5
Dassel 1,233	D5
Dawson 1,539	B6
Day 4,443	E5
Dayton 4,699	E5
Deephaven 3,853	G5
Deer Creek 328	C4
Deer River 903	E3
Deerwood 590	E4
Delano 3,837	E5
Delavan 223	D7
Dellwood 1,033	F5
Dent 192	C4
Detroit Lakes▲ 7,348	C4
Dilworth 3,001	B4
Dodge Center 2,226	F6
Donnelly 254	B5
Dover 438	F7
Dumont 122	B5
Dundas 547	E6
Dundee 102	C7
Dunnell 197	D7
Eagan 63,557	G6
Eagle Bend 595	D4
Eagle Lake 1,787	E6
East Bethel 10,941	E5
East Grand Forks 7,501	B3
East Gull Lake 978	D4
Easton 229	E7
Echo 278	C6
Eden Prairie 54,901	G6
Eden Valley 866	D5
Edgerton 1,033	B7
Edina 47,425	G5
Eitzen 229	G7
Elba 214	F6
Elbow Lake▲ 1,275	B5
Elgin 826	F6
Elizabeth 172	B4
Elko 472	E6
Elk River▲ 16,447	E5
Elkton 149	F7
Ellendale 590	E7
Ellsworth 540	C7
Elmore 735	D7
Elrosa 166	C5
Ely 3,724	G3
Elysian 486	E6
Emily 847	E4
Emmons 432	E7
Erhard 150	B4
Erskine 437	B3
Esko 500	F4
Evansville 566	C4
Eveleth 3,865	F3
Excelsior 2,393	G5
Eyota 1,644	F7
Fairfax 1,295	D6
Fairmont▲ 10,889	D7
Falcon Heights 5,572	G5
Faribault▲ 20,818	E6
Farmington 12,365	E6
Federal Dam 101	D3
Felton 216	B3
Fergus Falls▲ 13,471	B4
Fertile 893	C3
Fifty Lakes 392	D4
Finlayson 314	F4
Fisher 435	B3
Flensburg 244	D5
Floodwood 503	F4
Florenton 635	F3
Foley▲ 2,154	D5
Forada 197	C5
Forest Lake 6,798	F5
Foreston 389	E5
Fosston 1,575	F7
Fountain 343	F7
Foxhome 143	B4
Franklin 498	D6
Frazee 1,377	C4
Freeborn 305	E7
Freeport 451	D5
Fridley 27,449	G5
Frost 251	D7
Fulda 1,283	C7
Garfield 281	C5
Gary 215	B3
Gaylord▲ 2,279	D6
Geneva 449	E7
Ghent 315	C6
Gibbon 808	D6
Gilbert 1,847	F3
Gilman 215	E5
Glen• 442	E4
Glencoe▲ 5,453	D6
Glenville 720	E7
Glenwood▲ 2,594	C5
Glyndon 1,049	B4
Golden Valley 20,281	G5
Gonvick 294	C3
Goodhue 778	F6
Good Thunder 592	D6
Goodview 3,373	G6
Graceville 605	B5
Granada 317	D7
Grand Marais▲ 1,353	G2
Grand Meadow 945	F7
Grand Rapids▲ 7,764	E3
Granite Falls▲ 3,070	C6
Greenbush 784	B2
Greenfield 2,544	F5
Green Isle 334	E6
Greenwald 201	D5
Grey Eagle 335	D5
Grove City 608	D5
Grygla 228	C2
Hackensack 285	D4
Hallock▲ 1,196	A2
Halstad 622	B3
Ham Lake 12,710	E5
Hamburg 538	E6
Hamel 3,096	F5
Hammond 198	F6
Hampton 434	E6
Hancock 717	C5
Hanley Falls 323	C6
Hanover 1,355	E5
Hanska 443	D6
Hardwick 222	B7
Harmony 1,080	F7
Harris 1,121	F5
Hartland 288	E7
Hastings▲ 18,204	F6
Hawley 1,882	B4
Hayfield 1,283	F7
Hayward 246	E7
Hector 1,145	D6
Henderson 910	E6
Hendricks 725	B6
Hendrum 315	B3
Henning 719	C4
Herman 452	B5
Hermantown 7,448	F4
Heron Lake 768	C7

Agriculture, Industry and Resources

DOMINANT LAND USE

- Wheat, General Farming
- Dairy, Livestock
- Dairy, Hay, Potatoes
- Cattle Feed, Hogs
- Livestock, Cash Grain
- Forests
- Swampland, Limited Agriculture
- Urban Areas

MAJOR MINERAL OCCURRENCES

Cl	Clay	Gn	Granite
Fe	Iron Ore	Ls	Limestone
		Mn	Manganese

⚡ Water Power
▨ Major Industrial Areas

(continued on following page)

Topography

| Below Sea Level | 100 m. 328 ft. | 200 m. 656 ft. | 500 m. 1,640 ft. | 1,000 m. 3,281 ft. | 2,000 m. 6,562 ft. | 5,000 m. 16,404 ft. |

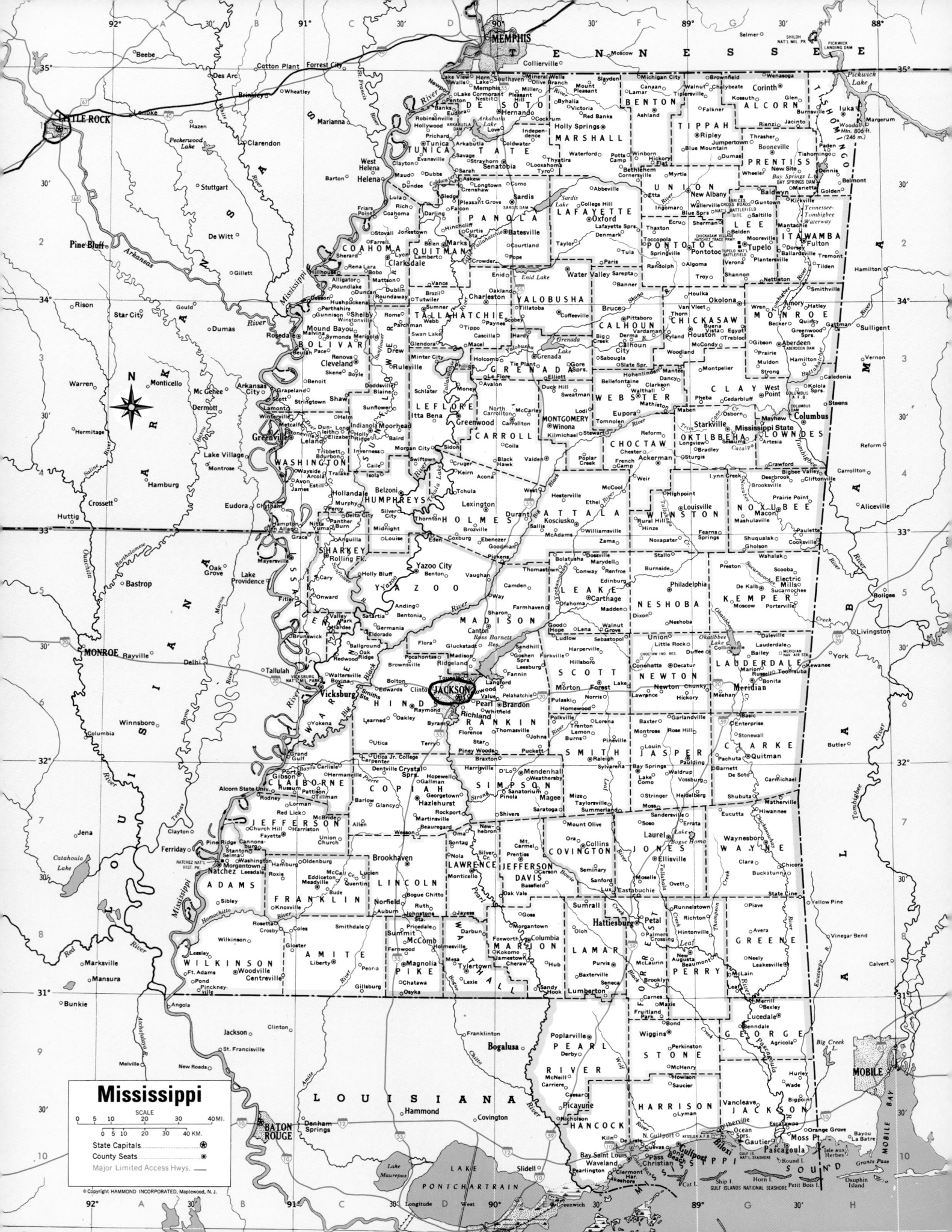

Mississippi

SCALE
0 5 10 20 30 40 MI.
0 5 10 20 30 40 KM.

State Capitals .... ⊛
County Seats .... ⊛
Major Limited Access Hwys. _____

© Copyright HAMMOND INCORPORATED, Maplewood, N.J.

COUNTIES

Adams 34,340 B8
Alcorn 34,558 G1
Amite 13,599 C8
Attala 19,661 E4
Benton 8,026 F1
Bolivar 40,633 C3
Calhoun 15,069 F3
Carroll 10,769 E4
Chickasaw 19,440 G3
Choctaw 9,758 F4
Claiborne 11,831 C7
Clarke 17,955 G6
Clay 21,979 G3
Coahoma 30,622 C2
Copiah 28,757 D7
Covington 19,407 E7
De Soto 107,199 E1
Forrest 72,604 F8
Franklin 8,448 C8
George 19,144 G9
Greene 13,299 G8
Grenada 23,263 E3
Hancock 42,967 E10
Harrison 189,601 F10
Hinds 250,800 D6
Holmes 21,609 D4
Humphreys 11,206 C4
Issaquena 2,274 B5
Itawamba 22,770 H2
Jackson 131,420 G9
Jasper 18,149 F6
Jefferson 9,740 B7
Jefferson Davis 13,962 E7
Jones 64,958 F7
Kemper 10,453 G5
Lafayette 38,744 E2
Lamar 39,070 E8
Lauderdale 78,161 G6
Lawrence 13,258 D7
Leake 20,940 E5
Lee 75,755 G2
Leflore 37,947 D3
Lincoln 33,166 D8
Lowndes 61,586 H4
Madison 74,674 D5
Marion 25,595 E8
Marshall 34,993 E1
Monroe 38,014 H3
Montgomery 12,189 E4
Neshoba 28,684 F5
Newton 21,838 F6
Noxubee 12,548 G4
Oktibbeha 42,902 G4
Panola 34,274 E2
Pearl River 48,621 E9
Perry 12,138 G8
Pike 38,940 D8
Pontotoc 26,726 F2
Prentiss 25,556 G1
Quitman 10,117 D2
Rankin 115,327 E6
Scott 28,423 E6
Sharkey 6,580 C5
Simpson 27,639 E7
Smith 16,182 E6
Stone 13,622 F9
Sunflower 34,369 C3
Tallahatchie 14,903 D3
Tate 25,370 E1
Tippah 20,826 G1
Tishomingo 19,163 H1
Tunica 9,227 D1
Union 25,362 F2
Walthall 15,156 D8
Warren 49,644 C6
Washington 62,977 C4
Wayne 21,216 G7
Webster 10,294 F3
Wilkinson 10,312 B8
Winston 20,160 F4
Yalobusha 13,051 E2
Yazoo 28,149 D5

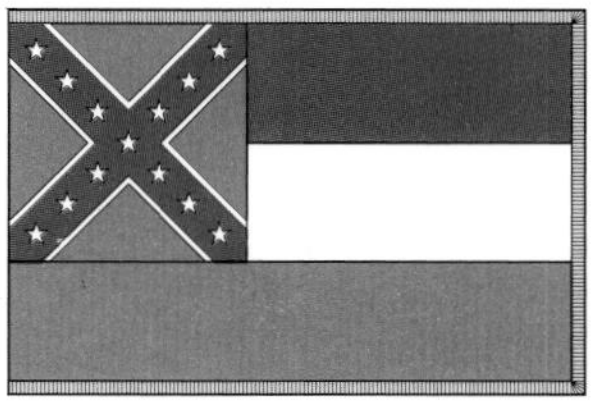

AREA 47,689 sq. mi. (123,515 sq. km.)
POPULATION 2,844,658
CAPITAL Jackson
LARGEST CITY Jackson
HIGHEST POINT Woodall Mtn. 806 ft. (246 m.)
SETTLED IN 1716
ADMITTED TO UNION December 10, 1817
POPULAR NAME Magnolia State
STATE FLOWER Magnolia
STATE BIRD Mockingbird

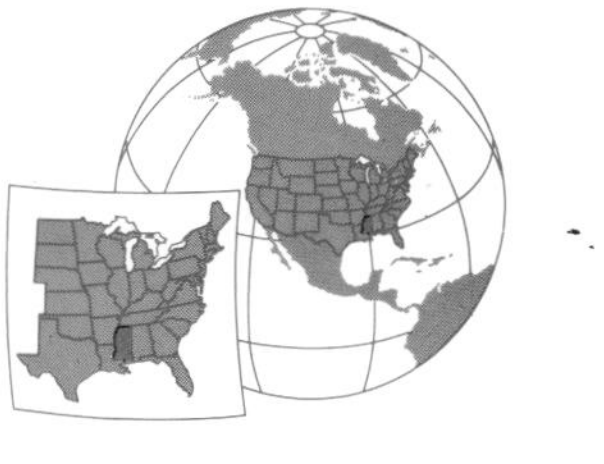

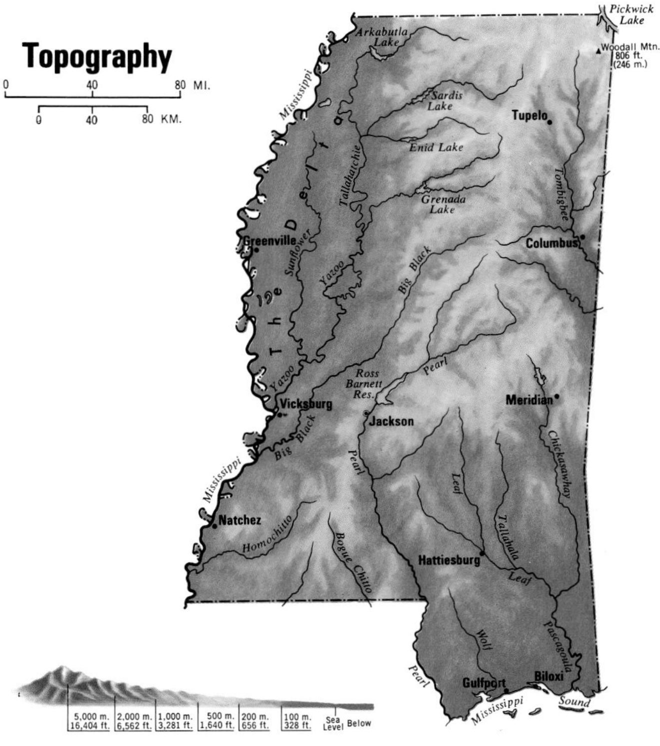

Topography
0 40 80 MI.
0 40 80 KM.
Pickwick Lake
Woodall Mtn. 806 ft. (246 m.)
Arkabutla Lake
Sardis Lake
Enid Lake
Grenada Lake
Tupelo
Columbus
Greenville
Vicksburg
Jackson
Meridian
Natchez
Hattiesburg
Gulfport
Biloxi
Ross Barnett Res.
Mississippi
Tallahatchie
Sunflower
Yazoo
Big Black
Pearl
Tombigbee
Chickasawhay
Leaf
Tallahala
Homochitto
Bogue Chitto
Wolf
Pascagoula
Sound
The Delta
5,000 m. 16,404 ft.
2,000 m. 6,562 ft.
1,000 m. 3,281 ft.
500 m. 1,640 ft.
200 m. 656 ft.
100 m. 328 ft.
Sea Level
Below

CITIES and TOWNS

Abbeville 423 F2
Aberdeen▲ 6,415 H3
Ackerman▲ 1,696 F4
Acona 200 D4
Agricola 200 G9
Alcorn State University B7
Algoma 508 G2
Alligator 220 C2
Amory 6,956 H3
Anguilla 907 C5
Arcola 563 C4
Arkabutla 400 D1
Artesia 498 G4
Ashland▲ 577 F1
Askew 300 D1
Auburn 500 C8
Avalon 100 D3
Avera 150 G8
Avon 400 B4
Bailey 320 G6
Baird 150 C4
Baldwyn 3,321 G2
Ballardsville 105 H2
Banks 100 D1
Banner 120 F2
Bassfield 315 E8
Batesville▲ 7,113 E2
Baxterville 100 E8
Bay Saint Louis▲ 8,209 F10
Bay Springs▲ 2,097 F7
Beaumont 977 G8
Beauregard 265 D7
Becker 350 G3
Belden 241 G2
Belen 400 D2
Bellefontaine 400 F3
Belmont 1,961 H1
Belzoni▲ 2,663 C4
Benndale 500 G9
Benoit 611 C3
Benton 390 D5
Bentonia 500 D5
Bethlehem 210 F1
Beulah 473 B3
Bexley 130 G9
Big Creek 127 F3
Bigbee Valley 370 H4
Bigpoint 115 H9
Biloxi 50,644 G10
Blue Mountain 670 G1
Blue Springs 144 G2
Bobo 200 C2
Bogue Chitto 533 D8
Bolatusha 87 E5
Bolton 629 D6
Bond 350 F9
Bonita 300 G6
Booneville▲ 8,625 G1
Bourbon 200 C4
Boyle 720 C3
Brandon▲ 16,436 E6
Braxton 181 D6
Brazil 229 D2
Brookhaven▲ 9,861 C7
Brooklyn 450 F8
Brooksville 1,182 G4
Brownfield 125 G1
Brownsville 200 D6
Brozville 150 D4
Bruce 2,097 F3
Brunswick 90 C5
Buckatunna 500 G7
Bude 1,037 C8
Burns 949 E6
Burnsville 1,034 H1
Byhalia 706 E1
Byram 7,386 D6
Caesar 80 E9
Caledonia 1,015 H3
Calhoun City 1,872 F3
Camden 150 E5
Canaan 200 F1
Cannonsburg 240 B7
Canton▲ 12,911 D5
Carlisle 425 C7
Carpenter 200 C6
Carriere 900 E9
Carrollton▲ 408 E4
Carson 400 E7
Carthage▲ 4,637 E5
Cary 427 C5
Cascilla 230 D3
Cedarbluff 175 G3
Centreville 1,680 B8
Chalybeate 150 G1
Charleston▲ 2,198 D2
Chatawa 300 D8
Chatham 150 B4
Cheraw 100 E8
Chunky 344 G6
Church Hill 350 B7
Clara 275 G7
Clarksdale▲ 20,645 D2
Clarkson 100 F3
Clermont Harbor 550 F10
Cleveland▲ 13,841 C3
Cliftonville 280 H4
Clinton 23,347 D6
Coahoma 325 C2
Cockrum 150 E1
Coffeeville▲ 930 E3
Coldwater 1,674 E1
Coles 150 C8
College Hill 150 E2
Collins▲ 2,683 E7
Collinsville 1,823 G6
Columbia▲ 6,603 E8
Columbus▲ 25,944 H3
Como 1,310 E1
Conehatta 997 F6
Corinth▲ 14,054 G1
Courtland 460 E2
Coxburg 300 D5
Crawford 655 G4
Crenshaw 916 D2
Crosby 360 B8
Crowder 766 D2
Cruger 449 D4
Crystal Springs 5,873 D7
Cuevas 200 F10
Curtis Station 350 D2
D'Iberville 7,608 G10
D'Lo 394 E7
Daleville 210 G5
Dancy 116 F3
Darbun 100 D8
Darling 275 D2
De Kalb▲ 972 G5
De Lisle 450 F10
De Soto 150 G7
Decatur▲ 1,426 F6
Delta City 310 C4
Dennis 150 H1
Dentville 175 C7
Derby 298 E9
Derma 1,023 F3
Dixon 125 F5
Doddsville 108 C3
Dorsey 100 H2
Drew 2,434 C3
Dublin 100 C2
Duck Hill 746 E3
Duffee 175 G6
Dumas 452 G1
Duncan 578 C2
Dundee 600 D1
Dunleith 140 C4
Durant 2,932 E4
Eastabuchie 200 F8
Ebenezer 200 D5
Ecru 947 F2
Eden 126 D5
Edinburg 200 F5
Edwards 1,347 C6
Egypt 100 G3
Electric Mills 100 G5
Elizabeth 500 C4
Elliott 200 E3
Ellisville▲ 3,465 F7
Enid E2
Enterprise 474 G6
Errata 85 F7
Escatawpa 3,566 G10
Estill 100 C4
Ethel 452 F4
Eudora 200 D1
Eupora 2,326 F3
Falcon 317 D2
Falkner 212 G1
Fannin 250 E6
Farrell 300 C2
Fayette▲ 2,242 B7
Fernwood 500 D8
Fitler 175 B5
Flora 1,546 D5
Florence 2,396 D6
Flowood 4,750 D6
Forest▲ 5,987 F6
Forkville 185 E6
Foxworth 800 E8
French Camp 393 F4
Friars Point 1,480 C2
Fulton▲ 3,882 H2
Gallman D7
Garlandville 150 F6
Gattman 114 H3
Gautier 11,681 G10
Georgetown 344 D7
Glen 286 H1
Glen Allan 650 B4
Glendora 285 D3
Gloster 1,073 B8
Gluckstadt 150 D5
Golden 201 H2
Good Hope 125 E5
Goodman 1,252 E5
Gore Springs 125 E3
Goshen Springs 100 E6
Goss 100 E8
Grace 325 C5
Grapeland 200 B3
Greenville▲ 41,633 B4
Greenwood▲ 18,425 D4
Greenwood Springs 170 H3
Grenada▲ 14,879 E3
Gulfport▲ 71,127 F10
Gunnison 633 C3
Guntown 1,183 G2
Hamburg 150 B7
Hamilton 500 H3
Hampton 200 B4
Hardee 100 C5
Harperville 200 E6
Harriston 500 C7
Harrisville 500 D7
Hatley 476 H3
Hattiesburg▲ 44,779 F8
Hazlehurst▲ 4,400 D2
Heidelberg 840 F7
Helm 80 C4
Hermanville 750 C7
Hernando▲ 6,812 E1
Hickory 499 F6
Hickory Flat 565 F1
Hillsboro 800 E6
Hintonville 300 F8
Hiwannee 250 G7
Hohenlinden 96 F3
Hollandale 3,437 C4
Holly Bluff 700 C5
Holly Ridge 350 C4
Holly Springs▲ 7,957 E1
Hollywood 80 D1
Hopewell 250 D7
Horn Lake 14,099 D1
Houlka 710 G2
Houston▲ 4,079 G3
Howison 300 F9
Hub 80 E8
Hurley 985 H9
Independence 150 E1
Indianola▲ 12,066 C4
Ingomar 150 F2
Inverness 1,153 C4
Isola 768 C4
Itta Bena 2,208 D4
Iuka▲ 3,059 H1
Jackson (cap.)▲ 184,256 D6
James 100 B4
Jayess 200 D8
Johns 90 E6
Jonestown 1,701 D2
Jumpertown 404 G1
Kewanee 250 H6
Kilmichael 830 E4
Kiln 2,040 F10
Kirkville 200 H2
Kokomo 250 D8
Kolola Springs 100 H3
Kosciusko▲ 7,372 E4
Kossuth 170 G1
Lafayette Springs 80 F2
Lake 408 F6
Lake Como 150 F7
Lake Cormorant 300 D1
Lakeshore 550 F10
Lake View 125 D1
Lamar 200 F1
Lambert 1,967 D2
Lamont 400 B3
Langford 100 E6
Lauderdale 600 G5
Laurel▲ 18,393 F7
Lawrence 250 F6
Le Flore 99 D3
Leaf 250 G8
Leakesville▲ 1,026 G8
Learned 50 C6
Leland 5,502 C4
Lemon 90 E6
Lena 167 E5
Lessley 100 B8
Lexington▲ 2,025 D4
Liberty▲ 633 C8
Long 15,804 C4
Long Beach 17,320 F10
Longtown 150 D1
Longview 800 G4
Looxahoma 200 E1
Lorena 90 F6
Lorman 350 B7
Louin 339 F6
Louise 315 C5
Louisville▲ 7,006 G4
Lucedale▲ 2,458 G9
Ludlow 350 E5
Lula 370 C2
Lumberton 2,228 E8
Lyman 1,081 F10
Lyon 418 D2
Maben 803 F3
Macon▲ 2,461 G4
Madden 450 F5
Madison 14,692 D6
Magee 4,200 E7
Magnolia▲ 2,071 D8
Malvina 100 C3
Mantachie 1,107 H2
Mantee 169 F3
Marietta 248 H2
Marion 1,305 G6
Marks▲ 1,551 D2
Marydell 99 F5
Mashulaville 227 G4
Matherville 150 G7
Mathiston 720 F3
Mattson 200 C2
Maxie 233 F9
Mayersville▲ 795 B5
Mayhew 150 G4
McAdams 350 E4
McCall Creek 250 C7
McCarley 250 E3
McComb 13,337 D8
McCondy 150 G3
McCool 182 F4
McHenry 660 F9
McLain 603 G8
McLaurin 100 F8
McNeill 800 E9
Meadville▲ 519 C8
Meehan 100 G6
Mendenhall▲ 2,555 E7
Meridian▲ 39,968 G6
Merigold 664 C3
Merrill 100 G9
Metcalfe 1,109 B4
Michigan City 350 F1
Midnight 500 C4
Mineral Wells 250 E1
Minter City 150 D3
Mississippi State G4
Mize 285 E7
Money 350 D3
Monticello▲ 1,726 D7
Montpelier 175 G3
Montrose 127 F6
Mooreville 200 G2
Moorhead 2,573 C4
Morgan City 305 D4
Morgantown 32,880 B7
Morgantown 325 E8
Morton 3,482 E6
Moselle 525 F8
Moss 17,837 F7
Moss Point 15,851 G10
Mound Bayou 2,102 C3
Mount Olive 893 E7
Mount Pleasant 250 E1
Murphy 100 C4
Myrtle 407 F1
Natchez▲ 18,464 B7
Neely 270 G8
Nesbit 366 D1
Neshoba 250 F5
Nettleton 1,932 G2
New Albany▲ 7,607 G2
New Augusta▲ 715 F8
Newhebron 447 D7
New Houlka (Houlka) 710 G2
New Site 100 H1
Newton 3,699 F6
Nicholson 400 E10
Nitta Yuma 150 C4
Nola 120 D7
North Carrollton 499 E3
North Gulfport 4,966 F10
Noxapater 419 F5
Oakland 586 E2
Oakley 133 D6
Oak Ridge 350 C6
Oak Vale E8
Ocean Springs 17,225 G10
Ofahoma 350 E5
Okolona▲ 3,056 G2
Olive Branch 21,054 E1
Oloh 93 E8
Oma 200 D7
Ora 15,676 E7
Orange Grove H10
Osyka 481 D8
Ovett 600 F8
Oxford▲ 11,756 F2
Pace 364 C3
Pachuta 245 G6
Paden 106 H1
Palmers Crossing 2,765 F8
Panther Burn 300 C4
Parchman 200 D3
Paris 253 F2
Pascagoula▲ 26,200 G10
Pass Christian 6,579 F10
Pattison 540 C7
Paulding▲ 630 F6
Paulette 230 H4
Paynes 100 D3
Pearl 21,961 D6
Pearlington 1,684 E10
Pelahatchie 1,461 E6
Penton 175 D1
Peoria 100 C8
Perkinston 950 F9
Petal 7,579 F8
Pheba 280 G3
Philadelphia▲ 7,303 F5
Philipp 975 D3
Piave 150 G8
Picayune 10,535 E9
Pickens 1,325 E5
Pine Ridge 175 B7
Pineville 80 F6
Piney Woods 450 D6
Pinola E7
Pittsboro▲ 212 F3
Plantersville 1,144 G2
Pleasant Grove 100 D2
Pleasant Hill 400 E1
Polkville 132 E6

(continued on following page)

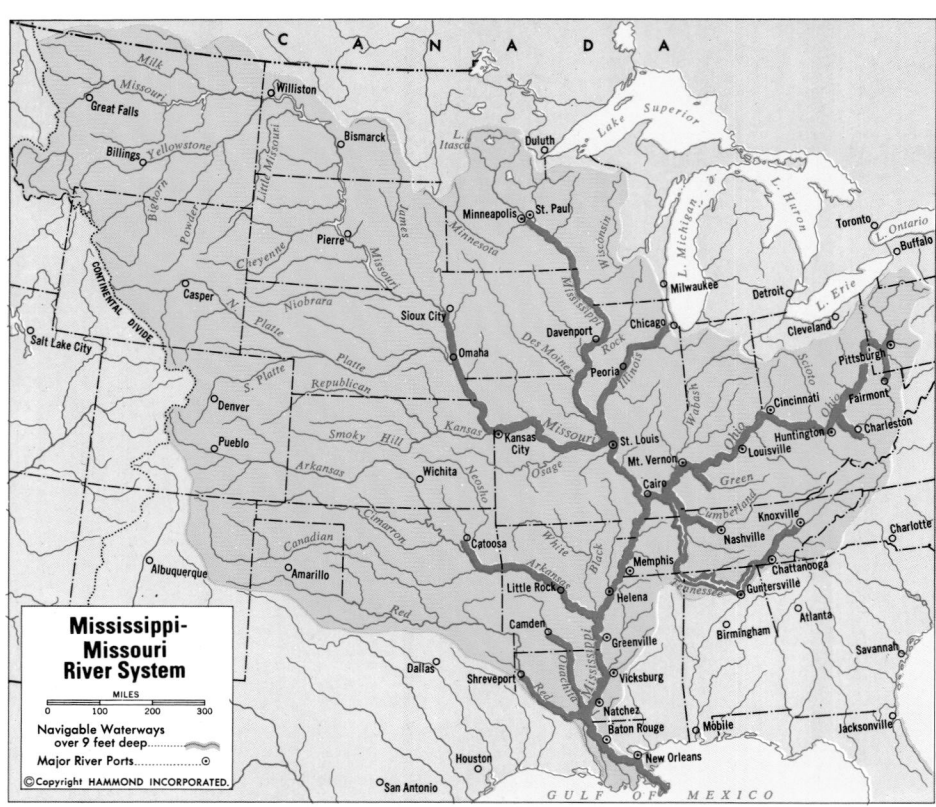

Mississippi-Missouri River System

MILES
0 100 200 300

Navigable Waterways over 9 feet deep........
Major River Ports...............⊚

©Copyright HAMMOND INCORPORATED.

Agriculture, Industry and Resources

DOMINANT LAND USE

- Specialized Cotton
- Cotton, Livestock
- Cotton, General Farming
- Cotton, Forest Products
- Truck and Mixed Farming
- Forests
- Swampland, Limited Agriculture

MAJOR MINERAL OCCURRENCES

- Cl Clay
- Fe Iron Ore
- G Natural Gas
- O Petroleum
- ⊘ Major Industrial Areas

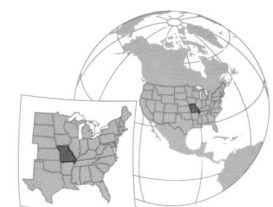

AREA 69,697 sq. mi. (180,515 sq. km.)
POPULATION 5,595,211
CAPITAL Jefferson City
LARGEST CITY St. Louis
HIGHEST POINT Taum Sauk Mtn. 1,772 ft. (540 m.)
SETTLED IN 1764
ADMITTED TO UNION August 10, 1821
POPULAR NAME Show Me State
STATE FLOWER Hawthorn
STATE BIRD Bluebird

COUNTIES

Adair 24,977	G2
Andrew 16,492	C3
Atchison 6,430	B2
Audrain 25,853	J4
Barry 34,010	E9
Barton 12,541	D7
Bates 16,653	D6
Benton 17,180	F6
Bollinger 12,029	M8
Boone 135,454	H4
Buchanan 85,998	C3
Butler 40,867	M9
Caldwell 8,969	E3
Callaway 40,766	J5
Camden 37,051	G6
Cape Girardeau 68,693	N8
Carroll 10,285	F4
Carter 5,941	L9
Cass 82,092	D5
Cedar 13,733	E7
Chariton 8,438	F3
Christian 54,285	F9
Clark 7,416	J2
Clay 184,006	D4
Clinton 18,979	D3
Cole 71,397	H6
Cooper 16,670	G5
Crawford 22,804	K7
Dade 7,923	E8
Dallas 15,661	F7
Daviess 8,016	E3
De Kalb 11,597	D3
Dent 14,927	J7
Douglas 13,084	G9
Dunklin 33,155	M1
Franklin 83,807	K6
Gasconade 15,342	J6
Gentry 6,861	D2
Greene 240,391	F8
Grundy 10,432	E2
Harrison 8,850	E2
Henry 21,997	E6
Hickory 8,940	F7
Holt 5,351	B2
Howard 10,212	G4
Howell 37,238	J9
Iron 10,697	L7
Jackson 654,880	R5
Jasper 104,686	D8
Jefferson 198,099	L6
Johnson 48,258	E5
Knox 4,361	H2
Laclede 32,513	G7
Lafayette 32,960	E4
Lawrence 35,204	E8
Lewis 10,494	J2
Lincoln 38,944	L4
Linn 13,754	F3
Livingston 14,558	E3
Macon 15,762	G3
Madison 11,800	M8
Maries 8,903	J6
Marion 28,289	J3
McDonald 21,681	D9
Mercer 3,757	E2
Miller 23,564	H6
Mississippi 13,427	O9
Moniteau 14,827	G5
Monroe 9,311	H3
Montgomery 12,136	K5
Morgan 19,309	G6
New Madrid 19,760	N9
Newton 52,636	D9
Nodaway 21,912	C2
Oregon 10,344	K9
Osage 13,062	J6
Ozark 9,542	H9
Pemiscot 20,047	N1
Perry 18,132	N7
Pettis 39,403	F5
Phelps 39,825	J7
Pike 18,351	K4
Platte 73,781	C4
Polk 26,992	F7
Pulaski 41,165	H7
Putnam 5,223	F2
Ralls 9,626	J3
Randolph 24,663	G3
Ray 23,354	E4
Reynolds 6,689	L8
Ripley 13,509	L9
Saint Charles 283,883	M2
Saint Clair 9,652	E6
Sainte Genevieve 17,842	M7
Saint Francois 55,641	M7
Saint Louis 1,016,315	O3
Saint Louis (city county) 348,189	P3
Saline 23,756	F4
Schuyler 4,170	G2
Scotland 4,983	H2
Scott 40,422	N8
Shannon 8,324	K8
Shelby 6,799	H3
Stoddard 29,705	N9
Stone 28,658	F9

CITIES and TOWNS

Sullivan 7,219	F2
Taney 39,703	F9
Texas 23,003	J8
Vernon 20,454	D7
Warren 24,525	K5
Washington 23,344	L7
Wayne 13,259	L8
Webster 31,045	G8
Worth 2,382	D2
Wright 17,955	H8
Adrian 1,780	D6
Advance 1,244	N8
Affton 20,535	P4
Agency 599	C3
Alba 588	D8
Albany▲ 1,937	D2
Allenton 800	M4
Alma 399	E4
Altamont 218	D3
Altenburg 309	O7
Alton▲ 668	K9
Amazonia 277	C3
Amoret 211	C6
Amsterdam 281	D6
Anderson 1,856	D9
Annapolis 310	L8
Anniston 285	O9
Appleton City 1,314	D6
Arbyrd 528	M10
Arcadia 567	L7
Archie 890	D5
Argyle 164	J6
Armstrong 287	G4
Arnold 19,965	M6
Asbury 218	C8
Ash Grove 1,430	E8
Ashland 1,869	H5
Atlanta 450	H3
Augusta 218	L5
Aurora 7,014	E9
Auxvasse 901	J4
Ava▲ 3,021	G9
Avondale 529	P5
Bakersfield 285	H9
Ballwin 31,283	N3
Baring 159	H2
Barnard 257	C2
Barnett 207	G6
Bates City 245	E5
Battlefield 2,385	F8
Beaufort 175	K6
Belgrade 350	L7
Bell City 461	N8
Bella Villa 687	R4
Belle 1,344	J6
Bellefontaine 10,922	N2
Bellefontaine Neighbors 11,271	R2
Bellflower 427	K4
Bel-Nor 1,598	P2
Bel-Ridge 3,082	P2
Belton 21,730	C5
Benton▲ 732	O8
Benton City 122	J4
Berkeley 10,063	P2
Bernie 1,777	M9
Bertrand 740	O9
Bethany▲ 3,087	E2
Bevier 723	G3
Billings 1,091	F8
Birch Tree 634	K9
Birmingham 214	R5
Bismarck 1,470	L7
Bixby 500	K7
Black 6,128	L7
Black Jack 6,792	R1
Blackburn 284	F4
Blackwater 1991	G5
Bland 565	J6
Blodgett 265	O8
Bloomfield▲ 1,952	M9
Bloomsdale 419	M6
Blue Springs 48,080	R6
Bogard 234	E4
Bolckow 234	C2
Bolivar▲ 9,143	F7
Bonne Terre 4,039	L7
Boonville▲ 8,202	G5
Bosworth 382	F4
Bourbon 1,348	K6
Bowling Green▲ 3,260	K4
Braggadocio 400	N10
Branson 6,050	F9
Brashear 280	H2
Braymer 910	E3
Breckenridge 454	E3
Breckenridge Hills 4,817	O2
Brentwood 7,693	P3
Bridgeton 15,550	O2
Bridgeton Terrace 334	O2
Bronaugh 245	C7
Brookfield 4,769	F3
Browning 317	F2
Brunswick 925	F4
Bucklin 524	G3

Buckner 2,725	R5
Buffalo▲ 2,781	F7
Bunceton 348	G5
Bunker 427	K8
Burlington Junction 632	B2
Butler▲ 4,209	D6
Butterfield 397	H8
Cabool 2,168	H8
Cainsville 370	E2
Cairo 293	H4
Caledonia 158	L7
Calhoun 491	E6
California▲ 4,005	H5
Callao 291	G3
Calverton Park 1,322	P2
Camden 209	D4
Camden Point 484	C4
Camdenton▲ 2,779	G6
Cameron 8,312	D3
Campbell 1,883	M9
Canalou 348	N8
Canton 2,557	J2
Cape Girardeau 35,349	O8
Cardwell 789	M10
Carl Junction 5,294	C8
Carrollton▲ 4,122	E4
Carterville 1,850	D8
Carthage▲ 12,668	D8
Caruthersville▲ 6,760	N10
Carytown 217	D8
Cassville▲ 2,890	E9
Catawissa 250	L6
Cedar City 427	H5
Cedar Hill Lakes 229	L6
Center 644	J3
Centertown 257	H5
Centerview 249	E5
Centerville▲ 171	L8
Centralia 3,774	H4
Chaffee 3,044	N8
Chamois 456	J5
Charlack 1,431	P2
Charleston▲ 4,732	O9
Chesterfield 37,991	N2
Chilhowee 329	E5
Chillicothe▲ 8,968	E3

Chula 198	F3
Circle City 154	N9
Clarence 915	H3
Clark 275	H4
Clarksburg 375	G5
Clarksdale 351	D3
Clarkson Valley 2,675	N3
Clarksville 490	K4
Clarkton 1,330	M10
Claycomo 1,267	P5
Clayton▲ 12,825	P3
Clearmont 191	C1
Cleveland 592	C5
Clever 1,010	F8
Clinton▲ 9,311	E6
Cobalt City 189	M7
Coffey 140	E2
Cole Camp 1,028	F6
Columbia 84,531	H5
Commerce 110	O8
Conception Junction 202	C2
Concord 16,689	P4
Concordia 2,360	E5
Conway 743	G7
Cool Valley 1,081	P2
Cooter 440	N10
Corder 427	E4
Cottleville 1,928	M2
Cottleville 4,247	M6
Country Club Village 1,846	C3
Cowgill 247	E3
Craig 309	B2
Crane 1,390	E9
Creighton 322	D6
Crestwood 11,863	O3
Creve Coeur 16,500	O2
Crocker 1,033	H7
Cross Timbers 185	F6
Crystal City 4,247	M6
Crystal Lake Park 457	O3
Cuba 3,230	K6
Curryville 251	K4
Dadeville 224	E8
Dearborn 529	C3
Deepwater 507	E6
De Kalb 257	C3
Dellwood 5,255	R2

Delta 517	N8
Des Arc 187	L8
Desloge 4,802	M7
De Soto 6,375	L6
Des Peres 8,592	O3
Dexter 7,356	N9
Diamond 807	D9
Diehlstadt 163	N9
Diggins 298	G8
Dixon 1,570	H6
Doniphan▲ 1,932	L9
Doolittle 644	J7
Downing 396	H2
Drexel 1,090	C6
Dudley 289	M9
Duenweg 1,034	D8
Duquesne 1,640	D8
East Lynne 300	D5
East Prairie 3,227	O9
Easton 258	C3
Edgar Springs 190	J7
Edgerton 533	C3
Edina▲ 1,233	H2
Edmundson 840	O2
Eldon 4,895	G6
El Dorado Springs 3,775	E7
Ellington 1,045	L8
Ellisville 9,104	M3
Ellsinore 363	L9
Elmo 166	B1
Elsberry 2,047	L4
Elvins 1,391	L7
Eminence▲ 548	K8
Emma 243	F5
Eolia 435	L4
Essex 524	N9
Esther 1,071	M7
Eugene 141	H6
Eureka 7,676	M4
Everton 322	E8
Ewing 464	J2
Excelsior Springs 10,847	R4
Exeter 707	E9
Fairfax 645	B2
Fair Grove 1,107	F8

Fair Play 418	E7
Fairview 395	D9
Farber 411	J4
Farley 226	O4
Farmington▲ 13,924	M7
Fayette▲ 2,793	G4
Fenton 4,360	O3
Ferguson 22,406	P2
Ferrelview 593	O4
Festus 9,660	M6
Fillmore 211	C2
Fisk 363	M9
Flat 4,823	J7
Flat River 4,443	M7
Fleming 130	E4
Flemington 141	F7
Flinthill 379	L5
Florissant 50,497	P1
Foley 178	L4
Fordland 523	G8
Forest City 338	B3
Forsyth▲ 1,686	F9
Fortuna 175	G6
Foster 130	D6
Frankford 351	K4
Franklin 112	G4
Fredericktown▲ 3,928	M7
Freeburg 423	J6
Freeman 521	C5
Freistatt 166	E8
Fremont 201	K9
Frohna 192	N7
Frontenac 3,483	O3
Fulton▲ 12,128	J5
Gainesville▲ 632	G9
Galena▲ 451	F9
Gallatin▲ 1,789	E3
Galt 275	F2
Garden City 1,500	D5
Gasconade 267	J5
Gerald 1,171	K6
Gideon 1,113	N10
Gilliam 229	F4
Gilman City 380	D2
Gladstone 26,365	P5
Glasgow 1,263	G4

Glenaire 553	R5
Glencoe 400	M3
Glendale 5,767	P3
Glenwood 203	G1
Golden City 884	D8
Goodman 1,183	C9
Gordonville 425	N8
Gower 1,399	C3
Graham 191	C2
Grain Valley 5,160	S6
Granby 2,121	D9
Grandin 236	L9
Grandview 24,881	P6
Grant City▲ 926	D2
Grantwood 883	O4
Gray Summit 2,640	L6
Green Castle 308	G2
Green City 688	F2
Greenfield▲ 1,358	E8
Green Ridge 445	F5
Greentop 427	H2
Greenville▲ 451	M8
Greenwood 3,952	R6
Hale 473	F3
Half Way 176	F7
Hallsville 978	H4
Halltown 161	E8
Hamilton 1,813	E3
Hanley Hills 2,124	P2
Hannibal 17,757	K3
Hardin 614	E4
Harrisburg 184	H4
Harrisonville▲ 8,946	D5
Hartville▲ 607	G8
Hawk Point 459	K5
Haywood City 239	N9
Hazelwood 26,206	P2
Henrietta 457	E4
Herculaneum 2,805	M6
Hermann▲ 2,674	K5
Hermitage▲ 406	F7
Higbee 623	H4
Hayti 3,207	N10
Hayti Heights 771	N10
High Hill 231	K5
Higginsville 4,682	E4

(continued on following page)

(continued on following page)

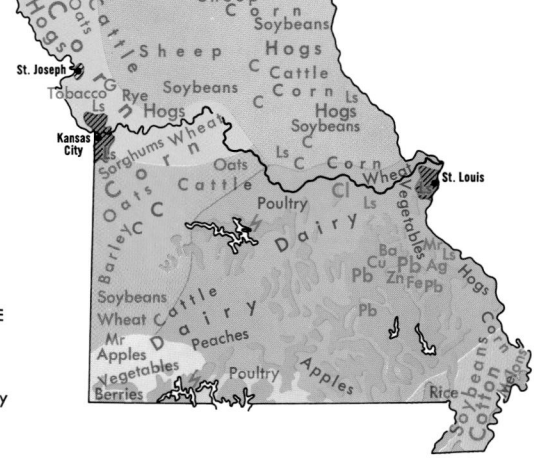

Agriculture, Industry and Resources

DOMINANT LAND USE

- Cattle Feed, Hogs
- Livestock, Cash Grain, Dairy
- Pasture Livestock
- Specialized Cotton
- General Farming, Dairy, Livestock, Poultry
- General Farming, Livestock, Truck Farming, Cotton
- Fruit and Mixed Farming
- Forests
- Urban Areas

MAJOR MINERAL OCCURRENCES

Ag	Silver	G	Natural Gas
Ba	Barite	Ls	Limestone
C	Coal	Mr	Marble
Cl	Clay	Pb	Lead
Cu	Copper	Zn	Zinc
Fe	Iron Ore		

⚡ Water Power　　▨ Major Industrial Areas

High Ridge 4,236 ....M6
Hillsboro▲ 1,675 ....L6
Hillsdale 1,477 ....R2
Holcomb 696 ....N10
Holden 2,510 ....E5
Holland 246 ....N10
Holliday 129 ....H3
Hollister 3,867 ....F9
Holt 337 ....D4
Holts Summit 2,935 ....H5
Homestown 181 ....M6
Hopkins 579 ....C1
Horine 923 ....M6
Hornersville 686 ....M10
Houston▲ 1,992 ....J8
Houstonia 275 ....F5
Houston Lake 284 ....O5
Howardville 342 ....N9
Hughesville 174 ....F5
Humansville 946 ....E7
Hume 237 ....C6
Hunnewell 227 ....J3
Huntleigh 323 ....O3
Huntsville▲ 1,553 ....H4
Hurdland 239 ....H2
Hurricane Deck 210 ....F7
Iberia 605 ....H6
Illmo 1,368 ....O8
Imperial 4,373 ....M6
Independence▲ 113,288 ....R5
Irondale 437 ....L7
Iron Gates 309 ....L7
Ironton▲ 1,471 ....L7
Jackson▲ 11,947 ....O8
Jameson 120 ....E2
Jamesport 505 ....E2
Jamestown 382 ....G5
Jasper 1,011 ....D8
Jefferson City (cap.)▲ 39,636 ....H5
Jennings 15,469 ....R2
Jerico Springs 259 ....E7
Jonesburg 695 ....K5
Joplin 45,504 ....C8
Junction City 319 ....M7
Kahoka▲ 2,241 ....J2
Kansas City 441,545 ....P5
Kearney 5,472 ....D4
Kelso 527 ....O8
Kennett▲ 11,260 ....M10
Keytesville▲ 533 ....G4
Kidder 271 ....D3
Kimberling City 2,253 ....F9
Kimmswick 94 ....M6
King City 1,012 ....D2
Kingston▲ 287 ....E3
Kingsville 257 ....D5
Kinloch 2,702 ....R2
Kirksville▲ 16,988 ....H2
Kirkwood 27,324 ....O3
Knob Noster 2,462 ....E5
Knox City 223 ....H2
Koshkonong 205 ....K6
Krakow 900 ....K6
La Belle 669 ....H2
Laddonia 620 ....J4
Ladue 8,645 ....O3
La Grange 1,000 ....K2
Lake Lotawana 1,872 ....R6
Lake Ozark 1,489 ....G6
Lake Saint Louis 10,169 ....L5
Lakeshire 1,375 ....P4
Lake Tapawingo 843 ....R6
Lakeview Heights 150 ....F6
Lake Waukomis 917 ....P5

Lake Winnebago 902 ....R6
Lamar▲ 4,425 ....D8
Lamar Heights 216 ....D8
La Monte 1,064 ....F5
Lanagan 411 ....C9
Lancaster▲ 737 ....H1
La Plata 1,486 ....H2
Laredo 250 ....E2
Lathrop 2,092 ....D3
Lawson 2,336 ....D4
Leadington 206 ....M7
Leadwood 1,160 ....L7
Leasburg 323 ....K6
Lebanon▲ 12,155 ....G7
Lee's Summit 70,700 ....R6
Leeton 619 ....E5
Lemay 17,215 ....R4
Lemons 200 ....F2
Lesterville 200 ....L8
Levasy 108 ....S5
Lewistown 595 ....J2
Lexington▲ 4,453 ....E4
Liberal 779 ....D7
Liberty▲ 26,232 ....R5
Licking 1,471 ....J8
Lilbourn 1,303 ....N9
Lincoln 1,026 ....F6
Linn▲ 1,354 ....J5
Linn Creek 280 ....G6
Linneus▲ 369 ....F3
Lockwood 989 ....E8
Lohman 168 ....H5
Lone Jack 528 ....S6
Louisiana 3,863 ....K4
Lowry City 728 ....E6
Ludlow 204 ....E3
Lutesville 865 ....M8
Mackenzie 137 ....P3
Macks Creek 267 ....G7
Macon▲ 5,538 ....H3
Madison 586 ....H4
Maitland 342 ....B2
Malden 4,782 ....M9
Malta Bend 249 ....F4
Manchester 19,161 ....O3
Mansfield 1,349 ....G8
Maplewood 9,228 ....P3
Marble Hill▲ 1,502 ....N8
Marceline 2,558 ....F3
Marionville 2,113 ....E8
Marlborough 2,235 ....P3
Marquand 251 ....M8
Marshall▲ 12,433 ....F4
Marshfield▲ 5,720 ....G8
Marston 610 ....N9
Marthasville 837 ....L5
Martinsburg 326 ....J4
Maryland Heights 25,756 ....O2
Maryville▲ 10,581 ....C2
Matthews 605 ....N9
Maysville▲ 1,212 ....D3
Mayview 294 ....E4
Maywood 300 ....J3
Meadville 457 ....F3
Mehlville 27,557 ....P4
Memphis▲ 2,061 ....H2
Mendon 208 ....F3
Mercer 342 ....F2
Meta 249 ....H6
Mexico▲ 11,320 ....J4
Miami 160 ....F4
Middletown 199 ....J4
Milan▲ 1,958 ....F2
Mill Spring 219 ....L8

Miller 753 ....E8
Mindenmines 409 ....C8
Mine La Motte 125 ....M7
Miner 1,056 ....N9
Mineral Point 363 ....L7
Missouri City 295 ....R5
Moberly 11,945 ....G4
Mokane 188 ....J5
Moline Acres 2,662 ....R2
Monett 7,396 ....E9
Monroe City 2,588 ....J3
Montgomery City▲ 2,442 ....K5
Monticello▲ 126 ....J2
Montrose 417 ....E6
Morehouse 1,015 ....N9
Morley 792 ....N8
Morrison 123 ....J5
Morrisville 344 ....F8
Mosby 242 ....R4
Moscow Mills 1,742 ....K5
Mound City 1,193 ....B2
Moundville 140 ....C7
Mountain Grove 4,574 ....H8
Mountain View 2,430 ....J8
Mount Vernon▲ 4,017 ....E8
Murphy 9,048 ....O4
Napoleon 208 ....E4
Naylor 610 ....L9
Neelyville 487 ....M9
Nelson 212 ....F4
Neosho▲ 10,505 ....D9
Nevada▲ 8,607 ....D7
New Bloomfield 599 ....J5
New Cambria 222 ....G3
New Florence 764 ....K5
New Franklin 1,145 ....G4
New Hampton 349 ....D2
New Haven 1,867 ....K5
New London 1,001 ....K3
New Madrid▲ 3,334 ....O9
New Market 164 ....L5
New Melle 124 ....L5
Newtonia 231 ....D9
Niangua 445 ....G8
Nixa 12,124 ....F8
Noel 1,480 ....D9
Norborne 805 ....E4
Normandy 5,153 ....R2
North Kansas City 4,714 ....P5
Northmoor 399 ....P5
Northwoods 4,643 ....R2
Norwood 552 ....H8
Novelty 119 ....H2
Novinger 534 ....G2
Oak Grove 382 ....H4
Oak Grove 5,535 ....S6
Oakland 1,540 ....P3
Oak Ridge 202 ....N7
Oakview 386 ....P5
Oakwood 197 ....P5
Oakwood Park 183 ....P5
Ocie 153 ....G9
Odessa 4,818 ....E4
O'Fallon 46,169 ....L5
Old Monroe 250 ....L5
Olivette 7,438 ....O2
Olympian Village 669 ....M6
Oran 1,264 ....N8
Oregon▲ 935 ....B2
Oronogo 976 ....D8
Orrick 889 ....E4
Osage Beach 3,662 ....G6
Osborn 455 ....D3
Osceola▲ 835 ....E6

Otterville 476 ....G5
Overland 16,838 ....O2
Owensville 2,500 ....K6
Oxly 152 ....L9
Ozark▲ 9,665 ....F8
Pacific 5,482 ....L5
Pagedale 3,616 ....P2
Palmyra▲ 3,467 ....J3
Paris▲ 1,529 ....J4
Parkville 4,059 ....O5
Parkway 280 ....L6
Parma 852 ....N9
Parnell 197 ....C2
Pascola 138 ....N10
Patton 414 ....M8
Pattonsburg 261 ....D2
Peculiar 2,604 ....D5
Perry 666 ....J4
Perryville▲ 7,667 ....N7
Pevely 3,768 ....M6
Philadelphia 200 ....J3
Pickering 154 ....C2
Piedmont▲ 1,992 ....L8
Pierce City 1,385 ....D9
Pilot Grove 723 ....G5
Pilot Knob 697 ....L7
Pine 4,204 ....K9
Pine Lawn 6,600 ....R2
Pineville▲ 768 ....D9
Platte City▲ 3,866 ....C4
Platte Woods 474 ....O5
Plattsburg▲ 2,354 ....D3
Pleasant Hill 5,582 ....D5
Pleasant Hope 548 ....F8
Pleasant Valley 3,321 ....R5
Polo 582 ....D3
Poplar Bluff▲ 16,651 ....L9
Portage Des Sioux 351 ....M5
Portageville 3,295 ....N10
Potosi▲ 2,662 ....L7
Prairie Home 220 ....G5
Princeton▲ 1,047 ....E2
Purcell 357 ....D8
Purdin 223 ....F3
Purdy 1,103 ....E9
Puxico 1,145 ....M9
Queen City 638 ....H2
Qulin 467 ....M9
Ravenwood 448 ....C2
Raymondville 442 ....J8
Raymore 11,146 ....D5
Raytown 30,388 ....P6
Rayville 204 ....E4
Reeds Spring 465 ....F9
Renick 251 ....H4
Republic 8,438 ....E8
Rhineland 158 ....J5
Rich Hill 1,461 ....D6
Richland 1,805 ....H7
Richmond▲ 6,116 ....D4
Richmond Heights 9,602 ....P3
Ridgeway 530 ....D2
Risco 392 ....N9
Rivermines 459 ....L7
Riverside 2,979 ....O5
Riverview 3,146 ....R2
Rocheport 208 ....H5
Rockaway Beach 577 ....F9
Rock Hill 4,765 ....P3
Rock Port▲ 1,395 ....B2
Rockville 162 ....D6
Rogersville 1,508 ....G8
Rolla▲ 16,367 ....J7
Rosebud 364 ....K6

Rosendale 180 ....C2
Rushville 280 ....B3
Russellville 758 ....H6
Saginaw 276 ....C8
Saint Ann 13,607 ....O2
Saint Charles▲ 60,321 ....N1
Saint Clair 4,390 ....K6
Sainte Genevieve▲ 4,476 ....M6
Saint Elizabeth 297 ....H6
Saint George 1,288 ....P4
Saint James 3,704 ....J6
Saint John 6,871 ....P2
Saint Joseph▲ 73,990 ....C3
Saint Louis▲ 348,189 ....R3
Saint Martins 1,023 ....H5
Saint Marys 377 ....M7
Saint Paul 1,634 ....L5
Saint Peters 51,381 ....M1
Saint Robert 2,760 ....H7
Saint Thomas 287 ....H6

Salem▲ 4,854 ....J7
Salisbury 1,726 ....G4
Sappington 7,287 ....P4
Sarcoxie 1,354 ....D8
Savannah▲ 4,763 ....C3
Schell City 286 ....D6
Scott City 4,591 ....O8
Sedalia▲ 20,339 ....F5
Seligman 877 ....D9
Senath 1,650 ....M10
Seneca 2,135 ....C9
Seymour 1,834 ....G8
Shelbina 1,943 ....H3
Shelbyville▲ 682 ....H3
Sheldon 529 ....D7
Sheridan 185 ....C1
Sherman 202 ....N3
Shrewsbury 6,644 ....P3
Sibley 347 ....S5
Sikeston 16,992 ....N9

Silex 206 ....K4
Skidmore 342 ....B2
Slater 2,083 ....F4
Smithton 510 ....F5
Smithville 5,514 ....D4
South West City 855 ....D9
Spanish Lake 21,337 ....R1
Sparta 1,144 ....F9
Spickard 315 ....F2
Springfield▲ 151,580 ....F8
Stanberry 1,243 ....C2
Stanton 400 ....K6
Steelville▲ 1,429 ....K7
Stewartsville 759 ....C3
Stockton▲ 1,960 ....E7
Stotts City 250 ....E8
Stoutland 177 ....G7
Stover 968 ....G6
Strafford 1,845 ....F8
Sturgeon 944 ....H4

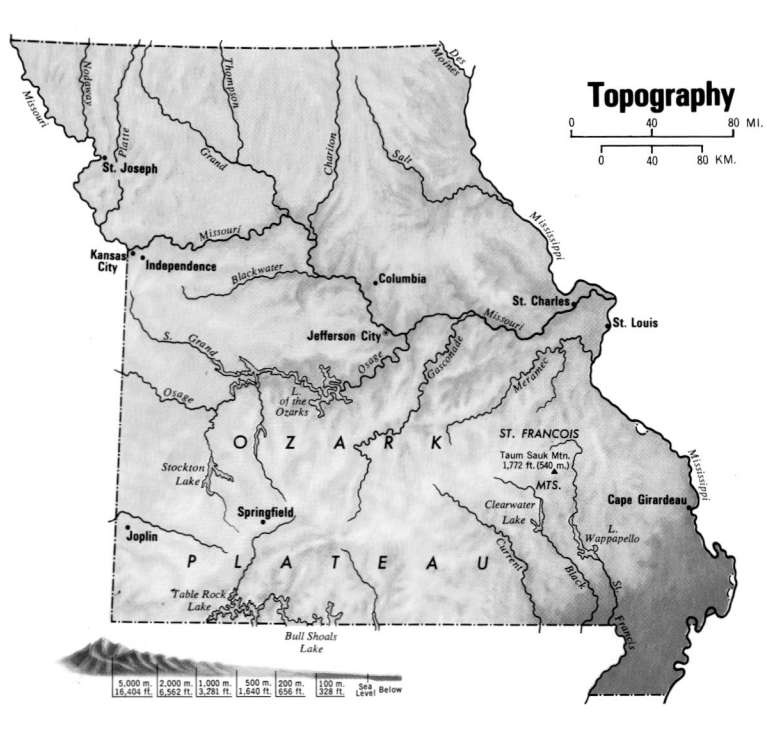

Topography

0 ——— 40 ——— 80 MI.

0 ——— 40 ——— 80 KM.

St. Joseph

Kansas City · Independence

Columbia

Jefferson City · St. Charles · St. Louis

O Z A R K   P L A T E A U

Springfield

Joplin

Stockton Lake

Table Rock Lake

Bull Shoals Lake

Clearwater Lake

L. Wappapello

Cape Girardeau

ST. FRANCOIS
Taum Sauk Mtn.
1,772 ft. (540 m.)
MTS.

L. of the Ozarks

5,000 m. | 2,000 m. | 1,000 m. | 500 m. | 200 m. | 100 m. | Sea Level | Below
16,404 ft. | 6,562 ft. | 3,281 ft. | 1,640 ft. | 656 ft. | 328 ft.

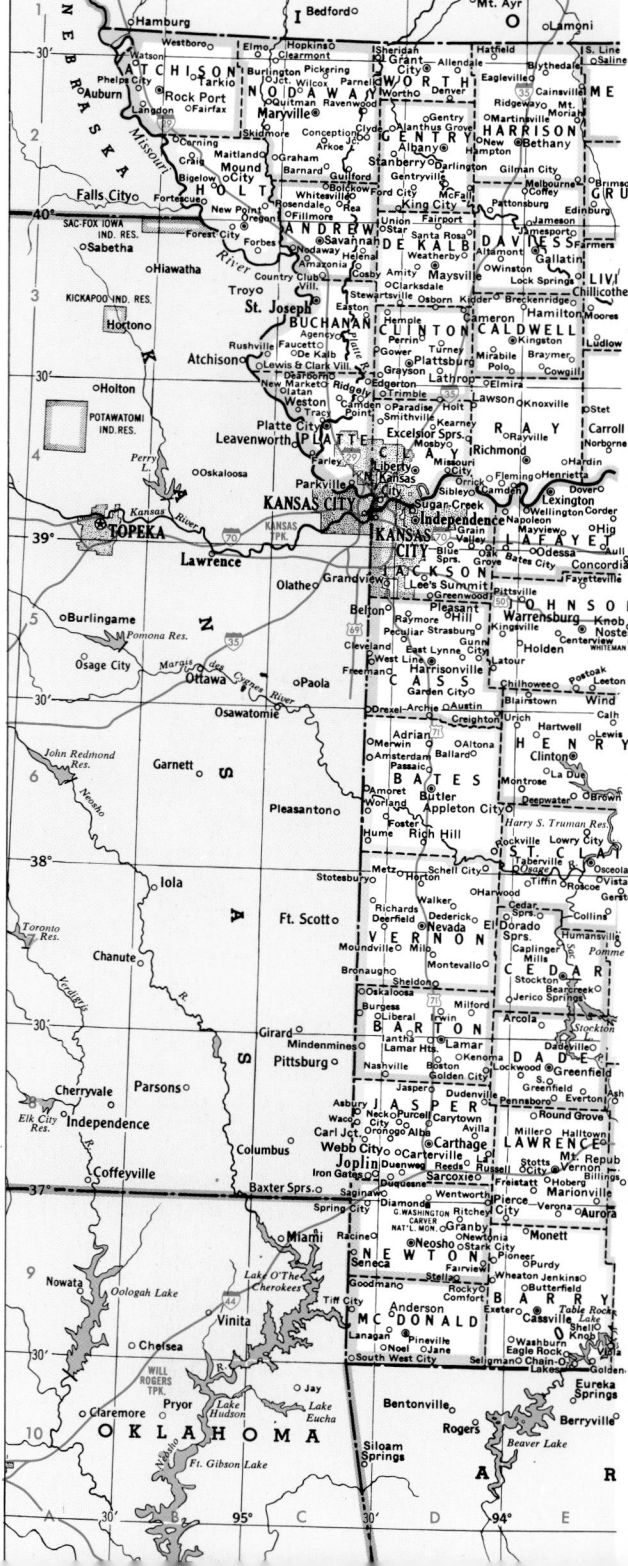

Sugar Creek 3,839 ... R5
Sullivan 6,351 ... K6
Summersville 544 ... J8
Sumner 142 ... F3
Sunset Hills 8,267 ... O4
Swedeborg 180 ... H7
Sweet Springs 1,628 ... F5
Syracuse 172 ... G5
Tallapoosa 204 ... N9
Taneyville 359 ... F9
Taos 870 ... H5
Tarkio 1,935 ... B2
Tarsney Lakes 329 ... R6
Thayer 2,201 ... J9
Theodosia 240 ... G9
Tina 193 ... F3
Tipton 3,261 ... G5
Town and Country 10,894 ... O3
Tracy 213 ... C4
Trenton▲ 6,216 ... E2
Trimble 451 ... D4
Troy▲ 6,737 ... L5
Truesdale 397 ... K5
Tuscumbia▲ 218 ... H6
Twin Oaks 362 ... N3
Union▲ 7,757 ... L6
Union Star 433 ... C2
Unionville▲ 2,041 ... G2
Unity Village 140 ... R6
University City 37,428 ... P3
Urbana 407 ... F7
Urich 499 ... E6
Utica 274 ... E3
Valley Park 6,518 ... O3
Van Buren▲ 845 ... L8
Vandalia 2,529 ... J4
Vanduser 217 ... N9
Velda 1,616 ... P2
Verona 714 ... E9
Versailles▲ 2,565 ... G6
Viburnum 825 ... K7
Vienna▲ 628 ... H6
Vinita Park 1,924 ... P2
Walker 275 ... D7
Walnut Grove 630 ... F8
Wardell 278 ... N10
Wardsville 976 ... H6
Warren 15,244 ... J3
Warrensburg▲ 16,340 ... E5
Warrenton▲ 5,281 ... K5
Warsaw▲ 2,070 ... F6
Warson Woods 1,983 ... O3
Washburn 448 ... E9
Washington 13,243 ... K5
Waverly 806 ... E4
Wayland 425 ... J2
Waynesville▲ 3,507 ... H7
Weatherby Lake 1,873 ... O5
Weaubleau 518 ... F7
Webb City 9,812 ... C8
Webster Groves 23,230 ... P3
Wellington 784 ... E4
Wellston 2,460 ... R2
Wellsville 1,423 ... K4
Wentzville 6,896 ... L5
West Alton 573 ... M5
Westboro 163 ... B1
Weston 1,631 ... C4
Westphalia 320 ... J6
West Plains▲ 10,866 ... J9
Westwood 284 ... O3
Wheatland 388 ... F7
Wheaton 721 ... E9
Wheeling 268 ... F3
Wilbur Park 475 ... P3
Willard 3,193 ... F8
Williamsville 379 ... L9
Willow Springs 2,147 ... H9
Wilson City 165 ... O9
Winchester 1,651 ... N3
Windsor 3,087 ... E5
Winfield 723 ... L5
Winona 1,290 ... K8
Winston 247 ... D3
Woods Heights 742 ... S4
Woodson Terrace 4,189 ... P2
Wright City 1,532 ... K5
Wyaconda 310 ... J2
Wyatt 364 ... O9

OTHER FEATURES

Bagnell (dam) ... G6
Big (riv.) ... L6
Black (riv.) ... L10
Bull Shoals (lake) ... G10
Chariton (riv.) ... G1
Clearwater (lake) ... L8
Cuivre (riv.) ... N2
Current (riv.) ... K8
Des Moines (riv.) ... J1
Fort Leonard Wood 13,666 ... H7
Gasconade (riv.) ... H7
George Washington Carver Nat'l Mon. ... D9
Grand (riv.) ... F3
Harry S Truman (res.) ... E6
Harry S Truman Nat'l Hist. Site ... R5
Jefferson Nat'l Expansion Mem. ... R3
Lake City Arsenal ... R5
Meramec (riv.) ... N3
Mississippi (riv.) ... L4
Missouri (riv.) ... H5
Norfork (lake) ... H10
Osage (riv.) ... E6
Ozark (plat.) ... F9
Ozark National Scenic Riverways ... K8
Ozarks, Lake of the (lake) ... F6
Platte (riv.) ... C3
Pomme de Terre (lake) ... E7
Richards-Gebaur A.F.B. ... P6
Sac (riv.) ... E7
Saint Francis (riv.) ... M9
Saint Francis (mts.) ... L7
Salt (riv.) ... J3
Stockton (lake) ... E7
Table Rock Lake (res.) ... E9
Taneycomo (lake) ... F9
Taum Sauk (mt.) ... L7
Wappapello (lake) ... L8
White (riv.) ... G10
Whiteman A.F.B. 3,814 ... E5
Wilson's Creek Nat'l Battlefield ... F8

▲County seat

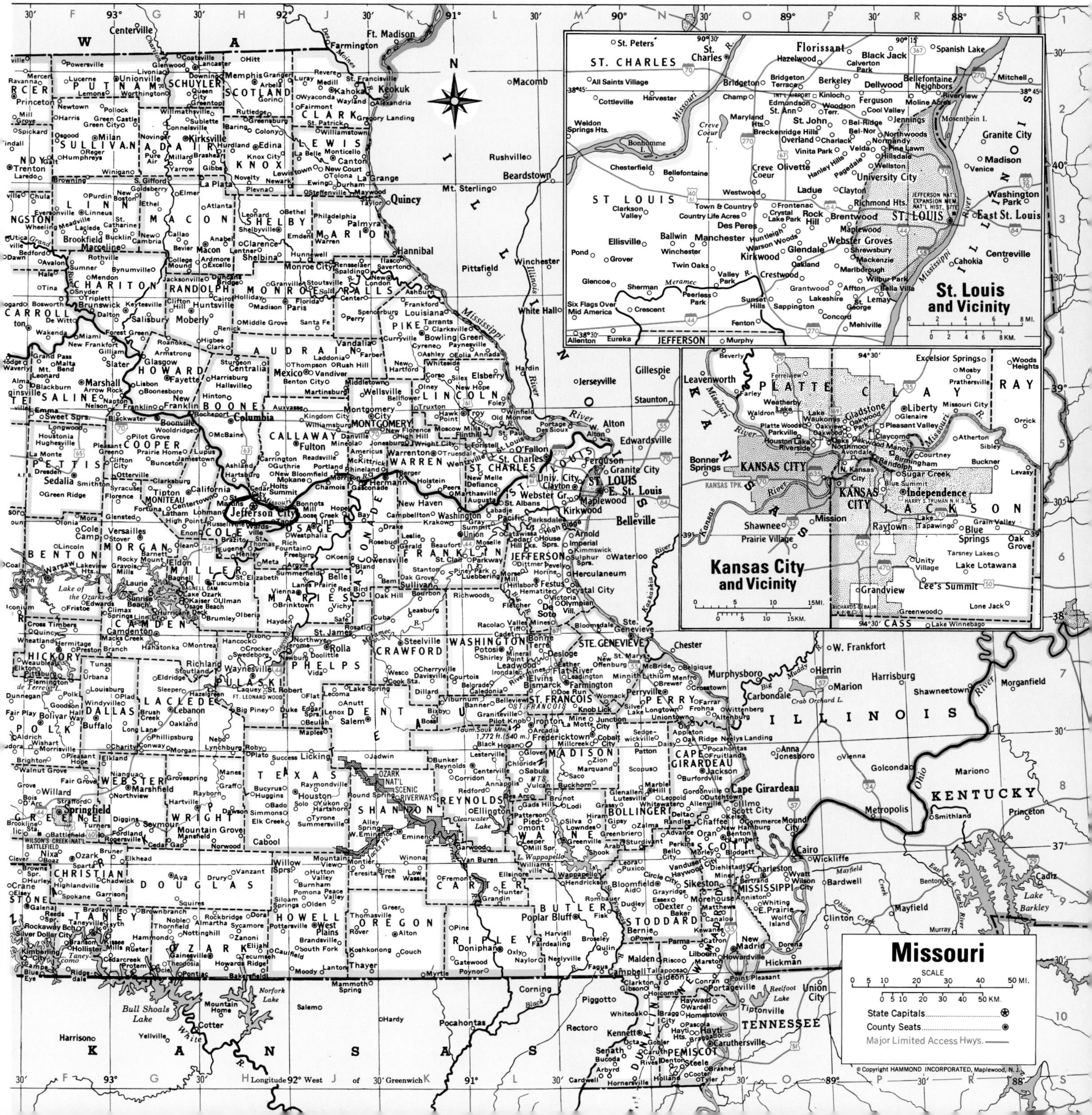
St. Louis and Vicinity
Kansas City and Vicinity
Missouri
SCALE
0 5 10 20 30 40 50 MI.
0 5 10 20 30 40 50 KM.
State Capitals
County Seats
Major Limited Access Hwys.
Longitude 92° West of Greenwich
ILLINOIS
KENTUCKY
TENNESSEE
KANSAS
ARKANSAS
Jefferson City
Kansas City
St. Louis
Springfield
Columbia
© Copyright HAMMOND INCORPORATED, Maplewood, N. J.

Agriculture, Industry and Resources

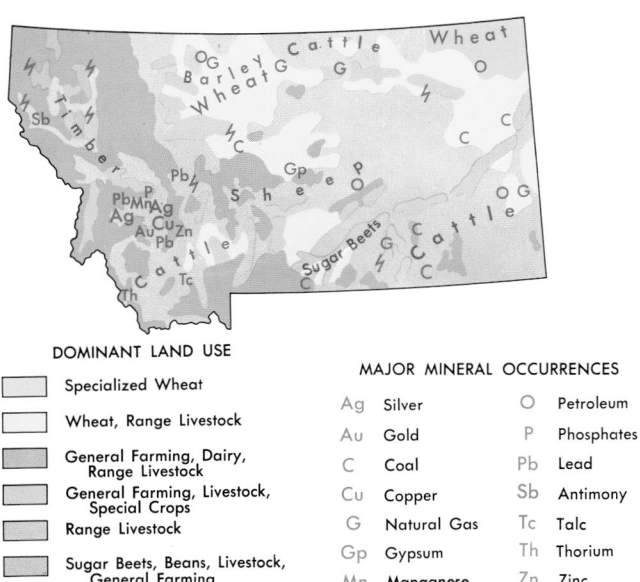

DOMINANT LAND USE

- Specialized Wheat
- Wheat, Range Livestock
- General Farming, Dairy, Range Livestock
- General Farming, Livestock, Special Crops
- Range Livestock
- Sugar Beets, Beans, Livestock, General Farming
- Forests

MAJOR MINERAL OCCURRENCES

Ag	Silver	O	Petroleum
Au	Gold	P	Phosphates
C	Coal	Pb	Lead
Cu	Copper	Sb	Antimony
G	Natural Gas	Tc	Talc
Gp	Gypsum	Th	Thorium
Mn	Manganese	Zn	Zinc

ϟ Water Power

COUNTIES

Beaverhead 9,202 .... C5
Big Horn 12,671 .... J5
Blaine 7,009 .... G2
Broadwater 4,385 .... E4
Carbon 9,552 .... G5
Carter 1,360 .... M5
Cascade 80,357 .... E3
Chouteau 5,970 .... F3
Custer 11,696 .... L4
Daniels 2,017 .... L2
Dawson 9,059 .... M3
Deer Lodge 9,417 .... C5
Fallon 2,837 .... M4
Fergus 11,893 .... G3
Flathead 74,471 .... B2
Gallatin 67,831 .... E5
Garfield 1,279 .... J3
Glacier 13,247 .... C2
Golden Valley 1,042 .... G4
Granite 2,830 .... C4
Hill 16,673 .... F2
Jefferson 10,049 .... D4
Judith Basin 2,329 .... F4
Lake 26,507 .... B3
Lewis and Clark 55,716 .... D3
Liberty 2,158 .... E2
Lincoln 18,837 .... A2
Madison 6,851 .... D5
McCone 1,977 .... L3
Meagher 1,932 .... F4
Mineral 3,884 .... B3
Missoula 95,802 .... C3
Musselshell 4,497 .... H4
Park 15,694 .... F5
Petroleum 493 .... H3
Phillips 4,601 .... J2
Pondera 6,424 .... D2
Powder River 1,858 .... L5
Powell 7,180 .... D4
Prairie 1,199 .... L4
Ravalli 36,070 .... B4
Richland 9,667 .... M3
Roosevelt 10,620 .... L2
Rosebud 9,383 .... K4
Sanders 10,227 .... A3
Sheridan 4,105 .... M2
Silver Bow 34,606 .... D5
Stillwater 8,195 .... G5
Sweet Grass 3,609 .... G5
Teton 6,445 .... D3
Toole 5,267 .... E2
Treasure 861 .... J4
Valley 7,675 .... K2
Wheatland 2,259 .... G4
Wibaux 1,068 .... M4
Yellowstone 129,352 .... H4

CITIES and TOWNS

Absarokee 1,234 .... G5
Acton 50 .... H5
Alberton 374 .... B3
Alder 116 .... D5
Alzada 52 .... M5
Amsterdam 727 .... E5
Anaconda-Deer Lodge County▲ 9,417 .... C4
Angela 50 .... K4
Antelope 43 .... M2
Apgar 25 .... B2
Arlee 602 .... B3
Armington 75 .... F3
Ashland 464 .... K5
Augusta 284 .... D3
Avon 124 .... D4
Babb 150 .... C2
Bainville 153 .... M2
Baker▲ 1,695 .... M4
Ballantine 346 .... J5
Bannack 2 .... C5
Basin 255 .... D4
Bearcreek 83 .... G5
Becket 35 .... G4
Belfry 219 .... H5
Belgrade 5,728 .... E5
Belt 633 .... E3
Biddle 28 .... L5
Big Arm 131 .... B3
Bigfork 1,421 .... C2
Big Sandy 703 .... G2
Big Sky 1,221 .... E5
Big Timber▲ 1,650 .... G5
Billings▲ 89,847 .... H5
Birney 108 .... K5
Black Eagle 914 .... E3
Blackfoot 100 .... D2
Bloomfield 28 .... M3
Bonner-West Riverside 1,693 .... C4
Boulder▲ 1,300 .... E4
Box Elder 794 .... F2
Boyd 32 .... G5
Bozeman▲ 27,509 .... E5
Brady 450 .... E2
Bridger 745 .... H5
Broadus▲ 451 .... L5
Broadview 150 .... H4
Brockton 245 .... M2
Brockway 55 .... L3
Browning 1,065 .... C2
Busby 695 .... J5
Butte-Silver Bow County▲ 34,606 .... D5
Bynum 49 .... D3
Camas Prairie 160 .... B3
Cameron 150 .... E5
Canyon Creek 100 .... D4
Canyon Ferry 100 .... E4
Cardwell 40 .... E5
Carter 62 .... E3
Cartersville 115 .... K4
Cascade 819 .... E3
Charlo 439 .... B3
Chester▲ 871 .... E2
Chinook▲ 1,386 .... G2
Choteau▲ 1,781 .... D3
Christina 60 .... G3
Circle▲ 644 .... L3
Clancy 1,406 .... E4
Clinton 549 .... C4
Clyde Park 310 .... F5
Coffee Creek 62 .... F3
Colstrip 2,346 .... K5
Columbia Falls 3,645 .... B2
Columbus▲ 1,748 .... G5
Condon 300 .... C3
Conner 420 .... B5
Conrad▲ 2,753 .... D2
Cooke City 140 .... G5
Coram 337 .... C2
Corvallis 443 .... C4
Craig 100 .... D3
Crane 163 .... M3
Creston 60 .... C2
Crow Agency 1,552 .... J5
Culbertson 716 .... M2
Custer 145 .... J4
Cut Bank▲ 3,105 .... D2
Dagmar 35 .... M2

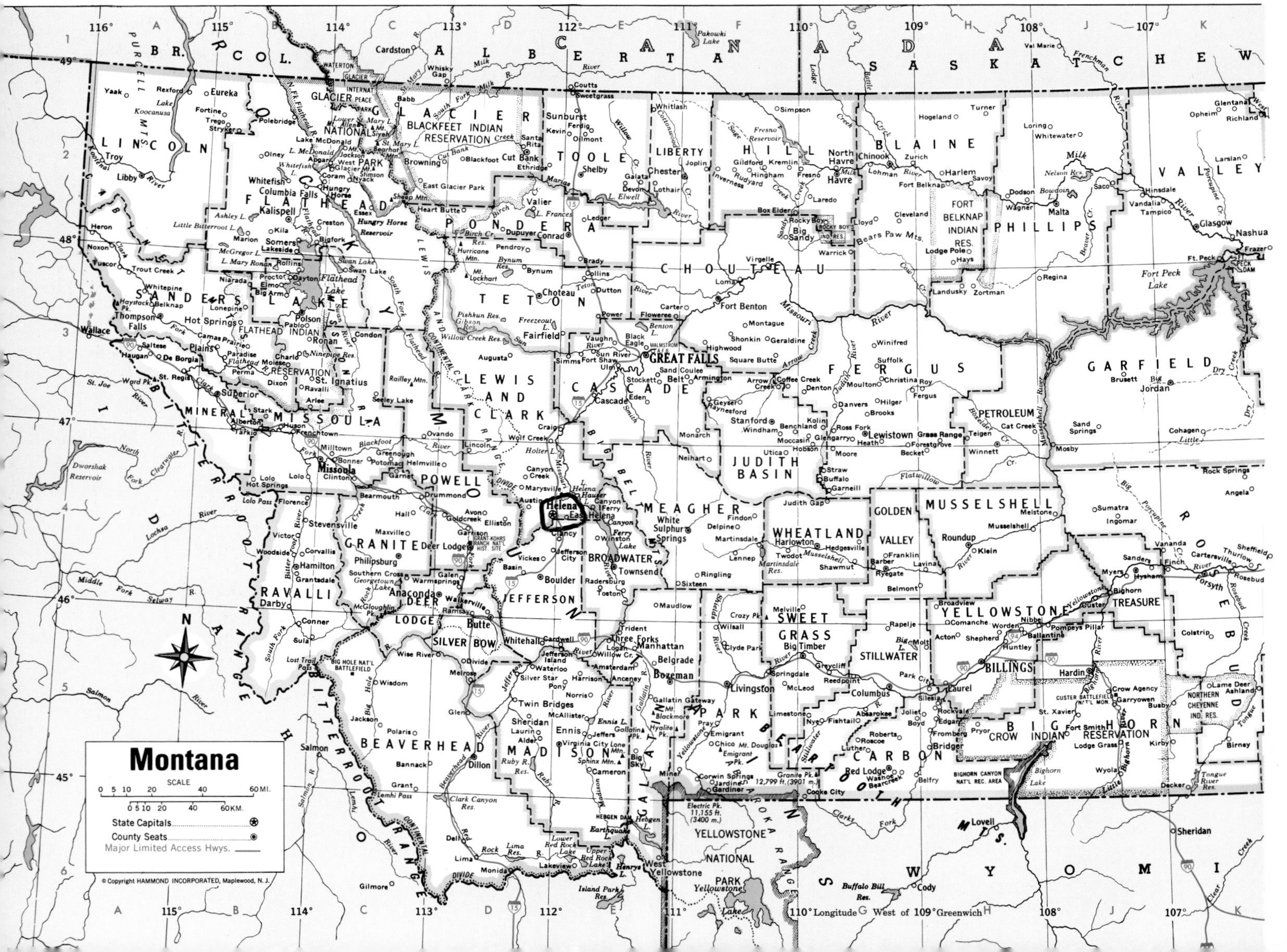

Topography

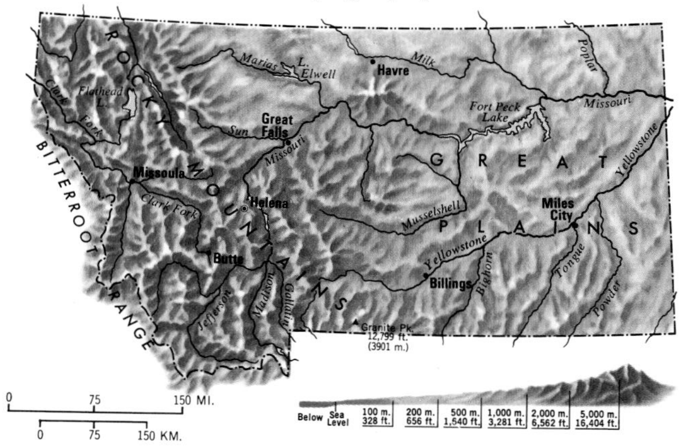
ROCKY MOUNTAINS
BITTERROOT RANGE
GREAT PLAINS
Havre
Great Falls
Missoula
Helena
Butte
Billings
Miles City
Flathead L.
Marias
L. Elwell
Milk
Missouri
Fort Peck Lake
Poplar
Yellowstone
Musselshell
Sun
Clark Fork
Jefferson
Madison
Gallatin
Bighorn
Tongue
Powder
Granite Pk. 12,799 ft. (3901 m.)
0 75 150 MI.
0 75 150 KM.
Below Sea Level
100 m. 328 ft.
200 m. 656 ft.
500 m. 1,540 ft.
1,000 m. 3,281 ft.
2,000 m. 6,562 ft.
5,000 m. 16,404 ft.

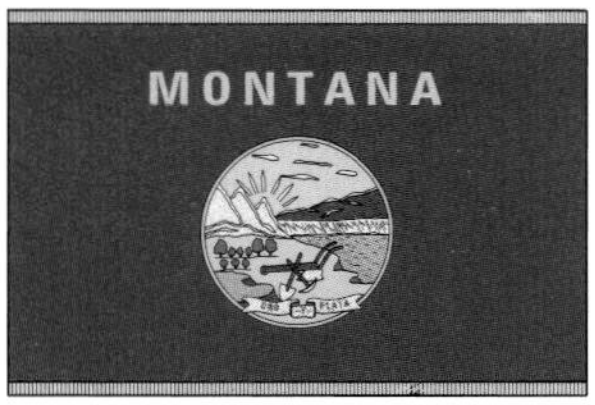
MONTANA

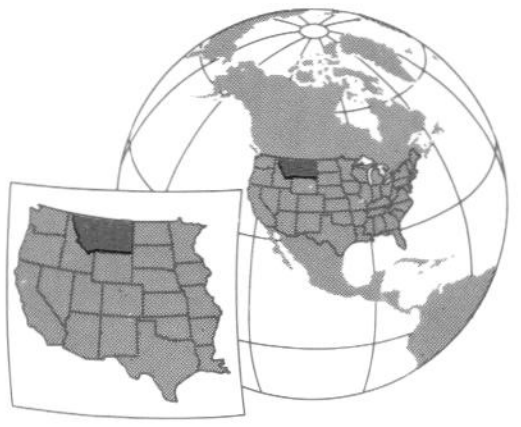

AREA 147,046 sq. mi. (380,849 sq. km.)
POPULATION 902,195
CAPITAL Helena
LARGEST CITY Billings
HIGHEST POINT Granite Pk. 12,799 ft. (3901 m.)
SETTLED IN 1809
ADMITTED TO UNION November 8, 1889
POPULAR NAME Treasure State; Big Sky Country
STATE FLOWER Bitterroot
STATE BIRD Western Meadowlark

Darby 710 B4
Dayton 95 B3
De Borgia 69 A3
Decker 150 K5
Deer Lodge▲ 3,421 D4
Dell 29 D6
Delpine 33 F4
Denton 301 G3
Dillon▲ 3,752 D5
Divide 275 D5
Dixon 216 B3
Dodson 122 H2
Drummond 318 D4
Dupuyer 105 D2
Dutton 389 E3
East Glacier Park 396 C2
East Helena 1,642 E4
Edgar 220 H5
Ekalaka▲ 410 M5
Elliston 225 D4
Elmo 143 B3
Emigrant 80 F5
Ennis 840 E5
Epsie 60 L5
Essex 48 C2
Eureka 1,017 B2
Fairfield 659 D3
Fairview 709 M3
Fallon 138 L4
Fishtail 46 G5
Flaxville 87 L2
Florence 901 B4
Floweree 48 E3
Forestgrove 100 H3
Forsyth▲ 1,944 K4
Fort Belknap 1,262 H2
Fort Benton▲ 1,594 F3
Fortine 169 A2
Fort Peck 240 K2
Fort Shaw 274 E3
Fort Smith 122 J5
Four Buttes 50 L2
Frazer 452 K2
Frenchtown 883 B3
Froid 195 M2
Fromberg 486 H5
Galata 100 E2
Galen 210 D4
Gallatin Gateway 600 E5
Gardiner 851 F5
Garneill 61 G4
Garrison 112 D4
Garryowen 200 J5
Geraldine 284 F3
Geyser 125 F3
Gildford 185 F2
Glasgow▲ 3,253 K2
Glen 4,802 D5
Glendive▲ 4,729 M3
Goldcreek 100 D4
Grant 25 C5
Grantsdale 500 B4
Grass Range 149 H3
Great Falls▲ 56,690 E3
Greenough 120 C4
Greycliff 56 G5
Hall 130 C4
Hamilton▲ 3,705 B4
Hardin▲ 3,384 J5
Harlem 848 H2
Harlowton▲ 1,062 F4
Harrison 162 E5
Hathaway 55 K4
Haugan 90 A3
Havre▲ 9,621 G2
Hays 702 H2
Heart Butte 698 C2
Helena (cap.)▲ 25,780 E4
Helmville 250 C4
Heron 149 A2
Highwood 189 F3
Hilger 38 G3
Hingham 157 F2
Hinsdale 260 K2
Hobson 244 G4
Hodges 50 M4
Hogeland 35 H2
Homestead 50 M2
Hot Springs 531 B3
Hungry Horse 934 C2
Huntley 411 H5
Huson 97 B3
Hysham▲ 330 J4
Ingomar 48 J4
Intake 60 M3
Inverness 103 F2
Jackson 210 C5
Jardine 30 F5
Jeffers 70 E5
Jefferson City 295 E4
Jefferson Island 25 E5
Joliet 575 G5
Joplin 210 F2
Jordan▲ 364 J3
Judith Gap 164 G4
Kalispell▲ 14,223 B2
Kevin 178 D2
Kila 350 B2
Kinsey 100 L4
Kirby 30 J5
Klein 188 H4
Kremlin 126 F2
Lakeside 1,679 B2
Lakeview 28 E6
Lambert 203 M3
Lame Deer 2,018 K5
Landusky 40 H3
Laurel 6,255 H5
Laurin 60 D5
Lavina 209 H4
Lewistown▲ 5,813 G3
Libby▲ 2,626 A2
Lima 242 D6
Lincoln 1,100 D4
Lindsay 50 L3
Livingston▲ 6,851 F5
Locate 55 L4
Lodge Grass 510 J5
Lodge Pole 214 H2
Logan 53 E5
Lohman 25 G2
Lolo 3,388 B4
Lolo Hot Springs 25 B4
Loma 92 F3
Lonepine 137 B3
Lothair 29 E2
Malta▲ 2,120 J2
Manhattan 1,396 E5
Marion 450 B2
Martinsdale 75 F4
Marysville 76 D4
Maxville 44 C4
McAllister 55 E5
McLeod 150 G5
Medicine Lake 269 M2
Melrose 350 D5
Melstone 136 H4
Melville 100 F4
Miles City▲ 8,487 L4
Mill Iron 66 M5
Milltown 300 C4
Missoula▲ 57,053 C4
Moccasin 57 F3
Molt 31 H5
Monarch 120 F3
Moore 186 G4
Musselshell 60 H4
Myers 120 J4
Nashua 325 K2
Neihart 91 F4
Nibbe 30 H4
Norris 55 E5
North Havre 1,230 G2
Noxon 230 A3
Nye 50 G5
Oilmont 50 E2
Olney 200 B2
Opheim 111 K2
Oswego 75 L2
Outlook 82 M2
Ovando 71 C3
Pablo 1,814 B3
Paradise 184 B3
Park City 870 H5
Peerless 110 L2
Pendroy 100 D2
Perma 50 B3
Philipsburg▲ 914 C4
Plains 1,126 B3
Plentywood▲ 2,061 M2
Plevna 138 M4
Polaris 53 C5
Polson▲ 4,041 B3
Pompeys Pillar 300 J5
Pony 130 E5
Poplar 911 L2
Potomac 80 C4
Power 171 E3
Pray 40 F5
Proctor 150 B3
Pryor 628 H5
Radersburg 70 E4
Ramsay 95 D4
Rapelje 50 G5
Ravalli 119 B3
Raymond 26 M2
Raynesford 35 F3
Red Lodge▲ 2,177 G5
Redstone 40 M2
Reedpoint 185 G5
Regina 83 J3
Reserve 37 M2
Rexford 151 A2
Richey 189 L3
Richland 48 K2
Ringling 102 F4
Roberts 312 G5
Rocky Boy 150 G2
Rollins 183 B3
Ronan 1,812 C3
Roscoe 40 G5
Rosebud 259 K4
Roundup▲ 1,931 H4
Roy 200 H3
Rudyard 275 F2
Ryegate▲ 268 G4
Saco 224 J2
Saint Ignatius 788 C3
Saint Regis 315 A3
Saint Xavier 67 J5
Saltese 90 A3
Sand Coulee 600 E3
Sanders 50 J4
Santa Rita 120 D2
Savage 300 M3
Scobey▲ 1,082 L2
Seeley Lake 1,436 C3
Shawmut 66 G4
Sheffield 49 K4
Shelby▲ 3,216 E2
Shepherd 193 H5
Sheridan 659 D5
Sidney▲ 4,774 M3
Silesia 90 H5
Silver Star 125 D5
Simms 373 E3
Simpson 70 F2
Somers 556 B2
Sonnette 42 L5
Springdale 45 F5
Square Butte 48 F3
Stanford▲ 454 F3
Stark 51 B3
Stevensville 1,553 C4
Stockett 500 E3
Stryker 96 B2
Suffolk 45 G3
Sula 200 B5
Sunburst 415 E2
Sun River 131 E3
Superior▲ 893 B3
Swan Lake 100 C3
Sweetgrass 250 E2
Terry▲ 611 L4
Thompson Falls▲ 1,321 A3
Three Forks 1,728 E5
Thurlow 84 K4
Toston 105 E4
Townsend▲ 1,867 E4
Trego 50 B2
Trident 50 E5
Trout Creek 261 A3
Troy 957 A2
Turner 150 H2
Twin Bridges 400 D5
Twodot 285 F4
Ulm 750 E3
Utica 30 F4
Valier 498 D2
Vananda 50 K4
Vandalia 35 J2
Vaughn 701 E3
Victor 859 B4
Vida 50 L3
Virgelle 28 F2
Virginia City▲ 130 E5
Volborg 125 L5
Wagner 32 H2
Walkerville 714 D4
Warmsprings 500 D4
Waterloo 102 D5
Westby 172 M2
West Glacier 150 C2
West Yellowstone 1,177 E6
Whitefish 5,032 B2
Whitehall 1,044 D5
White Sulphur Springs▲ 984 E4
Whitetail 150 L2
Whitewater 100 J2
Whitlash 50 E2
Wibaux▲ 567 M3
Wickes 60 D4
Willow Creek 209 E5
Wilsall 237 F5
Windham 63 F3
Winifred 156 G3
Winnett▲ 185 H4
Winston 73 E4
Wisdom 114 C5
Wise River 150 C5
Wolf Creek 500 D3
Wolf Point▲ 2,663 L2
Woodside 75 B4
Worden 506 H5
Wyola 186 J5
Zurich 60 G2

OTHER FEATURES

Absaroka (range) F5
Allen (mt.) C2
Arrow (creek) F3
Ashley (lake) B2
Battle (creek) G1
Bearhat (mt.) C2
Bears Paw (mts.) G2
Beartooth (mts.) G5
Beaver (creek) J2
Beaverhead (riv.) D5
Benton (lake) E3
Big (lake) G5
Big Belt (mts.) E4
Big Dry (creek) K3
Big Hole (riv.) C5
Big Hole Nat'l Battlefield C5
Bighorn (lake) H5
Bighorn (riv.) J5
Bighorn Canyon Nat'l Rec. Area H5
Big Muddy (riv.) M2
Big Porcupine (creek) J4
Birch (creek) D2
Birch Creek (res.) D2
Bitterroot (range) B4
Bitterroot (riv.) B4
Blackfeet Ind. Res. D2
Blackfoot (riv.) C4
Blackmore (mt.) F5
Bowdoin (lake) J2
Boxelder (creek) H3
Boxelder (creek) M5
Bynum (res.) D2
Cabinet (mts.) A2
Canyon Ferry (lake) E4
Clark Canyon (res.) D6
Clark Fork (riv.) A3
Clarks Fork, Yellowstone (riv.) G6
Cottonwood (creek) E2
Cow (creek) G2
Crazy peak) F4
Crow Ind. Res. H5
Cut Bank (creek) D2
Douglas (mt.) F5
Earthquake (lake) E6
Electric (peak) F6
Elwell (lake) E2
Emigrant (peak) F5
Ennis (lake) E5
Flathead (lake) C3
Flathead (riv.) B2
Flathead, North Fork (riv.) B2
Flathead, South Fork (riv.) C3
Flathead Ind. Res. B3
Flatwillow (creek) H4
Fort Belknap Ind. Res. H2
Fort Peck (lake) K3
Fort Union Trading Post Nat'l Hist. Site N2
Frances (lake) D2
Freezeout (lake) D3
Frenchman (riv.) J1
Fresno (res.) F2
Gallatin (peak) E5
Gallatin (riv.) E5
Georgetown (lake) C4
Gibson (res.) D3
Glacier Nat'l Park C2
Granite (peak) F5
Grant-Kohrs Ranch Nat'l Hist. Site D4
Hauser (lake) E4
Haystack (peak) A3
Hebgen (lake) E6
Helena (lake) E4
Holter (lake) D4
Hungry Horse (res.) C2
Hurricane (mt.) D2
Hyalite (peak) E5
Jackson (mt.) C2
Jefferson (riv.) D5
Judith (riv.) G3
Koocanusa (lake) A2
Kootenai (riv.) A2
Lemhi (pass) C6
Lewis and Clark (range) C3
Lima (res.) D6
Little Bighorn (riv.) J5
Little Bitterroot (lake) B2
Little Dry (creek) K3
Little Missouri (riv.) M5
Lockhart (mt.) D3
Lodge (creek) G1
Lolo (pass) B4
Lone (mt.) E5
Lost Trail (pass) B5
Lower Red Rock (lake) E6
Lower Saint Mary (lake) C2
Madison (riv.) E5
Malmstrom A.F.B. 4,544 E3
Marias (riv.) D2
Martinsdale (res.) F4
Mary Ronan (lake) B3
McDonald (lake) B2
McGloughlin (peak) C4
McGregor (lake) B3
Medicine (lake) M2
Milk (riv.) J2
Mission (range) C3
Missouri (riv.) L3
Musselshell (riv.) J3
Nelson (res.) J2
Ninepipe (res.) C3
Northern Cheyenne Indian Reservation K5
O'Fallon (creek) L4
Pishkun (res.) D3
Poplar (riv.) L2
Porcupine (creek) K2
Powder (riv.) L4
Purcell (mts.) A2
Railley (mt.) C3
Red Rock (lakes) E6
Red Rock (riv.) D6
Redwater (riv.) L3
Rock (creek) C4
Rocky (mts.) D4
Rocky Boy's Ind. Res. G2
Rosebud (creek) K4
Ruby (riv.) D5
Ruby River (res.) D5
Sage (creek) F2
Saint Mary (lake) C2
Saint Mary (riv.) C1
Sandy (creek) F2
Sheep (mt.) C2
Shields (riv.) F4
Siyeh (mt.) C2
Smith (riv.) E3
Sphinx (mt.) E5
Stillwater (riv.) G5
Stimson (mt.) C2
Sun (riv.) D3
Swan (lake) C3
Teton (riv.) E3
Tongue (riv.) K5
Upper Red Rock (lake) E6
Ward (peak) A3
Waterton-Glacier Int'l Peace Park C2
Whitefish (lake) B2
Willow (creek) E2
Willow Creek (res.) D3
Yellowstone (riv.) M3
Yellowstone Nat'l Park F6

▲County seat

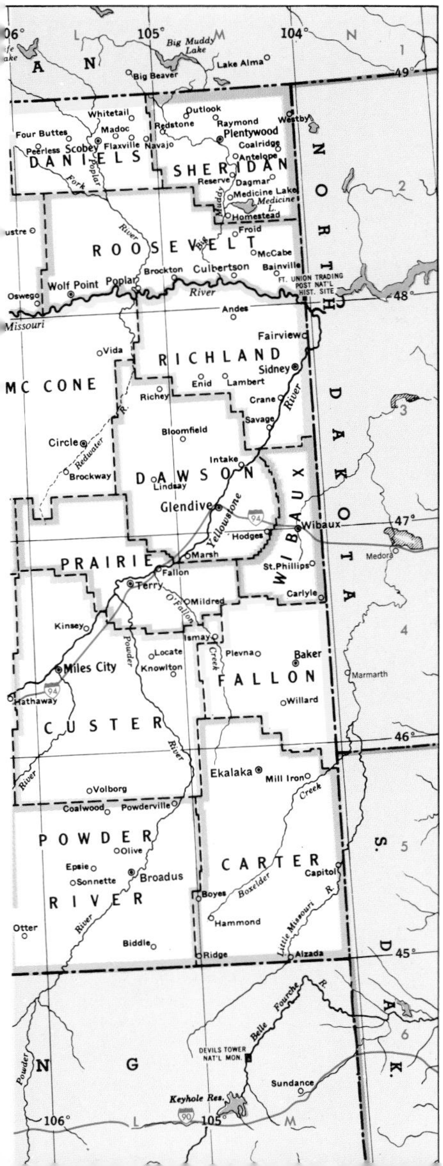
DANIELS
SHERIDAN
ROOSEVELT
RICHLAND
MC CONE
DAWSON
WIBAUX
PRAIRIE
FALLON
CUSTER
POWDER RIVER
CARTER
NORTH DAKOTA
S. DAKOTA
Big Muddy Lake
Lake Alma
Big Beaver
Whitetail
Outlook
Raymond
Plentywood
Westby
Coalridge
Antelope
Four Buttes
Madoc
Redstone
Peerless
Scobey
Flaxville
Navajo
Reserve
Dagmar
Medicine Lake
Homestead
Froid
McCabe
Brockton
Culbertson
Bainville
FT. UNION TRADING POST NAT'L HIST. SITE
Wolf Point
Poplar
Oswego
Andes
Fairview
Sidney
Vida
Enid
Lambert
Richey
Crane
Savage
Bloomfield
Circle
Intake
Brockway
Lindsay
Glendive
Wibaux
Hodges
Marsh
St.Phillips
Fallon
Terry
Mildred
Carlyle
Kinsey
Ismay
Plevna
Baker
Miles City
Locate
Knowlton
Marmarth
Hathaway
Willard
Ekalaka
Mill Iron
Volborg
Coalwood
Powderville
Olive
Epsie
Sonnette
Broadus
Capitol
Boyes
Hammond
Otter
Biddle
Ridge
Alzada
DEVILS TOWER NAT'L MON.
Sundance
Keyhole Res.
Missouri
River
Yellowstone
O'Fallon Creek
Little Missouri
Boxelder
Belle Fourche R.
Powder
49°
48°
47°
46°
45°
104°
105°
106°

COUNTIES

Adams 31,151 F4
Antelope 7,452 F2
Arthur 444 C3
Banner 819 A3
Blaine 583 E3
Boone 6,259 F3
Box Butte 12,158 A2
Boyd 2,438 F2
Brown 3,525 E2
Buffalo 42,259 E4
Burt 7,791 H3
Butler 8,767 G3
Cass 24,334 H4
Cedar 9,615 G2
Chase 4,068 C4
Cherry 6,148 C2
Cheyenne 9,830 A3
Clay 7,039 F4
Colfax 10,441 G3
Cuming 10,203 H3
Custer 11,793 E3
Dakota 20,253 H2
Dawes 9,060 A2
Dawson 24,365 E4
Deuel 2,098 B3
Dixon 6,339 H2
Dodge 36,160 H3
Douglas 463,585 H3
Dundy 2,292 C4
Fillmore 6,634 G4
Franklin 3,574 F4
Frontier 3,099 D4
Furnas 5,324 E4
Gage 22,993 H4
Garden 2,292 B3
Garfield 1,902 F3
Gosper 2,143 E4
Grant 747 C3
Greeley 2,714 F3
Hall 53,534 F4
Hamilton 9,403 F4
Harlan 3,786 E4
Hayes 1,068 C4
Hitchcock 3,111 C4
Holt 11,551 F2
Hooker 783 C3
Howard 6,567 F3
Jefferson 8,333 G4
Johnson 4,488 H4
Kearney 6,882 F4
Keith 8,875 C3
Keya Paha 983 E2
Kimball 4,089 A3
Knox 9,374 G2
Lancaster 250,291 H4
Lincoln 34,632 D4
Logan 774 D3
Loup 712 E3
Madison 35,226 G3
McPherson 533 C3
Merrick 8,204 F3
Morrill 5,440 A3
Nance 4,038 F3
Nemaha 7,576 J4
Nuckolls 5,057 F4
Otoe 15,396 H4
Pawnee 3,087 H4
Perkins 3,200 C4
Phelps 9,747 E4
Pierce 7,857 G2
Platte 31,662 G3
Polk 5,639 G3
Red Willow 11,448 D4
Richardson 9,531 J4
Rock 1,756 E2
Saline 13,843 G4
Sarpy 122,595 H3
Saunders 19,830 H3
Scotts Bluff 36,951 A3
Seward 16,496 G4
Sheridan 6,198 B2
Sherman 3,318 F3
Sioux 1,475 A2
Stanton 6,455 G3
Thayer 6,055 G4
Thomas 729 D3
Thurston 7,171 H2
Valley 4,647 E3
Washington 18,780 H3
Wayne 9,851 G2
Webster 4,061 F4
Wheeler 886 F3
York 14,598 G4

CITIES and TOWNS

Adams 489 H4
Ainsworth▲ 1,862 D2
Albion▲ 1,797 F3
Alda 652 F4
Alexandria 216 G4
Allen 411 H2
Alliance▲ 8,959 A2
Alma▲ 1,214 E4
Alvo 142 H4
Amherst 277 E4
Anselmo 159 E3
Ansley 520 E3
Arapahoe 1,028 E4
Arcadia 359 F3
Arlington 1,197 H3
Arnold 630 D3
Arthur▲ 145 C3
Ashland 2,262 H3
Ashton 237 F3
Atkinson 1,244 E2
Auburn▲ 3,350 J4
Aurora▲ 4,225 F4
Avoca 270 H4
Axtell 696 E4
Bancroft 520 H2
Bartlett▲ 128 F3
Bartley 355 D4
Bassett▲ 743 E2
Battle Creek 1,158 G3
Bayard 1,247 A3
Beatrice▲ 12,496 H4
Beaver City▲ 641 E4
Beaver Crossing 457 G4
Bee 223 H3
Beemer 773 H3
Belden 131 G2
Belgrade 134 G3
Bellevue 44,382 J3
Bellwood 446 G3
Benedict 278 G3
Benkelman▲ 1,006 C4
Bennet 570 H4
Bennington 937 H3
Bertrand 786 E4
Big Springs 418 B3
Bladen 291 F4
Blair▲ 7,512 H3
Bloomfield 1,126 G2
Blue Hill 867 F4
Blue Springs 383 H4
Boys Town 818 H3
Bradshaw 336 G4
Brady 366 D3
Brainard 351 G3
Brewster▲ 29 D3
Bridgeport▲ 1,594 A3
Broadwater 140 B3
Brock 162 H4
Broken Bow▲ 3,491 E3
Brownville 146 J4
Brule 372 C3
Bruning 300 G4
Bruno 112 G3
Brunswick 179 G2
Burwell▲ 1,130 E3
Butte▲ 366 F2
Cairo 790 F3
Callaway 637 D3
Cambridge 1,041 D4
Campbell 387 F4
Carleton 136 G4
Carroll 238 G2
Cedar Bluffs 615 H3
Cedar Creek 396 H3
Cedar Rapids 407 F3
Center▲ 90 G2
Central City▲ 2,998 F3
Ceresco 920 H3
Chadron▲ 5,634 B2
Chambers 333 F2
Chapman 341 F3
Chappell▲ 983 B3
Chester 294 G4
Clarks 361 G3
Clarkson 685 G3
Clatonia 275 H4
Clay Center▲ 861 F4
Clearwater 384 F2
Cody 149 C2
Coleridge 541 G2
Colon 138 H3
Columbus▲ 20,971 G3
Cook 322 H4
Cordova 127 G4
Cortland 488 H4
Cozad 4,163 E4
Craig 241 H3
Crawford 1,107 A2
Creighton 1,270 G2
Creston 215 G3
Crete 6,028 G4
Crofton 754 G2
Culbertson 594 C4
Curtis 832 D4
Dakota City▲ 1,821 H2
Dalton 332 B3
Dannebrog 352 F3
Davenport 339 G4
Davey 153 H4
David City▲ 2,597 G3
Dawson 209 J4
Daykin 177 G4
Decatur 618 H2
Denton 189 H4
Deshler 879 G4
De Witt 572 G4
Diller 287 H4
Dix 267 A3
Dodge 700 H3
Doniphan 763 F4
Dorchester 615 G4
Douglas 231 H4
Dunbar 237 J4
Duncan 359 G3
Dwight 259 G3
Eagle 1,105 H4
Edgar 539 F4
Edison 154 E4
Elba 243 F3
Elgin 735 F3
Elkhorn 6,062 H3
Elm Creek 894 E4
Elmwood 668 H4
Elsie 139 C4
Elwood▲ 761 E4
Emerson 817 H2
Endicott 139 G4
Eustis 464 D4
Ewing 433 F2
Exeter 712 G4
Fairbury▲ 4,262 G4
Fairfield 467 G4
Fairmont 691 G4
Falls City▲ 4,671 J4
Farnam 223 D4
Farwell 148 F3
Filley 174 H4
Firth 564 H4
Fordyce 182 G2
Fort Calhoun 856 J3
Franklin▲ 1,026 E4
Fremont▲ 25,174 H3
Friend 1,174 G4
Fullerton▲ 1,378 F3
Funk 204 E4
Garland 247 G4
Geneva▲ 2,226 G4
Genoa 981 G3
Gering▲ 7,751 A3
Gibbon 1,759 F4
Giltner 389 F4
Glenvil 332 F4
Goehner 186 G4
Gordon 1,756 B2
Gothenburg 3,619 D4
Grafton 152 G4
Grand Island▲ 42,940 F4
Grant▲ 1,225 C4
Greeley▲ 531 F3
Greenwood 544 H3
Gresham 270 G3
Gretna 2,355 H3
Guide Rock 245 F4
Gurley 228 B3
Hadar 312 G2
Haigler 211 C4
Hallam 276 H4
Hampton 439 G4
Hardy 179 G4
Harrisburg▲ 75 A3
Harrison▲ 279 A2
Hartington▲ 1,640 G2
Harvard 998 F4
Hastings▲ 24,064 F4
Hayes Center▲ 240 C4
Hay Springs 652 B2
Hebron▲ 1,565 G4
Hemingford 993 A2
Henderson 986 G4
Henry 162 A2
Herman 310 H3
Hershey 572 D3
Hickman 1,084 H4
Hildreth 370 E4
Holbrook 225 D4
Holdrege▲ 5,636 E4
Holmesville 100 H4
Holstein 229 F4
Homer 590 H2
Hooper 827 H3
Hoskins 283 G2
Howells 632 H3
Hubbard 234 H2
Humboldt 941 J4
Humphrey 786 G3
Hyannis▲ 287 C3
Imperial▲ 1,982 C4
Indianola 642 D4
Inglewood 382 H3
Inman 148 F2
Jackson 205 H2
Johnson 280 J4
Juniata 693 F4
Kearney▲ 27,431 E4
Kenesaw 873 F4
Kennard 371 H3
Kimball▲ 2,559 A3
Laurel 986 G2
La Vista 11,699 J3
Lawrence 312 F4
Leigh 442 G3
Lewellen 282 B3
Lexington▲ 10,011 E4
Lincoln (cap.)▲ 225,581 H4
Lindsay 276 G3
Litchfield 280 E3
Lodgepole 348 B3
Long Pine 341 E2
Loomis 397 E4
Louisville 1,046 H3
Loup City▲ 996 E3
Lyman 421 A3
Lynch 269 F2
Lyons 963 H3
Macy 956 H2
Madison▲ 2,367 G3
Madrid 265 C4
Malcolm 413 H4
Manley 191 H4
Marquette 282 G4
Mason City 178 E3
Max 285 C4
Maxwell 315 D3
Maywood 331 D4
McCook▲ 7,994 D4
McCool Junction 385 G4
Mead 564 H3
Meadow Grove 311 G2
Merna 391 E3
Merriman 118 C2
Milford 2,070 H4
Milligan 315 G4
Minatare 810 A3
Minden▲ 2,964 F4
Mitchell 1,831 A3
Monroe 307 G3
Morrill 957 A3
Mullen▲ 491 C2
Murdock 269 H4
Murray 481 J4
Nebraska City▲ 7,228 J4
Nehawka 232 H4
Neligh▲ 1,651 G2
Nelson▲ 587 F4
Nemaha 178 J4
Newcastle 299 H2
Newman Grove 797 G3
Newport 136 E2
Nickerson 431 H3
Niobrara 379 G2
Norfolk 23,516 G2
North Bend 1,213 H3
North Loup 339 F3
North Platte▲ 23,878 D3
Oakdale 345 F2
Oakland 1,367 H3
Oconto 141 E3
Odell 345 H4
Ogallala▲ 4,930 C3
Ohiowa 142 G4
Omaha▲ 390,007 J3
O'Neill▲ 3,733 F2
Orchard 391 F2
Ord▲ 2,269 F3
Orleans 425 E4
Osceola▲ 921 G3
Oshkosh▲ 887 B3
Osmond 796 G2
Otoe 217 H4
Overton 646 E4
Oxford 876 E4
Page 157 F2
Palisade 386 C4
Palmer 472 F3
Palmyra 546 H4
Panama 253 H4
Papillion▲ 16,363 J3
Pawnee City▲ 1,033 H4
Paxton 614 C3
Pender▲ 1,148 H2
Peru 569 J4
Petersburg 374 G3
Phillips 336 F4
Pickrell 182 H4
Pierce▲ 1,774 G2
Pilger 378 G2
Plainview 1,353 G2
Platte Center 359 G3
Plattsmouth▲ 6,887 J3
Pleasant Dale 245 G4
Pleasanton 360 E4
Plymouth 477 G4
Polk 322 G3
Ponca▲ 1,062 H2
Potter 390 A3
Prague 346 H3
Ralston 6,314 J3
Randolph 955 G2
Ravenna 1,341 F4
Raymond 186 H4
Red Cloud▲ 1,131 F4
Republican City 209 E4
Rising City 386 G3
Riverdale 213 E4
Riverton 145 F4
Rosalie 194 H2
Rose 247 E2
Roseland 242 F4
Rulo 226 J4
Rushville▲ 999 B2
Ruskin 195 G4
Saint Edward 796 G3
Saint Libory 325 F3
Saint Paul▲ 2,218 F3
Santee 302 G2
Sargent 649 E3
Schuyler▲ 5,371 G3
Scotia 308 F3
Scottsbluff 14,732 A3
Scribner 971 H3
Seward▲ 6,319 H4
Shelby 690 G3
Shelton 1,140 F4
Shickley 376 G4
Shubert 252 J4
Sidney▲ 6,282 B3
Silver Creek 441 G3
Snyder 318 H3
South Sioux City 11,925 H2
Spalding 537 F3
Spencer 541 F2
Sprague 146 H4
Springfield 1,450 H3
Springview▲ 244 E2
Stamford 202 E4
Stanton▲ 1,627 G3
Staplehurst 270 G4
Stapleton▲ 301 D3
Stella 220 J4
Sterling 507 H4
Stockville▲ 36 D4
Stratton 396 C4
Stromsburg 1,232 G3
Stuart 625 E2
Sumner 237 E4
Superior 2,055 F4
Sutherland 1,129 C3
Sutton 1,447 G4
Swanton 106 H4
Syracuse 1,762 H4
Table Rock 264 H4
Talmage 268 H4
Taylor▲ 207 E3
Tecumseh▲ 1,716 H4
Tekamah▲ 1,892 H3
Terrytown 646 A3
Thedford▲ 211 D3
Tilden 1,078 G2

Agriculture, Industry and Resources

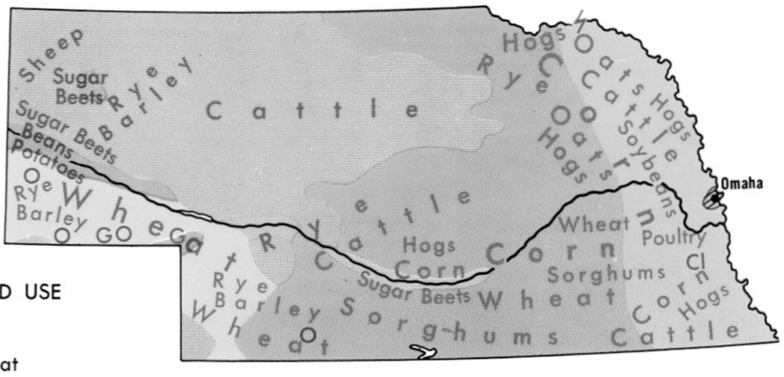
Sheep
Sugar Beets
Barley
Rye
Cattle
Hogs
Oats
Corn
Soybeans
Omaha
Sugar Beets
Beans
Potatoes
Rye
Wheat
Barley
Wheat
Poultry
Sorghums
Sugar Beets
Cl

DOMINANT LAND USE

- Specialized Wheat
- Cattle Feed, Hogs
- Livestock, Cash Grain
- General Farming, Livestock, Special Crops
- Sugar Beets, Dry Beans, Livestock, General Farming
- Range Livestock

MAJOR MINERAL OCCURRENCES

- Cl Clay
- G Natural Gas
- O Petroleum
- Water Power
- Major Industrial Areas

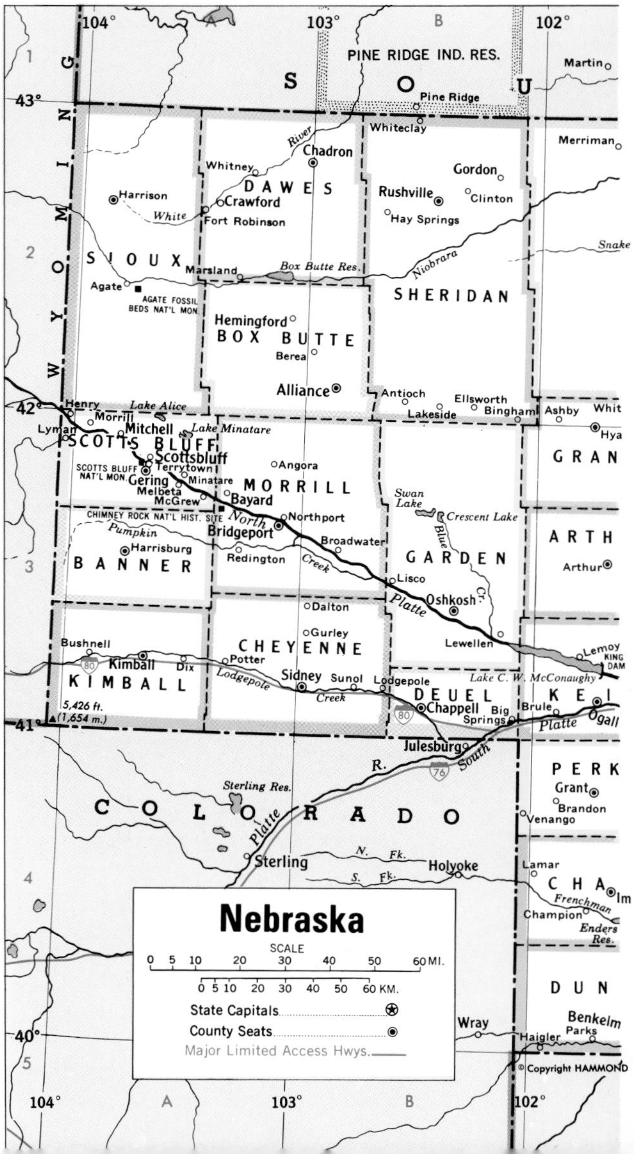
PINE RIDGE IND. RES.
Pine Ridge
Martin
Whiteclay
Merriman
Chadron
Whitney
DAWES
Crawford
Fort Robinson
Harrison
Gordon
Rushville
Clinton
Hay Springs
SIOUX
Marsland
Box Butte Res.
Agate
AGATE FOSSIL BEDS NAT'L MON.
SHERIDAN
Hemingford
BOX BUTTE
Berea
Alliance
Antioch
Ellsworth
Lakeside
Bingham
Ashby
Henry
Lake Alice
Morrill
Lyman
Mitchell
Lake Minatare
SCOTTS BLUFF
Scottsbluff
Terrytown
SCOTTS BLUFF NAT'L MON.
Gering
Minatare
Melbeta
McGrew
Angora
MORRILL
Bayard
Swan Lake
Crescent Lake
CHIMNEY ROCK NAT'L HIST. SITE
Northport
Bridgeport
Broadwater
Harrisburg
BANNER
Redington
GARDEN
Arthur
Lisco
Oshkosh
Dalton
Gurley
Lewellen
Bushnell
CHEYENNE
Potter
Kimball
Dix
KIMBALL
Sidney
Sunol
Lodgepole
Lake C. W. McConaughy
DEUEL
Chappell
Big Springs
Brule
5,426 ft. (1,654 m.)
Julesburg
PERK
Grant
Brandon
Venango
Sterling Res.
COLORADO
Sterling
Holyoke
Lamar
CHA
Champion
Enders Res.
Nebraska
SCALE
0 5 10 20 30 40 50 60 MI.
0 5 10 20 30 40 50 60 KM.
State Capitals
County Seats
Major Limited Access Hwys.
DUN
Benkelman
Parks
Wray
Haigler
© Copyright HAMMOND

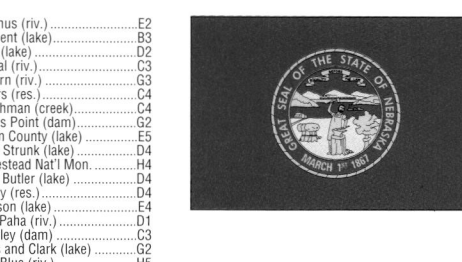

AREA 77,355 sq. mi. (200,349 sq. km.)
POPULATION 1,711,263
CAPITAL Lincoln
LARGEST CITY Omaha
HIGHEST POINT (Kimball Co.) 5,424 ft. (1654 m.)
SETTLED IN 1847
ADMITTED TO UNION March 1, 1867
POPULAR NAME Cornhusker State
STATE FLOWER Goldenrod
STATE BIRD Western Meadowlark

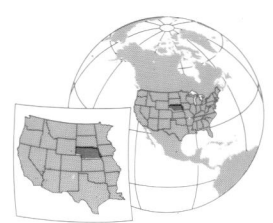

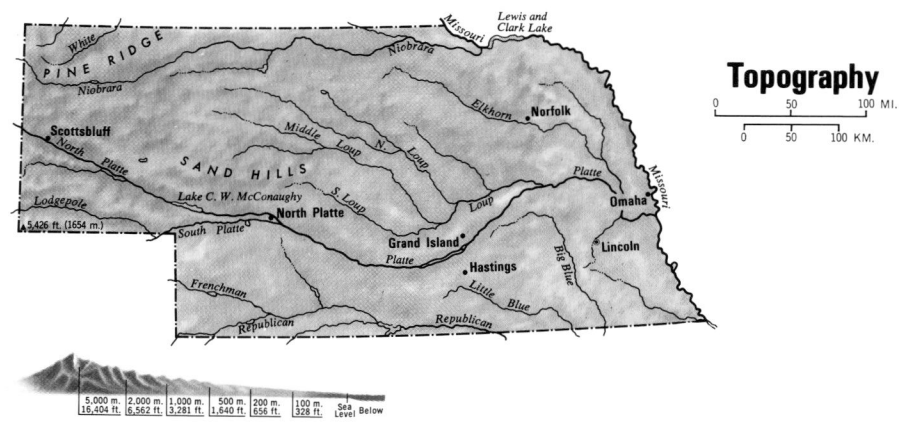

Topography

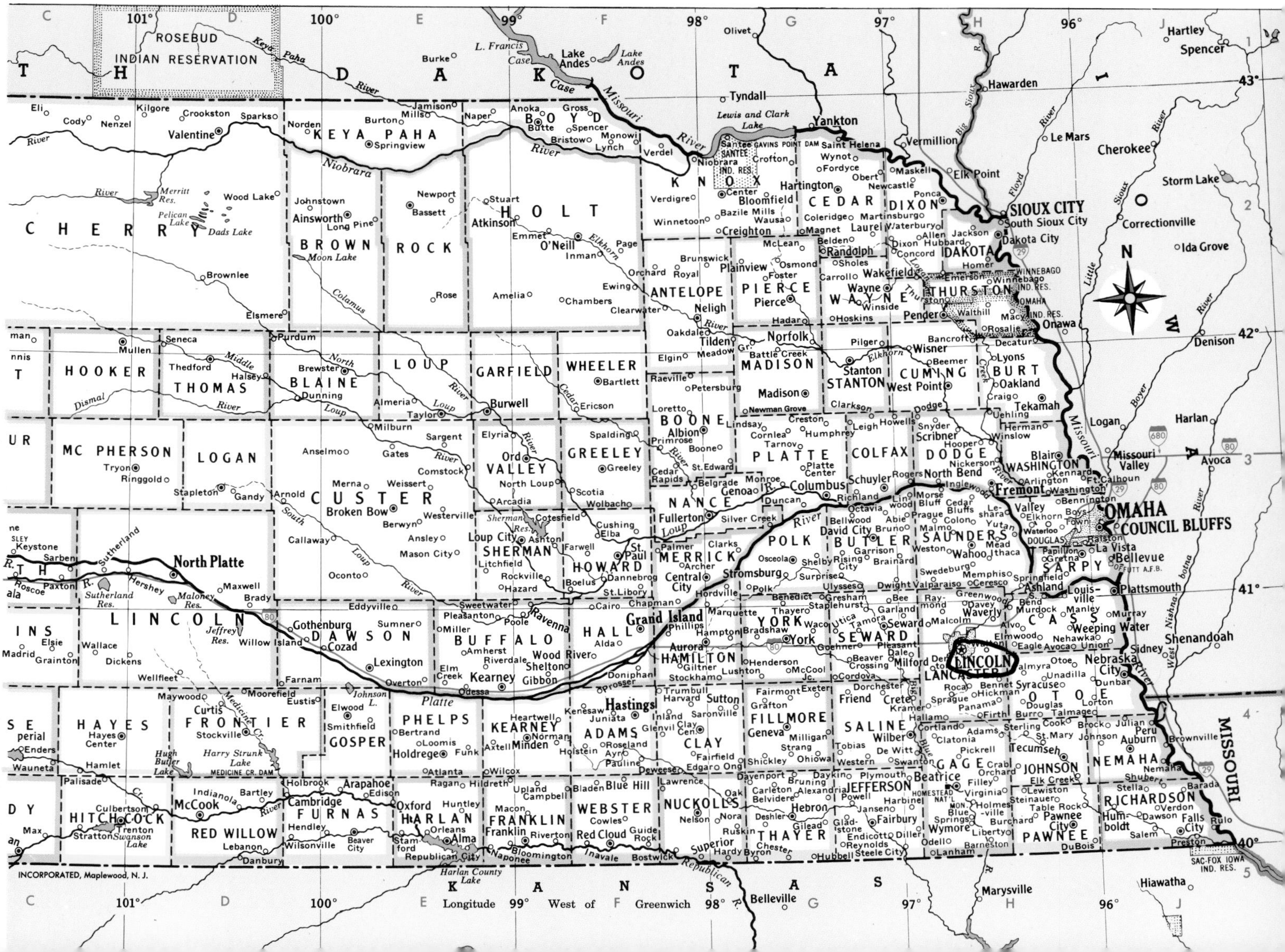

5,000 m. / 2,000 m. / 1,000 m. / 500 m. / 200 m. / 100 m. / Sea Level / Below
16,404 ft. / 6,562 ft. / 3,281 ft. / 1,640 ft. / 656 ft. / 328 ft.

INCORPORATED, Maplewood, N. J.

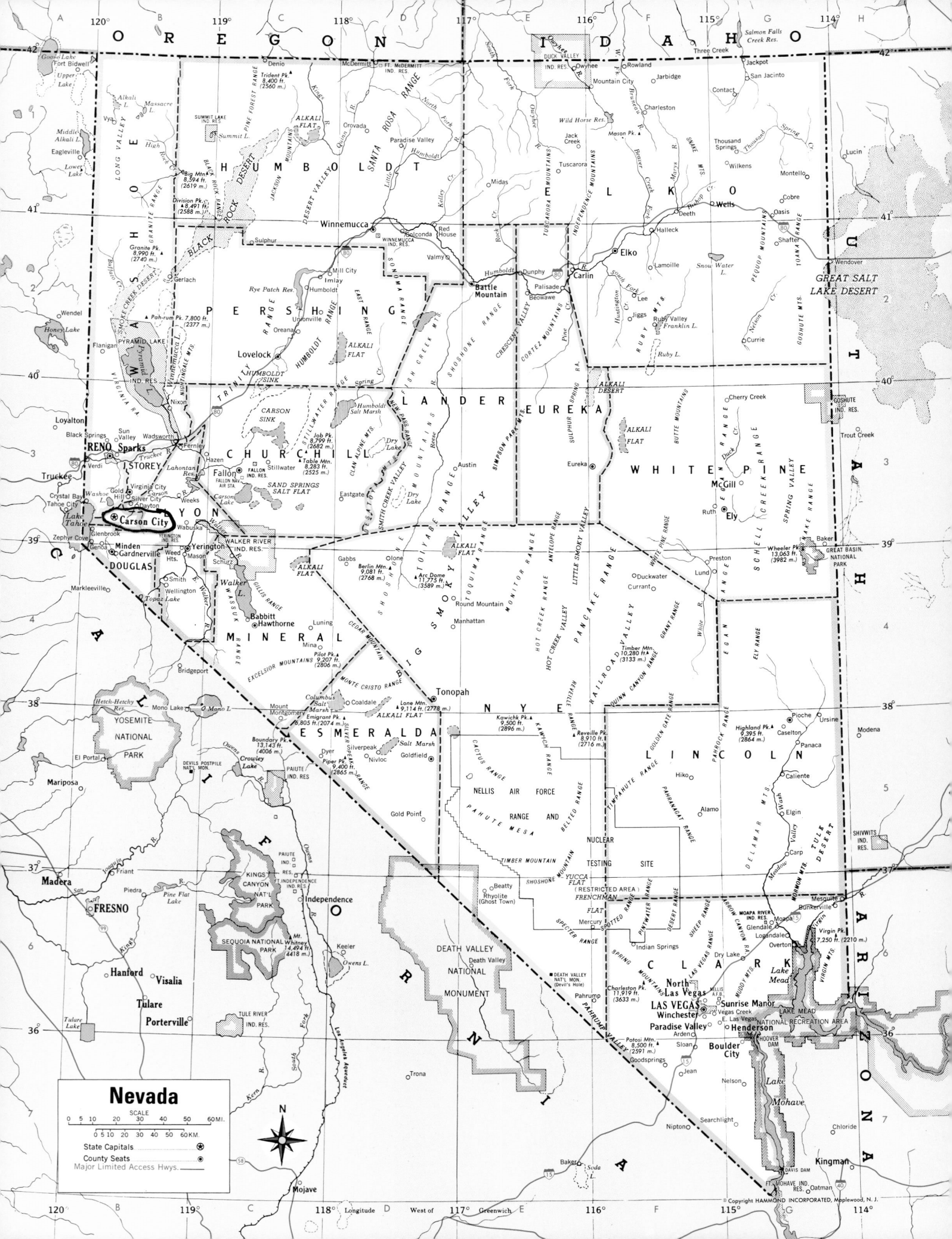

Nevada

SCALE

0 5 10   20   30   40   50   60 MI.

0 5 10   20   30   40   50   60 KM.

⊛ State Capitals

⊙ County Seats

Major Limited Access Hwys. ━━━

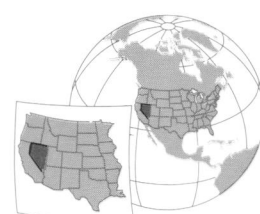

AREA 110,561 sq. mi. (286,353 sq. km.)
POPULATION 1,998,257
CAPITAL Carson City
LARGEST CITY Las Vegas
HIGHEST POINT Boundary Pk. 13,140 ft.
(4005 m.)
SETTLED IN 1850
ADMITTED TO UNION October 31, 1864
POPULAR NAME Silver State; Sagebrush State
STATE FLOWER Sagebrush
STATE BIRD Mountain Bluebird

MAJOR MINERAL OCCURRENCES

Ag Silver
Au Gold
Ba Barite
Cu Copper
Gp Gypsum
Hg Mercury
Lt Lithium
Mg Magnesium
Mo Molybdenum
Na Salt
O Petroleum
Pb Lead
S Sulfur
W Tungsten   ⚡ Water Power
Zn Zinc

DOMINANT LAND USE

General Farming, Dairy, Livestock
General Farming, Livestock, Special Crops
Range Livestock
Forests
Nonagricultural Land

Agriculture, Industry and Resources

Topography

0   60   120 MI.
0   60   120 KM.

5,000 m. 2,000 m. 1,000 m. 500 m. 200 m. 100 m. Sea Level Below
16,404 ft. 6,562 ft. 3,281 ft. 1,640 ft. 656 ft. 328 ft.

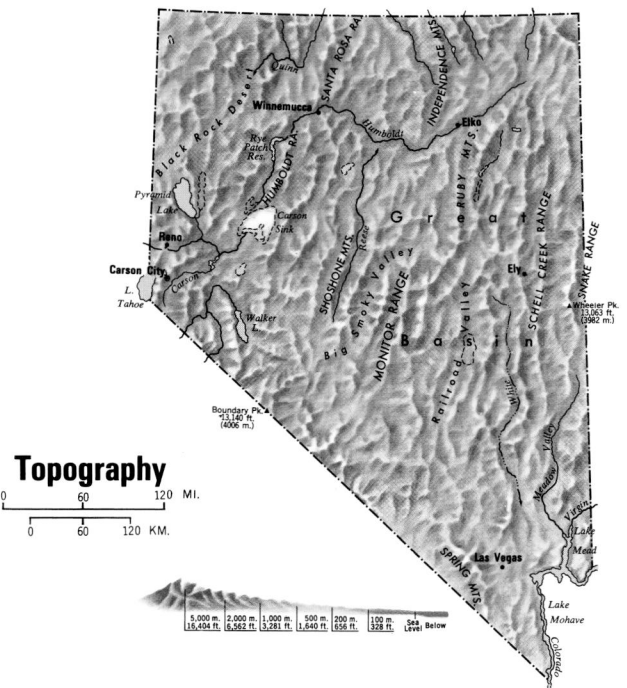

New Hampshire
and Vermont

SCALE

0 5 10 15 20 25 MI.

0 5 10 15 20 25 KM.

State Capitals .... ⊛
County Seats .... ◉
Major Limited Access Hwys. _____

© Copyright HAMMOND INCORPORATED, Maplewood, N.J.

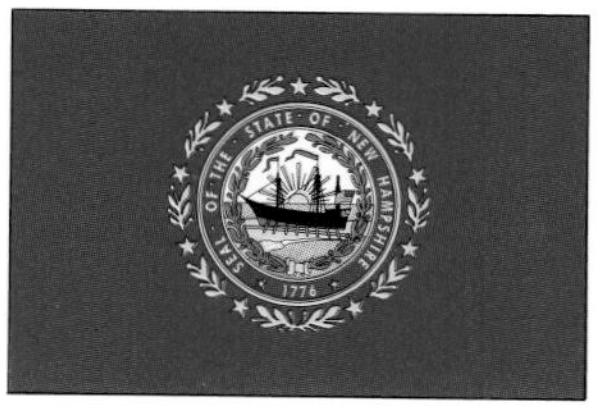

NEW HAMPSHIRE

AREA 9,279 sq. mi. (24,033 sq. km.)
POPULATION 1,235,786
CAPITAL Concord
LARGEST CITY Manchester
HIGHEST POINT Mt. Washington 6,288 ft. (1917 m.)
SETTLED IN 1623
ADMITTED TO UNION June 21, 1788
POPULAR NAME Granite State
STATE FLOWER Purple Lilac
STATE BIRD Purple Finch

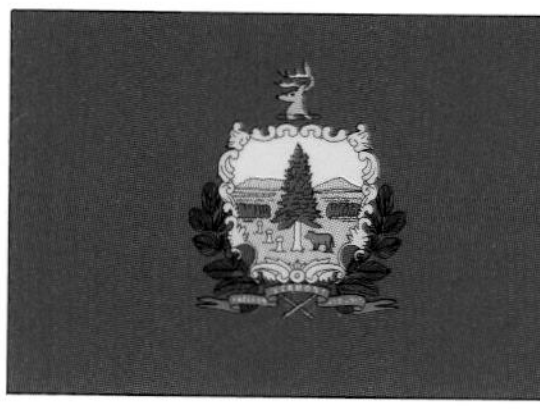

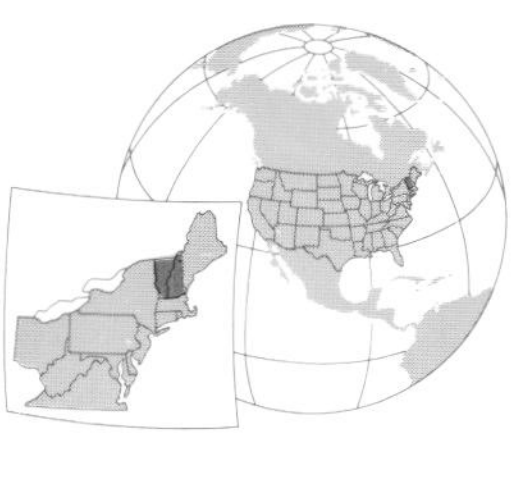

VERMONT

AREA 9,614 sq. mi. (24,910 sq. km.)
POPULATION 562,758
CAPITAL Montpelier
LARGEST CITY Burlington
HIGHEST POINT Mt. Mansfield 4,393 ft. (1339 m.)
SETTLED IN 1764
ADMITTED TO UNION March 4, 1791
POPULAR NAME Green Mountain State
STATE FLOWER Red Clover
STATE BIRD Hermit Thrush

Topography

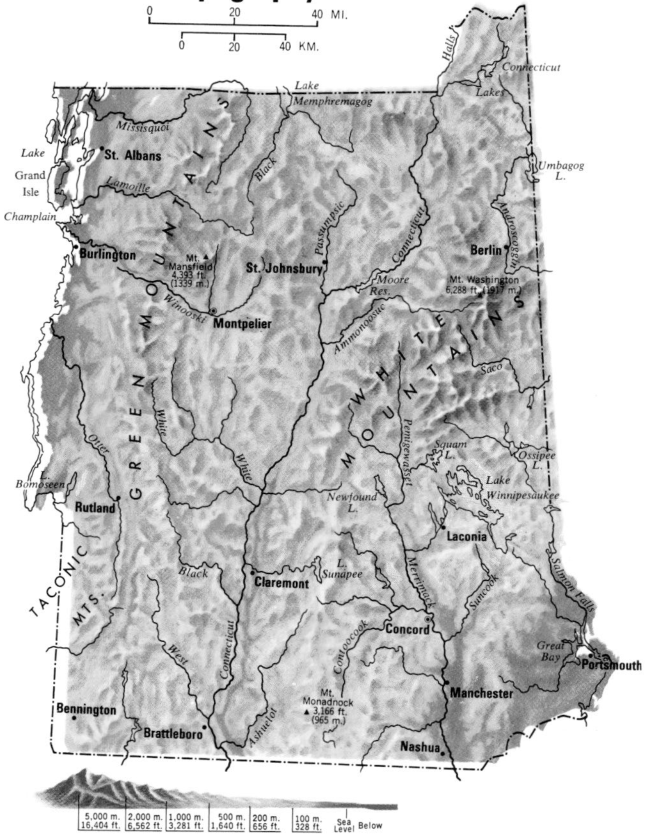
0 20 40 MI.
0 20 40 KM.
Lake Memphremagog
Missisquoi
Lake Champlain
Grand Isle
St. Albans
Lamoille
Black
Passumpsic
Connecticut
Connecticut Lakes
Halls
Umbagog L.
Androscoggin
Berlin
Burlington
Mt. Mansfield 4,393 ft. (1339 m.)
St. Johnsbury
Moore Res.
Mt. Washington 6,288 ft. (1917 m.)
Winooski
Montpelier
Ammonoosuc
GREEN MOUNTAINS
WHITE MOUNTAINS
Saco
White
Otter
Pemigewasset
Squam L.
Ossipee L.
Lake Winnipesaukee
Bomoseen
Rutland
Newfound L.
Laconia
TACONIC MTS.
Black
Claremont
L. Sunapee
Merrimack
Suncook
Salmon Falls
Concord
Contoocook
Great Bay
Portsmouth
West
Connecticut
Mt. Monadnock 3,166 ft. (965 m.)
Manchester
Bennington
Ashuelot
Brattleboro
Nashua
5,000 m. 16,404 ft.
2,000 m. 6,562 ft.
1,000 m. 3,281 ft.
500 m. 1,640 ft.
200 m. 656 ft.
100 m. 328 ft.
Sea Level
Below

NEW HAMPSHIRE

COUNTIES

Belknap 56,325 ... D4
Carroll 43,666 ... E4
Cheshire 73,825 ... C6
Coos 33,111 ... E2
Grafton 81,743 ... D4
Hillsborough 380,841 ... D6
Merrimack 136,225 ... D5
Rockingham 277,359 ... E5
Strafford 112,233 ... E5
Sullivan 40,458 ... C5

CITIES and TOWNS

Acworth• 836 ... C5
Albany• 654 ... E4
Alexandria• 1,329 ... D4
Allenstown• 4,843 ... E5
Alstead• 1,944 ... C5
Alton• 4,502 ... E5
Alton Bay 500 ... E5
Amherst• 10,769 ... D6
Andover• 2,109 ... D5
Antrim• 2,449 ... D5
Antrim 1,389 ... D5
Ashland• 1,955 ... D4
Ashland 1,915 ... D4
Ashuelot 810 ... C6
Atkinson• 6,178 ... E6
Auburn• 4,682 ... E5
Barnstead• 3,886 ... E5
Barrington• 7,475 ... E5
Bartlett• 2,705 ... E3
Bath• 893 ... D3
Bedford• 18,274 ... D6
Beebe River 355 ... D4
Belmont• 6,716 ... E5
Bennington• 1,401 ... D5
Benton• 314 ... D3
Berlin 10,331 ... E3
Bethlehem• 2,199 ... D3
Boscawen• 3,672 ... D5
Bow Mills 802 ... D5
Bradford• 1,454 ... D5
Brentwood• 3,197 ... E6
Bretton Woods 6 ... E3
Bridgewater• 974 ... D4
Bristol• 3,033 ... D4
Bristol 1,670 ... D4
Brookfield• 604 ... E4
Brookline• 4,181 ... D6
Campton• 2,719 ... D4
Canaan• 3,319 ... C4
Candia• 3,911 ... E5
Canobie Lake 500 ... E6
Canterbury• 1,979 ... D5
Carroll• 663 ... D3
Cascade 350 ... E3
Center Barnstead 400 ... E5
Center Conway 558 ... E4
Center Harbor• 996 ... E4
Center Ossipee• 800 ... E4
Charlestown• 4,749 ... C5
Charlestown 1,145 ... C5
Chatham• 260 ... E3
Chester• 3,792 ... E6
Chesterfield• 3,542 ... C6
Chichester• 2,236 ... E5
Chocorua 575 ... E4
Claremont 13,151 ... C5
Clarksville• 294 ... E1
Colebrook• 2,459 ... E2
Colebrook 2,321 ... E2
Concord (cap.)▲ 40,687 ... D5
Contoocook 1,444 ... D5
Conway• 8,604 ... E4
Conway 1,692 ... E4
Croydon• 661 ... C5
Crystal 75 ... E2
Dalton• 927 ... D3
Danbury• 1,071 ... D4
Danville• 4,023 ... E6
Deerfield• 3,678 ... E5
Deering• 1,875 ... D5
Derry• 34,021 ... E6
Derry 22,661 ... E6
Dixville Notch 10 ... E2
Dorchester• 353 ... D4
Dover▲• 26,884 ... F5
Dublin• 1,476 ... C6
Dummer• 309 ... E2
Durham• 12,664 ... F5
Durham 9,024 ... F5
East Andover 500 ... D5
East Hampstead 900 ... E6
East Kingston• 1,784 ... F6
East Lempster 300 ... C5
Easton• 256 ... D3
East Sullivan 300 ... C6
East Swanzey 500 ... C6
East Wolfeboro 400 ... F4
Eaton (Eaton Center) 362 ... F4
Ellsworth• 87 ... D4
Enfield• 4,618 ... C4
Enfield 1,698 ... C4
Epping• 5,476 ... E5
Epping 1,673 ... E5
Epsom• 4,021 ... E5
Errol• 298 ... E2
Etna 550 ... C4
Exeter• 14,058 ... F6
Exeter▲ 9,759 ... F6
Farmington• 5,774 ... E5
Farmington 3,468 ... E5
Fitzwilliam• 2,141 ... C6
Fitzwilliam Depot 350 ... C6
Francestown• 1,480 ... D6
Franconia• 924 ... D3
Franklin 8,405 ... D5
Freedom• 1,303 ... E4
Fremont• 3,510 ... E6
Georges Mills 375 ... C5
Gerrish 500 ... D5
Gilford• 6,803 ... E4
Gilmanton• 3,060 ... E5
Gilmanton Iron Works 300 ... E5
Gilsum• 777 ... C5
Glen 600 ... E3
Goffstown• 16,929 ... D5
Gorham• 2,895 ... E3
Gorham 1,773 ... E3
Goshen• 741 ... C5
Grafton• 1,138 ... D4
Grantham• 2,167 ... C5
Grasmere 400 ... E5
Greenfield• 1,657 ... D6
Greenland• 3,208 ... F5
Greenville• 2,224 ... D6
Greenville 1,131 ... D6
Groton• 456 ... D4
Groveton 1,197 ... D2
Guild 500 ... C5
Hampstead• 8,297 ... E6
Hampton• 14,937 ... F6
Hampton 9,126 ... F6
Hampton Beach 975 ... F6
Hampton Falls• 1,880 ... F6
Hancock• 1,739 ... C6
Hanover• 10,850 ... C4
Hanover 8,162 ... C4
Harrisville• 1,075 ... C6
Haverhill• 4,416 ... C3
Hebron• 459 ... D4
Henniker• 4,433 ... D5
Henniker 1,627 ... D5
Hill• 992 ... D4
Hillsboro• 4,928 ... D5
Hillsboro 1,842 ... D5
Hinsdale• 4,082 ... D6
Hinsdale 1,713 ... D6
Holderness• 1,930 ... D4
Hollis• 7,015 ... D6
Hooksett• 11,721 ... E5
Hooksett 3,609 ... E5
Hopkinton• 5,399 ... D5
Hudson• 22,928 ... E6
Hudson 7,814 ... E6
Intervale 725 ... E3
Jackson• 835 ... E3
Jaffrey• 5,476 ... C6
Jaffrey 2,802 ... C6
Jaffrey Center 340 ... C6
Jefferson• 1,006 ... D3
Kearsarge 350 ... F3
Keene▲ 22,563 ... C6
Kingston• 5,862 ... E6
Laconia▲• 16,411 ... E4
Lancaster• 3,280 ... D3
Lancaster▲ 1,695 ... D3
Landaff• 378 ... D3
Langdon• 586 ... C5
Lebanon 12,568 ... C4
Lee• 4,145 ... F5
Lempster• 971 ... C5
Lincoln• 1,271 ... D3
Lisbon• 1,587 ... D3
Lisbon 1,070 ... D3
Litchfield• 7,360 ... E6
Littleton• 5,845 ... D3
Littleton 4,431 ... D3
Lochmere 300 ... D5
Londonderry• 23,236 ... E6
Loudon• 4,481 ... E5
Lyman• 487 ... D3
Lyme• 1,679 ... C4
Lyndeborough• 1,585 ... D6
Madbury• 1,509 ... F5
Madison• 1,984 ... E4
Manchester 107,006 ... E6
Marlborough• 2,009 ... C6
Marlborough 1,089 ... C6
Marlow• 747 ... C5
Melvin Village 450 ... F4
Meredith• 5,943 ... D4
Meredith 1,739 ... D4
Meriden 800 ... C4
Merrimack• 25,119 ... D6
Middleton• 1,440 ... E5
Milan• 1,331 ... E2
Milford• 13,535 ... D6
Milford 8,293 ... D6
Milton• 3,910 ... F5
Milton Mills 450 ... F4
Mirror Lake 350 ... F4
Monroe• 759 ... C3
Mont Vernon• 2,034 ... D6
Moultonboro• 4,484 ... E4
Nashua▲ 86,605 ... D6
Nelson• 634 ... C5
New Boston• 4,138 ... D6
Newbury• 1,702 ... C5
New Castle• 1,010 ... F5
New Durham• 2,220 ... E5
Newfields• 1,551 ... F5
New Hampton• 1,950 ... D4
Newington• 775 ... F5
New Ipswich• 4,289 ... D6
New London• 4,116 ... D5
New London 2,935 ... D5
Newmarket• 8,027 ... F5
Newmarket 5,124 ... F5
Newport• 6,269 ... C5
Newport▲ 4,008 ... C5
Newton• 4,289 ... E6
Newton Junction 450 ... E6
North Chichester 450 ... E5
North Conway 2,069 ... E3
Northfield• 4,548 ... D5
Northfield-Tilton 3,081 ... D5
North Hampton• 4,259 ... F6
North Haverhill 400 ... D3
North Stratford 600 ... D2
Northumberland• 2,438 ... D2
North Walpole 950 ... C5
North Weare 400 ... D5
Northwood• 3,640 ... E5
Northwood Narrows 325 ... E5
North Woodstock 750 ... D3
Nottingham• 3,701 ... E5
Orange• 299 ... D4
Orford• 1,091 ... C4
Ossipee• 4,211 ... E4
Pelham• 10,914 ... E6
Pembroke• 6,897 ... D5
Peterborough• 5,883 ... D6
Peterborough 2,944 ... D6
Piermont• 709 ... C4
Pike 433 ... C3
Pittsburg• 867 ... E1
Pittsfield• 3,931 ... E5
Pittsfield 1,669 ... E5
Plainfield• 2,241 ... C4
Plaistow• 7,747 ... E6
Plymouth• 5,892 ... D4
Plymouth 3,528 ... D4
Portsmouth 20,784 ... F5
Randolph• 339 ... E3
Raymond• 9,674 ... E5
Raymond 2,839 ... E5
Redstone 300 ... E3
Richmond• 877 ... C6
Rindge• 5,451 ... C6
Rochester 28,461 ... E5
Roxbury• 237 ... C6
Rumney• 1,480 ... D4
Rye• 5,182 ... F5
Rye Beach 600 ... F6
Rye North Beach 700 ... F5
Salem• 28,112 ... E6
Salem Depot 975 ... E6
Salisbury• 1,137 ... D5
Salmon Falls 950 ... F5
Sanbornton• 2,581 ... D5
Sanbornville 750 ... F4
Sandown• 5,143 ... E6
Sandwich• 1,286 ... E4
Seabrook• 7,934 ... F6
Sharon• 360 ... D6
Shelburne• 379 ... E3
Shelburne 318 ... E3
Silver Lake 350 ... E4
Somersworth 11,477 ... F5
South Deerfield 500 ... E5
South Hampton• 844 ... F6
South Lyndeboro 300 ... D6
South Merrimack 650 ... D6
South Seabrook 500 ... F6
South Weare 400 ... D5
Spofford 750 ... C6
Springfield• 945 ... C4
Stark• 516 ... E2
Stewartstown• 1,012 ... E2
Stoddard• 928 ... C5
Strafford• 3,626 ... E5
Stratford• 942 ... D2
Stratham• 6,355 ... F5
Sugar Hill• 563 ... D3
Sullivan• 746 ... C5
Sunapee• 3,055 ... C5
Suncook 5,362 ... E5
Surry• 673 ... C5
Sutton• 1,544 ... D5
Swanzey• 6,800 ... C6
Tamworth• 2,510 ... E4
Temple• 1,297 ... D6
Thornton• 1,843 ... D4
Tilton• 3,477 ... D5
Tilton-Northfield 3,231 ... D5
Troy• 1,962 ... C6
Troy 2,097 ... C6
Tuftonboro• 2,148 ... E4
Twin Mountain 500 ... D3
Unity• 1,530 ... C5
Wakefield• 4,252 ... F4
Walpole• 3,594 ... C5
Warner• 2,760 ... D5
Warren• 873 ... D4
Washington• 895 ... C5
Waterville Valley• 257 ... D4
Weare• 7,776 ... D5
Webster• 1,579 ... D5
Wentworth• 798 ... D4
Wentworths Location• 44 ... E2
West Campton 400 ... D4
West Epping 400 ... E5
West Henniker 500 ... D5
West Lebanon ... C4
West Milan 350 ... E2
Westmoreland• 1,747 ... C6
West Rye 350 ... F6
West Stewartstown 700 ... D2
West Swanzey 1,118 ... C6
Westville 750 ... E6
Whitefield• 2,038 ... D3
Whitefield 1,089 ... D3
Wilmot• 1,144 ... D5
Wilmot Flat 450 ... D5
Wilton• 3,743 ... D6
Wilton 1,236 ... D6
Winchester• 4,144 ... C6
Windham• 10,709 ... E6
Winnisquam 500 ... E5
Wolfeboro• 6,083 ... E4
Wolfeboro 2,979 ... E4
Wolfeboro Falls 600 ... E4
Woodstock• 1,139 ... D4
Woodsville▲ 1,081 ... C3

OTHER FEATURES

Adams (mt.) ... E3
Ammonoosuc (riv.) ... D3
Androscoggin (riv.) ... E2
Ashuelot (riv.) ... C6
Back (lake) ... E1
Baker (riv.) ... D4
Bearcamp (riv.) ... E4
Beaver (brook) ... E6
Belknap (mt.) ... E5
Blackwater (res.) ... D5
Blue (mt.) ... E2
Bond (mt.) ... E3
Bow (lake) ... E5
Cabot (mt.) ... E2
Cannon (mt.) ... D3
Cardigan (mt.) ... D4
Carrigain (mt.) ... E3
Carter Dome (mt.) ... E3
Chocorua (mt.) ... E4
Cocheco (riv.) ... E5
Cold (riv.) ... C5
Comerford (dam) ... D3
Connecticut (riv.) ... B6
Contoocook (riv.) ... D6
Conway (lake) ... E4
Crawford Notch (pass) ... E3
Croydon (peak) ... C5
Croydon Branch, Sugar (riv.) ... C5
Crystal (lake) ... E5
Cube (mt.) ... D4
Dixville (peak) ... E2
Dixville Notch (pass) ... E2
Edward MacDowell (res.) ... D6
Ellis (riv.) ... E3
Everett (dam) ... D5
Exeter (riv.) ... E6
First Connecticut (lake) ... E1
Francis (lake) ... E1
Franconia Notch (pass) ... D3
Franklin Falls (res.) ... D4
Gale (riv.) ... D3
Great (bay) ... F5
Halls (stream) ... E1
Hancock (mt.) ... D3
Highland (lake) ... C5
Hutchins (mt.) ... E2
Indian (stream) ... E1
Jefferson (mt.) ... E3
Kearsarge (mt.) ... D5
Kinsman (mt.) ... D3
Kinsman Notch (pass) ... D3
Lafayette (mt.) ... D3
Lamprey (riv.) ... E5
Liberty (mt.) ... D3
Lincoln (mt.) ... D3
Long (mt.) ... E2
Mad (riv.) ... D4
Madison (mt.) ... E3
Mascoma (lake) ... C4
Massabesic (lake) ... E6
Merrimack (riv.) ... D5
Merrymeeting (lake) ... E5
Mohawk (riv.) ... E2
Monadnock (mt.) ... C6
Monroe (mt.) ... E3
Moore (dam) ... D3
Moore (res.) ... D3
Moosilauke (mt.) ... D3
Nash (stream) ... E2
Newfound (lake) ... D4
North Carter (mt.) ... E3
North Twin (mt.) ... D3
Nubanusit (lake) ... C5
Osceola (mt.) ... E3
Ossipee (lake) ... E4
Ossipee (mts.) ... E4
Ossipee (riv.) ... F4
Passaconaway (mt.) ... E4
Pawtuckaway (pond) ... E5
Pease A.F.B. ... F5
Pemigewasset (riv.) ... D4
Perry (stream) ... E1
Pine (riv.) ... E4
Pinkham Notch (pass) ... E3
Piscataqua (riv.) ... F5
Piscataquog (riv.) ... D5
Presidential (range) ... E3
Rice (mt.) ... E2
Saco (riv.) ... E3

(continued on following page)

Agriculture, Industry and Resources

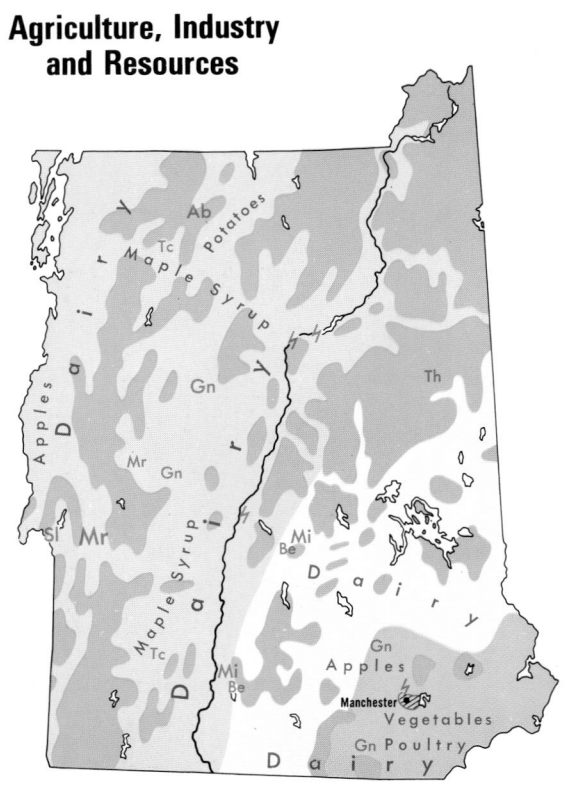

DOMINANT LAND USE

	Specialized Dairy
	Dairy, General Farming
	Dairy, Poultry, Mixed Farming
	Forests

⚡ Water Power

▨ Major Industrial Areas

MAJOR MINERAL OCCURRENCES

Ab	Asbestos	Mr	Marble
Be	Beryl	Sl	Slate
Gn	Granite	Tc	Talc
Mi	Mica	Th	Thorium

AREA 7,787 sq. mi. (20,168 sq. km.)
POPULATION 8,414,350
CAPITAL Trenton
LARGEST CITY Newark
HIGHEST POINT High Point 1,803 ft. (550 m.)
SETTLED IN 1617
ADMITTED TO UNION December 18, 1787
POPULAR NAME Garden State
STATE FLOWER Purple Violet
STATE BIRD Eastern Goldfinch

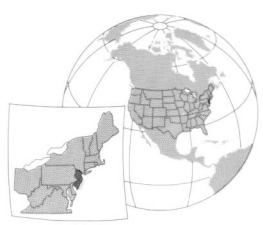

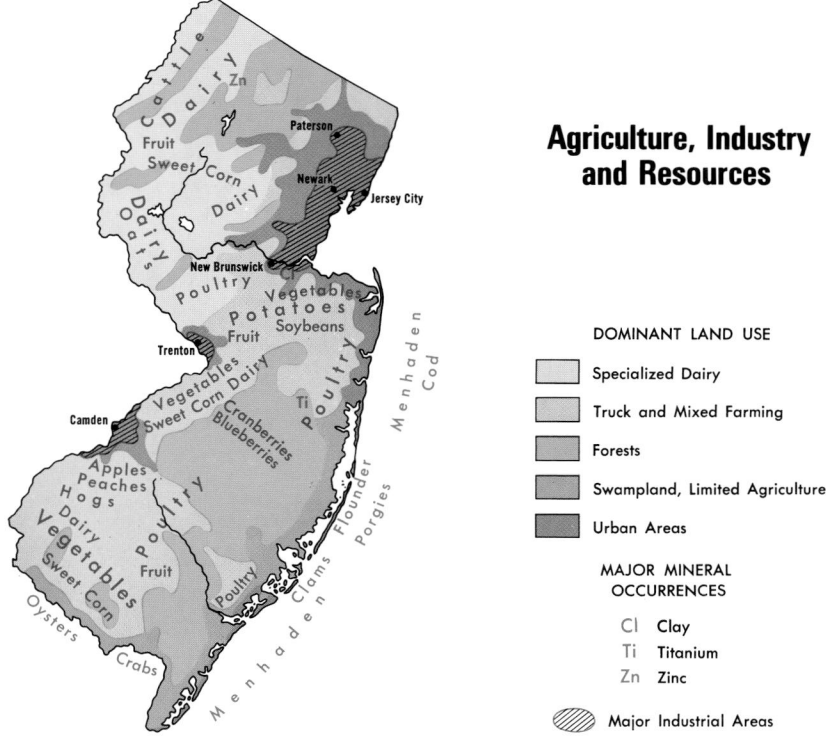

Agriculture, Industry and Resources

DOMINANT LAND USE

Specialized Dairy

Truck and Mixed Farming

Forests

Swampland, Limited Agriculture

Urban Areas

MAJOR MINERAL OCCURRENCES

Cl Clay

Ti Titanium

Zn Zinc

Major Industrial Areas

The Urban Northeast

Urbanized Areas

• Places with more than 10,000 inhabitants
• Places with 5,000-10,000 inhabitants
• Places with 2,500-5,000 inhabitants

© Copyright HAMMOND INCORPORATED, Maplewood, N. J.

COUNTIES

Atlantic 252,552	D5
Bergen 884,118	E2
Burlington 423,394	D4
Camden 508,932	D4
Cape May 102,326	D5
Cumberland 146,438	C5
Essex 793,633	E2
Gloucester 254,673	C4
Hudson 608,975	E2
Hunterdon 121,989	D2
Mercer 350,761	D3
Middlesex 750,162	E3
Monmouth 615,301	E3
Morris 470,212	D2
Ocean 510,916	E4
Passaic 489,049	E1
Salem 64,285	C4
Somerset 297,490	D2
Sussex 144,166	D1
Union 522,541	E2
Warren 102,437	C2

CITIES and TOWNS

Aberdeen • 17,454	E3
Absecon 7,638	D5
Allaire	E3
Allamuchy • 3,877	D2
Allendale 6,699	B1
Allenhurst 718	F3
Allentown 1,882	D3
Allenwood 935	E3
Alloway 1,128	C4
Alpha 2,482	C2
Alpine 2,183	C1
Andover 658	D2
Annandale 1,276	D2
Asbury Park 16,930	F3
Atlantic City▲ 40,517	E5
Atlantic Highlands 4,705	F3
Audubon 9,182	B3
Audubon Park 1,102	B3
Avalon 2,143	D5
Avenel 17,552	E2
Avon By The Sea 2,244	E3
Barnegat 1,690	E4
Barnegat Light 764	E4
Barrington 7,084	B3
Basking Ridge	D2
Batsto 19	D4
Bay Head 1,226	E3
Bayonne 61,842	B2
Bayville	E4
Beach Haven 1,278	E4
Beachwood 10,375	E4
Bedminster • 8,302	D2
Belle Mead	D3
Belleville 35,928	B2
Bellmawr 11,262	B3
Belmar 6,045	E3
Belvidere▲ 2,771	C2
Bergenfield 26,247	C1
Berkeley Heights • 13,407	E2
Berlin 6,149	D4
Bernardsville 7,345	D2
Beverly 2,661	D3
Bivalve 30	C5
Blackwood 4,692	C4
Blairstown • 5,747	C2
Bloomfield 47,683	B2
Bloomingdale 7,610	E1
Bloomsbury 886	C2
Bogota 8,249	B2
Boonton 8,496	E2
Bordentown 3,969	D3
Bound Brook 10,155	D2
Bradley Beach 4,793	F3
Branchville 845	D1
Breton Woods	E3
Brick • 76,119	E3
Bridgeport 750	C4
Bridgeton▲ 22,771	C5
Bridgewater • 42,940	D2
Brielle 4,893	E3
Brigantine 12,594	E5
Brooklawn 2,354	B3
Browns Mills 11,257	D4
Budd Lake 8,100	D2
Buena 3,873	D4
Burlington 9,736	D3
Butler 7,420	E2
Caldwell 7,584	B2
Califon 1,055	D2
Camden▲ 79,904	B3
Candlewood 6,750	E3
Cape May 4,034	D6
Cape May Court House▲ 4,704	D5
Cape May Point 241	D6
Carlstadt 5,917	B2
Carneys Point 6,914	C4
Carteret 20,709	E2
Cedar Brook 600	D4
Cedar Grove • 12,300	B2

Cedar Knolls	E2
Cedarville 793	C5
Chatham 8,460	E2
Chatsworth 700	D4
Cheesequake	E3
Cherry Hill • 69,965	B3
Chesilhurst 1,520	D4
Chester 1,635	D2
Chesterfield • 5,955	D3
Cinnaminson • 14,595	B3
Clark • 14,597	A3
Clarksburg 800	E3
Clayton 7,139	C4
Clementon 4,986	D4
Cliffside Park 23,007	C2
Clifton 78,672	B2
Clinton 2,632	D2
Closter 8,383	C1
Cold Spring 500	D6
Collingswood 14,326	B3
Cologne 800	D4
Colonia 17,811	E2
Colts Neck 12,331	E3
Columbia 600	C2
Columbus 800	D3
Convent Station	E2
Corbin City 468	D5
Cranberry Lake 500	D2
Cranbury 2,008	E3
Cranford • 22,578	A3
Cresskill 7,746	C1
Crosswicks 265	D3
Dayton 6,235	D3
Deal 1,070	F3
Delair	B3
Delanco • 3,237	D3
Delran • 15,536	B3
Demarest 4,845	C1
Dennisville 890	D5
Denville • 15,824	E2
Deptford • 26,763	B4
Dorothy 900	D5
Dover 18,188	D2
Dumont 17,503	C1
Dunellen 6,823	D2
East Brunswick • 46,756	E3
East Hanover • 11,393	E2
East Millstone 950	D3
East Newark 2,377	B2
East Orange 69,824	B2
East Rutherford 8,716	B2
Eatontown 14,008	E3
Echo Lake 300	E1
Edgewater 7,677	C2
Edgewater Park • 7,864	D3
Edison • 97,687	E2
Egg Harbor City 4,545	D4
Elberon	F3
Elizabeth▲ 120,568	B2
Elmer 1,384	C4
Elmwood Park 18,925	B2
Elwood 1,392	D4
Emerson 7,197	B1
Englewood 26,203	C2
Englewood Cliffs 5,322	C2
Englishtown 1,764	E3
Essex Fells 2,162	B2
Estell Manor 1,585	D5
Fairfield • 7,063	A2
Fair Haven 5,937	E3
Fair Lawn 31,637	B1
Fairton 2,253	C5
Fairview 13,255	C2
Fanwood 7,174	E2
Far Hills 859	D2
Farmingdale 1,587	E3
Flagtown 800	D2
Flanders	D2
Flemington▲ 4,200	D2
Florence-Roebling 8,200	D3
Florham Park 8,857	E2
Folsom 1,972	D4
Fords 15,032	E2
Forked River 4,914	E4
Fort Lee 35,461	C2
Franklin 5,160	D1
Franklin Lakes 10,422	B1
Franklin Park	D3
Franklinville	C4
Freehold▲ 10,976	E3
Frenchtown 1,488	C2
Garfield 29,786	B2
Garwood 4,153	E2
Gibbsboro 2,435	B4
Gibbstown 3,758	C4
Gilford Park 8,668	E4
Gillette	E2
Glassboro 19,068	C4
Glendora 4,907	B4
Glen Gardner 1,902	D2
Glen Ridge 7,271	B2
Glen Rock 11,546	B1
Gloucester City 11,484	B3
Green Brook • 5,654	D2
Green Pond 800	E1
Green Village 800	D2

(continued on following page)

Topography

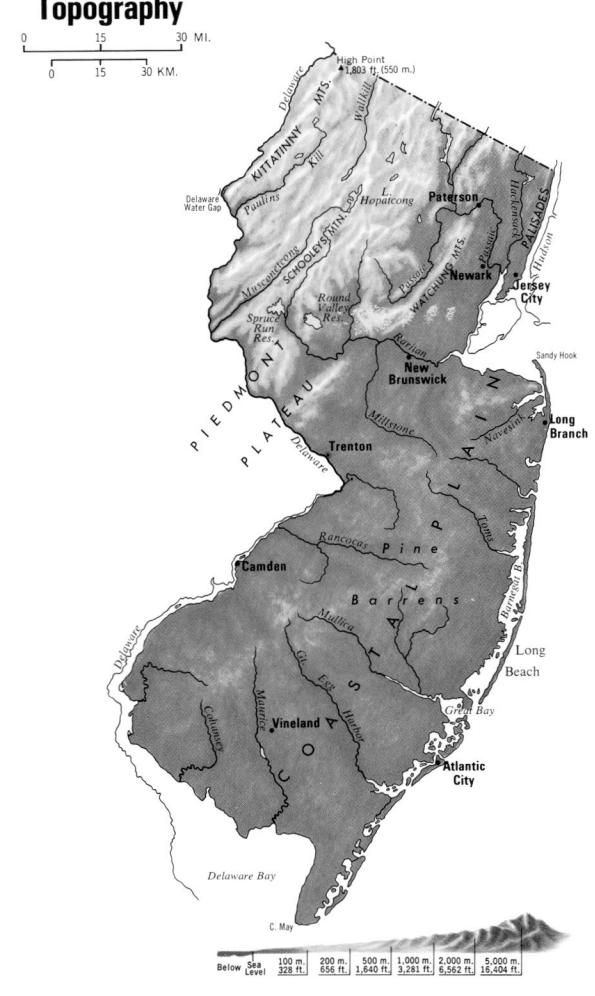

0   15   30 MI.

0   15   30 KM.

Below Sea Level / 100 m. 328 ft. / 200 m. 656 ft. / 500 m. 1,640 ft. / 1,000 m. 3,281 ft. / 2,000 m. 6,562 ft. / 5,000 m. 16,404 ft.

New Jersey

SCALE

0   5   10   15   20 MI.

0   5   10   15   20 KM.

State Capitals .... ⊛
County Seats .... ⊙
Canals ....
Major Limited Access Hwys. ....

Longitude 75° West of Greenwich

274 New Mexico

New Mexico

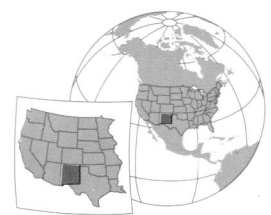

AREA 121,593 sq. mi. (314,926 sq. km.)
POPULATION 1,819,046
CAPITAL Santa Fe
LARGEST CITY Albuquerque
HIGHEST POINT Wheeler Pk. 13,161 ft.
(4011 m.)
SETTLED IN 1605
ADMITTED TO UNION January 6, 1912
POPULAR NAME Land of Enchantment
STATE FLOWER Yucca
STATE BIRD Road Runner

Topography

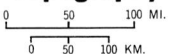

0   50   100 MI.

0   50   100 KM.

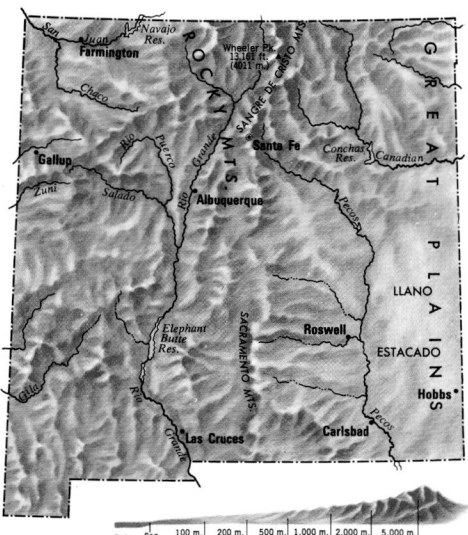

Below Sea Level | 100 m. 328 ft. | 200 m. 656 ft. | 500 m. 1,640 ft. | 1,000 m. 3,281 ft. | 2,000 m. 6,562 ft. | 5,000 m. 16,404 ft.

Agriculture, Industry and Resources

DOMINANT LAND USE

Wheat, Grain Sorghums, Range Livestock

General Farming, Livestock, Special Crops

General Farming, Livestock, Cash Grain

Dry Beans, General Farming

Cotton, Forest Products

Range Livestock

Forests

Nonagricultural Land

MAJOR MINERAL OCCURRENCES

Ag	Silver	Gp	Gypsum			U	Uranium
Au	Gold	K	Potash			V	Vanadium
C	Coal	Mo	Molybdenum			⚡	Water Power
Cu	Copper	Mr	Marble	O	Petroleum		
G	Natural Gas	Na	Salt	Pb	Lead	Zn	Zinc

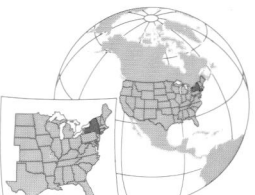

AREA 49,108 sq. mi. (127,190 sq. km.)
POPULATION 18,976,457
CAPITAL Albany
LARGEST CITY New York
HIGHEST POINT Mt. Marcy 5,344 ft. (1629 m.)
SETTLED IN 1614
ADMITTED TO UNION July 26, 1788
POPULAR NAME Empire State
STATE FLOWER Rose
STATE BIRD Bluebird

Topography

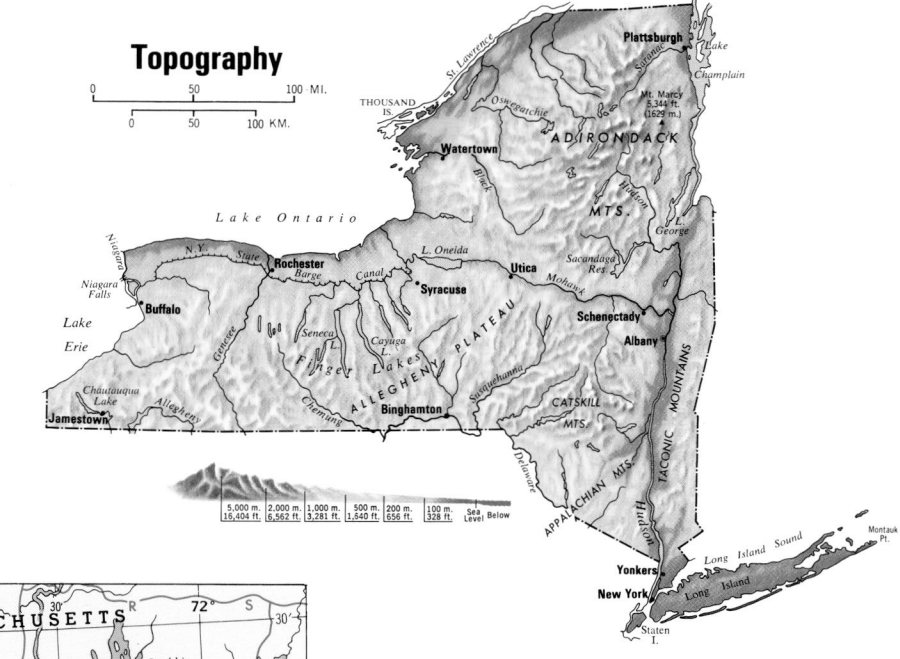

5,000 m. 2,000 m. 1,000 m. 500 m. 200 m. 100 m. Sea
16,404 ft. 6,562 ft. 3,281 ft. 1,640 ft. 656 ft. 328 ft. Level Below

(continued on following page)

© Copyright HAMMOND INCORPORATED, Maplewood, N.J.

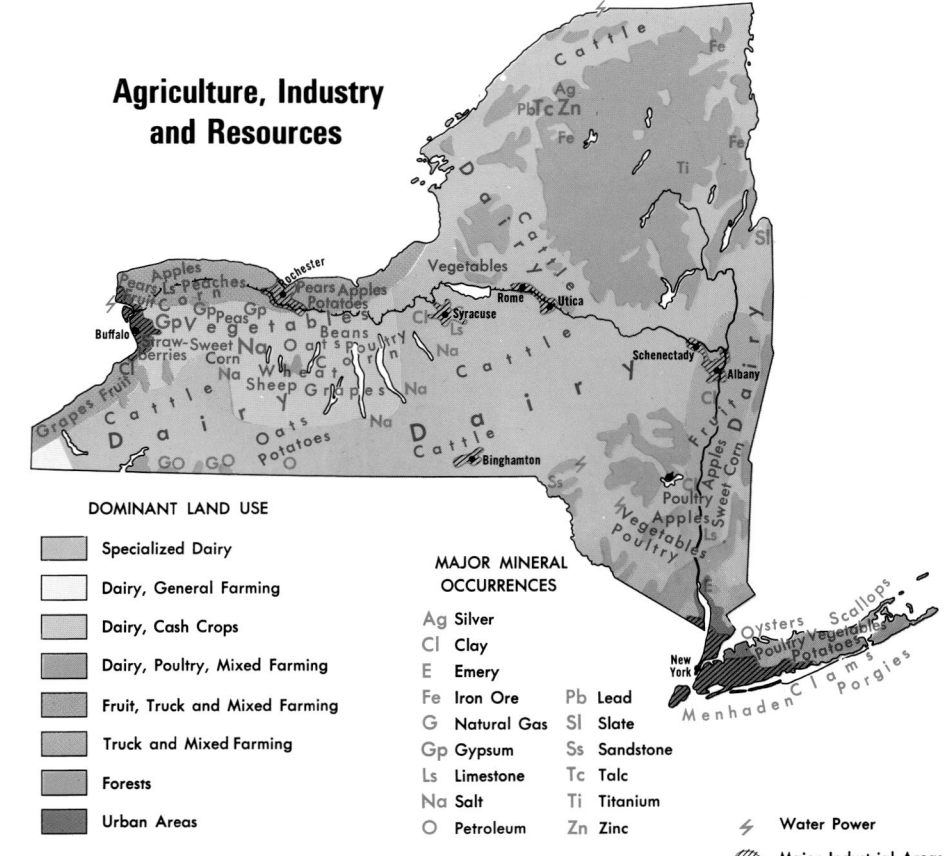

Agriculture, Industry and Resources

DOMINANT LAND USE

- Specialized Dairy
- Dairy, General Farming
- Dairy, Cash Crops
- Dairy, Poultry, Mixed Farming
- Fruit, Truck and Mixed Farming
- Truck and Mixed Farming
- Forests
- Urban Areas

MAJOR MINERAL OCCURRENCES

Ag Silver
Cl Clay
E Emery
Fe Iron Ore
G Natural Gas
Gp Gypsum
Ls Limestone
Na Salt
O Petroleum
Pb Lead
Sl Slate
Ss Sandstone
Tc Talc
Ti Titanium
Zn Zinc

⚡ Water Power
🖤 Major Industrial Areas

AREA 52,669 sq. mi. (136,413 sq. km.)
POPULATION 8,049,313
CAPITAL Raleigh
LARGEST CITY Charlotte
HIGHEST POINT Mt. Mitchell 6,684 ft. (2037 m.)
SETTLED IN 1650
ADMITTED TO UNION November 21, 1789
POPULAR NAME Tarheel State
STATE FLOWER Flowering Dogwood
STATE BIRD Cardinal

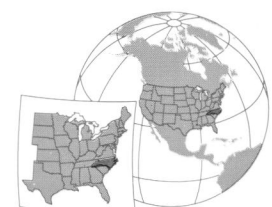

COUNTIES

Alamance 130,800 ....L3
Alexander 33,603..............G3
Alleghany 10,677...............G1
Anson 25,275...................J4
Ashe 24,384.....................F2
Avery 17,167....................F2
Beaufort 44,958................R4
Bertie 19,773...................P2
Bladen 32,278..................M5
Brunswick 73,143..............N6
Buncombe 206,330...........D3
Burke 89,148...................F3
Cabarrus 131,063.............H4
Caldwell 77,415................F3
Camden 6,885..................S2
Carteret 59,383................R5
Caswell 23,501................L2
Catawba 141,685.............G3
Chatham 49,329...............L3
Cherokee 24,298..............A4
Chowan 14,526................R2
Clay 8,775......................B4
Cleveland 96,287..............F4
Columbus 54,749.............M6
Craven 91,436.................P4
Cumberland 302,963.........M4
Currituck 18,190..............S2
Dare 29,967....................T3
Davidson 147,246............J3
Davie 34,835...................H3
Duplin 49,063..................O5
Durham 223,314..............M3
Edgecombe 55,606...........P3
Forsyth 306,067...............J2
Franklin 47,260................N2
Gaston 190,365................G4
Gates 10,516...................R2
Graham 7,993..................B4
Granville 48,498...............M2
Greene 18,974.................O3
Guilford 421,048..............K3
Halifax 57,370.................O2
Harnett 91,025................M4
Haywood 54,033..............C3
Henderson 89,173............D4
Hertford 22,601...............P2
Hoke 33,646...................L4
Hyde 5,826.....................S3
Iredell 122,660................H3
Jackson 33,121................C4
Johnston 121,965............N4
Jones 10,381..................P4

Lee 49,040.....................L4
Lenoir 59,648.................O4
Lincoln 63,780................G3
Macon 29,811.................B4
Madison 19,635..............D3
Martin 25,593.................P3
McDowell 42,151.............E3
Mecklenburg 695,454.......H4
Mitchell 15,687...............E2
Montgomery 26,822.........K4
Moore 74,769.................L4
Nash 87,420...................O2
New Hanover 160,307......O6
Northampton 22,086........P2
Onslow 150,355..............P5
Orange 118,227..............L2
Pamlico 12,934...............R4
Pasquotank 34,897..........S2
Pender 41,082.................O5
Perquimans 11,368..........S2
Person 35,623.................M2
Pitt 133,798....................P3
Polk 18,324....................E4
Randolph 130,454...........K3
Richmond 46,564............K4
Robeson 123,339.............L5
Rockingham 91,928.........K2
Rowan 130,340...............H3
Rutherford 62,899...........E4
Sampson 60,161..............N4
Scotland 35,998..............L5
Stanly 58,100.................J4
Stokes 44,711.................J2
Surry 71,219...................H2
Swain 12,968..................B3
Transylvania 29,334.........D4
Tyrrell 4,149...................S3
Union 123,677................J4
Vance 42,954..................N2
Wake 627,846.................M3
Warren 19,972................N2
Washington 13,723..........R3
Watauga 42,695..............F2
Wayne 113,329...............N4
Wilkes 65,632.................G2
Wilson 73,814.................N3
Yadkin 36,348.................H2
Yancey 17,774................E3

CITIES and TOWNS

Abbottsburg 425..............M5
Aberdeen 3,400...............L4
Acme.............................N6
Advance........................J3
Ahoskie 4,523.................P2
Alamance 310.................K2
Alarka 900......................C4
Albemarle▲ 15,680..........J4
Alexander Mills 662..........F4
Alliance 781....................R4
Altamahaw 996...............L2
Andrews 1,602................B4
Angier 3,419...................M4
Ansonville 636................J4
Apex 20,212...................M3
Arapahoe 436.................R4
Archdale 9,014................K3
Arlington 795..................H2
Ash 150.........................N6
Asheboro▲ 21,672..........K3
Asheville▲ 68,889...........D3
Askewville 180................R2
Atkinson 236..................N5
Atlantic 1,938.................S5
Atlantic Beach 1,781........R5

Aulander 888..................P2
Aurora 583.....................R4
Autryville 196.................M4
Avon 500........................U4
Avondale.......................
Ayden 4,622...................P4
Badin 1,154....................J4
Bahama 280...................M2
Bailey 670......................N3
Bakersville▲ 357.............E2
Balfour 1,200..................E4
Banner Elk 811................F2
Bannertown 1,028...........H1
Barco 325......................T2
Barker Heights 1,237........D4
Barco 325......................T2
Bat Cave 450..................E4
Bath 275........................R4
Battleboro 447................O2
Bayboro▲ 741.................R4
Bear Creek 500...............L3
Beargrass 53...................P3
Beaufort▲ 3,771.............R5

Belhaven 1,968...............R3
Bellarthur 350.................O3
Belmont 8,705................H4
Belvidere 275.................S2
Belville 285....................N6
Belwood 962...................F4
Benham 400...................G2
Bennett 254....................K3
Benson 2,923.................N4
Bessemer City 5,119........G4
Beta 500........................C4
Bethel 1,681...................P3
Beulaville 1,067..............O5
Biltmore Forest 1,440.......E3
Biscoe 1,700..................K4
Black Creek 714..............O3
Black Mountain 7,511.......D3
Bladenboro 1,718...........M5
Blowing Rock 1,418.........F2
Boardman 202................M6
Boger City 554................G4
Boiling Spring Lakes 2,972..N7

Boiling Springs 3,866.......F4
Bolivia▲ 148...................N6
Bolton 494.....................N6
Bonlee 300....................L3
Boomer 250...................G2
Boone▲ 1,138.................D2
Boonville 1,138...............H2
Brevard▲ 6,789..............D4
Bridgeton 328.................R4
Broadway 1,015..............M4
Brookford 434................G3
Browns Summit 500.........K2
Brunswick 360................M6
Bryson City▲ 1,411.........C4
Buies 2,085....................L5
Buies Creek 2,215...........M4
Bullock 525....................M2
Bunn 357.......................N3
Bunnlevel.......................M4
Burgaw▲ 3,337..............N5
Burlington 44,917............K2
Burnsville▲ 1,623...........E3
Buxton 700....................U4
Bynum 312.....................L3
Calabash 711..................M7
Calypso 410...................N4
Camden▲ 300................S2
Cameron 151..................L4
Candler 950....................D3
Candor 825....................K4
Canton 4,029..................D3
Cape Carteret 1,214........P5
Carolina Beach 4,701.......O6
Carrboro 16,782..............L3
Carthage▲ 1,871............L4
Cary 94,536....................M3
Casar 308.......................F3
Cashiers 196...................C4
Castalia 340....................N2
Castle Hayne 1,116..........O6
Caswell Beach 370...........N7
Catawba 698...................G3
Catharine Lake 500..........O5
Cedar Falls 400...............K3
Cedar Grove 250.............L2
Cedar Island 310.............S5
Cedar Mountain 250.........D4
Centerville 99..................N2
Cerro Gordo 244..............M6
Chadbourn 2,129.............M6
Chadwick Acres 15...........P6
Chapel Hill 48,715...........L3
Charlotte▲ 540,828........H4
Cherokee 975.................C4
Cherry 4,756...................R3
Cherryville 5,361.............G4
China Grove 3,616...........H3
Chinquapin 280...............O5
Chocowinity 733..............P4
Claremont 1,038..............G3
Clarendon 300................M6
Clark 739.......................P4
Clarkton 705...................M6
Clayton 6,973.................N3
Clemmons 13,827...........J2
Cleveland 808.................H3
Cliffside 950...................F4
Climax 475.....................K3
Clinton▲ 8,600...............N5
Clyde 1,324....................D3
Coats 1,845....................M4
Cofield 347.....................R2
Coinjock 650...................S2

Colerain 221...................R2
Collettsville 275...............F3
Columbia▲ 819...............S3
Columbus▲ 992..............E4
Comfort 325...................O5
Como 78.........................P1
Concord▲ 55,977...........H4
Conetoe 365...................O3
Connellys Springs 1,814....F3
Conover 6,604................G3
Conway 734....................P2
Cooleemee 905...............H3
Cornelius 11,969.............H4
Council..........................M6
Cove City 433..................P4
Cramerton 2,976.............G4
Creedmoor 2,232............M2
Creswell 278...................S3
Crisp 435........................O3
Crossnore 242.................F2
Cruso 800.......................D4
Culberson.......................A4
Cullowhee 3,579..............C4
Cumberland 400..............M5
Currie 294......................N6
Currituck▲ 700...............T2
Dallas 3,402....................G4
Dalton 400......................J2
Danbury▲ 108................J2
Davidson 7,139...............H4
Davis 612........................R5
Delco 450.......................N6
Denton 1,450..................J3
Dillsboro 205..................C4
Dobson▲ 1,457...............H2
Dortches 809..................O2
Dover 443.......................P4
Drexel 1,938...................F3
Dublin 250......................M5
Dudley...........................N4
Dulah 350.......................M6
Dundarrach.....................L5
Dunn 9,196.....................M4
Durham▲ 187,035..........M2
Dysartsville 950...............F3
Eagle Springs 280...........K4
Earl 234.........................F4
East Arcadia 524.............N6
East Bend 659.................H2
East Flat Rock 4,151........E4
East Laurinburg 295.........L5
East Marion 1,851............F3
East Spencer 1,755..........J3
Eden 15,908...................K1
Edenton▲ 5,394.............R2
Edward...........................R4
Efland 600......................L2
Elizabeth City▲ 17,188....S2
Elizabethtown▲ 3,698.....M5
Elk Park 459...................E2
Elkin 4,109.....................H2
Ellenboro 479.................F4
Ellerbe 1,021..................K4
Elm City 1,165................O3
Elon College 6,738...........L2
Emerald Isle 3,488...........P5
Enfield 2,347...................O2
Engelhard 500.................T3
Enka 5,567.....................D3
Ernul 350.......................P4
Erwin 4,537....................M4
Ether 425.......................K4
Etowah 2,766..................D4
Eure 282........................R2

(continued on following page)

Agriculture, Industry and Resources

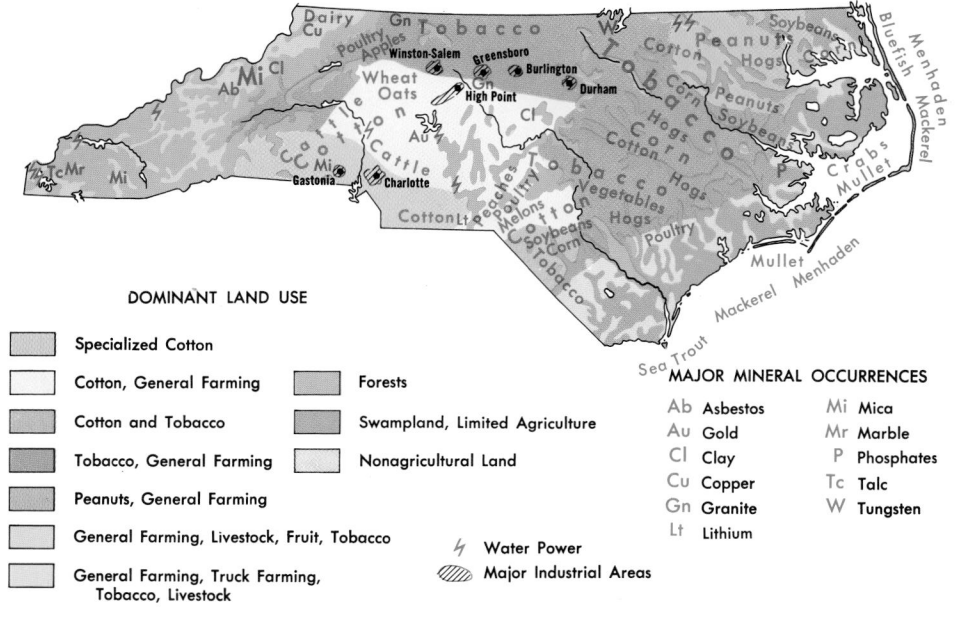

DOMINANT LAND USE

- Specialized Cotton
- Cotton, General Farming
- Cotton and Tobacco
- Tobacco, General Farming
- Peanuts, General Farming
- General Farming, Livestock, Fruit, Tobacco
- General Farming, Truck Farming, Tobacco, Livestock
- Forests
- Swampland, Limited Agriculture
- Nonagricultural Land

⚡ Water Power
▨ Major Industrial Areas

MAJOR MINERAL OCCURRENCES

Ab Asbestos
Au Gold
Cl Clay
Cu Copper
Gn Granite
Lt Lithium
Mi Mica
Mr Marble
P Phosphates
Tc Talc
W Tungsten

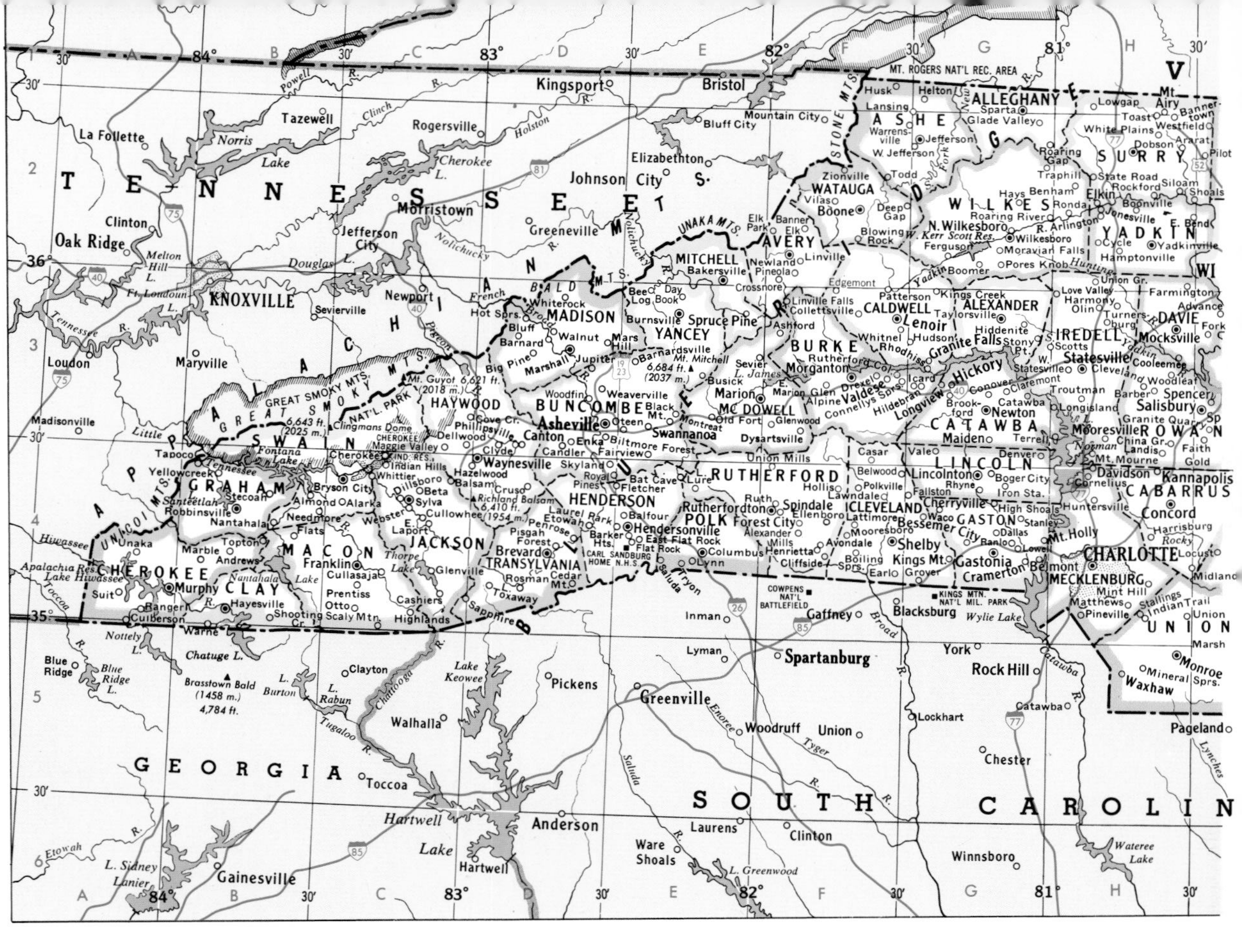

Topography

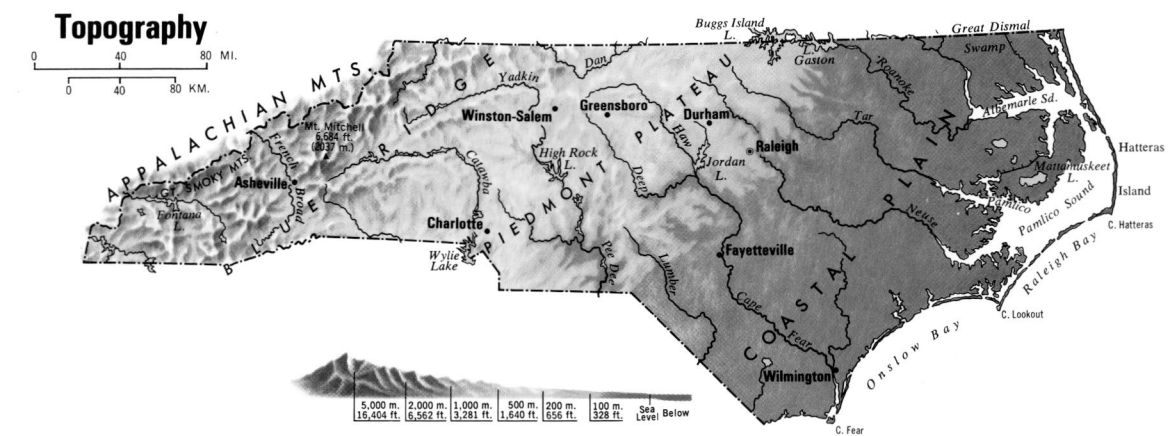

| 5,000 m. 16,404 ft. | 2,000 m. 6,562 ft. | 1,000 m. 3,281 ft. | 500 m. 1,640 ft. | 200 m. 656 ft. | 100 m. 328 ft. | Sea Level | Below |

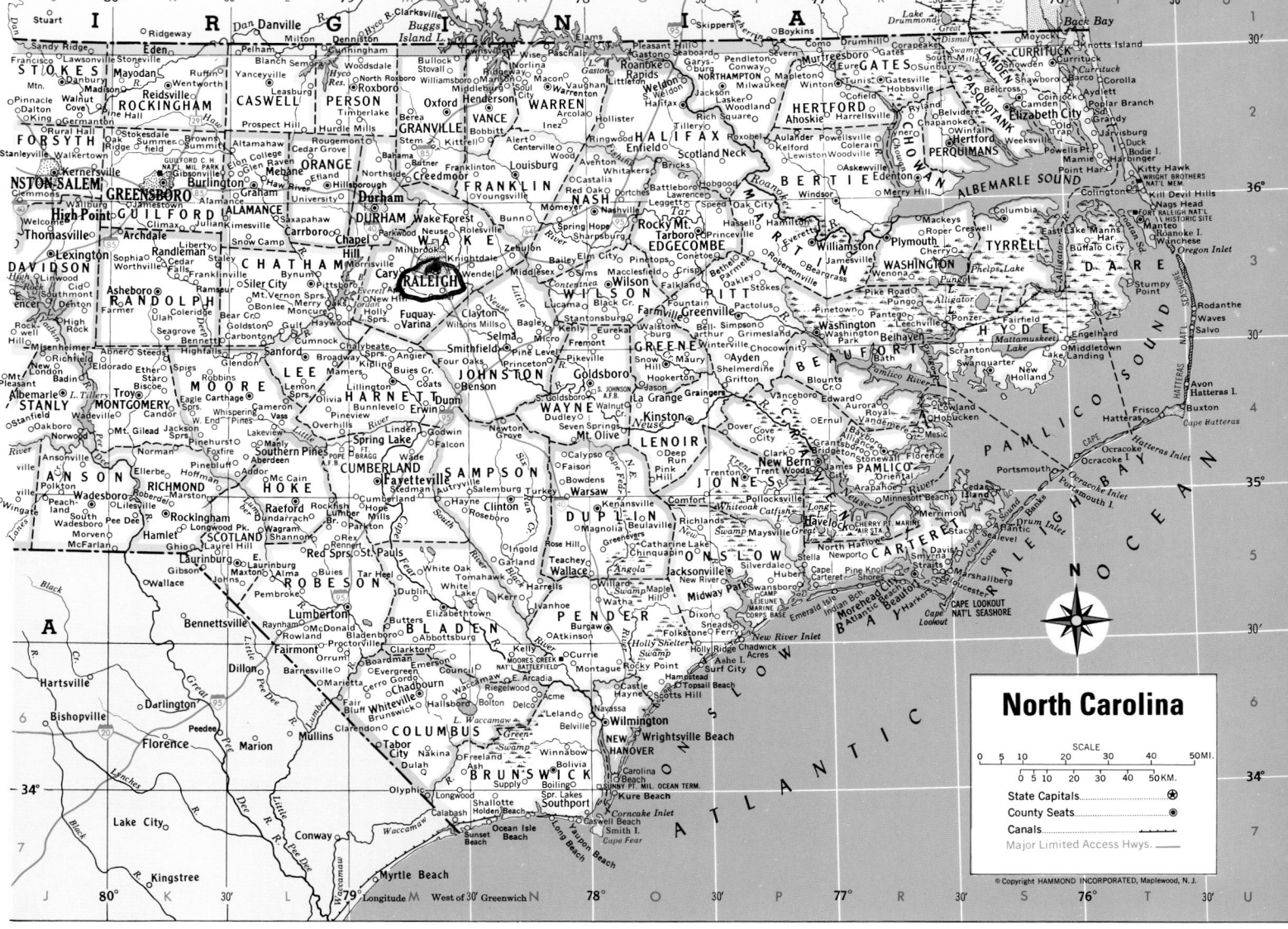

North Carolina

SCALE
0   5   10   20   30   40   50MI.
0 5 10   20   30   40   50KM.

State Capitals............................⊛
County Seats............................⊛
Canals............................↑——↑
Major Limited Access Hwys. ————

© Copyright HAMMOND INCORPORATED, Maplewood, N.J.

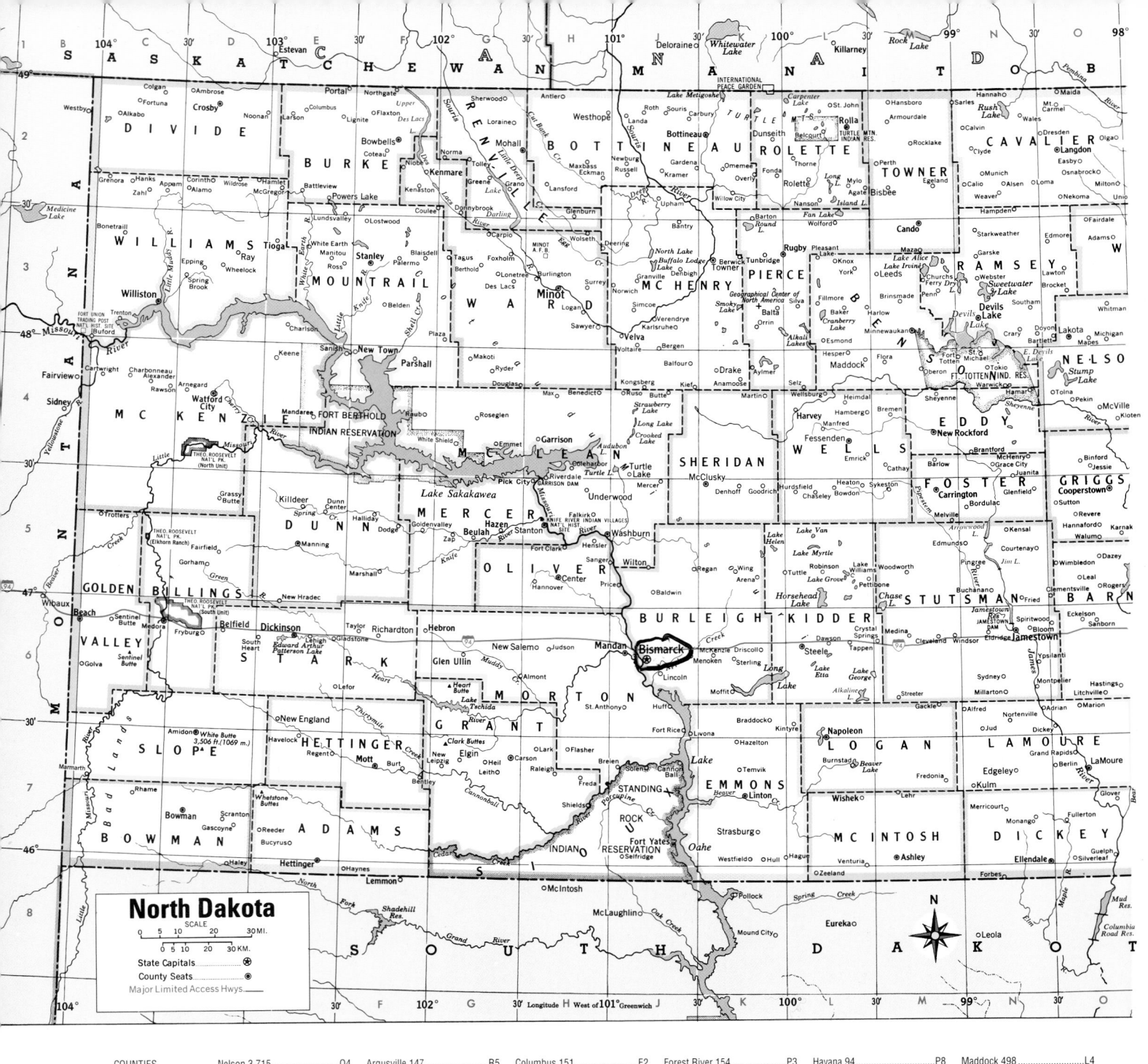

North Dakota

SCALE

0 5 10   20   30 MI.

0 5 10   20   30 KM.

State Capitals .... ⍟

County Seats .... ◉

Major Limited Access Hwys. _____

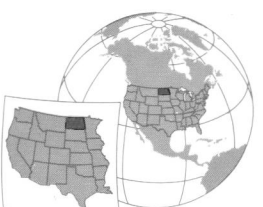

AREA 70,702 sq. mi. (183,118 sq. km.)
POPULATION 642,200
CAPITAL Bismarck
LARGEST CITY Fargo
HIGHEST POINT White Butte 3,506 ft. (1069 m.)
SETTLED IN 1780
ADMITTED TO UNION November 2, 1889
POPULAR NAME Flickertail State; Sioux State
STATE FLOWER Wild Prairie Rose
STATE BIRD Western Meadowlark

Topography

Turtle Lake 580	J4
Tuttle 106	L5
Underwood 812	H5
Upham 155	J2
Valley City▲ 6,826	P6
Velva 1,049	J3
Verona 108	O7
Wahpeton▲ 8,586	S7
Walcott 189	R6
Walhalla 1,057	P2
Washburn▲ 1,389	J5
Watford City▲ 1,435	D4
West Fargo 14,940	S6
Westhope 533	H2
White Shield 348	G4
Wildrose 129	D2
Williston▲ 12,512	C3
Willow City 221	K2
Wilton 807	J5
Wimbledon 237	O5
Wing 124	K5
Wishek 1,122	L7
Woodworth 80	M5
Wyndmere 533	R7
Ypsilanti 225	N6
Zap 231	G5
Zeeland 141	L8

OTHER FEATURES

Alkali (lakes)	L3
Alkaline (lake)	L6
Apple (creek)	J6
Arrowwood (lake)	N5
Ashtabula (Baldhill Res.) (lake)	P5
Audubon (lake)	H4
Bad Lands (reg.)	C7
Baldhill (Ashtabula Lake) (res.)	P5
Bear (creek)	O7
Beaver (creek)	K7
Beaver (creek)	B5
Beaver (creek)	L7
Buffalo Lodge (lake)	J3
Cannonball (riv.)	G7
Carpenter (lake)	L2
Cedar (creek)	G7
Chase (lake)	M5
Cherry (creek)	D4
Clark (buttes)	G7
Coteau du Missouri (plain)	G3
Cranberry (lake)	L3
Crooked (lake)	J4
Cut Bank (creek)	H2
Darling (lake)	G2
Deep (riv.)	J1
Des Lacs (riv.)	G3
Devils (lake)	N3
Dry (lake)	M3
East Devils (lake)	N4
Egg (creek)	H3
Elm (riv.)	N8
Elm (riv.)	R5
Etta (lake)	L6

Fan (lake)	L2
Forest (riv.)	P3
Fort Berthold Ind. Res.	E4
Fort Totten Ind. Res.	N4
Fort Union Trading Post Nat'l Hist. Site	B3
Garrison (dam)	H5
George (lake)	L6
Goose (riv.)	P4
Grand, North Fork (riv.)	F7
Grand Forks A.F.B. 4,832	R4
Green (riv.)	D5
Grove (lake)	M4
Heart (butte)	G6
Heart (riv.)	F6
Helen (lake)	K5
Horsehead (lake)	L5
International Peace Garden	K1
Irvine (lake)	M3
Island (lake)	L2
James (riv.)	N6
Jamestown (res.)	N6
Jim (riv.)	N5
Knife (riv.)	G5
Knife R. Indian Villages Nat'l Hist. Site	H5
Little Deep (creek)	G2
Little Knife (riv.)	F3

Little Missouri (riv.)	D4
Little Muddy (riv.)	C3
Long (lake)	J4
Long (lake)	K6
Long (lake)	L2
Maple (riv.)	O8
Maple (riv.)	R6
Metigoshe (lake)	K2
Minot A.F.B. 7,599	H3
Missouri (riv.)	H5
Muddy (creek)	G6
Myrtle (lake)	L5
North (lake)	J3
Oahe (lake)	J7
Oak (creek)	J8
Park (riv.)	R3
Patterson, Edward A. (lake)	E6
Pembina (riv.)	O1
Pipestem (riv.)	M5
Porcupine (creek)	J7
Red River of the North (riv.)	S4
Round (lake)	K3
Rush (lake)	N2
Rush (riv.)	R5
Sakakawea (lake)	G5
Sentinel (butte)	C6
Shell (creek)	F3
Sheyenne (riv.)	O6

Smoky (lake)	K3
Souris (riv.)	J2
Spring (creek)	E5
Standing Rock Ind. Res.	J7
Strawberry (lake)	J4
Stump (lake)	O4
Sweetwater (lake)	N3
Theodore Roosevelt Nat'l Park	C5
Theodore Roosevelt Nat'l Park	D4
Theodore Roosevelt Nat'l Park	D6
Thirty Mile (creek)	F6
Tongue (riv.)	P2
Tschida (lake)	G6
Turtle (lake)	H4
Turtle (mts.)	K2
Turtle Mountain Ind. Res.	K2
Upper Des Lacs (lake)	F2
Van (lake)	L5
Whetstone (buttes)	E7
White Butte (mt.)	D7
White Earth (riv.)	E3
Wild Rice (riv.)	R7
Yellowstone (riv.)	B4
▲County seat	

New England 555	E6
New Leipzig 274	G7
New Rockford▲ 1,463	N4
New Salem 938	G6
New Town 1,367	F4
Noonan 154	D2
Northwood 959	P4
Oakes 1,979	O7
Oriska 128	P6
Osnabrock 174	O2
Page 248	P5
Palermo 77	F3
Park River 1,535	P3
Parshall 981	F4
Pekin 80	O4
Pembina 642	R2
Petersburg 195	P3
Pick City 166	G5
Plaza 167	G3
Portal 131	E2
Portland 604	R5
Powers Lake 309	E2
Ray 534	D3
Reeder 181	E7
Regent 211	E7
Reile's Acres 254	S6
Reynolds 350	R4
Rhame 189	C7
Richardton 619	F6
Riverdale 273	H4
Riverside 465	S6
Rocklake 194	M2
Rolette 538	L2

Rolla▲ 1,417	L2
Rugby▲ 2,939	L3
Rutland 220	P7
Saint John 358	L2
Saint Thomas 447	R2
Sanborn 194	O6
Sanish	E4
Sawyer 377	H3
Scranton 304	D7
Selfridge 223	J7
Sharon 109	P4
Sheldon 135	P6
Sherwood 255	G2
Sheyenne 318	M4
Shields 125	H7
South Heart 307	D6
Stanley▲ 1,279	F3
Stanton▲ 345	H5
Starkweather 157	N3
Steele▲ 761	L6
Strasburg 549	K7
Streeter 172	M6
Surrey 917	H3
Sykeston 153	M5
Tappen 210	L6
Taylor 150	F6
Thompson 1,006	R4
Tioga 1,125	E3
Tokio 245	N4
Tolna 202	O4
Tower City 252	P6
Towner▲ 574	K3
Trenton 300	C3

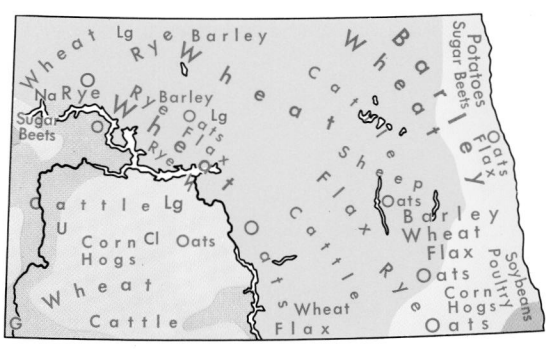

DOMINANT LAND USE

Agriculture, Industry and Resources

▢ Specialized Wheat	▢ Sugar Beets, Dry Beans, Livestock, General Farming
▢ Wheat, General Farming	▢ Range Livestock
▢ Wheat, Range Livestock	⚡ Water Power
▢ Livestock, Cash Grain	

MAJOR MINERAL OCCURRENCES

Cl	Clay
G	Natural Gas
Lg	Lignite
Na	Salt
O	Petroleum
U	Uranium

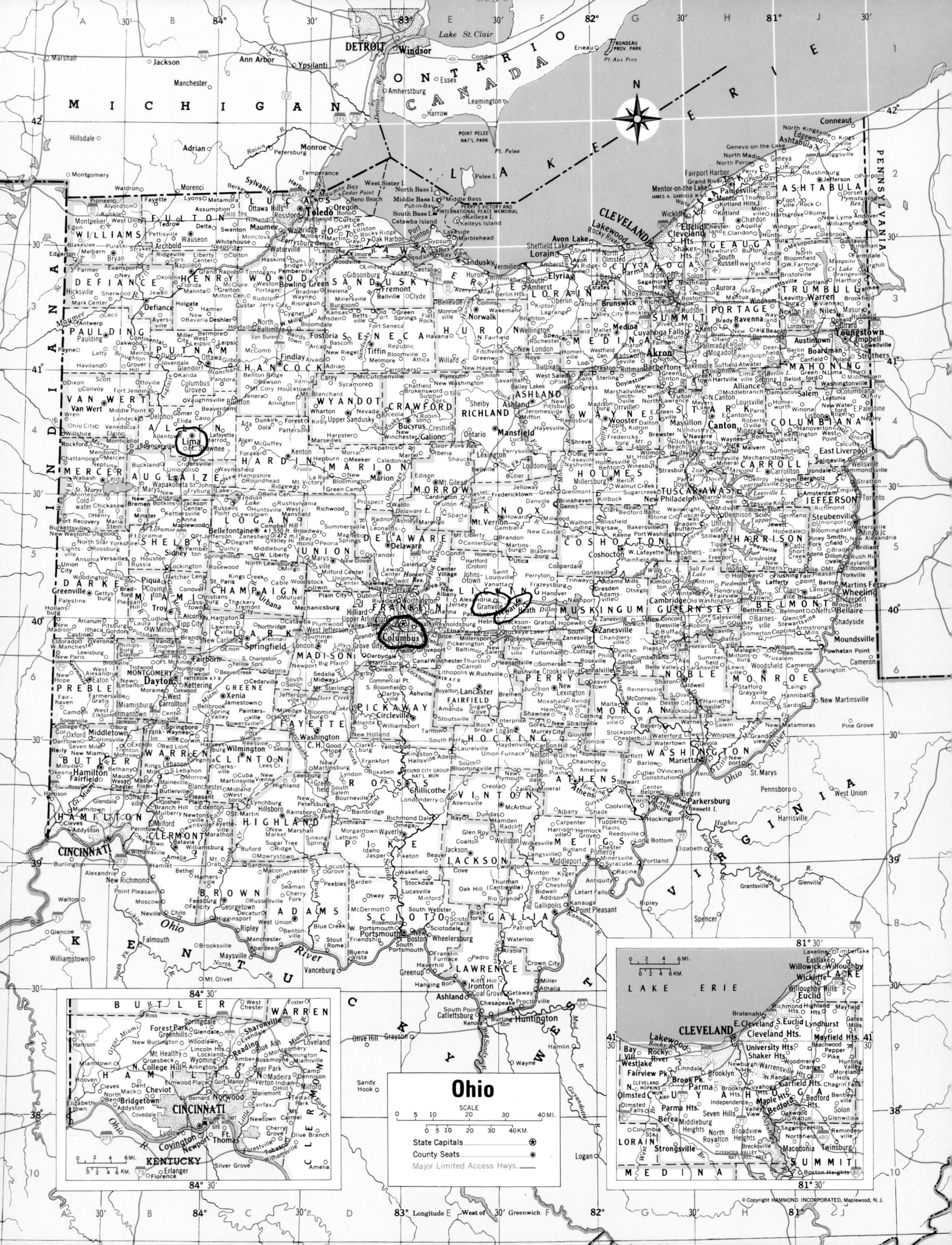

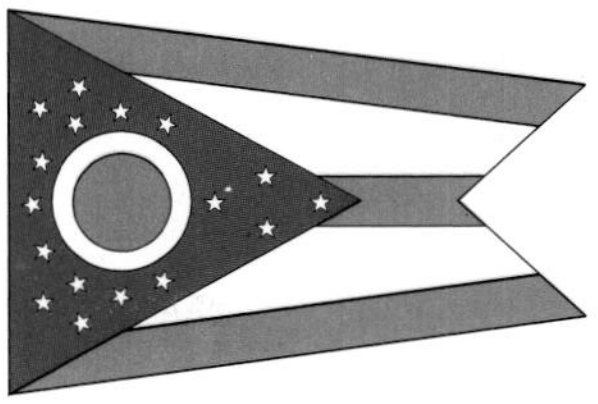

AREA 41,330 sq. mi. (107,045 sq. km.)
POPULATION 11,353,140
CAPITAL Columbus
LARGEST CITY Cleveland
HIGHEST POINT Campbell Hill 1,549 ft. (472 m.)
SETTLED IN 1788
ADMITTED TO UNION March 1, 1803
POPULAR NAME Buckeye State
STATE FLOWER Scarlet Carnation
STATE BIRD Cardinal

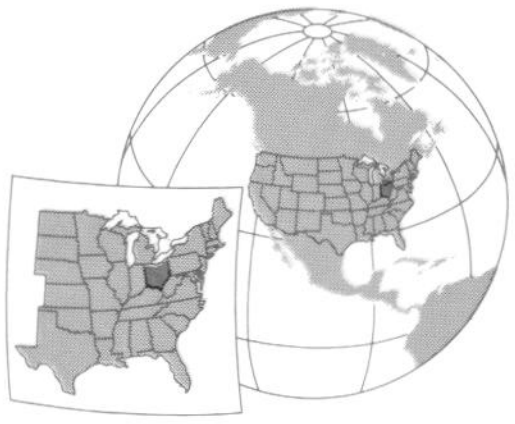

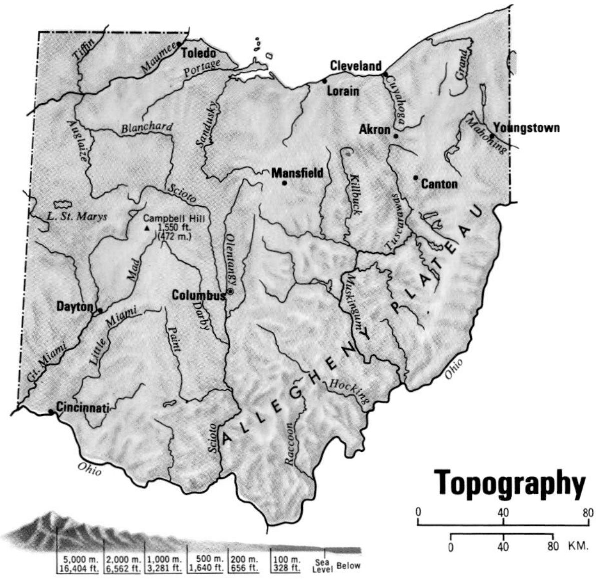

COUNTIES

Adams 27,330 D8
Allen 108,473 B4
Ashland 52,523 F4
Ashtabula 102,728 J2
Athens 62,223 F7
Auglaize 46,611 B4
Belmont 70,226 J5
Brown 42,285 C8
Butler 332,807 A7
Carroll 28,836 H4
Champaign 38,890 C5
Clark 144,742 C6
Clermont 177,977 B7
Clinton 40,543 C7
Columbiana 112,075 J4
Coshocton 36,655 G5
Crawford 46,966 E4
Cuyahoga 1,393,978 G3
Darke 53,309 A5
Defiance 39,500 A3
Delaware 109,989 D5
Erie 79,551 E3
Fairfield 122,759 E6
Fayette 28,433 D6
Franklin 1,068,978 E5
Fulton 42,084 B2
Gallia 31,069 F8
Geauga 90,895 H3
Greene 147,886 C6
Guernsey 40,792 H5
Hamilton 845,303 A7
Hancock 71,295 C3
Hardin 31,945 C4
Harrison 15,856 H5
Henry 29,210 B3
Highland 40,875 C7
Hocking 28,241 F6
Holmes 38,943 G4
Huron 59,487 E3
Jackson 32,641 E7
Jefferson 73,894 J5
Knox 54,500 F5
Lake 227,511 H2
Lawrence 62,319 E8
Licking 145,491 F5
Logan 46,005 C5
Lorain 284,664 F3
Lucas 455,054 C2
Madison 40,213 D6
Mahoning 257,555 J4
Marion 66,217 D4
Medina 151,095 G3
Meigs 23,072 F7
Mercer 40,924 A4
Miami 98,868 B5
Monroe 15,180 H6
Montgomery 559,062 B6
Morgan 14,897 G6
Morrow 31,628 E4
Muskingum 84,585 G5
Noble 14,058 G6
Ottawa 40,985 D2
Paulding 20,293 A3
Perry 34,078 F6
Pickaway 52,727 D6
Pike 27,695 D7
Portage 152,061 H3
Preble 42,337 A6
Putnam 34,726 B3
Richland 128,852 E4
Ross 73,345 D7
Sandusky 61,792 D3
Scioto 79,195 D8
Seneca 58,683 D3
Shelby 47,910 B5
Stark 378,098 H4
Summit 542,899 G3
Trumbull 225,116 J3
Tuscarawas 90,914 H5
Union 40,909 D5
Van Wert 29,659 A4
Vinton 12,806 E7
Warren 158,383 B7
Washington 63,251 H7
Wayne 111,564 G4
Williams 39,188 A2
Wood 121,065 C3
Wyandot 22,908 D4

CITIES and TOWNS

Aberdeen 1,603 C8
Ada 5,582 C4
Adamsville 151 G5
Addyston 1,010 B9
Adelphi 371 E7
Adena 815 J5
Akron▲ 217,074 G3
Albany 808 F7
Alexandria 85 E5
Alger 888 C4
Alliance 23,253 H4
Alvordton 298 A2
Amanda 707 E6
Amberley 3,425 C9
Amelia 2,752 D10
Amesville 250 F7
Amherst 11,797 F3
Amsterdam 568 J5
Andover 1,269 J2
Anna 1,319 B5
Ansonia 1,145 A5
Antioch 68 H6
Antwerp 1,740 A3
Apple Creek 999 G4
Aquilla 360 H2
Arcadia 537 D3
Arcanum 2,076 A6
Archbold 4,290 B2
Arlington 1,351 C4
Arlington Heights 899 C9
Ashland▲ 21,249 F4
Ashley 1,216 E5
Ashtabula 20,962 J2
Ashville 3,174 E6
Athalia 346 F8
Athens▲ 21,342 F7
Attica 955 E3
Aurora 13,556 H3
Austintown 31,627 J3
Avon 11,446 F3
Avon Lake 18,145 F2
Bailey Lakes 397 F4
Bainbridge 1,012 D7
Bairdstown 130 C3
Ballville 3,255 D3
Baltic 743 G5
Baltimore 2,881 E6
Barberton 27,899 G4
Barnesville 4,225 H6
Barnhill 364 H5
Barton 1,039 J5
Batavia▲ 1,617 B7
Batesville 95 H6
Bay View 692 E3
Bay Village 16,087 G9
Beach City 1,137 G4
Beachwood 12,186 J9
Beallsville 423 J6
Beaver 336 E7
Beavercreek 37,984 C6
Beaverdam 356 C4
Bedford 14,214 H9
Bedford Heights 11,375 J9
Bellaire 4,892 J5
Bellbrook 7,009 C6
Belle Center 807 C4
Bellefontaine▲ 13,069 C5
Belle Valley 263 G6
Bellevue 8,193 E3
Bellville 1,773 E4
Belmont 532 J5
Belmore 161 B3
Beloit 1,024 J4
Belpre 6,660 G7
Bentleyville 947 J9
Benton 351 G4
Benton Ridge 343 C4
Berea 18,970 G10
Bergholz 769 J4
Berkey 264 C2
Berlin 685 G4
Berlin Heights 756 F3
Bethel 2,637 B8
Bethesda 1,413 H5
Bettsville 784 D3
Beverly 1,282 G6
Bexley 13,203 E6
Blakeslee 128 A2
Blanchester 4,220 B7
Bloomdale 724 D3
Bloomingburg 874 D6
Bloomingdale 227 J5
Bloomville 1,045 D3
Blue Ash 12,513 C9
Bluffton 3,896 C4
Boardman 37,215 J3
Bolivar 894 G4
Boston Heights 1,186 J10
Botkins 1,205 B5
Bowerston 414 H5
Bowersville 225 C6
Bowling Green▲ 29,636 C3
Bradford 1,859 B5
Bradner 1,171 C3
Brady Lake 513 H3
Brecksville 13,382 H10
Bremen 1,265 F6
Brewster 2,324 G4
Brice 109 E6
Bridgeport 2,186 J5
Bridgetown 11,748 B9
Brilliant 1,672 J5
Brimfield 3,248 H3
Broadview Heights 15,967 H10
Brookfield 1,288 J3
Brooklyn 11,586 H9
Brooklyn Heights 1,558 H9
Brook Park 21,218 G9
Brookside 644 J5
Brookville 5,289 B6
Broughton 151 B3
Brunswick 33,388 G3
Bryan▲ 8,333 A3
Buchtel 574 F7
Buckeye Lake 3,049 F6
Buckland 239 B4
Bucyrus▲ 13,224 E4
Burbank 289 F4
Burgoon 224 D3
Burkettsville 268 A5
Burlington 2,794 F9
Burton 1,450 H3
Butler 921 F4
Butlerville 188 C7
Byesville 2,574 G6
Cadiz▲ 3,308 J5
Cairo 499 B4
Calcutta 3,491 J4
Caldwell▲ 1,956 G6
Caledonia 578 D4
Cambridge▲ 11,520 G5
Camden 2,302 A6
Campbell 9,460 J3
Canal Fulton 5,061 H4
Canal Winchester 4,478 E6
Canfield 7,374 J3
Canton▲ 80,806 H4
Cardington 1,849 E5
Carey 3,901 D4
Carlisle 5,121 B6
Carroll 488 E6
Carrollton▲ 3,190 J4
Casstown 246 B5
Castalia 935 E3
Castine 163 A6
Catawba 312 C6
Cecil 249 A3
Cedarville 3,828 C6
Celina▲ 10,303 A4
Centerburg 1,432 E5
Centerville 134 B6
Chagrin Falls 4,024 J9
Chardon▲ 5,156 H2
Chatfield 206 E4
Chauncey 1,067 F7
Cherry Fork 178 C8
Cherry Grove 4,555 C10
Chesapeake 842 E9
Cheshire 250 F8
Chester 309 G7
Chesterhill 305 G6
Chesterland 2,646 H2
Chesterville 286 E5
Cheviot 9,015 B9
Chickasaw 378 A5
Chillicothe▲ 21,796 E7
Chilo 130 B8
Christiansburg 553 C5
Cincinnati▲ 331,285 B9
Circleville▲ 13,485 D6
Clarington 444 J6
Clark 523 G5
Clarksburg 483 D7
Clarksville 497 C7
Clay Center 289 D2
Clayton 13,347 B6
Cleveland▲ 478,403 H9
Cleveland Heights 49,958 H9
Cleves 2,790 B9
Clinton 1,337 G4
Cloverdale 270 B3
Clyde 6,064 E3
Coal Grove 2,027 E9
Coalton 545 E7
Coldwater 4,482 A5
College Corner 379 A6
Columbiana 5,635 J4
Columbus (cap.)▲ 711,470 E6
Columbus Grove 2,200 B4
Commercial Point 405 D6
Conesville 420 G5
Congress 162 F4
Conneaut 12,485 J2
Continental 1,188 B3
Convoy 1,110 A4
Coolville 528 G7
Corning 593 F6
Cortland 6,830 J3
Corwin 225 B6
Coshocton▲ 11,682 G5
Cove 6,669 E8
Covedale 6,360 B10
Covington 2,559 B5
Craig Beach 1,254 H3
Crestline 5,088 E4
Creston 2,161 G3
Cridersville 1,817 B4
Crooksville 2,483 F6
Crown City 411 F8
Cumberland 402 G6
Custar 209 C3
Cuyahoga Falls 49,374 G3
Cuyahoga Heights 599 J9
Cygnet 564 C3
Dalton 1,605 G4
Danville 1,104 F5
Darbydale 825 D6
Darbyville 293 D6
Dayton▲ 166,179 B6
Deer Park 5,982 C9
Deersville 86 H5
Defiance▲ 16,465 B3
Degraff 1,212 C5
Delaware▲ 25,243 E5
Dellroy 314 H4
Delphos 6,944 B4
Delta 2,930 B2
Dennison 2,992 H5
Dent 7,612 B9
Deshler 1,831 C3
Devola 2,771 H7
Dexter City 161 G6
Dillonvale 3,716 J5
Dover 12,210 G4
Doylestown 2,799 G4
Dresden 1,423 G5
Dublin 31,392 D5
Dunkirk 952 C4
Dupont 279 B3
East Canton 1,629 H4
East Cleveland 27,217 H9
Eastlake 20,255 J8
East Liverpool 13,089 J4
East Palestine 4,917 J4
East Sparta 806 H4
Eaton▲ 8,133 A6
Eaton Estates 1,409 G3
Edgerton 2,117 A3
Edgewood 4,762 J2
Edison 437 E4
Edon 898 A2
Eldorado 543 A6
Elgin 71 B4
Elida 1,917 B4
Elmore 1,426 D3
Elmwood Place 2,681 B9
Elyria▲ 55,953 F3
Empire 300 J5
Englewood 12,235 B6
Evendale 3,090 C9
Fairborn 32,052 B6
Fairfax 1,938 C9
Fairfield 42,097 A7
Fairlawn 7,307 G3
Fairport Harbor 3,180 H2
Fairview Park 17,572 G9
Farmer 932 A3
Farmersville 980 A6
Fayette 1,340 B2
Fayetteville 372 C7
Felicity 922 B8
Findlay▲ 38,967 C3
Fletcher 510 B5
Florida 304 B3
Flushing 900 J5
Forest 1,488 C4
Forest Park 19,463 B9
Forestville 10,978 C10
Fort Jennings 432 B4
Fort Loramie 1,344 B5
Fort McKinley 3,989 B6
Fort Recovery 1,273 A5
Fort Shawnee 3,855 B4
Fostoria 13,931 D3
Frankfort 1,011 D7
Franklin 11,396 B6
Franklin Furnace 1,537 E8
Frazeysburg 1,201 F5
Fredericksburg 487 G4
Fredericktown 2,428 F5
Freeport 398 H5
Fremont▲ 17,375 D3
Fulton 325 E5
Fultonham 178 F6
Gahanna 32,636 E5
Galena 361 E5
Galion 11,341 E4
Gallipolis▲ 4,180 F8
Gambier 1,871 F5
Garfield Heights 30,734 J9
Garrettsville 2,262 H3
Gates Mills 2,493 J9
Geneva 6,595 J2
Geneva-on-the-Lake 1,545 H2
Genoa 2,230 D2
Georgetown▲ 3,691 C8
Germantown 4,884 B6
Gettysburg 558 A5
Gibsonburg 2,506 D3
Gilboa 208 C3
Girard 10,902 J3
Glandorf 919 B3
Glendale 2,188 C9
Glenford 208 F6
Glenmont 233 F4
Glenwillow 449 J10
Glouster 1,972 F6
Gnadenhutten 1,280 G5
Golf Manor 3,999 C9
Gordon 206 B6
Grafton 2,302 F3
Grand Rapids 1,002 C3
Grand River H2
Grandview 1,391 H7
Grandview Heights 6,695 D6
Granville 3,167 E5
Gratiot 195 F6
Gratis 934 A6
Green Camp 393 D4
Greenfield 4,906 D7
Greenhills 4,103 B9
Greensburg 3,306 G4
Green Springs 1,247 E3
Greentown 3,154 H4
Greenville▲ 13,294 A5
Greenwich 1,525 E3
Groesbeck 7,202 B9
Grove City 27,075 D6
Groveport 3,865 E6
Grover Hill 412 B3
Hamden 871 F7
Hamersville 515 C8
Hamilton▲ 60,690 A7
Hamler 650 B3
Hanging Rock 306 E8
Hanover 885 F5
Hanoverton 387 J4
Harbor View 122 C2
Harpster 233 D4
Harrisburg 340 D6
Harrison 7,487 A9
Harrisville 308 J5
Harrod 491 C4
Hartford 418 J3
Hartford 412 E5
Hartville 2,174 H4
Harveysburg 563 C7
Haskins 638 C3
Haviland 210 A3
Hayesville 348 F4
Heath 8,527 F5
Hebron 2,034 E6
Helena 267 D3
Hemlock 203 F6
Hicksville 3,649 A3
Higginsport 298 C8
Highland 275 C7
Highland Heights 8,082 J9
Hilliard 24,230 D5
Hillsboro▲ 6,368 C7
Hiram 1,242 H3
Holgate 1,194 B3
Holland 1,306 C2
Hollansburg 300 A5
Holloway 345 H5
Holmesville 386 G4
Hopedale 984 J5
Hoytville 296 C3
Hubbard 8,284 J3
Huber Heights 38,212 B6
Hudson 22,439 H3
Hunting Valley 735 J9
Huntsville 454 C5
Huron 7,958 E3
Independence 7,109 H9
Indian Hill 5,383 C9
Irondale 418 J4
Ironton▲ 11,211 E8
Ithaca 119 A6
Jackson▲ 6,184 E7
Jackson Center 1,369 B5
Jacksonville 544 F7
Jamestown 1,917 C6
Jefferson▲ 3,572 J2
Jefferson (West Jefferson) 4,331 D6
Jeffersonville 1,288 C6
Jenera 285 C4
Jeromesville 478 F4
Jerry City 453 C3
Jerusalem 144 H6
Jewett 784 H5
Johnstown 3,440 E5
Junction City 818 F6
Kalida 1,031 B4
Kelleys Island 367 E2
Kent 27,906 H3
Kenton▲ 8,336 C4
Kettering 57,502 B6
Kettlersville 194 B5
Killbuck 839 G5
Kimbolton 134 G5
Kingston 1,032 E7
Kingsville 1,243 J2
Kipton 265 F3
Kirby 155 D4
Kirkersville 520 E6
Kirtland 6,670 H2
Kirtland Hills 597 H2
Lafayette 304 C4
Lagrange 1,815 F3
Lakeline 210 J8
Lakemore 2,561 H3
Lakeview 1,074 C4
Lakewood 56,646 G9
Lancaster▲ 35,335 E6
La Rue 775 D4
Latty 205 A3
Laura 487 B6
Laurelville 533 E7
Lawrenceville 302 C6
Lebanon▲ 16,962 B7
Leesburg 1,253 D7
Leesville 156 H5
Leetonia 2,043 J4
Leipsic 2,236 C3
Lewisburg 1,798 A6
Lewisville 261 H6
Lexington 4,165 E4
Liberty Center 1,109 B3
Lima▲ 40,081 B4
Limaville 152 H4
Lincoln Heights 4,113 C9
Lindsey 504 D3
Linndale 159 G9
Lisbon▲ 2,788 J4
Lithopolis 600 E6
Lockbourne 173 E6
Lockington 214 B5
Lockland 3,707 C9
Lodi 3,061 F3
Logan▲ 6,704 F6
London▲ 8,771 C6
Lorain 68,652 F3
Lordstown 3,633 J3
Lore City 305 H6
Loudonville 2,906 F4
Louisville 8,904 H4
Loveland 11,677 D9
Lowell 628 H6
Lowellville 1,281 J3
Lower Salem 103 H6
Lucas 620 F4
Lucasville 1,588 E8
Luckey 998 D3
Ludlow Falls 300 B6
Lynchburg 1,350 C7
Lyndhurst 15,279 J9
Lyons 559 B2
Macedonia 9,224 J10
Mack 2,816 B9
Macksburg 218 G6
Madeira 8,923 C9
Madison 2,921 H2
Magnetic Springs 323 D5
Magnolia 931 H4
Maineville 359 C9
Malinta 294 B3

(continued on following page)

Agriculture, Industry and Resources

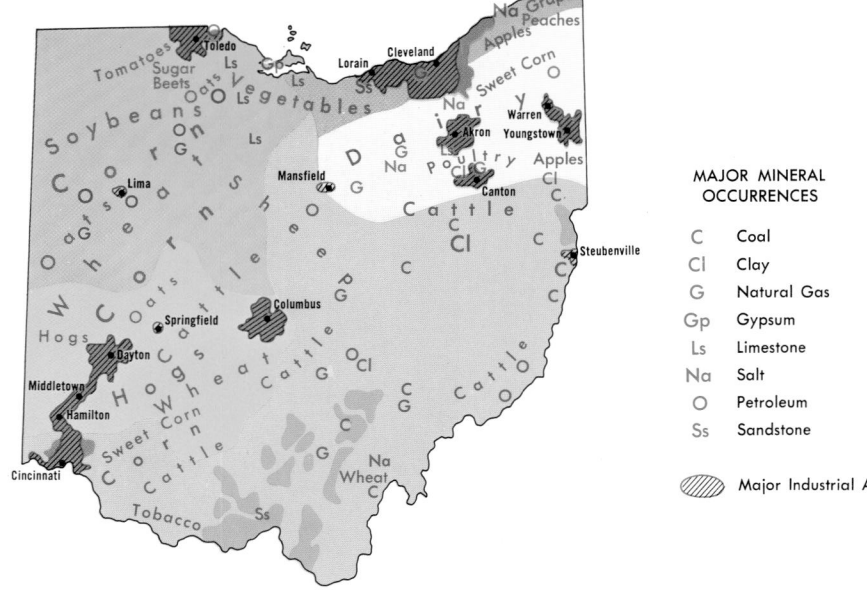

DOMINANT LAND USE

- Hogs, Soft Winter Wheat
- Livestock, Dairy, Soybeans, Cash Grain
- Dairy, General Farming
- General Farming, Livestock, Tobacco
- Fruit, Truck and Mixed Farming
- Forests
- Urban Areas

MAJOR MINERAL OCCURRENCES

- C   Coal
- Cl   Clay
- G   Natural Gas
- Gp   Gypsum
- Ls   Limestone
- Na   Salt
- O   Petroleum
- Ss   Sandstone

Major Industrial Areas

AREA 69,956 sq. mi. (181,186 sq. km.)
POPULATION 3,450,654
CAPITAL Oklahoma City
LARGEST CITY Oklahoma City
HIGHEST POINT Black Mesa 4,973 ft. (1516 m.)
SETTLED IN 1889
ADMITTED TO UNION November 16, 1907
POPULAR NAME Sooner State
STATE FLOWER Mistletoe
STATE BIRD Scissor-tailed Flycatcher

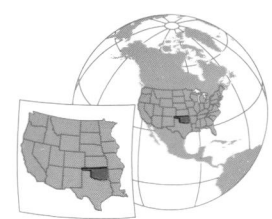

COUNTIES

Adair 21,038	S3
Alfalfa 6,105	K1
Atoka 13,879	O6
Beaver 5,857	E1
Beckham 19,799	G4
Blaine 11,976	K3
Bryan 36,534	O7
Caddo 30,150	K4
Canadian 87,697	K3
Carter 45,621	M6
Cherokee 42,521	R3
Choctaw 15,342	P6
Cimarron 3,148	A1
Cleveland 208,016	M4
Coal 6,031	O5
Comanche 114,996	K5
Cotton 6,614	K6
Craig 14,950	R1
Creek 67,367	O3
Custer 26,142	H3
Delaware 37,077	S2
Dewey 4,743	H2
Ellis 4,075	G2
Garfield 57,813	L2
Garvin 27,210	M5
Grady 45,516	L5
Grant 5,144	L1
Greer 6,061	G5
Harmon 3,283	G5
Harper 3,562	G1
Haskell 11,792	R4
Hughes 14,154	O4
Jackson 28,439	H5
Jefferson 6,818	L6
Johnston 10,513	N6
Kay 48,080	M1
Kingfisher 13,926	L3
Kiowa 10,227	J5
Latimer 10,692	R5
Le Flore 48,109	S5
Lincoln 32,080	N3
Logan 33,924	M3
Love 8,831	M7
Major 7,545	K2
Marshall 13,184	N6
Mayes 38,369	R2
McClain 27,740	L5

McCurtain 34,402	S6
McIntosh 19,456	P4
Murray 12,623	M6
Muskogee 69,451	R3
Noble 11,411	M2
Nowata 10,569	P1
Okfuskee 11,814	O3
Oklahoma 660,448	M3
Okmulgee 39,685	P3
Osage 44,437	O1
Ottawa 33,194	S1
Pawnee 16,612	N2
Payne 68,190	N2
Pittsburg 43,953	P5
Pontotoc 35,143	N5
Pottawatomie 65,521	N4
Pushmataha 11,667	R6
Roger Mills 3,436	G3
Rogers 70,641	P2
Seminole 24,894	N4
Sequoyah 38,972	S3
Stephens 43,182	L6
Texas 20,107	C1
Tillman 9,287	J6
Tulsa 563,299	P2
Wagoner 57,491	P3
Washington 48,996	P1
Washita 11,508	J4
Woods 9,089	J1
Woodward 18,486	H2

CITIES and TOWNS

Achille 506	O7
Ada▲ 15,691	N5
Adair 704	R2
Adams 150	D1
Adamson 150	P5
Addington 117	L6
Afton 1,118	S1
Agra 356	N3
Akins 449	S3
Albany 65	O7
Albert 100	K4
Albion 143	R5
Alderson 261	P5
Alex 635	L5
Alfalfa 70	J4
Aline 214	K1

Allen 951	O5
Altus▲ 21,447	H5
Alva▲ 5,288	J1
Amber 490	L4
Ames 199	K2
Amorita 44	K1
Anadarko▲ 6,645	K4
Antlers▲ 2,552	P6
Apache 1,616	K5
Apperson 30	N1
Aqua Park	R3
Arapaho▲ 748	H3
Arcadia 279	M3
Ardmore▲ 23,711	M6
Arkoma 2,180	T4
Arnett▲ 520	G2
Asher 419	N5
Ashland 53	O5
Atoka▲ 2,988	O6
Atwood 113	O5
Avant 372	O1
Avard 35	J1
Avery 35	N3
Bache 100	P5
Bacone 786	R3
Baker 70	D1
Balko 100	E1
Barnsdall 1,325	O1
Baron 300	S3
Bartlesville▲ 34,748	O1
Battiest 250	S6
Bearden 140	O4
Beaver▲ 1,570	E1
Beggs 1,364	P3
Belzoni 50	R6
Bengal 300	R5
Bennington 289	P7
Bentley 75	O6
Berlin 50	G4
Bernice 504	S1
Bessie 190	H4
Bethany 20,307	L3
Bethel 2,505	S6
Bethel Acres 2,735	M4
Big Cabin 293	R1
Billings 436	M1
Binger 708	K4
Bison 103	L2
Bixby 13,336	P3

Blackburn 102	N2
Blackgum 150	S3
Blackwell 7,668	M1
Blair 894	H5
Blanchard 2,816	L4
Blanco 215	P5
Blocker 135	P4
Blue 175	O7
Bluejacket 274	R1
Boggy Depot 100	O6
Boise City▲ 1,483	B1
Bokchito 564	O6
Bokhoma 35	S7
Bokoshe 450	S4
Boley 1,126	O4
Boswell 703	P6
Bowlegs 371	N4
Bowring 115	O1
Boyd 10	E1
Boynton 274	P3
Braden 15	S4
Bradley 182	L5
Braggs 301	R3
Braman 244	M1
Bray 1,035	L5
Breckinridge 239	L2
Briartown 55	R4
Bridgeport 109	K3
Brinkman	G4
Bristow 4,325	O3
Broken Arrow 74,859	P2
Broken Bow 4,230	S7
Bromide 163	N6
Brooksville 90	M4
Bryant 74	P4
Buffalo▲ 1,200	G1
Bunch 64	S3
Burbank 155	N1
Burlington 156	K1
Burneyville 150	M7
Burns Flat 1,782	H4
Butler 345	H3
Byars 280	N5
Byng 1,090	N5
Byron 45	K1
Cache 2,371	J5
Caddo 944	O6
Cairo 50	O5
Calera 1,739	O7

Calumet 535	K3
Calvin 279	O5
Camargo 115	H2
Cameron 312	T4
Canadian 239	P4
Canadian City	L4
Caney 199	O6
Canton 618	J2
Canute 524	H4
Capron 42	J1
Cardin 150	S1
Carmen 411	J1
Carnegie 1,637	J4
Carney 150	N3
Carrier 77	K2
Carter 254	H4
Cartersville 79	S4
Cashion 635	L3
Castle 122	O4
Catoosa 5,449	P2
Cement 530	K5
Center 100	N5
Centrahoma 110	O5
Centralia	R1
Chandler▲ 2,842	N3
Chattanooga 432	J6
Checotah 3,481	R4
Chelsea 2,136	P1
Cherokee▲ 1,630	K1
Chester 104	J2
Cheyenne▲ 778	G3
Chickasha▲ 15,850	L4
Chilocco 400	M1
Choctaw 9,377	M3
Chouteau 1,931	R2
Christie 166	S3
Cimarron 71	L3
Claremore▲ 15,873	R2
Clarita 72	O6
Clayton 719	R5
Clearview 56	O4
Clemscot 52	L6
Cleora 1,113	S1
Cleo Springs 326	K2
Cleveland 3,282	O2
Clinton 8,833	H3
Cloud Chief 12	J4
Cloudy 175	R6
Coalgate▲ 2,005	O5

Cogar 40	K4
Colbert 1,065	O7
Colcord 819	S2
Cold Springs 24	J5
Cole 473	L4
Coleman 200	N6
Collinsville 4,077	P2
Colony 147	J4
Comanche 1,556	L6
Commerce 2,645	R1
Concho 300	L3
Connerville 150	N6
Cooperton 20	J5
Copan 796	P1
Cordell▲ 2,903	H4
Corinne 100	R6
Corn 591	J4
Cornish 172	L6
Council Hill 129	P3
Countyline 550	L6
Courtney 12	L7
Covington 553	L2
Coweta 7,139	P3
Cowlington 133	S4
Cox City 285	L5
Coyle 337	M3
Crawford 53	G3
Crescent 1,281	L3
Cromwell 265	N4
Crowder 436	P4
Cumberland 100	N6
Curtis 30	H2
Cushing 8,371	N3
Custer City 393	J3
Cyril 1,168	K5
Dacoma 148	J1
Daisy 250	P5
Dale 160	M4
Darwin 50	P6
Davenport 881	N3
Davidson 375	J6
Davis 2,610	M5
Deer Creek 147	L1
Del City 22,128	L4
Dela 434	P6
Delaware 456	P1
Delhi 41	G4
Depew 564	O3
Devol 150	J6
Dewar 919	P4
Dewey 3,179	P1
Dibble 389	L4
Dickson 1,139	M6
Dill City 526	H4
Disney 226	S2
Dougherty 224	M6
Douglas 32	L2
Douthat 30	S1
Dover 367	L3
Dow 500	P5
Driftwood	K1
Drummond 405	L2
Drumright 2,905	O3
Duke (East Duke) 445	G5
Dunbar 10	P6
Duncan▲ 22,505	L5
Durham 30	G3
Dustin 452	O4
Eagle City 56	J3
Eagletown 650	S6
Eakly 276	K4
Earlsboro 633	N4
Edmond 68,315	M3
Eldorado 527	G6
Elgin 1,210	K5
Elk City 10,510	G4
Elmer 96	H6
Elmore City 756	M5
Elmwood 300	F1
Empire City 734	L6
Enid▲ 47,045	L2
Enterprise 130	R4
Erick 1,023	G4
Eucha 210	S2
Eufaula▲ 2,639	P4
Fair Oaks 122	P2
Fairfax 1,555	N1
Fairland 1,025	S1
Fairmont 147	L2
Fairview▲ 2,733	J2
Fallis 28	M3
Fanshawe 384	S5
Fargo 326	G2
Farris 100	P6
Faxon 134	J5
Fay 140	J3
Featherston 75	P4
Felt 120	A1
Fillmore 60	N6
Finley 350	R6
Fittstown 500	N5
Fitzhugh 204	N5
Fleetwood 12	L7
Fletcher 1,022	K5
Foraker 23	O1

Forest Park 1,066	M3
Forgan 532	E1
Fort Cobb 667	K4
Fort Gibson 4,054	R3
Fort Supply 328	G1
Fort Towson 611	R7
Foss 127	H4
Foster 100	M5
Fox 400	M6
Foyil 234	R2
Francis 332	N5
Frederick▲ 4,637	H6
Freedom 271	H1
Gage 429	G2
Gans 208	S4
Garber 845	M2
Garvin 143	S7
Gate 112	F1
Geary 1,258	K3
Gene Autry 99	N6
Geronimo 959	K6
Gerty 101	O5
Glencoe 583	M2
Glenpool 8,123	P3
Glover 244	S6
Golden 300	S6
Goldsby 1,204	L4
Goltry 268	K1
Goodwater 240	S7
Goodwell 1,192	C1
Gore 850	R3
Gotebo 272	J4
Gould 206	G5
Gowen 75	R5
Gracemont 336	K4
Grady 85	L6
Graham 200	M6
Grainola 31	N1
Grandfield 1,110	J6
Grand Lake Towne 65	S1
Granite 1,844	H5
Grant	R7
Gray Horse 60	N1
Grayson 134	P3
Greenfield 123	K3
Griggs 15	B1
Grove 5,131	S1
Guthrie▲ 9,925	M3
Guymon▲ 10,472	D1
Haileyville 891	P5
Hall Park 1,088	M4
Hallett 168	N2
Hammon 469	H3
Hanna 133	P4
Hanson 250	S4
Harden City 250	N5
Hardesty 277	D1
Hardy	N1
Harjo 35	N4
Harmon 27	G2
Harrah 4,719	M4
Harris 192	S7
Hartshorne 2,102	R5
Haskell 1,765	P3
Hastings 155	K6
Haworth 354	S7
Haywood 175	P5
Headrick 130	H5
Healdton 2,786	M6
Heavener 3,201	S5
Helena 443	K1
Hendrix 79	O7
Hennepin 300	M5
Hennessey 2,058	L2
Henryetta 6,096	O4
Herd 18	O1
Hess 29	H6
Hester 25	H5
Hickory 87	N5
Hillsdale 101	K1
Hinton 2,175	K4
Hitchcock 141	K3
Hitchita 113	P3
Hobart▲ 3,997	J5
Hockerville 125	S1
Hodgen 150	S5
Hoffman 148	P4
Holdenville▲ 4,732	O4
Hollis▲ 2,264	G5
Hollister 60	J6
Homestead 35	K2
Hominy 2,584	O2
Honobia 80	R5
Hooker 1,788	D1
Hoot Owl 42	R2
Hopeton 42	J1
Howe 697	S5
Hoyt 160	R4
Hugo▲ 5,536	P7
Hulah 50	O1
Hulbert 543	R3
Humphreys 68	H5
Hunter 173	L1
Hydro 1,060	J3
Idabel▲ 6,952	S7
Indiahoma 374	J5

Agriculture, Industry and Resources

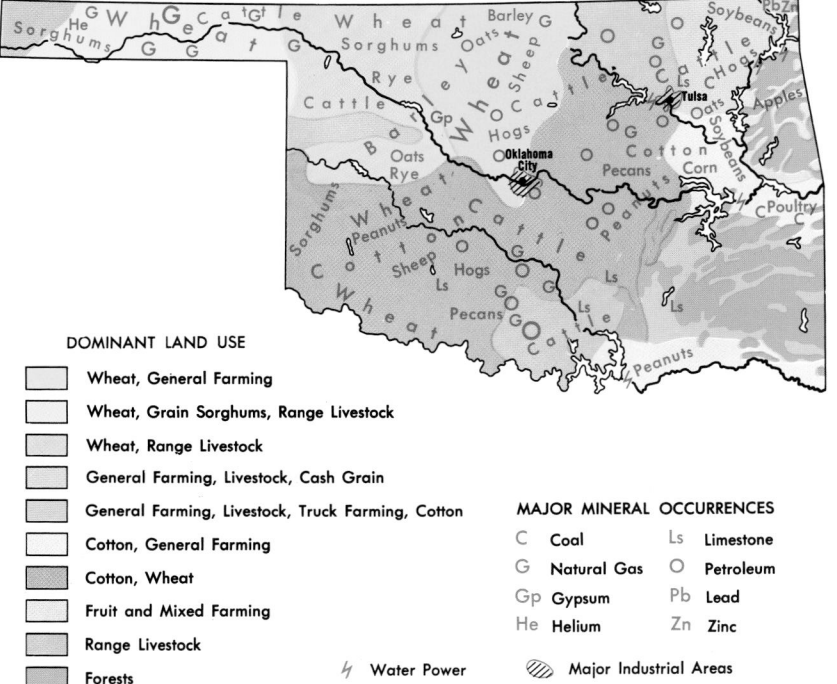

DOMINANT LAND USE

- Wheat, General Farming
- Wheat, Grain Sorghums, Range Livestock
- Wheat, Range Livestock
- General Farming, Livestock, Cash Grain
- General Farming, Livestock, Truck Farming, Cotton
- Cotton, General Farming
- Cotton, Wheat
- Fruit and Mixed Farming
- Range Livestock
- Forests

MAJOR MINERAL OCCURRENCES

C	Coal	Ls	Limestone
G	Natural Gas	O	Petroleum
Gp	Gypsum	Pb	Lead
He	Helium	Zn	Zinc

⚡ Water Power

▨ Major Industrial Areas

(continued on following page)

Oklahoma

SCALE
0 5 10 20 30 40 MI.
0 10 20 30 40 KM
State Capitals ....⊛
County Seats ....◉
Major Limited Access Hwys.

® Copyright HAMMOND INCORPORATED, Maplewood, N. J.

Topography
0   50   100 MI.
0   50   100 KM.

5,000 m. 2,000 m. 1,000 m. 500 m. 200 m. 100 m. Sea Below
16,404 ft. 6,562 ft. 3,281 ft. 1,640 ft. 656 ft. 328 ft. Level

OTHER FEATURES

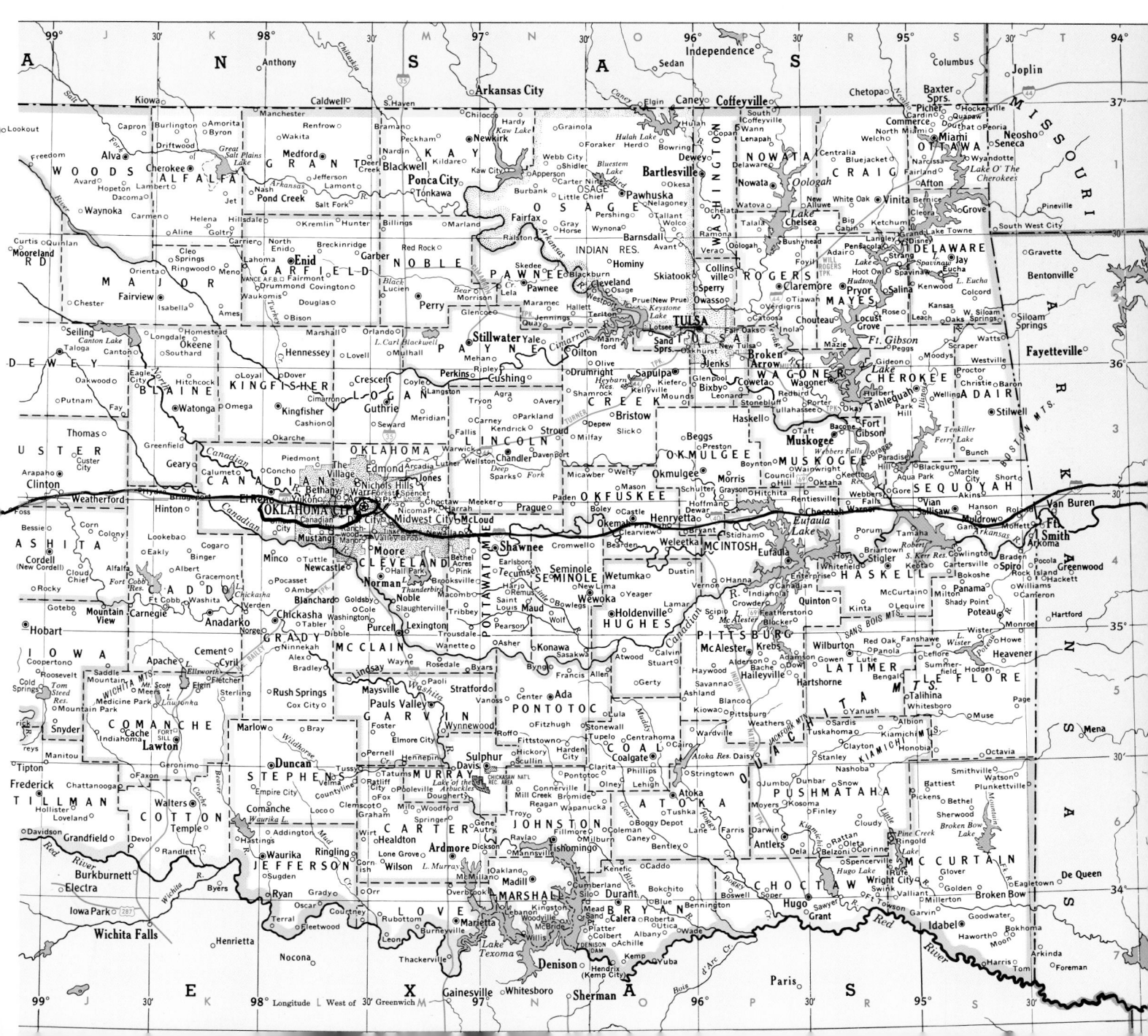

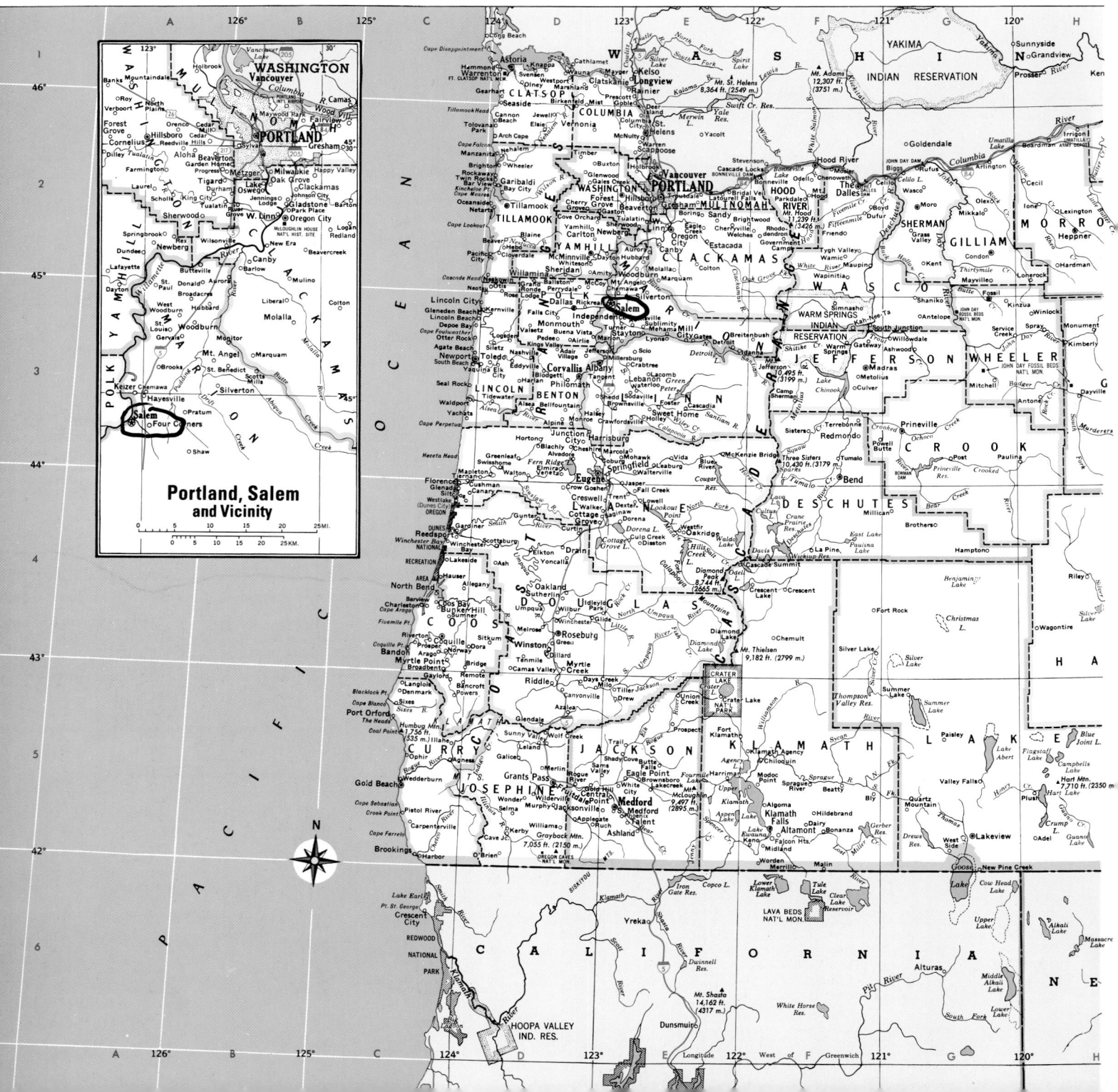

STATE OF OREGON

1859

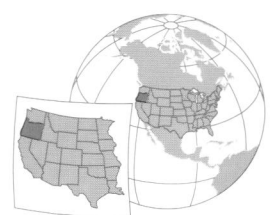

AREA 97,073 sq. mi. (251,419 sq. km.)
POPULATION 3,421,399
CAPITAL Salem
LARGEST CITY Portland
HIGHEST POINT Mt. Hood 11,239 ft. (3426 m.)
SETTLED IN 1810
ADMITTED TO UNION February 14, 1859
POPULAR NAME Beaver State
STATE FLOWER Oregon Grape
STATE BIRD Western Meadowlark

Topography

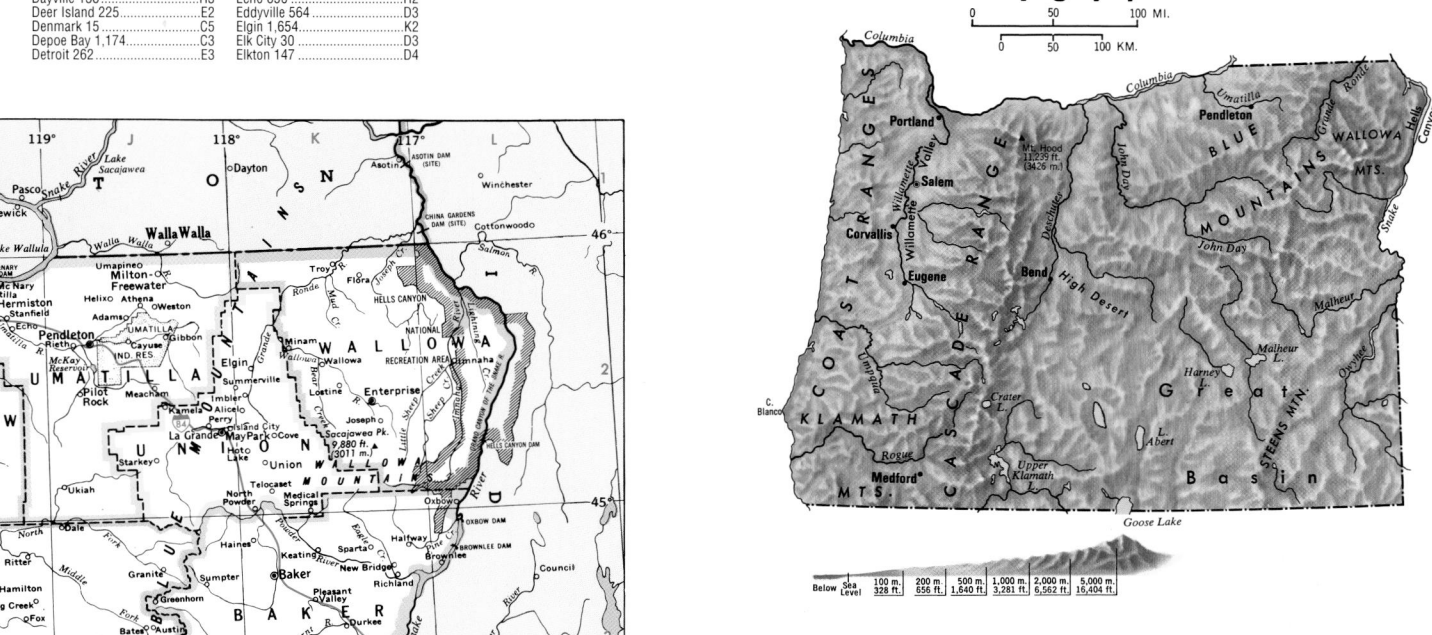

(continued on following page)

Oregon

SCALE
0 5 10 20 30 40 50 60 MI.
0 5 10 20 30 40 50 60KM.

State Capitals ⊛
County Seats ◉
Major Limited Access Hwys.

Agriculture, Industry and Resources

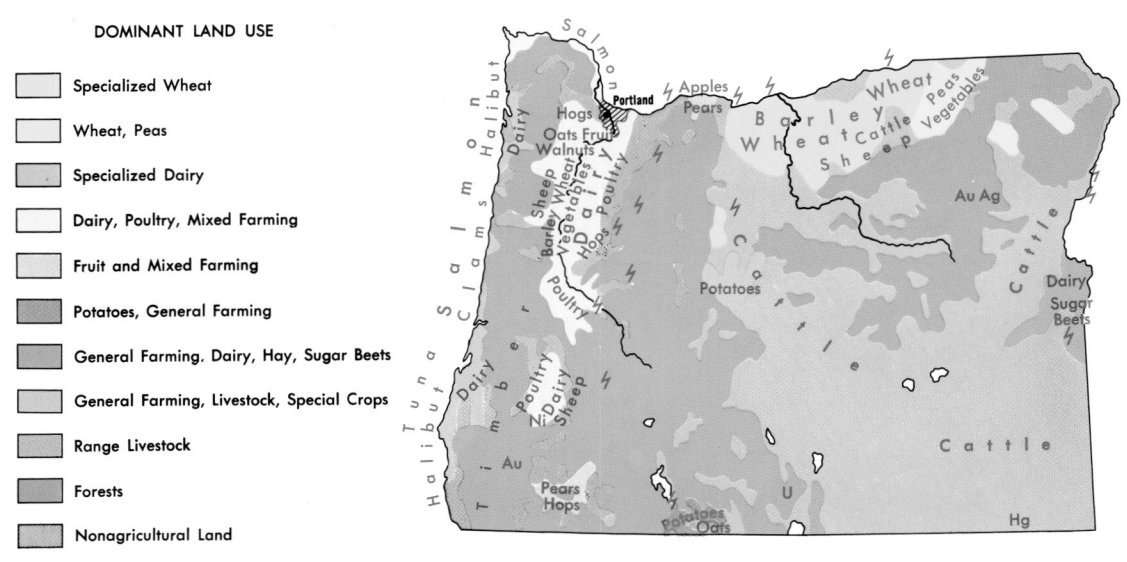

DOMINANT LAND USE

- Specialized Wheat
- Wheat, Peas
- Specialized Dairy
- Dairy, Poultry, Mixed Farming
- Fruit and Mixed Farming
- Potatoes, General Farming
- General Farming. Dairy, Hay, Sugar Beets
- General Farming, Livestock, Special Crops
- Range Livestock
- Forests
- Nonagricultural Land

MAJOR MINERAL OCCURRENCES

Ag Silver   Hg Mercury
Au Gold   Ni Nickel
U Uranium

⚡ Water Power
▨ Major Industrial Areas

DOMINANT LAND USE

- Specialized Dairy
- Dairy, General Farming
- Fruit and Mixed Farming
- Fruit, Truck and Mixed Farming
- General Farming, Livestock, Tobacco
- General Farming, Livestock, Fruit, Tobacco
- Forests
- Urban Areas

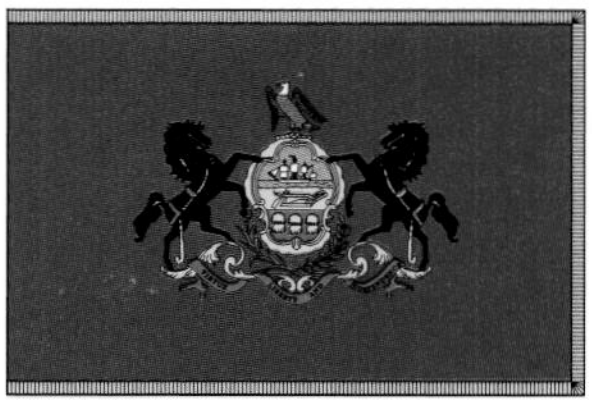

AREA 45,308 sq. mi. (117,348 sq. km.)
POPULATION 12,281,054
CAPITAL Harrisburg
LARGEST CITY Philadelphia
HIGHEST POINT Mt. Davis 3,213 ft. (979 m.)
SETTLED IN 1682
ADMITTED TO UNION December 12, 1787
POPULAR NAME Keystone State
STATE FLOWER Mountain Laurel
STATE BIRD Ruffed Grouse

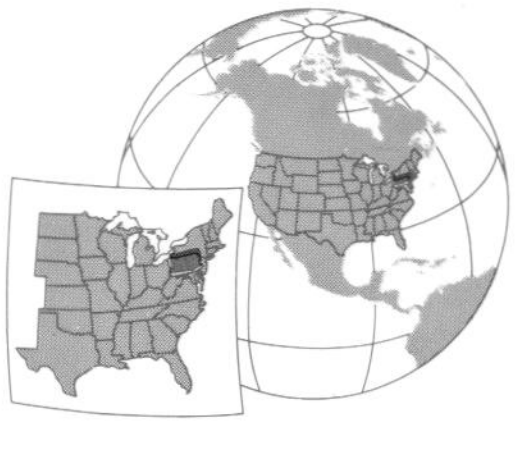

MAJOR MINERAL OCCURRENCES

C	Coal	G	Natural Gas	Sl	Slate
Cl	Clay	Ls	Limestone	Ss	Sandstone
Co	Cobalt	O	Petroleum	Zn	Zinc
Fe	Iron Ore				

Water Power
Major Industrial Areas

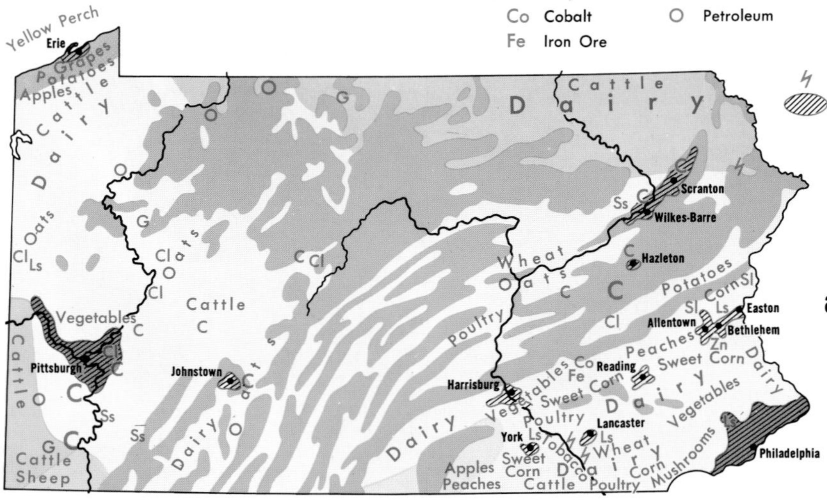

Agriculture, Industry and Resources

COUNTIES

Adams 91,292 H6
Allegheny 1,281,666 B5
Armstrong 72,392 D4
Beaver 181,412 B4
Bedford 49,984 E6
Berks 373,638 K5
Blair 129,144 F4
Bradford 62,761 J2
Bucks 597,635 M5
Butler 174,083 C4
Cambria 152,598 E4
Cameron 5,974 F3
Carbon 58,802 L4
Centre 135,758 G4
Chester 433,501 L6
Clarion 41,765 D3
Clearfield 83,382 F3
Clinton 37,914 G3
Columbia 64,151 K3
Crawford 90,366 B2
Cumberland 213,674 H5
Dauphin 251,798 J5
Delaware 550,864 M6
Elk 35,112 E3
Erie 280,843 B2
Fayette 148,644 C6
Forest 4,946 D2
Franklin 129,313 G6
Fulton 14,261 F6
Greene 40,672 B6
Huntingdon 45,586 F5
Indiana 89,605 D4
Jefferson 45,932 D3
Juniata 22,821 H4
Lackawanna 213,295 L3
Lancaster 470,658 K5
Lawrence 94,643 B4
Lebanon 120,327 K5
Lehigh 312,090 L4
Luzerne 319,250 L3
Lycoming 120,044 H3
McKean 45,936 E2
Mercer 120,293 B3
Mifflin 46,486 G4
Monroe 138,687 M3
Montgomery 750,097 M5
Montour 18,236 J3
Northampton 267,066 M4
Northumberland 94,556 J4
Perry 43,602 H5
Philadelphia (city county) 1,517,550 M6
Pike 46,302 M3
Potter 18,080 G2
Schuylkill 150,336 K4
Snyder 37,546 H4
Somerset 80,023 D6
Sullivan 6,556 J3
Susquehanna 42,238 L2
Tioga 41,373 H2
Union 41,624 H4
Venango 57,565 C3
Warren 43,863 D2
Washington 202,897 B5
Wayne 47,722 M2
Westmoreland 369,993 D5
Wyoming 28,080 K2
York 381,751 J6

CITIES and TOWNS

Abbottstown 905 J6
Abington• 56,103 M5
Adamstown 1,203 K5
Akron 4,046 K5
Albion 1,607 B2
Alburtis 2,117 L5
Aldan 4,313 M7
Alexandria 411 F4
Aliquippa 11,734 B4
Allentown▲ 106,632 L4
Allison Park 10,000 C4
Altoona 49,523 F4
Ambler 6,426 M5
Ambridge 7,769 B4
Annville 4,518 J5
Apollo 1,765 C4
Archbald 6,220 F6
Ardmore 12,616 M6
Arendtsville 848 H6
Arnold 5,667 C4
Ashland 3,283 K4
Ashley 2,866 E7
Aspinwall 2,960 C6
Atglen 1,217 K6
Athens 3,415 K2
Atlas 1,162 K4
Auburn 839 K4
Austin 623 F2
Avalon 5,294 B6
Avella 900 B5
Avis 1,492 H3
Avoca 2,851 F7
Avondale 1,108 L6
Avonmore 820 C4
Baden 4,377 B4
Bala-Cynwyd N6
Baldwin 19,999 B7
Bally 1,062 L5
Bangor 5,319 M4
Barnesboro 2,530 E4
Bath 2,678 M4
Beallsville 511 C5
Beaver▲ 4,775 B4
Beaverdale 1,230 E5
Beaver Falls 9,920 B4
Beaver Meadows 968 L4
Beavertown 870 H4
Bedford▲ 3,141 F5
Beech Creek 717 G3
Bellefonte▲ 6,395 G4
Belle Vernon 1,211 C5
Belleville 1,386 G4
Bellevue 8,770 B6
Bellwood 2,016 F4
Ben Avon 1,917 B6
Bendersville 576 H6
Bentleyville 2,502 B5
Benton 955 K3
Berlin 2,192 E6
Bernville 865 K5
Berrysburg 376 J4
Berwick 10,774 K3
Berwyn-Devon 5,019 L5
Bessemer 1,172 B4
Bethel Park 33,556 B7
Bethlehem 71,329 M4
Biglerville 1,101 H6
Big Run 686 E4
Birdsboro 5,064 L5
Black Lick 1,438 D4
Blairsville 3,607 D5
Blakely 7,027 F6
Blawnox 1,550 C6
Bloomfield (New Bloomfield)▲ 1,077 H5
Blooming Valley 391 B2
Bloomsburg▲• 12,375 J3
Blossburg 1,480 H2
Boalsburg 3,578 G4
Bobtown 1,008 B6
Boiling Springs 2,769 H5
Bolivar 501 D5
Boothwyn 5,206 L7
Boswell 1,364 E5
Bowmanstown 895 L4
Boyertown 3,940 L5
Brackenridge 3,543 C4
Braddock 2,912 C7
Bradford 9,175 E2
Brentwood 10,466 B7
Briar Creek 651 K3
Brickerville 1,287 K5
Bridgeport 4,371 M5
Bridgeville 5,341 B5
Bridgewater 739 B4
Brisbin 413 F4
Bristol 9,923 N5
Bristol• 55,521 N5
Broad Top 384 F5
Brockway 2,182 E3
Brodheadsville 1,637 M4
Brookhaven 7,985 M7
Brookville▲ 4,230 D3
Broomall 11,046 M6
Brownstown 883 K5
Brownsville 2,804 C5
Bruin 534 C3
Bryn Athyn 1,351 M5
Bryn Mawr 4,382 M5
Burgettstown 1,576 A5
Burlington 479 J2
Burnham 2,144 H4
Burnside 350 E4
Butler▲ 15,121 C4
Cadogan• 390 C4
Cairnbrook 1,081 E5
California 5,274 C5
Carlisle 420 C4
Cambridge Springs 2,363 C2
Camp Hill 7,636 H5
Canonsburg 8,607 B5
Canton 1,807 J2
Carbondale 9,804 L2
Carlisle▲ 17,970 H5
Carmichaels 556 B6
Carnegie 8,389 B7
Carrolltown 1,049 E4
Carroll Valley 3,291 H6
Castle Shannon 8,556 B7
Catasauqua 6,588 M4
Catawissa 1,589 K4
Centerville 3,390 B6
Central City 1,258 E5
Centre Hall 1,079 G4
Chalfont 3,900 M5
Chambersburg▲ 17,862 G6
Charleroi 4,871 C5
Cheltenham• 36,875 M5
Cherry Tree 431 E4
Chester 36,854 L7
Chester Heights 2,481 L7
Chester Hill 918 F4
Cheswick 1,899 C6
Chicora 1,021 C4
Christiana 1,124 K6
Churchill 3,566 C7
Clairton 8,491 C7
Clarendon 564 D2
Clarion▲ 6,185 D3
Clark (Clarksville) 633 B3
Clarks Green 1,630 F6
Clarks Summit 5,126 F6
Claysburg 1,503 F5
Claysville 724 B5
Clearfield▲ 6,631 F3
Clifton Heights 6,779 M7
Clintonville 528 C3
Clymer 1,547 E4
Coalport 490 E4
Coatesville 10,838 L5
Cochranton 1,148 B2
Codorus (Jefferson) 685 J6
Cokeburg 705 B5
Collegeville 8,032 M5
Collingdale 8,664 N7
Columbia 1,162 K5
Colver 1,035 E4
Colwyn 2,453 N7
Confluence 834 D6
Conneaut Lake 708 B2
Conneautville 848 A2
Connellsville 9,146 C5
Connoquenessing 564 B4
Conshohocken 7,589 M5
Conway 2,290 B4
Conyngham 1,958 K3
Coopersburg 2,582 M5
Cooperstown 460 C2
Coplay 3,387 L4
Coraopolis 6,131 B4
Cornwall 3,486 K5
Corry 6,834 C2
Coudersport▲ 2,650 G2
Crabtree 320 D5
Crafton 6,706 B7
Cranesville 600 B2
Cresson 1,631 E5
Cressona 1,635 K4
Cross Roads 322 J6
Curwensville 2,650 E4
Dale 1,503 E5
Dallas 2,557 E7
Dallastown 4,087 J6
Dalton 1,294 L2
Danville▲ 4,897 J4
Darby 10,299 M7
Dauphin 773 J5
Dayton 543 D4
Delaware Water Gap 744 M4
Delmont 2,497 D5
Delta 741 K6
Denver 3,332 K5
Derry 2,991 D5
Dickson City 6,205 F7
Dillsburg 2,063 J5
Donora 5,653 C5
Dormont 9,305 B7
Dover 1,815 J6
Downingtown 7,589 L5
Doylestown▲ 8,227 M5
Dravosburg 2,015 C7
Drexel Hill 29,364 M6
Drifton 1,786 L3
Dublin 2,083 M5
DuBois 8,123 E3
Duboistown 1,280 H3
Dunbar 1,219 C6
Duncannon 1,450 H5
Duncansville 1,508 F5
Dunmore 14,018 F7
Dupont 2,719 F7
Duquesne 7,332 C7
Duryea 4,634 F7
Dushore 663 K2
East Bangor 979 M4
East Berlin 1,365 J6
East Berwick 1,998 K3
East Brady 1,038 C3
East Butler 679 C4
East Conemaugh 1,291 E5
East Faxon 3,951 J3
East Greenville 3,103 L5
East Lansdowne 2,586 M7
Easton▲ 26,263 M4
East Petersburg 4,450 K5
East Pittsburgh 2,017 C7
East Prospect 678 J6
East Stroudsburg 9,888 M4
East Washington 1,930 B5
Eau Claire 371 C3
Ebensburg▲ 3,091 E5
Economy 9,363 B4
Eddystone 2,442 M7
Edgewood 3,311 B7
Edgeworth 1,730 B4
Edinboro 6,950 B2
Edwardsville 4,984 E7
Elderton 371 D4
Eldred 858 F2
Elizabeth 1,609 C5
Elizabethtown 11,887 J5
Elizabethville 1,344 J4
Elkland 1,786 H1
Ellsworth 1,083 B5
Ellwood City 8,688 B4
Elverson 959 L5
Elysburg 2,067 K4
Emigsville 2,467 J5
Emlenton 784 C3
Emmaus 11,313 M4
Emporium▲ 2,526 F2
Emsworth 2,598 B6
Enola 5,627 J5
Enon Valley 355 B4
Ephrata 13,213 K5
Erie▲ 103,717 B1
Ernest 492 D4
Espy 1,428 K4
Etna 3,924 B6
Etters (Goldsboro) 939 J5
Evans City 2,009 B4
Everett 1,905 F5
Everson 842 C5
Exeter 5,955 F7
Export 895 C5
Factoryville 1,144 L2
Fairchance 2,174 C6
Fairfield 486 H6
Fairless Hills 8,365 N5
Falls Creek 983 E3
Farrell 6,050 A3
Fawn Grove 489 J6
Fayette City 714 C5
Fayetteville 2,774 G6
Felton 438 J6
Ferndale 1,834 E5
Finleyville 446 B5
Fleetwood 4,018 L5
Fleming (Unionville) 361 G4
Flemington 1,319 G3
Folcroft 6,978 M7
Folsom 8,072 M7
Ford City 3,451 D4
Ford Cliff 412 D4
Forest City 1,855 L2
Forest Hills 6,831 C7
Forty Fort 4,579 F7
Fountain Hill 4,614 L4
Fox Chapel 5,436 C6
Frackville 4,361 K4
Franklin▲ 7,212 C3
Franklintown 373 H5
Fredericksburg 1,140 B2
Fredericktown 1,094 C6
Fredonia 652 B3
Freeburg 584 H4
Freedom 1,763 B4
Freeland 3,643 L3
Freemansburg 1,897 M4
Freeport 1,962 C4
Galeton 1,325 G2
Gallitzin 1,756 E4
Gap 1,611 L6
Garden View 2,679 H3
Garrett 449 D6
Geistown 2,555 E5
Gettysburg▲ 7,490 H6
Gilberton 867 K4
Girard 3,164 B2
Girardville 1,742 K4
Glassport 4,993 C7
Glen Lyon 1,881 E7
Glenolden 7,476 M7
Glen Rock 1,809 J6
Glenside 7,914 M5
Grampian 395 E4
Gratz 676 J4
Great Bend 700 L2
Greencastle 3,722 G6
Greensburg▲ 15,889 D5
Greentree 4,719 B7
Greenville 6,380 B3
Grove City 8,024 B3
Halifax 875 J5
Hallstead 1,216 L2
Hamburg 4,114 L4
Hanover 14,535 J6
Harmony 937 B4
Harrisburg (cap.)▲ 48,950 H5
Harrisville 883 B3
Harveys Lake 2,888 E7
Hastings 1,398 E4
Hatboro 7,393 M5
Hatfield 2,605 M5
Haverford• 48,498 M6
Havertown M6
Hawley 1,303 M3
Hawthorn 587 D3
Hazleton 23,329 L4
Heidelberg 1,225 B7
Hellam (Hallam) 1,532 J6
Hellertown 5,606 M4
Herndon 422 J4
Hershey 12,771 J5
Highland Park 1,446 H4
Highspire 2,720 J5
Hollidaysburg▲ 5,368 F5
Homer City 1,844 D4
Homestead 3,569 B7
Honesdale▲ 4,874 M2
Honey Brook 1,287 L5
Hooversville 779 E5
Hop Bottom 345 L2
Hopwood 2,006 C6
Houston 1,314 B5
Houtzdale 941 F4
Howard 699 G3
Hughestown 1,541 F7
Hughesville 2,220 J3
Hummelstown 4,360 J5
Huntingdon▲ 6,918 G5
Hyde 1,491 F4
Hyde Park 513 D4
Hydetown 605 C2
Hyndman 1,005 E6
Imperial-Enlow 3,514 B5
Indiana▲ 14,895 D4
Indian Lake 450 E5
Industry 1,921 B4
Ingram 3,712 B7
Irvona 680 E4
Irwin 4,366 C5
Jacobus 1,203 J6
Jamestown 636 A3
Jeannette 10,654 C5
Jenkintown 4,478 M5
Jennerstown 714 D5
Jermyn 2,287 L2
Jerome 1,068 D5
Jersey Shore 4,482 H3
Jessup 4,718 F6
Jim Thorpe (Mauch Chunk)▲ 4,804 L4

(continued on following page)

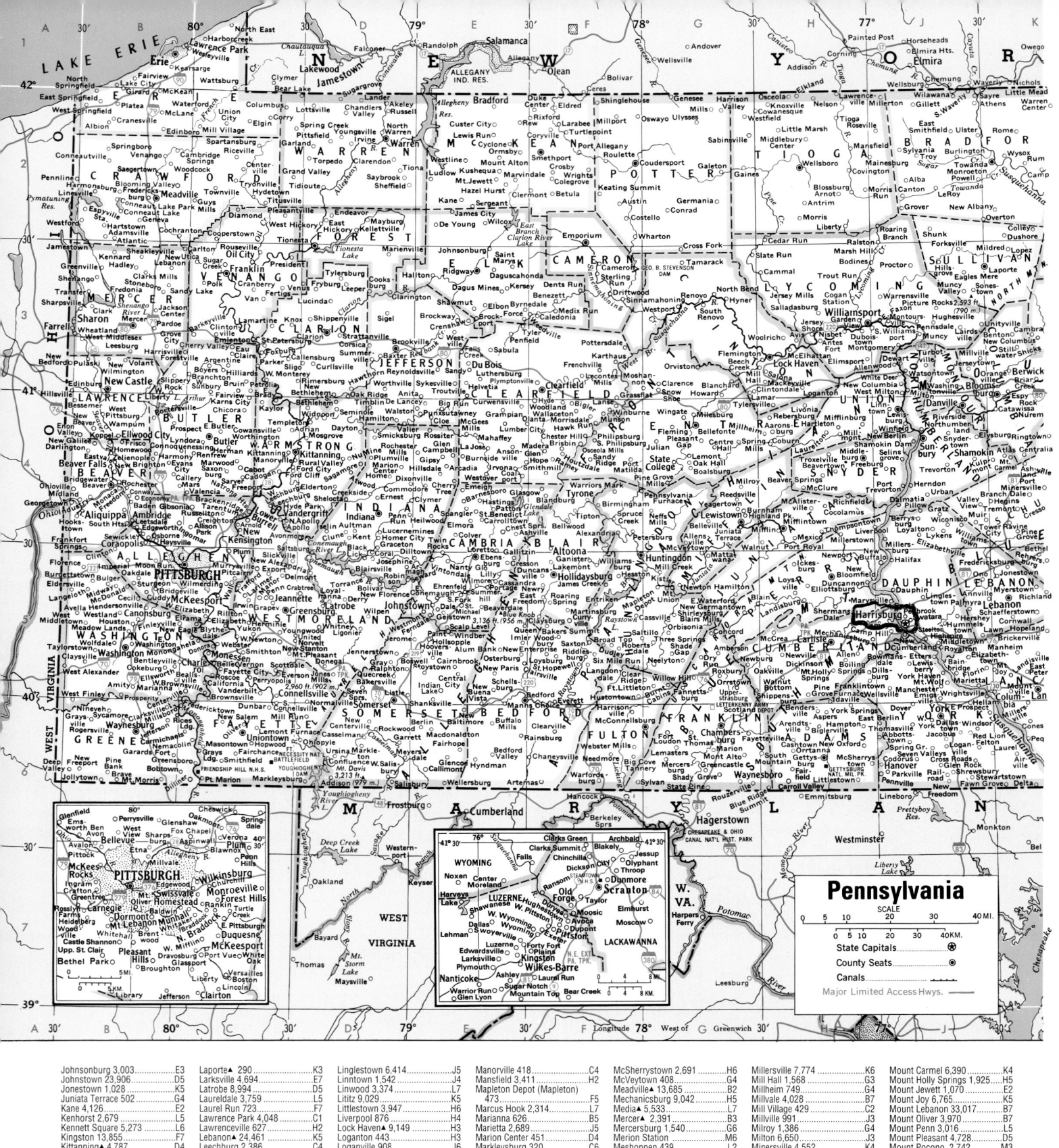

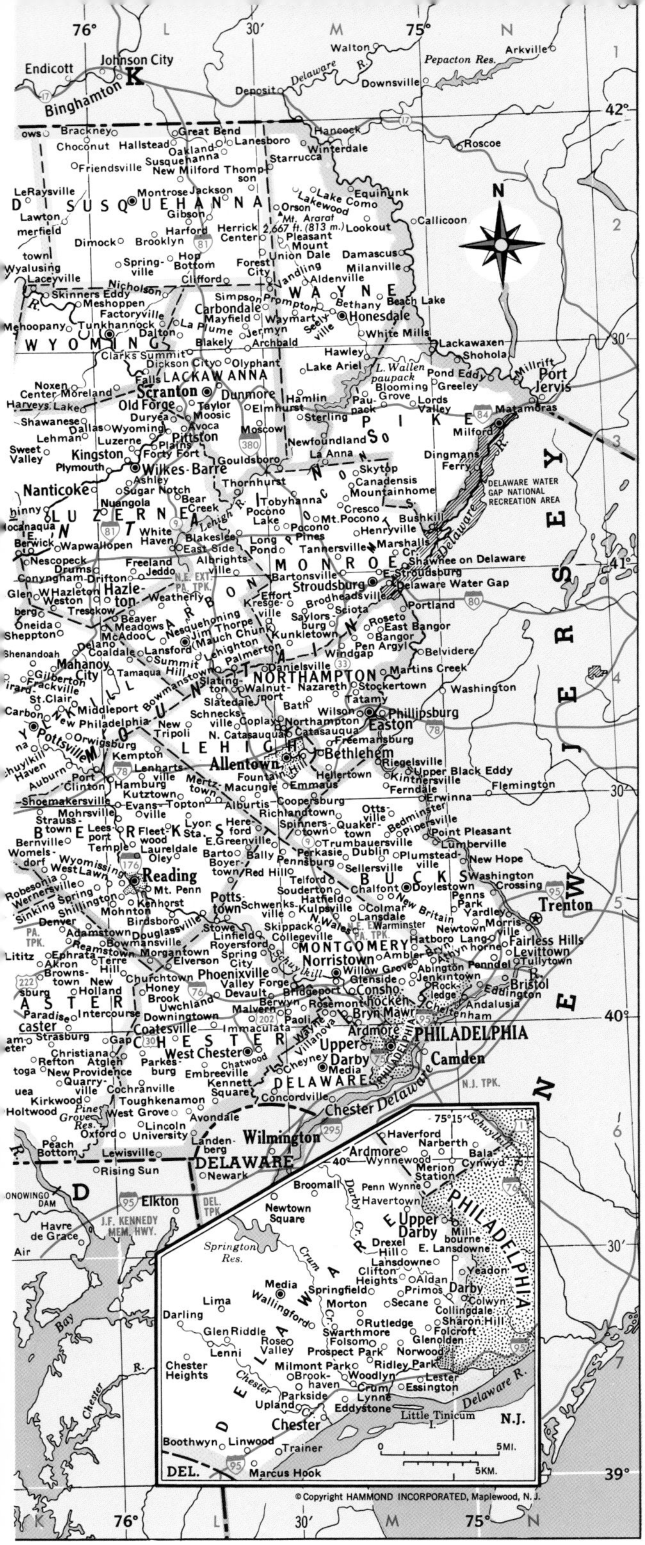

New Beaver 1,677 B4
New Berlin 838 J4
New Bethlehem 1,057 D3
New Bloomfield▲ 1,109 H5
New Brighton 6,641 B4
New Britain 3,125 M5
New Castle▲ 26,309 B3
New Cumberland 7,349 J5
New Eagle 2,262 B5
New Florence 784 D5
New Freedom 3,512 J6
New Galilee 424 A4
New Holland 5,092 K5
New Hope 2,252 N5
New Kensington 14,701 C4
New Milford 878 L2
New Oxford 1,696 H6
New Philadelphia 1,149 K4
Newport 1,506 H5
New Salem (Delmont) 648 D5
New Stanton 1,906 C5
Newtown 2,312 N5
Newtown Square• 11,775 L6
Newville 1,367 H5
New Wilmington 2,452 B3
Nicholson 713 L2
Norristown▲ 31,282 M5
Northampton 9,405 M4
North Apollo 1,426 D4
North Braddock 6,410 C7
North Catasauqua 2,814 L4
North East 4,601 C1
Northumberland 3,714 J4
North Wales 3,342 M5
North Warren 1,232 D2
Norvelt 2,541 D5
Norwood 5,985 M7
Nuangola 671 L3
Oakdale 1,551 B5
Oakland 622 L2
Oakmont 6,911 C6
Ohioville 3,759 B4
Oil City 11,504 C3
Old Forge 8,798 F7
Oliver 2,925 C6
Olyphant 4,978 F7
Orangeville 500 K3
Orbisonia 447 G5
Orwigsburg 3,106 K4
Osborne 566 B4
Osceola Mills 1,249 F4
Oxford 4,315 K6
Paint 1,103 E5
Palmerton 5,248 L4
Palmyra 7,096 J5
Paoli 5,425 M5
Paradise 1,028 K5
Parker 799 C3
Parkesburg 3,373 L6
Parkside 2,267 M7
Parkville 6,593 J6
Patton 2,023 E4
Pen Argyl 3,615 M4
Penbrook 3,044 J5
Penn 460 C5
Penndel 2,420 N5
Penn Hills 46,809 C7
Pennsburg 2,732 M5
Pennville 1,964 J6
Pen Wynne 5,382 M6
Perkasie 8,828 M5
Perryopolis 1,764 C5
Petersburg 469 G4
Philadelphia▲ 1,517,550 N6
Philipsburg 3,056 F4
Phoenixville 14,788 L5
Picture Rocks 693 J3
Pillow 341 J4
Pine Grove 2,154 K4
Pine Grove Mills 1,141 G4
Pitcairn 3,689 C5
Pittsburgh▲ 334,563 B7
Pittston 8,104 F7
Plains• 10,906 F7
Platea 467 B2
Pleasant Gap 1,611 G4
Pleasant Hills 8,397 B7
Plum 26,940 C5
Plumville 342 D4
Plymouth 6,507 E7
Plymptonville 1,040 E3
Pocono Pines 1,013 M3
Point Marion 1,333 C6
Polk 1,031 C3
Portage 2,837 E5
Port Allegany 2,355 F2
Port Carbon 2,019 K4
Portland 579 M4
Port Matilda 638 F4
Port Royal 977 H4
Port Vue 4,228 C7
Pottstown 21,859 L5
Pottsville▲ 15,549 K4
Prospect 1,234 B4
Prospect Park 6,594 M7
Punxsutawney 6,271 E4
Quakertown 8,931 M5
Quarryville 1,994 K6
Ramey 525 F4
Rankin 2,315 C7
Reading▲ 81,207 L5
Reamstown 3,498 K5
Red Hill 2,196 L5
Red Lion 6,149 J6
Reedsville 858 G4
Renovo 1,318 G3
Reynoldsville 2,710 D3
Rices Landing 457 C6
Richland 1,508 K5
Richlandtown 1,283 M5
Ridgway▲ 4,591 E3
Ridley Park 7,196 M7
Riegelsville 863 M4
Rimersburg 1,051 D3
Ringtown 826 K4
Riverside 1,861 J4
Roaring Spring 2,418 F5
Robesonia 2,036 K5
Rochester 4,014 B4
Rockledge 2,577 M5
Rockwood 954 D6
Rome 475 K2
Roscoe 848 C5
Rose Valley 944 L7
Roseto 1,555 M4
Rosslyn Farms 483 B7
Rouseville 472 C3
Rouzerville 862 G6
Royalton 963 J5
Royersford 4,246 L5
Rural Valley 922 D4
Russellton 1,530 C4
Rutledge 860 M7
Saegertown 1,071 B2
Saint Clair 3,254 K4
Saint Marys 14,502 E3
Saint Michael-Sidman 973 E5
Saint Petersburg 405 C3
Salisbury 878 D6
Saltillo 347 G5
Saltsburg 955 C4
Sandy 1,687 E3
Sandy Lake 743 B3
Saxonburg 1,629 C4
Saxton 803 F5
Sayre 5,813 K2
Scalp Level 851 E5
Schnecksville 1,989 L4
Schuylkill Haven 5,548 K4
Schwenksville 1,693 L5
Scottdale 4,772 C5
Scranton▲ 76,415 F7
Selinsgrove 5,383 J4
Sellersville 4,564 M5
Seven Valleys 483 J6
Seward 484 E5
Sewickley 3,902 B4
Shamokin 8,009 J4
Shamokin Dam 1,502 J4
Sharon 16,328 B3
Sharon Hill 5,468 N7
Sharpsburg 3,594 B6
Sharpsville 4,500 A3
Sheffield 1,268 D2
Shenandoah 5,624 K4
Shickshinny 959 K3
Shillington 5,059 K5
Shinglehouse 1,250 F2
Shippensburg 5,586 H5
Shippenville 505 D3
Shoemakersville 2,124 K4
Shrewsbury 3,378 J6
Sinking Spring 2,639 K5
Skippack 2,889 M5
Slatington 4,434 L4
Slickville 4,434 C5
Sligo 728 C3
Slippery Rock 3,068 B3
Smethport▲ 1,684 F2
Smithfield 854 C6
Smithton 388 C5
Snow Shoe 771 G3
Snydertown 357 J4
Somerset▲ 6,762 D6
Souderton 6,730 M5
South Bethlehem 444 D4
South Connellsville 2,281 C6
South Fork 1,138 E5
South Heights 542 B4
South Philipsburg 438 F4
South Renovo 557 G3
South Waverly 987 J2
South Williamsport 6,412 J3
Spangler 2,068 E4
Spartansburg 333 C2
Springboro 491 B2
Spring City 3,305 L5
Springdale 3,828 C6
Springfield 23,677 M7
Spring Grove 2,050 J6
State College 38,420 G4
State Line 1,253 G6
Steelton 5,858 J5
Stewartstown 1,752 K6
Stockertown 687 M4
Stoneboro 1,104 B3
Stowe 3,585 L5
Stoystown 428 E5
Strasburg 2,800 K6
Strattanville 542 D3
Strausstown 353 K5
Stroudsburg▲ 5,756 M4
Sturgeon 1,764 B5
Sugar Creek 5,331 C3
Sugargrove 613 D1
Sugar Notch 1,023 E7
Summerhill 521 E5
Summerville 525 D3
Summit Hill 2,974 L4
Sunbury▲ 10,610 J4
Susquehanna 1,690 L2
Swarthmore 6,170 M7
Swatara• 18,796 J5
Swissvale 9,653 C7
Swoyerville 5,157 E7
Sykesville 1,246 E3
Tamaqua 7,174 L4
Tarentum 5,674 C4
Tatamy 930 M4
Taylor 2,239 F7
Telford 4,680 M5
Temple 1,491 L5
Terre Hill 1,237 L5
Thompsontown 711 H4
Three Springs 445 G5
Throop 4,010 F7
Tidioute 792 D2
Tioga 622 H2
Tionesta▲ 615 C2
Tipton 1,225 F4
Titusville 6,146 C2
Topton 1,948 L5
Toughkenamon 1,375 L6
Towanda▲ 3,024 J2
Tower City 1,396 J4
Townville 358 C2
Trafford 3,236 C5
Trainer 1,901 L7
Tremont 1,784 K4
Tresckow 964 K4
Trevorton 2,010 J4
Troy 1,508 J2
Trumbauersville 1,059 M5
Tullytown 2,031 N5
Tunkhannock▲ 1,911 L2
Turbotville 691 J3
Turtle Creek 6,076 C7
Tyrone 5,528 F4
Ulysses (Lewisville) 684 G2
Union City 3,463 C2
Uniontown▲ 12,422 C6
Upland 2,977 L7
Upper Darby• 81,821 M6
Upper Saint Claire• 20,053 B7
Valencia 384 C4
Valley Forge 400 L5
Valley View 1,677 J4
Vanderbilt 553 C5
Vandergrift 5,455 D4
Vandling 738 M2
Verona 3,124 C6
Versailles 1,724 C7
Villanova M6
Vintondale 528 E5
Wall 727 C5
Walnutport 2,043 L4
Wampum 678 B4
Warren▲ 10,259 D2
Warrior Run 624 E7
Washington▲ 15,268 B5
Waterford 1,449 B2
Watsontown 2,255 J3
Wattsburg 378 C1
Waymart 1,429 M2
Wayne M6
Waynesboro 9,614 G6
Waynesburg▲ 4,184 B6
Weatherly 2,612 L4
Wellsboro▲ 3,328 H2
Wernersville 2,150 K5
Wesleyville 3,617 C1
West Brownsville 1,075 C5
West Chester▲ 17,861 L6
West Elizabeth 565 C5
Westfield 1,190 H2
West Grove 2,652 L6
West Hazleton 3,542 K4
West Kittanning 1,199 C4
West Lawn 1,597 K5
West Leechburg 1,290 C4
West Middlesex 929 B3
West Mifflin 22,464 C7
Westmont 5,523 D5
West Newton 3,083 C5
Westover 446 E4
West Pittsburg 1,133 B4
West Pittston 5,072 F7
West View 7,277 B6
West Wyoming 2,833 E7
West York 4,321 J6
Wheatland 748 B3
Whitaker 1,338 C7
Whitehall 14,444 B7
White Haven 1,182 L3
White Oak 8,437 C7
Wiconisco• 1,168 J4
Wilkes-Barre▲ 43,123 F7
Wilkinsburg 19,196 C7
Williamsburg 1,345 F5
Williamsport▲ 30,706 H3
Williamstown 1,433 J4
Willow Grove 16,234 M5
Wilmerding 2,145 C5
Wilson 7,682 M4
Windber 4,395 E5
Windgap 2,812 M4
Windsor 1,331 J6
Wolfdale 2,873 B5
Womelsdorf 2,599 K5
Woodlyn 10,036 M7
Worthington 778 C4
Wrightsville 2,223 J5
Wyalusing 564 K2
Wyoming 3,221 E7
Wyomissing 8,587 K5
Yardley 2,498 N5
Yeadon 11,762 N7
Yeagertown 1,035 G4
York▲ 40,862 J6
York Haven 809 J5
York Springs 574 H6
Youngsville 1,834 D2
Youngwood 4,138 D5
Zelienople 4,123 B4

OTHER FEATURES

Allegheny (res.) E2
Allegheny (riv.) D2
Allegheny Front (mts.) E5
Appalachian (mts.) H4
Ararat (mt.) M2
Arthur (lake) C4
Beaver (riv.) B4
Blue (mt.) G5
Blue Knob (mt.) E5
Casselman (riv.) D6
Clarion (riv.) D3
Conemaugh (riv.) D5
Conemaugh River (lake) D4
Conewango (creek) D1
Davis (mt.) D6
Delaware (riv.) N3
Delaware Water Gap Nat'l Rec. Area N3
Erie (lake) B1
Fort Necessity Nat'l Battlefield C6
George B. Stevenson (dam) G3
Gettysburg Nat'l Mil. Park H6
Glendale (lake) F4
Juniata (riv.) G5
Laurel Hill (mt.) D5
Lehigh (riv.) L3
Letterkenny Army Depot G6
Licking (creek) F6
Little Tinicum (isl.) M7
Lycoming (creek) H3
Monongahela (riv.) C6
North (mt.) K3
Ohio (riv.) A4
Oil (creek) C2
Pine (creek) H2
Pine Grove (res.) K6
Pocono (mts.) M3
Pymatuning (res.) A2
Redbank (creek) E3
Schuylkill (riv.) M5
Shenango River (lake) B3
Sinnemahoning (creek) F3
South (mt.) H6
Steamtown Nat'l Hist. Site F7
Susquehanna (riv.) K6
Tioga (riv.) H1
Tionesta Creek (lake) D3
Towanda (creek) J2
Tuscarora (mt.) G5
Wallenpaupack (lake) M3
Youghiogheny River (lake) D6

▲County seat
•Population of town or township

Topography

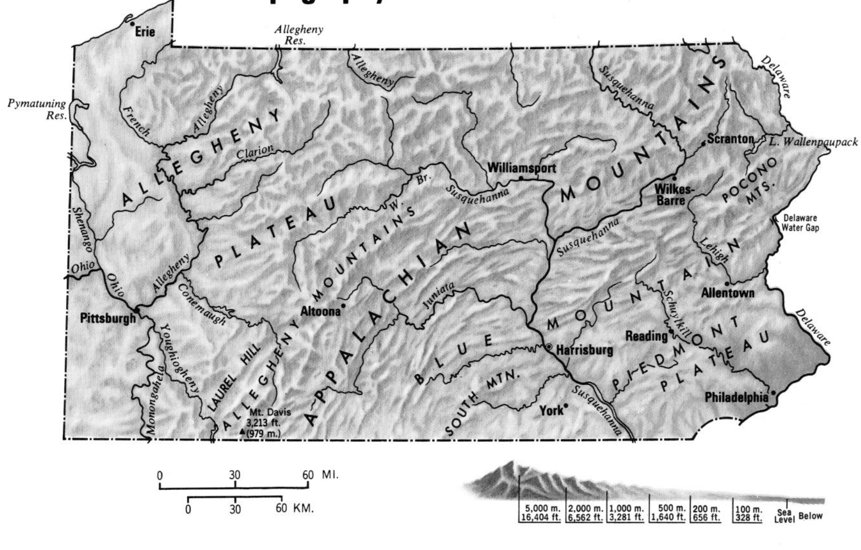

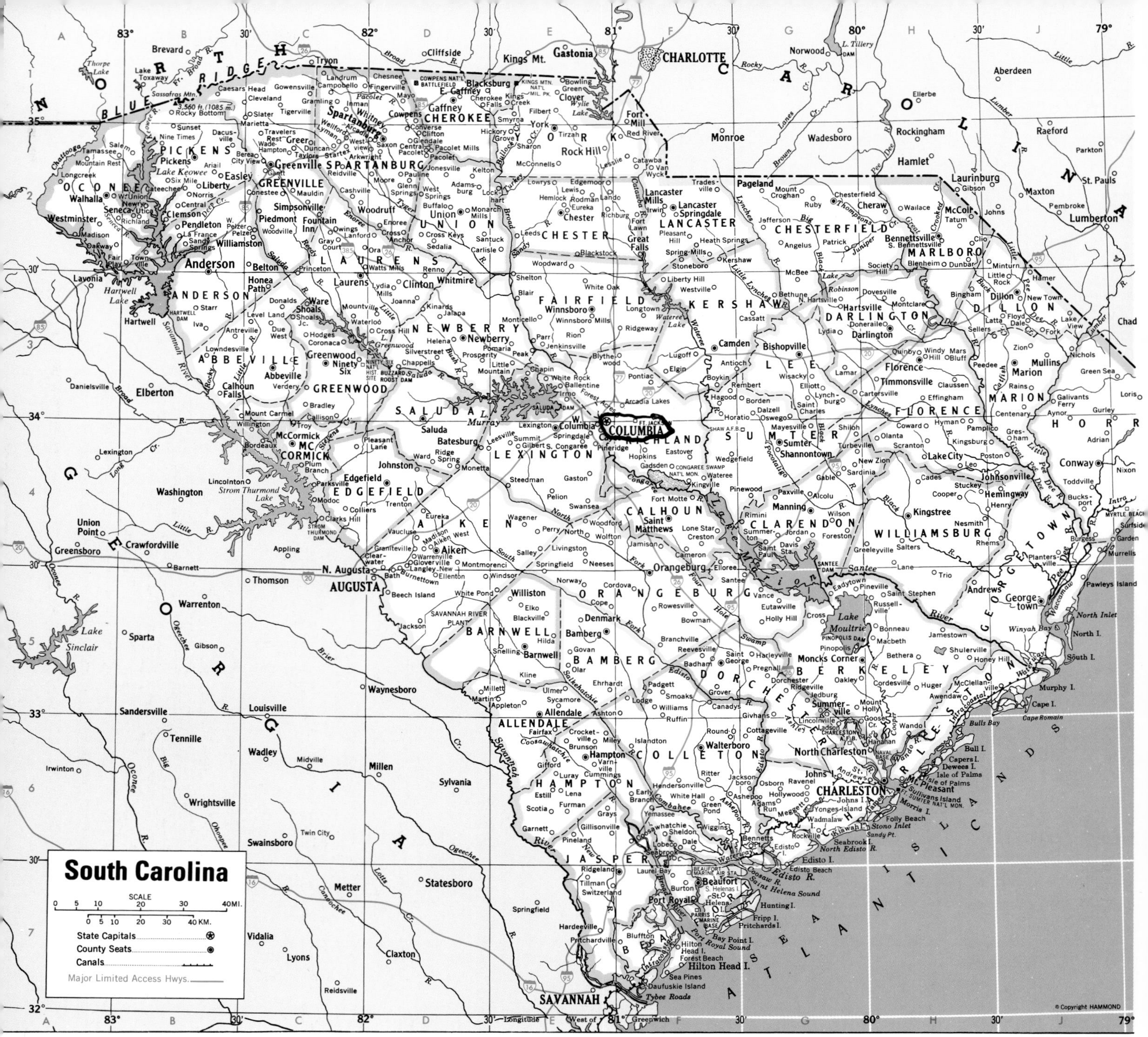

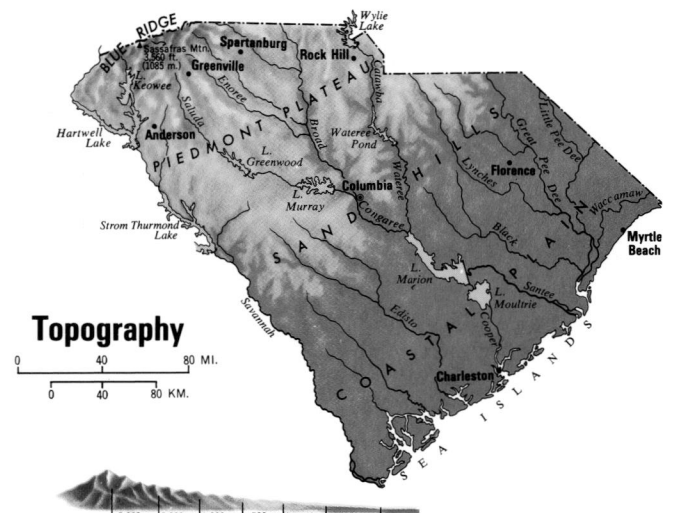

South Carolina

SCALE
0 5 10 20 30 40 MI.
0 5 10 20 30 40 KM.

State Capitals .... ⊛
County Seats .... •
Canals
Major Limited Access Hwys.

Topography
0   40   80 MI.
0   40   80 KM.

5,000 m. | 2,000 m. | 1,000 m. | 500 m. | 200 m. | 100 m. | Sea
16,404 ft. | 6,562 ft. | 3,281 ft. | 1,640 ft. | 656 ft. | 328 ft. | Level Below

COUNTIES

Abbeville 26,167	B3
Aiken 142,552	D4
Allendale 11,211	E6
Anderson 165,740	B2
Bamberg 16,658	E5
Barnwell 23,478	E5
Beaufort 120,937	F7
Berkeley 142,651	G5
Calhoun 15,185	E4
Charleston 309,969	H6
Cherokee 52,537	D1
Chester 34,068	E2
Chesterfield 42,768	G2
Clarendon 32,502	G4
Colleton 38,264	F6
Darlington 67,394	H3
Dillon 30,722	J2
Dorchester 96,413	G5
Edgefield 24,595	D4
Fairfield 23,454	E3
Florence 125,761	H3
Georgetown 55,797	J5
Greenville 379,616	C2
Greenwood 66,271	C3
Hampton 21,386	E6
Horry 196,629	J4
Jasper 20,678	E6
Kershaw 52,647	F3
Lancaster 61,351	F2
Laurens 69,567	D2
Lee 20,119	G3
Lexington 216,014	E4
Marion 35,466	J3
Marlboro 28,818	H2
McCormick 9,958	C4
Newberry 36,108	D3
Oconee 66,215	A2
Orangeburg 91,582	F5
Pickens 110,757	B2
Richland 320,677	F4
Saluda 19,181	D3
Spartanburg 253,791	D2
Sumter 104,646	G4
Union 29,881	D2
Williamsburg 37,217	H4
York 164,614	E2

CITIES and TOWNS

Abbeville▲ 5,840	C3
Adams Run 500	G6
Adrian 110	J4
Aiken▲ 25,337	D4
Aiken West 3,083	D4
Alcolu 600	G4
Allendale▲ 4,052	E5
Allsbrook 100	K3
Anderson 25,514	B2
Andrews 3,068	H5
Antioch 500	F3
Antreville 118	B3
Appleton 200	E5
Arcadia 899	C2
Arcadia Lakes 882	F3
Ariail 2,607	B2
Arkwright 2,623	C2
Atlantic Beach 351	K4
Awendaw 1,195	H5
Aynor 587	J3
Ballentine 550	E3
Bamberg▲ 3,733	E5
Barnwell▲ 5,035	E5
Batesburg 5,517	D4
Bath 2,242	D5
Beaufort▲ 12,950	F7
Beech Island 400	D5
Belton 4,461	C2
Bennettsville 9,425	H2
Berea 14,158	C2
Bethera 265	H5
Bethune 352	G3
Bingham 200	H3
Bishopville▲ 3,670	G3
Blacksburg 1,880	D1
Blackville 2,973	E5
Blenheim 137	H2
Blythewood 170	E3
Bonneau 354	H5
Bowman 1,198	F5
Boykin 350	F3
Branchville 1,083	F5
Brunson 589	E6
Bucksport 1,117	J4
Buffalo 1,426	D2
Burgess 250	J4
Burnettown 2,720	D5
Burton 7,180	F7
Calhoun Falls 2,303	B3
Camden▲ 6,682	F3
Cameron 449	F4
Campobello 449	C1
Canadys 130	F5
Carlisle 496	D2
Cashville 200	C2
Catawba 500	E2
Cateechee 225	B2
Cayce 12,150	E4
Centenary 700	J3
Central 3,522	B2
Central Pacolet 267	D2
Chapin 628	D3
Chappells 45	D3
Charleston▲ 96,650	G6
Cheraw 5,524	H2
Cherokee Falls 250	D1
Chesnee 1,003	D1
Chester▲ 6,476	E2
Chesterfield 1,318	G2
City View 1,254	C2
Clarks Hill 376	C4
Claussen 500	H3
Clearwater 4,199	D5
Clemson 11,939	B2
Cleveland 800	C1
Clifton 950	D2
Clinton 8,091	D3
Clio 774	H2
Clover 4,014	E1
Columbia (cap.)▲ 116,278	F4

© Copyright HAMMOND

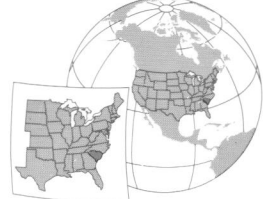

AREA 31,113 sq. mi. (80,583 sq. km.)
POPULATION 4,012,012
CAPITAL Columbia
LARGEST CITY Columbia
HIGHEST POINT Sassafras Mtn. 3,560 ft.
(1085 m.)
SETTLED IN 1623
ADMITTED TO UNION May 23, 1788
POPULAR NAME Palmetto State
STATE FLOWER Carolina (Yellow) Jessamine
STATE BIRD Carolina Wren

Edisto Beach 641	G7	Langley 1,714	D4	Port Royal 3,950	F7	Stuckey 263	H4	Woodford 196	E4	Kiawah (isl.)	G6
Edisto Island 900	G6	Latta 1,410	J3	Poston 250	J4	Sullivans Island 1,911	H6	Woodruff 4,229	D2	Kings Mountain
Effingham 300	H3	Laurel Bay 6,625	F7	Princeton 65	C2	Summerton 1,061	G4	Woodville		Nat'l Mil. Park	E1
Ehrhardt 614	E5	Laurens▲ 9,916	C3	Prosperity 1,047	D3	Summerville 27,752	G5	Yemassee 807	F6	Little (riv.)	C3
Elgin 806	F3	Leesville		Quinby 842	H3	Summit 219	E4	Yonges Island 500	G6	Little (riv.)	D3
Elko 212	E5	Lena 275	E6	Rains 450	J3	Sumter▲ 39,643	G4	York▲ 6,985	E1	Little Lynches (riv.)	G3
Elliott 500	G3	Lesslie 2,268	E1	Ravenel 2,214	G6	Surfside Beach 4,425	K4			Little Pee Dee (riv.)	J4
Elloree 742	F4	Level Land 100	C3	Red River	F2	Swansea 533	E4			Little River (inlet)	L4
Enoree 1,107	D2	Lexington▲ 9,793	E4	Reevesville 207	F5	Sycamore 185	E5	**OTHER FEATURES**		Lumber (riv.)	J3
Estill 2,425	E6	Liberty 3,009	B2	Reidville 478	C2	Tamassee 320	A2			Lynches (riv.)	H3
Eureka 1,738	E2	Lincolnville 904	G6	Rembert 406	G3	Tatum 69	H2	Ashepoo (riv.)	F6	Marion (lake)	G5
Eutawville 344	G5	Little Mountain 255	E3	Richburg 332	E2	Taylors 20,125	C2	Ashley (riv.)	G6	Morris (isl.)	H6
Fairfax 3,206	E6	Little River 7,027	K4	Ridgeland▲ 2,518	E7	Tigerville 975	C1	Bay Point (isl.)	F7	Moultrie (lake)	G5
Fair Play 500	A2	Little Rock 500	J3	Ridge Spring 823	D4	Tillman 225	E7	Beaufort Marine Air Sta.	F7	Murphy (isl.)	J5
Filbert 203	E1	Livingston 148	E4	Ridgeville 1,690	G5	Timmonsville 2,315	H3	Big Black (creek)	G2	Murray (lake)	D4
Fingerville 320	D1	Lobeco 345	F7	Ridgeway 328	F3	Toddville 200	H4	Black (riv.)	H4	Myrtle Beach A.F.B.	K4
Florence▲ 30,248	H3	Lockhart 39	D2	Rimini 525	G4	Townville 300	B2	Blue Ridge (mts.)	B1	Naval Base	H6
Floyd Dale 450	J3	Lodge 114	F5	Rion 300	E3	Tradesville 500	F2	Broad (riv.)	D2	New (riv.)	E6
Folly Beach 2,116	H6	Longcreek 200	A2	Ritter 300	F6	Travelers Rest 4,099	C2	Broad (riv.)	D3	Ninety Six Nat'l Hist. Site	C3
Forest Acres 10,558	E3	Longtown 400	F3	Rock Hill 49,765	E2	Trenton 226	D4	Buck (creek)	J3	North (inlet)	J5
Forest Beach 500	F7	Loris 2,079	K3	Rodman 300	E2	Trio 400	H5	Bull (isl.)	H6	North (isl.)	J5
Foreston 300	G4	Lowndesville 166	B3	Rowesville 378	F5	Troy 105	C4	Bullock (creek)	E2	North Edisto (riv.)	G6
Fort Lawn 864	F2	Lowrys 207	E2	Ruby 348	G2	Turbeville 602	G4	Bulls (bay)	H6	Pacolet (riv.)	D1
Fort Mill 7,587	E1	Lugoff 6,278	F3	Ruffin 400	F6	Ulmer 102	E5	Bush (riv.)	D3	Palms, Isle of (isl.)	H6
Fort Motte 700	F4	Luray 115	E6	Saint Andrews 21,814	G6	Union▲ 8,793	D2	Buzzard Roost (dam)	D3	Parris Island Marine Base	F7
Fountain Inn 6,017	C2	Lydia 500	G3	Saint George▲ 2,092	F5	Utica 1,322	B2	Cape (isl.)	J5	Pee Dee (riv.)	J4
Furman 286	E6	Lydia Mills 925	C2	Saint Matthews▲ 2,107	F4	Vance 208	G5	Capers (isl.)	H6	Pocotaligo (riv.)	G4
Gable 230	G4	Lyman 2,659	C2	Saint Paul 725	G4	Van Wyck 500	F2	Catawba (riv.)	F2	Port Royal (sound)	F7
Gadsden 500	F4	Lynchburg 588	G3	Saint Stephen 1,776	H5	Varnville 2,074	E6	Catfish (creek)	J3	Pritchards (isl.)	G7
Gaffney▲ 12,968	D1	Madison	A2	Salem 126	A2	Vaucluse 606	D7	Chattooga (riv.)	A2	Reedy (riv.)	C2
Gantt 13,962	C2	Madison 1,150	A2	Salley 410	E4	Wade-Hampton 20,458	C2	Combahee (riv.)	F6	Robinson (lake)	G3
Garden City Beach 300	K4	Manning▲ 4,025	G4	Salters 300	H4	Wagener 863	E4	Congaree (riv.)	E4	Romain (cape)	J6
Garnett 500	E7	Marietta-Slater 2,245	C1	Saluda▲ 3,066	D4	Walhalla▲ 3,801	A2	Congaree Nat'l Mon.	F4	Saint Helena (isl.)	F7
Gaston 1,304	E4	Marion▲ 7,042	J3	Sandy Springs 150	B2	Wallace 500	H2	Cooper (riv.)	H6	Saint Helena (sound)	G7
Georgetown▲ 8,950	J5	Mars Bluff 500	H3	Santee 740	F5	Walterboro▲ 5,153	F6	Coosaw (riv.)	G7	Salkehatchie (riv.)	E5
Gifford 370	E6	Mauldin 15,224	C2	Sardinia 225	G4	Wampee 200	K4	Coosawhatchie (riv.)	E6	Saluda (riv.)	D3
Gilbert 500	E4	Mayesville 1,001	G3	Saxon 3,707	D2	Wando 500	H6	Cowpens Nat'l Battlefield	D1	Sandy (pt.)	H6
Gillisonville 350	E6	Mayo 1,842	D1	Scotia 227	E6	Ward 111	D3	Crooked (creek)	H2	Sandy (riv.)	E2
Givhans 400	G5	McBee 714	G2	Scranton 942	H4	Ware Shoals 2,363	C3	Deep (creek)	B2	Santee (dam)	G4
Glendale 1,049	D2	McClellanville 459	H5	Seabrook 1,250	F6	Warrenville 1,029	D4	Dewees (isl.)	H6	Santee (riv.)	H5
Glenn Springs 350	D2	McColl 2,498	H2	Sea Pines 500	F7	Waterloo 203	C3	Donaldson A.F.B.	C2	Sassafras (mt.)	B1
Gloverville 2,805	D4	McConnells 287	E2	Sellers 277	H3	Watts Mill 1,479	C2	Edisto (isl.)	G6	Savannah (riv.)	E6
Goose Creek 29,208	H6	McCormick▲ 1,489	C4	Seneca 7,652	A2	Wedgefield 550	F4	Edisto (riv.)	G7	Savannah River Plant	D5
Govan 67	E5	Meggett 1,230	G6	Shannontown	G4	Wellford 2,030	C2	Enoree (riv.)	C2	Sea (isls.)	G7
Gowensville 200	C1	Modoc 256	C4	Sharon 421	E2	West Columbia 13,064	E4	Fort Jackson	F4	Seabrook (isl.)	G6
Gramling 400	C1	Monarch Mills 1,930	D2	Sheldon 256	F6	Westminster 2,743	A2	Fort Sumter Nat'l Mon.	H6	Seneca (riv.)	B2
Graniteville 1,158	D4	Moncks Corner▲ 5,952	H5	Shiloh 259	G4	West Pelzer 879	B2	Four Hole Swamp (creek)	F5	Shaw A.F.B.	F4
Gray Court 1,021	C2	Monetta 224	D4	Shulerville 500	H5	West Springs 500	D2	Fripp (isl.)	G7	South (isl.)	J5
Great Falls 2,194	F2	Montmorenci 500	D4	Silverstreet 216	D3	West Union 297	B2	Great Pee Dee (riv.)	J4	Stevens (creek)	C4
Greeleyville 452	H4	Moore 500	C2	Simpsonville 14,352	C2	Westview 1,999	C2	Greenwood (lake)	D3	Stono (inlet)	H6
Greenwood▲ 22,071	C3	Mountain Rest 500	A2	Six Mile 553	B2	Westville 440	F3	Hartwell (dam)	B3	Thompsons (creek)	G2
Greer 16,843	C2	Mount Carmel 237	C3	Slater-Marietta 2,228	C1	White Pond 200	D5	Hartwell (lake)	A3	Tugaloo (riv.)	A2
Gresham 350	J4	Mount Croghan 155	G2	Snelling 246	E5	White Rock 600	E3	Hilton Head (isl.)	F7	Turkey (creek)	E2
Gurley 425	J3	Mount Holly 200	H5	Society Hill 700	H2	Whitmire 1,512	D3	Hunting (isl.)	G7	Tybee Roads (chan.)	F7
Hamer 588	J3	Mount Pleasant 47,609	H6	South Bennettsville 1,065	H2	Whitney 4,052	D1	Intracoastal Waterway	H5	Tyger (riv.)	D2
Hampton▲ 2,837	E6	Mullins 5,029	J3	South Congaree 2,266	E4	Williams 116	G5	James (isl.)	H6	Waccamaw (riv.)	J5
Hanahan 12,937	H6	Murrells Inlet 5,519	K4	Spartanburg▲ 39,673	C1	Williamston 3,791	B2	J. Strom Thurmond		Wadmalaw (isl.)	G6
Hardeeville 1,793	E7	Myrtle Beach 22,759	K4	Springdale 2,864	F2	Williston 3,307	E5	(dam)	C4	Wando (riv.)	H6
Harleyville 594	G5	Neeses 413	E4	Springdale 2,877	E4	Windsor 127	E5	J. Strom Thurmond		Wateree (lake)	F3
Hartsville 7,556	G3	Nesmith 350	H4	Springfield 504	E4	Windy Hill 1,622	H3	(lake)	C4	Winyah (bay)	J5
Heath Springs 864	F2	Newberry▲ 10,580	D3	Spring Mills 1,419	F2	Winnsboro▲ 3,599	E3	Juniper (creek)	H2	Wylie (lake)	E1
Helena 500	D3	New Ellenton 2,250	D5	Starr 173	B3	Winnsboro Mills 2,263	E3	Keowee (lake)	B2	
Hemingway 573	J4	Newry 400	B2	Startex 988	C2	Wisacky 250	G3	Keowee (riv.)	B2	▲County seat
Hemlock (Eureka) 1,738	E2	New Town 950	J3							
Hickory Grove 337	E2	New Zion 200	H4							
Hilda 436	E5	Nichols 528	J3							
Hilton Head Island 33,862	F7	Ninety Six 1,936	C3							
Hodges 158	C3	Norris 847	B2							
Holly Hill 1,281	G5	North 813	E4							
Hollywood 3,946	G6	North Augusta 17,574	C5							
Honea Path 3,504	C3	North Charleston 79,641	G6							
Hopkins 300	F4	North Hartsville 3,136	G3							
Horatio 500	F3	North Myrtle Beach 10,974	K4							
Huger 500	H5	Norway 389	E5							
Inman 1,884	C1	Oakley 250	G5							
Irmo 11,039	D3	Olanta 613	H4							

Conestee 500	C2	Isle of Palms 4,583	H6
Converse 1,173	D2	Iva 1,156	B3
Conway▲ 11,788	J4	Jackson 1,625	D5
Coosawhatchie 250	F6	Jacksonboro 475	G6
Cope 107	E5	Jamestown 97	H5
Cordesville 300	H5	Jedburg 900	G5
Cordova 157	F5	Jefferson 704	G2
Coronaca 170	C3	Joanna 1,609	C3
Cottageville 707	G6	Johns Island 200	G6
Coward 650	H4	Johnsonville 1,418	J4
Cowpens 2,279	D1	Johnston 2,336	D4
Cross 469	G5	Jonesville 982	D2
Cross Anchor 350	D2	Kershaw 1,645	G2
Cross Hill 601	D3	Kinards 300	D3
Cross Keys 200	D2	Kingsburg 300	H4
Cummings 275	E6	Kingstree▲ 3,496	H4
Dacusville 350	C2	Kingville 500	F4
Dale 500	F6	Kline 238	E5
Dalzell 2,260	G3	Ladson 13,264	G6
Darlington▲ 6,720	H3	La France 875	B2
Davis Station 300	G4	Lake City 6,478	H4
Denmark 3,328	E5	Lake View 789	J3
Dillon▲ 6,316	J3	Lamar 1,015	G3
Donalds 354	C3	Lancaster▲ 8,177	F2
Doneraile 1,276	H3	Lancaster Mills 2,109	F2
Dorchester 400	G5	Lando 250	E2
Due West 1,209	C3	Landrum 2,472	C1
Duncan 2,870	C2	Lane 585	H5
Easley 17,754	C2		
East Gaffney 3,349	D1		
Eastover 830	F4		
Edgefield 4,449	C4		
Edgemoor 500	E2		

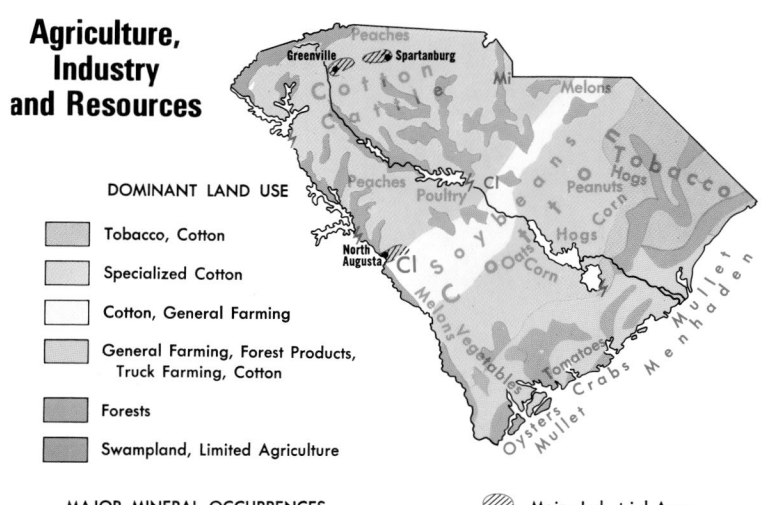

Agriculture, Industry and Resources

DOMINANT LAND USE

- Tobacco, Cotton
- Specialized Cotton
- Cotton, General Farming
- General Farming, Forest Products, Truck Farming, Cotton
- Forests
- Swampland, Limited Agriculture

MAJOR MINERAL OCCURRENCES

Cl Clay
Mi Mica

Major Industrial Areas
Water Power

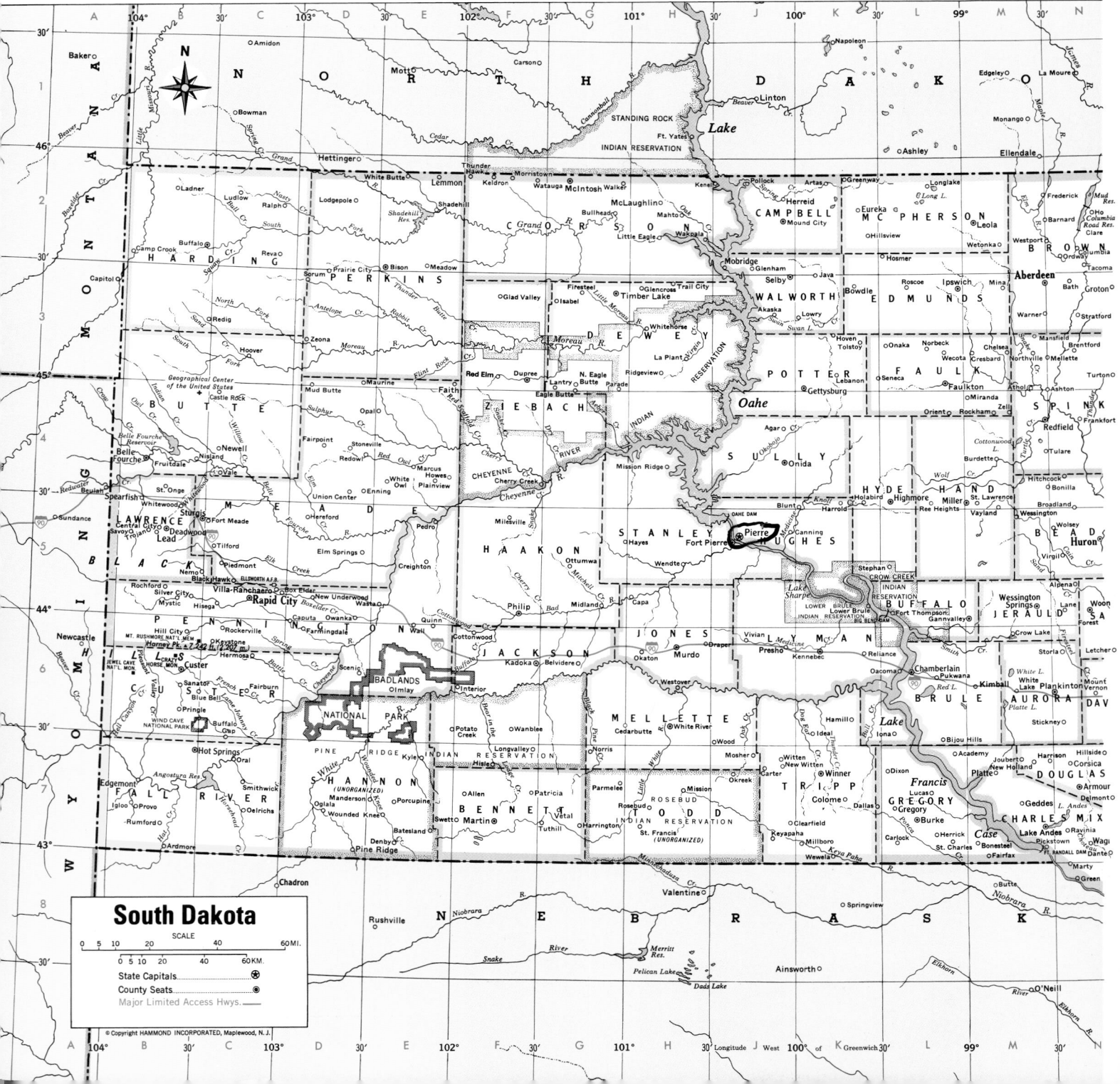

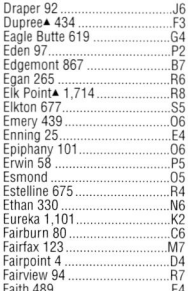

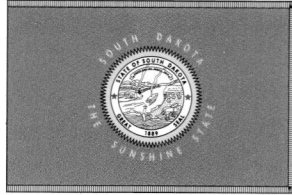

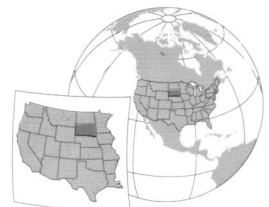

AREA 77,116 sq. mi. (199,730 sq. km.)
POPULATION 754,844
CAPITAL Pierre
LARGEST CITY Sioux Falls
HIGHEST POINT Harney Pk. 7,242 ft.
(2207 m.)
SETTLED IN 1856
ADMITTED TO UNION November 2, 1889
POPULAR NAME Coyote State;
Mt. Rushmore State
STATE FLOWER Pasqueflower
STATE BIRD Ring-necked Pheasant

(continued on following page)

Topography

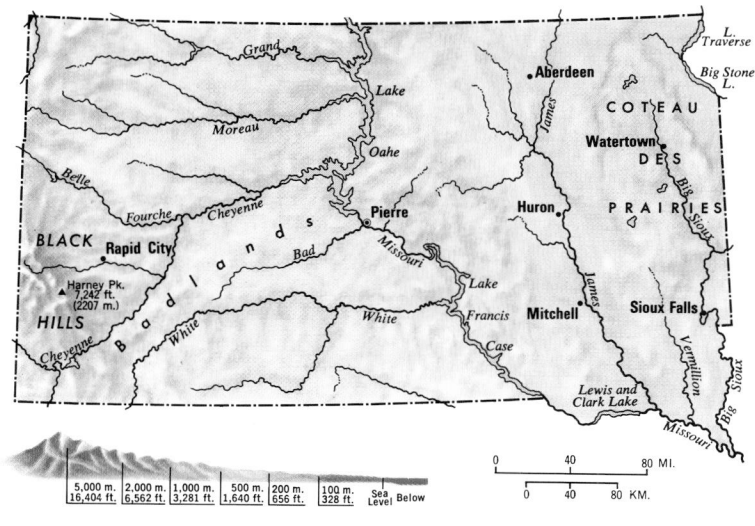

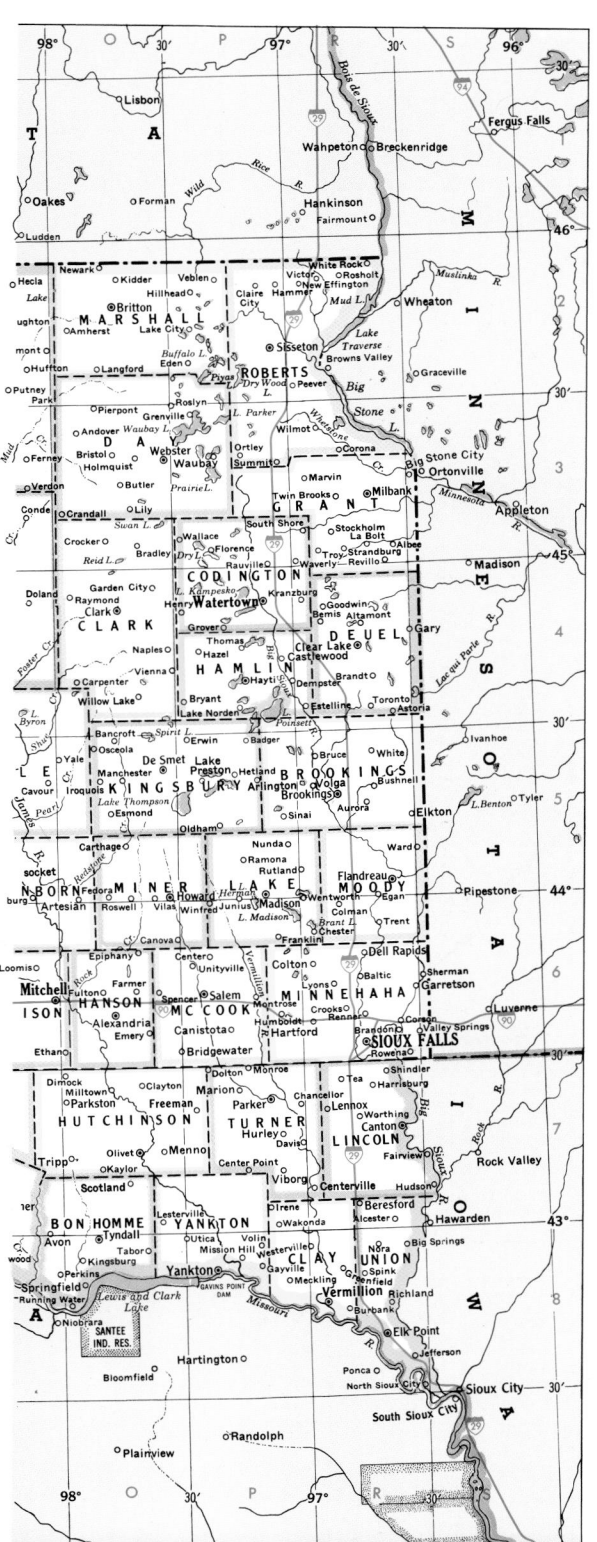

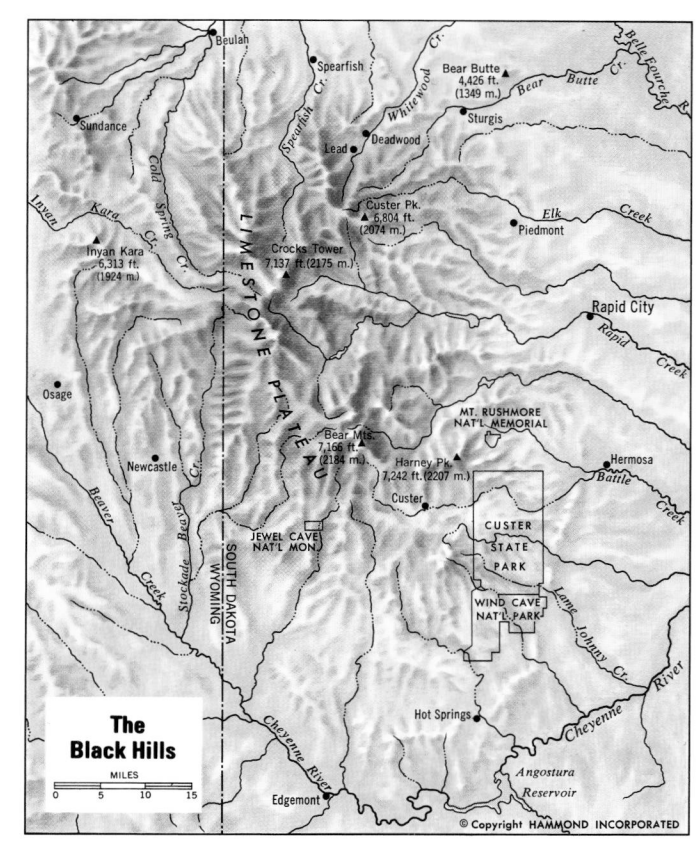

The Black Hills

MILES

Agriculture, Industry and Resources

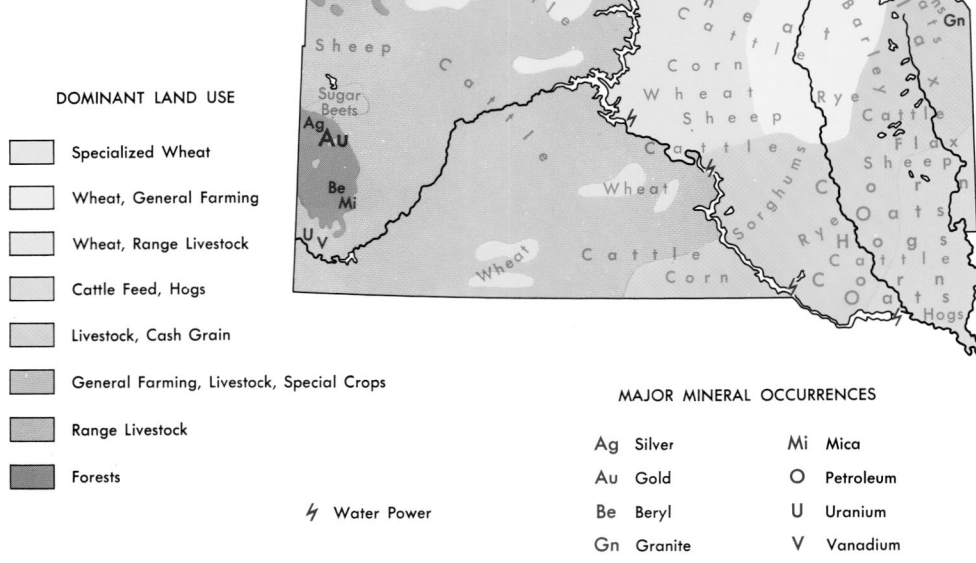

DOMINANT LAND USE

- Specialized Wheat
- Wheat, General Farming
- Wheat, Range Livestock
- Cattle Feed, Hogs
- Livestock, Cash Grain
- General Farming, Livestock, Special Crops
- Range Livestock
- Forests

⚡ Water Power

MAJOR MINERAL OCCURRENCES

Ag	Silver	Mi	Mica
Au	Gold	O	Petroleum
Be	Beryl	U	Uranium
Gn	Granite	V	Vanadium

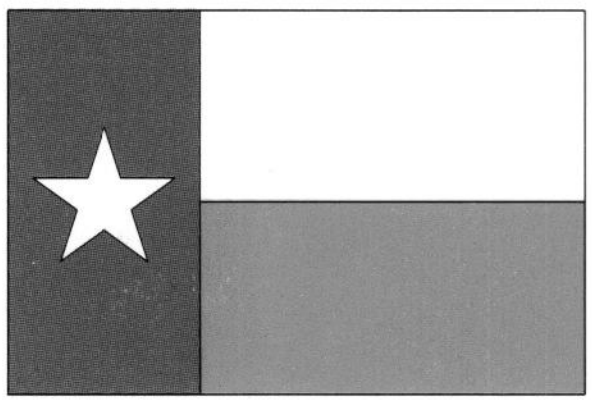

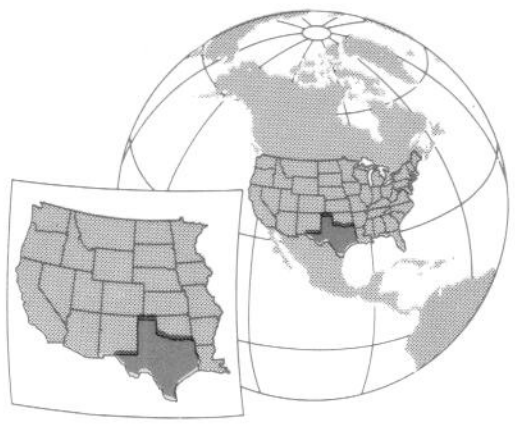

AREA 266,807 sq. mi. (691,030 sq. km.)
POPULATION 20,851,820
CAPITAL Austin
LARGEST CITY Houston
HIGHEST POINT Guadalupe Pk. 8,749 ft. (2667 m.)
SETTLED IN 1686
ADMITTED TO UNION December 29, 1845
POPULAR NAME Lone Star State
STATE FLOWER Bluebonnet
STATE BIRD Mockingbird

COUNTIES

Anderson 55,109....J6
Andrews 13,004....B5
Angelina 80,130....K6
Aransas 22,497....H10
Archer 8,854....F4
Armstrong 2,148....C3
Atascosa 38,628....F9
Austin 23,590....H8
Bailey 6,594....B3
Bandera 17,645....E8
Bastrop 57,733....G7
Baylor 4,093....E4
Bee 32,359....G9
Bell 237,974....G6
Bexar 1,392,931....F8
Blanco 8,418....F8
Borden 729....C5
Bosque 17,204....G6
Bowie 89,306....K4
Brazoria 241,767....J8
Brazos 152,415....H7
Brewster 8,866....A8
Briscoe 1,790....C3
Brooks 7,976....F11
Brown 37,674....F6
Burleson 16,470....H7
Burnet 34,147....F7
Caldwell 32,194....G8
Calhoun 20,647....H9
Callahan 12,905....E5
Cameron 335,227....G11
Camp 11,549....K5
Carson 6,516....C2
Cass 30,438....K4
Castro 8,285....B3
Chambers 26,031....K8
Cherokee 46,659....J6
Childress 7,688....D3
Clay 11,006....F4
Cochran 3,730....B4
Coke 3,864....D6
Coleman 9,235....E6
Collin 491,675....H4
Collingsworth 3,206....D3
Colorado 20,390....H8
Comal 78,021....F8
Comanche 14,026....F5
Concho 3,966....E6
Cooke 36,363....G4
Coryell 74,978....G6
Cottle 1,904....D3
Crane 3,996....B6
Crockett 4,099....C7
Crosby 7,072....C4
Culberson 2,975....C11
Dallam 6,222....B1
Dallas 2,218,899....H5
Dawson 14,985....C5
Deaf Smith 18,561....B3
Delta 5,327....J4
Denton 432,976....G4
De Witt 20,013....G9
Dickens 2,762....D4
Dimmit 10,248....E9
Donley 3,828....D2
Duval 13,120....F10
Eastland 18,297....F5
Ector 121,123....B6
Edwards 2,162....D7
Ellis 111,360....H5
El Paso 679,622....A10
Erath 33,001....F5
Falls 18,576....H6
Fannin 31,242....H4
Fayette 21,804....H8
Fisher 4,344....D5
Floyd 7,771....C3
Foard 1,622....E3
Fort Bend 354,452....J8
Franklin 9,458....J4
Freestone 17,867....H6
Frio 16,252....E9
Gaines 14,467....B5
Galveston 250,158....K8
Garza 4,872....C4
Gillespie 20,814....F7
Glasscock 1,406....C6
Goliad 6,928....G9
Gonzales 18,628....G8
Gray 22,744....D2
Grayson 110,595....H4
Gregg 111,379....K5
Grimes 23,552....J7
Guadalupe 89,023....G8
Hale 36,602....C3
Hall 3,782....D3
Hamilton 8,229....F6
Hansford 5,369....C1
Hardeman 4,724....E3
Hardin 48,073....K7
Harris 3,400,578....J8
Harrison 62,110....K5
Hartley 5,537....B2
Haskell 6,093....E4
Hays 97,589....F7
Hemphill 3,351....D2
Henderson 73,277....J5
Hidalgo 569,463....F11
Hill 32,321....G5
Hockley 22,716....B4
Hood 41,100....G5
Hopkins 31,960....J4
Houston 23,185....J6
Howard 33,627....C5
Hudspeth 3,344....B10
Hunt 76,596....H4
Hutchinson 23,857....C2
Irion 1,771....C6
Jack 8,763....F4
Jackson 14,391....H9
Jasper 35,604....K7
Jeff Davis 2,207....C11
Jefferson 252,051....K8
Jim Hogg 5,281....F11
Jim Wells 39,326....F10
Johnson 126,811....G5
Jones 20,785....E5
Karnes 15,446....G9
Kaufman 71,313....H5
Kendall 23,743....F8
Kenedy 414....G11
Kent 859....D4
Kerr 43,653....E7
Kimble 4,468....E7
King 356....D4
Kinney 3,379....D8
Kleberg 31,549....G10
Knox 4,253....E4
Lamar 48,499....J4
Lamb 14,709....B3
Lampasas 17,762....F6
La Salle 5,866....E9
Lavaca 19,210....H8
Lee 15,657....H7
Leon 15,335....J6
Liberty 70,154....K7
Limestone 22,051....H6
Lipscomb 3,057....D1
Live Oak 12,309....F9
Llano 17,044....F7
Loving 67....A6
Lubbock 242,628....C4
Lynn 6,550....C4
Madison 12,940....J6
Marion 10,941....K5
Martin 4,746....C5
Mason 3,738....E7
Matagorda 37,957....H9
Maverick 47,297....D9
McCulloch 8,205....E6
McLennan 213,517....G6
McMullen 851....F9
Medina 39,304....E8
Menard 2,360....E7
Midland 116,009....B6
Milam 24,238....H7
Mills 5,151....F6
Mitchell 9,698....D5
Montague 19,117....G4
Montgomery 293,768....J7
Moore 20,121....C2
Morris 13,048....K4
Motley 1,426....D3
Nacogdoches 59,203....K6
Navarro 45,124....H5
Newton 15,072....L7
Nolan 15,802....D5
Nueces 313,645....G10
Ochiltree 9,006....D1
Oldham 2,185....B2
Orange 84,966....L7
Palo Pinto 27,026....F5
Panola 22,756....K5
Parker 88,495....G5
Parmer 10,016....B3
Pecos 16,809....B7
Polk 41,133....K7
Potter 113,546....C2
Presidio 7,304....C12
Rains 9,139....J5
Randall 104,312....C2
Reagan 3,326....C6
Real 3,047....E8
Red River 14,314....J4
Reeves 13,137....D11
Refugio 7,828....G9
Roberts 887....D2
Robertson 16,000....H6
Rockwall 43,080....H5
Runnels 11,495....E6
Rusk 47,372....K5
Sabine 10,469....L6
San Augustine 8,946....K6
San Jacinto 22,246....J7
San Patricio 67,138....G10
San Saba 6,186....F6
Schleicher 2,935....D7
Scurry 16,361....D5
Shackelford 3,302....E5
Shelby 25,224....K6
Sherman 3,186....C1
Smith 174,706....J5
Somervell 6,809....G5
Starr 53,597....F11
Stephens 9,674....F5
Sterling 1,393....C6
Stonewall 1,693....D4
Sutton 4,077....D7
Swisher 8,378....C3
Tarrant 1,446,219....G5
Taylor 126,555....E5
Terrell 1,081....B7
Terry 12,761....B4
Throckmorton 1,850....E4
Titus 28,118....K4
Tom Green 104,010....D6
Travis 812,280....G7
Trinity 13,779....J6
Tyler 20,871....K7
Upshur 35,291....K5
Upton 3,404....B6
Uvalde 25,926....E8
Val Verde 44,856....C8
Van Zandt 48,140....J5
Victoria 84,088....H9
Walker 61,758....J7
Waller 32,663....J8
Ward 10,909....A6
Washington 30,373....H7
Webb 193,117....E10
Wharton 41,188....H8
Wheeler 5,284....D2
Wichita 131,664....F3
Wilbarger 14,676....E3
Willacy 20,082....G11
Williamson 249,967....G7
Wilson 32,408....F8
Winkler 7,173....A6
Wise 48,793....G4
Wood 36,752....J5
Yoakum 7,322....B4
Young 17,943....F4
Zapata 12,182....E11
Zavala 11,600....E9

CITIES and TOWNS

Abernathy 2,839....B4
Abilene▲ 115,930....E5
Addison 14,166....G2
Alamo 14,760....F11
Alamo Heights 7,319....K10
Albany▲ 1,921....E5
Alice▲ 19,010....F10
Allen 43,554....H1
Alpine▲ 5,786....D12
Alvarado 3,288....G5
Alvin 21,413....J3
Anahuac▲ 2,210....K8
Anderson▲ 257....J7
Andrews▲ 9,652....B5
Angleton▲ 18,130....J8
Anson▲ 2,556....E5
Anthony 3,850....A10
Aransas Pass 8,138....G10
Archer City▲ 1,848....F4
Arlington 332,969....F2
Aspermont▲ 1,021....D4
Athens▲ 11,297....J5
Atlanta 5,745....K4
Austin (cap.)▲ 656,562....G7
Azle 9,600....E2
Baird▲ 1,623....E5
Balch Springs 19,375....H2
Balcones Heights 3,016....J10
Ballinger▲ 4,243....E6
Bandera▲ 957....F8
Barrett 2,872....K1
Bastrop▲ 5,340....G7
Bay City▲ 18,667....H9
Baytown 66,430....L2
Beaumont▲ 113,866....K7
Bedford 47,152....F2
Beeville▲ 13,129....G9
Bellaire 15,642....J2
Bellmead 9,214....H6
Bellville▲ 3,794....H8
Belton▲ 14,623....G7
Benavides 1,788....F10
Benbrook 20,208....E2
Benjamin▲ 264....E4
Big Lake▲ 2,885....C6
Big Spring▲ 25,233....C5
Bishop 3,305....G10
Bloomington 1,888....H9
Blue Mound 2,388....E2
Boerne▲ 6,178....J10
Bonham▲ 9,990....H4
Borger 14,302....C2
Boston▲ 400....K4
Bowie 5,219....G4
Brackettville▲ 1,876....D8
Brady▲ 5,523....E6
Brazoria 2,787....J9
Breckinridge▲ 5,868....F5
Brenham▲ 13,507....H7
Briar 5,350....E1
Bridge City 8,651....L7
Bridgeport 4,309....G4
Brookshire 3,450....J8
Brownfield▲ 9,488....B4
Brownsville▲ 139,722....G12
Brownwood▲ 18,813....F6
Bryan▲ 65,660....H7
Buda 1,795....G7
Buna 2,269....L7
Bunker Hill Village 3,654....J1
Burkburnett 10,927....F3
Burleson 20,976....F3
Burnet▲ 4,735....F7
Caldwell▲ 3,449....H7
Cameron▲ 5,634....H7
Canadian▲ 2,233....D2
Canton▲ 3,292....J5
Canutillo 5,129....A10
Canyon▲ 12,875....C3
Carrizo Springs▲ 5,655....E9
Carrollton 109,576....G2
Carthage▲ 6,664....K5
Castle Hills 4,202....J10
Castroville 2,664....J11
Cedar Hill 32,093....G3
Cedar Park 26,049....G7
Center▲ 5,678....K6
Centerville▲ 903....H6
Channelview 29,685....K1
Channing▲ 356....B2
Childress▲ 6,778....D3
Cisco 3,851....E5
Clarendon▲ 1,974....C3
Clarksville▲ 3,883....K4
Claude▲ 1,313....C2
Clear Lake Shores 1,096....K2
Cleburne▲ 26,005....G5
Cleveland 7,605....K7
Clifton 3,542....G6
Clute 10,424....J9
Clyde 3,345....E5
Cockrell Hill 4,443....G2
Coldspring▲ 691....J7
Coleman▲ 5,127....E6
College Station▲ 67,890....H7
Colleyville 19,636....F2
Colorado City▲ 4,281....C5
Columbus▲ 3,916....H8
Comanche▲ 4,482....F6
Commerce 7,669....J4
Conroe▲ 36,811....J7
Converse 11,508....K11
Cooper▲ 2,150....J4
Coppell 35,958....G2
Copperas Cove 29,592....G6
Corpus Christi▲ 277,454....G10
Corsicana▲ 24,485....H5
Cotulla▲ 3,614....E9
Crane▲ 3,191....B6
Crockett▲ 7,141....J6
Crosby 1,811....J8
Crosbyton▲ 1,874....C4
Crowell▲ 1,141....E4
Crowley 7,467....E3
Crystal City▲ 7,190....E9
Cuero▲ 6,571....G8
Daingerfield▲ 2,517....K4
Dalhart▲ 7,237....B1
Dallas▲ 1,188,580....G2
Dalworthington Gardens 1,758....F2
Dayton 5,709....J7
Decatur▲ 5,201....G4
Deer Park 28,520....K2
De Kalb 1,976....K4
De Leon 2,433....F5
Del Rio▲ 33,867....D8
Denison 22,773....H4
Denton▲ 80,537....G4
Denver City 3,985....B4
De Soto 37,646....G3
Devine 4,140....E8
Diboll 5,470....K6
Dickens▲ 332....D4
Dickinson 17,093....K3
Dilley 3,674....E9
Dimmitt▲ 4,375....B3
Donna 14,768....F11
Double Oak 1,664....F1

(continued on following page)

Agriculture, Industry and Resources

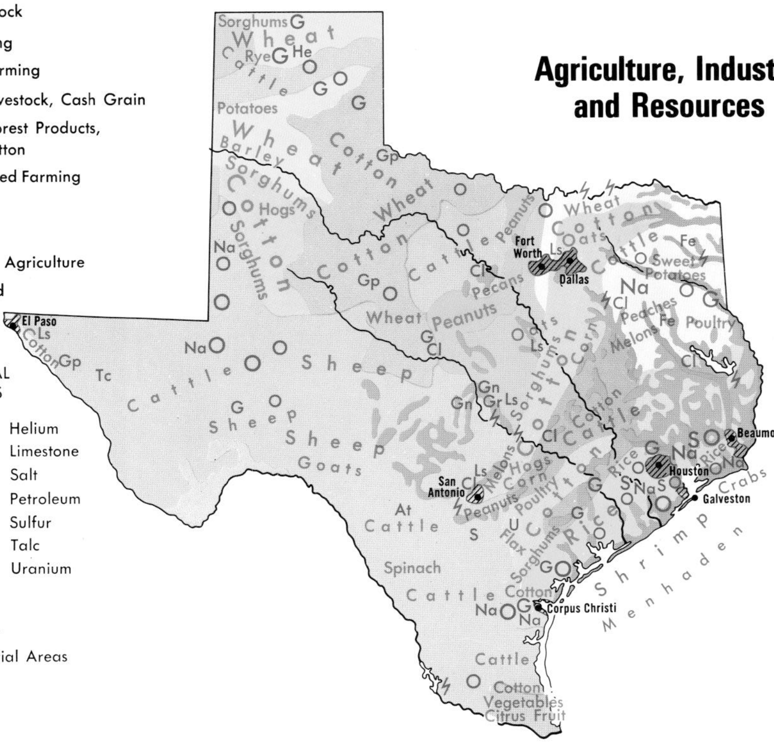

Sorghums
Wheat
Rye
He
Cattle
Potatoes
Barley
Cotton
Sorghums
Hogs
Gp
Na
Cattle
Peanuts
Fort Worth
Ls
Oats
Dallas
Fe
Sweet Potatoes
Cl
Pecans
Wheat
Peanuts
Peaches
Melons
Poultry
El Paso
Ls
Cotton
Gp
Tc
Sheep
Goats
Gn
Gr
Corn
Beaumont
Houston
Galveston
San Antonio
At
Rice
Flax
Crabs
Shrimp
Menhaden
Spinach
Corpus Christi
Vegetables
Citrus Fruit

DOMINANT LAND USE

- Wheat, Grain Sorghums, Range Livestock
- Cotton, Wheat
- Specialized Cotton
- Cotton, General Farming
- Cotton, Forest Products
- Cotton, Range Livestock
- Rice, General Farming
- Peanuts, General Farming
- General Farming, Livestock, Cash Grain
- General Farming, Forest Products, Truck Farming, Cotton
- Fruit, Truck and Mixed Farming
- Range Livestock
- Forests
- Swampland, Limited Agriculture
- Nonagricultural Land
- Urban Areas

MAJOR MINERAL OCCURRENCES

At	Asphalt	He	Helium
Cl	Clay	Ls	Limestone
Fe	Iron Ore	Na	Salt
G	Natural Gas	O	Petroleum
Gn	Granite	S	Sulfur
Gp	Gypsum	Tc	Talc
Gr	Graphite	U	Uranium

ϟ Water Power
Major Industrial Areas

Dublin 3,754 F5
Dumas▲ 13,747 C2
Duncanville 36,081 G3
Eagle Lake 3,664 H8
Eagle Pass▲ 22,413 D9
Eastland▲ 3,769 F5
Edcouch 3,342 G11
Edgecliff 2,550 E2
Edinburg▲ 48,465 F11
Edna▲ 5,899 H9
El Campo 10,945 H8
Eldorado▲ 1,951 D7
Electra 3,168 F4
Elgin 5,700 G7
El Lago 3,075 K2
El Paso▲ 563,662 A10
Elsa 5,549 G11
Emory▲ 1,021 J5
Ennis 16,045 H5
Euless 46,005 F2
Everman 5,836 F3
Fabens 8,043 B10
Fairfield▲ 3,094 H6
Falfurrias▲ 5,297 F10
Farmers Branch 27,508 G2
Farmersville 3,118 H4
Farwell▲ 1,364 A3
Ferris 2,175 H3
Floresville▲ 5,868 K11
Flower Mound 50,702 F1
Floydada▲ 3,676 C3
Forest Hill 12,949 F2
Forney 5,588 H5
Fort Davis▲ 1,050 D11
Fort Stockton▲ 7,846 A7
Fort Worth▲ 534,694 F2
Franklin▲ 1,470 H7
Fredericksburg▲ 8,911 E7
Fredonia 50 E7
Freeport 12,708 J9
Freer 3,241 F10
Fresno 6,603 J2
Friendswood 29,037 J2
Friona 3,854 B3
Frisco 33,714 H4
Fritch 2,235 C2
Gail▲ 171 C5
Gainesville▲ 15,538 G4
Galena Park 10,592 J1
Galveston▲ 57,247 L3
Ganado 1,701 H8
Garden City▲ 350 C6
Garland 215,768 H2
Gatesville▲ 15,591 G6
Georgetown▲ 28,339 G7
George West▲ 2,524 F9
Giddings▲ 5,105 H7
Gilmer▲ 4,799 J5
Gladewater 6,078 K5
Glenn Heights 7,224 G2
Glen Rose▲ 2,122 G5
Goldthwaite▲ 1,802 F6
Goliad▲ 1,975 G9
Gonzales▲ 7,202 G8
Graham▲ 8,716 F4
Granbury▲ 5,718 G5
Grand Prairie 127,427 G2
Grand Saline 3,028 J5
Grapevine 42,059 F2
Greenville▲ 23,960 H4
Groesbeck▲ 4,291 H6
Groves 15,733 L8
Groveton▲ 1,107 J7
Guthrie▲ 170 D4
Hale Center 2,263 C3
Hallettsville▲ 2,345 G8
Hallsville 2,772 K5
Haltom City 39,018 F2
Hamilton▲ 2,977 G6
Hamlin 2,248 E5
Harlingen 57,564 G11
Haskell▲ 3,106 E4
Hearne 4,690 H7
Hebbronville▲ 4,498 F10
Hedwig Village 2,334 H1
Hemphill▲ 1,106 L6
Hempstead▲ 4,691 J7
Henderson▲ 11,273 K5
Henrietta▲ 3,264 F4
Hereford▲ 14,597 B3
Hickory Creek 2,078 F1
Hidalgo 7,322 F11
Highland Park 8,842 G2
Highlands 7,089 K1
Highland Village 12,173 F1
Hillsboro▲ 8,232 G5
Hitchcock 6,386 K3
Hollywood Park 2,983 K10
Hondo▲ 7,897 E8
Honey Grove 1,746 J4
Hooks 2,973 K4
Houston▲ 1,953,631 J2
Howe 2,478 H4
Hughes Springs 1,856 K5
Humble 14,579 J7
Hunters Creek Village 4,374 J1
Huntington 2,068 K6
Huntsville▲ 35,078 J7
Hurst 36,273 F2
Hutchins 2,805 G3
Idalou 2,157 C4
Iowa Park 6,431 F4
Irving 191,615 G2
Italy 1,993 H5
Jacinto City 10,302 J1
Jacksboro▲ 4,533 F4
Jacksonville 13,868 J5
Jasper▲ 8,247 L7
Jayton▲ 513 D4
Jefferson▲ 2,024 K5
Jersey Village 6,880 J1
Johnson City▲ 1,191 F7
Jones Creek 2,130 J9
Jourdanton▲ 3,732 F9
Junction▲ 2,618 E7
Karnes City▲ 3,457 G9
Katy 11,775 J8
Kaufman▲ 6,490 H5
Keene 5,003 G5
Keller 27,345 F2
Kenedy 3,487 G9
Kennedale 5,850 F3
Kerens 1,702 H5
Kermit▲ 5,714 B6
Kerrville▲ 20,425 E7
Kilgore 11,301 K5
Killeen 86,911 G6
Kingsland 4,584 F7
Kingsville▲ 25,575 G10
Kirby 8,673 K11
Kirbyville 2,085 K7
Kountze▲ 2,115 K7
Kyle 5,314 G8
La Feria 6,115 G11
La Grange▲ 4,478 G8
La Joya 3,303 F11
Lake Dallas 6,166 G1
Lake Jackson 26,386 J8
Lake Worth 4,618 E2
La Marque 13,682 K3
Lamesa▲ 9,952 C5
Lampasas▲ 6,786 F6
Lancaster 25,894 G3
La Porte 31,880 K2
Laredo▲ 176,576 E10
League City 45,444 K2
Leakey▲ 387 E8
Leon Valley 9,239 J10
Leonard 1,744 H4
Levelland▲ 12,866 B4
Lewisville 77,737 G1
Liberty▲ 8,033 K7
Lindale 2,954 J5
Linden▲ 2,256 K4
Lipscomb▲ 44 D1
Littlefield▲ 6,507 B4
Live Oak 9,156 K10
Livingston▲ 5,433 K7
Llano▲ 3,325 F7
Lockhart▲ 11,615 G8
Lockney 2,056 C3
Lomax 2,991 K2
Longview▲ 73,344 K5
Los Fresnos 4,512 G11
Lubbock▲ 199,564 C4
Lucas 2,890 H1
Lufkin▲ 32,709 K6
Luling 5,080 G8
Lumberton 8,731 K7
Lyford 1,973 G11
Lytle 2,383 J11
Mabank 2,151 H5
Madisonville▲ 4,159 J7
Malakoff 2,257 H5
Mansfield 28,031 F3
Manvel 3,046 J3
Marble Falls 4,959 F7
Marfa▲ 2,121 C12
Marlin▲ 6,628 H6
Marshall▲ 23,935 K5
Mart 2,273 H6
Mason▲ 2,134 E7
Matador▲ 740 D3
Mathis 5,034 G9
McAllen 106,414 F11
McCamey 1,805 B6
McGregor 4,727 G6
McKinney▲ 54,369 H4
Memphis▲ 2,479 D3
Menard▲ 1,653 E7
Mentone▲ 50 D10
Mercedes 13,649 F12
Meridian▲ 1,491 G6
Merkel 2,637 E5
Mertzon▲ 839 C6
Mesquite 124,523 H2
Mexia 6,563 H6
Miami▲ 588 D2
Midland▲ 94,996 C6
Midlothian 7,480 G5
Mineola 4,550 J5
Mineral Wells 16,946 F5
Mission 45,408 F11
Missouri City 52,913 J2
Monahans▲ 6,821 B6
Montague▲ 1,253 G4
Morton▲ 2,249 B4
Mount Pleasant▲ 13,935 K4
Mount Vernon▲ 2,286 J4
Muleshoe▲ 4,530 B3
Nacogdoches▲ 29,914 J6
Nash 2,169 K4
Nassau Bay 4,170 K2
Navasota 6,789 J7
Nederland 17,422 K8
Needville 2,609 J8
New Boston 4,808 K4
New Braunfels▲ 36,494 K10
Newton▲ 2,459 L7
Nixon 2,186 G8
Nocona 3,198 G4
North Richland Hills 55,635 F2
Odessa▲ 90,943 B6
Olmos Park 2,343 K11
Olney 3,396 F4
Olton 2,288 B3
Orange▲ 18,643 L7
Overton 2,350 K5
Ovilla 3,405 G3
Ozona▲ 3,436 C7
Paducah▲ 1,498 D4
Paint Rock▲ 320 E6
Palacios 5,153 H9
Palestine▲ 17,598 J6
Palo Pinto▲ 350 F5
Pampa▲ 17,887 D2
Panhandle▲ 2,589 C2
Pantego 2,318 F2
Paris▲ 25,898 J4
Pasadena 141,674 J2
Pearland 37,640 J2
Pearsall▲ 7,157 E9
Pecos▲ 9,501 D10
Perryton▲ 7,774 D1

Topography

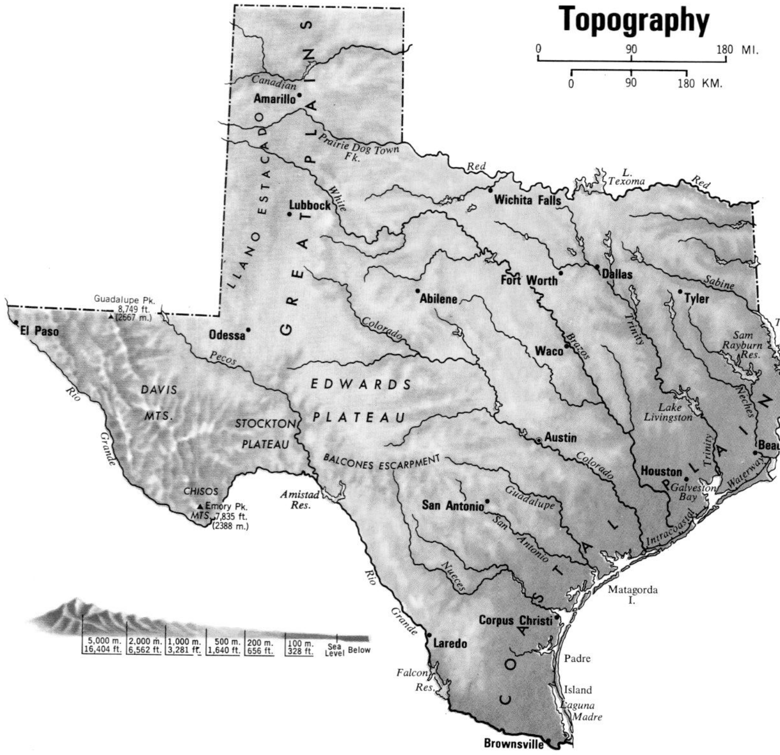

Pflugerville 16,335 G7
Pharr 46,660 F11
Pickton 1,729 C2
Pilot Point 3,538 H4
Piney Point Village 3,380 J1
Pittsburg▲ 4,347 J4
Plains▲ 1,450 B4
Plainview▲ 22,336 C3
Plano 222,030 G1
Pleasanton 8,266 F9
Port Aransas 3,370 H10
Port Arthur 57,755 K8
Port Isabel 4,865 G11
Portland 14,827 G10
Port Lavaca▲ 12,035 H9
Port Neches 13,601 K7
Post▲ 3,708 C4
Poteet 3,305 F8
Prairie View 4,410 J7
Premont 2,772 F10
Presidio 4,167 C12
Quanah▲ 3,022 E3
Queen City 1,613 L4
Quitman▲ 2,030 J5
Ralls 2,252 C4
Ranger 2,584 F5
Rankin▲ 800 B6
Raymondville▲ 9,733 G11
Red Oak 4,301 H5
Refugio▲ 2,941 G9
Reno 2,441 E2
Richardson 91,802 G2
Richland Hills 8,132 F2
Richmond▲ 11,081 J8
Rio Grande City▲ 11,923 F11
Rio Hondo 1,942 G11
River Oaks 6,985 E2
Robert Lee▲ 1,171 D6
Robstown 12,727 G10
Roby▲ 673 D5
Rockdale 5,439 G7
Rockport▲ 7,385 H9
Rocksprings▲ 1,285 D8
Rockwall▲ 17,976 H5
Roma-Los Saenz 3,384 E11
Rosenberg 24,043 J8
Rotan 1,913 D5
Round Rock 61,136 G7
Rowlett 44,503 H2
Royse City 2,957 H4
Rusk▲ 5,085 J6
Sachse 9,751 H2
Saginaw 12,374 E2
San Angelo▲ 88,439 D6
San Antonio▲ 1,144,646 J11
San Augustine▲ 2,475 K6
San Benito 23,444 G12
Sanderson▲ 861 B7
San Diego▲ 4,753 F10
San Elizario 11,046 A10
Sanger▲ 4,534 G4
San Juan 26,229 F11
San Leon 4,365 L2
San Marcos▲ 34,733 F8
San Saba▲ 2,637 F6
Sansom Park Village 4,181 E2
Santa Fe 9,548 K3
Sarita▲ 200 G10
Schertz 18,694 K10
Schulenburg 2,699 H8
Seabrook 9,443 K2
Seagoville 10,823 H3
Seagraves 2,334 B5
Sealy 5,248 H8
Seguin▲ 22,011 G8
Seminole▲ 5,910 K10
Seymour▲ 2,908 E4
Shamrock 2,029 D2
Shepherd 2,029 K7
Sherman▲ 35,082 H4
Shiner 2,070 G8
Sierra Blanca▲ 533 B11
Silsbee 6,393 K7
Silverton▲ 771 C3
Sinton▲ 5,676 G9
Slaton 6,109 C4
Smithville 3,901 G7
Snyder▲ 10,783 D5
Sonora▲ 2,924 D7
South Houston 15,833 J2
South Padre Island 1,677 F11
Spearman▲ 3,021 C1
Spring 36,385 J7
Spring Valley 3,611 J1
Stafford 15,681 J2
Stamford 3,636 E5
Stanton▲ 2,556 C5
Stephenville▲ 14,921 F5
Sterling City▲ 1,081 D6
Stinnett▲ 1,936 C2
Stratford▲ 1,991 C1
Sugar Land 63,328 J8
Sulphur Springs▲ 14,551 J4
Sundown 1,759 B4
Sunnyvale 2,693 H2
Sweeny 3,624 J8
Sweetwater▲ 11,415 D5
Taft 3,396 G9
Tahoka▲ 2,910 C4
Taylor 13,575 G7
Taylor Lake Village 3,694 K2
Teague 4,557 H6
Temple 54,514 G6
Terlingua 100 D12
Terrell 13,606 H5
Terrell Hills 5,019 K11
Texarkana 34,782 L4
Texas City 41,521 K3
Texhoma 291 C1
The Colony 26,531 G1
Three Rivers 1,878 F9
Throckmorton▲ 905 F4
Tilden▲ 450 F9
Tomball 9,089 J7
Trinity 2,721 J7
Tulia▲ 5,117 C3
Tyler▲ 83,650 J5
Universal City 14,849 K10
University Park 23,324 F2
Uvalde▲ 1,972 E8
Van 1,854 J5
Van Alstyne 2,502 H4
Van Horn▲ 2,435 C11
Vega▲ 936 B2
Vernon▲ 11,660 E3
Victoria▲ 60,603 H9
Vidor 11,440 L7
Waco▲ 113,726 G6
Wake Village 5,129 K4
Waskom 1,812 L5
Watauga 21,908 F2
Waxahachie▲ 21,426 H5
Weatherford▲ 19,000 G5
Webster 9,083 K2
Weimar 1,981 H8
Wellington▲ 2,275 D3
Weslaco 26,935 F11
West 2,692 G6
West Columbia 4,255 J8
West Orange 4,111 L7
West University Place 14,211 J2
Westworth 2,124 E2
Wharton▲ 9,237 J8
Wheeler▲ 1,378 D2
White Oak 5,624 K5
Whitesboro 3,760 H4
White Settlement 14,831 E2
Whitewright 1,713 H4
Wichita Falls▲ 104,197 F4
Willis 3,985 J7
Wills Point 3,496 J5
Wilmer 3,393 H3
Windcrest 5,105 K11
Winnie 2,914 K8
Winnsboro 3,584 J5
Winters 2,880 E6
Wolfforth 2,554 C4
Woodsboro 1,731 G9
Woodville▲ 2,415 K7
Wylie 15,132 H1
Yoakum 5,731 G8
Yorktown 2,271 G9
Zapata▲ 4,856 E11

OTHER FEATURES

Alibates Flint Quarries Nat'l Mon. C2
Amistad (res.) C8
Amistad Nat'l Rec. Area D8
Angelina (riv.) K6
Apache (mts.) C11
Aransas (passage) H10
Arlington (lake) F2
Baffin (bay) G10
Balcones Escarpment (plat.) E8
Benbrook (lake) E3
Bergstrom A.F.B. G7
Big Bend Nat'l Park A8
Big Thicket Nat'l Preserve K7
Bolivar (pen.) K8
Brazos (riv.) H7
Brooks A.F.B. K11
Brownwood (lake) E6
Buchanan (lake) F7
Caddo (lake) L5
Calaveras (lake) K11
Canadian (riv.) D1
Carrizo (creek) A1
Carswell A.F.B. E2
Cathedral (mt.) D12
Cavallo (passage) H9
Cedar (lake) B5
Cerro Alto (mt.) B10
Chamizal Nat'l Mem. A10
Chinati (mts.) C12
Chinati (peak) C12
Chisos (mts.) A8
Cibolo (creek) K11
Clear Fork, Brazos (riv.) D5
Coldwater (creek) B1
Colorado (riv.) F7
Copano (bay) G9
Corpus Christi (lake) F9
Corpus Christi N.A.S. G10
Cottonwood Draw (dry riv.) C10
Davis (mts.) C11
Deep (creek) C5
Delaware (creek) C10
Delaware (mts.) C10
Denison (dam) H4
Devils (riv.) D7
Double Mountain Fork, Brazos (riv.) C4
Dyess A.F.B. D5
Eagle (peak) C11
Eagle Mountain (lake) E2
Edwards (plat.) C7
Elephant (mt.) D12
Elm Fork, Trinity (riv.) G2
Emory (peak) A8
Falcon (res.) E11
Finlay (mts.) B10
Fort Bliss 8,264 A10
Fort Davis Nat'l Hist. Site D11
Fort Hood 35,580 G6
Fort Sam Houston K11
Frio (riv.) E8
Galveston (bay) L2
Galveston (isl.) K8
Glass (mts.) A7
Goodfellow A.F.B. D6
Grapevine (lake) F2
Guadalupe (mts.) C10
Guadalupe (peak) B10
Guadalupe (riv.) G8
Guadalupe Mountains Nat'l Park C10
Houston (lake) J8
Houston Ship (chan.) K2
Howard (creek) C7
Hubbard Creek (lake) F5
Hueco (mts.) B10
Intracoastal Waterway J9
Johnson Draw (dry riv.) C7
Kelly A.F.B. J11
Kemp (lake) E4
Kingsville N.A.S. G10
Kiowa (creek) D1
Lackland A.F.B. 7,123 J11
Lake Meredith Nat'l Rec. Area C2
Lampasas (riv.) G6
Laughlin A.F.B. 2,225 D8
Lavon (lake) H1
Leon (riv.) F6
Livermore (mt.) C11
Livingston (lake) K7
Llano (riv.) D7
Locke (mt.) D11
Los Olmos (creek) F10
Los Olmos (creek) F11
Lyndon B. Johnson Nat'l Hist. Park F7
Lyndon B. Johnson Space Center K2
Madre (lag.) G11
Maravillas (creek) A8
Matagorda (bay) H9
Matagorda (isl.) H9
Matagorda (pen.) H9
Medina (lake) E8
Medina (riv.) J11
Mexico (gulf) K9
Middle Concho (riv.) C6
Mountain Creek (lake) G2
Mustang (creek) A1
Mustang (isl.) G10
Mustang Draw (dry riv.) B5
Navasota (riv.) H7
Navidad (riv.) H8
Neches (riv.) K6
North Concho (riv.) C6
North Pease (riv.) D3
Nueces (riv.) F9
Padre (isl.) G10
Padre Island Nat'l Seashore G11
Palo Duro (creek) B2
Palo Duro (creek) C1
Pease (riv.) D3
Pecos (riv.) C7
Pedernales (riv.) F7
Possum Kingdom (lake) F5
Prairie Dog Town Fork, Red (riv.) C3
Quitman (mts.) B11
Randolph A.F.B. K10
Ray Hubbard (lake) H2
Red (riv.) F3
Red Bluff (lake) D10
Reese A.F.B. B4
Rio Grande (riv.) D9
Rita Blanca (creek) B2
Sabine (riv.) L7
Salt Fork, Red (riv.) D3
Sam Rayburn (res.) K6
San Antonio (bay) H9
San Antonio (mt.) B10
San Antonio Missions Nat'l Hist. Park J11
San Francisco (creek) B8
San Luis (passage) K8
San Martine Draw (dry riv.) C11
San Saba (riv.) D7
Santa Isabel (creek) E10
Santiago (mts.) A8
Santiago (peak) D12
Sheppard A.F.B. F3
Sierra Diablo (mts.) C10
Sierra Vieja (mts.) C11
Staked (plain) B4
Stamford (lake) E4
Stockton (plat.) B7
Sulphur (riv.) J4
Sulphur Draw (dry riv.) B4
Sulphur Springs (creek) B4
Tenmile (creek) G3
Terlingua (creek) D12
Texoma (lake) H3
Thomas (lake) C5
Tierra Blanca (creek) B3
Toledo Bend (res.) L6
Toyah (creek) D11
Toyah (lake) A6
Travis (lake) G7
Trinity (bay) L2
Trinity (riv.) H5
Trinity, West Fork (riv.) G2
Trujillo (creek) A2
Washita (riv.) D2
West (bay) K3
White (riv.) C3
White River (lake) C4
White Rock (creek) G2
Wichita (riv.) F4
Wolf (creek) D1
Worth (lake) E2
Wright Patman (lake) K4

▲County seat

Texas

State Capitals ⊛
County Seats ◉
Major Limited Access Hwys. ⎯

Western Part of Texas
Same scale as main map

© Copyright HAMMOND INCORPORATED, Maplewood, N. J.

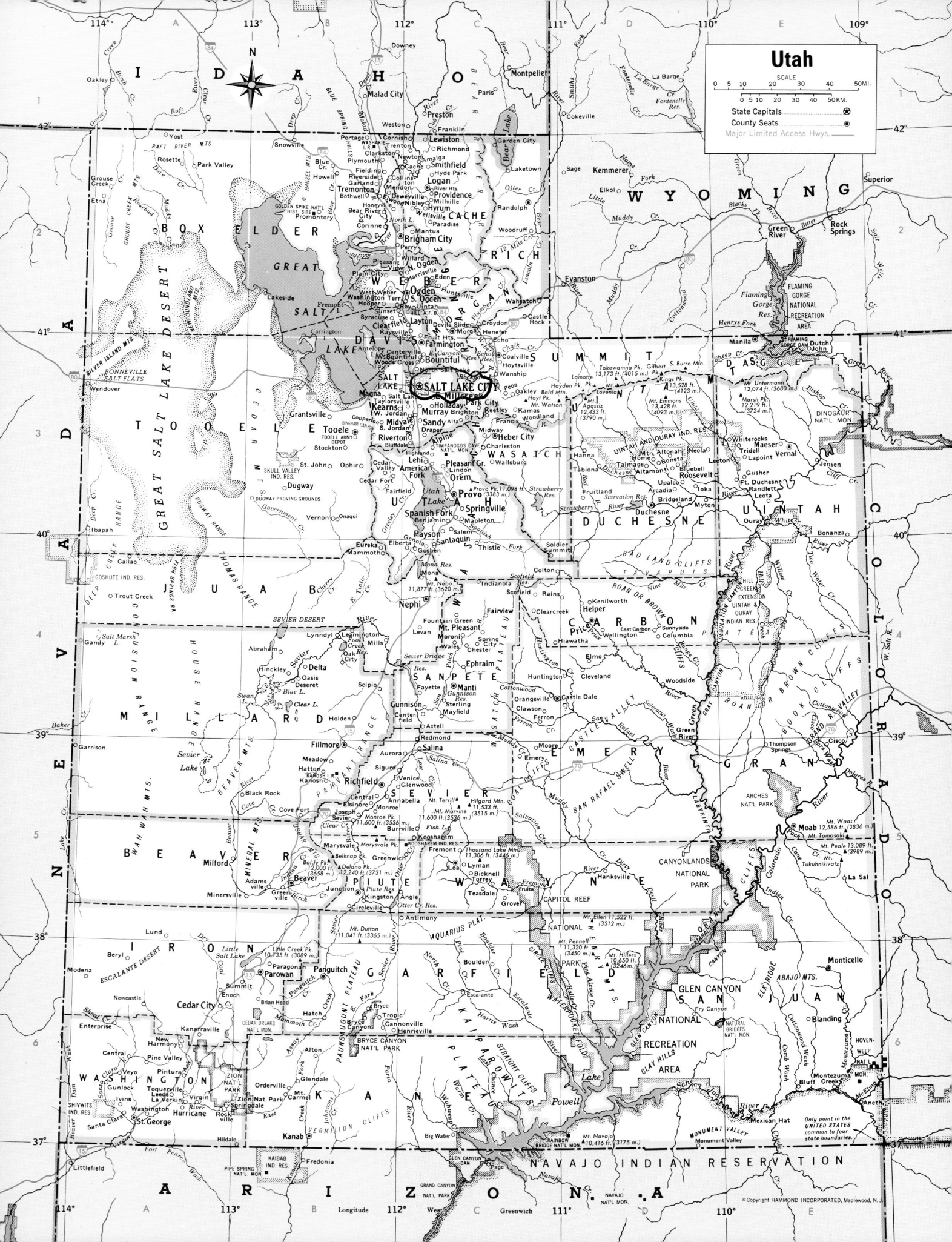

Utah

SCALE

State Capitals ⊛
County Seats ⊙
Major Limited Access Hwys.

© Copyright HAMMOND INCORPORATED, Maplewood, N.J.

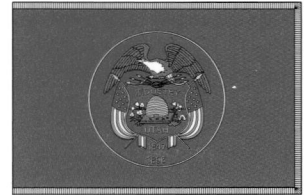

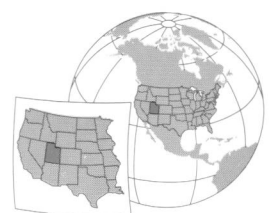

AREA 84,899 sq. mi. (219,888 sq. km.)
POPULATION 2,233,169
CAPITAL Salt Lake City
LARGEST CITY Salt Lake City
HIGHEST POINT Kings Pk. 13,528 ft. (4123 m.)
SETTLED IN 1847
ADMITTED TO UNION January 4, 1896
POPULAR NAME Beehive State
STATE FLOWER Sego Lily
STATE BIRD Sea Gull

COUNTIES

Beaver 6,005 A5
Box Elder 42,745 A2
Cache 91,391 C2
Carbon 20,422 D4
Daggett 921 E3
Davis 238,994 B3
Duchesne 14,371 D3
Emery 10,860 D4
Garfield 4,735 C6
Grand 8,485 E5
Iron 33,779 A6
Juab 8,238 A4
Kane 6,046 B6
Millard 12,405 A4
Morgan 7,129 C2
Piute 1,435 B5
Rich 1,961 C2
Salt Lake 898,387 B3
San Juan 14,413 E6
Sanpete 22,763 C4
Sevier 18,842 C5
Summit 29,736 C3
Tooele 40,735 A3
Uintah 25,224 E3
Utah 368,536 C3
Wasatch 15,215 C3
Washington 90,354 A6
Wayne 2,509 C5
Weber 196,533 B2

CITIES and TOWNS

Alpine 7,146 C3
Alta 370 C3
Amalga 427 C2
American Fork 21,941 C3
Annabella 487 B5
Aneth 598 E6
Aurora 947 B5
Bear River City 750 B2
Beaver▲ 2,454 B5
Benjamin 1,029 C3
Bicknell 353 C5
Big Water 417 C6
Blanding 3,162 E6
Bluebell 275 D3
Bluffdale 4,700 B3
Bothwell 410 B2
Bountiful 41,301 C3
Brigham City▲ 17,411 C2
Brighton 150 C3
Castle Dale▲ 1,657 D4
Castle Rock C2
Cedar City 20,527 A6
Cedar Fort 341 B3
Cedar Valley 286 B3
Centerfield 1,048 C4
Centerville 14,585 C3
Charleston 378 C3
Circleville 505 B5
Clarkston 688 B2
Clearfield 25,974 B2
Cleveland 508 D4
Coalville▲ 1,382 C3
Columbia 280 D4
Corinne 621 B2
Delta 3,209 B4
Deweyville 278 B2
Draper 25,220 C3
Duchesne▲ 1,408 D3
Dugway 2,016 B3
East Carbon 1,393 D4
East Millcreek 21,385 C3
Eden 421 C2
Elberta 278 B4
Elmo 368 D4
Elsinore 733 B5
Elwood 678 B2
Emery 308 C5
Enoch 3,467 A6
Enterprise 1,285 A6
Ephraim 4,505 C4
Escalante 818 C6
Eureka 766 B4
Fairview 1,160 C4
Farmington▲ 12,081 C4
Ferron 1,623 C4
Fielding 448 B2
Fillmore▲ 2,253 B5
Fort Duchesne 621 E3
Fountain Green 945 C4
Francis 698 C3
Fruit Heights 4,701 C2
Garden City 357 C2
Garland 1,943 B2
Genola 965 C4
Glendale 355 B6
Glenwood 437 C5
Goshen 874 C4
Grantsville 6,015 B3
Green River 973 D4
Gunnison 2,394 C4
Hanksville 250 D5
Harrisville 3,645 C2
Heber City▲ 7,291 C3
Helper 2,025 D4
Henefer 684 C2
Hiawatha 249 D4
Highland 8,172 C3
Hildale 1,895 A6
Hinckley 698 B4
Holden 400 B4
Holladay 14,561 C3
Honeyville 1,214 B2
Hooper 3,926 B2
Howell 221 B2
Hoytsville 500 C3
Huntington 2,131 C4
Huntsville 649 C2
Hurricane 8,250 A6
Hyde Park 2,955 C2
Hyrum 6,316 C2
Ivins 4,450 A6
Jensen 750 E3
Joseph 269 B5
Junction▲ 177 B5
Kamas 1,274 C3
Kanab▲ 3,564 B6
Kanarraville 311 A6
Kanosh 485 B5
Kaysville 20,351 B2
Kearns 33,659 B3
Kenilworth 335 D4
Koosharem 276 C5
Laketown 188 C2
Lapoint 430 E3
La Sal 339 E5
La Verkin 3,392 A6
Layton 58,474 C2
Leamington 217 B4
Leeds 547 A6
Lehi 19,028 C3
Levan 688 C4
Lewiston 1,877 C2
Lindon 8,363 C3
Loa▲ 525 C5
Logan▲ 42,670 C2
Lyman 234 C5
Maeser 2,855 E3
Magna 22,770 B3
Manila▲ 308 E3
Manti▲ 3,040 C4
Mantua 791 C2
Mapleton 5,809 C3
Marysvale 381 B5
Mayfield 420 C4
Meadow 254 B5
Mendon 898 B2
Mexican Hat 88 E6
Midvale 27,029 B3
Midway 2,121 A5
Milford 1,451 A5
Millville 1,507 C2
Minersville 817 A5
Moab▲ 4,779 E5
Mona 850 C4
Monroe 1,845 B5
Montezuma Creek 507 E6
Monticello▲ 1,958 E6
Morgan▲ 2,635 C2
Moroni 1,280 C4
Mount Pleasant 2,707 C4
Murray 34,024 C3
Myton 539 D3
Neola 533 D3
Nephi▲ 4,733 C4
Newcastle 350 A6
Newton 699 C2
Nibley 2,045 C2
North Ogden 15,026 C2
North Salt Lake 8,749 C3
Oak City 650 B4
Oakley 948 C3
Ogden▲ 77,226 C2
Orangeville 1,398 C4
Orderville 596 B6
Orem 84,324 C3
Panguitch▲ 1,623 B6
Paradise 759 C2
Paragonah 470 B6
Park City 7,371 C3
Parowan▲ 2,565 B6
Payson 12,716 C3
Perry 2,383 C2
Plain City 3,489 B2
Pleasant Grove 23,468 C3
Pleasant View 5,632 B2
Plymouth 328 B2
Price▲ 8,402 D4
Providence 4,377 C2
Provo▲ 105,166 C3
Randlett 224 E3
Randolph▲ 483 C2
Redmond 788 C4
Richfield▲ 6,847 B5
Richmond 2,051 C2
River Heights 1,496 C2
Riverside 678 B2
Riverton 25,011 B3
Roosevelt 4,299 D3
Roy 32,885 C2
Saint George▲ 49,663 A6
Saint John 350 B3
Salem 4,372 C3
Salina 2,393 C5
Salt Lake City (cap.)▲ 181,743 C3
Sandy 88,418 C3
Santa Clara 4,630 A6
Santaquin 4,834 C4
Scipio 290 B4
Sigurd 430 B5
Smithfield 7,261 C2
South Jordan 29,437 B3
South Ogden 14,377 C2
South Salt Lake 22,038 C3
Spanish Fork 20,246 C3
Spring City 956 C4
Springdale 457 B6
Springville 20,424 C3
Stockton 443 B3
Sunnyside 404 D4
Sunset 5,204 B2
Syracuse 9,398 B2
Taylorsville-Bennion 57,439 B3
Tooele▲ 22,502 B3
Toquerville 910 A6
Tremonton 5,592 B2
Trenton 449 B2
Tropic 508 B6
Uintah 1,127 C2
Upalco 280 D3
Vernal▲ 7,714 E3
Wallsburg 274 C3
Washington 8,186 A6
Washington Terrace 8,551 B2
Wellington 1,666 D4
Wellsville 2,728 C2
Wendover 1,537 A3
West Bountiful 4,484 B3
West Jordan 68,336 B3
West Weber 750 B2
Whiterocks 341 E3
Willard 1,630 C2
Woodland 335 C3
Woods Cross 6,419 B3

OTHER FEATURES

Abajo (mts.) E6
Agassiz (mt.) D3
Antelope (isl.) B3
Aquarius (plat.) C5
Arches Nat'l Park E5
Assay (creek) B6
Bad Land (cliffs) D4
Baldy (peak) B5
Bear (lake) C2
Bear (riv.) B2
Beaver (mts.) A5
Beaver (riv.) A5
Beaver Dam Wash (creek) A6
Birch (creek) B5
Blue (creek) B2
Bonneville (salt flats) A3
Book (cliffs) E4
Bryce Canyon Nat'l Park B6
Canyonlands Nat'l Park D5
Capitol Reef Nat'l Park C5
Castle (valley) D4
Cedar (mts.) B3
Cedar Breaks Nat'l Mon. B6
Chalk (creek) C3
Chinle (creek) E6
Clear (lake) B4
Cliff (creek) E3
Coal (cliffs) C5
Colorado (riv.) E5
Confusion (range) A4
Cottonwood (creek) C4
Cub (creek) C1
Deep (creek) B1
Deep Creek (range) A4
Delano (peak) B5
Desolation (canyon) E4
Dinosaur Nat'l Mon. E3
Dirty Devil (riv.) D5
Dolores (riv.) E5
Dry Coal (creek) A6
Duchesne (riv.) D3
Dugway (range) A3
Dugway Proving Grounds B3
Dutton (mt.) B5
East Canyon (res.) C3
Echo (res.) C3
Elk (ridge) E6
Ellen (mt.) D5
Emmons (mt.) D3
Escalante (des.) A6
Escalante (riv.) C6
Fish (lake) C5
Fish Springs (range) A4
Flaming Gorge (res.) E3
Flaming Gorge Nat'l Rec. Area E2
Fool Creek (res.) B4
Fremont (isl.) B2
Fremont (riv.) C5
Glen Canyon Nat'l Rec. Area D6
Golden Spike Nat'l Hist. Site B2
Goshute Ind. Res. A4
Government (creek) B3
Gray (canyon) D4
Great Salt (lake) B2
Great Salt Lake (des.) A3
Greeley (creek) B3
Green (riv.) D4
Grouse (creek) A2
Grouse Creek (mts.) A2
Gunnison (res.) C4
Henry (mts.) D6
Hilgard (mt.) C5
Hill (creek) E4
Hill A.F.B. C2
Hill Creek Extension, Uintah and Ouray Ind. Res. E4
Hillers (mt.) D6
House (range) A4
Hovenweep Nat'l Mon. E6
Hoyt (peak) C3
Huntington (creek) C4
Indian (creek) B5
Jordan (riv.) C3
Kaiparowits (plat.) C6
Kanab (creek) B7
Kanosh Ind. Res. B5
Kings (peak) D3
Koosharem Ind. Res. C5
Little Creek (peak) B6
Little Salt (lake) A6
Malad (riv.) B1
Marsh (peak) E3
Marvine (mt.) C5
Mineral (mts.) B5
Mona (res.) C4
Monroe (peak) B5
Montezuma (creek) E6
Monument (valley) D6
Muddy (creek) C4
Natural Bridges Nat'l Mon. E6
Navajo (mt.) D6
Navajo Ind. Res. D7
Nebo (mt.) C4
Newfoundland (mts.) A2
Nine Mile (creek) D4
North (lake) B2
Orange (cliffs) D5
Otter (creek) C5
Otter Creek (res.) C5
Paria (riv.) B6
Paunsaugunt (plat.) B6
Pahvant (range) B5
Peale (mt.) E5
Pennell (mt.) D6
Piute (res.) B5
Plumber (creek) C2
Powell (lake) D6
Price (riv.) D4
Provo (peak) C3
Provo (riv.) C3
Raft River (mts.) A2
Rainbow Bridge Nat'l Mon. C6
Roan (cliffs) E4
Rockport (lake) C3
Salvation (creek) C5
San Juan (riv.) D6
San Pitch (riv.) C4
San Rafael (riv.) D4
San Rafael Swell (mts.) D5
Santa Clara (riv.) A6
Sevier (des.) B4
Sevier (lake) A5
Sevier (riv.) B4
Sevier Bridge (res.) C4
Shivwits Ind. Res. A6
Silver Island (mts.) A3
Skull Valley Ind. Res. B3
Spanish Fork (riv.) C3
Strait (cliffs) C6
Strawberry (res.) C3
Strawberry (riv.) D3
Swan (lake) B4
Tavaputs (plat.) D4
Thomas (range) A4
Thousand Lake (mt.) C5
Timpanogos Cave Nat'l Mon. C3
Tokewamna (peak) D3
Tooele Army Depot B3
Two Water (creek) E4
Uinta (mts.) D3
Uinta (riv.) D3
Uintah and Ouray Ind. Res. D3
Utah (lake) C3
Virgin (riv.) A6
Waas (mt.) E5
Wah Wah (mts.) A5
Wasatch (range) C3
Washakie Ind. Res. B2
Waterpocket Fold (cliffs) D6
Weber (riv.) C3
White (riv.) E3
Willow (creek) E4
Zion Nat'l Park A6

▲County seat

Agriculture, Industry and Resources

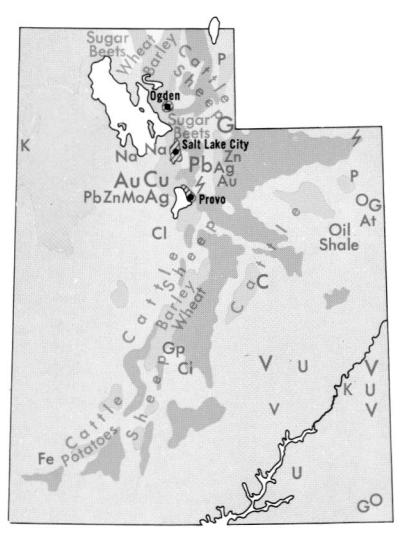

DOMINANT LAND USE

Wheat, General Farming

General Farming, Livestock, Special Crops

Range Livestock

Forests

Nonagricultural Land

MAJOR MINERAL OCCURRENCES

Ag	Silver	Fe	Iron Ore	O	Petroleum
At	Asphalt	G	Natural Gas	P	Phosphates
Au	Gold	Gp	Gypsum	Pb	Lead
C	Coal	K	Potash	U	Uranium
Cl	Clay	Mo	Molybdenum	V	Vanadium
Cu	Copper	Na	Salt	Zn	Zinc

Water Power

Major Industrial Areas

Topography

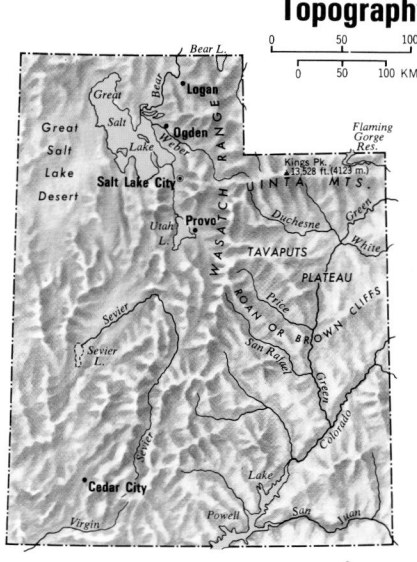

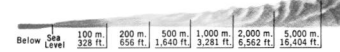

Topography

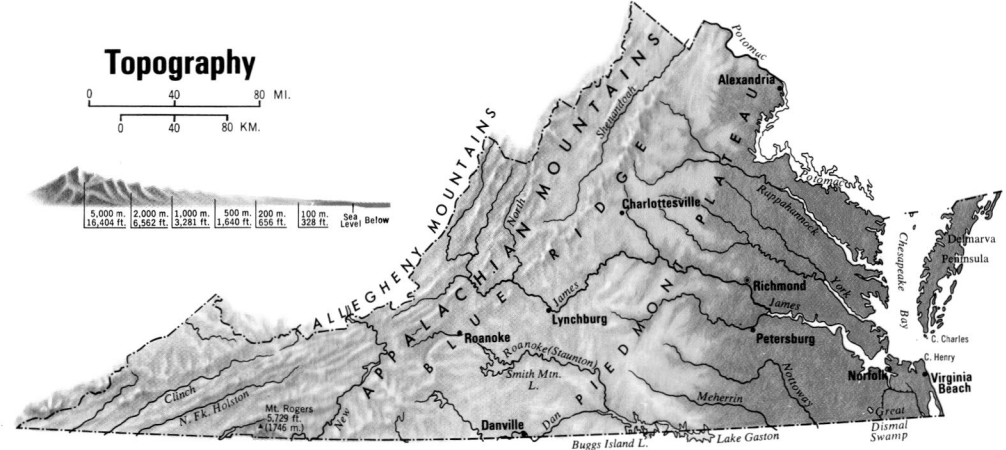

5,000 m. 16,404 ft.	2,000 m. 6,562 ft.	1,000 m. 3,281 ft.

500 m. 1,640 ft. 200 m. 656 ft. 100 m. 328 ft. Sea Level Below

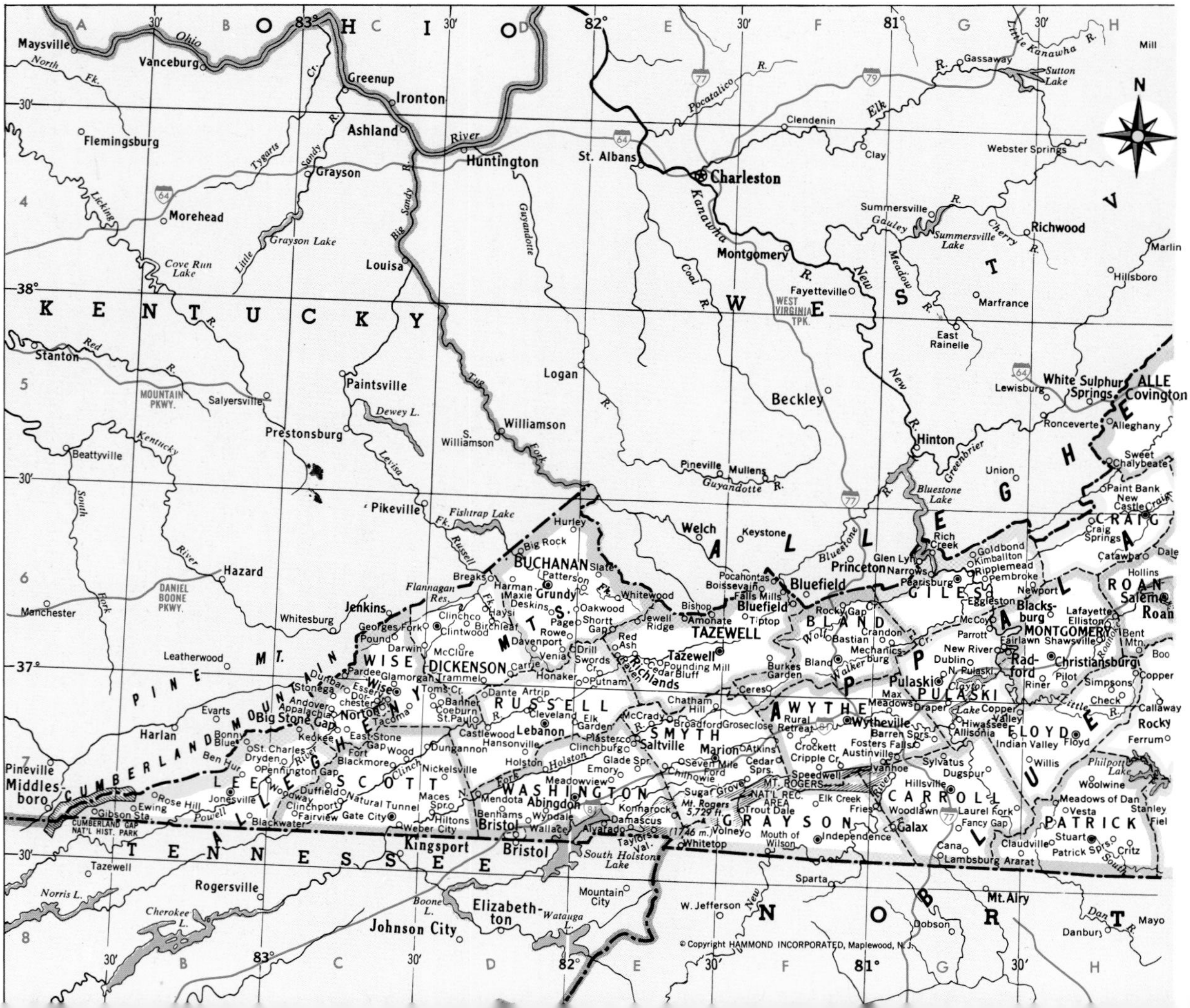

AREA 40,767 sq. mi. (105,587 sq. km.)
POPULATION 7,078,515
CAPITAL Richmond
LARGEST CITY Norfolk
HIGHEST POINT Mt. Rogers 5,729 ft. (1746 m.)
SETTLED IN 1607
ADMITTED TO UNION June 26, 1788
POPULAR NAME Old Dominion
STATE FLOWER Dogwood
STATE BIRD Cardinal

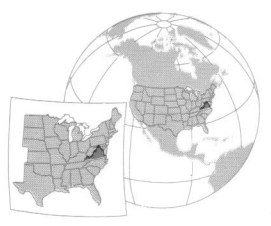

(continued on following page)

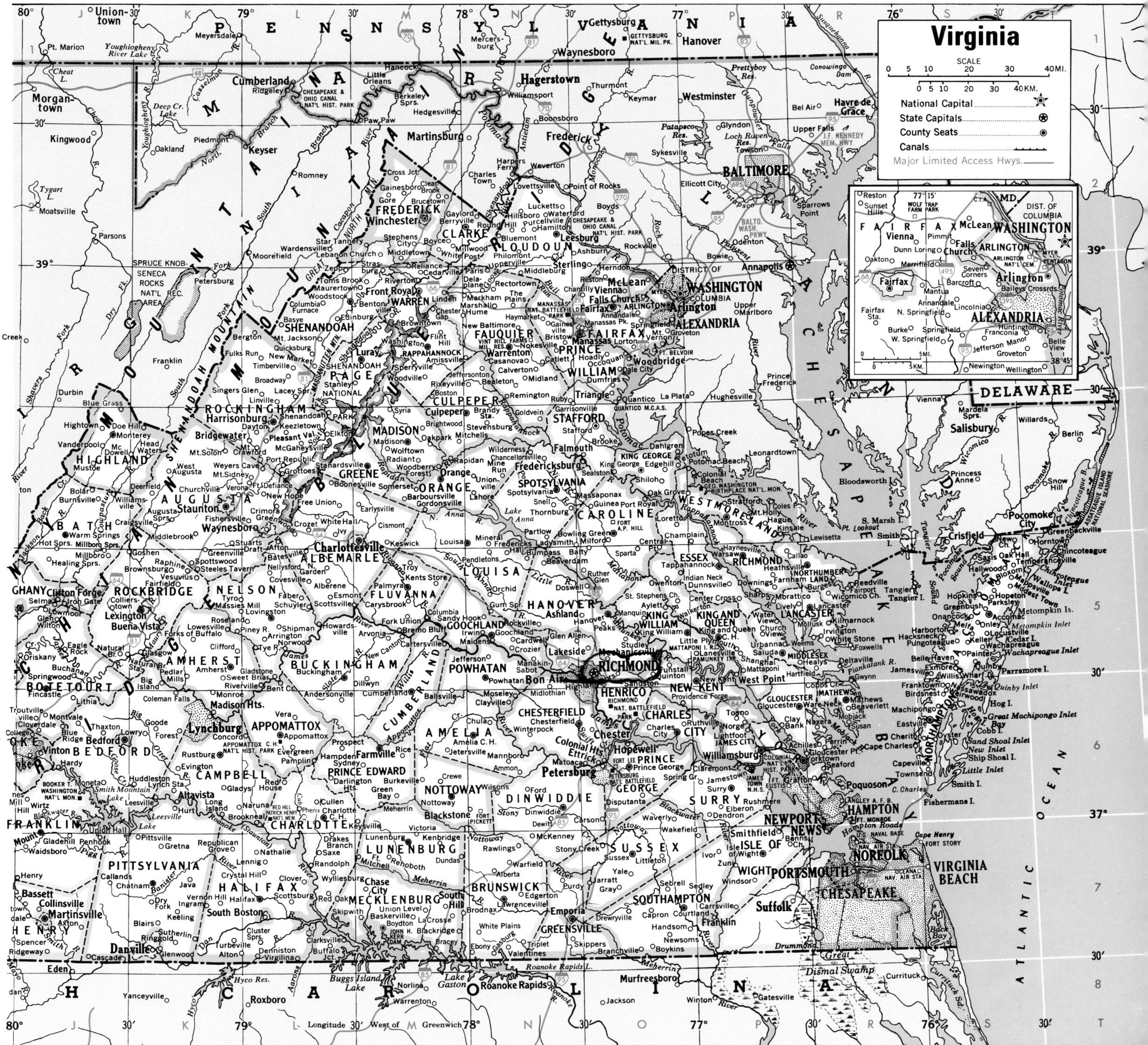

Virginia

SCALE
0 5 10 20 30 40MI.
0 5 10 20 30 40KM.

National Capital ....★
State Capitals ....⊛
County Seats ....●
Canals ....
Major Limited Access Hwys. ____

Agriculture, Industry and Resources

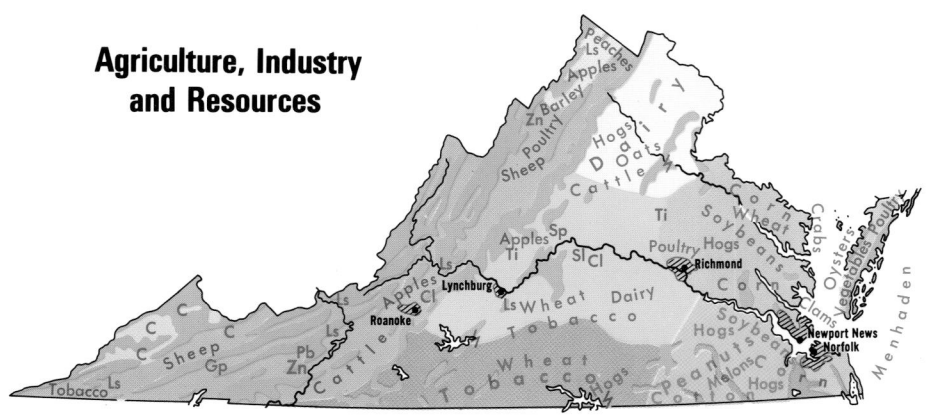

MAJOR MINERAL OCCURRENCES

C	Coal	Sl	Slate
Cl	Clay	Sp	Soapstone
Gp	Gypsum	Ti	Titanium
Ls	Limestone	Zn	Zinc
Pb	Lead		

⚡ Water Power
▨ Major Industrial Areas

DOMINANT LAND USE

- Dairy, General Farming
- General Farming, Livestock, Dairy
- General Farming, Livestock, Tobacco
- General Farming, Livestock, Fruit, Tobacco
- General Farming, Truck Farming, Tobacco, Livestock
- Tobacco, General Farming
- Peanuts, General Farming
- Fruit and Mixed Farming
- Truck and Mixed Farming
- Forests
- Swampland, Limited Agriculture

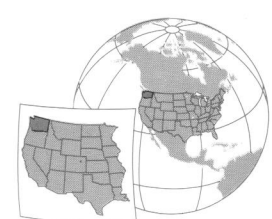

AREA 68,139 sq. mi. (176,480 sq. km.)
POPULATION 5,894,121
CAPITAL Olympia
LARGEST CITY Seattle
HIGHEST POINT Mt. Rainier 14,410 ft. (4392 m.)
SETTLED IN 1811
ADMITTED TO UNION November 11, 1889
POPULAR NAME Evergreen State
STATE FLOWER Western Rhododendron
STATE BIRD Willow Goldfinch

Agriculture, Industry and Resources

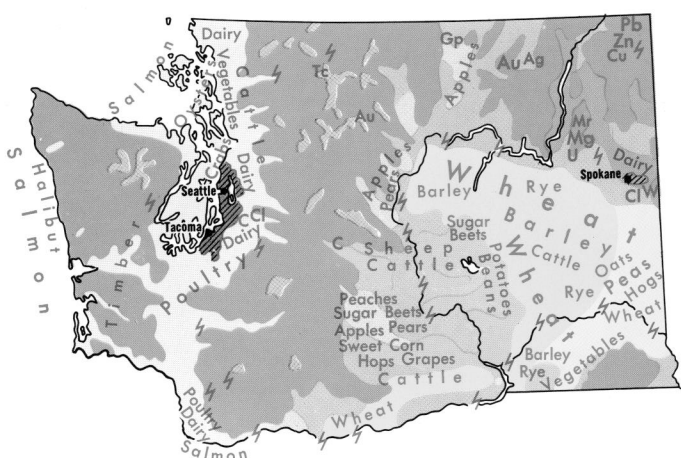

DOMINANT LAND USE

- Specialized Wheat
- Wheat, Peas
- Dairy, Poultry, Mixed Farming
- Fruit and Mixed Farming
- General Farming, Dairy, Range Livestock
- General Farming, Livestock, Special Crops
- Range Livestock
- Forests
- Urban Areas
- Nonagricultural Land

MAJOR MINERAL OCCURRENCES

Ag	Silver	Mr	Marble
Au	Gold	Pb	Lead
C	Coal	Tc	Talc
Cl	Clay	U	Uranium
Cu	Copper	W	Tungsten
Gp	Gypsum	Zn	Zinc
Mg	Magnesium		

⚡ Water Power

▨ Major Industrial Areas

(continued on following page)

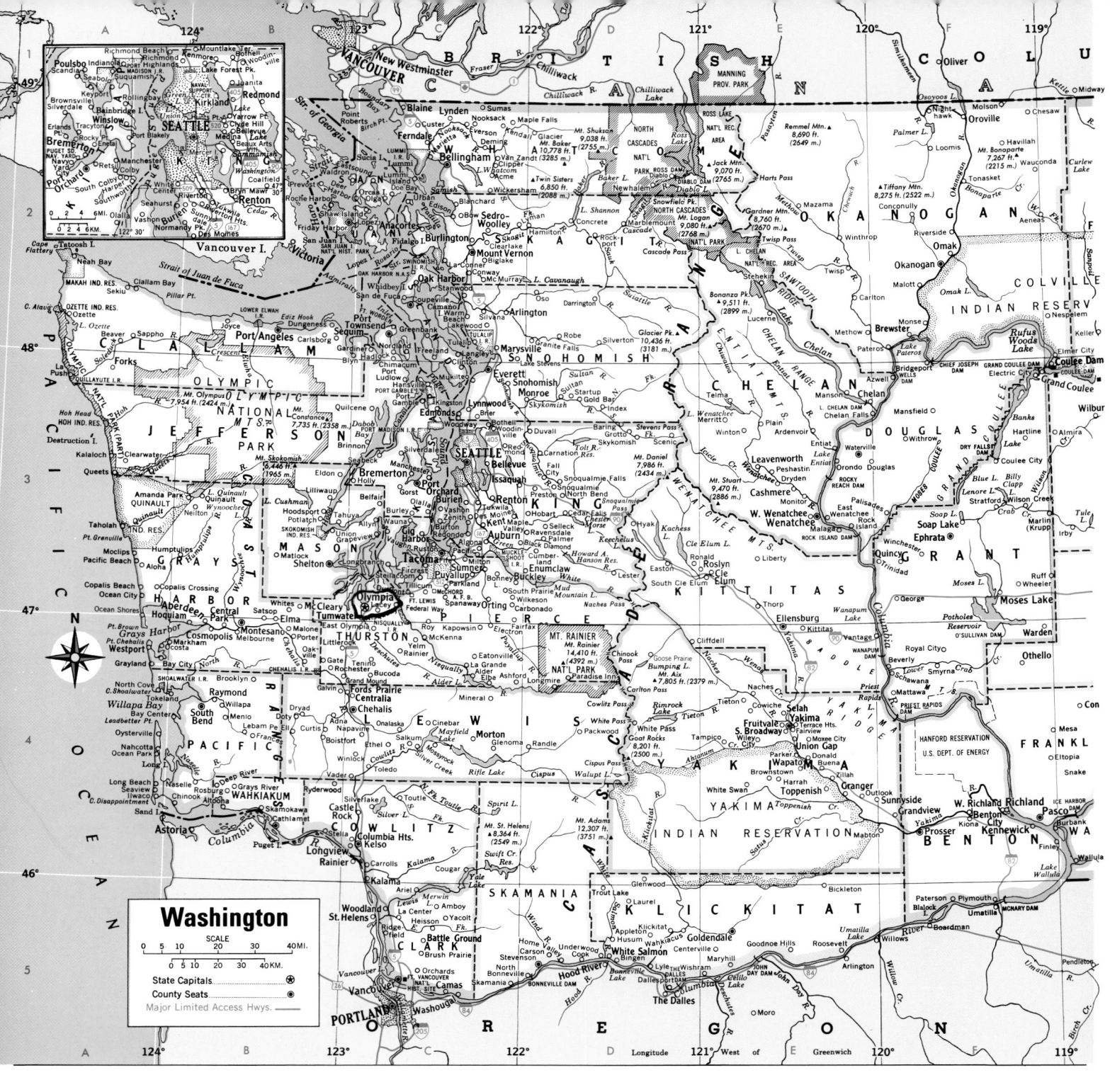

Washington

SCALE
0 5 10 20 30 40MI.
0 5 10 20 30 40KM.
State Capitals............⊛
County Seats.............◉
Major Limited Access Hwys.

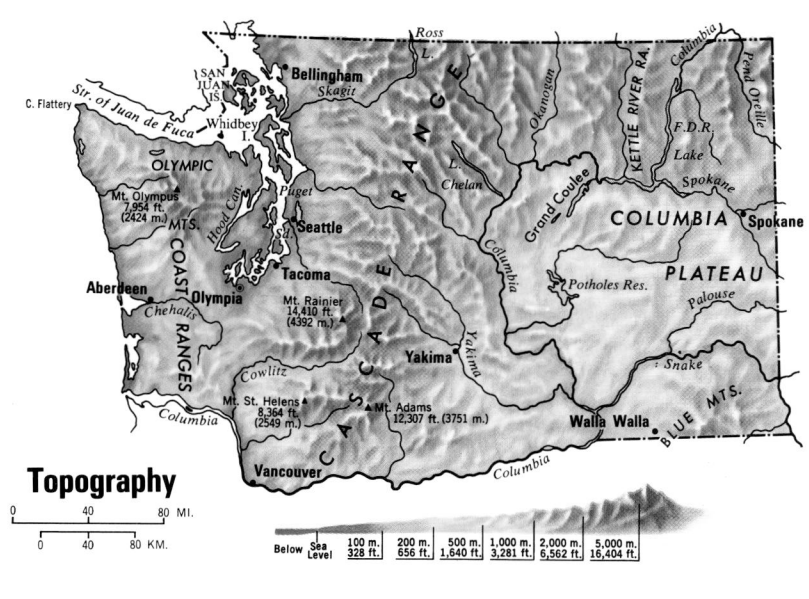

Topography

0   40   80 MI.

0   40   80 KM.

Below Sea Level	100 m. 328 ft.	200 m. 656 ft.	500 m. 1,640 ft.	1,000 m. 3,281 ft.	2,000 m. 6,562 ft.	5,000 m. 16,404 ft.

© Copyright HAMMOND INCORPORATED, Maplewood, N.J.

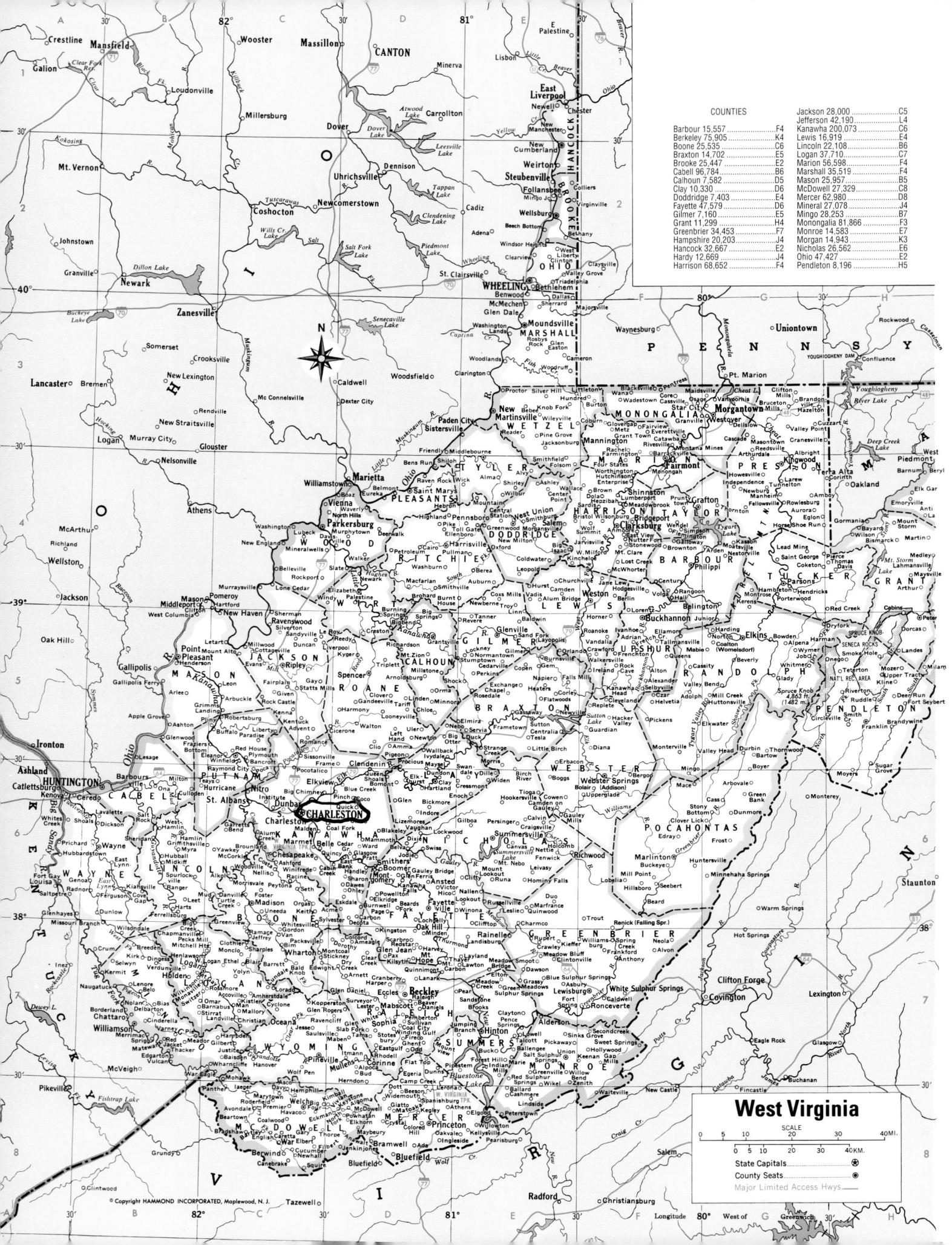

West Virginia

COUNTIES

County	Population	Grid
Barbour	15,557	F4
Berkeley	75,905	K4
Boone	25,535	C6
Braxton	14,702	E5
Brooke	25,447	E2
Cabell	96,784	B6
Calhoun	7,582	D5
Clay	10,330	D6
Doddridge	7,403	E4
Fayette	47,579	D6
Gilmer	7,160	E5
Grant	11,299	H4
Greenbrier	34,453	F7
Hampshire	20,203	J4
Hancock	32,667	E2
Hardy	12,669	J4
Harrison	68,652	F4
Jackson	28,000	C5
Jefferson	42,190	L4
Kanawha	200,073	C6
Lewis	16,919	E4
Lincoln	22,108	B6
Logan	37,710	C7
Marion	56,598	F4
Marshall	35,519	F4
Mason	25,957	B5
McDowell	27,329	C8
Mercer	62,980	D8
Mineral	27,078	J4
Mingo	28,253	B7
Monongalia	81,866	F3
Monroe	14,583	E7
Morgan	14,943	K3
Nicholas	26,562	E6
Ohio	47,427	E2
Pendleton	8,196	H5

SCALE

0 5 10 20 30 40 MI.

0 5 10 20 30 40 KM.

State Capitals ⊛

County Seats ◉

Major Limited Access Hwys.

AREA 24,231 sq. mi. (62,758 sq. km.)
POPULATION 1,808,344
CAPITAL Charleston
LARGEST CITY Charleston
HIGHEST POINT Spruce Knob 4,861 ft. (1482 m.)
SETTLED IN 1774
ADMITTED TO UNION June 20, 1863
POPULAR NAME Mountain State
STATE FLOWER Big Rhododendron
STATE BIRD Cardinal

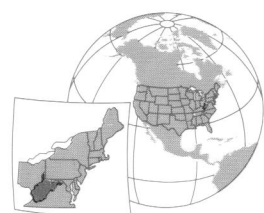

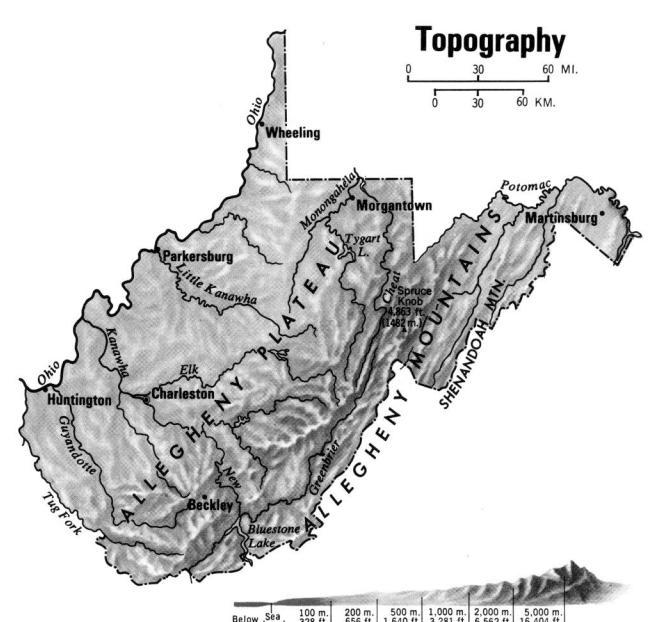

Topography

0   30   60 MI.

0   30   60 KM.

Below Sea Level | 100 m. 328 ft. | 200 m. 656 ft. | 500 m. 1,640 ft. | 1,000 m. 3,281 ft. | 2,000 m. 6,562 ft. | 5,000 m. 16,404 ft.

(continued on following page)

Agriculture, Industry and Resources

DOMINANT LAND USE

- Dairy, General Farming
- General Farming, Livestock, Dairy
- General Farming, Livestock, Tobacco
- General Farming, Livestock, Fruit, Tobacco
- Fruit and Mixed Farming
- Forests

MAJOR MINERAL OCCURRENCES

- C   Coal
- Cl   Clay
- G   Natural Gas
- Ls   Limestone
- Na   Salt
- O   Petroleum

- ⚡   Water Power
- Major Industrial Areas

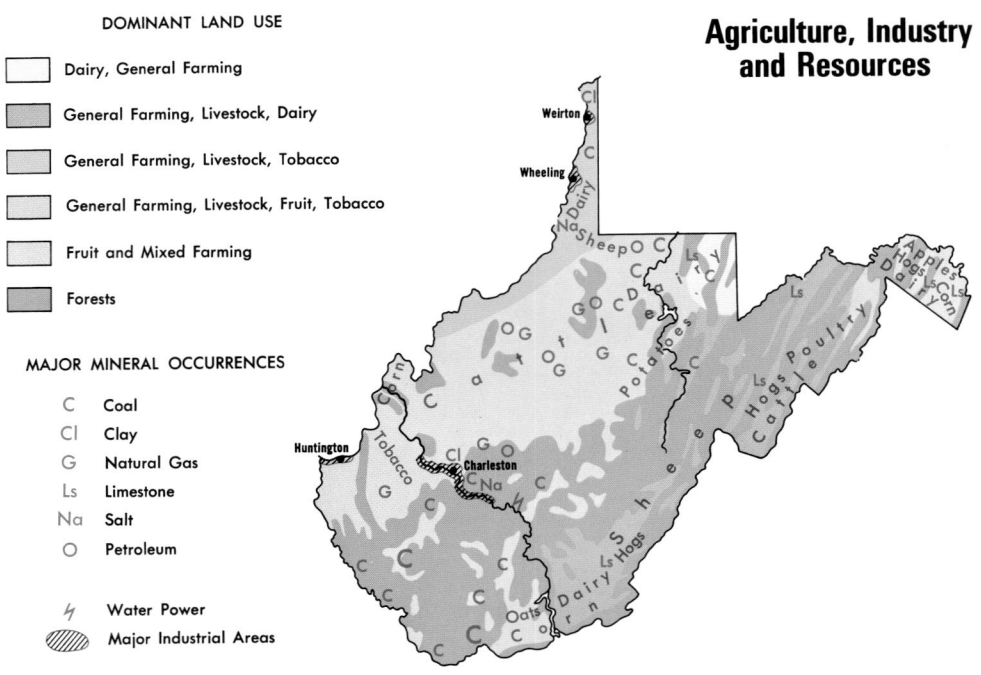

WISCONSIN
1848

AREA 56,153 sq. mi. (145,436 sq. km.)
POPULATION 5,363,675
CAPITAL Madison
LARGEST CITY Milwaukee
HIGHEST POINT Timms Hill 1,951 ft. (595 m.)
SETTLED IN 1670
ADMITTED TO UNION May 29, 1848
POPULAR NAME Badger State
STATE FLOWER Wood Violet
STATE BIRD Robin

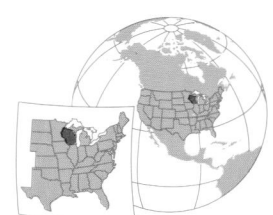

COUNTIES

Adams 18,643	G7
Ashland 16,866	E3
Barron 44,963	C5
Bayfield 15,013	D3
Brown 226,778	L7
Buffalo 13,804	C7
Burnett 15,674	B4
Calumet 40,631	K7
Chippewa 55,195	D5
Clark 33,557	E6
Columbia 52,468	H9
Crawford 17,243	E9
Dane 426,526	H9
Dodge 85,897	J9
Door 27,961	M6
Douglas 43,287	C3
Dunn 39,858	D6
Eau Claire 93,142	D6
Florence 5,088	K4
Fond du Lac 97,296	K8
Forest 10,024	J4
Grant 49,597	E10
Green 33,647	G10
Green Lake 19,105	H8
Iowa 22,780	F9
Iron 6,861	G3
Jackson 19,100	F3
Jefferson 74,021	J9
Juneau 24,316	F8
Kenosha 149,577	K10
Kewaunee 20,187	L6
La Crosse 107,120	D8
Lafayette 16,137	F10
Langlade 20,740	H5
Lincoln 29,641	G5
Manitowoc 82,887	L7
Marathon 125,834	G6
Marinette 43,384	K5
Marquette 15,832	H8
Menominee 4,562	J5
Milwaukee 940,164	L9
Monroe 40,899	E8
Oconto 35,634	K6
Oneida 36,776	G4
Outagamie 160,971	K7
Ozaukee 82,317	L9
Pepin 7,213	C6
Pierce 36,804	B6
Polk 41,319	B5
Portage 67,182	G6
Price 15,822	F4
Racine 188,831	K10
Richland 17,924	F9
Rock 152,307	H10
Rusk 15,347	D5
Saint Croix 63,155	B5
Sauk 55,225	G9
Sawyer 16,196	D4
Shawano 40,664	J6
Sheboygan 112,646	L8
Taylor 19,680	E5
Trempealeau 27,010	D7
Vernon 28,056	E8
Vilas 21,033	G3
Walworth 93,759	J10
Washburn 16,036	C4
Washington 117,493	K9
Waukesha 360,767	K9
Waupaca 51,731	J6
Waushara 23,154	H7
Winnebago 156,763	J8
Wood 75,555	F7

CITIES and TOWNS

Abbotsford 1,956	F6
Abrams 1,757	L6
Adams 1,914	G8
Adell 517	L8
Afton 225	H10
Albany 1,191	G10
Albion 1,093	H10
Algoma 3,357	M6
Allenton 915	K9
Allouez 15,443	L7
Alma▲ 942	C7
Alma Center 446	E7
Almena 720	B5
Almond 459	G7
Alto 1,103	J8
Altoona 6,698	C6
Alvin 186	J4
Amberg 854	K5
Amery 2,845	B5
Amherst 964	H7
Amherst Junction 305	H7
Angelica 1,635	K6
Angelo 100	E3
Aniwa 272	H6
Antigo▲ 8,560	H5
Appleton▲ 70,087	J7
Arbor Vitae 3,153	G4
Arcadia 2,402	D7
Arena 685	G9

Argonne 532	G9
Argyle 823	G10
Arkansaw 400	B6
Arlington 484	H9
Armstrong Creek 463	K4
Arpin 337	G6
Ashippun 2,308	H1
Ashland▲ 8,620	E2
Ashwaubenon 17,634	K7
Athens 1,095	G5
Auburndale 738	F6
Augusta 1,460	D6
Auroraville 250	H7
Avoca 608	F9
Avon 120	H10
Babcock 250	F7
Bagley 339	D10
Baileys Harbor 1,003	M5
Baldwin 2,667	B6
Balsam Lake▲ 950	B5
Bancroft 355	G7
Bangor 1,400	E8
Baraboo▲ 10,711	G9
Barnes 610	D3
Barneveld 1,088	F10
Barron▲ 3,248	C5
Barronett 405	B4
Batavia 125	K8
Bay City 465	B6
Bayfield 611	E2
Bayside 4,518	M1
Bear Creek 415	J6
Beaver 100	K5
Beaver Dam 15,169	J9
Beetown 150	E10
Beldenville 175	A6
Belgium 1,678	L8
Bell Center 116	E9
Belleville 1,908	G10
Belmont 871	F10
Beloit 35,775	H10
Bennett• 622	C3
Benton 976	F10
Berlin 5,305	H8
Bethel 210	F6
Bevent• 1,126	H6
Big Bend 1,278	K2
Birchwood 518	C4
Birnamwood 795	H6
Biron 915	G7
Black Creek 1,192	K7
Black Earth 1,320	G9
Black River Falls▲ 3,618	E7
Blackwell• 347	J4
Blair 1,273	D7
Blanchardville 806	G10
Bloom City 167	E8
Bloomer 3,347	D5
Bloomington 701	E10
Blue Mounds 708	G9
Blue River 429	E9
Boardman 100	A5
Boaz 131	E9
Bohners Lake 1,952	K10
Bonduel 1,416	K6
Boscobel 3,047	E9
Boulder Junction• 958	G3
Bowler 343	J6
Boyceville 1,043	C5
Boyd 680	E6
Brackett 150	D6
Bradley 100	G4
Branch 300	L7
Brandon 912	J8
Brantwood 500	F4
Bridgeport 946	D9
Briggsville 250	H8
Brighton 100	K3
Brill 200	C4
Brillion 2,937	L7
Brodhead 3,180	G10
Brokaw 107	G5
Brookfield 38,649	K1
Brooklyn 916	H10
Brooks 103	G8
Brothertown 100	K7
Brown Deer 12,170	L11
Brown's Lake 1,933	K3
Brownsville 570	J8
Browntown 252	G10
Bruce 787	D5
Brule• 591	C2
Brussels• 1,112	L6
Buffalo 1,040	C7
Burlington 9,936	K10
Burnett• 919	J9
Butler 1,881	K1
Butte Des Morts	J7
Butternut 407	E3
Cable• 836	D3
Cadott 1,345	D6
Caldwell 101	J2
Caledonia 100	L2
Cambria 792	H8
Cambridge 1,101	H9
Cameron 1,546	C5

Campbellsport 1,913	K8
Camp Douglas 592	F8
Camp Lake 3,255	K10
Canton 100	C5
Caroline 450	J6
Carter 100	J5
Cascade 666	K8
Casco 572	L6
Cashton 1,005	E8
Cassville 1,085	E10
Cataract 200	E7
Catawba 149	E4
Cazenovia 326	F8
Cecil 466	K6
Cedarburg 10,908	L9
Cedar Grove 1,887	L8
Centuria 865	A5
Chaseburg 306	D8
Chelsea• 719	F5
Chenequa 583	J1
Chetek 2,180	C5
Chili 185	F6
Chilton▲ 3,708	K7
Chippewa Falls▲ 12,925	D6
City Point 110	F7
Clam Lake 140	E3
Clayton 507	B5
Clear Lake 1,051	B5
Clearwater Lake 200	H4
Cleveland 1,361	L8
Clinton 2,162	J10
Clintonville 4,736	J6
Clyman 388	J9
Cobb 442	F10
Cochrane 435	C7
Colby 1,616	F6
Coleman 716	L5
Colfax 1,136	C6
Coloma 461	H7
Columbus 4,479	H9
Combined Locks 2,422	K7
Commonwealth• 419	K4
Como 1,870	K10
Comstock 160	C5
Concord• 2,023	H1
Conover• 1,137	H3
Conrath 92	E5
Coon Valley 714	E8
Cornell 1,466	D5
Cornucopia 250	D2
Couderay 92	D4
Crandon▲ 1,961	H4
Cream 120	C7
Crivitz 998	L5
Cross Plains 3,084	G9
Cuba City 2,156	F10
Cudahy 18,429	M2
Cumberland 2,280	C4
Curtiss 198	F6
Cushing 150	A4
Cylon 100	B5
Dale• 2,288	J7
Dallas 356	C5
Dalton 300	H8
Danbury 350	B3
Dane 799	G9
Darien 1,572	J10
Darlington▲ 2,418	F10
Deerfield 1,971	H9
Deer Park 227	B5
De Forest 7,368	H9
Delafield 6,472	J1
Delavan 7,956	J10
Delavan Lake 2,352	J10
Dellwood 120	G7
Denmark 1,958	L7
De Pere 20,559	K7
De Soto 366	D9
Dexterville 100	F7
Diamond Bluff 100	A6
Dickeyville 1,043	E10
Dodge• 414	D7
Dodgeville▲ 4,220	F10
Dorchester 827	F5
Dousman 1,584	J1
Downing 257	B5
Downsville 200	C6
Doylestown 328	H9
Draper 171	E4
Dresser 732	A5
Drummond• 541	D3
Dunbar 106	K4
Durand▲ 1,968	C6
Dyckesville 300	L6
Eagle 1,707	H2
Eagle River▲ 1,443	H4
Eastman 437	D9
Easton• 1,194	G8
East Troy 3,564	J2
Eau Claire▲ 61,704	D6
Eden 687	K8
Edgar 1,386	G6
Edgerton 4,933	H10
Egg Harbor 250	M5
Eland 251	H6
Elcho• 1,317	H5

Elderon 189	H6
Eldorado• 1,447	J8
Eleva 635	D6
Elkhart Lake 1,021	L8
Elkhorn▲ 7,305	J10
Elk Mound 785	C6
Ellison Bay 112	M5
Ellsworth▲ 2,909	A6
Elm Grove 6,249	K1
Elmwood 841	B6
Elmwood Park 474	M3
Elroy 1,578	F8
Elton 150	J5
Embarrass 399	J6
Emerald• 691	B5
Endeavor 440	G8
Ephraim 353	M5
Ettrick 521	D7
Evansville 4,039	H10
Fairchild 564	D6
Fair Water 350	J8
Fall Creek 1,236	D6
Fall River 1,097	H9

Fence• 231	K4
Fennimore 2,387	E9
Fenwood 174	F6
Ferryville 174	D9
Fifield• 989	F4
Fish Creek 119	M5
Florence▲ 2,319	K4
Fond du Lac▲ 42,203	K8
Fontana 1,754	J10
Footville 788	H10
Forest Junction 140	K7
Forestville 429	L6
Fort Atkinson 11,621	J10
Fountain City 983	C7
Foxboro 360	B2
Fox Lake 1,454	J8
Fox Point 7,012	M1
Francis Creek 681	L7
Franklin 29,494	L2
Franksville 1,789	M3
Frederic 1,262	B4
Fredonia 1,934	L8
Fremont 666	J7
Friendship▲ 698	G8

Friesland 298	H8
Galesville 1,427	D7
Galloway 200	H6
Gays Mills 625	E9
Genesee• 7,284	J2
Genesee Depot 350	J2
Genoa 263	D8
Genoa City 1,949	K11
Germantown 18,260	K1
Gibbsville 408	L8
Gillett 1,256	K6
Gilman 474	E5
Gilmanton• 474	C7
Gleason 200	G5
Glenbeulah 378	L8
Glendale 13,367	M1
Glen Flora 108	E4
Glen Haven 490	E10
Glenwood City 1,183	B5
Glidden 940	F3
Goodman 820	K4
Gordon• 645	C3
Gotham 250	F9
Grafton 10,312	L9

Grand Marsh 725	G8
Grand View• 483	D3
Granton 406	E6
Grantsburg▲ 1,369	A4
Gratiot 252	F10
Green Bay▲ 102,313	K6
Greendale 14,405	L2
Greenfield 35,476	L2
Green Lake▲ 1,100	H8
Greenleaf 300	L7
Green Valley 104	K6
Greenville• 6,844	J7
Greenwood 1,079	E6
Gresham 575	J6
Gurney 158	F3
Hager City 110	A6
Hales Corners 7,765	K2
Hallie	D6
Hamburg• 910	G5
Hammond 1,153	A6
Hancock 463	G7
Hartford 10,905	K9
Hartland 7,905	J1
Hatfield 500	E7

(continued on following page)

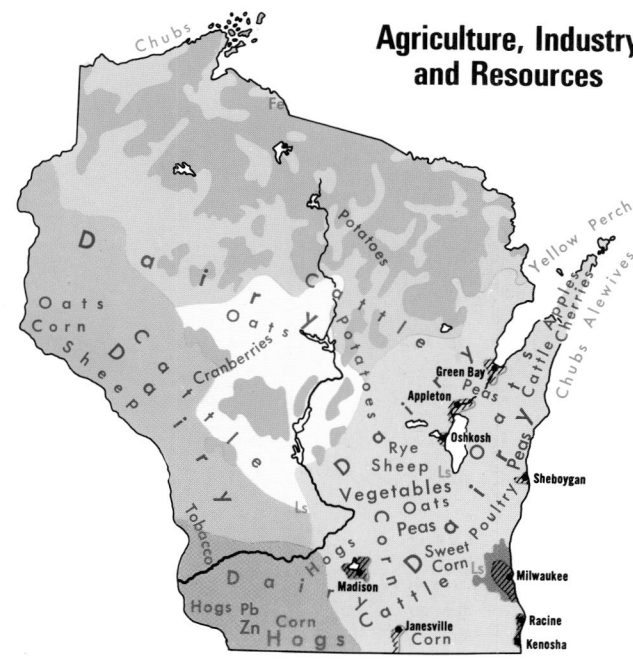

Agriculture, Industry and Resources

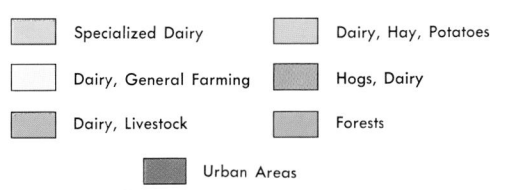

DOMINANT LAND USE

- Specialized Dairy
- Dairy, General Farming
- Dairy, Livestock
- Dairy, Hay, Potatoes
- Hogs, Dairy
- Forests
- Urban Areas

MAJOR MINERAL OCCURRENCES

Fe Iron Ore   Pb Lead
Ls Limestone   Zn Zinc

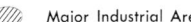 Major Industrial Areas

Topography

0 40 80 MI.
0 40 80 KM.

Legend: Below Sea Level | 100 m. 328 ft. | 200 m. 656 ft. | 500 m. 1,640 ft. | 1,000 m. 3,281 ft. | 2,000 m. 6,562 ft. | 5,000 m. 16,404 ft.

Map labels: APOSTLE ISLANDS, SUPERIOR, Superior, SUPERIOR UPLAND, St. Croix, Namekagon, Chippewa L., Chippewa, Flambeau, Flambeau Flowage, Timms Hill 1,951 ft (595 m.), Menominee, Red Cedar, Eau Claire, Wausau, Wisconsin, Wolf, Peshtigo, Washington I., Door Pen., Green Bay, Appleton, Lake Winnebago, Oshkosh, L. Poygan, Petenwell Lake, Castle Rock Lake, Sheboygan, La Crosse, The Dells, Black, Yellow, Kickapoo, Mississippi, Madison, Milwaukee, Rock, Janesville, Racine, Kenosha, Fox

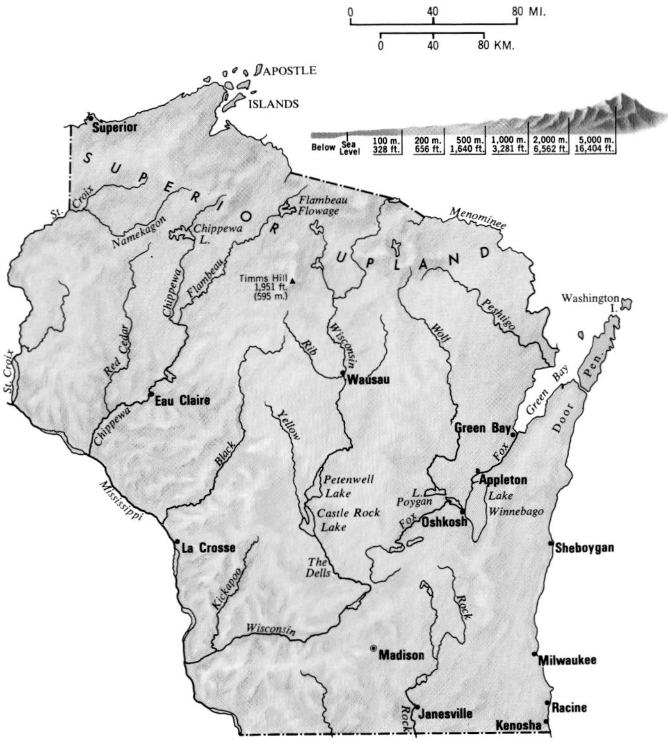

Wisconsin

SCALE
0 5 10   20   30   40 MI.
0 5 10   20   30   40 KM.

✳ State Capitals
◉ County Seats
Canals
Major Limited Access Hwys

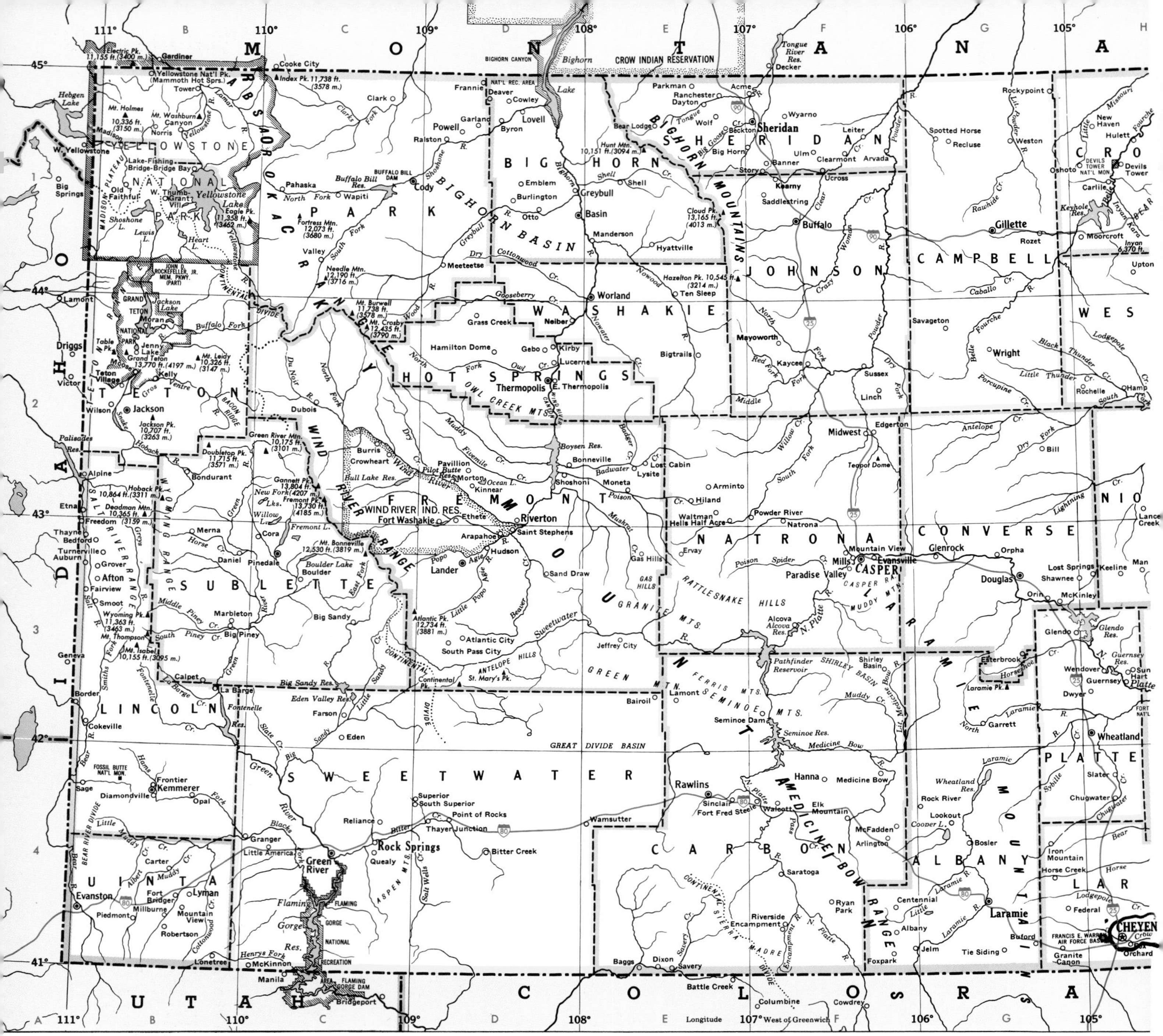

Agriculture, Industry and Resources

DOMINANT LAND USE

- Specialized Wheat
- Specialized Dairy
- General Farming, Livestock, Special Crops
- Sugar Beets, Dry Beans, Livestock, General Farming
- Range Livestock
- Forests
- Nonagricultural Land

MAJOR MINERAL OCCURRENCES

C	Coal	G	Natural Gas	So	Soda Ash
Cl	Clay	O	Petroleum	U	Uranium
Fe	Iron Ore	P	Phosphates	V	Vanadium
		ϟ	Water Power		

COUNTIES

CITIES and TOWNS

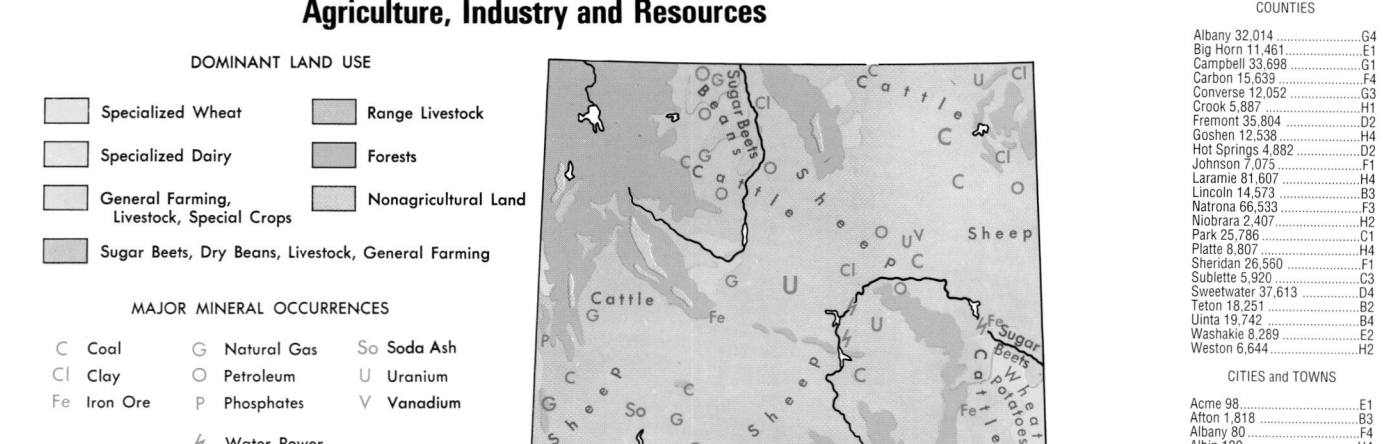

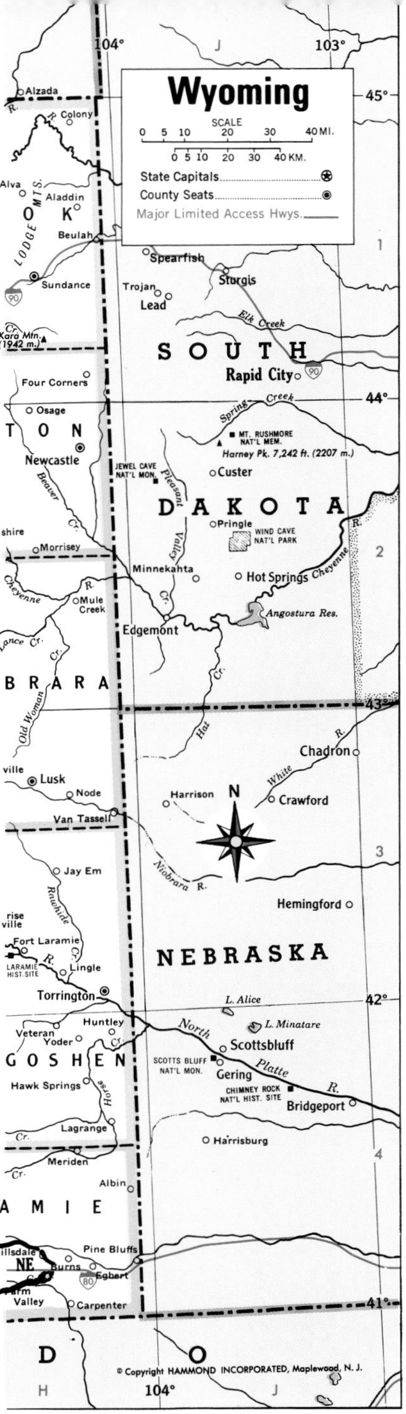
Wyoming
SCALE
0 5 10 20 30 40 MI.
0 5 10 20 30 40 KM.
State Capitals
County Seats
Major Limited Access Hwys.
SOUTH DAKOTA
NEBRASKA
Rapid City
Spearfish
Sturgis
Lead
Trojan
Sundance
Beulah
Aladdin
Alva
Colony
Alzada
Four Corners
Osage
Newcastle
Custer
Pringle
WIND CAVE NAT'L PARK
JEWEL CAVE NAT'L MON.
MT. RUSHMORE NAT'L MEM.
Harney Pk. 7,242 ft. (2207 m.)
Hot Springs
Minnekahta
Edgemont
Mule Creek
Morrisey
Angostura Res.
Chadron
Crawford
Harrison
Lusk
Node
Van Tassell
Jay Em
Hemingford
Fort Laramie
Lingle
LARAMIE HIST. SITE
Torrington
Huntley
Veteran
Yoder
L. Alice
L. Minatare
Scottsbluff
Gering
SCOTTS BLUFF NAT'L MON.
CHIMNEY ROCK NAT'L HIST. SITE
Bridgeport
Hawk Springs
Lagrange
Harrisburg
Meriden
Albin
Pine Bluffs
Hillsdale
Burns
Egbert
Carpenter
North Platte R.
Niobrara R.
Cheyenne R.
White R.
Kara Mtn. (1942 m.)
© Copyright HAMMOND INCORPORATED, Maplewood, N. J.
104°
103°
45°
44°
43°
42°
41°

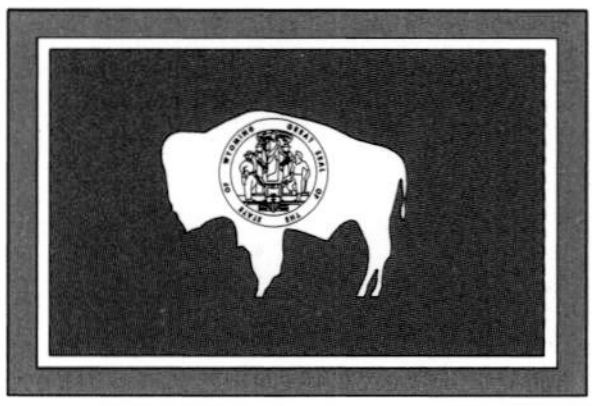

AREA 97,809 sq. mi. (253,325 sq. km.)
POPULATION 493,782
CAPITAL Cheyenne
LARGEST CITY Cheyenne
HIGHEST POINT Gannett Pk. 13,804 ft. (4207 m.)
SETTLED IN 1834
ADMITTED TO UNION July 10, 1890
POPULAR NAME Equality State
STATE FLOWER Indian Paintbrush
STATE BIRD Meadowlark

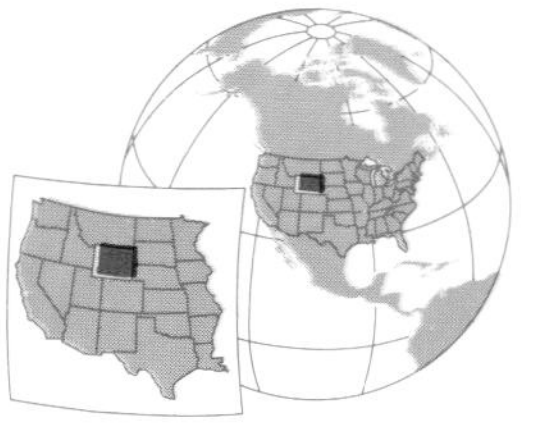

Topography

0 50 100 MI.
0 50 100 KM.

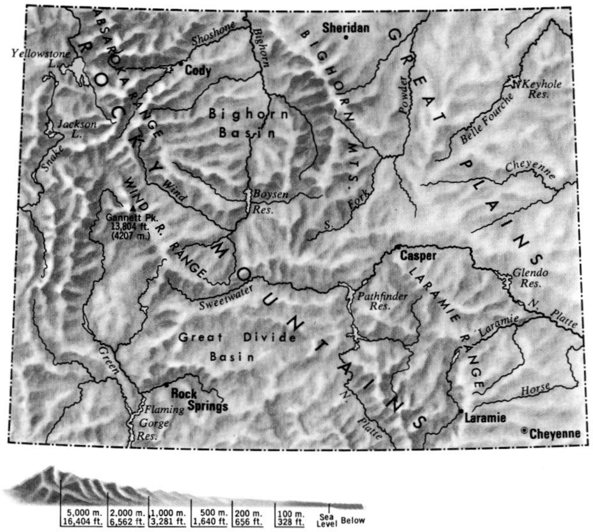
Yellowstone L.
Sheridan
Cody
Bighorn Basin
BIGHORN MTS.
GREAT PLAINS
ABSAROKA RANGE
ROCKY MOUNTAINS
WIND R. RANGE
Gannett Pk. 13,804 ft. (4207 m.)
Jackson L.
Boysen Res.
Keyhole Res.
Belle Fourche
Cheyenne
Casper
Glendo Res.
Pathfinder Res.
LARAMIE RANGE
Sweetwater
Great Divide Basin
Rock Springs
Flaming Gorge Res.
Laramie
Cheyenne
5,000 m. 16,404 ft.
2,000 m. 6,562 ft.
1,000 m. 3,281 ft.
500 m. 1,640 ft.
200 m. 656 ft.
100 m. 328 ft.
Sea Level
Below

Alcova 20 F3
Alpine 550 B2
Alva 50 H1
Arapahoe 1,766 D3
Arvada 33 F1
Atlantic City 39 D3
Auburn 276 A3
Baggs 348 E4
Bairoil 97 E3
Banner 40 F1
Basin▲ 1,238 E1
Beckton 110 E1
Bedford 169 A3
Beulah 184 H1
Big Horn 198 E1
Big Piney 408 B3
Bondurant 155 B2
Border 25 B3
Bosler 195 G4
Boulder 30 C3
Buffalo▲ 3,900 F1
Buford 36 G4
Burlington 250 D1
Burns 285 H4
Burris 30 C2
Byron 557 D1
Carpenter 75 H4
Casper▲ 49,644 F3
Centennial 191 G4
Cheyenne (cap.)▲ 53,011 H4
Chugwater 244 H4
Clark 25 C1
Clearmont 115 F1
Cody▲ 8,835 D1
Cokeville 506 B3
Colony 50 H1
Cowley 560 D1
Crowheart 163 C2
Daniel 89 B3
Dayton 678 E1
Deaver 177 D1
Devils Tower 28 H1
Diamondville 716 B4
Dixon 79 E4
Douglas▲ 5,288 G3
Dubois 962 C2
East Thermopolis 274 D2
Eden 388 C3
Edgerton 169 F2
Egbert 75 H4
Elk Mountain 192 F4
Encampment 611 F4
Ethete 1,455 D2
Etna 123 A2
Evanston▲ 11,507 B4
Evansville 2,255 F3
Fairview 277 B3
Farson 242 C3
Fort Bridger 400 B4
Fort Laramie 243 H3
Fort Washakie 1,477 C2
Fox Farm 3,272 H4
Foxpark 78 F4
Frannie 209 D1
Freedom 400 B3
Frontier 150 B4
Garland 95 D1
Gas Hills 150 E3
Gillette▲ 19,646 G1
Glendo 229 G3
Glenrock 2,231 G3
Granger 146 C4
Granite Canon 80 G4
Grass Creek 152 D2
Green River▲ 11,808 C4
Greybull 1,815 E1
Grover 137 B3
Guernsey 1,147 H3
Hamilton Dome 80 D2
Hanna 873 F4
Hartville 76 H3
Hawk Springs 69 H4
Hillsdale 160 H4
Horse Creek 225 G4
Hudson 407 D3
Hulett 408 H1
Huntley 21 H4
Hyattville 73 E1
Iron Mountain 45 G4
Jackson▲ 8,647 B2
Jeffrey City 106 E3
Jelm 29 G4
Kaycee 249 F2
Kearny 49 F1
Kelly 100 B2
Kemmerer▲ 2,651 B4
Kinnear 145 D2
Kirby 57 D2
La Barge 431 B3
Lagrange 332 H4
Lamont 30 E3
Lance Creek 51 H2
Lander▲ 6,867 D3
Laramie▲ 27,204 G4
Leiter 46 F1
Linch 187 F2
Lingle 510 H3
Little America 56 C4
Lost Cabin 25 E2
Lovell 2,281 D1
Lucerne 525 D2
Lusk▲ 1,447 H3
Lyman 1,938 B4
Lysite 175 E2
Mammoth Hot Springs (Yellowstone Nat'l Park) 350 B1
Manderson 104 E1
Manville 101 H3
Marbleton 720 B3
McFadden 47 F4
McKinnon 49 C4
Medicine Bow 274 F4
Meeteetse 351 D1
Meriden 55 H4
Merna 25 B3
Midwest 408 F2
Millburne 54 B4
Mills 2,591 F3
Moorcroft 807 H1
Moose 150 B2
Moran 200 B2
Morrisey 28 H2
Morton 35 D2
Mountain View 1,153 B4
Mountain View 76 F3
Newcastle▲ 3,065 H2
New Haven 35 H1
Old Faithful 75 B1
Opal 102 B4
Orchard Valley 3,327 H4
Osage 215 H2
Otto 50 D1
Pahaska 75 C1
Parkman 137 E1
Pavillion 165 D2
Piedmont 25 B4
Pine Bluffs 1,153 H4
Pinedale▲ 1,412 C3
Point of Rocks 425 D4
Powder River 51 F2
Powell 5,373 D1
Ralston 233 D1
Ranchester 701 E1
Rawlins▲ 8,538 E4
Recluse 225 G1
Reliance 665 C4
Riverside 59 F4
Riverton 9,310 D2
Robertson 59 B4
Rochelle 23 H2
Rock River 235 G4
Rock Springs 18,708 C4
Saddlestring 100 F1
Saint Stephens 80 D3
Sand Draw 40 D3
Saratoga 1,726 F4
Savageton 30 G2
Savery 29 E4
Shell 80 E1
Sheridan▲ 15,804 F1
Shirley Basin 400 F3
Shoshoni 635 D2
Sinclair 423 E4
Slater 82 H4
Smoot 182 B3
South Superior 586 D4
Story 887 F1
Sundance▲ 1,161 H1
Sunrise 29 H3
Superior 244 D4
Sussex 25 F2
Ten Sleep 304 E1
Thayne 267 A3
Thermopolis▲ 3,172 D2
Torrington▲ 5,776 H3
Turnerville 155 A3
Ucross 17 F1
Ulm 25 F1
Upton 872 H1
Valley 10 C1
Veteran 28 H4
Walcott 200 F4
Wamsutter 261 E4
Wapiti 130 C1
Wendover 17 H3
Wheatland▲ 3,548 H3
Wilson 1,294 B2
Worland▲ 5,250 E1
Wright 1,347 G2
Wyarno 101 G2
Yellowstone National Park 350 B1
Yoder 169 H4

OTHER FEATURES

Absaroka (range) C1
Antelope (creek) G2
Antelope (hills) D3
Aspen (mts.) C4
Atlantic (peak) D3
Bear (creek) H4
Bear (riv.) B4
Bear Lodge (mts.) H1
Bear River Divide (mts.) B4
Beaver (creek) D3
Beaver (creek) H2
Belle Fourche (riv.) H1
Big Goose (creek) E1
Bighorn (basin) D1
Bighorn (lake) D1
Bighorn (mts.) E1
Bighorn (riv.) D1
Bighorn Canyon Nat'l Rec. Area D1
Big Sandy (riv.) C3
Bitter (creek) C4
Blacks Fork, Green (riv.) C4
Black Thunder (creek) G2
Bonneville (mt.) C3
Boysen (res.) D2
Buffalo Bill (dam) C1
Buffalo Bill (res.) C1
Buffalo Fork, Snake (riv.) B2
Burwell (mt.) C2
Caballo (creek) G1
Casper (range) F3
Cheyenne (riv.) H2
Chugwater (creek) H4
Clarks Fork (riv.) C1
Clear (creek) F1
Cloud (peak) E1
Cottonwood (creek) B4
Crazy Woman (creek) F1
Crosby (mt.) C2
Crow (creek) H4
Deadman (mt.) B2
Devils Tower Nat'l Mon. H1
Doubletop (peak) B2
Dry (creek) C2
Dry Cottonwood (creek) D1
Eagle (peak) B1
Fivemile (creek) D2
Flaming Gorge (res.) C4
Flaming Gorge Nat'l Rec. Area C4
Fontenelle (creek) B3
Fontenelle (res.) B3
Fort Laramie Nat'l Hist. Site H3
Fortress (mt.) C1
Fossil Butte Nat'l Mon. B4
Francis E. Warren A.F.B. 3,832 G4
Fremont (lake) C3
Fremont (peak) C2
Gannett (peak) C2
Gas (hills) E3
Glendo (res.) H3
Gooseberry (creek) D1
Grand Teton (mt.) B2
Grand Teton Nat'l Park B2
Granite (mts.) E3
Great Divide (basin) E3
Green (mt.) E3
Green (riv.) C4
Green, East Fork (riv.) C3
Green River (mt.) C2
Greybull (riv.) D1
Greys (riv.) B3
Gros Ventre (riv.) B2
Guernsey (res.) H3
Hams Fork (riv.) B4
Hazelton (peak) E1
Henrys Fork, Green (riv.) C4
Hoback (peak) B2
Hoback (riv.) B2
Holmes (mt.) B1
Horse (creek) H4
Horseshoe (creek) G3
Hunt (mt.) E1
Index (peak) C1
Inyan Kara (creek) H1
Inyan Kara (mt.) H1
Isabel (mt.) B3
Jackson (lake) B2
Jackson (peak) B2
John D. Rockefeller, Jr., Mem. Pkwy. B1
Keyhole (res.) H1
Lamar (riv.) B1
Lance (creek) H2
Laramie (mts.) G3
Laramie (peak) G3
Laramie (riv.) G4
Leidy (mt.) B2
Lewis (lake) B1
Lightning (creek) G2
Little Missouri (riv.) H1
Little Muddy (creek) B4
Little Powder (riv.) G1
Little Sandy (creek) C3
Little Thunder (creek) G2
Lodgepole (creek) H2
Lodgepole (creek) H4
Madison (plat.) B1
Medicine Bow (range) F4
Medicine Bow (riv.) F3
Middle Piney (creek) B3
Muddy (creek) D2
Muskrat (creek) E2
Needle (mt.) C1
Niobrara (riv.) J3
North Laramie (riv.) G3
North Platte (riv.) H3
Nowater (creek) E2
Nowood (riv.) E1
Owl, North Fork (creek) D2
Owl Creek (mts.) D2
Palisades (res.) A2
Pass (creek) F4
Pathfinder (res.) F3
Poison (creek) E2
Poison Spider (creek) F3
Popo Agie (riv.) D3
Powder (riv.) F2
Rattlesnake (hills) E3
Rawhide (creek) G1
Rocky (mts.) C1
Salt (riv.) B3
Salt River (range) B3
Salt Wells (creek) D4
Seminoe (mts.) E3
Seminoe (res.) F3
Shell (creek) E1
Shirley (basin) F3
Shoshone (lake) B1
Shoshone (riv.) D1
Sierra Madre (mts.) E4
Slate (creek) C3
Smiths Fork (riv.) B3
Snake (riv.) B2
South Cheyenne (riv.) H2
South Piney (creek) B3
Sweetwater (riv.) D3
Sybille (creek) G4
Teapot Dome (mt.) F2
Teton (range) B2
Tongue (riv.) E1
Washburn (mt.) B1
Wheatland (res.) G4
Willow (creek) C2
Wind (riv.) C2
Wind River (canyon) D2
Wind River (range) C2
Wind River Ind. Res. C2
Wood (riv.) C2
Wyoming (peak) B3
Wyoming (range) B2
Yellowstone (lake) B1
Yellowstone (riv.) B1
Yellowstone Nat'l Park B1

▲County seat

Acquisitions of Territory

The United States in 1783 comprised the thirteen original states and included lands acquired by conquest during the Revolution and by the Treaty of 1783.

Rank by Area

Rank by Population
1990
Rank by Population

Rank by Area

MAP COLORS INDICATE RANKINGS IN GROUPS OF TEN

YEAR OF ADMISSION TO THE UNION

State	Year
DELAWARE ☆	1787
PENNSYLVANIA ☆	
NEW JERSEY ☆	
GEORGIA ☆	1788
CONNECTICUT ☆	
MASSACHUSETTS ☆	
MARYLAND ☆	
SOUTH CAROLINA ☆	
NEW HAMPSHIRE ☆	
VIRGINIA ☆	
NEW YORK ☆	
NORTH CAROLINA ☆	1789
RHODE ISLAND ☆	1790
VERMONT ☆	1791
KENTUCKY ☆	1792
TENNESSEE ☆	1796
OHIO ☆	1803
LOUISIANA ☆	1812
INDIANA ☆	1816
MISSISSIPPI ☆	1817
ILLINOIS ☆	1818
ALABAMA ☆	1819
MAINE ☆	1820
MISSOURI ☆	1821

1959 ☆HAWAII ☆ALASKA

1912 ☆ARIZONA ☆NEW MEXICO

1907 ☆OKLAHOMA

1896 UTAH
1890 WYOMING
1889 ☆NORTH DAKOTA ☆SOUTH DAKOTA ☆MONTANA ☆WASHINGTON
IDAHO
1876 COLORADO
1867 NEVADA
1864 ☆KANSAS
1863 WEST VIRGINIA
1861 OREGON
1859 MINNESOTA
1858 CALIFORNIA
1850 WISCONSIN
1848 IOWA
1846 FLORIDA
1845 TEXAS
1837 MICHIGAN
1836 ARKANSAS

© Copyright
HAMMOND INCORPORATED

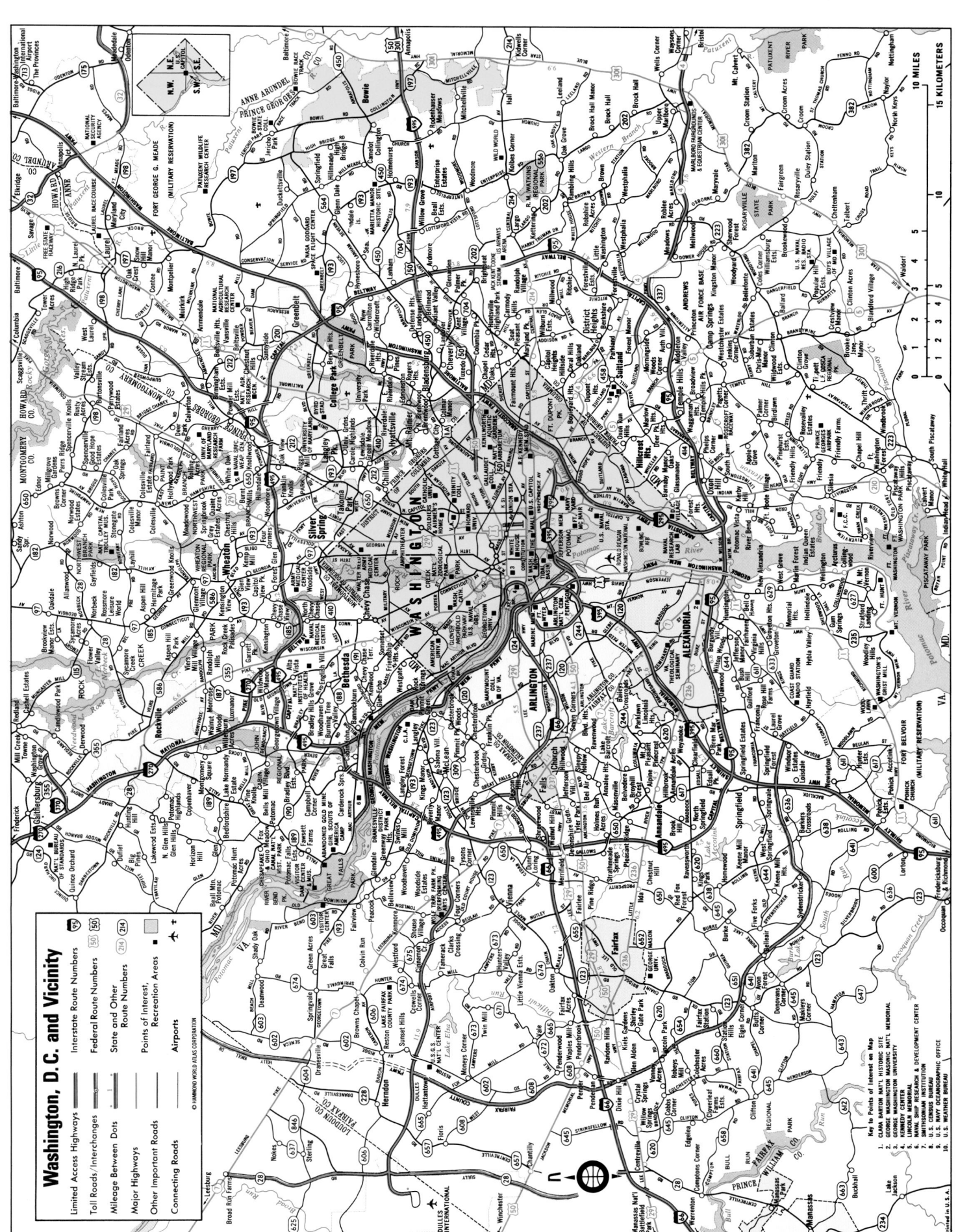

Washington, D.C. and Vicinity

Limited Access Highways

Toll Roads/Interchanges 95 50

Mileage Between Dots 56 214

Major Highways 56 214

Other Important Roads

Connecting Roads

Interstate Route Numbers

Federal Route Numbers

State and Other Route Numbers

Points of Interest

Recreation Areas

Airports

© HAMMOND WORLD ATLAS CORPORATION

Key to Points of Interest on Map
1. CLARA BARTON NAT'L HISTORIC SITE
2. GEORGE WASHINGTON MASONIC NAT'L MEMORIAL
3. GEORGE WASHINGTON UNIVERSITY
4. KENNEDY CENTER
5. LINCOLN MEMORIAL
6. NAVAL SHIP RESEARCH & DEVELOPMENT CENTER
7. SMITHSONIAN INSTITUTION
8. U.S. CENSUS BUREAU
9. U.S. NAVY OCEANOGRAPHIC OFFICE
10. U.S. WEATHER BUREAU

10 MILES

15 KILOMETERS

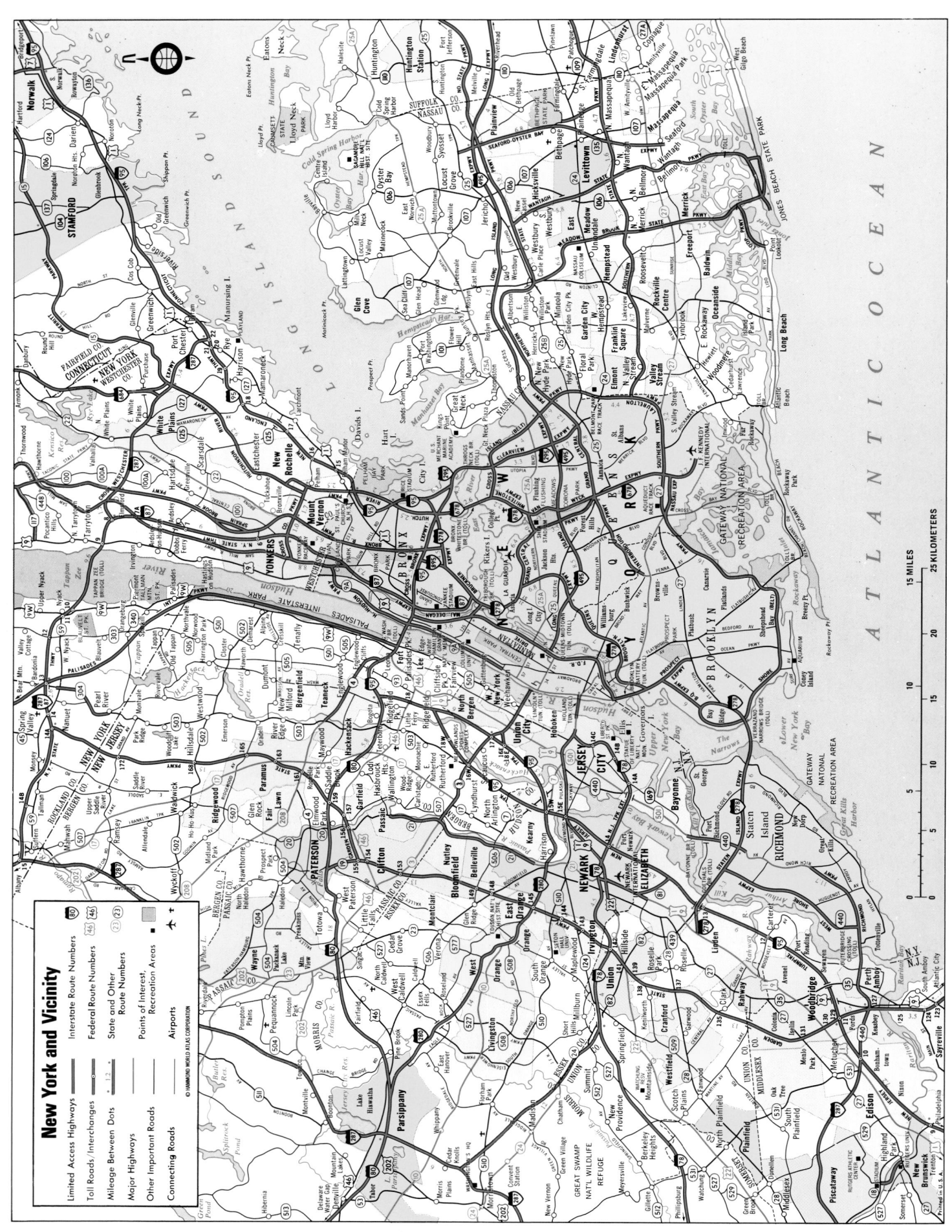

New York and Vicinity

Interstate Route Numbers
Federal Route Numbers
State and Other Route Numbers
Points of Interest, Recreation Areas
Airports

Limited Access Highways
Toll Roads/Interchanges
Mileage Between Dots
Major Highways
Other Important Roads
Connecting Roads

© HAMMOND WORLD ATLAS CORPORATION

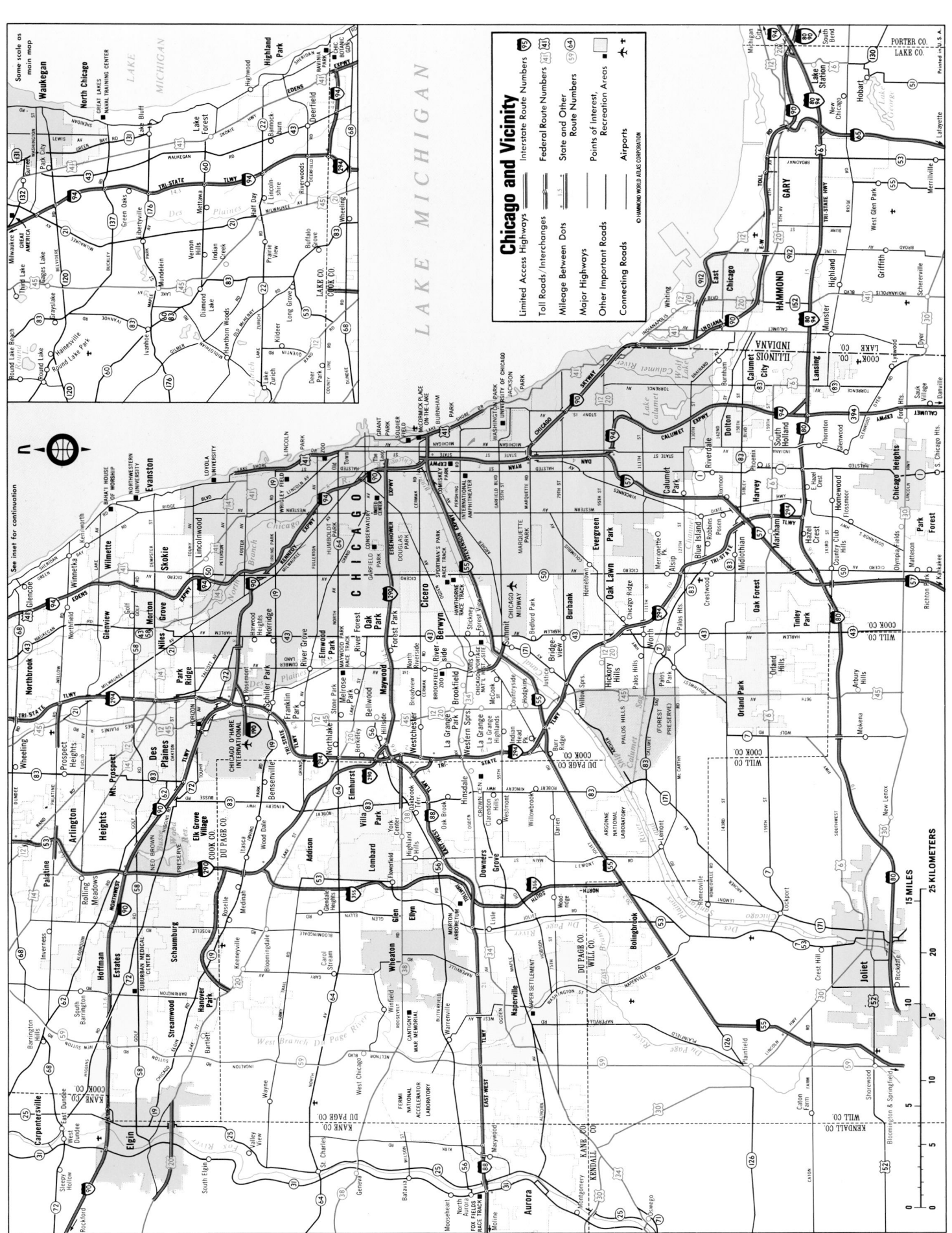

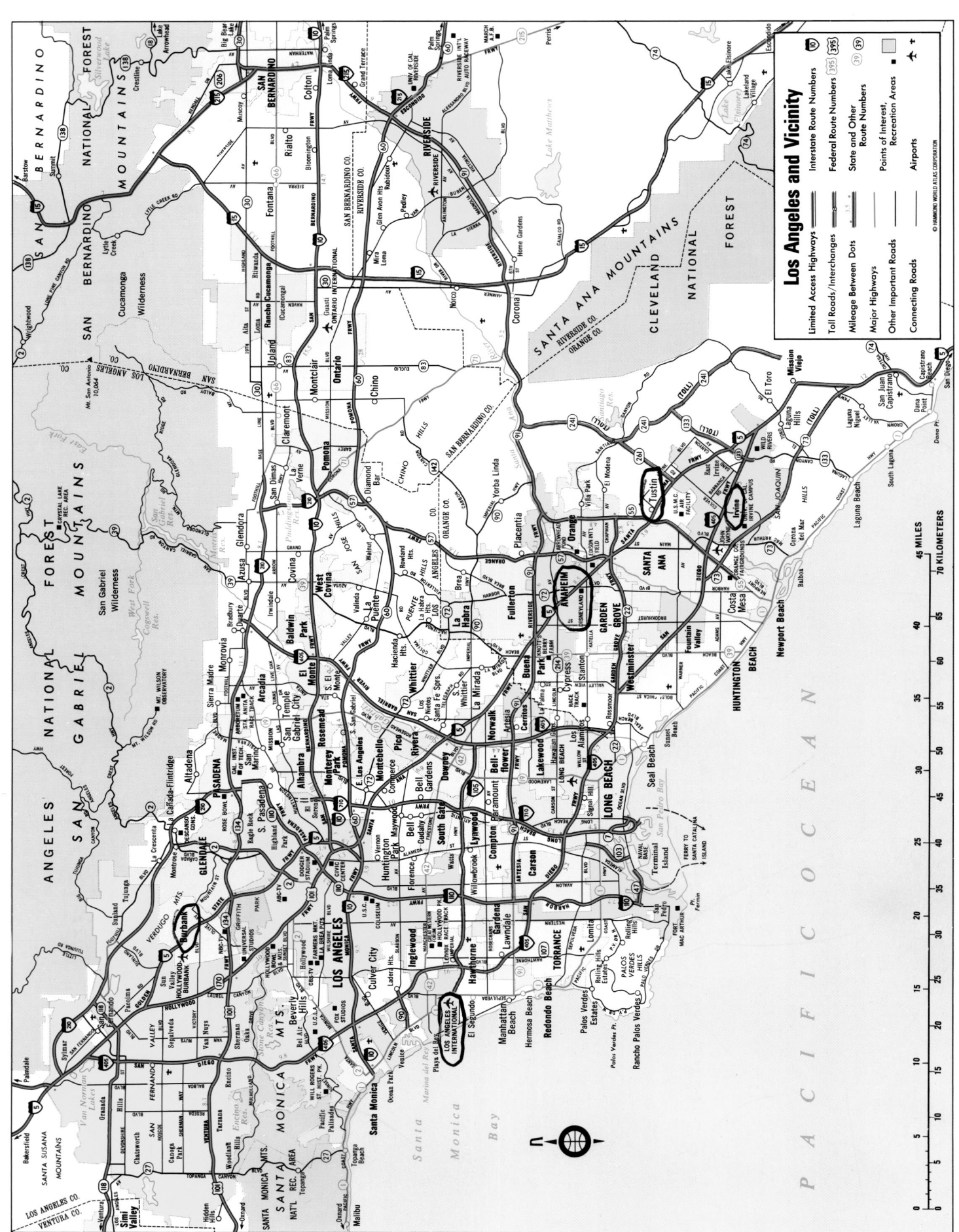

Los Angeles and Vicinity

GEOGRAPHICAL TERMS

A. = Arabic Burm. = Burmese Camb. = Cambodian Ch. = Chinese Czech. = Czechoslovakian Dan. = Danish Du. = Dutch Finn. = Finnish Fr. = French Ger. = German Ice. = Icelandic It. = Italian Jap. = Japanese Mong. = Mongol Nor. = Norwegian Per. = Persian Port. = Portuguese Russ. = Russian Sp. = Spanish Sw. = Swedish Turk. = Turkish

Å	Nor., Sw.	Stream
Aas	Dan., Nor	Hills
Abajo	Sp	Lower
Ada, Adasi	Turk	Island
Altipiano	It	Plateau
Altiplano	Sp	Plateau
Alv, Alf, Elf	Sw	River
Arrecife	Sp	Reef
Asa	Nor., Sw	Hill
Asaga	Turk	Lower
Austral	Sp	Southern
Baai	Du	Bay
Bab	Arabic	Gate or Strait
Bahia	Sp	Bay
Bahr	Arabic	Marsh, Lake, Sea, River
Baia	Port	Bay
Baie	Fr	Bay, Gulf
Baizo	Port	Low
Bakke	Dan	Hill
Bana	Jap	Cape
Bañados	Sp	Marshes
Band	Per	Mt. Range
Bandao	Ch	Peninsula
Bandar	Per	Harbor
Barra	Sp	Reef
Bel	Turk	Pass
Belt	Ger	Strait
Ben	Gaelic	Mountain
Bera	Du	Mountain
Berg	Ger., Du	Mountain
Bir	Arabic	Well
Boca	Sp	Gulf, Inlet
Boğhaz	Turk	Strait
Bolshoi, Bolshaya	Russ	Big
Bolson	Sp	Depression
Bong	Korean	Mountain
Boreal	Sp	Northern
Breen	Nor	Glacier
Bro	Dan., Nor., Sw.	Bridge
Bucht	Ger	Bay
Bugt	Dan	Bay
Bukhta	Russ	Bay
Bukit	Malay	Hill, Mountain
Bukt	Nor., Sw	Bay, Gulf
Burnu, Burun	Turk	Cape, Point
By	Dan., Nor., Sw	Town
Cabo	Port., Sp	Cape
Campos	Port	Plains
Canal	Port., Sp	Channel
Cap, Capo	Fr., It	Cape
Cataratas	Sp	Falls
Catena	It	Mt. Range
Catingas	Port	Open Woodlands
Cayos	Sp	Islands
Central, Centrale	Fr., It	Middle
Cerrito, Cerro	Sp	Hill
Cerros	Sp	Hills, Mountains
Chai	Turk	River
Chott	Arabic	Salt Lake
Ciénaga	Sp	Swamp
Ciudad	Sp	City
Col	Fr	Pass
Cordillera	Sp	Mt. Range, Mts.
Côte	Fr	Coast
Csatoria	Magyar	Canal
Cuchilla	Sp	Mt. Range
Curiche	Sp	Swamp
Dağ, Dağı	Turk	Mountain, Peak
Dağlari	Turk	Mt. Range
Dal	Nor., Sw	Valley
Dar	Arabic	Land
Dar'ya	Russ	River
Daryacheh	Per	Marshy Lake
Dasht	Per	Desert, Plain
Deniz, Denizi	Turk	Sea, Lake
Desierto	Sp	Desert
Détroit	Fr	Strait
Djeziret	Arabic, Turk	Island
Do	Korean	Island
Doi	Thai	Mountain
Eiland	Du	Island
Elv	Dan., Nor	River
Embalse	Sp	Reservoir
Emi	Berber	Mountain
Erg	Arabic	Dune, Desert
Eski	Turk	Old
Est, Este	Fr., Port., Sp	East
Estero	Sp	Estuary, Creek
Estrecho, Estreito	Sp., Port	Strait
Etang	Fr	Pond, Lagoon, Lake
Feng	Ch	Mountain
Fiume	It	River
Fjäll	Sw	Mountain
Fjeld, Fjell	Nor	Hills, Mountain
Fjord	Dan., Nor., Sw	Fiord
Fleuve	Fr	River
Fljót	Ice	Stream
Fluss	Ger	River
Fors	Sw	Waterfall
Fos, Foss	Dan., Nor	Waterfall
Gamla	Nor	Old
Gamle	Dan	Old
Gata	Jap	Lake
Gawa	Jap	River
Gebel	Arabic	Mountain
Gebergte	Du	Mt. Range
Gebirge	Ger	Mt. Range
Gobi	Mongol	Desert
Goe	Jap	Pass
Gol	Mongol, Turk	Lake, Stream
Golf	Ger., Du	Gulf
Golfe	Fr	Gulf
Golfo	Sp., It., Port	Gulf
Gölü	Turk	Lake
Gora	Russ	Mountain
Grand, Grande	Fr., Sp	Big
Groot	Du	Big
Gross	Ger	Big
Grosso	It., Port	Big
Guba	Russ	Bay, Gulf
Gunto	Jap	Archipelago
Gunung	Malay	Mountain
Hai	Ch	Sea
Haixia	Ch	Strait
Halbinsel	Ger	Peninsula
Hamáda, Hammada	Arabic	Rocky Plateau
Hamn	Sw	Harbor
Hamún	Per	Marsh
Hanto	Jap	Peninsula
Has, Hassi	Arabic	Well
Hav	Dan., Nor., Sw	Sea, Ocean
Havet	Nor	Bay
Havn	Dan., Nor	Harbor
Havre	Fr	Harbor
He	Ch	River, Stream
Higashi, Higasi	Jap	East
Hochebene	Ger	Plateau
Hoek	Du	Cape
Hoku	Jap	North
Holm	Dan., Nor., Sw	Island
Hory	Czech	Mountains
Hoved	Dan., Nor	Cape, Promontory
Hu	Ch	Lake
Huang	Ch	Yellow
Huk	Dan., Nor., Sw	Point
Hus, Huus	Dan., Nor., Sw	House
Idehan	Arabic	Desert
Ile	Fr	Island
Ilet	Fr	Islet
Ilot	Fr	Islet
Indre	Dan., Nor	Inner
Inferieur, Inferiore	Fr., It	Lower
Inner, Inre	Sw	Inner
Insel	Ger	Island
Irmak	Turk	River
Isla	Sp	Island
Isola	It	Island
Jabal, Jebel	Arabic	Mountains
Järvi	Finn	Lake
Jaure	Sw	Lake
Jiang	Ch	River, Stream
Jima	Jap	Island
Joki	Finn	River
Kaap	Du	Cape
Kabir, Kebir	Arabic	Big
Kai	Jap	Sea
Kaikyo	Jap	Strait
Kami	Turk	Upper
Kanaal	Du	Canal
Kanal	Russ., Ger	Canal, Channel
Kao	Thai	Mountain
Kap, Kapp	Nor., Sw., Ice	Cape
Kaupunki	Finn	Town
Kawa	Jap	River
Khao	Thai	Mountain
Khrebet	Russ	Mt. Range
Kita	Jap	North
Klein	Du., Ger	Small
Klint	Dan	Promontory
Kô	Jap	Lake
Ko	Thai	Island
Koh	Camb., Khmer	Island
Kop	Du	Peak, Head
Köping	Sw	Market, Borough
Körfez, Körfezi	Turk	Gulf
Kosa	Russ	Spit
Kosui	Jap	Lake
Kraal	Du	Native Village
Kuchuk	Turk	Small
Kuh, Kuhha	Per	Mt. Range, Mts.
Kul	Sinkiang Turki	Lake
Kum	Turk	Desert
Kuro	Jap	Black
Laag	Du	Low
Lac	Fr	Lake
Lago	Port., Sp., It	Lake
Lagoa	Port	Lagoon
Laguna	Sp	Lagoon
Lagune	Fr	Lagoon
Lahti	Finn	Bay, Bight
Län	Sw	County
Liedao	Ch	Islands, Archipelago
Lilla	Sw	Small
Lille	Dan., Nor	Small
Ling	Ch	Mountain
Llanos	Sp	Plains
Mae Nam	Thai	River
Mali, Malaya	Russ	Small
Man	Korean	Bay
Mar	Sp., Port	Sea
Mare	It	Sea
Medio	Sp	Middle
Meer	Du	Lake
Meer	Ger	Sea
Mer	Fr	Sea
Meridionale	It	Southern
Meseta	Sp	Plateau
Middelst, Midden	Du	Middle
Minami	Jap	Southern
Mis	Russ	Cape
Misaki	Jap	Cape
Mittel	Ger	Middle
Mont	Fr	Mountain
Montagne	Fr	Mountain
Montaña	Sp	Mountains
Monte	Sp., It., Port.	Mountain
More	Russ	Sea
Mörön	Mong	Stream
Morro	Port., Sp	Mountain, Promontory
Morue	Fr	Hill
Moyen	Fr	Middle
Muang	Siamese	Town
Mui	Vietnamese	Cape, Point
Mys	Russ	Cape
Nada	Jap	Sea
Naka	Jap	Middle
Nam	Burm., Lao	River
Namakzar	Per	Salt Waste
Nan	Jap	South
Nes	Nor	Cape, Point
Nevado	Sp	Snow-covered Peak
Nieder	Ger	Lower
Nishi, Nisi	Jap	West
Nizhni, Nizhnyaya	Russ	Lower
Njarga	Finn	Peninsula, Promontory
Nong	Thai	Lake
Noord	Du	North
Nord	Fr., Ger	North
Norte	Sp., It., Port	North
Nos	Russ	Cape
Novi, Novaya	Russ	New
Nur, Nuur	Ch., Mong	Lake
Nuruu	Mong	Mountains
Nusa	Malay	Island
Ny, Nya	Nor., Sw	New
O	Jap	Big
ö	Nor., Sw	Island
Ober	Ger	Upper
Occidental, Occidentale	Sp., It	Western
Odde	Dan	Point
Oeste	Port	West
Ooster	Du	Eastern
Opper, Over	Du	Upper
Oriental	Sp., Fr	Eastern
Orientale	It	Eastern
Orta	Turk	Middle
Ost	Ger	East
Ostrov	Russ	Island
Ouest	Fr	West
öy	Nor	Island
Ozero	Russ	Lake
Pampa	Sp	Plain
Pas	Fr	Channel, Strait
Paso	Sp	Pass
Passo	It., Port	Pass
Peña	Sp	Rock, Mountain
Pendi	Ch	Basin
Penisola	It	Peninsula
Pequeño	Sp	Small
Pereval	Russ	Pass
Peski	Russ	Desert
Petit, Petite	Fr	Small
Phu	Lao, Annamese	Mtn.
Pic	Fr	Mountain
Piccolo	It	Small
Pico	Port., Sp	Mountain, Peak
Pik	Russ	Mountain, Peak
Piton	Fr	Mountain, Peak
Planalto	Port	Plateau
Plato	Russ	Plateau
Pointe	Fr	Point
Poluostrov	Russ	Peninsula
Ponta	Port	Point
Presa	Sp	Reservoir
Presqu'île	Fr	Peninsula
Proliv	Russ	Strait
Pulou, Pulo	Malay	Island
Punt	Du	Point
Punta	Sp., It., Port	Point
Qiryat	Hebrew	City, Settlement
Qum	Turk	Desert
Qundao	Ch	Islands
Rada	Sp	Inlet
Rade	Fr	Bay, Inlet
Ras	Arabic	Cape
Reka	Russ	River
Retto	Jap	Archipelago
Ria	Sp	Estuary
Río	Sp	River
Rivier, Rivière	Du., Fr	River
Rud	Per	River
Sai	Jap	West
Saki	Jap	Cape
Salar, Salina	Sp	Salt Deposit
Salto	Sp., Port	Falls
San	Jap., Korean	Hill
Sanmaek	Korean	Mt. Range
Schiereiland	Du	Peninsula
Se	Camb., Khmer	River
See	Ger	Sea, Lake
Selvas	Sp., Port	Woods, Forest
Seno	Sp	Bay, Gulf
Serra	Port	Mts.
Serranía	Sp	Mts.
Seto	Jap	Strait
Settentrionale	It	Northern
Severni, Severnaya	Russ	North
Shamo	Ch	Desert
Shan	Ch., Jap	Hill, Mts.
Shankou	Ch	Pass
Shatt	Arabic	River
Shima	Jap	Island
Shimo	Jap	Lower
Shin	Jap	Land
Shiro	Jap	White
Shoto	Jap	Islands
Si	Ch	West
Sierra	Sp	Mt. Range, Mts.
Sjö	Nor., Sw	Lake, Sea
Sok, Suk, Souk	Arabic	Market
Song	Annamese	River
Sopka	Russ	Volcano
Spitze	Ger	Mt. Peak
Sredni, Srednyaya	Russ	Middle
Stad	Dan., Nor., Sw	City
Stari, Staraya	Russ	Old
Step	Russ	Treeless Plain
Straat	Du	Strait
Strasse	Ger	Strait
Stretto	It	Strait
Ström	Dan., Nor., Sw	Sound
Stung	Camb., Khmer	River
Su	Turk	River
Sud, Süd	Sp., Fr., Ger	South
Suido	Jap	Strait, Channel
Sul	Port	South
Sund	Dan., Nor., Sw	Sound
Sungei	Malay	River
Supérieur	Fr	Upper
Superior, Superiore	Sp., It	Upper
Sur	Sp	South
Suyu	Turk	River
Ta	Ch	Big
Tafelland	Du	Plateau
Tagh	Turk	Mt. Range
Take	Jap	Peak, Ridge
Takht	Arabic	Lower
Tal	Ger	Valley
Tanjung	Malay	Cape, Point
Tell	Arabic	Hill
Thale	Thai	Sea, Lake
Tind	Nor	Peak
Tō	Jap	East
To	Jap	Island
Toge	Jap	Pass
Trask	Finn	Lake
Tugh	Somali	Dry River
Ujung	Malay	Point
Umi	Jap	Bay
Unter	Ger	Lower
Ura	Jap	Inlet
Uul	Mong	Mountain
Val	Fr	Valley
Vatn	Nor	Lake
Vecchio	It	Old
Veld	Du	Plain, Field
Velho	Port	Old
Verkhni	Russ	Upper
Vesi	Finn	Lake
Viejo	Sp	Old
Vik	Nor., Sw	Bay
Vishni, Vishnyaya	Russ	High
Vodokhranilishche	Russ	Reservoir
Volcán	Sp	Volcano
Vostochni, Vostochnaya	Russ	East, Eastern
Wadi	Arabic	Dry River
Wald	Ger	Forest
Wan	Jap	Bay
Westersch	Du	Western
Wüste	Ger	Desert
Yama	Jap	Mountain
Yug, Yuzhni, Yuzhnaya	Russ	South, Southern
Zaki	Jap	Cape
Zaliv	Russ	Bay, Gulf
Zangbo	Tibetan	River, Stream
Zapadni, Zapadnaya	Russ	Western
Zee	Du	Sea
Zemlya	Russ	Land
Zizhiqu	Ch	Autonomous Region
Zuid	Du	South

WORLD STATISTICS

Elements of the Solar System

	Mean Distance from Sun: in Miles	in Kilometers	Period of Revolution around Sun	Period of Rotation on Axis	Equatorial Diameter in Miles	in Kilometers	Surface Gravity (Earth = 1)	Mass (Earth = 1)	Mean Density (Water = 1)	Number of Satellites
Mercury	35,990,000	57,900,000	87.97 days	58.7 days	3,032	4,880	0.38	0.055	5.4	0
Venus	67,240,000	108,200,000	224.70 days	243.7 days†	7,521	12,104	0.91	0.815	5.2	0
Earth	93,000,000	149,700,000	365.26 days	23h 56m	7,926	12,755	1.00	1.00	5.5	1
Mars	141,610,000	227,900,000	686.98 days	24h 37m	4,221	6,794	0.38	0.107	3.9	2
Jupiter	483,675,000	778,400,000	11.86 years	9h 55m	88,846	142,984	2.36	317.8	1.3	16
Saturn	886,572,000	1,426,800,000	29.46 years	10h 30m	74,898	120,536	0.92	95.2	0.7	18
Uranus	1,783,957,000	2,871,000,000	84.01 years	17h 14m†	31,763	51,118	0.89	14.5	1.3	15
Neptune	2,795,114,000	4,498,300,000	164.79 years	16h 6m	30,778	49,532	1.13	17.1	1.6	8
Pluto	3,670,000,000	5,906,400,000	247.70 years	6.4 days†	1,413	2,274	0.07	0.002	2.1	1

† Retrograde motion

Source: NASA, National Space Science Data Center

Dimensions of the Earth

	Area in: Sq. Miles	Sq. Kilometers
Superficial area	196,939,000	510,072,000
Land surface	57,506,000	148,940,000
Water surface	139,433,000	361,132,000

	Distance in: Miles	Kilometers
Equatorial circumference	24,902	40,075
Polar circumference	24,860	40,007
Equatorial diameter	7,926.4	12,756.4
Polar diameter	7,899.8	12,713.6
Equatorial radius	3,963.2	6,378.2
Polar radius	3,949.9	6,356.8

Volume of the Earth	2.6×10^{11} cubic miles	10.84×10^{11} cubic kilometers
Mass or weight	6.6×10^{21} short tons	6.0×10^{21} metric tons
Maximum distance from Sun	94,600,000 miles	152,000,000 kilometers
Minimum distance from Sun	91,300,000 miles	147,000,000 kilometers

Oceans and Major Seas

	Area in: Sq. Miles	Sq. Kms.	Greatest Depth in: Feet	Meters
Pacific Ocean	63,855,000	166,241,000	36,198	11,033
Atlantic Ocean	31,744,000	82,217,000	28,374	8,648
Indian Ocean	28,417,000	73,600,000	25,344	7,725
Arctic Ocean	5,427,000	14,056,000	17,880	5,450
Caribbean Sea	970,000	2,512,300	24,720	7,535
Mediterranean Sea	969,000	2,509,700	16,896	5,150
South China Sea	895,000	2,318,000	15,000	4,600
Bering Sea	875,000	2,266,250	15,800	4,800
Gulf of Mexico	600,000	1,554,000	12,300	3,750
Sea of Okhotsk	590,000	1,528,100	11,070	3,370
East China Sea	482,000	1,248,400	9,500	2,900
Yellow Sea	480,000	1,243,200	350	107
Sea of Japan	389,000	1,007,500	12,280	3,740
Hudson Bay	317,500	822,300	846	258
North Sea	222,000	575,000	2,200	670
Black Sea	185,000	479,150	7,365	2,245
Red Sea	169,000	437,700	7,200	2,195
Baltic Sea	163,000	422,170	1,506	459

The Continents

	Area in: Sq. Miles	Sq. Kms.	Percent of World's Land
Asia	17,128,500	44,362,815	29.5
Africa	11,707,000	30,321,130	20.2
North America	9,363,000	24,250,170	16.2
South America	6,879,725	17,818,505	11.9
Antarctica	5,405,000	14,000,000	9.4
Europe	4,057,000	10,507,630	7.0
Australia	2,967,893	7,686,850	5.1

Major Ship Canals

	Length in: Miles	Kms.	Minimum Depth in: Feet	Meters
Volga-Baltic, Russia	225	362	–	–
Baltic-White Sea, Russia	140	225	16	5
Suez, Egypt	100.76	162	42	13
Albert, Belgium	80	129	16.5	5
Moscow-Volga, Russia	80	129	18	6
Volga-Don, Russia	62	100	–	–
Göta, Sweden	54	87	10	3
Kiel (Nord-Ostsee), Germany	53.2	86	38	12
Panama Canal, Panama	50.72	82	41.6	13
Houston Ship, U.S.A.	50	81	36	11

Largest Islands

	Area in: Sq. Miles	Sq. Kms.
Greenland	840,000	2,175,600
New Guinea	305,000	789,950
Borneo	286,000	740,740
Madagascar	226,656	587,040
Baffin, Canada	195,928	507,454
Sumatra, Indonesia	164,000	424,760
Honshu, Japan	88,000	227,920
Great Britain	84,400	218,896
Victoria, Canada	83,896	217,290
Ellesmere, Canada	75,767	196,236
Celebes, Indonesia	72,986	189,034
South I., New Zealand	58,393	151,238
Java, Indonesia	48,842	126,501
North I., New Zealand	44,187	114,444
Cuba	42,803	110,860
Newfoundland, Canada	42,031	108,860
Luzon, Philippines	40,420	104,688
Iceland	39,768	103,000
Mindanao, Philippines	36,537	94,631
Ireland	32,589	84,406
Hokkaido, Japan	30,436	75,066
Sakhalin, Russia	29,500	76,405

	Area in: Sq. Miles	Sq. Kms.
Hispaniola, Haiti & Dom. Rep.	29,399	76,143
Banks, Canada	27,038	70,028
Ceylon, Sri Lanka	25,332	65,610
Tasmania, Australia	24,600	63,710
Svalbard, Norway	23,957	62,049
Devon, Canada	21,331	55,247
Novaya Zemlya (north isl.), Russia	18,600	48,200
Marajó, Brazil	17,991	46,597
Tierra del Fuego, Chile & Argentina	17,900	46,360
Alexander, Antarctica	16,700	43,250
Axel Heiberg, Canada	16,671	43,178
Melville, Canada	16,274	42,150
Southhampton, Canada	15,913	41,215
New Britain, Papua New Guinea	14,100	36,519
Taiwan, China	13,836	35,835
Kyushu, Japan	13,770	35,664
Hainan, China	13,127	33,999
Prince of Wales, Canada	12,872	33,338
Spitsbergen, Norway	12,355	31,999
Vancouver, Canada	12,079	31,285
Timor, Indonesia	11,527	29,855
Sicily, Italy	9,926	25,708

	Area in: Sq. Miles	Sq. Kms.
Somerset, Canada	9,570	24,786
Sardinia, Italy	9,301	24,090
Shikoku, Japan	6,860	17,767
New Caledonia, France	6,530	16,913
Nordaustlandet, Norway	6,409	16,599
Samar, Philippines	5,050	13,080
Negros, Philippines	4,906	12,707
Palawan, Philippines	4,550	11,785
Panay, Philippines	4,446	11,515
Jamaica	4,232	10,961
Hawaii, United States	4,038	10,458
Viti Levu, Fiji	4,010	10,386
Cape Breton, Canada	3,981	10,311
Mindoro, Philippines	3,759	9,736
Kodiak, Alaska, U.S.A.	3,670	9,505
Cyprus	3,572	9,251
Puerto Rico, U.S.A.	3,435	8,897
Corsica, France	3,352	8,682
New Ireland, Papua New Guinea	3,340	8,651
Crete, Greece	3,218	8,335
Anticosti, Canada	3,066	7,941
Wrangel, Russia	2,819	7,301

Principal Mountains

	Height in: Feet	Meters
Everest, Nepal-China	29,028	8,848
K2 (Godwin Austen), Pakistan-China	28,250	8,611
Kanchenjunga, Nepal-India	28,208	8,598
Lhotse, Nepal-China	27,923	8,511
Makalu, Nepal-China	27,789	8,470
Dhaulagiri, Nepal	26,810	8,172
Nanga Parbat, Pakistan	26,660	8,126
Annapurna, Nepal	26,504	8,078
Nanda Devi, India	25,645	7,817
Rakaposhi, Pakistan	25,550	7,788
Kongur Shan, China	25,325	7,719
Tirich Mir, Pakistan	25,230	7,690
Gongga Shan, China	24,790	7,556
Ismail Samani Peak, Tajikistan	24,590	7,495
Pobeda Peak, Kyrgyzstan	24,406	7,439
Chomo Lhari, Bhutan-China	23,997	7,314
Muztag, China	23,891	7,282
Cerro Aconcagua, Argentina	22,831	6,959
Ojos del Salado, Chile-Argentina	22,572	6,880
Bonete, Chile-Argentina	22,546	6,872
Tupungato, Chile-Argentina	22,310	6,800

	Height in: Feet	Meters
Pissis, Argentina	22,241	6,779
Mercedario, Argentina	22,211	6,770
Huascarán, Peru	22,205	6,768
Llullaillaco, Chile-Argentina	22,057	6,723
Nevada Ancohuma, Bolivia	21,489	6,550
Chimborazo, Ecuador	20,561	6,267
McKinley, Alaska	20,320	6,194
Logan, Yukon, Canada	19,524	5,951
Cotopaxi, Ecuador	19,347	5,897
Kilimanjaro, Tanzania	19,340	5,895
El Misti, Peru	19,101	5,822
Pico Cristóbal Colón, Colombia	18,947	5,775
Huila, Colombia	18,865	5,750
Citlaltépetl (Orizaba), Mexico	18,700	5,700
Damavand, Iran	18,605	5,671
El'brus, Russia	18,510	5,642
St. Elias, Alaska, U.S.A.-Yukon, Canada	18,008	5,489
Dykhtau, Russia	17,070	5,203
Kenya, Kenya	17,058	5,199
Ararat, Turkey	16,946	5,165
Vinson Massif, Antarctica	16,864	5,140

	Height in: Feet	Meters
Margherita, D.R. Congo-Uganda	16,795	5,119
Kazbek, Georgia-Russia	16,558	5,047
Puncak Jaya, Indonesia	16,503	5,030
Blanc, France	15,771	4,807
Klyuchevskaya Sopka, Russia	15,584	4,750
Fairweather, Br. Col., Canada	15,300	4,663
Dufourspitze, Italy-Switzerland	15,203	4,634
Ras Dashen, Ethiopia	15,157	4,620
Matterhorn, Switzerland	14,691	4,478
Whitney, California, U.S.A.	14,494	4,418
Elbert, Colorado, U.S.A.	14,433	4,399
Rainier, Washington, U.S.A.	14,410	4,392
Shasta, California, U.S.A.	14,162	4,317
Pikes Peak, Colorado, U.S.A.	14,110	4,301
Finsteraarhorn, Switzerland	14,022	4,274
Mauna Kea, Hawaii, U.S.A.	13,796	4,205
Mauna Loa, Hawaii, U.S.A.	13,677	4,169
Jungfrau, Switzerland	13,642	4,158
Grossglockner, Austria	12,457	3,797
Fuji, Japan	12,389	3,776
Cook, New Zealand	12,349	3,764

Longest Rivers

	Length in: Miles	Kms.
Nile, Africa	4,145	6,671
Amazon, S. America	4,007	6,448
Mississippi-Missouri-Red Rock, U.S.A.	3,710	5,971
Chang Jiang (Yangtze), China	3,500	5,633
Ob'-Irtysh, Russia-Kazakhstan	3,362	5,411
Yenisey-Angara, Russia	3,100	4,989
Huang He (Yellow), China	2,950	4,747
Congo, Africa	2,780	4,474
Amur-Shilka-Onon, Asia	2,744	4,416
Lena, Russia	2,734	4,400
Mackenzie-Peace-Finlay, Canada	2,635	4,241
Paraná-La Plata, S. America	2,630	4,232
Mekong, Asia	2,610	4,200
Niger, Africa	2,580	4,152
Missouri-Red Rock, U.S.A.	2,564	4,125
Yenisey, Russia	2,500	4,028
Mississippi, U.S.A.	2,348	3,778
Murray-Darling, Australia	2,310	3,718
Volga, Russia	2,290	3,685
Madeira, S. America	2,013	3,240
Purus, S. America	1,995	3,211
Yukon, Alaska-Canada	1,979	3,185
Zambezi, Africa	1,950	3,138
São Francisco, Brazil	1,930	3,106
St. Lawrence, Canada-U.S.A.	1,900	3,058

	Length in: Miles	Kms.
Rio Grande, Mexico-U.S.A.	1,885	3,034
Syrdar'ya-Naryn, Asia	1,859	2,992
Indus, Asia	1,800	2,897
Danube, Europe	1,775	2,857
Brahmaputra, Asia	1,700	2,736
Tocantins, Brazil	1,677	2,699
Salween, Asia	1,675	2,696
Euphrates, Asia	1,650	2,655
Xi Jiang, China	1,650	2,655
Amudar'ya, Asia	1,616	2,601
Nelson-Saskatchewan, Canada	1,600	2,575
Orinoco, S. America	1,600	2,575
Paraguay, S. America	1,584	2,549
Kolyma, Russia	1,562	2,514
Ganges, Asia	1,550	2,494
Ural, Russia-Kazakhstan	1,509	2,428
Japurá, S. America	1,500	2,414
Arkansas, U.S.A.	1,450	2,334
Colorado, U.S.A.-Mexico	1,450	2,334
Negro, S. America	1,400	2,253
Dnieper, Russia-Belarus-Ukraine	1,368	2,202
Orange, Africa	1,350	2,173
Irrawaddy, Burma	1,325	2,132
Brazos, U.S.A.	1,309	2,107
Ohio-Allegheny, U.S.A.	1,306	2,102

	Length in: Miles	Kms.
Kama, Russia	1,252	2,031
Don, Russia	1,222	1,967
Red, U.S.A.	1,222	1,966
Columbia, U.S.A.-Canada	1,214	1,953
Tigris, Asia	1,181	1,901
Darling, Australia	1,160	1,867
Angara, Russia	1,135	1,827
Songhua Jiang (Sungari), Asia	1,130	1,819
Pechora, Russia	1,124	1,809
Snake, U.S.A.	1,038	1,670
Churchill, Canada	1,000	1,609
Pilcomayo, S. America	1,000	1,609
Uruguay, S. America	994	1.600
Platte-N. Platte, U.S.A.	990	1,593
Ohio, U.S.A.	981	1,578
Magdalena, Colombia	956	1,538
Pecos, U.S.A.	926	1,490
Oka, Russia	918	1,477
Canadian, U.S.A.	906	1,458
Colorado, Texas, U.S.A.	894	1,439
Dniester, Ukraine-Moldova	876	1,410
Fraser, Canada	850	1,369
Rhine, Europe	820	1,319
Northern Dvina, Russia	809	1,302
Ottawa, Canada	790	1,271

Principal Natural Lakes

	Area in: Sq. Miles	Sq. Kms.	Max. Depth in: Feet	Meters
Caspian Sea, Asia	143,243	370,999	3,264	995
Lake Superior, U.S.A.-Canada	31,820	82,414	1,329	405
Lake Victoria, Africa	26,628	69,215	270	82
Lake Huron, U.S.A.-Canada	23,010	59,596	748	228
Lake Michigan, U.S.A.	22,400	58,016	923	281
Aral Sea, Kazakhstan-Uzbekistan	15,830	41,000	213	65
Lake Tanganyika, Africa	12,650	32,764	4,700	1,433
Lake Baykal, Russia	12,162	31,500	5,316	1,620
Great Bear Lake, Canada	12,096	31,328	1,356	413
Lake Nyasa (Malawi), Africa	11,555	29,928	2,320	707
Great Slave Lake, Canada	11,031	28,570	2,015	614
Lake Erie, U.S.A.-Canada	9,940	25,745	210	64
Lake Winnipeg, Canada	9,417	24,390	60	18
Lake Ontario, U.S.A.-Canada	7,540	19,529	775	244
Lake Balkhash, Kazakhstan	7,081	18,340	87	27
Lake Ladoga, Russia	6,900	17,871	738	225
Lake Maracaibo, Venezuela	5,120	13,261	100	31
Lake Chad, Africa*	10,000 – 4,000	25,900 – 10,360	25	8
Lake Onega, Russia	3,761	9,741	377	115

	Area in: Sq. Miles	Sq. Kms.	Max. Depth in: Feet	Meters
Lake Eyre, Australia*	3,500-0	9,065-0	–	–
Lake Titicaca, Peru-Bolivia	3,200	8,288	1,000	305
Lake Nicaragua, Nicaragua	3,100	8,029	230	70
Lake Athabasca, Canada	3,064	7,936	400	122
Reindeer Lake, Canada*	2,568	6,651	–	–
Lake Turkana (Rudolf), Africa	2,463	6,379	240	73
Issyk-Kul', Kyrgyzstan	2,425	6,281	2,303	702
Lake Torrens, Australia*	2,230	5,776	–	–
Vänern, Sweden	2,156	5,584	328	100
Nettilling Lake, Canada*	2,140	5,543	–	–
Lake Winnipegosis, Canada	2,075	5,374	38	12
Lake Mobutu Sese Seko (Albert), Africa	2,075	5,374	160	49
Lake Kariba, Zambia-Zimbabwe	2,050	5,310	295	90
Lake Nipigon, Canada	1,872	4,848	540	165
Lake Mweru, Dem. Rep. of the Congo-Zambia	1,800	4,662	60	18
Lake Manitoba, Canada	1,799	4,659	12	4
Lake Taymyr, Russia	1,737	4,499	85	26
Lake Khanka, China-Russia	1,700	4,403	33	10
Lake Kioga, Uganda	1,700	4,403	25	8
Lake of the Woods, U.S.A.-Canada	1,679	4,349	70	21

* Figures subject to great seasonal variations.

TABLES OF AIRLINE DISTANCES

ALL DISTANCES IN STATUTE MILES

BETWEEN PRINCIPAL CITIES OF THE WORLD

FROM/TO	AZORES	BAGHDAD	BERLIN	BOMBAY	BUENOS AIRES	CALLAO	CAIRO	CAPE TOWN	CHICAGO	ISTANBUL	GUAM	HONOLULU	JUNEAU	LONDON	LOS ANGELES	MELBOURNE	MEXICO CITY	MONTREAL	NEW ORLEANS	NEW YORK	PANAMA	PARIS	RIO DE JANEIRO	SAN FRANCISCO	SANTIAGO	SEATTLE	SHANGHAI	SINGAPORE	TOKYO	WELLINGTON
AZORES	....	3906	2118	5930	5385	4825	3325	5670	3305	2880	8985	7421	4715	1562	5034	12190	4584	2548	3718	2604	3918	1617	4312	5114	5718	4720	7324	8338	7370	11475
BAGHDAD	3906	....	2040	2022	8215	8618	785	4923	6490	1085	6380	8445	6180	2568	7695	8150	8155	5814	7212	6066	7807	2385	7012	7521	8876	6848	4468	4443	5242	9782
BERLIN	2148	2040	....	3947	7411	6937	1823	5949	4458	1068	7158	7384	4638	575	5849	9992	6119	3776	5182	4026	5902	540	6246	5744	7842	5121	5323	6226	5623	11384
BOMBAY	5930	2022	3947	....	9380	10530	2698	5133	8144	3043	4831	8172	6992	4526	8810	6140	9818	7582	8952	7875	9832	4391	8438	8523	10127	7830	3219	2425	4247	7752
BUENOS AIRES	5385	8215	7411	9380	....	1982	7428	4332	5598	7638	10516	7653	7964	6919	6148	7336	4609	5619	4902	5295	3319	6891	1230	6487	731	6956	12295	9940	11601	6341
CALLAO	4825	8618	6937	10530	1982	....	7870	6195	3765	7666	9760	5993	5806	6376	4155	8196	2619	3954	2990	3633	1450	6455	2490	4500	1548	4964	10760	11700	9740	6696
CAIRO	3325	785	1823	2698	7428	7870	....	4476	6231	780	7175	8925	6352	2218	7675	8720	7807	5502	6862	5701	7230	2020	6242	7554	8100	6915	5290	5152	6005	10360
CAPE TOWN	5670	4923	5949	5133	4332	6195	4476	....	8551	5210	8918	11655	10382	5975	10165	6510	8620	7975	8390	7845	7090	5732	3850	10340	5080	10305	8179	6025	9234	7149
CHICAGO	3305	6490	4458	8144	5598	3765	6231	8551	....	5530	7510	4315	2310	4015	1741	9837	1690	750	827	727	2320	4219	5320	1875	5325	1753	7155	9475	6410	8465
ISTANBUL	2880	1085	1068	3043	7638	7666	780	5210	5530	....	7015	8200	5665	1540	6895	9189	7160	4825	6220	5060	6797	1390	6420	6770	8230	6124	5084	5440	5649	10790
GUAM	8985	6380	7158	4831	10516	9760	7175	8918	7510	7015	....	3896	5225	7605	6253	3497	7690	7840	7895	8115	9220	7675	11710	5952	9946	5785	1945	2990	1596	4206
HONOLULU	7421	8445	7384	8172	7653	5993	8925	11655	4315	8200	3896	....	2825	7320	2620	3581	3846	4992	4305	5051	5347	7525	8400	2407	6935	2707	5009	6874	3940	4676
JUNEAU	4715	6180	4638	6992	7964	5806	6352	10382	2310	5665	5225	2825	....	4496	1835	8162	3210	2647	2860	2874	4456	4700	7611	1530	7320	870	4968	7375	4117	7501
LONDON	1562	2568	575	4526	6919	6376	2218	5975	4015	1540	7605	7320	4496	....	5496	10590	5605	3370	4656	3500	5310	210	5747	5440	7275	4850	5841	6818	6050	11790
LOS ANGELES	5034	7695	5849	8810	6148	4155	7675	10165	1741	6895	6255	2620	1835	5496	....	8098	1445	2468	1695	2466	3025	5711	6330	345	5595	961	6598	8955	5600	6806
MELBOURNE	12190	8150	9992	6140	7336	8196	8720	6510	9837	9189	3497	3581	8162	10590	8098	....	8599	10553	9455	10541	9211	10500	8340	7970	7130	8330	4967	3768	5172	1655
MEXICO CITY	4584	8155	6119	9818	4609	2619	7807	8620	1690	7160	7690	3846	3210	5605	1445	8599	....	2247	940	2110	1532	5800	4810	1870	4122	2339	8120	10495	7190	7003
MONTREAL	2548	5814	3776	7582	5619	3954	5502	7975	750	4825	7840	4992	2647	3370	2468	10553	2247	....	1390	340	2545	3490	5110	2557	5461	2309	7141	9280	6546	9206
NEW ORLEANS	3718	7212	5182	8952	4902	2990	6862	8390	827	6220	7895	4305	2860	4656	1695	9455	940	1390	....	1161	1600	4846	4798	1960	4553	2137	7830	10255	6993	7950
NEW YORK	2604	6066	4026	7875	5295	3633	5101	7845	727	5060	8115	5051	2874	3500	2466	10541	2110	340	1161	....	2211	3600	4810	2606	5134	2440	7460	9617	6846	9067
PANAMA	3918	7807	5902	9832	3319	1450	7230	7090	2320	6797	9220	5347	4456	5310	3025	9211	1532	2545	1600	2211	....	5440	3311	3349	3000	3680	9430	11800	8560	7580
PARIS	1617	2385	540	4391	6891	6455	2020	5762	4219	1390	7675	7525	4700	210	5711	10500	5800	3490	4846	3600	5440	....	5710	5680	7300	5080	5855	6730	6132	11865
RIO DE JANEIRO	4312	7012	6246	8438	1230	2400	6242	3850	5320	6420	11710	8400	7611	5747	6330	8340	4810	5110	4798	4810	3311	5710	....	6655	1852	11800	6730	9875	8440	10270
SAN FRANCISCO	5114	7521	5744	8523	6487	4500	7554	10340	1875	6770	5952	2407	1530	5440	345	7970	1870	2557	1960	2606	3349	5680	6655	....	5960	692	6245	8440	5250	6800
SANTIAGO	5718	8876	5902	10127	731	1548	8100	5080	5325	8230	9946	6935	7320	7275	5595	7130	4122	5461	4553	5134	3000	7300	1852	5960	....	6466	11850	10270	10850	5925
SEATTLE	4720	6848	5121	7830	6956	4964	6915	10305	1753	6124	5785	2707	870	4850	961	8330	2339	2309	2137	2440	3680	5080	6945	692	6466	....	5780	8200	4863	7310
SHANGHAI	7324	4468	5323	3219	12295	10760	5290	8179	7155	5084	1945	5009	4968	5841	6598	4967	8120	7141	7830	7460	9430	5855	11510	6245	11850	5780	....	2395	1095	6080
SINGAPORE	8338	4443	6226	2425	9940	11700	5152	6025	9475	5440	2990	6874	7375	6818	8955	3768	10495	9280	10255	9617	11800	6730	9875	8440	10270	8200	2395	....	3350	5360
TOKYO	7370	5242	5623	4247	11601	9740	6005	9234	6410	5649	1596	3940	4117	6050	5600	5172	7190	6546	6993	6846	8560	6132	11600	5250	10850	4863	1095	3350	....	5730
WELLINGTON	11475	9782	11384	7752	6341	6696	10360	7149	8465	10790	4206	4676	7501	11790	6806	1655	7003	9206	7950	9067	7580	11865	7510	6800	5925	7310	6080	5360	5730	....

BETWEEN PRINCIPAL CITIES OF EUROPE

FROM/TO	AMSTERDAM	ATHENS	BAKU	BARCELONA	BELGRADE	BERLIN	BRUSSELS	BUCHAREST	BUDAPEST	COLOGNE	COPENHAGEN	ISTANBUL	DRESDEN	DUBLIN	FRANKFURT	HAMBURG	ST. PETERSBURG	LISBON	LONDON	LYON	MADRID	MARSEILLES	MILAN	MOSCOW	MUNICH	OSLO	PARIS	RIGA	ROME	SOFIA	STOCKHOLM	TOULOUSE	WARSAW	VIENNA	ZURICH
AMSTERDAM	....	1340	2218	770	875	365	105	1100	710	128	381	1360	385	468	228	232	1090	1140	220	458	912	627	517	1325	415	568	257	820	808	1073	695	625	673	580	375
ATHENS	1340	....	1395	1160	500	1112	1292	460	698	1200	1320	350	1022	1765	1113	1250	1535	1770	1476	1100	1463	1025	900	1388	925	1610	1300	1310	650	335	1495	1215	990	795	1000
BAKU	2218	1395	....	2427	1487	1867	2240	1220	1562	2127	1980	1070	1837	2490	2055	2020	1570	3050	2435	2238	2742	2238	2028	1175	1912	2118	2335	1590	1900	1360	1862	2425	1555	1700	2050
BARCELONA	770	1160	2427	....	998	925	638	1210	924	692	1085	1380	860	919	665	910	1740	610	707	327	316	211	450	1852	648	1330	518	1140	530	1072	1410	156	1150	830	513
BELGRADE	875	500	1487	998	....	618	850	295	205	750	840	502	530	1327	652	760	1165	1555	1040	752	1235	750	540	1160	475	1112	890	855	440	231	1005	930	510	300	590
BERLIN	365	1112	1867	925	618	....	401	798	425	300	225	1068	95	815	268	165	815	1410	575	601	1149	730	570	995	310	520	540	520	730	810	503	815	320	322	410
BRUSSELS	105	1292	2240	638	850	401	....	1110	700	110	475	1345	407	480	198	301	1175	998	200	352	807	521	435	1392	372	672	170	900	730	945	793	515	720	568	312
BUCHAREST	1100	460	1220	1210	295	798	1110	....	295	982	970	272	725	1560	890	950	1080	1842	1285	1025	1518	1020	819	920	725	1245	1152	870	700	194	1080	1210	580	520	855
BUDAPEST	710	698	1562	924	205	425	700	295	....	590	629	650	345	1176	504	572	963	1515	900	680	1214	718	476	965	350	920	770	685	500	395	820	883	342	128	498
COLOGNE	128	1200	2127	692	750	300	110	982	590	....	400	1240	292	585	93	228	1090	1126	308	370	875	528	390	1285	282	635	250	805	675	945	722	875	602	460	259
COPENHAGEN	381	1320	1960	1085	840	225	475	970	629	400	....	1240	315	768	412	180	708	1520	590	760	1272	970	720	970	520	303	634	453	948	1010	330	962	415	538	595
ISTANBUL	1360	350	1070	1380	502	1068	1345	272	650	1240	1240	....	995	1830	1150	1222	1292	2003	1540	1238	1690	1205	1030	1180	975	1535	1390	1115	840	315	1340	1400	852	790	1090
DRESDEN	385	1022	1837	860	530	95	407	725	345	292	315	995	....	852	236	238	885	1380	592	540	1100	655	435	1200	227	620	523	585	630	730	598	762	325	235	342
DUBLIN	468	1765	2490	919	1327	815	480	1560	1176	585	768	1830	852	....	671	668	1440	1015	300	720	902	875	880	1728	655	786	480	1210	1175	1525	1010	761	1130	1040	768
FRANKFURT	228	1113	2055	665	652	268	198	890	504	93	412	1150	236	671	....	250	1075	1160	392	350	888	492	323	1240	193	675	295	780	698	860	730	560	550	370	193
HAMBURG	232	1250	2020	910	760	165	301	970	572	228	180	1222	238	668	250	....	880	1301	448	580	1098	730	570	1100	378	445	459	600	810	954	502	780	462	460	432
ST. PETERSBURG	1090	1535	1570	1740	1165	815	1175	1080	965	1090	708	1292	885	1440	1075	880	....	2235	1300	1420	1980	1540	1315	391	1100	610	1335	300	1440	1218	435	1635	640	975	1225
LISBON	1140	1770	3050	610	1555	1410	998	1842	1515	1126	1520	2005	1380	1015	1160	1301	2235	....	975	850	313	810	1350	430	1208	1690	890	1940	1150	1685	1848	640	1700	1415	1058
LONDON	220	1476	2435	707	1040	375	202	1285	900	308	590	1540	590	300	392	448	1300	975	....	455	777	620	595	1540	526	720	210	1035	890	1235	885	550	890	762	480
LYON	458	1090	2238	327	752	601	352	1025	890	370	760	1238	540	720	350	580	1420	850	455	....	577	170	210	1560	352	1005	248	1122	462	928	1080	228	850	562	206
MADRID	912	1463	2742	316	1235	1149	807	1518	1214	875	1272	1690	1100	902	888	1098	1980	313	777	557	....	394	728	2120	910	1474	645	1670	840	1385	1598	344	1410	1110	765
MARSEILLES	627	1025	2238	211	750	730	521	1020	718	528	906	1205	655	875	492	730	1540	810	620	170	394	....	238	1642	445	1165	410	1238	372	895	1225	196	950	620	318
MILAN	517	900	2028	450	540	570	435	819	476	390	720	1030	435	880	323	570	1315	1350	595	210	728	238	....	1408	215	1000	400	1010	295	715	1020	400	705	385	137
MOSCOW	1325	1388	1175	1852	1160	995	1392	920	965	1285	970	1180	1200	1728	1240	1100	391	430	1540	1560	2120	1642	1408	....	1220	1030	1538	500	1462	1100	770	1710	710	1020	1353
MUNICH	415	925	1912	648	475	310	372	725	350	282	520	975	227	855	193	378	1100	1208	526	352	910	445	215	1220	....	810	425	800	430	672	811	570	500	222	158
OSLO	568	1610	2118	1330	1112	520	672	1245	920	635	303	1505	620	786	675	445	610	1690	720	1005	1474	1165	1000	1030	810	....	830	531	1242	1295	267	1140	653	835	869
PARIS	257	1300	2335	518	890	540	170	1152	770	250	634	1390	523	480	295	459	1335	890	210	248	645	410	400	1538	425	830	....	1050	690	1080	950	431	845	770	295
RIGA	820	1310	1590	1440	855	520	900	870	685	805	453	1115	585	1210	780	600	300	1940	1035	1122	1670	1238	1010	520	800	531	1050	....	1155	985	276	1335	350	685	930
ROME	808	650	1900	530	440	730	730	700	500	675	948	840	630	1175	698	810	1440	890	1035	462	840	372	295	1462	430	1242	690	1155	....	545	1220	690	810	470	421
SOFIA	1073	335	1360	1072	231	810	945	194	395	945	1010	315	730	1525	860	954	1218	1685	1235	928	1385	895	715	1100	672	1295	1080	985	545	....	1170	1080	662	500	780
STOCKHOLM	695	1495	1862	1410	1005	503	793	1080	820	722	330	1340	598	1010	730	502	435	1848	885	1080	1598	1225	1020	770	811	267	950	276	1220	1170	....	1281	500	770	908
TOULOUSE	625	1215	2425	156	930	815	515	1210	883	875	962	1400	762	761	560	780	1635	640	550	228	344	196	400	1710	570	1140	431	1335	569	1080	1281	....	1062	725	425
WARSAW	673	990	1555	1150	510	320	720	580	342	602	415	852	325	1130	550	462	640	1700	890	850	1410	950	705	710	500	653	845	350	810	662	500	1062	....	345	640
VIENNA	580	795	1700	830	300	322	568	520	128	460	538	790	235	1040	370	460	975	1415	762	562	1110	620	385	1028	222	835	770	685	470	500	770	725	345	....	365
ZURICH	375	1000	2050	513	590	410	312	855	498	259	595	1090	342	768	193	432	1225	1058	480	206	765	318	137	1350	158	869	295	930	421	780	908	425	640	365	....